Grundlehren der mathematischen Wissenschaften 268

A Series of Comprehensive Studies in Mathematics

For further volumes:
http://www.springer.com/series/138

Enrico Arbarello • Maurizio Cornalba •
Phillip A. Griffiths

Geometry of Algebraic Curves

Volume II
with a contribution by Joseph Daniel Harris

Springer

Enrico Arbarello
Dipartimento di Matematica "Guido Castelnuovo"
Università di Roma La Sapienza
00185 Roma
Italy
ea@mat.uniroma1.it

Phillip A. Griffiths
Institute for Advanced Study
Einstein Drive
Princeton, NJ 08540
USA
pg@ias.edu

Maurizio Cornalba
Dipartimento di Matematica "Felice Casorati"
Università di Pavia
Via Ferrata 1
27100 Pavia
Italy
maurizio.cornalba@unipv.it

Contribution by:

Joseph D. Harris
Department of Mathematics
Harvard University
One Oxford Street
Cambridge, MA 02138
USA

ISSN 0072-7830
ISBN 978-3-662-50620-2 ISBN 978-3-540-69392-5 (eBook)
DOI 10.1007/978-3-540-69392-5
Springer Heidelberg Dordrecht London New York

Mathematics Subject Classification (2010): 14xx, 32xx, 30xx, 57xx, 05xx

Cover design: VTeX UAB, Lithuania

Printed on acid-free paper

Springer is part of Springer Science+Business Media (www.springer.com)

To the memory of Aldo Andreotti

Preface

This volume is devoted to the foundations of the theory of moduli of algebraic curves defined over the complex numbers. The first volume was almost exclusively concerned with the geometry on a fixed, smooth curve. At the time it was published, the local deformation theory of a smooth curve was well understood, but the study of the geometry of global moduli was in its early stages. This study has since undergone explosive development and continues to do so. There are two reasons for this; one predictable at the time of the first volume, the other not.

The predictable one was the intrinsic algebro-geometric interest in the moduli of curves; this has certainly turned out to be the case. The other is the external influence from physics. Because of this confluence, the subject has developed in ways that are incredibly richer than could have been imagined at the time of writing of Volume I.

When this volume, GAC II, was planned it was envisioned that the centerpiece would be the study of linear series on a general or variable curve, culminating in a proof of the Petri conjecture. This is still an important part of the present volume, but it is not the central aspect. Rather, the main purpose of the book is to provide comprehensive and detailed foundations for the theory of the moduli of algebraic curves. In addition, we feel that a very important, perhaps distinguishing, aspect of GAC II is the blending of the multiple perspectives—algebro-geometric, complex-analytic, topological, and combinatorial—that are used for the study of the moduli of curves.

It is perhaps keeping this aspect in mind that one can understand our somewhat unusual choice of topics and of the order in which they are presented. For instance, some readers might be surprised to see a purely algebraic proof of the projectivity of moduli spaces immediately followed by a detailed introduction to Teichmüller theory. And yet Teichmüller theory is needed for our subsequent discussion of smooth Galois covers of moduli, which in turn is immediately put to use in our approach to the theory of cycles on moduli spaces. Besides, all the above are essential tools in Kontsevich's proof of Witten's conjecture, which is presented in later chapters. Concerning this, the main motivation of our choice of presenting Kontsevich's original proof instead of one of the several more recent ones is—in addition to the great beauty of the proof itself—a desire to be as self-contained as possible. This same desire also motivates in part the presence, at the beginning of the book, of two introductory chapters on the Hilbert scheme and on deformation theory.

In the *Guide for the Reader* we will briefly go through the material we included in this volume. Among the topics we did not cover are the theory of Gromov–Witten invariants, the birational geometry of moduli spaces, the theory of moduli of vector bundles on a fixed curve, the theory of syzygies for the canonical curve, the various incarnations of the Schottky problem together with the related theory of theta function, and the theory of stable rational cohomology of moduli spaces of smooth curves. Some of these topics are covered by excellent publications like [14] for syzygies and [532] for the birational geometry of moduli spaces. On other topics, like the intersection theory of cycles or the theory of the ample cone of moduli spaces of stable curves, we limited ourselves to the foundational material.

Much of Volume I was devoted to the study of the relationship between an algebraic curve and its Jacobian variety. In this volume there is relatively little emphasis on the universal Jacobian or Picard variety and discussion of the moduli of abelian varieties. The latter is a vast and deep subject, especially in its arithmetic aspect, that goes well beyond the scope of this book.

In some instances, important topics, such as the Kodaira dimension of moduli spaces of stable curves, the theory of limit linear series, or the irreducibility of the Severi variety, have appeared elsewhere, specifically in the book *Moduli of Curves* by Joe Harris and Ian Morrison [352]. This is in fact a good opportunity to thank Joe and Ian for their kind words in the introduction of their book. We believe that our respective books complement each other, and we encourage our readers to benefit from their work.

In the bibliographical notes we try to point the reader to the most significant developments, not covered in this volume, of which we were aware at the time of writing. In fact, we view our bibliography and our bibliographical notes as, potentially, an ongoing project.

There is virtually no area in the theory of moduli of curves where the contribution of David Mumford has not been crucial. Our first debt of gratitude is therefore owed to him.

There is a long list of people to whom we would also like to express our gratitude. The first one is Joe Harris, whose generous contribution consists of approximately half of the exercises in this book.

During the long years of preparation of this volume, the following people have greatly contributed with ideas, comments, remarks, and corrections: Gilberto Bini, Alberto Canonaco, Alessandro Chiodo, Herb Clemens, Eduardo Esteves, Domenico Fiorenza, Claudio Fontanari, Jeffrey Giansiracusa, John Harer, Eduard Looijenga, Marco Manetti, Elena Martinengo, Gabriele Mondello, Riccardo Murri, Filippo Natoli, Giuseppe Pareschi, Gian Pietro Pirola, Marzia Polito, Giulia Saccà, Edoardo Sernesi, Roy Smith, Lidia Stoppino, Angelo Vistoli. To all of them we extend our heartfelt sense of gratitude.

We also wish to thank the students in the courses that we taught out of draft versions of parts of the book, who also offered a number of suggestions for improvements.

The first two authors are also grateful to several institutions which hosted them during the preparation of this volume, in particular the Courant Institute of New York University, Columbia University, the Italian Academy in New York, IMPA in Rio de Janeiro, the Institut Henri Poincaré in Paris, the Accademia dei Lincei in Rome, and above all the Institute for Advanced Study. Special thanks go to Enrico Bombieri, who was instrumental in arranging the first two authors' stays at the Institute. It was through his good offices that they were supported on one of these stays as "Sergio Serapioni, Honorary President, Società Trentina Lieviti – Trento (Italy) Members."

We gratefully acknowledge financial support provided by the PRIN projects "Spazi di moduli e teoria di Lie" funded by the Italian Ministry for Education and Research, and by the EAGER project funded by the European Union.

Rome, Pavia, Princeton, 2010

Contents

Guide for the Reader

The first four chapters of this volume, that is, Chapters IX, X, XI, and XII, are devoted to the construction of the moduli space $\overline{M}_{g,n}$ of stable n-pointed curves of genus g. The three main characters in these chapters are: nodal curves, deformation theory, and Kuranishi families.

Chapter IX gives a self-contained introduction to the Hilbert scheme, explaining the various implications of the concept of flatness and highlighting the case of curves as, for instance, in Mumford's example.

Nodal curves are studied in **Chapter X**. There, we establish the Stable Reduction Theorem (4.11), the theorem on isomorphisms of families of stable curves (5.1), and, in Section 6, the basic constructions of clutching, projection, and stabilization. All these results are fundamental in the construction of the moduli space of stable curves and in the study of its boundary.

The Kodaira–Spencer deformation theory is ubiquitous in this book. Its first appearance is in Section 5 of Chapter IX. It presents itself in its most classical guise as the study of the characteristic system which, in modern terms, translates into the study of the tangent space to the Hilbert scheme. The deformation theory of nodal curves, and in particular of stable ones, is the central theme of **Chapter XI**. There, (5.10) is the key exact sequence describing the tangent space to the local deformation space of a nodal curve. The concept of Kuranishi family is pivotal in the entire volume. The (bases of) Kuranishi families provide the analytic charts for the atlases of moduli stacks of curves. Kuranishi families are constructed by slicing the Hilbert scheme $H_{g,n,\nu}$ of ν-log-canonical embedded stable n-pointed curves of genus g, transversally with respect to the orbits of the natural projective group acting on $H_{g,n,\nu}$, and then restricting to these slices the universal family of curves over $H_{g,n,\nu}$ (see Theorem (6.5) and the key Definitions (6.7) and (6.8) in Chapter XI).

The moduli space $\overline{M}_{g,n}$ is then constructed in **Chapter XII**. We exhibit $\overline{M}_{g,n}$ first as an analytic space, then as an algebraic space, and finally as an orbifold and as a Deligne–Mumford stack. Actually, one of the purposes of this chapter, besides the construction of moduli spaces, and the study of the first properties of their boundary strata, is to give an utilitarian and essentially self-contained introduction to the theory of stacks. This is done in Sections 3–9.

Several topics treated in the first four chapters are not directly aimed at the construction of moduli spaces. Specifically:

- Section 9 of Chapter IX deals with the universal property of the Hilbert scheme with respect to *continuous* families of projective manifolds. Its natural continuation is Section 7 of Chapter XI, where it is shown that the universal property of the Kuranishi family holds also in this context of continuous families of Riemann surfaces. These results will be essential in our presentation of Teichmüller theory in Chapter XV.
- Section 9 of Chapter X is devoted to the Picard–Lefschetz theory of vanishing cycles describing the topological picture of a family of smooth curves degenerating to a nodal one.
- Section 8 of Chapter XI deals with the classical theory of the period map for Riemann surfaces and its infinitesimal behavior.
- In Section 9 of the same chapter we study the positivity properties of the Hodge bundle from the viewpoint of its curvature.
- In the final section of Chapter XI we present Kempf's study of deformations of the symmetric product of a curve leading to the proof of Green's theorem about quadrics passing through the canonical curve (cf. Theorem (4.1) in Chapter VI).

In **Chapter XIII** we present the theory of line bundles on moduli stacks of curves, developing the necessary theory of descent. In the first two sections we introduce the Hodge bundle, the point-bundles $\mathcal{L}_i$, the tangent bundle to the stack $\overline{\mathcal{M}}_{g,n}$, the canonical bundle, the stack divisors corresponding to the codimension one components of its boundary, and the normal bundles to the various boundary strata. The following Section 4 is devoted to the theory of the determinant of the cohomology. This theory is well suited to producing line bundles on moduli stacks, and, at the end of this section, we treat the boundary of moduli as a determinant, leading to important formulae of "restriction to the boundary" as in Lemma (4.22), Proposition (3.10), and formula (4.31). In Section 5 we present the theory of the Deligne pairing, we introduce Mumford's κ_1 class, and we give a concrete version, "without denominators," of the Riemann–Roch theorem for line bundles on families of nodal curves (cf. Theorem (5.31)). In Section 6 we compare the various notions of Picard group for moduli spaces of curves. Section 7 is devoted to Mumford's remarkable idea that the Grothendieck–Riemann–Roch theorem can be effectively used to produce relations among classes in the moduli spaces of curves. There we prove the key formula $\kappa_1 = 12\lambda + \psi - \delta$ for Mumford's class and the formula $K_{\overline{M}_{g,n}} = 13\lambda + \psi - 2\delta - \delta_{1,\emptyset}$ for the canonical class. In the final Section 8 we study the Picard group of the closure $\overline{H}_g \subset \overline{M}_g$ of the hyperelliptic locus.

The fact that $\overline{M}_{g,n}$ is a projective variety (and therefore a scheme) is established in **Chapter XIV**. To prove this we use a mixture of two techniques that are of independent interest. The first one is Mumford's geometric invariant theory. In Sections 2 and 3, we prove the Hilbert–Mumford criterion of stability (Proposition (2.2)), and we use this criterion to prove the stability of the ν-log-canonically embedded *smooth* curves, viewed as points in the appropriate Hilbert scheme. We then take a sharp turn and use stability of smooth curves to find numerical inequalities among cycles in moduli spaces and, consequently, positivity results. Using the same techniques, we then prove the ampleness of Mumford's class κ_1 and hence the projectivity of $\overline{M}_{g,n}$.

Chapter XV gives a self-contained treatment of Teichmüller space and of the modular group. The Teichmüller space $\mathcal{T}_S$ is constructed in Section 2, as a complex manifold, by patching together bases of Kuranishi families. We then examine the natural map $\Phi : B \to \mathcal{T}_S$ from the unit ball, in the space of quadratic differentials on the reference Riemann surface S, to the Teichmüller space $\mathcal{T}_S$. The continuity of this map is an immediate consequence of the results proved in Section 7 of Chapter XI about the universal property of Kuranishi families with respect to continuous families of Riemann surfaces. To prove the injectivity of Φ we first study, in Section 5, the geometry associated to quadratic differentials and then prove, in the following section, Teichmüller's Uniqueness Theorem. As we explain at the end of Section 4, the fact that Φ is a diffeomorphism follows readily from Teichmüller's Uniqueness Theorem and from the elementary theory of the Beltrami equation. In the last section of this chapter we introduce a bordification of Teichmüller space which is very close to the one defined in terms of Fenchel–Nielsen coordinates. Although this bordification is interesting in itself, its only use in our book is in Chapter XIX, where we present Kontsevich's combinatorial expression for the point-bundle classes ψ_i.

Teichmüller space can be thought of as the space representing a rigidification of the moduli functor in which each Riemann surface C comes equipped with a marking (i.e., the homotopy class of a diffeomorphism onto a fixed reference surface). This marking eliminates the automorphism group of C, with the result that Teichmüller space is smooth. The same process of rigidification of the moduli functor can be performed algebraically by considering, for example, pairs consisting of a Riemann surface and the group of points of order n in its Jacobian. More generally, one is looking for finite index normal subgroups Λ of the mapping class group $\Gamma_{g,n}$. Then $\mathcal{T}_{g,n}/\Lambda$ is a Galois cover of $M_{g,n}$ with Galois group $H = \Gamma_{g,n}/\Lambda$. In many instances $\mathcal{T}_{g,n}/\Lambda$ is smooth, so that $M_{g,n}$ can be represented as the quotient of a smooth variety by a finite group H. The main results in this circle of ideas are proved in the first two sections of **Chapter XVI**. When trying to naively push the same ideas to prove analogous results regarding $\overline{M}_{g,n}$, one encounters significant difficulties. These difficulties are addressed in Section 4, and the way to analyze them is to use the Picard–Lefschetz transformation. The problem of expressing $\overline{M}_{g,n}$ as a quotient X/H where X is a smooth variety and H a finite group was solved by Looijenga. In the remaining part of the chapter we present a variation of Looijenga construction due to Abramovich, Corti, and

Vistoli which exhibits X as a fine moduli space, in fact as a moduli space for admissible G-covers, where G is an appropriate finite group, and H is a quotient of the semidirect product $G^n \ltimes \mathrm{Aut}(G)$.

The fact that $\overline{M}_{g,n}$ can be expressed as the quotient X/H, with X a smooth variety and H a finite group, makes it relatively easy to talk about its Chow rings. The theory of cycles in $\overline{M}_{g,n}$ is the central subject of **Chapter XVII**. After presenting, in Section 2, the foundational material on the intersection theory of stacks of the form $[X/H]$ with X smooth and H a finite group, in Section 3 we introduce the tautological classes. These are the Mumford–Morita–Miller classes (i.e., the κ-classes), the point-bundle classes (i.e., the ψ-classes), the Hodge classes (i.e, the λ-classes), and the boundary classes (i.e., the δ-classes). In Section 4 we describe the behavior of these classes under push-forward and pullback via the projection morphism $\pi : \overline{M}_{g,n+1} \to \overline{M}_{g,n}$ and the clutching morphisms $\xi_\Gamma : \overline{M}_\Gamma \to \overline{M}_{g,n}$ from the various boundary strata. In Section 5, following Mumford, we use, on the one hand, Grothendieck's Riemann–Roch theorem to find relations between the Hodge classes and the κ classes, and on the other hand, using the flatness of the Gauss–Manin connection, we exhibit a set of generators for the tautological ring $R^\bullet(M_g)$ (i.e., the ring generated by the tautological classes). At the end of the section we discuss Deligne's canonical extension of the Gauss–Manin connection to the boundary of moduli. Section 6 offers a brief and informal discussion of the tautological ring, presenting two results, a nonvanishing theorem for the tautological class κ_{g-2} due to Faber, and a vanishing theorem for polynomials of degree greater than $g-2$ in the tautological classes, due to Looijenga. Both results are proved in subsequent parts of the book. In the last section we present Keel's result on the Chow ring of $\overline{M}_{0,n}$, and we give a direct computation of $A^1(\overline{M}_{0,n})$.

The fact that $M_{g,n}$ is a rational $K(\Gamma_{g,n}, 1)$ for the mapping class group $\Gamma_{g,n}$ hints to the possibility of studying $M_{g,n}$ from a combinatorial point of view. This is done in **Chapter XVIII**, where we introduce a $\Gamma_{g,n}$-invariant triangulation of the Teichmüller space $\mathcal{T}_{g,n}$. Loosely speaking, the complex structure on a Riemann surface determines (and is determined by) a graph embedded in the Riemann surface itself. This makes it possible to give a $\Gamma_{g,n}$-invariant cellular decomposition of $\mathcal{T}_{g,n}$ where the cells are labelled by these graphs. In the first two sections we introduce the arc system complex and, by duality, the ribbon graphs, which are the basic tools for the combinatorial description of $\mathcal{T}_{g,n}$. To prove that (a subcomplex of) the arc system complex gives a combinatorial model of $\mathcal{T}_{g,n}$, one may choose either the theory of Jenkins–Strebel differentials, or alternatively, via uniformization, the canonical hyperbolic metric on Riemann surfaces. We choose the latter since it more easily enables one to extend the cellular decomposition to the bordification of $\mathcal{T}_{g,n}$. After explaining, in Section 4, how hyperbolic geometry is used to obtain the cellular decomposition of $\mathcal{T}_{g,n}$ and after recalling, in Sections 5 and 6, some basic facts about the uniformization theorem and the Poincaré metric, in Sections 7 and 8 we give the construction of the cellular decomposition of $\mathcal{T}_{g,n}$. In this book the cellular decomposition of moduli spaces is used in two ways. First of all to give a simple and direct proof of the vanishing of the rational homology of $M_{g,n}$ in high degree. These applications are given in Chapter XIX. The second enters when computing the intersection number of tautological classes in Kontsevich's proof of Witten conjecture, which is given in Chapter XX. In fact, this last application requires that the cellular decomposition of $M_{g,n}$ be extended to a suitable compactification of moduli space. This task, which is technically more demanding, is carried out in Sections 9–12.

Chapter XIX discusses the first consequences of the cellular decomposition constructed in Chapter XVIII. We begin by computing the rational cohomology of $\overline{M}_{g,n}$ in degrees one and two. This computation can be performed by elementary methods by virtue of the vanishing of the high homology of $M_{g,n}$ which, in turn, is a direct consequence of the cellular decomposition. This is carried out in Sections 2, 3, and 4. In Section 5, after a very brief discussion of Harer's stability theorem and of the Madsen–Weiss and Tillmann theorems on the stable rational cohomology of $M_{g,n}$, we prove Harer's theorem on the second homology of $M_{g,n}$. This we do by using the knowledge of $H^2(\overline{M}_{g,n};\mathbb{Q})$ and Deligne's spectral sequence for the complement of a divisor with normal crossings. Further uses of the cellular decomposition are presented in Section 7, where we give Kontsevich's combinatorial expression for the point-bundle classes ψ, and in Section 8, where we give Kontsevich's combinatorial expression for an orientation form on $M_{g,n}$.

Chapter XX is almost entirely devoted to Kontsevich's proof of Witten's conjecture on the intersection numbers of the ψ-classes. The proof is self-contained, with the exception of an algebraic result by Itzykson for which there is a very clear and well-written reference. In the first two sections we review Witten's generating series for the intersection numbers of the ψ-classes, introduce the Virasoro operators, and describe their link with the KdV hierarchy. In Section 4 we prove Kontsevich's combinatorial formula expressing Witten's generating series as a sum over ribbon graphs. We then give a self-contained treatment of the Feynman diagram expansion of matrix integrals, and finally, in Section 6, we express Kontsevich's combinatorial sum as a matrix integral and, using this, conclude the proof of Witten's conjecture. As we show in Section 7, the knowledge of the intersection numbers of the ψ-classes can be used to prove the nonvanishing of the class κ_{g-2}. This result by Faber gives the threshold for the non-vanishing of the tautological ring of M_g. In fact, in Section 4 of Chapter XXI we prove a theorem by Looijenga stating that the ring of tautological classes on M_g vanishes in degree strictly larger than $g-2$. After recalling some basic facts about equivariant cohomology, in the last two sections of Chapter XX we present Harer and Zagier's computation of the virtual Euler–Poincaré characteristic of $M_{g,n}$.

The Brill–Noether theory is one the central themes of the first volume of this book. There we study the static aspect of this theory, namely the theory of special linear series on a fixed curve. In our final **Chapter XXI** we study the Brill–Noether theory for smooth curves moving with moduli. In the first few sections, aside for an intermission in which we prove the vanishing theorem of Looijenga we mentioned above, we construct the basic varieties of the Brill–Noether theory for smooth moving curves, and we describe their tangent spaces in terms of the fundamental homomorphisms $\mu_0 : H^0(C; L) \otimes H^0(C; \omega_C L^{-1}) \to H^0(\omega_C)$ and $\mu_1 : \ker \mu_0 \to H^0(\omega_C^2)$, where C is a smooth curve and L a line bundle on it. We also connect these maps to the normal sheaf relative to the morphism $\phi_L : C \to \mathbb{P}^r$, where $r = h^0(C, L) - 1$. In Section 7 we present Lazarsfeld's elegant proof of Petri's conjecture. In the remaining part of the chapter we concentrate mostly on the study of g^1_d's and g^2_d's on smooth curves. We revisit a number of classical results and present some nonclassical ones, related to, among others, the Hurwitz scheme, the Severi variety of plane curves of given degree and genus and the unirationality of M_g for small values of g.

Notational conventions and blanket assumptions

- Unless otherwise stated, all schemes are implicitly assumed to be of finite type over $\mathbb{C}$.
- If V is a vector space or a vector bundle, $\mathbb{P}V$ is the projective space, or projective bundle, of *lines* in V, or in the fibers of V.
- If $\varphi : X \to S$ is a morphism of schemes or of analytic spaces and T is a locally closed subscheme or subspace of S, we write X_T to denote the fiber product $X \times_S T$. Likewise, if s is a point of S, we write X_s to denote the fiber $\varphi^{-1}(s)$.
- We usually write $\operatorname{Sym}^q V$ to indicate the q-th symmetric product of the module or coherent sheaf V. Occasionally, we instead use the notation $S^q V$, especially when V is a vector space.

List of Symbols

$\mathcal{A}_X$	Sheaf of differentiable functions on X	
$\mathcal{A}_X^{p,q}$	Sheaf of smooth (p,q)-forms on the complex manifold X	
$\mathcal{A}^q(E)$	Sheaf of smooth q-forms with values in the vector bundle E	
$\mathcal{A}^q_{X/Y}$	Sheaf of relative smooth q-forms of a smooth fibration $X \to Y$	
$\mathbb{G}(k,n)$	Grassmannian of k-planes in $\mathbb{C}^n$	
$\mathbb{G}(k,V)$	Grassmannian of k-planes in the vector space V	
$\mathcal{I}_X$	Ideal sheaf of subscheme X	
K_X	Class of canonical bundle of X in the Picard group of X	
$\mathbb{P}V$	Projective space of *lines* in the vector space V, or in the fibers of the vector bundle V	
$S^m V$	mth symmetric power of module or coherent sheaf V	
$\operatorname{Sym}^m V$	mth symmetric power of module or coherent sheaf V	
T_X	Tangent bundle to X	
$T_x(X)$	Tangent space to X at x	
$V^\vee$	Dual of module or coherent sheaf V	
$< x_{\nu_1} x_{\nu_2} >$	Propagator	735
$< \tau_{d_1} \cdots \tau_{d_n} >$	Intersection number of point bundle classes	721
$< f >$	Expectation value	734
$< x_{ij} x_{kl} >$	Propagator in the matrix model	741
$\underset{t\to 0}{\asymp}$	Asymptotic expansion	736
$\mathbb{1}_X$	Unit graded line bundle	349
$\chi_{virt}(\Gamma)$	Virtual Euler–Poincaré characteristic of Γ	758
Δ_Γ	Boundary stratum of moduli space of stable curves attached to the graph Γ	312
$\Delta_{\mathcal{P}}$	Locus in moduli space parameterizing curves with at least one separating node of type $\mathcal{P}$	261
$\Delta_{a,A}$	Locus in moduli space parameterizing curves with at least one separating node of type (a,A)	262

Chapter IX.
The Hilbert Scheme

1. Introduction.

In this chapter we introduce the Hilbert scheme which, roughly speaking, parameterizes subschemes of a fixed projective space with a prescribed Hilbert polynomial. Our immediate motivation for discussing Hilbert schemes at this point is to be able to construct the moduli space of stable curves, first as an analytic space and then as an algebraic space and as a stack. The Hilbert scheme will continue to play a major role throughout the rest of the book.

In the first section we describe the basic results on families

$$\begin{array}{c} \mathcal{X} \subset \mathbb{P}^r \times S \\ f\big\downarrow \quad\quad\quad \\ S \quad\quad\quad\quad \end{array} \tag{1.1}$$

of closed subschemes of $\mathbb{P}^r$ parameterized by a scheme S. We give an informal introduction to Hilbert schemes, and we discuss a few elementary examples, including the Grassmannian which is, at the same time, the basic example of a Hilbert scheme and the ambient space where any Hilbert scheme is naturally embedded. The theorems announced in this section are proved in the following two. One of these results anticipates the meaning of flatness, which is the key notion in the deformation theory of schemes: if the family (1.1) is flat, then all of its fibers $X_s = f^{-1}(s)$ have the same Hilbert polynomial. In fact, on a reduced base S, this property can be taken as a characterization of flatness. Alternatively, the flatness of f is equivalent to the requirement that $f_*\mathcal{O}_{\mathcal{X}}(n)$ is locally free for large n. These results can be thought of as *global* consequences of flatness.

In the second section we prove what was announced in the first one. Specifically, using the theory of base change in cohomology, we show that, when the family (1.1) is flat, one gets a rather good control on how the cohomology of the fibers X_s varies, as the point s travels in S. At the end of this section we prove yet another property of flatness, which is an analogue of Sard's theorem, saying that, given a morphism $\varphi : Z \to T$, with T reduced, there is a Zariski dense open subset of T over which φ is flat.

E. Arbarello et al., *Geometry of Algebraic Curves*,
Grundlehren der mathematischen Wissenschaften 268,
DOI 10.1007/978-3-540-69392-5_1,

The third section is devoted to the actual construction of the Hilbert scheme and its universal family and to the realization of the Hilbert scheme as a subscheme of a Grassmannian. Fixing a positive integer r and a rational polynomial $p(t)$, the Hilbert scheme $\mathrm{Hilb}_r^{p(t)}$ represents the functor, from the category of schemes to the category of sets, which associates to a scheme S the set of flat families of projective subschemes of $\mathbb{P}^r$ parameterized by S and with Hilbert polynomial equal to $p(t)$. The closed points of $\mathrm{Hilb}_r^{p(t)}$ are just the projective subschemes of $\mathbb{P}^r$ with Hilbert polynomial equal to $p(t)$.

Apart from the use of flatness, which is ubiquitous in the construction of the Hilbert scheme, another tool is needed, namely the following strengthening of a classical results by Serre on projective varieties. Let r be a nonnegative integer, and let $p(t)$ be a rational polynomial. Then there exists an integer n_0, *depending only on* r *and* $p(t)$, such that, for any subscheme $X \subset \mathbb{P}^r$ with Hilbert polynomial $p(t)$ and for any $n \geq n_0$, the homomorphism

$$H^0(\mathbb{P}^r, \mathcal{O}_{\mathbb{P}^r}(n)) \xrightarrow{\varphi_X^n} H^0(X, \mathcal{O}_X(n))$$

is onto, and the scheme X is completely determined by its kernel. Moreover, the dimension of $\ker \varphi_X^n$ depends only on r and $p(t)$ and is given by

$$q(n) = \dim \ker \varphi_X^n = \binom{n+r}{r} - p(n).$$

Now fix $n \geq n_0$. The assignment $X \mapsto [\ker \varphi_X^n]$ gives then an injective set-theoretical map from $\mathrm{Hilb}_r^{p(t)}$ to the Grassmannian G of $q(n)$-planes in $H^0(\mathbb{P}^r, \mathcal{O}_{\mathbb{P}^r}(n)))$. It is in the Grassmannian G that one looks for equations defining the appropriate scheme structure of $\mathrm{Hilb}_r^{p(t)}$, and the universal property of $\mathrm{Hilb}_r^{p(t)}$ will be a reflection of the universal property of the Grassmannian.

As we will show in Section 6, a Hilbert scheme can be quite nasty. Of course, one of its advantages is that it represents a functor. Another one is that it is projective. We prove this result at the end of Section 4.

In Section 5, in order to compute the tangent space to the Hilbert scheme at a given point, we discuss another aspect of flatness. The basic picture that we analyze is now the *local* picture of a flat family

$$\begin{array}{c} \mathcal{X} \subset Y \times S \\ f \big\downarrow \\ S = \operatorname{Spec} \mathbb{C}[\varepsilon] \end{array} \tag{1.2}$$

of closed affine subschemes of an affine scheme $Y = \operatorname{Spec}(R)$ parameterized by $\operatorname{Spec} \mathbb{C}[\varepsilon]$ or, as one says, the picture of a first-order embedded deformation of the closed subscheme $X = X_{s_0} \subset Y \times \{s_0\}$. Here s_0 is the closed point of $\operatorname{Spec}\mathbb{C}[\varepsilon]$. Set $X = \operatorname{Spec}(R/I)$ and $\mathcal{X} = \operatorname{Spec}(R[\varepsilon]/J)$, where I is an ideal in R, and J is an ideal in $R[\varepsilon]$. It turns out that saying that the family (1.2) is flat is equivalent to saying that every presentation by generators and relations of the ideal I of the central fiber $X = \mathcal{X} \mod \varepsilon$ extends to a presentation of J. From this property it is not hard to see that the first-order embedded deformations of X are classified by $\operatorname{Hom}_{R/I}(I/I^2, R/I)$. All of this easily globalizes to show that the first-order embedded deformations of a subscheme $X \subset Y$ are in a one-to-one correspondence with the sections of the normal sheaf $N_{X/Y} = \mathcal{H}om_{\mathcal{O}_X}(\mathcal{I}/\mathcal{I}^2, \mathcal{O}_X)$, where $\mathcal{I}$ is the ideal sheaf of X in Y. This is the first instance of the Kodaira–Spencer theory, which in fact goes back to the classical notion of the characteristic system. In classical algebraic geometry, given, for example, an algebraic family of curves on a surface Y and a curve C in this family, the first-order approximation of this algebraic family was studied via the linear series cut out on C by the *infinitely near* curves belonging to the algebraic family. This linear series was called the characteristic linear series of the given algebraic family. The divisors belonging to the characteristic series are nothing but zero-divisors of sections of the normal bundle $N_{C/Y}$.

Having acquired the normal bundle description of the tangent spaces to the Hilbert scheme, in Section 6 we present Mumford's example which illustrates how bad a Hilbert scheme can be, even in innocent-looking cases. The case at hand is the one of a smooth connected space curve such that the corresponding point in the appropriate Hilbert scheme H lies in a component along which H is everywhere nonreduced.

Section 7 is rather technical but of central importance. There we study a number of variants of the Hilbert schemes. Let us very briefly present a few of these. Given a scheme S and a closed subscheme X of $\mathbb{P}^r \times S$, the first variant comes from the desire of parameterizing couples (s, Y), where s is a point of S, and Y is a closed subscheme of the fiber of $X \to S$ over s with given Hilbert polynomial. Another useful space that can be constructed is a scheme parameterizing nested pairs of subschemes in the fibers of a family of projective schemes. This construction will be routinely used when dealing with pointed curves. Another important variant is the *scheme* (not of finite type) $\mathbf{Isom}_S(X, Y)$, where $X \subset \mathbb{P}^r \times S$ and $Y \subset \mathbb{P}^t \times S$ are closed subschemes, and both X and Y are flat over S. This is a scheme representing the functor which associates to each scheme T over S the set of all isomorphisms, as schemes over T, from $X \times_S T$ to $Y \times_S T$. This scheme and its properties are essential in studying how the automorphism group of a curve varies as one varies the curve. Finally the Hilbert scheme of isomorphisms is one of the central objects in the theory of Deligne–Mumford stacks.

Section 8 contains tangent space computations for several of the Hilbert space variants introduced in the previous section.

The final section has a partially nonalgebraic flavor. The representability properties of the Hilbert functor can be effectively used also when dealing with *differentiable*, or even *continuous*, families of projective manifolds. From our perspective, this section introduces the necessary tools for proving analogous properties for Kuranishi families, which, in turn, will play an important role in our treatment of Teichmüller theory.

2. The idea of the Hilbert scheme.

We recall our general conventions. Unless otherwise stated, all schemes are implicitly assumed to be of finite type over $\mathbb{C}$.

Fix a projective space $\mathbb{P}^r$. In a very imprecise sense, we would like to parameterize closed subschemes of $\mathbb{P}^r$ with "fixed numerical characters." Naively, this means putting a scheme structure on the set H of all such subschemes which is, in some sense, natural. One way of making the adjective "natural" slightly more precise is to ask that the scheme H be the parameter space for a family $\{X_h\}_{h\in H}$, where X_h is the closed subscheme of $\mathbb{P}^r$ corresponding to h. Our first goal, then, is to formalize the notion of family of closed subschemes of $\mathbb{P}^r$ with fixed numerical characters. Clearly, given an ambient scheme Y, a *family of subschemes of Y parameterized by a scheme S* is nothing but a subscheme $X \subset Y \times S$, viewed as fibered over S via the natural projection

$$Y \times S \supset X \xrightarrow{f} S\,. \tag{2.1}$$

We shall write X_s to denote the fiber $f^{-1}(s)$. Now we fix our attention on the case where $Y = \mathbb{P}^r$ and f is proper, or, which is the same, X is closed in $\mathbb{P}^r \times S$. As most of the numerical invariants of a closed subscheme of $\mathbb{P}^r$, like the dimension, the degree, and so forth, are encoded in its Hilbert polynomial, we may implement the vague requirement that the numerical characters of the fibers X_s be fixed by asking that the Hilbert polynomial

$$h_{X_s}(t) = \sum_i (-1)^i \dim H^i(X_s, \mathcal{O}_{X_s}(t))$$

be independent of s. Now it is a basic result that, at least when the base S is reduced, the local constancy of the Hilbert polynomial is equivalent to the flatness of f.

To explain this, we begin by recalling a few basic definitions. A module M over a commutative ring R is said to be *flat* if tensoring with M is an exact functor. As is well known, each of the following conditions is equivalent to the flatness of M:

- $\operatorname{Tor}_q^R(M,H) = 0$ for any R-module H and any $q > 0$;
- $\operatorname{Tor}_q^R(M,H) = 0$ for any finitely generated R-module H and any $q > 0$;
- $\operatorname{Tor}_1^R(M,H) = 0$ for any R-module H;
- $\operatorname{Tor}_1^R(M,H) = 0$ for any finitely generated R-module H.

A morphism of schemes

$$\varphi : X \to S$$

is *flat* if for any $x \in X$, $\mathcal{O}_{X,x}$ is a flat $\mathcal{O}_{S,\varphi(x)}$-module. More generally, a coherent $\mathcal{O}_X$-module $\mathcal{F}$ is said to be flat over S if $\mathcal{F}_x$ is a flat $\mathcal{O}_{S,\varphi(x)}$-module for every $x \in X$. This is equivalent to saying that, for any pair of affine open subsets $U \subset X$ and $V \subset S$ such that $\varphi(U) \subset V$, $\Gamma(U, \mathcal{F})$ is a flat $\Gamma(V, \mathcal{O}_S)$-module. It is useful to notice that to check that $\mathcal{F}$ is flat over S, it suffices to verify that $\mathcal{F}_x$ is $\mathcal{O}_{S,\varphi(x)}$-flat for all *closed* points $x \in X$. We shall say that a family as in (2.1) is flat if f is.

The notion of flatness carries over, with the same definition, to the context of analytic spaces and coherent analytic sheaves over them. Moreover, the analytic notion of flatness and the algebraic one agree in the following sense. Let $\varphi : X \to S$ be a morphism of schemes, and let $\mathcal{F}$ be a coherent $\mathcal{O}_X$-module. Denote by $\varphi^{an} : X^{an} \to S^{an}$ and $\mathcal{F}^{an}$ the corresponding analytic objects. Then $\mathcal{F}$ is flat over S if and only if $\mathcal{F}^{an}$ is flat over S^{an}. The proof is based on a fundamental result of Serre [626], asserting that, if Z is a scheme of finite type over $\mathbb{C}$ and $z \in Z$ is a closed point, then $\mathcal{O}_{Z^{an},z}$ is flat over $\mathcal{O}_{Z,z}$. The rest of the argument is a simple application of the basic properties of tensor products. One may, for instance, apply Exercise 1 for this chapter, with $A = \mathcal{O}_{S,\varphi(x)}$, $A' = \mathcal{O}_{S^{an},\varphi^{an}(x)}$, $B = \mathcal{O}_{X,x}$, and $B' = \mathcal{O}_{X^{an},x}$. These observations will allow us, when talking about flatness, to go back and forth without risks between the algebraic and the analytic category.

If $\mathcal{F}$ is a coherent sheaf on $\mathbb{P}^r$, its Hilbert polynomial is

$$h_{\mathcal{F}}(n) = \sum_i (-1)^i h^i(\mathbb{P}^r, \mathcal{F}(n)) .$$

The Hilbert polynomial $h_X(t)$ of a projective scheme X is nothing but the Hilbert polynomial of $\mathcal{O}_X$. The link between flatness and Hilbert polynomials is formally expressed by the following result.

PROPOSITION (2.2). *Let*

$$\mathbb{P}^r \times S \supset X \xrightarrow{f} S$$

be a family of closed subschemes of $\mathbb{P}^r$. *Then*

i) *If f is flat, then the Hilbert polynomial $h_{X_s}(t)$ is locally constant as a function of $s \in S$,*
ii) *If S is reduced, then the converse of i) holds. In fact, it suffices to check the local constancy of $h_{X_s}(t)$ as s varies among the closed points of S.*

The proposition captures one of the main geometric aspects of flatness. We shall study some of the technical aspects of flatness in the next section, where the proposition will be proved.

In view of (2.2), it seems reasonable to take flatness as a good formalization of the notion of having "fixed numerical characters" for families of closed subschemes of projective space. Let us then fix a rational polynomial $p(t)$. According to our initial program, it is natural to consider the set H of all closed subschemes of $\mathbb{P}^r$ with Hilbert polynomial equal to $p(t)$ and to try and give the set H a scheme structure. Actually we shall construct a scheme $\mathrm{Hilb}_r^{p(t)}$ whose closed points are in one-to-one correspondence with the points of H and a flat family

$$\mathbb{P}^r \times \mathrm{Hilb}_r^{p(t)} \supset \mathcal{X} \xrightarrow{\pi} \mathrm{Hilb}_r^{p(t)}$$

of subschemes of $\mathbb{P}^r$ such that the identification between closed points of $\mathrm{Hilb}_r^{p(t)}$ and points of H is given by

$$\mathrm{Hilb}_r^{p(t)} \ni h \mapsto X_h = \pi^{-1}(h)\,.$$

Even more will turn out to be true. The family over $\mathrm{Hilb}_r^{p(t)}$ is universal in the sense that any flat family

$$\mathbb{P}^r \times S \supset X \xrightarrow{f} S$$

of closed subschemes of $\mathbb{P}^r$ with Hilbert polynomial $p(t)$ is the pullback

$$\mathcal{X} \times_{\mathrm{Hilb}_r^{p(t)}} S \to S$$

via a unique morphism $\alpha : S \to \mathrm{Hilb}_r^{p(t)}$. In other words, there is a commutative diagram of cartesian squares

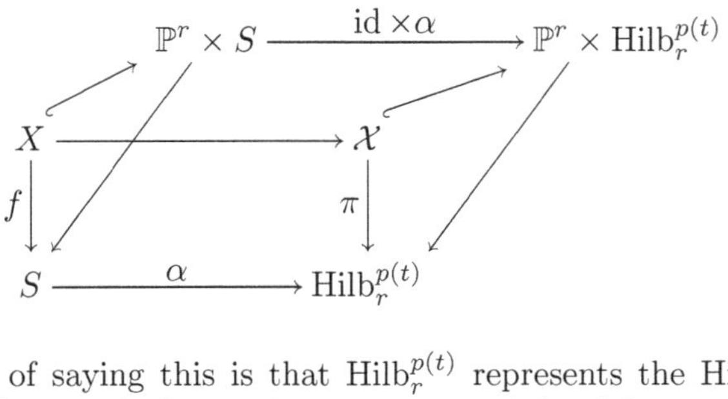

Another way of saying this is that $\mathrm{Hilb}_r^{p(t)}$ represents the Hilbert functor, that is, the functor $\mathbf{h}$ from schemes to sets defined by

$$\mathbf{h}(S) = \{\text{families of closed subschemes of } \mathbb{P}^r \text{ with Hilbert polynomial } p(t) \text{ parameterized by } S\}\,.$$

This means that there is an isomorphism of functors between $\mathbf{h}$ and the functor

$$S \mapsto \mathrm{Hom}(S, \mathrm{Hilb}_r^{p(t)}) .$$

The isomorphism is given as follows:

$$\begin{gathered} \mathrm{Hom}(S, \mathrm{Hilb}_r^{p(t)}) \longrightarrow \mathbf{h}(S) \\ \left(\alpha : S \to \mathrm{Hilb}_r^{p(t)}\right) \mapsto \left(\mathcal{X} \times_{\mathrm{Hilb}_r^{p(t)}} S \to S\right) . \end{gathered}$$

The scheme $\mathrm{Hilb}_r^{p(t)}$ is called the *Hilbert scheme of closed subschemes of* $\mathbb{P}^r$ *with Hilbert polynomial* $p(t)$. If Y is a subscheme of $\mathbb{P}^r$ with Hilbert polynomial $p(t)$, the corresponding point in $\mathrm{Hilb}_r^{p(t)}$ will usually be denoted by $[Y]$.

Let us immediately give a few elementary examples.

EXAMPLE (2.3). Maybe the simplest example is the one of hypersurfaces in $\mathbb{P}^r$ of a fixed degree d. Observe first that, given such a hypersurface X, its Hilbert polynomial is

$$p(n) = \binom{n+r}{r} - \binom{n-d+r}{r},$$

as can be seen by taking cohomology of the exact sequence

$$0 \to \mathcal{O}_{\mathbb{P}^r}(n-d) \to \mathcal{O}_{\mathbb{P}^r}(n) \to \mathcal{O}_X(n) \to 0 .$$

Notice that $p(t)$ does not depend on the particular X we are considering, but only on its degree, and that its leading term is

$$d\frac{t^{r-1}}{(r-1)!} .$$

Moreover, hypersurfaces of degree d are characterized, among all subschemes of $\mathbb{P}^r$, by having $p(t)$ as Hilbert polynomial. In fact, suppose that the Hilbert polynomial of $Y \subset \mathbb{P}^r$ is $p(t)$. Let $Y_1, \dots, Y_h$ be the irreducible components of Y of dimension $r-1$, and let $\mu_1, \dots, \mu_h$ be the multiplicities with which they occur in Y. The hypersurface $X = \sum \mu_i Y_i$ is a subscheme of Y, and the quotient $\mathcal{Q} = \mathcal{I}_X / \mathcal{I}_Y$ is supported on a subscheme of $\mathbb{P}^r$ of dimension at most $r-2$. Since, as is well known, the degree of the Hilbert polynomial of a coherent sheaf equals the dimension of its support, it follows that the Hilbert polynomial of $\mathcal{Q}$ has degree at most $r-2$. On the other hand,

$$p(t) = h_Y(t) = h_X(t) + h_{\mathcal{Q}}(t) ,$$

so it follows that the leading term of $h_X(t)$ is the same as that of $p(t)$, hence that X has degree d and that $h_X(t) = p(t)$. But then $h_{\mathcal{Q}}(t) = 0$, so $\mathcal{Q} = 0$, that is, $X = Y$.

Now set

$$N = \binom{d+r}{r} - 1 .$$

Denote by $x_0, \ldots, x_r$ homogeneous coordinates in $\mathbb{P}^r$ and by a_I, where I runs through all multiindices $(i_0, \ldots, i_r)$ such that $\sum i_k = d$, homogeneous coordinates in $\mathbb{P}^N$. Consider the hypersurface $\mathcal{X}$ in $\mathbb{P}^r \times \mathbb{P}^N$ defined by the equation

$$\sum_I a_I x^I = 0 . \tag{2.4}$$

Then $\mathcal{X}$, together with its projection π onto $\mathbb{P}^N$, is a family of hypersurfaces of degree d in $\mathbb{P}^r$. As the Hilbert polynomial of the fiber $\pi^{-1}(a)$ is independent of the point $a \in \mathbb{P}^N$, by part *ii)* of Proposition (2.2) this family is flat. It is essentially evident that $\mathbb{P}^N$ is the Hilbert scheme of hypersurfaces of degree d in $\mathbb{P}^r$ and that $\pi : \mathcal{X} \to \mathbb{P}^N$ is its universal family. To prove this, let

$$\mathbb{P}^r \times S \supset X \xrightarrow{f} S$$

be a flat family of hypersurfaces of degree d in $\mathbb{P}^r$. We wish to show that this family comes from a morphism $\alpha : S \to \mathbb{P}^N$. We shall rely on an elementary characterization of flatness which we now state and whose proof will be found in the next section.

PROPOSITION (2.5). *Let $\mathcal{F}$ be a coherent sheaf on $\mathbb{P}^r \times S$, and denote by ξ the projection of $\mathbb{P}^r \times S$ onto S. Then $\mathcal{F}$ is flat over S if and only if $\xi_*(\mathcal{F}(n))$ is locally free for any sufficiently large n.*

Now let $\mathcal{I}$ be the ideal sheaf of X in $\mathbb{P}^r \times S$. Consider the exact sequence

$$0 \to \mathcal{I}(n) \to \mathcal{O}_{\mathbb{P}^r \times S}(n) \to \mathcal{O}_X(n) \to 0 .$$

For large n, $R^1\xi_*\mathcal{I}(n)$ vanishes, so we obtain the exact sequence

$$0 \to \xi_*\mathcal{I}(n) \to \xi_*\mathcal{O}_{\mathbb{P}^r \times S}(n) \to \xi_*\mathcal{O}_X(n) \to 0 .$$

Clearly, $\xi_*\mathcal{O}_{\mathbb{P}^r \times S}(n) = H^0(\mathbb{P}^r, \mathcal{O}_{\mathbb{P}^r}(n)) \otimes \mathcal{O}_S$ is free, while, by the proposition, $\xi_*\mathcal{O}_X(n)$ is locally free, so that $\xi_*\mathcal{I}(n)$ is locally free as well. Again by the proposition, $\mathcal{I}$ is flat over S. Now consider $\xi_*(\mathcal{I}(d))$. Since, for any hypersurface Y of degree d in $\mathbb{P}^r$, $H^0(\mathbb{P}^r, \mathcal{I}_Y(d))$ is one-dimensional, the theory of base change (which we will very briefly review at the beginning of Section 3) tells us that $\xi_*\mathcal{I}(d)$ is a line bundle on S. Thus we can find a finite cover $\{U_i\}$ of S and a generator σ_i of $\xi_*\mathcal{I}(d)$

on each U_i. In concrete terms, we may think of σ_i as a homogeneous polynomial of degree d

$$\sum_I b_{i,I} x^I$$

whose coefficents are regular functions on U_i. In addition, for any $s \in U_i$, the equation of $f^{-1}(s)$ as a subscheme of $\mathbb{P}^r$ is precisely

$$\sum_I b_{i,I}(s) x^I = 0.$$

We may then define a map $\alpha_i : U_i \to \mathbb{P}^N$ which, from an intuitive point of view, is given by

$$s \mapsto [\cdots : b_{i,I}(s) : \cdots].$$

More exactly, the morphism α_i corresponds to the homomorphism from the homogeneous coordinate ring of $\mathbb{P}^N$ to the coordinate ring of U_i defined by

$$a_I \mapsto b_{i,I}.$$

The fact that σ_i and σ_j are local generators of the line bundle $\xi_* \mathcal{I}(d)$ implies that there is a unit u on $U_i \cap U_j$ such that $\sigma_j = u\sigma_i$. In other words, $b_{j,I} = u b_{i,I}$ for any multiindex I. This shows that the morphisms α_i patch together to define a global morphism $\alpha : S \to \mathbb{P}^N$, and it is obvious from the definitions that the family $f : X \to S$ is the pullback, via α, of the universal family $\pi : \mathcal{X} \to \mathbb{P}^N$.

The uniqueness of α comes from the fact that to say that there is a commutative diagram of cartesian squares

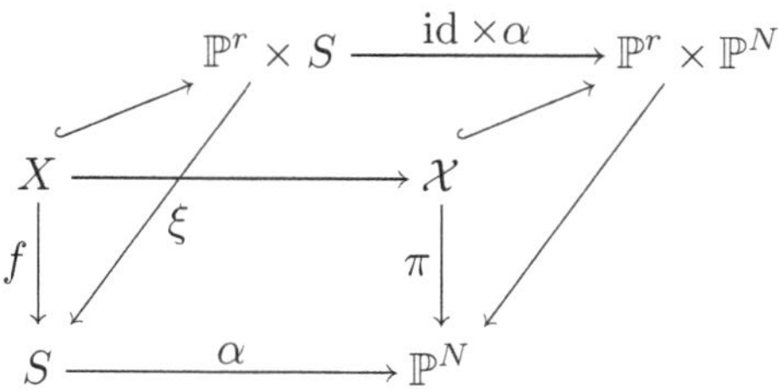

means that X, as a subscheme of $\mathbb{P}^r \times S$, is defined by the pullback, via α, of the universal equation (2.4). Since this equation has degree 1 in the a_I variables and degree d in the x variables, the pullback in question, in intrinsic terms, is nothing but a nowhere vanishing section of

$$\xi_* \mathcal{I}_X(d) \otimes \alpha^* \mathcal{O}_{\mathbb{P}^N}(1).$$

Any local trivialization of $\alpha^* \mathcal{O}_{\mathbb{P}^N}(1)$ gives rise to a local generator of $\xi_* \mathcal{I}_X(d)$. It is evident that applying to this section the construction performed in the existence part, one gets back α. This proves the uniqueness.

Example (2.6). The second elementary example is the Grassmannian $G = \mathbb{G}(k+1, r+1)$ of $(k+1)$-dimensional subspaces of $\mathbb{C}^{r+1}$. Let us show that G is the Hilbert scheme of k-dimensional linear subspaces of $\mathbb{P}^r$; of course, the relevant Hilbert polynomial is $p(t) = \binom{k+t}{k}$. Let $\mathcal{S}$ be the universal subbundle on G; its projectivization $\mathbb{P}\mathcal{S}$ yields a flat family of k-planes in $\mathbb{P}^r$ parameterized by G

$$\mathbb{P}^r \times G \supset \mathbb{P}\mathcal{S} \xrightarrow{\pi} G$$

This turns out to be the universal family on $G = \mathrm{Hilb}_r^{p(t)}$. In fact, given any flat family

$$\mathbb{P}^r \times S \supset X \xrightarrow{f} S$$

of k-planes in $\mathbb{P}^r$, we can attach to it the rank $k+1$ vector bundle on S defined by $f_*\mathcal{O}_X(1)$. We also have an inclusion

$$(2.7) \qquad (f_*\mathcal{O}_X(1))^\vee \hookrightarrow H^0(\mathbb{P}^r, \mathcal{O}_{\mathbb{P}^r}(1))^\vee \otimes \mathcal{O}_S = \mathbb{C}^{r+1} \otimes \mathcal{O}_S .$$

It is immediate to check that the projectivization of this inclusion can be identified with the inclusion of X in $\mathbb{P}^r \times S$. At this point we may use the standard universal property of the Grassmannian to get a unique morphism $\alpha : S \to G$ such that the inclusion (2.7) is the pullback, via α, of the inclusion of the universal subbundle $\mathcal{S}$ in the trivial bundle of rank $r+1$ on G. This proves that G coincides with the Hilbert scheme of k-planes in $\mathbb{P}^r$.

The preceding example can be viewed as a Hilbert-scheme theoretic interpretation of the universal property of the Grassmannian. As we shall see momentarily, this universal property is one of the key ingredients in the construction of general Hilbert schemes.

Example (2.8). For a zero-dimensional subscheme Z of $\mathbb{P}^r$, the Hilbert polynomial is the constant polynomial

$$d = \deg Z = \dim H^0(Z, \mathcal{O}_Z) .$$

Let H be the Hilbert scheme of degree d, zero-dimensional subschemes of $\mathbb{P}^r$. It is easy to give an explicit model for an open subset of H. Consider the d-fold symmetric product of $\mathbb{P}^r$ minus the big diagonal. This is a smooth variety S which is the parameter space of the (flat) family of all zero-dimensional subschemes of $\mathbb{P}^r$ consisting of d distinct points. By universality, S maps to H. In Section 5 we shall see that (as is intuitively evident) S actually embeds in H as an open subset; this will be just a tangent space computation. We shall see in Section 4 that Hilbert schemes are complete, as was clearly the case in the two preceding examples. Thus, in the case at hand, H is a compactification of S. It is worth pointing out that this compactification is not the d-fold symmetric product of $\mathbb{P}^r$, except, of course, for $d = 1$, when S is already complete. Even worse, for $r > 2$ and $d \geq 2$, H is not even irreducible, contrary to what one might intuitively expect (cf. [380])

EXAMPLE (2.9). Even innocent-looking subschemes of projective space may degenerate, inside the Hilbert scheme, to very ugly ones. As a simple example, consider curves of degree two in $\mathbb{P}^3$. The smooth ones come in two kinds, the smooth conics and the unions of two skew lines. These can be told apart by looking at their Hilbert polynomial; the one of a smooth conic is $Q(n) = 2n+1$, the one of a union of skew lines is $P(n) = 2n+2$. We shall look at the Hilbert scheme Hilb_3^P. Let $X_0, \dots, X_3$ be homogeneous coordinates in $\mathbb{P}^3$. For $t \neq 0$, let C_t be the union of the lines $\{X_1 = X_3 = 0\}$ and $\{X_2 = X_3 - tX_0 = 0\}$. The family $\{C_t\}$ extends across $\{t = 0\}$ to a flat family of subschemes of $\mathbb{P}^3$ as follows. Let $\mathcal{C} \subset \mathbb{P}^3 \times \mathbb{C}$ be the subscheme defined by the equations $X_1X_2 = X_1(X_3 - tX_0) = X_2X_3 = X_3(X_3 - tX_0) = 0$. The fiber of $\mathcal{C} \to \mathbb{C}$ over $t \neq 0$ is C_t, while the fiber C_0 over $t = 0$ is the nonreduced subscheme of $\mathbb{P}^3$ defined by the equations $X_1X_2 = X_1X_3 = X_2X_3 = X_3^2 = 0$. Informally, C_0 can be described as the reducible conic Γ defined by the equations $X_1X_2 = X_3 = 0$ plus the embedded component $X_1 = X_2 = X_3^2 = 0$. To see that $\mathcal{C} \to \mathbb{C}$ is flat, it suffices to notice that the Hilbert polynomial of C_0 is $P(n) = 2n+2$. In fact, the kernel of $\mathcal{O}_{C_0} \to \mathcal{O}_\Gamma$ is a one-dimensional complex vector space concentrated at $X_1 = X_2 = X_3 = 0$, while, as we observed, the Hilbert polynomial of Γ is $Q(n) = 2n+1$.

Before closing this section, let us give a quick preview of the construction of the Hilbert scheme.

Fix a rational polynomial $p(t)$. Let X be a subscheme of $\mathbb{P}^r$ with $p(t)$ as Hilbert polynomial. Look at the map

$$H^0(\mathbb{P}^r, \mathcal{O}_{\mathbb{P}^r}(n)) \xrightarrow{\varphi_n} H^0(X, \mathcal{O}_X(n)).$$

By Serre's theorems there is an integer n_0 such that, for all $n \geq n_0$, the following properties hold:

i) φ_n is surjective;
ii) $h^0(X, \mathcal{O}_X(n)) = p(n)$;
iii) X is completely determined by the kernel of φ_n, in the sense that the degree n hypersurfaces passing through X generate $H^0(\mathbb{P}^r, \mathcal{I}_X(m))$ for any $m \geq n$.

As we shall see in a crucial lemma, one can find an n_0 that *depends only on r and $p(t)$ but not on the particular X under consideration.* Set

$$q(n) = h^0(\mathbb{P}^r, \mathcal{O}_{\mathbb{P}^r}(n)) - p(n) = \binom{n+r}{r} - p(n).$$

For any fixed $n \geq n_0$, we may then associate to X the point

$$H^0(\mathbb{P}^r, \mathcal{I}_X(n)) = \ker \varphi_n \in G = \mathbb{G}(q(n), H^0(\mathbb{P}^r, \mathcal{O}_{\mathbb{P}^r}(n))).$$

If H is the set of all subschemes of $\mathbb{P}^r$ with Hilbert polynomial equal to $p(t)$, the above construction gives an injection

$$\begin{gathered} H \hookrightarrow G \\ X \mapsto [H^0(\mathbb{P}^r, \mathcal{I}_X(n)) \subset H^0(\mathbb{P}^r, \mathcal{O}_{\mathbb{P}^r}(n))] \, . \end{gathered}$$

Suppose that

$$\mathbb{P}^r \times S \supset Y \xrightarrow{f} S$$

is a flat family of subschemes of $\mathbb{P}^r$ with Hilbert polynomial $p(t)$. Then S is equipped with the vector subbundle

$$f_*(\mathcal{I}(n)) \subset H^0(\mathbb{P}^r, \mathcal{O}_{\mathbb{P}^r}(n)) \otimes \mathcal{O}_S \, ,$$

where $\mathcal{I}$ is the ideal sheaf of Y in $\mathbb{P}^r \times S$. Of course, we have

$$f_*(\mathcal{I}(n)) \otimes k(s) = H^0(Y_s, \mathcal{I}_{f^{-1}(s)}(n))$$

for any $s \in S$. Then the universal property of the Grassmannian provides a morphism

$$\alpha : S \to G$$

which, set theoretically, lands in H. Now that a certain measure of the required universal property is built into H, the remaining technical challenge will be to equip H with a scheme structure such that two conditions are satisfied. First of all, any α as above must land into H as a morphism of schemes. As for the second condition, look at the universal subbundle on G. It generates a sheaf of ideals on $\mathbb{P}^r \times G$ and hence determines a family of closed subschemes

$$\mathbb{P}^r \times G \supset \mathcal{Y} \xrightarrow{\eta} G \, .$$

The fibers of η are all possible subschemes of $\mathbb{P}^r$ defined by $q(n)$ linearly independent equations of degree n. What we will have to show is that the restriction $\mathcal{X} \xrightarrow{\pi} H$ of this family to H is flat. At this point H will be the Hilbert scheme $\mathrm{Hilb}_r^{p(t)}$ and $\mathcal{X} \xrightarrow{\pi} H$ its universal family.

3. Flatness.

In this section we have collected a few technical results about flatness that either have been announced in the previous section or will be needed in the construction of the Hilbert scheme.

One of the most important applications of flatness is to the theory of base change in cohomology (cf., for instance, Mumford's book [552], Section 5). Since this will be the main tool in this section, we shall briefly digress on it. A typical setup is the following. We have a scheme

S, a coherent sheaf $\mathcal{F}$ on $\mathbb{P}^r \times S$, and a morphism of schemes $f : T \to S$. Look at the diagram

$$\begin{array}{ccc} \mathbb{P}^r \times T & \xrightarrow{\;g\;} & \mathbb{P}^r \times S \\ \eta \big\downarrow & & \big\downarrow \xi \\ T & \xrightarrow{\;f\;} & S \end{array} \tag{3.1}$$

where ξ and η are the projections, and $g = \mathrm{id} \times f$. Given a point $s \in S$, we write $\mathcal{F}_s$ to denote $\mathcal{F} \otimes \mathcal{O}_{\xi^{-1}(s)}$, viewed as a sheaf on $\mathbb{P}^r = \xi^{-1}(s)$. The problem of base change is to compare $R^q\eta_* g^*\mathcal{F}$ with $f^* R^q \xi_* \mathcal{F}$ for any $q \geq 0$. In general, these two sheaves are far from being equal. When $\mathcal{F}$ is flat over S, we have better control on the situation. The theory of base change asserts that, under this assumption, any point of S has an open neighborhood $U \subset S$ over which there is a bounded complex

$$K^0 \to K^1 \to K^2 \to \cdots \tag{3.2}$$

of locally free coherent sheaves which calculates the direct images of $\mathcal{F}$, functorially under base change. Thus, for *any* base change $f : T \to S$ such that $f(T) \subset U$, the cohomology of the complex

$$f^*K^0 \to f^*K^1 \to f^*K^2 \to \cdots$$

is $R^\bullet \eta_* g^* \mathcal{F}$. As a corollary, one gets the following well-known result.

PROPOSITION (3.3). *Let S be a scheme, and $\mathcal{F}$ a coherent sheaf on $\mathbb{P}^r \times S$, flat over S. Let $\xi : \mathbb{P}^r \times S \to S$ be the projection. Then the following statements are equivalent:*

i) *$R^q \xi_* \mathcal{F} = 0$ for any $q > 0$,*
ii) *$R^q \eta_* g^* \mathcal{F} = 0$ for any base change (3.1) and any $q > 0$,*
iii) *$H^q(\mathbb{P}^r, \mathcal{F}_s) = 0$ for any $s \in S$ and any $q > 0$,*
iv) *$H^q(\mathbb{P}^r, \mathcal{F}_s) = 0$ for any closed point s of S and any $q > 0$.*

Moreover, if one of the above holds, $\xi_*\mathcal{F}$ is locally free, and the natural homomorphisms

$$f^*\xi_*\mathcal{F} \to \eta_* g^* \mathcal{F},$$
$$\xi_*\mathcal{F} \otimes k(s) \to H^0(\mathbb{P}^r, \mathcal{F}_s)$$

are isomorphisms for any base change (3.1) and any $s \in S$.

Proof. The statement is local on S, so we may as well assume that a complex (3.2) exists on all of S. Condition iii) is a special case of ii), and iv) a special case of iii). Likewise, that $\xi_*\mathcal{F} \otimes k(s) = H^0(\mathbb{P}^r, \mathcal{F}_s)$ for any $s \in S$ is a special case of the statement that $f^*\xi_*\mathcal{F} = \eta_* g^*\mathcal{F}$ for any base change (3.1).

Let us show that i) implies ii). Condition i) means that $K^\bullet$ is a locally free *resolution* of $\xi_*\mathcal{F}$. Therefore, $\xi_*\mathcal{F}$ is locally free as well. Put otherwise,

$$0 \to \xi_*\mathcal{F} \to K^0 \to K^1 \to \cdots$$

is an exact sequence of locally free, and hence flat, $\mathcal{O}_S$-modules. This implies that the pulled-back sequence

$$0 \to f^*\xi_*\mathcal{F} \to f^*K^0 \to f^*K^1 \to \cdots$$

is also exact. Thus $R^q\eta_*g^*\mathcal{F} = 0$ for $q > 0$, and $\eta_*g^*\mathcal{F} = f^*\xi_*\mathcal{F}$. It remains to show that iv) implies i). It follows from iv) that the complex

$$K^0 \otimes k(s) \to K^1 \otimes k(s) \to K^2 \otimes k(s) \to \cdots$$

is exact for any closed s, and what must be shown is that (3.2) is exact as well. More generally, we shall show, by induction on n, that any complex

$$L^0 \to L^1 \to \cdots \to L^n \to 0 \to \cdots$$

of locally free coherent sheaves on U is exact if $L^\bullet \otimes k(s)$ is exact for any s. This is obviously true when $n = 0$. For $n > 0$, the assumption says in particular that $L^{n-1} \otimes k(s) \to L^n \otimes k(s)$ is onto for any s, and Nakayama's lemma then implies that $L^{n-1} \to L^n$ is also onto. It follows that the kernel of this homomorphism, which we denote by Z^{n-1}, is locally free and that $Z^{n-1} \otimes k(s)$ equals the kernel of $L^{n-1} \otimes k(s) \to L^n \otimes k(s)$. In particular, tensoring

$$L^0 \to \cdots \to Z^{n-1} \to 0 \to \cdots \tag{3.4}$$

with $k(s)$ yields an exact complex for any closed point s. By induction hypothesis, this implies that (3.4) is exact, concluding the proof.

We now begin to prove the results we have announced in the previous section, starting with Proposition (2.5). We briefly recall its statement. We are given a coherent sheaf $\mathcal{F}$ on $\mathbb{P}^r \times S$, where S is a scheme, and denote by ξ the projection of $\mathbb{P}^r \times S$ onto S. The proposition says that $\mathcal{F}$ is flat over S if and only if $\xi_*(\mathcal{F}(n))$ is locally free for any sufficiently large n.

Suppose first that $\mathcal{F}$ is flat over S. If n is large enough, the higher direct images of $\mathcal{F}(n)$ vanish, and hence (3.3) tells us that $\xi_*(\mathcal{F}(n))$ is locally free. Conversely, assume that $\xi_*(\mathcal{F}(n))$ is locally free for all large n. To show that $\mathcal{F}$ is flat over S, we must show that, for any injection $\mathcal{G}_1 \hookrightarrow \mathcal{G}_2$ of coherent $\mathcal{O}_S$-modules, $\xi^*\mathcal{G}_1 \otimes \mathcal{F}$ injects into $\xi^*\mathcal{G}_2 \otimes \mathcal{F}$. Since the question is local on S, we may assume that S is affine; let A be its coordinate ring. Let $R = \oplus_{n\geq 0}R_n$ be the homogeneous coordinate ring of $\mathbb{P}^r \times S$, so that $R_0 = A$. We may write $\mathcal{G}_i = \widetilde{G}_i$, $i = 1, 2$, and $\mathcal{F} = \widetilde{F}$, where the G_i are finitely generated A-modules, and $F = \oplus_{n \geq n_0} F_n$ is a finitely generated graded R-module. To say that $\xi_*(\mathcal{F}(n))$ is locally free for large n is the same as saying that F_n is a projective A-module for large n. Since changing a finite number of summands of F does not change $\mathcal{F}$, we may then assume that F_n is projective, and hence in particular flat over A, for all $n \geq n_0$. That $\xi^*\mathcal{G}_1 \otimes \mathcal{F}$ injects into $\xi^*\mathcal{G}_2 \otimes \mathcal{F}$ then follows from the remark that

$$\xi^*\mathcal{G}_i \otimes \mathcal{F} = (\widetilde{G_i \otimes_A R}) \otimes_R F = \widetilde{G_i \otimes_A F}$$

and from the flatness of F over A. Proposition (2.5) is now fully proved.

We next turn to Proposition (2.2). It pays to prove the following, slightly more general, result.

PROPOSITION (3.5). *Let $\mathcal{F}$ be a coherent sheaf on $\mathbb{P}^r \times S$, where S is a scheme. Denote by ξ the projection of $\mathbb{P}^r \times S$ onto S. Then*

i) *if $\mathcal{F}$ is flat over S, then the Hilbert polynomial $h_{\mathcal{F}_s}(t)$ is locally constant as a function of $s \in S$;*
ii) *if S is reduced, then the converse of i) holds. In fact, it suffices to check the local constancy of $h_{\mathcal{F}_s}(t)$ as s varies among the closed points of S.*

Notice that ii) does not necessarily hold when S is not reduced. The simplest example is provided by the case $S = \operatorname{Spec}\mathbb{C}[\varepsilon]$, $r = 0$, and $\mathcal{F} = \mathbb{C}[\varepsilon]/(\varepsilon) \simeq \mathbb{C}$. That $\mathcal{F}$ is not flat can be seen by noticing that tensoring the injection of $\mathbb{C}[\varepsilon]$-modules $(\varepsilon) \hookrightarrow \mathbb{C}[\varepsilon]$ with $\mathcal{F} \simeq \mathbb{C}$ yields the zero homomorphism $\mathbb{C} \to \mathbb{C}$.

Proposition (3.5) is an almost immediate consequence of the following lemma.

LEMMA (3.6). *Let $\mathcal{F}$ be a coherent sheaf on $\mathbb{P}^r \times S$, and let ξ be the projection from $\mathbb{P}^r \times S$ to S. Then there is an integer n_0 such that, for any point $s \in S$ and any $n \geq n_0$,*

i) *the natural map $\xi_*\mathcal{F}(n) \otimes k(s) \to H^0(\mathbb{P}^r, \mathcal{F}_s(n))$ is an isomorphism,*
ii) *$H^q(\mathbb{P}^r, \mathcal{F}_s(n)) = 0$ for any $q > 0$.*

Assuming for the moment that the lemma has been proved, we deduce Proposition (3.5) from it. Lemma (3.6) implies that, for large n, $h_{\mathcal{F}_s}(n)$ equals the dimension of $\xi_*\mathcal{F}(n) \otimes k(s)$. If $\mathcal{F}$ is flat, Proposition (2.5) implies that this dimension is locally constant. Conversely, if S is reduced, it is well known that a coherent sheaf $\mathcal{G}$ such that the dimension of $\mathcal{G} \otimes k(s)$ is locally constant as s varies among the closed points of S, is locally free; this is the content of Lemma (3.7) below. Thus the local constancy of the dimension of $\xi_*\mathcal{F}(n) \otimes k(s)$ implies that $\xi_*\mathcal{F}(n)$ is locally free. That $\mathcal{F}$ is flat then follows from Proposition (2.5).

Recall that a commutative ring is said to be a *Jacobson ring* if every prime ideal is the intersection of maximal ones. As is well known, the general form of the Nullstellensatz asserts that a finitely generated algebra over a Jacobson ring is also Jacobson (cf., for instance, [194], Theorem 4.19), so that in particular finitely generated algebras over a field are Jacobson rings.

LEMMA (3.7). *Let A be a reduced commutative ring, and let M be a finitely generated A-module. For any prime ideal P in A, let $k(P)$ be the quotient of A_P modulo its maximal ideal. The following are equivalent:*

i) *M is projective;*
ii) *the dimension of $M \otimes k(P)$ over $k(P)$ is locally constant on* $\mathrm{Spec}(A)$*;*

When A is a Jacobson ring, i) *and* ii) *are equivalent to:*

iii) *the dimension of $M \otimes k(P)$ over $k(P)$ is locally constant on the maximal spectrum of A.*

That i) implies ii) and iii) is clear. Conversely, suppose ii) holds. Let Q be a prime ideal. We must show that M_f is a free A_f-module for a suitable $f \in A \smallsetminus Q$. Let $m_1, \dots, m_n$ be elements of M whose classes in $M \otimes k(Q)$ form a basis over $k(Q)$. Nakayama's lemma then implies that the classes of $m_1, \dots, m_n$ generate M_Q. Replacing A with A_f for a suitable $f \in A$, we may in fact assume that the classes of $m_1, \dots, m_n$ generate M_P for all $P \in \mathrm{Spec}(A)$ and hence that the homomorphism $A^n \to M$ sending the elements of the canonical basis to $m_1, \dots, m_n$ is onto; we may also assume that the dimension of $M \otimes k(P)$ is constant on $\mathrm{Spec}(A)$. We claim that M is a free A-module. To see this, let K be the kernel of $A^n \to M$, and tensor the exact sequence $0 \to K \to A^n \to M \to 0$ with $k(P)$ to get the exact sequence

$$K \otimes k(P) \to k(P)^n \to M \otimes k(P) \to 0.$$

The middle homomorphism is an isomorphism by assumption. This means that $K \subset PA^n = P^{\oplus n}$. Since this is true for all prime ideals P, it follows that $K \subset N^{\oplus n}$, where N is the nilradical of A. Since A is assumed to be reduced, $K = \{0\}$, and our claim is proved.

To prove that iii) also implies i) when A is a Jacobson ring, it suffices to observe that the argument above shows that K is contained in $I^{\oplus n}$, where I stands for the intersection of the maximal ideals of A, and that I equals the nilradical since A is Jacobson. Notice also that replacing A with A_f is legitimate since $A_f = A[1/f]$ is also a Jacobson ring. This concludes the proof of (3.7).

It remains to prove Lemma (3.6). The crux of the matter is to find an n_0 that works *uniformly* for *all* points $s \in S$. In fact, if the uniformity requirement were dropped, (3.6) would be a special case of the following variant of Serre's correspondence for coherent sheaves on projective space.

LEMMA (3.8). *Let $\mathcal{F}$ be a coherent sheaf on $\mathbb{P}^r \times S$, and let ξ be the projection of $\mathbb{P}^r \times S$ onto S. Let $f : T \to S$ be any morphism of schemes, and set $g = \mathrm{id} \times f : \mathbb{P}^r \times T \to \mathbb{P}^r \times S$. Then there is an n_0 such that, for any $n \geq n_0$,*

i) *the natural map $f^*\xi_*\mathcal{F}(n) \to \eta_*g^*\mathcal{F}(n)$ is an isomorphism,*
ii) *$R^q\eta_*g^*\mathcal{F}(n) = 0$ for any $q > 0$.*

Let us quickly prove this lemma. Part ii) is just Serre's vanishing theorem. Statement i) is local on the base, so we may assume that $S = \operatorname{Spec} A$, $T = \operatorname{Spec} B$. We set $R_A = A[x_0, \ldots, x_r]$, where $x_0, \ldots, x_r$ are indeterminates, and similarly for R_B. Notice that $R_B = R_A \otimes_A B$. Serre's theorem tells us that there is a graded R_A-module $F = \oplus F_h$ such that $\mathcal{F} = \widetilde{F}$, and that moreover

$$H^0(\mathbb{P}^r \times S, \mathcal{F}(h)) = F_h$$

for large h. The sheaf version of this statement is that

$$\xi_*(\mathcal{F}(h)) = \widetilde{F}_h\,,$$

always for large h. The pullback $g^*\mathcal{F}$ is nothing but $\widetilde{F \otimes_{R_A} R_B}$. On the other hand,

$$F \otimes_{R_A} R_B = F \otimes_{R_A} (R_A \otimes_A B) = F \otimes_A B\,.$$

In particular, for large h,

$$\eta_*g^*\mathcal{F}(h) = \widetilde{F_h \otimes_A B}\,.$$

But, again for large h, one has that

$$f^*\xi_*\mathcal{F}(h) = f^*\widetilde{F}_h = \widetilde{F_h \otimes_A B}\,,$$

proving the lemma.

Returning to the problem of finding a uniform n_0 in Lemma (3.8), that this is possible would be an immediate consequence of (3.3) if the sheaf $\mathcal{F}$ were flat over S. It would in fact suffice to take as n_0 an integer m such that $R^q\xi_*\mathcal{F}(n) = 0$ for any $n \geq m$ and $q > 0$.

Unfortunately, we cannot afford to limit ourselves to the flat case. To meet the uniformity requirement on n_0, we will have to use flatness in a more indirect way, combining it with (3.8). The key ingredient to do so is the following result, which is, in some sense, reminiscent of Sard's lemma from differential geometry.

Proposition (3.9) (Sard's lemma for flatness). *Let $\alpha : X \to Y$ be a morphism of schemes, and let $\mathcal{G}$ be a coherent sheaf on X. Assume that Y is reduced. Then there exists a Zariski dense open subset $U \subset Y$ such that $\mathcal{G}_{|\alpha^{-1}(U)}$ is flat over U.*

In proving (3.9) we may as well assume that Y is irreducible. Moreover we may assume that $X = \operatorname{Spec} B$ and $Y = \operatorname{Spec} A$. Thus $\mathcal{G}$ corresponds to a finitely generated B-module M. We will be done if we can show that there is $a \in A$ such that the localization M_a is a free A_a-module. Notice that, if M sits in an exact sequence

$$0 \to L \to M \to N \to 0,$$

and there are $a, a' \in A$ such that L_a is free over A_a and $N_{a'}$ is free over $A_{a'}$, then $M_{aa'}$ is free over $A_{aa'}$. Thus it suffices to deal with the quotients M_i/M_{i-1} of a composition series

$$0 = M_0 \subset M_1 \subset \cdots \subset M_n = M.$$

Since we may find a composition series whose successive quotients are isomorphic to modules of the form B/P, where P is a prime ideal, we are reduced to the case where B is a domain and $M = B$.

Let K be the quotient field of A; then $B_K = B \otimes_A K$ is a K-algebra of finite type. We argue by induction on the dimension d of $\operatorname{Spec}(B_K)$. If $d = -1$, that is, if $B_K = \{0\}$, then there is a nonzero element a in the kernel of $A \to B$. Thus $B_a = \{0\}$ is free over A_a. If $d \geq 0$, by Noether's normalization lemma we may find algebraically independent elements $b_1, \dots, b_d$ of B_K such that B_K is integral over $K[b_1, \dots, b_d]$. Thus there exists $c \in A$ such that $b_1, \dots, b_d \in B_c$ and B_c is integral over $A_c[b_1, \dots, b_d]$. It follows, in particular, that B_c is a finitely generated $A_c[b_1, \dots, b_d]$-module, so there is an exact sequence of $A_c[b_1, \dots, b_d]$-modules

$$0 \to A_c[b_1, \dots, b_d]^m \to B_c \to C \to 0$$

such that C is a torsion module. Since $A_c[b_1, \dots, b_d]^m$ is free over A_c, it suffices to show that there is $a \in A$ such that C_a is free over A_{ca}. As before, using a composition series for C, we may reduce to the case where C is an integral A_c-algebra. But in this case, since C is a torsion $A_c[b_1, \dots, b_d]$-module, the dimension of $\operatorname{Spec}(C_K)$ is strictly less than d, and we are done by induction.

We are now in a position to prove lemma (3.6). We first construct a sequence of subschemes of S. The first one is $S_0 = S_{red}$. By Sard's lemma for flatness the pullback of $\mathcal{F}$ to $\mathbb{P}^r \times S_0$ is flat over an open subset of S_0, whose complement we call S_1. Now we restrict $\mathcal{F}$ to $\mathbb{P}^r \times S_1$

and proceed in the same way to obtain S_2, and so on. The process terminates in a finite number of steps by noetherianity. We thus get a finite sequence

$$S_0 \supset S_1 \supset \cdots \supset S_N \supset S_{N+1} = \emptyset$$

of reduced closed subschemes of S with the property that, setting $T_i = S_i \smallsetminus S_{i+1}$, the pullback of $\mathcal{F}$ to $\mathbb{P}^r \times T_i$ is flat over T_i.

Let s be a point of S. Clearly, s belongs to a unique T_i. Let us apply Lemma (3.8) when $T = T_i$ and f is the inclusion of T_i in S. We get

$$f^*\xi_*\mathcal{F}(n) = \eta_* g^*\mathcal{F}(n)$$

for $n \geq \nu_i$. On the other hand,

$$\begin{aligned} (g^*\mathcal{F})_s &= \mathcal{F}_s\,, \\ \xi_*\mathcal{F}(n) \otimes k(s) &= f^*\xi_*\mathcal{F}(n) \otimes k(s)\,. \end{aligned}$$

Finally, since $g^*\mathcal{F}$ is flat over T_i, by the remark immediately preceding the statement of Sard's lemma for flatness there is an integer μ_i, not depending on $s \in T_i$, such that

$$\begin{aligned} H^0(\mathbb{P}^r, (g^*\mathcal{F})_s(n)) &= \eta_* g^*\mathcal{F}(n) \otimes k(s)\,, \\ H^q(\mathbb{P}^r, (g^*\mathcal{F})_s(n)) &= 0\,, \qquad q > 0\,, \end{aligned}$$

for $n \geq \mu_i$. Putting everything together, we conclude that

$$\begin{aligned} H^0(\mathbb{P}^r, \mathcal{F}_s(n)) &= \xi_*\mathcal{F}(n) \otimes k(s)\,, \\ H^q(\mathbb{P}^r, \mathcal{F}_s(n)) &= 0\,, \qquad q > 0\,, \end{aligned}$$

whenever $n \geq \nu_i$, $n \geq \mu_i$. Thus it suffices to take as n_0 the maximum among all the ν_i and μ_i. This concludes the proof of Lemma (3.6).

4. Construction of the Hilbert scheme.

Fix a projective space $\mathbb{P}^r$ and a rational polynomial $p(t)$. We are ready to construct the Hilbert scheme $\mathrm{Hilb}_r^{p(t)}$ parameterizing subschemes of $\mathbb{P}^r$ with Hilbert polynomial $p(t)$, following the strategy outlined at the end of Section 2. According to it, we wish to realize $\mathrm{Hilb}_r^{p(t)}$ as a subscheme of the Grassmannian $G = \mathbb{G}(q(n), H^0(\mathbb{P}^r, \mathcal{O}(n)))$, for large n, where

$$q(n) = h^0(\mathbb{P}^r, \mathcal{O}(n)) - p(n) = \binom{n+r}{r} - p(n)\,.$$

More exactly, given a subscheme $X \subset \mathbb{P}^r$ with Hilbert polynomial $p(t)$, the idea is to associate to X the point of G corresponding to the vector subspace $H^0(\mathbb{P}^r, \mathcal{I}_X(n)) \subset H^0(\mathbb{P}^r, \mathcal{O}(n))$, where n is so large that $h^0(\mathbb{P}^r, \mathcal{I}_X(n)) = q(n)$ and $H^0(\mathbb{P}^r, \mathcal{I}_X(n))$ generates the homogeneous ideal of X in degree greater than n. The first problem is to find an n that does not depend on X, but only on r and $p(t)$. That this is possible is a consequence of the following result.

LEMMA (4.1). *Let r be a nonnegative integer, and let $q(t)$ be a rational polynomial. Then there exists an integer n_0 such that, for any ideal sheaf $\mathcal{I} \subset \mathcal{O}_{\mathbb{P}^r}$ with Hilbert polynomial $q(t)$ and for any $n \geq n_0$,*

i) *$H^i(\mathbb{P}^r, \mathcal{I}(n)) = 0$ for every $i \geq 1$,*
ii) *the natural map $H^0(\mathbb{P}^r, \mathcal{I}(n)) \otimes H^0(\mathbb{P}^r, \mathcal{O}(1)) \longrightarrow H^0(\mathbb{P}^r, \mathcal{I}(n+1))$ is onto.*

It is well known that an integer n_0, possibly depending on $\mathcal{I}$, such that *i)* and *ii)* hold for $n \geq n_0$, exists. We shall see that one of the standard proofs of this fact actually yields (4.1), provided that a little more attention is paid to details. We argue by induction on r. For $r = 0$, there is nothing to prove. If $r > 0$, denote by X the projective scheme defined by $\mathcal{I}$, choose a hyperplane H not containing any of the components of X, including the embedded ones, and set $\mathcal{J} = \mathcal{I} \otimes \mathcal{O}_H$. Notice that $\mathcal{J}$ injects into $\mathcal{O}_H$ and hence can be viewed as a sheaf of ideals in $\mathcal{O}_H$; in fact, tensoring

$$(4.2) \qquad 0 \longrightarrow \mathcal{O}_{\mathbb{P}^r}(-1) \longrightarrow \mathcal{O}_{\mathbb{P}^r} \longrightarrow \mathcal{O}_H \longrightarrow 0$$

with $\mathcal{O}_X$, one gets the exact sequence

$$0 = \mathcal{T}or_1(\mathcal{O}_{\mathbb{P}^r}, \mathcal{O}_X) \longrightarrow \mathcal{T}or_1(\mathcal{O}_H, \mathcal{O}_X) \longrightarrow \mathcal{O}_X(-1) \xrightarrow{\alpha} \mathcal{O}_X \longrightarrow \cdots$$

By the choice of H, the homomorphism α is injective, so $\mathcal{T}or_1(\mathcal{O}_H, \mathcal{O}_X)$ vanishes. Tensoring

$$0 \longrightarrow \mathcal{I} \longrightarrow \mathcal{O}_{\mathbb{P}^r} \longrightarrow \mathcal{O}_X \longrightarrow 0$$

with $\mathcal{O}_H$, we find that $\mathcal{J} = \mathcal{I} \otimes \mathcal{O}_H$ injects in $\mathcal{O}_H$.

Tensoring (4.2) with $\mathcal{I}(m+1)$, one gets the exact sequence

$$(4.3) \qquad 0 \longrightarrow \mathcal{I}(m) \longrightarrow \mathcal{I}(m+1) \longrightarrow \mathcal{J}(m+1) \longrightarrow 0\,.$$

Thus the Hilbert polynomial of $\mathcal{J}$ satisfies the identity

$$h_{\mathcal{J}}(t) = h_{\mathcal{I}}(t) - h_{\mathcal{I}}(t-1) = q(t) - q(t-1)$$

and therefore depends only on $q(t)$ and not on $\mathcal{I}$ and H. By induction there exists n_1 such that *i)* and *ii)* are satisfied for $\mathcal{J}$ whenever $n \geq n_1$. In particular, it follows from (4.3) that, for such an n, the vector space $H^i(\mathbb{P}^r, \mathcal{I}(n))$ is isomorphic to $H^i(\mathbb{P}^r, \mathcal{I}(n+1))$ whenever $i \geq 2$; as $H^i(\mathbb{P}^r, \mathcal{I}(m))$ vanishes for very large m, it follows that $H^i(\mathbb{P}^r, \mathcal{I}(n))$ vanishes for $i \geq 2$. It remains to deal with $H^1(\mathbb{P}^r, \mathcal{I}(n))$. For $n \geq n_1$, there is an exact sequence

$$H^0(\mathbb{P}^r, \mathcal{I}(n+1)) \xrightarrow{\alpha_n} H^0(\mathbb{P}^r, \mathcal{J}(n+1)) \to H^1(\mathbb{P}^r, \mathcal{I}(n)) \to H^1(\mathbb{P}^r, \mathcal{I}(n+1)) \to 0\,.$$

Thus, either α_n is surjective, or $h^1(\mathbb{P}^r, \mathcal{I}(n+1))$ is strictly less than $h^1(\mathbb{P}^r, \mathcal{I}(n))$. Observe that if α_n is surjective, the same is true for α_{n+1}. In fact, the map

$$H^0(\mathbb{P}^r, \mathcal{I}(n+1)) \otimes H^0(\mathbb{P}^r, \mathcal{O}(1)) \longrightarrow H^0(H, \mathcal{J}(n+1)) \otimes H^0(H, \mathcal{O}(1))$$

is surjective because α_n is, and the map

$$H^0(H, \mathcal{J}(n+1)) \otimes H^0(H, \mathcal{O}(1)) \longrightarrow H^0(H, \mathcal{J}(n+2))$$

is surjective by induction, so the image in $H^0(\mathbb{P}^r, \mathcal{I}(n+2))$ of $H^0(\mathbb{P}^r, \mathcal{I}(n+1)) \otimes H^0(\mathbb{P}^r, \mathcal{O}(1))$ already surjects onto $H^0(H, \mathcal{J}(n+2))$. In conclusion, as n increases, $h^1(\mathbb{P}^r, \mathcal{I}(n))$ decreases strictly for a while and then stabilizes; since $h^1(\mathbb{P}^r, \mathcal{I}(m))$ is certainly zero for very large m, it stabilizes at zero. It follows, in particular, that $H^1(\mathbb{P}^r, \mathcal{I}(n)) = 0$ if $n \geq n_1 + h^1(\mathbb{P}^r, \mathcal{I}(n_1))$. On the other hand, there is an upper bound for $h^1(\mathbb{P}^r, \mathcal{I}(n_1))$ which is independent of $\mathcal{I}$; in fact,

$$h^1(\mathbb{P}^r, \mathcal{I}(n_1)) = h^0(\mathbb{P}^r, \mathcal{I}(n_1)) - q(n_1) \leq h^0(\mathbb{P}^r, \mathcal{O}(n_1)) - q(n_1) \, .$$

We claim that

$$n_0 = n_1 + h^0(\mathbb{P}^r, \mathcal{O}(n_1)) - q(n_1) + 1$$

will do. By what has just been said, *i)* certainly holds for any $n \geq n_0 - 1$. Now we turn to *ii)*. Look at the diagram

$$\begin{array}{ccccc} H^0(\mathbb{P}^r, \mathcal{I}(n)) & \xrightarrow{\eta} & H^0(\mathbb{P}^r, \mathcal{I}(n+1)) & \xrightarrow{\alpha_n} & H^0(\mathbb{P}^r, \mathcal{J}(n+1)) \\ & & \uparrow{\scriptstyle \beta} & & \uparrow{\scriptstyle \delta} \\ & & H^0(\mathbb{P}^r, \mathcal{I}(n)) \otimes H^0(\mathbb{P}^r, \mathcal{O}(1)) & \xrightarrow{\gamma} & H^0(\mathbb{P}^r, \mathcal{J}(n)) \otimes H^0(\mathbb{P}^r, \mathcal{O}_H(1)) \end{array}$$

The top row is exact, and γ is surjective for $n \geq n_0$, since *i)* holds for any $n \geq n_0 - 1$. As $n_0 > n_1$, δ is also surjective for $n \geq n_0$, by induction hypothesis. On the other hand, the image of η is contained in the image of β, and a diagram chase shows that β is onto. This ends the proof of the lemma.

REMARK (4.4). It may be of some use to observe that the result we just finished proving continues to hold, with the same proof, if in the statement $\mathcal{I}$ is replaced with a coherent subsheaf of a *fixed* coherent sheaf on $\mathbb{P}^r$. This is a crucial step in generalizing the construction of the Hilbert scheme to the one of the *Quot* scheme parameterizing coherent quotients of a fixed coherent sheaf.

In practice, we shall often use the following, seemingly more general, version of Lemma (4.1).

COROLLARY (4.5). *Let r be a nonnegative integer, and let $p(t)$ be a rational polynomial. Then there exists an integer n_0 with the following property. Let $\mathbb{P}^r \times S \supset X \to S$ be any flat family of subschemes of $\mathbb{P}^r$ with Hilbert polynomial $p(t)$, let $\psi : \mathbb{P}^r \times S \to S$ be the projection, and denote by $\mathcal{I}_X$ the ideal sheaf of X in $\mathbb{P}^r \times S$. Then, for any $n \geq n_0$, the following hold:*

i) $\psi_* \mathcal{I}_X(n)$ *is locally free of rank* $q(n)$;

ii) $R^i \psi_* \mathcal{I}_X(n) = 0\,, \quad i \geq 1$;

iii) the multiplication map $\psi_* \mathcal{I}_X(n) \otimes \psi_* \mathcal{O}_{\mathbb{P}^r \times S}(1) \to \psi_* \mathcal{I}_X(n+1)$ *is onto;*

iv) for any morphism $\alpha : T \to S$, *the natural homomorphism*

$$\alpha^* \psi_* \mathcal{I}_X(n) \to \varphi_* \mathcal{I}_Y(n)$$

is an isomorphism, where $Y = X \times_S T \subset \mathbb{P}^r \times T$, *and* $\varphi : \mathbb{P}^r \times T \to T$ *is the projection.*

We first observe that $\mathcal{I}_X$ is flat over S. This can be seen, for instance, by noticing that, for large m, the sequence

$$0 \to \psi_* \mathcal{I}_X(m) \to \psi_* \mathcal{O}_{\mathbb{P}^r \times S}(m) \to \psi_* \mathcal{O}_X(m) \to 0$$

is exact, and then that, by (2.5), $\psi_* \mathcal{O}_X(m)$ is locally free for large m; it then follows that $\psi_* \mathcal{I}_X(m)$ is also locally free for large m, and another application of (2.5) proves our claim.

We next let n_0 be the integer provided by Lemma (4.1), where $q(m) = \binom{m+r}{r} - p(m)$. Combining part i) of Lemma (4.1) with Proposition (3.3) proves i), ii), and also iv), since $(\mathrm{id} \times \alpha)^* \mathcal{I}_X(n) = \mathcal{I}_Y(n)$. Finally, the fact that

$$\psi_* \mathcal{I}_X(n) \otimes k(s) = H^0(X_s, \mathcal{I}_{X_s}(n))$$

reduces iii) to part ii) of (4.1).

We may now complete the construction of the Hilbert scheme $\mathrm{Hilb}_r^{p(t)}$. Let X be a closed subscheme of $\mathbb{P}^r$ with Hilbert polynomial $p(t)$. Let

$$q(n) = \binom{n+r}{r} - p(n)$$

be the Hilbert polynomial of the ideal sheaf $\mathcal{I}_X$ of X. Fix an integer n greater or equal to the integer n_0 provided by Corollary (4.5). We may then associate to $X \subset \mathbb{P}^r$ its nth *Hilbert point*, that is, the surjective homomorphism

$$\varphi_n : H^0(\mathbb{P}^r, \mathcal{O}_{\mathbb{P}^r}(n)) \to H^0(X, \mathcal{O}_X(n))\,,$$

or, dually, the kernel of φ_n, that is, the subspace

$$H^0(\mathbb{P}^r, \mathcal{I}_X(n)) \subset H^0(\mathbb{P}^r, \mathcal{O}_{\mathbb{P}^r}(n))\,.$$

The nth Hilbert point of $X \subset \mathbb{P}^r$ is a point of the Grassmannian

$$G = \mathbb{G}(q(n), H^0(\mathbb{P}^r, \mathcal{O}_{\mathbb{P}^r}(n)))$$

parameterizing $q(n)$-dimensional subspaces of $H^0(\mathbb{P}^r, \mathcal{O}_{\mathbb{P}^r}(n))$. The nth Hilbert point determines $X \subset \mathbb{P}^r$ completely; in fact, part *ii*) of (4.1) says that the homogeneous ideal of X is generated, in degree n or more, by $H^0(\mathbb{P}^r, \mathcal{I}_X(n))$. Moreover, by part *i*) of (4.1), the dimension of $H^0(\mathbb{P}^r, \mathcal{I}_X(m))$ equals $q(m)$ for any $m \geq n$ (in fact, for any $m \geq n_0$). In view of part *ii*) of (4.1), this means that the nth Hilbert point of X belongs to the subset H of the Grassmannian G whose points are the vector subspaces $V \subset H^0(\mathbb{P}^r, \mathcal{O}_{\mathbb{P}^r}(n))$ such that the image of the multiplication map

$$\rho_{m,V} : V \otimes H^0(\mathbb{P}^r, \mathcal{O}_{\mathbb{P}^r}(m-n)) \to H^0(\mathbb{P}^r, \mathcal{O}_{\mathbb{P}^r}(m))$$

has dimension $q(m)$. Conversely, suppose that V belongs to H. Then, by definition, V generates a homogeneous ideal whose Hilbert polynomial is $q(t)$ and therefore comes from a closed subscheme of $\mathbb{P}^r$ with Hilbert polynomial $p(t)$. Thus attaching to any $X \subset \mathbb{P}^r$ its nth Hilbert point establishes a one-to-one correspondence between closed subschemes of $\mathbb{P}^r$ with Hilbert polynomial $p(t)$ and points of H. The Hilbert scheme $\mathrm{Hilb}_r^{p(t)}$ will be exhibited as a subscheme of G whose set of closed points is H. To construct it, we begin by globalizing the multiplication maps $\rho_{h,V}$. We let

$$\xi : \mathbb{P}^r \times G \to G$$

be the projection, and

$$\mathcal{F} \hookrightarrow \xi_* \mathcal{O}_{\mathbb{P}^r \times G}(n)$$

the universal subsheaf on G. We then let

$$\rho_h : \mathcal{F} \otimes \xi_* \mathcal{O}_{\mathbb{P}^r \times G}(h-n) \to \xi_* \mathcal{O}_{\mathbb{P}^r \times G}(h)$$

be the multiplication map. We denote by Σ_h the determinantal locally closed subscheme of G defined by the condition

$$\operatorname{rank} \rho_h = q(h)\,.$$

To be more precise, Σ_h is the intersection of the open subset of G where at least one $q(h) \times q(h)$ minor of ρ_h does not vanish with the subscheme of G whose ideal sheaf is generated by the $(q(h)+1) \times (q(h)+1)$ minors of ρ_h. Clearly,

$$H = \bigcap_{h \geq n} (\Sigma_h)_{red}\,.$$

We would like to define $\mathrm{Hilb}_r^{p(t)}$ to be the scheme-theoretic intersection of all the Σ_h. The problem is that it is not clear that this makes sense, as we are dealing with an infinite intersection of locally closed, and not necessarily closed, subschemes of G. The way out is to show that the finite intersections

$$\Xi_k = \bigcap_{n \le h \le k} \Sigma_h$$

stabilize for large k. To do this, it suffices to show that the Ξ_k stabilize *as sets*. In fact, if this happens, then there is a k_0 such that, for k larger than k_0, $\{\Xi_k\}$ is a decreasing sequence of *closed* subschemes of Ξ_{k_0} and hence stabilizes by noetherianity.

Denote by $|\Xi_k|$ the set of closed points of Ξ_k. It is enough to show that the sequence $\{|\Xi_k|\}$ is constant for large k. To prove this, it suffices to find, for every closed point s in G, a closed subscheme Y_s of $\mathbb{P}^r$ and an integer $N \ge n$, independent of s, such that

$$\operatorname{rank} \rho_{h,s} = h_{\mathcal{I}_{Y_s}}(h) \qquad \text{for} \quad h \ge N\,, \tag{4.6}$$

where $\mathcal{I}_{Y_s}$ is the ideal sheaf of Y_s. Suppose in fact that this has been done. By definition, given a closed point $s \in \Xi_{N+r}$, we have

$$\operatorname{rank} \rho_{h,s} = q(h) \qquad \text{for} \quad n \le h \le N+r\,.$$

As both $q(h)$ and $h_{\mathcal{I}_{Y_s}}(h)$ are polynomials in h of degree not exceeding r, they must coincide. This shows that $\operatorname{rank} \rho_{h,s} = q(h)$ for $h \ge N$, or equivalently that $s \in \Xi_h$ for all $h \ge N$, proving the claim.

It is clear that a candidate for Y_s is the subscheme of $\mathbb{P}^r$ corresponding to the ideal generated by the subspace $V \subset H^0(\mathbb{P}^r, \mathcal{O}_{\mathbb{P}^r}(n))$, where $[V] = s \in G$. To prove the existence of an integer N such that (4.6) holds, it is best to exhibit Y_s as the fiber over s of a closed subscheme

$$\mathcal{Y} \subset \mathbb{P}^r \times G. \tag{4.7}$$

We will later realize that this subscheme, when restricted to $\mathrm{Hilb}_r^{p(t)}$, is nothing but the universal family over $\mathrm{Hilb}_r^{p(t)}$. We set

$$\mathcal{F}_m = \rho_m(\mathcal{F} \otimes \xi_* \mathcal{O}_{\mathbb{P}^r \times G}(m-n))\,.$$

Then $\oplus_j \mathcal{F}_j$ is a graded sheaf of ideals in $\oplus_j \xi_* \mathcal{O}_{\mathbb{P}^r \times G}(j)$. Hence there are a sheaf of ideals $\mathcal{J} \subset \mathcal{O}_{\mathbb{P}^r \times G}$ and an integer $N \ge n$ such that

$$\mathcal{F}_m = \xi_* \mathcal{J}(m) \qquad \text{for } m \ge N\,.$$

We denote by $\mathcal{Y}$ the subscheme of $\mathbb{P}^r \times G$ corresponding to $\mathcal{J}$. We may assume that

$$R^1 \xi_* \mathcal{J}(m) = 0$$

for $m \geq N$ and also, by Lemma (3.6), that

$$H^0(Y_s, \mathcal{O}_{Y_s}(m)) = \xi_* \mathcal{O}_{\mathcal{Y}}(m) \otimes k(s)\,, \tag{4.8}$$
$$H^i(Y_s, \mathcal{O}_{Y_s}(m)) = 0\,, \quad i > 0\,,$$

for $m \geq N$ and for any $s \in G$, where Y_s stands for the fiber of $\mathcal{Y} \to G$ over s. Then the sequence

$$\mathcal{F} \otimes \xi_* \mathcal{O}_{\mathbb{P}^r \times G}(m-n) \xrightarrow{\rho_m} \xi_* \mathcal{O}_{\mathbb{P}^r \times G}(m) \to \xi_* \mathcal{O}_{\mathcal{Y}}(m) \to 0$$

is exact for $m \geq N$, and tensoring it with $k(s)$ we get another exact sequence

$$(\mathcal{F} \otimes k(s)) \otimes H^0(\mathbb{P}^r, \mathcal{O}_{\mathbb{P}^r}(m-n)) \xrightarrow{\rho_{m,s}} H^0(\mathbb{P}^r, \mathcal{O}_{\mathbb{P}^r}(m)) \to H^0(Y_s, \mathcal{O}_{Y_s}(m)) \to 0.$$

It follows that the image of $\rho_{m,s}$ is $H^0(Y_s, \mathcal{I}_{Y_s}(m))$. On the other hand, this same exact sequence and (4.8) imply that $H^i(Y_s, \mathcal{I}_{Y_s}(m))$ vanishes for $i > 0$ and $m \geq N$. In conclusion, $\operatorname{rank} \rho_{m,s} = h_{\mathcal{I}_{Y_s}}(m)$ when $m \geq N$, as we had to show. This concludes the construction of the scheme structure of $\mathrm{Hilb}_r^{p(t)}$.

It remains to prove that $\mathrm{Hilb}_r^{p(t)}$ represents the Hilbert functor. There is already a natural candidate for the universal family on $\mathrm{Hilb}_r^{p(t)}$, namely the restriction to $\mathrm{Hilb}_r^{p(t)}$ of $\mathcal{Y} \to G$, which we denote by

$$\pi : \mathcal{X} \to \mathrm{Hilb}_r^{p(t)}\ .$$

This family is flat. To show this, we use the criterion (2.5). Let $j : \mathrm{Hilb}_r^{p(t)} \to G$ be the inclusion. Lemma (3.8) implies that, for large h,

$$\pi_* \mathcal{O}_{\mathcal{X}}(h) = j^* \xi_* \mathcal{O}_{\mathcal{Y}}(h) = j^*(\operatorname{coker}(\rho_h))\,.$$

On the other hand, since the maps ρ_h have constant rank $q(h)$ on $\mathrm{Hilb}_r^{p(t)}$, the sheaves $j^*(\operatorname{coker}(\rho_h))$ are locally free of rank $p(h)$ for any $h \geq n$. Hence $\mathcal{X}$ is flat over $\mathrm{Hilb}_r^{p(t)}$. As a consequence, any morphism $\alpha : S \to \mathrm{Hilb}_r^{p(t)}$ induces by pullback a flat family

$$\mathbb{P}^r \times S \supset X = \mathcal{X} \times_{\mathrm{Hilb}_r^{p(t)}} S \xrightarrow{f} S$$

of subschemes of $\mathbb{P}^r$ with Hilbert polynomial $p(t)$. Denote by φ and ψ the projections from $\mathbb{P}^r \times S$ and $\mathbb{P}^r \times \mathrm{Hilb}_r^{p(t)}$ to their second factors. Then, by dimension reasons, it follows from part i) of Corollary (4.5) that $\psi_* \mathcal{I}_{\mathcal{X}}(n) \subset \psi_* \mathcal{O}(n)$ is just the restriction to $\mathrm{Hilb}_r^{p(t)}$ of the universal subbundle on G. It also follows from part iv) of (4.5) that this pulls back, via α, to $\varphi_* \mathcal{I}_X(n) \subset \varphi_* \mathcal{O}_{\mathbb{P}^r \times S}(n)$.

Conversely, let $\mathbb{P}^r \times S \supset X \to S$ be any flat family of subschemes of $\mathbb{P}^r$ with Hilbert polynomial $p(t)$, and let $\varphi : \mathbb{P}^r \times S \to S$ be the projection. Again by Corollary (4.5), the sheaf $\varphi_* \mathcal{I}_X(m)$ is locally free of rank $q(m)$ for any $m \geq n$, and moreover the multiplication map

$$\varphi_* \mathcal{I}_X(n) \otimes \varphi_* \mathcal{O}_{\mathbb{P}^r \times S}(m-n) \to \varphi_* \mathcal{I}_X(m)$$

is onto. The universal property of the Grassmannian tells us that there is a *unique* morphism $\alpha : S \to G$ such that the universal subbundle on G pulls back to $\varphi_* \mathcal{I}_X(n)$. By what we have just observed, α factors through $\mathrm{Hilb}_r^{p(t)}$, and, by construction

$$X = \mathcal{X} \times_{\mathrm{Hilb}_r^{p(t)}} S\,.$$

This ends the proof that $\mathrm{Hilb}_r^{p(t)}$ represents the Hilbert functor. The construction of the Hilbert scheme is complete.

We end this section with a few remarks. The first is that the universal property of the Hilbert scheme holds also with respect to *analytic* families of proper subschemes of $\mathbb{P}^r$, that is, for analytic subspaces $X \subset \mathbb{P}^r \times S$, where S is an analytic space, and the projection from X to S is proper and flat. In other words, any such family is obtained, in a unique way, by pullback via an analytic map from S to an appropriate Hilbert scheme. The proof of this fact is essentially the same as in the algebraic case. Here is an immediale and useful consequence.

LEMMA (4.9). *Every flat analytic family of projective schemes is locally the pullback of an algebraic one.*

The second remark is the following.

PROPOSITION (4.10). *The Hilbert scheme* $\mathrm{Hilb}_r^{p(t)}$ *is projective.*

By construction, the Hilbert scheme is a subscheme of a Grassmannian G, which is projective; thus to prove projectivity, it suffices to prove properness. We will use the valuative criterion, in the following form. We will show that, given a smooth curve S and a closed point $s \in S$, any morphism $\psi : S^* = S \smallsetminus \{s\} \to \mathrm{Hilb}_r^{p(t)}$ extends to all of S. First of all, ψ extends to a morphism $\varphi : S \to G$. Recall, from the construction of the Hilbert scheme, that the universal family $\mathcal{X} \to \mathrm{Hilb}_r^{p(t)}$ is the restriction of the (nonflat) family $\mathcal{Y} \to G$ introduced in (4.7). By base change via φ, the latter pulls back to a family

$$\mathbb{P}^r \times S \supset \mathcal{Z} \xrightarrow{q} S.$$

Clearly, the restriction $\mathcal{Z}^* \to S^*$ to S^* of this family is just the pullback of $\mathcal{X} \to \mathrm{Hilb}_r^{p(t)}$ via ψ, and hence is flat. Let $\mathfrak{m}$ be the ideal sheaf of s

in $\mathcal{O}_S$, and denote by $\mathcal{T}$ the ideal sheaf in $\mathcal{O}_{\mathcal{Z}}$ consisting of the sections which are annihilated by $\mathfrak{m}$. It is evident that $\mathcal{T}$ is concentrated on $q^{-1}(s)$. Now let $\mathcal{J}$ be the inverse image of $\mathcal{T}$ via $\mathcal{O}_{\mathbb{P}^r \times S} \to \mathcal{O}_{\mathcal{Z}}$, and denote by $\mathcal{W}$ the subscheme of $\mathbb{P}^r \times S$ it defines. We claim that $\mathcal{W}$ is flat over S. Since $\mathcal{W}$ agrees with $\mathcal{Z}^*$ over S^*, this will prove the proposition, by virtue of the universal property of the Hilbert scheme. To prove flatness, we use the criterion (2.5). We must show that the pushforward $q_*\mathcal{O}_{\mathcal{W}}(n)$ is locally free for large n. Away from s, this follows from the flatness of $\mathcal{Z}^*$ over S^*. At s, we argue as follows. Let t be a local parameter at s, that is, a local generator for $\mathfrak{m}$. By construction, t is not a zero divisor in $\mathcal{O}_{\mathcal{W}}$. Thus, $q_*\mathcal{O}_{\mathcal{W}}(n)$ is torsion-free at s for all n. Since S is a smooth curve, this implies that $q_*\mathcal{O}_{\mathcal{W}}(n)$ is locally free at s for all n.

It may be worth noticing that, as shown by Hartshorne [354], the Hilbert scheme $\mathrm{Hilb}^{p(t)}_r$ is also connected.

5. The characteristic system.

In this section we will study the tangent spaces to the Hilbert scheme. We denote by $\mathbb{C}[\varepsilon]$ the ring of dual numbers, by Σ the scheme $\operatorname{Spec}\mathbb{C}[\varepsilon]$, and by s_0 its closed point. The tangent space to $H = \mathrm{Hilb}^{p(t)}_r$ at a point h corresponding to a subscheme $X_0 \subset \mathbb{P}^r$ is in one-to-one correspondence with the set of morphisms of pointed schemes

$$\mathrm{Hom}((\Sigma, s_0), (H, h)).$$

Since H represents the Hilbert functor, this can be identified with the set of all flat families

$$\mathbb{P}^r \times \Sigma \supset X \xrightarrow{f} \Sigma = \operatorname{Spec}\mathbb{C}[\varepsilon]$$

whose fiber over s_0 is $X_0 \subset \mathbb{P}^r$. More generally, given a scheme Y and a closed subscheme $X_0 \subset Y$, we shall study the *first-order embedded deformations of X_0 in Y*, meaning those flat families

$$Y \times \Sigma \supset X \xrightarrow{f} \Sigma = \operatorname{Spec}\mathbb{C}[\varepsilon]$$

whose fiber over s_0 is $X_0 \subset Y$

We first wish to explain the case where Y is affine. Let then R be the coordinate ring of Y so that $R[\varepsilon] = R \otimes_{\mathbb{C}} \mathbb{C}[\varepsilon]$ is the coordinate ring of $Y \times \Sigma$. A subscheme X of $Y \times \Sigma$ corresponds to an ideal $J \subset R[\varepsilon]$, and the flatness of the natural projection from X to Σ means that $R[\varepsilon]/J$ is a flat $\mathbb{C}[\varepsilon]$-module. We wish to show that the flatness of this family of subschemes of Y is equivalent to saying that every presentation by generators and relations of the ideal I of the central fiber $X_0 = X \bmod (\varepsilon)$ extends to a presentation of J. Formally, we shall prove the following more general result.

LEMMA (5.1). *Let $\varphi : A \to B$ be a homomorphism of noetherian commutative rings, with B flat over A. Assume either that A is a local artinian ring or that A and B are both local rings and φ is a local homomorphism. Let J be an ideal in B, and set $C = B/J$. Let k be the quotient of A modulo its maximal ideal $\mathfrak{m}_A$, and set $B_k = B \otimes_A k$, $C_k = C \otimes_A k$. Then the following statements are equivalent:*

i) *C is flat over A;*
ii) *every exact sequence*

$$B_k^l \to B_k^h \to B_k \to C_k \to 0 \tag{5.2}$$

is the reduction modulo $\mathfrak{m}_A$ of an exact sequence:

$$B^l \to B^h \to B \to C \to 0;$$

iii) *there are generators $F_1, \ldots, F_h$ of J such that, denoting by f_i the image of F_i in B_k, $i = 1, \ldots, h$, every relation among the f_i extends to a relation among the F_i.*

The case of first-order deformations of embedded affine subschemes introduced above is the special one in which $A = \mathbb{C}[\varepsilon]$ and $B = R[\varepsilon]$. The reader may find it helpful to follow the proof with this specific example in mind.

In proving Lemma (5.1) we shall rely on the following well-known results.

LEMMA (5.3). *Let R be a commutative ring, and let $0 \to E \to F \to G \to 0$ be an exact sequence of R-modules. If G is flat, then E is flat if and only if F is flat.*

Tensoring the exact sequence with an R-module H, we get an exact sequence of Tor's

$$\cdots \to \mathrm{Tor}_2^R(G,H) \to \mathrm{Tor}_1^R(E,H) \to \mathrm{Tor}_1^R(F,H) \to \mathrm{Tor}_1^R(G,H) \to \cdots.$$

The terms on the right and on the left vanish since G is flat, so the middle terms are isomorphic. Thus E is flat, that is, $\mathrm{Tor}_1^R(E,H)$ vanishes for any H, if and only if the same is true for F.

LEMMA (5.4) (Local criterion of flatness). *Let $\varphi : A \to B$ be a homomorphism of commutative rings. Assume either that A is a local artinian ring or that A and B are both local noetherian rings and φ is a local homomorphism. Then a finitely generated B-module M (or any B-module if A is artinian) is flat over A if and only if $\mathrm{Tor}_1^A(M,k) = 0$, where k stands for the quotient of A modulo its maximal ideal $\mathfrak{m}_A$.*

The *only if* part is obvious. For a proof of the converse in the case where φ is a local homomorphism of local noetherian rings, we refer to [194], Theorem 6.8. If A is artinian, we need to prove that $\mathrm{Tor}_1^A(M,N)$ vanishes for any finitely generated A-module N. There is an integer v such that $\mathfrak{m}_A^v = 0$, and it suffices to show that $\mathrm{Tor}_1^A(M,N_i)$ vanishes for each quotient $N_i = \mathfrak{m}_A^i N/\mathfrak{m}_A^{i+1}N$ of the composition series

$$N = \mathfrak{m}_A^0 N \supset \mathfrak{m}_A^1 N \supset \cdots \supset \mathfrak{m}_A^v N = 0\,.$$

That $\mathrm{Tor}_1^A(M,N_i)$ is zero follows immediately from the assumptions since N_i is a finite-dimensional vector space over k.

We now return to the proof of Lemma (5.1). We shall show that i) implies ii) and that iii) implies i); that iii) follows from ii) is obvious. We may write $C_k = B_k/I$, where I is an ideal. Since $C_k = B/(J+\mathfrak{m}_A B)$, the ideal I is isomorphic to $J/(J\cap \mathfrak{m}_A B)$. Suppose now that C is flat over A, and let an exact sequence of the form (5.2) be given. Lemma (5.3), applied to the exact sequence $0\to J\to B\to C\to 0$, implies that J is flat over A. Moreover, tensoring this sequence with k, we obtain the exact sequence

$$\cdots \to \mathrm{Tor}_1^A(C,k) \to J\otimes_A k \to B_k \to C_k \to 0\,.$$

The term on the left vanishes by the flatness of C, and hence $J\otimes_A k = I$, or, equivalently, $\mathfrak{m}_A J = J\cap \mathfrak{m}_A B$, since $J\otimes_A k = J/\mathfrak{m}_A J$. We can construct the commutative diagram

$$\begin{array}{ccccc} B_k^h & \xrightarrow{\beta} & I & \longrightarrow & 0 \\ \uparrow & & \uparrow & & \\ B^h & \xrightarrow{\alpha} & J & & \end{array}$$

We claim that α is onto. In fact if Q is the cokernel of α, we have that $Q\otimes_A k$ can be identified with the cokernel of β and hence is zero; this means that $Q = \mathfrak{m}_A Q$. If B is a local ring, this implies that $Q = \mathfrak{m}_B Q$, so $Q = 0$ by Nakayama's lemma. If A is artinian, there is an integer v such that $\mathfrak{m}_A^v = 0$, so $Q = \mathfrak{m}_A Q = \mathfrak{m}_A^2 Q = \cdots = \mathfrak{m}_A^v Q = 0$. In any case the conclusion is that Q is zero. At this stage we have extended the generators of I to generators of J. It remains to extend relations. Let N be the kernel of β, and M the kernel of α. We then have the commutative diagram

$$\begin{array}{ccccccccc} 0 & \longrightarrow & N & \longrightarrow & B_k^h & \xrightarrow{\beta} & I & \longrightarrow & 0 \\ & & \uparrow & & \uparrow & & \uparrow & & \\ 0 & \longrightarrow & M & \longrightarrow & B^h & \xrightarrow{\alpha} & J & \longrightarrow & 0 \end{array}$$

Tensoring the lower sequence with k and using the flatness of J, we find that $M \otimes_A k$ equals N. We can construct a commutative diagram

$$\begin{array}{ccccc} B_k^l & \longrightarrow & N & \longrightarrow & 0 \\ \uparrow & & \uparrow & & \\ B^l & \xrightarrow{\ \gamma\ } & M & & \end{array}$$

and arguing exactly as before, we see that γ is onto, so that

$$B^l \xrightarrow{\gamma} B^h \xrightarrow{\alpha} B \to C \to 0 \tag{5.5}$$

is an exact sequence extending (5.2).

This proves that i) implies ii). Now assume that iii) holds. The choice of the generators $F_1, \dots, F_h$ corresponds to a surjective homomorphism $B^h \to J$; likewise, $f_1, \dots, f_h$ give a surjective homomorphism $B_k^h \to I$. Denoting by M and N the kernels of these homomorphisms, condition iii) means that M maps onto N. Thus, if $B^l \to M$ is onto, then B_k^l maps surjectively onto N. In other words, tensoring the exact sequence

$$B^l \to B^h \to B \to C \to 0 \tag{5.6}$$

with k yields an exact sequence

$$B_k^l \to B_k^h \to B_k \to C_k \to 0\,.$$

We can complete (5.6) to a free resolution of C; tensoring with k, we obtain a complex whose homology calculates the groups $\mathrm{Tor}_i^A(C, k)$. On the other hand, we just noticed that this complex is exact in degree one, so $\mathrm{Tor}_1^A(C, k)$ vanishes. By Lemma (5.4), this is sufficient to infer that C is flat over A. This concludes the proof of the lemma.

A way of rephrasing the equivalence of i)–iii) in the preceding lemma is the following:

COROLLARY (5.7). *Let A and B be as in (5.1). Let $f_1, \dots, f_h$ be elements of $B/\mathfrak{m}_A B$, and, for each $i = 1, \dots, h$, let F_i be an element in B which reduces to f_i modulo $\mathfrak{m}_A B$. Then $B/(F_1, \dots, F_h)$ is flat over A if and only if every relation among the f_i extends to a relation among the F_i.*

Lemma (5.1) makes it possible to completely classify embedded first-order deformations of a subscheme $X_0 \subset Y$. We begin with the affine case.

LEMMA (5.8). *Let R be a commutative noetherian $\mathbb{C}$-algebra, and let I be an ideal in R. The first-order embedded deformations of $X_0 = \mathrm{Spec}(R/I)$ within $Y = \mathrm{Spec}(R)$ are in one-to-one correspondence with $\mathrm{Hom}_{R/I}(I/I^2, R/I) = \mathrm{Hom}_R(I, R/I)$.*

Proof. Given an element g of R, we write $[g]$ to indicate its class modulo I. We must classify the ideals $J \subset R[\varepsilon]$ such that $R[\varepsilon]/J$ is flat over $\mathbb{C}[\varepsilon]$ and $J/((\varepsilon) \cap J) = I$. Let one such J be given. Given $i \in I$, pick $j \in J$ whose reduction modulo (ε) is i. We can then write $j = i - \varepsilon h$, where h is an element of R which depends R-linearly on j and which is uniquely determined by i modulo I; in fact, if $i = 0$, then εh belongs both to J and to (ε), and, during the proof of Lemma (5.1), we observed that the flatness of $R[\varepsilon]/J$ implies, in particular, that $J \cap (\varepsilon) = \varepsilon J = \varepsilon I$, so that $h \in I$. Conversely, suppose we are given a homomorphism $\alpha : I \to R/I$. Choose generators $f_1, \ldots, f_n$ for I, write $\alpha(f_i) = [g_i]$, where $g_i \in R$, and set $F_i = f_i - \varepsilon g_i$ and $J = (F_1, \ldots, F_n)$. Clearly, $J/(J \cap (\varepsilon)) = I$. We wish to show that $R[\varepsilon]/J$ is flat over $\mathbb{C}[\varepsilon]$. By Corollary (5.7), this follows if we can show that any relation among the f_i is the reduction modulo (ε) of a relation among the F_i. Let then $\sum a_i f_i = 0$ be a relation and notice that

$$\sum a_i[g_i] = \alpha\left(\sum a_i f_i\right) = 0 ,$$

meaning that $\sum a_i g_i \in I$, so that we can write $\sum a_i g_i = \sum b_i f_i$ for some elements b_i in R. Thus,

$$\sum (a_i + \varepsilon b_i) F_i = \sum a_i f_i + \varepsilon \left(\sum b_i f_i - \sum a_i g_i\right) = 0$$

is a relation among the F_i which extends $\sum a_i f_i = 0$. Summing up, we have associated to each ideal J in $R[\varepsilon]$ extending I and such that $R[\varepsilon]/J$ is flat over $\mathbb{C}[\varepsilon]$, a homomorphism of R-modules from I to R/I, and conversely. It is then a trivial exercise to check that these two maps are inverse to each other, thus proving the lemma.

Let X be a closed subscheme of a fixed scheme Y, and let $\mathcal{I}$ be the ideal of X in Y. The sheaf $\mathcal{I}/\mathcal{I}^2$, or rather its restriction to X, is called the *conormal sheaf* of X in Y and denoted $\mathcal{C}_{X/Y}$. Its dual $\mathcal{H}om_{\mathcal{O}_X}(\mathcal{C}_{X/Y}, \mathcal{O}_X) = \mathcal{H}om_{\mathcal{O}_Y}(\mathcal{I}, \mathcal{O}_X)$ is the *normal sheaf* of X in Y and is denoted $N_{X/Y}$. From Lemma (5.8) one immediately obtains a description of all first-order embedded deformations of X in Y.

PROPOSITION (5.9). *Let X be a closed subscheme of Y, and let $\mathcal{I}$ be the ideal sheaf of X in Y. Then the first-order embedded deformations of X in Y are in a ono-to-one correspondence with* $H^0(X, N_{X/Y}) = \mathrm{Hom}_{\mathcal{O}_X}(\mathcal{C}_{X/Y}, \mathcal{O}_X) = \mathrm{Hom}_{\mathcal{O}_Y}(\mathcal{I}, \mathcal{O}_X)$.

This is a natural place to introduce, in the case at hand, a general machinery which will occur several times, in different contexts, throughout this volume. Let

$$Y \times B \supset \mathcal{X} \to B$$

be a flat family of subschemes of Y parameterized by a scheme B. Let $X = X_{b_0}$ for some closed point $b_0 \in B$. Let v be a tangent vector to B at b_0. Such a vector can be interpreted as a morphism of pointed schemes

$$v : \operatorname{Spec} \mathbb{C}[\varepsilon] \longrightarrow (B, b_0) .$$

Pulling back the family $\mathcal{X}$ via v yields a first-order embedded deformation of X in Y and hence, by Proposition (5.9), an element of $H^0(X, N_{X/Y})$. This assignment defines a map

$$T_{b_0}(B) \longrightarrow H^0(X, N_{X/Y}) . \tag{5.10}$$

Following a classical terminology, this map is called by Kodaira and Spencer the *characteristic map*. The map (5.10) is linear; a proof is outlined in exercises A-1 to A-8.

It is perhaps of some interest to relate Proposition (5.9) with the classical notion of characteristic system. The setting is the one of a family $\mathcal{X} = \{X_s\}_{s \in S}$ of divisors in a smooth projective variety Y, parameterized by a smooth variety S. Let $X = X_{s_0}$ be a reduced and irreducible member of this family. Then the *characteristic linear system of the family at* X is the linear system cut out on X by "infinitely near" members of the family. More precisely, suppose that $\mathcal{X}$ is defined locally by the equation

$$f(y_1, \dots, y_n, t_1, \dots, t_k) = 0 ,$$

where $y_1, \dots, y_n$ are local coordinates on Y, and $t_1, \dots, t_k$ are local coordinates on S centered at s_0, so that the equation of X in Y is

$$f(y_1, \dots, y_n, 0, \dots, 0) = 0 .$$

For each i, the equation

$$\left. \frac{\partial f}{\partial t_i}(y_1, \dots, y_n, t_1, \dots, t_k) \right|_{t_1 = \dots = t_k = 0} = 0$$

defines, locally, a divisor on X. Since f is uniquely determined up to multiplication by a unit g, this divisor does not depend on the choice of f. In fact,

$$\left. \frac{\partial g f}{\partial t_i} \right|_{t_1 = \dots = t_k = 0} \equiv g \left. \frac{\partial f}{\partial t_i} \right|_{t_1 = \dots = t_k = 0} \mod f|_{t_1 = \dots = t_k = 0} .$$

Thus $(\partial f / \partial t_i)|_{t_1 = \dots = t_k = 0}$ determines a global divisor D_i on X. The linear system generated by the D_i is the classical characteristic system. In our notation, $\{\partial / \partial t_i\}$ is a basis for the tangent space to S at s_0, and $(\partial f/\partial t_i)|_{t_1=\cdots=t_k=0}$ is the image in $H^0(X, N_{X/Y})$ of $\{\partial/\partial t_i\}$ via the characteristic map (5.10). Of course, in this particular case, $\mathcal{I} = \mathcal{O}_Y(-X)$, so that $N_{X/Y} = \mathcal{O}_Y(X) \otimes \mathcal{O}_X \simeq \mathcal{O}_X(D_i)$.

Let us go back to the Hilbert scheme $\mathrm{Hilb}_r^{p(t)}$. Let h be a (closed) point of $\mathrm{Hilb}_r^{p(t)}$, and let X be the corresponding subscheme of $\mathbb{P}^r$. Proposition (5.9) has the following important consequence.

COROLLARY (5.11). *The tangent space to* $\mathrm{Hilb}_r^{p(t)}$ *at* h *is given by*

$$T_h(\mathrm{Hilb}_r^{p(t)}) = H^0(X, N_{X/\mathbb{P}^r}) .$$

In particular, this result shows that $h^0(X, N_{X/\mathbb{P}^r})$ is an upper bound for the dimension of $\mathrm{Hilb}_r^{p(t)}$ at h; a lower bound for this same dimension is given by the following important proposition.

PROPOSITION (5.12). *Let* X *be a closed local complete intersection subscheme of* $\mathbb{P}^r$, *and let* h *be the corresponding point of* $\mathrm{Hilb}_r^{p(t)}$. *Then the dimension of every irreducible component of* $\mathrm{Hilb}_r^{p(t)}$ *at* h *is at least*

$$h^0(X, N_{X/\mathbb{P}^r}) - h^1(X, N_{X/\mathbb{P}^r}) .$$

We shall not prove this fundamental existence result but refer to [440] or [625] for a proof. Now let us give some applications.

EXAMPLE (5.13). We revisit Example (2.8). Let H be the Hilbert scheme of degree d zero-dimensional subschemes of $\mathbb{P}^r$ and consider a point of H corresponding to a subscheme $Z \subset \mathbb{P}^r$. If Z consists of d distinct points $p_1, \ldots, p_d$, then

$$H^0(Z, N_{Z/\mathbb{P}^r}) = \oplus_{i=1}^d T_{p_i}(\mathbb{P}^r) = T_{p_1+\cdots+p_d}(\mathrm{Sym}^d(\mathbb{P}^r)) ,$$

where $\mathrm{Sym}^d(\ )$ stands for d-fold symmetric product. This shows that the open subset of $\mathrm{Sym}^d(\mathbb{P}^r)$ consisting of d-uples of distinct points embeds in H as an open subset.

EXAMPLE (5.14). Consider hypersurfaces of degree d in $\mathbb{P}^r$. We know that the Hilbert scheme of these is isomorphic to a projective space of dimension

$$\binom{d+r}{r} - 1 .$$

In fact, if X is a degree d hypersurface in $\mathbb{P}^r$, then

$$h^0(X, N_{X/\mathbb{P}^r}) = h^0(X, \mathcal{O}_X(d)) = \binom{d+r}{r} - 1 .$$

EXAMPLE (5.15). Consider a smooth complete nondegenerate degree d curve C in $\mathbb{P}^r$, and let g be its genus. The curve C corresponds to a point $[C]$ in the Hilbert scheme $\mathrm{Hilb}_r^{p(t)}$, where $p(t) = dt+1-g$. We shall write H to denote this Hilbert scheme. We might naively expect that $\dim_{[C]} H = h^0(C, N_{C/\mathbb{P}^r}) = \chi(N_{C/\mathbb{P}^r})$, as is certainly the case for $r = 2$, by the preceding example. Actually, all we can say is that

$$\dim_{[C]} H \leq h^0(C, N_{C/\mathbb{P}^r}).$$

On the other hand, in the case under consideration the lower bound on the dimension of H given by (5.12) equals the Euler characteristic of $N_{X/\mathbb{P}^r}$, whence the inequality

$$\dim_{[C]} H \geq \chi(N_{X/\mathbb{P}^r}). \tag{5.16}$$

The Euler characteristic of $N_{X/\mathbb{P}^r}$ is readily calculated. From the sequence

$$0 \to T_C \to T_{\mathbb{P}^r} \otimes \mathcal{O}_C \to N_{C/\mathbb{P}^r} \to 0$$

we see that the degree of the normal bundle is

$$\deg N_{C/\mathbb{P}^r} = (r+1)d + 2g - 2,$$

and then, by Riemann–Roch, we have

$$\begin{aligned} \chi(N_{C/\mathbb{P}^r}) &= \deg(N_{C/\mathbb{P}^r}) - (r-1)(g-1) \\ &= (r+1)d - (r-3)(g-1). \end{aligned} \tag{5.17}$$

We now sketch an alternative proof of (5.16), referring to Section 8 of Chapter XXI, and specifically to Remark (8.20) therein, for more details. As we shall see in that chapter, in any family of line bundles of degree d on curves of genus g, the locus of those line bundles having $r+1$ or more sections has codimension at most $(r+1)(g-d+r)$ in the neighborhood of a line bundle with exactly $r+1$ sections. Applying this to the "family of all line bundles of degree d on all curves of genus g," one concludes that the family of all linear series of degree d and dimension r on curves of genus g has local dimension at least $4g-3-(r+1)(g-d+r)$ everywhere. Since such a linear series, when without base points, determines a map of a curve to $\mathbb{P}^r$ up to the (r^2+2r)-dimensional family $PGL(r+1, \mathbb{C})$ of automorphisms of $\mathbb{P}^r$, we may conclude that the dimension of H at $[C]$ is at least

$$\begin{aligned} \dim_{[C]} H &\geq 4g - 3 - (r+1)(g-d+r) + (r^2+2r) \\ &= (r+1)d - (r-3)(g-1), \end{aligned}$$

as desired.

Note one curious feature of the behavior of $\chi(N_{C/\mathbb{P}^r})$ as a function of g. When $r = 2$, it increases with g; when $r = 3$, it equals $4d$ and hence is independent of the genus, while for $r \geq 4$, it decreases with g.

There is one important case where the dimension of H is exactly equal to $\chi(N_{C/\mathbb{P}^r})$. Suppose that C is as above and that, in addition, $\mathcal{O}_C(1)$ is nonspecial. We claim that H is smooth at $[C]$ of dimension $\chi(N_{C/\mathbb{P}^r})$. To see this, it suffices to show that $h^0(C, N_{C/\mathbb{P}^r}) = \chi(N_{C/\mathbb{P}^r})$ or, equivalently, that $h^1(C, N_{C/\mathbb{P}^r}) = 0$. Look at the cohomology of the diagram

$$\begin{array}{ccccccccc} & & & & 0 & & \\ & & & & \downarrow & & \\ & & & & T_C & & \\ & & & & \downarrow & & \\ 0 \longrightarrow & \mathcal{O}_C & \longrightarrow & \mathcal{O}_C^{r+1}(1) & \longrightarrow & T_{\mathbb{P}^r} \otimes \mathcal{O}_C & \longrightarrow 0 \\ & & & & \downarrow & & \\ & & & & N_{C/\mathbb{P}^r} & & \\ & & & & \downarrow & & \\ & & & & 0 & & \end{array}$$

where the horizontal sequence is the restriction to C of the Euler sequence for the tangent bundle of projective space. We get, in particular,

$$\begin{array}{ccccc} H^1(C, \mathcal{O}_C(1))^{r+1} & \longrightarrow & H^1(C, T_{\mathbb{P}^r} \otimes \mathcal{O}_C) & \longrightarrow & 0 \\ & & \downarrow & & \\ & & H^1(C, N_{C/\mathbb{P}^r}) & & \\ & & \downarrow & & \\ & & 0 & & \end{array}$$

From the assumption that $\mathcal{O}_C(1)$ be nonspecial, it then follows that $H^1(C, T_{\mathbb{P}^r} \otimes \mathcal{O}_C)$, and hence $H^1(C, N_{C/\mathbb{P}^r})$, vanishes. We can summarize what has been proved in the following statement.

PROPOSITION (5.18). *Let C be a smooth, irreducible, nondegenerate curve of degree d and genus g in $\mathbb{P}^r$ such that $\mathcal{O}_C(1)$ is nonspecial. Set $p(t) = dt + 1 - g$. Then:*

i) $\dim_{[C]} \mathrm{Hilb}_r^{p(t)} = \dim T_{[C]}(\mathrm{Hilb}_r^{p(t)}) = \chi(N_{C/\mathbb{P}^r}) = (r+1)d-(r-3)(g-1)$,
ii) $H^1(C, N_{C/\mathbb{P}^r}) = 0$.

We close this section by discussing some other consequences of Lemma (5.1), centered around the permanence, under small deformations, of the property of being a local complete intersection. Recall that a *regular sequence* in a commutative ring R is a finite sequence $f_1, \dots, f_n$ of elements of R such that the ideal $(f_1, \dots, f_n)$ is proper and f_i is not a zero divisor in $R/(f_1, \dots, f_{i-1})$ for all i with $1 \leq i \leq n$. An embedding of schemes $Y \hookrightarrow X$ is said to be a *regular embedding* if every point of Y has an affine neighborhood $U = \operatorname{Spec} R$ in X such that the ideal of $U \cap Y$ in R is generated by a regular sequence; the length of the sequence, which is independent of the particular sequence chosen (cf., for instance, Lemma (5.22) below), is called the *codimension* of the embedding. When X is smooth, the notion of regularly embedded subscheme of X coincides with the one of local complete intersection subscheme of X (cf. [503], pp. 105 and 121).

The relations among the elements of a regular sequence are particularly easy to describe. The following elementary lemma describes the first step of what usually goes under the name of Koszul resolution.

LEMMA (5.19). *Let R be a commutative ring, let $f_1, \dots, f_h$ be a regular sequence in R, and let $a_1, \dots, a_h$ be elements of R. Then $\sum a_i f_i = 0$ if and only if there is an $h \times h$ skew-symmetric matrix (c_{ij}) with entries in R such that $a_i = \sum_j c_{ij} f_j$ for all i.*

The "if" part is obvious and does not require the assumption that $f_1, \dots, f_h$ be a regular sequence. The converse is also easy to prove, using induction on h. Suppose in fact that $\sum a_i f_i = 0$. If $h = 1$, then $a_1 = 0$, and we are done. If $h > 1$, we have instead that

$$a_h = \sum_{i<h} b_i f_i \tag{5.20}$$

for suitable b_i and hence that

$$\sum_{i=1}^{h-1} (a_i + b_i f_h) f_i = 0 .$$

By induction hypothesis,

$$a_i + b_i f_h = \sum_{j<h} c_{ij} f_j \tag{5.21}$$

for $i < h$, where (c_{ij}) is a suitable skew-symmetric matrix. If we complete (c_{ij}) to an $h \times h$ skew-symmetric matrix by setting $c_{ih} = -b_i$, $c_{hi} = b_i$, and $c_{hh} = 0$, then (5.21) and (5.20) translate into

$$a_i = \sum_{j=1}^{h} c_{ij} f_j$$

for $1 \leq i \leq h$.

The following standard result is a corollary of the preceding lemma.

LEMMA (5.22). *Let R be a noetherian local ring, let $f_1, \dots, f_n$ be a regular sequence in R, and set $I = (f_1, \dots, f_n)$.*

i) *If σ is a permutation of $\{1, \dots, n\}$, then $f_{\sigma(1)}, \dots, f_{\sigma(n)}$ is a regular sequence;*
ii) *if $b_1, \dots, b_m$ is a minimal system of generators of I, then $m = n$, and $b_1, \dots, b_m$ is a regular sequence.*

Proof. To prove i), it suffices to deal with the case where σ is a transposition $(i, i+1)$. All we have to do in this case is to show that $\times f_i : R/(f_1, \dots, f_{i-1}) \to R/(f_1, \dots, f_{i-1})$ and $\times f_{i+1} : R/(f_1, \dots, f_i) \to R/(f_1, \dots, f_i)$ are injective. Replacing R with $R/(f_1, \dots, f_{i-1})$, we may then assume that $n = 2$ and $i = 1$. Suppose that $f_2 a \in R f_1$, that is, that $0 = f_1 b + f_2 a$ for some $b \in R$. Then, by (5.19), a and b are of the form $a = f_1 d$ and $b = -f_2 d$ for some d; in particular, $a \in (f_1)$. Suppose instead that $f_1 a = 0$. Always by (5.19), there is $a_1 \in R$ such that $a = f_2 a_1$, $0 = f_1 a_1$. But then there is $a_2 \in R$ such that $a_1 = f_2 a_2$, $0 = f_1 a_2$, and so on. In conclusion, for any integer $k > 0$, we may write $a = f_2^k a_k$, so that a belongs to $\mathfrak{m}^k$, where $\mathfrak{m}$ is the maximal ideal of R. Since $\bigcap_{k>0} \mathfrak{m}^k = 0$, it follows that $a = 0$.

We now prove ii). Write $b_m = \sum c_i f_i$ and $f_i = \sum d_{i,j} b_j$. Substituting the latter relations into the first yields a relation among the b_i. If all the coefficients c_i did belong to $\mathfrak{m}$, the coefficient of b_m in this relation would be invertible, contradicting the minimality of $b_1, \dots, b_m$. Thus, possibly rearranging the f_i, we may assume that c_n is a unit. It is then immediate to check that $f_1, \dots, f_{n-1}, b_m$ is a regular sequence and that it generates I. Now write b_{m-1} as a linear combination of $f_1, \dots, f_{n-1}, b_m$. If the coefficients of $f_1, \dots, f_{n-1}$ all belonged to $\mathfrak{m}$, we could conclude, arguing as before, that $b_1, \dots, b_m$ is not a minimal system of generators of I. Possibly rearranging the f_i, we may thus suppose that the coefficient of f_{n-1} is a unit. Hence $f_1, \dots, f_{n-2}, b_{m-1}, b_m$ is a regular sequence generating I. This procedure can be repeated until either all the b_i have been used, or there are no more f_i to be replaced. The first alternative must occur, since otherwise $b_1, \dots, b_m$ would not be a minimal system of generators of I. On the other hand, the final regular sequence one obtains cannot contain elements of R other than $b_1, \dots, b_m$, since these already generate I. In conclusion, $m = n$, and $b_1, \dots, b_m$ is a regular sequence. Q.E.D.

Here is another consequence of (5.19).

LEMMA (5.23). *Let $\varphi : A \to B$ be a local homomorphism of noetherian local rings, with B flat over A. Let $J = (F_1, \dots, F_h)$ be an ideal in B, and let k be the quotient of A modulo its maximal ideal $\mathfrak{m}_A$. Denote by $f_1, \dots, f_h$ the images of $F_1, \dots, F_h$ in $B_k = B \otimes_A k$. Then, if $f_1 \dots, f_h$ is a regular sequence, $C = B/J$ is flat over A. Moreover, $F_1, \dots, F_h$ is a regular sequence.*

We first prove flatness. Let $\sum a_i f_i = 0$ be a relation among the f_i. We know, by Lemma (5.19), that there is a skew-symmetric matrix (c_{ij}) with entries in B_k such that $a_i = \sum_j c_{ij} f_j$ for every i. For each choice of i and j with $i > j$, pick an element d_{ij} of B mapping to c_{ij}, and set $d_{ji} = -d_{ij}$; also set $d_{ii} = 0$. Then $\sum b_i F_i = 0$, where $b_i = \sum_j d_{ij} F_j$, is a relation among the F_i extending $\sum a_i f_i = 0$. Summing up, any relation among the f_i extends to a relation among the F_i, and the conclusion follows from (5.1), or better from (5.7).

It remains to prove that $F_1, \ldots, F_h$ is a regular sequence. Set $J_i = (F_1, \ldots, F_i)$. Since B is A-flat by assumption, and B/J_i is A-flat for any i, as we just proved, applying Lemma (5.3) successively to the exact sequences of B-modules

$$\begin{gathered} 0 \to J_i \to B \to B/J_i \to 0, \\ 0 \to J_i/J_{i-1} \to B/J_{i-1} \to B/J_i \to 0, \\ 0 \to K_i \to B/J_{i-1} \xrightarrow{\times F_i} J_i/J_{i-1} \to 0 \end{gathered}$$

shows that J_i, J_i/J_{i-1} and K_i are A-flat, where K_i is the kernel of the multiplication map $\times F_i : B/J_{i-1} \to B/J_{i-1}$. It follows that $K_i \otimes_A k$ can be identified with the kernel of $\times f_i : B_k/(f_1, \ldots, f_{i-1}) \to B_k/(f_1, \ldots, f_{i-1})$ and hence vanishes. This means that $K_i = \mathfrak{m}_A K_i$. A fortiori, $K_i = \mathfrak{m}_B K_i$, so K_i vanishes by Nakayama's lemma. This concludes the proof of (5.23).

The following property of regular sequences is well known.

LEMMA (5.24). *Let $I = (f_1, \ldots, f_h)$ be a proper ideal in a local ring A. If $f_1, \ldots, f_h$ is a regular sequence, then I/I^2 is a free A/I-module of rank h.*

The proof of this result is immediate. It is clear that the classes of $f_1, \ldots, f_h$ generate I/I^2; all we have to do is show that they are independent. Suppose that $\sum a_i f_i$ belongs to I^2, i.e., that we may write $\sum a_i f_i = \sum b_i f_i$, where the b_i belong to I. By Lemma (5.19), the coefficients of the relation $\sum (a_i - b_i) f_i = 0$ must be of the form $a_i - b_i = \sum_j c_{ij} f_j$ for some skew-symmetic matrix (c_{ij}). This implies that $a_i \in I$ for each i, finishing the proof.

A consequence of Lemma (5.24) is that, if $Y \hookrightarrow X$ is a regular embedding, then the conormal sheaf $\mathcal{C}_{Y/X}$ to Y in X, and hence also the normal sheaf $N_{Y/X}$, are locally free $\mathcal{O}_Y$-modules.

LEMMA (5.25). *Let $\pi : X \to S$ be a flat morphism of schemes, and let Y be a subscheme of X, flat and proper over S. Suppose that $Y_{s_0} \hookrightarrow X_{s_0}$ is a regular embedding for some $s_0 \in S$. Then $Y_s \hookrightarrow X_s$ is a regular embedding for all s in a suitable Zariski neighborhood of s_0.*

Let x be a point of Y_{s_0}. There is an affine neighborhood of x in X_{s_0} over which the ideal of Y_{s_0} in X_{s_0} is generated by a regular sequence $f_1, \ldots, f_n$. By (5.1), the f_i extend to generators $F_1, \ldots, F_n$ of the ideal of Y on an affine neighborhood of x in X; by (5.23), we may assume that $F_1, \ldots, F_n$ is a regular sequence. To prove (5.25), we are thus reduced to proving the following result.

LEMMA (5.26). *Let $\varphi : A \to B$ be a homomorphism of noetherian rings, J an ideal in B, and P a prime ideal in B containing J. Set $Q = \varphi^{-1}(P)$. Suppose that B and B/J are flat over A and that J is generated by a regular sequence $F_1, \ldots, F_n$. Then the images of $F_1, \ldots, F_n$ in $B_P \otimes_{A_Q} A_Q/\mathfrak{m}_{A_Q}$ form a regular sequence.*

Set $J_h = (F_1, \ldots, F_h)$, $1 \le h \le n$, and denote by g_i the image of F_i in $B_P \otimes_{A_Q} A_Q/\mathfrak{m}_{A_Q}$. We know that

$$0 \to B/J_{h-1} \xrightarrow{\times F_h} B/J_{h-1} \to B/J_h \to 0$$

is exact. Hence,

$$0 \to (B/J_{h-1})_P \to (B/J_{h-1})_P \to (B/J_h)_P \to 0$$

is also exact. If $(B/J_h)_P$ were flat over A_Q,

$$\times F_h \otimes 1 : (B/J_{h-1})_P \otimes_{A_Q} A_Q/\mathfrak{m}_{A_Q} \to (B/J_{h-1})_P \otimes_{A_Q} A_Q/\mathfrak{m}_{A_Q}$$

would be injective. On the other hand,

$$(B/J_{h-1})_P \otimes_{A_Q} A_Q/\mathfrak{m}_{A_Q} = (B_P \otimes_{A_Q} A_Q/\mathfrak{m}_{A_Q})/(g_1, \ldots, g_{h-1}),$$

and $\times F_h \otimes 1$ is the multiplication by g_h, so we would conclude that $g_1, \ldots, g_n$ is a regular sequence. We must then show that $(B/J_h)_P$ is flat over A_Q for every h. This we will do by descending induction on h, starting from the case $h = n$, where we have flatness by assumption. In general, from

$$0 \to B/J_h \to B/J_h \to B/J_{h+1} \to 0$$

we obtain an exact sequence

$$\mathrm{Tor}_2^A(B/J_{h+1}, N) \to \mathrm{Tor}_1^A(B/J_h, N) \to \mathrm{Tor}_1^A(B/J_h, N) \to \mathrm{Tor}_1^A(B/J_{h+1}, N)$$

for any A-module N. Since

$$\mathrm{Tor}_i^A(B/J_{h+1}, N)_P = \mathrm{Tor}_i^{A_Q}((B/J_{h+1})_P, N_Q)$$

and the latter is zero by induction hypothesis, multiplication by F_{h+1} gives an isomorphism

$$\mathrm{Tor}_1^A(B/J_h, N)_P \xrightarrow{\sim} \mathrm{Tor}_1^A(B/J_h, N)_P .$$

This implies that

$$\operatorname{Tor}_1^A(B/J_h, N)_P = \mathfrak{m}_{B_P} \operatorname{Tor}_1^A(B/J_h, N)_P$$

and hence, by Nakayama's lemma, that

$$\operatorname{Tor}_1^{A_Q}((B/J_h)_P, N_Q) = \operatorname{Tor}_1^A(B/J_h, N)_P = 0$$

when N_Q is finitely generated over A_Q. This proves that $(B/J_h)_P$ is flat over A_Q and concludes the proof of Lemma (5.26).

6. Mumford's example.

As we have already hinted, Hilbert schemes can be quite pathological. In this section we shall give an example, due to Mumford, of a Hilbert scheme of space curves which is everywhere non-reduced along one of its components. The general point of the offending component is a curve liying on a smooth cubic surface in $\mathbb{P}^3$. In our discussion, we shall freely use some well-known facts about cubic surfaces; for a more thorough discussion of these and for proofs, we refer to [318].

Let F be a smooth cubic surface in $\mathbb{P}^3$. It is a classical fact that on F there are exactly 27 lines. Denote by L one of them, and by X a hyperplane section. Using the adjunction formula, we get that

$$\omega_F = \mathcal{O}(-X), \quad (X \cdot X) = 3, \quad (X \cdot L) = 1, \quad (L \cdot L) = -1.$$

It is clear that $H^0(F, \mathcal{O}(nX + mL))$ vanishes when $n < 0$ and $m \leq 0$. In fact, the assumption that m be nonpositive is not needed. This can be shown, inductively on m, using the exact cohomology sequence of

$$0 \to \mathcal{O}(nX + (m-1)L) \to \mathcal{O}(nX + mL) \to \mathcal{O}_L(nX + mL) \to 0.$$

This observation, coupled with Serre duality, implies that

$$\begin{cases} h^0(F, \mathcal{O}(nX + mL)) = 0, & n < 0, \\ h^2(F, \mathcal{O}(nX + mL)) = 0, & n \geq 0. \end{cases}$$

We can say something also on $H^1(F, \mathcal{O}(nX + mL))$. In fact, it is easily checked that $|nX + mL|$ has no base points if $n \geq m > 0$; since X is very ample, it follows that $nX + mL$ is very ample if $n > m \geq 0$. Therefore, by Kodaira vanishing and Serre duality, we get

$$\begin{cases} h^1(F, \mathcal{O}(nX + mL)) = 0, & n \geq m \geq 0, \\ h^1(F, \mathcal{O}(nX + mL)) = 0, & 0 \geq m > n. \end{cases}$$

If n and m are integers such that $n > m \geq 0$, and C is a general member of $|nX + mL|$, then C is a smooth curve of genus

$$g = 3\left(\frac{n^2 - n}{2}\right) - \frac{m^2 + m}{2} + nm + 1$$

and degree $d = 3n + m$. Moreover, by the Riemann–Roch theorem on F,

$$\dim |C| = 3\left(\frac{n^2+n}{2}\right) - \frac{m^2-m}{2} + nm\,.$$

From now on, we let $n = 4$ and $m = 2$, so that C is a smooth curve of genus 24 and degree 14, lying on a smooth cubic surface. Our purpose is to study the Hilbert scheme $\mathrm{Hilb}_3^{p(t)}$, where $p(t) = 14t - 23$ is the Hilbert polynomial of C. To do it, we are going to move the curve C in its linear system, and the cubic surface as well. We denote by $V \subset \mathrm{Hilb}_3^{p(t)}$ the locus corresponding to smooth curves C belonging to $|4X + 2L|$ for some smooth cubic surface F and some line L on F. Since cubic surfaces in $\mathbb{P}^3$ depend on 19 parameters, the dimension of $|C|$ is equal to 37, and each curve C is contained in a unique cubic by degree reasons, we have

$$\dim V = 56\,.$$

It is well known, and fairly easy to prove, that monodromy acts transitively on the set of all lines lying on a given smooth cubic surface (see for instance Exercise I-1). As a consequence, V is an irreducible, locally closed subscheme of $\mathrm{Hilb}_3^{p(t)}$. Let H be an irreducible component of $\mathrm{Hilb}_3^{p(t)}$ containing V. We will prove that

$$\dim H = \dim V = 56 \tag{6.1}$$

and that, for every point $x \in H$,

$$\dim T_x(H) = \dim H + 1, \tag{6.2}$$

so that H is everywhere nonreduced.

Consider the normal bundle $N_{C/\mathbb{P}^3}$, and let x be the point in H corresponding to C. We know that

$$T_x(H) = H^0(C, N_{C/\mathbb{P}^3})$$

and moreover, by (5.17), that

$$\chi(N_{C/\mathbb{P}^3}) = 4 \cdot 14 = 56\,.$$

We wish to show that $h^0(C, N_{C/\mathbb{P}^3}) = 57$ or, equivalently, that $h^1(C, N_{C/\mathbb{P}^3}) = 1$. It follows from the exact cohomology sequence of

$$0 \to \mathcal{O}_C(C) = N_{C/F} \to N_{C/\mathbb{P}^3} \to N_{F/\mathbb{P}^3}{}_{|C} = \mathcal{O}(3X) \otimes \mathcal{O}_C \to 0$$

that $h^1(C, N_{C/\mathbb{P}^3}) = h^1(C, \mathcal{O}(3X) \otimes \mathcal{O}_C)$. Looking at the exact sequence

$$0 \to \mathcal{O}(-X - 2L) \to \mathcal{O}(3X) \to \mathcal{O}(3X) \otimes \mathcal{O}_C \to 0\,,$$

one easily sees that $h^1(C, \mathcal{O}(3X) \otimes \mathcal{O}_C) = h^2(F, \mathcal{O}(-X - 2L)) = h^0(F, \mathcal{O}(2L)) = 1$, proving that the dimension of $H^0(C, N_{C/\mathbb{P}^3})$ is equal to 57. To prove (6.1) and (6.2), we must show that the dimension of H is equal to 56. Since H contains V, it suffices to show that

$$\dim H \leq 56\,. \tag{6.3}$$

Let Γ denote a curve corresponding to a general point of H. Since $H \supset V$, we know that Γ is a smooth connected curve of genus 24 and degree 14. We also know that Γ does not lie in a plane or in a quadric. Furthermore, we can assume that Γ does not lie on a cubic, otherwise H would coincide with the closure of V, and we would be done. In fact, as the Picard group of a smooth cubic surface is discrete, Γ would belong to $|4X + 2L|$ for some line L on the surface, since this is the case for the curves which correspond to points of V. Since $h^0(\mathbb{P}^3, \mathcal{O}(4)) = 35$ and $h^0(\Gamma, \mathcal{O}_\Gamma(4)) = 33$, the curve Γ lies in a pencil of quartics, all of which we may assume to be irreducible. The quartics of this pencil meet in Γ and in a residual conic Γ'. By Bertini's theorem, the singular points of a general member G of this pencil of quartics lie on $\Gamma \cup \Gamma'$. Since this reducible curve can have at most triple points, it follows that G is smooth. In fact, G is a K3 surface, and hence, denoting by $|\Gamma|_G$ the linear system *on* G to which Γ belongs, $\dim |\Gamma|_G = g(\Gamma) = 24$. Now let us consider the variety

$$Y = \left\{ (G, [\Gamma]) \ : \ \begin{array}{c} G \text{ a smooth quartic, } \Gamma \subset G \text{ a smooth curve} \\ \text{of degree 14 and genus 24, with } [\Gamma] \in H \end{array} \right\}.$$

We just proved that the projection $\pi : Y \to H$ is dominant and that its fiber dimension is at least equal to 1. To prove (6.3), it then suffices to show that $\dim Y \leq 57$. But this is obvious. In fact, the quartics involved in the definition of Y are quartics containing a conic, and these quartics depend on 34–9+8=33 parameters (there are ∞^8 conics in $\mathbb{P}^3$, and there are 9 conditions to impose for a quartic to contain one of them). Thus,

$$\dim Y \leq 33 + \dim |\Gamma|_G = 33 + 24 = 57\,.$$

This completes the proof of (6.1) and (6.2).

It is interesting to interpret the nonreducedness of H in terms of first-order deformations. Let us look at the exact diagram

$$\begin{array}{ccccccccc} 0 & \longrightarrow & H^0(C, N_{C/F}) & \longrightarrow & H^0(C, N_{C/\mathbb{P}^3}) & \xrightarrow{\alpha} & H^0(C, \mathcal{O}(3X) \otimes \mathcal{O}_C) & \longrightarrow & 0 \\ & & & & & & \uparrow{\scriptstyle \beta} & & \\ & & & & & & H^0(\mathbb{P}^3, \mathcal{O}(3)) & & \end{array}$$

The 37-dimensional space $H^0(C, N_{C/F})$ represents the first-order deformations of C in F. The 57-dimensional space $H^0(C, N_{C/\mathbb{P}^3})$ represents the first-order deformations of C in $\mathbb{P}^3$, i.e., is the tangent space to H at the point corresponding to C. The kernel of β is one-dimensional and generated by an equation of the cubic F containing C. Therefore, the image of β is 19-dimensional and represents the first-order deformations of F in $\mathbb{P}^3$. Any element in $H^0(C, N_{C/\mathbb{P}^3})$ whose image under α lies in $\beta(H^0(\mathbb{P}^3, \mathcal{O}(3))) \subset H^0(C, \mathcal{O}(3X) \otimes \mathcal{O}_C)$ represents a first-order deformation of C in $\mathbb{P}^3$ for which the deformed curve still lies on a cubic. On the other hand, any element in $H^0(C, N_{C/\mathbb{P}^3})$ whose image under α falls out of $\beta(H^0(\mathbb{P}^3, \mathcal{O}(3))) \subset H^0(C, \mathcal{O}(3X) \otimes \mathcal{O}_C)$ represents a first-order deformation of C in $\mathbb{P}^3$ for which the deformed curve does not lie on any cubic. These are the first-order deformations that cannot be integrated to an actual deformation. The corresponding tangent vectors to H are the ones that do not belong to $T_{[C]}(H_{red})$.

7. Variants of the Hilbert scheme.

The existence of the Hilbert schemes parameterizing subschemes of projective spaces makes it relatively easy to perform several related constructions. In this section we shall briefly sketch a few of these.

The first generalization that comes to mind is that of a Hilbert scheme $\mathrm{Hilb}_X^{p(t)}$ parameterizing subschemes with Hilbert polynomial $p(t)$ of a fixed projective scheme $X \subset \mathbb{P}^r$. More generally, given a scheme S and a closed subscheme X of $\mathbb{P}^r \times S$, it is possible to construct a Hilbert scheme $\mathrm{Hilb}_{X/S}^{p(t)}$ parameterizing couples (s, Y), where s is a point of S, and Y is a closed subscheme of the fiber of $X \to S$ over s, with Hilbert polynomial $p(t)$. Formally, the problem is to represent the functor $\mathbf{hilb}_{X/S}^{p(t)}$ from schemes over S to sets defined by

$$\mathbf{hilb}_{X/S}^{p(t)}(T/S) = \left\{ \begin{array}{c} \text{closed subschemes } Y \subset X \times_S T, \text{ flat over } T \\ \text{with fibers having } p(t) \text{ as a Hilbert polynomial} \end{array} \right\}.$$

Clearly, $\mathrm{Hilb}_{\mathbb{P}^r \times S/S}^{p(t)}$ is nothing but $H \times S$, where H stands for $\mathrm{Hilb}_r^{p(t)}$. In general, we shall realize $\mathrm{Hilb}_{X/S}^{p(t)}$ as a closed subscheme of $H \times S$. We let $\mathcal{X} \subset \mathbb{P}^r \times H$ be the universal family over H and set $\mathcal{X}' = \mathcal{X} \times S$. Likewise, we let X' be the inverse image of X under the projection from $\mathbb{P}^r \times H \times S$ to $\mathbb{P}^r \times S$. We claim that there is a closed subscheme Z of $H \times S$ such that a morphism $T \to H \times S$ factors through Z if and only if $\mathcal{X}' \times_{H \times S} T$ is a subscheme of $X' \times_{H \times S} T$. But then $\mathrm{Hilb}_{X/S}^{p(t)}$ is just Z, since for a morphism $T \to H \times S$ of schemes over S, we have that $\mathcal{X}' \times_{H \times S} T = \mathcal{X} \times_H T$ and $X' \times_{H \times S} T = X \times_S T$. That a subscheme Z as above exists is a special case of the following simple result.

LEMMA (7.1). *Let A be a scheme, and let B and C be closed subschemes of $\mathbb{P}^r \times A$. Assume that B is flat over A. Then there is a closed subscheme*

D of A such that any morphism $T \to A$ factors through D if and only if $B \times_A T$ is a subscheme of $C \times_A T$.

In down-to-earth terms, the lemma asserts the existence of a subscheme D of A parameterizing points a of A such that the fiber of B over a is a subscheme of the corresponding fiber of C.

Proof. We shall denote by π the projection of $\mathbb{P}^r \times A$ to A; for any scheme T, we shall also write $\mathcal{O}$ instead of $\mathcal{O}_{\mathbb{P}^r \times T}$ when no confusion is likely. Let $\varphi : \mathcal{F} \to \mathcal{G}$ be a homorphism of coherent sheaves on A. The regular functions on A which are locally of the form $\lambda(\varphi(s))$, where s is a local section of $\mathcal{F}$, and λ is a local section of the dual of $\mathcal{G}$, make up a sheaf of ideals in $\mathcal{O}_A$. We shall refer to the corresponding subscheme of A as the scheme of zeros of φ. We apply this construction to the homomorphisms

$$\varphi_n : \pi_*\mathcal{I}_C(n) \to \pi_*\mathcal{O}_B(n)$$

obtained by composing the inclusions $\pi_*\mathcal{I}_C(n) \to \pi_*\mathcal{O}(n)$, where $\mathcal{I}_C$ stands for the ideal sheaf of C, with the restriction homomorphisms $\pi_*\mathcal{O}(n) \to \pi_*\mathcal{O}_B(n)$. We claim that the subscheme of zeros of φ_n does not depend on n for large enough n; this will be the subscheme D we are looking for. Since the claim is local on A, in proving it we are allowed to take A as small as necessary. Now denote by $\mathcal{I}_n$ the ideal sheaf of the subscheme of zeros of φ_n and look at the diagram

$$\begin{array}{ccc} \pi_*\mathcal{O}(1) \otimes \pi_*\mathcal{I}_C(n) & \xrightarrow{\mathrm{id} \otimes \varphi_n} & \pi_*\mathcal{O}(1) \otimes \pi_*\mathcal{O}_B(n) \\ \Big\downarrow{\scriptstyle \mu_n} & & \Big\downarrow{\scriptstyle \nu_n} \\ \pi_*\mathcal{I}_C(n+1) & \xrightarrow{\varphi_{n+1}} & \pi_*\mathcal{O}_B(n+1) \end{array}$$

If λ is a local section of the dual of $\pi_*\mathcal{O}_B(n+1)$, then $\lambda \circ \nu_n$ is of the form $\sum \sigma_i \otimes \lambda_i$, where the σ_i are sections of the dual of $\pi_*\mathcal{O}(1)$ and the λ_i are sections of the dual of $\pi_*\mathcal{O}_B(n)$. For large n, the map μ_n is onto, and hence any section s of $\pi_*\mathcal{I}_C(n+1)$ is of the form $\mu_n(\sum t_j \otimes s_j)$, where the t_j are sections of $\pi_*\mathcal{O}(1)$ and the s_j are sections of $\pi_*\mathcal{I}_C(n)$. Thus,

$$\lambda(\varphi_{n+1}(s)) = \sum_{i,j} \sigma_i(t_j)\lambda_i(\varphi_n(s_j))$$

is a section of $\mathcal{I}_n$. This proves that $\mathcal{I}_{n+1} \subset \mathcal{I}_n$ as soon as μ_n is onto.

It remains to show that, conversely, $\mathcal{I}_{n+1} \supset \mathcal{I}_n$ for large n. If a is a point of A, we shall denote by B_a the fiber of $B \to A$ over a. Let σ be a linear form on $\mathbb{P}^r$, and H the corresponding hyperplane; set $H' = H \times A$. There is an integer n_0 such that, for any $n \geq n_0$, $\pi_*\mathcal{O}_B(n)$ is locally free, $R^1\pi_*\mathcal{O}_B(n) = 0$, and in addition $H^0(B_a \cap H, \mathcal{O}_{B_a \cap H}(n)) = \pi_*\mathcal{O}_{B\cap(H\times A)}(n) \otimes k(a)$, $H^j(B_a \cap H, \mathcal{O}_{B_a\cap H}(n)) = 0$ for any $j > 0$ and any $a \in A$. Now fix a closed point a_0 in A and choose σ so that H does not contain any components of B_{a_0}, including the embedded ones. Thus the homomorphism $H^0(B_{a_0}, \mathcal{O}_{B_{a_0}}(n)) \to H^0(B_{a_0}, \mathcal{O}_{B_{a_0}}(n+1))$ given by σ is injective for any n. Shrinking A, if necessary, we may then assume that $H^0(B_a, \mathcal{O}_{B_a}(n)) \to H^0(B_a, \mathcal{O}_{B_a}(n+1))$ is injective for any $a \in A$ and for $n = n_0, \ldots, n_0 + r$. By our choice of n_0, for these values of n, the sequence

$$0 \to H^0(B_a, \mathcal{O}_{B_a}(n)) \to H^0(B_a, \mathcal{O}_{B_a}(n+1)) \to H^0(B_a \cap H, \mathcal{O}_{B_a \cap H}(n+1)) \to 0$$

is exact for any a; it follows in particular that $h^0(B_a \cap H, \mathcal{O}_{B_a \cap H}(n+1))$ is independent of a. This means that the values of the Hilbert polynomial $h_{B_a \cap H}(t)$ at $t = n_0 + 1, \ldots, n_0 + r + 1$ do not depend on a. Hence, $h_{B_a \cap H}$ is independent of a, since its degree is not greater than r. In particular, $h^0(B_a \cap H, \mathcal{O}_{B_a \cap H}(n+1))$ does not depend on a for any $n \geq n_0$. By dimension reasons it then follows that $H^0(B_a, \mathcal{O}_{B_a}(n)) \to H^0(B_a, \mathcal{O}_{B_a}(n+1))$ is injective for any $a \in A$ and for any $n \geq n_0$.

What the preceding argument proves is that, when n is at least n_0, $\pi_*\mathcal{O}_B(n)$ is a subsheaf of $\pi_*\mathcal{O}_B(n+1)$, and the quotient $\pi_*\mathcal{O}_B(n+1)/\pi_*\mathcal{O}_B(n)$ is locally free. As a consequence, any section of the dual of $\pi_*\mathcal{O}_B(n)$ over a sufficiently small open set extends to a section of the dual of $\pi_*\mathcal{O}_B(n+1)$. Now look at the diagram

$$\begin{array}{ccc}
\pi_*\mathcal{I}_C(n) & \xrightarrow{\varphi_n} & \pi_*\mathcal{O}_B(n) \\
{\scriptstyle i_n}\downarrow & & \downarrow{\scriptstyle j_n} \\
\pi_*\mathcal{I}_C(n+1) & \xrightarrow{\varphi_{n+1}} & \pi_*\mathcal{O}_B(n+1)
\end{array}$$

where i_n and j_n are multiplication by σ. If s and λ are sections of $\pi_*\mathcal{I}_C(n)$ and of the dual of $\pi_*\mathcal{O}_B(n)$, respectively, then λ is locally of the form $\lambda' \circ j_n$ for some section λ' of the dual of $\pi_*\mathcal{O}_B(n+1)$. Hence, $\lambda(\varphi_n(s)) = \lambda'(j_n(\varphi_n(s))) = \lambda'(\varphi_{n+1}(i_n(s)))$ is a section of $\mathcal{I}_{n+1}$. This proves that $\mathcal{I}_n \subset \mathcal{I}_{n+1}$ for large n, as desired.

We may now complete the proof of the lemma. Let $f : T \to A$ be a morphism. Denote by η the projection of $\mathbb{P}^r \times T$ onto T and by F the morphism $\mathrm{id} \times f : \mathbb{P}^r \times T \to \mathbb{P}^r \times A$. Set $B' = B \times_A T$, $C' = C \times_A T$. Now look at the diagram

$$\begin{array}{ccccccccc}
 & & f^*\pi_*\mathcal{I}_C(n) & \longrightarrow & f^*\pi_*\mathcal{O}(n) & \longrightarrow & f^*\pi_*\mathcal{O}_C(n) & \longrightarrow & 0 \\
 & & {\scriptstyle u_n}\downarrow & & {\scriptstyle v_n}\downarrow & & {\scriptstyle w_n}\downarrow & & \\
0 & \longrightarrow & \eta_*\mathcal{I}_{C'}(n) & \longrightarrow & \eta_*\mathcal{O}(n) & \longrightarrow & \eta_*\mathcal{O}_{C'}(n) & \longrightarrow & 0
\end{array}$$

For large n, the two rows of the diagram are exact, and moreover v_n and w_n are isomorphisms, by Lemma (3.8) and by the remark that

$\mathcal{O}_{C'} = F^*\mathcal{O}_C$. Thus, u_n is onto for large n. Since $\eta_*\mathcal{O}_{B'}(n) = f^*\pi_*\mathcal{O}_B(n)$, this implies that, for n large enough, $f^*\pi_*\mathcal{I}_C(n) \to f^*\pi_*\mathcal{O}_B(n)$ is zero if and only if $\eta_*\mathcal{I}_{C'}(n) \to \eta_*\mathcal{O}_{B'}(n)$ is. This means that f lands in D if and only if $\eta_*\mathcal{I}_{C'}(n) \to \eta_*\mathcal{O}_{B'}(n)$ vanishes for all sufficiently large n, that is, if and only if $\mathcal{I}_{C'} \to \mathcal{O}_{B'}$ is zero or, put otherwise, $B' = B \times_A T$ is a subscheme of $C' = C \times_A T$. This concludes the proof of (7.1). Q.E.D.

An immediate consequence of (4.10) and of the construction of $\mathrm{Hilb}^p_{X/S}$ is the following.

LEMMA (7.2). *$\mathrm{Hilb}^p_{X/S} \to S$ is proper.*

It goes without saying that, for fixed p, the Hilbert scheme Hilb^p_X, or more generally $\mathrm{Hilb}^p_{X/S}$, depends on the projective embedding of X. What is independent of the embedding is the infinite union

$$\mathrm{Hilb}_{X/S} = \coprod_p \mathrm{Hilb}^p_{X/S},$$

which, however, is not of finite type. When S is a point, we shall write Hilb_X instead of $\mathrm{Hilb}_{X/S}$.

Contrary to what happens when $X = \mathbb{P}^r$, the Hilbert scheme Hilb^p_X is in general not connected. For instance, when $p(t) = t + 1$ and X is a smooth cubic surface in three-space, Hilb^p_X parameterizes lines lying on X. As we know, there are exactly 27 of these, so in this case, Hilb^p_X consists of 27 points; we shall presently see that it is also reduced. If X is any closed subscheme of $\mathbb{P}^r$ and $p(t)$ is a rational polynomial, Proposition (5.9) describes the tangent spaces to $\mathrm{Hilb}^{p(t)}_X$. If h is a point of $\mathrm{Hilb}^{p(t)}_X$ corresponding to a subscheme Y of X, then the tangent space to $\mathrm{Hilb}^{p(t)}_X$ at h is

$$T_h(\mathrm{Hilb}^{p(t)}_X) = H^0(Y, N_{Y/X}). \tag{7.3}$$

In the special case where X is a smooth cubic in $\mathbb{P}^3$ and Y is a line in X, the adjunction formula for Y, coupled with the fact that the canonical bundle on X is $\mathcal{O}_X(-1)$, shows that the self-intersection of Y equals -1. This means that the degree of $N_{Y/X} = \mathcal{O}_Y(Y)$ is -1, which, by (7.3), implies that $T_h(\mathrm{Hilb}^{p(t)}_X) = 0$. Thus, $\mathrm{Hilb}^{p(t)}_X$ is reduced at h.

The formation of the Hilbert scheme is compatible with base change, in the sense that, for any morphism $T \to S$, there is a canonical isomorphism

$$\mathrm{Hilb}_{X \times_S T/T} \simeq \mathrm{Hilb}_{X/S} \times_S T.$$

This follows at once from the universal property of the Hilbert scheme.

REMARK (7.4). The argument above shows that $\mathrm{Hilb}_{Z/T}$ exists also when $Z \to T$ is just an analytic family of projective schemes, at least when Z is flat over T. The question is local on T, so by (4.9) we may suppose that $Z = X \times_S T$ for some morphism $T \to S$, where $X \to S$ is an algebraic family. Then $\mathrm{Hilb}_{X/S} \times_S T$ is the sought-for space, since it clearly possesses the required universal property.

A very important construction that can be carried out thanks to the existence of general Hilbert schemes is the one of the *scheme* (not of finite type) $\mathbf{Hom}_S(X, Y)$, where $X \subset \mathbb{P}^r \times S$ and $Y \subset \mathbb{P}^t \times S$ are closed subschemes, and X is flat over S. This is a scheme representing the functor

$$\mathbf{h}(T/S) = \mathrm{Hom}_T(X \times_S T, Y \times_S T).$$

In fact, we shall see that there is an open subscheme of $\mathrm{Hilb}_{X \times_S Y/S}$ representing $\mathbf{h}$. We begin by observing that, given schemes Z and W over T, associating to a morphism its graph gives a one-to-one correspondence between morphisms $Z \to W$ of schemes over T and closed subschemes Γ of $Z \times_T W$ which project isomorphically onto Z. The next remark is that, for a subscheme of a fibered product, being a graph is an open condition. Formally, we have the following result.

LEMMA (7.5). *Let A, B, and C be schemes, and let $\alpha : A \to C$ and $\beta : B \to C$ be flat projective morphisms. Let $f : A \to B$ be a morphism of schemes over C. Assume that there is a closed point $c \in C$ such that $f_c : A_c \to B_c$ is an isomorphism. Then there is a neighborhood U of c such that f maps $\alpha^{-1}(U)$ isomorphically onto $\beta^{-1}(U)$.*

To prove the lemma, we may argue as follows. Let M be a line bundle on A that is very ample relative to f, and let b_0 be a point of B_c. Using (3.6) and replacing M with a power, if necessary, we may assume that, for any $m \geq 1$, the map $f_*M^m \otimes k(b) \to H^0(f^{-1}(b), M^m \otimes \mathcal{O}_{f^{-1}(b)})$ is an isomorphism for any $b \in B$ and the map $f_*M \otimes f_*M^m \to f_*M^{m+1}$ is onto. Since $A_c \to B_c$ is an isomorphism, $H^0(f^{-1}(b_0), M \otimes \mathcal{O}_{f^{-1}(b_0)})$ is one-dimensional, and hence f_*M has rank 1 at b_0. Let V be a neighborhood of b_0 such that the rank of f_*M is not greater than 1 at any point of V. Then, for any $b \in V$, the dimension of $H^0(f^{-1}(b), M^m \otimes \mathcal{O}_{f^{-1}(b)})$ is at most 1 for any $m \geq 1$, so $f^{-1}(b)$ consists of at most one point. By compactness it follows that, possibly after shrinking C, we may assume that $f^{-1}(b)$ consists of at most one point for any $b \in B$. Thus, if L is a line bundle on B that is very ample relative to β, its pullback to A is ample relatively to α; replacing L with a suitable power, we may even assume that f^*L is relatively very ample.

Since A and B are flat over C, we may find an integer n_0 such that, for $n \geq n_0$, $\beta_*(L^n)$ and $\alpha_*f^*(L^n)$ are locally free and in addition $\alpha_*f^*(L) \otimes \alpha_*f^*(L^n) \to \alpha_*f^*(L^{n+1})$ is onto. Since A_c and B_c are isomorphic, $\beta_*(L^n) \otimes k(c) \to \alpha_*f^*(L^n) \otimes k(c)$ is an isomorphism for any $n \geq n_0$; in particular $\beta_*(L^n)$ and $\alpha_* f^*(L^n)$ have the same rank. Another consequence is that there is a neighborhood U of c over which $\beta_*(L^{n_0}) \to \alpha_* f^*(L^{n_0})$ is an isomorphism. By our choice of n_0, $\beta_*(L^{hn}) \to \alpha_* f^*(L^{hn})$ is onto for any $h \geq 1$ and hence an isomorphism, by dimension reasons. If follows that $\alpha^{-1}(U)$ and $\beta^{-1}(U)$ are isomorphic.

Lemma (7.5) enables us to conclude that $\mathbf{h}$ is represented by the open subset of $\mathrm{Hilb}_{X\times_S Y/S}$ consisting of the couples (s, Γ), where $s \in S$ and Γ is the graph of a morphism from X_s to Y_s.

When not only X, but also Y, is flat over S, in addition to $\mathbf{Hom}_S(X,Y)$, there is also the scheme $\mathbf{Hom}_S(Y,X)$. The intersection of $\mathbf{Hom}_S(X,Y)$ and $\mathbf{Hom}_S(Y,X)$ inside $\mathrm{Hilb}_{X\times_S Y/S}$ parameterizes isomorphisms between fibers of $X \to S$ and the corresponding fibers of $Y \to S$; we shall denote it by $\mathbf{Isom}_S(X,Y)$. Clearly, $\mathbf{Isom}_S(X,Y)$ represents the functor which associates to each scheme T over S the set of all isomorphisms, as schemes over T, from $X \times_S T$ to $Y \times_S T$.

REMARK (7.6). By Remark (7.4), $\mathbf{Hom}_S(X,Y)$ and $\mathbf{Isom}_S(X,Y)$ exist also, as analytic spaces, when $X \to S$ and $Y \to S$ are flat analytic families of projective schemes.

Another useful space that can be constructed by means of the general Hilbert scheme parameterizes nested pairs of subschemes in the fibers of a family of projective schemes. Let S be a scheme, and let X be a closed subscheme of $\mathbb{P}^r \times S$. Let $p_1(t)$ and $p_2(t)$ be rational polynomials. Denote by H_1 the Hilbert scheme $\mathrm{Hilb}^{p_1(t)}_{X/S}$ and by $\mathcal{Y}_1 \subset \mathbb{P}^r \times H_1$ the universal family on it. Then $\mathrm{Hilb}^{p_2(t)}_{\mathcal{Y}_1/H_1}$ parameterizes pairs (Y_1, Y_2), where $Y_1 \supset Y_2$ are closed subschemes of a fiber of $X \to S$ and the Hilbert polynomial of Y_i is $p_i(t)$, $i = 1, 2$. More generally, given rational polynomials $p_1(t), \dots, p_n(t)$, iterating this construction yields a "flag Hilbert scheme" parameterizing n-tuples $(Y_1, Y_2, \dots, Y_n)$ such that $Y_1 \supset Y_2 \supset \cdots \supset Y_n$ are subschemes of a fiber of $X \to S$ with Hilbert polynomials $p_1(t), \dots, p_n(t)$. In a slightly different direction, the same kind of construction makes it possible, given $p_1(t), \dots, p_n(t)$ as above and an additional polynomial $p(t)$, to construct a Hilbert scheme parameterizing $(n+1)$-tuples $(Y; Z_1, \dots, Z_n)$, where Y is a subscheme of a fiber of $X \to S$ with Hilbert polynomial $p(t)$, and Z_i is a subscheme of Y with Hilbert polynomial $p_i(t)$ for $i = 1, \dots, n$.

Here is another construction that can be performed thanks to the existence of general Hilbert schemes. Let S, X, $p_1(t)$, $p_2(t)$, H_1, and $\mathcal{Y}_1$ be as above. Set $H_2 = \mathrm{Hilb}^{p_2(t)}_{X/S}$ and $H = \mathrm{Hilb}^{p_2(t)}_{\mathcal{Y}_1/H_1}$. There are natural morphisms $H \to H_i$. Let $Z \subset X \times S$ be a closed subscheme, flat over S and such that the Hilbert polynomial of every fiber of $Z \to S$ is $p_2(t)$. Let $\alpha : S \to H_2$ be the corresponding map. Then the fiber product $H \times_{H_2} S$ parameterizes couples consisting of a point s of S and of a subscheme of $\mathbb{P}^r$ with Hilbert polynomial $p_1(t)$, containing the fiber of $Z \to S$ over s and contained in the fiber of $X \to S$ over s. More precisely, $H \times_{H_2} S$ represents the functor

$$\mathbf{h}(T/S) = \left\{ \begin{array}{c} \text{closed subschemes } Y \text{ of } X \times_S T, \text{ flat over } T, \\ \text{containing } Z \times_S T, \text{ and such that the Hilbert} \\ \text{polynomial of every fiber of } Y \to S \text{ is } p_1(t) \end{array} \right\}.$$

As a special case, given a closed subscheme Z of $\mathbb{P}^r$ and a rational polynomial $p(t)$, this construction provides a Hilbert scheme parameterizing closed subschemes of $\mathbb{P}^r$ with Hilbert polynomial $p(t)$ and containing Z as a subscheme.

Exercise (7.7). Let X and Y be closed subschemes of $\mathbb{P}^r \times S$, flat over S. Let $Z_1, \dots, Z_n$ be closed subschemes of X, and $W_1, \dots, W_n$ closed subschemes of Y, all flat over S. Generalizing the construction of $\mathbf{Isom}_S(X, Y)$, show that there exists a Hilbert scheme $\mathbf{Isom}_S((X; Z_1, \dots, Z_n), (Y; W_1, \dots, W_n))$ parameterizing pairs (s, φ), where $s \in S$ and φ is an isomorphism $X_s \to Y_s$ carrying $Z_{i,s}$ isomorphically to $W_{i,s}$ for $i = 1, \dots, n$.

8. Tangent space computations.

As we observed, the formation of the Hilbert scheme is compatible with base change. In particular, this has the following implication. Let S be a scheme, and let X be a closed subscheme of $\mathbb{P}^r \times S$. If s is a closed point of S and, as usual, we write X_s for the fiber of $X \to S$ at s, and similarly for the one of $\mathrm{Hilb}_{X/S}$ at s, then

$$(\mathrm{Hilb}_{X/S})_s = \mathrm{Hilb}_{X_s}$$

It follows that, for any closed point $h = [Y] \in \mathrm{Hilb}_{X_s}$, the sequence of tangent spaces

$$0 \to T_h(\mathrm{Hilb}_{X_s}) \to T_h(\mathrm{Hilb}_{X/S}) \to T_s(S) \tag{8.1}$$

is exact. In particular,

$$\dim_h(\mathrm{Hilb}_{X/S}) \leq H^0(Y, N_{Y/X_s}) + \dim_s(S)\,. \tag{8.2}$$

We warn the reader that the rightmost homomorphism in (8.1) is generally not onto, even when $X \to S$ and Y are "nice" (see, for instance, Example (8.21) below).

We know from (5.9) that, when S is a point, $T_h(\mathrm{Hilb}_{X/S}) = T_h(\mathrm{Hilb}_X)$ is just $H^0(Y, N_{Y/X})$. We wish to show that it is possible to describe $T_h(\mathrm{Hilb}_{X/S})$ along similar lines for arbitrary S. As usual, we write Σ to indicate $\mathrm{Spec}\,\mathbb{C}[\varepsilon]$, where $\mathbb{C}[\varepsilon]$ is the ring of dual numbers. We

also denote by π the projection from X to S, and by Y the subscheme of X_s corresponding to h. An element of $T_h(\mathrm{Hilb}_{X/S})$ consists of:
(8.3)
- a morphism $\varphi : \Sigma \to (S, s)$;
- a subscheme Y' of $X \times_S \Sigma$, flat over Σ, extending Y.

Via the inclusion $X \times_S \Sigma \subset X \times \Sigma$, the subscheme Y' corresponds to an element of $H^0(Y, N_{Y/X}) = \mathrm{Hom}_{\mathcal{O}_X}(\mathcal{J}, \mathcal{O}_Y)$, where $\mathcal{J}$ is the ideal sheaf of Y in X. On the other hand, if we denote by $\mathfrak{m}$ the ideal sheaf of s in S, φ corresponds to an element of $\mathrm{Hom}(\mathfrak{m}, \mathcal{O}_S/\mathfrak{m}) = T_s(S)$.

LEMMA (8.4). *$T_h(\mathrm{Hilb}_{X/S})$ is the subspace of $\mathrm{Hom}_{\mathcal{O}_X}(\mathcal{J}, \mathcal{O}_Y) \oplus \mathrm{Hom}(\mathfrak{m}, \mathcal{O}_S/\mathfrak{m})$ consisting of all pairs (v, w) such that*

$$v(\pi^*(u)) = w(u) \quad \text{for all sections } u \text{ of } \mathfrak{m}. \tag{8.5}$$

We denote by φ the morphism $\Sigma \to (S, s)$ corresponding to w, and by Y' the subscheme of $X \times \Sigma$, flat over Σ and extending Y, corresponding to v. We must show that Y' is a subscheme of $X \times_S \Sigma$ if and only if (8.5) is valid. The ideal of $X \times_S \Sigma$ in $X \times \Sigma$ is generated by the functions

$$\pi^*(u) \otimes 1 - 1 \otimes \varphi^*(u)\,, \tag{8.6}$$

where u runs through all sections of $\mathfrak{m}$. Recall that $\varphi^*(u) = w(u)\varepsilon$ for any section u of $\mathfrak{m}$. Since $w(u)$ is a constant, (8.6) is equal to

$$\pi^*(u) \otimes 1 - w(u) \otimes \varepsilon\,.$$

On the other hand, the ideal of Y' consists of all functions of the form $f \otimes 1 - f' \otimes \varepsilon$ such that the reduction of f' modulo $\mathcal{J}$ is equal to $v(f)$. Thus (8.6) belongs to the ideal of Y' if and only if $v(\pi^*(u)) = w(u)$. This proves (8.4).

COROLLARY (8.7). *The projection $T_h(\mathrm{Hilb}_{X/S}) \to H^0(Y, N_{Y/X})$ is injective.*

It is not difficult to describe the tangent spaces to all the variants of the Hilbert scheme we have introduced in the previous section; here we limit ourselves to the one that we will most often encounter in the sequel. Let $\eta : X \to S$ be a projective morphism, and denote by H the Hilbert scheme parameterizing $(n+1)$-tuples $(Y; Z_1, \dots, Z_n)$ of subschemes of the fibers of η such that $Z_i \subset Y$ for $i = 1, \dots, n$. There are morphisms $\pi_i : H \to \mathrm{Hilb}_{X/S}$, $i = 0, \dots, n$, given by $\pi_0(Y; Z_1, \dots, Z_n) = Y$, $\pi_i(Y; Z_1, \dots, Z_n) = Z_i$ for $i = 1, \dots, n$, and $(\pi_0, \dots, \pi_n)$ identifies H to a closed subscheme of ${\mathrm{Hilb}_{X/S}}^{n+1}$. If W is a closed subscheme of X, we denote by $\mathcal{I}_W$ its ideal sheaf. Moreover, given a tangent vector v to $\mathrm{Hilb}_{X/S}$ at W, we denote by α_v the corresponding homomorphism of $\mathcal{O}_X$-modules from $\mathcal{I}_W/\mathcal{I}_W^2$ to $\mathcal{O}_W$. Given a point h of H, corresponding to an $(n+1)$-tuple $(Y; Z_1, \dots, Z_n)$, set $h_i = \pi_i(h)$ for $i = 0, \dots, n$. Then the tangent space to H at h is described by the following result, which is an immediate consequence of Lemma (8.4) and Corollary (8.7).

LEMMA (8.8). *The tangent space $T_h(H)$ is the subspace of $T_{h_0}(\mathrm{Hilb}_{X/S}) \oplus T_{h_1}(\mathrm{Hilb}_{X/S}) \oplus \cdots \oplus T_{h_n}(\mathrm{Hilb}_{X/S})$ consisting of the $(n+1)$-tuples $(u; v_1, \ldots, v_n)$ such that the diagrams*

$$\begin{array}{ccc} \mathcal{I}_Y/\mathcal{I}_Y^2 & \xrightarrow{\sigma_i} & \mathcal{I}_{Z_i}/\mathcal{I}_{Z_i}^2 \\ \alpha_u \downarrow & & \downarrow \alpha_{v_i} \\ \mathcal{O}_Y & \xrightarrow{\rho_i} & \mathcal{O}_{Z_i} \end{array}$$

commute for $i = 1, \ldots, n$, where $\sigma_i : \mathcal{I}_Y \to \mathcal{I}_{Z_i}$ and $\rho_i : \mathcal{O}_Y \to \mathcal{O}_{Z_i}$ are the obvious maps. In other words,

$$T_h(H) = \mathrm{Hom}(\mathcal{C}^\bullet, \mathcal{D}^\bullet),$$

where $\mathcal{C}^\bullet$ and $\mathcal{D}^\bullet$ are the complexes

$$\mathcal{C}^\bullet = (\mathcal{I}_Y/\mathcal{I}_Y^2 \to \oplus_i(\mathcal{I}_{Z_i}/\mathcal{I}_{Z_i}^2)), \quad \mathcal{D}^\bullet = (\mathcal{O}_Y \to \oplus_i \mathcal{O}_{Z_i}).$$

EXERCISE (8.9). Let H be the Hilbert scheme parameterizing $(n+1)$-tuples $(Y; p_1, \ldots, p_n)$, where Y is a closed subscheme of $\mathbb{P}^r$, and the p_i are points of Y. Let $h = (Y; p_1, \ldots, p_n)$ be a point of H, and let h_0 be the point of $\mathrm{Hilb}_{\mathbb{P}^r}$ corresponding to Y. Suppose the p_i are distinct.

i) Use Lemma (8.8) to show that there is an exact sequence

$$0 \to \oplus_i T_{p_i}(Y) \to T_h(H) \to T_{h_0}(\mathrm{Hilb}_{\mathbb{P}^r}).$$

ii) Show that, if the p_i are smooth points of Y, then the rightmost homomorphism in this sequence is onto.

iii) Show with an example that the conclusion of ii) may not be valid if at least one of the p_i is not a smooth point of Y.

We now return to sequence (8.1), with the goal of interpreting it in cohomological terms, under the assumption that X is flat over S. As above, we indicate by $\mathcal{J}$ the ideal sheaf of Y in $\mathcal{O}_X$, and by $\mathfrak{m}$ the ideal of s in $\mathcal{O}_S$. In addition, we write $\mathcal{I}$ for the ideal of X_s in $\mathcal{O}_X$, and $\mathcal{K} = \mathcal{J}/\mathcal{I}$ for the ideal of Y in $\mathcal{O}_{X_s}$. From the exact sequence of $\mathcal{O}_{X_s}$-modules

(8.10)
$$0 \to \mathcal{I}/\mathcal{I}\mathcal{J} \to \mathcal{J}/\mathcal{I}\mathcal{J} \to \mathcal{K} \to 0,$$

observing that $\mathrm{Hom}_{\mathcal{O}_{X_s}}(\mathcal{J}/\mathcal{I}\mathcal{J}, \mathcal{O}_Y) = \mathrm{Hom}_{\mathcal{O}_X}(\mathcal{J}, \mathcal{O}_Y)$ and $\mathrm{Hom}_{\mathcal{O}_{X_s}}(\mathcal{I}/\mathcal{I}\mathcal{J}, \mathcal{O}_Y) = \mathrm{Hom}_{\mathcal{O}_X}(\mathcal{I}, \mathcal{O}_Y)$, we get an exact sequence

(8.11)
$$0 \to \mathrm{Hom}_{\mathcal{O}_{X_s}}(\mathcal{K}, \mathcal{O}_Y) \to \mathrm{Hom}_{\mathcal{O}_X}(\mathcal{J}, \mathcal{O}_Y) \to \mathrm{Hom}_{\mathcal{O}_X}(\mathcal{I}, \mathcal{O}_Y) \to \mathrm{Ext}^1_{\mathcal{O}_{X_s}}(\mathcal{K}, \mathcal{O}_Y).$$

The second and third terms from the left in this sequence are $H^0(Y, N_{Y/X_s}) = T_h(\mathrm{Hilb}_{X_s})$ and $H^0(Y, N_{Y/X})$. Since X is flat over S, we have that $\mathcal{I} = \mathfrak{m} \otimes_{\mathcal{O}_S} \mathcal{O}_X$ and $\mathcal{I}^2 = \mathfrak{m}^2 \otimes_{\mathcal{O}_S} \mathcal{O}_X$, and hence that $\mathcal{I}/\mathcal{I}^2 = T_s(S)^\vee \otimes_{\mathbb{C}} \mathcal{O}_{X_s}$. It follows that

$$\mathrm{Hom}_{\mathcal{O}_X}(\mathcal{I}, \mathcal{O}_Y) = \mathrm{Hom}_{\mathcal{O}_{X_s}}(\mathcal{I}/\mathcal{I}^2, \mathcal{O}_Y) = T_s(S) \otimes_{\mathbb{C}} H^0(Y, \mathcal{O}_Y),$$

and (8.11) becomes

(8.12)

$$0 \to H^0(Y, N_{Y/X_s}) \to H^0(Y, N_{Y/X}) \to T_s(S) \otimes H^0(Y, \mathcal{O}_Y) \to \mathrm{Ext}^1_{\mathcal{O}_{X_s}}(\mathcal{K}, \mathcal{O}_Y).$$

By Lemma (8.7), $T_h(\mathrm{Hilb}_{X/S})$ is a subspace of $H^0(Y, N_{Y/X})$. Lemma (8.4) can then be rephrased as saying that an element of $H^0(Y, N_{Y/X})$ belongs to $T_h(\mathrm{Hilb}_{X/S})$ if and only if its image in $T_s(S) \otimes_{\mathbb{C}} H^0(Y, \mathcal{O}_Y)$ belongs to $T_s(S) \otimes 1 \simeq T_s(S)$. In particular, this says that $T_h(\mathrm{Hilb}_{X/S}) = H^0(Y, N_{Y/X})$ when Y is reduced and connected.

EXAMPLE (8.13). Here is a concrete example of an element of $H^0(Y, N_{Y/X})$ which does not come from a tangent vector to $\mathrm{Hilb}_{X/S}$. We choose as S the affine line with coordinate t, and the origin as s. We also set $X = \mathbb{P}^1 \times S$ and denote by π the projection of X to S. Finally, we let z be an affine coordinate on $\mathbb{P}^1$ (minus one point) and choose as Y the subscheme with the equations $t = 0$ and $z^2 = 0$. Let $Y' \subset X \times \Sigma$ be the subscheme defined by the equations $z^2 = t - \varepsilon z = 0$. For this choice of Y', we have that $\mathcal{J}/\mathcal{J}^2 = (z^2, t)/(z^4, z^2 t, t^2)$ is a free $\mathcal{O}_Y$-module generated by the classes $[t]$ and $[z^2]$, while

$$H^0(Y, N_{Y/X}) = \mathrm{Hom}(\mathcal{J}/\mathcal{J}^2, \mathcal{O}_Y) = H^0(Y, \mathcal{O}_Y)u + H^0(Y, \mathcal{O}_Y)v,$$

where u, v is the dual basis of $[t], [z^2]$. The image of Y' in $H^0(Y, N_{Y/X})$ under the characteristic map is zu, and its image in $T_s(S) \otimes H^0(Y, \mathcal{O}_Y)$ is $\frac{\partial}{\partial t} \otimes z$, which does not belong to $T_s(S) \otimes 1$.

As is clear from its construction, the homomorphism $T_h(\mathrm{Hilb}_{X/S}) \to T_s(S)$ in (8.1) agrees with the one obtained by restricting to $T_h(\mathrm{Hilb}_{X/S})$ the homomorphism $H^0(Y, N_{Y/X}) \to T_s(S) \otimes H^0(Y, \mathcal{O}_Y)$ in (8.12). We can thus summarize what we have proved in the following statement.

PROPOSITION (8.14). *Let X be flat and projective over S. Let s be a closed point of S, let Y be a closed subscheme of X_s, and denote by h the corresponding point of $\mathrm{Hilb}_{X/S}$. Then the exact sequence (8.1) extends to an exact sequence*

$$(8.15) \qquad 0 \to T_h(\mathrm{Hilb}_{X_s}) \to T_h(\mathrm{Hilb}_{X/S}) \to T_s(S) \to \mathrm{Ext}^1_{\mathcal{O}_{X_s}}(\mathcal{K}, \mathcal{O}_Y),$$

where $\mathcal{K}$ stands for the ideal sheaf of Y in X_s. Moreover, under the natural identifications and inclusions $T_h(\mathrm{Hilb}_{X_s}) = H^0(Y, N_{Y/X_s})$, $T_h(\mathrm{Hilb}_{X/S}) \subset H^0(Y, N_{Y/X})$, $T_s(S) \subset T_s(S) \otimes H^0(Y, \mathcal{O}_Y)$, the homomorphisms in this sequence agree with those in (8.12). Finally, when Y is reduced and connected, $T_h(\mathrm{Hilb}_{X/S}) = H^0(Y, N_{Y/X})$.

Let v be a tangent vector to S at s. We shall say that v is *obstructed at h* if it maps to a nonzero element of $\mathrm{Ext}^1_{\mathcal{O}_{X_s}}(\mathcal{K}, \mathcal{O}_Y)$. In other words, v is obstructed at h if it cannot be lifted to a $v' \in T_h(\mathrm{Hilb}_{X/S})$, that is, to a first-order family of subschemes in the fibers of $X \to S$.

The exact sequence (8.15) can be refined somewhat. Comparing the exact sequence of $\mathcal{O}_Y$-modules

$$0 \to \mathcal{I}/(\mathcal{I} \cap \mathcal{J}^2) \to \mathcal{J}/\mathcal{J}^2 \to \mathcal{K}/\mathcal{K}^2 \to 0$$

with (8.10), we get a commutative diagram with exact top and bottom rows

$$\begin{array}{ccccccc}
0 \to \mathrm{Hom}\left(\frac{\mathcal{K}}{\mathcal{K}^2}, \mathcal{O}_Y\right) & \to & \mathrm{Hom}\left(\frac{\mathcal{J}}{\mathcal{J}^2}, \mathcal{O}_Y\right) & \to & \mathrm{Hom}\left(\frac{\mathcal{I}}{\mathcal{I} \cap \mathcal{J}^2}, \mathcal{O}_Y\right) & \to & \mathrm{Ext}^1_{\mathcal{O}_Y}\left(\frac{\mathcal{K}}{\mathcal{K}^2}, \mathcal{O}_Y\right) \\
\| & & \| & & \downarrow & & \\
0 \to \mathrm{Hom}(\mathcal{K}, \mathcal{O}_Y) & \longrightarrow & \mathrm{Hom}(\mathcal{J}, \mathcal{O}_Y) & \longrightarrow & \mathrm{Hom}(\mathcal{I}, \mathcal{O}_Y) & \longrightarrow & \mathrm{Ext}^1_{\mathcal{O}_{X_s}}(\mathcal{K}, \mathcal{O}_Y)
\end{array}$$

whence an exact sequence

$$(8.16) \qquad 0 \to T_h(\mathrm{Hilb}_{X_s}) \to T_h(\mathrm{Hilb}_{X/S}) \to V \to \mathrm{Ext}^1_{\mathcal{O}_Y}(\mathcal{K}/\mathcal{K}^2, \mathcal{O}_Y),$$

where V stands for the intersection of $\mathrm{Hom}(\mathcal{I}/(\mathcal{I} \cap \mathcal{J}^2), \mathcal{O}_Y)$ and $T_s(S)$ inside $\mathrm{Hom}_{\mathcal{O}_X}(\mathcal{I}, \mathcal{O}_Y) = T_s(S) \otimes H^0(Y, \mathcal{O}_Y)$. It may well happen that an element of $T_s(S)$ is obstructed simply because it does not belong to V, as the following example shows.

EXAMPLE (8.17). Let $X \subset \mathbb{P}^3 \times \mathbb{C}$ be the locus $\{([x_0 : \cdots : x_3], t) : tx_0^2 + x_1^2 + x_2^2 + x_3^2 = 0\}$, and let π be the projection of X to $S = \mathbb{C}$. The fibers of π are all smooth quadrics, except for X_0 which is a cone with vertex p at $t = x_1 = x_2 = x_3 = 0$. We choose $\{p\}$ as Y_0, and, as usual, we denote by h the corresponding point of $\mathrm{Hilb}_{X/S}$. It is easy to show that every nonzero element of $T_s(S)$ is obstructed at h. To see this, we pass to the affine coordinates $y_i = x_i/x_0$, $i = 1, 2, 3$. In these coordinates, the equation defining X near p is $t + y_1^2 + y_2^2 + y_3^2 = 0$. Thus $\mathcal{J} = (y_1, y_2, y_3)$, and $t \in \mathcal{J}^2$. Since $\mathcal{I}$ is generated by t, it follows that $\mathcal{I} \subset \mathcal{J}^2$ and hence that $V = \{0\}$. This proves our claim.

When $Y \subset X_s$ is a regular embedding, sequence (8.16) is particularly nice, as the following refinement of (8.14) shows.

THEOREM (8.18). *Let X be flat and projective over S. Let s be a closed point of S, let Y be a regularly embedded closed subscheme of X_s, and denote by h the corresponding point of* $\mathrm{Hilb}_{X/S}$. *Then the exact sequence (8.1) extends to an exact sequence*

$$(8.19) \qquad 0 \to T_h(\mathrm{Hilb}_{X_s}) \to T_h(\mathrm{Hilb}_{X/S}) \to T_s(S) \to H^1(Y, N_{Y/X_s}).$$

The first step in proving the proposition is to observe that, under its assumptions, $\mathcal{I} \cap \mathcal{J}^2 = \mathcal{I}\mathcal{J}$, and hence $\mathrm{Hom}_{\mathcal{O}_Y}(\mathcal{I}/(\mathcal{I} \cap \mathcal{J}^2), \mathcal{O}_Y) = \mathrm{Hom}_{\mathcal{O}_X}(\mathcal{I}, \mathcal{O}_Y)$, so that $V = T_s(S)$. To this end, pick a point x of Y and write A for $\mathcal{O}_{S,s}$, B for $\mathcal{O}_{X,x}$, $\mathfrak{m}_A$ and $\mathfrak{m}_B$ for their maximal ideals, and I and J for $\mathcal{I}_x$ and $\mathcal{J}_x$. We must show that $I \cap J^2 = IJ$. By assumption, the ideal J/I in B/I is generated by a regular sequence $f_1, \dots, f_h$. For each i, choose $F_i \in B$ mapping to f_i, and set $L = (F_1, \dots, F_h)$; clearly, $J = I + L$. By Lemma (5.23), B/L is A-flat, and hence the sequence

$$0 \to L \otimes_A A/\mathfrak{m}_A \to B \otimes_A A/\mathfrak{m}_A \to B/L \otimes_A A/\mathfrak{m}_A \to 0$$

is exact. This means that $L/IL = L \otimes_A A/\mathfrak{m}_A$ is the kernel of $B/I \to B/J$. On the other hand, this kernel is also equal to $(I + L)/I \simeq L/(I \cap L)$, and hence $IL = I \cap L$. But then

$$I \cap J^2 = I \cap (IJ + L^2) = IJ + I \cap L^2 \subset IJ + I \cap L \subset IJ\,.$$

For the second step of the proof, recall that there is a spectral sequence abutting to $\mathrm{Ext}^{\bullet}_{\mathcal{O}_Y}(\mathcal{K}/\mathcal{K}^2, \mathcal{O}_Y)$ whose E_2 term is

$$E_2^{p,q} = H^p(Y, \mathcal{E}xt^q_{\mathcal{O}_Y}(\mathcal{K}/\mathcal{K}^2, \mathcal{O}_Y))\,.$$

On the other hand, since $Y \hookrightarrow X_s$ is a regular embedding, $\mathcal{K}/\mathcal{K}^2$ is a locally free $\mathcal{O}_Y$-module by (5.24), and hence the sheaves $\mathcal{E}xt^q_{\mathcal{O}_Y}(\mathcal{K}/\mathcal{K}^2, \mathcal{O}_Y)$ all vanish for $q > 0$. It follows that

$$\mathrm{Ext}^p_{\mathcal{O}_Y}(\mathcal{K}/\mathcal{K}^2, \mathcal{O}_Y) = H^p(Y, \mathcal{H}om_{\mathcal{O}_Y}(\mathcal{K}/\mathcal{K}^2, \mathcal{O}_Y)) = H^p(Y, N_{Y/X_s})$$

for all q. This concludes the proof of (8.18).

An upper bound for the dimension of $\mathrm{Hilb}_{X/S}$ is given by (8.2), while Theorem (8.18) gives the lower bound $h^0(N_{Y/X_s}) - h^1(N_{Y/X_s}) + \dim T_s S$ for the dimension of the *tangent space* to the Hilbert scheme at h, in the case of regular embeddings. The following powerful generalization to relative Hilbert schemes of the lower bound (5.12) shows that a similar bound holds also for the dimensions of the *spaces* involved.

THEOREM (8.20). *Let X be flat and projective over S, let s be a point of S, and assume that S is equidimensional at s. Let Y be a regularly embedded closed subscheme of X_s, and write h to indicate the corresponding point of* $\mathrm{Hilb}_{X/S}$. *Then every component of* $\mathrm{Hilb}_{X/S}$ *at h has dimension at least*

$$h^0(Y, N_{Y/X_s}) - h^1(Y, N_{Y/X_s}) + \dim_s S\,.$$

As we did for (5.12), we refer to [440] or [625] for a proof. Here we just want to give a simple example of an obstructed situation which shows, among other things, that this lower bound is sharp.

EXAMPLE (8.21). As we mentioned, the homomorphism $T_h(\mathrm{Hilb}_{X/S}) \to T_s(S)$ in (8.1) is in general not onto, even when the morphism $X \to S$ is very nice, say, for instance, smooth with S smooth, and the subscheme corresponding to h is also smooth. A slight variation on Example (8.17) provides what is probably the simplest instance of this phenomenon. We take as S the affine line with coordinate t and let X' be the locus in $\mathbb{P}^4 \times S$ defined by the two equations $x_1x_2 - x_3x_4 = 0$ and $x_3 + x_4 = tx_0$, where $x_0, \ldots, x_4$ are homogeneous coordinates in $\mathbb{P}^4$. Thus, $X' \to S$ is a family of quadrics in 3-space, all smooth except the one at $t = 0$, which is a cone over a smooth conic. The family $X \to S$ is obtained from $X' \to S$ by blowing up the vertex of this cone. Formally, let λ and μ be homogeneous coordinates in $\mathbb{P}^1$, and let X be defined, inside $\mathbb{P}^4 \times \mathbb{P}^1 \times S$, by the equations

$$x_1x_2 - x_3x_4 = 0\,, \quad \lambda x_1 = \mu x_3\,, \quad \mu x_2 = \lambda x_4\,, \quad x_3 + x_4 = tx_0\,.$$

It is immediate to check that $X \to S$ is smooth, and that X is isomorphic to X' except for the fact that the point $x_1 = x_2 = x_3 = x_4 = 0$ is replaced by a smooth rational curve E. Set $z_i = x_i/x_0$ for $i = 1, \ldots, 4$, $\zeta = \lambda/\mu$, and $\xi = \mu/\lambda$, and let U be a neighborhood of E. Then ζ, z_1, z_4 are local coordinates for X on the complement, inside U, of the locus $\mu = 0$. Likewise, ξ, z_2, z_3 are coordinates on the complement of $\lambda = 0$ in U. A generator for N_{E/X_0} away from $\mu = 0$ is $e_\mu = \partial/\partial z_1 - \zeta\partial/\partial z_4$, and one away from $\lambda = 0$ is $e_\lambda = \xi\partial/\partial z_3 - \partial/\partial z_2$. Since $e_\mu = \zeta^2 e_\lambda$, the normal bundle to E in X_0 is $\mathcal{O}_E(-2)$, and the self-intersection of E is -2. In particular, E cannot be "moved" to any nearby fiber since a smooth quadric does not contain curves with negative self-intersection. Neither can E be moved inside X_0, since $H^0(E, N_{E/X_0}) = 0$. Summing up, the point corresponding to E is isolated in $\mathrm{Hilb}_{X/S}$. In particular, in this case the lower bound given by (8.20) is an equality, since $h^1(E, N_{E/X_0}) = 1$. This, in itself, is not sufficient to conclude that $T_h(\mathrm{Hilb}_{X/S}) \to T_s(S)$ is not onto, since in principle E could still move to first order transversely to Hilb_{X_0}. To prove our claim, we must show that $T_h(\mathrm{Hilb}_{X/S}) = H^0(E, N_{E/X})$ is actually zero. For this, it suffices to look at the exact sequence (8.15), which in this case reduces to

$$0 \to H^0(E, N_{E/X}) \to H^0(E, \mathcal{O}_E) \xrightarrow{\delta} H^1(E, \mathcal{O}_E(-2)) = H^1(E, N_{E/X_0}),$$

since $H^0(E, N_{E/X_0})$ vanishes. All we have to show is that the coboundary map δ is not zero. We do this by explicit computation. The unit generator of $H^0(E, \mathcal{O}_E)$ corresponds to the tangent vector $\partial/\partial t$ at $0 \in S$, viewed as a normal vector field to X_0 along E. A lift of $\partial/\partial t$ to a normal field to E for $\mu \neq 0$ is $\partial/\partial z_4$; another lift for $\lambda \neq 0$ is $\partial/\partial z_3$. The difference between these liftings is a representative of the coboundary $\delta(\partial/\partial t)$ and equals

$$\xi\frac{\partial}{\partial z_1} - \frac{\partial}{\partial z_4} = \zeta^{-1}e_\mu = \xi^{-1}e_\lambda\,.$$

Thus $\delta(\partial/\partial t)$ is the standard generator of $H^1(E, N_{E/X_0}) = H^1(E, \mathcal{O}_E(-2))$.

9. C^n families of projective manifolds.

Up to now, we have concentrated on algebraic families of schemes or on analytic families of analytic spaces. However, in many contexts, it is useful to be able to deal also with families of complex manifolds that depend only continuously or differentiably, in an appropriate sense, on parameters. In particular, we will use continuous families of curves in our approach to Teichmüller's theorem. We begin by explaining what we mean by continuous families of complex manifolds and by establishing a few basic facts about them.

Let $\alpha : X \to B$ be a morphism of C^m manifolds, where m can be $0, 1, \ldots, \infty$. A *relative C^m atlas* for $\alpha : X \to B$ is a collection of open subsets V_i of X and C^m diffeomorphisms $\varphi_i : U_i \times S_i \to V_i$, where U_i is an open set in some $\mathbb{R}^k$ and S_i is an open subset of B, such that the V_i cover X and the φ_i are compatible with the projections to B. We shall often refer to the U_i-coordinates as the *vertical coordinates* on V_i and to the derivatives with respect of them as *vertical derivatives.* Consider relative C^m atlases having the additional property that the coordinate changes are C^∞ in the vertical coordinates and, moreover, all their vertical derivatives, of any order, are C^m as functions of *all* the coordinates. Any such atlas is contained in a unique maximal one. We will say that a maximal atlas defines a structure of C^m *family of differentiable manifolds* on $\alpha : X \to B$ and that $\alpha : X \to B$, together with the atlas, is a C^m family of differentiable manifolds. For brevity, the charts of the atlas will be said to be *adapted* (to the given structure of family of differentiable manifolds, of course).

If $\alpha : X \to B$ and $\alpha' : X' \to B'$ are two C^m families of differentiable manifolds, a morphism from the first to the second is a pair (f, F) of C^m morphisms fitting in a commutative square

$$\begin{array}{ccc} X & \xrightarrow{F} & X' \\ {\scriptstyle\alpha}\downarrow & & \downarrow{\scriptstyle\alpha'} \\ B & \xrightarrow{f} & B' \end{array} \tag{9.1}$$

such that derivatives of arbitrary order of the coordinates on X' with respect to vertical coordinates on X are C^m functions of all coordinates, vertical or not. In particular, taking $X' = \mathbb{R}$ and B' a point, we can speak of C^m families of differentiable functions on $\alpha : X \to B$; these families will also be called *adapted functions.* Clearly, we can also speak of adapted functions on open subsets of X. There are other objects for which the notion of being adapted makes sense, for example, relative forms; we will say that such a form is adapted when, written in adapted coordinates, it has components which are adapted functions. More generally, there is an obvious notion of C^m family of differentiable vector bundles on a C^m family of differentiable manifolds, and one can speak of adapted metrics on these and of adapted sections and vector-valued relative forms. When working with adapted "objects," the main fact to keep in mind is that the class of these is closed under the four operation, under functional composition, and, crucially, also under vertical differentiation of arbitrary order. In particular, the class of C^m families of differentiable vector bundles is closed under direct sum, tensor product, and pullback via morphisms of families. We shall denote the sheaf of adapted functions on X by $\widetilde{\mathcal{A}}_X$, or simply by $\widetilde{\mathcal{A}}$.

Global adapted objects can often be constructed by gluing together local ones by means of suitable partitions of unity. Let $\mathcal{W}$ be an open covering of a neighborhood of $\alpha^{-1}(b_0)$ for some point $b_0 \in B$. An *adapted partition of unity* on a neighborhood of $\alpha^{-1}(b_0)$ subordinated to $\mathcal{W}$ is a finite collection $\{\chi_i\}$ of compactly supported adapted functions on X such that $\sum \chi_i = 1$ on a neighborhood of $\alpha^{-1}(b_0)$ and that the support of each χ_i is contained in an element of $\mathcal{W}$.

LEMMA (9.2). *When $\alpha : X \to B$ is a proper map, for any open covering $\mathcal{W}$ of $\alpha^{-1}(b_0)$, there exist adapted partitions of unity on a neighborhood of $\alpha^{-1}(b_0)$ subordinated to $\mathcal{W}$.*

To prove the lemma, cover X, possibly after shrinking B, with finitely many adapted charts $\varphi_i : U_i \times B \to V_i$ such that each V_i is contained in one of the elements of $\mathcal{W}$. For each i, choose real-valued nonnegative C^∞ compactly supported functions μ_i' and μ_i'' on U_i and B, denote by μ_i the function $\mu_i' \times \mu_i''$ on $U_i \times B$, and call λ_i the function on X which equals $\mu_i \circ \varphi_i^{-1}$ on V_i and is zero on $X - V_i$. Clearly, λ_i is an adapted function. We may choose λ_i in such a way that $\sum \lambda_i$ is strictly positive on all of $\alpha^{-1}(b_0)$, and hence on $\alpha^{-1}(A)$, where A is a small enough ball centered at $0 \in B$. Choose another ball D centered at 0 whose closure is contained in A and a nonnegative C^∞ function σ on B which is zero on D and strictly positive outside A. If we set $\lambda = \sigma + \sum \lambda_i$, then $\chi_i = \lambda_i / \lambda$ will do.

There is a version of the inverse function theorem for adapted functions, that is, a version "with parameters."

PROPOSITION (9.3). *Let U be an open subset of $\mathbb{R}^n$, V an open subset of $\mathbb{R}^\ell$, $f : U \times V \to \mathbb{R}^n$ a continuous function, and m a nonnegative integer. Write $x = (x_1, \ldots, x_n)$ to indicate the standard coordinates in U and $t = (t_1, \ldots, t_\ell)$ to indicate the standard ones in V. Suppose that:*

i) *the function f is C^∞ in x for any fixed t;*
ii) *the function f and all its derivatives, of any order, with respect to the x variables are C^m functions of x and t;*
iii) *the Jacobian $\frac{\partial f}{\partial x}(x,t)$ is nonsingular at a point $(x_0, t_0) \in U \times V$.*

Set $F(x,t) = (f(x,t),t)$. Then there is an open neighborhood A of (x_0,t_0) such that $F(A)$ is open in $\mathbb{R}^n \times \mathbb{R}^\ell$ and F induces a homeomorphism from A to $F(A)$. Moreover, writing the inverse of this function under the form $(x,t) \mapsto (g(x,t),t)$, the function g is C^∞ in x, and g and all its derivatives, of any order, with respect to the x variables are C^m functions of x and t.

Proof. When $m > 0$, the Jacobian matrix $\frac{\partial F}{\partial(x,t)}$ is nonsingular at (x_0,t_0); hence the standard inverse function theorem shows that F has a local inverse of class C^m. To obtain the same result when $m = 0$, we must retrace the proof of the standard inverse function theorem. We can assume that $x_0 = 0 \in \mathbb{R}^n$ and $t_0 = 0 \in \mathbb{R}^\ell$. Shrinking U and V, we may suppose that $\frac{\partial f}{\partial x}$ is invertible and that $\left\| \left(\frac{\partial f}{\partial x}\right)^{-1} \right\|$ is bounded by some positive constant C on all of $U \times V$. We may also suppose that $U = B(0,r) \subset \mathbb{R}^n$. We first prove the theorem under the following additional assumptions:

(9.4)

$$\begin{cases} f(0,t) = 0 \text{ for all } t \in V; \\ \dfrac{\partial f}{\partial x}(0,t) = I \text{ for all } t \in V, \text{ where } I = I_n \text{ is the } n \times n \text{ identity matrix.} \end{cases}$$

Shrinking U and V, we may assume that

$$\left\| \frac{\partial f}{\partial x} - I \right\| < \frac{1}{4}$$

everywhere. We then set

$$T_y(x,t) = x + y - f(x,t)$$

and notice that $T_y(x) = x$ if and only if $f(x,t) = y$. Since

$$\frac{\partial}{\partial x} T_y = I - \frac{\partial f}{\partial x},$$

we find that

$$\|T_y(x,t) - T_y(x',t)\| < \frac{1}{4} \|x - x'\|.$$

As a special case, when $x' = 0$ and hence $T_y(x',t) = y$, we deduce that

$$\|T_y(x,t) - y\| < \frac{1}{4} \|x\|$$

and hence that

$$\|T_y(x,t)\| < \|y\| + \frac{1}{4} \|x\|.$$

Thus, if $y \in \overline{B(0, r/4)}$ and $x \in B(0, r)$,

$$\|T_y(x,t)\| < \|y\| + \frac{r}{4} < \frac{r}{2}.$$

In conclusion, when $y \in \overline{B(0, r/4)}$, $x \mapsto T_y(x,t)$ maps $B(0,r)$ to $B(0, \|y\| + r/4)$ and is a contraction. Hence, $x \mapsto T_y(x,t)$ has a unique fixed point in $B(0,r)$, which lies in $\overline{B(0, \|y\| + r/4)} \subset B(0, r/2)$. Put otherwise, for any $y \in \overline{B(0, r/4)}$, there is a unique $x \in B(0,r) = U$ such that $f(x,t) = y$, and moreover $x \in B(0, r/2)$. Now let W be an open neighborhood of the origin in $\mathbb{R}^\ell$ whose closure is compact and contained in V, and set $M = B(0, r/4) \times W$ and $E = F^{-1}(M)$. Then F maps E to M in a one-to-one fashion, and the same is true for $F^{-1}(\overline{M})$ and $\overline{M} = \overline{B(0, r/4)} \times \overline{W}$. On the other hand, $F^{-1}(\overline{M})$ is a closed subset of the compact set $\overline{B(0, r/2)} \times \overline{W}$ and hence is compact. Thus, $F : F^{-1}(\overline{M}) \to \overline{M}$ is a one-to-one continuous map from a compact space to a Hausdorff one and hence is a homeomorphism. It follows that $F : E \to M$ is also a homeomorphism.

Now we remove the assumption (9.4). If we set

$$\varphi(x,t) = \left(\frac{\partial f}{\partial x}(0,t)\right)^{-1} \cdot \left(f(x,t) - f(0,t)\right),$$

then φ satisfies (9.4); moreover, φ is C^∞ in x, and all its x-derivatives are continuous functions of x and t. By what we have proved, there is an open neighborhood E of the origin such that, writing $\Phi(x,t) = (\varphi(x,t), t)$, $\Phi(E)$ is open, and $E \to \Phi(E)$ is a homeomorphism. The map

$$H(x,t) = \left(\frac{\partial f}{\partial x}(0,t) \cdot (x + f(0,t)), t\right)$$

is clearly C^∞ in x, and all its x-derivatives are continuous functions of x and t. Moreover, H has an inverse with the same properties, and $F = H \circ \Phi$. A local inverse of F is thus obtained by composing H^{-1} with a local inverse of Φ.

Up to now we have shown that there exists an open neighborhood A of (x_0, t_0) such that $F(A)$ is open and $F : A \to F(A)$ has a C^m inverse. We denote this inverse by G; clearly, G is of the form

$$G(y,t) = (g(y,t), t).$$

The standard inverse function theorem says that g is C^∞ in the y variables and that

$$\frac{\partial g}{\partial y}(y,t) = \left(\frac{\partial f}{\partial x}(g(y,t), t)\right)^{-1}. \tag{9.5}$$

Since all the x-derivatives of f are C^m, and g is also C^m, the first y-derivatives of g are C^m functions of y and t. More generally, differentiating (9.5), we see that all order k derivatives of g with respect to the y variables are polynomials in the x-derivatives of the entries of $\left(\frac{\partial f}{\partial x}\right)^{-1}$, evaluated at $(g(y,t),t)$, and in the y-derivatives of g up to order $k-1$. This shows, inductively on k, that all y-derivatives of g are C^m functions of y and t, concluding the proof. Q.E.D.

An immediate application of (9.3) is the following observation. Suppose that the maps f and F in (9.1) give a morphism of C^m families of differentiable manifolds, that f is a C^m diffeomorphism, and that the restriction of F to each fiber of α is a diffeomorphism to the corresponding fiber of α'. Then the pair (f^{-1}, F^{-1}) is also a morphism of C^m families of differentiable manifolds. In other words, (f, F) is an isomorphism of C^m families, and (f^{-1}, F^{-1}) is its inverse.

Families of compact differentiable manifolds are always locally trivial.

LEMMA (9.6). *Let $\alpha : X \to B$ be a C^m family of compact differentiable manifolds, and let b_0 be a point of B. Then, if U is a sufficiently small neighborhood of b_0, there is an isomorphism of C^m families of differentiable manifolds between $\alpha^{-1}(U) \to U$ and the product family $\alpha^{-1}(b_0) \times U \to U$.*

This result is well known for C^∞ families, but its standard proof runs into difficulties when one tries to adapt it to the continuous case, for the reasons that we now illustrate. Suppose that B is an interval with coordinate t. The usual argument involves lifting the vector field $\partial/\partial t$ to a vector field v on X, obtained, for instance, by patching together local lifts by means of a partition of unity and then integrating v to trivialize the family. In the C^0 case, however, the very notion of lift of a vector field does not make good sense, so we are forced to adopt a different approach.

As a first step, we show that families of compact differentiable manifolds can be locally embedded in euclidean space.

LEMMA (9.7). *Let $\alpha : X \to B$ and b_0 be as above. Then, if U is a sufficiently small neighborhood of b_0, for sufficiently large N, there is a map $F : \alpha^{-1}(U) \to \mathbb{R}^N$ such that the pair $(\mathrm{id}_U, F \times \alpha)$ is a morphism of C^m families of differentiable manifolds from $\alpha^{-1}(U) \to U$ to $\mathbb{R}^N \times U \to U$ which is a fiberwise embedding.*

In proving Lemmas (9.7) and (9.6) we may assume that B is a ball of radius ε centered at the origin of $\mathbb{R}^\ell$ and that b_0 is the origin. Without loss of generality, we may assume that all fibers of α are connected and have the same dimension n. Denote by D_r the ball of radius r centered at the origin of $\mathbb{R}^n$. Shrinking B, if necessary, we may find finitely many adapted charts $(\varphi_i, \alpha) : U_i \to D_2 \times B$, where U_i is an open subset of X, such that the open sets $(\varphi_i, \alpha)^{-1}(D_1 \times B)$ cover X. Choose a nonnegative C^∞ function χ on D_2 which is identically equal to 1 on D_1 and vanishes identically on the complement of $D_{4/3}$. By pullback via φ_i we get a function on U_i, which we denote by χ_i. We denote by $\varphi_{i,j}$ the jth component of φ_i and write $\psi_{i,j}$ to indicate the function $\chi_i\varphi_{i,j}$ extended to zero on the complement of U_i in X; it is an adapted function on X. The $\psi_{i,j}$, collectively, give an adapted map $\psi : X \to \mathbb{R}^M$ whose restriction to each fiber of α is a local embedding. It follows in particular that there is an open neighborhood W of the diagonal Δ in $X \times_B X$ such that $\psi(x) \neq \psi(y)$ whenever $(x, y) \in W$. Shrinking B again, we may find finitely many adapted charts $(\xi_i, \alpha) : V_i \to D_2 \times B$, where V_i is an open subset of X, having the following property. There is a set I of pairs of indices such that $V_i \times_B V_j$ does not meet Δ if $(i, j) \in I$ and $X \times_B X \setminus W$ is covered by the open sets $(\xi_i, \alpha)^{-1}(D_1 \times B) \times_B (\xi_j, \alpha)^{-1}(D_1 \times B)$ as (i, j) varies in I. Denote by λ_i the pullback of χ via ξ_i, extended to zero on the complement of V_i. By construction, if $(x, y) \in X \times_B X \setminus W$, there is an index i such that $\lambda_i(x) \neq 0$ but $\lambda_i(y) = 0$. Thus, adding the λ_i to the $\psi_{i,j}$, we get an adapted map $\Psi : X \to \mathbb{R}^N$ whose restriction to each fiber of α is an embedding. This finishes the proof of (9.7).

We now prove (9.6). For each $b \in B$, we set $X_b = \alpha^{-1}(b)$. Using (9.7), we view X as embedded in $\mathbb{R}^N \times B$ and hence each X_b as embedded in $\mathbb{R}^N$. A neighborhood V of X_{b_0} in $\mathbb{R}^N$ is diffeomorphic to a neighborhood of the zero section of the normal bundle to X_{b_0} in $\mathbb{R}^N$, and projection to the zero section corresponds to a C^∞ map $\eta : V \to X_{b_0}$. On the other hand, X_b is contained in V for b close to b_0, and hence, possibly after shrinking B, η gives an adapted map $\beta : X \to X_{b_0}$. The fact that the embedding of X in $\mathbb{R}^N$ is adapted implies in particular that the tangent spaces to the X_b vary continuously. Therefore $\beta_{|X_b} : X_b \to X_{b_0}$ is a local diffeomorphism for b close to b_0; more precisely, the inverse function theorem (9.3) implies that every point of X_{b_0} has a neighborhood U such that β induces a diffeomorphism from $X_b \cap \beta^{-1}(U)$ to U for b close to b_0. Actually, $\beta_{|X_b}$ is a diffeomorphism, always for b close to b_0. To see this, it suffices to prove that $\beta_{|X_b}$ is injective. We argue by contradiction. Suppose that there are a sequence $\{b_n\}$ in B converging to b_0 and sequences of points $x_n, y_n \in X_{b_n}$ such that $x_n \neq y_n$ but $\beta(x_n) = \beta(y_n)$; we may assume that $\{x_n\}$ and $\{y_n\}$ converge, respectively, to points $x, y \in X_{b_0}$. Passing to the limit shows that $x = \beta(x) = \beta(y) = y$. Thus, if U is neighborhood of x as above, x_n and y_n belong to U for large n, and hence $\beta(x_n) \neq \beta(y_n)$, a contradiction. This shows that, after suitably shrinking B, $(\beta, \alpha) : X \to X_{b_0} \times B$ is an isomorphism of C^m families of differentiable manifolds, and concludes the proof of (9.6).

There is also an analogue of (9.6) for families of vector bundles.

LEMMA (9.8). *Let X be a compact differentiable manifold, B a C^m manifold, and b_0 a point of B. Denote by p and q the projections of*

$X \times B$ to B and X, respectively. Let F be a C^m family of differentiable vector bundles on $p : X \times B \to B$. Then, if U is a sufficiently small neighborhood of b_0, the restriction of F to $p^{-1}(U) \to U$ is isomorphic, as a C^m family of differentiable vector bundles, to $q^*(F_{b_0})$, where F_b stands for the restriction of F to $p^{-1}(b)$.

The proof is similar to the one of (9.6), and we only sketch it. First observe that, possibly after shrinking B, we may find sufficiently many adapted sections of F to embed F as a vector subbundle of a trivial bundle $\mathbb{R}^N \times X \times B$ for some large N. Orthogonal projection with respect to the standard Euclidean metric on $\mathbb{R}^N$ yields a surjective morphism of C^∞ vector bundles on X from $\mathbb{R}^N \times X$ to F_{b_0} and hence a surjective morphism $\beta : \mathbb{R}^N \times X \times B \to q^*(F_{b_0})$. Composing this with the inclusion of F in $\mathbb{R}^N \times X \times B$ gives a morphism of adapted vector bundles $F \to q^*(F_{b_0})$, which is an isomorphism when restricted to $p^{-1}(U)$ for any sufficiently small neighborhood U of b_0.

We will be interested in families of differentiable manifolds whose fibers are complex manifolds. Formally, a C^m $(m = 0, 1, \ldots, \infty, \omega)$ *family of compact complex manifolds* is a proper surjective C^m map $\alpha : X \to B$ of C^m manifolds whose fibers are compact complex manifolds, satisfying the following local triviality condition. For every $x \in X$, there is a C^m diffeomorphism $\varphi : U \times S \to V$, where S is a neighborhood of $\alpha(x)$ in B, V is a neighborhood of x in X, and U is a ball in $\mathbb{C}^h$ centered at 0 such that:

i) $\varphi(0, \alpha(x)) = x$;
ii) φ is compatible with the projections to B;
iii) the restriction of φ to $U \times \{s\}$, viewed as a map to $V \cap \alpha^{-1}(s)$, is biholomorphic for every $s \in S$.

A C^m family of compact complex manifolds has a natural structure of C^m family of differentiable manifolds, as a consequence of the following standard result.

LEMMA (9.9). *Let $f(z_1, \ldots, z_h, t_1, \ldots, t_\ell)$ be a C^m function of the complex variables $z_1, \ldots, z_h$ and of the real variables $t_1, \ldots, t_\ell$ which is holomorphic in $z_1, \ldots, z_h$. Then the partials of f, of any order, with respect to the variables $z_1, \ldots, z_h$ are C^m functions of $z_1, \ldots, z_h, t_1, \ldots, t_\ell$.*

The proof is an easy application of Cauchy's integral formula. In fact, given a multiindex $(k_1, \ldots, k_h)$, and setting $k = \sum k_i$, this says that

$$\frac{\partial^k f}{\partial z_1^{k_1} \ldots \partial z_h^{k_h}}(z_1, \ldots, z_h, t_1, \ldots, t_k)$$
$$= \frac{\prod k_i!}{(2\pi\sqrt{-1})^h} \int_{\substack{|\zeta_i - z_i| = \varepsilon \\ i=1,\ldots,h}} \frac{f(\zeta_1, \ldots, \zeta_h, t_1, \ldots, t_\ell)}{\prod(\zeta_i - z_i)^{k_i+1}} d\zeta_1 \ldots d\zeta_h$$

and the right-hand side can be continuously differentiated m times under the integral sign with respect to the variables t_i.

We shall denote the sheaf of adapted relative (0,1)-forms on X by $\widetilde{\mathcal{A}}^{0,1}$. We also denote by $\widetilde{\mathcal{O}}_X$ the sheaf on X of all C^m functions whose restrictions to the fibers are holomorphic.

A C^m family of projective varieties is a commutative diagram

$$\begin{array}{ccc} X & \subset & \mathbb{P}^r \times B \\ & \searrow^{\alpha} & \downarrow \pi \\ & & B \end{array} \tag{9.10}$$

such that $\alpha : X \to B$ is a C^m family of compact complex manifolds and $X \subset \mathbb{P}^r \times B$ is a C^m embedding whose restriction to each fiber is holomorphic. It follows from (9.9) that the inclusion of X in $\mathbb{P}^r \times B$ is a morphism of families of differentiable manifolds. We shall say that a map f from B to an analytic space Y is of class C^m at $b \in B$ if there is an embedding j of a neighborhood of $f(b)$ in a complex manifold Z such that $j \circ f$ is C^m.

PROPOSITION (9.11). *Consider a C^m family of projective varieties as in (9.10). Then the map ψ from B to the Hilbert scheme obtained by attaching to each $b \in B$ the Hilbert point of $\alpha^{-1}(b)$ is of class C^m. Moreover, $\alpha : X \to B$ is just the pullback via ψ of the universal family.*

The problem is local on B; hence in the proof we may assume that B is a ball in $\mathbb{R}^\ell$ centered at the origin; we will also feel free to shrink B where necessary. We write $\widetilde{\mathcal{O}}$ as a shorthand for $\widetilde{\mathcal{O}}_{\mathbb{P}^r \times B}$ and, as usual, set $X_b = \alpha^{-1}(b)$.

Since the Euler–Poincaré characteristic of a holomorphic line bundle depends only on the topological type of the bundle and on the one of the holomorphic tangent bundle of the underlying manifold, it follows from (9.6) and (9.8) that all the fibers of α have the same Hilbert polynomial. Then (4.1) says that there is an integer n_0 such that, for $n \geq n_0$ and all $b \in B$, the cohomology groups $H^i(X_b, \mathcal{I}_{X_b}(n))$, $i > 0$, vanish and the multiplication map

$$H^0(\mathbb{P}^r, \mathcal{O}(1)) \otimes H^0(X_b, \mathcal{I}_{X_b}(n)) \to H^0(X_b, \mathcal{I}_{X_b}(n+1))$$

is onto. This implies, among other things, that the dimension of $H^0(X_b, \mathcal{O}_{X_b}(n))$ is equal to $p(n)$ and that the map

$$H^0(\mathbb{P}^r, \mathcal{O}(n)) \to H^0(X_b, \mathcal{O}_{X_b}(n))$$

is onto for every $b \in B$.

LEMMA (9.12). *Suppose $n \geq n_0$. Then $\alpha_* \widetilde{\mathcal{O}}_X(n)$ is a locally free module over the sheaf of C^m functions on B, and $\rho_n : \pi_* \widetilde{\mathcal{O}}(n) \to \alpha_* \widetilde{\mathcal{O}}_X(n)$ is surjective as a homomorphism of vector bundles.*

Let b_0 be a point of B. Pick sections $s_1, \dots, s_{p(n)}$ of $\mathcal{O}_{\mathbb{P}^r}(n)$ which map to independent elements of $H^0(X_{b_0}, \mathcal{O}_{X_{b_0}}(n))$, and points $q_1, \dots, q_{p(n)}$ of X_{b_0} such that the matrix

$$\begin{pmatrix} s_1(q_1) & \dots & s_{p(n)}(q_1) \\ \vdots & & \vdots \\ s_1(q_{p(n)}) & \dots & s_{p(n)}(q_{p(n)}) \end{pmatrix}$$

is nonsingular. Choose local C^m sections $\sigma_1, \dots, \sigma_{p(n)}$ of $X \to B$ passing through the points $q_1, \dots, q_{p(n)}$. By continuity,

$$A(b) = \begin{pmatrix} s_1(\sigma_1(b)) & \dots & s_{p(n)}(\sigma_1(b)) \\ \vdots & & \vdots \\ s_1(\sigma_{p(n)}(b)) & \dots & s_{p(n)}(\sigma_{p(n)}(b)) \end{pmatrix}$$

stays nonsingular for b in a neighborhood of b_0. Therefore, if s is a section of $\widetilde{\mathcal{O}}(n)$ on a small neighborhood of X_{b_0}, we can write, in a unique way, $s = \sum a_i(b) s_i$. The a_i are given by

$$\begin{pmatrix} a_1(b) \\ \vdots \\ a_{p(n)}(b) \end{pmatrix} = A(b)^{-1} \begin{pmatrix} s(\sigma_1(b)) \\ \vdots \\ s(\sigma_{p(n)}(b)) \end{pmatrix}$$

and hence are C^m functions of b. This proves the lemma.

We may now conclude the proof of (9.11). Lemma (9.12) implies that $\ker \rho_n$ is a C^m vector subbundle of the trivial bundle $\pi_* \widetilde{\mathcal{O}}(n)$ whose fiber over $b \in B$ is $H^0(X_b, \mathcal{I}_{X_b}(n))$. Hence the map to the Grassmannian $\mathbb{G}(\binom{n+r}{r} - p(n), H^0(\mathbb{P}^r, \mathcal{O}(n)))$ it defines is of class C^m. Since this is the composition of ψ with the inclusion of the Hilbert scheme $\mathrm{Hilb}_r^{p(t)}$ in the Grassmannian, the result follows.

10. Bibliographical notes and further reading.

The Hilbert scheme was introduced and constructed by Grothendieck in [330]. Further general references on the Hilbert scheme are Chapter 5 of [243] by Nitsure, Sernesi [624,625], Kollár [439], and Huybrechts–Lehn [379]. Our treatment here is largely patterned on the one in [624]. The connectedness of the Hilbert scheme of projective space is due to Hartshorne [354].

The infinitesimal theory of the Hilbert scheme, in the form of the theory of the characteristic series, goes back at least to the early days of the Italian school of algebraic geometry. The problem of the completeness of the characteristic series, i.e., the one of deciding whether an embedded scheme admits a deformation over a smooth base whose characteristic map is onto, has been a guiding theme throughout much of the development of algebraic geometry. The Italian school extensively studied the problem and struggled with it without coming to a satisfactory conclusion (see, for instance, Enriques [212], Severi [630,632,635,636,637], Segre [620]; another classical source is the book by Coolidge [129]). A very extensive discussion of the problem is contained in Chapter V of Zariski's book [690] and is updated in Mumford's appendix to the same chapter. A satisfactory criterion for the completeness of the characteristic system was finally found by Kodaira and Spencer [436] (for weaker results, see also Kodaira [430,431]). A standard reference for all these questions is Mumford's book [550] (see, in particular, Lecture 22).

Hilbert schemes exhibit a bewildering variety of pathological behaviors. That Hilbert schemes of points are in general reducible was first proved by Iarrobino [380]; an overview of the theory of Hilbert schemes of points can be found in Iarrobino [381]. Mumford's example of a nonreduced component of the Hilbert scheme first appeared in [547]; of course, this same example provides an instance of non-completeness of the characteristic series. Further pathological phenomena were discovered by others, in particular, by Martin-Deschamps and Perrin [501] and by Vakil [667]; this last paper provides an overview of the subject.

Variants of the Hilbert scheme are commonly used in the literature, and are discussed in many of the general references given above. Flag Hilbert schemes were first studied by Kleppe [424].

C^n families of projective manifolds are treated by the first two authors in [29].

11. Exercises.

1. Let

$$\begin{array}{ccc} A & \longrightarrow & A' \\ \downarrow & & \downarrow \\ B & \longrightarrow & B' \end{array}$$

be a commutative diagram of homomorphisms of commutative rings, and let M be a B-module.

i) Show that, if M is A-flat and B' is B-flat, then $M \otimes_B B'$ is A'-flat (Hint: observe that, if P is any A-module, then $(M \otimes_B B') \otimes_{A'} (P \otimes_A A') = B' \otimes_B (M \otimes_A P)$).

ii) Assume that $B \to B'$ is a local homomorphism of local rings, that B is noetherian, and that M is finitely generated as a B-module. Show that, if $M \otimes_B B'$ is A'-flat, A' is A-flat, and B' is B-flat, then M is A-flat.

A. Linearity of the characteristic map

The following set of exercises sketches a proof of the linearity of the characteristic map (5.10). We treat the affine case first. The line of reasoning closely parallels the first half of the proof of (5.8). We let R and A be noetherian $\mathbb{C}$-algebras, $\mathfrak{m}$ a maximal ideal in A, and we wish to look at a flat family of subschemes of $\operatorname{Spec} R$ parameterized by $\operatorname{Spec} A$. This corresponds to an ideal $J \subset A \otimes R$ such that $(A \otimes R)/J$ is a flat A-module; here and in what follows, when we omit mention of the ring over which we are tensoring, we mean that we are tensoring over $\mathbb{C}$. The fiber of this family over the point corresponding to $\mathfrak{m}$ is $\operatorname{Spec}(R/I)$, where

$$R/I = A/\mathfrak{m} \otimes_A (A \otimes R)/J\,.$$

A-1. Show that I is naturally isomorphic to $J/(J \cap (M \otimes R))$

A-2. Show that J is flat over A (hint: look at the Tor sequence obtained by tensoring

$$(11.1) \qquad 0 \to J \to A \otimes R \to (A \otimes R)/J \to 0$$

with an arbitrary A-module and use the flatness of $A \otimes R$ and $(A \otimes R)/J$).

A-3. Show that $J \cap (\mathfrak{m} \otimes R) = \mathfrak{m}J$ (hint: tensoring (11.1) with $A/\mathfrak{m}$, show that $I \simeq J/\mathfrak{m}J$).

A-4. Given $i \in I$, let $\varphi(i)$ be the class of $j - 1 \otimes i$ modulo $\mathfrak{m}J$, where $j \in J$ is a lift of i. Show that this defines a homomorphism of R-modules

$$\varphi : I \to (\mathfrak{m} \otimes R)/\mathfrak{m}J = \mathfrak{m} \otimes_A (A \otimes R)/J\,.$$

A-5. Show that

$$\varphi(I^2) \subset \mathfrak{m}(\mathfrak{m} \otimes_A (A \otimes R)/J)\,,$$

so that φ defines an R-module homomorphism

$$\psi : I/I^2 \to \mathfrak{m}/\mathfrak{m}^2 \otimes_A (A \otimes R)/J\,.$$

A-6. Show that the homomorphism ψ is functorial under base change and that, when $A = \operatorname{Spec} \mathbb{C}[\varepsilon]$, it agrees with the homomorphism $I/I^2 \to R/I$ constructed in the proof of (5.8).

A-7. Let $v \in \operatorname{Hom}(\mathfrak{m}/\mathfrak{m}^2, \mathbb{C})$ be a tangent vector to $\operatorname{Spec} A$ at the point corresponding to $\mathfrak{m}$. Let $A \to \operatorname{Spec} \mathbb{C}[\varepsilon]$ be the corresponding ring homomorphism. Pull back to $\operatorname{Spec} \mathbb{C}[\varepsilon]$ the family of subschemes of

Spec R defined by J. Applying the construction in Lemma (5.8) to the pulled-back family yields a homomorphism

$$\psi' : I/I^2 \to R/I .$$

Show that ψ' is the composition of ψ with the homomorphism

$$v \otimes 1 : \mathfrak{m}/\mathfrak{m}^2 \otimes_A (A \otimes R)/J \to \mathbb{C} \otimes_A (A \otimes R)/J = R/J .$$

A-8. Show that the characteristic map (5.10) is linear.

B. **Conics in $\mathbb{P}^3$**

In this series of exercises we describe the Hilbert scheme parameterizing plane conics in $\mathbb{P}^3$ and use this description to answer an enumerative problem.

B-1. Let $X \subset \mathbb{P}^3$ be any subscheme of dimension 1 and degree 2, that is, with Hilbert polynomial $p(t) = 2t + c$ for some c. Show that $c \geq 1$.

B-2. Now let $X \subset \mathbb{P}^3$ be any subscheme with Hilbert polynomial $p(t) = 2t + 1$. Show that the Hilbert function $q_X(t) = h^0(X, \mathcal{O}(t))$ of X is equal to $p(t)$ and hence that X is a complete intersection of a plane and a quadric.

B-3. Using the preceding result, show that the Hilbert scheme $\mathcal{H} = \mathrm{Hilb}_3^{2t+1}$ is a $\mathbb{P}^5$-bundle over the dual projective space ${\mathbb{P}^3}^\vee$; specifically, it is the projectivization of the symmetric square $\mathrm{Sym}^2(S^\vee)$ of the dual of the universal subbundle S on ${\mathbb{P}^3}^\vee$.

B-4. Using the results of Chapter VII of Volume I, calculate the cohomology ring of the Hilbert scheme $\mathcal{H}$.

B-5. Let $L \subset \mathbb{P}^3$ be a line, and let $D_L \subset \mathcal{H}$ be the locus of conics C such that $C \cap L \neq \emptyset$. Show that $D_L \subset \mathcal{H}$ is a divisor and find its fundamental class $\alpha \in H^2(\mathcal{H}, \mathbb{Z})$.

B-6. Combining the results of the last two exercises, find the number of smooth conic curves meeting each of eight general lines $L_1, \ldots, L_8 \subset \mathbb{P}^3$. (This involves three things: calculating the eighth power $\alpha^8 \in H^{16}(\mathcal{H}, \mathbb{Z}) \cong \mathbb{Z}$, showing that, for general $L_1, \ldots, L_8$, the divisors $D_{L_i} \subset \mathcal{H}$ intersect transversely and that all these points of intersection correspond to smooth conics.)

B-7. Now let $H \subset \mathbb{P}^3$ be a plane, and let $E_H \subset \mathcal{H}$ be the closure of the locus of smooth conics $C \subset \mathbb{P}^3$ tangent to H. Show that this is a divisor and find its fundamental class $\beta \in H^2(\mathcal{H}, \mathbb{Z})$.

B-8. Find the number of smooth conics in $\mathbb{P}^3$ meeting each of seven general lines $L_1, \ldots, L_7 \subset \mathbb{P}^3$ and tangent to a general plane $H \subset \mathbb{P}^3$. More generally, find the number of smooth conics in $\mathbb{P}^3$ meeting each of $8-k$ general lines $L_1, \ldots, L_{8-k} \subset \mathbb{P}^3$ and tangent to k general planes $H_1, \ldots, H_k \subset \mathbb{P}^3$ for $k = 1, 2$, and 3.

B-9. Why do not the methods developed here work to calculate the number of smooth conics in $\mathbb{P}^3$ meeting each of $8-k$ general lines $L_1, \ldots, L_{8-k} \subset \mathbb{P}^3$ and tangent to k general planes $H_1, \ldots, H_k \subset \mathbb{P}^3$ for $k \geq 4$? What can you do to find these numbers?

C. Twisted cubics in $\mathbb{P}^3$

In this series of exercises we will look at the geometry of the Hilbert scheme Hilb_3^{3t+1}, one component of which parameterizes twisted cubic curves.

C-1. Let $X \subset \mathbb{P}^3$ be any subscheme of dimension 1 and degree 3. Show that the Hilbert polynomial of X is of the form $3t + c$ with $c \geq 0$ and that $c = 0$ if and only if X is a plane cubic curve.

C-2. Let $\mathcal{H} = \mathrm{Hilb}_3^p$ be the Hilbert scheme parameterizing subschemes of $\mathbb{P}^3$ with Hilbert polynomial $p(t) = 3t + 1$. Show that $\mathcal{H}$ has two irreducible components: one component $\mathcal{H}_0$ whose general point corresponds to a twisted cubic curve and one component $\mathcal{H}_1$ whose general point corresponds to the union of a plane cubic and a point.

For the next few exercises, consider the following three subschemes of $\mathbb{P}^3$:

a. the subscheme defined by the square of the ideal of a line, for example,
$$C_1 = V(X^2, XY, Y^2);$$

b. a planar triple line with a spatial embedded point of multiplicity 1, for example,
$$C_2 = V(X^2, XY, XZ, Y^3);$$

c. a planar triple line with a planar embedded point, for example,
$$C_3 = V(X, Y^3 Z, Y^4).$$

(Note that the tangent space to C_2 at the embedded point $[0,0,0,1]$ is three-dimensional, whereas in the case of C_3 it is two-dimensional; hence the names spatial and planar embedded point.)

C-3. Show that all three have Hilbert polynomial $p(t) = 3t + 1$; that is, they correspond to points of the Hilbert scheme $\mathcal{H}$.

C-4. Find the Hilbert function $q_{C_i}(t) = h^0(C_i, \mathcal{O}(t))$ of all three and show that the Hilbert function of C_3 is minimal, that is, if X is any subscheme of $\mathbb{P}^3$ with Hilbert polynomial $3t+1$, we have

$$q_X(t) \geq q_{C_3}(t)$$

for all m.

C-5. For each $i = 1, 2$ and 3, does the corresponding point $[C_i] \in \mathcal{H}$ lie in $\mathcal{H}_0$, $\mathcal{H}_1$, or both?

D. Dimension of the Hilbert schemes of twisted cubics and other curves

For the following, let $U \subset \mathrm{Hilb}_3^{3t+1}$ be the open subset parameterizing twisted cubics.

D-1. Prove that U is irreducible of dimension 12 by using the fact that a twisted cubic is residual to a line in an intersection of two quadric surfaces.

D-2. Prove that U is irreducible of dimension 12 by using the fact that a twisted cubic is the image of $\mathbb{P}^1$ under a Veronese map.

D-3. Prove that if $p_1, \ldots p_6 \in \mathbb{P}^3$ are any six points with no four coplanar, there exists a unique twisted cubic curve passing through all six; and use this to prove that U is irreducible of dimension 12.

D-4. Let $A(X)$ be a 2×3 matrix whose entries are general linear forms $L_{i,j}(X)$ on $\mathbb{P}^3$. Show that the locus

$$\{X \in \mathbb{P}^3 : \mathrm{rank}(A(X)) = 1\}$$

is a twisted cubic and use this to prove yet again that U is irreducible of dimension 12.

For the following, we define the *restricted Hilbert scheme* to be the union of those components of the Hilbert scheme whose general point corresponds to an irreducible and nondegenerate variety.

D-5. Using the techniques introduced in the above problems (mainly in problems D-1 and D-2), classify the irreducible components of the restricted Hilbert scheme parameterizing curves of degree 4 in $\mathbb{P}^3$ and find their dimensions.

D-6. Similarly, classify the irreducible components of the restricted Hilbert scheme parameterizing curves of degree 5 in $\mathbb{P}^3$ and find their dimensions.

D-7. What is the smallest degree d such that the restricted Hilbert scheme parameterizing curves of degree d in $\mathbb{P}^3$ has a component of dimension other than $4d$?

D-8. What is the smallest degree d such that the restricted Hilbert scheme parameterizing curves of degree d in $\mathbb{P}^3$ has two or more irreducible components whose general points correspond to curves of the same genus?

D-9. Are there any components of the restricted Hilbert scheme parameterizing curves in $\mathbb{P}^r$ whose general point corresponds to a singular curve?

E. Chow varieties

The *Chow variety* $C_{n,k,d}$ is a variety parameterizing subvarieties $X \subset \mathbb{P}^n$ of dimension k and degree d; it has a natural compactification parameterizing *cycles* of dimension k and degree d, that is, formal linear combinations $\sum a_i X_i$ with $X_i \subset \mathbb{P}^n$ irreducible of dimension k, $a_i \in \mathbb{N}$, and $\sum a_i \deg(X_i) = d$. It is cruder than the Hilbert scheme—it does not see scheme structures or components of dimension less than k—but sometimes that's a good thing. While it has been largely superseded by the Hilbert scheme, it is still useful in some contexts; for example, the construction of moduli spaces: taking the quotient of the Chow variety by $PGL(n+1)$ often yields a different result from taking the quotient of the corresponding components of the Hilbert scheme.

The following exercises will establish some of the basic facts about the Chow construction. For the following, $\mathbb{G}$ will stand for the Grassmannian $\mathbb{G}(n-k, n+1)$ of $(n-k-1)$-planes in $\mathbb{P}^n$, and $\mathbb{P}^N = \mathbb{P}H^0(\mathbb{G}, \mathcal{O}_{\mathbb{G}}(d))$ will denote the projective space parameterizing hypersurfaces of degree d in $\mathbb{G}$.

E-1. Let $X \subset \mathbb{P}^n$ be a variety of pure dimension k and degree d. Show that the subvariety

$$\Phi_X = \{\Lambda : \Lambda \cap X \neq \emptyset\} \subset \mathbb{G}$$

is a hypersurface in $\mathbb{G}$.

E-2. Let $X \subset \mathbb{P}^n$ and $\Phi_X \subset \mathbb{G}$ be as above. Show that Φ_X is a hypersurface of degree d, that is, as a divisor on $\mathbb{G}$, it is linearly equivalent to d times the hyperplane class on $\mathbb{G}$.

The *open Chow variety* is defined to be the locus $\tilde{C}_{n,k,d} \subset |\mathcal{O}_{\mathbb{G}}(d)| = \mathbb{P}^N$ of hypersurfaces arising in this way; the Chow variety is its closure in $\mathbb{P}^N$. The next few exercises will be devoted to a proof that $\tilde{C}$ is locally closed in $\mathbb{P}^N$.

E-3. Show that the following are closed subvarieties:

$$\{(x, \Lambda) : x \in \Lambda\} \subset \mathbb{P}^n \times \mathbb{G}$$

and

$$\{(\Lambda, \Phi) : \Lambda \in \Phi\} \subset \mathbb{G} \times \mathbb{P}^N.$$

E-4. For any point $p \in \mathbb{P}^n$, let $\mathbb{G}_p \subset \mathbb{G}$ be the sub-Grassmannian of planes containing p. Now let $\Phi \subset \mathbb{G}$ be any hypersurface of degree d. Show that

$$X_\Phi = \{p \in \mathbb{P}^n : \mathbb{G}_p \subset \Phi\}$$

is a closed subvariety of $\mathbb{P}^n$ of dimension at most k. Show moreover that if Φ is reduced and irreducible, and $\dim X_\Phi = k$, then $\deg X_\Phi = d$.

E-5. Show that

$$\Theta = \{(p, \Phi) : \mathbb{G}_p \subset \Phi\}$$

is a closed subvariety.

E-6. Let $U \subset \mathbb{P}^N$ be the open subset of reduced and irreducible hypersurfaces. Using the preceding problems, show that the locus of Chow forms of irreducible varieties $X \subset \mathbb{P}^n$ is closed in U. (A similar argument works to show that $C_{n,k,d}$ is locally closed in all of $\mathbb{P}^N$, but it is notationally messier.)

E-7. Let B be a smooth, one-dimensional variety, and let $X \subset B \times \mathbb{P}^n$ be a family whose general fiber $X_t \subset \mathbb{P}^n$ is a reduced and irreducible variety of dimension k and degree d. Suppose that the scheme-theoretic special fiber X_0 of the family is a union $X_0 = \cup Y_i$ of irreducible components Y_i, with Y_i a scheme of multiplicity M_i. Show that the limit of the Chow forms F_{X_t} of the general fiber is

$$\lim_{t \to 0} F_{X_t} = \prod F_{Y_i}^{m_i}.$$

(Hint: reduce to the case $k = 0$.)

E-8. We would like to make a statement to the effect that if $X \subset B \times \mathbb{P}^n$ is a closed subvariety such that the fibers $X_b = X \cap \{b\} \times \mathbb{P}^n$ are varieties of dimension k and degree d, the map

$$B \to C_{n,k,d}$$

obtained by sending $b \in B$ to the Chow form of X_b is a regular map. What hypotheses do we need on X?

F. Punctual Hilbert schemes.

These exercises deal with the Hilbert schemes Hilb_n^d parameterizing zero-dimensional subschemes of $\mathbb{P}^n$ of degree d (that is, schemes with Hilbert polynomial the constant d). This Hilbert scheme always contains one irreducible component $\mathcal{H}_0$, of dimension dn, whose general point $[X]$ corresponds to a reduced scheme $X \subset \mathbb{P}^n$, that is, d general points. The point of these exercises is to show that, in general, Hilb_n^d may contain many other components as well. The method in each case is to exhibit families of such schemes of dimension strictly greater than dn.

F-1. Start with $\mathbb{P}^3$ and consider, for any point $p \in \mathbb{P}^3$, the family of schemes X whose ideals are sandwiched between successive powers of the maximal ideal $\mathfrak{m}$ at p, that is, for some k and ℓ, we have

$$\mathfrak{m}^{k+1} \subset \mathcal{I}_X \subset \mathfrak{m}^k$$

with $\dim(\mathfrak{m}^k/\mathcal{I}_X) = \ell$. Calculate the degree d of such a scheme X and the dimension of the family of such schemes; conclude that for k large enough and $\ell \sim k^2/2$, the Hilbert scheme Hilb_3^d is reducible.

F-2. What is actually the smallest d for which the above argument shows that Hilb_3^d is reducible?

F-3. On to $\mathbb{P}^n$, and this time consider specifically schemes X of degree $n+1+\delta$ with

$$\mathfrak{m}^3 \subset \mathcal{I}_X \subset \mathfrak{m}^2.$$

Show that for n large enough, the Hilbert scheme Hilb_n^{n+4} is reducible. What is the smallest n for which you can prove this?

F-4. Going even further, consider now subschemes $X \subset \mathbb{P}^n$ of degree d that are supported at a point p, contained in a subspace $\Lambda = \mathbb{P}^r$, and whose ideals *in that* $\mathbb{P}^r$ are sandwiched between the schemes defined by the square and the cube of the maximal ideal at p, that is, such that

$$\mathcal{I}_\Lambda + \mathfrak{m}^3 \subset \mathcal{I}_X \subset \mathcal{I}_\Lambda + \mathfrak{m}^2.$$

Using these, show that the schemes Hilb_n^d may be reducible even when $d < n$.

We should say that the actual number of components of Hilb_n^d, or their dimensions, is completely unknown; nor is it known exactly for which pairs (d,n) the scheme Hilb_n^d is reducible. Even worse, given a zero-dimensional subscheme $X \subset \mathbb{P}^n$ of degree d, we know of no effective way of determining whether or not it belongs to the principal component $\mathcal{H}_0$ of Hilb_n^d, that is, whether it is a limit of d distinct points. Moreover, this uncertainty about punctual Hilbert schemes spills over into Hilbert schemes of curves, since these will in general have components whose general member is the disjoint union of a curve and a zero-dimensional scheme. Thus, for example, it is impossible to say (except in very special cases) how many components the Hilbert scheme Hilb_3^{dt-g+1} parameterizing curves of degree d and genus g will have, or what their dimensions might be.

G. **Sections of the Hilbert scheme**

Note: for these problems, we are concerned with sections of the universal family of rational normal curves; since we are only interested in rational sections, it does not matter if we take the Chow variety or Hilbert scheme as our parameter space.

G-1. Let Hilb_2^{2t+1} be the Hilbert scheme of plane conics; let $X \subset B \times \mathbb{P}^2 \to B$ be the universal family over B. Show that $X \to B$ has no rational section.

G-2. Let $B \subset \mathrm{Hilb}_3^{3t+1}$ be the locus, in the Hilbert scheme, of twisted cubic curves; let $X \subset B \times \mathbb{P}^3 \to B$ be the universal family over B. Show that $X \to B$ *does* have a rational section and show how to construct one.

G-3. More generally, for any n, let $B \subset \mathrm{Hilb}_n^{nt+1}$ be the locus of rational normal curves in $\mathbb{P}^n$, and again let $X \subset B \times \mathbb{P}^n \to B$ be the universal family over B. For what n does $X \to B$ have a rational section?

H. **Special classes of curves**

These exercises deal with various classes of curves in $\mathbb{P}^3$. In each case, the degree and arithmetic genus of the curves in question can be calculated readily; in addition, it can be shown that the curves in question form a single irreducible family, whose dimension also can be calculated. The problem then is to say whether the locus of such curves is open in the Hilbert scheme or, in other words, whether they are dense in an irreducible component of $\mathcal{H}$. Note: series (iii) is a direct generalization of series (i); series (iv) is somewhat harder than the first three; and Problem 6 is open.

i. **Complete Intersections.**

A curve $C \subset \mathbb{P}^3$ is called a *complete intersection* of type (d, e) if it is scheme-theoretically the intersection of a surface S of degree d and a surface T of degree e; by Noether's $AF + BG$ theorem, this says that the homogeneous ideal I_C is generated by the polynomials defining S and T.

H-1. Find the degree and genus of a complete intersection $C \subset \mathbb{P}^3$ of type (d, e) by applying Bézout's theorem and the adjunction formula.

H-2. Now find the degree k and genus g of C by calculating the Hilbert polynomial.

H-3. Assuming $d \leq e$, find the dimension of the family of such curves and show that the family is irreducible.

H-4. Now let Hilb_3^{kt-g+1} be the Hilbert scheme parameterizing curves of degree k and genus g in $\mathbb{P}^3$, and let $U \subset \mathrm{Hilb}_3^{kt-g+1}$ be the locus of complete intersection curves of type (d, e). Show that $U \subset \mathrm{Hilb}_3^{kt-g+1}$ is open.

H-5. Show by example that it is not closed.

H-6. Now let $\mathcal{H}^s \subset \mathcal{H}$ be an open subset of smooth curves, and let $U^s = U \cap \mathcal{H}^s$ be the locus of smooth complete intersection curves of type (d, e). Is U^s closed in $\mathcal{H}^s$?

ii. Curves on quadrics

In this series, we deal with *curves of type* (a, b) *on a quadric*, that is, curves $C \subset Q \subset \mathbb{P}^3$ that lie on a smooth quadric Q and are linearly equivalent on Q to a lines of one ruling plus b lines of the other (equivalently, the zero locus of a bihomogeneous polynomial of bidegree (a, b) on $Q \cong \mathbb{P}^1 \times \mathbb{P}^1$).

H-7. Find the degree k and genus g of a complete intersection $C \subset \mathbb{P}^3$ by applying the adjunction formula.

H-8. Now find the degree and genus of C by calculating the Hilbert polynomial.

H-9. Assuming that $a \leq b$, find the dimension of the family of all curves of type (a, b) on quadrics and show that it is irreducible. (Note that the quadric is not fixed!)

H-10. Show that, in the Hilbert scheme Hilb_3^{kt-g+1}, the locus $V \subset \mathrm{Hilb}_3^{kt-g+1}$ of curves of type (a, b) on quadrics is open if a and b are both at least 3, but not otherwise.

H-11. Analogously to problem H-6 above, let $\mathcal{H}^s \subset \mathcal{H}$ be the open subset of smooth curves, and let $V^s = V \cap \mathcal{H}^s$ be the locus of smooth curves of type (a, b) on quadrics. Show that V^s is closed in $\mathcal{H}^s$ if $|b - a| > 1$, but not otherwise.

iii. Quasi-Complete Intersections

We call a curve $C \subset \mathbb{P}^3$ a *quasi-complete intersection* of type (d, e, m) if it is residual to a plane curve of degree m in a complete intersection of type (d, e). For our purposes, we will take that to mean that there is a plane curve D of degree m whose support has no component in common with the support of C such that

$$C \cup D = S \cap T$$

for surfaces S and T of degrees d and e (though the notion of two curves C and D being residual in a complete intersection can be extended to the case where their supports may have components in common).

H-12. Find the degree and genus of a quasi-complete intersection $C \subset \mathbb{P}^3$ of type (d, e, m) by applying Bézout's theorem and the adjunction formula.

H-13. Now find the degree and genus of C by calculating the Hilbert polynomial.

H-14. Assuming $d \leq e$, find the dimension of the family of such curves and show that the family is irreducible.

H-15. Show that the locus of quasi-complete intersection curves of type (d, e, m) is open in the Hilbert scheme.

iv. Determinantal Curves

To keep the notation from getting completely out of hand, we will restrict ourselves here and say that a curve $C \subset \mathbb{P}^3$ is *determinantal of type* n if it is the rank $n-1$ locus of an $n \times (n+1)$ matrix A of linear forms on $\mathbb{P}^3$; that is, its ideal is generated by the maximal minors of A.

H-16. Find the degree and genus of a determinantal curve of type n.

H-17. Find the dimension of the locus in $\mathcal{H}$ of determinantal curves of type n.

H-18. Is this locus open in the Hilbert scheme?

I. More on Mumford's example

In this series of exercises, we will go through an analysis of the Hilbert scheme Hilb_3^{14t-23} parameterizing curves of degree 14 and genus 24 in $\mathbb{P}^3$, describing all the components of $\mathcal{H}$. We will consider only the restricted Hilbert scheme, that is, the union $\mathcal{H}$ of those components of Hilb_3^{14t-23} whose general point corresponds to a smooth irreducible curve.

For the following, we will assume throughout that $[C]$ is a general point of its component of $\mathcal{H}$.

I-1. Let F be a smooth cubic surface in $\mathbb{P}^3$. The following are well known (cf., for instance, [318]):

i) If $\{E_1, \ldots, E_6\}$ is any set of six disjoint lines on F, then there are six points $\{p_1, \ldots, p_6\} \in \mathbb{P}^2$ not lying on a conic, no three of which are collinear, such that there is an isomorphism between F and the blow up of $\mathbb{P}^2$ at $\{p_1, \ldots, p_6\}$ under which the lines $\{E_1, \ldots, E_6\}$ correspond to the exceptional curves of the blow-up.

ii) Let $\{p_1, \ldots, p_6\}$ and $\{E_1, \ldots, E_6\}$ be as above. Denote by $H_{i,j}$ the proper transform of the line through p_i and p_j and by G_i the proper transform of the conic through $p_1, \ldots, \widehat{p_i}, \ldots, p_6$, where the hat indicates omission. Then the set consisting of all the E_i, all the $H_{i,j}$, and all the G_i is the set of all lines lying on F. Moreover, the sets of six disjoint lines on F are the following:

(1) $\{E_1, \ldots, E_6\}$;
(2) $\{G_1, \ldots, G_6\}$;
(3) all the sets of the form $\{E_i, E_j, E_h, H_{k,l}, H_{k,m}, H_{l,m}\}$, where i, j, k, l, m, n are distinct;
(4) all the sets of the form $\{G_i, G_j, G_h, H_{k,l}, H_{k,m}, H_{l,m}\}$, where i, j, k, l, m, n are distinct;
(5) all the sets of the form $\{E_i, G_i, H_{j,h}, H_{j,k}, H_{j,l}, H_{j,m}\}$, where i, j, k, l, m, n are distinct.

Show that the monodromy group of the family of all smooth cubics in $\mathbb{P}^3$ acts transitively on the set of lines lying on F.

I-2. Show that C cannot lie in a plane, a quadric surface, a cubic cone, or a cubic with a double line, because there simply do not exist smooth curves of degree 14 and genus 24 on any of these surfaces.

I-3. Suppose now that C lies on a cubic surface S and that S is either smooth or has isolated double points. Show that the linear system $|C|$ on S has dimension exactly 37. Deduce that C cannot lie on a cubic surface with isolated double points, that is, S must be smooth, because such a curve is a specialization of a curve on a smooth cubic surface and C was taken to be general.

I-4. Now suppose that C lies on a smooth cubic surface S. Show that C will also lie on a sextic surface T not containing S; we can then write the intersection $S \cap T$ as the union of C with a curve D of degree 4. Verify in order the following intersection numbers (of divisors on S):

$$(C \cdot \omega_S) = -14;$$
$$(C \cdot C) = 60;$$
$$(C \cdot D) = 24;$$
$$(D \cdot D) = 0.$$

Deduce that D is a curve of arithmetic genus -1 and hence that D is linearly equivalent either to the union of two disjoint conics or to the disjoint union of a line and a twisted cubic; in these cases we will say that C is of type 1 or of type 2, respectively.

I-5. Continuing the preceding problem, show that curves C of type 1 form an irreducible, 56-dimensional family, and the same for curves of type 2.

I-6. With the conventions of the last two problems, show that $h^0(N_{C/\mathbb{P}^3}) = 57$ if C is of type 1, while $h^0(N_{C/\mathbb{P}^3}) = 56$ if C is of type 2.

I-7. Suppose now that C lies on no cubic surface. We know C lies on a pencil of quartic surfaces; argue (using Bertini's and Bézout's theorems) that the general such quartic is smooth along C. Then, using a series of calculations analogous to those of Problem 3 above, deduce that C is residual to a plane conic in a complete intersection of quartic surfaces. (We will say such a curve C is of type 3.)

I-8. Finally, show that curves of type 3 form an irreducible family of dimension 56; deduce that curves of type 1 are dense in a component of the Hilbert scheme and that this component is everywhere nonreduced.

I-9. One last note: if you tried to do the dimension count of the preceding problem by looking at the map from the locus of curves of type 3 to the space $\mathbb{P}^{34}$ of quartic surfaces and estimating the fiber dimension, you got the wrong answer. Why?

Chapter X.
Nodal curves

1. Introduction.

A nodal curve is a complete algebraic curve such that every one of its points is either smooth or is locally complex-analytically isomorphic to a neighborhood of the origin in the locus with equation $xy = 0$ in $\mathbb{C}^2$. Of course, we are interested in families of nodal curves. There are two ways of defining what a family of nodal curves over a base S is, and they are both very useful. The first one is to say that it is a proper surjective morphism of schemes, or analytic spaces, $\varphi : \mathcal{C} \to S$ such that φ is flat and every fiber of φ is a nodal curve. The second is to say that it is a proper surjective morphism $\varphi : \mathcal{C} \to S$ of analytic spaces such that for any $p \in \mathcal{C}$, either φ is smooth at p with one-dimensional fibers, or else, setting $s = \varphi(p)$, there is a neighborhood of p which is isomorphic, as a space over S, to a neighborhood of $(0, s)$ in the analytic subspace of $\mathbb{C}^2 \times S$ with equation $xy = f$, where f is a function on a neighborhood of s in S whose germ at s belongs to the maximal ideal of $\mathcal{O}_{S,s}$. The equivalence of these definitions is somewhat subtle, and establishing it takes a good part of Section 2. The rest of the section is used to recall the elementary theory of nodal curves, including the definition of dualizing sheaf and the statement of the Riemann–Roch theorem. We conclude the section by discussing relative Kähler differentials and the relative dualizing sheaf for a family of nodal curves.

An n-pointed nodal curve consists of the datum $(C; p_1, \dots, p_n)$ of a nodal curve C together with n distinct smooth point of C. We set $D = p_1 + \dots + p_n$. In Section 3 we introduce the concept of n-pointed stable curve. This is an n-pointed nodal curve whose automorphism group, as a pointed curve, is finite. It turns out that this is equivalent to requiring that the pullback of the log-canonical sheaf $\omega_C(D)$ be positive on any component of the normalization of C. An n-pointed nodal curve is said to be semistable if it satisfies a condition similar to the one expressed by the last sentence, but with "nonnegative" substituted for "positive." The main object of study in this book, to be formally introduced in Chapter XII, is the moduli space of n-pointed stable curve of given genus g. It is denoted by $\overline{M}_{g,n}$ and, as a set, is simply the set of isomorphism classes of n-pointed genus g stable curves. We also denote by $M_{g,n}$ the set of isomorphism classes of smooth n-pointed genus g stable curves. It will turn out that $M_{g,n}$ has the structure of a $(3g - 3 + n)$-dimensional

E. Arbarello et al., *Geometry of Algebraic Curves*,
Grundlehren der mathematischen Wissenschaften 268,
DOI 10.1007/978-3-540-69392-5_2, © Springer-Verlag Berlin Heidelberg 2011

quasi-projective variety and that $\overline{M}_{g,n}$ is a projective compactification of it. The fact that $M_{g,n}$ is not compact depends on the fact that a family of smooth curves over a noncomplete base S cannot, in general, be extended to a family of smooth curves over a completion of S. This remains true even if finite base changes are permitted. To complete the family, one has to allow singular fibers of some sort. It is a remarkable fact that one can get by with fibers having at worst nodes. This is a consequence of the stable reduction theorem (4.11), which is the main result proved in Section 4. The compactness of $\overline{M}_{g,n}$ follows directly from this theorem.

In Section 5 we study how the automorphism group of a stable pointed curve varies as the curve moves in a family. More generally, we look at two families of stable pointed curves $\alpha : X \to U$ and $\beta : Y \to U$, and prove that $\mathbf{Isom}_U(X, Y)$ is proper over U. This result will be instrumental in proving that $\overline{M}_{g,n}$ is Hausdorff. We also prove that $\mathbf{Isom}_U(X, Y)$ is unramified over U; as we will see in Chapter XII, this will turn out to be one of the properties that makes the stack counterpart of $\overline{M}_{g,n}$ a Deligne–Mumford stack.

In Section 6 and in the following two, we discuss a number of constructions involving families of nodal curves which will be continuously used throughout the book.

The first construction goes under the name of passing to the stable model. Let g and n be such that stable n-pointed genus g curves exist. This means that $2g - 2 + n > 0$. Let $(C; x_1, \dots, x_n)$ be a semistable n-pointed curve of genus g. Then there is a canonical way of constructing, out of this semistable curve, a stable one, which is called the stable model of $(C; x_1, \dots, x_n)$. For this, one notices that the irreducible components which prevent $(C; x_1, \dots, x_n)$ from being stable are precisely those components which are smooth rational and contain just two points which are either marked or nodes. The connected components of their union are chains of smooth rational curves, the so-called exceptional chains. The n-pointed nodal curve $(C'; x'_1, \dots, x'_n)$ obtained by collapsing to a point each exceptional chain is clearly stable and is, by definition, the stable model of $(C; x_1, \dots, x_n)$. Our aim in this section is to show that the procedure we just described, producing the stable model of a semistable curve, can be performed simultaneously and consistently for all fibers of any family of semistable curves.

The stable model makes it possible to easily perform two other operations, which usually go under the name of contraction and projection. Let us start with projection. Let $(C; x_1, \dots, x_n)$ be a stable n-pointed curve. Remove from it, say, the nth point. The resulting $(n - 1)$-pointed curve is semistable but may not be stable. If we pass to its stable model, the result is a stable $(n - 1)$-pointed curve. By our general construction, this operation can be performed in families, and this leads to the so-called projection map

$$\pi : \overline{M}_{g,n} \to \overline{M}_{g,n-1}\,.$$

The operation of contraction is strictly linked to the one of projection. It basically consists in keeping track of the nth point while performing the projection. Starting from a family of stable n-pointed genus g curves, this procedure produces a family of stable $(n-1)$-pointed, genus g curves, plus an extra section which, however, may meet one or more of the marked sections, or go through singular points of fibers. In Section 8 we show that the contraction operation has an inverse which, of course, can be consistently performed on families. This operation will be used in Chapter XI in constructing Kuranishi families of n-pointed curves out of Kuranishi families of $(n-1)$-pointed ones.

In Section 7 we describe the so-called clutching operation. Here we give the two main examples. In the first one we start from a pair consisting of a stable $(a+1)$-pointed genus p curve and a stable $(b+1)$-pointed genus q curve, and we produce a stable $(a+b)$-pointed curve of genus $p+q$ by identifying the $(a+1)$st marked point of the first curve with the $(b+1)$st marked point of the second one. In the second example we start from a stable $(n+2)$-pointed curve of genus $g-1$, and we produce a stable n-pointed genus g curve by identifying the $(n+1)$st and $(n+2)$nd point of the given curve. Of course, one may imagine more complicated operations. The important fact is that, again, all these operations can be consistently performed in families. The clutching operation is crucial in describing the boundary of moduli space $\partial M_{g,n} = \overline{M}_{g,n} \smallsetminus M_{g,n}$. For example, the two simple clutching operations we described above will lead to two fundamental boundary morphisms

$$\begin{aligned} &\xi_{p,a} : \overline{M}_{p,a+1} \times \overline{M}_{q,b+1} \to \overline{M}_{g,n}\,, \quad g = p+q,\ n = a+b;\\ &\xi_{irr} : \overline{M}_{g-1,n+2} \to \overline{M}_{g,n}. \end{aligned}$$

The last section is devoted to the so-called Picard–Lefschetz transformation. To illustrate it in simple terms, we consider a family $\mathcal{C} \to \Delta$ of genus g curves, parameterized by the disk $\Delta = \{t \in \mathbb{C} : |t| < 1\}$, and whose fibers are smooth except for the central one, which has a single node. Now look at a simple loop $\gamma : [0,1] \to \Delta \smallsetminus \{0\}$ based at the point $a \in \Delta$. We pull back the given family of curves to the interval $[0,1]$ and choose a C^∞ trivialization $F : \gamma^*\mathcal{C} \to C_a \times [0,1]$ of this pulled-back family. We then get a diffeomorphism $\varphi_\gamma = F_0 F_1^{-1} : C_a \to C_a$. One can verify that the isotopy class of φ_γ only depends on the homotopy class of γ. This leads to the following general picture. Suppose that $\pi : \mathcal{X} \to B$ is an algebraic or an analytic family of stable genus g curves. Let $B^* \subset B$ be the locus of points in B corresponding to smooth fibers of π. Given $a \in B^*$, there is a natural group homomorphism

$$\begin{aligned} \mathrm{PL} : \pi_1(B^*, a) &\to \mathrm{Diff}(C_a)/\,\mathrm{Diff}_0(C_a)\\ [\gamma] &\mapsto [\varphi_\gamma] \end{aligned}$$

where $\mathrm{Diff}_0(C_a)$ is the group of diffeomorphisms of C_a that are isotopic to the identity. The group $\mathrm{Diff}(C_a)/\mathrm{Diff}_0(C_a)$ is denoted by Γ_g and is called the Teichmüller modular group, while the homomorphism PL is called the Picard–Lefschetz transformation. If there were a universal family over M_g one could apply this procedure to get a map from the fundamental group of M_g to the Teichmüller modular group and even hope that this is an isomorphism. It actually turns out that, as a space, M_g is simply connected. However, we will learn that it is more useful to think of M_g as a stack (which we denote by $\mathcal{M}_g$). It then turns out that the fundamental group of the stack $\mathcal{M}_g$ is exactly Γ_g. In fact, in Chapter XV we will construct a contractible space $\mathcal{T}_g$, the Teichmüller space, which is the parameter space for a family of smooth genus g curves, on which Γ_g acts properly discontinuously, with finite stabilizers, and such that the moduli map $\mathcal{T}_g \to M_g$ is the quotient of $\mathcal{T}_g$ by Γ_g. The fact that M_g is the quotient of a contractible space modulo the action of a discrete group Γ_g, acting with finite stabilizers, tells us that the rational cohomology of Γ_g is the same as the rational cohomology of M_g, and one traditionally refers to this picture by saying that M_g is a rational $K(\Gamma_g, 1)$. Another way of describing the same picture is to say that, as a stack, $\mathcal{T}_g$ is the universal cover of M_g.

Whatever the fundamental group of the stack $\mathcal{M}_g$ may be, it is intuitively clear that the Picard–Lefschetz transformation gives a homomorphism from this group to Γ_g. But it is far from clear why this homomorphism should be an isomorphism and, in particular, why it should be surjective. The reason why it is possible to give an answer to this question is that the Picard–Lefschetz transformation can be described in very explicit terms. Let us go back to the family $\mathcal{C} \to \Delta$ of smooth curves acquiring a node at the central fiber C_0. One can think that the node of C_0 is produced by contracting a smooth simple closed curve c on a smooth fiber C_a. The cycle $[c] \in H_1(C_a; \mathbb{Z})$ is called the vanishing cycle of the family. Now, one can explicitly compute the diffeomorphism $\varphi_\gamma : C_a \to C_a$ and verify that it is a Dehn twist around the cycle c. By this we mean the homeomorphism δ_c (well defined up to isotopy) obtained by choosing an orientation on c, cutting the surface C_a along c, rotating the right edge c' of c by 180° in the positive direction, rotating the left edge c'' of c by 180° in the negative direction, and gluing the two edges together again. It is a theorem in the topology of surfaces that the group Γ_g is generated by Dehn twists, so that any element in Γ_g is in the image of a Picard–Lefschetz transformation. This is why the fundamental group of the stack $\mathcal{M}_g$ surjects onto Γ_g.

We can then identify $\mathrm{Sp}(H_1(C_a, \mathbb{Z}))$ with $Sp_{2g}(\mathbb{Z})$, and we have a homomorphism

$$\begin{aligned} \chi : \Gamma_g &\longrightarrow Sp_{2g}(\mathbb{Z}) \\ h &\mapsto h_* \end{aligned} \tag{1.1}$$

As we shall see in Chapter XV, this homomorphism is surjective. It is instructive to observe that the image of a Dehn twist under this homomorphism has a very simple form, namely that $\delta_{c*}(d) = d + (d \cdot c)c$, where $(d \cdot c)$ is the intersection number of the two cycles d and c.

Another aspect of the Picard–Lefschetz theory has to do with vanishing cycles. Let us go back to a family of stable curves $\pi : \mathcal{X} \to B$, and let us assume that B is a "small enough" polydisc. We prove that the total family $\mathcal{X}$ contracts to the central fiber X_0 and that the injection of a smooth fiber X_a into $\mathcal{X}$ induces a surjective homomorphism $H_1(X_a, \mathbb{Z}) \to H_1(\mathcal{X}, \mathbb{Z}) = H_1(X_0, \mathbb{Z})$ whose kernel is generated by the vanishing cycles. This result is intuitively clear, but its proof requires some computation. In fact we will give explicit formulae describing the retraction from $\mathcal{X}$ to X_0. To do so, it will be convenient to make use of real blow-ups. For example, let us look at the one-dimensional family $\mathcal{C} \to \Delta$ and blow up the disk at the origin. In the blown-up disk $\widetilde{\Delta}$, the origin is substituted with a copy of S^1. Now look at C_0 and normalize it. In the normalization N there are two points, s and t, corresponding to the node of C_0. Blow up N at s and t. The resulting surface $\widetilde{N}$ has a boundary formed by two copies of S^1, call them S and T. Fix a point p in S and a point q in T. One can construct a (real-analytic) family of smooth surfaces $\widetilde{\mathcal{X}} \to \widetilde{\Delta}$ which coincides with the original family over $\Delta \smallsetminus \{0\}$ and such that, over each point ϑ of the exceptional S^1 in $\widetilde{\Delta}$, there lies the surface obtained from $\widetilde{N}$ by identifying S and T in such a way that the angle between p and q is equal to ϑ. All of this can be done explicitly, and one then sees that the family $\widetilde{\mathcal{C}} \to \widetilde{\Delta}$ contracts to its restriction to S^1. Blowing everything down gives the desired retraction of $\mathcal{C}$ to C_0. Passing from a disk to a polydisc is notationally a bit more involved but presents no conceptual difficulty.

2. Elementary theory of nodal curves.

For future use, it is convenient to treat the elementary theory of nodal curves both from the analytic and the algebraic point of view.

One says that a singular point of a one-dimensional analytic space is a *node* if it has a neighborhood which is complex-analytically isomorphic to a neighborhood of the origin in the locus with equation $xy = 0$ in $\mathbb{C}^2$. A *nodal curve* is a complete algebraic curve such that every one of its points is either smooth or a node. More generally, *a family of nodal curves over a base* Y is a proper surjective morphism of schemes or analytic spaces

$$\varphi : \mathcal{C} \to Y$$

such that:

i) φ is flat;
ii) every geometric fiber of φ is a nodal curve.

An equivalent, and often more manageable, characterization of families of nodal curves is provided by the following result.

PROPOSITION (2.1). *A proper surjective morphism $\pi : X \to S$ of analytic spaces is a family of nodal curves if and only if the following holds. For any $p \in X$, either π is smooth at p with one-dimensional fibers, or else, setting $s = \pi(p)$, there is a neighborhood of p which is isomorphic, as a space over S, to a neighborhood of $(0, s)$ in the analytic subspace of $\mathbb{C}^2 \times S$ with equation*

$$xy = f\,, \tag{2.2}$$

where f is a function on a neighborhood of s in S whose germ at s belongs to the maximal ideal of $\mathcal{O}_{S,s}$.

To prove the proposition, we first need a differential characterization of nodes.

LEMMA (2.3). *Let f be a holomorphic function on a neighborhood of $0 \in \mathbb{C}^2$, and suppose that $f(0) = 0$. Then the analytic space defined by f has a node at the origin if and only if the first-order partials of f vanish at the origin and the Hessian of f is nonsingular at the origin.*

One implication is obvious; if $f = zw$ in suitable coordinates z and w, then the partials of f vanish for $z = w = 0$, and the Hessian is nonsingular. To prove the converse, notice that, if we choose coordinates x, y on $\mathbb{C}^2$ and assume that $\partial f/\partial x$ and $\partial f/\partial y$ vanish at 0 and that the Hessian is non-singular there, then on a neighborhood of the origin we may write

$$f(x, y) = a(x, y)x^2 + 2b(x, y)xy + c(x, y)y^2\,,$$

where a, b, and c are holomorphic, and $ac - b^2$ does not vanish for $x = y = 0$. We may assume, possibly after a linear change of coordinates, that $a(0, 0) \neq 0$. We then set $x_1 = x + (b/a)y$, $y_1 = y$, and notice that

$$f = a_1 x_1^2 + c_1 y_1^2$$

with $a_1(0, 0)$ and $c_1(0, 0)$ different from zero. Choose new coordinates x_2 and y_2 by setting $x_2 = x_1\alpha$, $y_2 = y_1\gamma$, where α and γ are square roots of a_1 and c_1, respectively. Thus,

$$f = x_2^2 + y_2^2\,.$$

The last step consists in choosing as final coordinates $x_3 = x_2 + \sqrt{-1}y_2$ and $y_3 = x_2 - \sqrt{-1}y_2$. With this choice of coordinates,

$$f = x_3 y_3\,.$$

This finishes the proof of the lemma.

Now we can prove (2.1). For any $s \in S$, we write X_s to denote $\pi^{-1}(s)$. Suppose first that $\pi : X \to S$ is a family of nodal curves, and let p be a point of X_s. We may realize an open neighborhood of p in X as a locally closed subspace of $\mathbb{C}^r \times S$ in such a way that the restriction of the projection to the second factor agrees with π and that p corresponds to $(0, s)$. The Zariski tangent space to X_s at p has dimension h, where h can equal 1 or 2. Choose a linear projection $\mathbb{C}^r \to \mathbb{C}^h$ in such a way that the induced map $T_p(X_s) \to \mathbb{C}^h$ is an isomorphism. Since the tangent space to X at p sits in an exact sequence

$$0 \to T_p(X_s) \to T_p(X) \to T_s(S)\,,$$

the composite map $T_p(X) \to \mathbb{C}^r \oplus T_s(S) \to \mathbb{C}^h \oplus T_s(S)$ is injective. This means that there is a neighborhood U of p in X such that $U \to \mathbb{C}^r \times S \to \mathbb{C}^h \times S$ identifies U to a locally closed subspace of $\mathbb{C}^h \times S$. In other words, we may choose r equal to h. If $h = 1$, this shows that π is smooth at p. If $h = 2$, we are essentially dealing with a family of plane curves. In this case we may view X_s, or rather a neighborhood of p in it, as a subspace of $\mathbb{C}^2$; under this identification, p corresponds to $0 \in \mathbb{C}^2$. Since X_s has no embedded components, it is locally defined by a single equation $f = 0$. Since we are assuming that X is flat over S, the local ring $\mathcal{O}_{X,p}$ is the quotient of $\mathcal{O}_{\mathbb{C}^2 \times S,(0,s)}$ modulo a principal ideal with a generator F which reduces to f on $\mathbb{C}^2 \times \{s\} = \mathbb{C}^2$. Thus X is locally defined, inside $\mathbb{C}^2 \times S$, by the single equation $F = 0$. We may realize a neighborhood of s in S as an analytic subspace of some $\mathbb{C}^k$ with s identified to the origin. Let $z = (z_1, \dots, z_k)$ be the coordinates in $\mathbb{C}^k$, and extend F to an element of the local ring of $\mathbb{C}^2 \times \mathbb{C}^k$ at the origin, which we will also denote by F. Our goal is to put F in as simple a form as possible, by a judicious change of coordinates. The argument we shall use is just a version "with parameters" of the proof of Lemma (2.3).

We begin by considering the pair of functions $\partial F/\partial x$, $\partial F/\partial y$. Their Jacobian matrix is just the Hessian of F with respect to x and y, which by Lemma (2.3) is nonsingular for $x = y = 0$ and $z = 0$. The implicit function theorem then says that there are functions $\varphi(z)$, $\psi(z)$ such that the locus of points where $\partial F/\partial x$ and $\partial F/\partial y$ vanish is given, near the origin of $\mathbb{C}^2 \times \mathbb{C}^k$, by $x = \varphi(z)$, $y = \psi(z)$. Replacing the coordinates x and y with $x - \varphi(z)$ and $y - \psi(z)$, we may then suppose that the partials $\partial F/\partial x$ and $\partial F/\partial y$ vanish identically for $x = y = 0$. This means that we may write

$$F(x, y, z) = -f(z) + a(x, y, z)x^2 + 2b(x, y, z)xy + c(x, y, z)y^2\,,$$

where a, b, and c are holomorphic functions such that $ac - b^2$ does not vanish for $x = y = 0$ and $z = 0$. We will operate on

$$Q = ax^2 + 2bxy + cy^2\,,$$

mimicking what we would do to diagonalize a quadratic form in x and y. Possibly after a linear change of coordinates, we may assume that $a(0,0,0) \neq 0$. We then set $x_1 = x + (b/a)y$, $y_1 = y$, and notice that

$$Q = a_1 x_1^2 + c_1 y_1^2$$

with $a_1(0,0,0)$ and $c_1(0,0,0)$ different from zero. Choose new coordinates x_2 and y_2 by setting $x_2 = x_1\alpha$, $y_2 = y_1\gamma$, where α and γ are square roots of a_1 and c_1, respectively. Thus,

$$Q = x_2^2 + y_2^2 .$$

Then set $x_3 = x_2 + \sqrt{-1}y_2$ and $y_3 = x_2 - \sqrt{-1}y_2$. With this choice of coordinates,

$$F = x_3 y_3 - f(z) .$$

This proves the "only if" part of the proposition.

The proof of the converse is simpler. First of all, if π is smooth at p, then it is also flat at p, and $X_{\pi(p)}$ is smooth. On the other hand, any space of the form (2.2) is flat over S. To prove this, it suffices to deal with the special case where S is the affine line, and $f = t$ is a linear coordinate on it, since the general case can be obtained by base change from the special one. Away from $x = y = t = 0$, the projection from $xy = t$ to the t-coordinate is smooth. At $x = y = t = 0$, the generator xy of the ideal of the central fiber in $\mathcal{O}_{\mathbb{C}^2,0}$ extends by definition to the generator $xy - t$ of the ideal of X in $\mathcal{O}_{\mathbb{C}^2\times\mathbb{C},0}$. Moreover, any relation among generators of the ideal of the central fiber extends to a relation among generators of the ideal of X, since there are no relations at all. The conclusion follows from Lemma (5.1) in Chapter IX.

Recall that a morphism $f : X \to Y$ of schemes or of analytic spaces is said to be a *local complete intersection morphism* (a *l.c.i. morphism* for short) if it factors, locally on X, as a regular embedding followed by a smooth morphism. Clearly, from the analytic point of view, a family of nodal curves is a l.c.i. morphism. We wish to show that the same is true in the scheme setup. More generally, we shall prove the following.

LEMMA (2.4). *Let $f : X \to Y$ be a morphism of schemes (of finite type over $\mathbb{C}$). Then f is a l.c.i. morphism if and only if the corresponding morphism of analytic spaces is l.c.i.*

In the proof, we shall rely on standard properties of l.c.i. morphisms, for which we refer to Section 6.3 of [480]; the results we shall use are stated and proved there only for morphisms of schemes, but they remain true, with essentially the same proof, for morphisms of analytic spaces. The next lemma summarizes the standard property we need.

LEMMA (2.5). *Let $f : X \to Y$ be a l.c.i. morphism. Suppose that f is the composition of an embedding $\alpha : X \to Z$ and a smooth morphism $Z \to Y$. Then α is a regular embedding.*

To prove (2.4), we just have to show that, if f is l.c.i. analytically, then it is l.c.i. also scheme-theoretically. Since the statement is of a local nature, we may assume that f factors into an embedding $\alpha : X \to Y \times \mathbb{C}^N$ followed by the projection of $Y \times \mathbb{C}^N$ to Y. Lemma (2.5) says that α is a regular embedding of analytic spaces. We must show that it is also a regular embedding of schemes. In other words, we are reduced to proving the following.

LEMMA (2.6). *Let $h : W \to Z$ be an embedding of schemes of finite type over $\mathbb{C}$. Then h is a regular embedding if and only if the corresponding morphism of analytic spaces is a regular embedding.*

Let z be a closed point of W. We denote by A the local ring of Z at z, and by I the ideal of W in A. We also denote by B the local ring at z of the analytic space underlying Z. Clearly, A is a subring of B. The ideal in B of the analytic subspace corresponding to W is just BI. Now, one of the main technical results of Serre's GAGA [626] is that (A, B) is a flat pair, so that in particular B is A-flat. As the inclusion $A \hookrightarrow B$ is a local homomorphism, B is actually faithfully flat over A. Recall in fact that a module M over a commutative ring R is said to be *faithfully flat* when any complex $C^\bullet$ of R-modules is exact if and only if $C^\bullet \otimes_R M$ is exact; equivalently, when M is flat and, for any R-module N, $N \otimes_R M$ is zero if and only if N is. Faithful flatness of B over A is then a consequence of the following standard result.

LEMMA (2.7). *Let $R \to R'$ be a local homomorphism of local rings. If R' is R-flat, then it is faithfully flat.*

The proof is straightforward. Let M be a nonzero R-module, and let L be a submodule of M generated by a nonzero element; thus $L \simeq R/J$ for some proper ideal J. By flatness, $L \otimes_R R'$ is a submodule of $M \otimes_R R'$; on the other hand, $L \otimes_R R'$ does not vanish since, again by flatness, $R'/J \otimes_R R' \cong R'/JR'$, and JR' is contained in the maximal ideal of R'.

Returning to the proof of (2.6), let $b_1, \dots, b_n$ be a minimal system of generators of I. Obviously, the b_i also generate BI; we claim that, in fact, they are a minimal system of generators. For each fixed j, denote by I_j the ideal of A generated by all the b_i except b_j and look at the exact sequence

$$I_j \to I \to Q \to 0\,,$$

where $Q = I/I_j$. Since $b_1, \dots, b_n$ is a minimal system of generators of I, $Q \neq 0$. Tensoring with B gives an exact sequence

$$BI_j \to BI \to Q \otimes_A B \to 0\,,$$

since $I \otimes_A B = BI$ and $I_j \otimes_A B = BI_j$, by flatness. On the other hand, $Q \otimes_A B$ cannot vanish, by faithful flatness. This proves that $b_1, \ldots, b_n$ is a minimal system of generators of BI; by assumption and by Lemma (5.22) it is thus a regular sequence in B. To conclude the proof, we must show that it is a regular sequence in A as well, that is, that the sequence

$$(b_1, \ldots, b_i) \hookrightarrow A \xrightarrow{\times b_{i+1}} A/(b_1, \ldots, b_i)$$

is exact for all i. Tensoring with B gives an exact sequence since $b_1, \ldots, b_n$ is a regular sequence in B; hence the conclusion follows from faithful flatness. This concludes the proof of Lemma (2.6) and hence of Lemma (2.4).

Given a nodal curve C and a set S of nodes of C, let $\beta : X \to C$ be the partial normalization of C at S. We may associate to this setup a graph. There is one vertex for each connected component of X, and the half-edges issuing from a given vertex are the points of the corresponding component of X mapping to nodes of C belonging to S. A pair $\{q, q'\}$ of distinct half-edges constitutes an edge when q and q' map to the same node of C. To each vertex v we assign an integer weight g_v, the arithmetic genus of the corresponding component of X. We shall sometimes write $\mathrm{Graph}^S(C)$ to indicate the graph we have just described, equipped with the genus weight function. When S is the set of all nodes of C, we shall write $\mathrm{Graph}(C)$ for $\mathrm{Graph}^S(C)$; this is the so-called *dual graph* of C.

Since β is finite, for any coherent sheaf $\mathcal{F}$ on X, the higher direct images $R^i\beta_*\mathcal{F}$ vanish, and hence, by the Leray spectral sequence, $H^i(X, \mathcal{F}) = H^i(C, \beta_*\mathcal{F})$ for any i. It then follows from the exact sequence

$$0 \to \mathcal{O}_C \to \beta_*\mathcal{O}_X \to \bigoplus_{p \in S} \mathbb{C}_p \to 0$$

that the genera of X and C are related by

$$p_a(C) = p_a(X) + |S| .$$

If $X_1, \ldots, X_v$ are the connected components of X, the above formula can also be rewritten in this form:

$$\begin{aligned} p_a(C) &= \sum p_a(X_i) + 1 - v + |S| \\ &= \sum p_a(X_i) + 1 - \chi(\mathrm{Graph}^S(C)) . \end{aligned} \tag{2.8}$$

We denote by

$$\alpha : N \to C$$

the normalization of C, by $C_1, \ldots, C_t$ the irreducible components of C, and by ν the number of nodes of C; we shall also denote by N_i the normalization of C_i. As we have just seen, the genera of N and C are related by

$$\begin{aligned} p_a(C) &= p_a(N) + \nu \\ &= \sum p_a(N_i) + 1 - t + \nu \,. \end{aligned}$$

We now discuss the group $\mathrm{Pic}(C) = H^1(C, \mathcal{O}_C^\times)$ of isomorphism classes of line bundles on C. An essentially complete description of this group can be extracted from the exact sequence

$$1 \to \mathcal{O}_C^\times \to \alpha_* \mathcal{O}_N^\times \xrightarrow{e} \prod_{p \in C_{\mathrm{sing}}} \mathbb{C}_p^\times \to 1 \,,$$

where e is defined as follows. For each singular point p of C, let $\alpha^{-1}(p) = \{q, r\}$. Then for any section f of $\alpha_* \mathcal{O}_N^\times$, the $\mathbb{C}_p$-component of $e(f)$ is $f(q)/f(r)$. Of course, this definition depends on the choice of an ordering of the points of $\alpha^{-1}(p)$. The exact cohomology sequence associated to the exact sheaf sequence above is

$$1 \to (\mathbb{C}^\times)^s \to (\mathbb{C}^\times)^t \to (\mathbb{C}^\times)^\nu \to \mathrm{Pic}(C) \xrightarrow{\alpha^*} \mathrm{Pic}(N) \to 0 \,,$$

where s is the number of connected components of C, and α^* is the pullback map. The sequence may be interpreted as saying:

i) To give a line bundle L on C, we have to specify its pullback $\tilde{L} = \alpha^* L$ to N, plus "descent data," that is, we have to specify when a section of $\tilde{L}$ is the pullback of a section of L. More precisely, for any node p of C, we have to give an identification $\varphi_p : \tilde{L}_q \tilde{\to} \tilde{L}_r$, where $\alpha^{-1}(p) = \{q, r\}$.

ii) When $\tilde{L}$ is trivial, a choice of trivialization identifies each φ_p with a well-defined nonzero complex number.

iii) Given a trivial $\tilde{L}$, two choices of descent data define the same element of $\mathrm{Pic}(C)$ exactly when they are obtained one from another by a change of trivialization for $\tilde{L}$, that is, multiplying the given trivialization by a nonzero constant on each component.

If L is a line bundle on C, the *multidegree* of L is the t-tuple $(d_1, \ldots, d_t)$, where $d_i = \deg_{C_i}(L)$ is the degree of the pullback of L to N_i; the *total degree* (or simply degree) of L is the sum of the d_i's and is written $\deg(L)$. If $\mathbf{d} = (d_1, \ldots, d_t)$ and $\mathbf{e} = (e_1, \ldots, e_t)$ are multidegrees, we shall write $\mathbf{d} \geq \mathbf{e}$ to mean that $d_i \geq e_i$ for every i, and $\mathbf{d} > \mathbf{e}$ to mean that $\mathbf{d} \geq \mathbf{e}$ and $d_i > e_i$ for at least one i. The Jacobian of C, which we shall denote by $\mathrm{Pic}^0(C)$ or $J(C)$, is the subgroup of $\mathrm{Pic}(C)$ consisting of line bundles with zero multidegree, so that we have the exact sequence

$$1 \to (\mathbb{C}^\times)^{\nu - t + s} \to J(C) \to J(N) \to 0 \,.$$

Thus the Jacobian of C is a $\mathbb{C}^\times$-extension of an abelian variety. From this point of view a rational irreducible g-nodal curve represents one extreme in the sense that $J(N) = 0$, while $J(C) = (\mathbb{C}^\times)^g$; at the other extreme, we have curves C satisfying any one of the following equivalent conditions:

a) $J(C)$ is compact.
b) The sum of the geometric genera of the components of C is g.
c) Every component of C is smooth, and the dual graph of C is a tree.

We will call such a curve *of compact type*.

In complete analogy with the smooth case, the Riemann–Roch theorem states that, for any line bundle L on C, one has

$$\chi(L) = \deg(L) + 1 - p_a(C)\,.$$

This formula is a straightforward consequence of the Riemann–Roch theorem for the line bundle $\alpha^*(L)$ on N and of the exact sequence

$$0 \to L \to \alpha_*\alpha^*(L) \to \bigoplus_{p\in C_{\mathrm{sing}}} \mathbb{C}_p \to 0\ .$$

The Serre duality theorem (cf., for instance, Hartshorne [356]) says, in our case, that there exist on C a *dualizing sheaf* ω_C and a "trace" homomorphism

$$\xi_C : H^1(C, \omega_C) \to \mathbb{C} \tag{2.9}$$

with the property that, for any coherent sheaf $\mathcal{F}$ on C, the pairing

$$H^1(C, \mathcal{F}) \times \mathrm{Hom}(\mathcal{F}, \omega_C) \to H^1(C, \omega_C) \xrightarrow{\xi_C} \mathbb{C} \tag{2.10}$$

is a duality. Furthermore, since C is, locally, a complete intersection, it is also the case that ω_C is invertible, and, for any coherent sheaf $\mathcal{F}$, the pairing

$$H^0(C, \mathcal{F}) \times \mathrm{Ext}^1(\mathcal{F}, \omega_C) \to \mathrm{Ext}^1(\mathcal{O}_C, \omega_C) \cong H^1(C, \omega_C) \xrightarrow{\xi_C} \mathbb{C} \tag{2.11}$$

is a duality. That $\mathrm{Ext}^1(\mathcal{O}_C, \omega_C)$ is isomorphic to $H^1(C, \omega_C)$ is a special case of the following simple observation. Let $\mathcal{F}$ be a coherent sheaf on a scheme X. There is a spectral sequence abutting to $\mathrm{Ext}^\bullet(\mathcal{O}_X, \mathcal{F})$ whose E_2 term is $E_2^{p,q} = H^p(X, \mathcal{E}xt^q(\mathcal{O}_X, \mathcal{F}))$. When $\mathcal{F}$ is locally free, the higher sheaf Ext's $\mathcal{E}xt^q(\mathcal{O}_X, \mathcal{F})$, $q > 0$, all vanish, and hence

$$\mathrm{Ext}^p(\mathcal{O}_X, \mathcal{F}) \cong H^p(X, \mathcal{H}om(\mathcal{O}_X, \mathcal{F})) \cong H^p(X, \mathcal{F})$$

for all p. It follows in particular that, for a locally free $\mathcal{F}$ on a nodal curve C, the duality pairings (2.10) and (2.11) yield a duality between $H^q(C,\mathcal{F})$ and $H^{1-q}(\omega_C \otimes \mathcal{F}^\vee)$ for $q = 0, 1$, where $\mathcal{F}^\vee = \mathcal{H}om(\mathcal{F}, \mathcal{O}_C)$ is the dual of $\mathcal{F}$.

In the case at hand, ω_C and ξ_C can be described quite explicitly. Let $\{q_1, q_1'\}, \ldots, \{q_\nu, q_\nu'\}$ be the preimages in N of the nodes $p_1, \ldots, p_\nu$ of C. Then ω_C is the invertible subsheaf

$$\omega_C \subset \alpha_*\left(\omega_N\left(\sum(q_i + q_i')\right)\right)$$

defined by the following prescription. A section φ of $\alpha_*(\omega_N(\sum(q_i + q_i')))$, viewed as a section of $\omega_N(\sum(q_i + q_i'))$, is a section of ω_C if and only if

$$\text{(2.12)} \qquad \operatorname{Res}_{q_i}(\varphi) + \operatorname{Res}_{q_i'}(\varphi) = 0\,, \quad i = 1, \ldots, \nu\,.$$

In particular, when C is smooth, ω_C is nothing but the canonical sheaf. The residue theorem is still valid on C in the following version. Let φ be a meromorphic section of ω_C which is *holomorphic at the nodes of* C; then the sum of the residues of φ vanishes. Taking into account (2.12), this is an immediate consequence of the ordinary residue theorem applied to the form $\alpha^*\varphi$ on N.

To define ξ_C, choose a divisor $D = r_1 + \cdots + r_h$ consisting of h distinct smooth points of C, with the property that any component of C contains at least one of the r_i's. We shall see in a moment that $H^1(C, \omega_C(D))$ vanishes. Assuming this and looking at the exact cohomology sequence

$$\cdots \to H^0(C, \omega_C(D)) \to H^0(C, \omega_C(D)/\omega_C) \to H^1(C, \omega_C) \to H^1(C, \omega_C(D))\,,$$

we find that any class φ in $H^1(C, \omega_C)$ lifts to a section $\overline{\varphi}$ of $\omega_C(D)/\omega_C$. Set

$$\xi_C(\varphi) = 2\pi\sqrt{-1}\sum \operatorname{Res}_{r_i}\overline{\varphi}\,.$$

Notice that the right-hand side does not depend on the choice of the lifting $\overline{\varphi}$, nor on the choice of D, by the residue theorem. It remains to show that $H^1(C, \omega_C(D))$ is zero. Let $\beta : X \to C$ be the partial normalization of C at the nodes $p_{i_1}, \ldots, p_{i_l}$. It is a useful general observation that, by the very definition of ω_C, one has two exact sequences on C:

$$\text{(2.13)} \qquad \begin{gathered} 0 \to \beta_*\omega_X \to \omega_C \to \mathbb{C}_{p_{i_1}} \oplus \cdots \oplus \mathbb{C}_{p_{i_l}} \to 0\,, \\ 0 \to \omega_C \to \beta_*\omega_X\left(\sum(q_{i_j} + q_{i_j}')\right) \to \mathbb{C}_{p_{i_1}} \oplus \cdots \oplus \mathbb{C}_{p_{i_l}} \to 0\,. \end{gathered}$$

We apply this to the full normalization $\alpha : N \to C$. Tensoring the first exact sequence with $\mathcal{O}_C(D)$ and passing to cohomology, we get a surjection

$$H^1(C, \alpha_*\omega_N(D)) \twoheadrightarrow H^1(C, \omega_C(D))\,.$$

The cohomology group on the left equals $H^1(N, \omega_N(D))$, where D is viewed as a divisor on N. On the other hand, this group vanishes by degree considerations since every component of N meets D.

Suppose that C is smooth and view the elements of $H^1(C, \omega_C)$ as represented by (1,1)-forms via the Dolbeault isomorphism. It is a simple exercise, which we leave to the reader, to check that in this special case ξ_C is just integration over C. It follows in particular that for a smooth C, the duality pairings (2.10) and (2.11) agree with the ones that we have been using so far.

As is the case for smooth curves, the duality theorem for nodal curves has several useful consequences. Here we record only one.

LEMMA (2.14). *Let C be a connected nodal curve, and let M be a line bundle on C. Let* $\mathbf{d}$ *be the multidegree of M, and* $\mathbf{e}$ *the one of ω_C. If* $\mathbf{d} > \mathbf{e}$, *then* $H^1(C, M) = 0$.

In fact, duality says that the vanishing of $H^1(C, M)$ is equivalent to the one of $H^0(C, \omega_C M^{-1})$. On the other hand, by assumption, the degrees of $\omega_C M^{-1}$ on the various components of C are all nonpositive, and one among them is strictly negative. Thus, any section of $\omega_C M^{-1}$ vanishes identically on one component and is at best constant on the remaining ones. Since C is connected, it must then vanish identically.

In analogy with what happens for smooth curves, ampleness of a line bundle on a nodal curve depends only on the multidegree.

LEMMA (2.15). *Let C be a nodal curve, and let M be a line bundle on C. Then M is ample if and only if* $\deg_D M > 0$ *for every irreducible component D of C.*

The proof is an easy exercise using (2.14) and is left to the reader. Alternatively, the lemma can be viewed as the simplest case of Seshadri's ampleness criterion (Proposition (9.11) in Chapter XI).

The partial normalization X of a nodal curve at a set S of nodes carries a distinguished finite set of smooth points, consisting of those points of X which map to nodes in S. It is important, from a technical point of view, to consider not just nodal curves, but also objects of this sort, that is, pairs $(C; D)$ consisting of a nodal curve C plus a finite set D of smooth points of C (often referred to as the *marked* points). We shall call such a pair a *nodal curve with marked points*, and we shall often view D as a divisor on C. The invertible sheaf $\omega_C(D)$ will be referred to as the *log-canonical sheaf* of $(C; D)$. As is the case for ordinary curves, we can associate a graph to a nodal curve with marked points. This seems a good place to specify what kind of graphs we shall be dealing with and to establish some terminology and notation.

DEFINITION (2.16). *A graph Γ is the datum of:*

- *a finite nonempty set $V = V(\Gamma)$ (the set of vertices);*
- *a finite set $L = L(\Gamma)$ (the set of half-edges);*
- *an involution ι of L;*
- *a partition of L indexed by V, that is, the assignment to each $v \in V$ of a (possibly empty) subset L_v of L such that $L = \bigcup_{v\in V} L_v$ and $L_v \cap L_w = \emptyset$ if $v \neq w$.*

A pair of distinct elements of L interchanged by the involution is called an edge of the graph. A fixed point of the involution is called a leg of the graph. The set of edges of Γ is denoted by $E(\Gamma)$. A dual graph is the datum of a graph together with the assignment of a nonnegative integer weight g_v to each vertex v. The genus of a dual graph Γ is defined to be

$$g = \sum_{v\in V(\Gamma)} g_v + 1 - \chi(\Gamma)\,.$$

A graph (or a dual graph) endowed with a one-to-one correspondence between a finite set P and the set of its legs will be said to be P-marked, or numbered if P is of the form $\{1,\dots,n\}$ for some nonnegative integer n.

As the reader will have noticed, we allow free half-edges, that is, half-edges which are not part of an edge; these are the legs. It is also important to observe that there may be multiple edges connecting two given vertices, and there may be loops, that is, edges going from one vertex to itself. Moreover, each edge $\{\ell,\ell'\}$ carries two possible orientations corresponding to the ordered pairs (ℓ,ℓ') and (ℓ',ℓ).

The notion of isomorphism between graphs (or dual graphs, or P-marked graphs) is the obvious one. An isomorphism is the datum of a bijection between vertices and a bijection between half-edges respecting the relevant structure.

The geometric realization $|\Gamma|$ of a graph Γ is the one-dimensional CW-complex obtained by realizing each half edge $\ell \in L_v$ as a segment, denoted by $|\ell|$, having a point $|v|$ as one of its endpoints, and then identifying the free endpoint of a segment $|\ell|$ with the free endpoint of a segment $|\ell'|$ if and only if $\iota(\ell) = \ell'$; the resulting segment is then called an edge of $|\Gamma|$. The set of edges of $|\Gamma|$ is denoted by $E(|\Gamma|)$ and is usually identified with $E(\Gamma)$. A graph Γ and its realization $|\Gamma|$ determine each another. A graph Γ is said to be connected if its realization $|\Gamma|$ is. Notice that $\chi(\Gamma) = \chi(|\Gamma|)$. Clearly, an isomorphism of graphs induces an isomorphism of the respective geometric realizations.

Let us go back to curves with marked points. Let C be a nodal curve, and let D be a finite set of smooth points of C. To associate a dual graph $\mathrm{Graph}(C;D)$ to $(C;D)$, we proceed essentially as for ordinary curves. As in that case, there is a vertex for each component of the normalization of C, and its weight is the genus of the component. The half-edges issuing from a vertex are the points of the corresponding component which map either to a node of C or to a marked point. As for plain curves, the edges of the graph are the pairs of half-edges mapping to the same node of C. The legs are the half-edges coming from the marked points. It follows from formula (2.8) that the genus of the graph $\mathrm{Graph}(C; D)$ as defined above is just the genus of C.

Most often, when considering curves with marked points, we will want the latter to be ordered. If n is a nonnegative integer, we then define an *n-pointed nodal curve* to be the datum of a nodal curve C together with an ordered set of n smooth points of C. More generally, if P is a finite set, a *P-pointed nodal curve* is the datum of a nodal curve C together with an injective map from P to the smooth locus of C; when $P = \{1, \dots, n\}$, we recover the notion of n-pointed nodal curve. The dual graph of a P-pointed curve $(C; \{x_p\}_{p\in P})$ is naturally P-marked and will be denoted $\mathrm{Graph}(C; \{x_p\}_{p\in P})$. Figure 1 below illustrates a 3-pointed genus 8 nodal curve and its graph; vertices are represented by small circles bearing in the middle the genus of the normalization of the corresponding component.

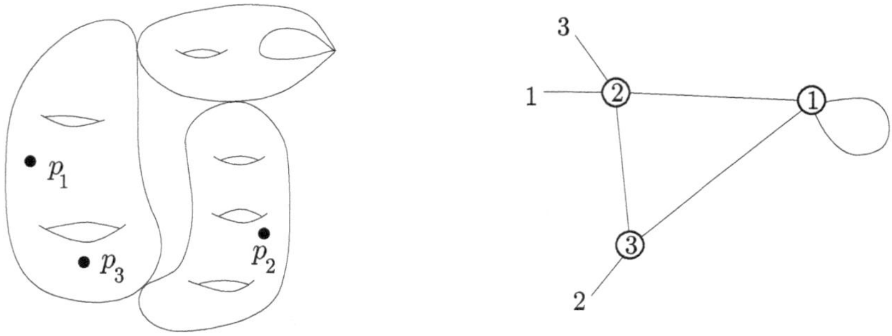

Figure 1. A 3-pointed curve (left) and its graph (right)

As is the case with unpointed curves, we can also associate a P-marked dual graph $\mathrm{Graph}^S(C; \{x_p\}_{p\in P})$ to the datum of a P-pointed nodal curve $(C; \{x_p\}_{p\in P})$ and a set S of nodes of C. The vertices of this graph are the connected components of the partial normalization X of C at S, the weight g_v of a vertex is the arithmetic genus of the corresponding component of X, the edges correspond to the nodes in S, and the half-edges are the marked points or the points of X mapping to nodes in S. The genus of $\mathrm{Graph}^S(C; \{x_p\}_{p\in P})$ is equal to the genus of C.

It is possible to classify the nodes of a connected P-pointed nodal curve $(C; \{x_p\}_{p\in P})$ of genus g according to the combinatorics of the partial normalization of $(C; \{x_p\}_{p\in P})$ at the node. Let q be a singular point of C, and let e be the corresponding edge of the dual graph $\Gamma = \mathrm{Graph}(C; \{x_p\}_{p\in P})$. We shall say that q is a *nonseparating node* or that e is a *nondisconnecting edge of* Γ if the partial normalization of C at q is connected, i.e., if the graph obtained from Γ by removing e is connected. In this case the partial normalization of $(C; \{x_p\}_{p\in P})$ at q is a connected nodal curve of genus $g-1$ with $|P|+2$ marked points, all but the two coming from the normalization process indexed by P. The other possibility is that the normalization of C at q has two connected components C_1 and C_2 of genera a and b adding to g. In this case the set P is partitioned in two complementary subsets A and B indexing, respectively, marked points on C_1 and marked points of C_2. We shall then say that q is a *separating node* (or that e is a *disconnecting edge*) *of type* $\mathcal{P}$, where $\mathcal{P} = \{(a,A),(b,B)\}$. We shall sometimes refer to such a $\mathcal{P}$ as a *bipartition* of (g,P). In practice, we shall usually say that q is a separating node of type (a,A) (or, equivalently, a separating node of type $(b,B) = (g-a, P \smallsetminus A)$).

Naturally, we can also speak of families of P-pointed curves. Formally, a *family of P-pointed nodal curves* is the datum of a family $\alpha : X \to S$ of nodal curves, plus disjoint sections $\sigma_p : S \to X$, $p \in P$, which do not meet the singular locus of any one of the fibers of α. If $\beta : X' \to S'$, $\{\sigma'_p : p \in P\}$ is another family of P-pointed nodal curves, a *morphism* from the first family to the second is a cartesian square

$$\begin{array}{ccc} X & \xrightarrow{H} & X' \\ {\scriptstyle\alpha}\downarrow & & \downarrow{\scriptstyle\alpha'} \\ S & \xrightarrow{h} & S' \end{array}$$

such that $H \circ \sigma_p = \sigma'_p \circ h$ for every $p \in P$. The composition of morphisms is the composition of cartesian squares.

When dealing with families of P-pointed curves, we shall sometimes use the symbols we use to denote *sections* also to denote the corresponding *divisors*. For instance, we shall often write $\mathcal{O}(\sum \sigma_p)$ instead of $\mathcal{O}(\sum \sigma_p(S))$.

The notion of dualizing sheaf, in a sense, provides a generalization of what, for smooth curves, is the sheaf of holomorphic 1-forms. There is, however, another more direct generalization, given by the notion of Kähler differentials. Recall that, for any morphism of schemes, or of analytic spaces,

$$\psi : X \to Y\,,$$

the sheaf of relative Kähler differentials $\Omega^1_{X/Y}$ (or Ω^1_ψ) is defined to be the pullback to X, via the diagonal map, of the ideal sheaf I of the diagonal in $X \times_Y X$. Denoting by π_1 and π_2 the projections of $X \times_Y X$ onto the two factors, the differentiation operator

$$d : \mathcal{O}_X \to \Omega^1_{X/Y}$$

defined by

$$d(h) = \pi_1^*(h) - \pi_2^*(h) \qquad (\text{mod. } I^2)$$

is $\mathcal{O}_Y$-linear and satisfies Leibniz' rule. The sheaf Ω^1_X of Kähler differentials on X is nothing but the sheaf $\Omega^1_{X/Y}$ for Y a point. If x, y are points of X and Y such that $\psi(x) = y$, the stalk of $\Omega^1_{X/Y}$ over x is the module of differentials $\Omega_{\mathcal{O}_x/\mathcal{O}_y}$. More generally, for any morphism $Y' \to Y$,

$$\Omega^1_{X'/Y'} \simeq p^*\Omega^1_{X/Y}\,,$$

where $X' = X \times_Y Y'$, and $p : X' \to X$ is the projection to the first factor. We refer to Matsmura's book [503] or to Chapter 6 of Qing Liu's book [480] for the properties of modules of differentials. Here we wish only to record, for future reference, just a few of their consequences. The first is that to any commutative diagram

$$\begin{array}{ccc} X & \xrightarrow{\quad\psi\quad} & Y \\ & \searrow \quad \swarrow & \\ & S & \end{array}$$

of schemes over S there is associated a canonical sheaf homomorphism

$$\psi^*\Omega^1_{Y/S} \to \Omega^1_{X/S}$$

fitting into an exact sequence

$$\psi^*\Omega^1_{Y/S} \to \Omega^1_{X/S} \to \Omega^1_{X/Y} \to 0\,.$$

In particular, when S is just a point, this reduces to

$$\psi^*\Omega^1_Y \to \Omega^1_X \to \Omega^1_{X/Y} \to 0\,. \tag{2.17}$$

A second consequence is as follows. Suppose Z is a closed subscheme (or analytic subspace) of X, and J is its sheaf of ideals. Then

$$\Omega^1_{Z/Y} = \Omega^1_{X/Y}/M\,,$$

where M stands for the sub-$\mathcal{O}_X$-module generated by $J\Omega^1_{X/Y}$ and by the differentials of sections of J. In other words, there is a canonical exact sequence

$$J/J^2 \xrightarrow{d} \Omega^1_{X/Y} \otimes_{\mathcal{O}_X} \mathcal{O}_Z \to \Omega^1_{Z/Y} \to 0.$$

As an example of application, suppose that

$$\varphi : \mathcal{C} \to Y \tag{2.18}$$

is a family of nodal curves, locally of the form $xy = f$; it then follows that $\Omega^1_{\mathcal{C}}$ is locally generated by dx, dy, and $\varphi^*\Omega^1_Y$, subject to the only relation

$$ydx + xdy = df\,,$$

while $\Omega^1_{\mathcal{C}/Y}$ is generated by dx and dy, modulo the relation

$$ydx + xdy = 0\,.$$

Like the sheaf of Kähler differentials, the dualizing sheaf also has a relative counterpart $\omega_{\mathcal{C}/Y}$ (or ω_φ) for a family of nodal curves (2.18). In fact, any l.c.i. morphism $h : X \to Y$ of schemes or analytic spaces admits a *relative dualizing sheaf*, an invertible $\mathcal{O}_X$-module denoted ω_h or $\omega_{X/Y}$. A convenient reference for the construction and elementary properties of the relative dualizing sheaf is Section 6.4 of [480]. Here we just want to recall what ω_h looks like locally. Let U be an open subset of X over which h factors into the composition of a regular embedding $\iota : U \to Z$ and a smooth morphism $\pi : Z \to Y$. The sheaf of relative Kähler differentials Ω^1_π is locally free of rank equal to the fiber dimension r of π, while the normal sheaf $N_{U/Z}$ is locally free of rank equal to the codimension k of U in Z. Then there is a canonical isomorphism

$$\omega_{h|U} \simeq \wedge^r \iota^*(\Omega^1_\pi) \otimes \wedge^k N_{U/Z}\,. \tag{2.19}$$

For a family of nodal curves $\varphi : \mathcal{C} \to Y$, this means the following. Near a smooth point of a fiber, φ is smooth, so (2.19) just says that $\omega_{\mathcal{C}/Y}$ is the same as $\Omega^1_{\mathcal{C}/Y}$. Near a singular point p of a fiber, instead, $\mathcal{C}$ is locally of the form $xy = f$. Using our previous notation, a neighborhood U of p in C is viewed as the closed set $F = 0$ in $Z = \mathbb{C}^2 \times Y$, where $F = xy - f$, and the normal bundle of U in Z is nothing but $\mathcal{O}_U((F))$. Hence a local generator of ω_φ is the class of $F^{-1}dx\wedge dy$ modulo F, where $F = xy - f$. The identification between $\omega_{\mathcal{C}/Y}$ and $\Omega^1_{\mathcal{C}/Y}$ away from the singularities of the fibers extends to a homomorphism $\rho : \Omega^1_{\mathcal{C}/Y} \to \omega_{\mathcal{C}/Y}$, given locally near p by

$$\rho(\alpha) = \text{class of } F^{-1}\widetilde{\alpha} \wedge dF \text{ modulo } F,$$

where $\widetilde{\alpha}$ is any relative differential on $\mathbb{C}^2 \times Y \to Y$ restricting to α; keep in mind that here the exterior differentiation symbol stands for relative differentiation $/Y$, so that for instance $dF = xdy + ydx$. In particular,

$$\rho(dx) = x\frac{dx \wedge dy}{F}\,, \quad \rho(dy) = -y\frac{dx \wedge dy}{F}\,.$$

In a sense, then, one may interpret the standard local generator $F^{-1}dx\wedge dy$ as $\frac{dx}{x}$ or $-\frac{dy}{y}$, and say that $\omega_{\mathcal{C}/Y}$ is locally generated by $\frac{dx}{x}$ and $\frac{dy}{y}$, subject to the relation

$$\frac{dx}{x} + \frac{dy}{y} = 0\,.$$

The formation of the relative dualizing sheaf is compatible with base change. In particular, taking into account the local description above, the restriction of ω_φ to a fiber $\varphi^{-1}(y)$ is just the dualizing sheaf $\omega_{\varphi^{-1}(y)}$, as described in the previous section.

Look at the exact sequence

$$0 \to \ker\rho \to \Omega^1_{\mathcal{C}/Y} \xrightarrow{\rho} \omega_{\mathcal{C}/Y} \to \operatorname{coker}\rho \to 0. \tag{2.20}$$

Clearly, both $\ker\rho$ and $\operatorname{coker}\rho$ are concentrated along the locus of those points of $\mathcal{C}$ which are singular in their fiber. Write $\rho(\Omega^1_{\mathcal{C}/Y}) = \mathcal{I}\omega_{\mathcal{C}/Y}$, where $\mathcal{I}$ is an ideal in $\mathcal{O}_\mathcal{C}$, and denote by S the corresponding subspace of $\mathcal{C}$. With this notation, $\operatorname{coker}\rho = \omega_{\mathcal{C}/Y} \otimes \mathcal{O}_S$. In a local representation $xy = f$ of $\mathcal{C}$, the image of ρ is generated by $x\,F^{-1}dx\wedge dy$ and $y\,F^{-1}dx\wedge dy$, so $\mathcal{I} = \mathcal{I}_S$ is locally generated by x and y.

It is not as straightforward to describe the kernel of ρ. Look at a point where $\mathcal{C}$ is of the form $xy = f$. Without being too pedantic about it, we shall sketch what the possibilities are for $\ker\rho$. Let us start with the easy case where Y is a smooth curve and $f = t$, where t is a local parameter on Y. An element $\alpha dx + \beta dy$ belongs to the kernel of ρ if and only if $\alpha x - \beta y = 0$. Since x and y are local parameters on the smooth surface $\mathcal{C}$, it follows that $\alpha = y\tilde\alpha$, $\beta = x\tilde\beta$, and that $xy(\tilde\alpha - \tilde\beta) = t(\tilde\alpha - \tilde\beta) = 0$. As $\mathcal{C}$ is smooth, it follows that $\tilde\alpha = \tilde\beta$ and therefore that $\alpha dx + \beta dy$ is zero in $\Omega^1_{\mathcal{C}/Y}$.

The above argument suggests how to produce nonzero elements in $\ker\rho$ whenever there exists a nonzero $g \in \mathcal{O}_Y$ such that $fg = 0$; one such element is, for instance, $hgydx$, where h is any function on $\mathcal{C}$. This situation may present itself in two cases only, when f is a zero divisor in $\mathcal{O}_Y$ or when f is zero. This last case corresponds to the situation where the family is locally a product near the singularity under examination (so that, in particular, the generic curve of the family is singular). When Y is a single point, i.e., when we are dealing with a single nodal curve, the kernel of ρ is the one-dimensional complex vector space generated by the class of $xdy = -ydx$. As a consequence of this discussion, we may assert that, when the general fiber of φ is smooth and Y is reduced and irreducible, ρ is injective.

We conclude this section with a quick glimpse at a very particular case of relative duality. We recall that a key ingredient of Serre duality on a nodal curve is the trace homomorphism (2.9). When dealing with a family $\varphi : \mathcal{C} \to Y$ of nodal curves, this globalizes to a sheaf homomorphism

$$\xi : R^1\varphi_*\omega_{\mathcal{C}/Y} \to \mathcal{O}_Y\,.$$

When the fibers of φ are connected, ξ is an isomorphism.

Suppose now that $\mathcal{L}$ is a line bundle on $\mathcal{C}$ and suppose further that both $\varphi_*\mathcal{L}$ and $R^1\varphi_*\mathcal{L}$ are vector bundles on Y. Then relative duality asserts that the pairing

$$\varphi_*\mathcal{L} \otimes R^1\varphi_*\left(\omega_{\mathcal{C}/Y} \otimes \mathcal{L}^{-1}\right) \to R^1\varphi_*\omega_{\mathcal{C}/Y} \xrightarrow{\xi} \mathcal{O}_Y$$

identifies $R^1\varphi_*\left(\omega_{C/Y} \otimes \mathcal{L}^{-1}\right)$ with the dual of $\varphi_*\mathcal{L}$.

EXERCISE (2.21). Let $C \subset \mathbb{P}^r$ be a nodal curve in projective space, and let $\mathcal{I}$ be its ideal sheaf. Show that the sequence

$$0 \to \mathcal{I}/\mathcal{I}^2 \xrightarrow{d} \Omega^1_{\mathbb{P}^r} \otimes \mathcal{O}_C \to \Omega^1_C \to 0$$

is exact.

3. Stable curves.

As we explained in the introduction to this chapter, singular curves of some sort are necessary if we want to take "limits" of families of smooth curves over noncomplete schemes. In the next two sections we shall prove that nodal curves suffice. What is immediately clear is that, on the other hand, nodal curves are "too many." Consider in fact the following setup. Let $f : X \to S$ be a family of smooth curves over, say, a smooth curve S. We blow up the surface X at a point p. The result is a new family which is identical to the old one except for the fiber at $f(p)$, which is now singular. In short, limits are not unique if arbitrary nodal curves are allowed. Put otherwise, "moduli spaces" constructed using unrestricted nodal curves are of necessity highly nonseparated. To restore uniqueness of the limit, one must restrict the class of nodal curves used. Let $(C; D)$ be a connected nodal curve with n marked points. One says that $(C; D)$ is *stable* if it has a finite automorphism group. In the same way, when P is a finite set, one defines the notion of stable P-pointed curve.

Attached to a P-pointed stable curve (C, D) is the *log-canonical sheaf* $\omega_C(D)$. This sheaf is intrinsic in the sense that, given an isomorphism $\varphi : (C, D) \to (C', D')$ between P-pointed stable nodal curves, there is a canonical isomorphism between $\omega'_C(D')$ and $\varphi^*\omega_C(D)$.

There is a simple combinatorial characterization of stability which is often easier to work with. To state it, let Γ be a connected genus g graph with n legs, and, as customary, for any vertex v, let g_v and l_v be the weight (the "genus") of v and the number of half-edges emanating from v. One says that Γ is a *stable graph* if $2g_v - 2 + l_v > 0$ for every vertex v.

LEMMA (3.1). *The following are equivalent:*

a) *$(C; D)$ is stable;*
b) *the graph of $(C; D)$ is stable;*
c) *for any component Y of the normalization of C, the pullback of the log-canonical sheaf $\omega_C(D)$ to Y has strictly positive degree.*

The proof is straightforward. There is a natural homomorphim from $\mathrm{Aut}(C; D)$ to $\mathrm{Aut}(\Gamma)$, whose kernel we denote by K. Since $\mathrm{Aut}(\Gamma)$ is finite, the finiteness of $\mathrm{Aut}(C;D)$ is equivalent to the one of K. On the other hand, K is the direct product of the groups K_v, where v ranges over all vertices of Γ, and K_v stands for the automorphism group, as an l_v-pointed curve, of the component of the normalization of C corresponding to v. Now, if $(X;E)$ is a smooth curve of genus g with n marked points, the group $\mathrm{Aut}(X;E)$ is finite unless $g=0$ and $n \leq 2$, or $g=1$ and $n=0$, i.e., unless $2g-2+n \leq 0$. Applying this remark to all the components of the normalization of C proves the equivalence of a) and b). As for condition c), it is just a rephrasing of b). In fact, if Y is the component of the normalization of C corresponding to vertex v of the graph of $(C;D)$, then the degree of the pullback of $\omega_C(D)$ to Y is exactly equal to $2g_v - 2 + l_v$.

An immediate consequence of the lemma is that n-pointed stable curves of genus g exist if and only if $g > 1$, or $g = 1$ and $n \geq 1$, or else $g = 0$ and $n \geq 3$.

As we explained in Section 2, the nodes of a stable P-pointed curve of genus g come in various flavors, nonseparating or separating of type $\mathcal{P}$, for some bipartition $\mathcal{P} = \{(a,A),(b,B)\}$ of (g,P). However, not all bipartitions can occur, but only the "stable" ones, that is, those allowed by the stability condition. To see what these are, recall that the partial normalization at a separating node produces two nodal curves, one of genus a with $|A|+1$ marked points and one of genus b with $|B|+1$ marked points, which must both be stable. Thus the bipartitions which are not "stable," and hence cannot occur, are those such that $a=0$ and $|A|<2$, or $b=0$ and $|B|<2$.

A useful generalization of the notion of stability is the one of semistability. We shall say that a connected graph Γ with n legs as above is *semistable* if $2g_v - 2 + l_v \geq 0$ for every vertex v. Likewise, a nodal curve with n marked points $(C;D)$ is said to be *semistable* if its graph is. In practice, a curve is semistable if and only if at least one of the following conditions is satisfied: either it is irreducible of genus one with no marked points, or else every one of its smooth rational components contains at least two points which are either singular or marked. The following characterization of semistability needs no proof.

LEMMA (3.2). *The following are equivalent:*

a) *$(C;D)$ is semistable;*
b) *for any component Y of the normalization of C, the pullback of $\omega_C(D)$ to Y has nonnegative degree.*

To be more precise, the pullback of $\omega_C(D)$ to any component of the normalization of a semistable C which is smooth rational and contains just two marked points is the trivial line bundle, while the pullback to any other component has strictly positive degree, with one exception: $\omega_C(D)$ is trivial if $n = 0$ and C is smooth of genus 1.

REMARK (3.3). Let $(C; D)$ be a nodal curve with n marked points, and let $\nu : N \to C$ be the partial normalization of C at t nodes. Let E be the set of those points of N which map either to marked points of C or to nodes where the normalization has taken place. Then $(N; E)$ is a nodal curve with $(n + 2t)$ marked points. It is clear from the description of the dualizing sheaf of a nodal curve that

$$\omega_N(E) = \nu^*(\omega_C(D)).$$

It is then an immediate consequence of (3.1) and (3.2) that, when $(C; D)$ is connected, it is stable (resp., semistable) if and only if every connected component of $(N; E)$ is.

There is an obvious extension of the notion of stability (or semistability) to families. A family of n-pointed nodal curves is said to be a *family of stable* (resp., *semistable*) *P-pointed curves* if every one of its fibers is stable (resp., semistable). A useful consequence of Lemma (3.1) is that, in families of nodal curves, stability of fibers is an open condition. Incidentally, this is true also for semistability, but this will be proved later (cf. Corollary (6.6)).

LEMMA (3.4). *Let $f : X \to S$, $\sigma_p : S \to X$, $p \in P$, be a family of P-pointed nodal curves. Then the set of $s \in S$ such that $(X_s; \{\sigma_p(s)\}_{p \in P})$ is stable is Zariski open in S.*

In proving the lemma we may assume, without loss of generality, that the fibers of f are connected. The proof is immediate. It suffices to notice that, by Lemma (3.1), stability is equivalent to the ampleness of the log-canonical sheaf, and that ampleness itself is an open condition.

It is possible to generalize to the stable case the notion of hyperelliptic curve. We would like to call hyperelliptic those stable curves of genus $g > 1$ which are "limits" of smooth hyperelliptic ones, that is, those stable curves which appear as fibers of families of stable curves over irreducible bases whose general fibers are smooth hyperelliptic. As this definition is a bit unwieldy and indirect, one of our first tasks will be to replace it with an equivalent one which is more intrinsic. This is modeled on the standard characterization of smooth hyperelliptic curves as double covers of $\mathbb{P}^1$. In the stable case the projective line is replaced by a genus zero nodal curve, i.e., a nodal curve whose graph is a tree and whose irreducible components are all copies of $\mathbb{P}^1$. We shall say that a genus $g > 1$ stable curve C is a *hyperelliptic stable curve* if there is an order two automorphism σ of C with isolated fixed points such that the quotient $P = C/\langle\sigma\rangle$ is a genus zero nodal curve. The requirement that the fixed points of σ be isolated insures that the quotient morphism

$C \to P$ has degree two above each component of P. The automorphism σ is the analogue, in the stable case, of the hyperelliptic involution, and our main goal here is to show that, as in the smooth case, it is unique. A proof that hyperelliptic stable curves, as defined above, are indeed the "limits" of ordinary hyperelliptic curves will be given in Lemmas (6.14) and (6.15) of Chapter XI (see also Exercise (4.14) in the same chapter).

Before we prove the uniqueness of the hyperelliptic involution, let us remark that the nodes of C can be classified into different types, depending on their behavior under σ. A node p of C can be fixed under σ, or not. In the first case, σ may or may not interchange the branches of C at p. If it does, then p maps to a smooth point of P and is not a separating node. In fact, if it were separating, C would consist of two copies of P joined at p, contradicting the stability of C. We shall say that a node of this kind is a *node of type* η_0. If instead σ does not interchange the two branches of C at p, then p maps to a node q of P, and the projection from each branch of C at p to the corresponding branch of P at q has degree two. Furthermore, since p is the only point mapping to q, and q is separating, p is also separating. Such a node divides C in two pieces of genera i and $j \geq i$, with $i + j = g$; in the terminology of Section 2, p is a separating node of type $(i, \emptyset)$. We shall also say that p is a *node of type* δ_i. There remain the nodes which are not fixed under σ. If p is such a node, it maps to a node q of P, and the only other point of C mapping to q is $\sigma(p)$. Normalizing C at $\{p, \sigma(p)\}$ breaks it into two curves C_1 and C_2. Observe that C_1 and C_2 must be connected, for otherwise they would be the union of two copies of a subcurve of P, contradicting the stability of C. In particular, p and $\sigma(p)$ are nonseparating. The genera i and j of C_1 and C_2 add to $g-1$ and are strictly positive, since otherwise C would not be stable. If h is the minimum between i and j, we shall say that $\{p, \sigma(p)\}$ is a *pair of nodes of type* η_i.

Summing up, while the classification of separating nodes is the same for hyperelliptic curves as for general ones, in the hyperelliptic case the nonseparating nodes fall into several finer subclasses. We now prove the uniqueness of the hyperelliptic involution.

LEMMA (3.5). *Let C be a stable curve of genus $g > 1$. There exists at most one order two automorphism σ of C which has isolated fixed points and is such that $C/\langle\sigma\rangle$ is a genus zero nodal curve.*

Proof. Suppose that C admits an automorphism σ as in the statement of the lemma. Set $P = C/\langle\sigma\rangle$, and let $\pi : C \to P$ be the quotient morphism. Let τ be another order two automorphism with isolated fixed points and with $C/\langle\tau\rangle$ a genus zero nodal curve. We shall show that $\tau = \sigma$, arguing by induction on the number of irreducible components of P.

We first deal with the initial case of the induction. We thus assume that P is a $\mathbb{P}^1$. If C is irreducible, $C/\langle\tau\rangle$ is also irreducible and hence isomorphic to a projective line. It then follows from the classification of nodes that those of C are fixed under τ and that their branches get interchanged by τ. We wish to show that the same conclusion holds if C is reducible. In this case, C consists of two copies of P joined at three or more points. Then τ interchanges the two components of C, since otherwise $C/\langle\tau\rangle$ would have positive genus. It follows that $C/\langle\tau\rangle$ is irreducible, and the same argument as above goes through. Now let N be the normalization of C, and let $\{p_1, p_1'\}, \dots, \{p_n, p_n'\}$ be the pairs of points or N mapping to the n nodes of C. The automorphism τ lifts to N and interchanges p_i and p_i' for each i. We may distinguish several cases:

i) N is irreducible of genus strictly larger than 1. Then $\sigma = \tau$ by the uniqueness of the hyperelliptic involution in the smooth case.
ii) N is irreducible of genus 1. Then $n \geq 1$, since $g \geq 2$. In this case there is a unique automorphism of N interchanging p_1 and p_1' (the symmetry about p_1 followed by translation of p_1 to p_1'). This shows that $\sigma = \tau$.
iii) N is irreducible of genus 0. Then $n \geq 2$. There is a unique automorphism of N interchanging p_1 with p_1' and p_2 with p_2' ($z \mapsto a/z$ if we normalize things so that $p_1 = 0$, $p_1' = \infty$, $p_2 = 1$, and $p_2' = a$). Thus $\sigma = \tau$.
iv) N is reducible. As we observed above, N is the disjoint union of two copies N_1 and N_2 of P, and C is obtained by identifying points $p_i \in N_1$ and $p_i' \in N_2$ for $i = 1, \dots, n$, $n \geq 3$. Clearly, there is a unique automorphism of N interchanging p_i with p_i' for $i = 1, \dots, 3$. Hence $\sigma = \tau$ also in this last case.

There is a variant of the initial case of the induction that we will also need. Let (E, e) be a stable 1-pointed curve of genus 1. Then there is a unique order two automorphism of (E, e). If E is smooth, this is the symmetry about e. If E is obtained from $\mathbb{P}^1$ by identifying 0 and ∞ and if $e = 1$, then the automorphism comes from $z \mapsto 1/z$.

Now we turn to the inductive step. Let P' be an irreducible component of P, and set $C' = \pi^{-1}(P')$. If C' is connected and has strictly positive genus, it must be carried to itself by τ, since otherwise $C/\langle\tau\rangle$ would contain a subcurve of strictly positive genus. Now suppose that P' is a leaf of P, i.e., that it meets the union P'' of the other components of P at a single point. It follows, in particular, that P'' is connected. Set $C'' = \pi^{-1}(P'')$. As we observed while classifying singularities of hyperelliptic stable curves, C' and C'' are connected and moreover have strictly positive genus. It follows that τ carries C' to itself and hence also C'' to itself. If C' and C'' meet at a single point, this is fixed under τ. In this case we set $E' = C'$, $E'' = C''$. If instead C' and C'' meet at a pair $\{p_1, p_2\}$ of nonseparating nodes of type η_i, then we claim that τ interchanges p_1 and p_2. In fact, the only other possibility is that p_1 and p_2 are both fixed under τ. But, as they are nonseparating nodes, τ would have to interchange the two branches of both, which is not the case since τ maps C' and C'' to themselves. In this case we let E' and E'' be the curves obtained from C' and C'', respectively, by identifying p_1 and p_2. It is clear that σ descends to automorphisms σ' and σ'' of E' and E''. By what we have said, τ also descends to automorphisms τ' and τ'' of E' and E'' such that $E'/\langle\tau'\rangle \cong \mathbb{P}^1$ and $E''/\langle\tau''\rangle$ is a rational nodal curve. By the initial case of the induction and its variant, and by induction hypothesis, $\tau' = \sigma'$ and $\tau'' = \sigma''$. Thus $\tau = \sigma$, as claimed.

EXERCISE (3.6). Let C be a nodal curve, and let γ be an involution on C with isolated fixed points. Let $\pi : C \to P$ be the quotient map. Show that every point of P has an open neighborhood U, in the ordinary topology, such that one of the following occurs:

i) $\pi^{-1}(U)$ is a disjoint union $U_1 \coprod U_2$ of open sets each of which maps isomorphically to U;
ii) U is isomorphic to a neighborhood of the origin in the complex z-plane, $\pi^{-1}(U)$ is isomorphic to a neighborhood of the origin in the complex t-plane, and π is locally given by $t = z^2$;
iii) U is isomorphic to a neighborhood of the origin in the complex t-plane, $\pi^{-1}(U)$ is isomorphic to a neighborhood of the origin in $xy = 0$, and π is locally given by $t = x + y$;
iv) U is isomorphic to a neighborhood of the origin in $zw = 0$, $\pi^{-1}(U)$ is isomorphic to a neighborhood of the origin in $xy = 0$, and π is locally given by $z = x^2$, $w = y^2$.

Q.E.D.

4. Stable reduction.

In the introduction to this chapter and in the previous section, we observed that the most naive approach to compactifying the moduli space of smooth genus g curves, that is, throwing in the isomorphism classes of all possible singular curves of arithmetic genus g, fails because it unavoidably produces highly nonseparated spaces.

As we announced, one remedy to this is to strongly limit the class of singular curves one considers, allowing only stable curves. The main technical tools to prove that this does indeed work will be given in this section and in the next one.

Before proceeding, we give a very brief overview of those facts concerning the moduli space of stable curves which can serve as a motivation for what we will be doing. We shall limit ourselves to the case of unpointed curves. First of all, the moduli space $\overline{M}_g$ of stable genus g curves is, set-theoretically, just the set of isomorphism classes of stable curves of genus g. It will be proved in Chapters XII and XIV that $\overline{M}_g$ has a natural structure of normal algebraic variety of dimension $3g-3$. Given a stable curve C of genus g, we shall denote by $[C] \in \overline{M}_g$ the isomorphism class of C. It is of course natural to ask whether $\overline{M}_g$ is the parameter space of a family

$$\pi : \mathcal{C} \longrightarrow \overline{M}_g$$

of stable curves such that $[\pi^{-1}(x)] = x$ for any $x \in \overline{M}_g$. It turns out that such a family exists only on the Zariski open subset of $\overline{M}_g$ corresponding to automorphism-free stable curves and that the presence of automorphisms prevents this family from existing over the whole of $\overline{M}_g$. On the other hand, what turns out to be true (cf. Theorem (2.9) of Chapter XII) is that there exist a normal algebraic variety Z, a *finite* morphism

$$m : Z \longrightarrow \overline{M}_g \ , \tag{4.1}$$

and a family of stable curves of genus g

$$\eta : \mathcal{X} \longrightarrow Z \tag{4.2}$$

such that, for every $z \in Z$, $[\eta^{-1}(z)] = m(z)$.

One of the main results of the first five chapters of this volume, to be proved in Chapter XIV, is that $\overline{M}_g$ is a projective variety. A basic step in the proof will be to show that $\overline{M}_g$ is complete. This will be checked by using the valuative criterion for completeness.

Informally speaking, the valuative criterion for completeness asserts that an algebraic variety X is complete whenever any holomorphic map from a punctured disc to X which is meromorphic at the origin can be extended across the puncture. Clearly this criterion can be somewhat softened by only requiring that the extension be possible after a base change $z = \zeta^k$ on the punctured disc.

In applying the valuative criterion of completeness to the moduli space $\overline{M}_g$, we are thus confronted with maps from a punctured disc $\dot{\Delta} = \{z \in \mathbb{C} : 0 < |z| < \varepsilon\}$ to $\overline{M}_g$ which are meromorphic at the origin. In view of the existence of the finite map (4.1), after a harmless base change, we can lift any such map to a map f from $\dot{\Delta}$ to Z. Then pulling back via f the family of stable curves (4.2) produces a family of stable curves

$$\dot{\pi} : \dot{\mathcal{C}} \to \dot{\Delta} \tag{4.3}$$

From this point of view, extending the original map from $\dot{\Delta}$ into moduli space over the puncture corresponds to filling in the family (4.3), possibly after a base change, with a stable curve as central fiber. We notice first that we can fill in the family with a possibly very nasty central fiber. This is a consequence of the meromorphicity of the map

$$f : \dot{\Delta} \to Z \ ,$$

as we now explain. We begin by completing Z to a normal variety $\overline{Z}$, so that f extends to a morphism

$$\overline{f} : \Delta = \{z \in \mathbb{C} | \, |z| < \varepsilon\} \longrightarrow \overline{Z} \ .$$

We now look at the family (4.2) over Z, take a completion $\mathcal{X}'$ of $\mathcal{X}$, and let $\overline{\mathcal{X}}$ be the closure of $\mathcal{X}$ in $\mathcal{X}' \times \overline{Z}$. Clearly the natural projection $\overline{\eta}$ from $\overline{\mathcal{X}}$ to $\overline{Z}$ extends the family (4.2). But then a completion of the family (4.3) is

$$\pi : \mathcal{C} = \overline{\mathcal{X}} \times_{\overline{Z}} \Delta \longrightarrow \Delta \ .$$

From this point on we may forget how the above family was constructed and deal exclusively with the abstract situation of a proper morphism

$$\pi : \mathcal{C} \longrightarrow \Delta \tag{4.4}$$

having the property of being a family of stable curves away from the origin of Δ. We wish to show that after a base change on Δ of the form $z = \zeta^k$, the central fiber of π can be replaced so as to get a family of stable curves over Δ. This process goes under the name of *stable reduction*. We begin by proving the slightly weaker statement that the central fiber can be replaced, after a finite base change, so as to get a family of nodal curves. We shall first deal with the case where the fibers of π over $\dot{\Delta}$ are smooth. Thus $\mathcal{C}$ is a surface whose singularities are concentrated along the central fiber. By successive blow-ups of points in the central fiber and normalization, we may then assume that $\mathcal{C}$ is smooth and that the central fiber of $\mathcal{C}$ is a divisor with normal crossings. It is important to realize that this does not mean that the central fiber is a nodal curve, since it can very well be nonreduced. In fact, to say that $\pi^{-1}(0)$ is a divisor with normal crossing means that, given any point p on $\pi^{-1}(0)$, one can choose local analytic coordinates x, y on $\mathcal{C}$ centered at p in such a way that, in a neighborhood of p, the map π is given by either

$$z = x^c$$

or

$$z = x^a y^b$$

for suitable positive integers a, b, and c.

By compactness we may cover the central fiber with a finite number of coordinate patches for $\mathcal{C}$ in which π is given as above. We then perform a base change $z = \zeta^k$, where k is chosen to be a multiple of c and of ab, for all these patches. So we obtain a new family

$$\pi' : \mathcal{C}' \to \Delta \ .$$

Finally we normalize $\mathcal{C}'$ and get yet another family

$$\pi'' : \mathcal{C}'' \longrightarrow \Delta \, .$$

We claim that the central fiber at π'' is now a nodal curve. To show this, we work on each of the patches of $\mathcal{C}$ separately. Suppose first that we are on a patch where π looks like $z = x^c$. Then, since k is a multiple of c, we can write $k = ch$, so that $\mathcal{C}'$ is given locally by the equation

$$0 = x^c - \zeta^{ch} = \prod_{\omega^c = 1} (x - \omega\zeta^h) \, .$$

Normalizing breaks up $\mathcal{C}'$ into the disjoint union of the smooth branches $\omega\zeta^h = x$. The other case is more interesting. In this case, π looks like $z = x^a y^b$, and we can write $k = rsuv$, where r and s are relatively prime, $a = ru$, and $b = su$. Hence, locally, the equation of $\mathcal{C}'$ is

$$0 = x^a y^b - \zeta^k = \prod_{\omega^u = 1} (x^r y^s - \omega\zeta^{vrs}) \ .$$

We can think of the normalization of $\mathcal{C}'$ to happen, locally, in two stages. The first one consists in breaking up $\mathcal{C}'$ into the disjoint union of the branches

$$\omega\zeta^{vrs} = x^r y^s \ .$$

The second one consists in normalizing each branch separately. As one sees replacing ζ with $\zeta' = \sqrt[vrs]{\omega}\zeta$, each of the branches of $\mathcal{C}'$ is a copy of $\zeta^{vrs} = x^r y^s$. We claim that the normalization of this branch is the surface in $\mathbb{C}^3$ with equation

$$\zeta^v = \alpha\beta \tag{4.5}$$

and that the normalization map is given by

$$x = \alpha^s \quad , \quad y = \beta^r \, . \tag{4.6}$$

Checking this involves verifying two properties. The first is that equation (4.5) defines a normal surface. This is certainly so by Proposition (5.4) of Chapter II of Volume 1, since the only singular point of (4.5) is at the origin, and the surface in question is a hypersurface in affine 3-space. The second point to be checked is that the map (4.6) is birational, that is, generically one-to-one and onto. To see that it is generically one-to-one, assume that (α, β, ζ) and $(\alpha', \beta', \zeta')$ map to the same point. This translates into

$$\zeta = \zeta' \ , \ \alpha^s = \alpha'^s \ , \ \beta^r = \beta'^r \ .$$

We can then write $\alpha' = \sigma\alpha$, where σ is an sth root of unity and $\beta' = \tau\beta$, where τ is an rth root of unity. Since $\alpha\beta = \alpha'\beta'$, unless α or β vanishes, we get that $\sigma\tau = 1$; since r and s are relatively prime, this implies that $\sigma = \tau = 1$. To check the surjectivity of the map (4.6), all we have to do is the following. Given complex numbers x, y, ζ satisfying the equation $x^r y^s = \zeta^{vrs}$, we must find α and β such that $\alpha\beta = \zeta^v$ and $x = \alpha^s$, $y = \beta^r$. We can certainly find α and β satisfying the last two conditions. But then

$$(\alpha\beta)^{rs} = x^r y^s = \zeta^{vrs} \ ,$$

so that $\alpha\beta = \xi\zeta^v$, where $\xi^{rs} = 1$. Since r and s are relatively prime, we may write $1 = mr + ns$ and replace the original α and β with $\alpha\xi^{-mr}$ and $\beta\xi^{-ns}$. These obviously solve our problem.

We have thus proved that the surface $\mathcal{C}''$ is locally either of the form $\zeta^h = x$ or of the form $\zeta^v = \alpha\beta$, showing that $\mathcal{C}'' \to \Delta$ is a family of nodal curves.

The preceding argument has been carried out under the simplifying hypothesis that the general fiber of π is smooth. Let us now remove this unnecessary assumption. Thus

$$\pi : \mathcal{C} \to \Delta$$

is a morphism with one-dimensional fibers such that all fibers other than the central one $\pi^{-1}(0)$ are nodal curves. Denote by $Z \subset \mathcal{C}$ the closure of the locus of nodes in the fibers of $\pi^{-1}(z)$ for $z \neq 0$. Possibly after shrinking Δ we may assume that Z has no zero-dimensional components. Possibly after a base change of the form $z = \zeta^k$ we may also assume that Z is a union of sections $Z_1, \dots, Z_\delta$. Now normalize the surface $\mathcal{C}$. The resulting surface $\mathcal{C}'$ is the disjoint union of components $\mathcal{C}_1, \dots, \mathcal{C}_c$ and is fibered over Δ via a morphism $\pi' : \mathcal{C}' \to \Delta$. Possibly after a further base change, we may assume that the preimage of each Z_i in $\mathcal{C}'$ is the disjoint union of two sections X_i and Y_i: these may either lie in the same component of $\mathcal{C}'$ or in different ones.

Since all fibers of π', except the central one, are smooth, by what has already been proved, after a finite base change one may transform the family $\pi' : \mathcal{C}' \to \Delta$ into a family of nodal curves by replacing the central fiber. Let $\pi'' : \mathcal{C}'' \to \Delta$ be the resulting family.

By abuse of language we continue to denote by X_i and Y_i the sections of π'' corresponding to the sections of π' with the same names. The idea now is to glue X_i to Y_i for each i and thus to get a family of nodal curves over Δ, which, by construction, will be obtained from

$\pi : \mathcal{C} \to \Delta$, away from the central fiber, via a base change. There are however a couple of small details to be taken care of. In fact, it could very well happen that one or more of the sections X_i, Y_i pass through a node of the central fiber of π'' or that two of these sections meet on the central fiber; in either case the gluing procedure produces a central fiber which is not a nodal curve. We shall deal with the two problems separately. We will show below that we can assume $\mathcal{C}''$ to be smooth. This will solve the first problem. In fact, if a section of π'' went through a singular point of the central fiber, it would have intersection number at least equal to two with this same fiber and hence with any other fiber; since the general fiber is smooth, this is absurd. Once we can assume $\mathcal{C}''$ to be smooth, the second problem can be immediately solved as well by repeatedly blowing up at the intersection points of sections.

Our task then is to show that the singularities of $\mathcal{C}''$, if any, can be resolved by successive blow-ups without destroying the property of the central fiber being reduced. To see that this is a point to worry about, just notice that blowing up the origin in a family which locally looks like $xy = z$ produces a fiber over $z = 0$ which has a multiple component, namely the exceptional curve counted twice.

By what we have shown, any singularity that $\mathcal{C}''$ may have is of the form

$$xy = z^{n+1} \tag{4.7}$$

or, as one says, is an A_n singularity. To blow up such a singularity, one may proceed as follows. Let U be a neighborhood of the origin in the space of the variables x, y, and let ξ, η, ζ be homogeneous coordinates in $\mathbb{P}^2$; denote by X the subvariety of U defined by (4.7) and by f its projection to the z factor. The blow-up of U at the origin is the subvariety of $U \times \mathbb{P}^2$ defined by the equations

$$\begin{aligned} x\eta &= y\xi \ , \\ x\zeta &= z\xi \ , \\ y\zeta &= z\eta \ . \end{aligned}$$

To get the total transform of X, one must add to these equation (4.7). The blow-up of X at the origin, which we denote by $\hat{X}$, is nothing but the proper transform of X. To see what it looks like, we shall examine it separately in the open subsets $\{\xi \neq 0\}$, $\{\eta \neq 0\}$, and $\{\zeta \neq 0\}$ of $U \times \mathbb{P}^2$. Let us begin with $\{\zeta \neq 0\}$. On this open set one can take as local coordinates $z, \xi/\zeta, \eta/\zeta$ and write

$$x = z\xi/\zeta \quad , \quad y = z\eta/\zeta \ .$$

In these coordinates, the equation of the total transform of X then becomes

$$z^2 \frac{\xi}{\zeta}\frac{\eta}{\zeta} = z^{n+1} \ ,$$

so that the equation of $\hat{X}$ is

$$\frac{\xi}{\zeta}\frac{\eta}{\zeta} = z^{n-1} . \tag{4.8}$$

In other words, $\hat{X} \cap \{\zeta \neq 0\}$ is smooth except for an A_{n-2} singularity at $x = y = z = \xi = \eta = 0$ when $n > 2$ and smooth when $n = 1, 2$. We next look at what takes place inside $\{\xi \neq 0\}$. On this open set one can take as local coordinates $x, \eta/\xi$, ζ/ξ, and write

$$y = x\eta/\xi \quad , \quad z = x\zeta/\xi . \tag{4.9}$$

In these coordinates the equation of the total transform of X is

$$x^2\eta/\xi = x^{n+1}(\zeta/\xi)^{n+1} ,$$

so the equation of $\hat{X}$ is

$$\eta/\xi = x^{n-1}(\zeta/\xi)^{n+1} .$$

In particular, we find that $\hat{X} \cap \{\xi \neq 0\}$ is smooth. Interchanging x with y and ξ with η shows that the same happens in $\{\eta \neq 0\}$. If we denote by $\hat{f}$ the composition of the natural map $\hat{X} \to X$ with f, formulas (4.8) and (4.9) also show that the fiber $\hat{f}^{-1}(0)$ is reduced with nodes as its only singularities. More exactly, for $n > 1$, it consists of the four components

a) $x = z = \xi = \zeta = 0$,

b) $y = z = \eta = \zeta = 0$,

c) $x = y = z = \xi = 0$,

d) $x = y = z = \eta = 0$.

The union of components a) and b) is the proper transform of $f^{-1}(0)$. Components c) and d) are projective lines whose union is the exceptional divisor of $\hat{X} \to X$. When $n = 1$, the picture is slightly different: $\hat{f}^{-1}(0)$ consists of the two components a) and b), plus the exceptional divisor of $\hat{X} \to X$, which is the smooth conic

$$x = y = z = \xi\eta - \zeta^2 = 0 .$$

When $n = 1, 2$, one blow-up resolves the singularity of X. In general, one must blow up $\left[\frac{n+1}{2}\right]$ times to desingularize X; denote by $\tilde{X}$ the resulting manifold and by $\tilde{f}$ the composition of f with $\tilde{X} \to X$. In any case the fiber $\tilde{f}^{-1}(0)$ is reduced, has only nodes as singularities, and consists of the two branches of the normalization of $f^{-1}(0)$, joined by a chain $E_1 + \cdots + E_n$ of n smooth rational curves

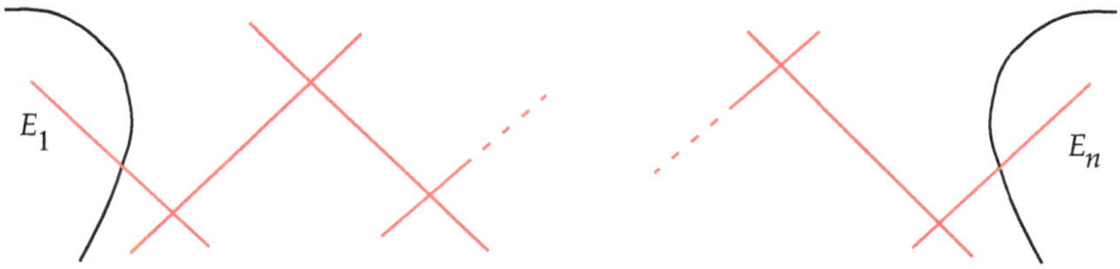

Figure 2.

The exceptional divisor of $\tilde{X} \to X$ is precisely $E_1 + \cdots + E_n$.

Returning to the family

$$\pi'' : \mathcal{C}'' \to \Delta \ ,$$

the analysis of the minimal resolution of an A_n singularity we just completed shows that we can assume $\mathcal{C}''$ to be smooth, and, as we already remarked, this insures that, possibly after further blow-ups at smooth points of the central fiber, gluing each section X_i to the corresponding section Y_i produces a family of nodal curves

$$\overline{\pi} : \overline{\mathcal{C}} \to \Delta \,. \tag{4.10}$$

To sum up, what we have done so far is to replace the central fiber of the family (4.4) so as to get a family of nodal curves, possibly after a finite base change. What remains to be done is to show that one can blow down suitable smooth rational components in the central fiber so as to end up with a family of stable curves. This is a special case of a general procedure, which goes under the name of "passing to the stable model" and will be described in detail in Section 6. In the case at hand, we give a direct argument.

Denote the central fiber of $\overline{\pi} : \overline{\mathcal{C}} \to \Delta$ by C; if it fails to be stable, this is because it contains smooth rational components which meet the rest of C in no more than two points. Let Γ be a connected component of the union of all these rational curves. Of necessity Γ is a chain of smooth rational curves which may either be attached to the rest of C at both ends, as in Fig. 2 above, or else be attached to the rest of C at one end only:

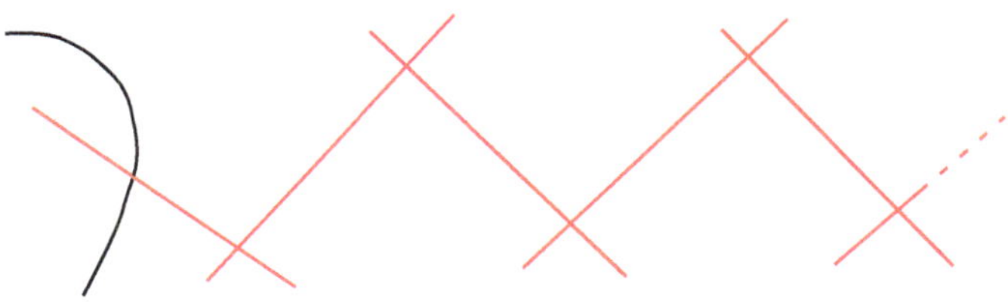

Figure 3.

We shall refer to such a Γ as an exceptional chain.

Denote by C' the nodal curve obtained by contracting all exceptional chains to points. By construction C' is stable. Moreover the genus of C and the genus of C' are equal. This follows immediately from formula (2.8). Now pick sections $D_1, \dots, D_d$ of $\overline{\pi}$ such that $D_1 + \dots + D_d$ meets all components of C save those lying in exceptional chains. We claim that, possibly after shrinking Δ, the divisor $D_1 + \dots + D_d$ meets every component of any fiber different from the central one. To prove this, it is convenient to recall that $\overline{\mathcal{C}}$ was obtained from $\mathcal{C}''$ by gluing together sections X_i, Y_i for $i = 1, \dots, \delta$ and to lift $D_1, \dots, D_d$ to sections $E_1, \dots, E_d$ of $\mathcal{C}''$. If our claim is false, there must be an entire component $\mathcal{D}$ of $\mathcal{C}''$ which does not contain any one of the sections $E_1, \dots, E_d$. By construction this means that the central fiber of $\mathcal{D} \to \Delta$ is a connected piece Γ_0 of an exceptional chain of C; thus the arithmetic genus $p_a(\Gamma_0)$ vanishes. Since the arithmetic genus is constant in flat families, the fibers of $\mathcal{D} \to \Delta$ other than the central one are smooth and rational. Since the fibers of $\overline{\mathcal{C}} \to \Delta$, except the central one, are stable, there must be at least three among the section X_i, Y_i which are contained in $\mathcal{D}$. But then Γ_0 cannot be a piece of an exceptional chain.

We now denote by L the line bundle $\mathcal{O}(D_1 + \dots + D_d)$ on $\overline{\mathcal{C}}$. We just proved that, if $z \neq 0$, the restriction $L|_{\overline{\pi}^{-1}(z)}$ has positive degree on every component of $\overline{\pi}^{-1}(z)$ and hence is ample. On the other hand, denoting by λ the natural map from C to C', the restriction $L|_C$ is of the form $\lambda^* M$ for a suitable line bundle M; by construction, M is ample on C'.

For any large enough k and any $z \neq 0$, $L^k|_{\overline{\pi}^{-1}(z)}$ is very ample, and moreover

$$H^1(\overline{\pi}^{-1}(z), L^k) = 0,$$

$$\dim H^0(\overline{\pi}^{-1}(z), L^k) = kd + 1 - g,$$

where g stands for the genus of the fibers of $\overline{\pi}$. Likewise, for large k, M^k is very ample, and

$$\dim H^0(\overline{\pi}^{-1}(0), L^k) = \dim H^0(C', M^k) = kd + 1 - g \ ;$$

thus, by Riemann–Roch,

$$H^1(\overline{\pi}^{-1}(0), L^k) = 0 \ .$$

By the theory of base change in cohomology the direct image $\overline{\pi}_* L^k$ is free. This implies that we can choose a frame $\sigma_0, \dots, \sigma_N$ of $\overline{\pi}_* L^k$; this yields a well-defined map $\varphi : \overline{\mathcal{C}} \to \mathbb{P}^N$. Now denote by $\tilde{\mathcal{C}}$ the image of $\overline{\mathcal{C}}$ in $\mathbb{P}^N \times \Delta$ via $(\varphi, \overline{\pi})$ and denote by $\tilde{\pi}$ the projection to Δ.

Then $\tilde{\pi} : \tilde{\mathcal{C}} \to \Delta$ agrees with $\overline{\pi} : \overline{\mathcal{C}} \to \Delta$, except that now the central fiber is not C but C'. On the other hand, by what has been proved, the Hilbert polynomial of $\tilde{\pi}^{-1}(z)$ is independent of z. This, as we observed in Section 2 of Chapter IX, implies that $\tilde{\pi}$ is a flat morphism. Since the fibers of $\tilde{\pi}$ are all stable, $\tilde{\pi} : \tilde{\mathcal{C}} \to A$ is a family of stable curves. The proof of stable reduction is now complete.

The argument we have given contains all the ingredients needed to prove also the following stable reduction theorem for pointed curves.

THEOREM (4.11) (STABLE REDUCTION). *Let*

$$\mathcal{C} \xrightarrow{\pi} \Delta = \{z \in \mathbb{C} |\ |z| < \varepsilon\}$$

be a proper morphism whose fibers are (possibly nonreduced) complete curves. Let

$$\sigma_i : \Delta \to \mathcal{C}, \quad i = 1, \dots, n,$$

be sections of π. *Assume that the restriction of the family* π, *together with the sections* $\sigma_1, \dots, \sigma_n$, *to the pointed disc* $\dot{\Delta} = \{z \in \Delta |\ |z| > 0\}$, *is a family of* n*-pointed stable curves. Then there exist an integer* k *and a family of stable* n*-pointed curves*

$$\overline{\pi} : \overline{\mathcal{C}} \to \Delta \ , \ \overline{\sigma}_i : \Delta \to \overline{\mathcal{C}} \ , \quad i = 1, \dots, n,$$

whose restriction to the pointed disc $\dot{\Delta}$ *is the pullback of the family* π *via the base change* $z = \zeta^k$.

We leave to the reader to fill in the missing details in the pointed case. To this end, it may be of help to notice that the construction of family (4.10) only uses the fact that all fibers of π, except the central one, are nodal, and not the full strength of the stability assumption, which is used just in the blowing-down procedure.

5. Isomorphisms of families of stable curves.

In Section 4 we proved the stable reduction theorem which, very loosely speaking, asserts that any family of stable curves over a punctured disc can be filled in with a stable curve. The main result of this section is that such a fill-in is unique. In the previous section we also explained that the stable reduction theorem is essentially equivalent to the assertion that the moduli space of stable curves is complete. Likewise, the uniqueness of stable reduction is essentially equivalent to the assertion that the moduli space of stable curves is separated. Consider two families

$$\alpha : X \to S \, , \ \sigma_1, \dots, \sigma_n : S \to X \quad \text{and} \quad \beta : Y \to S \, , \ \tau_1, \dots, \tau_n : S \to Y$$

of stable n-pointed curves over the same base S. We may then construct a scheme or analytic space $\mathbf{Isom}_S((X; \sigma_1, \dots, \sigma_n), (Y; \tau_1, \dots, \tau_n))$ parameterizing pairs (s, φ), where $s \in S$ and φ is an isomorphism $X_s \to Y_s$ carrying $\sigma_i(s)$ to $\tau_i(s)$ for $i = 1, \dots, n$ (cf. Exercise (7.7) in Chapter IX). The main result we have in mind is the following.

THEOREM (5.1). *Let*

$$\alpha : X \to S\,,\ \sigma_1, \dots, \sigma_n : S \to X \quad \text{and} \quad \beta : Y \to S\,,\ \tau_1, \dots, \tau_n : S \to Y$$

be two families of stable n-pointed curves over S, where S is a scheme or an analytic space. Then $\mathbf{Isom}_S((X; \sigma_1, \dots, \sigma_n), (Y; \tau_1, \dots, \tau_n))$ *is finite and unramified over S.*

Proof. To keep things simple, throughout the proof we shall omit any mention of the sections in the notation for families of n-pointed curves; thus, for instance, we shall write $\mathbf{Isom}_S(X,Y)$ instead of $\mathbf{Isom}_S((X; \sigma_1, \dots, \sigma_n), (Y; \tau_1, \dots, \tau_n))$. In the proof, we may limit ourselves to the case where the two families involved are algebraic, since any flat analytic family of projective schemes is locally the pullback of an algebraic one, by Lemma (4.9) in Chapter IX, and stability is an open condition in families. The morphism $\mathbf{Isom}_S(X,Y) \to S$ is quasi-finite, since a stable curve has a finite automorphism group; to prove that it is a finite morphism, we must show that it is proper. By the valuative criterion, properness is a consequence of the following statement:

(5.2). *Let $\alpha : W \to \Delta$, $\beta : Z \to \Delta$ be two families of stable n-pointed curves over the disk $\Delta = \{t \in \mathbb{C} : |t| < \varepsilon\}$, and let $W^* \to \Delta^*$, $Z^* \to \Delta^*$ be their restrictions to the pointed disk $\Delta^* = \Delta \smallsetminus \{0\}$. Let $\gamma : W^* \to Z^*$ be an isomorphism of families of n-pointed curves over Δ^*. Suppose that γ is meromorphic as a map from W to Z. Then γ extends uniquely to an isomorphism $\xi : W \to Z$ of families of n-pointed curves over Δ.*

For the sake of simplicity, in proving (5.2) we shall deal only with the essential case in which the general fibers of the families involved are smooth; the general case follows by the same cut-and-paste procedures used in the previous section. The uniqueness of ξ is clear, as Z is separated; what needs to be proved is its existence. The singularities of W and Z are, in suitable coordinates, of the form $xy = t^{n+1}$, that is, they are A_n singularities. In particular they are normal and have a minimal resolution in which the exceptional divisor is a chain of n smooth rational curves $E_1, \dots, E_n$, as we explained in the previous section (cf. Fig. 2). Notice that the self-intersection of E_i equals -2 for each i.

If we resolve all the singular points of W and Z situated on the central fibres, we may replace our original families with new ones

$$\alpha' : W' \to \Delta\ ,\ \beta' : Z' \to \Delta,$$

where W' and Z' are smooth along their central fibres. The price we have to pay is that now the central fibres are no longer stable, but only semistable. Notice however that the only destabilizing components are smooth rational with self-intersection equal to -2 and that they do not meet any of the marked sections; furthermore, any exceptional curve of the first kind in the central fibers meets at least two marked sections. Let $\Gamma \subset W^* \times_{\Delta^*} Z^*$ be the graph of γ. By meromorphicity, Γ extends to an analytic subspace $\overline{\Gamma} \subset W' \times_\Delta Z'$, proper over Δ. Let $\widetilde{\Gamma}$ be a desingularization of $\overline{\Gamma}$. By the structure theorem for morphisms between smooth surfaces, $\widetilde{\Gamma} \to W'$ is a finite sequence of blow-ups. Summing up, there is a smooth surface $\widetilde{W}$, obtained from W' by blowing up finitely many times at points of the central fiber, such that γ extends to a morphism $\widetilde{\gamma} : \widetilde{W} \to Z'$: we choose $\widetilde{W}$ minimal with respect to this property. If $\widetilde{W}$ is different from W', it contains an exceptional curve of the first kind E which is not contracted by $\widetilde{\gamma}$ but gets contracted in $\widetilde{W} \to W'$. It follows in particular that E does not meet marked sections, so the same is true of its image in W', which we denote by E'; as a consequence, E' is not an exceptional curve of the first kind. We know that the inverse of $\widetilde{\gamma}$ is a sequence of monoidal transformations. The proper transform of E' under any one of these has self-intersection not greater than the one of E'; thus the self-intersection of E' is greater than or equal to -1. Since E' is rational and is not an exceptional curve of the first kind, it must be singular. But then, since E is smooth, one of the above monoidal transformations must be centered at a singular point of E' so that the self-intersection of E' is at least equal to the self-intersection of E plus three. This leads to the absurd conclusion that Z' contains curves with positive self-intersection. This contradiction establishes that γ extends to a morphism

$$\gamma' : W' \to Z' \ .$$

The same argument establishes that ${\gamma'}^{-1}$ also extends to a morphism from Z' to W'. Hence γ' must be an isomorphism. Clearly γ' sends chains of self-intersection -2 rational curves to chains of self-intersection -2 rational curves. This shows that γ extends to a morphism from W_{reg} to Z_{reg} and therefore, by Hartogs' theorem, to a morphism ξ from W to Z. As the same argument applies to γ^{-1}, it follows that ξ is an isomorphism. This concludes the proof of (5.2) and hence shows that $\pi : \mathbf{Isom}_S(X, Y) \to S$ is proper.

We now show that $\pi : \mathbf{Isom}_S(X, Y) \to S$ is unramified. This means that, for any point $\gamma \in \mathbf{Isom}_S(X, Y)$, the map of Zariski tangent spaces

$$d\pi : T_\gamma(\mathbf{Isom}_S(X, Y)) \to T_{\pi(\gamma)}(S)$$

is injective. To show that this is true, set $s = \pi(\gamma)$ and observe that an element v of $T_\gamma(\mathbf{Isom}_S(X, Y))$ consists of a morphism $\varphi : \operatorname{Spec} \mathbb{C}[\varepsilon] = \Sigma \to (S, s)$ plus an isomorphism $\overline{\gamma} : \varphi^* X \to \varphi^* Y$ of families of n-pointed curves over Σ extending γ; moreover, $d\pi(v)$ is just φ. Thus, if $d\pi(v) = 0$, then $\varphi^* X$ and $\varphi^* Y$ are trivial families $X_s \times \Sigma$ and $Y_s \times \Sigma$, where $X_s = \alpha^{-1}(s)$ and $Y_s = \beta^{-1}(s)$. If we identify X_s and Y_s via γ, then $\overline{\gamma}$ gets identified to a Σ-automorphism of $X_s \times \Sigma$ restricting to the identity on the central fiber and on the marked sections, that is, to what is usually referred to as an infinitesimal automorphism of X_s. Formally, an *infinitesimal automorphism* of an n-pointed curve $(C; p_1, \dots, p_n)$ is a Σ-automorphism of $C \times \Sigma$ which restricts to the identity on the central fiber and on the marked sections $\{p_i\} \times \Sigma$. The result is then a consequence of the following simple remark.

LEMMA (5.3). *A stable n-pointed curve has no nontrivial infinitesimal automorphisms.*

To prove the lemma, denote by g the genus and by N the normalization of C, viewed as an $(n + 2\delta)$-pointed curve, where δ is the number of singular points of C, and let $N = \cup N_i$ be the decomposition of N into irreducible components; recall that each of the N_i is stable. Then observe that any infinitesimal automorphism of $(C; p_1, \dots, p_n)$ lifts to an infinitesimal automorphism of N. This reduces us to the case where C is smooth; an infinitesimal automorphism is then just a vector field vanishing at the marked points, that is, an element of $H^0(C, T_C(-\sum p_i))$. On the other hand, the stability assumption means that $2g - 2 + n > 0$, that is, that the degree of $T_C(-\sum p_i)$ is negative. This concludes the proof of the lemma and hence of Theorem (5.1). Q.E.D.

An immediate corollary of Theorem (5.1) is that one can do away with the meromorphicity assumption in (5.2). More precisely, the following result holds.

COROLLARY (5.4). *Let*

$$\alpha : X \to S\,, \quad \sigma_1, \dots, \sigma_n : S \to X \quad \text{and} \quad \beta : Y \to S\,, \quad \tau_1, \dots, \tau_n : S \to Y$$

be families of stable n-pointed curves as in Theorem (5.1), where S is a reduced and irreducible normal scheme or analytic space. Let S^* be the complement in S of a closed proper subscheme or analytic subspace, and set $X^* = \alpha^{-1}(S^*)$, $Y^* = \beta^{-1}(S^*)$. Let $\gamma : X^* \to Y^*$ be an isomorphism of families of n-pointed curves over S^*. Then γ extends uniquely to an isomorphism $\xi : X \to Y$ of families of n-pointed curves over S.

Here is another immediate consequence of Theorem (5.1).

COROLLARY (5.5). *Let*

$$\alpha : X \to S\,, \ \sigma_1, \dots, \sigma_n : S \to X \quad \text{and} \quad \beta : Y \to S\,, \ \tau_1, \dots, \tau_n : S \to Y$$

be families of stable n-pointed curves as in Theorem (5.1), and let $\varphi, \psi : X \to Y$ be isomorphisms of families of n-pointed curves over S. Suppose that the induced isomorphisms φ_s and ψ_s between $X_s = \alpha^{-1}(s)$ and $Y_s = \beta^{-1}(s)$ are equal for each (closed) point s of S. Then $\varphi = \psi$.

Proof. Clearly the corollary is trivial when S is reduced. Set

$$Z = \mathbf{Isom}_S((X;\sigma_1,\dots,\sigma_n),(Y;\tau_1,\dots,\tau_n))$$

and let $\pi : Z \to S$ be the natural projection. Then φ and ψ can be viewed as sections $a, b : S \to Z$. Let s be a (closed) point of S, set $z = a(s) = b(s)$, and denote by $\mathfrak{m}_z$ and $\mathfrak{m}_s$ the maximal ideals of $\mathcal{O}_{Z,z}$ and $\mathcal{O}_{S,s}$. Pullback via π gives a homomorphism $f : \mathcal{O}_{S,s} \to \mathcal{O}_{Z,z}$, and pullback via a and b homomorphisms $u, v : \mathcal{O}_{Z,z} \to \mathcal{O}_{S,s}$. By construction, u and v are left inverses to f. We let

$$f_h : \mathfrak{m}_s/\mathfrak{m}_s^h \to \mathfrak{m}_z/\mathfrak{m}_z^h\,, \qquad u_h, v_h : \mathfrak{m}_z/\mathfrak{m}_z^h \to \mathfrak{m}_s/\mathfrak{m}_s^h$$

be the homomorphisms induced by f, u, and v. Clearly, u_h and v_h are left inverses to f_h; in particular, f_h is always injective. Theorem (5.1) asserts that π is unramified and hence that f_2 is onto. One then easily shows, inductively on h, that f_h is onto for every $h \geq 2$. The upshot is that the homomorphism of completions

$$\widehat{f} : \widehat{\mathcal{O}}_{S,s} \to \widehat{\mathcal{O}}_{Z,z} \tag{5.6}$$

is an isomorphism, i.e., that the projection π is étale at z, and that $\widehat{u}$ and $\widehat{v}$ are inverses of $\widehat{f}$ and hence are equal. It follows that u and v are also equal. Q.E.D.

For future reference, we record here another useful fact concerning isomorphisms of families of nodal curves, which is an immediate consequence of Lemma (7.5) in Chapter IX.

LEMMA (5.7). *Let $f : X \to S$ and $f' : X' \to S$ be families of nodal curves, and let $\pi : X \to X'$ be an S-morphism. Suppose that, for each point $s \in S$, the induced morphism $X_s \to X'_s$ between fibers is an isomorphism. Then π is also an isomorphism.*

6. The stable model, contraction, and projection.

In this section and in the following two, we shall discuss a number of constructions involving families of nodal curves which will be crucial in the rest of the book.

The first construction we wish to present is the one of the stable model, a special instance of which we already encountered in Section 4. Let g and n be such that stable n-pointed genus g curves exist, i.e., such that $2g - 2 + n > 0$. Then there is a canonical way of attaching a stable n-pointed genus g curve to a semistable one $(C; x_1, \dots, x_n)$. To see why this is the case, recall that the irreducible components which prevent $(C; x_1, \dots, x_n)$ from being stable are precisely those components which are smooth rational and contain just two points which are either marked or nodes. The connected components of their union are chains of smooth rational curves, and, as we did in Section 4, we refer to them as exceptional chains. The n-pointed nodal curve $(C'; x'_1, \dots, x'_n)$ obtained by collapsing to a point each exceptional chain is clearly stable and is called the *stable model* of $(C; x_1, \dots, x_n)$. Let $\alpha : C \to C'$ be the collapsing map. It is important to notice that, since ω_C is trivial on each exceptional chain, $\alpha^*(\omega_{C'}(\sum x'_i))$ is canonically isomorphic to $\omega_C(\sum x_i)$. Exceptional chains come in two flavors, depending on whether they contain no marked points or one marked point. The figure below illustrates the two kinds of chains, drawn in red, and the corresponding collapsing maps.

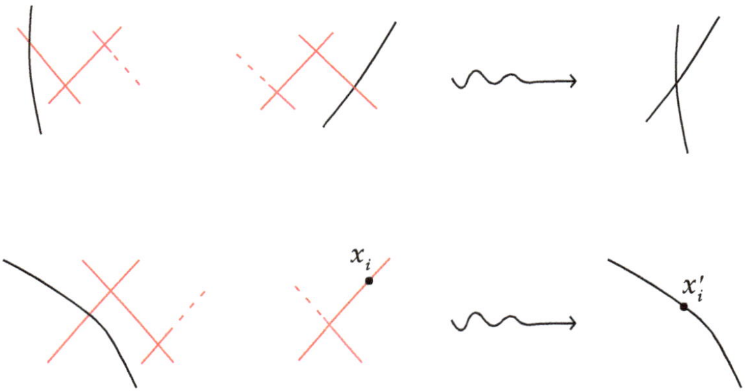

Figure 4.

As we shall presently see, the collapsing operation which produces the stable model of a semistable curve can be performed simultaneously and consistently for all fibers of any family of semistable curves. We shall use the following simple result, which generalizes a well-known property of smooth (unpointed) curves.

LEMMA (6.1). *Let $(C; D)$ be a semistable curve of genus g with n marked points, and let V be the set of vertices of its graph. Suppose that $2g - 2 + n > 0$. For each $v \in V$, denote by C_v the corresponding component of C, by g_v the genus of its normalization, and by l_v the number of the half-edges issuing from v. Let L be a line bundle on C and set $d_v = \deg_{C_v} L$. Then*

i) *$H^1(C, L) = 0$ if $d_v \geq 2(2g_v - 2 + l_v)$ for all v;*
ii) *L is base-point-free if $d_v \geq 2(2g_v - 2 + l_v)$ for all v;*
iii) *L is very ample if C is stable and $d_v \geq 3(2g_v - 2 + l_v)$ for all v.*

Proof. The assumptions imply that $d_v \geq 0$ for every v and in fact that $d_v > 0$ unless C_v is a smooth rational curve and $l_v = 2$, which of course can happen only if C is not stable. Thus, if $d_v = 0$, L is trivial on C_v. But then, denoting by $(C^\sharp; D^\sharp)$ the curve obtained by contracting C_v to a point and by $\varphi : C \to C^\sharp$ the contraction map, $L = \varphi^* L^\sharp$ for some line bundle $L^\sharp$ on $C^\sharp$. The curve $(C^\sharp; D^\sharp)$ and the line bundle $L^\sharp$ still satisfy the assumptions of the lemma, and moreover

$$H^i(C, L) = H^i(C^\sharp, L^\sharp), \quad i = 0, 1.$$

Iterating the contraction procedure, we are thus reduced to proving the lemma under the additional assumption that $d_v > 0$ for every vertex v. Property i) follows from (2.14), since $d_v > \deg_{C_v} \omega_C$ for every v. In fact, $d_v \geq 2(2g_v - 2 + l_v) > \deg_{C_v} \omega_C$ when $2g_v - 2 + l_v > 0$, while $d_v > 0 = \deg_{C_v} \omega_C$ when $2g_v - 2 + l_v = 0$.

To prove ii), we must show that $H^1(C, I_x L)$ vanishes for every $x \in C$, where I_x stands for the ideal sheaf of x. We distinguish two cases. Suppose first that x is not a singular point of C, so that $I_x L$ is a line bundle. If C_v does not contain x, then $\deg_{C_v} I_x L = d_v > \deg_{C_v} \omega_C$. If $x \in C_v$, then

$$\deg_{C_v} I_x L = d_v - 1 \geq \deg_{C_v} \omega_C\,. \tag{6.2}$$

If equality holds, we are in one of the following cases. Either $2g_v - 2 + l_v = 0$, and hence $g_v = 0$, $l_v = 2$, or else $2g_v - 2 + l_v = 1$, and C_v does not contain marked points. The latter can happen only for $g_v = 0$, $l_v = 3$ and for $g_v = 1$, $l_v = 1$. In all cases, C_v cannot be the only component of C. Thus (2.14) applies to $I_x L$.

Suppose instead that x is a singular point of C. We let $\alpha : C' \to C$ be the partial normalization at x. If C_v is a component of C, we denote by C'_v the corresponding component of C'. The sheaf $I_x L$ is of the form $\alpha_* L'$, where L' is a line bundle on C', and hence $H^1(C, I_x L) = H^1(C', L')$. We also recall that $\alpha_* \omega_{C'} = I_x \omega_C$. If C_v is any component of C, then

$$\deg_{C_v} L - \deg_{C'_v} L' = \deg_{C_v} \omega_C - \deg_{C'_v} \omega_{C'}\,. \tag{6.3}$$

Thus,

$$\begin{aligned} \deg_{C'_v} L' &\geq \deg_{C'_v} \omega_{C'} - \deg_{C_v} \omega_C + 2(2g_v - 2 + l_v) \\ &\geq \deg_{C'_v} \omega_{C'} + 2g_v - 2 + l_v > \deg_{C'_v} \omega_{C'} \end{aligned}$$

when $2g_v - 2 + l_v > 0$, while $\deg_{C'_v} L' \geq d_v - 1 > -1 = \deg_{C'_v} \omega_{C'}$ when $2g_v - 2 + l_v = 0$, and (2.14) again proves the vanishing of $H^1(C', L') = H^1(C, I_x L)$.

It remains to prove iii). We must show that, for any choice of points $x, y \in C$, the group $H^1(C, I_x I_y L)$ vanishes. We distinguish various cases. First suppose that x and y are not singular points of C. Then $I_x I_y L$ is a line bundle, and its degree on C_v is at least $3(2g_v - 2 + l_v) - 2 \geq 2g_v - 2 + l_v \geq \deg_{C_v} \omega_C$, since C is stable. Equality occurs only when both x and y belong to C_v, $2g_v - 2 + l_v = 1$, and C_v

does not contain marked points. This can happen only for $g_v = 0$, $l_v = 3$ and for $g_v = 1$, $l_v = 1$. In both cases, C_v cannot be the only component of C and, if C_w is any other component, the degree of I_xI_yL on it is strictly greater than $\deg_{C_w}\omega_C$, so that (2.14) applies.

In the remaining cases, at least one among x and y is singular. Let $\alpha : C' \to C$ be the partial normalization at those points of $\{x, y\}$ which are nodes of C. If C_v is a component of C, we denote by C'_v the corresponding component of C'. The sheaf I_xI_yL is of the form α_*L', where L' is a line bundle on C', and hence $H^1(C, I_xI_yL) = H^1(C', L')$. The second case we consider is the one where $x \neq y$ and x, y are both singular. One proceeds as in the second part of the proof of ii). Formula (6.3) is valid and implies that

$$\deg_{C'_v} L' \geq \deg_{C'_v}\omega_{C'} + 2(2g_v - 2 + l_v) > \deg_{C'_v}\omega_{C'}\,.$$

We conclude by appealing to (2.14) again.

A similar argument covers the case where x is singular and y is not. If $y \notin C_v$, formula (6.3) is valid, and one can argue exactly as in the previous case; if $y \in C_v$, formula (6.3) gets replaced by $\deg_{C_v} L - \deg_{C'_v} L' = \deg_{C_v}\omega_C - \deg_{C'_v}\omega_{C'} + 1$, which implies

$$\deg_{C'_v} L' \geq \deg_{C'_v}\omega_{C'} + 2(2g_v - 2 + l_v) - 1 > \deg_{C'_v}\omega_{C'}\,.$$

The final case to be considered is the one where $x = y$ is singular. Instead of (6.3), we get that $\deg_{C_v} L - \deg_{C'_v} L' = 2\deg_{C_v}\omega_C - 2\deg_{C'_v}\omega_{C'}$, which gives

$$\deg_{C'_v} L' \geq 2\deg_{C'_v}\omega_{C'} - 2\deg_{C_v}\omega_C + 3(2g_v - 2 + l_v) \geq 2\deg_{C'_v}\omega_{C'} + 2g_v - 2 + l_v.$$

Moreover, if these inequalities are equalities, then $\deg_{C_v}\omega_C = 2g_v - 2 + l_v$; in other words, C_v does not contain marked points. Observe that $\deg_{C'_v}\omega_{C'} = \deg_{C_v}\omega_C - h_v$, where h_v equals 2 if C_v is the only component containing x, 1 if x belongs to C_v and to another component, and 0 if $x \notin C_v$; moreover, $h_v \leq l_v$. Hence, since $2g_v - 2 + l_v > 0$ by stability, $\deg_{C'_v}\omega_{C'} + 2g_v - 2 + l_v$ is nonnegative, and vanishes if and only if $g_v = 0$, $l_v = 3$, and $h_v = 2$. Thus $\deg_{C'_v} L' \geq \deg_{C'_v}\omega_{C'}$, and if we have equality, then C_v contains no marked points, $g_v = 0$, $l_v = 3$, and $h_v = 2$. If this happens, C' is connected and has components other than C'_v, as $l_v - h_v = 1$ and C_v contains no marked points. Since the degree of L' on any component of C' different from C'_v is strictly greater than the one of $\omega_{C'}$, (2.14) shows that $H^1(C', L') = H^1(C, I_x^2 L) = 0$. This concludes the proof. Q.E.D.

COROLLARY (6.4). *Let $(C; D)$ be as in (6.1). Set $M = \omega_C(D)$. Then*

$$H^0(C, M^2) \otimes H^0(C, M^k) \to H^0(C, M^{k+2})$$

is onto for every $k \geq 4$.

This can be proved by a standard argument. Notice that M^h satisfies the assumptions of parts i) and ii) of (6.1) for $h \geq 2$, and those of part iii) for $h \geq 3$. In particular, we can choose sections s and t of M^2 which have no common zeros. Let E be the divisor of s. By part i) of (6.1), from the exact sequence

$$0 \to M^{i-2} \xrightarrow{\times s} M^i \to M^i_{|E} \to 0$$

for $i = k+2, k$ we deduce the exact sequence

$$\text{(6.5)} \qquad H^0(C, M^k) \xrightarrow{\times s} H^0(C, M^{k+2}) \to H^0(E, M^{k+2}_{|E}) \to 0$$

and the surjectivity of $H^0(C, M^k) \to H^0(E, M^k_{|E})$. Thus the composite mapping

$$H^0(C, M^k) \to H^0(E, M^k_{|E}) \xrightarrow{\times t} H^0(E, M^{k+2}_{|E})$$

is onto, since s and t have no common zeros. But then $s \otimes H^0(C, M^k) + t \otimes H^0(C, M^k)$ maps onto $H^0(C, M^{k+2})$, by (6.5). This concludes the proof.

COROLLARY (6.6). *Let $f : X \to S$, $\sigma_p : S \to X$, $p \in P$, be a family of P-pointed nodal curves. Then the set of $s \in S$ such that $(X_s; \{\sigma_p(s)\}_{p \in P})$ is semistable is Zariski open in S.*

It does no harm to assume that f has connected fibers. Set $\mathcal{L} = \omega_f(\sum \sigma_p)$. Suppose that the fiber of f at $s_0 \in S$ is semistable, and set $x_p = \sigma_p(s_0)$. Let $(X'_{s_0}; \{x'_p\}_{p \in P})$ be the stable model of $(X_{s_0}; \{x_p\}_{p \in P})$. Then $\mathcal{L}_{s_0} = \omega_{X_{s_0}}(\sum x_p)$ is the pullback of $\mathcal{L}' = \omega_{X'_{s_0}}(\sum x'_p)$. Since the fibers of $X_{s_0} \to X'_{s_0}$ are points or chains of $\mathbb{P}^1$'s, a Leray spectral sequence argument shows that

$$H^q(X_{s_0}, \mathcal{L}^k_{s_0}) = H^q(X'_{s_0}, \mathcal{L}'^k_{s_0})$$

for all $q \geq 0$ and all k. Combining this with Lemma (6.1) shows that, for high enough k, the linear system $|\mathcal{L}^k_{s_0}|$ has no base points, and $H^1(X_{s_0}, \mathcal{L}^k_{s_0})$ vanishes. The theory of base change in cohomology (cf. Proposition (3.3) in Chapter IX) then implies, in particular, that every section of $\mathcal{L}^k_{s_0}$ extends to a section of $\mathcal{L}^k$ over a neighborhood of X_{s_0}. It follows that $|\mathcal{L}^k_s|$ has no base points for all s in a neighborhood of s_0. But then Lemma (3.2) implies that, when s belongs to this neighborhood, $(X_s; \{\sigma_p(s)\}_{p \in P})$ is semistable.

We now return to the problem of simultaneously constructing the stable models of all fibers in a family of semistable curves. Here is what we shall prove.

Proposition (6.7). *Consider a family F of semistable P-pointed curves of genus g*

$$f : X \to S\,, \quad \sigma_p : S \to X\,, p \in P.$$

Suppose that $2g - 2 + |P| > 0$. Then there exist a family F' of stable P-pointed genus g curves

$$f' : X' \to S\,, \quad \sigma'_p : S \to X'\,, p \in P,$$

and a morphism $\pi : X \to X'$ such that:

i) $f' \circ \pi = f$;
ii) $\sigma'_p = \pi \circ \sigma_p$ *for every* $p \in P$;
iii) *for each closed point s of S, the fiber X'_s is the stable model of X_s, and $\pi_s : X_s \to X'_s$ is the collapsing map.*

The pair (F', π) is unique up to a unique isomorphism. Set $\mathcal{L} = \omega_{X/S}(\sum \sigma_p)$, $\mathcal{L}' = \omega_{X'/S}(\sum \sigma'_p)$. Then:

a) *for every $k > 0$, there are canonical isomorphisms $\pi^* \mathcal{L}'^k \xrightarrow{\sim} \mathcal{L}^k$ and $\mathcal{L}'^k \xrightarrow{\sim} \pi_* \mathcal{L}^k$;*
b) $R^q \pi_* \mathcal{L}^k = 0$ *for all $q > 0$ and all $k \geq 0$;*
c) $R^q f'_* \mathcal{L}'^k$ *and $R^q f_* \mathcal{L}^k$ are canonically isomorphic for every $q \geq 0$ and every $k \geq 0$.*

We shall only prove existence and uniqueness of F' and π. For a proof of properties a), b) and c) the reader is referred to [426]; here we simply observe that these statements are essentially obvious when S is reduced. We begin with the proof of existence. As usual, for any $s \in S$, we write $\mathcal{L}_s$ for the pullback of $\mathcal{L}$ to the fiber X_s. Set $x_p = \sigma_p(s)$, and let $(X'_s; \{x'_p\}_{p \in P})$ be the stable model of $(X_s; \{x_p\}_{p \in P})$. Clearly, $\mathcal{L}_s$ is the pullback to X_s of $\mathcal{L}'_s = \omega_{X'_s}(\sum x'_p)$, and, as we also observed in the proof of Corollary (6.6),

$$H^q(X_s, \mathcal{L}_s^k) = H^q(X'_s, \mathcal{L}'^k_s)$$

for all $q \geq 0$ and all k. Lemma (6.1) and the theory of base change in cohomology imply that $R^1 f_* \mathcal{L}^k = 0$ for $k \geq 2$ and that $f_* \mathcal{L}^k$ is locally free for all k. Moreover, Corollary (6.4) implies that

$$f_* \mathcal{L}^2 \otimes f_* \mathcal{L}^k \to f_* \mathcal{L}^{k+2}$$

is onto as soon as $k \geq 4$; in particular, $\oplus_{k \geq 0} f_* \mathcal{L}^{4k}$ is a locally finitely generated graded $\mathcal{O}_S$-algebra. We set $\mathcal{F} = \mathcal{L}^4$ and

$$X' = \operatorname{Proj}(\oplus_{k \geq 0} f_* \mathcal{F}^k)\,.$$

The family $f' : X' \to S$ is flat since, for large k, $f'_*(\mathcal{O}_{X'}(k))$ equals $f_*\mathcal{F}^k$, which is locally free, and by part ii) of Lemma (6.1) there is a surjective map π fitting in a commutative diagram

$$\begin{array}{ccc} X & \xrightarrow{\ \pi\ } & X' \\ & {\scriptstyle f}\searrow \quad \swarrow{\scriptstyle f'} & \\ & S & \end{array}$$

We let σ'_p be the composition of σ_p and π. Then, by part iii) of Lemma (6.1), the fiber of $f' : X' \to S$, $\{\sigma'_p\}_{p\in P}$ at s is just the stable model $(X'_s; \{\sigma'_p(s)\}_{p\in P})$ of $(X_s; \{\sigma_p(s)\}_{p\in P})$, and the fiber of π at s is just the collapsing map.

We now come to uniqueness. The homomorphism

$$\mathcal{O}_{X'} \to \pi_*\mathcal{O}_X \tag{6.8}$$

is clearly injective. We claim that, in fact, it is an isomorphism. It suffices to treat the algebraic case, since any flat analytic family of projective schemes is locally the pullback of an algebraic one, by Lemma (4.9) in Chapter IX, and semistability is an open condition. Let s be a closed point of S, and let $q_1, \dots, q_\ell \in X'_s$ be the images of the exceptional chains of X_s. Choose an affine open subset $Y = \operatorname{Spec} A$ of X' containing $q_1, \dots, q_\ell$. Denote by $\mathfrak{m}_i \subset A$ the maximal ideal corresponding to q_i, and let T be the complement of $\mathfrak{m}_1 \cup \cdots \cup \mathfrak{m}_\ell$ in A. We also set $B = \Gamma(\pi^{-1}(U), \mathcal{O}_X)$. It will suffice to show that $T^{-1}A \to T^{-1}B$ is an isomorphism. Let u be a function on an open neighborhood U of $\pi^{-1}(q_1, \dots, q_\ell)$. Possibly after shrinking, we may assume that U is the complement of a Cartier divisor in $f^{-1}(V)$, where V is a suitable open neighborhood of s. We may thus view u as a section over $f^{-1}(V)$ of $\mathcal{O}_{f^{-1}(V)}(D)$, where D is a Cartier divisor not meeting $\pi^{-1}(q_1, \dots, q_\ell)$. The line bundle $\mathcal{F}$ has strictly positive degree on all components of X_s except those which belong to exceptional chains. On the latter, which are smooth rational, it has degree zero. Thus, for large enough k, the line bundle $\mathcal{F}^k(-D) \otimes \mathcal{O}_{X_s}$ satisfies the assumptions of parts i), ii) of Lemma (6.1); it follows that it is base-point-free and that its higher cohomology groups vanish. By the theory of base change, then, possibly after shrinking V, the direct image via f of $\mathcal{F}^k(-D)$ is locally free over V, and its fiber at s is $H^0(X_s, \mathcal{F}^k(-D) \otimes \mathcal{O}_{X_s})$; it follows that there is a section v of $\mathcal{F}^k(-D)$ over $f^{-1}(V)$ which does not vanish at any point of $\pi^{-1}(q_1, \dots, q_\ell)$. We may view uv as a section of $\mathcal{F}^k$ over $f^{-1}(V)$. If we also regard v as a section of $\mathcal{F}^k$, the quotient uv/v is a regular function on a neighborhood of $\{q_1, \dots, q_\ell\}$ which pulls back to u via π. This shows that $T^{-1}A \to T^{-1}B$ is onto, proving that (6.8) is an isomorphism.

Now suppose that $f' : X'' \to S$, $\{\sigma''_p\}_{p\in P}$, is another family of stable curves and that $\pi' : X \to X''$ is another collapsing map sharing with F

and π properties i), ii), and iii). Clearly, there is a bijection of sets $\alpha : X' \to X''$ such that $\pi' = \alpha \circ \pi$. Since (6.8) is an isomorphism, this map actually comes from a morphism. By Lemma (5.7), this morphism is an isomorphism. This concludes the proof of uniqueness in (6.7).

Summing up, we have associated to the family F a new family

$$\mathrm{StMd}(F) \quad = \quad \begin{array}{c} X' \\ f' \Big\downarrow \Big\uparrow \sigma'_p, p \in P \\ S \end{array}$$

plus the collapsing map $\mathrm{Coll}_F : X \to X'$. The family $\mathrm{StMd}(F)$ is called the *stable model* of F.

REMARK (6.9). It follows immediately from the functoriality of $\omega_{X/S}(\sum \sigma_p)$ under morphisms of families of P-pointed nodal curves that $\mathrm{StMd}(F)$ and Coll_F depend functorially on F in the following sense. Suppose we are given two families F_1 and F_2 of semistable P-pointed genus g curves and a morphism

$$\mathbf{H} \quad = \quad \begin{array}{ccc} X_1 & \xrightarrow{H} & X_2 \\ \Big\downarrow & & \Big\downarrow \\ S_1 & \xrightarrow{h} & S_2 \end{array}$$

between them. Then there is a natural morphism from $\mathrm{StMd}(F_1)$ to $\mathrm{StMd}(F_2)$

$$\mathrm{StMd}(\mathbf{H}) \quad = \quad \begin{array}{ccc} X'_1 & \xrightarrow{H'} & X'_2 \\ \Big\downarrow & & \Big\downarrow \\ S_1 & \xrightarrow{h'} & S_2 \end{array}$$

such that $H' \circ \mathrm{Coll}_{F_1} = \mathrm{Coll}_{F_2} \circ H$, and moreover $\mathrm{StMd}(\mathbf{HK}) = \mathrm{StMd}(\mathbf{H})\,\mathrm{StMd}(\mathbf{K})$ whenever $\mathbf{HK}$ is defined.

In particular, the stable model is a functor from the category of families of semistable P-pointed genus g curves to the category of families of stable ones.

The stable model makes it possible to easily perform two other operations, which usually go under the name of contraction and projection, for reasons that will become clear in Chapter XII. To explain what these are, start with a stable P-pointed curve, where P is a finite nonempty set, and remove (or better, unmark) one of the marked points, say the one labeled by $p \in P$. The result is a $(P \smallsetminus \{p\})$-pointed curve, which may not be stable but is certainly semistable. If we pass to its stable model, the result is a stable $(P \smallsetminus \{p\})$-pointed curve. This same operation can be performed in families. Given a family

$$F = \begin{array}{c} X \\ f \Big\downarrow \Big\uparrow \sigma_p, p \in P \\ S \end{array}$$

of stable P-pointed genus g curves, we get a family of semistable $(P \smallsetminus \{p\})$-pointed ones by ignoring the pth section. We call this new family F' and pass to its stable model $\mathrm{StMd}(F')$, which we also denote by $\mathrm{Pr}_p(F)$. By Remark (6.9), Pr_p is a functor from families of stable P-pointed genus g curves to families of stable $(P \smallsetminus \{p\})$-pointed ones, which we shall call the pth *projection*. We may also elect to keep track of the pth section of F; composing it with the collapsing map gives an additional section δ of $\mathrm{Pr}(F)$, which however may meet one or more of the marked sections or go through singular points of fibers. Associating to F the pair $\mathrm{Contr}_p(F) = (\mathrm{Pr}_p(F), \delta)$ gives a morphism from families of stable P-pointed genus g curves to pairs consisting of a family of stable $(P \smallsetminus \{p\})$-pointed ones plus an extra section. This will be called the pth *contraction*.

LEMMA (6.10). *If p and q are distinct points of P, then Pr_p and Pr_q commute. In other words, $\mathrm{Pr}_p \mathrm{Pr}_q(F)$ and $\mathrm{Pr}_q \mathrm{Pr}_p(F)$ are canonically isomorphic. The same applies to Contr_p and Contr_q.*

The lemma is obviously true when S is a single point. In the general case, denote by X' the total space of $\mathrm{Pr}_p(F)$, by X'' the one of $\mathrm{Pr}_q(F)$, and by X''' the one of $\mathrm{Pr}_p \mathrm{Pr}_q(F)$. There are a diagram of collapsing morphisms

$$\begin{array}{ccc} X & \xrightarrow{\alpha} & X' \\ \beta \Big\downarrow & & \\ X'' & \xrightarrow{\gamma} & X''' \end{array}$$

and a set-theoretic map $\eta : X' \to X'''$ whose restriction to each fiber of $X' \to S$ is the collapsing morphism associated to the projection Pr_q. Set-theoretically, $\eta\alpha = \gamma\beta$. As we observed in the proof of (6.7), the homomorphism $\mathcal{O}_{X'} \to \alpha_* \mathcal{O}_X$ is an isomorphism. We thus get a sheaf homomorphism (actually, an isomorphism) $\mathcal{O}_{X'''} \to \eta_* \mathcal{O}_{X'}$ by composing

$$\mathcal{O}_{X'''} \to \gamma_* \mathcal{O}_{X''} \to \gamma_* \beta_* \mathcal{O}_X \simeq \eta_* \alpha_* \mathcal{O}_X \simeq \eta_* \mathcal{O}_{X'}.$$

Thus η comes from a morphism. The lemma now follows from the uniqueness of the stable model.

7. Clutching.

Another construction, which is related to passage to the stable model and to projection, is the so-called clutching operation. We first describe it for a single curve and then explain how it can be generalized to families. Let P be a finite set, and let Γ be a P-marked dual graph; thus to each $p \in P$ there is attached a leg ℓ_p. Let V be the set of vertices of Γ. As usual, for each vertex v, we denote by L_v the set of half-edges issuing from v. Suppose that we are given an L_v-pointed genus g_v nodal curve C_v for each vertex v. We may then construct a new P-pointed curve C by identifying two points of $\coprod C_v$ if and only if they are marked points labeled by the two halves of an edge of Γ, and by labeling with $p \in P$ the point originally labeled by the leg ℓ_p.

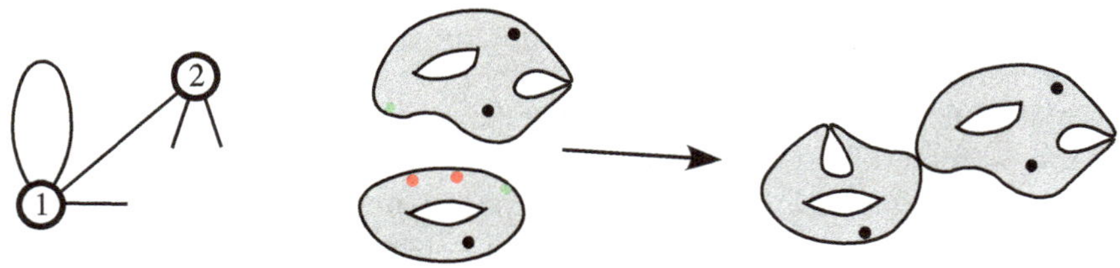

Figure 5. A graph and a clutching associated to it.

Clearly, the curve $\coprod C_v$ is the partial normalization of C at the nodes introduced by the identifications. Hence the process we have just described is a sort of inverse to partial normalization. If Γ is connected and all the C_v are connected, then C is connected as well. Conversely, the connectedness of C implies the connectedness of Γ. From now on, we assume that Γ and all the C_v are connected. The genus g of C is given by formula (2.8), which can be rewritten as

$$g = \sum_{v \in V} g_v + 1 - \chi(\Gamma) = \sum_{v \in V} g_v + h^1(\Gamma)\,.$$

Since one passes from C to $\coprod C_v$ by a process of partial normalization, C is stable if and only if all the C_v are (cf. Remark (3.3)). Stability is thus equivalent to

$$2g_v - 2 + l_v > 0\,,$$

where $l_v = |L_v|$.

The curve C we have constructed is said to be obtained from the C_v by *clutching along the graph* Γ. The clutching procedure is readily generalized to families. Let Γ be as above, and denote by E the set of its edges. Suppose that for each $v \in V$, we are given a family

$$F_v \;=\; \begin{array}{c} X_v \\ f_v\Big\downarrow\Big\uparrow\sigma_\ell, \;\; \ell \in L_v \\ S \end{array}$$

of stable L_v-pointed genus g_v curves, all over the same base S. We let X be the disjoint union of the X_v, and $f : X \to S$ the morphism whose restriction to X_v is f_v. This gives an L-pointed family of nodal curves over S, which we denote by F. If $m \in L$, taking residues along σ_m yields a well-defined surjective homomorphism

$$\omega_f(\sum \sigma_\ell) \to \mathcal{O}_{\sigma_m(S)}\,,$$

whence surjective homomorphisms

$$\omega_f^k(k\sum \sigma_\ell) \to \mathcal{O}_{\sigma_m(S)}$$

for all $k > 0$. These drop down to homomorphisms

$$R_\ell^{(k)} : f_*\left(\omega_f^k(k\sum \sigma_\ell)\right) \to \mathcal{O}_S\,,$$

which are surjective for $k \geq 2$, by Lemma (6.1). We may then construct homomorphisms

$$R^{(k)} : f_*\left(\omega_f^k(k\sum \sigma_\ell)\right) \to \mathcal{O}_S^E$$

by defining the component of $R^{(k)}$ indexed by edge $\{\ell, \ell'\}$ to be $R_\ell^{(k)} + (-1)^{k-1}R_{\ell'}^{(k)}$ (here we have arbitrarily chosen an orientation on $\{\ell, \ell'\}$). The homomorphism $R^{(k)}$ has constant rank; in fact, the kernel of its fiber at $s \in S$ can be identified with $H^0(X'_s, \omega_{X_s}^k(k\sum_{p\in P}\sigma_{\ell_p}(s)))$, where X'_s is the curve obtained from X_s by clutching along Γ, and hence its dimension is independent of s. It follows that the kernel of $R^{(k)}$, which we denote by $\mathcal{S}_k$, is locally free. It also follows from (6.4) that the graded $\mathcal{O}_S$-algebra $\oplus_{k\geq 0}\mathcal{S}_k$ is locally finitely generated. We set

$$X' = \mathrm{Proj}(\oplus_{k\geq 0}\mathcal{S}_k)\,.$$

It is a consequence of (6.1) that the fiber of $X' \to S$ at s is precisely the curve obtained from X_s by clutching along Γ. For each $p \in P$, we define $\sigma'_p : S \to X'$ to be the composition of σ_{ℓ_p} with the natural morphism $X \to X'$. The family

$$F' = \begin{array}{c} X' \\ f' \Big\downarrow \Big\uparrow \sigma'_p, \; p \in P \\ S \end{array}$$

is said to be obtained from F by clutching along Γ. As was the case for the stable model, it is clear from the construction that clutching is functorial.

8. Stabilization.

The contraction operation produces, out of a family of stable n-pointed curves, a family of $(n-1)$-pointed ones plus an extra section.

We shall show that this operation has an inverse, which goes under the name of *stabilization*. Suppose that we are given a family of P-pointed nodal curves, consisting of a family

$$\varphi : X \to S \tag{8.1}$$

of nodal curves plus sections σ_p, $p \in P$, and, in addition, a further section

$$\delta : S \to X\,.$$

Set $Q = P \cup \{q\}$, where q does not belong to P. Adding δ to the σ_p and labeling it with q does not, in general, produce a family of Q-pointed nodal curves, for two reasons; the first is that δ may well meet the other sections, and the second is that it may go through nodes of the fibers of f. As we shall see, there is however a canonical way of modifying X so as to obtain a family of Q-pointed nodal curves. If one starts with a family of stable P-pointed curves, the result is a family of stable Q-pointed ones; in addition, applying to this family the operation Pr_q gives back the family of P-pointed curves one started with.

It is useful, though not strictly necessary, to remark that it suffices to perform the construction in the special case in which the family is the projection of a fiber product to one of the factors and the extra section is the diagonal. Consider in fact the commutative diagram

$$\begin{array}{ccc} X & \xrightarrow{\;\overline{\delta}\;} & X \times_S X = Y \\ {\scriptstyle\varphi}\big\downarrow & & \big\downarrow{\scriptstyle\pi_2} \\ S & \xrightarrow{\;\delta\;} & X \end{array} \tag{8.2}$$

where $\overline{\delta} = (\mathrm{id}, \delta)$, and $\pi_i : X \times_S X \to X$ stands for projection to the ith factor. The diagram is cartesian, and the sections σ_p are the pullbacks of the sections $\tau_p : (\sigma_p, \mathrm{id}) : X \to X \times_S X$. Hence diagram (8.2) is a morphism of families of P-pointed nodal curves. Furthermore, the extra section δ is the pullback of the diagonal section $\Delta : X \to X \times_S X$. Thus, if the problem we have posed can be solved for $Y \to X$ and the sections τ_p, $p \in P$, and Δ, then a solution to our original problem can be simply obtained by pullback.

We begin by describing the construction in rather informal and naive terms, starting with the "universal" case of $Y \to X$ and of the diagonal section Δ. What we have to do is find a morphism $Y' \to Y$ and lifts of the sections τ_p and Δ to sections of $Y' \to X$ which make the latter into a family of Q-pointed nodal curves. Clearly, nothing needs to be done to Y except where a section crosses the diagonal, or where a section meets a node.

Suppose first that τ_p crosses the diagonal, i.e., that δ meets σ_p somewhere. Since φ is smooth at the crossing point, by the very definition of family of P-pointed curves, from the analytic point of view, $X \times_S X$ can be locally represented as $U \times V$, where U is an open neighborhood of the origin in $\mathbb{C}^2$, and V an open subset of S, with π_2 the projection to the product of the second factor of $\mathbb{C}^2$ times S. Furthermore, if x, x' are the standard coordinates in $\mathbb{C}^2$, we may set things up so that section τ_p is $\{x = 0\}$, and Δ is $\{x = x'\}$. It is then clear that we just need to perform a blow-up in the x, x' coordinates and replace τ_p and Δ by their proper transforms. In formulas, this amounts to the following. Let λ, μ be homogeneous coordinates in $\mathbb{P}^1$. Then the sought-for modification of $Y = X \times_S X$ is locally the projection

$$\{((x, x'), s, [\lambda : \mu]) \in U \times V \times \mathbb{P}^1 : x\mu = x'\lambda\} \to U \times V,$$

and the proper transforms of the sections τ_p and Δ are, respectively,

$$\lambda = 0 \quad \text{and} \quad \lambda = \mu.$$

Notice that this construction does not affect the fibers away from the locus $x' = 0$; on the other hand, its effect on a fiber at a point of X where x' vanishes is to add a $\mathbb{P}^1$ (the red line in Fig. 6 below) meeting the rest of the fiber at a single point and crossed by the proper transforms of τ_p and Δ at distinct points.

Figure 6.

We now turn to the case where Δ (or, which is the same, δ) meets a node of a fiber. Say the fiber lies over $s_0 \in S$. Near the node, X can be analytically represented as the locus with equation $xy = f$, where f is a function on an open neighborhood V of s_0, vanishing at s_0, and hence $Y = X \times_S X$ can be locally realized as the locus

$$W = \{((x, y, x', y'), s) \in U \times V : xy = f = x'y'\},$$

where U is a neighborhood of the origin in $\mathbb{C}^4$, and Δ as the locus wih equations

$$x = x', \ \ y = y'.$$

In this case also the solution is simple and consists in replacing W with

$$W' = \{((x,y,x',y'),s,[\lambda:\mu]) \in U \times V \times \mathbb{P}^1 : xy = f = x'y', \lambda x' = \mu x, \lambda y = \mu y'\},$$

and Δ with the section Δ' corresponding to the locus

$$\lambda = \mu .$$

The net effect on the fiber at $x' = y' = 0$, $s = s_0$, is to replace the node with a $\mathbb{P}^1$ (the red line in Fig. 7 below), meeting once each of the two branches of the former node, and crossed by Δ' at a further point.

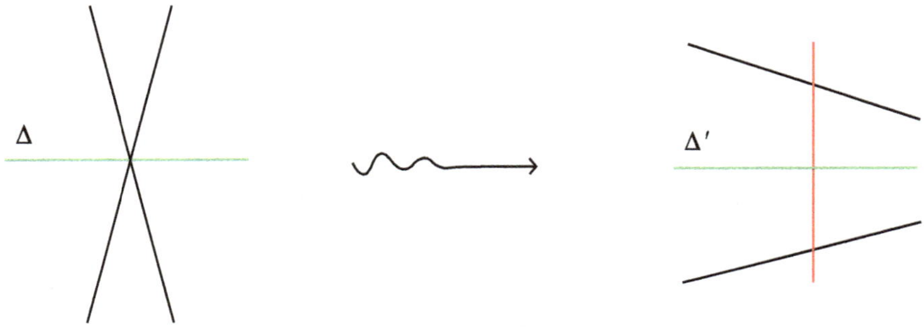

Figure 7.

We now go back to our original family (8.1). We want to find a morphism $\pi : X' \to X$, plus sections σ'_p, $p \in P$ and $\delta' = \sigma'_q$ of $\varphi' = \varphi \circ \pi : X' \to S$ which make the latter into a family of Q-pointed nodal curves. As we have observed, a solution can be obtained from $Y' \to X$, $\{\tau'_p\}_{p \in P}$ and Δ' by base change via $\delta : S \to X$. To describe the answer we distinguish two cases, as above.

Case 1. Suppose δ meets σ_p somewhere. Analytically, X can be locally represented as $U \times V$, where U is an open neighborhood of the origin in $\mathbb{C}$ and V an open subset of S, and φ as the projection to the second factor. Furthermore, if x is the standard coordinate in $\mathbb{C}$, we can arrange things so that section σ_p is $\{x = 0\}$ and δ is $\{x = \xi\}$, where ξ is a function on V. Choose homogeneous coordinates λ and μ on $\mathbb{P}^1$. Locally, X' is the subspace of the product $U \times V \times \mathbb{P}^1$ with equation

$$x\mu = \xi\lambda, \tag{8.3}$$

and the equations of sections σ'_p and δ' are

$$\lambda = 0 \quad \text{and} \quad \lambda = \mu, \tag{8.4}$$

respectively.

Case 2. Suppose δ meets a node of a fiber, say of $\varphi^{-1}(s_0)$. Analytically, a neighborhood of the node is isomorphic to the subspace $\{xy = f\}$ of $U \times V$, where V is a neighborhood of s_0 in S, f is a function on V, vanishing at s_0, and U is a neighborhood of the origin in the $\mathbb{C}^2$ with coordinates x, y. The section δ is defined by $x = \xi$, $y = \eta$, where ξ, η are functions on V such that $\xi\eta = f$. Then, letting again λ, μ be homogeneous coordinates in $\mathbb{P}^1$, X' is locally the locus in $U \times V \times \mathbb{P}^1$ with equations

$$xy = f \ , \quad \lambda\xi = \mu x \ , \quad \lambda y = \mu\eta \, , \tag{8.5}$$

while the lift δ' of δ is the locus

$$\lambda = \mu \, . \tag{8.6}$$

We are now confronted with two main difficulties. The first is that it must be shown that the various local constructions we have performed fit together. The second is that the constructions are analytic, and it is not a priori clear that, if we start with an algebraic family, the result will be algebraic. We shall address these two problems simultaneously. We denote by D the subspace of X corresponding to δ, by $\mathcal{I}$ its ideal sheaf, and by $\mathcal{I}^\vee = \mathcal{H}om_{\mathcal{O}_X}(\mathcal{I}, \mathcal{O}_X)$ the dual of $\mathcal{I}$. There is a natural homomorphism $\mathcal{O}_X \to \mathcal{I}^\vee$ which is injective since $\mathcal{I}$ contains elements which are not zero divisors, as is clear from the local analysis carried out above. We consider the "diagonal" homomorphism $h \mapsto (h, h)$

$$\mathcal{O}_X \to \mathcal{I}^\vee \oplus \mathcal{O}_X(\sum \sigma_i) \, ,$$

and denote by $\mathcal{K}$ its cokernel. We then set

$$X' = \operatorname{Proj}(\oplus_{k \geq 0} \operatorname{Sym}^k \mathcal{K}) \, .$$

To define the lifts of the sections σ_p to sections of $X' \to S$, consider the natural surjective homomorphism $\mathcal{K} \to \mathcal{O}_X(\sum \sigma_i)/\mathcal{O}_X$, and its pullback

$$\sigma_p^*(\mathcal{K}) \to \sigma_p^*(\mathcal{O}_X(\sum \sigma_i)/\mathcal{O}_X)$$

via σ_p. Since $\sigma_p^*(\mathcal{O}_X(\sum \sigma_i)/\mathcal{O}_X)$ is invertible, this defines a section of $X' \to X$ along σ_p. We let the lifting σ_p' be the composition of it with σ_p. To define the lifting δ' we shall proceed along the same lines, using the following result.

Lemma (8.7). *$(\mathcal{I}^\vee/\mathcal{O}_X) \otimes_{\mathcal{O}_X} \mathcal{O}_D$ is an invertible $\mathcal{O}_D$-module.*

Granting the lemma, the surjective homomorphism

$$\mathcal{K} \otimes_{\mathcal{O}_X} \mathcal{O}_D \to (\mathcal{I}^\vee/\mathcal{O}_X) \otimes_{\mathcal{O}_X} \mathcal{O}_D \tag{8.8}$$

defines a section of $X' \to X$ along D. Composing it with δ gives the required lifting.

We will now show that this new definition of X' and δ' agrees, locally, with the previous constructions and, at the same time, we will prove (8.7). Notice, to begin with, that D is a Cartier divisor except where δ goes through a node of a fiber; away from these points, $\mathcal{I}^\vee/\mathcal{O}_X = \mathcal{O}_X(D)/\mathcal{O}_X = \mathcal{O}_X(D) \otimes_{\mathcal{O}_X} \mathcal{O}_D$, so the lemma is trivial. In proving our contention, we shall give separate arguments for cases 1 and 2 described above, keeping the notation introduced there, and working in the analytic setup. Things are very simple in case 1. We view $1/(\xi - x)$ as a section of $\mathcal{I}^\vee$, and $1/x$ as a section of $\mathcal{O}_X(\sum \sigma_i)$, and denote by α and β' their classes in $\mathcal{K}$. Clearly, α and β' locally generate $\mathcal{K}$, and the relations among them are generated by $(x-\xi)\alpha = x\beta'$. Thus, if we set $\beta = \alpha - \beta'$, the classes α and β also generate $\mathcal{K}$ locally, and the relations among them are generated by

$$\xi\alpha = x\beta\,. \tag{8.9}$$

It is now essentially obvious that X' is locally just (8.3). In fact, set $W = U \times V$, denote by W' the portion of X' lying above W, and consider the surjective sheaf homomorphism $\mathcal{O}_W^2 \to \mathcal{K}_{|W}$ sending (f,g) to $f\alpha + g\beta$. We can view W' as the subspace of $\mathbb{P}^1 \times W = \operatorname{Proj}(\oplus \operatorname{Sym}^k \mathcal{O}_W^2)$ whose homogeneous ideal sheaf is the kernel of

$$\operatorname{Proj}(\oplus_{k\geq 0} \operatorname{Sym}^k \mathcal{O}_W^2) \to \operatorname{Proj}(\oplus_{k\geq 0} \operatorname{Sym}^k \mathcal{K}_{|W})\,.$$

Let λ be the section $(1,0)$ of $\mathcal{O}_W^2$, and μ the section $(0,1)$. These two sections give homogeneous coordinates on the first factor of $\mathbb{P}^1 \times W$, and map, respectively, to α and β. It then follows from (8.9) that the homogeneous ideal sheaf of W' is generated by $\xi\lambda - x\mu$, as claimed. Furthermore, α and β have the same image under under (8.8), since β' maps to zero. This shows that δ' is indeed described, in coordinates, by the second equation in (8.4). One similarly checks that the first of these equations describes σ'_p.

Case 2 requires more work. Set $\chi = \xi - x$, $\rho = y - \eta$, and notice that $\mathcal{K}$ locally agrees with $\mathcal{I}^\vee$. What needs to be proved is the following.

LEMMA (8.10). *The ideal $\mathcal{I}$ is locally generated by χ and ρ. The relations between χ and ρ are generated by*

$$\eta\chi = x\rho\,; \quad y\chi = \xi\rho\,.$$

The $\mathcal{O}_X$-module $\mathcal{I}^\vee$ is locally generated by sections α and β such that

$$\alpha(\chi)=x\,;\quad \alpha(\rho)=\eta\,;\quad \beta(\chi)=\xi\,;\quad \beta(\rho)=y\,. \tag{8.11}$$

The relations between α and β are generated by

$$y\alpha=\eta\beta\,;\quad \xi\alpha=x\beta\,.$$

Granting the lemma, and arguing as in case 1, the last two statements show that, indeed, X' is locally as described by (8.5).

Some parts of the lemma are obvious. For instance, it is clear that χ and ρ generate $\mathcal{I}$. Assuming that the relations between them are as described in the lemma, it is also clear that (8.11) can be taken as the definition of α and β. The real work will be in showing that α and β generate $\mathcal{I}^\vee$, and in determining the relations between χ and ρ, and between α and β.

The lemma asserts the exactness, for any point z of D, of the two sequences

$$\begin{gathered}\mathcal{O}^2_{X,z}\xrightarrow{G}\mathcal{O}^2_{X,z}\xrightarrow{\Phi}\mathcal{I}_z\to 0,\\ \mathcal{O}^2_{X,z}\xrightarrow{F}\mathcal{O}^2_{X,z}\xrightarrow{\Psi}\mathcal{I}^\vee_z\to 0,\end{gathered} \tag{8.12}$$

where F and G are the matrices

$$F=\begin{pmatrix}\xi & y\\ -x & -\eta\end{pmatrix}\,,\qquad G=\begin{pmatrix}\eta & y\\ -x & -\xi\end{pmatrix},$$

and Φ and Ψ are given by

$$\Phi\begin{pmatrix}u\\ v\end{pmatrix}=u\chi+v\rho\;,\qquad \Psi\begin{pmatrix}u\\ v\end{pmatrix}=u\alpha+v\beta\,.$$

Since the completion of a local ring R is faithfully flat over R, it suffices to prove the exactness of the analogues of (8.12) with $\mathcal{O}_{X,z}$ and $\mathcal{I}_z$ replaced by their completions. In other words, we must prove the exactness of

$$\begin{gathered}B^2\xrightarrow{G}B^2\xrightarrow{\Phi}I\to 0,\\ B^2\xrightarrow{F}B^2\xrightarrow{\Psi}I^\vee\to 0,\end{gathered} \tag{8.13}$$

where B stands for the completion of the local ring of X at z, and I for the ideal in B generated by χ and ρ.

Denoting by A the completion of the local ring of S at $\varphi(z)$, we have that

$$B=A[[x,y]]/(xy-\xi\eta).$$

Set

$$H = \begin{pmatrix} 0 & 1 \\ -1 & 0 \end{pmatrix} , \quad K = \begin{pmatrix} 0 & -1 \\ 1 & 0 \end{pmatrix} .$$

LEMMA (8.14). *The diagram*

$$\begin{array}{ccccccccccccc}
\cdots & \longrightarrow & B^2 & \xrightarrow{F} & B^2 & \xrightarrow{G} & B^2 & \xrightarrow{F} & B^2 & \xrightarrow{G} & B^2 & \longrightarrow & \cdots \\
 & & \downarrow{\scriptstyle H} & & \downarrow{\scriptstyle K} & & \downarrow{\scriptstyle H} & & \downarrow{\scriptstyle K} & & \downarrow{\scriptstyle H} & & \\
\cdots & \longrightarrow & B^2 & \xrightarrow{{}^tG} & B^2 & \xrightarrow{{}^tF} & B^2 & \xrightarrow{{}^tG} & B^2 & \xrightarrow{{}^tF} & B^2 & \longrightarrow & \cdots
\end{array}$$

is commutative with exact rows.

The commutativity is straightforward, as is the fact that $GF = FG = 0$. To prove the exactness, we shall construct an operator $\Lambda : B^2 \to B^2$ such that

$$\Lambda F + G\Lambda = \mathrm{id} = \Lambda G + F\Lambda . \tag{8.15}$$

This operator, however, will not be B-linear, but just A-linear. Notice that any element of B can be written uniquely in the form

$$\sum_{i\geq 0} a_i x^i + \sum_{i<0} a_i y^{-i} ,$$

where the a_i belong to A, and that conversely any such series comes from an element of B. Thus, the elements of B can be viewed as infinite column vectors

$$\begin{pmatrix} \cdot \\ \cdot \\ a_1 \\ a_0 \\ a_{-1} \\ \cdot \\ \cdot \end{pmatrix}$$

with entries in A. Left multiplication of such vectors by an infinite matrix with entries in A does not necessarily make sense, but this problem certainly does not occur with matrices having the property that every row contains only finitely many nonzero elements. Such a matrix thus defines an A-linear map of B to itself. For instance, multiplication by x (resp., by y) corresponds to the first (resp., the second) of the matrices

$$\begin{pmatrix} \ddots & & & & & & \\ & 0 & 1 & & & & \\ & & 0 & 1 & & & \\ & & & \boxed{0} & \xi\eta & & \\ & & & & 0 & \xi\eta & \\ & & & & & 0 & \\ & & & & & & \ddots \end{pmatrix}, \quad \begin{pmatrix} \ddots & & & & & & \\ & 0 & & & & & \\ & \xi\eta & 0 & & & & \\ & & \xi\eta & \boxed{0} & & & \\ & & & 1 & 0 & & \\ & & & & 1 & 0 & \\ & & & & & & \ddots \end{pmatrix}.$$

Here and elsewhere, when dealing with infinite matrices, we single out with a box the position with indices $0, 0$. On the other hand, multiplication by ξ (resp., by η) clearly corresponds to the diagonal matrix with ξ (resp., η) on the diagonal.

We then let Λ be given by the block matrix

$$\begin{pmatrix} 0 & U \\ V & 0 \end{pmatrix},$$

where

$$U = \begin{pmatrix} \ddots & & & & & & \\ & 0 & & & & & \\ & -1 & 0 & & & & \\ & & -1 & \boxed{0} & & & \\ & & & 0 & 0 & & \\ & & & & 0 & 0 & \\ & & & & & & \ddots \end{pmatrix},$$

$$V = \begin{pmatrix} \ddots & & & & & & \\ & 0 & 0 & & & & \\ & & 0 & 0 & & & \\ & & & \boxed{0} & 1 & & \\ & & & & 0 & 1 & \\ & & & & & 0 & \\ & & & & & & \ddots \end{pmatrix}.$$

It is straightforward to check that (8.15) holds. This finishes the proof of Lemma (8.14).

We denote by J the image of F. From the top row of the diagram in (8.14) we get in particular exact sequences

(8.16)
$$B^2 \xrightarrow{G} B^2 \to J \to 0,$$
$$0 \to J \to B^2 \xrightarrow{G} B^2.$$

Dualizing them gives exact sequences

$$0 \to J^\vee \xrightarrow{a} B^2 \xrightarrow{{}^tG} B^2,$$
$$B^2 \xrightarrow{{}^tG} B^2 \xrightarrow{b} J^\vee,$$

where $a \circ b = {}^tF$. The exactness of the bottom row of the diagram in (8.14) then implies that b is onto, and the commutativity of the diagram yields an exact sequence

$$B^2 \xrightarrow{F} B^2 \to J^\vee \to 0. \tag{8.17}$$

LEMMA (8.18). *The homomorphism* $\begin{pmatrix} u \\ v \end{pmatrix} \mapsto u+v$ *from* B^2 *to* B *induces an isomorphism from* J *to* I.

Clearly, J is generated by $\begin{pmatrix} \xi \\ -x \end{pmatrix}$ and $\begin{pmatrix} y \\ -\eta \end{pmatrix}$, whose images are precisely χ and ρ. Thus J maps surjectively to I, and all we have to show is the injectivity. To prove it, consider the block matrix $M = (M_1\ M_2)$, where

$$M_1 = \begin{pmatrix} \ddots & & & & & & \\ & 1 & 0 & & & & \\ & & 1 & 0 & & & \\ & & & \boxed{0} & -\eta & & \\ & & & & 0 & -\eta & \\ & & & & & 0 & \ddots \\ & & & & & & \ddots \end{pmatrix},$$

$$M_2 = \begin{pmatrix} \ddots & & & & & \\ & 0 & & & & \\ & \xi & 0 & & & \\ & & \xi & 0 & & \\ & & & 0 & \boxed{-1} & \\ & & & & 0 & -1 \\ & & & & & \ddots \end{pmatrix}.$$

A straightforward calculation shows that $MF = 0$. This says, in particular, that M maps J to zero. If $\begin{pmatrix} u \\ v \end{pmatrix} \in J$ goes to zero in B, that is, if $v = -u$, this means that

$$(M_1 - M_2)u = 0\,.$$

Writing

$$u = \begin{pmatrix} \cdot \\ \cdot \\ u_1 \\ u_0 \\ u_{-1} \\ \cdot \\ \cdot \end{pmatrix}$$

and keeping in mind the explicit form of M_1 and M_2, this translates into

$$\begin{aligned} &\dots \\ &u_2 = \xi u_3, \\ &u_1 = \xi u_2, \\ &u_0 = \eta u_{-1}, \\ &u_{-1} = \eta u_{-2}, \\ &\dots \end{aligned}$$

which in turn implies that, for any $n \geq 0$, u_i belongs to (ξ^n) when $i > 0$ and to (η^n) when $i \leq 0$. Since ξ and η belong to the maximal ideal of A, this shows that $u_i = 0$ for any i, finishing the proof of Lemma (8.18).

Combining (8.18) with (8.16) and (8.17) proves the exactness of the sequences (8.13), modulo checking that the homomorphisms $B^2 \to I$ and $B^2 \to I^\vee$ one obtains do indeed agree, up to sign, with Φ and Ψ. This simple verification is left to the reader. Lemma (8.10) is now fully proved.

Corollary (8.19). *$\mathcal{I}^\vee$ is $\mathcal{O}_S$-flat.*

By faithful flatness, to prove the corollary, it suffices to show that $I^\vee$ is A-flat. By Lemmas (8.14) and (8.18),

$$\text{(8.20)} \qquad \cdots \to B^2 \xrightarrow{F} B^2 \xrightarrow{G} B^2 \xrightarrow{F} B^2 \xrightarrow{G} B^2 \xrightarrow{F} B^2 \to I^\vee \to 0$$

is a resolution of $I^\vee$ by A-flat modules. On the other hand, we know that there is an A-linear operator $\Lambda : B^2 \to B^2$ satisfying (8.15). This shows that tensoring (8.20) with any A-module yields an exact sequence. This proves that $I^\vee$ is A-flat.

We now finish the proof of (8.7). In the notation of Lemma (8.10), one immediately sees that

$$\beta = \alpha + 1\,,$$

meaning that the difference between the homomorphisms β and α acts as multiplication by the function 1. This implies, in particular, that α

and β are the same modulo $\mathcal{O}_X$. From the second of the exact sequences (8.12) we deduce the other exact sequence

$$\mathcal{O}^2_{X,z} \xrightarrow{\overline{F}} \mathcal{O}^2_{X,z}/\mathcal{O}_{X,z} \xrightarrow{\overline{\Psi}} \mathcal{I}^\vee_z/\mathcal{O}_{X,z} \to 0,$$

where the inclusion $\mathcal{O}_{X,z} \hookrightarrow \mathcal{O}^2_{X,z}$ is given by

$$u \mapsto \begin{pmatrix} u \\ -u \end{pmatrix},$$

and $\overline{F}$ and $\overline{\Psi}$ are obtained from F and Ψ by passage to the quotient. If we restrict this last exact sequence to D, $\overline{F}$ pulls back to the zero map, by the explicit form of F. The outcome is an isomorphism between $\mathcal{I}^\vee_z/\mathcal{O}_{X,z}\otimes_{\mathcal{O}_{X,z}}\mathcal{O}_{D,z}$ and $\mathcal{O}^2_{D,z}/\mathcal{O}_{D,z}$, which in turn is isomorphic to $\mathcal{O}_{D,z}$. This completes the proof of (8.7). In the course of it, we observed that α and β have the same image in $\mathcal{I}^\vee_z/\mathcal{O}_{X,z}$. This shows that δ' is given, in coordinates, by (8.6). The construction of $X' \to S$ and of δ' is now complete.

The stabilization procedure functorially associates to a pair

$$F = (\text{family of stable } P\text{-pointed genus } g \text{ curves, extra section})$$

a new family $\mathrm{Stab}(F)$ of stable $(P \cup \{q\})$-pointed genus g curves, the section labeled with q coming from the extra section. As we loosely announced, the stabilization construction is the inverse of contraction. To make this precise, and anticipating notation that will be more formally introduced in Chapter XII, we denote by $\overline{\mathcal{M}}_{g,P\cup\{q\}}$ the category of families of stable $(P \cup \{q\})$-pointed genus g curves, and by $\overline{\mathcal{C}}_{g,P}$ the one of families of stable P-pointed genus g curves plus an extra section.

THEOREM (8.21). *The contraction and stabilization functors*

$$\mathrm{Contr}_q : \overline{\mathcal{M}}_{g,P\cup\{q\}} \to \overline{\mathcal{C}}_{g,P}\,, \qquad \mathrm{Stab} : \overline{\mathcal{C}}_{g,P} \to \overline{\mathcal{M}}_{g,P\cup\{q\}}$$

are inverse to each other up to isomorphism of functors. In particular, they are equivalences of categories.

The proof uses the following general result and its corollary.

LEMMA (8.22). *Let $f : Z \to W$ be a proper morphism of schemes or analytic spaces, $\mathcal{G}$ a coherent sheaf on Z, and w a point of W. Denote by $\mathfrak{m}_w$ the maximal ideal in $\mathcal{O}_{W,w}$. Suppose that the dimension of $f^{-1}(w)$ is at most 1 and that $H^1(f^{-1}(w), \mathcal{G}\otimes_{\mathcal{O}_W} k(w)) = 0$. Then:*

i) *there is an open neighborhood U of w such that $R^1f_*\mathcal{G}|_U = 0$;*
ii) *for each $h \geq 0$, the homomorphism*

$$(f_*\mathcal{G})_w \to H^0(f^{-1}(w), \mathcal{G}\otimes_{\mathcal{O}_W} \mathcal{O}_{W,w}/\mathfrak{m}^h_w)$$

is onto.

Proof. Since $R^1 f_* \mathcal{G}$ is coherent, i) is equivalent to the assertion that the stalk of $R^1 f_* \mathcal{G}$ at w is zero. In turn, by the theorem on formal functions (see [356], Theorem 11.1 or [324], (4.2.1)), this will follow if we can show that

$$H^1(f^{-1}(w), \mathcal{G} \otimes_{\mathcal{O}_W} \mathcal{O}_{W,w}/\mathfrak{m}_w^h) = 0 \quad \text{for } h \geq 1 .$$

This assertion reduces to one of the assumptions for $h = 1$, and we shall prove it in general by induction on h. Look at the exact cohomology sequence of

$$0 \to \mathfrak{m}_w^h \mathcal{G}/\mathfrak{m}_w^{h+1}\mathcal{G} \to \mathcal{G} \otimes_{\mathcal{O}_W} \mathcal{O}_{W,w}/\mathfrak{m}_w^{h+1} \to \mathcal{G} \otimes_{\mathcal{O}_W} \mathcal{O}_{W,w}/\mathfrak{m}_w^h \to 0.$$

It is clear that, to do the inductive step, what must be shown is that

$$H^1(f^{-1}(w), \mathfrak{m}_w^h \mathcal{G}/\mathfrak{m}_w^{h+1}\mathcal{G}) = 0 . \tag{8.23}$$

The natural homomorphism of $\mathcal{O}_{f^{-1}(w)}$-modules

$$\mathcal{G} \otimes_{k(w)} \mathfrak{m}_w^h/\mathfrak{m}_w^{h+1} \to \mathfrak{m}_w^h \mathcal{G}/\mathfrak{m}_w^{h+1}\mathcal{G}$$

is onto, and the cohomology of its kernel vanishes in degree greater than 1 for dimension reasons. Passing to cohomology, this gives a surjection

$$H^1(f^{-1}(w), \mathcal{G} \otimes_{k(w)} \mathfrak{m}_w^h/\mathfrak{m}_w^{h+1}) \to H^1(f^{-1}(w), \mathfrak{m}_w^h \mathcal{G}/\mathfrak{m}_w^{h+1}\mathcal{G}) .$$

On the other hand, the left-hand side is a direct sum of copies of $H^1(f^{-1}(w), \mathcal{G} \otimes_{\mathcal{O}_W} k(w))$, which is zero by assumption, and hence we are done.

We now turn to ii). Let $\mathcal{I}$ be the ideal sheaf of w. A piece of the higher direct image exact sequence of

$$0 \to \mathcal{I}^h \mathcal{G} \to \mathcal{G} \to \mathcal{G}/\mathcal{I}^h \mathcal{G} \to 0$$

is

$$f_* \mathcal{G} \to f_*(\mathcal{G}/\mathcal{I}^h \mathcal{G}) \to R^1 f_*(\mathcal{I}^h \mathcal{G}).$$

Since $f_*(\mathcal{G}/\mathcal{I}^h \mathcal{G})$ is a skyscraper sheaf whose stalk at w is just $H^0(f^{-1}(w), \mathcal{G} \otimes_{\mathcal{O}_W} \mathcal{O}_{W,w}/\mathfrak{m}_w^h)$, to prove ii), it suffices to show that $R^1 f_*(\mathcal{I}^h \mathcal{G})$ vanishes in a neighborhood of w. Clearly,

$$H^1(f^{-1}(w), \mathcal{I}^h \mathcal{G} \otimes_{\mathcal{O}_W} k(w)) = H^1(f^{-1}(w), \mathfrak{m}_w^h \mathcal{G}/\mathfrak{m}_w^{h+1}\mathcal{G}) ,$$

so (8.23) shows that the sheaf $\mathcal{I}^h \mathcal{G}$ satisfies the assumptions of the lemma, whose part i) then gives the required local vanishing of $R^1 f_*(\mathcal{I}^h \mathcal{G})$.

Q.E.D.

Corollary (8.24). *Let Z, W be schemes or analytic spaces over S, and let $f : Z \to W$ be a proper S-morphism whose fibers have dimension at most equal to one. Let $\mathcal{G}$ be a coherent sheaf on Z, flat over S. Consider the following assumptions on $\mathcal{G}$:*

a) $H^1(f^{-1}(w), \mathcal{G} \otimes_{\mathcal{O}_W} k(w)) = 0$ *for each (closed) point w of W;*
b) $\mathcal{G} \otimes_{\mathcal{O}_W} k(w)$ *is generated by global sections for each (closed) point $w \in W$.*

If a) holds, then:

i) $R^1 f_* \mathcal{G} = 0$;
ii) $f_* \mathcal{G}$ *is $\mathcal{O}_S$-flat;*
iii) *for any morphism $\alpha : T \to S$, there is a canonical isomorphism*

$$f_* \mathcal{G} \otimes_{\mathcal{O}_S} \mathcal{O}_T \to (f \times \mathrm{id})_* (\mathcal{G} \otimes_{\mathcal{O}_S} \mathcal{O}_T)\,;$$

If, in addition, b) is also satisfied, then

iv) $f^* f_* \mathcal{G} \to \mathcal{G}$ *is onto.*

We shall give the proof only for schemes, leaving to the reader the task of adapting it to the analytic context. Property i) follows immediately from part i) of Lemma (8.22). Without loss of generality, we may suppose that S and W are affine. Choose a finite cover $\mathcal{U}$ of Z with affine open sets, and denote by $\mathcal{C}^q(\mathcal{U}, \mathcal{G})$ the sheaf of alternating q-cochains with values in $\mathcal{G}$, relative to the cover $\mathcal{U}$; thus, $\mathcal{C}^\bullet(\mathcal{U}, \mathcal{G})$ is a finite resolution of $\mathcal{G}$. Then $f_*\mathcal{C}^\bullet(\mathcal{U}, \mathcal{G})$ is a resolution of $f_*\mathcal{G}$, because of i) and since $R^i f_* \mathcal{G}$ vanishes when $i > 1$ for dimension reasons. Since each $f_*\mathcal{C}^q(\mathcal{U}, \mathcal{G})$ is a finite direct sum of $\mathcal{O}_S$-flat sheaves, it follows that $f_*\mathcal{G}$ is $\mathcal{O}_S$-flat. In proving iii), we may suppose that T is affine. Denote by $\mathcal{V}$ the cover of $Z \times_S T$ consisting of the affine open sets of the form $U \times_S T$, where U belongs to $\mathcal{U}$. Then

$$\mathcal{C}^\bullet(\mathcal{V}, \mathcal{G} \otimes_{\mathcal{O}_S} \mathcal{O}_T) = \mathcal{C}^\bullet(\mathcal{U}, \mathcal{G}) \otimes_{\mathcal{O}_S} \mathcal{O}_T\,;$$
$$(f \times \mathrm{id})_* \mathcal{C}^\bullet(\mathcal{V}, \mathcal{G} \otimes_{\mathcal{O}_S} \mathcal{O}_T) = (f_* \mathcal{C}^\bullet(\mathcal{U}, \mathcal{G})) \otimes_{\mathcal{O}_S} \mathcal{O}_T\,.$$

The left-hand side of the second equality is a resolution of $(f \times \mathrm{id})_*(\mathcal{G} \otimes_{\mathcal{O}_S} \mathcal{O}_T)$, while the right-hand side is a resolution of $f_*\mathcal{G} \otimes_{\mathcal{O}_S} \mathcal{O}_T$ by flatness. Property iii) follows.

Now suppose that $\mathcal{G} \otimes_{\mathcal{O}_W} k(w)$ is generated by global sections for each $w \in W$. This assumption, together with part ii) of Lemma (8.22), implies that

$$f^* f_* \mathcal{G} \to \mathcal{G} \otimes_{\mathcal{O}_Z} k(z)$$

is onto for any $z \in Z$. Property iv) then follows from Nakayama's lemma, and the proof of the corollary is complete.

We will also need this elementary result.

LEMMA (8.25). *Let $f : X \to S$ be a morphism of schemes or analytic spaces, let $\mathcal{E}$ and $\mathcal{F}$ be coherent sheaves on X, and let $\alpha : \mathcal{E} \to \mathcal{F}$ be an $\mathcal{O}_X$-module homomorphism. Suppose that, for each (closed) point $s \in S$, the induced homomorphism $\mathcal{E} \otimes_{\mathcal{O}_S} k(s) \to \mathcal{F} \otimes_{\mathcal{O}_S} k(s)$ is an isomorphism. If $\mathcal{F}$ is $\mathcal{O}_S$-flat, α is an isomorphism.*

Proof. Let $\mathcal{Q}$ be the cokernel of α. The sequence

$$\mathcal{E} \otimes_{\mathcal{O}_S} k(s) \to \mathcal{F} \otimes_{\mathcal{O}_S} k(s) \to \mathcal{Q} \otimes_{\mathcal{O}_S} k(s) \to 0$$

is exact, and hence $\mathcal{Q}_x = \mathfrak{m}_s \mathcal{Q}_x$ for any (closed) point x above s, where $\mathfrak{m}_s$ indicates the maximal ideal in $\mathcal{O}_{S,s}$. It follows that $\mathcal{Q}_x = \mathfrak{m}_x \mathcal{Q}_x$ for any (closed) point x of X and hence that $\mathcal{Q}$ is zero by Nakayama's lemma. Now let $\mathcal{K}$ be the kernel of α. By the flatness assumption, $\mathrm{Tor}_1^{\mathcal{O}_{S,s}}(\mathcal{F}_x, k(s)) = 0$ for any x above s, and hence,

$$0 \to \mathcal{K} \otimes_{\mathcal{O}_S} k(s) \to \mathcal{E} \otimes_{\mathcal{O}_S} k(s) \to \mathcal{F} \otimes_{\mathcal{O}_S} k(s) \to 0$$

is exact. Arguing as above, it follows that $\mathcal{K}$ vanishes, which proves the lemma.

Q.E.D.

Proof of Theorem (8.21). It follows from the uniqueness of the stable model (cf. Proposition (6.7)) that applying stabilization and then contraction to an object in $\overline{\mathcal{C}}_{g,P}$ produces another object which is canonically isomorphic to the original one. Conversely, suppose that we are given a family of stable $(P \cup \{q\})$-pointed curves consisting of a family

$$\psi : Y \to S$$

of nodal curves plus sections τ_t, $t \in P \cup \{q\}$. Applying the qth contraction yields a family of stable P-pointed curves consisting of a family

$$\varphi : X \to S$$

and sections σ_p, $p \in P$, plus an extra section δ coming from τ_q. We may then apply to this setup the stabilization procedure and get a new family of stable $(P \cup \{q\})$-pointed curves consisting of a family

$$\varphi' : X' \to S$$

of nodal curves plus sections σ'_t, $t \in P \cup \{q\}$. We shall show that there is a unique isomorphism $\gamma : Y \to X'$ making the diagram

$$\begin{array}{ccc} Y & \xrightarrow{\gamma} & X' \\ & {\scriptstyle\pi}\searrow \quad \swarrow{\scriptstyle\pi'} & \\ & X & \end{array} \qquad (8.26)$$

commute, where π and π' are the collapsing morphisms, and carrying section τ_t to σ'_t for each t. In the proof we shall freely use the notation introduced earlier in this section while constructing $X' \to X$. Several simple, but somewhat tedious, verifications will be left to the reader.

We begin by proving existence. The sheaf $\mathcal{O}_Y(-\sum_{t\in P}\tau_t)$ satisfies assumption a) in Corollary (8.24) relative to the morphism π. In fact, the positive-dimensional fibers of π are projective lines, and each one of them meets at most one of the τ_t with $t \in P$. It then follows from Corollary (8.24) that $\pi_*\mathcal{O}_Y(-\sum_{t\in P}\tau_t)$ is S-flat and that its formation commutes with base change; in particular, its pullback to any fiber $X_s = \varphi^{-1}(s)$ is the pushforward via $Y_s \to X_s$ of the pullback of $\mathcal{O}_Y(-\sum_{t\in P}\tau_t)$ to Y_s. By direct inspection, one easily checks that the restriction to X_s of the pullback homomorphism

$$\mathcal{O}_X(-\sum_{t\in P}\sigma_t) \to \pi_*\mathcal{O}_Y(-\sum_{t\in P}\tau_t) \tag{8.27}$$

is an isomorphism for any s. By the S-flatness of $\pi_*\mathcal{O}_Y(-\sum_{t\in P}\tau_t)$, it then follows from Lemma (8.25) that (8.27) is an isomorphism. Pullback via π gives a homomorphism $\mathcal{I} \to \pi_*\mathcal{O}_Y(-\tau_q)$, and hence a pairing

$$\begin{aligned}\pi_*\mathcal{O}_Y(\tau_q - \sum_{t\in P}\tau_t) \otimes \mathcal{I} &\to \pi_*\mathcal{O}_Y(\tau_q - \sum_{t\in P}\tau_t) \otimes \pi_*\mathcal{O}_Y(-\tau_q)\\ &\to \pi_*\mathcal{O}_Y(-\sum_{t\in P}\tau_t) \cong \mathcal{O}_X(-\sum_{t\in P}\sigma_t)\end{aligned}$$

i.e., a homomorphism

$$\pi_*\mathcal{O}_Y(\tau_q - \sum_{t\in P}\tau_t) \to \mathcal{I}^\vee(-\sum_{t\in P}\sigma_t). \tag{8.28}$$

We leave to the reader the straightforward task of checking, using the explicit presentation of $\mathcal{I}^\vee$ given by Lemma (8.10), that the restriction of this homomorphism to each fiber X_s is an isomorphism. Since $\mathcal{I}^\vee$ is S-flat by Lemma (8.19), Lemma (8.25) then implies that (8.28) is an isomorphism.

The sequence

$$0 \to \mathcal{O}_Y(-\sum_{t\in P}\tau_t) \xrightarrow{u} \mathcal{O}_Y(\tau_q - \sum_{t\in P}\tau_t) \oplus \mathcal{O}_Y \xrightarrow{v} \mathcal{O}_Y(\tau_q) \to 0\,, \tag{8.29}$$

where $u(a) = (a,a)$ and $v(a,b) = a - b$, is exact. By Corollary (8.24), $R^1\pi_*\mathcal{O}_Y(-\sum_{t\in P}\tau_t)$ vanishes, and hence the pushforward via π of (8.29) is also exact. We thus get a commutative diagram with exact rows

$$\begin{array}{ccccccccc} 0 & \longrightarrow & \mathcal{O}_X(-\sum_{t\in P}\sigma_t) & \longrightarrow & \mathcal{I}^\vee(-\sum_{t\in P}\sigma_t) \oplus \mathcal{O}_X & \longrightarrow & \mathcal{K}(-\sum_{t\in P}\sigma_t) & \longrightarrow & 0\\ & & \downarrow\cong & & \downarrow\cong & & & & \\ 0 & \longrightarrow & \pi_*\mathcal{O}_Y(-\sum_{t\in P}\tau_t) & \longrightarrow & \pi_*\mathcal{O}_Y(\tau_q - \sum_{t\in P}\tau_t) \oplus \pi_*\mathcal{O}_Y & \longrightarrow & \pi_*\mathcal{O}_Y(\tau_q) & \longrightarrow & 0 \end{array}$$

whence an isomorphism $\mathcal{K}(-\sum_{t\in P}\sigma_t) \xrightarrow{\cong} \pi_*\mathcal{O}_Y(\tau_q)$. Pulling this back to Y, we get a homomorphism

$$\pi^*\mathcal{K}(-\sum_{t\in P}\sigma_t) \xrightarrow{\cong} \pi^*\pi_*\mathcal{O}_Y(\tau_q) \to \mathcal{O}_Y(\tau_q) \tag{8.30}$$

This homomorphism is onto since $\mathcal{O}_Y(\tau_q)$ satisfies the assumptions of Corollary (8.24). On the other hand,

$$X' = \mathrm{Proj}(\oplus_{k\geq 0}\,\mathrm{Sym}^k\,\mathcal{K}) \cong \mathrm{Proj}(\oplus_{k\geq 0}\,\mathrm{Sym}^k\,\mathcal{K}')\,,$$

where $\mathcal{K}' = \mathcal{K}(-\sum_{t\in P}\sigma_t)$. Hence (8.30) defines an X-morphism $\gamma : Y \to X'$. The induced morphism $Y_s \to X'_s$ is an isomorphism for any $s \in S$, and hence γ is an isomorphism by Lemma (5.7). Clearly, it maps τ_t to σ'_t for each $t \in P$; we leave it to the reader to check that it carries τ_q to σ'_q.

Having proved the existence, we turn to the uniqueness. We must show that, if ψ is an automorphism of Y over X carrying each τ_t to itself, then ψ is the identity. This is certainly true set-theoretically. In fact, the fibers of $Y \to X$ which do not consist of a single point are projective lines, and ψ fixes three points on each of them. That ψ is the identity morphism then follows Lemma (5.5). Q.E.D.

9. Vanishing cycles and the Picard–Lefschetz transformation.

In this section we shall discuss the topology of families of curves. Suppose that we are given a family of nodal curves parameterized by a complex space and that the general fiber of the family is smooth. We shall be concerned with the following question, which is the central one of Picard–Lefschetz theory: how is the topology of the family of smooth curves reflected in the nature of the singular fibers? For our purposes, this is a very relevant question. In fact, in the next chapters, when compactifying moduli spaces, we will often be able to analyze the nature of boundary points (e.g., their smoothness) only by "going around them."

The fundamental notion in this circle of ideas is the one of Picard–Lefschetz transformation, which we now introduce. Let

$$\alpha : \mathcal{C} \to B \quad \sigma_i : B \to \mathcal{C},\ i = 1, \dots, n, \tag{9.1}$$

be a family of smooth n-pointed genus g curves parameterized by a complex space B. Let

$$\eta : I = [0,1] \to B \tag{9.2}$$

be a path, and set $C_t = \alpha^{-1}(\eta(t))$, $\rho_i(t) = \sigma_i(\eta(t))$. Consider the pullback of $\mathcal{C}$ to I and take any topological trivialization of it

$$F : \eta^*\mathcal{C} \xrightarrow{\cong} C_0 \times I \tag{9.3}$$

as a family of n-pointed curves, that is, a trivialization which maps the sections ρ_i to constant sections. For each choice of $r, s \in I$, this trivialization defines a homeomorphism $F_{r,s} : C_r \to C_s$ between the fibers of $\eta^*\mathcal{C}$ at r and s carrying $\rho_i(r)$ to $\rho_i(s)$ for every i. In particular, we get a homeomorphism $F_{0,1} : C_0 \to C_1$. The isotopy class of $F_{0,1}$ relative to $(\rho_1(0), \dots, \rho_n(0))$, that is, the class modulo isotopies which do not move the image of $\rho_i(0)$ for every i, is independent of the choice of trivialization. In fact, if G is another trivialization of $\eta^*\mathcal{C}$, an isotopy between $F_{0,1}$ and $G_{0,1}$ is, for instance,

$$H_t = F_{t,1} \circ G_{1,t} \circ G_{0,1} .$$

The isotopy class of $F_{0,1}$ relative to $(\rho_1(0), \dots, \rho_n(0))$ is also independent of the path η, within each homotopy class with fixed endpoints. To see this, let $\vartheta : I \times I \to B$ be a homotopy with fixed endpoints between η and another path ξ; thus, $\vartheta(t, 0) = \eta(t)$, and $\vartheta(t, 1) = \xi(t)$. We can trivialize the pullback of $\mathcal{C}$ via ϑ as a family of n-pointed curves; by restriction this induces trivializations of the pullbacks via η and ξ. Trivializing $\vartheta^*\mathcal{C}$ determines a homeomorphism

$$M_{(t_1,s_1),(t_2,s_2)} : \alpha^{-1}(\vartheta(t_1, s_1)) \to \alpha^{-1}(\vartheta(t_2, s_2))$$

for any pair of points (t_1, s_1) and (t_2, s_2) of $I \times I$. Then

$$K_t = M_{(t,1),(1,1)} \circ M_{(t,0),(t,1)} \circ M_{(0,0),(t,0)}$$

is an isotopy between $M_{(0,1),(1,1)} \circ M_{(0,0),(0,1)}$ and $M_{(1,0),(1,1)} \circ M_{(0,0),(1,0)}$. On the other hand, by the independence on the choice of trivialization, $M_{(0,0),(0,1)}$ and $M_{(1,0),(1,1)}$ are, respectively, isotopic to the identity on C_0 and to the identity on C_1. Thus, $M_{(0,1),(1,1)}$ and $M_{(0,0),(1,0)}$, which are, respectively, the homeomorphism from C_0 to C_1 given by the trivialization of $\xi^*\mathcal{C}$ and the one given by the trivialization of $\eta^*\mathcal{C}$, are isotopic.

The most important instance of the above construction is the following. Let b_0 be a point of B, let η be a loop based at b_0, and set $C = \alpha^{-1}(b_0)$, $p_i = \sigma_i(b_0)$. Then the construction associates to the homotopy class with fixed endpoints of η the isotopy class relative to $(p_1, \dots, p_n)$ of an oriented homeomorphism of $(C; p_1, \dots, p_n)$ to itself, the so-called *Picard–Lefschetz transformation* associated to η.

To put this into perspective, we introduce the notion of mapping class group. If S is a compact, connected, and oriented topological surface, the *mapping class group* Γ_S, also called *Teichmüller modular group*, is the group of all isotopy classes of orientation-preserving homeomorphism of S to itself. More generally, if $q_1, \dots, q_n$ are distinct points of S, the *mapping class group of* $(S; q_1, \dots, q_n)$ is the group of all isotopy classes relative to $(q_1, \dots, q_n)$ of orientation-preserving homeomorphism of $(S; q_1, \dots, q_n)$ to itself, and we denote it by $\Gamma_{(S;q_1,\dots,q_n)}$, or simply by $\Gamma_{g,n}$, where g is the genus of S; we shall usually write Γ_g for $\Gamma_{g,0}$. Associating to the homotopy class of a loop based at b_0 the corresponding Picard–Lefschetz transformation thus yields a group homomorphism

$$PL : \pi_1(B, b_0) \to \Gamma_{(C;p_1,\dots,p_n)} \tag{9.4}$$

called the *Picard–Lefschetz representation.*

If we are given a homomorphism from the mapping class group $\Gamma_{(C;p_1,\dots,p_n)}$ to some other group G, the basic Picard–Lefschetz representation (9.4) induces a representation

$$\pi_1(B, b_0) \to G,$$

which often also goes under the name of Picard–Lefschetz representation. The premier example of this situation is the one in which G is the group of automorphisms of $H_1(C, R)$, where R is any coefficient group, and $\Gamma_C \to G$ sends φ to φ_*.

We now introduce certain basic elements of the mapping class group, called Dehn twists. Informally, the *Dehn twist* δ_c around a smooth simple closed curve c on an oriented smooth surface S is the homeomorphism (well defined up to isotopy) obtained by choosing an orientation on c, cutting the surface S along c, rotating the right edge c' of c by 180° in the positive direction, rotating the left edge c'' of c by 180° in the negative direction, and gluing the two edges together again.

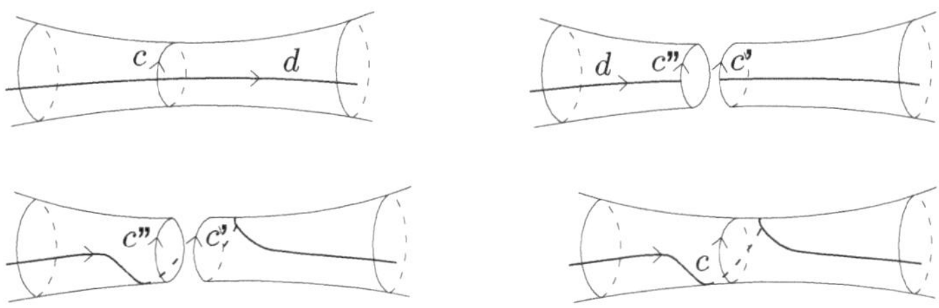

Figure 8. A Dehn twist

In formulas, δ_c can be defined as follows. Choose a tubular neighborhood U of c, and assume that it is parameterized by a positively oriented system of smooth coordinates y, ϑ, where $-1 < y < 1$, c is the locus $y = 0$, and the parameterization is periodic of period 2π in ϑ. Let ψ be a nonincreasing C^∞ function such that $\psi(x) = 0$ for all $x \le -1/2$, $\psi(0) = -1$, and $\psi(x) = -2$ for all $x \ge 1/2$. Then δ_c is defined to be the identity outside of U and

$$(y, \vartheta) \mapsto (y, \vartheta + \psi(y)\pi) \tag{9.5}$$

inside U. The definition of δ_c is clearly independent of the orientation we choose on c. It follows from formula (9.5) that δ_c is a diffeomorphism.

It is easy to check that the smooth isotopy class of δ_c does not depend on the choices made (i.e., on the choice of a tubular neighborhood, of its coordinatization, and of a function ψ). It is also important to notice, and equally easy to check, that if c is smoothly isotopic to another smooth simple closed curve $\tilde{c}$, then the Dehn twists attached to c and $\tilde{c}$ are smoothly isotopic. For details, we refer the reader to Chapter XV, where a more thorough discussion of the mapping class group can be found.

It is useful to record how Dehn twists act on homology. Looking at Fig. 8, it is immediate to see that, for a Dehn twist δ_c, the homomorphism $\delta_{c*} : H_1(S, \mathbb{Z}) \to H_1(S, \mathbb{Z})$ is given by

$$\delta_{c*}(d) = d + (d \cdot c)c\,, \tag{9.6}$$

where $(d \cdot c)$ is the intersection number of the two cycles d and c.

Dehn twists and Picard–Lefschetz transformation are intimately related through the concept of vanishing cycle, which we now introduce informally. The essential picture to have in mind is the one of a one-parameter family of smooth curves degenerating to a curve with a node p. Let $\pi : X \to \Delta$ be such a family, where $\Delta = \{t \in \mathbb{C} : |t| < R\}$, and the singular fiber is the one over the origin. We assume that X is smooth.

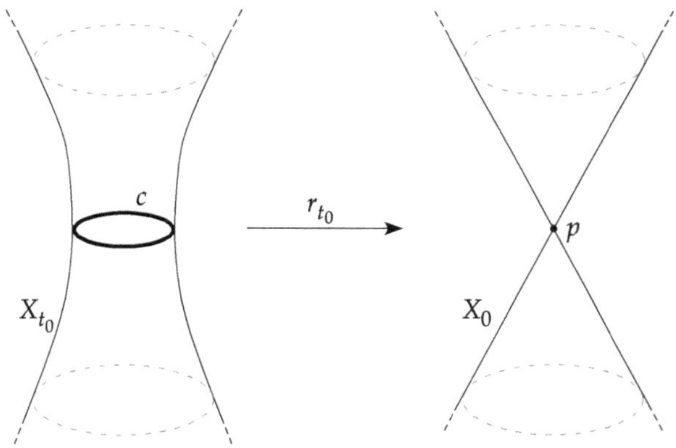

Figure 9.

The fundamental topological observations about this setup, whose proofs will be quickly given below, are the following. First of all, the central fiber X_0 is a deformation retract of X. Composing the retraction with the inclusion $X_{t_0} \hookrightarrow X$, where $t_0 \neq 0$, gives a map $r_{t_0} : X_{t_0} \to X_0$, and one can in fact set things up so that r_{t_0} is a homeomorphism everywhere except along a smoothly embedded circle c which gets collapsed to the node of X_0; one refers to c (or to any simple closed curve isotopic to it) as the vanishing cycle on X_{t_0}, relative to the given family. The second basic observation is that the Picard–Lefschetz representation

$$PL : \pi_1(\Delta \smallsetminus \{0\}, t_0) \to \Gamma_{X_{t_0}} \tag{9.7}$$

sends the positively oriented generator to the Dehn twist δ_c along the vanishing cycle. If X_0 is stable, or even just semistable, the vanishing cycle is homotopically nontrivial in X_{t_0}, and this easily implies that δ_c is also nontrivial. A remarkable consequence is that the restriction of $\pi : X \to \Delta$ to the punctured disk is topologically not a product family. Put otherwise, in general one cannot get rid of singular fibers because of purely topological reasons.

We now proceed to prove the two claims above. In proving the existence of a deformation retraction onto X_0 notice, first of all, that we may shrink Δ at will, since the family is differentiably locally trivial away from the central fiber, where all topological complications occur. We may then assume, possibly after shrinking the base of the family and rescaling the t coordinate, that there is an open neighborhood V of the node of the central fiber which is biholomorphic to $\{(x, y) \in \mathbb{C}^2 : |x| < 2, |y| < 2\}$ and that π is locally given by $t = xy$. Set

$$U = \{(x, y) \in V : |xy| < 1, \big||x|^2 - |y|^2\big| < 1\}.$$

The locus $|x| = |y|$ is fibered in circles above Δ, except over the origin, where the fiber is a single point. It will turn that these shrinking circles are the vanishing cycles in their respective fibers; the name "vanishing cycle" originates from this picture. The complement of $|x| = |y|$ in U decomposes in the two connected components $U_+ = \{|x| > |y|\}$ and $U_- = \{|y| > |x|\}$. We shall express each one of these as a product of $\{|t| < 1\}$ with a suitable open set in the central fiber. Explicitly, a diffeomorphism between $U_+ \cup U_-$ and $(X_0 \cap (U_+ \cup U_-)) \times \{|t| < 1\}$ is given by the prescription

$$\begin{aligned}(9.8)\qquad &(x, y) \mapsto \left(\frac{x}{|x|}(|x|^2 - |y|^2), 0, xy\right) \quad \text{when } (x, y) \in U_+\,,\\ &(x, y) \mapsto \left(0, \frac{y}{|y|}(|y|^2 - |x|^2), xy\right) \quad \text{when } (x, y) \in U_-\,.\end{aligned}$$

An inverse to the first of these transformations is given by

$$(\xi, 0, t) \mapsto (x, y)\,,$$

where

$$\begin{aligned}(9.9)\qquad x &= \frac{\xi}{|\xi|}\sqrt{\frac{|\xi| + \sqrt{|\xi|^2 + 4|t|^2}}{2}}\,,\\ y &= t/x\,,\end{aligned}$$

and a similar expression is valid for the second one. Write $t = \sigma + \sqrt{-1}\tau$. The product representation (9.8) provides canonical liftings u_0 and v_0 of the vector fields $\partial/\partial\sigma$ and $\partial/\partial\tau$. By a partition of unity argument, we can in fact find liftings u and v to all of X which agree with u_0 and v_0, respectively, on a neighborhood of $\{|x| = |y|\}$. Integrating u and v yields a diffeomorphism $H : (X \smallsetminus \{|x| = |y|\}) \to (X_0 \smallsetminus \{p\}) \times \{|t| < 1\}$ of fiber spaces over $\{|t| < 1\}$.

The trivialization we just described gives in particular a deformation retraction $\{h_s\}_{s\in[0,1]}$ of $X \smallsetminus \{|x| = |y|\}$ to $X_0 \smallsetminus \{p\}$ lifting the radial deformation retraction of Δ to the origin. As one can easily check using the coordinate change (9.8) and its inverse (9.9), the retraction can be continuously extended to all of X by setting $h_s(x,y) = \sqrt{s}(x,y)$ for $|x| = |y|$. This concludes the proof that X_0 is a deformation retract of X.

We now show that the Picard–Lefschetz transformation attached to the positively oriented generator of $\pi_1(\Delta \smallsetminus \{0\}, t_0)$ is just the Dehn twist δ_c along the vanishing cycle. It is clear that we may assume that t_0 is as close to 0 as we wish. Taking as generator of $\pi_1(\Delta \smallsetminus \{0\}, t_0)$ the loop

$$\lambda(\theta) = t_0 e^{2\pi\sqrt{-1}\theta}, \quad \theta \in I = [0,1],$$

we shall describe, for each $\theta \in I$, a specific homeomorphism k_θ, depending continuously on θ, between X_{t_0} and $X_{\lambda(\theta)}$. To this end, we choose an odd nondecreasing C^∞ function χ such that $\chi(s) = -1$ for all $s \le -1/2$ and $\chi(s) = 1$ for all $s \ge 1/2$. Outside the region U, k_θ is simply defined to be the map given by the trivialization H described above. Inside U, instead, we set

$$k_\theta(x,y) = \left(e^{\sqrt{-1}\pi(1+\chi(s))\theta}x, e^{\sqrt{-1}\pi(1-\chi(s))\theta}y\right), \tag{9.10}$$

where s is essentially the logarithm of $|x|$, or, more exactly,

$$s = \frac{\log(|x|/\sqrt{|t_0|})}{\log(1/\sqrt{|t_0|})}.$$

In terms of the coordinates s and $\varphi = \arg x$, the map k_1 is given by

$$k_1(s,\varphi) = (s, \varphi + (1+\chi(s))\pi), \tag{9.11}$$

while it is the identity away from the vicinity of the vanishing cycle. By formula (9.5) in Chapter XV, this says that k_1 is the Dehn twist associated to the vanishing cycle c, as was to be shown.

Another useful way of visualizing a curve acquiring a node and the corresponding vanishing cycle, is to look at the map

$$\begin{aligned}\varphi : V_t = \{(x,y,t) : xy = t, |x| < 1, |y| < 1\} &\longrightarrow \{z : |z| < 2\},\\ (x,y,t) &\longmapsto x + y,\end{aligned}$$

which exhibits V_t as a two-sheeted ramified cover branched at the points $\pm 2\sqrt{t}$. In fact, φ is the quotient map via the involution $(x, y, t) \mapsto (y, x, t)$. The vanishing cycle is the preimage, under the covering map, of the segment $[-2\sqrt{t}, 2\sqrt{t}]$. This segment, and hence the vanishing cycle above it, shrinks to a point as $t \to 0$.

As we observed, the nontriviality of the Picard–Lefschetz representation makes it impossible to replace the central fiber of $\pi : X \to \Delta$ with a topological surface. On the other hand, nothing prevents us from performing this operation if we restrict to a real ray issuing from the origin. The new fiber over the origin is obtained by normalizing at the node, substituting each of the points mapping to the node with a circle and gluing the two circles together; the specific gluing is dictated by the direction of the ray. This dependence on the ray is the obstruction to performing the construction over all of Δ. On the other hand, this observation suggests that the obstruction disappears if one substitutes the origin of Δ with a circle parameterizing all real directions through it. This calls for a digression on real blow-ups, which will come in handy later on, when we will need to work with a bordification of Teichmüller space while presenting Kontsevich's proof of the Witten conjecture.

Real blow-ups.

We only need to introduce real blow-ups in a special situation. Let X be a complex manifold, and let $D \subset X$ be a reduced divisor with normal crossings. As the reader will recall, this means that, for any point p of X, there are a neighborhood U of p and local coordinates $t_1, \dots, t_n$ on U, centered at p, such that $D \cap U$ is of the form $t_1 t_2 \cdots t_k = 0$, with k possibly equal to 0. We are going to define the *real (oriented) blow-up* of X along D, which we will denote $\mathrm{Bl}_D(X)$. We shall do the construction locally and then show how the various local patches fit together. Let p, U, and $t_1, \dots, t_n$ be as above. We then define $\mathrm{Bl}_D(U) \subset U \times (S^1)^k$ to be the locus

$$\mathrm{Bl}_D(U) = \{(x, \tau_1, \dots, \tau_k) : t_i(x) = |t_i(x)|\tau_i, \ \ i = 1, \dots, k\},$$

where we view S^1 as the unit circle in the complex plane. Equivalently, one can describe $\mathrm{Bl}_D(U)$ as the region

$$\mathrm{Re}(t_i \tau_i^{-1}) \geq 0, \qquad i = 1, \dots, k,$$

inside the locus with equations

$$\mathrm{Im}(t_i \tau_i^{-1}) = 0, \qquad i = 1, \dots, k. \tag{9.12}$$

Using the Jacobian criterion, it is easy to check that equations (9.12) define a real-analytic submanifold of $U \times (S^1)^k$ and that the projection

$$\pi_U : \mathrm{Bl}_D(U) \to U$$

is a real-analytic isomorphism away from $\pi_U^{-1}(D \cap U)$. The structure of $\mathrm{Bl}_D(U)$ can be made even more explicit if U is the polycylinder $\{|t_i| < \varepsilon,\ i = 1, \dots, n\}$; we can always reduce to this case by shrinking U. Then $\mathrm{Bl}_D(U)$ is real-analytically isomorphic to the product of the real k-torus $(S^1)^k$, of k copies of the half-closed interval $[0, \varepsilon)$, and $n-k$ ε-disks. In fact, writing Δ_ε for the ε-disk centered at the origin of the complex plane,

$$(S^1)^k \times [0,\varepsilon)^k \times \Delta_\varepsilon^{n-k} \longrightarrow \mathrm{Bl}_D(U),$$
$$(\tau_1, \dots, \tau_k, r_1, \dots, r_k, t_{k+1}, \dots, t_n) \mapsto (r_1\tau_1, \dots, r_k\tau_k, t_{k+1}, \dots, t_n, \tau_1, \dots, \tau_k),$$

is a real-analytic isomorphism whose inverse is

$$\mathrm{Bl}_D(U) \longrightarrow (S^1)^k \times [0,\varepsilon)^k \times \Delta_\varepsilon^{n-k},$$
$$(t_1, \dots, t_n, \tau_1, \dots, \tau_k) \mapsto (\tau_1, \dots, \tau_k, |t_1|, \dots, |t_k|, t_{k+1}, \dots, t_n).$$

Notice that these isomorphisms are compatible with the natural projection $\mathrm{Bl}_D(U) \to U = \Delta_\varepsilon^n$ and with the "polar coordinates" map $(S^1)^k \times [0,\varepsilon)^k \times \Delta_\varepsilon^{n-k} \to \Delta_\varepsilon^n$ given by $(\tau_1, \dots, \tau_k, r_1, \dots, r_k, t_{k+1}, \dots, t_n) \mapsto (r_1\tau_1, \dots, r_k\tau_k, t_{k+1}, \dots, t_n)$. It is clear from this description that the fiber of π_U at a point q of U is an h-dimensional real torus, where h is the number of coordinates t_i, $i = 1, \dots, k$, vanishing at q. In particular, the fiber over the central point p is a k-dimensional real torus

$$\mathbb{T} = \pi_U^{-1}(p) \cong (S^1)^k. \tag{9.13}$$

We next check the independence of the definition of real blow-up on the choice of coordinates. Let q be a point of U, and $V \subset U$ a neighborhood of q. Pick coordinates $s_1, \dots, s_n$ on V in such a way that $D \cap V = \{s_1 \cdots s_h = 0\}$, where $h \le k$. Possibly after renumbering, there are nowhere vanishing holomorphic functions $a_1, \dots, a_h$ on V such that $s_i = a_i t_i$, $i = 1, \dots, h$. We define a real-analytic map $\alpha : \mathrm{Bl}_D(V) \to \mathrm{Bl}_D(U)$ by setting

$$\alpha(x, \sigma_1, \dots, \sigma_h) = \left(x, \frac{|a_1(x)|}{a_1(x)}\sigma_1, \dots, \frac{|a_h(x)|}{a_h(x)}\sigma_h, \frac{t_{h+1}(x)}{|t_{h+1}(x)|}, \dots, \frac{t_k(x)}{|t_k(x)|} \right).$$

Clearly, $\pi_U \circ \alpha = \pi_V$, and α is the only continuous map with this property. Moreover, α has a real-analytic inverse on $\alpha(\mathrm{Bl}_D(V))$ given by

$$(x, \tau_1, \dots, \tau_k) \mapsto \left(x, \frac{a_1(x)}{|a_1(x)|}\tau_1, \dots, \frac{a_h(x)}{|a_h(x)|}\tau_h \right).$$

The existence and uniqueness of α imply that the various local blow-ups $\mathrm{Bl}_D(U)$ patch together to yield a well-defined real-analytic manifold with corners $\mathrm{Bl}_D(X)$, together with a projection $\pi : \mathrm{Bl}_D(X) \to X$. It is clear from the construction that the fiber of π at a point $x \in X$ is a real k-dimensional torus, where k is the number of components of the germ of D at x.

Let B be the polydisc $\{(t_1, \dots, t_n) \in \mathbb{C}^n : |t_i| < \varepsilon,\ i = 1, \dots, n\}$, and let D be the divisor $t_1 \cdots t_k = 0$. Using polar coordinates, it is easy to describe the universal cover $\widetilde{B}$ of $\mathrm{Bl}_D(B)$. In fact,

$$\widetilde{B} = \{(r_1, \dots, r_k, \vartheta_1, \dots, \vartheta_k, t_{k+1}, \dots, t_n) \in \mathbb{R}^{2k} \times \mathbb{C}^{n-k} : 0 \le r_i \le \varepsilon, |t_i| \le \varepsilon\},$$

and the covering map $\rho : \widetilde{B} \to \mathrm{Bl}_D(B)$ is

$$(r_1, \dots, r_k, \vartheta_1, \dots, \vartheta_k, t_{k+1}, \dots, t_n) \mapsto \\ (r_1 e^{2\pi\sqrt{-1}\vartheta_1}, \dots, r_k e^{2\pi\sqrt{-1}\vartheta_k}, t_{k+1}, \dots, t_n, e^{2\pi\sqrt{-1}\vartheta_1}, \dots, e^{2\pi\sqrt{-1}\vartheta_k}).$$

Having dealt with real blow-ups, we turn to families of nodal curves. First, we consider the case of a single nodal curve C of genus g, and we let

$$C_{\text{sing}} = \{p_1, \dots, p_k\}.$$

We denote by $\nu : N \to C$ the normalization map, and we set

$$\nu^{-1}(p_i) = \{r_i, s_i\},\ i = 1, \dots, k \qquad \text{and} \quad R = \{r_1, s_1, \dots, r_k, s_k\}.$$

Consider the real oriented blow-up of N along R,

$$\tau : \mathrm{Bl}_R(N) \longrightarrow N.$$

By definition, the fiber of τ over a point $r \in R$ is a copy of S^1. Hence, $\mathrm{Bl}_R(N)$ is a, possibly disconnected, Riemann surface with boundary (see Fig. 10 below). We may construct an oriented topological surface Σ of genus g by giving, for each point $y \in C_{\text{sing}}$, an identification between the two boundary components $\tau^{-1}(r) \cong S^1$ and $\tau^{-1}(s) \cong S^1$, where $\{r, s\} = \nu^{-1}(y)$. Of course, in order to get an oriented surface, this identification should carry the orientation of $\tau^{-1}(r)$ into the opposite orientation of $\tau^{-1}(s)$. By construction, the surface Σ comes equipped with a system of simple closed curves $\{\gamma_1, \dots, \gamma_k\}$. We denote by

$$h : \mathrm{Bl}_R(N) \to \Sigma$$

the quotient map and by

$$\xi : \Sigma \to C \tag{9.14}$$

the natural contraction map defined by the property: $\xi h = \nu\tau$.

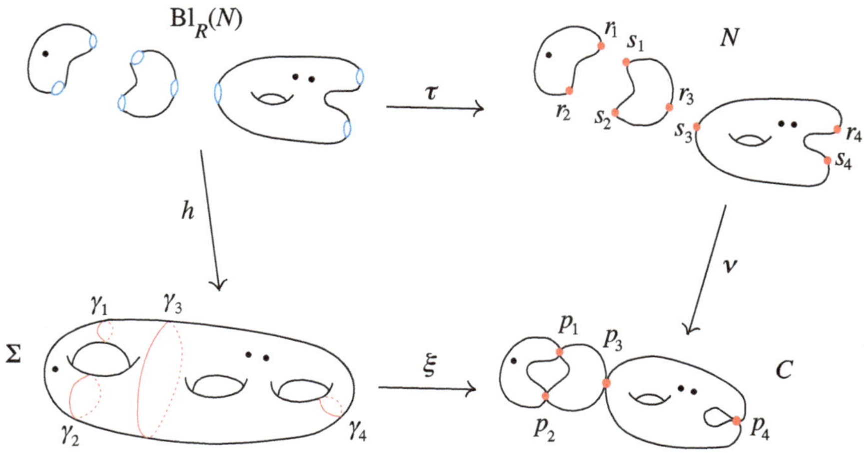

Figure 10.

Given a family $f : X \to B$ of nodal curves, we would like to perform the same operation simultaneously on all singular fibers of f. We should of course expect this process to destroy complex analyticity. Worse still, since the gluing of the circles $\tau^{-1}(r)$ and $\tau^{-1}(s)$ is noncanonical, we will be forced to base-change from B to $\mathrm{Bl}_D(B)$.

In carrying out this procedure, we shall limit ourselves to families $f : X \to B$ which are *transverse*, in the following sense. Let b be any point of B, and let $p_1, \ldots, p_k$ be the nodes of the fiber $f^{-1}(b)$. Locally near p_i, the family is of the form $xy = t_i$, where t_i is a function on a neighborhood of b. When the functions $t_1, \ldots, t_k$ are part of a system of local coordinates $t_1, \ldots, t_\ell$ on a neighborhood of b, we shall say that the family $f : X \to B$ is *transverse at* b and that $t_1, \ldots, t_\ell$ is a *standard system of parameters*. A family will be said to be transverse if it is transverse at every point of the base. Notice that the base and the general fiber of a transverse family are automatically smooth. As we shall see, when we discuss moduli spaces of stable curves in Chapter XII, the attribute "transverse" is justified by the fact that families of this sort give rise to "subvarieties" of moduli space which are transverse to all strata of the stratification of moduli space by topological type.

Let then $f : X \to B$ be a transverse family of nodal curves. We denote by D the locus in B parameterizing singular curves. Clearly, in a neighborhood of a point $b \in B$, a local equation for D is $t_1 \cdots t_k = 0$, where $t_1, \cdots, t_\ell$ is a standard system of parameters at b. Hence D is a divisor with normal crossings. We will define a new family

$$h : Z \to \mathrm{Bl}_D(B)$$

of compact orientable differentiable surfaces which agrees with f away from D. As was the case for real blow-ups, it will suffice to do the construction locally on B; we leave it to the reader to check that the local families we obtain patch together. We then assume that B is the polydisc $\{(t_1,\dots,t_\ell) : |t_i| < \varepsilon^2,\ i = 1,\dots,\ell\}$, that the fiber $f^{-1}(0)$ has k nodes $p_1,\dots,p_k$, that the family is of the form $xy = t_i$ in a neighborhood U_i of p_i, and that it is smooth outside the U_i. Thus D is the locus $t_1 \cdots t_k = 0$. We shall construct $Z \to \mathrm{Bl}_D(B)$ by suitably modifying the family $f' : X' \to \mathrm{Bl}_D(B)$ obtained by pullback from $f : X \to B$. To keep notation simple, we will assume that $k = 1$, and we will write t for t_1 and p for p_1. The real blow-up of B is the locus in $B \times S^1$ with equation $t = |t|\tau$. On the other hand, X is locally isomorphic, near p, to the locus with equation $xy = t$ in $\mathbb{C}^2 \times B$, and hence X' is locally isomorphic to a neighborhood of $\{x = y = 0\}$ in

$$W' = \left\{(x, y, t_2, \dots, t_\ell, \tau) \in \mathbb{C}^{\ell+1} \times S^1 \,\middle|\, \begin{matrix} xy = |xy|\tau,\ |x| < \varepsilon, \\ |y| < \varepsilon, |t_i| < \varepsilon^2 \end{matrix} \right\},$$

whereas the projection of X' to $\mathrm{Bl}_D(B)$ can be identified with

$$(x, y, t_2, \dots, t_\ell, \tau) \mapsto (xy, t_2, \dots, t_\ell, \tau)\,.$$

Now look at

(9.15)
$$W = \left\{(x, y, t_2, \dots, t_\ell, \xi, \eta) \in \mathbb{C}^{\ell+1} \times (S^1)^2 \,\middle|\, \begin{matrix} x = |x|\xi,\ y = |y|\eta, \\ |x| < \varepsilon, |y| < \varepsilon, |t_i| < \varepsilon^2 \end{matrix} \right\}.$$

Since $|xy|\xi\eta = xy$,

$$(x, y, t_2, \dots, t_\ell, \xi, \eta) \mapsto (x, y, t_2, \dots, t_\ell, \xi\eta)$$

defines a map β from W to W'. The map is one-to-one except where $x = y = 0$; the fiber above $(0, 0, t_2, \dots, t_\ell, \tau)$ is a circle, namely the locus of those points $(0, 0, t_2, \dots, t_\ell, \xi, \eta)$ such that $\xi\eta = \tau$. To construct Z, we just graft onto $X' \smallsetminus \{x = y = 0\}$, via the attaching map $\beta_{|\{x\neq 0, y\neq 0\}}$, a neighborhood of the locus $\{x = y = 0\}$ in W. This procedure does to a singular fiber of f exactly what we did above to a single nodal curve. First one normalizes the fiber at the node $x = y = 0$, then one performs real blow-ups at the two resulting points of the normalized curve, and finally one identifies the two circles thus obtained, keeping track of orientations; how the identification is made depends on the value of the angle parameter τ. When the number of nodes of $f^{-1}(0)$ is greater than 1, all one has to do to construct $h : Z \to \mathrm{Bl}_D(B)$ is to perform the procedure described above at every node. By construction, the fibers of h are all homeomorphic topological surfaces.

The composite map $W \to W' \to \mathrm{Bl}_D(B)$ is not a smooth fibration, but we are going to show that one can put on Z a structure of real-analytic manifold with corners in such a way that $Z \to \mathrm{Bl}_D(B)$ is a real-analytic fibration. While this is not strictly necessary in the sequel, it does simplify certain proofs. In the considerations that follow, for the sake of simplicity, we ignore the variables $t_2, \dots, t_\ell$, which behave as harmless parameters. In $Z \smallsetminus \{x = y = 0\}$, we keep the real-analytic structure coming from the one of X'. Inside W, we proceed as follows. Recall that, in our case,

$$W = \left\{ (x, y, \xi, \eta) \,\middle|\, \begin{matrix} x = |x|\xi,\ y = |y|\eta \\ |x| < \varepsilon,\ |y| < \varepsilon \end{matrix} \right\} \subset \mathbb{C}^2 \times (S^1)^2.$$

In other words, W is just the product of two blown-up disks. Set

$$V = \{(s, r, \xi, \eta) : s, r \in \mathbb{R},\ \xi, \eta \in S^1,\ 0 \le r < \varepsilon^2,\ |s| < \varepsilon - r/\varepsilon\}.$$

Define the map $\varphi : W \to V$ by

$$(x, y, \xi, \eta) \mapsto (|x| - |y|, |xy|, \xi, \eta)$$

and the map $\psi : V \to W$ by

$$(s, r, \xi, \eta) \mapsto \left(\frac{s + \sqrt{s^2 + 4r}}{2}\xi, \frac{-s + \sqrt{s^2 + 4r}}{2}\eta, \xi, \eta \right).$$

We leave to the reader the easy verification that these maps are inverse to each other. We transplant to W the real-analytic structure of V, and we check that it is compatible with the one of $Z \smallsetminus \{x = y = 0\}$. One immediately sees that the map

$$(x, r, \eta) \mapsto \left(x, \frac{r\eta}{|x|}, \frac{x}{|x|}, \eta \right)$$

gives a real-analytic diffeomorphism from

$$U = \{(x, r, \eta) \in \mathbb{C} \times \mathbb{R} \times S^1 : 0 \le r < \varepsilon^2,\ r/\varepsilon < |x| < \varepsilon\}$$

to $W \smallsetminus \{x = 0\}$, endowed with the real-analytic structure induced by the one of $Z \smallsetminus \{x = y = 0\}$. It is equally easy to see that the two maps

$$(x, r, \eta) \mapsto \left(|x| - \frac{r}{|x|}, r, \frac{x}{|x|}, \eta \right),$$

$$(s, r, \xi, \eta) \mapsto \left(\frac{s + \sqrt{s^2 + 4r}}{2}\xi, r, \eta \right)$$

between U and $V \smallsetminus \varphi(\{x=0\})$ are inverse of each other. They are also real-analytic. In fact, the only point to worry about is the possibility of s^2+4r vanishing; but this occurs only when $s=r=0$, that is, for $x=y=0$. One similarly deals with the open set $W \smallsetminus \{y=0\}$. This completes the construction of the real-analytic structure on Z. It is obvious from the definitions that the map $h: Z \to \mathrm{Bl}_D(B)$ is a real-analytic fibration. Its fibers are compact orientable real-analytic surfaces. We may summarize much of what has been proved in the following statement.

PROPOSITION (9.16). *Let $f: X \to B$ be a transverse family of nodal curves, and let $D \subset B$ be the locus parameterizing singular fibers. Then there are a real-analytic fibration $h: Z \to \mathrm{Bl}_D(B)$ and a surjective continuous map $\lambda: Z \to X \times_B \mathrm{Bl}_D(B)$ such that the diagram*

$$\begin{array}{ccccc} Z & \xrightarrow{\lambda} & X \times_B \mathrm{Bl}_D(B) & \longrightarrow & X \\ {\scriptstyle h}\downarrow & & \downarrow & & \downarrow{\scriptstyle f} \\ \mathrm{Bl}_D(B) & \xrightarrow{\mathrm{id}} & \mathrm{Bl}_D(B) & \xrightarrow{\pi} & B \end{array}$$

is commutative. Furthermore, denoting by $\Sigma \subset X$ the locus of singular points in the fibers of f, the map λ restricts to a real-analytic diffeomorphism $Z \smallsetminus \lambda^{-1}(\Sigma \times_B \mathrm{Bl}_D(B)) \to X \smallsetminus \Sigma \times_B \mathrm{Bl}_D(B)$, and $\lambda^{-1}(\Sigma \times_B \mathrm{Bl}_D(B))$ is a real-analytic submanifold of Z, fibered in circles over $\Sigma \times_B \mathrm{Bl}_D(B)$.

REMARK (9.17). In fact, the construction of h gives more than (9.16) says. It is particularly important to keep track of the detailed structure of Z above the loci of the form $\pi^{-1}(b_0)$, where b_0 is a point of B. Of course, this is of interest only when b_0 belongs to D, that is, when $X_{b_0} = f^{-1}(b_0)$ is singular. Let $p_1, \dots, p_k$ be the nodes of X_{b_0}, and let x_i, y_i be local coordinates on the two branches of X_{b_0} at p_i, centered at the singular point. The locus $\pi^{-1}(b_0)$ is a real k-torus and can be viewed as the set of k-tuples $\tau = (\tau_1, \dots, \tau_k)$ of complex numbers of absolute value one. The fiber $h^{-1}(\tau_1, \dots, \tau_k)$ is constructed as follows. The first step is to normalize X_{b_0} to obtain a smooth curve Y. We then perform real blow-ups at all point of Y mapping to nodes of X_{b_0}. In other words, we replace ε-neighborhoods of the origin in the coordinates x_i and y_i with the manifolds with boundary $M_i = \{(x_i, \xi_i) \in \mathbb{C}^2 : |x_i| < \varepsilon, |\xi_i| = 1, x_i = |x_i|\xi_i\}$ and $N_i = \{(y_i, \eta_i) \in \mathbb{C}^2 : |y_i| < \varepsilon, |\eta_i| = 1, y_i = |y_i|\eta_i\}$, respectively. The final step is to glue together the boundaries of M_i and N_i, for each i, by identifying $(0, \xi_i)$ and $(0, \eta_i)$ when $\xi_i \eta_i = \tau_i$. One thus obtains a topological surface Y_τ. We denote by A_i the union of the images in Y_τ of M_i and N_i, and by γ_i the common image of their boundaries, a circle. Loosely speaking, one can say that Y_τ has been obtained from X_{b_0} by replacing the node p_i with the circle γ_i for each i. The fiber of h at the point of $\pi^{-1}(b_0)$ corresponding to τ is Y_τ, endowed with a suitable real-analytic structure. This structure coincides with the one of X_{b_0} away from the circles γ_i, that is, from the point of view of X_{b_0}, away from the nodes. In A_i, the real-analytic structure is obtained by transplanting the natural one of $(-\varepsilon, \varepsilon) \times S^1$ via the homeomorphism $(-\varepsilon, \varepsilon) \times S^1 \to A_i$ given by

$$\begin{aligned} &(s, \rho) \mapsto (s\rho, \rho) \in M_i && \text{when } s \geq 0, \\ &(s, \rho) \mapsto (-s\tau_i\rho^{-1}, \tau_i\rho^{-1}) \in N_i && \text{when } s \leq 0. \end{aligned}$$

COROLLARY (9.18). *Let $f : X \to B$, Z, h, and π be as in (9.16). Let b_0 be a point of B, set $\mathbb{T} = \pi^{-1}(b_0)$, and let k and ℓ be, respectively, the number of nodes of $f^{-1}(b_0)$ and the dimension of B at b_0. Denote by Δ the unit disk in the complex plane. Then there are arbitrarily small open neighborhoods U of b_0 such that there exist diffeomorphisms $\alpha : \pi^{-1}(U) \to \mathbb{T} \times [0,1)^k \times \Delta^{\ell-k}$ and $\beta : (\pi \circ h)^{-1}(U) \to h^{-1}(\mathbb{T}) \times [0,1)^k \times \Delta^{\ell-k}$ with the property that the diagram*

$$\begin{array}{ccc} (\pi \circ h)^{-1}(U) & \xrightarrow{\ \beta\ } & h^{-1}(\mathbb{T}) \times [0,1)^k \times \Delta^{\ell-k} \\ \Big\downarrow h & & \Big\downarrow h \times \mathrm{id} \times \mathrm{id} \\ \pi^{-1}(U) & \xrightarrow{\ \alpha\ } & \mathbb{T} \times [0,1)^k \times \Delta^{\ell-k} \end{array}$$

is commutative. In particular, there are compatible deformation retractions or $\pi^{-1}(U)$ to $\mathbb{T}$ and of $(\pi \circ h)^{-1}(U)$ to $h^{-1}(\mathbb{T})$. Furthermore, α, β, and the retractions can be chosen so that the latter descend to compatible deformation retractions of U to b_0 and of $f^{-1}(U)$ to the fiber $f^{-1}(b_0)$. Finally, if $\sigma_1, \dots, \sigma_n$ are sections of $f : X \to B$ making it into a family of n-pointed nodal curves, the product decomposition of $(\pi \circ h)^{-1}(U)$ and the retractions can be chosen so as to be compatible with the sections $\sigma_1, \dots, \sigma_n$.

For the proof, we may suppose that B is the polydisc $\{(t_1, \dots, t_\ell) \in \mathbb{C}^\ell : |t_i| < 1, i = 1, \dots, \ell\}$, that b_0 is the origin of $\mathbb{C}^\ell$, and that D is the locus $t_1 \cdots t_k = 0$. We may also take $U = B$, so that $\pi^{-1}(U) = \mathrm{Bl}_D(B)$ and $(\pi \circ h)^{-1}(U) = Z$. We know that $\mathrm{Bl}_D(B)$ is real-analytically diffeomorphic to the product of $\mathbb{T}$ and $[0,1)^k \times \Delta^{\ell-k}$, where $\mathbb{T}$ is the set of all k-tuples $(\tau_1, \dots, \tau_k)$ of complex numbers of absolute value one. We indicate the ith coordinate in $[0,1)^k$ by r_i. As in the statement of (9.16), we denote by Σ the locus of singular points in the fibers of f and by λ the natural map from Z to $X \times_B \mathrm{Bl}_D(B)$. Clearly, $\lambda^{-1}(\Sigma)$ decomposes into connected components $F_1, \dots, F_k$, and F_i is fibered in circles over E_i, the locus in $\mathrm{Bl}_D(B) \cong \mathbb{T} \times [0,1)^k \times \Delta^{\ell-k}$ defined by the vanishing of r_i. The required diffeomorphism between Z and $h^{-1}(\mathbb{T}) \times [0,1)^k \times \Delta^{\ell-k}$ can be obtained by lifting to Z the vector fields on $\mathbb{T} \times [0,1)^k \times \Delta^{\ell-k}$ corresponding to differentiation in the t_i and r_i variables, and successively integrating them. A deformation retraction of $Z \cong h^{-1}(\mathbb{T}) \times [0,1)^k \times \Delta^{\ell-k}$ to $h^{-1}(\mathbb{T})$ is then, for instance, $\rho_s(x,r,t) = (x, sr, st)$, $0 \leq s \leq 1$; this lifts a similarly defined deformation retraction of $\mathrm{Bl}_D(B)$ to $\mathbb{T}$, which in turn lifts the radial deformation retraction of B to b_0. The trouble is that, in general, this does not descend to a deformation retraction of X, since ρ_s does not necessarily map each submanifold F_i to itself. In other words, F_i does not necessarily correspond, in $h^{-1}(\mathbb{T}) \times [0,1)^k \times \Delta^{\ell-k}$, to the product of a submanifold of $h^{-1}(\mathbb{T})$ and of the locus $r_i = 0$ in $[0,1)^k \times \Delta^{\ell-k}$. The remedy is to be more careful in the choice of the liftings of the coordinate vector fields. More exactly, we choose the liftings of $\partial/\partial t_i$ and $\partial/\partial \bar{t}_i$ to be everywhere tangent to all the F_j, and the lifting of $\partial/\partial r_i$ to be everywhere tangent to all the F_j except F_i. As the reader will readily verify, this yields a product decomposition of Z in which all the F_i appear as products of submanifolds of the factors. This in turn leads to a deformation retraction of Z to $h^{-1}(\mathbb{T})$ which descends to one of X to $f^{-1}(b_0)$. Finally, to make the product decomposition of $(\pi \circ h)^{-1}(U)$ and the deformation retractions compatible with the given sections, it suffices to choose the liftings of the coordinate vector field in such a way that they are tangent to the sections.

Picard–Lefschetz revisited.

We now go back to the Picard–Lefschetz transformation. Our goal is to study it in the context of a general transverse family of nodal curves $f : X \to B$, using the constructions we have just introduced. For future reference, we wish to highlight the main consequence of Corollary (9.18).

LEMMA (9.19). *Let $f : X \to B$ be a transverse family of nodal curves. Let b_0 be a point of B. Set $C_0 = f^{-1}(b_0)$. Then there are arbitrary small neighbourhoods U of b_0 such that there exist compatible retractions $r : U \to b_0$ and $R : X_{|U} \to C_0$. Moreover, if $\sigma_1, \dots, \sigma_n$ are sections of $f : X \to B$ making it into a family of n-pointed nodal curves, the retractions can be chosen so as to be compatible with the sections $\sigma_1, \dots, \sigma_n$.*

If $p_1, \dots, p_k$ are the nodes of $C_0 = f^{-1}(b_0)$, and $C = f^{-1}(b)$, $b \in U$, is a smooth fiber, then $R^{-1}(p_i) \cap C$ is a circle c_i. Going back to Corollary (9.18), an alternative description of c_i is as follows. There is a unique point of $\pi^{-1}(U)$ mapping to b. Suppose that, under the diffeomorphism $\pi^{-1}(U) \to \mathbb{T} \times [0,1)^k \times \Delta^{\ell-k}$, this point corresponds to $(\tau_1, \dots, \tau_k, r_1, \dots, r_k, t_{k+1}, \dots, t_\ell)$. Then the diffeomorphism $(\pi \circ h)^{-1}(U) \to h^{-1}(\mathbb{T}) \times [0,1)^k \times \Delta^{\ell-k}$ maps c_i to $\gamma_i \times \{(r_1, \dots, r_k, t_{k+1}, \dots, t_\ell)\}$, where γ_i is the fiber over the node p_i of the contraction map $h^{-1}(\tau_1, \dots, \tau_k, 0, \dots, 0) \to f^{-1}(b_0)$ (see also Remark (9.17), where Y_τ, $\tau = (\tau_1, \dots, \tau_k)$, corresponds to $h^{-1}(\tau_1, \dots, \tau_k, 0, \dots, 0)$).

In accordance with the discussion at the beginning of this section, by *vanishing cycle* on C we mean (any simple closed curve isotopic to) one of the circles $c_1, \dots, c_k$. Occasionally, we will use the words "vanishing cycle" also to designate the homology class of one of the c_i.

As we know, vanishing cycles and Dehn twists are closely linked. We wish to go over this link again, in the standard situation of a transverse family of nodal curves

$$f : X \to B = \{(t_1, \dots, t_\ell) : |t_i| < 1, i = 1, \dots, \ell\}, \tag{9.20}$$

where $t_1, \dots, t_\ell$ is a standard system of parameters. Thus, if $p_1, \dots, p_k$ are the singular points of the central fiber X_0, then, near p_i, X is of the form $xy = t_i$. The singular fibers of the family are parameterized by $D = \{t_1 \cdots t_k = 0\} \subset B$, and hence by restriction to $B^* = B \setminus D$ we get a family of smooth curves $X^* \to B^*$. We fix a base point $b_0 \in B^*$, set $C = X_{b_0}$, and look at the Picard–Lefschetz representation

$$PL : \pi_1(B^*, b_0) \to \Gamma_C \, .$$

The fundamental group of B^* is abelian and freely generated by k loops $\lambda_1, \dots, \lambda_k$, where each λ_i winds simply around the divisor $t_i = 0$ in the positive direction. In other words, using additive notation,

$$\pi_1(B^*, b_0) = \mathbb{Z}\lambda_1 \oplus \cdots \oplus \mathbb{Z}\lambda_k \, .$$

For each singular point p_i of the central fiber of f, let c_i be the corresponding vanishing cycle in C. Let us show again that the Picard–Lefschetz representation assigns to each λ_i the Dehn twist around c_i, i.e., that

$$PL(\lambda_i) = \delta_{c_i} \, . \tag{9.21}$$

Consider the family $h : Z \to \mathrm{Bl}_D(B)$ constructed in Proposition (9.16). Corollary (9.18) asserts in particular that, from a real-analytic point of view, this family is just

$$h \times \mathrm{id} \times \mathrm{id} : h^{-1}(\mathbb{T}) \times [0,1)^k \times \Delta^{\ell-k} \to \mathbb{T} \times [0,1)^k \times \Delta^{\ell-k} \, , \tag{9.22}$$

where $\mathbb{T}$ is the k-dimensional real torus which replaces the origin of B in the blow-up. Restricting to $\mathbb{T} \times (0,1)^k \times \Delta^{\ell-k}$, we get back the family $X^* \to B^*$, at least real-analytically. The point b_0 is of the form $(\overline{\tau}_1, \dots, \overline{\tau}_k, r_1, \dots, r_k, t_{k+1}, \dots, t_\ell)$ with $r_i > 0$ for every i. An explicit form of the loop λ_i is then

$$\lambda_i(t) = (\overline{\tau}_1, \dots, \overline{\tau}_i e^{2\pi\sqrt{-1}t}, \dots, \overline{\tau}_k, r_1, \dots, r_k, t_{k+1}, \dots, t_\ell) \, .$$

In view of the product representation (9.22), to prove (9.21), it suffices to prove its analogue for the real family $h^{-1}(\mathbb{T}) \to \mathbb{T}$. Recall that, in the notation of Remark (9.17), the fiber of this family at $\tau = (\tau_1, \dots, \tau_k)$ is denoted by Y_τ. We take $\overline{\tau} = (\overline{\tau}_1, \dots, \overline{\tau}_k)$ as base point and consider the loop $\mu_i : [0,1] \to \mathbb{T}$ given by

$$\mu_i(t) = (\overline{\tau}_1, \dots, \overline{\tau}_i e^{2\pi\sqrt{-1}t}, \dots, \overline{\tau}_k).$$

We shall describe an explicit trivialization of the family over the unit interval obtained from $h^{-1}(\mathbb{T}) \to \mathbb{T}$ by pullback via μ_i, by giving explicit diffeomorphisms $F_t : Y_{\overline{\tau}} \to Y_{\mu_i(t)}$ depending smoothly on t. By definition, the Picard–Lefschetz transformation corresponding to the loop μ_i will be (the isotopy class of) F_1. Recall from Remark (9.17) that Y_τ is constructed starting from a fixed surface Y with $2k$ boundary components by identifying the boundary components pairwise; what varies with $\tau = (\tau_1, \dots, \tau_k)$ is just how these identifications are made. More specifically, the boundary components have neighborhoods $M_j = \{(x_j, \xi_j) \in \mathbb{C}^2 : |x_j| < \varepsilon, |\xi_j| = 1, x_j = |x_j|\xi_j\}$ and $N_j = \{(y_j, \eta_j) \in \mathbb{C}^2 : |y_j| < \varepsilon, |\eta_j| = 1, y_j = |y_j|\eta_j\}$, $1 \le j \le k$, which are glued together by identifying $(0, \xi_j) \in M_j$ and $(0, \eta_j) \in N_j$ whenever $\xi_j \eta_j = \tau_j$. In our situation, all components of $\mu_i(t)$ are independent of t, except the ith one; thus, all the gluings are independent of t, except the one between M_i and N_i. Hence, we may regard $Y_{\mu_i(t)}$ as being obtained from a *fixed* curve Y' with two boundary components by gluing the neighborhoods M_i and N_i of these as specified above. We are now in a position to describe F_t. We choose an even smooth function $\chi : (-\varepsilon, \varepsilon) \to \mathbb{R}$ which vanishes outside $[-\varepsilon/2, \varepsilon/2]$, is identically equal to 1 on a neighborhood of 0, and is nonincreasing for positive values of s. We then define F_t to be the identity outside M_i and N_i, and

$$\begin{aligned} (x_i, \xi_i) &\mapsto \left(x_i e^{\pi\sqrt{-1}\chi(|x_i|)t}, \xi_i e^{\pi\sqrt{-1}\chi(|x_i|)t} \right) \qquad \text{in } M_i\,, \\ (y_i, \eta_i) &\mapsto \left(y_i e^{\pi\sqrt{-1}\chi(|y_i|)t}, \eta_i e^{\pi\sqrt{-1}\chi(|y_i|)t} \right) \qquad \text{in } N_i\,. \end{aligned}$$

This is a good definition since it respects the gluings. In fact, $\xi_i e^{\pi\sqrt{-1}\chi(|0|)t} \eta_i e^{\pi\sqrt{-1}\chi(|0|)t} = \xi_i \eta_i e^{2\pi\sqrt{-1}t} = \overline{\tau}_i e^{2\pi\sqrt{-1}t}$. In Remark (9.17) we have introduced explicit real-analytic coordinates s, ρ on the open set A_i of $Y_{\overline{\tau}}$ obtained from M_i and N_i by gluing their boundaries. In terms of these coordinates, F_1 is given by

$$(s, \rho) \mapsto \begin{cases} (s, \rho e^{\pi\sqrt{-1}\chi(s)}) & \text{if } s \ge 0\,, \\ (s, \rho e^{-\pi\sqrt{-1}\chi(s)}) & \text{if } s \le 0\,. \end{cases}$$

These formulas transform into (9.5) if we take as ψ the function which equals $-\chi$ for negative values of its argument and $\chi - 2$ for positive ones. Therefore, F_1 is just the Dehn twist δ_{γ_i}, where γ_i is the simple closed curve in $Y_{\overline{T}}$ which is the image of the boundaries of M_i and N_i. On the other hand, as we observed earlier, the curve γ_i corresponds to the vanishing cycle c_i in $C = X_{b_0}$ under the product decomposition (9.22). This proves (9.21).

Our next goal is to compare the homology of a smooth fiber $C = X_{b_0}$ of the family (9.20) with the one of the central fiber X_0. Let R be any commutative ring with unit. The inclusion $j : C \hookrightarrow X$ induces a homomorphism $H_1(C, R) \to H_1(X, R) = H_1(X_0, R)$. This map can be identified with

$$r_* : H_1(C, R) \to H_1(X_0, R)\,, \tag{9.23}$$

where $r : C \to X_0$ is the composition of j with the retraction map from X to X_0 given by (9.18). The map r is a homeomorphism, except for the fact that it contracts each vanishing cycle c_i to the corresponding node p_i. Let Γ be the dual graph of X_0. We claim that the homomorphism (9.23) fits into an exact sequence

$$0 \to H^1(|\Gamma|, R) \to H_1(C, R) \xrightarrow{r_*} H_1(X_0, R) \to 0 \tag{9.24}$$

and that the image of $H^1(|\Gamma|, R)$ in $H_1(C, R)$ is the subgroup generated by the homology classes of the vanishing cycles. We prove this by comparing the long exact homology sequences of the pairs $(C, \cup c_i)$ and $(X_0, \cup\{p_i\})$. Parts of the two sequences give the commutative diagram with exact rows

$$\begin{array}{ccccccccc}
H_2(C, \cup c_i) & \xrightarrow{\partial} & H_1(\cup c_i) & \longrightarrow & H_1(C) & \longrightarrow & H_1(C, \cup c_i) & \longrightarrow & H_0(\cup c_i) \\
 & & \downarrow & & \downarrow{\scriptstyle r_*} & & \| & & \| \\
 & & 0 & \longrightarrow & H_1(X_0) & \longrightarrow & H_1(X_0, \cup\{p_i\}) & \longrightarrow & H_0(\cup\{p_i\})
\end{array}$$

from which one immediately sees that r_* is onto and that $H_1(\cup c_i)$ maps surjectively onto its kernel. All that remains to be done is to identify the cokernel of ∂ with $H^1(|\Gamma|)$. Now, $H_2(C, \cup c_i)$ is freely generated by the fundamental classes of the closures of the connected components of $C \smallsetminus \cup c_i$. Denote these components by $C_1, C_2, \ldots$ and the respective fundamental classes by $[C_1], [C_2]$, and so on. We denote by v_i the vertex of Γ corresponding to C_i and by e_i the edge corresponding to c_i. Choose an orientation on each c_i, and denote by $[c_i]$ the generator of $H_1(c_i)$ corresponding to this orientation. Then $\partial[C_i] = \sum \varepsilon_{ij}[c_j]$, where ε_{ij} may be 1, –1, or 0. Denote by $v_1^*, v_2^*, \ldots$ and by $e_1^*, e_2^*, \ldots$ the bases of $C^0(|\Gamma|)$ and $C^1(|\Gamma|)$ which are dual to $v_1, v_2, \ldots$ and $e_1, e_2, \ldots$, respectively. We put an orientation on $|\Gamma|$ by defining the incidence number $[e_j : v_i]$ to equal ε_{ij}; this amounts to requiring that $v_i^*(\partial e_j) = \varepsilon_{ij}$. It is now a simple routine check to verify that the assignments $[C_i] \mapsto v_i^*$ and $[c_i] \mapsto e_i^*$ give an isomorphism of complexes between $H_2(C, \cup c_i) \xrightarrow{\partial} H_1(\cup c_i)$ and

$C^0(|\Gamma|) \xrightarrow{\delta} C^1(|\Gamma|)$. The construction of (9.24) and the proof of its exactness are now complete.

10. Bibliographical notes and further reading.

The notion of stable curve is due to Alan Mayer and David Mumford, and first appeared in the notes of a seminar by Mumford [548], included in the unpublished notes of the 1964 Woods Hole Summer Institute. In the same seminar the stable reduction theorem for curves is stated (Lemma A). A proof of the stable reduction theorem similar to the one given in the present book can be found in Alan Mayer's seminar notes [505]. As a side remark, we may observe that stable reduction for curves is a key ingredient in de Jong's theory of alterations [395]. In [398] Oort and de Jong prove the following extension result. *Let D be a divisor with normal crossings on a scheme S. Let $p : \mathcal{C} \to U = S \smallsetminus D$ be a family of stable curves with locally constant topological type. Assume that p extends, as a stable curve, to the generic points of D. Then p extends to a family of stable curves over S.*

The basic source for the foundations of the theory of stable n-pointed curves, including in particular the key constructions of projection, clutching, and stabilization, is Knudsen's paper [426].

The classical reference for the Picard–Lefschetz theory is Lefschetz's book *L'analysis situs et la géométrie algébrique* [471]. A nice treatment of the theory is given by Clemens in his seminar notes [125] and also by Barth, Hulek, Peters, and Van de Ven in [52]. A thorough treatment of real oriented blow-ups can be found in the paper [373] by Hubbard, Papadopol, and Veselov.

11. Exercises.

A. Stable reduction I.

In the following exercise the reader is led, step by step, through the process of semi-stable reduction for the family of elliptic curves with the affine equation given by

$$y^2 = x^3 - t, \tag{11.1}$$

where t is the parameter. Equation (11.1) defines a smooth surface. Let $C_0 : y^2 = x^3$ be the central fiber of this family.

A-1. First reduce the central fiber to a divisor with normal crossing. For this, one does three successive blow-ups at $(x, y) = (0, 0)$ and gets a central fiber $\widehat{C}_0$ with normal crossing and four components: $\widehat{C}_0 = C + 2E + 3F + 6G$, where C is the proper transform of C_0, and $C \cong E \cong F \cong G \cong \mathbb{P}^1$, with $E \cdot F = E \cdot C = F \cdot C = 0$, $G \cdot E = G \cdot F = G \cdot E = 1$.

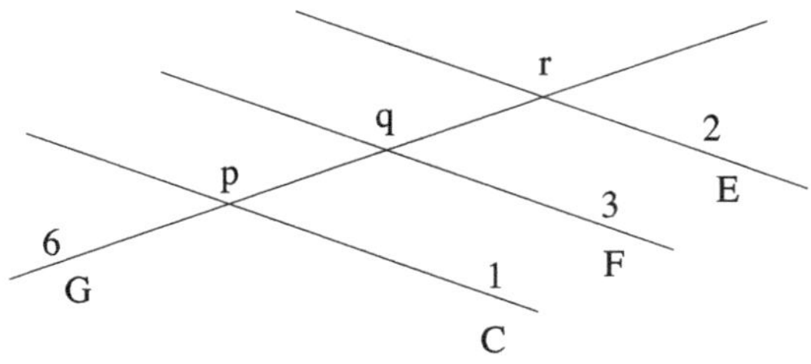

Figure 11.

A-2. Set $p = C \cap G$, $q = F \cap G$, $r = E \cap G$. In a neighborhood of q the family looks like $t = x^6y^3$. Perform the base change $t \mapsto t^6$ to get $(t^2 - x^2y)(t^2\varepsilon - x^2y)(t^2\varepsilon^2 - x^2y) = 0$ with ε a primitive 3rd root of 1. Normalize each one of the three components via $(x, y, t) \mapsto (v, u^2, c_iuv)$ with $i = 1, 2, 3$ and c_i a suitable constant. The result consists, locally, in 3 copies of a curve with a node. In each copy one branch of the node maps 2–1 over G with simple ramification over q, while the other branch maps, locally, 1–1 to F (see Fig. 11).

A-3. With the same base change and normalization, analyze what happens near p and r. Show that, over p, one gets a single node with one branch mapping 6–1 to G with total ramification over p and the other branch mapping 1–1 to C, while, over r, one gets two nodes, and in each of these, one branch maps 3–1 to G with total ramification over r, and the other branch maps 1–1 to E (see Fig. 12).

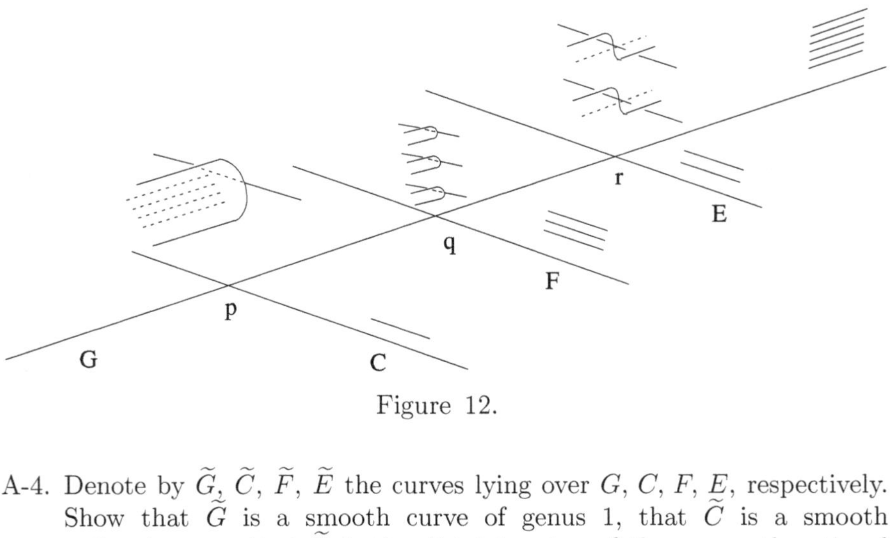

Figure 12.

A-4. Denote by $\widetilde{G}$, $\widetilde{C}$, $\widetilde{F}$, $\widetilde{E}$ the curves lying over G, C, F, E, respectively. Show that $\widetilde{G}$ is a smooth curve of genus 1, that $\widetilde{C}$ is a smooth rational curve, that $\widetilde{F}$ is the disjoint union of three smooth rational curves, and that $\widetilde{E}$ is the disjoint union of two smooth rational curves. If p_1 is the point lying over p, if q_1, q_2, q_3 are the ones lying over q and r_1, r_2 are the ones lying over r, then the central fiber looks like this:

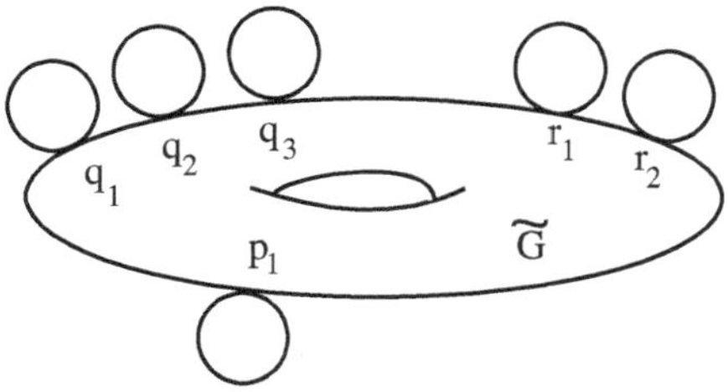

Figure 13.

A-5. Show that all the nonsingular fibers of the original family (11.1) are mutually isomorphic.

A-6. Show that each rational component in Fig. 13 above has self-intersection equal to -1 in the resulting family. Show that by blowing down these curves one gets a trivial family of elliptic curves.

B. **Stable reduction II.**

In each of the following problems, we ask you to find the stable limit of a family of curves $\{C_t\}$ parameterized by a disc or spectrum of a discrete valuation ring with parameter t.

B-1. C_t is smooth for $t \neq 0$; the total space $\mathcal{C}$ is smooth; C_0 has a tacnode, e.g.,

$$(11.2) \qquad \mathcal{C} = \{([X, Y, Z], t) \in \mathbb{P}^2 \times U \ : \ Z^2Y^2 = X^4 + t(Y^4 + Z^4)\}.$$

B-2. C_t is smooth for $t \neq 0$; the total space $\mathcal{C}$ is smooth; C_0 has a planar triple point.

B-3. C_t is smooth for $t \neq 0$; C_0 has a spatial triple point. (This is not necessary, but you can assume that the total space $\mathcal{C}$ has the minimal singularity possible, an ordinary double point.)

C. **Stable reduction III.**

Here again we ask you to find the limit of families of curves $\{C_t\}$ parameterized by a disc. The difference is that in the preceding batch in each case the general member of the family was smooth; here the general member is singular. One note: problems 5–9 may be easier to do after you have done the following series.

C-1. C_t has one node p_t for $t \neq 0$; C_0 has a cusp at $p_0 = \lim p_t$.

C-2. C_t has one node p_t for $t \neq 0$; C_0 has a tacnode at $p_0 = \lim p_t$.

C-3. C_t has one node p_t for $t \neq 0$; C_0 has a planar triple point at $p_0 = \lim p_t$.

C-4. C_t has two nodes p_t, q_t for $t \neq 0$; C_0 has a cusp at $p_0 = \lim p_t = \lim q_t$.

C-5. C_t has three nodes p_t, q_t, r_t for $t \neq 0$; C_0 has a cusp at $p_0 = \lim p_t = \lim q_t = \lim r_t$.

C-6. Let B be a smooth curve of genus $g-1$, and $p(t), q(t)$ two arcs on C with $p(0) = q(0)$ but $p(t) \neq q(t)$ for $t \neq 0$. Let $C_t = B/p(t) \sim q(t)$ be the curve obtained from B by identifying $p(t)$ and $q(t)$. What is the stable limit of the curves C_t as $t \to 0$?

C-7. Let B be a stable curve of genus $g-1$, p a node of B, and q a smooth point of B. Let $p(t)$ be an arc on C with $p(0) = p$. Let $C_t = B/p(t) \sim q$ be the curve obtained from B by identifying $p(t)$ and q. What is the stable limit of the curves C_t as $t \to 0$?

C-8. Let B be a stable curve of genus $g-1$, p, q two nodes of B, and $p(t), q(t)$ arcs on C with $p(0) = p$ and $q(0) = q$. Let $C_t = B/p(t) \sim q(t)$ be the curve obtained from B by identifying $p(t)$ and $q(t)$. What is the stable limit of the curves C_t as $t \to 0$?

C-9. Finally, let B be a smooth curve of genus $g-2$, and $p(t), q(t), r(t), s(t)$ four arcs on C with $p(0) = q(0) = r(0) = s(0)$, but with $p(t), q(t), r(t), s(t)$ distinct for $t \neq 0$. Let $C_t = B/p(t) \sim q(t), r(t) \sim s(t)$ be the curve obtained from B by identifying $p(t)$ with $q(t)$ and $r(t)$ with $s(t)$. What are *all the possible* stable limits of the curves C_t as $t \to 0$?

D. Stable reduction IV.

Now we ask the reader to find the limits of families of pointed curves $\{(C_t; p_1(t), \ldots, p_n(t))\}$.

D-1. Let C be a smooth curve of genus g, and $p(t), q(t)$ two arcs on C with $p(0) = q(0)$ but $p(t) \neq q(t)$ for $t \neq 0$. What is the stable limit of the pointed curves $(C; p(t), q(t))$ as $t \to 0$?

D-2. Let C be a stable curve of genus g, p a node of C, and $p(t)$ an arc on C with $p(0) = p$. What is the stable limit of the pointed curves $(C; p(t))$ as $t \to 0$?

D-3. Let C be a smooth curve of genus g, and $p(t), q(t), r(t)$ three arcs on C with $p(0) = q(0) = r(0)$. What are *all the possible* stable limits of the pointed curves $(C; p(t), q(t), r(t))$ as $t \to 0$?

D-4. Suppose that C is a curve with a node and $p, q \in C$ are two distinct smooth points of C. Describe the limit of the stable pointed curve (C, p, q) as p and q approach the node

a. along different branches of the node; and
b. along the same branch of the node.

Note that in one of these cases the stable limit is determined by the information given, while in the other case it is not!

E. Other limits.

E-1. Let $\{C_t = V(F^2 + tG) \subset \mathbb{P}^2\}$ be a general pencil of quartic plane curves specializing to a double conic. What is the stable limit of the curves C_t?

E-2. Let $\{C_t\}$ be as above, and let $p_1, \ldots, p_{24}$ be the flex points of the quartic C_t for $t \neq 0$. What are the limits of the points p_i as $t \to 0$?

E-3. Again, let $\{C_t\}$ be as above, and let $L_1, \ldots, L_{28}$ be the bitangent lines of the quartic C_t for $t \neq 0$. What are the limits of the lines L_i as $t \to 0$?

E-4. Now let $\{C_t = V(F^2 + tG)\}$ be a general pencil of sextic plane curves specializing to a double cubic, or more generally still a general pencil of plane curves of degree $2d$ specializing to the double of a curve of degree d. We ask the same questions as above: what is the stable limit of the curves C_t; what are the limits of their flex points, and what are the limits of their bitangent lines.

F. Miscellaneous exercises on nodal curves.

F-1. It is a classical fact that a smooth curve of genus $g \geq 2$ has at most $84(g-1)$ automorphisms. Is this true for stable curves as well?

F-2. Show that for $g \geq 4$, the singular locus of M_g is exactly the locus of curves with nontrivial automorphisms. (Note: this is not true for $g = 2$ and 3. Why not?)

F-3. Show that it is not the case for any g that the singular locus of $\overline{M}_g$ is the locus of stable curves with nontrivial automorphisms.

F-4. Describe $\Omega_{\mathcal{C}/U}$ and $\omega_{\mathcal{C}/U}$, where

$$\mathcal{C} = \{([X,Y,Z],t) \in \mathbb{P}^2 \times U : (X^2+Y^2+Z^2)(X^2-Y^2-Z^2)-t(X^4+Y^4+Z^4)\}.$$

F-5. Describe Picard–Lefschetz for the families in Exercises B-1 and B-2.

Chapter XI.

Elementary deformation theory and some applications

1. Introduction.

In this chapter we shall begin our study of moduli of curves by looking at the local picture. To be just a little more specific, we shall fix a stable curve and try to classify the stable curves that are "small perturbations" of it. This will be achieved, in a precise sense, in Section 6, via the construction of the so-called Kuranishi family. The notion of Kuranishi family is central to this book and is the main building block in the construction of all the moduli spaces we will consider. In turn, our construction of the Kuranishi family will rely on the Hilbert scheme and its universal property.

A deformation of a complete scheme (or of a compact analytic space) X, parameterized by a pointed scheme (or by a pointed analytic space) (Y, y_0), consists of a proper flat morphism $\varphi : \mathcal{X} \to Y$ together with an identification of X with the fiber X_{y_0} of φ over y_0. An infinitesimal deformation of X is just a deformation parameterized by the spectrum of the ring of dual numbers. In the first section we start by describing infinitesimal deformations for a compact complex manifold X of arbitrary dimension. We show that the isomorphism classes of infinitesimal deformations of X are in one-to-one correspondence with the elements of the first cohomology group of X with coefficients in the tangent sheaf T_X. One of the central points of this infinitesimal study will be the notion of Kodaira–Spencer map. Given a deformation $\varphi : \mathcal{X} \to (Y, y_0)$, there is a homomorphism $T_{Y,y_0} \to H^1(X; T_X)$ assigning to each tangent vector $v : \operatorname{Spec} \mathbb{C}[\varepsilon] \to (Y, y_0)$ the class in $H^1(X; T_X)$ corresponding to the infinitesimal deformation of X obtained pulling back via v the family φ to a family over $\operatorname{Spec} \mathbb{C}[\varepsilon]$.[1]

---

[1]One of the authors, PG, was a graduate student working under the supervision of Don Spencer at Princeton University in the late 1950s and early 1960s. Much of Spencer's early work in the 1940s had been on the moduli of (not necessarily compact) Riemann surfaces, especially Teichmüller theory. He remarked that, for many years, he had wanted to do deformation theory for higher-dimensional compact complex manifolds but was stuck because he did not know what in higher dimensions should replace quadratic differentials on Riemann surfaces. Spencer said that it was when Kodaira, who was in

E. Arbarello et al., *Geometry of Algebraic Curves*,
Grundlehren der mathematischen Wissenschaften 268,
DOI 10.1007/978-3-540-69392-5_3,

We then turn our attention to smooth curves. Let $(C, p_1, \dots, p_n)$ be a smooth, n-pointed curve of genus g and set $D = p_1 + \cdots + p_n$. The infinitesimal deformations of $(C, p_1, \dots, p_n)$ are parameterized by $H^1(C, T_C(-D))$. Using Schiffer variations, we construct an explicit deformation

$$\pi : \mathcal{C} \to (B, 0) \tag{1.1}$$

of $(C, p_1, \dots, p_n)$, parameterized by a polydisc of dimension $3g - 3 + n + h^0(C, T_C(-D))$, and we verify that the Kodaira–Spencer map of this deformation is an isomorphism. Schiffer variations are a useful tool in deformation theory; they consist in changing the complex structure of a Riemann surface around a single point leaving the structure unchanged outside a neighborhood of the chosen point. The explicit deformation we construct is obtained by choosing $h^1(C, T_C(-D))$ general points on C and simultaneously performing independent Schiffer variations around them. Essentially by construction, the resulting deformation has the property that its Kodaira–Spencer map $T_0(B) \to H^1(C, T_C(-D))$ is an isomorphism.

In the second section we extend the previous considerations to the case of nodal pointed curves. As a first step, we show that the infinitesimal deformations of an n-pointed nodal curve $(C, p_1, \dots, p_n)$ are parameterized by $\mathrm{Ext}^1(\Omega^1_C, \mathcal{O}_C(-D))$. To realize why an infinitesimal deformation $\varphi : \mathcal{Y} \to S = \operatorname{Spec} \mathbb{C}[\varepsilon]$ of an n-pointed nodal curve yields an element of $\mathrm{Ext}^1(\Omega^1_C, \mathcal{O}_C(-D))$, let us look, for simplicity, at the unpointed case. A natural extension of Ω^1_C by $\mathcal{O}_C$ is given by the exact sequence

$$0 \to \mathcal{O}_C \cong \varphi^*\Omega^1_S \xrightarrow{\alpha} \Omega^1_{\mathcal{Y}} \otimes \mathcal{O}_C \to \Omega^1_C \to 0\,.$$

This is the extension associated to the given infinitesimal deformation. As in the smoth case, given a deformation $\varphi : \mathcal{C} \to (Y, y_0)$ of $(C, p_1, \dots, p_n)$, one can define a Kodaira–Spencer map $T_{Y,y_0} \to \mathrm{Ext}^1(\Omega^1_C, \mathcal{O}_C(-D))$. In order to construct an explicit deformation of $(C, p_1, \dots, p_n)$ parameterized by a polydisc and whose Kodaira-Spencer map is an isomorphism, one would like to mimick what was done in the smooth case using Schiffer variations. To do this, it is necessary to better understand the vector space $\mathrm{Ext}^1(\Omega^1_C, \mathcal{O}_C(-D))$. Again, we look, for simplicity, at the unpointed

---

Princeton at that time, proved what is now called Kodaira–Serre duality giving the isomorphism

$$H^0(C, \omega_C^{\otimes 2}) \cong H^1(C, T_C)^\vee$$

for a smooth curve, that he and Kodaira realized what first-order deformations of a complex structure should be in general. With this insight, Kodaira and Spencer were off and running to lay the foundations for modern deformation theory.

case. The interpretation of the elements in $\mathrm{Ext}^1(\Omega^1_C, \mathcal{O}_C)$ is provided by the "local-to-global" spectral sequence of Ext's which gives the short exact sequence

(1.2)

$$0 \to H^1(C, \mathcal{H}om(\Omega^1_C, \mathcal{O}_C)) \to \mathrm{Ext}^1(\Omega^1_C, \mathcal{O}_C) \to H^0(C, \mathcal{E}xt^1(\Omega^1_C, \mathcal{O}_C)) \to 0\,.$$

We first look at the right-hand term of this sequence. The sheaf $\mathcal{E}xt^1(\Omega^1_C, \mathcal{O}_C)$ is concentrated at the nodes of C, so that $H^0(C, \mathcal{E}xt^1(\Omega^1_C, \mathcal{O}_C)) \cong \oplus_{p \in \mathrm{Sing}(C)} \mathbb{C}_p$. Look at a node p. In suitable analytic coordinates, C is of the form $xy = 0$ near p. It turns out that the infinitesimal deformation corresponding to a generator of $\mathbb{C}_p$ is obtained by pasting together the deformation $xy = \epsilon$ around p with the trivial deformation outside a neighborhood of p. This is the "Schiffer variation" centered at p. It consists in smoothing the node p. Moreover, this construction shows at the same time that this infinitesimal deformation can be easily integrated. We now turn to the left-hand term of (1.2) If $\alpha : N \to C$ is the normalization map, and $r_1, q_1, \dots, r_\delta, q_\delta$ are the points of N mapping to the nodes $\alpha(r_1) = \alpha(q_1), \dots, \alpha(r_\delta) = \alpha(q_\delta)$ of C, then we will check that

$$\mathcal{H}om(\Omega^1_C, \mathcal{O}_C) = \alpha_* T_N(-\textstyle\sum(r_i + q_i))\,. \tag{1.3}$$

The space

$$H^1(C, \mathcal{H}om(\Omega^1_C, \mathcal{O}_C)) = H^1(N, T_N(-\textstyle\sum(r_i + q_i)))\,,$$

as we have seen, parameterizes first-order deformations of N together with the 2δ marked points $r_1, q_1, \dots, r_\delta, q_\delta$. Thus, the elements of $H^1(C, \mathcal{H}om(\Omega^1_C, \mathcal{O}_C))$ parameterize infinitesimal deformations which are locally trivial at the nodes of C or, equivalently, infinitesimal deformations of the pointed curve $(N; r_1, q_1, \dots, r_\delta, q_\delta)$. The same analysis can be carried out starting from a pointed nodal curve $(C, p_1, \dots, p_n)$. We now proceed as in the smooth case. We can simultaneously integrate Schiffer variations at nodes and at smooth points of C to get a deformation $\pi : \mathcal{C} \to B$, parameterized by a polydisc of dimension $3g - 3 + n + h^0(C, T_C(-D))$ and check that the Kodaira–Spencer map of this deformation is an isomorphism.

In the third section we introduce the concept of Kuranishi family for a nodal pointed curve $(C; p_1, \dots, p_n)$. This is a local universal deformation of $(C; p_1, \dots, p_n)$,

$$\pi : \mathcal{C} \to (B, b_0)\,, \qquad \sigma_i : B \to \mathcal{C}\,, \quad i = 1, \dots, n,$$
$$\chi : (C; p_1, \dots, p_n) \cong (\pi^{-1}(0); \sigma_1(0), \dots, \sigma_n(0))\,.$$

It turns out that a Kuranishi family for $(C; p_1, \dots, p_n)$ exists if and only if $(C; p_1, \dots, p_n)$ is stable. This existence theorem is proved in the fourth and fifth sections of the chapter. In the third one we prove a number of formal consequences of the basic properties characterizing Kuranishi families. The first important remark is that any deformation of a stable pointed curve whose associated Kodaira–Spencer map is an isomorphism is automatically a Kuranishi family. As a consequence, the explicit deformations constructed in the previous sections via Schiffer variations are indeed Kuranishi families. Another important remark is the following. If one changes the identification χ between $(C; p_1, \dots, p_n)$ and the central fiber of one of its Kuranishi families, the result is passing from the given Kuranishi family to another one. By the universal property, the two Kuranishi families are induced, one from the other, by an automorphism of the base. Thus, in the final analysis, the automorphism group G of the central fiber $(C; p_1, \dots, p_n)$ acts on (B, b_0) and on the total space $\mathcal{C}$. We then proceed to analyze the action of G on the tangent space to B at b_0 and prove that this action is faithful, except when $g = n = 1$ or $g = 2$, $n = 0$.

Let $(C; p_1, \dots, p_n)$ be a stable n-pointed curve of genus g. Set $D = p_1 + \cdots + p_n$. Fix $\nu \geq 3$. The linear series $|(\omega_C(D))^\nu|$ embeds C in $\mathbb{P}^r$, where $r = (2\nu - 1)(g - 1) + \nu n - 1$. This embedding is the so-called ν-fold log-canonical embedding of C. We consider C as a point of the Hilbert scheme $\mathrm{Hilb}_r^{p(t)}$, where $p(t) = (2\nu t - 1)(g - 1) + \nu n t$, and we look at the locus $\mathrm{H}_{\nu,g,n}$ whose points represent ν-fold log-canonically embedded stable n-pointed curves of genus g. In Section 5 we prove that $\mathrm{H}_{\nu,g,n}$ is a smooth $(3g - 3 + n + (r + 1)^2 - 1)$-dimensional subvariety of the Hilbert scheme.

In Section 6 we show that Kuranishi families exist in the strongest sense. We start from the Hilbert scheme $\mathrm{H}_{\nu,g,n}$. This is a smooth $(3g - 3 + n + (r + 1)^2 - 1)$-dimensional variety acted on by the group $G = PGL(r + 1)$. The set-theoretical quotient $\mathrm{H}_{\nu,g,n} / G$ is the set $\overline{M}_{g,n}$ of isomorphism classes of stable n-pointed curves of genus g. Given a point $x_0 \in \mathrm{H}_{\nu,g,n}$ corresponding to a ν-fold log-canonically embedded stable n-pointed curve $(C; p_1, \dots, p_n)$ of genus g, the stabilizer G_{x_0} of x_0 can be identified with the automorphism group of $(C; p_1, \dots, p_n)$. Now we take in $\mathrm{H}_{\nu,g,n}$ a "slice" X through x_0 which is transversal to the orbits of G. We can do it in such a way that the following conditions are satisfied:

a) X is affine;
b) X is $\mathbb{G}_{x_0}$-invariant;
c) for every $y \in X$, the stabilizer $\mathbb{G}_y$ of y is contained in $\mathbb{G}_{x_0}$;
d) for every $y \in X$, there is a $\mathbb{G}_y$-invariant neighborhood U of y in X, for the analytic topology, such that $\{\gamma \in \mathbb{G} : \gamma U \cap U \neq \emptyset\} = \mathbb{G}_y$.

Now, the restriction $\pi : \mathcal{C} \to (X, x_0)$ has the property of being locally, in the analytic topology, a Kuranishi family for each one of its fibers and in particular for its central fiber $(C; p_1, \dots, p_n)$. The Kuranishi universal property is a direct consequence of the universal property of $H_{\nu,g,n}$. We close the section by studying the subloci of the bases of Kuranishi families parameterizing curves with specified automorphisms; in particular, we analyze the loci parameterizing hyperelliptic stable curves.

In Section 7, building on our discussion of continuous families of Riemann surfaces begun in Section 9 of Chapter IX, we prove that Kuranishi families satisfy a universal property also with respect to continuous deformations of smooth pointed curves. This property will be essential in our treatement of Teichmüller space in Chapter XV.

As Riemann understood, the local perturbation of the complex structure on a compact Riemann surface can be read in the variation of its period matrix. In Section 8 we discuss these ideas from the point of view of deformations. Starting from a family of Riemann surfaces $\pi : \mathcal{C} \to B$, which we may assume to be a Kuranishi family, the period map Z from B to the Siegel upper half-space $\mathbb{H}_g$ assigns to each point b in B the period matrix

$$\Omega(b) = \left(\int_{\gamma_{i,b}} \omega_{\alpha,b} \right)_{\alpha=1,\dots,g\,;\,i=1,\dots,2g},$$

where $\{\omega_{\alpha,b}\}$ is a continuously varying basis for $H^0(C_b, \mathbb{C})$, and $\{\gamma_{i,b}\}$ is a continuously varying symplectic system of generators for $H^1(C_b, \mathbb{Z})$. We carry out the local study of the period map introducing the so-called Gauss–Manin connection. We prove that the period map is holomorphic, and we also prove the so-called local Torelli theorem which amounts to the injectivity of the differential of the period map (at nonhyperelliptic curves). The dual of the differential dZ is identified with Max Noether's map

$$S^2 H^0(C, \omega_C) \to H^0(C, \omega_C^2),$$

whose surjectivity, for a nonhyperelliptic curve C, was proved in the first volume.

The introduction of the Gauss–Manin connection leads us, in Section 9, to a digression on the curvature properties of the Hodge bundles. This, although not strictly necessary in the sequel of the book, is quite interesting in its own right and could lead, if pursued further, to an alternative proof of the projectivity of the moduli space of stable curves which is quite different from the one that will be given in Chapter XIV.

In Section 10, the last one of the chapter, we present a theorem of Kempf on infinitesimal deformations of symmetric products of smooth curves which we already used in the first volume of this work as an essential step in the proof of Mark Green's theorem on quadrics through the canonical curve (cf. Chapter VI, Theorem (4.1)).

2. Deformations of manifolds.

We shall begin by discussing the rudiments of the Kodaira–Spencer deformation theory.

Let X be a complete scheme (or a compact analytic space). A *deformation of X parameterized by a pointed scheme* (resp., *a pointed analytic space*) (Y, y_0), $y_0 \in Y$, is a proper flat morphism

$$\varphi : \mathcal{X} \longrightarrow (Y, y_0) \tag{2.1}$$

plus a given isomorphism between X and the central fiber $\varphi^{-1}(y_0)$. It should be kept in mind that it is absolutely crucial to consider deformations with the same φ, but with different identifications between X and $\varphi^{-1}(y_0)$, as distinct.

A *morphism of deformations* between (2.1) and another deformation $\varphi' : \mathcal{X}' \to (Y', y_0')$ of X is a cartesian diagram

$$\begin{array}{ccc} \mathcal{X} & \xrightarrow{\ \alpha\ } & \mathcal{X}' \\ {\scriptstyle\varphi}\downarrow & & \downarrow{\scriptstyle\varphi'} \\ (Y, y_0) & \xrightarrow{\ \beta\ } & (Y', y_0') \end{array} \tag{2.2}$$

where α and β are morphisms inducing the identity on X. The notion of *isomorphism of deformations* is the obvious one. An *equivalence of deformations* is an isomorphism (2.2) such that $Y = Y'$ and β is the identity.

A *first-order deformation* of X is simply a deformation of X parameterized by $\operatorname{Spec} \mathbb{C}[\varepsilon]$, where $\mathbb{C}[\varepsilon]$ is the ring of dual numbers: we shall begin by classifying these in the simplest case, the one in which X is a complex manifold.

We shall work in the analytic category. By the GAGA comparison theorems of Serre the results will also apply without change to the algebraic case.

For brevity, we shall denote $\operatorname{Spec} \mathbb{C}[\varepsilon]$ by the letter S. We may imagine X as being given by transition data $\{U_\alpha, z_\alpha, f_{\alpha\beta}(z_\beta)\}$, where

$$\begin{cases} \mathcal{U} = \{U_\alpha\} \text{ is a finite cover of } X\,, \\ z_\alpha = {}^t(z_\alpha^1, \ldots, z_\alpha^n) \text{ are holomorphic coordinates in } U_\alpha\,, \\ z_\alpha = f_{\alpha\beta}(z_\beta) \text{ in } U_\alpha \cap U_\beta\,. \end{cases}$$

In triple intersections $U_\alpha \cap U_\beta \cap U_\gamma$ the cocycle rule

$$f_{\alpha\beta}(f_{\beta\gamma}(z_\gamma)) = f_{\alpha\gamma}(z_\gamma)$$

holds. Similarly, the total space $\mathcal{X}$ of a first-order deformation

$$\varphi : \mathcal{X} \longrightarrow S$$

of X may be thought of as being given by gluing the $U_\alpha \times S$ via the identifications

$$z_\alpha = \tilde{f}_{\alpha\beta}(z_\beta, \varepsilon) = f_{\alpha\beta}(z_\beta) + \varepsilon b_{\alpha\beta}(z_\beta)\,, \tag{2.3}$$

while φ is given, locally, by $\varphi(z_\alpha, \varepsilon) = \varepsilon$. The $\tilde{f}_{\alpha\beta}$ also have to satisfy the cocycle rule

$$\tilde{f}_{\alpha\beta}(\tilde{f}_{\beta\gamma}(z_\gamma, \varepsilon), \varepsilon) = \tilde{f}_{\alpha\gamma}(z_\gamma, \varepsilon)\,,$$

which reduces to the cocycle rule for the $f_{\alpha\beta}$ plus

$$b_{\alpha\beta} + \frac{\partial f_{\alpha\beta}}{\partial z_\beta} b_{\beta\gamma} = b_{\alpha\gamma}\,. \tag{2.4}$$

If $\frac{\partial}{\partial z_\alpha}$ stands for the row vector

$$\frac{\partial}{\partial z_\alpha} = \left(\frac{\partial}{\partial z_\alpha^1}, \dots, \frac{\partial}{\partial z_\alpha^n}\right),$$

then (2.4) just says that the 1-cochain $\vartheta = \{\vartheta_{\alpha\beta}\} \in C^1(\mathcal{U}, T_X)$ given by

$$\vartheta_{\alpha\beta} = {}^t b_{\alpha\beta}\ {}^t\!\left(\frac{\partial}{\partial z_\alpha}\right)$$

is a cocycle. The class

$$[\vartheta] \in H^1(X, T_X)$$

it defines is called the *Kodaira–Spencer class* of the first-order deformation φ. There is an exact sequence of $\mathcal{O}_X$-modules

$$0 \to T_X \to T_{\mathcal{X}} \to \varphi^* T_S \to 0\,.$$

Passing to the cohomology sequence, it is clear from the definitions that the Kodaira–Spencer class is just the coboundary of

$$\varphi^*\left(\frac{\partial}{\partial \varepsilon}\right) \in H^0(X, \varphi^* T_S)\,.$$

Therefore, $[\vartheta]$ depends only on the equivalence class of the first-order deformation $\mathcal{X} \to S$. Conversely, every 1-cocycle with coefficients in T_X defines via (2.3) a first-order deformation of X. Suppose now that

$$\vartheta = \left\{ {}^t b_{\alpha\beta} \cdot {}^t\!\left(\frac{\partial}{\partial z_\alpha}\right)\right\},$$

$$\vartheta' = \left\{ {}^t b'_{\alpha\beta} \cdot {}^t\!\left(\frac{\partial}{\partial z_\alpha}\right)\right\}$$

are cohomologous, i.e., that there are holomorphic vector-valued functions

$$c_\alpha \in \Gamma(U_\alpha, \mathcal{O}_X)^n$$

such that

$$\frac{\partial f_{\alpha\beta}}{\partial z_\beta} c_\beta - c_\alpha = b_{\alpha\beta} - b'_{\alpha\beta}\,.$$

Let

$$\mathcal{X} \xrightarrow{\varphi} S\,, \quad \mathcal{X}' \xrightarrow{\varphi'} S'$$

be the first-order deformations of X corresponding to ϑ, ϑ'. We may think of $\varphi : \mathcal{X} \to S$ as being described by (2.3) and of $\varphi' : \mathcal{X}' \to S'$ as being described by

$$z'_\alpha = \tilde{f}'_{\alpha\beta}(z'_\beta, \varepsilon) = f_{\alpha\beta}(z'_\beta) + \varepsilon b'_{\alpha\beta}(z'_\beta)\,.$$

Now set

$$(z'_\alpha, \varepsilon) = \Phi_\alpha(z_\alpha, \varepsilon) = (z_\alpha + \varepsilon c_\alpha, \varepsilon)\,.$$

It is straightforward to check that the Φ_α define an equivalence between the two first-order deformations. This sets up a 1–1 correspondence between $H^1(X, T_X)$ and the set of equivalence classes of first-order deformations of X.

We may observe, and we shall need this observation, that the notion of first-order deformation makes sense also for an open complex manifold X. One simply has to drop the properness assumption in the definition. More importantly, the proof that equivalence classes of first-order deformations of X are classified by $H^1(X, T_X)$ carries over to this case verbatim.

Let us next consider an arbitrary deformation

$$\varphi : \mathcal{X} \longrightarrow (Y, y_0)$$

of a compact complex manifold X. For any morphism

$$f : (Z, z_0) \to (Y, y_0),$$

the fiber product

$$f^*\varphi : \mathcal{X} \times_Y Z \to (Z, z_0)$$

is a deformation of X which is called the *pullback* of φ. Apply this construction to the case where Z is $S = \operatorname{Spec} \mathbb{C}[\varepsilon]$. Then the tangent space to Y at y_0 is the set of all morphisms f, i.e.,

$$T_{Y,y_0} = \operatorname{Hom}(S, (Y, y_0)),$$

and $f^*\varphi$ is nothing but the first-order approximation of φ in the direction of the tangent vector corresponding to f. We then get a map

$$\rho : T_{Y,y_0} \longrightarrow H^1(X, T_X)$$

associating to each f in $\mathrm{Hom}(S,(Y,y_0))$ the Kodaira–Spencer class of the first-order deformation $f^*\varphi$. The map ρ is easily seen to be linear and is called the *Kodaira–Spencer homomorphism*.

We now specialize to the case where $X = C$ is a smooth curve of genus g. The first remark to be made is that the vector space $H^1(C, T_C)$ parameterizing first-order deformations of C has dimension $3g-3$ unless $g = 0, 1$, in which case it has dimension 0 and 1, respectively (this, of course, follows from the Riemann–Roch theorem). In this space, there live certain distinguished classes that go under the name of *Schiffer variations*. Up to multiplicative constants, there is one of these for each point p on C, namely the image δ_p of a generator of

$$H^0(C, T_C(p) \otimes \mathbb{C}_p) \cong \mathbb{C}$$

under the coboundary map of

$$0 \to T_C \to T_C(p) \to T_C(p) \otimes \mathbb{C}_p \to 0\,.$$

If U is a small neighborhood of p, z a coordinate on U, and V is the complement of p, then, with respect to the cover $\{U, V\}$, a cocycle representing δ_p is $\frac{1}{z}\frac{\partial}{\partial z}$. Let $p_1, \dots, p_h$ be distinct points of C. Since $H^1(C, T_C(\sum p_i))$ vanishes for large enough h, the coboundary homomorphism

$$\bigoplus_i T_C(p_i) \otimes \mathbb{C}_{p_i} \to H^1(C, T_C)$$

deduced from

$$0 \to T_C \to T_C(\textstyle\sum p_i) \to \bigoplus_i T_C(p_i) \otimes \mathbb{C}_{p_i} \to 0$$

is onto, showing that $H^1(C, T_C)$ is generated by Schiffer variations. Thus, the Schiffer variations based at $h^1(C, T_C)$ general points of C form a basis of $H^1(C, T_C)$. One of the advantages of Schiffer variations is that they can be easily integrated, that is, for each point p of C and for sufficiently small η, one can find a deformation

$$\varphi : \mathcal{D} \to \Delta = \{t \in \mathbb{C} : |t| \le \eta\}$$

of C parameterized by a disc which to first-order is δ_p. Let $z : U \to \{z \in \mathbb{C} :\ |z| < b\}$ be a local coordinate centered at p. Choose a positive number $a < b$, and set

$$A = \{w \in \mathbb{C} :\ |w| < a\}\,.$$

The total space $\mathcal{D}$ can be constructed by pasting together

$$C - \left\{z : |z| \le \frac{a}{2}\right\} \times \Delta \quad \text{and} \quad A \times \Delta$$

by means of

$$w = z + \frac{t}{z} \tag{2.5}$$

(of course, η depends on a and b). Sometimes we shall refer to this procedure as *performing a Schiffer variation at p*. Now perform Schiffer variations simultaneously at $h = h^1(C, T_C)$ general points of C, using independent deformation parameters $t_1, \dots, t_h$. This yields a deformation

$$\varphi : \mathcal{C} \to (B, b_0) \tag{2.6}$$

of C, parameterized by an $h^1(C, T_C)$-dimensional polydisc, having the remarkable property that the Kodaira–Spencer map

$$\rho : T_{B,b_0} \to H^1(C, T_C) \tag{2.7}$$

is an isomorphism. This follows at once from the remark that Schiffer variations generate $H^1(C, T_C)$.

All these considerations extend, with minor formal modifications, to deformations of n-pointed smooth curves. A deformation of an n-pointed smooth curve $(C; q_1, \dots, q_n)$ consists of a deformation $\psi : \mathcal{X} \to T$ of C together with n disjoint section of ψ passing through the points of the central fiber corresponding to $q_1, \dots, q_n$. The notions of morphism, isomorphism, and equivalence of deformations of n-pointed curves are the obvious analogues of the corresponding notions for unpointed curves. When dealing with deformations of n-pointed curves, we shall often omit mention of the sections, when this is unlikely to cause confusion. Equivalence classes of first-order deformations of $(C; q_1, \dots, q_n)$ are classified by $H^1\left(C, T_C\left(-\sum_{i=1}^n q_i\right)\right)$; the easy verification of this fact is left to the reader (see Exercise (2.13)). One can construct a deformation

$$\varphi : \mathcal{C} \to (B, b_0)\,, \qquad \sigma_i : B \to \mathcal{C}\,, \;\; i = 1, \dots, n, \tag{2.8}$$

for $(C; q_1, \dots, q_n)$, parameterized by a polydisc, which is a direct analogue of (2.6). As before, we proceed by integrating Schiffer variations. Some of these are the ordinary Schiffer variations associated to points of C which are distinct from $q_1, \dots, q_n$. The remaining Schiffer variations η_i, $i = 1, \dots, n$, correspond to first-order deformations moving one marked point and leaving C and the remaining ones fixed. They are defined as follows. One looks at the exact sequence

$$0 \to T_C(-\textstyle\sum q_i) \to T_C \to \bigoplus T_{C,q_i} \to 0 \tag{2.9}$$

and chooses a coordinate z_i centered at q_i for each i. Then η_i is the image of $\partial/\partial z_i \in T_{C,q_i}$ under the coboundary map

$$\bigoplus T_{C,q_i} \to H^1(C, T_C(-\sum q_i))$$

and can be integrated exactly as one does for an ordinary Schiffer variation, using as gluing,

$$w = z_i + t \tag{2.10}$$

instead of (2.5). The same argument used in the unpointed case shows that a basis for $H^1(C, T_C(-\sum q_i))$ is given by the Schiffer variations based at $h^1(C, T_C(-\sum q_i))$ general points of C. In fact, one can do a little better. Looking at the exact cohomology sequence of (2.9), one easily sees that we can get it by the Schiffer variations based at $n - h^0(T_C) + h^0(T_C(-\sum q_i))$ of the marked points, plus $h^1(C, T_C)$ Schiffer variations based at general points of C. In any case, integrating these variations yields a family (2.8) such that the Kodaira–Spencer map

$$T_{B,b_0} \to H^1(C, T_C(-\sum q_i)) \tag{2.11}$$

is an isomorphism. We can summarize what has been achieved in the following statement.

THEOREM (2.12). *Let $(C; q_1, \ldots, q_n)$ be a smooth n-pointed genus g curve. There exists a deformation*

$$\begin{array}{c} \mathcal{C} \\ \varphi \Big\downarrow \Big\uparrow \ \sigma_i, \ i = 1, \ldots, n \\ (B, b_0) \end{array} \qquad \chi : (C; q_1, \ldots, q_n) \xrightarrow{\cong} (\varphi^{-1}(b_0); \sigma_1(b_0), \ldots, \sigma_n(b_0))$$

of $(C; q_1, \ldots, q_n)$ such that the Kodaira–Spencer map

$$\rho : T_{b_0}(B) \to H^1(C, T_C(-\sum q_i))$$

is an isomorphism and B is a polydisc of dimension $3g - 3 + n + h^0(C, T_C(-\sum q_i))$. When $(C; q_1, \ldots, q_n)$ is stable, i.e., when $2g - 2 + n > 0$, the dimension of B is $3g - 3 + n$.

EXERCISE (2.13). Fill in the details of the proof of Theorem (2.12)

EXERCISE (2.14). Let D be an effective divisor, possibly with multiple points, on a smooth curve C. Interpret $H^1(C, T_C(-D))$ from the point of view of infinitesimal deformations.

3. Deformations of nodal curves.

We now wish to study deformations of nodal curves. As in the previous section, we will generally work in the analytic category. Needless to say, the definitions of deformations and first-order deformations of nodal curves (and open subsets thereof) are as in the smooth case.

We shall begin by showing that the equivalence classes of first-order deformations of an open subset C of a nodal curve are in 1–1 correspondence with the elements of $\mathrm{Ext}^1_{\mathcal{O}_C}(\Omega^1_C, \mathcal{O}_C)$. Let

$$\varphi : \mathcal{Y} \to S = \operatorname{Spec} \mathbb{C}[\varepsilon] \tag{3.1}$$

be a first-order deformation of C. In the case at hand the exact sequence (2.17) of Chapter X restricts on C to the following one:

$$0 \to \mathcal{O}_C \cong \varphi^*\Omega^1_S \xrightarrow{\alpha} \Omega^1_{\mathcal{Y}} \otimes \mathcal{O}_C \to \Omega^1_C \to 0\,. \tag{3.2}$$

We only have to show that α is injective. In fact, $\varphi^*\Omega^1_S$ is generated by $d\varepsilon$, and $\alpha(d\varepsilon)$ certainly is not zero at smooth points of C, where $\mathcal{Y}$ is locally a product, and φ a projection. Since C is reduced, α must be injective.

We have thus associated to any first-order deformation of C an extension of Ω^1_C by $\mathcal{O}_C$, hence an element of $\mathrm{Ext}^1_{\mathcal{O}_C}(\Omega^1_C, \mathcal{O}_C)$. Exactly as in the case of smooth curves, we may then conclude that any deformation $\psi : \mathcal{X} \to Y$ of a nodal curve $C = \psi^{-1}(y_0)$ defines a Kodaira–Spencer map

$$\rho : T_{y_0}(Y) \to \mathrm{Ext}^1_{\mathcal{O}_C}(\Omega^1_C, \mathcal{O}_C)\,.$$

We next observe that, when two first-order deformations

$$\varphi : \mathcal{Y} \to S \qquad \text{and} \qquad \varphi' : \mathcal{Y}' \to S$$

define the same extension class, they are equivalent. We must produce a sheaf isomorphism

$$\beta : \mathcal{O}_{\mathcal{Y}} \to \mathcal{O}_{\mathcal{Y}'}$$

inducing the identity on $\mathcal{O}_C$ and commuting with projections, assuming that there is a commutative diagram

$$\begin{array}{ccccc} & & \Omega^1_{\mathcal{Y}} \otimes \mathcal{O}_C & & \\ & \nearrow & & \searrow & \\ \mathcal{O}_C & & \downarrow \gamma & & \Omega^1_C \\ & \searrow & & \nearrow & \\ & & \Omega^1_{\mathcal{Y}'} \otimes \mathcal{O}_C & & \end{array}$$

Indeed, we shall construct a β that induces the equivalence γ of extensions. We shall do so by showing that, for any function h on $\mathcal{Y}$, there is a unique function $\beta(h)$ on $\mathcal{Y}'$ such that

$$h|_C = \beta(h)|_C \qquad \text{and} \qquad \tilde{d}\beta(h) = \gamma(\tilde{d}h),$$

where $\tilde{d}$ stands for the composition of d with the natural map $\Omega^1_{\mathcal{Y}} \to \Omega^1_{\mathcal{Y}} \otimes \mathcal{O}_C$. Uniqueness is obvious: if f is a function on $\mathcal{Y}'$ such that $f|_C = 0$, then f is of the form εg, and $\tilde{d}f = gd\varepsilon$ in $\Omega^1_{\mathcal{Y}'} \otimes \mathcal{O}_C$, so that, if $\tilde{d}f = 0$, then g, and hence f, vanish. Having proved the uniqueness, the existence is a local problem. For any h, the restriction $h|_C$ locally extends to a function $\tilde{h}$ on $\mathcal{Y}'$. But the difference between $\tilde{d}\tilde{h}$ and $\gamma(\tilde{d}h)$ restricts to zero on C, so that

$$\tilde{d}\tilde{h} - \gamma(\tilde{d}h) = gd\varepsilon\,,$$

and we may set $\beta(h) = \tilde{h} - \varepsilon g$. It remains to show that β is a ring homomorphism. By Leibniz' rule,

$$\begin{aligned}\tilde{d}\beta(hg) &= \gamma(\tilde{d}(hg))\\ &= g|_C\gamma(\tilde{d}h) + h|_C\gamma(\tilde{d}g)\\ &= \tilde{d}(\beta(h)\beta(g))\,,\end{aligned}$$

so that, by uniqueness, $\beta(hg)$ indeed equals $\beta(h)\beta(g)$. A similar argument shows that $\beta(h+g)$ is equal to $\beta(h)+\beta(g)$. This ends the construction of β. It may be observed that the arguments developed so far apply, without changes, to deformations of a reduced scheme or analytic space, and not only to deformations of a nodal curve. First-order deformations of a reduced scheme or analytic space X are thus parameterized by a subspace of $\operatorname{Ext}^1_{\mathcal{O}_X}(\Omega^1_X, \mathcal{O}_X)$.

Before showing that any element of $\operatorname{Ext}^1(\Omega^1_C, \mathcal{O}_C)$ comes from a first-order deformation of C, we need a few preliminary observations. The "local-to-global" spectral sequence of Ext's (cf. [318], for example) gives a short exact sequence

$$(3.3)\qquad \begin{aligned}&0 \to H^1(C, \mathcal{H}om_{\mathcal{O}_C}(\Omega^1_C, \mathcal{O}_C)) \to \operatorname{Ext}^1_{\mathcal{O}_C}(\Omega^1_C, \mathcal{O}_C) \to\\ &\qquad H^0(C, \mathcal{E}xt^1_{\mathcal{O}_C}(\Omega^1_C, \mathcal{O}_C)) \to 0\,.\end{aligned}$$

The sheaf $\mathcal{E}xt^1(\Omega^1_C, \mathcal{O}_C)$ is concentrated at the nodes of C, and hence the sequence (3.3) can also be rewritten as

$$(3.4)\qquad \begin{aligned}&0 \to H^1(C, \mathcal{H}om_{\mathcal{O}_C}(\Omega^1_C, \mathcal{O}_C)) \to \operatorname{Ext}^1_{\mathcal{O}_C}(\Omega^1_C, \mathcal{O}_C) \to\\ &\qquad \bigoplus_{p\in \operatorname{Sing}(C)} \operatorname{Ext}^1_{\mathcal{O}_{C,p}}(\Omega^1_{C,p}, \mathcal{O}_{C,p}) \to 0\,.\end{aligned}$$

To compute it at a node p, we embed a small neighborhood of p as a closed subspace of an open subset V of $\mathbb{C}^2$, so that in suitable coordinates, C is of the form $xy = 0$. From the conormal sequence

$$0 \to \mathcal{O}_{V,p}(-C) \otimes \mathcal{O}_{C,p} \xrightarrow{d} \Omega^1_{V,p} \otimes \mathcal{O}_{C,p} \to \Omega^1_{C,p} \to 0$$

we get an exact sequence

$$
\begin{aligned}
(3.5)\qquad &\operatorname{Hom}(\Omega^1_{V,p}\otimes\mathcal{O}_{C,p},\mathcal{O}_{C,p}) \xrightarrow{\eta} \operatorname{Hom}(\mathcal{O}_{V,p}(-C)\otimes\mathcal{O}_{C,p},\mathcal{O}_{C,p}) \to \\
&\quad \operatorname{Ext}^1(\Omega^1_{C,p},\mathcal{O}_{C,p}) \to 0
\end{aligned}
$$

since $\Omega^1_V\otimes\mathcal{O}_C$ is locally free, and hence $\operatorname{Ext}^1(\Omega^1_{V,p}\otimes\mathcal{O}_{C,p},\mathcal{O}_{C,p})$ vanishes. Let φ be a homomorphism from $\Omega^1_{V,p}\otimes\mathcal{O}_{C,p}$ to $\mathcal{O}_{C,p}$. Then $\eta(\varphi)(xy)=\varphi(d(xy))=x\varphi(dy)+y\varphi(dx)$. Thus, the image of η is $\mathfrak{m}_p\operatorname{Hom}(\mathcal{O}_{V,p}(-C)\otimes\mathcal{O}_{C,p},\mathcal{O}_{C,p})$, where $\mathfrak{m}_p$ is the maximal ideal in $\mathcal{O}_{C,p}$, and hence (3.5) gives an isomorphism

$$
(3.6)\qquad \operatorname{Ext}^1(\Omega^1_{C,p},\mathcal{O}_{C,p})\cong(\mathcal{O}_{V,p}(-C)\otimes\mathbb{C})^\vee\,.
$$

It follows in particular that $\mathcal{E}xt^1(\Omega^1_C,\mathcal{O}_C)$ is (noncanonically) isomorphic to $\bigoplus_{p\in C_{\text{sing}}}\mathbb{C}_p$. It is useful to give an alternate intrinsic interpretation of $\operatorname{Ext}^1(\Omega^1_{C,p},\mathcal{O}_{C,p})$. Let τ be the set consisting of the two possible orderings of the branches of C at p, and let $\mu_2=\{\pm1\}$ act nontrivially on it. We claim that there is a canonical isomorphism

$$
(3.7)\qquad \operatorname{Ext}^1_{\mathcal{O}_{C,p}}(\Omega^1_{C,p},\mathcal{O}_{C,p})\cong\textstyle\bigwedge^2(\mathfrak{m}_p/\mathfrak{m}_p^2)^\vee\otimes_{\mu_2}\tau\,.
$$

This is obtained by composing (3.6) with the transpose of an isomorphism ϑ between $\bigwedge^2(\mathfrak{m}_p/\mathfrak{m}_p^2)\otimes_{\mu_2}\tau$ and $\mathcal{O}_{V,p}(-C)\otimes\mathbb{C}$ which we now describe. Let $\mathcal{I}$ be the maximal ideal of $\mathcal{O}_{V,p}$. For any element ξ of $\mathcal{O}_{V,p}$ (resp., of $\mathcal{O}_{V,p}(-C)$), we denote by $[\xi]$ its class in $\mathcal{O}_{C,p}/\mathfrak{m}_p$ (resp., in $\mathcal{O}_{V,p}(-C)/\mathcal{I}\mathcal{O}_{V,p}(-C)$). Let x and y be local equations for the two branches of C at p. Denote by b_x (resp., b_y) the branch defined by the vanishing of x (resp., of y). One then sets

$$
\vartheta\big([x]\wedge[y]\otimes(b_x,b_y)\big)=[xy]\,.
$$

It is immediate to see that this is well defined and linear. At first sight the isomorphism (3.7) seems to depend on the choice of an embedding of a neighborhood of $p\in C$ in an open subset of $\mathbb{C}^2$, but it is easy to show that this is not the case. In fact, given another embedding in an open subset $V'\subset\mathbb{C}^2$, possibly after shrinking V and V', there is an isomorphism $f:V\to V'$ inducing the identity on C. This yields a commutative diagram

$$
\begin{array}{ccccc}
 & \overset{\cong}{\nearrow} & (\mathcal{O}_{V,p}(-C)\otimes\mathbb{C})^\vee & \overset{{}^t\vartheta}{\searrow} & \\
\operatorname{Ext}^1(\Omega^1_{C,p},\mathcal{O}_{C,p}) & & \Big\downarrow{\scriptstyle f_*} & & \bigwedge^2(\mathfrak{m}_p/\mathfrak{m}_p^2)^\vee\otimes_{\mu_2}\tau \\
 & \underset{\cong}{\searrow} & (\mathcal{O}_{V',p}(-C)\otimes\mathbb{C})^\vee & \underset{{}^t\vartheta'}{\nearrow} &
\end{array}
$$

showing that (3.7) is independent of the embedding. Another version of (3.7) can be obtained as follows. Let N be the normalization of C at

p, and denote by p_1, p_2 the points of N mapping to p. Then

$$(\mathfrak{m}_p/\mathfrak{m}_p^2)^\vee = T_{C,p} = T_{N,p_1} \oplus T_{N,p_2} \,,$$

whence an identification

$$\text{(3.8)} \qquad \operatorname{Ext}^1_{\mathcal{O}_C}(\Omega^1_{C,p}, \mathcal{O}_{C,p}) \cong T_{N,p_1} \otimes T_{N,p_2} \,.$$

In fact, the identification between $\bigwedge^2(T_{N,p_1} \oplus T_{N,p_2})$ and $T_{N,p_1} \otimes T_{N,p_2}$ depends on the choice of an ordering of the two summands, i.e., of an ordering of the branches of the node, and changes sign if the ordering is reversed; tensoring with τ exactly compensates for this ambiguity.

We are now in a position to construct a first-order deformation of C corresponding to a given class in $\operatorname{Ext}^1(\Omega^1_C, \mathcal{O}_C)$. We shall first deal with the case where C is a neighborhood of the origin in $xy = 0$. Since C is Stein, in this case

$$\operatorname{Ext}^1(\Omega^1_C, \mathcal{O}_C) = H^0(C, \mathcal{E}xt^1(\Omega^1_C, \mathcal{O}_C)) \cong \mathbb{C} \,.$$

The family $xy = a\varepsilon$, where a is a complex number, is a first-order deformation $\mathcal{Y}$ of C which can be regarded as the reduction modulo t^2 of the surface X with equation $xy = at$. As this surface is smooth for $a \neq 0$, the sheaf

$$\Omega^1_{\mathcal{Y}} \otimes \mathcal{O}_C = \Omega^1_{X|C}$$

is, in this case, locally free and hence a nontrivial extension of Ω^1_C by $\mathcal{O}_C$. If we denote by ζ_a the class of this extension, a straightforward computation, based, for instance, on (3.6), shows that $\zeta_a = a\zeta_1$. It follows that, in the special case at hand, any class in $\operatorname{Ext}^1(\Omega^1_C, \mathcal{O}_C)$ can be realized as a first-order deformation of C.

In treating the general case we shall assume, for simplicity of notation, that C has a single node at p. Let ξ be an element of $\operatorname{Ext}^1(\Omega^1_C, \mathcal{O}_C)$. Let U be a small neighborhood of p. We have shown that there exists a first-order deformation $\mathcal{U}$ of U inducing $\xi|_U$. On the other hand, since $C \smallsetminus \{p\}$ is smooth and affine, the exact sequence (3.3) shows that $\xi|_{C \smallsetminus \{p\}}$ is trivial. Thus, ξ can be obtained by gluing the trivial extension on $C \smallsetminus \{p\}$ with $\xi|_U$ along $U \smallsetminus \{p\}$ by means of an equivalence of extensions of $\Omega^1_{U \smallsetminus \{p\}}$ by $\mathcal{O}_{U \smallsetminus \{p\}}$. As we have shown, over $U \smallsetminus \{p\}$ there is, between the first-order deformation $\mathcal{U}$ of U and the trivial deformation $\mathcal{D}$ of $C \smallsetminus \{p\}$, an equivalence β that induces γ. Gluing $\mathcal{U}$ and $\mathcal{D}$ by means of β yields a deformation $\mathcal{Y}$ of C whose class in $\operatorname{Ext}^1(\Omega^1_C, \mathcal{O}_C)$ is precisely ξ.

The construction we have performed provides an interpretation of the two extreme terms in the sequence (3.3). The term on the right corresponds to deformations that smooth one or more of the nodes of C, while the term on the left corresponds to deformations that are locally trivial, that is, locally products. Another way of explaining this last assertion is to notice that, if $\alpha : N \to C$ is the normalization map, and $r_1, q_1, \dots, r_\delta, q_\delta$ are the points of N mapping to the nodes $\alpha(r_1) = \alpha(q_1), \dots, \alpha(r_\delta) = \alpha(q_\delta)$ of C, then, as we shall see in a moment, one has

$$\mathcal{H}om(\Omega^1_C, \mathcal{O}_C) = \alpha_* T_N(-\textstyle\sum(r_i + q_i)) . \tag{3.9}$$

The space

$$H^1(C, \mathcal{H}om(\Omega^1_C, \mathcal{O}_C)) = H^1(N, T_N(-\textstyle\sum(r_i + q_i))) ,$$

as we have seen, parameterizes first-order deformations of N together with the 2δ marked points $r_1, q_1, \dots, r_\delta, q_\delta$. It remains to justify (3.9). Observe first that the equality is trivially true away from the nodes of C. Let then w be a node, and z_1, z_2 the points of N mapping to it. We will be done if we can show that

$$\mathrm{Hom}(\Omega^1_{C,w}, \mathcal{O}_{C,w}) = \mathrm{Hom}(\mathcal{I}\omega_{C,w}, \mathcal{O}_{C,w}) = \bigoplus_{i=1,2} \mathrm{Hom}(\omega_{N,z_i}, \mathcal{O}_{N,z_i}(-z_i)) ,$$

where $\mathcal{I}$ is the ideal of w. The equality on the left follows from the fact that the natural homomorphism $\Omega^1_{C,w} \to \omega_{C,w}$ sends $\Omega^1_{C,w}$ onto $\mathcal{I}\omega_{C,w}$ and has a one-dimensional vector space as kernel. As for the other equality, let φ be a homomorphism from $\mathcal{I}\omega_{C,w} = \omega_{N,z_1} \oplus \omega_{N,z_2}$ to $\mathcal{O}_{C,w}$. Clearly, $\varphi(\omega_{N,z_i})$ vanishes on one of the branches of the node, and hence is contained in $\mathcal{I} = \mathcal{O}_{N,z_1}(-z_1) \oplus \mathcal{O}_{N,z_2}(-z_2)$. Thus $\mathrm{Hom}(\mathcal{I}\omega_{C,w}, \mathcal{O}_{C,w})$ equals

$$\begin{aligned}\mathrm{Hom}_{\mathcal{O}_{C,w}} &\left(\omega_{N,z_1} \oplus \omega_{N,z_2}, \mathcal{O}_{N,z_1}(-z_1) \oplus \mathcal{O}_{N,z_2}(-z_2)\right) \\ &= \mathrm{Hom}(\omega_{N,z_1}, \mathcal{O}_{N,z_1}(-z_1)) \oplus \mathrm{Hom}(\omega_{N,z_2}, \mathcal{O}_{N,z_2}(-z_2))\end{aligned}$$

since $\mathrm{Hom}_{\mathcal{O}_{C,w}}(\omega_{N,z_i}, \mathcal{O}_{N,z_j}(-z_j))$ vanishes when $i \neq j$.

It goes almost without saying that one can develop a deformation theory for n-pointed nodal curves paralleling the deformation theory for nodal curves. Recall that an n-pointed nodal curve is the datum of a nodal curve C plus distinct smooth points $p_1, \dots, p_n$ of C. Look at a first-order deformation of $(C; p_1, \dots, p_n)$ consisting of a family (3.1) plus sections $\rho_1, \dots, \rho_n$. Set $D = p_1 + \dots + p_n$ and $\mathcal{R} = \rho_1 + \dots + \rho_n$. The extension (3.2) can be enriched to the commutative diagram

$$\begin{array}{ccccccccc} 0 & \longrightarrow & \mathcal{O}_C & \longrightarrow & \Omega^1_{\mathcal{Y}} \otimes \mathcal{O}_C & \longrightarrow & \Omega^1_C & \longrightarrow & 0 \\ & & \downarrow & & \downarrow & & \downarrow & & \\ 0 & \longrightarrow & \mathcal{O}_D & \xrightarrow{\simeq} & \Omega^1_{\mathcal{R}} \otimes \mathcal{O}_D & \longrightarrow & 0 & \longrightarrow & 0 \end{array} \tag{3.10}$$

This diagram can be regarded as an extension of the complex $\mathcal{A}^\bullet = (\mathcal{A}^0 \to \mathcal{A}^1) = (\Omega^1_C \to 0)$ by the complex $\mathcal{D}^\bullet = (\mathcal{D}^0 \to \mathcal{D}^1) = (\mathcal{O}_C \to \mathcal{O}_D)$, that is, as an element of $\mathrm{Ext}^1_{\mathcal{O}_C}(\mathcal{A}^\bullet, \mathcal{D}^\bullet)$. We leave it to the reader to show that this sets up a one-to-one correspondence between isomorphism classes of first-order deformations of $(C; p_1, \dots, p_n)$ and $\mathrm{Ext}^1_{\mathcal{O}_C}(\mathcal{A}^\bullet, \mathcal{D}^\bullet)$, extending the proof we have given in the unpointed case. Notice that

$$\mathrm{Ext}^1_{\mathcal{O}_C}(\mathcal{A}^\bullet, \mathcal{D}^\bullet) \cong \mathrm{Ext}^1_{\mathcal{O}_C}(\Omega^1_C, \mathcal{O}_C(-D)). \tag{3.11}$$

In fact, the elements of $\mathrm{Ext}^1_{\mathcal{O}_C}(\mathcal{A}^\bullet, \mathcal{D}^\bullet)$ are (isomorphism classes of) commutative diagrams of extensions

$$\begin{array}{ccccccccc}
0 & \longrightarrow & \mathcal{D}^0 & \longrightarrow & \mathcal{E}^0 & \longrightarrow & \mathcal{A}^0 & \longrightarrow & 0 \\
 & & \downarrow & & \downarrow & & \downarrow & & \\
0 & \longrightarrow & \mathcal{D}^1 & \longrightarrow & \mathcal{E}^1 & \longrightarrow & \mathcal{A}^1 & \longrightarrow & 0
\end{array}$$

one such being trivial if and only if the two rows admit compatible splittings. Taking kernels of the vertical arrows and keeping in mind that $\mathcal{A}^1$ vanishes and $\mathcal{D}^0 \to \mathcal{D}^1$ is onto, we get a commutative diagram with exact rows and columns,

$$\begin{array}{ccccccccc}
 & & 0 & & 0 & & 0 & & \\
 & & \downarrow & & \downarrow & & \downarrow & & \\
0 & \longrightarrow & \mathcal{K} & \longrightarrow & \mathcal{E} & \longrightarrow & \mathcal{A}^0 & \longrightarrow & 0 \\
 & & \downarrow & & \downarrow & & \downarrow & & \\
0 & \longrightarrow & \mathcal{D}^0 & \longrightarrow & \mathcal{E}^0 & \longrightarrow & \mathcal{A}^0 & \longrightarrow & 0 \\
 & & \downarrow & & \downarrow & & \downarrow & & \\
0 & \longrightarrow & \mathcal{D}^1 & \longrightarrow & \mathcal{E}^1 & \longrightarrow & 0 & & \\
 & & \downarrow & & \downarrow & & & & \\
 & & 0 & & 0 & & & &
\end{array} \tag{3.12}$$

The top row is an extension of $\mathcal{A}^0 = \Omega^1_C$ by $\mathcal{K} = \mathcal{O}_C(-D)$, and it is clear that if this extension is trivial, that is, if $\mathcal{E} \to \mathcal{A}^0$ has a right inverse, then $\mathcal{E}^0 \to \mathcal{A}^0$ has a right inverse whose composition with $\mathcal{E}^0 \to \mathcal{E}^1$ is zero, and hence the original diagram of extensions is trivial. This shows that $\mathrm{Ext}^1_{\mathcal{O}_C}(\mathcal{A}^\bullet, \mathcal{D}^\bullet) \subset \mathrm{Ext}^1_{\mathcal{O}_C}(\Omega^1_C, \mathcal{O}_C(-D))$. Conversely, any extension

$$0 \to \mathcal{K} \to \mathcal{E} \to \mathcal{A}^0 \to 0$$

can be completed to a diagram (3.12) by setting

$$\mathcal{E}^0 = \frac{\mathcal{D}^0 \oplus \mathcal{E}}{\sim}, \quad \mathcal{E}^1 = \frac{\mathcal{E}^0}{\mathcal{E}},$$

where $(d, e) \sim 0$ if and only if d and $-e$ come from the same element of $\mathcal{K}$. In conclusion, we have a natural identification

$$\left\{ \begin{array}{c} \text{first-order deformations} \\ \text{of } (C; p_1, \dots, p_n) \end{array} \right\} \Big/ \text{isomorphism} \;\cong\; \mathrm{Ext}^1_{\mathcal{O}_C}(\Omega^1_C, \mathcal{O}_C(-\textstyle\sum p_i)). \tag{3.13}$$

Clearly, there is an exact sequence

$$\begin{aligned}(3.14)\qquad & 0 \to H^1(C, \mathcal{H}om(\Omega^1_C, \mathcal{O}_C(-D))) \to \mathrm{Ext}^1(\Omega^1_C, \mathcal{O}_C(-D)) \to \\ & H^0(C, \mathcal{E}xt^1(\Omega^1_C, \mathcal{O}_C)) \to 0\end{aligned}$$

which generalizes (3.3) to the n-pointed case and admits a similar interpretation. The term on the left classifies first-order deformations which are locally trivial, and the one on the right classifies first-order smoothings of the nodes. Alternatively, one can view the term on the left as classifying first-order deformations of the normalization of C, pointed by the counterimages of D and of the nodes.

Finally, one can extend to nodal curves the construction that led to Theorem (2.12). We begin with the following observation. The realization of $\mathrm{Ext}^1(\Omega^1_C, \mathcal{O}_C(-\sum p_i))$ as the space of first-order deformations of the n-pointed nodal curve $(C; p_1, \dots, p_n)$ was largely based on the possibility of infinitesimally smoothing a nodal curve. We want to show that the same can be done in finite terms, namely that, given a node p on C, one can find a deformation

$$\varphi : \mathcal{C} \to \Delta, \qquad \rho_i : \Delta \to \mathcal{C}$$

of $(C; p_1, \dots, p_n)$, where Δ is the disk $\{t \in \mathbb{C} : |t| < 1\}$, and, in a neighborhood of p, the fibers $\varphi^{-1}(t)$ are smooth for $t \neq 0$. In fact, near p, $\mathcal{C}$ will look like $xy = t$. The construction is as follows. There are neighborhoods U of p, not containing any of the p_i, which may be thought of as being obtained from two discs

$$\begin{aligned} V &= \{z \in \mathbb{C} : |z| < 1\}, \\ W &= \{w \in \mathbb{C} : |w| < 1\} \end{aligned}$$

via identification of the origins. Consider the regions

$$\begin{aligned} A &= \{x, y, t) \in \mathbb{C}^3 : |x| < 1, \ \ |y| < 1, \ \ |t| < 1, \ \ xy = t\}, \\ B &= ((C - U) \times \Delta) \cup B', \end{aligned}$$

where

$$B' = \{(z, t) \in V \times \Delta : |z| > \sqrt{|t|}\} \cup \{(w, t) \in W \times \Delta : |w| > \sqrt{|t|}\} \subset C \times \Delta .$$

Then $\mathcal{C}$ is the complex manifold obtained from A and B by identifying B' to $A' = A - \{(x, y, t) : |x| = |y|\}$ by means of

$$\begin{aligned}(3.15)\qquad & (z, t) \mapsto \left(z, \frac{t}{z}, t\right), \\ & (w, t) \mapsto \left(\frac{t}{w}, w, t\right),\end{aligned}$$

and φ is the projection to the last factor. As for the sections, we simply set $\rho_i(t) = (p_i, t)$ in $(C - U) \times \Delta \subset \mathcal{C}$.

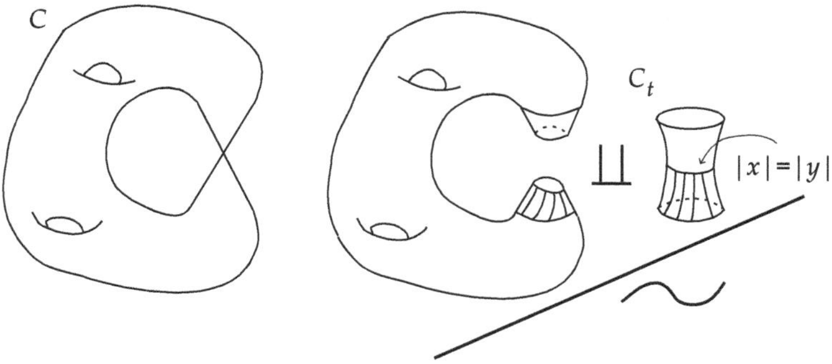

Figure 1.

To construct the analogue of the family in (2.12), we shall think of $(C; p_1, \ldots, p_n)$ as being obtained from its pointed normalization $(N; r_1, q_1, \ldots, r_\delta, q_\delta, p_1, \ldots, p_n)$ by identifying points $r_1, q_1, \ldots, r_\delta, q_\delta$ in pairs. The family

$$\psi : \mathcal{N} \to W$$

obtained by applying Theorem (2.12) to the pointed normalization comes equipped with disjoint sections

$$R_i : W \to \mathcal{N} \ ; \quad Q_i : W \to \mathcal{N} \ , \quad i = 1, \ldots, \delta \, ,$$
$$P_i : W \to \mathcal{N} \ , \quad i = 1, \ldots, n \, ,$$

passing, respectively, through the r_i, q_i, and p_i. Identifying R_i and Q_i for each i yields a locally trivial deformation

$$\varphi' : \mathcal{C}' \to W \ , \qquad P_i : W \to \mathcal{C}' \ , \quad i = 1, \ldots, n \, ,$$

of $(C; p_1, \ldots, p_n)$. The Kodaira–Spencer map provides an identification between the tangent space to W at w and the group

$$H^1(N, T_N(-\textstyle\sum(r_i + q_i) - \sum p_j)) = H^1(C, \mathcal{H}om(\Omega^1_C, \mathcal{O}_C(-\sum p_j))) \, .$$

The smoothing of a node which we performed on a fixed curve can be done "with parameters" on a locally trivial family such as φ'. Doing this independently at every node of C gives a deformation

(3.16) $$\varphi : \mathcal{C} \to B = W \times \Delta \, , \qquad \sigma_i : B \to \mathcal{C} \, , \quad i = 1, \ldots, n,$$

of $(C; p_1, \ldots, p_n)$, where Δ is a δ-dimensional polydisc. Moreover, we may choose coordinates $t_1, \ldots, t_\delta$ on Δ in such a way that, near the node of C arising from the identification of r_i with q_i, $\mathcal{C}$ is of the form $xy = t_i$.

It may be observed that the Kodaira–Spencer map provides an identification of the exact sequence

$$0 \to T_w(W) \to T_{(w,0)}(B) \to T_0(\Delta) \to 0$$

with (3.14). Furthermore, the locus in B parameterizing singular curves is the union of the hyperplanes $t_i = 0$, $i = 1, \ldots, \delta$. More precisely, the restriction of the family (3.16) to $t_i = 0$ parameterizes deformations of C which are locally trivial at the ith node. Summing up, here is what has been shown.

THEOREM (3.17). *Let $(C; p_1, \ldots, p_n)$ be an n-pointed nodal curve of genus g. There exists a deformation*

$$\begin{array}{c} \mathcal{C} \\ \varphi \Big\downarrow \Big\uparrow \sigma_i, \; i = 1, \ldots, n \\ (B, b_0) \end{array} \qquad \chi : (C; p_1, \ldots, p_n) \xrightarrow{\cong} (\varphi^{-1}(b_0); \sigma_1(b_0), \ldots, \sigma_n(b_0))$$

of $(C; p_1, \ldots, p_n)$ such that the Kodaira–Spencer map

$$\rho : T_{b_0}(B) \to \mathrm{Ext}^1_{\mathcal{O}_C}(\Omega^1_C, \mathcal{O}_C(-\textstyle\sum p_i))$$

is an isomorphism, and B is a polydisc of dimension $3g - 3 + n + \dim \mathrm{Hom}(\Omega^1_C, \mathcal{O}_C)$. In particular, when $(C; p_1, \ldots, p_n)$ is stable,

$$\dim \mathrm{Ext}^1(\Omega^1_C, \mathcal{O}_C(-\textstyle\sum p_i)) = \dim_{b_0}(B) = 3g - 3 + n\,.$$

Finally, if δ is the number of nodes of C, one can choose coordinates $t_1, \ldots, t_\delta, \ldots$ on B, vanishing at b_0, in such a way that the locus parameterizing deformations which are locally trivial at the ith node is $t_i = 0$; in particular, the locus parameterizing singular curves is $t_1 \cdots t_\delta = 0$.

We close with a useful generalization of the sequence (3.14). Let $(C; p_1, \ldots, p_n)$ be an n-pointed nodal curve, and let $W = \{w_1, \ldots, w_\ell\}$ be a set of nodes of C. Let $\alpha : \widetilde{C} \to C$ be the partial normalization at these nodes, and denote by r_i, q_i the two points of $\widetilde{C}$ mapping to w_i. We set $D = \sum p_i$ and $E = \sum (r_i + q_i)$, and write $\widetilde{D}$ also to designate the preimage of D in $\widetilde{C}$. Then (3.9) immediately generalizes to

$$\mathcal{H}om(\Omega^1_C, \mathcal{O}_C(-D)) = \alpha_* \mathcal{H}om(\Omega^1_{\widetilde{C}}, \mathcal{O}_{\widetilde{C}}(-\widetilde{D} - E))\,,$$

so that, in particular,

$$H^1(\widetilde{C}, \mathcal{H}om(\Omega^1_{\widetilde{C}}, \mathcal{O}_{\widetilde{C}}(-\widetilde{D} - E))) = H^1(C, \mathcal{H}om(\Omega^1_C, \mathcal{O}_C(-D)))\,.$$

The exact sequence (3.14) and its analogue for $\widetilde{C}$ then fit in a commutative diagram

$$
\begin{array}{ccccccc}
0 \to H^1(\mathcal{H}om(\Omega^1_{\widetilde{C}}, \mathcal{O}_{\widetilde{C}}(-\widetilde{D} - E))) & \to & \operatorname{Ext}^1(\Omega^1_{\widetilde{C}}, \mathcal{O}_{\widetilde{C}}(-\widetilde{D} - E)) & \to & \bigoplus\limits_{\substack{w \in C_{\mathrm{sing}} \\ w \notin W}} \operatorname{Ext}^1(\Omega^1_{C,w}, \mathcal{O}_{C,w}) \to 0 \\
\Vert & & \downarrow & & \downarrow \\
0 \longrightarrow H^1(\mathcal{H}om(\Omega^1_C, \mathcal{O}_C(-D))) & \longrightarrow & \operatorname{Ext}^1(\Omega^1_C, \mathcal{O}_C(-D)) & \longrightarrow & \bigoplus\limits_{w \in C_{\mathrm{sing}}} \operatorname{Ext}^1(\Omega^1_{C,w}, \mathcal{O}_{C,w}) \to 0
\end{array}
$$

whence an exact sequence

$$
\begin{aligned}
0 \to \operatorname{Ext}^1(\Omega^1_{\widetilde{C}}, \mathcal{O}_{\widetilde{C}}(-\widetilde{D} - E)) \to \operatorname{Ext}^1(\Omega^1_C, \mathcal{O}_C(-D)) \to \\
\bigoplus_{w \in W} \operatorname{Ext}^1(\Omega^1_{C,w}, \mathcal{O}_{C,w}) \to 0
\end{aligned}
\tag{3.18}
$$

Of course, the term on the left classifies first-order deformations which are locally trivial at the nodes belonging to W, and the one on the right classifies first-order smoothings of these nodes. It is useful to notice that, by (3.8),

$$
\bigoplus_{w \in W} \operatorname{Ext}^1(\Omega^1_{C,w}, \mathcal{O}_{C,w}) = \bigoplus_{i=1}^{\ell} T_{\widetilde{C}, r_i} \otimes T_{\widetilde{C}, q_i} \,. \tag{3.19}
$$

When W is the set of all the nodes of C, the sequence (3.18) reduces to (3.14). Its first two terms parameterize first-order deformations of $\widetilde{C}$ (as an $(n + 2\ell)$-pointed curve) and of $(C; p_1, \dots, p_n)$, respectively. The homomorphism connecting them is given by the clutching operation described in Section 7 of Chapter X.

Exercise (3.20). Construct an example of family $X \to S$ of n-pointed nodal curves over a connected base such that the dimension of $\operatorname{Ext}^1(\Omega^1_{X_s}, \mathcal{O}_{X_s}(-\sum p_i(s)))$ is not constant as a function of s.

4. The concept of Kuranishi family.

One of the central notions in the theory of deformations and moduli is the one of Kuranishi family. We will need it only for curves, and we limit ourselves to this case. We work in the analytic category. Let $(C; p_1, \dots, p_n)$ be an n-pointed connected nodal curve. A deformation

$$
\begin{array}{c}
\mathcal{C} \\
\varphi \Big\downarrow \Big\uparrow \ \sigma_i, \ i = 1, \dots, n \\
(B, b_0)
\end{array}
\qquad \chi : (C; p_1, \dots, p_n) \xrightarrow{\simeq} (\varphi^{-1}(b_0); \sigma_1(b_0), \dots, \sigma_n(b_0))
\tag{4.1}
$$

of $(C; p_1, \dots, p_n)$ is said to be a *Kuranishi family for* $(C; p_1, \dots, p_n)$ if it satisfies the following condition:

K) For any deformation $\mathcal{D} \xrightarrow{\psi} (E, e_0)$ of $(C; p_1, \dots, p_n)$ and for any sufficiently small connected neighborhood U of e_0, there is a *unique* morphism of deformations of n-pointed curves

$$\begin{array}{ccc} \mathcal{D}_U & \xrightarrow{F} & \mathcal{C} \\ \downarrow & & \downarrow{\scriptstyle\varphi} \\ (U, e_0) & \xrightarrow{f} & (B, b_0) \end{array} \tag{4.2}$$

Kuranishi families need not exist in general. In the next two sections we shall prove the following basic result.

THEOREM (4.3). *A Kuranishi family for* $(C; p_1, \dots, p_n)$ *exists if and only if* $(C; p_1, \dots, p_n)$ *is stable.*

The mere existence of a Kuranishi family has several important formal consequences, which we now detail. Most of them need no proof.

COROLLARY (4.4). *Kuranishi families are essentially unique, in the sense that any two Kuranishi families for* $(C; p_1, \dots, p_n)$, *up to restricting to sufficiently small connected neighborhoods of the base points, are isomorphic via a unique isomorphism.*

COROLLARY (4.5). *The Kodaira–Spencer map of a Kuranishi family at the base point is an isomorphism.*

COROLLARY (4.6). *Let there be given a deformation of a stable n-pointed curve* $(C; p_1, \dots, p_n)$ *over the pointed analytic space* (E, e_0). *Suppose that its Kodaira–Spencer map at* e_0 *is an isomorphism and that* E *is smooth at* e_0. *Then the deformation is a Kuranishi family for* $(C; p_1, \dots, p_n)$.

The proof is immediate. In fact, the universal property of the Kuranishi family asserts the existence of a diagram (4.2). The Kodaira–Spencer map in question can be identified with the differential of f, which is thus a local isomorphism at e_0.

COROLLARY (4.7). *When* $(C, p_1, \dots, p_n)$ *is stable, the deformation constructed in (3.17) is a Kuranishi family for* $(C, p_1, \dots, p_n)$.

COROLLARY (4.8). *The base of the Kuranishi family of a stable n-pointed curve* $(C, p_1, \dots, p_n)$ *of genus* g *is smooth of dimension* $3g - 3 + n$.

A family $X \to S$ of stable n-pointed curves can be viewed as a deformation of a fiber X_{s_0}, taking as identification between X_{s_0} and itself the identity.

COROLLARY (4.9). *Let* $X \to S$ *be a family of stable n-pointed curves, and* s_0 *a point of* S. *If* $X \to S$ *is a Kuranishi family for* X_{s_0}, *then it is a Kuranishi family for* X_s, *for all* s *in an open neighborhood* U *of* s_0. *If* $X \to S$ *is an algebraic family, then we can take* U *to be Zariski open.*

It follows from the previous results that $X \to S$ is Kuranishi for X_s if and only if s is a smooth point of S and the Kodaira–Spencer map at s is an isomorphism. The first of these conditions is clearly an open one, both in the ordinary and in the Zariski topologies. Since the dimension of $\mathrm{Ext}^1(\Omega^1_{X_s}, \mathcal{O}_{X_s}(-\sum \sigma_i(s)))$ is independent of s, the second condition translates into a rank condition for a map between vector bundles and hence is open as well. The corollary follows.

COROLLARY (4.10). *Let (4.1) be a Kuranishi family for* $(C; p_1, \dots, p_n)$. *Denote by* G *the automorphism group of* $(C; p_1, \dots, p_n)$. *Then there are arbitrarily small neighborhoods* V *of* b_0 *such that the action of* G *on* $(C; p_1, \dots, p_n)$ *extends to compatible actions on* V *and on* $\mathcal{C}_V$ *taking each distinguished section to itself.*

Proof. As usual, we write $\mathcal{C}_b$ to denote the fiber $\varphi^{-1}(b)$, and so on. Let γ be an element of G. Composing γ^{-1} with $\chi : C \to \mathcal{C}_{b_0}$, we get another identification of $(C; p_1, \dots, p_n)$ with the central fiber of the family and hence a new deformation of $(C; p_1, \dots, p_n)$. The universal property of the Kuranishi family, applied to this deformation, then gives a commutative diagram

$$\begin{array}{ccccc}
C & \xrightarrow{\ \gamma\ } & C & \xrightarrow{\ \mathrm{id}_C\ } & C \\
{\scriptstyle\chi}\downarrow & & \downarrow{\scriptstyle\chi} & & \downarrow{\scriptstyle\chi} \\
\mathcal{C}_{V_\gamma} & \xrightarrow{\ F_\gamma\ } & \mathcal{C}_{U_\gamma} & \hookrightarrow & \mathcal{C} \\
{\scriptstyle\varphi}\downarrow & & \downarrow{\scriptstyle\varphi} & & \downarrow{\scriptstyle\varphi} \\
V_\gamma & \xrightarrow{\ f_\gamma\ } & U_\gamma & \hookrightarrow & B
\end{array}$$

where V_γ, U_γ are suitable neighborhoods of b_0, and f_γ, F_γ are isomorphisms. Replacing each V_γ with $\cap\{V_\gamma : \gamma \in G\}$, we may suppose that all the V_γ are equal to the same neighborhood V of b_0. By uniqueness, we may also suppose that $f_\gamma f_\eta = f_{\gamma\eta}$ and $F_\gamma F_\eta = F_{\gamma\eta}$ where defined and that f_γ and F_γ are the identities when γ is the identity. Further replacing V with $\cap\{U_\eta : \eta \in G\}$, we may also assume that $U_\gamma = f_\gamma(V) = V$ for all $\gamma \in G$. The required actions of G on V and $\mathcal{C}_V$ are given, respectively, by $\gamma \mapsto f_\gamma$ and $\gamma \mapsto F_\gamma$. Q.E.D.

The action of G on $\mathcal{C}_V$ given by (4.10) is clearly faithful, while the same is not necessarily true of the action on V. This action is faithful if and only if the corresponding action on $T_{b_0}(B)$ is. Notice that, under the isomorphism $T_{b_0}(B) \cong \mathrm{Ext}^1_{\mathcal{O}_C}(\Omega^1_C, \mathcal{O}_C(-\sum p_i))$, the action of G on $T_{b_0}(B)$ corresponds to the natural one on $\mathrm{Ext}^1_{\mathcal{O}_C}(\Omega^1_C, \mathcal{O}_C(-\sum p_i))$. The situations in which this action is not faithful can be completely classified.

PROPOSITION (4.11). *Let* $(C; p_1, \dots, p_n)$ *be a stable n-pointed curve of genus* g, *and let* γ *be a nontrivial automorphism of* $(C; p_1, \dots, p_n)$ *which acts trivially on* $\mathrm{Ext}^1_{\mathcal{O}_C}(\Omega^1_C, \mathcal{O}_C(-\sum p_i))$. *Then one of the following occurs:*

i) $g = 1$, $n = 1$, *and* γ *is the symmetry about the marked point;*
ii) $g = 2$, $n = 0$, *and* γ *is the hyperelliptic involution.*

Proof. We begin by illustrating the main ingredient of the proof in the case of a smooth C. First of all, a smooth stable n-pointed curve of genus zero has no nontrivial automorphisms, so we may assume that $g \geq 1$. The dual of $\mathrm{Ext}^1_{\mathcal{O}_C}(\Omega^1_C, \mathcal{O}_C(-\sum p_i)) = H^1(C, T_C(-\sum p_i))$ is $H^0(C, \omega_C^2(\sum p_i))$. On the other hand,

$$\deg \omega_C^2(\textstyle\sum p_i) \geq \deg \omega_C + 3,$$

unless one of the following occurs:

- $g = 1$, $n \leq 2$;
- $g = 2$, $n = 0$.

Then, outside of these exceptions, $|\omega_C^2(\sum p_i)|$ embeds C in $\mathbb{P}H^1(C, T_C(-\sum p_i))$, and hence if γ acts trivially on $H^1(C, T_C(-\sum p_i))$, it must act trivially on its projectivization and therefore on C. It remains to examine the exceptional cases. The first of them will be dealt with when we treat the case of a possibly singular C. The second is taken care of by the argument we have just used. In fact, in genus two the bicanonical mapping is the composition of the hyperelliptic double covering and of a Veronese embedding, so that an automorphism of C acting trivially on $H^1(C, T_C)$ must be the identity or the hyperelliptic involution.

We now move on to the general case. Set $D = \sum p_i$. The vector space $\mathrm{Ext}^1(\Omega^1_C, \mathcal{O}_C(-D))$ sits in an exact sequence

$$0 \to H^1(C, \mathcal{H}om\,(\Omega^1_C, \mathcal{O}_C(-D))) \to \mathrm{Ext}^1(\Omega^1_C, \mathcal{O}_C(-D)) \to \bigoplus_{p \in C_{\mathrm{sing}}} \mathbb{C}_p \to 0$$

whose dual is

$$(4.12) \qquad 0 \to \bigoplus_{p \in C_{\mathrm{sing}}} \mathbb{C}_p \to H^0(C, \Omega^1_C \otimes \omega_C(D)) \to H^0(C, \mathcal{J}\omega_C^2(D)) \to 0,$$

where $\mathcal{J}$ stands for the ideal of the singular locus of C. Let γ be an automorphism of $(C; p_1, \dots, p_n)$ that acts as the identity on $H^0(C, \Omega^1_C \otimes \omega_C(D))$. Since γ acts trivially on the leftmost term in (4.12), it leaves every node of C fixed. As a consequence, γ carries every singular component of C into itself. It is also obvious that γ carries every component containing a marked point to itself. Now let $C = \cup C_i$ be the decomposition of C in irreducible components, let $\nu_i : N_i \to C_i$ be the normalization map, denote by E_i the divisor of all points of N_i mapping to singular points of C, and by D_i the divisor of all points of N_i mapping to one of the p_j. Clearly,

$$H^0(C, \mathcal{J}\omega_C^2(D)) = \bigoplus_i H^0(N_i, \omega_{N_i}^2(E_i + D_i)).$$

Each summand is nonzero, except those for which N_i is rational and $E_i + D_i$ has degree three. In particular, for any i such that N_i is not of this kind, $\mathcal{J}\omega_C^2(D)$ has nonzero sections vanishing on all components other than C_i. Since γ acts trivially on $H^0(C, \mathcal{J}\omega_C^2(D))$, it follows from this remark and the previous ones that γ carries every component of C into itself, except possibly those smooth rational components which meet the rest of the curve in three points and do not contain marked points. Since, however, the singular points of C are left fixed, the only case where there can be an interchange of components is where C is the union of two smooth rational curves meeting at three points (hence a curve of genus two), $n = 0$, and γ is the hyperelliptic involution. We may then assume that each component of C is carried into itself by γ. It follows in particular that γ must be the identity on each smooth rational component of C. If C_i is not one of these components, one immediately sees that

$$\deg \omega_{N_i}^2(E_i + D_i) \geq \deg \omega_{N_i} + 3$$

with the following four exceptions:

a) C_i is smooth of genus 1 and contains exactly two points which are either marked or points of contact with the rest of C;
b) C_i is a curve of genus 1 with one node and contains exactly two points which are either marked or points of contact with the rest of C;
c) $C_i = C$ has genus 1, $n = 1$;
d) C_i has genus 1, does not contain marked points, and meets the rest of C at one point;
e) $C_i = C$ has genus two, and $n = 0$.

Except in these cases, we then conclude that $|\omega_{N_i}^2(E_i + D_i)|$ embeds N_i in projective space, and since γ acts trivially on $H^0(N_i, \omega_{N_i}^2(E_i + D_i))$, it must act trivally on C_i as well.

We now examine cases a), b), c), d), e) above, in this order. In cases a) and b), we denote by p and q the points which are either marked or points of contact with the rest of C. In case a), $|\omega_{N_i}^2(E_i + D_i)|$ yields a degree two map of $C_i = N_i$ to $\mathbb{P}^1$, so that the restriction of γ to C_i can only be the identity or the sheet interchange. This last possibility, however, cannot occur, since p and q are mapped to the same point of $\mathbb{P}^1$, and we know that singular points and marked points of C are left fixed by γ. In case b), if z is a suitable local parameter centered at p, γ must be, locally, of the form

$$z \mapsto \zeta z\,, \tag{4.13}$$

where ζ is a root of unity. The space $H^0(N_i, \omega_{N_i}^2(E_i + D_i))$ has dimension 1, and a nonzero section is, locally near p, of the form $a(dz)^2/z$, where a is a nonvanishing function. By γ-invariance, ζ must be equal to 1, i.e., γ must be the identity on C_i.

In cases c) and d), we denote by p the unique point which is either marked or a point of contact with the rest of C. As before, γ is locally of the form (4.13) at p. Proceeding as above, one sees that a nonzero section of $H^0(N_i, \omega^2_{N_i}(E_i + D_i))$ is of the form $a(dz)^2$ at p, where a is a nonvanishing function. Such a section can be invariant under γ only if $\zeta^2 = 1$. In case c), this corresponds to the exceptional case i) in the statement of the proposition. In case d), suppose that γ does not act trivally on C_i, i.e., that $\zeta = -1$. Then γ cannot act as the identity on the component of C meeting C_i. Assume in fact that this is the case. Letting $zw = 0$ be the equation of C at p, if a belongs to $\mathbb{C}_p$ and $zw = a\varepsilon$ is the corresponding first-order deformation, γ clearly sends it to $zw = -a\varepsilon$, that is, γ acts on $\mathbb{C}_p$ as multiplication by -1, contrary to our assumption. Thus, if C_j is the component of C meeting C_i, by our previous considerations C_j must also fall under case d). Hence, C equals $C_i \cup C_j$ and has genus two. Moreover, γ is the symmetry about the point of attachment, both in C_i and in C_j, and therefore is the hyperelliptic involution. It remains to examine case e). When C is smooth, we have already seen at the beginning of the proof that the only automorphisms acting trivially on $\mathrm{Ext}^1(\Omega^1_C, \mathcal{O}_C)$ are the identity and the hyperelliptic involution. Suppose $C = C_i$ is singular. If C has one node, $|\omega^2_{N_i}(E_i)|$ maps N_i in two-to-one fashion to $\mathbb{P}^1$, so that γ must be the identity or the hyperelliptic involution. If C has two nodes, its only nontrivial automorphism fixing the two nodes is the hyperelliptic involution. Q.E.D.

While there exist no Kuranishi families for a nonstable nodal curve $(C; p_1, \dots, p_n)$, it is possible to construct what is called a *versal family.* By this one means a deformation (4.1) which satisfies condition **K)** except for uniqueness, which is replaced by the weaker property that the Kodaira–Spencer map at the central fiber be an isomorphism. It is a consequence of the definition that, modulo shrinking the base, a versal deformation of an n-pointed nodal curve is unique up to an isomorphism, which however need not be unique.

The existence of versal families follows easily from the one of Kuranishi families for stable curves. We add to the marked points of C the minimum number of additional ones needed to get a stable curve $(C; p_1, \dots, p_n, p_{n+1}, \dots)$ and look at its Kuranishi family. If one ignores the marked sections that go through the added points, one gets a deformation of $(C; p_1, \dots, p_n)$ which is the required versal family. In fact, given a deformation of $(C; p_1, \dots, p_n)$, by suitably adding sections and possibly shrinking the base, one obtains a deformation of $(C; p_1, \dots, p_n, p_{n+1}, \dots)$, which comes by pullback from the Kuranishi family. As for the second characteristic property of a versal family, it suffices to observe that the exact sequence (3.14) coincides with the corresponding sequence for $(C; p_1, \dots, p_n, p_{n+1}, \dots)$.

EXERCISE (4.14). Let C be a stable hyperelliptic curve. Use the cut-

and-paste methods employed in the construction of (3.16) to show that there is a family of stable curves over a disk whose central fiber is C and whose remaining fibers are smooth hyperelliptic.

EXERCISE (4.15). Consider a family of stable n-pointed curves

$$\mathcal{C} \to B\,, \quad \sigma_i : B \to \mathcal{C}, \ i = 1, \dots, n\,.$$

Assume that it is Kuranishi for any one of its fibers. Consider the family

$$\pi_2 : \mathcal{C} \times_B \mathcal{C} \to \mathcal{C}, \quad \tau_i, \ i = 1, \dots, n\,,$$

where π_2 is the projection to the second factor, and $\tau_i(x) = (\sigma_i(x), x)$. Apply to this family and to the diagonal the stabilization procedure described in Section 8 of Chapter X. Show that the resulting family is a Kuranishi family for any one of its fibers.

5. The Hilbert scheme of ν-canonical curves.

Consider the subset of the appropriate Hilbert scheme consisting of all stable curves of genus g embedded by the ν-canonical system for sufficiently large ν. The main aim of this section is to show that this subset is in fact a smooth subscheme of the Hilbert scheme and to compute its tangent space and dimension. We will also generalize this result to the case of stable n-pointed curves. We first need to show that for a flat family of curves, the condition of being nodal is an open one in the Zariski topology.

PROPOSITION (5.1). *Let $\varphi : X \to S$ be a flat proper morphism of schemes or of analytic spaces. Then the set of all $s \in S$ such that $X_s = \varphi^{-1}(s)$ is not a connected nodal curve is Zariski-closed in S. If, in addition, n sections $\sigma_1, \dots, \sigma_n$ of φ are given, then the set of all $s \in S$ such that $(X_s; \sigma_1(s), \dots, \sigma_n(s))$ is not a connected n-pointed nodal curve is Zariski-closed in S.*

We shall prove the proposition for a morphism of schemes, the proof for a morphism of analytic spaces being essentially identical, except for terminological changes, but simpler. Without loss of generality we may assume that S is connected and that at least one fiber of φ (and hence any fiber, by flatness) has dimension 1. Observe that, for any (closed) point s of S, the dimension of $H^0(X_s, \mathcal{O}_{X_s})$ is 1 when X_s is connected and reduced. Furthermore, if

$$K^0 \xrightarrow{\alpha} K^1 \to K^2 \to \cdots$$

is a complex of free sheaves on an open subset U of S which functorially calculates the higher direct images of $\mathcal{O}_X$ (cf. Section 3 in Chapter IX), then the locus of the points $s \in U$ such that the dimension of

$H^0(X_s, \mathcal{O}_{X_s})$ is strictly greater than 1 is the locus where the rank of α is $\operatorname{rank}(K^0) - 2$ or less. As such, it is Zariski-closed. This means that it suffices to prove the proposition under the additional assumption that $H^0(X_s, \mathcal{O}_{X_s})$ has dimension 1 for all s. In particular, we may assume that all the fibers of φ are connected and do not have embedded components.

Since φ is proper, to prove the proposition, it suffices to show that, for a point of X, the condition of being neither smooth nor a node in its fiber is a closed one. Possibly after shrinking S, we may embed X in a product $\mathbb{P}^r \times S$ in such a way that φ is the restriction to X of the projection to the second factor. Now suppose that a fiber X_s is a nodal curve. Then X_s is a local complete intersection in $\mathbb{P}^r$. In fact, a nodal curve is obviously a local complete intersection from the anaytic point of view; on the other hand, a subscheme of $\mathbb{C}^r$ is a local complete intersection if and only if it is such as an analytic subspace of $\mathbb{C}^r$. As we have seen (cf. Lemma (5.25) in Chapter IX), the property of being a local complete intersection is an open one. Therefore we may suppose that all the fibers of φ are local complete intersections. Furthermore, a nodal fiber of φ is locally the complete intersection of $r-2$ smooth hypersurfaces meeting transversely and a further hypersurface. This also is an open property, and we may therefore suppose that it is enjoyed by all fibers of φ. More precisely, we may suppose that $\mathbb{P}^r \times S$ is covered with a finite number of open sets U isomorphic to open subsets of $\mathbb{C}^r \times S$ and that $X \cap U$ is defined by equations $F, F_3, \dots, F_r$, where the intersection of the subscheme $F_3 = \dots = F_r = 0$ with $U \cap (\mathbb{P}^r \times \{s\})$ is smooth for any s. We denote by $x_1, \dots, x_r$ linear coordinates on $\mathbb{C}^r$ and regard $U_s = U \cap (\mathbb{P}^r \times \{s\})$ as an open subset of $\mathbb{C}^r$ and hence $X_s \cap U$ as a locally closed subscheme of $\mathbb{C}^r$. We also regard $F, F_3, \dots, F_r$ as functions on (varying) open subsets of $\mathbb{C}^r$ depending on parameters in S. Let s_0 be a point of S, and p_0 a point of X_{s_0}. After a linear change of coordinates in $\mathbb{C}^r$ we may suppose that $x_1, x_2, F_3, \dots, F_r$ give local coordinates on U_{s_0} at p_0, and hence, possibly after shrinking U and S, at every point of U_s for every $s \in S$. We set $y_i = x_i$ for $i = 1, 2$ and $y_i = F_i$ for $i > 3$. Clearly, p is a smooth point of X_s if and only if one of the two derivatives $\partial F / \partial y_i$, $i = 1, 2$, does not vanish at p. On the other hand, by Lemma (2.3) in Chapter X, p is a node if and only if these same derivatives vanish at p, but the Hessian determinant

$$\det \begin{pmatrix} \frac{\partial^2 F}{\partial y_1^2} & \frac{\partial^2 F}{\partial y_1 \partial y_2} \\ \frac{\partial^2 F}{\partial y_1 \partial y_2} & \frac{\partial^2 F}{\partial y_2^2} \end{pmatrix}$$

does not. The proof will be complete if we can show that these are algebraic conditions and not merely analytic ones. That this is the case follows from the observation that all partial derivatives with respect to the y variables of a rational function of $x_1, \dots, x_r$ are in fact rational functions of the x variables. To prove this, we may clearly restrict to first derivatives, since the statement for higher derivatives then follows immediately by induction. Let then G be a rational function, and write

$$\frac{\partial G}{\partial y_i} = \sum_j \frac{\partial G}{\partial x_j}\frac{\partial x_j}{\partial y_i}\,.$$

The functions $\partial G/\partial x_j$ are rational functions of the x variables. As for the functions $\partial x_j/\partial y_i$, it suffices to notice that, by Cramer's rule, the entries of the Jacobian matrix $\frac{\partial x}{\partial y}$ are rational functions of the entries of $\frac{\partial y}{\partial x}$, which are themselves rational functions of the x variables. This finishes the proof of the proposition in the unpointed case. The proof of the last statement is simple and left to the reader.

REMARK (5.2). The final part of the argument we used to prove Proposition (5.1) also shows that, in an algebraic family $X \to S$ of nodal curves, the locus of points which are singular in their fiber is a closed subscheme of X.

We now return to the central subject of this section. Consider a stable n-pointed genus g curve $(C; p_1, \dots, p_n)$, and set $D = \sum p_i$. It follows, for instance, from part iii) of Lemma (6.1) in Chapter X that, for $\nu \geq 3$, the ν-fold log-canonical sheaf $(\omega_C(D))^\nu$ is very ample and embeds C in $\mathbb{P}^r$, where $r = (2\nu-1)(g-1)+\nu n-1$. Its Hilbert polynomial is

$$p_\nu(t) = (2\nu t - 1)(g-1) + \nu n t\,. \tag{5.3}$$

We denote by H the Hilbert scheme of $(n+1)$-tuples $(Y; z_1, \dots, z_n)$, where Y is subscheme of $\mathbb{P}^r$ with Hilbert polynomial $p_\nu(t)$, and $z_1, \dots, z_n$ are points of Y (cf. the end of Chapter IX, Section 7). We just proved that the (nonempty) subset U of H parameterizing connected n-pointed nodal curves is Zariski open. Let

$$\begin{array}{c} \mathcal{Y} \\ \pi \Big\downarrow \Big\uparrow \sigma_i, \quad i = 1, \dots, n, \\ U \end{array}$$

be the restriction of the universal family. Of course, a general point of U does not correspond to an n-pointed curve embedded by the ν-fold log-canonical sheaf. We wish to construct a subscheme $\mathrm{H}_{\nu,g,n}$ of U parameterizing such curves. To be more precise, we would like $\mathrm{H}_{\nu,g,n}$ to satisfy the following universal property. Write L for the pullback to $\mathcal{Y}$ of the hyperplane bundle on $\mathbb{P}^r$ and observe that $\mathcal{F} = (\omega_\pi(\sum \sigma_i))^\nu L^{-1}$ has zero relative degree. Consider a map $\alpha : X \to U$ and the corresponding cartesian diagram

$$
\begin{array}{ccc}
\mathcal{Y} \times_U X & \xrightarrow{\beta} & \mathcal{Y} \\
\eta \big\downarrow & & \big\downarrow \pi \\
X & \xrightarrow{\alpha} & U
\end{array}
$$

Suppose that $\beta^* \mathcal{F}$ is isomorphic to $\eta^* G$ for some line bundle G on X. Then α factors through $\mathrm{H}_{\nu,g,n} \hookrightarrow U$.

We denote by W the subscheme of U defined by the sum of the gth Fitting ideal of $R^1\pi_*(\mathcal{F})$ and of the gth Fitting ideal of $R^1\pi_*(\mathcal{F}^{-1})$, and let $\pi' : \mathcal{Y}' \to W$ be the restriction of $\mathcal{Y} \to U$. By construction, the rank of $\pi'_*\mathcal{F}$ and $\pi'_*\mathcal{F}^{-1}$ is everywhere at least equal to 1. The scheme $\mathrm{H}_{\nu,g,n}$ is defined as the open subset of W where the rank of the multiplication map

$$\pi'_*\mathcal{F} \otimes \pi'_*\mathcal{F}^{-1} \to \pi'_*\mathcal{O}_{\mathcal{Y}'} = \mathcal{O}_W$$

is equal to one. We shall sometimes refer to $\mathrm{H}_{\nu,g,n}$ as the *Hilbert scheme of ν-log-canonically embedded, stable, n-pointed, genus g curves.* We end this section by showing that $\mathrm{H}_{\nu,g,n}$ is smooth of dimension $3g - 3 + n + (r+1)^2 - 1$, where $r = (2\nu - 1)(g-1) + n\nu - 1$ (recall that $\nu \geq 3$). We shall do this by explicitly computing the tangent space to $\mathrm{H}_{\nu,g,n}$.

Let then $C \subset \mathbb{P}^m$ be a nodal curve, and let $p_1, \dots, p_n$ be distinct smooth points of C. Set $D = \sum p_i$. Let $\mathcal{I}_C$ be the ideal sheaf of C in $\mathbb{P}^m$, and $\mathcal{I}_D$ the one of D. Look at the commutative diagram

$$
\begin{array}{ccccccccc}
0 & \longrightarrow & \mathcal{I}_C/\mathcal{I}_C^2 & \xrightarrow{d} & \Omega^1_{\mathbb{P}^r} \otimes \mathcal{O}_C & \longrightarrow & \Omega^1_C & \longrightarrow & 0 \\
 & & \big\downarrow & & \big\downarrow & & \big\downarrow & & \\
0 & \longrightarrow & \mathcal{I}_D/\mathcal{I}_D^2 & \longrightarrow & \Omega^1_{\mathbb{P}^r} \otimes \mathcal{O}_D & \longrightarrow & 0 & &
\end{array}
$$

Both rows are exact (cf. Exercise (2.21)). Hence we can view the diagram as an exact sequence

$$0 \to \mathcal{C}^\bullet \to \mathcal{B}^\bullet \to \mathcal{A}^\bullet \to 0$$

involving the complexes, all concentrated in degrees 0 and 1,

(5.4)

$$\mathcal{C}^\bullet = (\mathcal{I}_C/\mathcal{I}_C^2 \to \mathcal{I}_D/\mathcal{I}_D^2), \quad \mathcal{B}^\bullet = (\Omega^1_{\mathbb{P}^r} \otimes \mathcal{O}_C \to \Omega^1_{\mathbb{P}^r} \otimes \mathcal{O}_D), \quad \mathcal{A}^\bullet = (\Omega^1_C \to 0).$$

We define another complex $\mathcal{D}^\bullet = (\mathcal{D}^0 \to \mathcal{D}^1)$ by setting $\mathcal{D}^0 = \mathcal{O}_C$, $\mathcal{D}^1 = \mathcal{O}_D$. From (5.4) we get another exact sequence

(5.5)

$$0 \to \mathrm{Hom}_{\mathcal{O}_C}(\mathcal{A}^\bullet, \mathcal{D}^\bullet) \to \mathrm{Hom}_{\mathcal{O}_C}(\mathcal{B}^\bullet, \mathcal{D}^\bullet) \to \mathrm{Hom}_{\mathcal{O}_C}(\mathcal{C}^\bullet, \mathcal{D}^\bullet) \to \mathrm{Ext}^1_{\mathcal{O}_C}(\mathcal{A}^\bullet, \mathcal{D}^\bullet),$$

and our next goal is to understand the terms appearing in it. We begin from the left. Since $\mathcal{A}^1 = 0$, the homomorphisms from $\mathcal{A}^\bullet$ to $\mathcal{D}^\bullet$ are just the homomorphisms from $\mathcal{A}^0 = \Omega^1_C$ to the kernel of $\mathcal{O}_C \to \mathcal{O}_D$, that is, to $\mathcal{O}_C(-D)$. Passing to the next term, it is clear that a homomorphism from $\mathcal{B}^\bullet$ to $\mathcal{D}^\bullet$ is completely determined by its degree zero part $\Omega^1_{\mathbb{P}^r} \otimes \mathcal{O}_C \to \mathcal{O}_C$; hence this term is just $\mathrm{Hom}_{\mathcal{O}_C}(\Omega^1_{\mathbb{P}^r} \otimes \mathcal{O}_C, \mathcal{O}_C)$. The next term is the tangent space at $h = (C; p_1, \dots, p_n)$ to the Hilbert scheme H of $(n+1)$-tuples $(Y; q_1, \dots, q_n)$, where Y is a subscheme of $\mathbb{P}^m$ and $q_1, \dots, q_n$ points on it (cf. Lemma (8.8)). As for $\mathrm{Ext}^1_{\mathcal{O}_C}(\mathcal{A}^\bullet, \mathcal{D}^\bullet)$, we know from formula (3.11) that it is isomorphic to $\mathrm{Ext}^1_{\mathcal{O}_C}(\Omega^1_C, \mathcal{O}_C(-D))$ and from (3.13) that it classifies isomorphism classes of first-order deformations of $(C; p_1, \dots, p_n)$. Summing up, we get from (5.5) the exact sequence

(5.6)
$$\begin{gathered} 0 \to \mathrm{Hom}_{\mathcal{O}_C}(\Omega^1_C, \mathcal{O}_C(-D)) \to \mathrm{Hom}_{\mathcal{O}_{\mathbb{P}^m}}(\Omega^1_{\mathbb{P}^m}, \mathcal{O}_C) \to \\ T_h(H) \xrightarrow{\beta} \mathrm{Ext}^1_{\mathcal{O}_C}(\Omega^1_C, \mathcal{O}_C(-D)) \end{gathered}$$

Moreover, the homomorphism β is just the Kodaira–Spencer map at h associated to the universal family over H.

Now assume that C is connected and not contained in any hyperplane and that the linear system cut out on C by hyperplanes is complete. Combining the exact cohomology sequence of the Euler sequence

(5.7)
$$0 \to \mathcal{O}_C \to \mathcal{O}_C(1)^{m+1} \to \mathcal{H}om_{\mathcal{O}_{\mathbb{P}^m}}(\Omega^1_{\mathbb{P}^m}, \mathcal{O}_C) \to 0$$

with (5.6), we get the diagram of exact sequences

$$\begin{array}{ccccccc}
 & & & & 0 & & \\
 & & & & \downarrow & & \\
 & & & & \mathrm{Hom}(\Omega^1_C, \mathcal{O}_C(-D)) & & \\
 & & & & \downarrow & & \\
0 \to H^0(C, \mathcal{O}_C) & \to & H^0(C, \mathcal{O}_C(1))^{\oplus(m+1)} & \to & \mathrm{Hom}(\Omega^1_{\mathbb{P}^m}, \mathcal{O}_C) & \xrightarrow{\delta} & H^1(C, \mathcal{O}_C) \\
 & & & & \downarrow\gamma & & \\
 & & & & \mathrm{Hom}(\mathcal{C}^\bullet, \mathcal{D}^\bullet) & = & T_h(H) \\
 & & & & \downarrow\beta & & \\
 & & & & \mathrm{Ext}^1(\Omega^1_C, \mathcal{O}_C(-D)) & &
\end{array}$$

As we have observed, the map β associates to every first-order embedded deformation of $(C; p_1, \dots, p_n)$ the corresponding abstract deformation. The elements of $\mathrm{Hom}(\Omega^1_{\mathbb{P}^m}, \mathcal{O}_C)$ correspond to fiber space maps $C \times S \overset{j}{\hookrightarrow} \mathbb{P}^m \times S$, where $S = \mathrm{Spec}\,\mathbb{C}[\varepsilon]$, extending the inclusion of C in $\mathbb{P}^m$. The map δ associates to any such object the infinitesimal deformation of line bundles on C given by $j^*(\mathcal{O}_{\mathbb{P}^m}(1) \otimes \mathcal{O}_S)(\mathcal{O}_C(1) \otimes \mathcal{O}_S)^{-1}$. Finally, $H^0(C, \mathcal{O}_C(1))^{\oplus(m+1)}/H^0(C, \mathcal{O}_C)$ is the tangent space to the projective linear group $PGL(m+1)$. All these statements should by now be familiar, except perhaps those concerning $\mathrm{Hom}(\Omega^1_{\mathbb{P}^m}, \mathcal{O}_C)$ and δ. We shall return to these later. For the moment, we go on with the proof that $\mathrm{H}_{\nu,g,n}$ is smooth of dimension $3g - 3 + n + (r+1)^2 - 1$, specializing the above considerations to the case where $(C; p_1, \dots, p_n)$ is stable and $C \subset \mathbb{P}^m$ is embedded by the ν-fold log-canonical system (so that $m = r$). Pick an element v of $T_h(H)$ which is tangent to $\mathrm{H}_{\nu,g,n}$, and suppose in addition that it comes via γ from an element α of $\mathrm{Hom}(\Omega^1_{\mathbb{P}^m}, \mathcal{O}_C)$, i.e., from a map of fiber spaces $j : C \times S \to \mathbb{P}^m \times S$, such that $\mathcal{O}_{\mathbb{P}^m}(1) \otimes \mathcal{O}_S$ pulls back to $(\omega_C(D))^\nu \otimes \mathcal{O}_S$. Then $\delta(\alpha) = 0$ by the definition of δ. This means that α belongs to $H^0(C, \mathcal{O}_C(1))^{\oplus(r+1)}/H^0(C, \mathcal{O}_C)$. Conversely, it is clear that the image of $H^0(C, \mathcal{O}_C(1))^{\oplus(r+1)}/H^0(C, \mathcal{O}_C)$ in $T_h(H)$ is contained in the tangent space to $\mathrm{H}_{\nu,g,n}$ at h. At this point we need the following result, to be proved later.

LEMMA (5.8). *Let $(C; p_1, \dots, p_n)$ be a stable n-pointed curve. Then*

$$\mathrm{Hom}_{\mathcal{O}_C}(\Omega^1_C, \mathcal{O}_C(-\textstyle\sum p_i)) = 0\,.$$

Assuming the lemma, we get an exact sequence

$$0 \to H^0(C, \mathcal{O}_C(1))^{\oplus(r+1)}/H^0(C, \mathcal{O}_C) \to T_h(\mathrm{H}_{\nu,g,n}) \xrightarrow{\lambda} \mathrm{Ext}^1(\Omega^1_C, \mathcal{O}_C(-D))\,,$$

where λ stands for the restriction of β to $T_h(\mathrm{H}_{\nu,g,n})$. On the other hand, recall that $\mathrm{Ext}^1(\Omega^1_C, \mathcal{O}_C(-D))$ classifies abstract infinitesimal deformations of $(C; p_1, \dots, p_n)$. Any such can be embedded via the ν-fold log-canonical system, so the map λ is surjective. In conclusion, there is an exact sequence

(5.9)

$$0 \to H^0(C, \mathcal{O}_C(1))^{\oplus(r+1)}/H^0(C, \mathcal{O}_C) \to T_h(\mathrm{H}_{\nu,g,n}) \to \mathrm{Ext}^1(\Omega^1_C, \mathcal{O}_C(-D)) \to 0\,.$$

In particular,

$$\dim T_h(\mathrm{H}_{\nu,g,n}) = 3g - 3 + n + (r+1)^2 - 1\,,$$

and hence this quantity is an upper bound for the dimension of $\mathrm{H}_{\nu,g,n}$. If we can show that it is also a lower bound, we will have proved that $\mathrm{H}_{\nu,g,n}$ is smooth of dimension $3g - 3 + n + (r+1)^2 - 1$, as announced. To this end, consider the $(3g-3+n)$-dimensional deformation $\varphi : \mathcal{C} \to (B, b_0)$ given by Theorem (3.17). We shall write $\mathcal{C}_b$ for $\varphi^{-1}(b)$ and D_b for the divisor $\sum \sigma_i(b)$. Set $G = PGL(r+1)$ and consider the principal G-bundle over B defined by

$$\mathcal{B} = \{(b, F) |\, b \in B,\ F \text{ a basis for } H^0(\mathcal{C}_b, (\omega_{\mathcal{C}_b}(D_b))^\nu),\ \text{modulo homotheties}\}.$$

Let F_0 be the basis corresponding to the embedding $C \subset \mathbb{P}^r$. On the pulled back family

$$\mathcal{X} = \mathcal{B} \times_B \mathcal{C} \xrightarrow{\psi} \mathcal{B}\,, \qquad \tau_i : \mathcal{B} \to \mathcal{X}\,,\ i = 1, \dots, n,$$

there is a canonical projective frame for $\psi_*((\omega_{\mathcal{X}/\mathcal{B}}(\sum \tau_i))^\nu)$ which can be used to give a fiber space embedding

$$\begin{array}{ccc} \mathcal{X} & \hookrightarrow & \mathbb{P}^r \times \mathcal{B} \\ & \searrow \quad \swarrow & \\ & \mathcal{B} & \end{array}$$

By universality this family induces a map ξ from $\mathcal{B}$ to $\mathrm{H}_{\nu,g,n}$. There is a commutative diagram with exact rows

$$\begin{array}{ccccccc} 0 \longrightarrow & T_e(G) & \longrightarrow & T_{(b_0,F_0)}(\mathcal{B}) & \longrightarrow & T_{b_0}(B) & \longrightarrow 0 \\ & \| & & \downarrow d\xi & & \downarrow \rho & \\ & H^0(C,\mathcal{O}_C(1))^{\oplus(r+1)}/H^0(C,\mathcal{O}_C) & \longrightarrow & T_h(\mathrm{H}_{\nu,g,n}) & \longrightarrow & \mathrm{Ext}^1(\Omega^1_C,\mathcal{O}_C(-D)) & \longrightarrow 0 \end{array}$$

where ρ is the Kodaira–Spencer map. Since ρ is an isomorphism by construction of the family $\mathcal{C} \to (B, b_0)$, we get that $d\xi$ is an isomorphism. This shows that ξ is a local isomorphism at (b_0, F_0). Since $\mathcal{B}$ has the expected dimension $3g - 3 + n + (r+1)^2 - 1$, we get

$$\dim_h \mathrm{H}_{\nu,g,n} = \dim T_h(\mathrm{H}_{\nu,g,n}) = 3g - 3 + n + (r+1)^2 - 1\,,$$

proving our claim.

We now come to the proof of Lemma (5.8). Set $D = \sum p_i$, and consider the normalization map $\alpha : N \to C$. The exact sequence (2.20)

$$0 \to P \to \Omega^1_C \to \omega_C \to Q \to 0\,,$$

splits into two short exact sequences

$$0 \to P \to \Omega^1_C \to \alpha_*\omega_N \to 0\,, \quad 0 \to \alpha_*\omega_N \to \omega_C \to Q \to 0\,.$$

Since P is concentrated at the nodes, the first one of these tells us that

$$\mathrm{Hom}(\Omega^1_C, \mathcal{O}_C(-D)) = \mathrm{Hom}(\alpha_*(\omega_N), \mathcal{O}_C(-D))\,.$$

At a node, $\alpha_*(\omega_N)$ is generated by two sections, one vanishing identically on a branch of the node and one vanishing identically on the other. The image of either one under a homomorphism to $\mathcal{O}_C(-D)$ vanishes identically on one of the branches and hence at the node. Thus, denoting by $\mathcal{J}$ the ideal of the singular locus of C and by D' the divisor consisting of all the points of N mapping to nodes of C,

$$\begin{aligned} \mathrm{Hom}(\alpha_*(\omega_N), \mathcal{O}_C(-D)) &= \mathrm{Hom}(\alpha_*(\omega_N), \mathcal{J}\mathcal{O}_C(-D)) \\ &= \mathrm{Hom}(\alpha_*(\omega_N), \alpha_*(\mathcal{O}_N(-\alpha^*(D) - D'))) \\ &= H^0(N, \omega_N^{-1}(-\alpha^*(D) - D')). \end{aligned}$$

The last group vanishes, since $\omega_N^{-1}(-\alpha^*(D)-D')$ is the dual of the log-canonical sheaf of N, which has positive degree on every component of N, by stability.

Finally, as we promised, we return to the deformation-theoretic interpretation of the group $\mathrm{Hom}(\Omega^1_{\mathbb{P}^m}, \mathcal{O}_C)$ and of the coboundary map $\delta : \mathrm{Hom}(\Omega^1_{\mathbb{P}^m}, \mathcal{O}_C) \to H^1(C, \mathcal{O}_C)$ in the long cohomology sequence of (5.7). We recall that here $C \subset \mathbb{P}^m$ is a nodal curve. We set $S = \mathrm{Spec}\,\mathbb{C}[\varepsilon]$. Suppose we are given a map

$$C \times S \overset{j}{\hookrightarrow} \mathbb{P}^m \times S \tag{5.10}$$

of fiber spaces over S extending the inclusion $C \subset \mathbb{P}^m$. If f is a function or a form on $\mathbb{P}^m$, or on C, when no confusion is likely, we shall denote by the same symbol its pullback to $\mathbb{P}^m \times S$, or to $C \times S$, via the projection to the first factor. Given a section φ of $\Omega^1_{\mathbb{P}^m}$, pull it back to $C \times S$ via the composition of j with the projection of $\mathbb{P}^m \times S$ to $\mathbb{P}^m$. The pulled-back form can be written as $\psi + f d\varepsilon$, where ψ is a form on C, and f is a function on C. Associating f to φ yields a well-defined element of $\mathrm{Hom}(\Omega^1_{\mathbb{P}^m}, \mathcal{O}_C)$. We wish to show that any homomorphism $\alpha : \Omega^1_{\mathbb{P}^m} \to \mathcal{O}_C$ comes, via the procedure we have just described, from a unique morphism as in (5.10). What we have to find is a homomorphism $\eta : \mathcal{O}_{\mathbb{P}^m \times S} \to \mathcal{O}_{C \times S}$ extending the restriction homomorphism $\mathcal{O}_{\mathbb{P}^m} \to \mathcal{O}_C$. For any such η, given a function $f + \varepsilon h$ on $\mathbb{P}^m \times S$, where f and h are functions on $\mathbb{P}^m$, we must have

$$\eta(f + \varepsilon h) = \eta(f) + \varepsilon h\big|_C$$

and

$$\eta(f) = f\big|_C + \varepsilon\sigma$$

for a suitable function σ on C. Thus,

$$\eta^*(df) = d\eta(f) = d(f\big|_C + \varepsilon\sigma) = df\big|_C + \varepsilon d\sigma + \sigma d\varepsilon\,.$$

If α comes from η, we must then have

$$\sigma = \alpha(df)\,.$$

This proves that α comes from at most one η. To conclude the proof that $\mathrm{Hom}(\Omega^1_{\mathbb{P}^m}, \mathcal{O}_C)$ classifies maps of fiber spaces as in (5.10) extending $C \subset \mathbb{P}^m$, it remains to show that

$$\eta(f + \varepsilon h) = f\big|_C + \varepsilon(\alpha(df) + h\big|_C) \tag{5.11}$$

defines a homomorphism $\mathcal{O}_{\mathbb{P}^m \times S} \to \mathcal{O}_{C \times S}$ extending $\mathcal{O}_{\mathbb{P}^m} \to \mathcal{O}_C$. The only thing that needs some verification is that such an η is compatible with multiplication of functions. This is a straightforward calculation which is left to the reader. We also leave it to the reader to check that the interpretation of $\mathrm{Hom}(\Omega^1_{\mathbb{P}^m}, \mathcal{O}_C)$ as the set of fiber space maps $C \times S \hookrightarrow \mathbb{P}^m \times S$ extending $\mathcal{O}_{\mathbb{P}^m} \to \mathcal{O}_C$ is compatible, via γ, with the interpretation of $\mathrm{Hom}(\mathcal{C}^\bullet, \mathcal{D}^\bullet)$ as the tangent space to the Hilbert scheme.

We now come to the homomorphism

$$\delta : \mathrm{Hom}(\Omega^1_{\mathbb{P}^m}, \mathcal{O}_C) \to H^1(C, \mathcal{O}_C) .$$

What we claim is that δ associates to any map j as in (5.10) the infinitesimal deformation of line bundles on C given by $j^*(\mathcal{O}_{\mathbb{P}^m}(1) \otimes \mathcal{O}_S)(\mathcal{O}_C(1) \otimes \mathcal{O}_S)^{-1}$. To see this, we begin by choosing homogeneous coordinates $x_0, \dots, x_n$ on $\mathbb{P}^m$; for each i, we also let U_i be the open set in $\mathbb{P}^m$ (or in $\mathbb{P}^m \times S$) where x_i does not vanish. Next, let $\alpha \in \mathrm{Hom}(\Omega^1_{\mathbb{P}^m}, \mathcal{O}_C)$ be the homomorphism corresponding to j, and recall that the pullback homomorphism $\eta = j^* : \mathcal{O}_{\mathbb{P}^m \times S} \to \mathcal{O}_{C \times S}$ is given by formula (5.11). For brevity, we write L to denote the line bundle $\mathcal{O}_{\mathbb{P}^m}(1) \otimes \mathcal{O}_S$ on $\mathbb{P}^m \times S$ and M to denote $\mathcal{O}_C(1) \otimes \mathcal{O}_S$. Relative to the cover $\{U_h\}$, the line bundle L is given by the transition functions $\varphi_{hk} = x_h/x_k$, so $j^*(L)$ is given by the transition functions $\psi_{hk} = \eta(\varphi_{hk}) = (x_h/x_k)\big|_C + \varepsilon\alpha(d(x_h/x_k))$. On the other hand,

$$d\left(\frac{x_h}{x_k}\right) = \frac{dx_h}{x_k} - \frac{1}{x_k^2} x_h dx_k = \frac{x_h}{x_k}\frac{dx_h}{x_h} - \frac{x_k}{x_h}\frac{dx_k}{x_k} ;$$

In conclusion, the transition functions for $j^*(L)$ are

$$\psi_{hk} = \left(\frac{x_h}{x_k}\right)\bigg|_C \left(1 + \varepsilon\alpha\left[\frac{dx_h}{x_h} - \frac{dx_k}{x_k}\right]\right) .$$

Thus the Kodaira–Spencer class of the infinitesimal deformation of $\mathcal{O}_C$ given by $j^*(L) \otimes M^{-1}$ is the class in $H^1(C, \mathcal{O}_C)$ of the cocycle

$$u_{hk} = \alpha\left(\frac{dx_h}{x_h} - \frac{dx_k}{x_k}\right) .$$

We must show that this same cocycle represents the coboundary $\delta(\alpha)$. On each open set U_h there are sections ξ_i^h of $\mathcal{O}(1)$ such that α corresponds to contraction with the vector field

$$\sum_i \xi_i^h \frac{\partial}{\partial x_i} .$$

On the overlap between U_h and U_k we have that $\xi_i^k = \xi_i^h + x_i f_{hk}$, where the f_{hk} are holomorphic; the coboundary $\delta(\alpha)$ is the class in $H^1(C, \mathcal{O}_C)$

of the cocycle $\{f_{hk}\}$. On the other hand,

$$\begin{aligned} u_{hk} &= \alpha\left(\frac{dx_h}{x_h} - \frac{dx_k}{x_k}\right) \\ &= \sum_i \xi_i^k \left[\frac{\partial}{\partial x_i}\left(\frac{dx_h}{x_h}\right) - \frac{\partial}{\partial x_i}\left(\frac{dx_k}{x_k}\right)\right] \\ &= \sum_i \xi_i^k \left(\frac{\delta_{ih}}{x_h} - \frac{\delta_{ik}}{x_k}\right) \\ &= \frac{\xi_h^k}{x_h} - \frac{\xi_k^k}{x_k} \\ &= \frac{\xi_h^h}{x_h} - \frac{\xi_k^k}{x_k} + f_{hk} \,. \end{aligned}$$

Thus the two cocycles $\{u_{hk}\}$ and $\{f_{hk}\}$ differ by the coboundary of $\{\xi_h^h/x_h\}$. This completes the deformation-theoretic description of the homomorphism δ.

The following statement summarizes what has been proved in this section.

PROPOSITION (5.12). *Let g and n be nonnegative integers such that $2g-2+n > 0$, and let $\nu \geq 3$ be an integer. Set $r = (2\nu-1)(g-1)+\nu n-1$. Then the Hilbert scheme $\mathrm{H}_{\nu,g,n}$ parameterizing stable n-pointed genus g curves embedded in $\mathbb{P}^r$ via the ν-fold log-canonical system is smooth, quasi-projective, and of dimension*

$$\dim(\mathrm{H}_{\nu,g,n}) = 3g - 3 + n + (r+1)^2 - 1\,.$$

Moreover, the tangent space to $\mathrm{H}_{\nu,g,n}$ at a point h fits into an exact sequence (5.9), where λ is the Kodaira–Spencer map at h of the universal family on $\mathrm{H}_{\nu,g,n}$.

REMARK (5.13). It is useful to observe that $\mathrm{H}_{\nu,g,n}$ is naturally a smooth locally closed subscheme of the product

$$\mathrm{Hilb}_r^{p_\nu(t)} \times (\mathbb{P}^r)^n\,,$$

where $p_\nu(t)$ is given by (5.3). The natural action of $\mathbb{G} = PGL(r+1)$ on this product restricts to an action on $\mathrm{H}_{\nu,g,n}$, and the set-theoretical quotient $\mathrm{H}_{\nu,g,n}/\mathbb{G}$ is just the set of isomorphism classes of stable n-pointed genus g curves. Recall that the explicit construction of the Hilbert scheme we have given in Chapter IX exhibits $\mathrm{Hilb}_r^{p_\nu(t)}$ as a closed subscheme of a big projective space $\mathbb{P}^M = \mathbb{P}\wedge^k V$, where

$$V = H^0(\mathbb{P}^r, \mathcal{O}_{\mathbb{P}^r}(N))^\vee$$

for some large N, and where

$$k = \binom{N+r}{r} - p_\nu(N)\,.$$

Hence,

$$\mathrm{H}_{\nu,g,n} \subset \mathbb{P}^M \times (\mathbb{P}^r)^n \subset \mathbb{P}^K\,,$$

where the last inclusion is the Segre embedding. Of course, the action of $\mathbb{G}$ on $\mathrm{H}_{\nu,g,n}$ is given by $\mathbb{G} = PGL(r+1)$ acting on $\mathbb{P}^K$ via the obvious representation.

6. Construction of Kuranishi families.

We shall now use the Hilbert scheme $\mathrm{H}_{\nu,g,n}$ introduced in preceding section to construct Kuranishi families. Actually, we shall construct Kuranishi families endowed with several additional properties (e.g., they will be algebraic). We will be amply repaid for the additional effort involved when we construct moduli spaces in Chapter XII. Pick a stable n-pointed curve $(C; p_1, \ldots, p_n)$, fix an integer $\nu \geq 3$ and a ν-fold log-canonical embedding

$$\varphi : C \to \mathbb{P}^r\,, \quad r = (2\nu - 1)(g-1) - 1 + \nu n\,.$$

We identify C with its image via φ and denote by x_0 the corresponding point in $\mathrm{H}_{\nu,g,n}$ (which from now on we will simply denote by H). We let

$$\pi : \mathcal{Y} \to H\,, \quad \sigma_i : H \to \mathcal{Y}\,,\ i = 1, \ldots, n\,, \tag{6.1}$$

be the universal family on H, and for any subscheme or analytic subspace W of H, we shall write $\pi_W : \mathcal{Y}_W \to W$ to indicate its restriction to W; we will often omit mentioning the sections, and in any case their restrictions will always be denoted by the same symbols $\sigma_1, \ldots, \sigma_n$. Following Remark (5.13), we regard H as a smooth locally closed subscheme of a large projective space $\mathbb{P}^K$, acted on by $\mathbb{G} = PGL(r+1) \hookrightarrow PGL(K+1)$. For any point $x \in H$, let $\mathbb{G}_x \subset \mathbb{G}$ be the stabilizer of x. Since $\mathbb{G}_x$ can be identified with the automorphism group of the corresponding stable curve, it is a finite group. Consider the orbit $O(x_0) \subset H$ of x_0 under $\mathbb{G}$; this is a smooth subvariety of H of dimension $(r+1)^2 - 1$ passing through x_0. Since the linear subspace T of $\mathbb{P}^M$ tangent to $O(x_0)$ at x_0 is obviously $\mathbb{G}_{x_0}$-invariant, there is a $\mathbb{G}_{x_0}$-invariant linear subspace L of $\mathbb{P}^M$ of complementary dimension such that $L \cap T = \{x_0\}$. We now intersect H with L and claim that we can find a Zariski-open neighborhood X of x_0 in $H \cap L$ such that

i) X is affine

ii) X intersects $O(y)$ transversely at y for every $y \in X$,
iii) X is $\mathbb{G}_{x_0}$-invariant,
iv) for every $y \in X$, $\mathbb{G}_y \subset \mathbb{G}_{x_0}$.
v) for every $y \in X$, there is a $\mathbb{G}_y$-invariant neighborhood U of y in X, for the analytic topology, such that $\{\gamma \in \mathbb{G} : \gamma U \cap U \neq \emptyset\} = \mathbb{G}_y$.

That the first two properties can be satisfied is obvious, as is obvious that, once i), ii), iv, and v) are met, iii) can be satisfied as well. We next prove v), arguing by contradiction. Suppose there are sequences x_n and y_n of points of X converging to y, and elements $\gamma_n \in \mathbb{G} \setminus \mathbb{G}_y$ such that $\gamma_n x_n = y_n$. By Theorem (5.1) of Chapter X, possibly after passing to a subsequence, there is a $\gamma \in \mathbb{G}$ such that $\lim_{n\to\infty} \gamma_n = \gamma$. Clearly $\gamma \in G_y$, so that, replacing γ_n with $\gamma_n\gamma^{-1}$ and y_n with $\gamma^{-1}y_n$, we may assume that γ is the identity element e of G. Now consider the multiplication map $F : \mathbb{G} \times X \to H$ given by $F(\eta, x) = \eta x$. The transversality condition ii) tells us that F is a local biholomorphism at the point (e, y). On the other hand, we just constructed two sequences (e, y_n) and (γ_n, x_n), with $\gamma_n \neq e$ for every n, both converging to (e, y), and such that $F(e, y_n) = F(\gamma_n, x_n)$. This is absurd. In conclusion, one may find a neighborhood U of y in X such that the set of elements γ in $\mathbb{G}$ having the property that $\gamma U \cap U \neq \emptyset$ is equal to $\mathbb{G}_y$. By the finiteness of $\mathbb{G}_y$ the neighborhood U can be chosen to be $\mathbb{G}_y$-invariant.

To show that iv) can be met, we argue as follows. Set

$$I = \{(y, \gamma) | y \in X, \gamma \in \mathbb{G}_y\}.$$

Notice that $I = \mathbf{Isom}_X(\mathcal{Y}_X, \mathcal{Y}_X)$, and hence I is proper over X, by Theorem (5.1) of Chapter X. Therefore, removing from X the projections of those components of I which do not meet $\{x_0\} \times \mathbb{G}_{x_0}$, we can assume that, given any component I' of I, we have $I' \cap (\{x_0\} \times \mathbb{G}_{x_0}) \neq \emptyset$. Set $J = \{(y, \gamma) \in I' |\ \gamma \in \mathbb{G}_{x_0}\}$. Clearly, J is a nonempty Zariski-closed subset of I'. We must show that $I' = J$. To do this, it suffices to show that J contains a subset of I' which is open in the ordinary topology. This is an immediate consequence of v) applied to $y = x_0$. This concludes the proof that one can construct an $X \subset H \cap L$ satisfying i), ii), iii), iv), and v).

Now consider the restriction of the universal family (6.1) to X. We shall show that

$$\text{(6.2)} \qquad \begin{array}{c} \mathcal{Y}_X \\ \downarrow \pi \\ (X, x_0) \end{array} \qquad \begin{array}{l} \sigma_i : X \to \mathcal{Y}_X\,, \quad i = 1, \dots, n\,, \\ \varphi : (C; p_1, \dots, p_n) \xrightarrow{\cong} (\pi^{-1}(x_0); \sigma_1(x_0), \dots, \sigma_n(x_0)) \end{array}$$

is a Kuranishi family for $(C; p_1, \ldots, p_n)$. Let

$$(6.3) \qquad \begin{array}{c} \mathcal{D} \\ \big\downarrow \eta \\ (S, s_0) \end{array} \qquad \begin{array}{r} \tau_i : S \to \mathcal{D}, \ i = 1, \ldots, n, \\ \psi : (C; p_1, \ldots, p_n) \xrightarrow{\cong} (\eta^{-1}(s_0); \tau_1(x_0), \ldots, \tau_n(x_0)) \end{array}$$

be another deformation of $(C; p_1, \ldots, p_n)$. As we already observed, the transversality condition ii) shows that there are a neighborhood B of x_0 in X and a neighborhood V of e in $\mathbb{G}$ such that the multiplication map

$$\begin{aligned} F : V \times B &\longrightarrow H \\ (\gamma, b) &\mapsto \gamma b \end{aligned}$$

is a biholomorphism onto an open neighborhood W of x_0 in H. The inclusion $\varphi : C \hookrightarrow \mathbb{P}^r$ gives a canonical basis for $H^0(C, \omega_C^\nu(\nu \sum p_i))$. Transplant this basis to $\eta^{-1}(s_0)$ via the isomorphism $\psi : C \to \eta^{-1}(s_0)$ and extend it to a frame of $\eta_*(\omega^\nu_{\mathcal{D}/S}(\nu \sum \tau_i))$ over a neighborhood A of s_0. This frame can be used to realize $\eta : \mathcal{D}_A \to A$ as a family of subschemes of $\mathbb{P}^r$:

$$\begin{array}{ccc} \mathcal{D}_A & \hookrightarrow & A \times \mathbb{P}^r \\ \big\downarrow \eta & \swarrow & \\ A & & \end{array}$$

By the universal property of the Hilbert scheme this family of n-pointed curves is induced by a unique holomorphic map $f : A \to H$. Shrinking A, if necessary, we may assume that $f(A) \subset W$. We then get a cartesian diagram

$$(6.4) \qquad \begin{array}{ccc} \mathcal{D}_A & \xrightarrow{\tilde{f}} & \mathcal{Y}_W \subset \mathcal{Y} \\ \big\downarrow \eta & & \big\downarrow \pi \\ A & \xrightarrow{f} & W \subset H \end{array}$$

of families of n-pointed curves. The situation is illustrated by the picture below.

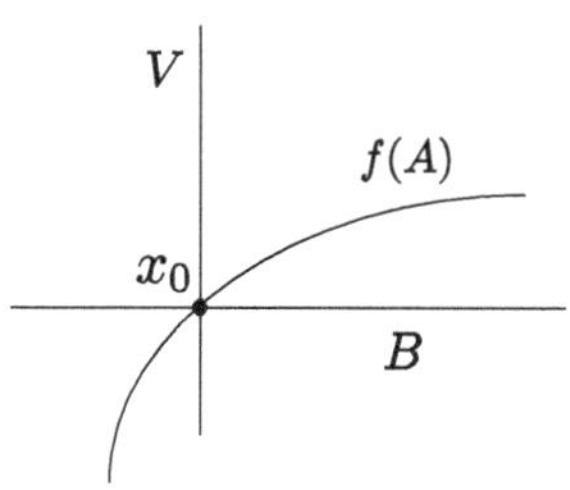

We now define $\alpha : A \to B \subset X$ by setting

$$\alpha = p_B F^{-1} f ,$$

where $p_B : V \times B \to B$ is the projection. We also set $\beta = p_V F^{-1} f$. We wish to show that the deformation

$$\begin{array}{c} \mathcal{Y}_B \\ \Big\downarrow \pi \\ (B, x_0) \end{array} \qquad \sigma_i : B \to \mathcal{Y}_B\,,\ i = 1, \dots, n\,, \qquad \varphi : C \to \pi^{-1}(x_0)$$

induces, via α, the restriction of the deformation (6.3) to A. Look at the cartesian diagram (6.4). The group $\mathbb{G}$ acts equivariantly on $\mathcal{Y} \to H$; therefore, if we replace f and $\tilde{f}$ with f' and $\tilde{f}'$ defined by

$$f'(a) = \beta(a)^{-1} f(a)\ ,\ \ \tilde{f}'(q) = \beta(\eta(q))^{-1} \tilde{f}(q)\,,$$

we get the other cartesian diagram

$$\begin{array}{ccc} \mathcal{D}_A & \xrightarrow{\ \tilde{f}'\ } & \mathcal{Y}_W \\ \Big\downarrow \eta & & \Big\downarrow \pi \\ A & \xrightarrow{\ f'\ } & W \end{array}$$

On the other hand, by definition,

$$f(a) = F(\beta(a), \alpha(a)) = \beta(a)\alpha(a)$$

so that $f' = \alpha$, and the cartesian diagram in question reduces to

$$\begin{array}{ccc} \mathcal{D}_A & \xrightarrow{\ \tilde{f}'\ } & \mathcal{Y}_B \\ \Big\downarrow \eta & & \Big\downarrow \pi \\ A & \xrightarrow{\ \alpha\ } & B \subset W \end{array}$$

Since $\beta(s_0) = e \in \mathbb{G}$, the identification φ of C with $\pi^{-1}(x_0)$ pulls back via $\tilde{f}'$ to the identification $\psi : C \to \eta^{-1}(s_0)$. This shows that (6.2) is a Kuranishi family for $(C; p_1, \dots, p_n)$. Notice that, among properties i)–v) of X, the only one we used in proving the universal property of (6.2) is the transversality condition ii). In particular it follows that (6.2) is a Kuranishi family for any one of its fibers.

We may summarize what we have done in the following:

Theorem (6.5). *Let $\nu \geq 3$ be an integer. Let $(C; p_1, \ldots, p_n) \subset \mathbb{P}^r$ be a stable n-pointed genus g curve, embedded in $\mathbb{P}^r$, $r = (2\nu-1)(g-1)+\nu n-1$, via the ν-fold log-canonical system. Let $x_0 \in \mathrm{H}_{\nu,g,n}$ be the corresponding Hilbert point, and let $\mathrm{Aut}(C; p_1, \ldots, p_n) = \mathbb{G}_{x_0} \subset \mathbb{G} = PGL(r+1)$ be the stabilizer of x_0. Then there is a locally closed $(3g-3+n)$-dimensional smooth subscheme X of $\mathrm{H}_{\nu,g,n}$ passing through x_0 such that the restriction to X of the universal family over $\mathrm{H}_{\nu,g,n}$ is a Kuranishi family for all of its fibers and hence, in particular, a Kuranishi family for $(C; p_1, \ldots, p_n)$. In addition, one can choose an X with the following properties:*

a) *X is affine;*
b) *X is $\mathbb{G}_{x_0}$-invariant;*
c) *for every $y \in X$, the stabilizer $\mathbb{G}_y$ of y is contained in $\mathbb{G}_{x_0}$;*
d) *for every $y \in X$, there is a $\mathbb{G}_y$-invariant neighborhood U of y in X, for the analytic topology, such that $\{\gamma \in \mathbb{G} : \gamma U \cap U \neq \emptyset\} = \mathbb{G}_y$.*

Now let

$$\begin{array}{c} \mathcal{C} \\ \big\downarrow \pi \\ (X, x_0) \end{array} \qquad \sigma_i : X \to \mathcal{C}, \ i = 1, \ldots, n, \qquad C = \pi^{-1}(x_0) \tag{6.6}$$

be the Kuranishi family constructed in the theorem. Since the projective group $PGL(r+1)$ acts equivariantly on the Hilbert scheme $\mathrm{H}_{\nu,g,n}$ and on the universal family over it, the $\mathbb{G}_{x_0}$-invariance of X implies that the finite group $\mathrm{Aut}(C; p_1, \ldots, p_n) = \mathbb{G}_{x_0}$ acts equivariantly on $\mathcal{C} \to X$. Moreover, the action of $\mathrm{Aut}(C; p_1, \ldots, p_n)$ on the central fiber $\pi^{-1}(x_0) = C$ is the natural one. This allows us to interpret Theorem (6.5) as the assertion that there exists a standard algebraic Kuranishi family for $(C; p_1, \ldots, p_n)$ in the sense of the following definition.

Definition (6.7) (Standard algebraic Kuranishi family). *Let $(C; p_1, \ldots, p_n)$ be a stable n-pointed curve and set $G = \mathrm{Aut}(C; p_1, \ldots, p_n)$. We will say that a Kuranishi family (6.6) for $(C; p_1, \ldots, p_n)$ is a standard algebraic Kuranishi family if the following conditions are satisfied. Denote by C_y the fiber of π over y and let G_y be the automorphism group of $(C_y; \sigma_1(y), \ldots, \sigma_n(y))$. Then*

a) *X is affine;*
b) *the family is Kuranishi at every point of X;*
c) *the action of the group G_{x_0} on the central fiber extends to compatible actions on $\mathcal{C}$ and X;*
d) *for every $y \in X$, the automorphism group G_y is equal to the stabilizer of y in G_{x_0}. In particular, G_y is a subgroup of G_{x_0};*
e) *for every $y \in X$, there is a G_y-invariant neighborhood U of y in X, for the analytic topology, such that any isomorphism (of n-pointed curves) between fibers over U is induced by an element of G_y.*

The only property that may need some explanation is e). But this follows at once from part d) of the theorem. In fact, any isomorphism between fibers over points of U is induced by a projectivity, that is, by an element of $\mathbb{G}$.

The following is a local counterpart of the notion of standard algebraic Kuranishi family.

DEFINITION (6.8) (STANDARD KURANISHI FAMILY). *Let $(C; p_1, \dots, p_n)$ be a stable n-pointed curve and set $G = \mathrm{Aut}(C; p_1, \dots, p_n)$. We will say that a Kuranishi family*

$$\mathcal{X} \to (B, b_0) \qquad \tau_i : B \to \mathcal{X}\,, \ \ i = 1, \dots, n\,, \qquad \varphi : C \xrightarrow{\simeq} \mathcal{X}_{b_0}$$

for $(C; p_1, \dots, p_n)$ is a standard Kuranishi family if the following conditions are satisfied:

i) *B is a connected complex manifold;*
ii) *the family is Kuranishi at every point of B;*
iii) *the action of G on the central fiber extends to compatible actions on $\mathcal{X}$ and B;*
iv) *any isomorphism (of n-pointed curves) between fibers is induced by an element of G.*

Standard Kuranishi families always exist. In fact, given any Kuranishi family, there is a neighborhood of the base point such that the restriction of the family to this neighborhood is standard. By the uniqueness of the Kuranishi family (cf. (4.4)), it suffices to notice that this is true for a standard algebraic Kuranishi family, as is implicit in the definitions.

REMARK (6.9). The universal property of the Kuranishi family (6.6), which has been stated in the analytic setup, has an exact counterpart in the algebraic category, provided that one works in the étale topology. What is true is that, given an *algebraic* deformation (6.3), there is an étale base change $(S', s'_0) \to (S, s_0)$, with S' connected, such that there is a unique morphism of deformations from the pulled-back family to (6.6).

REMARK (6.10). A useful application of the existence of Kuranishi families is that any family of nodal curves can be locally embedded in a family of nodal curves with a reduced, or even smooth, base. To see this, start with an analytic family of nodal curves $\eta : X \to S$ and let s_0 be a point of S. Let $\pi : \mathcal{X} \to (B, b_0)$ be a versal deformation for $\eta^{-1}(s_0)$. Then, possibly after shrinking S, there are a closed embedding $S \hookrightarrow T$, where T is smooth, and a cartesian square

$$\begin{array}{ccc} X & \xrightarrow{\ \beta\ } & \mathcal{X} \\ {\scriptstyle\eta}\big\downarrow & & \big\downarrow{\scriptstyle\pi} \\ S & \xrightarrow{\ \alpha\ } & B \end{array}$$

with $b_0 = \alpha(s_0)$. Out of this we construct another diagram of cartesian squares

$$\begin{array}{ccccc} X & \xrightarrow{(\eta,\beta)} & S \times \mathcal{X} & \longrightarrow & T \times \mathcal{X} \\ \eta \downarrow & & \downarrow (\mathrm{id}_S, \pi) & & \downarrow (\mathrm{id}_T, \pi) \\ S & \xrightarrow{(\mathrm{id}_S, \alpha)} & S \times B & \longrightarrow & T \times B \end{array}$$

Clearly, $T \times B$ is smooth, and $S \to T \times B$ is an embedding.

What we have just proved holds also in the algebraic setup, provided that we interpret "locally" as meaning "locally in the étale topology," and hence "shrinking S" as meaning that we base change $\eta : X \to S$ via a suitable étale morphism $S' \to S$ where S' is affine, and replace S with S'. All the rest goes through unchanged.

We close this section by describing the local structure of the Hilbert scheme $\mathbf{Aut}_B(\mathcal{X}) = \mathbf{Isom}_B(\mathcal{X}, \mathcal{X})$ parameterizing automorphisms of the fibers of a standard Kuranishi family

$$\alpha : \mathcal{X} \to (B, b_0)$$

for the stable n-pointed genus g curve $(C; p_1, \ldots, p_n)$. We already know, from Section 5 of Chapter X, that $\mathbf{Aut}_B(\mathcal{X})$ is finite and unramified over B. We denote by G the automorphism group of $(C; p_1, \ldots, p_n)$. It follows from property iv) of standard Kuranishi families that there is a closed immersion

$$\mathbf{Aut}_B(\mathcal{X}) \hookrightarrow G \times B\,.$$

LEMMA (6.11). *Write* $\mathbf{Aut}_B(\mathcal{X}) = \coprod A_\gamma$, *where* A_γ *stands for the intersection of* $\mathbf{Aut}_B(\mathcal{X})$ *with* $\{\gamma\} \times B$, $\gamma \in G = \mathrm{Aut}(C; p_1, \ldots, p_n)$. *For each* γ, *let* B^γ *be the fixed subspace of* B *under the action of* γ. *Then* B^γ *is smooth, and* A_γ *maps isomorphically to* B^γ *under the natural projection* $\mathbf{Aut}_B(\mathcal{X}) \to B$. *The tangent space to* B^γ *at* b_0 *is the space of* γ*-invariants* $T_{b_0}(B)^\gamma = \mathrm{Ext}^1_{\mathcal{O}_C}(\Omega^1_C, \mathcal{O}_C(-\sum p_i))^\gamma$. *Finally, possibly after shrinking* B, *in suitable coordinates centered at* b_0, *each of the* B^γ *is defined by linear equations.*

There is not much here that needs proof. That the B^γ are defined by linear equations is a consequence of the following well-known observation of Henri Cartan [106].

LEMMA (6.12). *Let* G *be a finite group acting on a complex manifold* U *and fixing a point* $u \in U$. *Then, on a suitable neighborhood of* u *and in suitable coordinates centered at* u, *the action of* G *is linear.*

Proof. Choose a system of coordinates centered at u. For each element $\gamma \in G$, denote by γ' its tangent transformation in the chosen coordinates. Set

$$\alpha = \frac{1}{|G|} \sum_{\gamma \in G} (\gamma')^{-1} \gamma.$$

This is a biholomorphism on a neighborhood of u, since its Jacobian at u is the identity. Moreover, one immediately checks that $\alpha\gamma = \gamma'\alpha$ for every $g \in G$. It follows that G acts linearly in the system of coordinates obtained from the original ones via α.

Q.E.D.

Lemma (6.12) proves, in particular, the smoothness of B^γ. It also identifies the action of G on B with the one on $T_{b_0}(B)$, proving the statement about the tangent space of B^γ. Finally, notice that $A_\gamma \to B^\gamma$ is a bijection since G acts equivariantly on $\mathcal{X}$ and B, and hence any γ fixing $b \in B$ restricts to an automorphism of $\alpha^{-1}(b)$; that $A_\gamma \to B^\gamma$ is an isomorphism then follows from Theorem (5.1) in Chapter X. This finishes the proof of Lemma (6.11).

REMARK (6.13). A straightforward consequence of Lemma (6.11) is that, if H is a subgroup of $G = \mathrm{Aut}(C; p_1, \dots, p_n)$, then the locus B^H of those points of B which are fixed under the action of H is smooth and its tangent space at b_0 can be identified with the space of invariants $T_{b_0}(B)^H = \mathrm{Ext}^1_{\mathcal{O}_C}(\Omega^1_C, \mathcal{O}_C(-\sum p_i))^H$. Furthermore, H is clearly a subgroup of $\mathrm{Aut}(\pi^{-1}(b))$ for any $b \in B^H$.

We next apply Lemma (6.11) to the case where C is a hyperelliptic stable curve and γ is its hyperelliptic involution. We first need a general remark. Let $X \to S$ be a proper flat morphism of analytic spaces and suppose that a finite group G acts on X, compatibly with π. We claim that $X/G \to S$ is flat. In fact, if p is a point of X, q its image in $W = X/G$, and s its image in S, then $\mathcal{O}_{W,q}$ is the subring of H-invariants of $\mathcal{O}_{X,p}$, where $H \subset G$ is the stabilizer of p. As such, when viewed as an $\mathcal{O}_{S,s}$-module, it is a direct summand of $\mathcal{O}_{X,p}$. The flatness of $\mathcal{O}_{W,q}$ over $\mathcal{O}_{S,s}$ thus follows from the one of $\mathcal{O}_{X,p}$.

LEMMA (6.14). *Let C be a stable curve of genus $g > 1$, and let*

$$\alpha : X \to (B, b_0), \qquad C \cong \alpha^{-1}(b_0),$$

be a standard Kuranishi family for C. Suppose that there are points $b \in B$ arbitrarily close to b_0 such that $\alpha^{-1}(b)$ is smooth hyperelliptic. Then C is hyperelliptic.

Proof. Possibly after shrinking B, we may choose coordinates on it, centered at b_0, in which the action of $\mathrm{Aut}(C)$ is linear. Pick a point b such that $\alpha^{-1}(b)$ is smooth hyperelliptic. By property iv) of standard Kuranishi families, there is an automorphism γ of C which fixes b and acts on $\alpha^{-1}(b)$ as the hyperelliptic involution. Clearly, γ has order two. Let $L \subset B^\gamma$ be the (piece of) line joining b_0 and b, and set $Y = \alpha^{-1}(L)$. Then $Y \to L$ is a family of stable curves, and γ acts on Y fiberwise. The fixed locus of γ does not contain components of fibers. In fact, if all points of a component E of a fiber were fixed, by Lemma (6.12) γ

would have to act nontrivially on the normal space to E at a general point and hence would have to act nontrivially on L, a contradiction. In particular, γ acts on C with isolated fixed points. We know that $Z = Y/\gamma$ is flat over L. It is easily seen that the fibers of $\beta : Z \to L$ are nodal curves. On the other hand, since $\alpha^{-1}(b)$ is smooth hyperelliptic and γ induces on it the hyperelliptic involution, the general fiber of β is a $\mathbb{P}^1$. Since the genus of $\beta^{-1}(h)$ is locally constant as a function of h by flatness, it follows that C/γ is a nodal curve of genus zero.

Q.E.D.

LEMMA (6.15). *Let C be a hyperelliptic stable curve of genus $g > 1$, and let*

$$\alpha : X \to (B, b_0)\,, \qquad C \cong \alpha^{-1}(b_0)\,,$$

be a standard Kuranishi family for C. Set $H = \{b \in B : \alpha^{-1}(b)$ is hyperelliptic$\}$ and let W be any component of the locus in B parameterizing singular curves. Then:

i) *H is a smooth complex submanifold of B of dimension $2g - 1$;*
ii) *H and W are transverse at b_0;*
iii) *if b is a general point of H, then $\alpha^{-1}(b)$ is smooth.*

Proof. Since $X \to B$ is a Kuranishi family at any point of B, in the proof we may shrink B at will. We may thus assume that, in suitable global coordinates on B, the automorphism group of C acts linearly on B. We begin by proving ii) and iii). Look at the exact sequence (3.4) and rewrite the term on the right, that is, $H^0(C, \mathcal{E}xt^1(\Omega^1_C, \mathcal{O}_C))$, grouping its summands according to node type, as the direct sum of the three terms

$$T_1 = \bigoplus_{i>0} \bigoplus_{\substack{p \text{ a node} \\ \text{of type } \delta_i}} \operatorname{Ext}^1(\Omega^1_{C,p}, \mathcal{O}_{C,p}),$$

$$T_2 = \bigoplus_{\substack{p \text{ a node} \\ \text{of type } \eta_0}} \operatorname{Ext}^1(\Omega^1_{C,p}, \mathcal{O}_{C,p}),$$

$$T_3 = \bigoplus_{i>0} \bigoplus_{\substack{\{p,q\} \text{ a pair of} \\ \text{nodes of type } \eta_i}} \left(\operatorname{Ext}^1(\Omega^1_{C,p}, \mathcal{O}_{C,p}) \oplus \operatorname{Ext}^1(\Omega^1_{C,q}, \mathcal{O}_{C,q})\right).$$

The hyperelliptic involution γ acts on all terms of (3.4). More in detail, we claim that it acts on the term on the right via the trivial action on T_1 and T_2, and by interchanging the summands in each term $\operatorname{Ext}^1(\Omega^1_{C,p}, \mathcal{O}_{C,p}) \oplus \operatorname{Ext}^1(\Omega^1_{C,q}, \mathcal{O}_{C,q})$ of T_3. This last assertion is clear. As for T_1 and T_2, it is convenient to use the isomorphism (cf. (3.8))

$$\operatorname{Ext}^1(\Omega^1_{C,p}, \mathcal{O}_{C,p}) \cong T_{N,p_1} \otimes T_{N,p_2}\,,$$

where N stands for the partial normalization of C at p, and p_1, p_2 are the points of N mapping to p. If p is a node of type δ_i, γ acts as multiplication by -1 on T_{N,p_1} and T_{N,p_2}, and hence acts trivially on their tensor product. If p is of type η_0, the action of γ is the interchange of the two factors and hence is again trivial. This proves our claim. What we have shown implies that there are γ-invariant elements of $H^0(C, \mathcal{E}xt^1(\Omega^1_C, \mathcal{O}_C))$ which map to a nonzero element of $\operatorname{Ext}^1(\Omega^1_{C,p}, \mathcal{O}_{C,p})$ for every $p \in \operatorname{Sing}(C)$. Any such element of $H^0(C, \mathcal{E}xt^1(\Omega^1_C, \mathcal{O}_C))$ can be lifted to a γ-invariant element $v \in \operatorname{Ext}^1_{\mathcal{O}_C}(\Omega^1_C, \mathcal{O}_C) = T_{b_0}(B)$. In other words, v is tangent to B^γ. On the other hand, $T_{b_0}(W)$ is a codimension 1 vector subspace of $T_{b_0}(B)$ whose image in $H^0(C, \mathcal{E}xt^1(\Omega^1_C, \mathcal{O}_C))$ is the direct sum of all the vector spaces $\operatorname{Ext}^1(\Omega^1_{C,p}, \mathcal{O}_{C,p})$, with p a node of C, except one. This proves that $T_{b_0}(W)$ and $T_{b_0}(B^\gamma)$ are transverse.

By what we have just shown, there are vectors $v \in T_{b_0}(B^\gamma)$ which are transverse to all components of the locus in B parameterizing singular curves. Thus, any deformation of C with v as Kodaira–Spencer class has the effect of smoothing all the nodes of C. In conclusion, the fiber above a general point of B^γ is smooth. On the other hand, every fiber $\alpha^{-1}(b)$ with $b \in B^\gamma$ is hyperelliptic, and γ acts on it as the hyperelliptic involution. To see this, one argues essentially as in the proof of Lemma (6.14). Set $Y = \alpha^{-1}(B^\gamma)$, $Z = Y/\gamma$, and consider the family of nodal curves $\beta : Z \to B^\gamma$. Since the fiber $\beta^{-1}(b_0)$ has genus zero by assumption, it follows that every fiber of β has genus zero. This shows that $B^\gamma \subset H$, proving ii), and at the same time proves iii).

It remains to prove i). It is implicit in the proof of (6.14) that H is contained in B^γ, so that in fact $H = B^\gamma$. Thus H is smooth. It remains to prove the dimension statement. Since H is smooth and the fiber of α at its general point is smooth, it suffices to prove it under the additional assumption that C is smooth. What we must show is that $H^1(C, T_C)^\gamma$ has dimension $2g-1$; dually, we must see that this is the dimension of the cotangent space to H at b_0, that is, of the space of (co)invariants $H^0(C, \omega_C^2)^\gamma$. Denote by $f : C \to \mathbb{P}^1$ the quotient map modulo the hyperelliptic involution and by D its branch locus. An elementary local calculation shows that any invariant quadratic differential on C is the pullback via f of a quadratic differential on $\mathbb{P}^1$ with simple poles along D, and conversely. Thus, $H^0(C, \omega_C^2)^\gamma$ can be identified with $H^0(\mathbb{P}^1, \omega_{\mathbb{P}^1}^2(D)) \cong H^0(\mathbb{P}^1, \mathcal{O}(2g-2))$ and hence has dimension $2g-1$, as claimed.

Q.E.D.

EXERCISE (6.16). Give a complete proof of the claim in Remark (6.9), mimicking the proof we have given in the analytic case.

7. The Kuranishi family and continuous deformations.

In Section 9 of Chapter IX we have introduced the notion of continuous, or differentiable, family of compact complex manifolds. We can therefore also speak about continuous, or differentiable, deformations of compact complex manifolds, and in particular of smooth curves. In this short section we wish to show that the universal property of the Kuranishi family also holds in this context. For simplicity, we shall limit ourselves to deformations of unpointed curves, but the arguments go through with virtually no change in the n-pointed case. Here is what we wish to prove.

PROPOSITION (7.1). *Let $\alpha : Y \to B$ be a C^m family of compact Riemann surfaces, and let b_0 be a point of B. Let $\pi : \mathcal{C} \to (X, x_0)$, $\pi^{-1}(x_0) \simeq Y_{b_0}$, be a Kuranishi family for Y_{b_0}. Then, for any sufficiently small neighborhood A of b_0, there is a cartesian diagram*

$$\begin{array}{ccc} \alpha^{-1}(A) & \xrightarrow{F} & \mathcal{C} \\ {\scriptstyle\alpha}\downarrow & & \downarrow{\scriptstyle\pi} \\ A & \xrightarrow{f} & X \end{array}$$

where $f(b_0) = x_0$, and (f, F) is a morphism of C^m families of complex manifolds such that F induces the identity on Y_{b_0}.

We claim that, possibly after shrinking B, we can find a cartesian diagram

(7.2)
$$\begin{array}{ccc} Y & \xrightarrow{H} & Y' \\ {\scriptstyle\alpha}\downarrow & & \downarrow{\scriptstyle\alpha'} \\ B & \xrightarrow{h} & B' \end{array}$$

where (h, H) is a morphism of C^m families of complex manifolds, and $Y' \to B'$ is a family of complex manifolds in the ordinary sense. The result then follows from the standard universal property of the Kuranishi family, applied to $Y' \to B'$.

In proving the existence of (7.2) we distinguish two cases. Suppose first that $Y \to B$ is real-analytic. We may assume that B is a ball in $\mathbb{R}^\ell$ centered at the origin, and that b_0 coincides with the origin. The family $Y \to B$ is given by real-analytic transition functions which are holomorphic in the vertical coordinates. The power series expansions of these make perfect sense even if the B-coordinates are allowed to take complex values with sufficiently small imaginary part, and define a complex-analytic family $\alpha' : Y' \to B'$, where B' is a neighborhood of B in $\mathbb{C}^\ell \supset \mathbb{R}^\ell$, whose restriction to B is just $\alpha : Y \to B$.

If m is a nonnegative integer or ∞, we have to argue differently. Let $\beta : \mathcal{Y} \to S$ be a C^m family of compact complex manifolds. Recall from Section 9 of Chapter IX that $\widetilde{\mathcal{O}}_{\mathcal{Y}}$, $\widetilde{\mathcal{A}}^{0,1}_{\mathcal{Y}}$, and $\widetilde{\mathcal{A}}_{\mathcal{Y}}$ denote, respectively, the sheaf on $\mathcal{Y}$ of all C^m functions whose restrictions to the fibers are holomorphic, the sheaf of adapted relative (0,1)-forms on $\mathcal{Y}$, and the sheaf of adapted functions on $\mathcal{Y}$. If $\mathcal{G}$ is a locally free $\widetilde{\mathcal{O}}_{\mathcal{Y}}$-module, for any $s \in S$, we denote by $\mathcal{G}_s$ the locally free $\mathcal{O}_{\mathcal{Y}_s}$-module $\mathcal{G} \otimes_{\widetilde{\mathcal{O}}_{\mathcal{Y}}} \mathcal{O}_{\mathcal{Y}_s}$.

LEMMA (7.3). *Let $\beta : \mathcal{Y} \to S$ be a C^m family of compact complex manifolds, and let $\mathcal{G}$ be a locally free $\widetilde{\mathcal{O}}_{\mathcal{Y}}$-module. Let s_0 be a point of S, and let σ be a section of $\mathcal{G}_{s_0}$. Assume that $H^1(\mathcal{Y}_s, \mathcal{G}_s) = 0$ for all s close to s_0. Then σ extends to a section of $\mathcal{G}$ over a neighborhood of $\mathcal{Y}_{s_0}$.*

Before proving the lemma, we show how it can be used to complete the proof of (7.1). We may of course shrink B if necessary; in particular, we may assume that α has a C^m section D. For any $n > 2g-2$, $L = \widetilde{\mathcal{O}}(nD)$ is an invertible $\widetilde{\mathcal{O}}_Y$-module which satisfies the assumptions of (7.3). Fix some $n > 2g$, let $s_1, \dots, s_h$ be a basis of $H^0(Y_{b_0}, L_{b_0})$, and let $\widetilde{s_1}, \dots, \widetilde{s_h}$ be sections of L which extend them to a neighborhood of Y_{b_0}; these give a C^m embedding of fiber spaces

$$\begin{array}{ccc} \alpha^{-1}(A) & \hookrightarrow & \mathbb{P}^{h-1} \times A \\ & \searrow & \downarrow \\ & & A \end{array}$$

where A is any sufficiently small neighborhood of b_0 in B. By (9.11) in Chapter IX, the family $\alpha^{-1}(A) \to A$ is the pullback, via a C^m map, of the universal family on $\mathrm{Hilb}^{p(t)}_{h-1}$, where $p(t) = tn + 1 - g$. This shows the existence of a diagram (7.2) and hence concludes the proof.

It remains to prove (7.3). When $\mathcal{Y} \to S$ is real-analytic, we already know that it comes locally by restriction from a complex-analytic family, so the conclusion follows from the theory of base change in cohomology. In the remaining cases, choose a covering $\mathcal{W}$ of a neighborhood of $\mathcal{Y}_{s_0}$ by means of adapted coordinate sets and choose an adapted partition of unity $\{\chi_i\}$ subordinated to $\mathcal{W}$; for each i, pick an element W_i of $\mathcal{W}$ containing the support of χ_i. The restriction of σ to $W_i \cap \mathcal{Y}_{s_0}$ extends to a section σ_i of $\mathcal{G}$ over W_i. Then $\tau = \sum \chi_i \sigma_i$ is a section of $\widetilde{\mathcal{A}}(\mathcal{G}) = \widetilde{\mathcal{A}} \otimes_{\widetilde{\mathcal{O}}_{\mathcal{Y}}} \mathcal{G}$ that extends σ. Let $\overline{\partial}_\beta$ be the relative $\overline{\partial}$ operator on $\mathcal{G}$ along the fibers of β and set $\eta = \overline{\partial}_\beta \tau$. We will show that one can solve the equation $\eta = \overline{\partial}_\beta u$ with an adapted u that vanishes identically on $\mathcal{Y}_{s_0}$. Then $\tau - u$ will be the sought-for extension of σ. We set $\widetilde{\mathcal{A}}^{0,1}(\mathcal{G}) = \widetilde{\mathcal{A}}^{0,1} \otimes_{\widetilde{\mathcal{O}}_{\mathcal{Y}}} \mathcal{G}$. Using an appropriate adapted partition of unity, we can put on $\widetilde{\mathcal{A}}^{0,1}$ and on $\mathcal{G}$ adapted hermitian metrics. We denote by ϑ_β the formal adjoint of $\overline{\partial}_\beta : \widetilde{\mathcal{A}}(\mathcal{G}) \to \widetilde{\mathcal{A}}^{0,1}(\mathcal{G})$ with respect to these metrics and form the Laplace–Beltrami operator

$$\square_\beta = \overline{\partial}_\beta \vartheta_\beta + \vartheta_\beta \overline{\partial}_\beta \tag{7.4}$$

operating on relative $\mathcal{G}$-valued (0,1)-forms. The equation $\Box_\beta v = \eta$ has a unique solution since, by the assumption that $H^1(\mathcal{Y}_s, \mathcal{G}_s) = 0$ for all s, there are no nonzero $\Box$–harmonic $\mathcal{G}$-valued (0,1)-forms on the fibers of β. In particular, v vanishes identically on $\mathcal{Y}_{s_0}$. Furthermore, $\overline{\partial}_\beta v = 0$. We then set $u = \vartheta_\beta v$. It remains to show that v is adapted.

The result we need about the Laplace–Beltrami operator follows from a theorem of Kodaira and Spencer [437] concerning the differentiability properties of solutions of differential equations $Ev = \eta$, where E is a C^m family of self-adjoint, strongly elliptic, differential operators, and η is an adapted section.

To set up, let $\alpha: \mathcal{Z} \to T$ be a C^m family of compact differentiable manifolds, and let $\mathcal{F}$ be a C^m family of differentiable vector bundles on it. Given two C^m families $\mathcal{F}$ and $\mathcal{F}'$ of differentiable vector bundles on $\alpha: \mathcal{Z} \to T$, a linear differential operator carrying sections of $\mathcal{F}$ to sections of $\mathcal{F}'$ will be said to be a C^m *family of linear differential operators* on $\alpha: \mathcal{Z} \to T$ if, when written in adapted coordinates and relative to adapted local trivializations of $\mathcal{F}$ and $\mathcal{F}'$, it involves only differentiation with respect to vertical coordinates, and its coefficients are adapted functions. For brevity, we will also say that such a differential operator is adapted. Thus, an adapted linear differential operator carries adapted sections to adapted sections. We denote by $\widetilde{A}(\mathcal{F})$ the vector space of adapted sections of $\mathcal{F}$. Let $E: \widetilde{A}(\mathcal{F}) \to \widetilde{A}(\mathcal{F})$ be a C^m family of linear differential operators. A metric on $\mathcal{F}$ will be said to be adapted if the inner product of any pair of adapted sections is an adapted function. Adapted metrics always exist and can, for instance, be constructed by gluing together flat local metrics by means of a partition of unity made up of adapted functions. Suppose an adapted metric is given on $\mathcal{F}$, and one on the relative tangent bundle to $\mathcal{Z} \to T$. We denote by $\langle\ ,\ \rangle_t$ the inner product on $\mathcal{F}_t$ and by dV_t the volume form on $\mathcal{Z}_t$ coming from the metric. Consider the inner product

$$(u, v) = \int_{\mathcal{Z}_t} \langle u, v \rangle_t \, dV_t \tag{7.5}$$

on $A(\mathcal{F}_t)$, the vector space of C^∞ sections of $\mathcal{F}_t$. We will say that E is a *family of formally self-adjoint, strongly elliptic differential operators* if each E_t is self-adjoint with respect to the inner product (7.5) and strongly elliptic. Under these circumstances the kernel of E_t is finite-dimensional, and there are linear operators

$$F_t, G_t: A(\mathcal{F}_t) \to A(\mathcal{F}_t),$$

where F_t is the orthogonal projection onto the kernel of E_t, and

$$u = F_t u + E_t G_t u \tag{7.6}$$

for any $u \in A(\mathcal{F}_t)$. We shall refer to F_t and G_t, respectively, as the *harmonic projector* and *Green operator* associated to L_t and to the chosen metrics. The theorem of Kodaira and Spencer reads as follows.

THEOREM (7.7) (cf. [437], Theorem 5). *Let m be a nonnegative integer or ∞. Let $\alpha: \mathcal{Z} \to T$ be a C^m family of compact differentiable manifolds, and let $\mathcal{F}$ be a C^m family of differentiable vector bundles on $\alpha: \mathcal{Z} \to T$. Suppose that $\mathcal{F}$ and the relative tangent bundle to $\mathcal{Z} \to T$ are endowed with adapted metrics. Let $E: \widetilde{A}(\mathcal{F}) \to \widetilde{A}(\mathcal{F})$ be a C^m family of formally self-adjoint, strongly elliptic linear differential operators. Suppose that the dimension of the kernel of E_t is independent of t. Then the family of harmonic projectors $F = \{F_t\}_{t \in T}$ and the family of Green operators $G = \{G_t\}_{t \in T}$ are of class C^m, in the sense that Fu and Gu are adapted for any adapted u.*

To be precise, the statement proved by Kodaira and Spencer is slightly less general than the one we have given, in two respects. First of all, they deal only with families of the form $Z_0 \times T \to T$, where Z_0 is a compact manifold. More importantly, they treat only the case $m = \infty$. These, however, are not serious difficulties. Since the statement of Theorem (7.7) is local on S, the first is resolved by Lemma (9.8) in Chapter IX. The second difficulty is also nonexistent, since the proof given by Kodaira and Spencer for their theorem, and in particular their crucial Proposition 1, work equally well, and virtually without changes, in our context, provided that we substitute Lemma (9.8) in Chapter IX, and in particular its second part, for their Lemma 1.

We now go back to the Laplace–Beltrami operator (7.4). Notice that, according to our definition, $\square_\beta$ is a family of strongly elliptic, formally self-adjoint linear differential operators. Since, by assumption, there are no nonzero $\square$-harmonic $\mathcal{G}$-valued (0,1)-forms on the fibers of β, the harmonic projector F is zero, so that $\square_\beta G\eta = \eta$, where G is the Green operator. But then Theorem (7.7) asserts that $u = G\eta$ is adapted whenever η is.

8. The period map and the local Torelli theorem.

This section is a utilitarian introduction to the period mapping in the case of 1-forms. Roughly speaking, the period map describes how the Hodge structure on the cohomology groups of a variety depends on the complex structure. In the case of compact Riemann surfaces, the period map associates to a compact Riemann surface C its normalized period matrix, up to the action of the integral symplectic group, and this, in turn, determines the Jacobian of C as a principally polarized abelian variety. This assignment is the Torelli map. The main result about it is Torelli's theorem, stating that it is injective (cf. Vol. I, Chapter VI, Section 3).

In this section we shall study the period map from an infinitesimal point of view, introducing the Gauss–Manin connection and proving the so called *local Torelli theorem*, which states that the differential of the period map is nondegenerate at nonhyperelliptic curves.

In the next section we shall discuss the curvature properties of the Hodge bundles, proving their semipositivity for families of curves.

To begin with, consider a family

$$\varphi : \mathcal{C} \to B$$

of compact, genus g Riemann surfaces parameterized by a polydisc B, and let $C = \varphi^{-1}(0)$ be its central fiber. From a C^∞ point of view this family can be trivialized. Fix then a C^∞ trivialization

$$f :\ C \times B \longrightarrow \mathcal{C}\ ,$$

and, for $b \in B$, set $C_b = \varphi^{-1}(b)$ and $f_b = f|_{C\times\{b\}}$. As usual, denote by $\Omega^1_{\mathcal{C}/B}$ the sheaf of relative holomorphic differentials. Choose a frame for $\varphi_*\Omega^1_{\mathcal{C}/B}$, that is, a basis $\eta_1(b), \dots, \eta_g(b)$ for $H^0(C_b, \omega_{C_b})$, varying holomorphically with b. Also let $\gamma_1, \dots, \gamma_{2g}$ be a symplectic basis for $H^0(C, \mathbb{Z})$. We may assume that our frame is normalized in such a way that the varying period matrix

$$\Omega(b) = \left(\int_{(f_b)_*(\gamma_i)} \eta_\alpha(b) \right)_{\alpha=1,\dots,g\ ;\ i=1,\dots,2g}$$

has the form

$$\Omega(b) = (I, Z(b))\,,$$

where I is the identity $g\times g$ matrix, and $Z(b)$ a point in the Siegel upper half-space $\mathbb{H}_g$ (cf. Chapter I, Section 3). The map

$$Z :\ B \longrightarrow \mathbb{H}_g$$

is called the *period map*.

We shall prove first of all that Z is holomorphic and then that the differential of Z fits into a commutative diagram

$$\text{(8.1)}\qquad \begin{array}{ccc} T_{b_0}(B) & \xrightarrow{\ dZ\ } & T_{Z(b_0)}(\mathbb{H}_g) \subset T_{Z(b_0)}(\mathbb{G}(g,2g)) \\ {\scriptstyle\rho}\big\downarrow & & \big\downarrow \\ H^1(C, T_C) & \xrightarrow{\ \ \nu\ \ } & \operatorname{Hom}(H^{1,0}(C), H^{0,1}(C)) \end{array}$$

where ρ is the Kodaira–Spencer map, and ν is defined by cup-product.

Before proving the statement we have announced, we shall reinterpret Z as a map into a Grassmannian, as is already suggested in the above diagram. Intrinsically, the map Z associates to each point b the point

$$H^{1,0}(C_b) \in \mathbb{G}(g, H^1(C, \mathbb{C})) = \mathbb{G}(g, 2g)\,.$$

In fact, the rows of the period matrix $\Omega(b)$ are simply the coordinates (relative to the basis of $H^1(C,\mathbb{C})$ which is dual to $\gamma_1,\dots,\gamma_{2g}$) of the holomorphic differentials $\eta_1(b),\dots,\eta_g(b)$. In the diagram above we used the well-known identification between the tangent space to $\mathbb{G}(g,2g)$ at the point $H^{1,0}(C)$ and

$$\operatorname{Hom}(H^{1,0}(C), H^1(C,\mathbb{C})/H^{1,0}(C)) \cong \operatorname{Hom}(H^{1,0}(C), H^{0,1}(C))\,.$$

To prove the holomorphicity of Z and the commutativity of the diagram above, it is convenient to put ourselves in a more general setting, since what has to be proved has nothing to do with the fact that we are dealing with a family of curves, but rather with the fact that we are dealing with a family of compact Kähler manifolds. Let

$$\varphi : \mathcal{X} \longrightarrow \Delta$$

be a smooth proper morphism with fibers compact Kähler manifolds and base a polydisc. Set $X_t = \varphi^{-1}(t)$, $X = X_0$. As before, we have a period map

$$Z : \Delta \longrightarrow \mathbb{G}(q, H^1(X,\mathbb{C}))\,,$$

where $q = h^{1,0}(X)$. We will show that Z is holomorphic and that the natural diagram

$$\text{(8.2)} \qquad \begin{array}{ccc} T_0(\Delta) & \xrightarrow{\ dZ\ } & \operatorname{Hom}(H^{1,0}(X), H^{0,1}(X)) \\ {\scriptstyle\rho}\big\downarrow & \nearrow{\scriptstyle\nu} & \\ H^1(X, T_X) & & \end{array}$$

commutes.

We shall need the following general fact about mappings into Grassmannians. Let

$$H : \Delta \longrightarrow \mathbb{G}(k, V)$$

be a C^∞ map. Let $t = (t_1,\dots,t_n)$ be coordinates on Δ. Let $w_1(t),\dots,w_k(t)$ be a frame for $H(t)$, varying smoothly with t. Suppose that $\partial w_i/\partial\bar t_j \in H(t)$ for all t, i, and j. We claim that H is holomorphic. It suffices to do this when $n=1$. By hypothesis, then

$$\frac{\partial w_i}{\partial \bar t} = \sum g_{i,j}(t) w_j(t)\,.$$

The system of differential equations in the $f_{i,j}$

$$\frac{\partial}{\partial \bar t}(\sum f_{i,j} w_j) + \sum g_{i,j} w_j = 0$$

can be solved uniquely with initial conditions $f_{i,j} = 0$, since by our assumption it can be rewritten in the form

$$\frac{\partial f_{i,j}}{\partial \bar{t}} + \sum_h f_{i,h} g_{h,j} + g_{i,j} = 0 \ , \quad i,j = 1,\ldots,k\,.$$

Then the vectors

$$v_i(t) = \sum f_{i,j}(t) w_j(t) + w_i(t)$$

provide, at least near zero, a new frame for $H(t)$ varying holomorphically with t.

Assuming now that H is holomorphic, we recall how to compute its differential

$$dH_0 : T_0(\Delta) \to \mathrm{Hom}(H(0), V/H(0)) = T_{H(0)}(\mathbb{G}(k,V))\,.$$

Let w be an element of $H(0)$, and $w(t)$ a path in V such that $w(t) \in H(t)$ and $w(0) = w$. Then

$$dH_0\left(\frac{\partial}{\partial t_i}\right)(w) = \text{projection of } \left.\frac{\partial w}{\partial t_i}\right|_{t=0} \text{ in } V/H\,. \tag{8.3}$$

Let us return to our original problem. Let η be a C^∞ closed relative 1-form on $\mathcal{X}$, i.e., a section of forms $\mathcal{A}^1_{\mathcal{X}/\Delta}$ such that $d_\varphi \eta = 0$, where d_φ stands for exterior differentiation along the fibers. Via the de Rham theorem, the 1-form η determines a section $[\eta]$ of $R^1\varphi_*\mathbb{C} \otimes \mathcal{A}_\Delta$. Now fix a C^∞ trivialization

$$f : X \times \Delta \longrightarrow \mathcal{X} \tag{8.4}$$

and denote by p the projection from $X \times \Delta$ to X. Pick closed 1-forms $\varphi_1, \ldots, \varphi_{2q}$ on X whose de Rham classes constitute a basis $[\varphi_1], \ldots, [\varphi_{2q}]$ of $H^1(X, \mathbb{C})$. Via the trivialization f, these elements give rise to relative closed 1-forms $\Phi_1, \ldots, \Phi_{2q}$ on $\mathcal{X}$, whose relative cohomology classes $[\Phi_1], \ldots, [\Phi_{2q}]$ give a frame for $R^1\varphi_*\mathbb{C} \otimes \mathcal{A}_\Delta$. By this we mean that

$$\Phi_\nu = \mathrm{rel}(f_* p^* \varphi_\nu)\,,$$

where rel is the homomorphism from differential forms on $\mathcal{X}$ to relative differential forms $\mathcal{A}^*_{\mathcal{X}/\Delta}$. Thus we can write

$$[\eta] = \sum a_\nu [\Phi_\nu]\,, \tag{8.5}$$

where the a_ν are C^∞ functions on Δ. Alternatively, using (8.4), one may view $[\eta]$ as a family $[\eta(t)]$ of classes in $H^1(X, \mathbb{C})$ depending on t. From this point of view, the above expression for η translates into

$$[\eta(t)] = \sum a_\nu(t) [\varphi_\nu]\,. \tag{8.6}$$

It now makes sense to differentiate $[\eta(t)]$ by differentiating its coefficients; the result is an element of the fixed vector space $H^1(X,\mathbb{C})$.

In order to show that the period mapping Z is holomorphic, we use the holomorphicity criterion for mappings into Grassmannians we discussed above. Therefore we must show that when η is of type $(1,0)$, then $\partial[\eta(t)]/\partial\bar{t}_i|_{t=0}$ is of type $(1,0)$ for every i as well. Once the holomorphicity has been established, to show the commutativity of diagram (8.2), we shall explicitly calculate the differential of the period mapping by evaluating the projection of $\partial[\eta(t)]/\partial t_i|_{t=0}$ in $H^1(X,\mathbb{C})/H^{1,0}(X)$, as required by formula (8.3).

To carry out this program, it is convenient to rephrase everything in terms of the so-called Gauss–Manin connection, which we now introduce. This is a connection ∇ on the bundle $R^1\varphi_*\mathbb{C}\otimes\mathcal{A}_\Delta$. For a brief discussion on connections and their curvatures, the reader may want to look at the beginning of the next section. Given a C^∞ section $[\eta]$ of $R^1\varphi_*\mathbb{C}\otimes\mathcal{A}_\Delta$ as in (8.5) and a C^∞ vector field v on Δ, the Gauss–Manin connection is defined by

$$\nabla_v[\eta] = \sum v(a_\nu)[\Phi_\nu]\,.$$

It is clear that this is a connection. When, as in (8.6), we think of $[\eta]$ as a family $[\eta(t)]$ of classes in $H^1(X,\mathbb{C})$ depending on t, we may write

$$\frac{\partial[\eta(t)]}{\partial t_i} = \nabla_{\frac{\partial}{\partial t_i}}[\eta]\,, \qquad \frac{\partial[\eta(t)]}{\partial\bar{t}_i} = \nabla_{\frac{\partial}{\partial\bar{t}_i}}[\eta]\,,$$

where again we think of $\nabla_{\frac{\partial}{\partial t_i}}[\eta]$ and $\nabla_{\frac{\partial}{\partial\bar{t}_i}}[\eta]$ as families of cohomology classes on X. In order to compute these derivatives in a form which is better suited to our goal of showing the holomorphicity of Z and calculating its differential, we shall express the Gauss–Manin connection in a different way. We shall show that

$$\nabla_v[\eta] = [\mathrm{rel}(\iota_{\tilde{v}}(d\tilde{\eta}))]\,, \tag{8.7}$$

where $\tilde{v}$ is any lifting of v to a vector field on $\mathcal{X}$, $\tilde{\eta}$ is any 1-form on $\mathcal{X}$ such that $\mathrm{rel}(\tilde{\eta}) = \eta$, and ι stands for interior product.

Assuming for a moment that (8.7) has been proved, we next indicate how one can conclude. To show the holomorphicity of Z, we must prove that $\nabla_{\frac{\partial}{\partial\bar{t}_i}}[\eta]$ is of type $(1,0)$ for every i whenever η is. Since φ is holomorphic, if η is of type $(1,0)$, we may choose $\tilde{\eta}$ to be of type $(1,0)$, and the lifting $\tilde{v}$ of $v = \partial/\partial\bar{t}_i$ to be of type $(0,1)$. The fact that $\nabla_{\frac{\partial}{\partial\bar{t}_i}}[\eta]$ is of type $(1,0)$ follows, by type considerations, by looking at the expression $[\mathrm{rel}(\iota_{\tilde{v}}(d\tilde{\eta}))]$. This proves the holomorphicity of Z.

We now show the commutativity of diagram (8.2). Explicitly, in view of (8.3), we must show that

$$\frac{\partial[\eta(t)]}{\partial t_i} \equiv \rho\left(\frac{\partial}{\partial t_i}\right) \cup [\eta] \qquad \mathrm{mod}\ H^{1,0}(X)$$

for every i or, equivalently, that

$$\left(\nabla_{\frac{\partial}{\partial t_i}}[\eta]\right)^{(0,1)} = \rho\left(\frac{\partial}{\partial t_i}\right) \cup [\eta] .$$

We may choose a lifting $\tilde{v}$ of $\partial/\partial t_i$ which is of type $(1,0)$, and, again by type considerations, we then see that

$$(\mathrm{rel}(\iota_{\tilde{v}}(d\tilde{\eta})))^{(0,1)} = \mathrm{rel}(\iota_{\tilde{v}}(\bar{\partial}\tilde{\eta})) .$$

On the other hand, writing locally

$$\tilde{v} = \frac{\partial}{\partial t_i} + \sum c_j \frac{\partial}{\partial z_j} ,$$

$$\tilde{\eta} = \sum_h a_h dt_h + \sum_j b_j dz_j ,$$

we get:

$$\begin{aligned}
\iota_{\tilde{v}}(\bar{\partial}\tilde{\eta}) &= -\bar{\partial} a_i - \sum c_j \bar{\partial} b_j , \\
\bar{\partial}\tilde{v} \wedge \tilde{\eta} &= \left(\sum_j \bar{\partial} c_j \otimes \frac{\partial}{\partial z_j}\right) \wedge \left(\sum_h a_h dt_h + \sum_j b_j dz_j\right) \\
&= \sum b_j \bar{\partial} c_j , \\
\bar{\partial}(\iota_{\tilde{v}}(\tilde{\eta})) &= \bar{\partial} a_i + \sum c_j \bar{\partial} b_j + \sum b_j \bar{\partial} c_j \\
&= \bar{\partial}\tilde{v} \wedge \tilde{\eta} - \iota_{\tilde{v}}(\bar{\partial}\tilde{\eta}) ,
\end{aligned}$$

so that:

$$\begin{aligned}
\left(\nabla_{\frac{\partial}{\partial t_i}}[\eta]\right)^{(0,1)} &= [\mathrm{rel}(\partial\tilde{v} \wedge \tilde{\eta})] \\
&= [\mathrm{rel}(\partial\tilde{v}) \wedge \eta] \qquad (8.8) \\
&= \rho\left(\frac{\partial}{\partial t_i}\right) \cup [\eta] ,
\end{aligned}$$

where the equality $\mathrm{rel}(\partial\tilde{v}) = \rho(\frac{\partial}{\partial t_i})$ is left as an exercise in the Dolbeault isomorphism for $H^1(X, T_X)$. This proves the commutativity of diagram (8.2).

It remains to prove (8.7). We begin by showing that the right-hand side does not depend on the choice of liftings for v and η. To do this, we must prove two facts. First of all, that

$$[\mathrm{rel}(\iota_w(d\tilde{\eta}))] = 0 \qquad (8.9)$$

whenever w is a vector field along the fibers of φ. Secondly, that

$$[\mathrm{rel}(\iota_{\tilde{v}}(d\psi))] = 0 \tag{8.10}$$

whenever $\mathrm{rel}(\psi) = 0$. We prove (8.9):

$$\begin{aligned}[\mathrm{rel}(\iota_w(d\tilde{\eta}))] &= [\iota_w(\mathrm{rel}(d\tilde{\eta}))] \\ &= [\iota_w(d_\varphi(\mathrm{rel}(\tilde{\eta})))] \\ &= [\iota_w(d_\varphi(\eta))] \\ &= 0\,,\end{aligned}$$

since η is d_φ-closed. We next prove (8.10). Clearly, ψ is of the form $\psi = \sum \alpha_i dt_i + \sum \beta_i d\bar{t}_i$. We have

$$\begin{aligned}[\mathrm{rel}(\iota_{\tilde{v}}(d\psi))] &= \left[\mathrm{rel}\left(\iota_{\tilde{v}} \sum (d\alpha_i \wedge dt_i + d\beta_i \wedge d\bar{t}_i)\right)\right] \\ &= -\sum [\mathrm{rel}(\iota_{\tilde{v}}(dt_i)d\alpha_i + \iota_{\tilde{v}}(d\bar{t}_i)d\beta_i)] \\ &= -\sum [\mathrm{rel}\, d(\iota_{\tilde{v}}(dt_i)\alpha_i + \iota_{\tilde{v}}(d\bar{t}_i)\beta_i)] \\ &= -\sum [d_\varphi\, \mathrm{rel}(\iota_{\tilde{v}}(dt_i)\alpha_i + \iota_{\tilde{v}}(d\bar{t}_i)\beta_i)] \\ &= 0\,,\end{aligned}$$

the second equality being a consequence of the fact that both $\iota_{\tilde{v}}(dt_i)$ and $\iota_{\tilde{v}}(d\bar{t}_i)$ are constant along the fibers of φ and can hence be moved under the sign of relative differentiation. This concludes the proof that the right-hand side of (8.7) is independent of the choice of liftings.

Let us denote temporarily this right-hand side by $\tilde{\nabla}_v[\eta]$, that is, let us set

$$\tilde{\nabla}_v[\eta] = [\mathrm{rel}(\iota_{\tilde{v}}(d\tilde{\eta}))]\,.$$

The operator $\tilde{\nabla}$ is a connection. Indeed, if g is a C^∞ function on Δ, then

$$\begin{aligned}\tilde{\nabla}_v(g[\eta]) &= [\mathrm{rel}(\iota_{\tilde{v}} d(g\tilde{\eta}))] \\ &= [\mathrm{rel}(\iota_{\tilde{v}}(dg \wedge \tilde{\eta} + g d\tilde{\eta}))] \\ &= [\mathrm{rel}(\tilde{v}(g)\tilde{\eta}) + g\, \mathrm{rel}(\iota_{\tilde{v}}(d\tilde{\eta}))] \\ &= v(g)[\eta] + g\tilde{\nabla}_v[\eta]\,.\end{aligned}$$

To prove (8.7), that is, to show that $\tilde{\nabla}$ is in fact the Gauss–Manin connection, it then suffices to show that

$$\nabla_v[\Phi_\nu] = \tilde{\nabla}_v[\Phi_\nu]$$

for every v and every ν. But now the left-hand side is zero by definition, and so is the right-hand side

$$[\mathrm{rel}(i_{\tilde{v}}(d\tilde{\Phi}_\nu))]\,,$$

since we may choose $\tilde{\Phi}_\nu$ to be $f_* p^* \varphi_\nu$, which is obviously closed.

The computations that we have just carried out show, among other things, that the Gauss–Manin connection is intrinsic and in particular does not depend on the choice of the trivialization (8.4) of $\mathcal{X} \to \Delta$ that we used to define the frame $\{[\Phi_\nu]\}$.

As we have already announced, one of the applications that we have in mind for the machinery of the period map and its differential is to the local Torelli theorem. This states that, when C is a smooth nonhyperelliptic curve of genus g and $\varphi : \mathcal{C} \to (B, b_0)$ is a standard Kuranishi family for C, then the differential of the period map

$$dZ : T_{b_0}(B) \to T_{Z(b_0)}(\mathbb{H}_g)$$

is injective. To show this, look at diagram (8.1). Under our assumption, the Kodaira–Spencer map is an isomorphism, so that it suffices to show that ν is injective. But the transpose of ν is the cup-product map

$$\nu^* : H^0(C, \omega_C) \otimes H^0(C, \omega_C) \to H^0(C, \omega_C^2)\,, \tag{8.11}$$

which is surjective for nonhyperelliptic C by Max Noether's theorem (cf. Chapter III, Section 2).

The local Torelli theorem fails at hyperelliptic curves of genus $g > 2$; in fact, if C is such a curve, the cup-product map (8.11) is not onto. However, the local Torelli theorem fails in directions which are transverse to the hyperelliptic locus, and not in those directions which are tangent to it. To explain what we mean by this, recall that the subvariety H of B parameterizing those fibers of π which are hyperelliptic has dimension $2g - 1$ and that its tangent space at b_0 is $H^1(C, T_C)^\gamma$, where γ is the hyperelliptic involution of C. Dually, the cotangent space to H at b_0 is $H^0(C, \omega_C^2)^\gamma$. Recall also from the proof of Lemma (6.15) that the elements of $H^0(C, \omega_C^2)^\gamma$ are precisely the pullbacks via the hyperelliptic double covering $f : C \to \mathbb{P}^1$ of the differentials in $H^0(\mathbb{P}^1, \omega_{\mathbb{P}^1}^2(D)) \cong H^0(\mathbb{P}^1, \mathcal{O}(2g-2))$, where D is the branch locus of f. Next, recall that $H^0(C, \omega_C)$ consists entirely of anti-invariants, since a nonzero invariant would descend to a nonzero abelian differential on $C/\gamma = \mathbb{P}^1$. Thus, the cup-product (8.11) lands inside $H^0(C, \omega_C^2)^\gamma$. We claim that, in fact, its image is all of $H^0(C, \omega_C^2)^\gamma$. To see this, notice that one may identify $H^0(C, \omega_C)$ with $H^0(\mathbb{P}^1, \mathcal{O}(g-1))$, and the cup product map

$$H^0(C, \omega_C) \times H^0(C, \omega_C) \to H^0(C, \omega_C^2)^{\langle\gamma\rangle} \subset H^0(C, \omega_C^2)$$

with

$$H^0(\mathbb{P}^1, \mathcal{O}(g-1)) \times H^0(\mathbb{P}^1, \mathcal{O}(g-1)) \to H^0(\mathbb{P}^1, \mathcal{O}(2g-2))\,,$$

which is onto. This proves our claim. Dually, what all of this shows is that the map

$$T_{b_0}(H) \to \mathrm{Hom}(H^0(C,\omega_C), H^1(C,\mathcal{O}_C)),$$

that is, the differential of the restriction to H of the period map, is injective and that its image equals the image of the differential

$$T_{b_0}(B) \to \mathrm{Hom}(H^0(C,\omega_C), H^1(C,\mathcal{O}_C)),$$

of the full period map. We shall refer to this result as the *local Torelli theorem for hyperelliptic curves.*

9. Curvature of the Hodge bundles.

As we mentioned at the beginning of this chapter, the introduction of the Gauss–Manin connection leads naturally to the study of the curvature properties of the Hodge bundles. This will be carried out in this section, although it will not be needed in any other part of this book. The reader will find an indication of some of the many applications of this study in the bibliographical notes.

Before talking about Hodge bundles, it is convenient to discuss connections and their curvatures on complex vector bundles in general. Let X be a complex manifold and E a complex vector bundle on it. Denote by $\mathcal{A}^i(E)$ the sheaf of E-valued C^∞ i-forms. Recall that a *connection* ∇ *on* E is a linear differential operator

$$\nabla : \mathcal{A}^0(E) \longrightarrow \mathcal{A}^1(E)$$

satisfying

$$\nabla(fs) = s \otimes df + f\nabla s\,.$$

If v is a tangent vector to X, one writes $\nabla_v s$ to denote the interior product $\iota_v(\nabla s)$. The ∇ operator can be extended to operators

$$\nabla : \mathcal{A}^k(E) \longrightarrow \mathcal{A}^{k+1}(E)$$

by setting

$$\nabla(s \otimes \omega) = s \otimes d\omega + \nabla s \wedge \omega\,.$$

The operator

$$R_E = \nabla^2 : \mathcal{A}^0(E) \longrightarrow \mathcal{A}^2(E)$$

turns out to be a tensor and, more precisely, a global section of $\mathcal{A}^2(E^* \otimes E)$. This tensor is called the *curvature form* of the connection ∇. Suppose that E is equipped with a nondegenerate (not necessarily definite) hermitian product $\langle\ ,\ \rangle$ and extend this product to E-valued forms by setting

$$\langle s \otimes \omega, t \otimes \varphi\rangle = \langle s, t\rangle \omega \wedge \bar{\varphi}\,.$$

Then one says that ∇ *is compatible with the hermitian product* if, given sections s and t of E, one has

$$d\langle s,t\rangle = \langle \nabla s,t\rangle + \langle s,\nabla t\rangle\,, \tag{9.1}$$

so that at the level of E-valued forms one has

$$d\langle \sigma,\tau\rangle = \langle \nabla\sigma,\tau\rangle + (-1)^{\deg\sigma}\langle \sigma,\nabla\tau\rangle\,. \tag{9.2}$$

If, in addition, E is holomorphic, one says that ∇ is a *hermitian connection* if it is compatible with the hermitian product and its (0,1) part is $\bar\partial_E$. We then write

$$\nabla = \nabla' + \bar\partial_E\,,$$

where ∇' is the $(1,0)$-part of ∇.

Suppose ∇ is hermitian. We observe that its curvature form is of type $(1,1)$. In fact, since R_E is a tensor, it suffices to calculate $R_E(s)$ when s is a holomorphic section of E. Thus,

$$\begin{aligned} R_E(s) &= (\nabla' + \bar\partial)(\nabla' + \bar\partial)s \\ &= \nabla'\nabla's + \bar\partial\nabla's\,. \end{aligned}$$

On the other hand, we claim that $\nabla'\nabla's = 0$. To show this, it suffices to prove that $\langle\nabla'\nabla's,t\rangle = 0$ for any local holomorphic section t of E. Breaking up (9.1) into pure-type components, we find that

$$\begin{aligned} \langle\nabla'\nabla's,t\rangle &= \partial\langle\nabla's,t\rangle + \langle\nabla's,\bar\partial t\rangle \\ &= \partial^2\langle s,t\rangle - \partial\langle s,\bar\partial t\rangle + \langle\nabla's,\bar\partial t\rangle \\ &= 0\,, \end{aligned}$$

since t is holomorphic and $\partial^2 = 0$. In conclusion,

$$R_E = \bar\partial\nabla'\,, \tag{9.3}$$

which is clearly of type $(1,1)$. It is also useful to observe that $\sqrt{-1}\langle R_E(s),s\rangle$ is a *real* $(1,1)$ form. In fact,

$$\begin{aligned} \langle R_E(s),s\rangle &= \langle\bar\partial\nabla's,s\rangle \\ &= \bar\partial\langle\nabla's,s\rangle + \langle\nabla's,\nabla's\rangle \\ &= \bar\partial\partial\langle s,s\rangle - \bar\partial\langle s,\bar\partial s\rangle + \langle\nabla's,\nabla's\rangle \\ &= \bar\partial\partial\langle s,s\rangle + \langle\nabla's,\nabla's\rangle\,, \end{aligned}$$

which is purely imaginary.

Now suppose that $F\subset E$ is a holomorphic subbundle on which the hermitian metric is nondegenerate. One can then define a connection D

on F by setting

$$Ds = p_F(\nabla s)\,,$$

where p_F stands for the orthogonal projection to F.

We leave to the reader the easy verification that the connection D is compatible with the hermitian product. As for its $(0,1)$ part, since F is holomorphic, it equals $p_F(\overline{\partial}_E) = \overline{\partial}_F$. Thus D is hermitian. For any section s of F, set

$$\sigma s = \nabla s - Ds\,.$$

Since, for any C^∞ function f,

$$\begin{aligned}\sigma(fs) &= s \otimes df + f\nabla s - s \otimes df + fDs \\ &= f\sigma(s)\,,\end{aligned}$$

σ is a tensor, and it clearly has type $(1,0)$. We now compute the curvature of D. Let s and t be holomorphic sections of F. Then

$$\begin{aligned}\langle R_E(s), t\rangle &= \langle \overline{\partial}\nabla' s, t\rangle \\ &= \langle \overline{\partial}D' s, t\rangle + \langle \overline{\partial}\sigma s, t\rangle \\ &= \langle R_F(s), t\rangle + \langle \overline{\partial}\sigma s, t\rangle\,.\end{aligned}$$

On the other hand,

$$\begin{aligned}\langle \overline{\partial}\sigma s, t\rangle &= \overline{\partial}\langle \sigma s, t\rangle + \langle \sigma s, \nabla' t\rangle \\ &= \overline{\partial}\langle \sigma s, t\rangle + \langle \sigma s, D' t\rangle + \langle \sigma s, \sigma t\rangle \\ &= \langle \sigma s, \sigma t\rangle\,,\end{aligned}$$

since σs is orthogonal to F. In conclusion,

$$\langle R_F(s), t\rangle = \langle R_E(s), t\rangle - \langle \sigma s, \sigma t\rangle\,. \tag{9.4}$$

Suppose for a moment that the hermitian product is positive. Recall that $\sqrt{-1}\langle R_E(s), s\rangle$ and $\sqrt{-1}\langle R_F(s), s\rangle$ are real. Then the above formula expresses the principle that *curvature decreases on holomorphic subbundles*, where, of course, we adopt the convention of saying that ∇ has positive curvature when $\sqrt{-1}\langle R_E(s), s\rangle$ is positive.

We now come to the central theme of this section, that is, to the positivity properties of the so-called Hodge bundles. Fix a family of compact Kähler manifolds

$$\pi : \mathcal{X} \to B$$

parameterized by a complex manifold B. The *Hodge bundles* are the direct images $\pi_* \omega^k_{\mathcal{X}/B}$, $k \geq 1$.

We consider the Gauss–Manin connection ∇ acting on sections of the rank $2q$ complex vector bundle E associated to the locally free sheaf

$R^1\pi_*\mathbb{C} \otimes \mathcal{A}_B$. We first observe that E has a complex structure which comes by writing

$$R^1\pi_*\mathbb{C} \otimes \mathcal{A}_B = (R^1\pi_*\mathbb{C} \otimes \mathcal{O}_B) \otimes_{\mathcal{O}_B} \mathcal{A}_B .$$

Secondly, the horizontal sections of E for the Gauss–Manin connection are exactly the sections of $R^1\pi_*\mathbb{C}$. Since locally holomorphic sections of E are sums of terms of the form $s = f\sigma$, where f is holomorphic and σ is a horizontal section, it follows that the $(0,1)$-part of ∇ is just $\overline{\partial}_E$.

Since E is locally generated by horizontal sections, the Gauss–Manin connection is clearly *flat*, that is, its curvature form vanishes.

The bundle E has a natural complex rank q subbundle F whose fiber over $b \in B$ is $H^{1,0}(X_b)$. As a C^∞ bundle, F coincides with the Hodge bundle $\pi_*\omega_{\mathcal{X}/B}$: as such, it carries a holomorphic structure. There is another holomorphic structure which F inherits from the holomorphic structure of E. In fact, when proving that the period mapping is holomorphic, we showed that, given local coordinates $t_1, \dots, t_n$ on B, for every section $[\psi]$ of F, i.e., for every closed relative $(1,0)$-form along the fibers of π, the covariant derivatives $\nabla_{\frac{\partial}{\partial \bar{t}_i}}[\psi]$ are still of type $(1,0)$. Thus F is a holomorphic subbundle of E. The $\overline{\partial}$ operator of this holomorphic structure is given locally by

$$\sum_i \nabla_{\frac{\partial}{\partial \bar{t}_i}} \otimes d\bar{t}_i .$$

Luckily, the two holomorphic structures on F turn out to be the same. To see this, let ψ be a section of $\pi_*\omega_{\mathcal{X}/B}$; in other words, let ψ be a holomorphic relative $(1,0)$ form. View $[\psi]$ as a C^∞ section of F. We must show that

$$\nabla_{\frac{\partial}{\partial \bar{t}_i}}[\psi] = 0 \qquad \text{for every } i .$$

Recall that

$$\nabla_{\frac{\partial}{\partial \bar{t}_i}}[\psi] = [\mathrm{rel}\,(\iota_v(d\tilde{\psi}))] ,$$

where v is a lifting of $\frac{\partial}{\partial \bar{t}_i}$, which we may choose to be of type $(1,0)$, and $\tilde{\psi}$ is a $(1,0)$ form such that $\psi = \mathrm{rel}(\tilde{\psi})$. In local coordinates,

$$v = \frac{\partial}{\partial \bar{t}_i} + \sum c_j \frac{\partial}{\partial \bar{z}_j} ,$$
$$\tilde{\psi} = \sum b_h dt_h + \sum a_j dz_j .$$

To say that ψ is holomorphic amounts to saying that the a_j are holomorphic. Thus,

$$\begin{aligned} \mathrm{rel}(\iota_v(d\tilde{\psi})) &= \mathrm{rel}(\iota_v(\overline{\partial}\tilde{\psi})) \\ &= \mathrm{rel}(\iota_v(\textstyle\sum \overline{\partial} b_h \wedge dt_h)) \\ &= \mathrm{rel}(\textstyle\sum \iota_v(\overline{\partial} b_h) \wedge dt_h) \\ &= 0 . \end{aligned}$$

From now on we restrict our study to the case where the fibers of π are one-dimensional, i.e., compact genus g Riemann surfaces. There is defined on E a natural, nondegenerate hermitian form. Given a point $b \in B$ and cohomology classes $[\eta]$, $[\xi] \in E_b = H^1(X_b, \mathbb{C})$, we set

$$\langle [\eta], [\xi] \rangle = \sqrt{-1} \int_{X_b} \eta \wedge \bar{\xi} \,. \tag{9.5}$$

Let us verify that the Gauss–Manin connection is hermitian. Given sections $[\omega]$ and $[\psi]$ of E and writing them in terms of a local horizontal frame $[\phi_1], \dots, [\phi_{2g}]$ around $b_0 \in B$ as introduced in the previous section,

$$[\omega] = \sum a_i [\Phi_i] \,,$$
$$[\psi] = \sum b_i [\Phi_i] \,,$$

we get

$$\begin{aligned} d\langle [\omega], [\psi] \rangle &= \sum d(a_i \bar{b}_j) \int_{X_{b_0}} \Phi_i \wedge \overline{\Phi}_j \\ &= \langle \nabla[\omega], [\psi] \rangle + \langle [\omega], \nabla[\psi] \rangle \,. \end{aligned}$$

Clearly, the hermitian product $\langle \;,\; \rangle$ on E is not definite; more specifically, it is positive definite on F and negative definite on the subbundle $F^\perp$, whose fiber over $b \in B$ can be canonically identified with $H^{0,1}(X_b)$. We compute the curvature of the hermitian connection on F. For this, we use formula (9.4) and get

$$\langle R_F \psi, \eta \rangle = -\langle \sigma\psi, \sigma\eta \rangle \,. \tag{9.6}$$

Since σ equals the difference between ∇ and its projection to F, $\sigma\psi$ is an $F^\perp$-valued $(1,0)$ form, and since the inner product is negative definite on $F^\perp$, it follows that

$$\sqrt{-1} \langle R_F \psi, \psi \rangle \geq 0 \tag{9.7}$$

for any ψ, that is, F has nonnegative curvature everywhere on B. One may be more explicit about the value of σ. In fact, we already computed in the previous section (cf. (8.8)) that for any tangent vector v to B,

$$\iota_v(\sigma\psi) = \rho(v) \cup \psi, \tag{9.8}$$

where ρ is the Kodaira–Spencer map.

Formula (9.7) is a first positivity property satisfied by a Hodge bundle. We would like to be able to talk about positivity properties of the Hodge bundles even for families of algebraic varieties containing singular fibers and specifically for families of nodal curves. There is no problem in defining the Hodge bundles in this situation. The trouble is that the natural metrics they carry are singular at degenerate fibers. One is therefore compelled to look for a positivity notion which is not metric in nature, but rather algebro-geometric. This is the notion of semipositivity, on which we now digress.

A locally free sheaf $\mathcal{F}$ over a complete scheme X is said to be *semipositive* if for every morphism

$$f : C \longrightarrow X ,$$

where C is a smooth complete curve, the pullback $f^*\mathcal{F}$ has no line bundle quotient of negative degree (notice that, when $\mathcal{F}$ is a line bundle, this amounts to saying that $\mathcal{F}$ is nef).

There are a number of standard constructions for locally free sheaves which yield semipositive sheaves when applied to semipositive ones. It is possible to give a general criterion to discriminate between those constructions which preserve semipositivity and those which do not. Here, however, we shall content ourselves with the following proposition, which is all we shall need in the sequel.

PROPOSITION (9.9). *Let $\mathcal{F}$ and $\mathcal{G}$ be semipositive locally free sheaves on a complete scheme X. Then:*

a) *any locally free quotient of $\mathcal{F}$ is semipositive;*
b) *if*

$$0 \to \mathcal{F} \to \mathcal{E} \to \mathcal{G} \to 0$$

is an exact sequence of locally free sheaves, then $\mathcal{E}$ is semipositive;
c) *$\mathcal{F} \otimes \mathcal{G}$ is semipositive.*

An immediate consequence of this proposition is the following:

COROLLARY (9.10). *Tensor powers of semipositive sheaves and their quotients, such as symmetric or exterior powers of semipositive sheaves, are semipositive.*

Property a) in Proposition (9.9) is immediate from the definition. As for b), suppose that L is a quotient line bundle of $f^*\mathcal{E}$ for some morphism f from a smooth complete curve C to X. Then, either the composite map from $f^*\mathcal{F}$ to L in nonzero, in which case L contains a line bundle quotient of $f^*\mathcal{F}$, so $\deg L \geq 0$, or else L is a quotient of $f^*\mathcal{G}$, and hence $\deg L \geq 0$ anyway. Proving part c) of Proposition (9.9) will not be as easy. Our strategy will be to rely on the notion of ampleness, which is closely related to the one of semipositivity, proving first the analogue of (9.9) with "ample" substituted for "semipositive" throughout and then deducing (9.9) as a corollary.

Let then $\mathcal{F}$ be a locally free sheaf on a complete scheme X, and consider the tautological line bundle $\mathcal{O}(1) = \mathcal{O}_{\mathbb{P}(\mathcal{F}^\vee)}(1)$ on $\mathbb{P}(\mathcal{F}^\vee)$. We recall that $\mathcal{F}$ is said to be *ample* if $\mathcal{O}(1)$ is ample.

Notice that ampleness implies semipositivity. In fact, any line bundle quotient of an ample locally free sheaf $\mathcal{F}$ on a smooth complete curve has a strictly positive degree. To see this, notice that a line bundle quotient L of $\mathcal{F}$ corresponds to a section Γ of $\mathbb{P}(\mathcal{F}^\vee)$ and that the degree of L is just the intersection number of Γ with $\mathcal{O}_{\mathbb{P}(\mathcal{F}^\vee)}(1)$, which is positive when $\mathcal{F}$ is ample.

It is a well-known fact that, loosely speaking, *nef line bundles* are the limits of ample (fractional) line bundles. We shall see in a moment that essentially the same relation intercurs between semipositive and ample locally free sheaves.

The fact that nef line bundles are limits of ample ones derives from the following fundamental ampleness criterion, due to Seshadri (cf. [355]).

PROPOSITION (9.11) (SESHADRI'S CRITERION). *A line bundle L on a complete scheme X is ample if and only if there is a positive constant k such that, for any irreducible reduced complete curve C in X,*

$$\deg(L_{|C}) \geq k\mu(C),$$

where $\mu(C)$ stands for the maximum of the multiplicities of the points of C.

Suppose then that L is ample, so that Seshadri's criterion is satisfied. If M is nef, then for any positive integer n,

$$\deg(L \otimes M^n)_{|C} \geq k\mu(C),$$

so that $L \otimes M^n$ is ample. Conversely, suppose that $L \otimes M^n$ is ample for any positive integer n and some ample line bundle L. Then, clearly,

$$n \deg(M_{|C}) + \deg(L_{|C}) = \deg(L \otimes M^n_{|C}) > 0.$$

Passing to the limit as n goes to infinity shows that $\deg M_{|C}$ is nonnegative, i.e., that M is nef. In particular, M is nef if and only if, for any ample line bundle L, the line bundle $L \otimes M^n$ is ample, so that if one is willing to deal with fractional line bundles, one sees that nef line bundles are limits of ample ones.

An analogue of this characterization of nef line bundles is the following characterization of semipositivity.

PROPOSITION (9.12). *Let $\mathcal{F}$ be a locally free sheaf on a complete scheme X. The following conditions are equivalent:*

i) *$\mathcal{F}$ is semipositive.*
ii) *$\mathcal{O}_{\mathbb{P}\mathcal{F}^\vee}(1)$ is nef.*
iii) *For any morphism $f : C \to X$, where C is a smooth complete curve, and any ample line bundle L on C, the locally free sheaf $L \otimes f^*\mathcal{F}$ is ample.*

To prove this, suppose first that $\mathcal{F}$ is semipositive. Let Γ be any irreducible complete curve in $\mathbb{P} = \mathbb{P}(\mathcal{F}^\vee)$. If Γ is contained in a fiber of the projection from $\mathbb{P}$ to X, then $\mathcal{O}_\mathbb{P}(1)$ has positive degree on Γ. If, on the other hand, Γ is not contained in a fiber, and we denote by $\tilde{\Gamma}$ its normalization and by k the natural morphism from $\tilde{\Gamma}$ to X, then $k^*\mathbb{P} \xrightarrow{h} \tilde{\Gamma}$ has a tautological section Γ' mapping to Γ, and moreover

$$\deg\left(\mathcal{O}_\mathbb{P}(1)_{|\Gamma}\right) = \deg\left(\mathcal{O}_{k^*\mathbb{P}}(1)_{|\Gamma'}\right).$$

On the other hand, one has a surjective restriction morphism

$$k^*\mathcal{F} = h_*\mathcal{O}_{k^*\mathbb{P}}(1) \longrightarrow h_*(\mathcal{O}_{k^*\mathbb{P}}(1)_{|\Gamma'}).$$

Since $\mathcal{F}$ is assumed to be semipositive, it follows that

$$\deg\left(\mathcal{O}_\mathbb{P}(1)_{|\Gamma}\right) \geq 0.$$

This shows that $\mathcal{O}_\mathbb{P}(1)$ is nef, proving that i) implies ii). Now suppose that ii) holds and let C, f, and L be as in iii). Consider any irreducible complete curve D in $\mathbb{P}f^*\mathcal{F}$ and let $\pi : P = \mathbb{P}f^*\mathcal{F}^\vee \to C$ be the projection. Then

$$\deg\left(\pi^*L \otimes \mathcal{O}_P(1)\right)_{|D} \geq \mu(D).$$

This is obvious if D is contained in a fiber. Otherwise, since $\mathcal{O}_\mathbb{P}(1)$ is nef, the left-hand side is larger than $\deg(\pi^*L_{|D})$, which is at least $(\deg(L))\cdot\mu(D)$. Then, by Seshadri's criterion, $\pi^*L\otimes\mathcal{O}_P(1) = \mathcal{O}_{\mathbb{P}(L^\vee\otimes f^*\mathcal{F}^\vee)}$ is ample, i.e., $L \otimes f^*\mathcal{F}$ is ample. We now prove that iii) implies i). Let f be a morphism from a complete smooth curve C to X, let M be a line bundle quotient of $f^*\mathcal{F}$, and let L be a line bundle of degree one on C. Then $M \otimes L$, being a line bundle quotient of the ample locally free sheaf $f^*\mathcal{F}\otimes L$, has positive degree; thus $\deg M \geq 0$. This shows that $\mathcal{F}$ is semipositive.

Let $\mathcal{F}$ be a locally free sheaf on a complete scheme X, and let $\varphi : Y \to X$ be a finite morphism. A simple remark that will be useful later is that, as is the case for line bundles, $\mathcal{F}$ is ample if and only if $f^*\mathcal{F}$ is. In fact, there is an obvious commutative diagram

$$\begin{array}{ccc}
\mathbb{P}\varphi^*\mathcal{F}^\vee & \xrightarrow{\psi} & \mathbb{P}\mathcal{F}^\vee \\
\xi\big\downarrow & & \big\downarrow\eta \\
Y & \xrightarrow{\varphi} & X
\end{array}$$

where ψ is finite. Moreover, $\psi^*\mathcal{O}_{\mathbb{P}\mathcal{F}^\vee}(1) = \mathcal{O}_{\mathbb{P}\varphi^*\mathcal{F}^\vee}(1)$, so that our claim follows from the line bundle case.

We are now ready to prove, in several steps, the analogue of Proposition (9.9) for ample sheaves. Actually, the proof will be complete only for locally free sheaves over smooth curves, but this is all we shall need.

LEMMA (9.13). *Let $\mathcal{F}$ and $\mathcal{G}$ be ample locally free sheaves on a complete scheme X. Then:*

a) *any locally free quotient of $\mathcal{F}$ is ample.*
b) *If*

$$0 \to \mathcal{F} \to \mathcal{E} \to \mathcal{G} \to 0$$

is exact, then $\mathcal{E}$ is ample.

We first prove a). Let $\mathcal{F} \to \mathcal{H}$ be a surjective map onto a locally free sheaf. Let $j : \mathbb{P}\mathcal{H}^\vee \to \mathbb{P}\mathcal{F}^\vee$ be the corresponding inclusion. Then $\mathcal{O}_{\mathbb{P}\mathcal{H}^\vee}(1) = j^*\mathcal{O}_{\mathbb{P}\mathcal{F}^\vee}(1)$, and the ampleness of $\mathcal{H}$ follows.

We shall prove b) only in case X is a smooth curve. Let H be an ample line bundle on X, and let $\eta : \mathbb{P}\mathcal{F}^\vee \to X$, $\xi : \mathbb{P}\mathcal{G}^\vee \to X$ be the natural projections. There is an integer $h > 0$ such that $\mathcal{O}_{\mathbb{P}\mathcal{F}^\vee}(h) \otimes \eta^* H^{-1}$ and $\mathcal{O}_{\mathbb{P}\mathcal{G}^\vee}(h) \otimes \eta^* H^{-1}$ are ample. Pick a degree h covering $\varphi : X' \to X$ and let $\psi : \mathbb{P}\varphi^*\mathcal{F}^\vee \to \mathbb{P}\mathcal{F}^\vee$ be the corresponding morphism. Then on X' there is a line bundle H' such that $H'^h = \varphi^* H$; clearly, H' is ample. Set $\mathcal{F}' = \varphi^*\mathcal{F}$, let η' be the projection $\mathbb{P}\mathcal{F}'^\vee \to X'$, and so on. Then

$$\mathcal{O}_{\mathbb{P}(\mathcal{F}' \otimes H'^{-1})^\vee}(1) = \eta'^*(H'^{-1}) \otimes \mathcal{O}_{\mathbb{P}\mathcal{F}^\vee}(1)\,,$$

so

$$\begin{aligned} \mathcal{O}_{\mathbb{P}(\mathcal{F}' \otimes H'^{-1})^\vee}(h) &= \eta'^* \varphi^*(H^{-1}) \otimes \mathcal{O}_{\mathbb{P}\mathcal{F}'^\vee}(h) \\ &= \psi^*(\eta^*(H^{-1}) \otimes \mathcal{O}_{\mathbb{P}\mathcal{F}^\vee}(h)) \end{aligned}$$

is ample because it is the pullback of an ample line bundle under a finite morphism. Thus, $\mathcal{O}_{\mathbb{P}(\mathcal{F}' \otimes H'^{-1})^\vee}(1)$ is ample, and the same is true, by the same argument, for $\mathcal{O}_{\mathbb{P}(\mathcal{G}' \otimes H'^{-1})^\vee}(1)$. By part b) of Proposition (9.9), $\mathcal{E}' \otimes H'^{-1}$ is semipositive. By Proposition (9.12), $\mathcal{E}'$ is ample. Since, as we already observed, ampleness is insensitive to finite base changes, $\mathcal{E}$ is ample as well. This concludes the proof of Lemma (9.13).

LEMMA (9.14). *Let $\mathcal{F}$ be an ample locally free sheaf on a complete scheme X. Then, for any $h > 0$, $S^h\mathcal{F}$ is ample.*

We begin by showing that the conclusion of the lemma holds for large enough h. To prove this, we let $\eta : \mathbb{P}\mathcal{F}^\vee \to X$ be the natural projection, and let $\mathcal{G}$ be any coherent sheaf on X. Then, if $h \geq 0$,

$$H^p(X, R^q\eta_*(\mathcal{O}_{\mathbb{P}\mathcal{F}^\vee}(h) \otimes \eta^*\mathcal{G})) = H^p(X, \mathcal{G} \otimes R^q\eta_*(\mathcal{O}_{\mathbb{P}\mathcal{F}^\vee}(h)))$$

vanishes for $q > 0$ and equals $H^p(X, \mathcal{G} \otimes S^h\mathcal{F})$ for $q = 0$. Thus, by a Leray spectral sequence argument,

$$H^p(X, \mathcal{G} \otimes S^h\mathcal{F}) = H^p(\mathbb{P}\mathcal{F}^\vee, \eta^*\mathcal{G}(h))$$

for any p and $h \geq 0$. Since $\mathcal{F}$ is ample, the right-hand side vanishes for $p > 0$ and large enough h. This implies that, for large enough h, $S^h\mathcal{F}$ is generated by its sections, i.e., that there is a surjective morphism

$$\mathcal{O}_X^{\oplus N} \longrightarrow S^h\mathcal{F} .$$

Thus, there is a surjective morphism

$$\mathcal{F}^{\oplus N} \longrightarrow S^{h+1}\mathcal{F} .$$

By Lemma (9.13), $S^{h+1}\mathcal{F}$ is ample, proving our claim.

Now let h be arbitrary. By what we have just proved, $S^{hk}\mathcal{F}$ is ample for large k. Consider the kth power morphism

$$\sigma : \mathbb{P}S^h\mathcal{F}^\vee \longrightarrow \mathbb{P}S^{hk}\mathcal{F}^\vee .$$

This is finite-to-one, and $\sigma^*\mathcal{O}_{\mathbb{P}S^{hk}\mathcal{F}^\vee}(1) = \mathcal{O}_{\mathbb{P}S^h\mathcal{F}^\vee}(k)$. Since $\mathcal{O}_{\mathbb{P}S^{hk}\mathcal{F}^\vee}(1)$ is ample, $\mathcal{O}_{\mathbb{P}S^h\mathcal{F}^\vee}(k)$ is also ample. This concludes the proof of the lemma.

COROLLARY (9.15). *Let $\mathcal{F}$ and $\mathcal{G}$ be ample locally free sheaves on a complete scheme X. Then $\mathcal{F} \otimes \mathcal{G}$ is ample.*

To see this, it suffices to notice that $\mathcal{F} \otimes \mathcal{G}$ is a direct summand of

$$S^2(\mathcal{F} \oplus \mathcal{G}) = S^2\mathcal{F} \oplus S^2\mathcal{G} \oplus \mathcal{F} \otimes \mathcal{G} ,$$

which is ample by lemmas (9.13) and (9.14).

The analogue of Proposition (9.9) for ample sheaves is now fully proved, at least when the base is a smooth curve. Notice that the only place where this extra assumption has been used is in the proof of part b) of Lemma (9.13).

To conclude the proof of Proposition (9.9), we argue as follows. By part iii) of (9.12) we must show that, for any morphism $f : C \to X$, where C is a smooth complete curve, and any ample line bundle H on C, $H \otimes f^*(\mathcal{F} \otimes \mathcal{G})$ is ample. Up to a finite base change, we may assume that H is the tensor product of two ample line bundles H_1 and H_2. But then

$$H \otimes f^*(\mathcal{F} \otimes \mathcal{G}) = (H_1 \otimes f^*\mathcal{F}) \otimes (H_2 \otimes f^*\mathcal{G}) ,$$

which is the tensor product of two ample sheaves and hence ample. Proposition (9.9) is then fully proved.

We now state the main result of this section (see, for instance, Kollár's paper [439]).

THEOREM (9.16). *Let $f : \mathcal{X} \to S$ be a family of nodal curves over a complete base S. Then the locally free sheaves $f_*\omega^k_{\mathcal{X}/S}$ are semipositive for any $k \geq 1$.*

Of course, it is enough to prove the theorem when S is a smooth complete curve. The crucial case is the one where $k = 1$ and the general fiber of f is a smooth curve. This is the case that we shall fully prove. At the end of the proof we shall briefly discuss the general case.

Until further notice we then consider the situation where $k = 1$, S is a smooth complete curve, and the generic fiber of f is smooth. By successive blow-ups we may also assume that $\mathcal{X}$ is a smooth surface. The proof that $f_*\omega_{\mathcal{X}/S}$ is semipositive will rest on curvature considerations. We recall that, away from singular fibers, $f_*\omega_{\mathcal{X}/S}$ has an intrinsic metric defined as follows. If η and ξ are C^∞ sections of $f_*\omega_{\mathcal{X}/S}$, the hermitian product between η and ξ is defined by

$$\langle \eta, \xi \rangle_s = \sqrt{-1} \int_{X_s} \eta \wedge \bar{\xi} .$$

We have shown that the curvature of this metric is nonnegative, away from singular fibers. At singular fibers the curvature has singularities. To prove the semipositivity, we are required to prove that quotient line bundles of $f_*\omega_{\mathcal{X}/S}$ have nonnegative degree. Therefore the idea of the proof is to use the principle that curvature increases on quotient bundles and to show that the singularities of the metric are mild enough to make it possible to compute degrees as integrals of the curvature. Control over the singularities of the metric is achieved by means of the following general criterion.

LEMMA (9.17). *Let L be a holomorphic line bundle over a smooth complete connected curve S, and let $p_1, \dots, p_n$ be distinct points of S. For each i, denote by z_i a local coordinate centered at p_i and by l_i a nonvanishing section of L on a neighborhood of p_i. Suppose that L is endowed with a hermitian metric $\| \;\|$ on $S \smallsetminus \{p_1, \dots, p_n\}$ and denote by $\mathcal{H}$ the corresponding Chern form. Let $\mu(x)$ be a real-valued positive function defined for large enough x such that*

$$\lim_{x \to \infty} \frac{\mu(x)}{x^N} = 0$$

for any positive exponent N.

Suppose that $\mathcal{H}$ is nonnegative on $S \smallsetminus \{p_1, \dots, p_n\}$, that $\log \|l_i\|$ is summable for each i, and that there are constants b_i such that, for every i,

i) $$\|l_i\|^{-1} = O(|z_i|^{b_i} \mu(|z_i|^{-1})) .$$

Then

$$\deg L \geq \int_{S \smallsetminus \{p_1, \dots, p_n\}} \mathcal{H} + \sum b_i .$$

In particular, when $\sum b_i \geq 0$, *one has that* $\deg L \geq 0$. *If, in addition, there are constants* a_i *such that, for every* i,

ii) $$\|l_i\| = O(|z_i|^{-a_i}\mu(|z_i|^{-1})),$$

then

$$\deg L \leq \int_{S \smallsetminus \{p_1,\dots,p_n\}} \mathcal{H} + \sum a_i \, .$$

In practical applications, $\mu(x)$ most often equals a power of $\log(x)$, and $a_i = b_i$ for each i. In this situation, then

$$\deg L = \int_{S \smallsetminus \{p_1,\dots,p_n\}} \mathcal{H} + \sum a_i \, .$$

It should also be observed that, if we assume both i) and ii), the summability of $\log \|l_i\|$ follows.

We now prove Lemma (9.17). For simplicity, and without any real loss of generality, we shall suppose that $n = 1$ and write $p = p_1$, $z = z_1$, and so on. Since $\log \|l\|$ is summable, $\partial\bar{\partial} \log \|l\|$ is a well-defined current on a neighborhood of p. Therefore, although the hermitian metric $\| \ \|$ has singularities, it has a well-defined Chern *current* $\hat{\mathcal{H}}$ whose local expression near the puncture is

$$\frac{1}{\pi\sqrt{-1}} \partial\bar{\partial} \log \|l\| \, .$$

Let χ be a nonnegative C^∞ function on S which is equal to 1 on a neighborhood of p and has support in a neighborhood U of p. Then

$$\deg L = \int_S \hat{\mathcal{H}} = \int_S \chi\hat{\mathcal{H}} + \int_S (1-\chi)\mathcal{H} \, .$$

On the other hand, if U_ε is a disc of radius ε centered at p, then, for small enough ε, χ is constant on U_ε, so that repeated applications of Stokes' theorem show that

$$\begin{aligned}
\pi\sqrt{-1} \int_S \chi\hat{\mathcal{H}} &= \int_U \chi\partial\bar{\partial} \log \|l\| \\
&= \int_U \bar{\partial} \log \|l\| \wedge \partial\chi \\
&= \int_{U \smallsetminus U_\varepsilon} \bar{\partial} \log \|l\| \wedge \partial\chi \\
&= \int_{U \smallsetminus U_\varepsilon} \chi\partial\bar{\partial} \log \|l\| + \int_{\partial U_\varepsilon} \chi\bar{\partial} \log \|l\| \\
&= \int_{U \smallsetminus U_\varepsilon} \chi\partial\bar{\partial} \log \|l\| + \int_{\partial U_\varepsilon} \bar{\partial} \log \|l\| \, .
\end{aligned}$$

Summing up, for small enough ε,

$$\deg L = \int_{S\smallsetminus U_\varepsilon} \mathcal{H} + \frac{1}{\pi\sqrt{-1}} \int_{\partial U_\varepsilon} \overline{\partial} \log \|l\| \,.$$

Passing to polar coordinates $z = re^{\sqrt{-1}\vartheta}$, one easily sees that, for any function f,

$$\overline{\partial} f \Big|_{\partial U_\varepsilon} = \left(-\frac{\sqrt{-1}}{2} r \frac{\partial f}{\partial r} - \frac{1}{2}\frac{\partial f}{\partial \vartheta} \right) d\vartheta \,.$$

In particular, it follows that

$$\int_{\partial U_\varepsilon} \overline{\partial} \log \|l\| = \frac{1}{2\sqrt{-1}} \int_{\partial U_\varepsilon} r \frac{\partial \log \|l\|}{\partial r} d\vartheta \,.$$

Hence, we can rewrite the degree of L as

$$\deg L = \int_{S\smallsetminus U_\varepsilon} \mathcal{H} - \frac{1}{2\pi} \int_{\partial U_\varepsilon} r \frac{\partial \log \|l\|}{\partial r} d\vartheta \,.$$

In the remainder of the proof we shall often write $F(\varepsilon)$ for $\frac{1}{2\pi}\int_{\partial U_\varepsilon} r \frac{\partial \log \|l\|}{\partial r} d\vartheta$. The positivity assumption on $\mathcal{H}$ implies that F is nonincreasing. Thus, for any $E \geq \varepsilon > 0$,

$$\begin{aligned} F(E)(\log E - \log \varepsilon) &= \int_\varepsilon^E F(E) \frac{dr}{r} \\ &\leq \int_\varepsilon^E F(r) \frac{dr}{r} \\ &= \frac{1}{2\pi} \int_\varepsilon^E \int_0^{2\pi} \frac{\partial \log \|l\|}{\partial r} dr d\vartheta \\ &= \frac{1}{2\pi} \int_0^{2\pi} (\log \|l\| \big|_{r=E} - \log \|l\| \big|_{r=\varepsilon}) d\vartheta \,. \end{aligned}$$

Keeping E fixed, letting ε vary, and using i), we thus find that

$$F(E) \log(\varepsilon^{-1}) \leq C + \log(\mu(\varepsilon^{-1})) + b \log \varepsilon$$

for a suitable constant C or, exponentiating, that

$$e^{-C} \leq \frac{\mu(\varepsilon^{-1})}{\varepsilon^{-(F(E)+b)}}$$

for any ε such that $0 < \varepsilon \leq E$. If $F(E) + b > 0$, this contradicts the assumption that $\frac{\mu(x)}{x^N} \to 0$ as $x \to \infty$ for any positive N. The conclusion is that $F + b$ is nonpositive or, in other terms, that

$$\int_{S\smallsetminus U_\varepsilon} \mathcal{H} + b \leq \deg L$$

for any small ε. This proves the first part of the lemma. To prove the second part, one uses a similar argument. Write

$$\deg L = \int_{S \setminus \{p\}} \mathcal{H} + K\,.$$

Then, for any r, $-F(r) \geq K$, so that, for any $E \geq \varepsilon > 0$, one has

$$\begin{aligned} K(\log E - \log \varepsilon) &\leq -\int_{\varepsilon}^{E} F(r)\frac{dr}{r} \\ &= \frac{1}{2\pi}\int_0^{2\pi} (\log \|l\|\big|_{r=\varepsilon} - \log \|l\|\big|_{r=E})d\vartheta\,. \end{aligned}$$

Keeping E fixed and letting ε vary, condition ii) then implies that

$$K \log(\varepsilon^{-1}) \leq C + a\log(\varepsilon^{-1}) + \log(\mu(\varepsilon^{-1}))$$

or, which is the same,

$$e^{-C} \leq \frac{\mu(\varepsilon^{-1})}{(\varepsilon^{-1})^{K-a}}\,.$$

This implies that $K - a \leq 0$, that is, that

$$\deg L \leq \int_{S \setminus \{p\}} \mathcal{H} + a\,,$$

as desired. This proves Lemma (9.17). We now complete the basic step in the proof of the semipositivity theorem for the Hodge bundles.

THEOREM (9.18). *Let $f : \mathcal{X} \to S$ be a family of nodal curves parameterized by a smooth complete curve. Suppose furthermore that the generic fiber of f is smooth and that $\mathcal{X}$ is smooth as well. Then $f_*\omega_{\mathcal{X}/S}$ is semipositive.*

Proof. By what we have seen, the curvature of $f_*\omega_{\mathcal{X}/S}$ is nonnegative away from singular fibers. If L is any line bundle quotient of $f_*\omega_{\mathcal{X}/S}$, the same is true for L by the principle that curvature increases on quotient bundles. In order to prove that the degree of L is nonnegative, we shall use Lemma (9.17), and thus what we need to do is to check conditions i) and ii) for L.

We begin by analyzing the singularities of the canonical metric of $f_*\omega_{\mathcal{X}/S}$. The problem being local, we focus our attention on

$$f : X \longrightarrow D = \{t \in \mathbb{C} |\ |t| < 1\},$$

where D is a disc in S, $X = f^{-1}(D)$, and $f^{-1}(t)$ is smooth for $t \neq 0$ and singular for $t = 0$. Denote by C the central fiber and let $N \xrightarrow{\pi} C$ be its normalization. Also denote by $p_1, \ldots, p_h$ the nodes of C. Then there is an exact sequence

$$0 \to H^0(N, \omega_N) \to H^0(C, \omega_C) \to \oplus \mathbb{C}_{p_i} .$$

Let φ and ψ be sections of $f_* \omega_{X/D}$ that do not restrict to zero in $H^0(C, \omega_C)$. We wish to estimate

$$\langle \varphi, \psi \rangle_t = \sqrt{-1} \int_{f^{-1}(t)} \varphi \wedge \overline{\psi}$$

as t goes to 0. We shall do this first when C has only one singular point p. We may choose local coordinate x and y centered at p on X in such a way that f is given in local coordinates by $t = xy$. Set

$$\begin{aligned} A &= \{(x,y) | \ |x| < 1, |y| < |\}, \\ B &= X \smallsetminus A, \\ A_t &= A \cap f^{-1}(t), \\ B_t &= B \cap f^{-1}(t). \end{aligned}$$

We have

$$\int_{f^{-1}(t)} \varphi \wedge \overline{\psi} = \int_{A_t} \varphi \wedge \overline{\psi} + \int_{B_t} \varphi \wedge \overline{\psi} ,$$

and the second summand is a C^∞ function of t. We now evaluate the first summand. We may write

$$\begin{aligned} A_t &= \{(x,y) \ | \ |t| < |x| < 1 \ , \ xy = t\} \\ &= \{(x,y) \ | \ |t| < |y| < 1 \ , \ xy = t\} \end{aligned}$$

and

$$\begin{aligned} \varphi &= (c + a_1 x + a_2 y) \frac{dx}{x}, \\ \psi &= (d + b_1 x + b_2 y) \frac{dx}{x}, \end{aligned}$$

where c and d are constants, and a_1, a_2, b_1, b_2 are holomorphic functions of x and y. We also set

$$\begin{aligned} \alpha &= (a_1 x + a_2 y) \frac{dx}{x}, \\ \beta &= (b_1 x + b_2 y) \frac{dx}{x} . \end{aligned}$$

In dealing with these expressions we should keep in mind that, on any fiber,

$$\frac{dx}{x} = - \frac{dy}{y} .$$

Of course, on the central fiber, we use the expression on the right for $x = 0$, and the one on the left for $y = 0$. We have

$$
(9.19)\qquad
\begin{aligned}
\int_{A_t} \alpha \wedge \overline{\beta} &= \int_{|t|<|x|<1} a_1 \overline{b}_1 dx \wedge d\overline{x} + \int_{|t|<|y|<1} a_2 \overline{b}_2 dy \wedge d\overline{y} \\
&+ \int_{|t|<|x|<1} a_1 \overline{b}_2 \frac{\overline{y}}{\overline{x}} dx \wedge d\overline{x} + \int_{|t|<|x|<1} a_2 \overline{b}_1 \frac{y}{x} dx \wedge d\overline{x}\,.
\end{aligned}
$$

The first two summands are bounded, and their limits for t going to zero are:

$$
\int_{A_0 \cap \{y=0\}} \alpha \wedge \overline{\beta}\ , \quad \int_{A_0 \cap \{x=0\}} \alpha \wedge \overline{\beta}\,.
$$

On the other hand,

$$
\begin{aligned}
&\left| \int_{|t|<|x|<1} a_1 \overline{b}_2 \frac{\overline{y}}{\overline{x}} dx \wedge d\overline{x} \right| = \left| \overline{t} \int_{|t|<|x|<1} a_1 \overline{b}_2 \frac{dx \wedge d\overline{x}}{\overline{x}^2} \right| \\
&\qquad \le |t| k \int_0^{2\pi} \int_{|t|}^1 \frac{r dr d\vartheta}{r^2} = 2\pi k |t| \,|\log|t|\,|\,,
\end{aligned}
$$

where k is a positive constant. Clearly, this quantity goes to zero as t tends to zero. The same argument applies to the last summand in (9.19). This computation shows that, when $\varphi|_C$ and $\psi|_C$ belong to $H^0(N, \omega_N)$, that is, when the constants c and d vanish, then $\langle \varphi, \psi \rangle_t$ is a continuous function of t and that

$$
\langle \varphi, \psi \rangle_0 = \sqrt{-1} \int_N \pi^* \varphi \wedge \pi^* \psi\,.
$$

Now we evaluate:

$$
\begin{aligned}
\int_{A_t} \alpha \wedge \overline{\left(\frac{dx}{x} \right)} &= \int_{|t|<|x|<1} \left(a_1 dx \wedge \frac{d\overline{x}}{\overline{x}} + a_2 \frac{y}{|x|^2} dx \wedge d\overline{x} \right) \\
&= \int_{|t|<|x|<1} \left(a_1 \frac{dx \wedge d\overline{x}}{\overline{x}} \right) + \int_{|t|<|x|<1} \left(t \frac{a_2}{x|x|^2} dx \wedge d\overline{x} \right)\,.
\end{aligned}
$$

The first integral is bounded, while the absolute value of the second can be bounded above by

$$
k|t| \int_0^{2\pi} \int_{|t|}^1 \frac{r dr d\vartheta}{r^3}\,,
$$

where k is a positive constant, and hence by

$$
2\pi k |t| \left(\frac{1}{|t|} - 1 \right) = 2\pi k (1 - |t|)\,.
$$

It follows that $\int_{A_t}\alpha\wedge\overline{\left(\frac{dx}{x}\right)}$ is a bounded function of t. We finally evaluate

$$\int_{A_t}\frac{dx}{x}\wedge\overline{\left(\frac{dx}{x}\right)}.$$

One has

$$\begin{aligned}\int_{A_t}\frac{dx}{x}\wedge\overline{\left(\frac{dx}{x}\right)} &= \int_{|t|<|x|<1}\frac{dx\wedge d\overline{x}}{|x|^2}\\ &= \frac{2}{\sqrt{-1}}\int_0^{2\pi}\int_{|t|}^1\frac{rdrd\vartheta}{r^2}\\ &= \frac{4\pi}{\sqrt{-1}}\,|\log|t||\,.\end{aligned}$$

Summing up, we obtain

$$\langle\varphi,\psi\rangle = 4\pi\mathrm{Res}_p(\varphi)\overline{\mathrm{Res}_p(\psi)}\,|\log|t||+ \text{ bounded terms},$$

where the residues are taken by restricting φ and ψ to (say) the branch $y=0$. In general, that is, when the central fiber has h singular points $p_1,\dots,p_h$, we get

$$\langle\varphi,\psi\rangle = 4\pi\sum\mathrm{Res}_{p_i}(\varphi)\overline{\mathrm{Res}_{p_i}(\psi)}\,|\log|t||+ \text{ bounded terms},$$

where again the residues are computed by restricting φ and ψ to one of the branches of each node. It is important to notice that, when $\mathrm{Res}_{p_i}\varphi=\mathrm{Res}_{p_i}\psi=0$ for every i, then $\langle\varphi,\psi\rangle$ is a continuous function whose value at $t=0$ is $\langle\pi^*\varphi,\pi^*\psi\rangle$.

Now let F be a rank m vector subbundle of $f_*\omega_{X/D}$. We may find a frame $s_1,\dots,s_m$ for F such that the restrictions of $s_1,\dots,s_l$ to C are independent modulo $H^0(N,\omega_N)$, while the restrictions of $s_{l+1}\dots,s_m$ to C belong to $H^0(N,\omega_N)$. Such a frame will be called *distinguished*. Then, writing

$$\begin{aligned}\sigma_i &= |\log|t||^{-\frac{1}{2}}s_i\ , \quad i=1,\dots,l\,,\\ \sigma_j &= s_j, \qquad j>l\,,\end{aligned}$$

we get

$$(\langle\sigma_i,\sigma_j\rangle)_{i,j=1\dots m} = \begin{pmatrix}P & 0\\ 0 & Q\end{pmatrix}+o(t)\,,$$

where

$$\begin{aligned}P &= \left(4\pi\sum_{v=1}^{h}\mathrm{Res}_{p_v}(s_i)\overline{\mathrm{Res}_{p_v}(s_j)}\right)_{i,j=1,\dots,l}\,,\\ Q &= (\langle\pi^*s_i,\pi^*s_j\rangle)_{i,j=l+1,\dots,m}\,.\end{aligned}$$

We obtain, in particular,

$$(\langle s_i,s_j\rangle)_{i,j=1,\dots,l} = H(t)(\langle\sigma_i,\sigma_j\rangle)_{i,j=1,\dots,l}\,H(t)\,,$$

where

$$H(t)=\begin{pmatrix} |\log|t||^{\frac{1}{2}} & & & & & 0 \\ & \ddots & & & & \\ & & |\log|t||^{\frac{1}{2}} & & & \\ & & & 1 & & \\ & 0 & & & \ddots & \\ & & & & & 1 \end{pmatrix}.$$

Since P and Q are both nonsingular, this implies that

$$\det(\langle s_i, s_j\rangle) \sim |\log|t||^l .$$

Now pick F to be the kernel of the morphism from $f_*\omega_{\mathcal{X}/S}$ to L. Let s be a section of $f_*\omega_{\mathcal{X}/S}$ such that its reduction $\overline{s}$ modulo F is a local generator for L. We can find a distinguished frame $s_1, \ldots, s_g$ for $f_*\omega_{\mathcal{X}/S}$ such that $s = s_n$ for some n. We may also set things up so that $s_1, \ldots, \hat{s}_n, \ldots, s_g$ is a distinguished frame for F. Then, denoting by pr_F the orthogonal projection onto F, we get

$$\begin{aligned} \|\overline{s}\|^2 = \|s - pr_F s\|^2 &= \frac{\det(\langle s_i, s_j\rangle)}{\det(\langle s_i, s_j\rangle)_{i,j\neq n}} \\ &\sim \begin{cases} 1 & \text{if } s\big|_C \in H^0(N, \omega_N) \\ |\log|t|| & \text{otherwise.} \end{cases} \end{aligned}$$

This shows that all the hypotheses of Lemma (9.17) are satisfied by our L. The proof of Theorem (9.18) is thus completed.

To prove Theorem (9.16) in general, two further steps are necessary. The first one is to prove the analogue of Theorem (9.18) for $f_*\omega^k_{\mathcal{X}/S}$. Without noticing, we have almost done this already. Suppose in fact that the general fiber of f is not hyperelliptic. Then, by Max Noether's theorem, the image $\mathcal{F}$ of the sheaf homomorphism

$$\underbrace{f_*\omega_{\mathcal{X}/S} \otimes \cdots \otimes f_*\omega_{\mathcal{X}/S}}_{k \text{ times}} \to f_*\omega^k_{\mathcal{X}/S}$$

equals $f_*\omega^k_{\mathcal{X}/S}$ at all but a finite number of points of S. Moreover, $\mathcal{F}$ is semipositive by parts a) and c) of Proposition (9.9). On the other hand, any line bundle quotient of $f_*\omega^k_{\mathcal{X}/S}$ contains as a subsheaf a line bundle quotient of $\mathcal{F}$, so $f_*\omega^k_{\mathcal{X}/S}$ is semipositive as well. This argument does not cover the case where all the fibers are hyperelliptic. To handle this, two strategies are possible. One is to proceed by direct explicit computation. The other, which is not restricted to the case where all the fibers are hyperelliptic, is to mimick the proof of (9.18), i.e., to put a natural metric on $f_*\omega^k_{\mathcal{X}/S}$, compute its curvature, and conclude using (9.17).

The last step would be to prove an analogue of (9.18) for $f_*\omega^k_{\mathcal{X}/S}$ and for families all of whose fibers are singular. In other words, we are dealing here with families $f : \mathcal{X} \to S$ such that the locus of all singular points in the fibers contains one-dimensional components. Modulo a finite base change, we can assume that these are all sections of f, and blowing them up yields a family of curves whose general fiber is smooth but possibly not connected. Although this is not completely correct, in a sense, this reduces us to the previous step. It is worth mentioning that a similar procedure will be used in Chapter XIV in the proof of the projectivity of moduli space.

10. Deformations of symmetric products.

In this section we give an interesting application of the infinitesimal deformation theory of curves and of the cohomological interpretation of the differential of the period mapping. Let

$$\varphi : \mathcal{C} \to (B, b_0) \ , \quad \varphi^{-1}(b_0) \cong C$$

be a deformation of a smooth genus g curve C. The symmetric group $\mathfrak{S}_d$ on d letters acts on the d-fold fiber product of $\mathcal{C}$ over B. We denote by $\mathcal{C}_d$ the quotient

$$\mathcal{C}_d = (\mathcal{C} \times_B \cdots \times_B \mathcal{C})/\mathfrak{S}_d$$

and by φ_d its projection onto B. Thus,

$$\varphi_d : \mathcal{C}_d \to B$$

is a deformation of the d-fold symmetric product C_d, which is clearly functorial in B. In particular, when $\varphi : \mathcal{C} \to B$ is the Kuranishi family for C, we get a linear map

$$\gamma : H^1(C, T_C) = T_{b_0}(B) \xrightarrow{\rho} H^1(C_d, T_{C_d})$$

associating to the Kodaira–Spencer class of a first-order deformation of C the Kodaira–Spencer class of the corresponding deformation of C_d. We may then ask how γ behaves. The answer was found by Kempf, who proved the following:

THEOREM (10.1). *If C is not hyperelliptic, γ is an isomorphism.*

To begin with, we shall show that γ is injective. Consider the diagram

$$\begin{array}{ccc} H^1(C_d, T_{C_d}) & \longrightarrow & \mathrm{Hom}(H^{1,0}(C_d), H^{0,1}(C_d)) \\ \uparrow{\scriptstyle\gamma} & & \uparrow{\scriptstyle\cong} \\ H^1(C, T_C) & \xrightarrow{\ \nu\ } & \mathrm{Hom}(H^{1,0}(C), H^{0,1}(C)) \end{array} \tag{10.2}$$

where the horizontal mappings are induced by cup-product (and hence are the differentials of the period mappings for the Kuranishi family $\varphi : \mathcal{C} \to B$ for C and its d-fold symmetric product over B), and where the vertical maps are induced by the diagonal map of C into C_d. Since the diagonal mapping induces identifications

$$H^1(C,\mathbb{C}) = H^1(C_d,\mathbb{C}),$$

$$H^{1,0}(C) = H^{1,0}(C_d),$$

the two period mappings into the Grassmannian

$$\mathbb{G}(g, H^1(C,\mathbb{C})) = \mathbb{G}(g, H^1(C_d,\mathbb{C}))$$

coincide. It follows that the diagram above is commutative. Since the local Torelli theorem, which we proved in Section 8, exactly says that ν is injective, γ must be injective as well.

The rest of this section will be devoted to the proof that $H^1(C, T_C)$ and $H^1(C_d, T_{C_d})$ have the same dimension; this will end the proof of Theorem (10.1). In addition to the deformation theory of curves, the argument makes an essential use of the classical theory of characteristic systems for divisors, which we discussed in Section 5 of Chapter IX. Observe that, given a relative divisor

$$\mathcal{D} \subset X \times Y$$

parameterized by a smooth Y, the characteristic homomorphisms of the fibers globalize to a characteristic bundle homomorphism

$$\alpha : T_Y \longrightarrow p_*\mathcal{O}_{\mathcal{D}}(\mathcal{D}),$$

where p is the projection of $X \times Y$ onto Y. The construction of α is as follows. Any vector field v on an open subset U of Y lifts canonically to a vector field $\tilde{v}$ on $X \times U$. If $\{U_i\}$ is an open cover of $X \times U$ and F_i a local equation of $\mathcal{E}$ in U_i, the meromorphic functions $\tilde{v}(F_i)/F_i$, when restricted to $\mathcal{D}$, give a section $\alpha(v)$ of $p_*\mathcal{O}_{\mathcal{D}}(\mathcal{D})$ over U. The reader will notice that this construction generalizes the construction of the bundle map ψ appearing in Lemma (2.30) of Chapter IV in Volume I.

We now return to the proof of Theorem (10.1). We denote by λ and π the projections of $C \times C_d$ onto C and C_d, respectively. Recall from Chapter IV, Section 2, that there exists a universal divisor $\Delta = \Delta_d$ in $C \times C_d$. This is the isomorphic image of the product of C with C_{d-1} under the map

$$\begin{aligned}(\lambda, \sigma) : C \times C_{d-1} &\longrightarrow C \times C_d \\ (x, D) &\mapsto (x, \sigma(x, D)) = (x, x + D).\end{aligned}$$

For future reference, we want to show that there is a canonical isomorphism

$$(10.3) \qquad (\lambda,\sigma)^*(\mathcal{O}(\Delta_d)) \cong \lambda^* T_C(\Delta_{d-1})$$

for any $d > 0$. To prove this, tensor both sides with

$$\omega_{C\times C_{d-1}} \cong \lambda^*\omega_C \otimes \pi^*\omega_{C_{d-1}}$$

and observe that since

$$\sigma : C \times C_{d-1} \longrightarrow C_d$$

ramifies simply along Δ_{d-1}, we have

$$\omega_{C\times C_{d-1}} \cong (\sigma^*\omega_{C_d})(\Delta_{d-1}).$$

This reduces the isomorphism to be proved to

$$\sigma^*\omega_{C_d} \otimes (\lambda,\sigma)^*\mathcal{O}(\Delta_d) \cong \pi^*\omega_{C_{d-1}}.$$

Now observe that, by adjunction,

$$(\lambda,\sigma)^*\omega_{C\times C_d}(\Delta_d) \cong \omega_{C\times C_{d-1}}.$$

Hence, formula (10.3) follows from the obvious identity

$$\sigma^*\omega_{C_d} \otimes \omega_{C\times C_{d-1}} = (\lambda,\sigma)^*\omega_{C\times C_d} \otimes \pi^*\omega_{C_{d-1}}.$$

Since $\Delta \subset C\times C_d$ is a relative divisor with respect to both projections, it defines characteristic homomorphisms

$$\alpha : T_C \longrightarrow \lambda_*\mathcal{O}_\Delta(\Delta),$$
$$\psi : T_{C_d} \longrightarrow \pi_*\mathcal{O}_\Delta(\Delta).$$

As we already remarked (cf. Lemma (2.30) of Chapter IV), ψ is an isomorphism. The same is true for α.

Lemma (10.4). *α is an isomorphism.*

Proof. We must show that

$$\alpha_x : T_x(C) \longrightarrow H^0(C_d, \mathcal{O}_{x+C_{d-1}}(x + C_{d-1}))$$

is an isomorphism for any $x \in C$. To show injectivity, we just notice that if $\sum x_i$, $x = x_1$, is a point of C_d which is not on the diagonals, we can take as local coordinates on C_d the d-tuple $(z_1, \ldots, z_d)$, z_i being a local coordinate centered at x_i. Let z be another name for z_1, viewed now as a coordinate on the first factor of $C \times C_d$. Thus, $(z, z_1, \ldots, z_d)$ are local coordinates on $C \times C_d$ in a neighborhood of $(x, \sum x_i)$, and a local equation for Δ is $z - z_1 = 0$. Now

$$\begin{aligned} \alpha_x\left(\frac{\partial}{\partial z}\right) &= \left.\frac{\frac{\partial}{\partial z}(z-z_1)}{z-z_1}\right|_{x+C_{d-1}} \\ &= \left.\frac{1}{z-z_1}\right|_{x+C_{d-1}} = -\frac{1}{z_1} \neq 0\,. \end{aligned} \tag{10.5}$$

To prove the lemma, it is now enough to show that

$$h^0(C_d,\ \mathcal{O}_{x+C_{d-1}}(x + C_{d-1})) = 1\,. \tag{10.6}$$

Since formula (10.3), when restricted to $x \times C_{d-1}$, tells us that

$$\mathcal{O}_{x+C_{d-1}}(x + C_{d-1}) \cong T_{C,x} \otimes \mathcal{O}_{C_{d-1}}(x + C_{d-2}))\,,$$

this is equivalent to showing that any meromorphic function on C_{d-1} with simple poles along $x + C_{d-2}$ is constant. The pullback of such a function to C^{d-1} is a meromorphic function having simple poles along the divisor

$$\{x\} \times C \times \cdots \times C + C \times \{x\} \times \cdots \times C + \cdots$$

and hence, by the Künneth formula, is an element of $H^0(C, \mathcal{O}(x))^{\otimes d-1} = \mathbb{C}$.

Q.E.D.

As we have said, to prove Theorem (10.1), we have to show that $H^1(C, T_C)$ and $H^1(C_d, T_{C_d})$ have the same dimension. Since T_{C_d} is isomorphic to $\pi_* \mathcal{O}_\Delta(\Delta)$, and the projection from Δ to C_d is finite, by the Leray spectral sequence for $\pi|_\Delta$ we get, for any $i \geq 0$, an isomorphism

$$H^i(C_d, T_{C_d}) \cong H^i(\Delta, \mathcal{O}_\Delta(\Delta))\,.$$

As we just proved that T_C is isomorphic to $\lambda_* \mathcal{O}_\Delta(\Delta)$, by pullback we get a map

$$H^1(C, T_C) \cong H^1(C, \lambda_* \mathcal{O}_\Delta(\Delta)) \xrightarrow{\beta} H^1(\Delta, \mathcal{O}_\Delta(\Delta)) \cong H^1(C_d, T_{C_d})\,.$$

Since C is a curve, the Leray spectral sequence for $\mathcal{O}_\Delta(\Delta)$ relative to λ degenerates at E_2 and yields an exact sequence

$$0 \to H^1(C, \lambda_* \mathcal{O}_\Delta(\Delta)) \xrightarrow{\beta} H^1(\Delta, \mathcal{O}_\Delta(\Delta)) \to H^0(C, R^1 \lambda_* \mathcal{O}_\Delta(\Delta))\,.$$

To show that $H^1(C, T_C)$ and $H^1(C_d, T_{C_d})$ are isomorphic, it then suffices to prove that $H^0(C, R^1 \lambda_* \mathcal{O}_\Delta(\Delta))$ is zero. By the isomorphism (10.3) we get

$$\begin{aligned} R^1 \lambda_* \mathcal{O}_{\Delta_d}(\Delta_d) &\cong R^1 \lambda_*(\lambda^* T_C(\Delta_{d-1})) \\ &\cong T_C \otimes R^1 \lambda_* \mathcal{O}_{C \times C_{d-1}}(\Delta_{d-1})\,. \end{aligned} \tag{10.7}$$

The conclusion now follows from the following:

LEMMA (10.8). *If C is a smooth genus curve of genus $g \geq 3$, then, for any $n > 0$,*

$$H^0(C, T_C^n \otimes R^1\lambda_*\mathcal{O}_{C\times C_d}(\Delta_d)) = 0\,.$$

If C is a smooth curve of genus 2, the same is true for any $n > 1$.

Proof. Consider the exact sequence on $C \times C_d$

$$0 \to \mathcal{O} \to \mathcal{O}(\Delta_d) \to \mathcal{O}_{\Delta_d}(\Delta_d) \to 0\,.$$

Taking into account (10.6) and (10.7), we get an exact sheaf sequence

$$\begin{aligned} 0 \to \lambda_*\mathcal{O}_{\Delta_d}(\Delta_d) &\xrightarrow{\delta} H^1(C_d, \mathcal{O}_{C_d}) \otimes \mathcal{O}_C \to \\ R^1\lambda_*\mathcal{O}(\Delta_d) &\to R^1\lambda_*\mathcal{O}_{\Delta_d}(\Delta_d) \cong T_C \otimes R^1\lambda_*\mathcal{O}_{C\times C_{d-1}}(\Delta_{d-1})\,. \end{aligned} \tag{10.9}$$

We now examine the map δ, and we claim that it fits into a commutative diagram

$$\begin{array}{ccc} T_C & \xrightarrow{\quad\eta\quad} & H^1(C, \mathcal{O}_C) \otimes \mathcal{O}_C \\ {\scriptstyle\alpha}\downarrow & & \uparrow{\scriptstyle\iota} \\ \lambda_*\mathcal{O}_{\Delta_d}(\Delta_d) & \xrightarrow{\quad\delta\quad} & H^1(C_d, \mathcal{O}_{C_d}) \otimes \mathcal{O}_C \end{array} \tag{10.10}$$

where α is the characteristic isomorphism of Lemma (10.4), η is the composition of α and δ in the case $d = 1$, and ι is induced by the map

$$\begin{aligned} &C \xrightarrow{\tau_D} C_d \\ &x \mapsto x + D \quad \text{for a fixed } D \in C_{d-1}\,. \end{aligned}$$

We immediately observe that ι is an isomorphism independent of D. The fact that it is an isomorphism has been proved in Section 4 of Chapter VI. Independence of D follows from the fact that the composite isomorphism

$$H^1(J(C), \mathcal{O}) \xrightarrow{u^*} H^1(C_d, \mathcal{O}) \xrightarrow{\tau_D^*} H^1(C, \mathcal{O})$$

is itself independent of D.

We prove commutativity of (10.10) over a point $x \in C$, that is, we show the commutativity of

$$\begin{array}{ccc} T_x(C) & \xrightarrow{\quad\eta_x\quad} & H^1(C, \mathcal{O}_C) \\ {\scriptstyle\alpha_x}\downarrow & & \uparrow{\scriptstyle\iota} \\ H^0(C_d, \mathcal{O}_{x+C_{d-1}}(x + C_{d-1})) & \xrightarrow{\quad\delta_x\quad} & H^1(C_d, \mathcal{O}_{C_d}) \end{array}$$

We fix a divisor $\sum x_i \in C_d$ not belonging to the diagonals, with $x_1 = x$. We set $D = \sum_{i>1} x_i$ and think of ι as induced by τ_D. We also fix coordinates $z, z_1, \ldots, z_d$ as in the proof of Lemma (10.4). We then have, by (10.5),

$$\left.\begin{aligned} \alpha_x\left(\frac{\partial}{\partial z}\right) &= -\frac{1}{z_1} \\ \delta_x\alpha_x\left(\frac{\partial}{\partial z}\right) &= \left\{-\frac{1}{z_1}\right\} \end{aligned}\right\} \quad \text{near } \sum x_i\,.$$

For the same reason, by the definition of η,

$$\eta_x\left(\frac{\partial}{\partial z}\right) = \left\{-\frac{1}{z_1}\right\}.$$

Since, in local coordinates, τ_D is just

$$z_1 \to (z_1, 0, \ldots, 0)\,,$$

the commutativity follows.

Now we can go back to the exact sequence (10.9), which we can rewrite as

$$0 \to \operatorname{coker}(\eta) \to R^1\lambda_*\mathcal{O}(\Delta_d) \to T_C \otimes R^1\lambda_*\mathcal{O}_{C\times C_{d-1}}(\Delta_{d-1})\,.$$

Tensoring with T_C^n, taking global sections, and assuming the lemma for C_{d-1}, we get an isomorphism

$$H^0(C, \operatorname{coker}(\eta) \otimes T_C^n) \cong H^0(C, T_C \otimes R^1\lambda_*\mathcal{O}(\Delta_d))\,.$$

If can prove that the left-hand-side vanishes, we will be done by induction, since, for $d = 0$, the statement of the lemma reduces to

$$H^0(C, T_C^n) = 0\ , \quad n > 0\,.$$

If we consider the exact sequence

$$0 \to T_C^{n+1} \to H^1(C, \mathcal{O}_C) \otimes T_C^n \to \operatorname{coker}(\eta) \otimes T_C^n \to 0\,,$$

it suffices to show that, for $n > 0$, the map

$$H^1(C, T_C^{n+1}) \xrightarrow{\xi} H^1(C, \mathcal{O}_C) \otimes H^1(C, T_C^n)$$

is injective. A moment of reflection will convince the reader that the dual of ξ is the cup-product homomorphism

$$H^0(C, \omega_C) \otimes H^0(C, \omega_C^{n+1}) \longrightarrow H^0(C, \omega_C^{n+2})\,.$$

On the other hand, this map is surjective for $g \geq 3$ and $n > 0$, or for $g = 2$ and $n > 1$. This is a consequence of Noether's theorem when C is not hyperelliptic and follows from an explicit computation when C is hyperelliptic. The proof of the lemma, and hence of Theorem (10.1), is now complete.

We end this section by observing that all the missing pieces in the proof of Green's theorem on quadrics through the canonical curves (Theorem (4.1) in Chapter VI) are now available. In fact, all we needed was to define γ, show that it is an isomorphism, and prove the commutativity of the leftmost square in diagram (4.4) of Chapter VI, that is, diagram (10.2).

11. Bibliographical notes and further reading.

The deformation theory of manifolds was founded by Kodaira and Spencer [435,437]; an introduction to their work can be found in Kodaira's book [434]. The theory was formalized and generalized by Grothendieck, in a series of Bourbaki seminars that go collectively under the name *Fondements de la Géometrie Algebrique*, and particularly in [327] and [328]; Grothendieck's seminars are revisited in [243]. A key step in the development of Grothendieck's program was the proof of the pro-representability of the deformation functor by Schlessinger [615]. There are several exhaustive accounts of the deformation theory of schemes. Among these, we may mention the books by Sernesi [625] and Hartshorne [357] and the papers by Fantechi and Manetti [244] and Manetti [496].

The existence of a Kuranishi family for a compact complex manifold M was first proved by Kodaira, Nirenberg, and Spencer [438] under the restrictive assumption that the second cohomology group of the tangent bundle to M vanish. This assumption was later removed by Kuranishi, who proved the existence of Kuranishi families for arbitrary compact complex manifolds [452,453,454]. The existence of Kuranishi families for compact, possibly nonreduced, analytic spaces is due to Grauert [311].

An introduction to the theory of the period map is provided by [317] and [149]. The earliest instance a local Torelli theorem is to be found in Kodaira [432], where it appears as Theorem 17 and is attributed to Andreotti and Weil. The observation that the local Torelli theorem for curves is equivalent to Noether's theorem appears in [315]. A refinement of the local Torelli theorem for curves is due to Oort and Steenbrink [579].

Our treatment of semipositivity follows in part Kollár's paper [439], where a substantially more general form of the semipositivity of direct images of powers of the relative dualizing sheaf is proved.

Kempf's theorem on deformations of symmetric products of curves can be found in [414]. Green's theorem on quadrics through the canonical curve appears in [312]; an alternate proof of the theorem is to be found in Smith and Varley [641].

Chapter XII.
The moduli space of stable curves

1. Introduction.

In this chapter we shall construct the moduli space $\overline{M}_{g,n}$ of stable n-pointed curves of genus g and look at its structure from various points of view. As a set, this space consists of all isomorphism classes of stable n-pointed curves of genus g. First we will put on $\overline{M}_{g,n}$ a structure of analytic space, and then we will see that this analytic space has a natural structure of an algebraic space. Only in Chapter XIV we will prove that $\overline{M}_{g,n}$ is, indeed, a projective variety. Finally we shall show that $\overline{M}_{g,n}$ is just a coarse reflection of a more fundamental object, the moduli stack $\overline{\mathcal{M}}_{g,n}$ of stable n-pointed curves of genus g.

Throughout the chapter we will make constant use of the construction of algebraic Kuranishi families, carried out in Chapter XI, which we now recall. Let $(C; p_1, \dots, p_n)$ be a stable, n-pointed curve of genus g. Then there exists an *algebraic* deformation

$$\begin{array}{c} \mathcal{C} \\ \big\downarrow \pi \\ (X, x_0) \end{array} \qquad \sigma_i : X \to \mathcal{C}\,, \ \ i = 1, \dots, n\,, \qquad C = \pi^{-1}(x_0) \tag{1.1}$$

of $(C; p_1, \dots, p_n)$ having the following properties. Denote by C_y the fiber of π over y and let G_y be the automorphism group of $(C_y; \sigma_1(y), \dots, \sigma_n(y))$. Then

a) X is affine;
b) the family is Kuranishi at every point of X;
c) the action of the group G_{x_0} on the central fiber extends to compatible actions on $\mathcal{C}$ and X;
d) for every $y \in X$, the automorphism group G_y is equal to the stabilizer of y in G_{x_0}. In particular, G_y is a subgroup of G_{x_0};
e) for every $y \in X$, there is a G_y-invariant neighborhood U of y in X, for the analytic topology, such that any isomorphism (of n-pointed curves) between fibers over U is induced by an element of G_y.

A family with the above properties is called a standard algebraic Kuranishi family, while its restriction to the analytic neighborhood U is simply called a standard Kuranishi family.

E. Arbarello et al., *Geometry of Algebraic Curves*,
Grundlehren der mathematischen Wissenschaften 268,
DOI 10.1007/978-3-540-69392-5_4, © Springer-Verlag Berlin Heidelberg 2011

In Section 2 we begin by putting on $\overline{M}_{g,n}$ a natural structure of normal analytic space of dimension $3g-3+n$. This is done by patching together quotients of bases of standard Kuranishi families modulo the action of the automorphism groups of the central fibers. This patching procedure is based on the universal property of Kuranishi families. The local analytic neighborhoods in $\overline{M}_{g,n}$ are therefore of the type B/G, where B is a bounded simply connected domain in $\mathbb{C}^{3g-3+n}$, and G is a finite group acting linearly on it. Built in the definition of the analytic structure of $\overline{M}_{g,n}$ is its versal property: for every analytic family $\mathcal{Y} \to T$ of n-pointed, stable curves of genus g, the moduli map $t \mapsto [X_t]$ is an analytic morphism from T to $\overline{M}_{g,n}$. Using the fact that there exist stable curves with nontrivial automorphism group, we then proceed to show that there cannot exist a universal family of curves over $\overline{M}_{g,n}$. There is a surrogate which, in several practical applications, is almost as good as the nonexistent universal family. It consists in a family of stable n-pointed genus g curves

$$\eta : \mathcal{X} \to Z\,,$$

parameterized by a normal scheme Z, whose moduli map

$$m : Z \to \overline{M}_{g,n} \tag{1.2}$$

is finite and surjective. A first application of the existence of this family is the following. Using the valuative criterion for properness together with stable reduction, we prove that Z, and therefore $\overline{M}_{g,n}$, is compact. Another important application will come in Chapter XIV. There, in order to prove that the analytic space $\overline{M}_{g,n}$ is a projective variety, we will use the scheme Z as an intermediary. Indeed, using Seshadri's criterion and a small amount of geometric invariant theory, we will show that the line bundle $\eta_*(c_1(\omega_\eta)^2)$ is the first Chern class of an ample line bundle, proving that Z, and therefore $\overline{M}_{g,n}$, is projective.

In order to construct the family (1.2) and in particular to introduce the scheme Z, we need to make a digression on algebraic spaces. We do this in Section 3. Indeed, implicit in the algebraic nature of the Kuranishi families is the fact that $\overline{M}_{g,n}$ has a natural structure of algebraic space. To see this, recall that, as a set, $\overline{M}_{g,n}$ is the quotient of the Hilbert scheme $\mathrm{H}_{\nu,g,n}$ modulo the action of a projective group G and that the base of an algebraic Kuranishi family was constructed by taking slices in $\mathrm{H}_{\nu,g,n}$ that are transversal to the orbits of G. By compactness one can cover $\mathrm{H}_{\nu,g,n}$ with the images of finitely many sets of the type $G \times X_i$, $i = 1, \dots, N$, where X_i is the base of an algebraic Kuranishi family $\pi_i : \mathcal{C}_i \to X_i$. Let G_i be the automorphism group of the central fiber of π_i. The properties of algebraic Kuranishi families imply that the natural map $X_i/G_i \to \overline{M}_{g,n}$ is étale. Set

$$Y_i = X_i/G_i\,, \quad i = 1, \dots, N, \qquad X = \coprod_{i=1}^{N} X_i, \qquad Y = \coprod_{i=1}^{N} Y_i. \tag{1.3}$$

Then the étale map

$$\varphi : Y \longrightarrow \overline{M}_{g,n} \tag{1.4}$$

is surjective. Now, by definition, a separated algebraic space is an étale morphism $\psi : S \to M$ from an affine scheme to an analytic space such that $S \times_M S$ is a closed subscheme of $S \times S$. The scheme S should be considered as a sort of algebraic atlas for M, while the subscheme $S \times_M S$ should be regarded as the set of compatibility conditions for this atlas. In order to prove that $\overline{M}_{g,n}$ is a separated algebraic space, one then needs to show that $R = Y \times_{\overline{M}_{g,n}} Y$ is Zariski-closed in $Y \times Y$. This turns out to be an immediate consequence of the properness of the natural projection $q : \mathbf{I} \longrightarrow X \times X$, where

$$\mathbf{I} = \{(x, x', g) : x, x' \in X, \ \ g \in G, \ \ x' = gx\},$$

which is a consequence of Theorem (5.1) of Chapter X.

The construction of moduli spaces of curves as algebraic spaces shows implicitly that moduli spaces like $\overline{M}_{g,n}$ are also orbifolds. Orbifolds are the differential-geometric counterparts of stacks. Essentially, an orbifold is an analogue of a differentiable variety in which local charts are not open immersions, but rather quotients of open subsets of $\mathbb{R}^n$ by the actions of finite groups. In Section 4 we give the basic definitions and examples of the theory of orbifolds. Furthermore, we show how de Rham theory naturally extends to orbifolds. This is important in view of the fact that, when developing the intersection theory of $\overline{M}_{g,n}$, it will be useful to express intersection numbers as integrals of top degree differential forms over $\overline{M}_{g,n}$.

Section 5 contains a utilitarian introduction to stacks, closely motivated by the case of moduli of curves.[2] In studying Kuranishi families or moduli spaces of curves, we constantly have to deal with the automorphism group of a curve. It is the presence of curves with a nontrivial automorphism group that prevents the moduli space from being smooth and a universal family from existing. Using stacks is a way of effectively keeping track of these automorphism groups. When thinking of moduli spaces as stacks, the automorphism groups become an essential part of the definition.

A stack is, first of all, a groupoid. A groupoid is a pair $\mathcal{M} = (\mathcal{C}, p)$, where $\mathcal{C}$ is a category, and

$$p : \mathcal{C} \to Sch/S$$

[2]Warning: in Sections 5 through 8 we deviate from our general convention that "scheme" stands for "scheme of finite type over $\mathbb{C}$," and allow general schemes.

is a functor. The "fibers" of p are supposed to be categories in which all morphisms are isomorphisms. The fiber over a scheme T is denoted by $\mathcal{M}(T)$. To better understand the properties a stack is asked to satisfy, one should keep in mind the example of the moduli stack $\overline{\mathcal{M}}_{g,n}$ of stable, n-pointed, genus g curves. In this case we take as $\mathcal{C}$ the category in which the objects are the families

$$\begin{array}{c} \mathcal{X} \\ \downarrow{\scriptstyle \xi} \\ T \end{array}$$

of stable n-pointed curves of genus g and in which a morphism

$$\varphi : \xi' \to \xi$$

between a family $\xi' : \mathcal{X}' \to T'$ and a family $\xi : \mathcal{X} \to T$ is a commutative diagram

$$\begin{array}{ccc} \mathcal{X}' & \longrightarrow & \mathcal{X} \\ {\scriptstyle \xi'}\downarrow & & \downarrow{\scriptstyle \xi} \\ T' & \xrightarrow{\ f\ } & T \end{array}$$

inducing an isomorphism $\mathcal{X}' \cong T' \times_T \mathcal{X}$. The functor p assigns to a family $\xi : \mathcal{X} \to T$ its parameter space T:

$$p(\xi) = T .$$

In the case of $\overline{\mathcal{M}}_{g,n}$, given a scheme T, the category $\overline{\mathcal{M}}_{g,n}(T)$ is simply the category of families of stable n-pointed curves of genus g parameterized by T, in which a morphism

$$\varphi : \xi' \to \xi$$

between a family $\xi' : \mathcal{X}' \to T$ and a family $\xi : \mathcal{X} \to T$ is an isomorphism of schemes over T from $\mathcal{X}$ to $\mathcal{X}'$. This is how automorphisms of curves are encoded in the stack definition of moduli spaces. Any scheme M can be considered as a groupoid $M = (\mathcal{C}_M, p_M)$. Here, the objects of $\mathcal{C}_M$ are pairs (T, f) where $f : T \to M$ is a morphism of schemes. The morphisms $\varphi : (T, f) \to (T', f')$ are the morphisms $h : T \to T'$ with $f'h = f$. Finally, $p_M((T, f)) = T$. A groupoid $(\mathcal{C}, p)$ is (represented by) a scheme if, for some scheme M, there exists an isomorphism of groupoids

$$(\mathcal{C}_M, p_M) \cong (\mathcal{C}, p).$$

It follows from the definitions that, if a universal family $\mathcal{X} \to \overline{M}_{g,n}$ existed, then the groupoid $\overline{\mathcal{M}}_{g,n}$ would be represented by the scheme $\overline{M}_{g,n}$.

One of the advantages of the category of groupoids is that in this category quotients always exist. For example, if a group scheme G acts on a scheme X, then one can form a quotient stack $[X/G]$. As a first result, we prove that

$$\overline{\mathcal{M}}_{g,n} = [\mathrm{H}_{\nu,g,n} / PGL(N)] , \qquad N = (2\nu - 1)(g-1) + \nu n .$$

In Section 6 we come to the second ingredient in the definition of a stack. This involves descent theory. Suppose that we are given a groupoid $\mathcal{M} = (\mathcal{C}, p)$, an étale surjective morphism of schemes $U \to T$, and an object ξ in $\mathcal{M}(U)$. The question is: when does ξ descend to T? In other words, when does there exists $\eta \in \mathcal{M}(T)$ with $f^*(\eta) \simeq \xi$? When $\mathcal{M} = \overline{\mathcal{M}}_{g,n}$, what we are given is a family of curves $\xi : X \to U$, and we look for conditions insuring the existence of a family $\eta : Y \to T$ with $f^*(\eta) \simeq \xi$. To understand these conditions, we consider the analogy between ordinary topology and étale topology. Instead of the étale map $U \to T$, we consider an open cover $\mathcal{U} = \{U_i\}$ of T. The collection of pairwise intersections $\{U_i \cap U_j\}$ is the topological counterpart of the fiber product $U \times_T U$, while the collection of triple intersections $\{U_i \cap U_j \cap U_k\}$ is the counterpart of the triple fiber product $U \times_T U \times_T U$. The datum of an object ξ_i on each U_i corresponds to the datum of an object ξ on U. An isomorphism φ_{ij} from $\xi_i|U_i \cap U_j$ to $\xi_i|U_i \cap U_j$ is translated into an isomorphism $\varphi : p_1^*\xi \to p_2^*\xi$, where p_1 and p_2 are the natural projections from $U \times_T U$ to U. The compatibility condition $\varphi_{ij}\varphi_{jk} = \varphi_{ik}$ on $\{U_i \cap U_j \cap U_k\}$ is translated into an appropriate "cocycle" condition for the isomorphism φ on $U \times_T U \times_T U$. If this cocycle condition is satisfied, (ξ, φ) are said to be a descent datum. However, this datum is not necessarily effective, meaning that the compatibility conditions are not always sufficient to make the object ξ "descend" from U to T. The first condition for a a groupoid to be a stack is that every étale descent datum is effective. To check this condition for the groupoid $\overline{\mathcal{M}}_{g,n}$, one has to use Grothendieck's descent theory for quasi-coherent sheaves, which we review in this same section.

In Section 7 we come to the third ingredient in the definition of a stack. This condition is almost automatically satisfied by the groupoid $\overline{\mathcal{M}}_{g,n}$, and it basically requires that a natural functor that can be concocted in terms of the *Isom* functor should, in fact, be a sheaf. Leaving the category of schemes to enter the category of stacks presents several advantages. Here are a few. First, as we observed, in the category of stacks one can take quotients. Secondly, looking at the case of curves, it makes sense to talk about a universal family of curves $\mathcal{C} \to \overline{\mathcal{M}}_{g,n}$ *over the stack* $\overline{\mathcal{M}}_{g,n}$. Moreover, as a stack, $\overline{\mathcal{M}}_{g,n}$ is smooth. In other words, in the category of stacks, modding out by finite groups destroys neither smoothness nor the property for a morphism to be étale. As another example, let us go back to (1.3) and (1.4). Look at the smooth variety X which is the disjoint union of bases $X_1, \ldots, X_N$ of Kuranishi families

$\mathcal{C}_i \to X_i$ and consider the surjective moduli map $m : X \to \overline{M}_{g,n}$. As a map of schemes, m is not étale. To make it étale, one has to divide each X_i by the automorphism group of the central fiber of π_i. One of the advantages of replacing $\overline{M}_{g,n}$ with the stack $\overline{\mathcal{M}}_{g,n}$ is that, as a morphism of stacks, $m : X \to \overline{\mathcal{M}}_{g,n}$ is étale.

There is a class of stacks which is particularly manageable and which includes the moduli stacks $\overline{\mathcal{M}}_{g,n}$. This is the class of Deligne–Mumford stacks. We discuss these in Section 8. The first property that characterizes a Deligne–Mumford stack $\mathcal{M}$ is the existence of an étale surjective morphism $m : X \to \mathcal{M}$, where X is a scheme. The other requirement is that the diagonal morphism $\Delta : \mathcal{M} \to \mathcal{M} \times \mathcal{M}$ should be representable. This last condition, for the stack $\overline{\mathcal{M}}_{g,n}$, translates into the following property which is the content of Proposition (3.10). Consider the family $\xi : \mathcal{C} \to X$ and the two projections $p_1, p_2 : X \times X \to X$. Then the natural projection

$$\mathbf{Isom}_{X\times X}(p_1^*\xi, p_2^*\xi) \longrightarrow X \tag{1.5}$$

is étale and surjective.

In Section 9, after digressing on Zariski's main theorem in the context of algebraic spaces, we state the basic result that, given a reduced, separated algebraic space X, there exists a scheme Z which is a finite Galois cover of X. A variant of the proof then yields the family (1.2).

The final section is devoted to the description of various natural morphisms between moduli stacks of curves. Building on the work done in Section 6 of Chapter X, we construct the universal curve

$$\overline{\mathcal{C}}_{g,n} \to \overline{\mathcal{M}}_{g,n}\,,$$

the basic projection morphisms

$$\overline{\mathcal{M}}_{g,n+1} \longrightarrow \overline{\mathcal{M}}_{g,n}\,,$$

and the basic clutching maps

$$\xi_\Gamma : \overline{\mathcal{M}}_\Gamma \to \overline{\mathcal{M}}_{g,n}\,,$$

where we adopt the notation introduced at the beginning of Section 7 of Chapter X. We also show that $\overline{\mathcal{C}}_{g,n}$ is naturally isomorphic to $\overline{\mathcal{M}}_{g,n+1}$.

We end this introduction by recalling some classical facts about elliptic curves that may be helpful to keep in mind in what follows.

Let $\mathbb{H}$ denote the upper half-plane. For $\tau \in \mathbb{H}$, we denote by E_τ the elliptic curve $\mathbb{C}/\Lambda_\tau$, where the lattice $\Lambda_\tau \cong \mathbb{Z}^2$ generated by 1 and τ acts by

$$(m_1, m_2) \cdot \tau = z + m_1 + m_2\tau.$$

It is well known that E_τ is embedded in $\mathbb{P}^2$ as a smooth cubic with affine equation

$$y^2 = 4x^3 + g_2(\tau)x + g_3(\tau)$$

by the map

$$\begin{cases} x = \mathcal{P}(z,\tau), \\ y = \mathcal{P}'(z,\tau), \end{cases}$$

where $\mathcal{P}(z,\tau)$ is the Weierstrass $\mathcal{P}$-function. This construction yields a family of smooth cubics

$$\alpha : \mathcal{C} \to \mathbb{H}. \tag{1.6}$$

This family, which is highly transcendental, has two remarkable properties. First of all, every elliptic curve appears in it, up to isomorphism, and, secondly, the family is everywhere Kuranishi. From g_2 and g_3 one constructs the j-function

$$j(\tau) = \frac{1728 g_2(\tau)^3}{g_2(\tau)^3 - 27 g_3(\tau)^2}.$$

Letting $\Gamma = SL_2(\mathbb{Z})$ denote the modular group acting as usual on $\mathbb{H}$ by

$$\begin{pmatrix} a & b \\ c & d \end{pmatrix} \cdot \tau = \frac{a\tau + b}{c\tau + d},$$

it is well known that $j(\tau)$ is Γ-invariant. Moreover, setting

$$\widehat{\mathbb{H}} = \mathbb{H} \cup \mathbb{Q} \cup \{\infty\},$$

one has a natural extension of the action of Γ to $\widehat{\mathbb{H}}$. The points of $\mathbb{Q} \cup \{\infty\}$ are called cusps and are permuted transitively by Γ. There are bijections

$$\begin{array}{ccc} j : \Gamma\backslash\widehat{\mathbb{H}} & \xrightarrow{\cong} & \mathbb{P}^1 \\ \cup & & \cup \\ \Gamma\backslash\mathbb{H} & \xrightarrow{\cong} & \mathbb{P}^1 \setminus \{\infty\} \end{array} \tag{1.7}$$

Anticipating notation to be used below, we set

$$\begin{cases} Y(1) = \Gamma\backslash\mathbb{H}, \\ \overline{Y(1)} = \Gamma\backslash\widehat{\mathbb{H}}. \end{cases}$$

One may take a slightly different approach to the family (1.6). The semi-direct product $\Gamma \ltimes \mathbb{Z}^2$ acts on $\mathbb{H} \times \mathbb{C}$ by

$$\left(\begin{pmatrix} a & b \\ c & d \end{pmatrix}, (m_1, m_2) \right) \cdot (\tau, z) = \left(\frac{a\tau + b}{c\tau + d}, \frac{z + m_1 + m_2\tau}{c\tau + d} \right).$$

Taking quotients only modulo the second factor $\mathbb{Z}^2$, we obtain a family over $\mathbb{H}$ which is just (1.6). Taking instead the full quotient, we get a variety $\mathbf{E}^*$ fibered over $Y(1)$:

$$\pi : \mathbf{E}^* \to Y(1).$$

As is well known, for τ and τ' in $\mathbb{H}$, E_τ is isomorphic to $E_{\tau'}$ if and only if $j(\tau) = j(\tau')$. It follows that we may identify $M_{1,1}$ with $Y(1)$. One's first guess might be that then $\mathbf{E}^* \to M_{1,1}$ is the universal elliptic curve. However, this is spectacularly not correct. The point is that one is in particular dividing by

$$\begin{pmatrix} -1 & 0 \\ 0 & -1 \end{pmatrix} \in \Gamma\,.$$

This automorphism acts trivially on $\mathbb{H}$ but gives the -1 involution on each E_τ, so that the fibers of π are in fact $\mathbb{P}^1$'s. In addition, two fibers of π are special in that they correspond to elliptic curves with automorphisms other than ± 1. These are precisely

$$\begin{array}{llll} E_{\tau_0}\,, & \tau_0 = e^{\pi\sqrt{-1}}\,, & j(\tau_0) = 1728\,, & |\operatorname{Aut}(E_0)| = 4; \\ E_{\tau_1}\,, & \tau_1 = e^{2\pi\sqrt{-1}/3}\,, & j(\tau_1) = 0\,, & |\operatorname{Aut}(E_1)| = 6. \end{array}$$

Although $\mathbf{E}^* \to M_{1,1}$ is not a universal family of elliptic curves, if we consider the stack $\mathcal{M}_{1,1}$ which, roughly speaking, refines $M_{1,1}$ by adding the data $\operatorname{Aut}(E_\tau)$, then, as will be explained below, there is a universal family $\mathcal{E} \to \mathcal{M}_{1,1}$ of elliptic curves. Thus, in this case enlarging our concept of variety to include stacks, one resolves the issue of having a universal elliptic curve. It is worth noticing that, as will be seen in the following chapters, one may do enumerative geometry in a stack context. Perhaps the first instance of this is due to Mumford [549], who showed that $\operatorname{Pic}(\mathcal{M}_{1,1}) \cong \mathbb{Z}/12\mathbb{Z}$, where the left-hand side is the Picard group of the stack $\mathcal{M}_{1,1}$.

A complementary approach to the issues raised above is the one of trying to *rigidify* the family of elliptic curves by adding additional data that kill the automorphisms groups of the E_s. This will be the central theme of Chapter XVI. Essentially, the additional data consists in considering the finite group of points of order N in each E_τ. For this, one sets

$$\Gamma(N) = \ker(SL_2(\mathbb{Z}) \to SL_2(\mathbb{Z}/N\mathbb{Z})).$$

Then the semi-direct product $\Gamma(N) \ltimes \mathbb{Z}^2$ acts on $\mathbb{H} \times \mathbb{C}$ as above. When $N \geq 3$, this action is free, so that the quotient $\mathbf{E}^*(N)$ is smooth. Moreover, the action of $\Gamma(N)$ on $\mathbb{H}$ extends to a free action on $\widehat{\mathbb{H}}$, giving rise to an open inclusion of smooth varieties $Y(N) = \Gamma(N)\backslash\mathbb{H} \subset \Gamma(N)\backslash\widehat{\mathbb{H}} =$

$\overline{Y(N)}$. The family of elliptic curves $\pi_N : \mathbf{E}^*(N) \to Y(N)$ can be completed to a family of nodal curves over $\overline{Y(N)}$, so that one gets a diagram

$$\begin{array}{ccc} \mathbf{E}^*(N) & \subset & \mathbf{E}(N) \\ {\scriptstyle \pi_N}\downarrow & & \downarrow{\scriptstyle \overline{\pi}_N} \\ Y(N) & \subset & \overline{Y(N)} \end{array}$$

Finally, setting $G_N = SL_2(\mathbb{Z}/N\mathbb{Z})$, we see that G_N acts naturally on the diagram

$$\begin{array}{ccc} Y(N) & \subset & \overline{Y(N)} \\ \downarrow & & \downarrow \\ M_{1,1} & \subset & \overline{M}_{1,1} \end{array}$$

and, consequently, $M_{1,1}$ and $\overline{M}_{1,1}$ are each represented as quotients of smooth varieties by finite groups.

2. Construction of moduli space as an analytic space.

Our goal in this section is to put a structure of analytic space on the set of isomorphism classes of stable P-pointed curves of genus g, where P is a finite set. The resulting space is called *the moduli space of stable P-pointed curves of genus g* and is denoted by $\overline{M}_{g,P}$. When $P = \{1, \ldots, n\}$, one writes $\overline{M}_{g,n}$ for $\overline{M}_{g,P}$. The construction relies on the existence of standard Kuranishi families, as defined in Chapter XI, definition (6.8), proved in the same chapter. We shall need the following well-known elementary result, due to Henri Cartan [106].

LEMMA (2.1). *Let G be a finite group acting on a complex manifold U. Then there is a unique structure of normal analytic space on the quotient U/G such that $U \to U/G$ is holomorphic.*

Without loss of generality, we may assume that the action of G is effective. Let u be a point of U, let p be its image in U/G, and denote by H the stabilizer of u. All sufficiently small open neighborhoods of p are of the form V/H, where V is a sufficiently small H-invariant open neighborhood of u, and conversely. Suppose that there exists a complex structure on U/G satisfying the requirements of the lemma. Then, if V and H are as above, a holomorphic function on V/H gives, by pullback, an H-invariant holomorphic function on V. Conversely, an H-invariant holomorphic function on V descends to a holomorphic function on V'/H, where V' is the open subset of V where the action of H is free, and hence to V/H, by Riemann's extension theorem. This proves the uniqueness.

In view of the uniqueness, to prove the existence, it suffices to put a structure of normal analytic space on all open sets of the form V/H, where H is the stabilizer of a point $u \in U$, and V is a sufficiently small H-invariant open neighborhood of u. So we may also assume that G

fixes a point $u \in U$ and that $V = U$. By Lemma (6.12) in Chapter XI, we may choose a system of coordinates centered at u in which G acts by linear transformations. We are thus reduced to the case where G acts linearly on $U = \mathbb{C}^n$. This is a special instance of the following more general result, whose full strength, in any case, we will later need.

LEMMA (2.2). *Let* Spec(A) *be a normal affine variety acted on algebraically by a finite group* Γ. *Then the ring* A^Γ *of* Γ*-invariant elements in* A *is an integrally closed finitely generated* $\mathbb{C}$*-algebra. Moreover, if* X *is the set of closed points of* Spec(A), *then the set of closed points of* Spec(A^Γ) *can be identified with* X/Γ.

Proof. Set $B = A^\Gamma$. Let $a_1, \dots, a_n$ be generators for A as a $\mathbb{C}$-algebra, so that $A = \mathbb{C}[a_1, \dots, a_n]$. Consider the polynomials

$$p_i(X) = \prod_{\gamma \in \Gamma} (X - \gamma a_i) \ , \qquad i = 1, \dots, n \, .$$

The coefficients of these polynomials are invariant under Γ, i.e., they belong to B. Let all these coefficients be $c_1, \dots, c_N$ and set $C = \mathbb{C}[c_1, \dots, c_N]$. Obviously, C is a subring of B, while A is integral over C since $p_i(a_i) = 0$. Moreover, A is a finitely generated C-module. To see this, given any polynomial $\alpha = p(a_1, \dots, a_n)$ with complex coefficients, one can use the integrality relations $p_i(a_i) = 0$ to recursively reduce α to a polynomial in $a_1, \dots, a_n$ with coefficients in C and degree in each variable bounded by the order of Γ minus one. Now, since C is noetherian and B is a C-submodule of the finitely generated C-module A, the C-module B is finitely generated as well; let $b_1, \dots, b_k$ be a set of generators of B over C. Then, clearly,

$$B = \mathbb{C}[c_1, \dots, c_N, b_1, \dots, b_k] \, .$$

We now show that B is integrally closed. Let K and L be the quotient fields of A and B. Suppose that $f \in L$ is integral over B. Since A is assumed to be integrally closed, f belongs to A. Since f is Γ-invariant, it belongs to B. It remains to show that the set of closed points of Spec B is X/Γ. We introduce the Reynolds operator

$$R : A \to B$$

defined by

$$R(a) = \frac{1}{|\Gamma|} \sum_{\gamma \in \Gamma} \gamma a \, .$$

The Reynolds operator has the following elementary properties:

1) R is the identity on B;
2) if $a \in A$ and $b \in B$, then $R(ba) = bR(a)$.

An easy consequence of these properties is that, for any ideal $I \subset B$,

$$AI \cap B = I\,. \tag{2.3}$$

In fact, let $\sum a_i f_i$ be invariant under Γ, where $f_i \in I$, $a_i \in A$. Then, applying the Reynolds operator and property 2), we get

$$\sum a_i f_i = \sum R(a_i) f_i \in I\,.$$

It then follows that the map

$$\begin{aligned} X &\longrightarrow \max(B) \\ J &\mapsto J \cap B \end{aligned}$$

is surjective. In fact, given any maximal ideal I in B, by (2.3) we have that $AI \neq A$, so that, if J is any maximal ideal in A containing AI, then $J \cap B = I$. On the other hand, let $J \neq J'$ be elements of X. If $J' = \gamma J$ for some $\gamma \in \Gamma$, then clearly $J \cap B = J' \cap B$. Conversely, if J and J' belong to different orbits under Γ, we may pick an element f in A such that

$$\begin{aligned} &f \notin J \\ &f \in \left(\bigcap_{\gamma J \neq J} \gamma J \right) \cap \left(\bigcap_{\gamma \in \Gamma} \gamma J' \right). \end{aligned}$$

But then

$$\begin{aligned} &Rf \in J' \cap B\,, \\ &Rf \notin J \cap B\,. \end{aligned}$$

In fact, the first summand of the right-hand side of the identity

$$R(f) = \frac{1}{|\Gamma|} \sum_{\gamma J \neq J} \gamma f + \frac{1}{|\Gamma|} \sum_{\gamma J = J} \gamma f$$

belongs to J by construction, while the second summand does not, as follows from the remark that the isotropy group Γ' of J acts trivially on $A/J = \mathbb{C}$, so that

$$\frac{1}{|\Gamma|} \sum_{\gamma J = J} \gamma f \equiv \frac{|\Gamma'|}{|\Gamma|} f \qquad \mod J\,.$$

Q.E.D.

Given any stable P-pointed genus g curve $(C; \{x_p\}_{p \in P})$, we shall write $[(C; \{x_p\}_{p \in P})]$ to indicate its isomorphism class. Consider a standard Kuranishi family

$$\mathcal{Y} \to (U, u_0) \qquad \tau_p : U \to \mathcal{Y}\,, \;\; p \in P\,, \qquad \varphi : C \xrightarrow{\simeq} \mathcal{Y}_{u_0}\,,$$

where of course φ is an isomorphism of P-pointed curves. We may suppose in addition that U is a bounded subset of $\mathbb{C}^{3g-3+n}$, $n = |P|$. Set $G = \mathrm{Aut}(C; \{x_p\}_{p \in P})$. There is a natural map of sets

$$\psi : U/G \hookrightarrow \overline{M}_{g,P} ,$$

whose injectivity is a consequence of property iv) in the definition of standard Kuranishi family (cf. (6.8) in Chapter XI). By Lemma (2.1), U/G is normal. Since the point $[(C; \{x_p\}_{p\in P})] \in \overline{M}_{g,P}$ is entirely arbitrary, one can cover $\overline{M}_{g,P}$ by "charts" of this kind. This will put a structure of analytic space on $\overline{M}_{g,P}$ if one can show that the "changes of coordinates" are analytic. More precisely, suppose that

$$\eta : U'/G' \hookrightarrow \overline{M}_{g,P}$$

is another chart obtained with the above procedure and suppose that

$$A = \psi(U/G) \cap \eta(U'/G') \neq \emptyset .$$

First observe that the preimage of A under ψ or η is open. This follows from the fact that the families of curves we are dealing with are Kuranishi families at any point of their respective parameter spaces and from the universal property characterizing Kuranishi families. Now we have to prove that $\psi^{-1}\eta$ is analytic.

Clearly, it suffices to deal with the case where U' is "sufficiently small," in particular, where $\eta(U'/G')$ is contained in $\psi(U/G)$. Again, since our families are Kuranishi at every point, we get by universality a commutative diagram

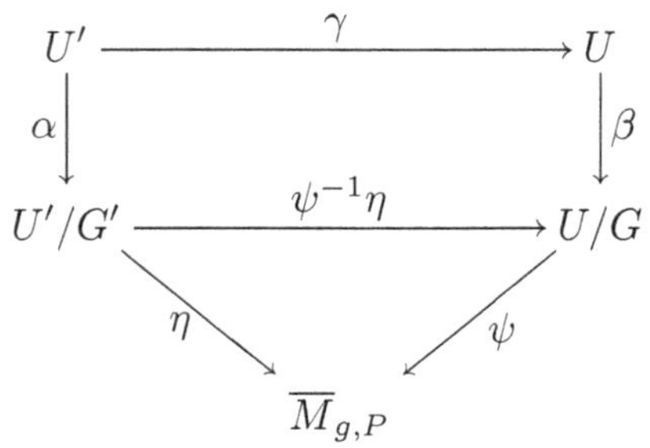

Since α is finite and holomorphic and β, γ are holomorphic, the map $\psi^{-1}\eta$ is holomorphic off the branch locus of α. Since U'/G' is normal and U/G can be realized as a bounded analytic subset of some $\mathbb{C}^N$, by Riemann's extension theorem $\psi^{-1}\eta$ is holomorphic everywhere.

This completes the construction of an analytic space structure on $\overline{M}_{g,P}$. What is already clear is that $\overline{M}_{g,P}$ is normal, since all the local patches U/G are. The results of Chapter X easily imply that $\overline{M}_{g,P}$ is separated and first countable, as we shall see in Section 3.

It follows from the construction that the analytic structure on $\overline{M}_{g,P}$ is natural in the following sense. Let

$$\psi : \mathcal{X} \to Z$$

be a family of stable P-pointed curves of genus g; then there is a morphism

$$m_\psi : Z \to \overline{M}_{g,P}$$

functorially attached to ψ such that, set-theoretically,

$$m_\psi(z) = \text{ isomorphism class of the } P\text{-pointed curve } \psi^{-1}(z)\,.$$

In addition, $\overline{M}_{g,P}$ dominates any variety having the above property. The map m_ψ is called the *moduli map of the family* ψ.

One denotes by $M_{g,P}$ the locus in $\overline{M}_{g,P}$ parameterizing smooth curves. It is an open subset of $\overline{M}_{g,P}$ since small deformations of smooth curves are smooth. Its complement, which parameterizes singular stable curves, is called the *boundary* of moduli space and is denoted by $\partial M_{g,P}$. Let x be a point of the boundary; it corresponds to a stable P-pointed genus g curve C with $\delta > 0$ nodes. Let U be the base of a (small) Kuranishi family for C. We know that, in suitable coordinates, the locus S in U parameterizing singular curves is the union of δ coordinate hyperplanes. This locus is obviously invariant under the action of $G = \mathrm{Aut}(C)$, so that, locally near x, the boundary $\partial M_{g,P}$ is just

$$S/G \subset U/G \subset \overline{M}_{g,P}\,. \tag{2.4}$$

As such, $\partial M_{g,P}$ is a closed codimension one analytic subvariety of $\overline{M}_{g,P}$. Of course, when $P = \{1, \ldots, n\}$, we write $M_{g,n}$ for $M_{g,P}$ and $\partial M_{g,n}$ for $\partial M_{g,P}$.

We know from the explicit description of Kuranishi families that curves with two or more singular points occur in codimension two in $\overline{M}_{g,P}$. Thus, a general point of any component of $\partial M_{g,P}$ corresponds to a curve with a single node. On the other hand, we know (cf. Section 2 of Chapter X) that nodes come in different flavors. There are nonseparating nodes and separating ones; moreover, the latter are classified in different types, indexed by the different stable bipartitions of (g, P). It is obvious that for a node being separating or nonseparating, and its type as a separating node, are deformation invariants. Thus the locus Δ_{irr} in $\overline{M}_{g,P}$ parameterizing curves with at least one nonseparating node is a closed analytic subset of $\partial M_{g,P}$, and the same can be said of the locus $\Delta_{\mathcal{P}}$ parameterizing curves with at least one separating node of type $\mathcal{P}$, where $\mathcal{P}$ is a stable bipartition of (g, P). In Chapter XV, and again in Chapter XXI, we shall prove that $\overline{M}_{g,P}$ is always irreducible; an immediate consequence (see Section 10) is that Δ_{irr} and the $\Delta_{\mathcal{P}}$ are all irreducible. Thus,

$$\partial M_{g,P} = \Delta_{irr} \cup \left(\bigcup \Delta_{\mathcal{P}} \right),$$

where $\mathcal{P}$ runs through all stable bipartitions, is the decomposition of the boundary of $\overline{M}_{g,P}$ in irreducible components. In the sequel, following the conventions introduced in Section 2 of Chapter X, given a stable bipartition $\mathcal{P} = \{(a, A), (b, B)\}$ of (g, P), we shall normally write $\Delta_{a,A}$ or, equivalently, $\Delta_{b,B}$ to indicate $\Delta_{\mathcal{P}}$.

Let $(C; x_1, \ldots, x_n)$ be a stable n-pointed curve of genus g, and let m be the corresponding point of $\overline{M}_{g,n}$. We know that a small neighborhood of m looks like U/G, where U is the base of a standard Kuranishi family for $(C; x_1, \ldots, x_n)$, and G is the automorphism group of $(C; x_1, \ldots, x_n)$. The point m can therefore be singular only if G is nontrivial. More precisely, m is a smooth point of $\overline{M}_{g,n}$ only in two cases. Either G acts trivially on U, or its fixed locus in U is a (smooth) divisor, in which case G is a cyclic group. The cases in which the first alternative occurs are implicitly described by Proposition(4.11) in Chapter XI: either G is trivial, or $g = 2$, $n = 0$ (resp., $g = 1$, $n = 1$), and the only nontrivial element of G is the hyperelliptic involution (resp., the symmetry about the marked point). To decide when the second alternative occurs, it is necessary to describe the divisor components of the locus Σ in $\overline{M}_{g,n}$ parameterizing curves with extra automorphisms; this locus is a closed analytic subspace of $\overline{M}_{g,n}$, as follows, for instance, from Lemma (6.11) in Chapter XI. A first result in this direction is the following.

PROPOSITION (2.5). *Let Σ be the closed analytic subspace of $\overline{M}_{g,n}$ parameterizing curves with nontrivial automorphism group. Then*

i) $\Sigma = \emptyset$ *if and only if* $g = 0$.
ii) $\Sigma = \overline{M}_{g,n}$ *if and only if* $g = 2$, $n = 0$ *or* $g = 1$, $n = 1$.
iii) *In the remaining cases the divisor components of* Σ *are:*

   a) *the closure in* $\overline{M}_{1,2}$ *of the locus parameterizing triples* $(C; x_1, x_2)$ *such that* C *is smooth and* $2(x_1 - x_2)$ *is linearly equivalent to zero;*
   b) *the closure in* $\overline{M}_{2,1}$ *of the locus parameterizing pairs* $(C; x)$ *such that* C *is smooth and* x *is a Weierstrass point of* C*;*
   c) *the closure in* $\overline{M}_3$ *of the locus parameterizing smooth hyperelliptic curves;*
   d) *for any* $g \geq 1$ *and any* n, *the locus* $\Delta_{1,\emptyset}$ *in* $\overline{M}_{g,n}$.

Proof. If $(C; x_1, \ldots, x_n)$ has a nontrivial automorphism group, it has an automorphism γ of prime order p. We have to classify the cases where γ propagates to all of $\overline{M}_{g,n}$ or along a codimension one subspace of $\overline{M}_{g,n}$. We first assume that C is smooth. Set $C' = C/\langle\gamma\rangle$. The quotient morphism $C \to C'$ is totally ramified at h points $x_1, \dots, x_h$. As the notation suggests, these include the marked points of C, since the latter are fixed for γ; thus $h \geq n$. Moreover, since C is stable, $h \geq 1$ if $g' = 1$ and $h \geq 3$ if $g' = 0$. We let $y_1, \dots, y_h$ be the images of $x_1, \dots, x_h$ in C'. Denoting by g' the genus of C', the Riemann–Hurwitz formula gives

$$2g - 2 = p(2g' - 2) + h(p-1). \tag{2.6}$$

According to Lemma (6.11) in Chapter XI, the dimension of the subspace of moduli along which γ propagates is $\dim H^1(C, T_C(x_1 + \cdots + x_n))^\gamma$. The key to calculating this dimension is the observation that there is a natural isomorphism

$$H^1(C, T_C(-x_1 - \cdots - x_n))^\gamma \cong H^1(C', T_{C'}(-y_1 - \cdots - y_h)).$$

To show this, it is convenient to prove the dual statement, namely that the pullback of forms induces an isomorphism

$$H^0(C', \omega_{C'}^2(y_1 + \cdots + y_h)) \xrightarrow{\simeq} H^0(C, \omega_C^2(x_1 + \cdots + x_n))^\gamma. \tag{2.7}$$

We may choose local coordinates such that, near x_i, the morphism $C \to C'$ is of the form $z \mapsto z^p$. The automorphism γ acts on z by multiplication by a nontrivial pth root of unity ζ. The pullback of an element of $H^0(C', \omega_{C'}^2(y_1 + \cdots + y_h))$ is locally of the form

$$f(z^p)d(z^p)^2 = p^2 z^{2p-2} f(z^p) dz^2 ,$$

where f has at most a simple pole at the origin, and hence is actually a holomorphic quadratic differential on C. Conversely, let η be a γ-invariant element of $H^0(C, \omega_C^2(x_1 + \cdots + x_n))$ and write it locally as $a(z)dz^2$, where $a(z) = \sum a_i z^i$ has at most a simple pole at the origin. By γ-invariance we must have

$$a(z)dz^2 = a(\zeta z)d(\zeta z)^2 = \zeta^2 a(\zeta z)dz^2,$$

Thus a_i can be nonzero only when $i \equiv -2$ modulo p, and hence $a(z) = z^{p-2}b(z^p)$ for some holomorphic function b. It follows that

$$\eta = \frac{1}{p^2} z^{-p} b(z^p) d(z^p)^2$$

and therefore that η is the pullback of an element of $H^0(C', \omega_{C'}^2(y_1 + \cdots + y_h))$, proving (2.7). A consequence of (2.7) is that

$$d = \dim H^1(C, T_C(x_1 + \cdots + x_n))^\gamma = 3g' - 3 + h. \tag{2.8}$$

We now assume that $\dim \overline{M}_{g,n} - d$ does not exceed 1 and classify the cases where this occurs. Using (2.6), we find that

$$\dim \overline{M}_{g,n} - d = (p-1)(3g' - 3) + h\frac{3p-5}{2} + n.$$

Thus, g' can only be equal to 1 or 0. If $g' = 1$, then $h \neq 0$ and $n = 0$, and the only possibility is that $p = h = 2$, so that $g = 2$. What this computation says, in effect, is that the locus of those genus 2 curves which are double coverings of an elliptic curve has dimension 2. Now suppose that $g' = 0$. It is straightforward to check that the only possibilities are:

(1) $p = 2,\ h = 4,\ n = 1$;
(2) $p = 2,\ h = 4,\ n = 2$;
(3) $p = 2,\ h = 6,\ n = 0$;
(4) $p = 2,\ h = 6,\ n = 1$;
(5) $p = 2,\ h = 8,\ n = 0$;
(6) $p = 3,\ h = 3,\ n = 1$.

The difference $\dim \overline{M}_{g,n} - d$ is zero only in the first and third cases, which correspond, respectively, to $g = n = 1$ and to $g = 2$, $n = 0$. This proves part ii) of the proposition. Cases (2), (4), and (5) correspond, respectively, to cases a), b), and c) in part iii) of the proposition. As for (6), it just says that there is a unique elliptic curve with an automorphism of order three (the quotient of $\mathbb{C}$ modulo the lattice generated by 1 and by a primitive third root ot unity).

We now turn to the divisor components of Σ which are entirely contained in the boundary $\partial M_{g,n}$. These may only be components of $\partial M_{g,n}$ itself. We know that a general member C of such a component has a single node, which is thus fixed for any automorphism of C. The automorphisms of C therefore induce automorphisms of its normalization N. However, by the analysis we carried out in the smooth case, it is immediate to check that, in general, N has no nontrivial automorphisms, with one exception: when N is the disjoint union of a 1-pointed genus 1 curve C_1 and an $(n+1)$-pointed genus $g-1$ curve C_2, it always has the nontrivial automorphism which restricts to the symmetry about the marked point on C_1 and to the identity on C_2. This takes care of case d) in part iii) of the proposition.

It remains to prove i). To do this, just observe that in genus $g \geq 1$ there are always curves with extra automorphisms for any value of n. We leave the construction of these curves, starting, for instance, from hyperelliptic ones, as an exercise for the reader.

Q.E.D.

It is in general very difficult to give a concrete description of $\overline{M}_{g,n}$, and this has been done only in a few low genus cases. Here are some of the simplest ones. Let us start with genus zero. In this case a stable curve has no nontrivial automorphisms, since an automorphism of $\mathbb{P}^1$ fixing three or more points is the identity. Thus, we expect a universal family to exist, and indeed we can easily construct one. Clearly, $\overline{M}_{0,3}$ is just a point, since any triple of distinct points on $\mathbb{P}^1$ is projectively equivalent, in a unique way, to the triple $(0, 1, \infty)$. We now turn to

$\overline{M}_{0,4}$. Consider the product $\mathbb{P}^1 \times \mathbb{P}^1$, let π' be the projection to the second factor, and denote by D_1', D_2', D_3' the horizontal sections of π' corresponding to the points $0, 1, \infty$ on the first factor and by Δ the diagonal.

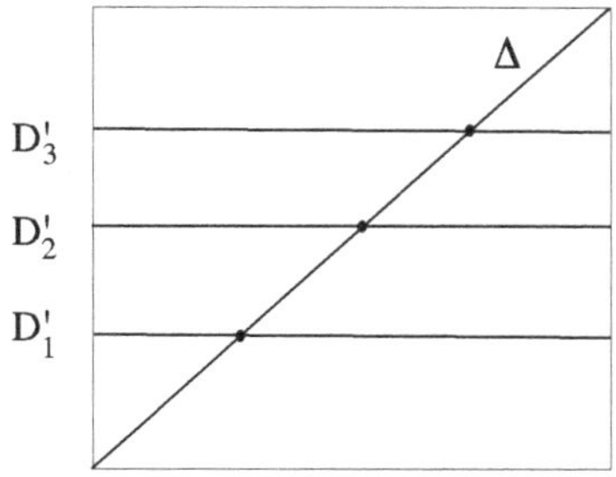

Figure 1.

Then blow up the three points where Δ meets D_1', D_2' or D_3', denote by X the resulting surface, and by $\pi : X \to \mathbb{P}^1$ the composition of the contraction map $X \to \mathbb{P}^1 \times \mathbb{P}^1$ with π'. Clearly, $\pi : X \to \mathbb{P}^1$ has four distinguished nonintersecting sections D_1, D_2, D_3, and D_4, which are the proper transforms of D_1', D_2', D_3', and Δ, respectively. This makes it into a family of 4-pointed genus zero curves. The fiber $\pi^{-1}(0)$ consists of two copies of $\mathbb{P}^1$ joined at one point, with the marked points labelled by 2 and 3 on one component, and those labelled by 1 and 4 on the other; in particular, it is a stable 4-pointed curve. The fibers $\pi^{-1}(1)$, $\pi^{-1}(\infty)$ can be similarly described. Then the moduli map $\mathbb{P}^1 \to \overline{M}_{0,4}$ attached to the family consisting of $\pi : X \to \mathbb{P}^1$ together with the sections D_1, D_2, D_3, D_4 is an isomorphism, and the family is a universal family.

The construction we have just carried out is the prototype of a general procedure which inductively constructs $\overline{M}_{0,n+1}$ out of the universal family on $\overline{M}_{0,n}$, as well as a universal family on it. We exemplify this for $\overline{M}_{0,5}$. Let $p_2 : X \times_{\overline{M}_{0,4}} X \to X$ be the projection to the second factor. Then $D_i \times_{\overline{M}_{0,4}} X$, $i = 1, \dots, 4$, are sections of p_2 and, together with it, constitute a family of stable 4-pointed curves of genus zero. We may apply to this family and to the diagonal Δ the general stabilization procedure described in Section 8 of Chapter X. The result is a family $Y \to X$ of stable 5-pointed curves of genus zero. Again, the moduli map $X \to \overline{M}_{0,5}$ is an isomorphism, and $Y \to X$ a universal family. This procedure shows that $\overline{M}_{0,n}$ is a smooth projective variety and that it carries a universal family which can be identified with the map $\pi : \overline{M}_{0,n+1} \to \overline{M}_{0,n}$ obtained by "forgetting" the $(n+1)$st point. If one takes this point of view, the ith section of π takes each n-pointed genus zero curve to the $(n+1)$-pointed curve obtained from it by attaching to the ith marked point a $\mathbb{P}^1$ with two marked points labelled with i and $n+1$, as illustrated in Fig. 2 below.

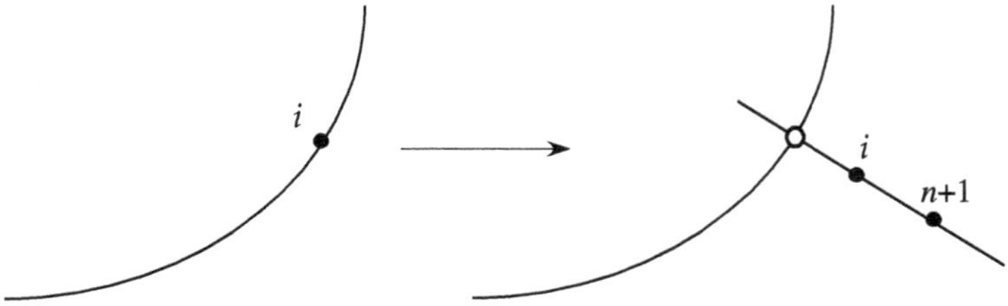

Figure 2.

We may apply what we have learned about $\overline{M}_{0,n}$ to construct other moduli spaces. The first of these is $\overline{M}_{1,1}$. Let $(E;x)$ be any smooth 1-pointed elliptic curve. Consider the group operation on E having x as origin, and let ι be the symmetry about x. Then the fixed points of the involution ι are precisely the four 2-torsion points of E, and the quotient E/ι is a $\mathbb{P}^1$. Conversely, given a stable 4-pointed curve of genus zero $(C;p_1,\ldots,p_4)$, the double covering of C branched at $p_1,\ldots,p_4$ is a genus 1 nodal curve E, which comes with four distinguished points $q_1,\ldots,q_4$, the inverse images of the marked points of C. This gives us a stable 4-pointed elliptic curve. We forget about the labeling of q_1,q_2,q_3, and keep q_4 as a marked point on E. This directly gives us a 1-pointed elliptic curve when E is smooth. When E is not smooth, to get a stable 1-pointed elliptic curve, we also have to contract the component not containing q_4.

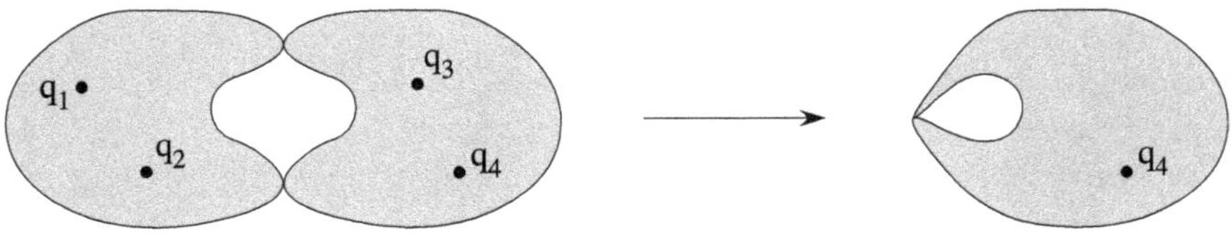

Figure 3.

By the previous discussion, we may get in this way all stable 1-pointed elliptic curves. This procedure defines a finite surjective morphism $\overline{M}_{0,4} \to \overline{M}_{1,1}$, and exhibits $\overline{M}_{1,1}$ as the quotient $\overline{M}_{0,4}/\mathfrak{S}_3$, where the symmetric group $\mathfrak{S}_3$ acts by changing the labeling of q_1, q_2, and q_3. An entirely similar construction shows that $\overline{M}_2$ can be identified with $\overline{M}_{0,6}/\mathfrak{S}_6$. In particular, $\overline{M}_{1,1}$ and $\overline{M}_2$ are both projective varieties; this is actually true for all moduli spaces $\overline{M}_{g,n}$, and Chapter XIV will be entirely devoted to proving it. As we explained above, neither $\overline{M}_{1,1}$ nor $\overline{M}_2$ carry a universal family.

Ideally, one would like to have, over $\overline{M}_{g,n}$, a universal family of stable curves, that is, one with the property that any family

$$\psi : \mathcal{X} \to Z$$

of stable curves is induced by the universal one via the map m_ψ. This, however, is impossible. In fact, let $(C; x_1, \ldots, x_n)$ be a smooth n-pointed curve of genus g, and p the corresponding point in $\overline{M}_{g,n}$. If a universal family existed, it would locally be a pullback of a standard Kuranishi family

$$\mathcal{Y} \to (U, u_0)$$

for $(C; x_1, \ldots, x_n)$, via a morphism

$$\alpha : V \to U\,,$$

where V is a neighborhood of p. Conversely, the Kuranishi family would be induced by the family on $\overline{M}_{g,n}$ via

$$\beta : U \to \overline{M}_{g,n}\,.$$

We claim that β has to be injective. In fact the composition $\alpha \circ \beta$ induces on (a neighborhood of u_0 in) U a deformation of $(C; x_1, \ldots, x_n)$ which differs from the Kuranishi family we started with at most because the identification between the central fiber and $(C; x_1, \ldots, x_n)$ has been changed by an automorphism of $(C; x_1, \ldots, x_n)$. A deformation of this nature is induced via an automorphism of U. By uniqueness, $\alpha \circ \beta$ must coincide with this automorphism: thus β is injective. When $(C; x_1, \ldots, x_n)$ has nontrivial automorphisms (or, in genus two, has nontrivial automorphisms other than the hyperelliptic involution), we reach a contradiction, since β maps any orbit of $\mathrm{Aut}(C; x_1, \ldots, x_n)$ to one point of $\overline{M}_{g,n}$ and, by Proposition (4.11) in Chapter XI, $\mathrm{Aut}(C; x_1, \ldots, x_n)$ acts nontrivially on U.

From the very construction of the analytic structure on $\overline{M}_{g,n}$ it follows that a universal family does exist on the open subset $\overline{M}^{\,0}_{g,n}$ of $\overline{M}_{g,n}$ whose points correspond to automorphism-free curves.

Something which is almost as good, in several practical applications, as a universal family on moduli, is the existence of such a family on a ramified covering of $\overline{M}_{g,n}$. In addition, the parameter space for this family may be taken to be a scheme rather than a mere analytic space. Formally, in Section 9, when we will have at our disposal some essential tools from the theory of stacks and algebraic spaces, we will prove the following result.

THEOREM (2.9). *There exists a family of stable n-pointed genus g curves $\eta : \mathcal{X} \to Z$, parameterized by a normal scheme Z, whose moduli map*

$$m : Z \to \overline{M}_{g,n}$$

is finite and surjective.

As we said, we will not prove this theorem in the current section. Here we shall only give one of its simplest consequences.

THEOREM (2.10). *$\overline{M}_{g,n}$ is compact.*

Proof. It suffices to show that the scheme Z in the statement of Theorem (2.9) is complete. For this, we use the valuative criterion of properness. Given an analytic map from $\Delta^* = \{z \in \mathbb{C} : 0 < |z| < 1\}$ to Z,

$$f : \Delta^* \to Z,$$

which is meromorphic at the origin, we must show that f extends across the puncture, possibly after a base change on Δ of the form $z = \zeta^k$. Look at the pullback, via f, of the family $\mathcal{X} \to Z$. As explained in Section 4 of Chapter X, stable reduction implies that, after a base change, the induced family can be extended across the puncture to a family of stable curves over Δ. Therefore, if not yet f, at least the composition $m \circ f$ extends to a map from Δ to $\overline{M}_{g,n}$. By the finiteness of m this can be lifted, after another base change, to the required extension of f. Q.E.D.

As we mentioned, in Chapter XIV we shall prove that $\overline{M}_{g,n}$ is a projective variety. In our proof of this result, Theorem (2.9) will be essential. In fact, since $Z \to \overline{M}_{g,n}$ is finite and surjective, the projectivity of $\overline{M}_{g,n}$ follows from the one of Z. But on Z we have the great advantage of being able to work with the family $\eta : \mathcal{X} \to Z$.

3. Moduli spaces as algebraic spaces.

Our main goal in this section is to show that a small variant of the constructions carried out in the previous one puts a structure of algebraic space on $\overline{M}_{g,n}$. For this, we first need to digress on the theory of algebraic spaces. In this book we will use only foundational facts of this theory. Whenever, for a given property of algebraic spaces, a clear reference exists and is easily attainable, we will point the reader to it. When this is not the case, we will provide the necessary proofs.

Let Y be a set. We will say that a set R, together with a pair of maps

$$R \underset{\beta}{\overset{\alpha}{\rightrightarrows}} Y$$

is an equivalence relation on Y if $(\alpha, \beta) : R \to Y \times Y$ is injective and $(\alpha, \beta)(R) \subset Y \times Y$ is an equivalence relation in the ordinary sense.

Now let B be a scheme, and Y a B-scheme. A *schematic equivalence relation*, or simply an *equivalence relation on* Y, is a B-scheme R together with a pair of morphisms

$$R \underset{t}{\overset{s}{\rightrightarrows}} Y$$

over B such that, for every B-scheme S,

$$\mathrm{Hom}_B(S,R) \underset{t_*}{\overset{s_*}{\rightrightarrows}} \mathrm{Hom}_B(S,Y)$$

is a set-theoretic equivalence relation. In particular, this implies that, for a schematic equivalence relation, $(s_*,t_*) : \mathrm{Hom}_B(S,R) \to \mathrm{Hom}_B(S,Y) \times \mathrm{Hom}_B(S,Y) = \mathrm{Hom}_B(S, Y \times_B Y)$ is injective for any S, i.e., that

$$(s,t) : R \to Y \times_B Y$$

is a *monomorphism*.

We will denote by $\eta : Y \times_B Y \to Y \times_B Y$ the involution interchanging the two factors, and by $\Delta : Y \to Y \times_B Y$ the diagonal morphism. Finally, we denote by $R\,{}_s{\times}_t R$ the fiber product induced by s and t. The following result follows from the definitions.

Proposition (3.1). *Let R and Y be B-schemes, and let $s,t : R \to Y$ be morphisms over B. Then $R \rightrightarrows Y$ is an equivalence relation if and only if $(s,t) : R \to Y \times_B Y$ is a monomorphism and there exist B-morphisms $u : Y \to R$, $i : R \to R$, and $m : R\,{}_s{\times}_t R \to Y$ such that the following diagrams commute:*

Reflexivity:

$$\begin{array}{ccc} Y & \xrightarrow{\ \Delta\ } & Y \times_B Y \\ & {}_{u}\searrow & \uparrow{\scriptstyle (s,t)} \\ & & R \end{array}$$

Symmetry:

$$\begin{array}{ccc} R & \xrightarrow{(s,t)} & Y \times_B Y \\ {\scriptstyle i}\downarrow & & \downarrow{\scriptstyle \eta} \\ R & \xrightarrow[(s,t)]{} & Y \times_B Y \end{array}$$

Transitivity:

$$\begin{array}{ccc} R\,{}_s{\times}_t R & \xrightarrow{t \times s} & Y \times_B Y \\ & {}_{m}\searrow & \uparrow{\scriptstyle (t,s)} \\ & & R \end{array}$$

In formulae:

(3.2) $$(s,t)u = \Delta\,, \quad (s,t)i = \eta(s,t)\,, \quad (t,s)m = t \times s\,.$$

The points of R are sometimes called "arrows," and, given an arrow a, $s(a)$ stands for its "source" and $t(a)$ for its "target." One often refers to m as the "composition" of arrows and writes ab for $m(a,b)$.

We will say that a morphism of schemes $\pi : Y \to X$ is a *quotient* of an equivalence relation $s, t : R \rightrightarrows Y$ if it has the following properties:

1) $\pi s = \pi t$;
2) every morphism $f : Y \to Z$ such that $fs = ft$ is of the form $h\pi$ for a unique morphism $h : X \to Z$.

We will say that $\pi : Y \to X$ is an *effective quotient* of $R \rightrightarrows Y$ if, in addition,

3) the induced morphism $R \to Y \times_X Y$ is an isomorphism.

Sometimes, we shall write $X = Y/R$ to mean that X is an effective quotient of the equivalence relation. In the category of schemes, effective quotients seldom exist, so one has to enlarge the category to accommodate them. One candidate for this enlargement is the category of algebraic spaces.

The definition of algebraic space follows a simple philosophy: if you cannot beat them, join them. A practical, though slightly incorrect, way of defining an *algebraic space* is simply to say that it is an étale equivalence relation, in other words, an equivalence relation

$$R \underset{t}{\overset{s}{\rightrightarrows}} Y \tag{3.3}$$

where s and t are both étale morphisms.

However, defining an algebraic space in this way is like defining a manifold via an atlas. In this case, the atlas is Y, the scheme R corresponds to the disjoint union of the pairwise intersections of charts, and the two morphisms s and t dictate how the charts of the atlas are patched together along their mutual intersections.

Exactly as in the case of manifolds, one can free the definition of algebraic space from the specific choice of an atlas. We will do this in Section 9, after introducing algebraic stacks, by interpreting algebraic spaces as a particular class of stacks. At that point we will also explain what one means by morphism between two algebraic spaces. In this section we really do not need any of these notions.

A typical example of an algebraic space is provided by the case of a finite group G acting on a scheme Y. In this case we set $R = G \times Y$, and we let $s : R \to Y$ and $t : R \to Y$ be, respectively, the projection and the action. This algebraic space is simply denoted by Y/G.

In practice, it does no harm to assume that the scheme Y in (3.3) is affine. An algebraic space $R \rightrightarrows Y$ is said to be *separated* if the map $(s,t) : R \to Y \times Y$ is a closed immersion. Since, over the complex numbers, an étale map is a local isomorphism for the underlying analytic structures, it is evident that, given a separated algebraic space $R \rightrightarrows Y$, an effective quotient M always exists in the analytic category. We say that M is the *underlying analytic space of the algebraic space* $R \rightrightarrows Y$, and we denote by $\pi : Y \to M$ the étale quotient map. Sometimes, we write $M = (Y/R)_{an}$.

We digress a moment to remark that the analytic space underlying a separated algebraic space is Hausdorff. For this, let $\{m_n\}$ be a sequence in M converging to points α and β. Write $\alpha = \pi(a)$, $\beta = \pi(b)$. Since π is étale, one can lift $\{m_n\}$ to sequences $\{x_n\}$, $\{y_n\}$ in Y such that $\{x_n\}$ converges to a and $\{y_n\}$ to b. By construction (x_n, y_n) belongs to $Y \times_M Y$. Since this is closed in $Y \times Y$, the point (a, b) belongs to $Y \times_M Y$, too. Thus $\alpha = \beta$.

Having said that, an alternative way of giving a *separated algebraic space over* $\mathbb{C}$ is to say that any one such space is defined by the datum of an étale, surjective, analytic map $\pi : Y \to M$, where M is an analytic variety, Y is an affine scheme, and $Y \times_M Y$ a closed subscheme of $Y \times Y$. This is what we shall mean when we will say that $\pi : Y \to M$ is a separated algebraic space. In particular, M will be a scheme exactly when the equivalence relation $Y \times_M Y \rightrightarrows Y$ is effective. Finally, we will say that the algebraic space $\pi : Y \to M$ is reduced, respectively irreducible, normal, complete if the underlying analytic space M is. In the next section we will use this simple way of viewing an algebraic space to see that the moduli space on n-pointed stable curves of given genus is indeed an algebraic space.

An elementary result about algebraic spaces that we will often use is the following.

THEOREM (3.4). *Let $X = (R \rightrightarrows Y)$ be an algebraic space. Then there exists an affine open dense subset $V \subset Y$ such that the induced relation $R_V \rightrightarrows V$ has an effective quotient. Hence, if M is the underlying analytic space of $R \rightrightarrows Y$, the scheme V/R_V is (isomorphic to) a dense open subset in M.*

A proof of this theorem can be found in [428], Prop. 5.19, p. 89, or [38], Prop. 4.5, p. 107.

Let X be an irreducible algebraic space. With the notation of Theorem (3.4), one defines *the field $K(X)$ of rational functions on X* by setting

$$K(X) = K(V/R_V)\,.$$

We will now prove the following result.

PROPOSITION (3.5). *$\overline{M}_{g,n}$ is a separated, normal, complete algebraic space.*

Proof. Consider a standard algebraic Kuranishi family

$$\xi : \mathcal{C} \longrightarrow (X_0, x_0)$$

as defined in Chapter XI, (6.7). Consider the natural map

$$\psi : X_0/G_{x_0} \longrightarrow \overline{M}_{g,n},$$

where G_{x_0} is the automorphism group of the central fiber of the family. By properties d) and e) in the definition of standard algebraic Kuranishi family, the map ψ is an étale morphism from the *affine scheme* X_0/G_{x_0} to the analytic space $\overline{M}_{g,n}$. The idea is then to take as Y a *finite* union of schemes of the form X_0/G_{x_0}. The first thing to show is that a finite number of these suffices to cover $\overline{M}_{g,n}$. We may argue as follows. Consider the natural map

$$m : \mathrm{H}_{\nu,g,n} \longrightarrow \overline{M}_{g,n}$$

and notice that it is obviously surjective. For each point $x \in \mathrm{H}_{\nu,g,n}$, let X_x be the parameter space for the (standard algebraic) Kuranishi family at x constructed in Proposition (6.5) of Chapter XI, and let G_x be the isotropy group of x. Recall that $\mathrm{H}_{\nu,g,n}$ is acted on algebraically by the projective group G and that X_x is a locally closed algebraic subvariety of $\mathrm{H}_{\nu,g,n}$ which is transverse to the orbits of G. It follows that $G \cdot X_x$ contains a Zariski-open subset of $\mathrm{H}_{\nu,g,n}$. By compactness we can cover $\mathrm{H}_{\nu,g,n}$ with finitely many sets of the type $G \times X_i$, $i = 1, \dots, N$. Set

$$Y_i = X_i/G_i\,, \quad i = 1, \dots, N, \qquad Y = Y = \coprod_{i=1}^{N} Y_i. \tag{3.6}$$

Then the étale map

$$\varphi : Y \longrightarrow \overline{M}_{g,n} \tag{3.7}$$

is surjective. Denote by φ_i the restriction of φ to Y_i. We now wish to show that

$$R = Y \times_{\overline{M}_{g,n}} Y$$

is Zariski-closed in $Y \times Y$. Set

$$X = \coprod_{i=1}^{N} X_i \tag{3.8}$$

and consider on X the family of curves $\xi : \mathcal{C} \to X$ induced by the universal family on $\mathrm{H}_{\nu,g,n}$. Denote by p_1 and p_2 the two projections from $X \times X$ to X. Look at the scheme

$$\mathbf{I} = \mathbf{Isom}_{X \times X}(p_1^*(\xi), p_2^*(\xi))\,. \tag{3.9}$$

Set-theoretically, $\mathbf{I}$ is nothing but the incidence correspondence in $X \times X \times G$ defined by

$$\mathbf{I} = \{(x, x', g) \mid x' = gx\},$$

where the X_i are viewed as embedded in the Hilbert scheme $\mathrm{H}_{\nu,g,n}$ on which the projective group G acts. By Theorem (5.1) of Chapter X, the natural projection

$$q : \mathbf{I} \longrightarrow X \times X$$

is finite. Thus, the composition η of this map with finite morphism

$$X \times X \longrightarrow Y \times Y$$

is also finite, so that $\eta(\mathbf{I})$ is a closed subscheme of $Y \times Y$. On the other hand, $\eta(\mathbf{I})$ is clearly equal to R.

This shows that $\overline{M}_{g,n}$ is a separated normal algebraic space. In particular, as we already observed, this means that the analytic space $\overline{M}_{g,n}$ is Hausdorff. The completeness of $\overline{M}_{g,n}$ is Theorem (2.10), which however depends on Theorem (2.9), to be proved later. Q.E.D.

In the sections of this chapter dealing with algebraic stacks, the scheme $\mathbf{I}$ in (3.9) will play a fundamental role. This is a good occasion to prove the following result.

PROPOSITION (3.10). *The natural projection $q_1 : \mathbf{I} \to X$ is étale and surjective.*

Proof. Recall, from point e) of Definition (6.7) in Chapter XI, that every point y in X possesses a G_y-invariant neighborhood U such that $\{\gamma \in G \mid \gamma U \cap U \neq \emptyset\} \subset G_y = \mathrm{Aut}(\mathcal{C}_y)$. Let $\alpha : \mathcal{C}_U \to U$ be the restriction to U of the family ξ over X. The following simple lemma gives a local description of the two maps q and q_1.

LEMMA (3.11). *Consider the Kuranishi family $\alpha : \mathcal{C}_U \to U$. Let p_1 and p_2 be the two projections from $U \times U$ to U. Consider the natural diagram*

$$\begin{array}{ccc} \mathbf{Isom}_{U\times U}(p_1^*\alpha, p_2^*\alpha) & \xrightarrow{q_1} & U \\ \downarrow{\scriptstyle q} & & \\ U \times U & & \end{array}$$

Let $C = C_{u_0}$ be the central fiber of α. Let $H = \mathrm{Aut}(C)$. Then, there is an isomorphism $\chi : H \times U \to \mathbf{Isom}_{U\times U}(p_1^*\alpha, p_2^*\alpha)$ such that $q_1\chi(g, u) = u$ and $q\chi(g, u) = (gu, u)$. In particular, q_1 is étale and surjective.

Proof. Set $\mathbf{I} = \mathbf{Isom}_{U\times U}(p_1^*\alpha, p_2^*\alpha)$. Define $\chi : H \times U \to \mathbf{I}$ by setting

$$\chi(g, u) = \{g^{-1} : C_{gu} \to C_u\}.$$

Since every isomorphism between two fibers of α is uniquely induced by an element of H, the morphism χ is, set-theoretically, a bijection. Set $k = |H|$. We then have a decomposition of $\mathbf{I}$ into irreducible components

$$\mathbf{I} = I_1 \cup \cdots \cup I_k .$$

We also have induced bijective morphisms $\chi : U \to I_i$ having the property that $q_1 \chi = \mathrm{id}_U$. But then χ is unramified. Thus, I_i must be smooth, and χ is an isomorphism.

Q.E.D.

Let us finish the proof of Proposition (3.10). Consider a point in $\mathbf{Isom}_{X\times X}(p_1^*\xi, p_2^*\xi)$ corresponding to an isomorphism $\varphi : \mathcal{C}_x \to \mathcal{C}_y$, where $(x,y) \in X \times X$. As an analytic neighborhood for φ, we take $\mathbf{Isom}_{U\times W}(p_1^*(\xi_{|U}), p_2^*(\xi_{|W}))$, where U (resp. W) is a neighborhood of x (resp. y), as described in point e) of Definition (6.7). It is then sufficient to show that

$$\mathbf{Isom}_{U\times W}(p_1^*(\xi_{|U}), p_2^*(\xi_{|W})) \longrightarrow W \tag{3.12}$$

is étale and surjective. Since $\xi_{|U} : \mathcal{C}_{|U} \to U$ and $\xi_{|W} : \mathcal{C}_{|W} \to W$ are Kuranishi families with isomorphic central fibers, we may assume that there is an isomorphism $\gamma : U \to W$ such that $\gamma^*(\xi_{|W}) = \xi_{|U}$. But then we are reduced to proving that

$$\mathbf{Isom}_{U\times U}(p_1^*(\xi_{|U}), p_2^*(\xi_{|U})) \longrightarrow U \tag{3.13}$$

is étale and surjective, which is the content of Lemma (3.11). Q.E.D.

In all of what we did so far, the presence of curves with nontrivial automorphism groups always appears as an impediment obstructing the existence of good quotients, of universal families, and so on. In the next sections, we will enter the realms of orbifolds and algebraic stacks, where the automorphism groups will no longer appear as a nuisance but rather as a physiological aspect of the structure.

4. The moduli space of curves as an orbifold.

The path we followed in putting an analytic structure on $\overline{M}_{g,n}$, by patching together quotients modulo finite groups of bases of Kuranishi families, suggests that $\overline{M}_{g,n}$ is just a shadow of a richer geometric structure. One way of formalizing what this consists of is by using the notion of orbifold, which we now introduce.

Let M be a Hausdorff topological space. A *V-cover for M* is a set $\mathcal{U}$ of connected open subsets of M such that

1) $M = \bigcup_{U\in\mathcal{U}} U$.

2) Given U and U' in $\mathcal{U}$ and $x \in U \cap U'$, there exists $W \in \mathcal{U}$ with $x \in W \subset U \cap U'$.
3) Each U comes equipped with an *orbifold local chart for* U, i.e., a triple (B, G, m), where B is a ball in $\mathbb{R}^n$, G is a finite group acting smoothly on B, and $m : B \to U$ is a continuous map inducing a homeomorphism between B/G and U.

In contrast with the usual definition of an orbifold local chart (see [612,614] or [514]), we are not asking that the action of G on B be effective. The example to keep in mind is the one of a Kuranishi family for a genus 2 curve C. If $\xi : \mathcal{C} \to B$ is a Kuranishi family for C, the group $\mathrm{Aut}(C)$ acts equivariantly on B and $\mathcal{C}$. But, while the hyperelliptic involution $\iota \in \mathrm{Aut}(C)$ acts nontrivially on $\mathcal{C}$, it instead acts trivially on B. In taking the quotient $B/G \subset M_2$, we want to retain a memory of this trivial action. This is the orbifold analogue of a feature which is an essential part of the stack definition of moduli spaces, where the relevant action of $\mathrm{Aut}(C)$ is the one on the family ξ and not on the base B.

We could continue in this vein to give a definition of an orbifold atlas on a space M by giving compatibility conditions between the various charts, but the notation quickly gets out of hand, and a more indirect approach is advisable.

A *Lie groupoid* $\mathbb{X}$ consists of the datum of two smooth manifolds X_0 and X_1 and five smooth structure maps

$$
\begin{aligned}
&s : X_1 \to X_0 \,, \\
&t : X_1 \to X_0 \,, \\
&m : X_1 \,{}_s\!\times_t X_1 \to X_1 \,, \\
&u : X_0 \to X_1 \,, \\
&i : X_1 \to X_1 \,,
\end{aligned}
\tag{4.1}
$$

satisfying formal properties that will be detailed below. Points of X_1 are called *arrows*, the map s is called the *source*, the map t the *target*, the map m the *composition*, the map u the *unit*, and the map i the *inverse*. Given arrows f and g with $t(f) = s(g)$, the composition $m(f, g)$ is also written fg. One also writes $i(g) = g^{-1}$. The following conditions must be satisfied:

i) $su(x) = x = tu(x)$, $\forall x \in X_0$.
ii) $ti(g) = s(g)$, $s(gh) = s(h)$, $t(gh) = t(g)$, whenever $t(h) = s(g)$.
iii) If $s(g) = x$, $t(g) = y$, then $gu(x) = g = u(y)g$, $g^{-1}g = u(x)$, $gg^{-1} = u(y)$.
iv) The composition m is associative.

Of course, all these properties can be expressed as the commutativity of a certain number of diagrams, including the ones in the statement of

Proposition (3.1). With the notation of that proposition, the equalities expressing these commutativity relations are the following:

$$
\begin{aligned}
&(s,t)u = \Delta\,, \quad (s,t)i = \eta(s,t)\,, \quad (t,s)m = (t \times s)\,, \\
&m \circ (ut, \mathrm{id}_{X_1}) = \mathrm{id}_{X_1}\,, \quad m \circ (\mathrm{id}_{X_1}, us) = \mathrm{id}_{X_1}\,, \\
&m \circ (\mathrm{id}_{X_1}, m) = m \circ (m, \mathrm{id}_{X_1})\,,
\end{aligned} \tag{4.2}
$$

where in the first row we rewrote equalities (3.2). Also notice that, for the last equality to make sense, we are using the identification

$$
\begin{aligned}
(X_1 \,{}_s\!\times_t X_1)\,{}_{sm}\!\times_t X_1 &= (X_1 \,{}_s\!\times_t X_1)\,{}_{p_2}\!\times_{p_1} (X_1 \,{}_s\!\times_t X_1) \\
&= X_1 \,{}_s\!\times_{tm} (X_1 \,{}_s\!\times_t X_1),
\end{aligned} \tag{4.3}
$$

where p_1 and p_2 are the two projections from $X_1 \,{}_s\!\times_t X_1$ to X_1.

A *proper étale Lie groupoid*, or simply an *orbifold groupoid*, is a Lie groupoid such that the two maps s and t are local diffeomorphisms while the map

$$(s,t) : X_1 \to X_0 \times X_0$$

is proper.

It is worth observing that the notion of Lie groupoid is a generalization of the one of equivalence relation. In fact, if we add to the axioms for an Lie groupoid the requirement that the map $(s,t) : X_1 \to X_0 \times X_0$ be a monomorphism, we obtain just the notion of equivalence relation, as the last two rows of (4.2) are then consequences of the first one.

In a sense, conditions i), ii), and iii) allow one to view X_1 as a generalization of a group acting on X_0, where each arrow acts by sending its source point to its target point.

If x is a point of X_0, one can easily verify that the set

$$G_x = \{g \in X_1 \mid s(g) = t(g) = x\} \tag{4.4}$$

is a finite group, which is called the *isotropy group at* x. The set $ts^{-1}(x)$ is called the *orbit* of x, and the orbit space $|\mathbb{X}|$ of $\mathbb{X}$ is the quotient

$$|\mathbb{X}| = X_0/\sim$$

where $x \sim y$ if and only if x and y belong to the same orbit. For $x \in X_0$, we will denote by $\overline{x}$ its class in $|\mathbb{X}|$.

A *morphism* $\varphi : \mathbb{X} \to \mathbb{Y}$ *of orbifold groupoids* consists of two smooth maps $\varphi_0 : X_0 \to Y_0$ and $\varphi_1 : X_1 \to Y_1$ commuting with all the structure maps defining the two orbifold groupoids.

Of course, one could also give the notion of *topological orbifold groupoid* by requiring that X_0 and X_1 are merely topological manifolds and by relaxing the C^∞ condition on the structure maps. At the other extreme, one could introduce the notion of *complex orbifold groupoid* by insisting that X_0 and X_1 be complex manifolds and that the structure maps be analytic.

An *orbifold structure* on a paracompact Hausdorff space M consists of an orbifold groupoid $\mathbb{X}$ and a homeomorphism $f : |\mathbb{X}| \to M$. An orbifold structure should be thought of as the analogue of a specific atlas on a manifold. As in the case of manifolds, it is seldom the case that different atlases can be compared directly, so one passes to a common refinement, and, in this way, one gets an atlas-free definition of a manifold. One can imitate this refinement procedure in the world of orbifold structures and thereby get to the notion of orbifold (freed from the choice of a specific atlas). For this and related matters, we refer to Chapter 1 in [3]. In the present book orbifolds will normally appear equipped with a specific orbifold structure. In Exercises A-1 and A-2 the reader will find the definition of the *orbifold quotient* $[M/G]$ *of a manifold* M *acted on by a finite group* G and will understand that, given an orbifold structure on a space M, every point in M has a neighborhood with an orbifold structure of the form $[B/G_x]$, where $x \in X_0$, and B is a chart around x, reconciling the notion of V-cover with the one of Lie groupoid.

Let us now equip the moduli space $\overline{M}_{g,n}$ with an orbifold structure. Let X be as in (3.8), so that $X = \coprod_i^N X_i$, each X_i is the (smooth) basis of a Kuranishi family, and the moduli map $m : X \to \overline{M}_{g,n}$ is surjective. Let $\mathcal{C} \to X$ be the total family over X. We define an orbifold groupoid $\overline{\mathbb{M}}_{g,n}$ in the following way. Consider the two projections p_1 and p_2 from $X \times X$ to X and set

$$\mathbf{I} = \mathbf{Isom}_{X \times X}(p_1^*\mathcal{C}, p_2^*\mathcal{C}) .$$

Since $X \times X$ is smooth, it follows from Theorem (5.1) in Chapter X that $\mathbf{I}$ is smooth as well. We then let

$$(\overline{\mathbb{M}}_{g,n})_0 = X , \quad (\overline{\mathbb{M}}_{g,n})_1 = \mathbf{I},$$

and we define s and t to be the natural projections from $\mathbf{I}$ onto the first and second factors of $X \times X$, respectively. The composition rule, the unit, and the inverse are the obvious ones. It is then an exercise to verify that there is a homeomorphism $\overline{m} : |\overline{\mathbb{M}}_{g,n}| \to \overline{M}_{g,n}$, giving $\overline{M}_{g,n}$ an orbifold structure.

As usual, enlarging a category (e.g., passing from manifolds to orbifolds) has the advantage of accommodating, inside the new category, operations that were not allowed in the old one, such as taking quotients. Let us then consider a *finite* group G acting on an orbifold $\mathbb{X}$. By this we mean that there are actions of G on X_0 and X_1 that are compatible with all structure morphisms. We then define an orbifold groupoid $[\mathbb{X}/G] = (Y_0, Y_1)$ in the following way:

$$Y_0 = X_0\,, \quad Y_1 = G \times X_1, \tag{4.5}$$

$$\begin{array}{ll} s_G : Y_1 \to Y_0 & t_G : Y_1 \to Y_0 \\ (\sigma, \varphi) \mapsto s(\varphi) & (\sigma, \varphi) \mapsto t(\sigma\varphi) \end{array} \tag{4.6}$$

The composition in Y_1 is given by

$$m_G((\sigma, \varphi), (\tau, \psi)) = (\sigma\tau, \tau^{-1} m(\varphi, \psi)). \tag{4.7}$$

The unit and the inverse are the obvious ones. It is then easy to verify that, under these rules, $[\mathbb{X}/G]$ is an orbifold groupoid. In Section 10 we will see an example of this situation when looking at the boundary of $\overline{M}_{g,n}$.

Next, let us say two words about the cohomology of orbifolds. We start with the de Rham complex. Let $\mathbb{X}$ be an orbifold groupoid. Set

$$\mathcal{A}^p(\mathbb{X}) = \{\varphi \in \mathcal{A}^p(X_0) \mid s^*\varphi = t^*\varphi\}.$$

For obvious reasons, the elements in $\mathcal{A}^p(\mathbb{X})$ are called *invariant forms.* Indeed, as we know, every point $\overline{x} \in |\mathbb{X}|$ has a neighborhood of the form B/G_x, where B is a local chart around x. Then the restrictions to B of the forms in $\mathcal{A}^p(\mathbb{X})$ are just the G_x-invariant forms on B. The differential $d : \mathcal{A}^p(\mathbb{X}) \to \mathcal{A}^{p+1}(\mathbb{X})$ is defined in the usual way, and the resulting cohomology groups are denoted with the symbol $H^p_{dR}(\mathbb{X})$. Satake proved that there is an isomorphism

$$H^*_{dR}(\mathbb{X}) \cong H^*(|\mathbb{X}|, \mathbb{R}), \tag{4.8}$$

where the right-hand-side denotes singular cohomology. When $\mathbb{X}$ is a complex orbifold groupoid, one can define in a similar fashion the vector space $\mathcal{A}^{p,q}(\mathbb{X})$ of invariant (p,q)-forms and the vector space $\Omega^p(\mathbb{X})$ of holomorphic p-forms.

Integration of forms requires some care. First of all, given B and G_x, as above, and a top-degree invariant form φ on $\mathbb{X}$, one defines

$$\int_{B/G_x} \varphi = \frac{1}{|G_x|} \int_B \varphi.$$

Now fix a locally finite cover of $\mathcal{U} = \{U_\alpha\}$ of $|\mathbb{X}|$, where U_α is of the form B_α/G_α, and let $\{\rho_\alpha\}$ be a partition of unit for $\underset{\alpha}{\cup} B_\alpha$ Then define

$$\int_{\mathbb{X}} \varphi = \sum_\alpha \int_{U_\alpha} \rho_\alpha \varphi .$$

One can then prove that, when $|\mathbb{X}|$ is compact, Poincaré duality holds in the sense that the pairing

$$\begin{aligned} H^p(\mathbb{X}) \times H^{n-p}(\mathbb{X}) &\longrightarrow \mathbb{R} \\ (\varphi, \psi) &\mapsto \int_{\mathbb{X}} \varphi \wedge \psi \end{aligned}$$

is nondegenerate.

We end this section by introducing the notion of divisor with normal crossings in a complex orbifold $\mathbb{X}$, presented as an orbifold groupoid (X_0, X_1). By definition, a divisor with normal crossings in $\mathbb{X}$ is just a divisor with normal crossings in X_0 which is X_1-invariant. The boundary $\partial M_{g,n}$ is a normal crossings divisor in $\overline{M}_{g,n}$, by the local description (2.4).

5. The moduli space of curves as a stack, I.

It is often useful to regard the moduli spaces $M_{g,n}$ and $\overline{M}_{g,n}$ as Deligne–Mumford stacks. The idea of stack is modeled on moduli spaces and on quotient spaces. In this brief introduction to stacks, we will continuously go back and forth between the abstract categorical concepts and their geometrical origins. In our treatement we will closely follow [167], [671], [190], and [94]. As already announced in the introduction to this chapter, in this section and in the following three, we deviate from our general convention that "scheme" stands for "scheme of finite type over $\mathbb{C}$" and allow general schemes.

As is generally done, we will introduce Deligne–Mumford stacks in three stages. First, we will introduce categories fibered in groupoids, then stacks, and, finally, algebraic stacks and Deligne–Mumford stacks.

Let S be a scheme and consider the category Sch/S of schemes over S. In what follows we will mostly consider the case $S = \operatorname{Spec} \mathbb{C}$.

A *category fibered in groupoids over Sch/S* or, more simply, a *groupoid over S*, is a pair $\mathcal{M} = (\mathcal{C}_\mathcal{M}, p_\mathcal{M})$, where $\mathcal{C}_\mathcal{M}$ is a category, and

$$p_\mathcal{M} : \mathcal{C}_\mathcal{M} \to Sch/S$$

is a functor satisfying the following two conditions:

A) Let $f : T' \to T$ be a morphism in Sch/S, and let η be an object in $\mathcal{C}_{\mathcal{M}}$ such that $p_{\mathcal{M}}(\eta) = T$. Then there exist an object ξ in $\mathcal{C}_{\mathcal{M}}$ and a morphism $\varphi : \xi \to \eta$ in $\mathcal{C}_{\mathcal{M}}$ with $p_{\mathcal{M}}(\varphi) = f$.

B) Every morphism $\varphi : \xi \to \eta$ in $\mathcal{C}_{\mathcal{M}}$ is *cartesian* in the following sense. Given any other arrow $\varphi' : \xi' \to \eta$ and a morphism $h : p_{\mathcal{M}}(\xi) \to p_{\mathcal{M}}(\xi')$ such that $p_{\mathcal{M}}(\varphi')h = p_{\mathcal{M}}(\varphi)$, there exists a unique morphism $\psi : \xi \to \xi'$ such that $p_{\mathcal{M}}(\psi) = h$ and $\varphi'\psi = \varphi$.

By abuse of language, we will refer to the objects of $\mathcal{C}_{\mathcal{M}}$ as to the objects of the groupoid $\mathcal{M}$, and given objects ξ and ξ' in $\mathcal{C}_{\mathcal{M}}$, we will write $\mathrm{Hom}_{\mathcal{M}}(\xi, \xi')$, instead of $\mathrm{Hom}_{\mathcal{C}_{\mathcal{M}}}(\xi, \xi')$.

A *morphism* $\alpha : \mathcal{M} \longrightarrow \mathcal{M}'$ *of groupoids over* Sch/S is a functor (also denoted by) $\alpha : \mathcal{C}_{\mathcal{M}} \to \mathcal{C}_{\mathcal{M}'}$ such that $p_{\mathcal{M}'} = \alpha p_{\mathcal{M}}$. The morphisms between $\mathcal{M}$ and $\mathcal{M}'$ form themselves a category $\mathrm{Hom}(\mathcal{M}, \mathcal{M}') = \mathrm{Hom}_{Sch/S}(\mathcal{C}_{\mathcal{M}}, \mathcal{C}_{\mathcal{M}'})$, whose arrows are the natural transformations between functors. Technically, one says that groupoids over Sch/S constitute a 2-category. We will not elaborate further on this notion, except to notice that some care must be exercised when dealing with commutativity question having to do with morphisms of groupoids, or, more generally, morphisms in a 2-category: one says that a diagram of groupoids and morphisms of groupoids over Sch/S

$$\begin{array}{ccc} \mathcal{F} & \xrightarrow{a} & \mathcal{G} \\ & \searrow^{c} & \downarrow b \\ & & \mathcal{E} \end{array}$$

is *commutative* if there is given an isomorphism of functors between c and ba. This is in keeping with the fact that the good notion of "being essentially the same" for categories is the one of equivalence. When a morphism of groupoids is an equivalence of categories, we shall sometimes improperly say that it is an *isomorphism of groupoids*.

As we already mentioned, in this book, we will mostly be concerned with groupoids over $Sch/\mathbb{C}$. We simply call these *groupoids*. The definition of groupoid looks modeled after the following example. Take as $\mathcal{C}$ the category in which the objects are the families

$$\begin{array}{c} \mathcal{X} \\ \downarrow \xi \\ T \end{array}$$

of smooth (resp. stable, n-pointed) curves of genus g and in which a morphism

$$\varphi : \xi' \to \xi$$

between a family $\xi' : \mathcal{X}' \to T'$ and a family $\xi : \mathcal{X} \to T$ is a commutative diagram

$$\begin{array}{ccc} \mathcal{X}' & \longrightarrow & \mathcal{X} \\ {\scriptstyle \xi'}\big\downarrow & & \big\downarrow{\scriptstyle \xi} \\ T' & \xrightarrow{\ f\ } & T \end{array}$$

inducing an isomorphism $\mathcal{X}' \cong T' \times_T \mathcal{X}$. The functor p assigns to a family $\xi : \mathcal{X} \to T$ its parameter space T:

$$p(\xi) = T\,.$$

With regard to morphisms, using the above notation, we set

$$p(\varphi) = f.$$

It is an easy exercise for the reader to prove that properties A) and B) are satisfied for the pair $(\mathcal{C}, p)$.

The groupoid of smooth, n-pointed, genus g curves is denoted with the symbol $\mathcal{M}_{g,n}$. The one of stable, n-pointed, genus g curves is denoted with the symbol $\overline{\mathcal{M}}_{g,n}$.

As suggested by the preceding examples, whenever a groupoid $\mathcal{M} = (\mathcal{C}, p)$ is given, it could help to think of an object $\xi \in \mathcal{C}$ as a "family over $p(\xi)$." The term groupoid has the following origin. Given a groupoid $\mathcal{M} = (\mathcal{C}, p)$, denote by $\mathcal{M}(T)$ the category whose objects are objects $\xi \in \mathcal{C}$ with $p(\xi) = T$ (i.e., the "families" over T) and whose morphisms are morphisms φ in $\mathcal{C}$ with $p(\varphi) = id$. Axiom B) tells us that a morphism φ in $\mathcal{C}$ is an isomorphism if, and only if, $p(\varphi)$ is. It follows that $\mathcal{M}(T)$ is a groupoid in the (more usual) sense that all morphisms in $\mathcal{M}(T)$ are isomorphisms. So one can view the functor $p : \mathcal{C} \to Sch/S$ as a "fibration" having groupoids as fibers:

$$\mathcal{M}(T) = p^{-1}(T).$$

The category $\mathcal{M}(T)$ is also called *the category of sections of $\mathcal{M}$ over T.*

Notice that, by axiom B), the object ξ in axiom A) is unique up to a unique isomorphism. We shall refer to this object, or rather to $\xi \to \eta$, as *a pullback of η to T'*. It is tempting to write $f^*(\eta)$ for ξ. The trouble is that pullbacks are generally not unique, and there is no way of singling out one which is "nicer" than the others. However, it is always possible to choose, for each object η in $\mathcal{C}$ and each arrow $T' \to T = p(\eta)$, a specific pullback $\xi \to \eta$. Technically, such a choice is called a *cleavage.* Once a cleavage has been chosen, it makes sense to write $f^*(\eta)$ for ξ, but one has to remember that the arrow $f^*(\eta) \to \eta$ is an essential part of the notion of pullback.

It is important to decide when two groupoids $\mathcal{M}$ and $\mathcal{M}'$ are isomorphic. As we already mentioned, for this to happen, there must exist an equivalence of categories $F : \mathcal{C}_{\mathcal{M}} \to \mathcal{C}_{\mathcal{M}'}$ such that $p_{\mathcal{M}} = p_{\mathcal{M}'}F$. It is well known, and easy to prove, that for a functor F to be an equivalence, it is necessary and sufficient that:

i) F is *fully faithful,* meaning that, for every pair of objects ξ and ξ' in $\mathcal{M}$, the induced map

$$\mathrm{Hom}_{\mathcal{M}}(\xi', \xi) \longrightarrow \mathrm{Hom}_{\mathcal{M}'}(F(\xi'), F(\xi))$$

is bijective, and

ii) F is *essentially surjective,* meaning that every object η in $\mathcal{M}'$ is isomorphic to $F(\xi)$ for some object ξ in $\mathcal{M}$.

For groupoids, the following lemma holds.

LEMMA (5.1). *A morphism $F : \mathcal{M} \to \mathcal{M}'$ of groupoids over Sch/S is an isomorphism if, and only if, for every T in Sch/S, the induced functor on fibers $F_T : \mathcal{M}(T) \to \mathcal{M}'(T)$ is an equivalence of categories.*

Proof. The only two nontrivial assertions hidden in this lemma are the following.

a) If F_T is fully faithful for every T, then F is fully faithful.
b) If F is essentially surjective, so is F_T for every T.

To prove a), it pays to use the following notation. Given a morphism $f : T' \to T$ and objects ξ' and ξ in $\mathcal{M}(T')$ and $\mathcal{M}(T)$, respectively, we set

$$\mathrm{Hom}^f_{\mathcal{M}}(\xi', \xi) = \{\varphi \in \mathrm{Hom}_{\mathcal{M}}(\xi', \xi) \mid p_{\mathcal{M}}(\varphi) = f\}.$$

Write $p = p_{\mathcal{M}}$ and $p' = p_{\mathcal{M}'}$. Since $p = p'F$, to prove a), it suffices to prove that F induces a bijection

$$\mathrm{Hom}^f_{\mathcal{M}}(\xi', \xi) \longrightarrow \mathrm{Hom}^f_{\mathcal{M}'}(F(\xi'), F(\xi)). \tag{5.2}$$

Consider the morphism $\varphi : f^*(\xi) \to \xi$. We have a commutative diagram

$$\begin{array}{ccc}
\mathrm{Hom}_{\mathcal{M}(T)}(\xi', f^*(\xi)) & \xrightarrow{\ F_T\ } & \mathrm{Hom}_{\mathcal{M}'(T)}(F(\xi'), F(f^*(\xi))) \\
\varphi\circ \downarrow & & \downarrow F(\varphi)\circ \\
\mathrm{Hom}^f_{\mathcal{M}}(\xi', \xi) & \xrightarrow{\ \ F\ \ } & \mathrm{Hom}^f_{\mathcal{M}'}(F(\xi'), F(\xi))
\end{array}$$

By the definition of groupoid, the two vertical arrows are bijective, so that the bijectivity of F_T implies the bijectivity of (5.2). As far as b) is concerned, let η be an object in $\mathcal{M}'(T)$. By hypothesis, there are an object ξ' in $\mathcal{M}$ and an isomorphism $\varphi : \eta \to F(\xi')$. Set $f = p'(\varphi) : T \to T'$ and let $\xi = f^*(\xi')$. We have an isomorphism $\psi : \xi = f^*(\xi') \to \xi'$ for which $p'(F(\psi)) = p'(\varphi) = f$. By property B), there exists an isomorphism $\sigma : F(\xi) \to \eta$ with $p'(\sigma) = id_T$. But then ξ is an object of $\mathcal{M}(T)$, and σ is an isomorphism in $\mathcal{M}'(T)$ between η and $F_T(\xi)$. Q.E.D.

It is important to notice that any scheme X and any contravariant functor $F : Sch \to Sets$ can be considered as groupoids.

Let us start with the case of schemes. We will consider a scheme X as a groupoid $X = (\mathcal{C}_X, p_X)$, where the objects of $\mathcal{C}_X$ are pairs (T, f) with $f : T \to X$ a morphism of schemes. The morphisms $\varphi : (T, f) \to (T', f')$ are the morphisms $h : T \to T'$ with $f'h = f$. Finally, the projection p_X is defined by $p_X(T, f) = T$. A groupoid $\mathcal{M}$ is *(represented by) a scheme* X if there exists an *isomorphism* of groupoids $\alpha : X \xrightarrow{\sim} \mathcal{M}$. This condition is equivalent to the existence of an object ξ_X in $\mathcal{M}(X)$ having the following universal property: for every object ξ in $\mathcal{M}$, there exists a unique morphism $f : \xi \to \xi_X$. Of course, given the equivalence α, we have $\xi_X = \alpha(X, \mathrm{id}_X)$. If, in the examples above, we limit ourselves to families of smooth (stable) *automorphism-free* curves, then the corresponding groupoids are indeed (represented by) smooth schemes. But, as we know, this is not the case for the groupoid of smooth (resp. stable) n-pointed curves. The lack of a universal family over $\overline{M}_{g,n}$ can be rephrased by saying that, although any family of n-pointed, genus g stable curves $\xi : \mathcal{C} \to S$ induces a moduli map $m_\xi : S \to \overline{M}_{g,n}$, not every map $S \to \overline{M}_{g,n}$ induces a family over S. Let us see how, inherent in the concept of groupoid, is the cure for this asymmetry.

Let $\mathcal{M}$ be a groupoid over Sch. Given an object ξ in $\mathcal{M}(S)$, we think of S as a groupoid, and we define an induced morphism of groupoids

$$m_\xi : S \to \mathcal{M}\,,$$

by associating to every object in $S(T)$, i.e., to every arrow $f : T \to S$, a pullback $f^*(\xi)$ in $\mathcal{M}(T)$, and proceeding similarly for morphisms. By consonance, one might call m_ξ a moduli map of ξ; of course, m_ξ is not unique but depends on the choice of pullbacks. But now, conversely, given a morphism $\mu : S \to \mathcal{M}$, one gets an object ξ in $\mathcal{M}(S)$ by setting $\xi = \mu(\mathrm{id}_S)$. As we shall presently see, this sets up an equivalence of categories between $\mathcal{M}(S)$ and $\mathrm{Hom}(S, \mathcal{M})$, and the symmetry is reestablished.

Actually, we shall prove something slightly more general. Let $\mathcal{M}$ be a groupoid over Sch. Consider the category $\widetilde{\mathcal{C}}$ whose objects are the morphisms of groupoids $S \to \mathcal{M}$, where S is a scheme, and whose arrows are commutative triangles. More precisely, an arrow from $\beta : T \to \mathcal{M}$ to $\alpha : S \to \mathcal{M}$ is a pair consisting of a morphism $f : T \to S$ and an isomorphism of functors between β and αf. As the reader will easily check, the functor $\widetilde{p} : \widetilde{\mathcal{C}} \to Sch$ which attaches S to $\alpha : S \to \mathcal{M}$ makes $\widetilde{\mathcal{M}} = (\widetilde{\mathcal{C}}, \widetilde{p})$ into a category fibered in groupoids over Sch. By definition we have $\widetilde{\mathcal{M}}(S) = \mathrm{Hom}(S, \mathcal{M})$. Moreover, $\widetilde{\mathcal{M}}$ is endowed with a canonical cleavage: given a morphism $f : T \to S$, one can take as pullback of an object $\alpha : S \to \mathcal{M}$ simply the composition $\alpha f : T \to \mathcal{M}$ mapping to α via the pair consisting of f and the identity isomorphism of functors. In fact, this cleavage is a *splitting*, meaning that it contains all the identities and is closed under composition. There is an obvious functor $F : \widetilde{\mathcal{C}} \to \mathcal{C}$. As above, we associate to $\alpha : S \to \mathcal{M}$ the object $F(\alpha) = \alpha(\mathrm{id}_S)$ in $\mathcal{M}(S)$. Similarly, given an arrow φ in $\widetilde{\mathcal{C}}$, consisting of a morphism of schemes $f : T \to S$ and of an isomorphism of functors $\beta \cong \alpha f$, we first get an isomorphism $\beta(\mathrm{id}_T) \cong \alpha f(\mathrm{id}_T) = \alpha(f)$. On the other hand, since α is a functor, it gives an arrow $\alpha(f) \to \alpha(\mathrm{id}_S)$. Composing the two, we get an arrow $F(\varphi) : F(\beta) = \beta(\mathrm{id}_T) \to \alpha(\mathrm{id}_S) = F(\alpha)$. We leave it to the reader to check that what we have defined is indeed a functor. It is clear that $\widetilde{p} = pF$, and hence that F can be viewed as a morphism $\widetilde{\mathcal{M}} \to \mathcal{M}$. The next result is a special instance of the *2-categorical Yoneda lemma.*

LEMMA (5.3). *The morphism $F : \widetilde{\mathcal{M}} \to \mathcal{M}$ is an equivalence of categories fibered in groupoids over Sch.*

Here is a sketch of the proof. First, we define a functor $G : \mathcal{C} \to \widetilde{\mathcal{C}}$. Suppose that ξ is an object in $\mathcal{M}(S)$ and $f : T \to S$ is a morphism of schemes. As above, we set $m_\xi(f) = f^*\xi$. If $\varphi : U \to T$ is a morphism of schemes, and $h = f\varphi$, then by cartesianness there is a unique arrow $m_\xi(\varphi) : m_\xi(h) \to m_\xi(f)$ lying above φ and making the diagram

$$\begin{array}{ccc} m_\xi(h) & & \\ {\scriptstyle m_\xi(\varphi)}\Big\downarrow & \searrow & \\ m_\xi(f) & \longrightarrow & \xi \end{array}$$

commute. Cartesianness also immediately shows that m_ξ is a functor $S \to \mathcal{M}$. We set $G(\xi) = m_\xi$. Now let $\alpha : \eta \to \xi$ be an arrow in $\mathcal{C}$ lying above a morphism of schemes $a : T \to S$. Let $b : U \to T$ be a morphism of schemes and set $c = ab$. Then $m_\eta(b)$ and $m_\xi(c)$ are both pullbacks of ξ to U and hence are canonically isomorphic. As $U \to T$ varies, these isomorphisms give an isomorphism of functors between m_η and $m_\xi a$, and hence an arrow $G(\alpha) : G(\eta) \to G(\xi)$. We leave to the reader the easy task of checking that G is a functor.

We claim that FG is isomorphic to the identity functor on $\mathcal{M}$, and GF to the identity on $\widetilde{\mathcal{M}}$. First of all, when ξ is an object in $\mathcal{M}(S)$, $FG(\xi)$ is just $\mathrm{id}_S^*(\xi)$, which is canonically isomorphic to ξ; in fact, in the definition of G, we can arrange things so that $\mathrm{id}_S^*(\xi)$ is just ξ, and $\mathrm{id}_S^*(\xi) \to \xi$ the identity. If we do this, FG turns out to be the identity functor.

Now we turn to GF. Let $\alpha : S \to \mathcal{M}$ be a morphism where S is a scheme. In other words, α is a base-preserving functor from the category of morphisms of schemes $T \to S$ to $\mathcal{C}$. This simply means that, for each morphism $a : T \to S$, we are given an object $\alpha(a)$ in $\mathcal{M}(T)$ and that, for any morphism $b : U \to T$, $\alpha(ab)$ is a pullback of $\alpha(a)$ via b. By the essential uniqueness of pullbacks, the objects $\alpha(a)$ are determined, up to a unique isomorphism, by $\alpha(\mathrm{id}_S) = F(\alpha)$. By the definition of G, this sets up a canonical isomorphism between α and $GF(\alpha)$. It is not difficult, but tedious, to check that this defines an isomorphism of functors between GF and the identity, ending the proof of (5.3).

As we have announced, a contravariant functor $F : Sch \to Sets$ can also be considered as a groupoid. For this groupoid, also denoted by F, the objects of $\mathcal{C}_F$ are pairs (T, ξ), where T is a scheme and $\xi \in F(T)$. A morphism $(T, \xi) \to (T', \xi')$ is a morphism $f : T \to T'$ such that $F(f)(\xi') = \xi$. The symbol $F(T)$ unambiguosly denotes both the set $F(T)$ and the fiber over T of F considered as a groupoid. When we choose as F the functor of points $h_X = \mathrm{Hom}(-, X)$ of a scheme X, the groupoids associated to h_X and X coincide. A contravariant functor $F : Sch \to Sets$ is said to be *representable* if there exists a scheme X and a groupoid isomorphism between F an h_X.

We now make an important remark regarding $\overline{\mathcal{M}}_{g,n}$ and $\mathcal{M}_{g,n}$. In Section 3 we constructed the spaces $M_{g,n}$ and $\overline{M}_{g,n}$ as algebraic spaces. We will see in Chapter XIV that they are actually schemes, and as such we will treat them now. Let us concentrate our attention on stable curves, the case of smooth curves being completely similar. Consider the scheme $\overline{M}_{g,n}$ and the groupoid $\overline{\mathcal{M}}_{g,n}$. A third object is linked to n-pointed, stable curves of genus g, namely the contravariant functor

$$\overline{F}_{g,n} : Sch/\mathbb{C} \longrightarrow Sets$$

defined as follows. For every scheme T, we set

$$\overline{F}_{g,n}(T) = \left\{ \begin{matrix} \text{Families of } n\text{-pointed genus } g \\ \text{stable curves parameterized by } T \end{matrix} \right\} \Big/ \text{isomorphisms.}$$

An element of $\overline{F}_{g,n}(T)$ is denoted by $[\xi : \mathcal{X} \to T]$, and for every morphism $f : T \to T'$ and every element $[\xi' : \mathcal{X}' \to T']$ in $\overline{F}_{g,n}(T')$, we set $F(f) = [\mathcal{X}' \times_{T'} T \to T]$. We now have three groupoids and two obvious morphisms:

$$\overline{\mathcal{M}}_{g,n} \xrightarrow{\alpha} \overline{F}_{g,n} \xrightarrow{\beta} \overline{M}_{g,n}\,. \tag{5.4}$$

In the rightmost groupoid we are looking at moduli as a scheme (or as an algebraic space). In the central one we look at moduli as a functor. In the first one we consider moduli as a bona fide groupoid. In a sense, we can regard α and β as forgetful functors. The functor α associates to the object $\xi : \mathcal{X} \to T$ of $\overline{\mathcal{M}}_{g,n}$ the object $[\xi]$ of $\overline{F}_{g,n}$. The functor β associates to the object $[\xi]$ the object (T, f) of $\overline{M}_{g,n}$, where $f : T \to \overline{M}_{g,n}$ is the morphism induced by the family ξ. Neither α nor β is an isomorphism of groupoids. The fact that β is not is a restatement of the fact that $\overline{F}_{g,n}$ is not representable or, which is the same, that there is no universal family of curves over $\overline{M}_{g,n}$ or, yet in other words, that there is no object in $\mathcal{C}_{\overline{F}_{g,n}}$ mapping, via β, to $(\overline{M}_{g,n}, \mathrm{id}_{\overline{M}_{g,n}})$. Clearly, α also fails to be an isomorphism of groupoids. Indeed, given an object $\xi : \mathcal{X} \to T$ in $\overline{\mathcal{M}}_{g,n}(T)$, we have

$$\mathrm{Hom}_{\overline{\mathcal{M}}_{g,n}(T)}(\xi, \xi) = \{\text{isomorphisms} \quad \varphi : \mathcal{X} \to \mathcal{X} \mid \xi = \varphi\xi\},$$

while

$$\mathrm{Hom}_{\overline{F}_{g,n}(T)}([\xi], [\xi]) = \{\mathrm{id}_{\mathcal{X}}\}.$$

In particular, when $T = \{pt\}$ is a single point, and $\mathcal{X} = X$ a stable curve,

$$\mathbf{Isom}_{\overline{\mathcal{M}}_{g,n}(pt)}(X, X) = \mathrm{Aut}(X)\,, \qquad \mathbf{Isom}_{\overline{F}_{g,n}(\{pt\})}([X], [X]) = \{\mathrm{id}_X\}.$$

The presence of stable curves with nontrivial automorphism group, which is the cause for the nonrepresentability of the moduli functor $\overline{F}_{g,n}$, is actually the distinctive feature of the geometric fibers of the moduli groupoid $\overline{\mathcal{M}}_{g,n}$.

A very important example of groupoid is the following. Suppose that a group scheme G acts on a scheme X. Then one can form the *quotient groupoid*

$$[X/G] = (\mathcal{P}_{G,X}, p)\,, \tag{5.5}$$

where $\mathcal{P}_{G,X}$ is the category whose objects are pairs (π, σ_π), where $\pi : E \to T$ is a principal G-bundle, and $\sigma_\pi : E \to X$ is a G-equivariant map. A morphism between (π, σ_π) and (π', σ'_π) is a pair of commutative diagrams

$$\begin{array}{ccc} E' & \xrightarrow{\varphi} & E \\ {\scriptstyle \pi'}\downarrow & & \downarrow{\scriptstyle \pi} \\ T' & \xrightarrow{f} & T \end{array} \qquad\qquad \begin{array}{ccc} E' & \xrightarrow{\varphi} & E \\ & {\scriptstyle \sigma_{\pi'}}\searrow & \downarrow{\scriptstyle \sigma_\pi} \\ & & X \end{array}$$

the first one of which is cartesian. Finally, the projection

$$p : \mathcal{P}_{G,X} \longrightarrow Sch$$

is given by

$$(\pi\,, \sigma_\pi) \quad \mapsto \quad T\,,$$

so that the fiber $\mathcal{P}_{G,X}(T)$ is the category of principal G-bundles over T, equipped with a G-equivariant map from their total space to X. Notice that only when G acts freely on X and the quotient X/G exists as a scheme, the groupoid $[X/G]$ is represented by X/G. Indeed, in this case, X is a principal G-bundle over X/G and every principal G-bundle $\pi : E \to T$, equipped with a G-equivariant morphism $\sigma_\pi : E \to X$, is isomorphic to the pullback bundle $X \times_f T$, via a unique map $f : T \to X/G$:

$$\begin{array}{ccc} X \times_f T \cong E & \xrightarrow{\sigma_\pi} & X \\ {\scriptstyle\pi}\downarrow & & \downarrow \\ T & \xrightarrow{f} & X/G \end{array}$$

On the other hand, if the action of G is not free, X/G may well exist as a scheme without the groupoid $[X/G]$ being representable. The case is the one where $X = \{pt\}$ is a single point. Then, for obvious reasons, one sets

$$[\{pt\}/G] = BG\,.$$

The geometric example we have in mind is of course the Hilbert scheme $\mathrm{H}_{\nu,g,n}$ of ν-log-canonically embedded n-pointed stable curves of genus g (where $\nu \geq 3$). As we saw in Section 5 of Chapter XI, $\mathrm{H}_{\nu,g,n}$ is acted on by $PGL(N)$, where $N = (2\nu-1)(g-1)+\nu n$.

THEOREM (5.6). *The moduli groupoid $\overline{\mathcal{M}}_{g,n}$ is isomorphic to the quotient groupoid $[\mathrm{H}_{\nu,g,n}/PGL(N)]$.*

Proof. Let us define a morphism

$$\Phi : \overline{\mathcal{M}}_{g,n} \longrightarrow [\mathrm{H}_{\nu,g,n}/PGL(N)]\,.$$

Given an object in $\mathcal{C}$, that is, a family $\xi : \mathcal{X} \to T$ of stable n-pointed genus g curves, $\Phi(\xi)$ must consist of a G-bundle $\pi : E \to T$ and a G-equivariant map $\sigma_\pi : E \to \mathrm{H}_{\nu,g,n}$. As far as the bundle is concerned, we let $\pi : E \to T$ be the principal G-bundle associated to the projective bundle $\mathbb{P}_\xi = \mathbb{P}(\xi_*(\omega_\xi^\nu(\nu D))) \to T$, where D is the divisor of the canonical sections of ξ. Consider the canonically trivialized G-bundle

$$\pi^*\mathbb{P}_\xi \to E$$

and the pulled-back family

$$\eta : \mathcal{Z} = \mathcal{X} \times_\pi E \to E\,.$$

There is a canonical isomorphism

$$\mathbb{P}_\eta \cong \pi^*\mathbb{P}_\xi.$$

The canonical trivialization of $\mathbb{P}_\eta$ exhibits $\mathcal{Z} \to E$ as a family of ν-log-canonically embedded curves and therefore gives a G-equivariant morphism

$$\sigma_\pi : E \longrightarrow \mathrm{H}_{\nu,g,n} \; .$$

The definition of Φ on objects is now completed. The definition of Φ on morphisms and the proof that Φ is indeed a morphism of groupoids are straightforward and are left to the reader. Let us show that Φ is an isomorphism. We use Lemma (5.1). We must show that, for every scheme T, the functor Φ_T is fully faithful and essentially surjective. For the first point, we must prove that Φ_T induces a bijection

$$\mathrm{Hom}_{\mathcal{C}(T)}(\xi,\xi) \xrightarrow{\sim} \mathrm{Hom}_{\mathcal{P}(T)}(\Phi(\xi),\Phi(\xi)).$$

Equivalently, we must show that if $\xi : \mathcal{X} \to T$ is a family of stable, n-pointed curves of genus g, then the automorphisms of this family and the automorphisms of the projective bundle $P_\xi \to T$ determine each other. Looking at the fiberwise ν-canonical embedding

$$\begin{array}{ccc} \mathcal{X} & \xrightarrow{\varphi} & \mathbb{P}^*_\xi \\ {\scriptstyle\pi}\downarrow & \swarrow & \\ T & & \end{array}$$

it is clear that the only thing to prove is that any automorphism γ of the family $\xi : \mathcal{X} \to T$ is induced by one of the bundles $\mathbb{P}_\xi$. This is certainly true locally, where the projective bundle can be trivialized. But then it is also true globally because, on each fiber $\mathcal{X}_t$, the automorphism γ_t is uniquely induced, via φ_t, by a projective automorphism of $(\mathbb{P}_\xi)_t$. We now address the essential surjectivity of Φ_T. Let then (π, σ_π) be an object of $\mathcal{P}$, so that $\pi : E \to T$ is a principal G-bundle, and $\sigma_\pi : E \to \mathrm{H}_{\nu,g,n}$ is a G-equivariant map. Look at the universal family $\mathcal{Y} \to \mathrm{H}_{\nu,g,n}$ and form the cartesian diagram

$$\begin{array}{ccc} \mathcal{Z} & \longrightarrow & \mathcal{Y} \\ {\scriptstyle\eta}\downarrow & & \downarrow \\ E & \xrightarrow{\sigma_\pi} & \mathrm{H}_{\nu,g,n} \end{array}$$

The group G acts equivariantly and *freely* on E and $\mathcal{Z}$. We can then form the quotient family $\xi : \mathcal{Z}/G = \mathcal{X} \to T = E/G$. It is now an exercise to prove that, indeed, $\Phi_T(\xi)$ is isomorphic to (π, σ_π). Q.E.D

6. The classical theory of descent for quasi-coherent sheaves.

In this section we recall the simplest instance of Grothendieck's descent theory, namely faithfully flat descent for quasicoherent sheaves. To explain what this means, consider a morphism of schemes $X \to Y$, and a quasicoherent $\mathcal{O}_X$-module $\mathcal{F}$. Via the two projections p_1 and p_2 of $X \times_Y X$ to the two factors, $\mathcal{F}$ pulls back to $\mathcal{F}_1 = p_1^*\mathcal{F}$ and $\mathcal{F}_2 = p_2^*\mathcal{F}$. Write p_{12}, p_{13}, and p_{23} to indicate the projections of $X \times_Y X \times_Y X$ to $X \times_Y X$ obtained by omitting the third, second, and first components, respectively, and q_1, q_2, q_3 to indicate the three projections of $X \times_Y X \times_Y X$ to X. We have the usual simplicial diagram

$$X \times_Y X \times_Y X \;\substack{\longrightarrow\\ \longrightarrow\\ \longrightarrow}\; X \times_Y X \;\substack{\longrightarrow\\ \longrightarrow}\; X \longrightarrow Y$$

Notice that $p_1 p_{12} = q_1 = p_1 p_{13}$, $p_2 p_{12} = q_2 = p_1 p_{23}$, and $p_2 p_{13} = q_3 = p_2 p_{23}$. By *descent data* for $\mathcal{F}$ relative to $X \to Y$ we mean an isomorphism $\varphi : \mathcal{F}_1 \to \mathcal{F}_2$ such that the following "benzene" diagram commutes:

$$\text{(6.1)}\qquad \begin{array}{ccccccc} & & p_{12}^*\mathcal{F}_2 & = & p_{23}^*\mathcal{F}_1 & & \\ & \overset{p_{12}^*\varphi}{\nearrow} & & & & \overset{p_{23}^*\varphi}{\searrow} & \\ p_{12}^*\mathcal{F}_1 & & & & & & p_{23}^*\mathcal{F}_2 \\ & \diagdown\!\diagdown & & & & /\!/ & \\ & & p_{13}^*\mathcal{F}_1 & \xrightarrow{p_{13}^*\varphi} & p_{13}^*\mathcal{F}_2 & & \end{array}$$

We will refer to this condition as the *cocycle condition.* Very roughly speaking, the existence of φ tells us that $\mathcal{F}$ "looks the same" at points belonging to the same fiber of $X \to Y$, and the cocycle condition guarantees that the ensuing identifications are consistent.

When $\mathcal{F}$ is the pullback of a quasicoherent $\mathcal{O}_Y$-module, there is a canonical isomorphism between $\mathcal{F}_1$ and $\mathcal{F}_2$ which provides $\mathcal{F}$ with canonical descent data. The problem of descent is to decide whether this process can be inverted, that is, whether a quasicoherent $\mathcal{O}_X$-module with descent data comes from an $\mathcal{O}_Y$-module. One may ask a similar question for morphisms. There is an obvious notion of morphism of quasicoherent $\mathcal{O}_X$-modules with descent data, and one may wonder whether morphisms between modules with descent data which arise by pullback from $\mathcal{O}_Y$-modules $\mathcal{F}$ and $\mathcal{G}$ do descend to morphisms between $\mathcal{F}$ and $\mathcal{G}$. The answer to both questions is yes if one assumes that $X \to Y$ be *faithfully flat*, that is, flat and surjective. Actually, one also has to assume that $X \to Y$ is quasi-compact. This is automatic when one deals only with schemes of finite type over a field, as we do in the rest of the book.

THEOREM (6.2). *Let $\pi : X \to Y$ be a faithfully flat and quasi-compact morphism of schemes. Then the pullback functor*

$$\{\textit{quasicoherent } \mathcal{O}_Y\textit{-modules}\} \to \left\{ \begin{array}{c} \textit{quasicoherent } \mathcal{O}_X\textit{-modules with} \\ \textit{descent data relative to } \pi \end{array} \right\}$$

is an equivalence of categories.

We postpone the proof for a moment, except for the following simple observation.

REMARK (6.3). There is one case in which the conclusion of Theorem (6.2) holds true, even without faithful flatness hypothesis, and this is the one when π admits a section, that is, a right inverse. Call this σ and consider the two morphisms $\tau_1 = (\mathrm{id}, \sigma)$ and $\tau_2 = (\sigma, \mathrm{id})$ from X to $X \times_Y X$:

$$Y \xrightarrow{\ \sigma\ } X \underset{\tau_2}{\overset{\tau_1}{\rightrightarrows}} X \times_Y X$$

Clearly,

$$\pi\sigma = \mathrm{id}_Y\,, \quad p_1\tau_1 = p_2\tau_2 = \mathrm{id}_X\,, \quad p_1\tau_2 = p_2\tau_1 = \sigma\pi\,.$$

In particular, if a quasicoherent $\mathcal{O}_X$-module $\mathcal{F}$ is of the form $\pi^*\mathcal{H}$ for some $\mathcal{H}$, then $\mathcal{H} = \sigma^*\mathcal{F}$. Similarly, any morphism $\mathcal{H} \to \mathcal{K}$ of quasicoherent $\mathcal{O}_Y$-modules can be recovered from $\pi^*\mathcal{H} \to \pi^*\mathcal{K}$ as $\sigma^*\pi^*\mathcal{H} \to \sigma^*\pi^*\mathcal{K}$. Thus, $\mathrm{Hom}(\mathcal{H}, \mathcal{K}) \to \mathrm{Hom}(\pi^*\mathcal{H}, \pi^*\mathcal{K})$ is injective.

Now suppose that $\mathcal{F}$ is a quasicoherent $\mathcal{O}_X$-module with descent data $p_1^*\mathcal{F} \to p_2^*\mathcal{F}$ and pull back these via τ_1. What we get is an isomorphism

$$\mathcal{F} = \tau_1^* p_1^* \mathcal{F} \to \tau_1^* p_2^* \mathcal{F} = \pi^*\sigma^*\mathcal{F}\,. \tag{6.4}$$

Thus, $\mathcal{F}$ is the pullback via π of a quasicoherent sheaf on Y, namely $\sigma^*\mathcal{F}$. The isomorphism between $\mathcal{F}$ and $\pi^*\sigma^*\mathcal{F}$ holds also if we take descent data into account. This is a direct consequence of the cocycle relation. Define a morphism $\tau_{12} : X \times_Y X \to X \times_Y X \times_Y X$ by setting $\tau_{12} = (p_1, p_2, \sigma\pi p_1) = (p_1, p_2, \sigma\pi p_2)$. Then one immediately checks that

$$p_{12}\tau_{12} = \mathrm{id}\,, \quad p_{13}\tau_{12} = \tau_1 p_1\,, \quad p_{23}\tau_{12} = \tau_1 p_2\,.$$

Since (6.1) commutes, pulling it back by means of τ_{12}, we get another commutative diagram, which, taking into account the above identities and recalling that $p_2\tau_1 = \sigma\pi$, reduces to

$$\begin{array}{ccc} p_1^*\mathcal{F} & \xrightarrow{\qquad\qquad\qquad} & p_2^*\mathcal{F} \\ \downarrow & & \downarrow \\ p_1^*\tau_1^*p_2^*\mathcal{F} = p_1^*\pi^*\sigma^*\mathcal{F} & = & p_2^*\pi^*\sigma^*\mathcal{F} = p_2^*\tau_1^*p_2^*\mathcal{F} \end{array}$$

This shows that the descent data for $\mathcal{F}$ and those for $\pi^*\sigma^*\mathcal{F}$ correspond to each other via the isomorphism $\mathcal{F} \to \pi^*\sigma^*\mathcal{F}$ given by (6.4).

Finally, let $\mathcal{F}$ and $\mathcal{G}$ be quasicoherent $\mathcal{O}_X$-modules with descent data and suppose that α is a morphism between them. In other words, $\alpha : \mathcal{F} \to \mathcal{G}$ is such that

$$\begin{array}{ccc} p_1^*\mathcal{F} & \xrightarrow{p_1^*\alpha} & p_1^*\mathcal{G} \\ \downarrow & & \downarrow \\ p_2^*\mathcal{F} & \xrightarrow{p_2^*\alpha} & p_2^*\mathcal{G} \end{array}$$

commutes. Pulling back this diagram via τ_1 yields another commutative diagram

$$\begin{array}{ccc} \mathcal{F} & \xrightarrow{\quad\alpha\quad} & \mathcal{G} \\ \downarrow & & \downarrow \\ \pi^*\sigma^*\mathcal{F} & \xrightarrow{\pi^*\sigma^*\alpha} & \pi^*\sigma^*\mathcal{G} \end{array}$$

This means that α is the pullback of the morphism of $\mathcal{O}_Y$-modules $\sigma^*\alpha : \sigma^*\mathcal{F} \to \sigma^*\mathcal{G}$.

We now prove (6.2). Let us recapitulate what must be shown. Recall that a diagram of mappings of sets

$$A \xrightarrow{\ a\ } B \overset{b_1}{\underset{b_2}{\rightrightarrows}} C$$

is said to be *exact* if a is injective and its image is the *equalizer* of the pair of mappings b_1, b_2, that is, the set of those element of B which map to the same element of C via b_1 and b_2. First of all, we must prove that, if $\mathcal{F}$ and $\mathcal{G}$ are quasicoherent $\mathcal{O}_Y$-modules and we set $\mathcal{F}' = \pi^*\mathcal{F}$, $\mathcal{F}'' = p_1^*\mathcal{F}' = p_2^*\mathcal{F}'$, then the diagram

$$\text{(6.5)} \qquad \mathrm{Hom}_{\mathcal{O}_Y}(\mathcal{F},\mathcal{G}) \xrightarrow{\ \pi^*\ } \mathrm{Hom}_{\mathcal{O}_X}(\mathcal{F}',\mathcal{G}') \overset{p_1^*}{\underset{p_2^*}{\rightrightarrows}} \mathrm{Hom}_{\mathcal{O}_{X\times_Y X}}(\mathcal{F}'',\mathcal{G}'')$$

is exact. Then we must show that any quasicoherent $\mathcal{O}_X$-module with descent data comes by pullback from a quasicoherent $\mathcal{O}_Y$-module; the latter is then automatically unique, up to a unique isomorphism, by the exactness of (6.5). By Remark (6.3), the conclusion of the theorem is valid if π has a section. The idea is to reduce to this case via the base change $X \to Y$. Once this has been done, however, we must push down from X to Y what we have obtained. Here is where the faithful flatness assumption comes into play. Recall in fact that a module M over a commutative ring A is said to be *faithfully flat* if the exactness of any sequence of homomorphisms of A-modules is equivalent to the exactness of the sequence obtained by tensoring it with M. It is an elementary result (see, for instance, [503]) that a commutative A-algebra B is faithfully flat if and only if it is flat and $f : \mathrm{Spec}\, B \to \mathrm{Spec}\, A$ is onto, that is, if and only if f is a faithfully flat morphism of schemes. It is then at least plausible that questions having to do with exactness of sequences on Y could be decided by examining the corresponding questions on X. Before we explain this in detail, however, it is best to perform a couple of reductions. First of all, we claim that it suffices to prove (6.2) when Y is affine. In fact, the injectivity of π^* in (6.5), that is, uniqueness of descent for homomorphisms, is a local property on Y. On the other hand, if a homomorphism between $\mathcal{F}'$ and $\mathcal{G}'$ descends locally to a homomorphism between $\mathcal{F}$ and $\mathcal{G}$, the compatibility of the descended homomorphisms can be checked locally and follows from uniqueness if the theorem is known to hold when Y is affine. Likewise, if a quasicoherent sheaf $\mathcal{G}$ with descent data on X descends to a quasicoherent sheaf $\mathcal{F}_i$ on each open set Y_i of a cover of Y, then the $\mathcal{F}_i$ are isomorphic, via a unique isomorphism, on any affine contained in any one of the overlaps of the Y_i and hence patch together to yield a quasicoherent sheaf $\mathcal{F}$ which lifts to $\mathcal{G}$.

The second reduction is that we may suppose that X is also affine. In fact, if Y is affine, then X is the union of finitely many affines X_i. We write X' for the disjoint union of the X_i, which is then affine. Clearly, $X' \to Y$ is faithfully flat and factors through $X \to Y$. Equally clearly, if the theorem holds for $X' \to Y$, it holds a fortiori for $X \to Y$. In conclusion, in proving (6.2) we may assume that both X and Y are affine.

We therefore assume that $Y = \operatorname{Spec} A$ and $X = \operatorname{Spec} A'$, that π corresponds to a ring homomorphism $\alpha : A \to A'$, and that A' is faithfully flat over A. We set $A'' = A' \otimes_A A'$; moreover, for any A-module M, we set $M' = M \otimes_A A'$, $M'' = M \otimes_A A''$. There are two natural homomorphisms β_1 and β_2 from A' to A'', given by $b \mapsto b \otimes 1$ and $b \mapsto 1 \otimes b$, corresponding to the two projections $X \times_Y X \to X$. For any A'-module H, these give rise to two tensor products $H \otimes_{A'} A''$, which we denote by H_1 and H_2. In the language of rings and modules, descent data for the quasicoherent $\mathcal{O}_X$-module $\widetilde{H}$ correspond to a homomorphism $H_1 \to H_2$ of A''-modules, satisfying an obvious cocycle condition. We shall refer to H, equipped with such a homomorphism, as a *module with descent data*. An A'-module of the form M' comes equipped with natural descent data. For any pair M, N of A-modules, the homomorphisms β_1 and β_2 give two distinct homomorphisms from $\operatorname{Hom}_{A'}(M', N')$ to $\operatorname{Hom}_{A''}(M'', N'')$. Rephrased in terms of modules over rings, what we have to prove is then:

i) for any pair M, N of A-modules, the diagram

$$\operatorname{Hom}_A(M, N) \longrightarrow \operatorname{Hom}_{A'}(M', N') \rightrightarrows \operatorname{Hom}_{A''}(M'', N'')$$

is exact;

ii) Any A'-module with descent data is isomorphic to one of the form M', for some A-module M.

The key to proving i) is the following special case of i) itself.

LEMMA (6.6). *For any A-module N, the diagram*

$$N \longrightarrow N' \rightrightarrows N'' \tag{6.7}$$

is exact.

The proof is based on the idea, mentioned earlier, of performing a faithfully flat base change $A \to B$ such that, setting $B' = B \otimes_A A'$, $B'' = B' \otimes_B B'$, the ring homomorphism $B' \to B''$ has a left inverse (i.e., such that $\operatorname{Spec} B'' \to \operatorname{Spec} B'$ has a section). One then deduces exactness from the existence of the left inverse, and finally exactness can be "pushed down" to A by faithful flatness. The required base change is provided to us for free; it suffices to take A' as B, since $\beta_i : A' \to A''$ has a left inverse, namely the multiplication homomorphism $A'' = A' \otimes_A A' \to A'$.

We now show that this idea can be made into an actual proof. We claim that the lemma is true if $\alpha : A \to A'$ has a left inverse. Actually, we have already proved this in Remark (6.3), but let us do it again here. Let ρ be a left inverse of α. We define homomorphisms $\sigma_i : A'' \to A'$, $i = 1, 2$, by $\sigma_1(b \otimes b') = \rho(b')b$, $\sigma_2(b \otimes b') = \rho(b)b'$. Then $\sigma_i\beta_i = \mathrm{id}$ and $\sigma_i\beta_j = \alpha\rho$ when $i \neq j$. Now tensor with N and call with the same names the resulting homomorphisms. We see that $N \to N'$ has a left inverse, and hence is injective. On the other hand, if $\beta_1(n') = \beta_2(n')$, then $n' = \sigma_1\beta_1(n') = \alpha\rho(n')$. This proves the claim.

Now observe that, if B is an A-algebra, the diagram obtained by tensoring (6.7) with B is just

$$N \otimes_A B \longrightarrow (N \otimes_A B) \otimes_B B' \rightrightarrows (N \otimes_A B) \otimes_B B''$$

as can be easily checked. Thus, when $B \to B'$ has a left inverse, this diagram is exact. If, in addition, B is a faithfully flat A-algebra, then the original diagram (6.7) is also exact. This proves the lemma.

Our next task is to deduce i) from the lemma. The latter says in particular that M is an A-submodule of M', which is an A'-submodule of M'', and similarly for N, N', and N''. Since N injects into N', the homomorphism $\operatorname{Hom}_A(M, N) \to \operatorname{Hom}_{A'}(M', N')$ is injective. Now let $\xi : M' \to N'$ be a homomorphism which belongs to the kernel of the pair of homomorphisms from $\operatorname{Hom}_{A'}(M', N')$ to $\operatorname{Hom}_{A''}(M'', N'')$. To show that ξ comes from $\operatorname{Hom}_A(M, N)$, it suffices to show that $\xi(m \otimes 1) \in N \otimes 1$ for any $m \in M$. But $\xi(m \otimes 1)$ is an element of the kernel of the pair of homomorphisms from N' to N'', so by the lemma it belongs to $N \otimes 1$. This proves i).

To prove ii), we may argue as follows. Let H be an A'-module with descent data $v : H_1 \to H_2$, and let $\gamma_i : H \to H_i$, $i = 1, 2$, be the natural homomorphisms. We must find an A-module M such that H is isomorphic, as a module with descent data, to M'. There is a natural candidate for M. Suppose in fact that an M exists. Then both H_1 and H_2 can be identified with M'', and the composition of the two identifications is just v. But then Lemma (6.6) says that M injects in H and gets identified with

$$N = \{h \in H : v\gamma_1(h) = \gamma_2(h)\}.$$

Thus, all that needs to be done is to prove that the homomorphism $N \otimes_A A' \to H$ is an isomorphism. We resort to the same trick used to prove i). Let B be a faithfully flat A-algebra such that $B \to B'$ has a left inverse (for instance, $B = A'$). Change ring by tensoring everything with B. Write $\overline{N}$ for $N \otimes_A B$, $\overline{H}$ for $H \otimes_A B$, $\overline{v}$ for $v \otimes \mathrm{id} : \overline{H}_1 \to \overline{H}_2$, and so on. Since B is A-flat, $\overline{N}$ is the set of all $h \in \overline{H}$ such that $\overline{v}\,\overline{\gamma}_1(h) = \overline{\gamma}_2(h)$. Since ii) is true for $B \to B'$, by Remark (6.3), we know that $\overline{N} \otimes_B B' \to \overline{H}$ is an isomorphism. On the other hand, this isomorphism can be obtained by tensoring with B the homomorphism $N \otimes_A A' \to H$, which is then also an isomorphism, by faithful flatness. This completes the proof of ii) and of (6.2).

The category $QCoh$ of quasi-coherent sheaves over schemes of finite type is an example of a fibered category (endowed with a cleavage). The theory of descent is best formalized in the framework of these categories. In this language, Grothendieck's theorem of descent for quasi-coherent sheaves can be stated by saying that *the fibered category $QCoh$ is a stack for the faithfully flat, quasi-compact topology.* We will not introduce general fibered categories in this book. As far as moduli spaces are concerned, it will suffice to restrict our attention to the particular case of categories fibered in groupoids ($QCoh$ is not one such), and for these categories, we will formally treat descent only in the étale topology. This is what we are doing in the next section.

7. The moduli space of curves as a stack, II.

Let $\mathcal{M} = (\mathcal{C}, p)$, with $p : \mathcal{C} \to Sch$, be a category fibered in groupoids or, briefly, a groupoid. Let T be a scheme, let ξ be an object of $\mathcal{M}(U)$, and let $f : U \to T$ be an étale surjective morphism. We consider the two projections p_1 and p_2 from $U \times_T U$ to the two factors. As before, we write p_{12}, p_{13}, and p_{23} to indicate the projections of $U \times_T U \times_T U$ to $U \times_T U$ obtained by omitting the third, second, and first component, respectively, and q_1, q_2, q_3 to indicate the three projections of $U \times_T U \times_T U$ to X, so that $p_1 p_{12} = q_1 = p_1 p_{13}$, $p_2 p_{12} = q_2 = p_1 p_{23}$, and $p_2 p_{13} = q_3 = p_2 p_{23}$. A *descent datum for ξ, relative to $f : U \to T$,* is an isomorphism $\varphi : p_1^* \xi \to p_2^* \xi$ such that the following diagram commutes:

$$(7.1)\qquad \begin{array}{ccccc} & p_{12}^*p_2^*\xi & = & p_{23}^*p_1^*\xi & \\ {\scriptstyle p_{12}^*\varphi}\nearrow & & & & \searrow{\scriptstyle p_{23}^*\varphi} \\ p_{12}^*p_1^*\xi & & & & p_{23}^*p_2^*\xi \\ \| & & & & \| \\ & p_{13}^*p_1^*\xi & \xrightarrow{\;p_{13}^*\varphi\;} & p_{13}^*p_2^*\xi & \end{array}$$

In other words, we must have

$$(7.2)\qquad p_{23}^*\varphi \circ p_{12}^*\varphi = p_{13}^*\varphi : q_1^*\xi \longrightarrow q_3^*\xi \,.$$

A descent datum for ξ, relative to f, is said to be *effective* if there exist an object $\eta \in \mathcal{M}(T)$ and an isomorphism $\psi : f^*(\eta) \to \xi$ such that

$$\varphi = (p_2^*\psi) \circ (p_1^*\psi)^{-1} . \tag{7.3}$$

This last condition just says that ψ identifies the descent data for $f^*(\eta)$ with those of ξ. The intuitive meaning of these definitions is clear when we translate the language of the usual topology into the language the étale topology. An open cover $\mathcal{U} = \{U_i\}$ of T is translated into a surjective étale map $U \to T$, the collection of pairwise intersections $\{U_i \cap U_j\}$ is translated into the fiber product $U \times_T U$, while the collection of triple intersections $\{U_i \cap U_j \cap U_k\}$ is translated into the triple fiber product $U \times_T U \times_T U$. The datum of an object ξ_i on each U_i corresponds to the datum of an object ξ on U. An isomorphism φ_{ij} from $\xi_i|U_i \cap U_j$ to $\xi_i|U_i \cap U_j$ is translated into the descent datum $\varphi : p_1^*\xi \to p_2^*\xi$. The compatibility condition $\varphi_{ij}\varphi_{jk} = \varphi_{ik}$ on $\{U_i \cap U_j \cap U_k\}$ is translated into the cocycle condition (7.2).

We are now ready to define what we mean by a *stack in groupoids for the étale topology* or, as we will usually say for the sake of brevity, a *stack*. Such an object is a groupoid $\mathcal{M} = (\mathcal{C}, p)$ having the following two properties.

1) Every (étale) descent datum is effective.
2) Given a scheme S and objects ξ and η in $\mathcal{M}(S)$, the functor

$$Isom_S(\xi, \eta) : Sch/S \longrightarrow Sets$$

which associates to a morphism $f : T \to S$ the set of isomorphisms in $\mathcal{M}(T)$ between $f^*\xi$ and $f^*\eta$ is a sheaf in the étale topology.

We recall that a contravariant functor $F : Sch/S \to Sets$ is a sheaf in the étale topology if, for every étale surjective morphism $\pi : X \to Y$ of S-schemes, the diagram

$$F(Y) \xrightarrow{F(\pi)} F(X) \underset{F(p_2)}{\overset{F(p_1)}{\rightrightarrows}} F(X \times_Y X)$$

is exact. Notice that, when condition 2) is satisfied, (7.3) implies that η is unique up to a unique isomorphism.

The following fundamental theorem is due to Grothendieck.

THEOREM (7.4). *Let S be a scheme. Let $F : Sch/S \to Sets$ be a contravariant, representable functor. Then F is a sheaf for the étale topology.*

Proof. Assume that F is represented by a scheme Z, so that $F = \mathrm{Hom}(-, Z)$. Given an étale surjective morphism $\pi : X \to Y$ of S-schemes, we must prove the exactness of the diagram

$$\mathrm{Hom}(Y, Z) \xrightarrow{\pi^*} \mathrm{Hom}(X, Z) \overset{p_1^*}{\underset{p_2^*}{\rightrightarrows}} \mathrm{Hom}(X \times_Y X, Z) \cdot$$

Following, almost word by word, the arguments we used to prove the exactness of (6.5), we are reduced to the case in which $Y = \mathrm{Spec}\, A$ and $X = \mathrm{Spec}\, A'$. In an analogous way, we may furthermore assume that $Z = \mathrm{Spec}\, B$. The above sequence may then be identified with

$$(7.5) \qquad \mathrm{Hom}(B, A) \longrightarrow \mathrm{Hom}(B, A') \rightrightarrows \mathrm{Hom}(B, A' \otimes_A A') \cdot$$

Recall that étale surjective morphisms are faithfully flat, so that A' is a faithfully flat A-algebra. Therefore the exactness of (7.5) follows from the exactness of the sequence of rings

$$A \longrightarrow A' \rightrightarrows A' \otimes_A A' \cdot$$

Q.E.D.

A *morphism between two stacks* $\mathcal{M}$ *and* $\mathcal{M}'$ is just a morphism between the underlying groupoids. As we already mentioned, we may consider a scheme S as a groupoid, and it is easily seen that this groupoid is indeed a stack.

THEOREM (7.6). *The groupoids* $\mathcal{M}_{g,n}$ *and* $\overline{\mathcal{M}}_{g,n}$ *are stacks (in groupoids in the étale topology).*

In view of Theorem (5.6), this is a consequence of the following general result.

THEOREM (7.7). *Given a group scheme* G *acting on a scheme* X, *the quotient groupoid* $[X/G]$ *is a stack.*

To prove (7.7), one needs to know more about descent than we have explained. Therefore we shall follow a different path. A sketch of proof of (7.7) can be found in [190], Proposition 2.1.

Proof. We only deal with $\overline{\mathcal{M}}_{g,n}$, the case of $\mathcal{M}_{g,n}$ being completely analogous. Property 2) in the definition of stack is easily checked. In fact, given families of stable curves $\xi : X \to S$ and $\eta : Y \to S$, that is, objects in $\overline{\mathcal{M}}_{g,n}(S)$, the functor $Isom_S(\xi, \eta)$ is represented by the scheme $\mathbf{Isom}_S(X, Y)$, introduced in Section 7 of Chapter IX. Thus, property 2) follows from Theorem (7.4). We turn to property 1). Let then

$$T \longrightarrow T'$$

be a surjective étale morphism, and let

$$\xi : X \longrightarrow T$$

be a family of stable curves endowed with descent data $\varphi : p_1^*(\xi) \to p_2^*(\xi)$,

$$\begin{array}{ccccc} p_1^*(X) = T \times_{T'} X & & \xrightarrow{\ \varphi\ } & & X \times_{T'} T = p_2^*(X) \\ & \searrow^{p_1^*(\xi)} & & \swarrow_{p_2^*(\xi)} & \\ & & T \times_{T'} T & & \end{array} \tag{7.8}$$

To check property 1), we must produce a family of stable curves $\eta : Y \to T'$ such that $\xi = \pi^*(\eta)$. The construction of the family η is a typical descent construction and, as is often the case, will be reduced to the theory of descent of quasi-coherent sheaves. Let us illustrate the two basic steps of this reduction.

We need some notation. Consider the family $\xi : X \to T$. Denote by L_ξ the line bundle $(\omega_\xi(D))^3$, where D is the divisor of marked points in the fibers of ξ, and consider the dual direct image bundle $E_\xi = \xi_*(L_\xi)^\vee$. The total space X of the family ξ can be viewed as embedded in $\mathbb{P}(E_\xi)$:

$$\begin{array}{l} X \subset \mathbb{P}(E_\xi) \\ \downarrow \xi \\ T \end{array}$$

Step 1. From the descent data for ξ we deduce descent data for the vector bundle (or better, locally free sheaf) E_ξ. From the theory of descent on $QCoh$, we get a vector bundle E' over T' with $\pi^*(E') = E$.

Step 2. At this stage there is a diagram

$$\begin{array}{ccc} \mathbb{P} = \mathbb{P}(E_\xi) & \xrightarrow{\ q\ } & \mathbb{P}(E') = \mathbb{P}' \\ \downarrow & & \downarrow \\ T & \xrightarrow{\ \pi\ } & T' \end{array} \tag{7.9}$$

Look at the étale morphism $q : \mathbb{P} \to \mathbb{P}'$. The descent data for $\xi : X \to T$ relative to the étale cover $\pi : T \to T'$ determine descent data for the ideal sheaf $\mathcal{I}_X \subset \mathcal{O}_\mathbb{P}$ with respect to the étale cover $q : \mathbb{P} \to \mathbb{P}'$. Using again the theory of descent for $QCoh$, we get the subscheme $Y \subset \mathbb{P}'$ and the family $\eta : Y \to T'$.

Before embarking on the actual proof of the two steps above, we need some preparation. Given a morphism $U \to U'$ and a cartesian diagram of families of stable curves

$$\begin{array}{ccc} Z & \xrightarrow{h} & Z' \\ {\scriptstyle\alpha}\downarrow & & \downarrow{\scriptstyle\beta} \\ U & \xrightarrow{f} & U' \end{array}$$

we have canonical isomorphisms:

$$\sigma_{h,f} : h^*(L_\beta) \xrightarrow{\sim} L_\alpha\,, \qquad \tau_{h,f} : f^*(E_\beta) \xrightarrow{\sim} E_\alpha\,. \tag{7.10}$$

Given two composable cartesian squares of families of stable curves

$$\begin{array}{ccccc} Z & \xrightarrow{h} & Z' & \xrightarrow{k} & Z'' \\ {\scriptstyle\alpha}\downarrow & & \downarrow{\scriptstyle\beta} & & \downarrow{\scriptstyle\gamma} \\ U & \xrightarrow{f} & U' & \xrightarrow{g} & U'' \end{array}$$

one can easily check the equalities

$$\begin{aligned} \sigma_{kh,gf} &= \sigma_{h,f} h^*(\sigma_{k,g}) : \ (kh)^*(L_\gamma) \xrightarrow{\sim} L_\alpha, \\ \tau_{kh,gf} &= \tau_{h,f} f^*(\tau_{k,g}) \ : \ (gf)^*(E_\gamma) \xrightarrow{\sim} E_\alpha. \end{aligned} \tag{7.11}$$

We return to the étale cover $\pi : T \to T'$ and to the descent datum $\varphi : p_1^*(X) \to p_2^*(X)$ for the family $\xi : X \to T$. Look at the diagram

$$\begin{array}{ccccccc} X & \xleftarrow{p_1} & p_1^*(X) & \xrightarrow{\varphi} & p_2^*(X) & \xrightarrow{p_2} & X \\ {\scriptstyle\xi}\downarrow & & \downarrow{\scriptstyle p_1^*(\xi)} & & \downarrow{\scriptstyle p_2^*(\xi)} & & \downarrow{\scriptstyle\xi} \\ T & \xleftarrow{p_1} & T\times_{T'} T & = & T\times_{T'} T & \xrightarrow{p_2} & T \end{array}$$

Using (7.11) and some patience, the reader will see that the isomorphism

$$\varphi_\xi = \tau^{-1}_{p_2,p_2} \tau^{-1}_{\varphi,id} \tau_{p_1,p_1} \ : \ p_1^* E_\xi \longrightarrow p_2^* E_\xi$$

satisfies the cocycle condition for the étale cover $\pi : T \to T'$, thus defining descent data for the coherent $\mathcal{O}_T$-module E_ξ. From the theory of descent on $QCoh$ we get a quasi-coherent $\mathcal{O}_{T'}$-module E' such that $E_\xi = \pi^*(E')$. Since π is étale, E' also is locally free. In particular $E = \pi^*(E')$. This concludes the first step of the proof.

Now we have the cartesian diagram (7.9). As we already mentioned, the descent data for $\xi : X \to T$, relative to the étale cover $\pi : T \to T'$, determine descent data for the ideal sheaf $\mathcal{I}_X \subset \mathcal{O}_\mathbb{P}$ with respect to the étale cover $q : \mathbb{P} \to \mathbb{P}'$. Again by descent in $QCoh$, we get an $\mathcal{O}_{\mathbb{P}'}$-module $\mathcal{G}$ such that $q^*(\mathcal{G}) = \mathcal{I}_X$. As q is étale, and hence faithfully flat, it follows that $\mathcal{G}$ is a sheaf of ideals in $\mathcal{O}_{\mathbb{P}'}$. This sheaf of ideals defines a subscheme $Y \subset \mathbb{P}'$ such that $X \cong q^*(Y) = Y \times_{\mathbb{P}'} \mathbb{P}$. But then

$$X \cong Y \times_{\mathbb{P}'} \mathbb{P} = Y \times_{\mathbb{P}'} \mathbb{P}' \times_{T'} T \cong Y \times_{T'} T\,.$$

We then get a cartesian square

$$\begin{array}{ccc} X & \longrightarrow & Y \\ {\scriptstyle\xi}\downarrow & & \downarrow{\scriptstyle\eta} \\ T & \xrightarrow{\ \pi\ } & T' \end{array}$$

Since π is étale, $\eta : Y \to T'$ is a family of stable curves, as desired.

Q.E.D.

We end this section by saying a few words about *fiber products* of stacks. These are defined in the following way. Suppose that $\alpha : \mathcal{M} \to \mathcal{P}$ and $\beta : \mathcal{N} \to \mathcal{P}$ are morphisms of stacks. Then $\mathcal{M} \times_{\mathcal{P}} \mathcal{N}$ is the groupoid whose objects are defined by

$$(\mathcal{M} \times_{\mathcal{P}} \mathcal{N})(T) = \{(\xi, \eta, \varphi) : (\xi, \eta) \in \mathcal{M}(T) \times \mathcal{N}(T),\ \ \varphi \in \mathrm{Isom}_T(\alpha(\xi), \beta(\eta))\}\,,$$

for every scheme T. A morphism in the category $\mathcal{M} \times_{\mathcal{P}} \mathcal{N}$ between two objects (ξ, η, φ) and (ξ', η', φ') is a pair (ψ_1, ψ_2), where $\psi_1 : \xi \to \xi'$ is a morphism in $\mathcal{M}$ and $\psi_2 : \eta \to \eta'$ is a morphism in $\mathcal{N}$, with $p_{\mathcal{M}}(\psi_1) = p_{\mathcal{N}}(\psi_2)$ and $\varphi' \alpha(\psi_1) = \beta(\psi_1)\varphi$. It can be easily checked that such a groupoid is indeed a stack. As an exercise in the language of 2-categories, and more specifically in the definition of a commutative diagram in the category of stacks, the reader is encouraged to state the universal property satisfied by the fiber product of a stacks as just defined.

8. Deligne–Mumford stacks.

As we already mentioned, we may view a scheme S as a stack, by considering the stack associated to the functor of points of S. It is in this sense that we will talk about morphisms between schemes and stacks. As we already observed, a morphism f from a scheme S to a stack $\mathcal{M}$ is equivalent to the datum of an object ξ in $\mathcal{M}(S)$; indeed, $\xi = f(\mathrm{id}_S)$. When a stack is isomorphic to a scheme, we will say that it is *represented* by this scheme. We also observed that, when talking about representable groupoids, the word isomorphism is crucial. As an exercise in subtleties, and using Lemma (5.1), the reader should give a detailed proof of the fact that, given a groupoid $\mathcal{M}$ and a morphism $S \to \mathcal{M}$, the groupoid $\mathcal{M} \times_{\mathcal{M}} S$ is represented by S.

A morphism of stacks $f : \mathcal{M} \to \mathcal{N}$ is said to be *representable* if, for every scheme S and every morphism $S \to \mathcal{N}$, the fiber product $\mathcal{M} \times_{\mathcal{N}} S$ is a scheme. Given a stack $\mathcal{M}$, let us consider the diagonal morphism $\Delta : \mathcal{M} \to \mathcal{M} \times \mathcal{M}$ (which is defined in the obvious way). The following lemma explains what the representability of Δ means.

Lemma (8.1). *$\Delta : \mathcal{M} \to \mathcal{M} \times \mathcal{M}$ is representable if and only if every morphism from a scheme S to $\mathcal{M}$ is.*

Proof. Let $f : S \to \mathcal{M}$ and $g : T \to \mathcal{M}$ be morphisms from schemes to $\mathcal{M}$. Look at $(f, g) : S \times T \to \mathcal{M} \times \mathcal{M}$. From the universal property of the fiber product we get an isomorphism

$$S \times_{\mathcal{M}} T \cong \mathcal{M} \times_{\mathcal{M} \times \mathcal{M}} (S \times T). \tag{8.2}$$

Now suppose that Δ is representable. The right-hand side of (8.2) is then a scheme. Since this is true for every f and g, this means that f (and g) is representable. Conversely, assume that every morphism $S \to \mathcal{M}$ is representable. Let $h : S \to \mathcal{M} \times \mathcal{M}$ be a morphism. We must show that $\mathcal{M} \times_{\mathcal{M} \times \mathcal{M}} S$ is a scheme. Write $h = (f, g) \circ \Delta_S$, where $(f, g) : S \times S \to \mathcal{M} \times \mathcal{M}$. Then

$$\mathcal{M} \times_{\mathcal{M} \times \mathcal{M}} S = (\mathcal{M} \times_{\mathcal{M} \times \mathcal{M}} (S \times S)) \times_{S \times S} S \cong (S \times_{\mathcal{M}} S) \times_{S \times S} S,$$

and the stack on the right-hand side is a scheme, since $S \to \mathcal{M}$ is representable. Q.E.D.

Let **P** be a property of morphisms of schemes which is stable under base change as, for example, being surjective, flat, faithfully flat, étale, unramified, quasi-compact, separated, or of finite type. Then, by definition, a *representable morphism* $f : \mathcal{M} \to \mathcal{N}$ *satisfies* **P** if, for every morphism $S \to \mathcal{M}$, where S is a scheme, the morphism of schemes $\mathcal{M} \times_{\mathcal{N}} S \to S$ satisfies **P**.

A *Deligne–Mumford stack* is a stack $\mathcal{M}$ having the following two properties.

1) The diagonal $\Delta : \mathcal{M} \to \mathcal{M} \times \mathcal{M}$ is representable, quasi-compact, and separated.
2) There exist a scheme X and an étale surjective morphism $\alpha : X \to \mathcal{M}$.

The morphism α is also called *an atlas for* $\mathcal{M}$. As far as terminology is concerned, the reader should be aware of the fact that the original name given by Deligne and Mumford to what we now call a Deligne–Mumford stack is "algebraic stack." On the other hand, an *Artin stack* is a stack satifying 1) and 2) with the word "étale" substituted with "smooth." We now prove the following theorem.

Theorem (8.3). *$\mathcal{M}_{g,n}$ and $\overline{\mathcal{M}}_{g,n}$ are Deligne–Mumford stacks.*

Proof. We prove the theorem for $\overline{\mathcal{M}}_{g,n}$, the proof for $\mathcal{M}_{g,n}$ being completely analogous. Set $\mathcal{M} = \overline{\mathcal{M}}_{g,n}$. The representabilty of $\Delta : \mathcal{M} \to \mathcal{M} \times \mathcal{M}$ is straightforward. Let $h : S \to \mathcal{M} \times \mathcal{M}$ be a morphism. The datum of h is equivalent to the datum of two families of stable pointed curves $\xi : X \to S$ and $\eta : Y \to S$ in $\mathcal{M}(S)$. From the representability results of Section 7 of Chapter IX we get:

(8.4)
$$\begin{aligned}(\mathcal{M} \times_{\mathcal{M}\times\mathcal{M}} S)(T) &= \{(f,\alpha) \mid f : T \to S\,,\ \alpha \in \mathrm{Isom}_T(f^*\xi, f^*\eta)\} \\ &= \{(f,\beta) \mid f : T \to S\,,\ \beta \in \mathrm{Hom}_S(T, \mathbf{Isom}_S(\xi,\eta))\} \\ &= \mathrm{Hom}(T, \mathbf{Isom}_S(\xi,\eta)).\end{aligned}$$

Therefore, $\mathcal{M} \times_{\mathcal{M}\times\mathcal{M}} S$ is represented by $\mathbf{Isom}_S(\xi,\eta)$, which is separated and quasi-compact. This proves property 1). Before proving the second property, let us observe that, given morphisms $f : S \to \mathcal{M}$ and $g : T \to \mathcal{M}$, where S and T are schemes, or equivalently given two families of stable curves, $\xi : X \to S$ in $\mathcal{M}(S)$ and $\eta : Y \to T$ in $\mathcal{M}(T)$, proceeding as in (8.4), we get

$$S \times_{\mathcal{M}} T = \mathbf{Isom}_{S\times T}(p_1^*\xi, p_2^*\eta), \tag{8.5}$$

where $p_1 : S \times T \to S$ and $p_2 : S \times T \to T$ are the two projections. We now proceed to prove property 2). Let us go back to the smooth variety X defined in (3.8). Recall that X is the disjoint union of a finite number of "slices", $X_1, \dots, X_N$, in the Hilbert scheme $\mathrm{H}_{\nu,g,n}$. Each one of these slices is a smooth affine $(3g-3+n)$-dimensional subvariety of $\mathrm{H}_{\nu,g,n}$ which is transversal to the orbits of $G = PGL(N)$ and satisfies all the properties listed in definition (6.7) of Chapter XI. The restriction to X of the universal family over $\mathrm{H}_{\nu,g,n}$ yields a family of stable curves $\xi : \mathcal{C} \to X$ and hence a morphism

$$\alpha : X \to \mathcal{M}\,. \tag{8.6}$$

We wish to prove that α is étale and surjective. For this, we must prove that, for every morphism f from a scheme S to $\mathcal{M}$, the induced morphism

$$X \times_{\mathcal{M}} S = \mathbf{Isom}_{X\times S}(p_1^*\xi, p_2^*\eta) \longrightarrow S \tag{8.7}$$

is étale and surjective. Let $\eta : \mathcal{X} \to S$ be the family corresponding to the morphism $f : S \to \mathcal{M}$. Since being étale is a local property and since, locally on S, the family η is the pullback of the family $\xi : \mathcal{C} \to X$, we are reduced to showing that the natural projections

$$X \times_{\mathcal{M}} X = \mathbf{Isom}_{X\times X}(p_1^*\xi, p_2^*\xi) \longrightarrow X \tag{8.8}$$

are étale and surjective, but this is exactly the content of Proposition (3.10).

Q.E.D.

Let us go back to the étale map

$$\varphi : Y \to \overline{M}_{g,n}$$

defined in (3.7). The scheme Y is the disjoint union of schemes $Y_1, \dots, Y_N$, and each of these is a quotient $Y_i = X_i/G_i$, where X_i is a Kuranishi family obtained as a slice of $\mathrm{H}_{\nu,g,n}$. Certainly, the composite map

$$\beta : X = \coprod X_i \to \overline{M}_{g,n}$$

is not étale. The scheme Y has the advantage of mapping surjectively and in an étale manner onto $\overline{M}_{g,n}$. The scheme X has the advantage of being the basis of family of stable curves, which in turn determines the moduli map β. These two advantages cannot be reconciled in the world of algebraic spaces or in the world of schemes. In proving that

$$\alpha : X \to \overline{\mathcal{M}}_{g,n}$$

is étale and surjective, we showed that this reconciliation is possible in the world of algebraic stacks.

The coarse moduli space of a stack

From now on all the stacks and schemes we consider will be over $\mathbb{C}$. We start with a definition and a general remark. A *geometric point of a stack* $\mathcal{M}$ is, by definition, a connected component of the groupoid $\mathcal{M}(\mathrm{Spec}(\mathbb{C}))$, that is, an isomorphism class of objects in $\mathcal{M}(\mathrm{Spec}(\mathbb{C}))$. Let S be a scheme, and $f : \mathcal{M} \to S$ a morphism. Then, from the definition of morphism it follows that if ξ and ξ' belong to the same connected component of $\mathcal{M}(\mathrm{Spec}(\mathbb{C}))$, then $f(\xi) = f(\xi')$.

A *coarse moduli space for a stack* $\mathcal{M}$ is a scheme M together with a morphism $m : \mathcal{M} \to M$ inducing a bijection on geometric points and such that every morphism from $\mathcal{M}$ to a scheme factors through M.

In the next chapter we will prove that the analytic space $\overline{M}_{g,n}$ is a projective variety. We are going to use this property to show that $\overline{M}_{g,n}$ is a coarse moduli space for $\overline{\mathcal{M}}_{g,n}$. The analogous statement regarding $M_{g,n}$ and $\mathcal{M}_{g,n}$ immediately follows from this. The morphism $m : \overline{\mathcal{M}}_{g,n} \to \overline{M}_{g,n}$ is readily described. If T is a scheme, the functor

$$m_T : \overline{\mathcal{M}}_{g,n}(T) \to \overline{M}_{g,n}(T)$$

is defined as follows. Given a family $\xi : X \to T$, $m_T(\xi)$ is nothing but the moduli map $m_T(\xi) : T \to \overline{M}_{g,n}$. The statement about geometric points is an immediate consequence of the general remark we made at the beginning of this subsection. Finally, let S be a scheme, and $f : \overline{\mathcal{M}}_{g,n} \to S$ a morphism. Consider the Kuranishi étale cover (8.6)

$$\alpha : X \to \overline{\mathcal{M}}_{g,n}.$$

We get morphisms $m\alpha : X \to \overline{M}_{g,n}$ and $f\alpha : X \to S$ with the property that $f\alpha$ factors, set-theoretically, through $m\alpha$ via a map $h : \overline{M}_{g,n} \to S$. Since $m\alpha$ is finite, h is necessarily analytic. We now use the fact that $\overline{M}_{g,n}$ is projective, together with Proposition 15 of [626], to conclude that h is algebraic.

An orbifold-like definition of Deligne–Mumford stacks

Let $\mathcal{M}$ be a Deligne–Mumford stack, and let $X_0 \to \mathcal{M}$ be an étale surjective morphism, where X_0 is a scheme. The fiber product $X_1 = X_0 \times_{\mathcal{M}} X_0$ is a scheme, and the projections to the two factors are étale. There are natural morphisms $s, t : X_1 \to X_0$ (the projections to the two factors), $u : X_0 \to X_1$ (the diagonal), $i : X_1 \to X_1$ (interchanging the factors), plus a composition morphism $m : X_1 {}_s\times_t X_1 \to X_1$ defined as the projection onto the first and third factors in

$$X_1 \times_{X_0} X_1 = X_0 \times_{\mathcal{M}} X_0 \times_{\mathcal{M}} X_0 \to X_0 \times_{\mathcal{M}} X_0 = X_1 .$$

These morphisms satisfy the scheme-theoretic analogues of the equalities (4.2) defining an orbifold structure, which, for convenience, we state again:

$$\begin{aligned} &(s,t)u = \Delta\,, \quad (s,t)i = \eta(s,t)\,, \quad (t,s)m = (t \times s)\,,\\ &m \circ (ut, \mathrm{id}_{X_1}) = \mathrm{id}_{X_1}\,, \quad m \circ (\mathrm{id}_{X_1}, us) = \mathrm{id}_{X_1}\,,\\ &m \circ (\mathrm{id}_{X_1}, m) = m \circ (m, \mathrm{id}_{X_1})\,. \end{aligned} \tag{8.9}$$

In addition, the two projections s and t are étale, as we observed. It can be shown that the datum of the schemes X_0 and X_1, together with the morphisms s, t, u, i, m, completely determines the stack $\mathcal{M}$.

Conversely, suppose that we are given schemes and étale morphisms

$$X_1 \underset{t}{\overset{s}{\rightrightarrows}} X_0\,, \tag{8.10}$$

plus morphisms s, t, u, i, m subject to conditions (8.9), such that $(s,t) : X_1 \to X_0 \times X_0$ is quasicompact and separated. Then one can define a "quotient Deligne–Mumford stack" $\mathcal{M}$ with an étale surjective map $X_0 \to \mathcal{M}$ such that $X_1 = X_0 \times_{\mathcal{M}} X_0$. Proving these facts would require more descent theory than is at our disposal. For a proof, we therefore refer to [298], or to [671], p. 668, and references therein. We may notice, however, that a "baby" version of the argument is the proof of Proposition (2.9) in Chapter XIII and that the latter proposition is in fact sufficient to fully prove the existence of the stack $\mathcal{M}$ in special instances (cf. specifically Remark (2.10) in Chapter XIII). Here we shall limit ourselves to giving an idea of a possible construction of the stack

$\mathcal{M}$ starting from (8.10). We will just describe the category $\mathcal{M}(T)$ for a given scheme T. Objects of this category should correspond to morphisms from T to $\mathcal{M}$. Thus, it is natural to look for "morphisms from T to $X_1 \rightrightarrows X_0$." To put T and $X_1 \rightrightarrows X_0$ on the same footing, we consider a groupoid presentation of T, that is, a surjective étale map $T' \to T$. We then define a category

$$\mathcal{M}(T' \to T) = \mathrm{Hom}\,(T' \times_T T' \rightrightarrows T', X_1 \rightrightarrows X_0)\,,$$

where of course an object of $\mathcal{M}(T' \to T)$ consists of a pair of morphisms $\varphi : T' \to X_0$ and $\Phi : T' \times_T T' \to X_1$ satisfying obvious compatibility conditions. Freeing T from the choice of a groupoid presentation, one arrives at the following definition of $\mathcal{M}(T)$:

$$\mathcal{M}(T) = \varinjlim_{\substack{T' \to T \\ \text{ét, surj}}} \mathcal{M}(T' \to T).$$

The datum of (8.10) and of the morphisms s, t, u, i, m is called a *groupoid presentation of the Deligne–Mumford stack* $\mathcal{M}$.

Among other things, the axioms (8.9) give, for each point $x \in X_0$, a group structure to the fiber

$$G_x = (s,t)^{-1}(x,x)\,, \tag{8.11}$$

where $(s,t) : X_1 \to X_0 \times X_0$. This is the isotropy group which we encountered in (4.4), in the orbifold context.

Groupoid presentations come particularly handy in performing various constructions on stacks and particularly in defining the notions of normalization of a stack and of quotient of a stack modulo the action of a finite group.

EXAMPLE (8.12) (Substacks). Let $\mathcal{M}$ be a Deligne–Mumford stack. A representable morphism $\mathcal{N} \to \mathcal{M}$ is a *closed* (resp., *open*) *immersion* if, for any morphism $S \to \mathcal{M}$ with S a scheme, $S \times_{\mathcal{M}} \mathcal{N} \to S$ is a closed (resp., open) immersion of schemes. It can be shown that under these circumstances $\mathcal{N}$ is necessarily a Deligne–Mumford stack; the easy proof is left to the reader. A *closed* (resp., *open*) *substack* of $\mathcal{M}$ is an equivalence class of closed (resp., open) immersions in $\mathcal{M}$ modulo isomorphism over $\mathcal{M}$. In other words, two immersions $\mathcal{N} \to \mathcal{M}$ and $\mathcal{A} \to \mathcal{M}$ define the same substack if and only if there is an isomorphism $\mathcal{N} \to \mathcal{A}$ such that the diagram

$$\begin{array}{ccc} \mathcal{N} & \xrightarrow{\;\cong\;} & \mathcal{A} \\ & \searrow \quad \swarrow & \\ & \mathcal{M} & \end{array}$$

commutes. Let $X \to \mathcal{M}$ be an atlas for the Deligne–Mumford stack $\mathcal{M}$, and let $\pi_1, \pi_2 : X \times_{\mathcal{M}} X \to X$ be the projections to the two factors. By definition, a closed (resp., open) substack of $\mathcal{M}$ yields a closed (resp., open) subscheme Y of X with the property that $\pi_1^{-1}(Y) = \pi_2^{-1}(Y)$. In fact, it can be shown (cf. Remark (2.10) in Chapter XIII) that giving a closed (resp., open) substack of $\mathcal{M}$ is equivalent to giving a closed (resp., open) subscheme of X with this property. More precisely, if we are given such a subscheme Y, the substack corresponding to it is the one defined by the groupoid presentation

$$\pi_1^{-1}(Y) = \pi_2^{-1}(Y) \overset{\pi_1}{\underset{\pi_2}{\rightrightarrows}} Y .$$

Example (8.13) (Normalization). Let $\mathcal{M}$ be a Deligne–Mumford stack with groupoid presentation (8.10). The normalization of $\mathcal{M}$ is obtained by normalizing both "space" (i.e., X_0) and "relations" (i.e., X_1). If $\widehat{X}_1$ and $\widehat{X}_0$ denote the normalizations of X_1 and X_0, respectively, the structure maps s, t, u, i, m lift to morphisms $\widehat{s}, \widehat{t} : \widehat{X}_1 \to \widehat{X}_0$ $\widehat{u} : \widehat{X}_0 \to \widehat{X}_1$, $\widehat{i} : \widehat{X}_1 \to \widehat{X}_1$, and $\widehat{m} : \widehat{X}_1 \, {}_s\times_t \widehat{X}_1 \to \widehat{X}_1$. These liftings satisfy the identities (4.2) necessary to make

$$\widehat{X}_1 \overset{\widehat{s}}{\underset{\widehat{t}}{\rightrightarrows}} \widehat{X}_0 \tag{8.14}$$

into a groupoid presentation of a Deligne–Mumford stack, since these identities are satisfied on open dense subsets of the various domains of definition. Moreover, the diagonal $\widehat{X}_1 \to \widehat{X}_0 \times \widehat{X}_0$ is separated. Thus, (8.14) indeed defines a Deligne–Mumford stack $\widehat{\mathcal{M}}$ which, by definition, is *the normalization of* $\mathcal{M}$. Of course, a posteriori, we have that

$$\widehat{X}_1 = \widehat{X}_0 \times_{\widehat{\mathcal{M}}} \widehat{X}_0 .$$

Example (8.15) (Quotient modulo a finite group action). We consider actions of a finite group G on a Deligne–Mumford stack $\mathcal{M} = (\mathcal{C}, p)$ in the following restrictive sense. We assume that to each element of G there corresponds a morphism of $\mathcal{M}$ into itself and that the following properties are satisfied. First of all, the identity element of G corresponds to the identity on $\mathcal{M}$. Secondly, the product in G corresponds to the composition of morphisms of $\mathcal{M}$, *in the strict sense*, and not just up to isomorphism of functors. From now on we assume that the category $\mathcal{C}$ has coproducts. To form the quotient Deligne–Mumford stack $[\mathcal{M}/G]$, we proceed as follows. First assume that there is a groupoid presentaton of the stack $\mathcal{M}$ such that the action of G lifts to G-actions μ_0 on X_0 and μ_1 on X_1 for which all structural morphisms are equivariant. In this case we proceed exactly as in the orbifold case, that is, we define the stack $[\mathcal{M}/G]$ as the one given by the groupoid presentation

$$Y_1 \underset{t_G}{\overset{s_G}{\rightrightarrows}} Y_0, \tag{8.16}$$

where

$$\begin{gathered} Y_0 = X_0\,, \quad Y_1 = G \times X_1\,, \\ s_G = s \circ pr_{X_1}\,, \quad t_G = t \circ \mu_1\,, \end{gathered}$$

the unit u_G and the inverse i_G are the obvious ones, and the composition map m_G is defined, in the scheme-theoretic context, exactly as we did in (4.7) in the orbifold context. The remaining problem is then to find a groupoid presentation of $\mathcal{M}$ to which we can lift the given G-action on $\mathcal{M}$. For this, start with an arbitrary presentation $Z_1 \underset{t}{\overset{s}{\rightrightarrows}} Z_0$ of $\mathcal{M}$. The étale morphism $f : Z_0 \to \mathcal{M}$ corresponds to an object ξ in $\mathcal{M}(Z_0)$. As G acts on $\mathcal{M}$, for every $\sigma \in G$, we may consider the object ξ^σ in $\mathcal{M}(Z_0)$ and its moduli morphism $f^\sigma : Z_0 \to \mathcal{M}$ which is also étale (cf. Exercise A-4). We then set

$$X_0 = \coprod_{\sigma \in G} Z_0 \cong G \times Z_0 \tag{8.17}$$

and define a new étale surjective morphism

$$f^G = \coprod_{\sigma \in G} f^\sigma \ : X_0 \to \mathcal{M}. \tag{8.18}$$

Then we set

$$X_1 = X_0 \times_{\mathcal{M}} X_0\,. \tag{8.19}$$

As an exercise, the reader should verify that $X_1 \rightrightarrows X_0$ is a groupoid presentation of $\mathcal{M}$ to which the action of G lifts.

Remark (8.20). It is important to remark that, by the way in which the quotient of a Deligne–Mumford stack $\mathcal{M}$ modulo a finite group G is defined, and in sharp contrast with the case, say, of affine schemes, the quotient morphism $\mathcal{M} \to [\mathcal{M}/G]$ is always étale. In a sense, since isotropy groups of points are part of the structure, the action of a finite group is always free.

As we have observed, the definition of Deligne–Mumford stack via groupoid presentations is the translation, in the algebraic context, of the definition of orbifold via orbifold groupoids. In fact, given a groupoid presentation $X_1 \rightrightarrows X_0$ of a *smooth* Deligne–Mumford stack, the underlying analytic object $(X_1)_{an} \rightrightarrows (X_0)_{an}$ is clearly an analytic orbifold groupoid. This is the sense in which we shall sometimes treat smooth Deligne–Mumford stacks, and in particular moduli stacks, as orbifolds.

9. Back to algebraic spaces.

The definition of Deligne–Mumford stack via groupoid presentations is obviously very close to the one of algebraic space. In fact, for a presentation $X_1 \rightrightarrows X_0$ to define a separated algebraic space, it suffices to add the requirement that $X_1 \to X_0 \times X_0$ be a closed immersion. Therefore, separated algebraic spaces can be regarded as a particular kind of Deligne–Mumford stacks. It turns out that separated algebraic spaces are precisely those Deligne–Mumford stacks $\mathcal{A}$ such that $\mathcal{A}(T)$ is a set for every scheme T (meaning that the only morphisms in the category $\mathcal{A}(T)$ are the identities). In this case, $T \mapsto \mathcal{A}(T)$ becomes a functor $Sch \to Sets$, and $\mathcal{A}$ is the stack associated to this functor. This is just the translation in the language of stacks of Artin's original definition of (separated) algebraic space.

Regarding an algebraic space as a stack has the considerable advantage of freeing us from relying on an accidental presentation and in particular makes it possible to define a morphism of algebraic spaces to be simply a morphism of stacks. In this sense we can make sense out of the assertion that a morphism of algebraic spaces is an open immersion, finite, proper, étale, and so on. On the other hand, as in the case of stacks, the groupoid presentation of an algebraic space comes in handy when defining the concept of normalization. It is clear that the normalization of an algebraic space, as defined in the previous section, is an algebraic space.

In this section we are going to give a proof of the following result.

THEOREM (9.1). *Let X be a reduced, separated algebraic space. Then there exist a normal scheme Z and a finite group G acting on Z such that X is isomorphic to the quotient Z/G.*

As an application, we shall then prove Theorem (2.9), which we restate here for the convenience of the reader.

THEOREM (9.2). *There exists a family of stable n-pointed genus g curves $\eta : \mathcal{X} \to Z$, parameterized by a normal scheme Z, whose moduli map*

$$m : Z \to \overline{M}_{g,n}$$

is finite and surjective. Moreover, we may choose Z so that $\overline{M}_{g,n}$ is the quotient of Z modulo the action of a finite group.

The proof of Theorem (9.1) relies on the concept of normalization of an irreducible algebraic space X in a finite field extension L of its field of rational function $K(X)$. To explain how this is defined, we need a basic fact about algebraic spaces, namely Zariski's connectedness theorem, whose proof can be found in [428], p. 233, Theorem 4.1.

THEOREM (9.3) (ZARISKI'S CONNECTEDNESS THEOREM). *Let $f: Y \to X$ be a proper morphism of separated algebraic spaces. Then there exists a factorization*

$$\begin{array}{ccc} Y & \xrightarrow{\ g\ } & T \\ & \searrow^{f} & \downarrow{h} \\ & & X \end{array} \tag{9.4}$$

where h is a finite morphism, and g is a proper morphism with connected fibers.

Now let X be an irreducible separated algebraic space, and let L be a finite extension of $K(X)$. We want to define the *normalization X^L of X in L*. We proceed as follows. As we know, there exists an open subspace U in X which is an affine scheme (cf. (3.4)). Let U^L be the normalization of U in L, and let V be a completion of U^L. Denote by Y the closure in $V \times X$ of the graph of $U^L \to X$. There is a proper morphism $f: Y \to X$, to which we apply Zariski's connectedness theorem to get a diagram (9.4). The morphism $U^L \to T$ is birational. We define X^L to be the normalization of T. Notice that the injective morphism $U \to X$ lifts to an open immersion $U^L \hookrightarrow X^L$, since both spaces are normal analytic varieties. The same argument proves the following slightly more general fact.

LEMMA (9.5). *Let X be an irreducible separated algebraic space. Let U be an irreducible scheme, let $f: U \to X$ be a quasi-finite dominant morphism, and let L be a finite extension of $K = K(U)$. Then f lifts to an open immersion $U^L \hookrightarrow X^L$.*

Now let us prove (9.1). We may as well assume that X is irreducible. Let $\coprod U_i \to X$ be an étale cover, where the U_i are irreducible and affine. Let L be a finite Galois extension of $K = K(X)$ containing all the $K(U_i)$. Let G be the Galois group of L over K. By the preceding lemma, the schemes U_i^L are openly immersed in X^L. We can use the action of G on X^L to move the U_i^L around; the result is a covering of X^L by open affine schemes. Thus, $Z = X^L$ is a scheme, and $X = Z/G$. This proves Theorem (9.1).

We now turn to Theorem (9.2). As we already mentioned, this result will be crucial in our proof of the projectivity of $\overline{M}_{g,n}$, and this is the only place where it will be used. We already know that $\overline{M}_{1,1}$ and $\overline{M}_2$

are projective, and hence, for our purposes, it is sufficient to give a simplified proof of Theorem (2.9) which applies to all cases except those of $\overline{M}_{1,1}$ and $\overline{M}_2$. The simplification is made possible by the fact that, as shown by Proposition (2.5), a general stable n-pointed curve of genus g has no nontrivial automorphisms except when $g = n = 1$, or $g = 2$, $n = 0$.

We then assume that $(g,n) \neq (2,0),(1,1)$. The argument is a slight variant of the proof of (9.1). Notice that, in the latter, one does not need the full strength of the fact that the U_i are étale over X, since all we need is that thery are quasi-finite over it. We then consider a finite number of standard algebraic Kuranishi families $\mathcal{Y}_i \to X_i$ such that $\coprod X_i \to \overline{M}_{g,n}$ is onto. We let L be a Galois extension of $K = K(\overline{M}_{g,n})$ containing all the $K(X_i)$, and we set $Z = X^L$. As in the proof of (9.1), we can say that each one of the X_i embeds as an open subset in Z. Moreover, the translates of the X_i under the action of $\mathrm{Gal}(L/K)$ cover Z. Let $U_1, \ldots, U_N$ be this cover. Each one of these open sets carries a family $\eta_i : \mathcal{W}_i \to U_i$ of stable n-pointed curves whose moduli map is the composition of $U_i \hookrightarrow Z$ and $Z \to \overline{M}_{g,n}$. We wish to patch together the various families $\eta_i : \mathcal{W}_i \to U_i$ to form the desired family $\eta : \mathcal{X} \to Z$. Since we are assuming that $(g,n) \neq (2,0),(1,1)$, this patching can be easily performed. In fact, it follows from Proposition (2.5) that, in this case, the curves equipped with nontrivial automorphisms are parameterized, in any Kuranishi family, by a proper analytic subset. In fact, the locus in question is just the projection of the Hilbert scheme of automorphisms of fibers, which, as we know from Theorem (5.1) in Chapter X, is proper over the base. Now suppose that $U_i \cap U_j \neq \emptyset$. For u outside a proper analytic subvariety Σ of $U_i \cap U_j$, there is a unique isomorphism between $\eta_i^{-1}(u)$ and $\eta_j^{-1}(u)$; this yields a canonical identification

$$\begin{array}{ccc} \mathcal{W}_i|_{U_i \cap U_j \smallsetminus \Sigma} & \xrightarrow{\ \sim\ } & \mathcal{W}_j|_{U_i \cap U_j \smallsetminus \Sigma} \\ & \searrow \quad \swarrow & \\ & U_i \cap U_j \smallsetminus \Sigma & \end{array}$$

Theorem (5.1) of Chapter X shows that this identification extends uniquely to all of $U_i \cap U_j$. The family $\eta : \mathcal{X} \to Z$ is thus constructed.

Exercise (9.6). Modify the proof of Theorem (2.9) so that it also covers the cases $(g,n) = (1,1)$ and $(g,n) = (2,0)$. (Hint: replace the original Z with a suitable finite cover.)

10. The universal curve, projections and clutchings.

In this section we rephrase the geometrical constructions discussed in Sections 6, 7, and 8 of Chapter X in the language of stacks and orbifolds.

The universal curve

We denote by $\overline{\mathcal{C}}_{g,P}$ the category whose objects are families of stable P-pointed curves of genus g

$$\xi : X \to S\,, \qquad \sigma_p : S \to X\,, \quad p \in P\,,$$

which are equipped with an extra section

$$\delta : S \to X$$

(on which we make no extra requirements, as, for instance, not intersecting the other sections or not hitting the nodes of the fibers). The morphisms in $\overline{\mathcal{C}}_{g,P}$ are defined in the usual way as cartesian squares

$$\begin{array}{ccc} X' & \xrightarrow{F} & X \\ {\scriptstyle \xi'}\downarrow & & \downarrow{\scriptstyle \xi} \\ S & \xrightarrow{f} & S' \end{array}$$

such that $F \circ \sigma'_y = \sigma_y \circ f$, $y \in P$, $F \circ \delta' = \delta \circ f$, the notation being self-explanatory. As usual, a functor $q : \overline{\mathcal{C}}_{g,P} \to Sch/\mathbb{C}$ is defined by assigning to a family of curves its parameter space. In this way the pair $(\overline{\mathcal{C}}_{g,P}, q)$ becomes a groupoid, which, for brevity, we simply denote with the symbol $\overline{\mathcal{C}}_{g,P}$. Theorem (8.21) in Chapter X shows that there is an isomorphism of groupoids

$$\lambda : \overline{\mathcal{C}}_{g,P} \to \overline{\mathcal{M}}_{g,P\cup\{x\}}\,,$$

so that, in particular, $\overline{\mathcal{C}}_{g,P}$ is a Deligne–Mumford stack. Explicitly, the definition of λ is as follows. Let $(\xi : X \to S,\ \sigma_p,\ p \in P,\ \delta)$ be an object in $\overline{\mathcal{C}}_{g,P}$. The stabilization procedure described in the above-mentioned section yields, in a functorial way, a family $\xi' : X' \to S$ of stable $(P\cup\{x\})$-pointed genus g curves. The assignement $\xi \mapsto \xi'$ defines λ. Theorem (8.21) in Chapter X says that assigning to a family $(\eta : Y \to S, \{\sigma_q : q \in P \cup \{x\}\})$ its xth contraction gives an inverse of λ up to isomorphism of functors.

The Deligne–Mumford stack $\overline{\mathcal{C}}_{g,P}$ has a scheme incarnation or, as one says, a coarse moduli space $\overline{C}_{g,P}$. As a set, $\overline{C}_{g,P}$ is the set of isomorphism classes of triples $(C, \{x_p\}_{p\in P}, x)$, where $(C, \{x_p\}_{p\in P})$ is a stable P-pointed curve of genus g, and x is a point on C. The analytic structure of $\overline{C}_{g,P}$ is given as follows. Let $[C, \{x_p\}_{p\in P}, x] \in \overline{C}_{g,P}$, and denote by G the automorphism group of $(C, \{x_p\}_{p\in P})$. Let $\xi : X \to B$ be a Kuranishi family for C. The group G acts on X. If U is a G-invariant neighborhood of $x \in X$, then a local patch for the analytic structure of $\overline{C}_{g,P}$ near $[C, \{x_p\}_{p\in P}, x]$ is given by U/G. By universality, the isomorphism λ drops to an isomorphism between moduli spaces

$$\lambda : \overline{C}_{g,P} \to \overline{M}_{g,P\cup\{x\}}.$$

The projectivity of $\overline{M}_{g,P\cup\{x\}}$ implies, in particular, that $\overline{C}_{g,P}$ is a scheme and that λ is an isomorphism of schemes.

Projections

In the language of stacks, the projection operation described in Section 6 of Chapter X gives morphisms

$$\mathrm{Pr}_x : \overline{\mathcal{M}}_{g,P\cup\{x\}} \longrightarrow \overline{\mathcal{M}}_{g,P}$$

for any finite set P and any $x \notin P$. This morphism is naturally called the xth *projection morphism.* As we observed in Lemma (6.10) of Chapter X, it makes good sense to ignore any number of sections so that, given a finite set L disjoint from P, we may define a projection morphism

$$\mathrm{Pr}_L : \overline{\mathcal{M}}_{g,P\cup L} \longrightarrow \overline{\mathcal{M}}_{g,P}.$$

An important property of the projection morphisms is the following.

LEMMA (10.1). *$\mathrm{Pr}_x : \overline{\mathcal{M}}_{g,P\cup\{x\}} \longrightarrow \overline{\mathcal{M}}_{g,P}$ is representable.*

Proof. To see why the lemma is true, it is convenient to identify the morphism Pr_x with

$$\overline{\mathcal{C}}_{g,P} \to \overline{\mathcal{M}}_{g,P}.$$

We must show that, given a scheme S and a morphism $\alpha : S \to \overline{\mathcal{M}}_{g,P}$, the stack $\overline{\mathcal{C}}_{g,P} \times_{\overline{\mathcal{M}}_{g,P}} S$ is (isomorphic to) a scheme. The morphism α corresponds to a family $X \to S$ of P-pointed stable curves. It is then essentially obvious that $\overline{\mathcal{C}}_{g,P} \times_{\overline{\mathcal{M}}_{g,P}} S$ is just X. Q.E.D.

The projection morphisms Pr_L can also be defined at the level of moduli spaces. In fact, since $\overline{M}_{g,P\cup L}$ is a coarse moduli space for $\overline{\mathcal{M}}_{g,P\cup L}$, the composition of Pr_L with the moduli map $m_L : \overline{\mathcal{M}}_{g,P} \to \overline{M}_{g,P}$ necessarily factors through the moduli map $m : \overline{\mathcal{M}}_{g,P\cup L} \to \overline{M}_{g,P\cup L}$. We denote the factoring morphism again with the symbol Pr_L:

$$\mathrm{Pr}_L : \overline{M}_{g,P\cup L} \longrightarrow \overline{M}_{g,P}$$

Clutchings

Our next objective is to describe the boundary $\partial M_{g,P} = \overline{M}_{g,P} \smallsetminus M_{g,P}$ of $\overline{M}_{g,P}$. Fix a stable P-marked, genus g dual graph Γ. We adopt the notation introduced at the beginning of Section 7 of Chapter X and in Definition (2.16) in the same chapter. We set

$$\overline{\mathcal{M}}_\Gamma = \prod_{v\in V} \overline{\mathcal{M}}_{g_v,L_v}, \qquad (10.2)$$

and we define a morphism

$$\xi_\Gamma : \overline{\mathcal{M}}_\Gamma \to \overline{\mathcal{M}}_{g,P} \tag{10.3}$$

as follows. An object η in $\overline{\mathcal{M}}_\Gamma(S)$ is the datum of a family

$$\eta_v : X_v \to S \tag{10.4}$$

of stable L_v-pointed curves of genus g_v for each $v \in V$. The morphisms in $\overline{\mathcal{M}}_\Gamma(S)$ are the isomorphisms between these families. The groupoid $\overline{\mathcal{M}}_\Gamma$ is obviously a Deligne–Mumford stack. The clutching procedure described in Section 7 of Chapter X yields a family

$$\xi_\Gamma(\eta) : X' \to S$$

of stable P-pointed genus g curves. The functoriality of this construction exactly says that ξ_Γ is a morphism of Deligne–Mumford stacks. It is in fact a finite morphism. We may also introduce a closed substack

$$\mathcal{D}_\Gamma \subset \overline{\mathcal{M}}_{g,P} \tag{10.5}$$

parameterizing the curves which are in the image of $\overline{\mathcal{M}}_\Gamma$ under ξ_Γ. An object in $\mathcal{D}_\Gamma(S)$ is the datum of a family

$$\sigma : X \to S$$

of stable P-pointed curves of genus g whose fibers have dual graphs which are specializations of Γ, in a sense to be made precise below. It is implicit in (8.12) that a closed substack of a Deligne–Mumford stack is itself a Deligne–Mumford stack. The codimension of $\mathcal{D}_\Gamma$ in $\overline{\mathcal{M}}_{g,P}$ is equal to the number $|E(\Gamma)|$ of edges of Γ. We denote by

$$\Delta_\Gamma \subset \overline{M}_{g,P} \tag{10.6}$$

the coarse moduli space of $\mathcal{D}_\Gamma$. If we assume for a moment what is going to be proved in the next chapter, namely that $\overline{M}_{g,P}$ is a projective scheme, then we see that Δ_Γ is a closed subscheme of $\overline{M}_{g,P}$. It will follow from the irreducibility of $\overline{M}_{g,n}$ (cf. Corollary (4.2) in Chapter XV or Corollary (11.9) in Chapter XXI) that Δ_Γ is always irreducible.

We shall often refer to the $\mathcal{D}_\Gamma$ (or the Δ_Γ) as the *boundary strata* of $\overline{\mathcal{M}}_{g,P}$ (or of $\overline{M}_{g,P}$). The simplest boundary strata are those of codimension 1, which correspond to the stable graphs with a single edge. These are of two kinds and are illustrated in Figure 4. First of all, there is the graph Γ_{irr} with only one vertex and one edge. In addition to this, there are graphs $\Gamma_\mathcal{P}$ attached to stable bipartitions $\mathcal{P} = \{(a, A), (b, B)\}$

of (g, P). These have two vertices, one of genus a and A-marked, the other of genus $b = g - a$ and B-marked, where $B = P \setminus A$.

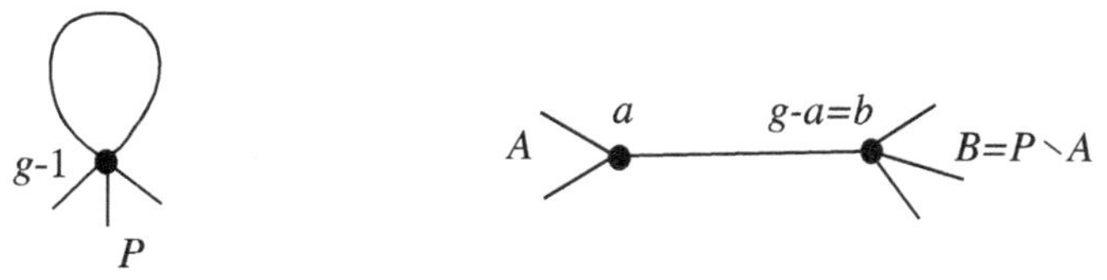

Figure 4.

In accordance with the conventions established in Section 2 of Chapter X, we usually write $\Gamma_{a,A}$ for $\Gamma_{\mathcal{P}}$. We normally write $\mathcal{D}_{irr}$ for $\mathcal{D}_{\Gamma_{irr}}$ and $\mathcal{D}_{a,A}$ for $\mathcal{D}_{\Gamma_{a,A}}$. The coarse counterparts of these substacks are the divisors Δ_{irr} and $\Delta_{a,A}$ introduced in Section 2. The clutching morphisms $\xi_{\Gamma_{irr}}$ and $\xi_{\Gamma_{a,A}}$ are usually written ξ_{irr} and $\xi_{a,A}$, respectively:

(10.7)

$$\xi_{irr} : \overline{\mathcal{M}}_{g-1,P\cup\{x,y\}} \to \overline{\mathcal{M}}_{g,P}\,, \qquad \xi_{a,A} : \overline{\mathcal{M}}_{a,A\cup\{x\}} \times \overline{\mathcal{M}}_{g-a,A^c\cup\{y\}} \to \overline{\mathcal{M}}_{g,P}\,.$$

We shall use the same symbols also to denote the corresponding morphisms between coarse moduli spaces.

We need to introduce some notions regarding dual graphs of stable curves and prove a number of auxiliary results about them. The reader should go back to the terminology introduced in the definition (2.16) of Chapter X.

Definition (10.8). *Let Γ be a graph. A subgraph $\Gamma' \subset \Gamma$ is a graph Γ' such that $L(\Gamma') \subset L(\Gamma)$, $V(\Gamma') \subset V(\Gamma)$, $\iota_{\Gamma'} = \iota_{\Gamma|L(\Gamma')}$, and $L_v(\Gamma') = L_v(\Gamma) \cap L(\Gamma')$ for every $v \in V(\Gamma')$.*

Suppose that we are given a P-marked dual graph Γ and a subgraph $I \subset \Gamma$ having no legs and containing all the vertices of Γ. Equivalently, we may think of I as being obtained from Γ by removing a subset from $E(\Gamma)$, together with all the legs. We want to construct a new graph Γ_I which is obtained from Γ by contracting to a point each connected component of I. Formally, we proceed as follows. Let W be the set of connected components of I. When we want to view $w \in W$ as a subgraph of Γ, we denote it by I_w; thus $I_w = w$. We let Γ_I be the P-marked dual graph defined as follows:

$$L(\Gamma_I) = L(\Gamma) \setminus L(I)\,, \quad V(\Gamma_I) = W\,, \quad L_w(\Gamma_I) = \bigcup_{v\in V(I_w)} (L_v(\Gamma) \setminus L_v(I_w))\,,$$

$$\iota_{\Gamma_I} = \iota_{\Gamma|L(\Gamma_I)}\,, \quad g_w(\Gamma_I) = g(I_w) \qquad \text{for} \quad w \in W\,,$$

while the indexing of the legs by P is the same as for Γ. Clearly, there is a continuous map $c_I : |\Gamma| \to |\Gamma_I|$ which contracts $|I_w|$ to the vertex $|w|$

and which induces a homeomorphism between $|\Gamma| \setminus |I|$ and $|\Gamma_I| \setminus W$, and we have

$$c_I^{-1}(E(|\Gamma_I|)) = E(|\Gamma|) \setminus E(|I|).$$

A P-marked dual graph Γ' is said to be a *specialization of* Γ if Γ is isomorphic to Γ'_I, for some $I \subset \Gamma'$. We call an *I-contraction* or simply a *contraction* any map $c : |\Gamma'| \to |\Gamma|$ which is the composition of c_I with the map $|\Gamma'_I| \to |\Gamma|$ induced by an isomorphism of graphs. In the following figure the graph I which is contracted is the one with bold edges.

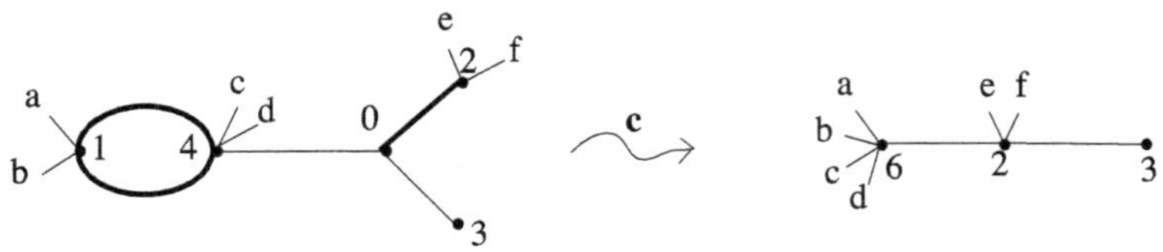

Figure 5.

Let Γ and Γ' be P-marked dual graphs and suppose that Γ is isomorphic to Γ'_I for some subgraph I of Γ'. Let σ' be an automorphism of Γ' which induces an automorphism of I. Then σ' descends to an automorphism σ of Γ, and we say that *σ' is a specialization of σ*.

We now come to stable curves. Let

(10.9) $$\pi : \mathcal{C} \to S\,, \quad \{\tau_p : S \to X\}_{p \in P}$$

be a family of stable P-pointed genus g curves parameterized by a scheme S. In what follows, to keep the notation simple, we will usually not mention the marked sections. Suppose that there exists a subvariety

(10.10) $$\Sigma \subset \mathrm{Sing}(\mathcal{C})$$

which is proper and étale over S. This subvariety cuts on each fiber C_s of π a finite set of nodes Σ_s, and we may form the graph $\mathrm{Graph}^{\Sigma_s}(C_s)$ associated to the P-pointed curve C_s and to the set Σ_s of nodes, as explained in Chapter X below definition (2.16). Let Γ be a fixed P-marked genus g graph. We shall say that Σ is a *weak* Γ*-marking* if each graph $\mathrm{Graph}^{\Sigma_s}(C_s)$ is isomorphic to Γ.

Families of stable P-pointed, genus g curves endowed with a weak Γ-marking form a stack, which we denote by $\mathcal{E}_\Gamma$.

Now suppose that the family (10.9) has been obtained from an object $X \to S$ of $\overline{\mathcal{M}}_\Gamma$ via clutching and that Σ is precisely the locus of nodes produced by the clutching operation. Then, by construction, the family (10.9) comes with a weak Γ-marking. However, in this case, there is more structure. First of all, the locus Σ is clearly a union of sections, and the same is true for its preimage in X, that is, in the partial normalization of $\mathcal{C}$ along Σ. At this point, one can attach to the family a graph $\mathrm{Graph}^{\Sigma}(\mathcal{C})$ repeating, word by word, the construction that led to the graph associated to a single curve and a choice of nodes on it. In effect, here we are treating $\mathcal{C}$ as a single curve and Σ as a set of nodes. Notice that there is a canonical isomorphism between Γ and $\mathrm{Graph}^{\Sigma}(\mathcal{C})$ (and also between $\mathrm{Graph}^{\Sigma}(\mathcal{C})$ and $\mathrm{Graph}^{\Sigma_s}(\mathcal{C}_s)$ for each s in S). One expresses all of this by saying that the family (10.9) is endowed with a Γ-*marking*. It is evident that giving an object in $\mathcal{M}_\Gamma(S)$ is equivalent to giving a family of stable P-pointed genus g curves over S together with a Γ-marking.

By what we have said, the morphism $\xi_\Gamma : \overline{\mathcal{M}}_\Gamma \to \overline{\mathcal{M}}_{g,P}$ can be viewed as the composition of two forgetful morphisms

$$\overline{\mathcal{M}}_\Gamma \to \mathcal{E}_\Gamma \to \mathcal{D}_\Gamma \subset \overline{\mathcal{M}}_{g,P},$$

where the first one forgets the Γ-marking while still keeping the corresponding weak one, and the second forgets the extra Γ-structure altogether.

It is also clear that the automorphism group $\mathrm{Aut}(\Gamma)$ acts on $\overline{\mathcal{M}}_\Gamma$ in the sense of Example (8.15). In the next proposition we will see that the morphism $\overline{\mathcal{M}}_\Gamma \to \mathcal{E}_\Gamma$ comes from taking the quotient of $\overline{\mathcal{M}}_\Gamma$ by $\mathrm{Aut}(\Gamma)$, while $\mathcal{E}_\Gamma \to \mathcal{D}_\Gamma$ is the normalization morphism.

PROPOSITION (10.11).

i) *$\mathcal{E}_\Gamma$ is the normalization of the substack $\mathcal{D}_\Gamma \subset \overline{\mathcal{M}}_{g,P}$.*
ii) *The morphism $\overline{\mathcal{M}}_\Gamma \to \mathcal{E}_\Gamma$ can be identified with the quotient morphism $\overline{\mathcal{M}}_\Gamma \to [\overline{\mathcal{M}}_\Gamma/\mathrm{Aut}(\Gamma)]$.*

Before proving this proposition, in order to form an intuition of what is going on, let us consider the clutching morphism at the more concrete level of moduli spaces. Since

$$\overline{M}_\Gamma = \prod_{v\in V} \overline{M}_{g_v,L_v}$$

is a coarse moduli space for $\overline{\mathcal{M}}_\Gamma$, the same reasoning we used in the context of the projection morphisms tells us that the clutching morphisms ξ_Γ descend to scheme morphisms

$$\xi_\Gamma : \overline{M}_\Gamma \to \overline{M}_{g,P} \tag{10.12}$$

between the corresponding coarse moduli spaces and that the image of $\overline{M}_\Gamma$ under ξ_Γ is Δ_Γ. In general, the morphisms ξ_Γ are not injective. Let us look at Figure 6 below. In the first row we consider

$$\xi_\Gamma : \overline{M}_{g-1,\{a,b\}} \to \overline{M}_g,$$

and we see that $\xi_\Gamma[C;x_a,x_b]=\xi_\Gamma[C;x_b,x_a]$. Therefore, in this case, ξ_Γ is generically two-to-one. But, as the second row shows, the cardinality of the fiber may jump. In fact,

$$\xi_\Gamma[C;x_a,x_b]=\xi_\Gamma[C';x_a,x_b]=\xi_\Gamma[C;x_b,x_a]=\xi_\Gamma[C';x_b,x_a],$$

where $(C;x_a,x_b)$ (resp. $(C';x'_a,x'_b)$) is obtained from a curve $(C_0;x_a,x_b,x'_a,x'_b)$ identifying x'_a with x'_b (resp x_a with x_b). We next consider the map

$$\xi_\Gamma:\ \overline{M}_{h,\{a\}}\times\overline{M}_{h+1,\{b\}}\to\overline{M}_{2h+1},$$

where Γ consists of two vertices joined by one edge. This morphism is generically injective, but, as we see in the third row of Figure 6, it is not injective. In fact, if E is elliptic and D is the curve obtained from C and E by identifying x_c with x_d, and if D' is the curve obtained from C' and E by identifying x_a with x_b, we get

$$\xi_\Gamma([D,x_a],[C',x_b])=\xi_\Gamma([D',x_d],[C,x_c])\,.$$

What these pictures suggest, and what the reader will easily verify, is that the morphism ξ_Γ factors through a generically injective morphism

$$\overline{\xi_\Gamma}:\overline{M}_\Gamma/\operatorname{Aut}(\Gamma)\longrightarrow\overline{M}_{g,P}\,. \tag{10.13}$$

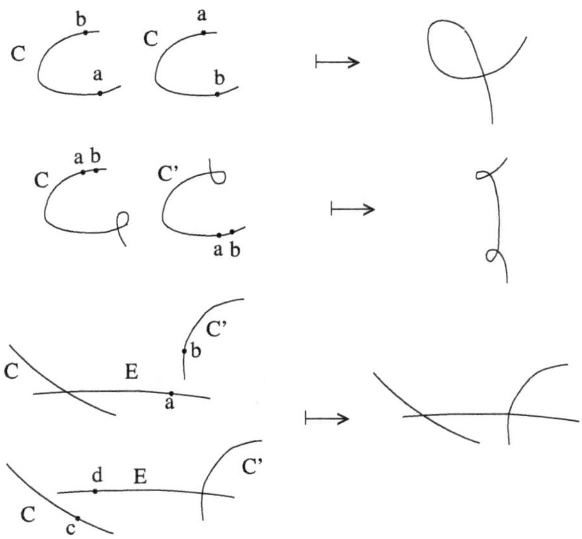

Figure 6.

In the above two examples we see that the subvariety Δ_Γ folds into itself forming a double point p.

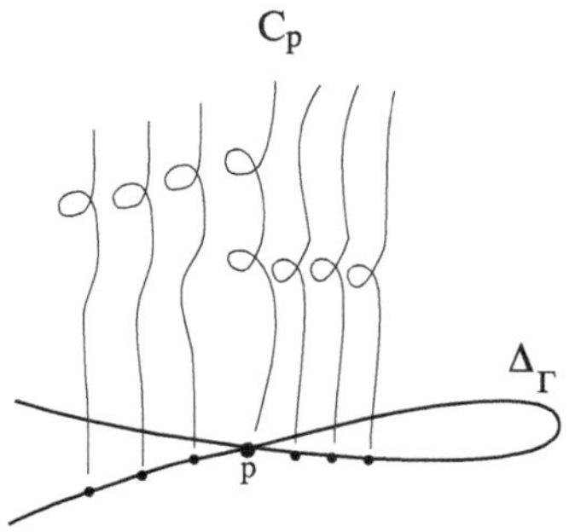

Figure 7.

The two branches of Δ_Γ at p correspond to the smoothing of one of the two nodes of C_p.

Proof of Proposition (10.11). Let us start by examining the substack $\mathcal{D}_\Gamma \subset \overline{\mathcal{M}}_{g,P}$. To construct an étale surjective morphism $X_\Gamma \to \mathcal{D}_\Gamma$ from a scheme X_Γ (i.e., an atlas for $\mathcal{D}_\Gamma$), we start from an atlas $X \to \overline{\mathcal{M}}_{g,P}$ consisting of the union of smooth bases of algebraic Kuranishi families (the "slices" of the Hilbert scheme $\mathrm{H}_{\nu,g,n}$ of ν-log-canonical stable P-pointed curves of genus g), and we look at the corresponding family of curves $\pi : \mathcal{C} \to X$. We then let X_Γ be the subvariety of X defined by

$$X_\Gamma = \{s \in X : \mathrm{Graph}(C_s) \text{ is a specialization of } \Gamma\},$$

and we denote by $\eta : \mathcal{C}_\Gamma \to X_\Gamma$ the restriction of π to X_Γ. We then get a surjective morphism $X_\Gamma \to \mathcal{D}_\Gamma$. Let us look more closely at the morphism

$$X_\Gamma \times_{\mathcal{D}_\Gamma} X_\Gamma = \mathbf{Isom}_{X_\Gamma \times X_\Gamma}(p_1^*\eta, p_2^*\eta) \longrightarrow X_\Gamma, \tag{10.14}$$

where p_1 and p_2 are the natural projections. The reason why this is an étale cover stems from the analogue, in the present setting, of Proposition (3.10) and therefore, in the final analysis, from the analogue of Lemma (3.11). Using the notation of that lemma, where $U \subset X$ denotes the basis of a local Kuranishi family, we set $U_\Gamma = U \cap X_\Gamma$ and let $\alpha : \mathcal{C}_{U_\Gamma} \to U_\Gamma$ be the restriction of the family η to U_Γ. If C is the central fiber of this family and $H = \mathrm{Aut}(C)$, then, also in this case, H acts on U_Γ, and we have an isomorphism between $H \times U_\Gamma$ and $\mathbf{Isom}_{U_\Gamma \times U_\Gamma}(p_1^*\alpha, p_2^*\alpha)$, under which the natural projection

$$U_\Gamma \times_{\mathcal{D}_\Gamma} U_\Gamma = \mathbf{Isom}_{U_\Gamma \times U_\Gamma}(p_1^*\alpha, p_2^*\alpha) \longrightarrow U_\Gamma \tag{10.15}$$

is just the projection

$$H \times U_\Gamma \to U_\Gamma\,. \tag{10.16}$$

This is the reason why (10.14) and, therefore, $X_\Gamma \to \mathcal{D}_\Gamma$ are étale. We next describe the normalization $\widehat{\mathcal{D}}_\Gamma$ of $\mathcal{D}_\Gamma$. Let us go back to Section 8 and in particular to the subsection on the orbifold-like definition of Deligne–Mumford stacks, where we explained what is the normalization of a stack. According to that definition, $\widehat{\mathcal{D}}_\Gamma$ is given by the groupoid presentation

$$X_\Gamma \widehat{\times_{\mathcal{D}_\Gamma}} X_\Gamma \rightrightarrows \widehat{X}_\Gamma\,,$$

where the hat stands for normalization. To analyze this normalization, the relevant local picture is given by the normalization of (10.15), that is, the normalization of (10.16). But the picture of U_Γ is the one of a certain number k of $(3g-3+n-\delta)$-dimensional linear subspaces in $\mathbb{C}^{3g-3+n}$ meeting transversally at the origin. If $|\operatorname{Sing}(C)| = \delta'$, the number k is the number of subsets I of $E(\operatorname{Graph}(C))$ consisting of $\delta' - \delta$ edges, contracting which, we obtain a graph isomorphic to Γ. Let then

$$U_\Gamma = U_1 \cup \cdots \cup U_k \tag{10.17}$$

be the decomposition of U_Γ in linear branches. The normalization $\widehat{U}_\Gamma$ is just the disjoint union of $U_1, \dots, U_k$, so that the normalization of (10.15), that is, the normalization of (10.16), is nothing but the projection

$$H \times \widehat{U}_\Gamma \to \widehat{U}_\Gamma\,. \tag{10.18}$$

But now, by virtue of the local description of $\widehat{U}_\Gamma$, the pullback $\widehat{\eta}$ to $\widehat{X}_\Gamma$ of the family η over X_Γ is a family of curves with weak Γ-marking, and we have

$$U_\Gamma \widehat{\times_{\mathcal{D}_\Gamma}} U_\Gamma = \widehat{H \times U_\Gamma} = H \times \widehat{U}_\Gamma = \mathbf{Isom}^{\mathcal{E}_\Gamma}_{\widehat{U}_\Gamma \times \widehat{U}_\Gamma}(p_1^*\widehat{\alpha}, p_2^*\widehat{\alpha}) \longrightarrow \widehat{U}_\Gamma, \tag{10.19}$$

where $\widehat{\alpha}$ is the pullback of α to $\widehat{U}_\Gamma$, and $\mathbf{Isom}^{\mathcal{E}_\Gamma}$ stands for isomorphisms respecting the weak Γ-marking. This tells us that the two projections

$$\begin{aligned} &X_\Gamma \widehat{\times_{\mathcal{D}_\Gamma}} X_\Gamma \longrightarrow \widehat{X}_\Gamma\,, \\ &\widehat{X}_\Gamma \times_{\mathcal{E}_\Gamma} \widehat{X}_\Gamma = \mathbf{Isom}^{\mathcal{E}_\Gamma}_{\widehat{X}_\Gamma \times \widehat{X}_\Gamma}(p_1^*\widehat{\eta}, p_2^*\widehat{\eta}) \longrightarrow \widehat{X}_\Gamma \end{aligned} \tag{10.20}$$

may be identified. The fact that the second projection is étale tells us that the morphism $\widehat{X}_\Gamma \to \mathcal{E}_\Gamma$ is étale. On the other hand, the fact that the two projections in (10.20) can be identified tells us that $\mathcal{E}_\Gamma$ and the normalization $\widehat{\mathcal{D}}_\Gamma$ of $\mathcal{D}_\Gamma$ have the same groupoid presentation and are therefore isomorphic.

It now remains to show that $[\overline{\mathcal{M}}_\Gamma/\operatorname{Aut}(\Gamma)] \cong \mathcal{E}_\Gamma$. We set $G = \operatorname{Aut}(\Gamma)$. We will prove that $[\overline{\mathcal{M}}_\Gamma/G]$ is isomorphic to $\mathcal{E}_\Gamma$ by showing that $[\overline{\mathcal{M}}_\Gamma/G]$

and $\mathcal{E}_\Gamma$ have a common groupoid presentation. By the definition of $\overline{\mathcal{M}}_\Gamma$, there is a Γ-marked family of curves

$$\eta : \mathcal{C} \to Y_0 \tag{10.21}$$

for which the moduli morphism $m : Y_0 \to \overline{\mathcal{M}}_\Gamma$ is étale and surjective. As we explained in Section 8, we may assume that it is G-equivariant, and we get a groupoid presentation $Y_1 \rightrightarrows Y_0$ of $[\overline{\mathcal{M}}_\Gamma/G]$ by setting

$$Y_1 = G \times Y_0 \times_{\overline{\mathcal{M}}_\Gamma} Y_0 .$$

On the other hand, we have

$$G \times Y_0 \times_{\overline{\mathcal{M}}_\Gamma} Y_0 = G \times \mathbf{Isom}^\Gamma_{Y_0 \times Y_0}(p_1^*\eta, p_2^*\eta) ,$$

where p_1 and p_2 are the natural projections, and $\mathbf{Isom}^\Gamma$ stands for isomorphisms respecting the Γ-marking. We now look at $\mathcal{E}_\Gamma$. Remembering only the weak Γ-marking, the family (10.21) may be considered as an object of $\mathcal{E}_\Gamma(Y_0)$, yielding a morphism $Y_0 \to \mathcal{E}_\Gamma$, which is readily seen to be étale. In fact, locally, Y_0 looks like a product $\prod_{v \in V} U_v$, where U_v is the basis of a standard Kuranishi family for a curve C_v in $\overline{\mathcal{M}}_{g_v, L_v}$, and this product, in turn, is isomorphic to one of the branches U_i of $U_\Gamma \subset U$, where U is the basis for a Kuranishi family of the curve C obtained via the clutching ξ_Γ, starting from the C_v. Finally, we have

$$Y_0 \times_{\mathcal{E}_\Gamma} Y_0 = \mathbf{Isom}^{\mathcal{E}_\Gamma}_{Y_0 \times Y_0}(p_1^*\eta, p_2^*\eta) = G \times \mathbf{Isom}^\Gamma_{Y_0 \times Y_0}(p_1^*\eta, p_2^*\eta) .$$

Q.E.D.

COROLLARY (10.22). *Let Γ be a stable P-marked dual graph of genus g. Assume that* $\mathrm{Aut}(\Gamma) = \{\mathrm{id}_\Gamma\}$. *Furthermore, assume that, for every graph Γ' which is a specialization of Γ, all the elements in* $\mathrm{Aut}(\Gamma')$ *are specializations of* id_Γ. *Then $\xi_\Gamma : \overline{\mathcal{M}}_\Gamma \to \overline{\mathcal{M}}_{g,P}$ is a closed immersion.*

Proof. The fact that $\mathrm{Aut}(\Gamma) = \{\mathrm{id}_\Gamma\}$ tells us that $\overline{\mathcal{M}}_\Gamma$ is the normalization of the closed substack $\mathcal{D}_\Gamma$. Let us then prove that $\mathcal{D}_\Gamma$ is a smooth Deligne–Mumford stack. Going back to the local picture (10.17), we must prove that $k = 1$. But now the various branches U_i of U_Γ correspond to the various contractions $c_I : |\mathrm{Graph}(C)| \to |\Gamma|$, where $[C] \in \overline{M}_\Gamma$. As one can easily check, any two contractions c_I and c_J are linked by an automorphism $\sigma \in \mathrm{Graph}(C)$. Since, by assumption, σ is a specialization of id_Γ, we must have $I = J$.

Q.E.D.

In studying the geometry of $\overline{\mathcal{M}}_{g,P}$ it is important to describe how the various boundary strata intersect, not only the divisorial ones. We must therefore understand cartesian diagrams of the type

$$\begin{array}{ccc}
\overline{\mathcal{M}}_\Gamma \times_{\mathcal{M}_{g,P}} \overline{\mathcal{M}}_{\Gamma'} & \longrightarrow & \overline{\mathcal{M}}_\Gamma \\
\big\downarrow & & \big\downarrow \xi_\Gamma \\
\overline{\mathcal{M}}_{\Gamma'} & \xrightarrow{\ \xi_{\Gamma'}\ } & \mathcal{M}_{g,P}
\end{array}$$

Set $\overline{\mathcal{M}}_{\Gamma\Gamma'} = \overline{\mathcal{M}}_\Gamma \times_{\mathcal{M}_{g,P}} \overline{\mathcal{M}}_{\Gamma'}$. We let $\mathcal{G}_{\Gamma\Gamma'}$ denote a set of representatives for the isomorphism classes of triples (Λ, c, c') where Λ is a P-marked, genus g, dual graph, and $c : |\Lambda| \to |\Gamma|$ and $c' : |\Lambda| \to |\Gamma'|$ are contractions. We also insist that

$$E(|\Lambda|) = c^{-1}(E(|\Gamma|) \cup {c'}^{-1}(E(|\Gamma'|)\,. \tag{10.23}$$

This simply means that, given a curve C with dual graph equal to Λ, smoothing the nodes corresponding to the edges of ${c'}^{-1}E(|\Gamma'|) \smallsetminus c^{-1}(E(|\Gamma|)$ produces a curve whose graph is Γ, and similarly when the roles of Γ and Γ' are reversed. An isomorphism between triples (Λ, c, c') and $(\Lambda_1, c_1, {c'}_1)$ is an isomorphism between Λ and Λ' commuting with the contractions. Figure 8 below describes $\mathcal{G}_{\Gamma\Gamma'}$ in four examples, where Γ and Γ' are two unpointed graphs (we are assuming that $a, b, g > 1$ and that $a \neq b$).

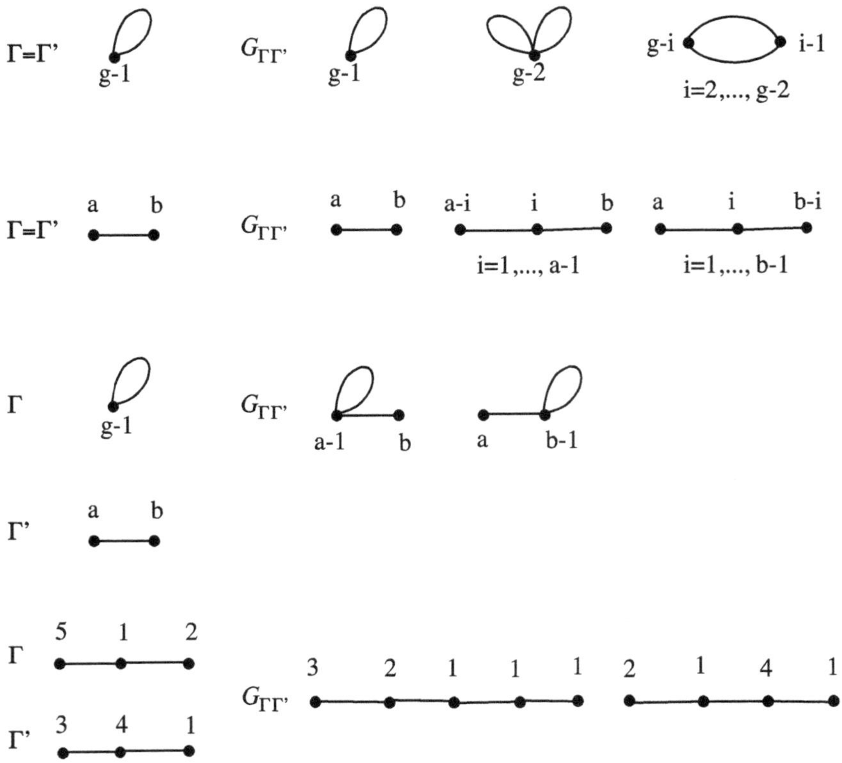

Figure 8.

We will now show how an intersection of two boundary strata decomposes into a disjoint union of boundary strata.

PROPOSITION (10.24). *There is an isomorphism*

$$\overline{\mathcal{M}}_{\Gamma\Gamma'} \cong \coprod_{\Lambda\in\mathcal{G}_{\Gamma\Gamma'}} \overline{\mathcal{M}}_{\Lambda}.$$

Proof. It suffices to give isomorphisms between $\overline{\mathcal{M}}_{\Gamma\Gamma'}(T)$ and $\coprod_{\Lambda\in\mathcal{G}_{\Gamma\Gamma'}} \overline{\mathcal{M}}_{\Lambda}(T)$, where T is a scheme. An object of $\overline{\mathcal{M}}_{\Gamma\Gamma'}(T)$ is a triple (ξ,ξ',φ), where ξ (resp. ξ') is a family of Γ-marked (resp. Γ'-marked) stable P-pointed genus g curves, parameterized by T, and $\varphi : \xi \to \xi'$ is a T-isomorphism. Now suppose $\xi : \mathcal{C} \to T$ is an object in $\overline{\mathcal{M}}_{\Lambda}(T)$. This means that we are given a subvariety Σ of $\mathrm{Sing}(\mathcal{C})$, proper and étale over T, whose inverse image in the partial normalization along Σ itself is a union of sections, plus an isomorphism $\gamma : \mathrm{Graph}^{\Sigma}(\mathcal{C}) \xrightarrow{\sim} \Lambda$. Composing γ with the given contractions $c : \Lambda \to \Gamma$ and $c' : \Lambda \to \Gamma'$ provides two subsets

$$\Sigma_1 = (c\gamma)^{-1}(E(\Gamma)) \quad \text{and} \quad \Sigma_2 = (c'\gamma)^{-1}(E(\Gamma'))$$

such that $\Sigma = \Sigma_1 \cup \Sigma_2$, and isomorphisms $\gamma_1 : \mathrm{Graph}^{\Sigma_1}(\mathcal{C}) \xrightarrow{\sim} \Gamma$ and $\gamma_2 : \mathrm{Graph}^{\Sigma_2}(\mathcal{C}) \xrightarrow{\sim} \Gamma'$. This exhibits ξ as an object in both $\overline{\mathcal{M}}_{\Gamma}$ and $\overline{\mathcal{M}}_{\Gamma'}$ and therefore as an object of $\overline{\mathcal{M}}_{\Gamma\Gamma'}$. Conversely, given an object (ξ,ξ',φ) in $\overline{\mathcal{M}}_{\Gamma\Gamma'}(T)$, the family ξ is endowed with a Γ-marking and, via φ, with a Γ'-marking, that is, with isomorphisms $\gamma : \mathrm{Graph}^{\Sigma_1}(\mathcal{C}) \xrightarrow{\sim} \Gamma$ and $\gamma' : \mathrm{Graph}^{\Sigma_2}(\mathcal{C}) \xrightarrow{\sim} \Gamma'$. From these one gets contractions $\bar{c} : \mathrm{Graph}^{\Sigma_1\cup\Sigma_2}(\mathcal{C}) \to \Gamma$ and $\bar{c}' : \mathrm{Graph}^{\Sigma_1\cup\Sigma_2}(\mathcal{C}) \to \Gamma'$, and therefore a unique element $(\Lambda, c, c') \in \mathcal{G}_{\Gamma\Gamma'}$ and a unique isomorphism $\theta : \mathrm{Graph}^{\Sigma_1\cup\Sigma_2}(\mathcal{C}) \to \Lambda$ such that $c\theta = \bar{c}$ and $c'\theta = \bar{c}'$. The reader will check that these associations carry over to morphisms, thus establishing the desired isomorphism between $\overline{\mathcal{M}}_{\Gamma\Gamma'}(T)$ and $\coprod_{\Lambda\in\mathcal{G}_{\Gamma\Gamma'}} \overline{\mathcal{M}}_{\Lambda}(T)$. Q.E.D.

Looking at Figure 8, in the first two lines we see the decomposition in strata for the self-intersection of the boundary divisors (in the unpointed case). There we see one "excess intersection" component, while the remaining components are "transverse." In the third example we see a bona fide transverse intersection of two boundary divisors. In the fourth example we are intersecting two codimension 2 boundary strata in $\overline{\mathcal{M}}_8$, and we get a transversal codimension 4 component and an "excess intersection" component of codimension 3.

We end this section by proving the following result.

PROPOSITION (10.25). *The clutching morphisms $\xi_\Gamma : \overline{\mathcal{M}}_{\Gamma} \to \overline{\mathcal{M}}_{g,P}$ are representable.*

Proof. First of all, we are going to show that any clutching morphism can be factored as the composition of a closed embedding with a projection map. Let Γ be the dual graph of a stable P-pointed genus g curve. As usual, we denote by $E = E(\Gamma)$ the set of edges of Γ. We construct a new graph $\widehat{\Gamma}$ in the following way. Fix an edge $\mathbf{l} = \{l, l'\} \in E$. Consider two graphs Γ_l and $\Gamma_{l'}$ as in Figure 9. Split $\mathbf{l}$ in the two halves l and l', then join l with l_∞, l' with l'_∞ and l_0 with l'_0.

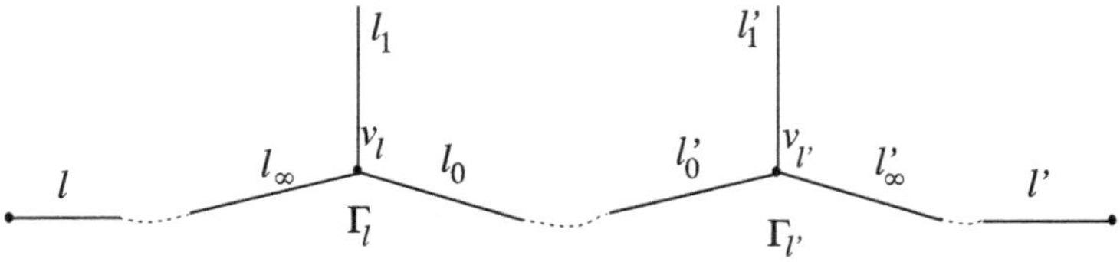

Figure 9.

Repeat this operation for every edge of Γ to obtain $\widehat{\Gamma}$ (see Figure 10 for an example of how to pass from Γ to $\widehat{\Gamma}$).

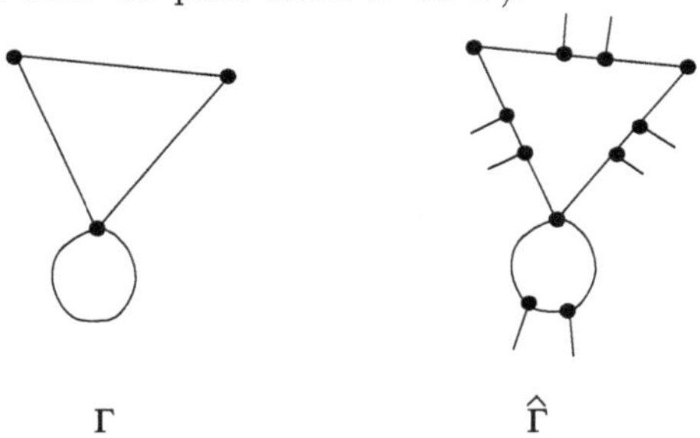

Figure 10.

By definition, $\widehat{\Gamma}$ is marked by the set $P \cup H$, where $H = H(\Gamma)$ is the set of half-edges of Γ which are not legs. From the construction it follows that $\mathrm{Aut}(\widehat{\Gamma}) = \{\mathrm{id}_{\widehat{\Gamma}}\}$. We have a decomposition

$$\xi_\Gamma = \pi_H \circ \xi_{\widehat{\Gamma}} \circ \iota_\Gamma, \tag{10.26}$$

where

$$\iota_\Gamma : \overline{\mathcal{M}}_\Gamma = \prod_{v \in V} \overline{\mathcal{M}}_{g_v, L_v} \longrightarrow \overline{\mathcal{M}}_{\widehat{\Gamma}} = \prod_{v \in V} \overline{\mathcal{M}}_{g_v, L_v} \times \prod_{\{l,l'\} \in E} \left(\overline{\mathcal{M}}_{0,\{l_0,l_1,l_\infty\}} \times \overline{\mathcal{M}}_{0,\{l'_0,l'_1,l'_\infty\}} \right)$$

is the natural isomorphism,

$$\xi_{\widehat{\Gamma}} : \overline{\mathcal{M}}_{\widehat{\Gamma}} = \prod_{v \in V} \overline{\mathcal{M}}_{g_v, L_v} \times \prod_{\{l,l'\} \in E} \left(\overline{\mathcal{M}}_{0,\{l_0,l_1,l_\infty\}} \times \overline{\mathcal{M}}_{0,\{l'_0,l'_1,l'_\infty\}} \right) \longrightarrow \overline{\mathcal{M}}_{g, P \cup H}$$

is the morphism defined by $\widehat{\Gamma}$, and π_H is the natural projection from $\overline{\mathcal{M}}_{g,P\cup H}$ to $\overline{\mathcal{M}}_{g,P}$.

Since ι_Γ is an isomorphism, and since we already proved that projections are representable, it is enough to prove that $\xi_{\widehat{\Gamma}}$ is representable. This follows from Corollary (10.22), which shows that $\xi_{\widehat{\Gamma}}$ is a closed immersion and hence is representable.

Q.E.D.

11. Bibliographical notes and further reading.

Our basic references for algebraic spaces are Artin [37,38] and Knutson [428]. A proof of Theorem (9.1) can be found in Laumon and Moret-Bailly [462], Corollaire 16.6.2.

A good introduction to the theory of orbifold is Adem, Leida, and Ruan [3]. Satake's seminal papers on the subject are [612,614].

The theory of stacks was initiated by Deligne and Mumford [167], based in part on ideas of Grothendieck. Good general references are Gillet [298], Vistoli [671], Edidin [190], Laumon and Moret-Bailly [462], Fantechi [242], Canonaco [94], Fantechi–Göttsche–Illusie–Kleiman–Nitsure–Vistoli [243], and the online Stacks Project [397]

A good reference for the theory of descent can be found in [243]. Our treatment of descent for quasicoherent modules is taken directly from Grothendieck's (cf. [326], exposé VIII).

Theorem (2.9) is a special instance of a more general result found in Kollár [439], whose treatment we follow in the proof. A vast generalization of our method of construction of the moduli space of curves as an algebraic space is given in Keel and Mori [410].

As already explained in the bibliographical notes to Chapter X, the projection, clutching, and stabilization constructions were first studied in Knudsen [426]. Our treatment of the intersections of boundary strata is inspired by the one by Graber and Pandharipande in Appendix A of [307].

12. Exercises.

A. **Orbifolds and stacks**

A-1. Let G be a finite group acting holomorphically on a complex manifold M. Suppose that M/G is Hausdorff. Set $X_0 = M$ and $X_1 = G\times M$. Let $s : X_1 = G\times M \to M$ be the projection, and $t : X_1 = G\times M \to M$ be the action. The composition rule, the unit, and the inverse are the obvious ones. Verify that this defines an orbifold structure on M/G. We call the orbifold described in this way, the *quotient orbifold of M modulo G*, and we shall denote it by $[M/G]$.

A-2. Let $f : |\mathbb{X}| \to M$ be an orbifold structure on a complex manifold M. Show that every point in M has a neighborhood with an orbifold structure of the form $[B/G_x]$ where $x \in X_0$, and B is a chart around x.

A-3. Using Lemma (5.1), give a detailed proof of the fact that, given a groupoid (or a Deligne–Mumford stack) $\mathcal{M}$ and a morphism $S \to \mathcal{M}$, the stack $\mathcal{M} \times_{\mathcal{M}} S$ is represented by S.

A-4. Let $h : \mathcal{N} \to \mathcal{N}'$ be an isomorphism of DM stacks. Show that if a representable morphism of Deligne–Mumford stacks $f : \mathcal{M} \to \mathcal{N}$ satisfies a property **P** of morphisms of schemes which is stable under base change, then also fh satisfies **P**.

A-5. Verify that G_x given in (4.4) and in (8.11) is indeed a group.

A-6. Show that (8.17), (8.18), and (8.19), together with appropriate structural maps s_G, t_G, u_G, i_G, and m_G, define a groupoid presentation of $\mathcal{M}$ to which the action of G lifts.

A-7. Show that a closed substack of a Deligne–Mumford stack is a Deligne–Mumford stack.

B. **Genus 0 and 1.**

B-1. Show that $M_{0,n}$ can be described as follows. Let X_n be the n-fold product of $\mathbb{P}^1$, minus the big diagonal. The group $PGL(2)$ acts naturally on X_n and $M_{0,n}$ is just the quotient.

B-2. Show that $\overline{M}_{0,n}$ is smooth.

B-3. Construct $\overline{M}_{0,n+1}$ by blowing up the diagonal in $\overline{M}_{0,n} \times_{\overline{M}_{0,n-1}} \overline{M}_{0,n}$.

B-4. Show that $M_{1,1}$ can be naturally identified, via the period map, with the quotient of the upper half-plane $\mathbb{H}$ modulo the action of $SL(2, \mathbb{Z})$. Deduce that $M_{1,1} = \mathbb{C}$ and $\overline{M}_{1,1} = \mathbb{P}^1$.

B-5. Construct explicitly Kuranishi families for curves in $M_{1,1}$, carefully describing the automorphism groups acting on them.

B-6. Describe the stack $\overline{\mathcal{M}}_{1,1}$.

C. **Hyperelliptic curves**

C-1. Let $H_g \subset M_g$ be the locus of hyperelliptic curves of genus g.

a) Show that $H_g \subset M_g$ is a $(2g-1)$-dimensional subvariety of M_g.
b) Show that H_g can be identified with the quotient of $M_{0,2g+2}$ modulo the action of the symmetric group $\mathfrak{S}_{2g+2}$.

C-2. Define the stack $\mathcal{H}_g$ of smooth hyperelliptic curves of genus g. Show that the hyperelliptic involution defines a nontrivial $\mathbb{Z}_2$-action on $\mathcal{H}_g$ and that H_g is a coarse moduli space for both $\mathcal{H}_g$ and $\mathcal{H}_g/\mathbb{Z}_2$.

D. Strata of moduli spaces.

In this series, we will ask the reader simply to list the strata of various moduli spaces $\overline{M}_{g,n}$. In each case we ask to make a chart listing the various possible topological types of stable n-pointed curves of genus g, the dimensions of the corresponding loci in $\overline{M}_{g,n}$, and the inclusion relations among their closures. We would suggest putting the open stratum (smooth curves with n distinct points) at the top; the various codimension 1 strata on a line below that, the codimension 2 strata on the next line, and so on; indicate the specialization relationships by vertical or diagonal lines.

D-1. Describe all the boundary strata of
$\overline{M}_{0,n}$ for $3 \le n \le 6$, $\overline{M}_2$, $\overline{M}_3$, $\overline{M}_{1,2}$, $\overline{M}_{2,1}$, $\overline{M}_{2,2}$, $\overline{M}_{3,1}$, $\overline{M}_{3,2}$, $\overline{M}_{3,3}$.

D-2. Consider the projection $\pi : \overline{M}_{g,n+1} \to \overline{M}_{g,n}$.

a) Describe $\pi^{-1}(\partial M_{g,n})$.
b) Describe $\pi^{-1}(M_{g,n}) \cap \partial M_{g,n+1}$.

D-3. Describe the intersection of the codimension one boundary components $\Delta_{a,A} \cap \Delta_{irr}$ and $\Delta_{a,A} \cap \Delta_{a',A'}$ in $\overline{M}_{g,P}$

D-4. How many boundary divisors are there in $\overline{M}_g$? In $\overline{M}_{g,n}$?

D-5. Do every pair of boundary divisors in $\overline{M}_g$ intersect? How about in $\overline{M}_{g,n}$?

D-6. Give an example of a pair of boundary divisors in $\overline{M}_g$ whose intersection is reducible. Can you find an example where the intersection is disconnected?

D-7. Let C be a stable curve of genus g with δ nodes. Show that $\delta \le 3g-3$. Similarly, show that a stable n-pointed curve of genus g has at most $3g-3+n$ nodes.

D-8. Show that the set $R_g \subset \overline{M}_g$ of stable curves with $3g-3$ nodes is finite. What is its cardinality for $g = 2$, 3, and 4?

D-9. Similarly, count the stable n-pointed curves of genus g having $3g-3+n$ nodes for $(g,n) = (2,1), (2,2), (3,1)$, and $(3,2)$.

D-10. Consider now the locus in $\overline{M}_g$ of stable curves of genus g having $3g-4$ nodes. Show that every component of this locus is a rational curve.

E. Curves in moduli spaces $\overline{M}_{g,n}$.

In what follows—in particular, when we discuss line bundles on moduli spaces in the next chapter—it will be useful to have some explicitly given curves in $\overline{M}_{g,n}$. These will be given by one-parameter families of stable curves of genus g. In these exercises, we ask you to verify that all members of the specified family are in fact stable.

E-1. A general pencil of plane curves of degree d, that is, we let F and G be general polynomials of degree d and consider curves C_t defined by linear combinations $t_0F + t_1G$ of the two.

E-2. Let $S \subset \mathbb{P}^3$ be a general surface of degree d, and $C_t = S \cap H_t$ a general pencil of plane sections of S.

E-3. Let $S \subset \mathbb{P}^3$ be an arbitrary smooth surface of degree d, and $C_t = S \cap H_t$ a general pencil of plane sections of S. (Note: This is much harder than the preceding problem and in particular requires a hypothesis of characteristic 0)

E-4. Let B be a smooth curve of genus $g - 1 > 0$, $p \in B$ any point, $\{E_t \subset \mathbb{P}^2\}$ a general pencil of plane cubics, and q a base point of the pencil. Let C_t be the curve obtained from $B \cup E_t$ by identifying p with q.

E-5. Let $p_1, \ldots, p_{2g+1} \in \mathbb{P}^1$ be distinct points, and for $p \in \mathbb{P}^1$, let C_p be the hyperelliptic curve given as a double cover of $\mathbb{P}^1$ with branch divisor $p + p_1 + \cdots + p_{2g+1}$.

F. Unirationality of moduli spaces $M_{g,n}$

We say that a variety X is *unirational* if there exists a dominant rational map $\mathbb{P}^n \to X$ from a projective (or affine) space to X. In particular, if there is an open subset $U \subset \mathbb{P}^n$ and a family of stable curves of genus g over U whose associated map $U \to M_g$ is dominant, we may conclude that M_g is unirational. (Of course, the converse need not be true—a priori, there might be a dominant rational map $\mathbb{P}^n \to M_g$, but not one arising from a family of stable curves—but in practice the only way we have ever shown a space M_g to be unirational is by exhibiting such a family.) In these exercises, we ask you to prove the unirationality of a particular moduli space $M_{g,n}$ by showing that the specified rationally parameterized family of curves is dominant, that is, the general member of the family is indeed stable, and the general curve of genus g does appear among the member of the family.

F-1. M_2: consider the family of curves given by $y^2 = f(x)$, where f ranges over all sextic polynomials.

F-2. M_3: consider the family of all plane quartics.

F-3. $M_{3,n}$ for $n \leq 14$: consider the family of pointed curves $(C, p_1, \ldots, p_n)$ with $p_i \in \mathbb{P}^2$ and C a plane quartic passing through $p_1, \ldots, p_n$. What is the largest n for which you can say that $M_{3,n}$ is unirational?

F-4. $M_{2,n}$: What is the largest n for which you can say that $M_{2,n}$ is unirational?

F-5. M_4: consider the family of curves of bidegree $(3,3)$ on $\mathbb{P}^1 \times \mathbb{P}^1$.

F-6. M_5: consider the family of complete intersections of three quadrics in $\mathbb{P}^4$.

F-7. M_6: let $S \subset \mathbb{P}^5$ be a (fixed) del Pezzo surface and consider the family of all intersections $S \cap Q$ of S with quadric hypersurfaces in $\mathbb{P}^5$.

F-8. Using Brill–Noether theory and plane sextics curves, show that M_6 is unirational.

F-9. In the same spirit of the preceding exercise, try $g = 7$.

F-10. Prove that the moduli H_g space of hyperelliptic curves is unirational.

G. Miscellaneous exercises

G-1. Find n_0 such that $M_{g,n}$ is smooth for $n \geq n_0$. Is there a similar lower bound for the moduli space of stable curves?

G-2. Let $\overline{M}_g^n$ be the set of isomorphism classes of triples $(C, \mathbf{x}, \mathbf{v})$, where C is a smooth genus g curve, $\mathbf{x} = (x_1, \ldots, x_n)$ is an n-tuple of points of C, and $\mathbf{v} = (v_1, \ldots, v_n)$ is an n-tuple of nonzero tangent vectors with $v_i \in T_{x_i}(C)$, $i = 1, \ldots, n$.

a) Show that $\overline{M}_g^n$ has a natural structure of smooth algebraic variety and compute its dimension.
b) Express the tangent spaces at points of $\overline{M}_g^n$ in cohomological terms.

G-3. Let $\overline{M}_{g,n}^0$ be the locus of automorphism-free, stable n-pointed curves of genus g. Show that $\overline{M}_{g,n}^0$ is an open smooth subset of $\overline{M}_{g,n}$. Set $V_g = \overline{M}_{g,n} \smallsetminus \overline{M}_{g,n}^0$

a) Describe the codimension-one components of $V_{g,n}$. Can any one of these be contained in the singular locus of $\overline{M}_{g,n}$?
b) Show that if Y is component of $V_{g,n}$, which is not of codimension equal to one, then

$$\dim Y \leq 2g - 1 \,. \tag{12.1}$$

c) Show that equality in (12.1) holds if and only if Y is the hyperelliptic locus.
d) Show that the singular locus of $\overline{M}_{g,n}$ consists precisely of those components of $V_{g,n}$ which are not divisors in $\overline{M}_{g,n}$.

G-4. Describe the infinitesimal behavior of projections and clutching maps.

G-5. Show that, given a point $\{[C_v; \mathbf{x}_v]\}_{v \in V}$ of $\overline{M}_\Gamma$ and denoting by $[C; \mathbf{x}]$ its image via ξ_Γ, there is an exact sequence

$$1 \to \prod_{v \in V} \mathrm{Aut}(C_v; \mathbf{x}_v) \to \mathrm{Aut}(C; \mathbf{x}) \stackrel{\alpha}{\to} \mathrm{Aut}(\Gamma)\,. \tag{12.2}$$

G-6. Look at the exact sequence (12.2). Give examples of stable curves for which

a) α is surjective,
a) α is injective,
a) α is neither injective nor surjective.

G-7. Give examples of dual graphs Γ' and Γ such that Γ' is a specialization of Γ, but not every element in $\mathrm{Aut}(\Gamma')$ is the specialization of an element in $\mathrm{Aut}(\Gamma)$.

Chapter XIII.
Line bundles on moduli

1. Introduction.

This chapter is devoted to the study of quasi-coherent sheaves, and in particular invertible ones, on the moduli stacks of curves. There are several equivalent notions of quasi-coherent sheaf on a stack $\mathcal{M}$. According to the basic one, such an animal consists of the datum, for any scheme S and any object ξ in $\mathcal{M}(S)$, of a quasi-coherent sheaf $\mathcal{F}_\xi$ on S, plus suitable compatibility conditions between the various $\mathcal{F}_\xi$. When $\mathcal{M}$ is a scheme X, this is equivalent to the usual definition of quasi-coherent sheaf on X. For an object $\xi \in \mathcal{M}(S)$, that is, for a morphism $\xi : S \to X$, the sheaf $\mathcal{F}_\xi$ is just the pullback $\xi^*\mathcal{F}$ of $\mathcal{F} = \mathcal{F}_{\mathrm{id}_X}$.

When $\mathcal{M}$ is a Deligne–Mumford stack, an alternative approach comes from the following observation. Giving a quasi-coherent sheaf on a scheme Z is equivalent to giving its restriction to every open subset of Z, plus compatibility conditions, that is, a description of how the various restrictions fit together. The analogues of open sets for a Deligne–Mumford stack $\mathcal{M}$ are schemes étale over $\mathcal{M}$, and one might define a quasi-coherent sheaf on $\mathcal{M}$ to be a quasi-coherent sheaf on each one of these, plus suitable compatibility conditions. A variant of this definition would be to look at an atlas $X \to \mathcal{M}$ and to define a quasi-coherent sheaf on $\mathcal{M}$ to be a quasi-coherent sheaf on X endowed with suitable "descent data." In Section 2 we show that on a Deligne–Mumford stack the above three notions of quasi-coherent sheaf are equivalent. The proof is a simple application of descent.

For a Deligne–Mumford stack, there is a well-defined notion of Picard group, and in Section 2 we begin to study the Picard groups of the moduli stacks $\overline{\mathcal{M}}_{g,n}$ of pointed stable curves. We first introduce some of the natural or, as one says, tautological line bundles and vector bundles on $\overline{\mathcal{M}}_{g,n}$. The first one is the Hodge vector bundle $\mathbb{E}$ whose "fiber" at a curve C is the vector space of abelian differentials $H^0(C, \omega_C)$. The Hodge line bundle is the top exterior power of $\mathbb{E}$; its class in the Picard group of $\overline{\mathcal{M}}_{g,n}$ is commonly designated by λ. Further natural line bundles are the point bundles $\mathcal{L}_i$, $i = 1, \ldots, n$. The fiber of $\mathcal{L}_i$ at a stable n-pointed curve $(C; x_1, \ldots, x_n)$ is the cotangent space to C at x_i, and the class of $\mathcal{L}_i$ in the Picard group of $\overline{\mathcal{M}}_{g,n}$ is denoted by ψ_i. Finally, one may attach a line bundle $\mathcal{O}(\mathcal{D})$ to any Cartier divisor $\mathcal{D}$ on $\overline{\mathcal{M}}_{g,n}$; doing this for the boundary or for its irreducible components produces other natural line

E. Arbarello et al., *Geometry of Algebraic Curves*,
Grundlehren der mathematischen Wissenschaften 268,
DOI 10.1007/978-3-540-69392-5_5, © Springer-Verlag Berlin Heidelberg 2011

bundles on moduli; the class of the line bundle attached to the boundary is usually denoted by δ. In general, none of these line bundles descends to a line bundle on the coarse moduli space $\overline{M}_{g,n}$, the reason being that the automorphism group of $(C; x_1, \ldots, x_n)$, when nontrivial, acts nontrivially on the fiber of the relevant line bundle at $(C; x_1, \ldots, x_n)$. The section closes with an alternative description of quasi-coherent sheaves on the quotient stack $[H/G]$ of a scheme H modulo the action of an algebraic group G. We show that giving a quasi-coherent sheaf on $[H/G]$ is the same as giving a G-equivariant quasi-coherent sheaf on H. Roughly speaking, such an object consists of a quasi-coherent sheaf $\mathcal{F}$ on H together with a lifting to $\mathcal{F}$ of the action of G on H. In particular, this shows that $\mathrm{Pic}([H/G])$ coincides with the group $\mathrm{Pic}(H, G)$ which parameterizes isomorphism classes of G-equivariant line bundles on H.

In Section 3 we discuss the tangent and cotangent bundles to the moduli stack $\overline{\mathcal{M}}_{g,P}$, we study the normal sheaf N_{ξ_Γ} to a clutching morphism $\xi_\Gamma : \overline{\mathcal{M}}_\Gamma \to \overline{\mathcal{M}}_{g,P}$, and we compute the "excess intersection bundles" for the intersections of two boundary strata.

One way of producing line bundles on moduli spaces is via the determinant construction. This is what we discuss in Section 4. In general, given a (connected) scheme S and a vector bundle F, the determinant of F is a $\mathbb{Z}/2\mathbb{Z}$-graded line bundle $\det F = (\wedge^{\max} F, \mathrm{rank}\, F)$, where the rank is to be taken modulo 2. The introduction of a grading is essentially forced when comparing $\wedge^{\max}(F \oplus G)$ with $\wedge^{\max}(G \oplus F)$. The notion of determinant naturally extends to a finite complex $F^\bullet$ of vector bundles over S. The determinant of $F^\bullet$ is defined to be

$$\det(F^\bullet) = \cdots \otimes (\det F^q)^{(-1)^q} \otimes (\det F^{q-1})^{(-1)^{q-1}} \otimes \cdots .$$

Now if $\pi : X \to S$ is a family of nodal curves, and $\mathcal{F}$ a coherent sheaf on X, one may attempt to define the determinant of the cohomology of $\mathcal{F}$ as[3]

$$\mathrm{d}_\pi(\mathcal{F}) = \det(R\pi_* \mathcal{F}).$$

This means that, when $R\pi_*\mathcal{F}$ is quasi-isomorphic to a finite complex $C^\bullet$ of vector bundles, one sets $\mathrm{d}_\pi(\mathcal{F}) = \det(C^\bullet)$. The properties of determinants ensure that this is independent of the particular complex $C^\bullet$ chosen. More generally, suppose that S can be covered with open sets U such that the restriction of $R\pi_*\mathcal{F}$ to U is quasi-isomorphic to a finite complex $C^\bullet_U$ of vector bundles. Then the line bundles $\det(C^\bullet_U)$ patch together on the overlaps of the open sets U, yielding a line bundle on S, which we again call determinant of the cohomology of $\mathcal{F}$ and denote by $\mathrm{d}_\pi(\mathcal{F})$. In practice, in defining the determinant of the cohomology we

[3]The standard notation for the determinant of the cohomology of $\mathcal{F}$ is $\lambda_\pi(\mathcal{F})$; in this book we depart from this convention since it conflicts with the ubiquitous notation for the Hodge class.

will not follow this path but will rely on a somewhat ad hoc construction which avoids some of the technicalities involved in the above general setup.

When $\pi_*\mathcal{F}$ and $R^1\pi_*\mathcal{F}$ are locally free, then $\mathrm{d}_\pi(\mathcal{F}) \cong \det(R^1\pi_*\mathcal{F})^{-1} \otimes \det(\pi_*\mathcal{F})$. The determinant of the cohomology construction also interacts with relative duality; given a vector bundle $\mathcal{F}$ on X, there is a canonical isomorphism

$$\mathrm{d}_\pi(\omega_\pi \otimes \mathcal{F}^\vee) \cong \mathrm{d}_\pi(\mathcal{F})\,.$$

From this point of view the Hodge line bundle can be viewed both as $\mathrm{d}_\pi(\omega_\pi)$ and as $\mathrm{d}_\pi(\mathcal{O}_X)$. The notion of determinant of the cohomology can be generalized to hypercohomology. Given a family of nodal curves $\pi : X \to S$ and a finite complex of coherent sheaves $\mathcal{F}^\bullet$ on X, one can define the determinant $\mathrm{d}_\pi(\mathcal{F}^\bullet)$ of the hypercohomology of the complex $\mathcal{F}^\bullet$. A significant application of this construction is the description of the boundary divisor $\mathcal{D} \subset \overline{\mathcal{M}}_{g,P}$ as the determinant

$$\mathcal{O}(\mathcal{D}) = \mathrm{d}_\pi(\Omega^1_\pi \to \omega_\pi).$$

This expression is particularly useful when dealing with intersection questions involving the boundary, as we show in a number of examples.

The classical Weil reciprocity for a smooth curve, which we discussed in Section 2, Appendix B of Chapter VI, can be generalized to a family of nodal curves, yielding the so-called Deligne pairing. Given a family of nodal curves $\pi : X \to S$, the Deligne pairing produces, out of line bundles L and M on X, a new line bundle on S denoted by $\langle L, M\rangle_\pi$. The Deligne pairing is compatible with base change and bilinear, with respect to tensor product, in both entries; furthermore, there is a canonical isomorphism between $\langle L, M\rangle_\pi$ and $\langle M, L\rangle_\pi$.

What makes the Deligne pairing a central object in the theory of line bundles on $\overline{\mathcal{M}}_{g,P}$ is its expression

$$\langle L, M\rangle_\pi \simeq \mathrm{d}_\pi(L \otimes M) \otimes \mathrm{d}_\pi(L)^{-1} \otimes \mathrm{d}_\pi(M)^{-1} \otimes \mathrm{d}_\pi(\mathcal{O}_X)$$

in terms of the determinant of the cohomology. Using this formula, we prove a very useful version of the Riemann–Roch theorem stating that, given a family of nodal curves $\pi : X \to S$ and a line bundle L on X, there is a canonical functorial isomorphism of line bundles (not just an equality of Chern classes)

$$\mathrm{d}_\pi(L)^2 \simeq \langle L, L \otimes \omega_\pi^{-1}\rangle_\pi \otimes \mathrm{d}_\pi(\mathcal{O}_X)^2.$$

The bridge between this "concrete" version of the Riemann–Roch theorem and the usual one is the formula

$$c_1(\langle L, M\rangle_\pi) = \pi_*(c_1(L) \cdot c_1(M)).$$

Finally, via the Deligne pairing, we define the first Mumford class κ_1. Specifically, given a family $\pi : X \to S$ of stable pointed curves, we set

$$\kappa_1 = c_1(\langle \omega_\pi(D), \omega_\pi(D) \rangle),$$

where $\omega_\pi(D)$ is the relative log-canonical sheaf, and we then proceed to prove a remarkable formula expressing the restriction of this class to the boundary of $\overline{\mathcal{M}}_{g,P}$.

Having introduced the notion of line bundle on a moduli stack, in Section 6 we discuss various natural notions of Picard groups. In particular, we compare the Picard group of the stack $\overline{\mathcal{M}}_{g,P}$, the Picard group of the space $\overline{M}_{g,P}$, and the group of $PGL(r+1)$)-invariant isomorphism classes of line bundles on the Hilbert scheme $\mathrm{H}_{\nu,g,n}$ of stable n-pointed genus g curves embedded in $\mathbb{P}^r$ by the νth power of the log-canonical sheaf, for some fixed $\nu \geq 3$, and $r = (2\nu-1)(g-1)+\nu n-1$.

Section 7 is devoted to Mumford's formulae relating various classes defined on $\overline{\mathcal{M}}_{g,P}$, the most notable of which are

$$\begin{aligned} \kappa_1 &= 12\lambda + \psi - \delta\,, \\ K_{\overline{\mathcal{M}}_{g,P}} &= 13\lambda + \psi - 2\delta\,, \end{aligned}$$

where δ is the class of the boundary, $K_{\overline{\mathcal{M}}_{g,P}}$ is the canonical class, and $\psi = \sum_{p\in P}\psi_p$ is the sum of the classes of the point bundles.

In the final Section 8 we study the stack $\overline{\mathcal{H}}_g$ of stable hyperelliptic curves, proving that it is indeed a closed substack of $\overline{\mathcal{M}}_g$ and describing the boundary $\partial\mathcal{H}_g = \overline{\mathcal{H}}_g \smallsetminus \mathcal{H}_g$. We then compute the Picard group $\mathrm{Pic}(\overline{\mathcal{H}}_g) \otimes \mathbb{Q}$ for $g \geq 2$, describe natural classes in it, and compute the relations between these classes.

2. Line bundles on the moduli stack of stable curves.

There are several equivalent ways of defining the notion of line bundle, or more generally of quasi-coherent sheaf, on a Deligne–Mumford stack. One of the most manageable ones comes from the observation that giving a quasi-coherent sheaf on a scheme Z is equivalent to giving its pullback via every morphism $S \to Z$, together with compatibility information among the various pullbacks. Now let $\mathcal{M} = (\mathcal{C}, p)$ be a Deligne–Mumford stack or, more generally, a category fibered in groupoids over $Sch/\mathbb{C}$. Here we keep the notation introduced in Section 5 of Chapter XII. Since morphisms from a scheme S to $\mathcal{M}$ correspond to objects $\xi \in \mathcal{M}(S)$, in a sense which is made precise by the 2-categorical Yoneda lemma (5.3) in Chapter XII, it is tempting to define a quasi-coherent sheaf on $\mathcal{M}$ to be the datum of an ordinary quasi-coherent sheaf on $p(\xi)$ for any object ξ of $\mathcal{M}$, plus suitable compatibility information. We shall return to this analogy later in this section. To make our intuition into a formal definition, we adopt the following conventions. Given an object ξ in $\mathcal{M}$, we sometimes write S_ξ to indicate $p(\xi)$. Moreover, we use uppercase letters to indicate morphisms between objects of $\mathcal{M}$ and the corresponding lowercase letters to indicate the corresponding morphisms between the bases. In other words, if ξ and η are objects in $\mathcal{M}$, and H is a morphism from ξ to η, then the corresponding morphism $p(\xi) \to p(\eta)$ will be designated by h.

DEFINITION (2.1). *Let $\mathcal{M} = (\mathcal{C}, p)$ be a category fibered in groupoids over $Sch/\mathbb{C}$. A quasi-coherent sheaf $\mathcal{F}$ on $\mathcal{M}$ is the datum of:*

i) *a quasi-coherent sheaf $\mathcal{F}_\xi$ on $p(\xi)$ for any object ξ of $\mathcal{M}$,*
ii) *an isomorphism*

$$\rho_H : h^*(\mathcal{F}_\eta) \xrightarrow{\simeq} \mathcal{F}_\xi$$

for any morphism $H : \xi \to \eta$ of objects of $\mathcal{M}$,

satisfying the cocycle condition:

iii) *for any pair of morphisms $H_1 : \xi_1 \to \xi_2$ and $H_2 : \xi_2 \to \xi_3$ of objects of $\mathcal{M}$, the diagram*

$$\begin{array}{ccc} h_1^* h_2^* \mathcal{F}_{\xi_3} & \overset{\sim}{=\!=\!=} & (h_2 h_1)^* \mathcal{F}_{\xi_3} \\ {\scriptstyle h_1^*(\rho_{H_2})} \downarrow & & \downarrow {\scriptstyle \rho_{H_2 \circ H_1}} \\ h_1^* \mathcal{F}_{\xi_2} & \xrightarrow{\rho_{H_1}} & \mathcal{F}_{\xi_1} \end{array}$$

of isomorphisms of sheaves over $p(F_1)$ commutes.

The sheaf $\mathcal{F}$ will be said to be coherent if each one of the $\mathcal{F}_F$ is.

A homomorphism from a quasi-coherent sheaf $\mathcal{F}$ to another quasi-coherent sheaf $\mathcal{G}$ is a collection of sheaf homomorphisms $\mathcal{F}_\xi \to \mathcal{G}_\xi$, where ξ varies over all objects of $\mathcal{M}$, which is compatible with the homomorphisms ρ_H.

The same kind of definition, with obvious terminological changes, makes sense in the context of "analytic stacks" and gives a notion of coherent (analytic) sheaf on these objects.

We shall say that a sheaf $\mathcal{F}$ on $\mathcal{M}$ is locally free or, by abuse of language, that it is a vector bundle if all the $\mathcal{F}_\xi$ are locally free. As usual, an invertible sheaf, or line bundle, is a locally free sheaf of rank one. It makes sense to perform on quasi-coherent sheaves on $\mathcal{M}$ the usual sheaf operations such as, for instance, direct sum, tensor product, exterior and symmetric products, kernel and cokernel, by performing them on the individual $\mathcal{F}_\xi$.

At least when $\mathcal{M}$ is a Deligne–Mumford stack, the isomorphism classes of line bundles on $\mathcal{M}$ form a group under the tensor product operation; this is called the *Picard group* of the stack and denoted

$\mathrm{Pic}(\mathcal{M})$. The usual convention that we will follow throughout the book is to write the operation in the Picard group additively. If L is a line bundle on $\mathcal{M}$, we shall sometimes write $c_1(L)$ to indicate the class of L in $\mathrm{Pic}(\mathcal{M})$.

We now specialize some of these considerations to the moduli stack $\overline{\mathcal{M}}_{g,n}$ of n-pointed stable curves of genus g. In this case an object of $\overline{\mathcal{M}}_{g,n}$ is just a family of n-pointed stable curves of genus g parameterized by a scheme S. Such a family also determines a morphism from S to moduli space. A quasi-coherent sheaf on $\overline{M}_{g,n}$ thus determines by pullback a quasi-coherent sheaf on $\overline{\mathcal{M}}_{g,n}$. However, there are plenty of quasi-coherent sheaves on the moduli stack which do not come by pullback from ordinary quasi-coherent sheaves on $\overline{M}_{g,n}$.

The premier example of quasi-coherent sheaf on moduli (actually, a vector bundle of rank g) is the *Hodge bundle* $\mathbb{E}$, which is defined to be the direct image of the relative dualizing sheaf. More precisely, for any family $\xi = (\pi : X \to S)$, one sets $\mathbb{E}_\xi = \pi_*(\omega_\pi)$; that this defines a quasi-coherent sheaf follows at once from the fact that the relative dualizing sheaf is functorial with respect to morphisms of families. The *Hodge line bundle* is the determinant bundle $\det \mathbb{E} = \wedge^g \mathbb{E}$. As we will show later in this section, the Hodge bundle and its determinant do *not* come from ordinary bundles on $\overline{M}_{g,n}$, except when $g = 0$, in which case they are trivial.

It is customary to denote by λ the class of the determinant of the Hodge bundle in the Picard group of $\overline{\mathcal{M}}_{g,n}$. Thus,

$$\lambda = \text{class of } \wedge^g \mathbb{E} \text{ in } \mathrm{Pic}(\overline{\mathcal{M}}_{g,n}) . \tag{2.2}$$

The construction of λ can be generalized in several directions. On one hand, we could take higher Chern classes of the Hodge bundle; this will be done in Chapter XVII. On the other hand, we can look at the virtual bundle $\pi_! \omega_\pi^\nu$, where ν is any integer, and set $\Lambda(\nu)_F = (\wedge^{\max} R^1 \pi_* \omega_\pi^\nu)^{-1} \otimes \wedge^{\max} \pi_* \omega_\pi^\nu$. This defines a new line bundle $\Lambda(\nu)$ on $\overline{\mathcal{M}}_{g,n}$. We then set

$$\lambda(\nu) = \text{class of } \Lambda(\nu) \text{ in } \mathrm{Pic}(\overline{\mathcal{M}}_{g,n}) .$$

When $\nu = 1$, this definition gives back λ. In fact, $R^1 \pi_* \omega_\pi$ is canonically trivial, and hence $\Lambda(1)_F$ and $\wedge^g \pi_* \omega_\pi$ are canonically isomorphic.

Another example of line bundle on the moduli stack of n-pointed stable curves is provided by the so-called *point bundles*, which are defined only when $n > 0$. Informally, the fiber of the ith point bundle at an n-pointed curve $(C; p_1, \dots, p_n)$ is just the cotangent space to C at p_i. A more formal definition of this bundle, which we denote by $\mathcal{L}_i$, is as follows. Consider a family ξ of n-pointed stable genus g curves, consisting of a family of curves $X \to S$ plus sections σ_i, $i = 1, \dots, n$. Then

$$(\mathcal{L}_i)_\xi = \sigma_i^* \omega_{X/S} . \tag{2.3}$$

One usually sets

$$(2.4)\qquad \begin{aligned} \psi_i &= \text{class of } \mathcal{L}_i \text{ in } \operatorname{Pic}(\overline{\mathcal{M}}_{g,n}), \\ \psi &= \sum_i \psi_i . \end{aligned}$$

As was the case for the Hodge line bundle, in general, $\mathcal{L}_i$ does not descend to a line bundle on moduli space.

The usual correspondence between Cartier divisors and invertible sheaves holds also for Deligne–Mumford stacks, although this is not immediate from the definition of line bundle we have given. There are, however, other equivalent definitions of quasi-coherent sheaf and line bundle which are better adapted to dealing with divisors. Before we explain what these are, we need some preliminary remarks. Let $\mathcal{M}$ be a category fibered in groupoids over $Sch/\mathbb{C}$. We denote by $\widetilde{\mathcal{M}}$ the category of schemes over $\mathcal{M}$. Recall that morphisms in this category are commutative triangles. More precisely, given objects $\alpha : S \to \mathcal{M}$ and $\beta : T \to \mathcal{M}$ in $\widetilde{\mathcal{M}}$, where S and T are schemes, a morphism from the first to the second consists of a morphism of schemes $f : S \to T$ together with an isomorphism of functors between βf and α. The content of the 2-categorical Yoneda lemma (5.3) in Chapter XII is that the functor which assigns to each object $\alpha : S \to \mathcal{M}$ of $\widetilde{\mathcal{M}}$ the object $\alpha(\mathrm{id}_S)$ of $\mathcal{M}$ is an equivalence of categories. A consequence is that the categories of quasi-coherent sheaves over $\mathcal{M}$ and $\widetilde{\mathcal{M}}$ are equivalent. In other words, giving a quasi-coherent sheaf on $\mathcal{M}$ amounts to giving

i') a quasi-coherent sheaf $\mathcal{F}_\alpha$ on S for any morphism $\alpha : S \to \mathcal{M}$, where S is a scheme,

ii') an isomorphism

$$\rho_H : h^*(\mathcal{F}_\beta) \xrightarrow{\simeq} \mathcal{F}_\alpha$$

for any morphism $H : \alpha \to \beta$ of schemes over $\mathcal{M}$,

satisfying the cocycle condition:

iii') for any pair of morphisms $H_1 : \alpha_1 \to \alpha_2$ and $H_2 : \alpha_2 \to \alpha_3$, where $\alpha_i : S_i \to \mathcal{M}$, $i = 1, 2, 3$, is a scheme over $\mathcal{M}$, the diagram

$$\begin{array}{ccc} h_1^*(h_2^*(\mathcal{F}_{\alpha_3})) & \overset{\sim}{=\!=\!=} & (h_2 \circ h_1)^*(\mathcal{F}_{\alpha_3}) \\ {\scriptstyle h_1^*(\rho_{H_2})}\big\downarrow & & \big\downarrow{\scriptstyle \rho_{H_2 \circ H_1}} \\ h_1^*(\mathcal{F}_{\alpha_2}) & \xrightarrow{\quad \rho_{H_1} \quad} & \mathcal{F}_{\alpha_1} \end{array}$$

of isomorphisms of sheaves over S_1 commutes.

In what follows, if $\mathcal{F}$ is a quasi-coherent sheaf on $\mathcal{M}$, S is a scheme, and $\alpha : S \to \mathcal{M}$ is a morphism, we shall sometimes write $\mathcal{F}_S$ for $\mathcal{F}_\alpha$, when no confusion is likely.

We are now in a position to discuss alternative definitions of the notion of quasi-coherent sheaf on a Deligne–Mumford stack $\mathcal{M}$. One of the most immediate generalizations of the usual definition of quasi-coherent sheaf on a scheme comes from the following observation. Giving such a sheaf on a scheme Z is equivalent to giving its restriction to every open subset of Z, plus "descent data," that is, a description of how the various restrictions are to be fitted together. For a Deligne–Mumford stack $\mathcal{M}$, the analogues of open sets are schemes étale over $\mathcal{M}$, and one may imagine that to give a quasi-coherent sheaf on $\mathcal{M}$, it should suffice to specify a quasi-coherent sheaf on each one of these, plus suitable descent data. We shall show that this is indeed the case. Formally, suppose that we are given data i'), ii') as above, where α and β are restricted to be étale morphisms, satisfying condition iii'). We will provisionally say that these data constitute an *extended quasi-coherent sheaf* on the Deligne–Mumford stack $\mathcal{M}$. Provisionally because, as announced, we will immediately show that this new notion is entirely equivalent to the one of quasi-coherent sheaf over $\mathcal{M}$. From that point on, we will simply speak of quasi-coherent sheaves, using one definition of the other according to need. There is an obvious functor

$$(2.5) \qquad \tau : \left\{ \begin{matrix} \text{quasi-coherent} \\ \text{sheaves on } \mathcal{M} \end{matrix} \right\} \to \left\{ \begin{matrix} \text{extended quasi-coherent} \\ \text{sheaves on } \mathcal{M} \end{matrix} \right\}.$$

We will show is that τ is an equivalence of categories. In the analytic setting, that τ is an equivalence is a triviality, mostly because locally, from the analytic point of view, étale morphisms are open embeddings; the reader is encouraged to provide a proof by himself. In the scheme setup, the proof, although not difficult, is more delicate and uses faithfully flat descent for quasi-coherent sheaves, i.e., Theorem (6.2) in Chapter XII.

We now show how to deduce from the above-mentioned theorem the fact that τ is an equivalence. Actually, we will find it convenient to introduce yet another equivalent definition of quasi-coherent sheaf on $\mathcal{M}$. Fix an atlas $X \to \mathcal{M}$, where X is a scheme. We write XX and XXX as shorthand for $X \times_{\mathcal{M}} X$ and $X \times_{\mathcal{M}} X \times_{\mathcal{M}} X$. We also let $p_i : XX \to X$ and $q_i : XXX \to X$ be the projections to the ith factor, and $p_{ij} : XXX \to XX$, $i \neq j$, the projection to the product of the ith and jth factors. Then we can speak of quasi-coherent sheaves on X with descent data relative to $X \to \mathcal{M}$ just as we did for morphisms of schemes.

If $\mathcal{F}$ is an extended quasi-coherent sheaf on $\mathcal{M}$, there are given isomorphisms $p_1^*\mathcal{F}_X \to \mathcal{F}_{XX}$ and $p_2^*\mathcal{F}_X \to \mathcal{F}_{XX}$. Composing the first with the inverse of the second, one gets an isomorphism $\varphi : p_1^*\mathcal{F}_X \to p_2^*\mathcal{F}_X$. This satisfies the cocycle condition, as one can see by looking at the diagram of isomorphisms

(2.6)

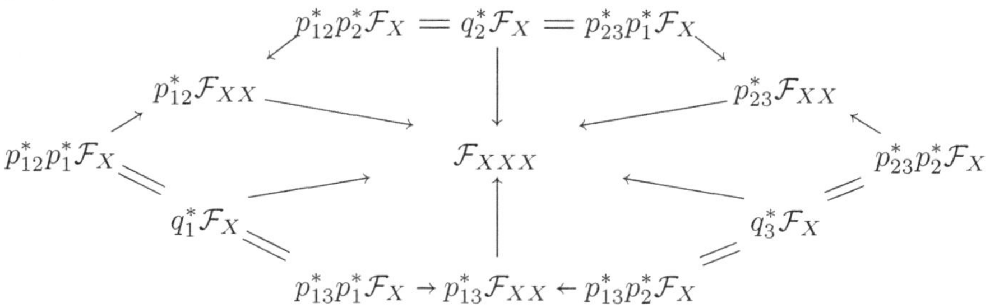

All the quadrilaterals in the diagram are commutative by condition iii') in the definition of extended quasi-coherent sheaf. Hence, the outer hexagon, with arrows inverted where needed, is also commutative. This is just the cocycle condition. Thus, to any $\mathcal{F}$ we can associate a quasi-coherent sheaf on X, namely $\mathcal{F}_X$, with descent data relative to $X \to \mathcal{M}$. As one readily checks, this gives a functor

$$(2.7)\quad \chi : \left\{ \begin{matrix} \text{extended quasi-coherent} \\ \text{sheaves on } \mathcal{M} \end{matrix} \right\} \to \left\{ \begin{matrix} \text{quasi-coherent sheaves} \\ \text{on } X \text{ with descent data} \\ \text{relative to } X \to \mathcal{M} \end{matrix} \right\}.$$

Now suppose we are given a quasi-coherent sheaf $\mathcal{G}$ on X, plus descent data ψ. Let ξ be an object in $\mathcal{M}(Z)$, where Z is a scheme. Then ξ corresponds to a morphism $Z \to \mathcal{M}$, well defined up to isomorphism of functors. Set $Y = Z \times_{\mathcal{M}} X$. Pulling back $\mathcal{G}$ and ψ to Y, we get a quasi-coherent sheaf plus descent data relative to $Y \to Z$; moreover, this construction is clearly functorial under base change. But now $Y \to Z$ is étale and surjective, because $X \to \mathcal{M}$ is, and hence in particular faithfully flat. Thus, Theorem (6.2) in Chapter XII applies and produces a quasi-coherent sheaf on Z, again functorial under base change; this simply means that we get a quasi-coherent sheaf on $\mathcal{M}$. In other words, descent gives another functor

$$(2.8)\qquad \zeta : \left\{ \begin{matrix} \text{quasi-coherent sheaves} \\ \text{on } X \text{ with descent data} \\ \text{relative to } X \to \mathcal{M} \end{matrix} \right\} \to \left\{ \begin{matrix} \text{quasi-coherent} \\ \text{sheaves on } \mathcal{M} \end{matrix} \right\}.$$

We then have the following result.

PROPOSITION (2.9). *Let $\mathcal{M}$ be a Deligne–Mumford stack, and let $X \to \mathcal{M}$ be an atlas for $\mathcal{M}$. The functors τ, χ, and ζ in (2.5), (2.7), and (2.8) are equivalences of categories. Their composition, in any one of the three possible orders, is isomorphic to the appropriate identity functor.*

The proposition is a more or less formal consequence of Theorem (6.2) in Chapter XII. What one has to find is isomorphisms between the compositions of τ, χ, and ζ, in each of the three possible orders, and the relevant identity functors. The proof consists of a series of standard verifications; we give a small sample of them, leaving the others to the reader.

Suppose first that we are given a quasi-coherent sheaf or an extended quasi-coherent sheaf, on $\mathcal{M}$, and denote it by $\mathcal{F}$. Let Z be a scheme étale over $\mathcal{M}$ or, more generally, a scheme over $\mathcal{M}$. From $\mathcal{F}$ we construct descent data φ for $\mathcal{F}_X$ relative to $X \to \mathcal{M}$. Write $Y = Z \times_{\mathcal{M}} X$. By pullback we get a quasi-coherent sheaf on Y plus descent data for it relative to $Y \to Z$. We wish to show that the sheaf on Z one then gets by descent is naturally isomorphic to $\mathcal{F}_Z$. For this, it is enough to give a natural isomorphism between the corresponding quasi-coherent sheaves with descent data on Y. We write π to indicate the projection $Y \to Z$, q_1 and q_2 to indicate the projections of $Y \times_Z Y$ to the two factors, α for the projection $Y \to X$, and β for the induced morphism $Y \times_Z Y \to X \times_{\mathcal{M}} X$. We also set $r = \pi \circ q_1 = \pi \circ q_2$. By the definition of quasi-coherent sheaf on $\mathcal{M}$, there are canonical isomorphisms $\alpha^*\mathcal{F}_X \to \mathcal{F}_Y \leftarrow \pi^*\mathcal{F}_Z$. We then get a diagram of isomorphisms

$$\begin{array}{ccccc}
q_1^*\alpha^*\mathcal{F}_X & \longrightarrow & \beta^*\mathcal{F}_{X\times_{\mathcal{M}}X} & \longleftarrow & q_2^*\alpha^*\mathcal{F}_X \\
\downarrow & & \downarrow & & \downarrow \\
q_1^*\mathcal{F}_Y & \longrightarrow & \mathcal{F}_{Y\times_Z Y} & \longleftarrow & q_2^*\mathcal{F}_Y \\
\uparrow & & \uparrow & & \uparrow \\
q_1^*\pi^*\mathcal{F}_Z & \longrightarrow & r^*\mathcal{F}_Z & \longleftarrow & q_2^*\pi^*\mathcal{F}_Z
\end{array}$$

The four squares are commutative because of property iii) in the definition of quasi-coherent sheaf on $\mathcal{M}$. Hence, the descent data for $\alpha^*\mathcal{F}_X$, which are given by the isomorphisms in the upper row, and those for $\pi^*\mathcal{F}_Z$, which are given by the isomorphisms in the lower row, correspond to each other under the isomorphism $\alpha^*\mathcal{F}_X \to \pi^*\mathcal{F}_Z$. It can be checked that what we have described is a natural transformation from the functor $\zeta\chi\tau$ to the identity, or from the functor $\tau\zeta\chi$ to the identity, as the case may be.

Now start instead with a quasi-coherent sheaf $\mathcal{G}$ on X together with descent data ψ relative to $X \to \mathcal{M}$. Applying ζ and τ, we get a quasi-coherent sheaf $\mathcal{F}$ on $\mathcal{M}$, in particular, also a quasi-coherent sheaf on X and, as we explained, descent data relative to $X \to \mathcal{M}$. We wish to show that $\mathcal{F}_X$ and $\mathcal{G}$, with their respective descent data, are naturally isomorphic. We first observe that there is a more direct procedure for obtaining $\mathcal{F}_X$ or, more generally, $\mathcal{F}_Z$ when $Z \to \mathcal{M}$ factors through $X \to \mathcal{M}$. Let $Z \to \mathcal{M}$ be the composition of $\alpha : Z \to X$ and $X \to \mathcal{M}$. Then the projection $r_1 : Z \times_{\mathcal{M}} X \to Z$ has a section, namely

$\sigma = (\mathrm{id}, \alpha)$, and, as we observed, under these circumstances, the pullback of $\mathcal{G}$ via $r_2 : Z \times_{\mathcal{M}} X \to X$, together with its descent data, is just $r_1^* \sigma^* r_2^* \mathcal{G} = r_1^* \alpha^* \mathcal{G}$. In other words, $\mathcal{F}_Z$ is canonically isomorphic to $\alpha^* \mathcal{G}$. This applies in particular to $X \to \mathcal{M}$, giving an isomorphism $\mathcal{F}_X \cong \mathcal{G}$, and to $X \times_{\mathcal{M}} X \to \mathcal{M}$. Thus we get isomorphisms $p_1^* \mathcal{G} \cong \mathcal{F}_{X \times_{\mathcal{M}} X} \cong p_2^* \mathcal{G}$. We spare the reader another diagram chase showing that the composite isomorphism $p_1^* \mathcal{G} \to p_2^* \mathcal{G}$ is just ψ and completing the proof that indeed $\mathcal{F}_X$ and $\mathcal{G}$, with their descent data, are isomorphic.

As we said earlier, we leave to the reader the remaining checks which are needed to finish the proof of (2.9).

The proposition we have just proved says that, given a Deligne–Mumford stack $\mathcal{M}$ and an atlas $X \to \mathcal{M}$, the notions of quasi-coherent sheaf on $\mathcal{M}$, extended quasi-coherent sheaf on $\mathcal{M}$, and quasi-coherent sheaf on X with descent data relative to $X \to \mathcal{M}$ are all equivalent. From now on we shall use the words "quasi-coherent sheaf on $\mathcal{M}$" to refer to each one of them, indifferently. The category of quasi-coherent sheaves on $\mathcal{M}$ will be denoted $QCoh(\mathcal{M})$.

We now have all the ingredients needed to discuss Cartier divisors and line bundles on Deligne–Mumford stacks. A Cartier divisor on a Deligne–Mumford stack $\mathcal{M}$ is the datum of a Cartier divisor on each scheme étale over $\mathcal{M}$, compatible with pullbacks under morphisms of schemes over $\mathcal{M}$. To each Cartier divisor D one can attach an invertible sheaf $\mathcal{O}(D)$ on $\mathcal{M}$, by performing the usual construction on S for each étale $S \to \mathcal{M}$. For instance, the boundary $\mathcal{D} = \partial \mathcal{M}_{g,n}$ "is" a stack divisor, and the same is true of its components $\mathcal{D}_{irr}$ and $\mathcal{D}_{\mathcal{P}}$ for each stable bipartition $\mathcal{P}$ of $(g, \{1, \dots, n\})$ (cf. Section 2 of Chapter X for the terminology). Clearly,

$$\mathcal{D} = \mathcal{D}_{irr} + \sum_{\mathcal{P}} \mathcal{D}_{\mathcal{P}} \,,$$

where the sum runs through all stable bipartitions of $(g, \{1, \dots, n\})$. These divisors are called *boundary divisors.* Naturally enough, the class of $\mathcal{O}(\mathcal{D}_{irr})$ in $\mathrm{Pic}(\overline{\mathcal{M}}_{g,n})$ is denoted by δ_{irr}, and the one of $\mathcal{O}(\mathcal{D}_{\mathcal{P}})$ by $\delta_{\mathcal{P}}$. As we said in Section 2 of Chapter X, one normally indicates a stable bipartition $\mathcal{P} = \{(a, A), (b, B)\}$ by specifying just "one half" of it, that is, (a, A) or (b, B). In accordance with this convention, one usually writes $\delta_{a,A}$ or, equivalently, $\delta_{b,B}$ to indicate $\delta_{\mathcal{P}}$, and $\mathcal{D}_{a,A}$ or, equivalently, $\mathcal{D}_{b,B}$, instead of $\mathcal{D}_{\mathcal{P}}$.

REMARK (2.10). As we mentioned in Section 8 of Chapter XII, the content of Proposition (2.9) is closely related to the statement that the standard definition of Deligne–Mumford stack is equivalent to the one via groupoid presentations. In fact, in certain special instances, Proposition (2.9) provides a proof of the equivalence of the two definitions of Deligne–Mumford stack. Here we shall sketch how one can deal with the case of substacks. Consider then a Deligne–Mumford stack $\mathcal{M}$, an atlas $X \to \mathcal{M}$ for it, and let $\pi_1, \pi_2 : X \times_{\mathcal{M}} X \to X$ be the projections to the two factors. Any closed immersion $\mathcal{N} \to \mathcal{M}$ gives a closed subscheme Y of X such that $\pi_1^{-1}(Y)$ and $\pi_2^{-1}(Y)$ are equal. We want to invert this operation. Given a subscheme Y of X, to say that $\pi_1^{-1}(Y) = \pi_2^{-1}(Y)$ means that the coherent sheaf $\mathcal{O}_Y$ on X and the surjective homomorphism $\mathcal{O}_X \to \mathcal{O}_Y$ come with descent data relative to $X \to \mathcal{M}$. Then Proposition (2.9) attaches to $\mathcal{O}_Y$ a coherent sheaf $\mathcal{F}$ on $\mathcal{M}$, and to $\mathcal{O}_X \to \mathcal{O}_Y$ a surjective homomorphism $\mathcal{O}_{\mathcal{M}} \to \mathcal{F}$. It is essentially evident, and easy to check, that one gets a new stack $\mathcal{Q}$ by considering $\mathcal{M}$ together with $\mathcal{O}_{\mathcal{M}} \to \mathcal{F}$, that the obvious forgetful morphism $\mathcal{Q} \to \mathcal{M}$ is a closed immersion, and that $Y \to \mathcal{Q}$ is an atlas for the Deligne–Mumford stack $\mathcal{Q}$. It is equally easy to check that, in case Y comes from a closed immersion $\mathcal{N} \to \mathcal{M}$, there is an equivalence of stacks over $\mathcal{M}$ between $\mathcal{Q}$ and $\mathcal{N}$.

The arguments we have used to prove Proposition (2.9) are valid in somewhat greater generality. Specifically, the construction of the functor ζ uses faithfully flat descent applied to morphisms of schemes which, in the case examined, happen to be étale. However, for the definition and for the subsequent arguments to go through, just faithful flatness is required. We shall use this observation to describe quasi-coherent sheaves on quotient stacks modulo group actions.

Consider an algebraic group G acting on a scheme H. A *G-equivariant quasi-coherent sheaf on H* is a quasi-coherent sheaf $\mathcal{F}$ on H plus a lifting to $\mathcal{F}$ of the action of G or, as one says, a *G-linearization* of $\mathcal{F}$. The intuitive notion of G-linearization is simple enough; one is merely asking for a lifting of the action of G on H to an action on the "total space" of $\mathcal{F}$ which is linear on "fibers." The precise formalization of this naive idea is somewhat more forbidding. Let $\alpha : G \times H \to H$ be the action, $p : G \times H \to H$ the projection to the second factor, $q : G \times G \times H \to G \times H$ the projection to the product of the second and third factors, and $\mu : G \times G \to G$ the product. Notice that $\alpha \circ q = p \circ (\mathrm{id} \times \alpha)$, $p \circ q = p \circ (\mu \times \mathrm{id})$, and $\alpha \circ (\mathrm{id} \times \alpha) = \alpha \circ (\mu \times \mathrm{id})$ since α is an action. Then a G-linearization of $\mathcal{F}$ is an isomorphism $\rho : p^*\mathcal{F} \to \alpha^*\mathcal{F}$ such that the following diagram of sheaves and sheaf isomorphisms on $G \times G \times H$ commutes:

$$
\begin{array}{ccccc}
 & q^*\alpha^*\mathcal{F} & \xrightarrow{\quad = \quad} & (\mathrm{id} \times \alpha)^* p^*\mathcal{F} & \\
{\scriptstyle q^*(\rho)} \nearrow & & & & \searrow {\scriptstyle (\mathrm{id} \times \alpha)^*(\rho)} \\
q^*p^*\mathcal{F} & & & & (\mathrm{id} \times \alpha)^*\alpha^*\mathcal{F} \\
{\scriptstyle =} \searrow & & & & \nearrow {\scriptstyle =} \\
 & (\mu \times \mathrm{id})^* p^*\mathcal{F} & \xrightarrow{(\mu \times \mathrm{id})^*(\rho)} & (\mu \times \mathrm{id})^*\alpha^*\mathcal{F} &
\end{array}
$$

There is an obvious notion of morphism between G-equivariant quasi-coherent sheaves on H; we write $QCoh(H, G)$ to denote the resulting category. We will show that there is an equivalence of categories between $QCoh(H,G)$ and $QCoh([H/G])$. Recall from (5.5) in Chapter XII that a section of the stack quotient $[H/G]$ over a base scheme T is a diagram

$$\begin{array}{ccc} P & \xrightarrow{\sigma} & H \\ {\scriptstyle f}\downarrow & & \\ T & & \end{array} \tag{2.11}$$

where P is a principal G-bundle over T, and σ is G-equivariant. A basic example is given by the trivial bundle $G \times H$ over H mapping to H via the action morphism α:

$$\begin{array}{ccc} G \times H & \xrightarrow{\alpha} & H \\ {\scriptstyle p}\downarrow & & \\ H & & \end{array} \tag{2.12}$$

This section corresponds to the quotient morphism $\pi : H \to [H/G]$.

LEMMA (2.13). *Let $\eta : T \to [H/G]$ be the morphism attached to (2.11). The diagram*

$$\begin{array}{ccc} P & \xrightarrow{\sigma} & H \\ {\scriptstyle f}\downarrow & & \downarrow{\scriptstyle \pi} \\ T & \xrightarrow{\eta} & [H/G] \end{array} \tag{2.14}$$

is commutative and induces an isomorphism between P and the fiber product $T \times_{[H/G]} H$.

First, the commutativity bit. We begin by recalling how one composes a morphism of schemes with a morphism to $[H/G]$, for example, a morphism $u : S \to T$ with $\eta : T \to [H/G]$. One forms the diagram

$$\begin{array}{ccccc} u^*P & \xrightarrow{\widetilde{u}} & P & \xrightarrow{\sigma} & H \\ {\scriptstyle \widetilde{f}}\downarrow & & {\scriptstyle f}\downarrow & & \\ S & \xrightarrow{u} & T & & \end{array} \tag{2.15}$$

and the composite morphism corresponds to the section of $[H/G]$ over S consisting of the principal G-bundle $u^*P \to S$ together with the G-equivariant map $\sigma\widetilde{u} : u^*P \to H$.

Now let $\gamma : S \to P$ be a morphism of schemes. The composition of this morphism with σ and π can be calculated using the recipe we just gave, by looking at the composite diagram

$$\begin{array}{ccccccc} G \times S & \xrightarrow{\mathrm{id} \times \gamma} & G \times P & \xrightarrow{\mathrm{id} \times \sigma} & G \times H & \xrightarrow{\alpha} & H \\ \downarrow & & \downarrow & & \downarrow & & \\ S & \xrightarrow{\gamma} & P & \xrightarrow{\sigma} & H & & \end{array} \tag{2.16}$$

where the vertical maps are all projections to the second factor. Similarly, the composition of γ with f and η can be obtained from

$$
(2.17)\qquad
\begin{array}{ccccccc}
\gamma^*f^*P & \longrightarrow & f^*P & \longrightarrow & P & \xrightarrow{\ \sigma\ } & H \\
\downarrow & & \downarrow & & \downarrow{\scriptstyle f} & & \\
S & \xrightarrow{\ \gamma\ } & P & \xrightarrow{\ f\ } & T & &
\end{array}
$$

Next, notice that the morphism $G\times P\to P\times_T P$ given by $(g,x)\mapsto(x,g\cdot x)$ is an isomorphism and that, under this isomorphism, the projection of $G\times P$ to the second factor and the action morphism $\beta: G\times P\to P$ get identified to the projections of $P\times_T P$ to the first and second factors, respectively. Observe also that this is a special instance of (2.13). The identification between $G\times P$ and $f^*P = P\times_T P$, and the resulting one between $G\times S$ and γ^*f^*P, transform (2.17) into

$$
(2.18)\qquad
\begin{array}{ccccccc}
G\times S & \xrightarrow{\ \mathrm{id}\times\gamma\ } & G\times P & \xrightarrow{\ \beta\ } & P & \xrightarrow{\ \sigma\ } & H \\
\downarrow & & \downarrow & & \downarrow{\scriptstyle f} & & \\
S & \xrightarrow{\ \gamma\ } & P & \xrightarrow{\ f\ } & T & &
\end{array}
$$

That the isomorphism between $G\times S$ and γ^*f^*P gives rise to an isomorphism of sections of $[H/G]$ over S follows from the fact that σ is G-equivariant, and hence $\alpha\circ(\mathrm{id}\times\sigma) = \sigma\circ\beta$. This proves the commutativity of diagram (2.14) and, more precisely, shows how to attach to a morphism $S\to P$ a specific section of $T\times_{[H/G]} H$ over S.

To complete the proof of (2.13), we must reverse the construction we just outlined. Let $u: S\to T$ and $v: S\to H$ be morphisms of schemes. Consider the diagrams (2.15) and

$$
\begin{array}{ccccc}
G\times S & \xrightarrow{\ \mathrm{id}\times v\ } & G\times H & \xrightarrow{\ \alpha\ } & H \\
\downarrow & & {\scriptstyle p}\downarrow & & \\
S & \xrightarrow{\ v\ } & H & &
\end{array}
$$

and suppose that there is an isomorphism $\varphi: G\times S\to u^*P$ of principal bundles over S such that $\alpha\circ(\mathrm{id}\times v) = \sigma\circ\widetilde{u}\circ\varphi$. Set $j=(1_G,\mathrm{id}): S\to G\times S$, where 1_G stands for the constant morphism mapping S to the identity element of G. Then we can attach to u, v, and φ the morphism $\widetilde{u}\circ\varphi\circ j: S\to P$. We claim that this provides an inverse to the construction described above. Suppose first that u, v, and φ come from a morphism $\gamma: S\to P$. Then $\widetilde{u}\circ\varphi\circ j$ can also be described as the composition of j with $\mathrm{id}\times\gamma$ and β in diagram (2.18) and therefore is equal to γ. Conversely, start with u, v, and φ and set $\gamma=\widetilde{u}\circ\varphi\circ j$. Then $f\gamma = u\widetilde{f\varphi j} = u$ since φ is an isomorphism of fiber spaces. On the other hand, $\sigma\gamma = \sigma\widetilde{u}\varphi j = \alpha(\mathrm{id}\times v)j = v$. Finally, the isomorphism

$\psi : G \times S \to \gamma^* f^* P = S \times_T P$ is given explicitly by $\psi(g,s) = (s, g \cdot \gamma(s))$. On the other hand,

$$(s, g\gamma(s)) = (s, g \cdot \widetilde{u}\varphi j(s)) = (s, \widetilde{u}\varphi(g,s)) = \varphi(g,s)\,.$$

This concludes the proof of Lemma (2.13).

We are now ready to compare $QCoh([H/G])$ and $QCoh(H,G)$. Lemma (2.13), applied to the basic diagram (2.12), yields an isomorphism between $G \times H$ and $H \times_{[H/G]} H$. The key observation is that, as can be immediately checked, under this isomorphism, a G-linearization of a quasi-coherent sheaf $\mathcal{F}$ on H becomes descent data for $\mathcal{F}$ relative to $\pi : H \to [H/G]$. A quasi-coherent sheaf on $[H/G]$ gives, by pullback, a quasi-coherent sheaf on H endowed with descent data, that is, a G-equivariant quasi-coherent sheaf. This defines a functor from $QCoh([H/G])$ and $QCoh(H,G)$.

Proposition (2.19). *Let G be an algebraic group acting on a scheme H. Then $QCoh([H/G]) \to QCoh(H,G)$ is an equivalence of categories.*

The proof is formally identical to that of Proposition (2.9). Suppose that $\mathcal{F}$ is a quasi-coherent sheaf on H endowed with descent data relative to $H \to [H/G]$. For any section (2.11) of $[H/G]$, pullback via σ produces a quasi-coherent sheaf on P with descent data relative to $P \to T$. Since the latter morphism is faithfully flat, these descent data are effective and produce a quasi-coherent sheaf on T. This yields a functor from $QCoh(H,G)$ to $QCoh([H/G])$. To check that this is an inverse to the one we defined above, one proceeds exactly as with the corresponding statement in (2.9).

Given a quotient stack $[H/G]$, we may introduce, in addition to the Picard group $\mathrm{Pic}([H/G])$, a Picard group $\mathrm{Pic}(H,G)$ consisting of the isomorphism classes of G-equivariant line bundles on H, with the group operation induced by tensor product. Pullback via $H \to [H/G]$ yields a homomorphism from $\mathrm{Pic}([H/G])$ to $\mathrm{Pic}(H,G)$. An immediate corollary of (2.19) is then the following:

Corollary (2.20). $\mathrm{Pic}([H/G]) \to \mathrm{Pic}(H,G)$ *is a group isomorphism.*

We now return to more concrete questions having to do with line bundles on the moduli spaces of curves. As a first application of the equivalence between the various notions of coherent sheaf, or of line bundle, on $\overline{\mathcal{M}}_{g,n}$, we shall show, as promised earlier, that the Hodge bundle and its determinant do not descend to coherent sheaves on the moduli space $\overline{M}_{g,n}$, except in genus zero. To see this, it clearly suffices to deal with the Hodge line bundle. Consider a family ξ of curves over a point, that is, a single stable n-pointed curve $(C; p_1, \ldots, p_n)$. The sheaf $\mathbb{E}_\xi$ is just the vector space $H^0(C, \omega_C)$, and an automorphism of the family is just an automorphism of $(C; p_1, \ldots, p_n)$. If $\det \mathbb{E}$ did come from $\overline{M}_{g,n}$, any automorphism of $(C; p_1, \ldots, p_n)$ would act trivially on $\wedge^g H^0(C, \omega_C)$; it then suffices to find a curve $(C; p_1, \ldots, p_n)$ and an automorphism of it which acts nontrivially. When g is odd, we may choose as C any hyperelliptic curve (any elliptic curve when $g = 1$), as automorphism the hyperelliptic involution (the -1 involution around a point of C when $g = 1$) since this acts as multiplication by -1 on $H^0(C, \omega_C)$, and as $\{p_1, \ldots, p_n\}$ any set of n points which is invariant under the involution. When g is even, we choose as C a ramified double covering of an elliptic curve and as $\{p_1, \ldots, p_n\}$ any set of n points which is invariant under the covering involution. The eigenvalues of the covering involution acting on $H^0(C, \omega_C)$ are 1 with multiplicity 1 and -1 with multiplicity $g - 1$. Thus, the covering involution acts as -1 on $\wedge^g H^0(C, \omega_C)$. This proves our claim.

REMARK (2.21). It is important to observe that the Hodge line bundle and the point bundles have been defined not only on the stacks of stable nodal curves but, more generally, also for arbitrary families of nodal curves, and that they behave graciously under base change. That one can do the same also for the boundary line bundles is true but not so obvious, as we will explain in Section 4.

3. The tangent bundle to moduli and related constructions.

Given a smooth Deligne–Mumford stack $\mathcal{M}$, it makes perfect sense to speak of its tangent bundle $\mathcal{T} = T_{\mathcal{M}}$. In fact, given a section F in $\mathcal{M}(S)$ such that S is étale over $\mathcal{M}$, one defines $\mathcal{T}_F$ to be the tangent bundle to S, while, if $G \to F$ is a morphism lying over $\varphi : S' \to S$, the isomorphism $\mathcal{T}_G \to \varphi^* \mathcal{T}_F$ is provided by the differential of φ. In particular, one may take as $\mathcal{M}$ the moduli stack of stable P-pointed genus g curves $\overline{\mathcal{M}}_{g,P}$. In this case the tangent bundle $\mathcal{T}$ can be described somewhat more explicitly using the results of Chapter XI. Consider in fact a stable P-pointed curve $(C; \{x_p\}_{p \in P})$ and let $\mathcal{X} \to (U, u_0)$ be its Kuranishi family. Then, as we know, the tangent space to U at u_0 is canonically isomorphic to $\mathrm{Ext}^1_{\mathcal{O}_C}(\Omega^1_C, \mathcal{O}_C(-\sum x_p))$, which in turn is the dual of $H^0(C, \Omega^1_C \otimes \omega_C(\sum x_p))$. Thus, for any family $F = (f : X \to S, \{\sigma_p\}_{p \in P})$ of stable P-pointed genus g curves, the bundle $\mathcal{T}_F$ is just $f_*(\Omega^1_f \otimes \omega_f(D))^\vee$, where $D = \sum \sigma_p(S)$ stands for the divisor of sections. Likewise, the cotangent bundle $\mathcal{T}^\vee$ to $\overline{\mathcal{M}}_{g,P}$ is given by $\mathcal{T}^\vee_F = f_*(\Omega^1_f \otimes \omega_f(D))$. The canonical bundle of $\overline{\mathcal{M}}_{g,P}$ is the top exterior power of the cotangent bundle; we shall denote by $K_{\overline{\mathcal{M}}_{g,P}}$ its class in the Picard group of $\overline{\mathcal{M}}_{g,P}$. Later in this chapter (Theorem (7.15)) we shall see that $K_{\overline{\mathcal{M}}_{g,P}}$ can be expressed as a linear combination of the Hodge, point, and boundary classes.

Recall that the *normal sheaf to a morphism* of smooth schemes $f : X \to Y$ is, by definition, the quotient $N_f = f^*T_Y/T_X$. This definition has a straightforward extension to morphisms of smooth Deligne–Mumford stacks. We shall describe the normal sheaf to a clutching morphism

$$\xi_\Gamma : \overline{\mathcal{M}}_\Gamma \to \overline{\mathcal{M}}_{g,P}\,, \tag{3.1}$$

where Γ is a stable graph (cf. Section 10 of Chapter XII). In this particular case, N_{ξ_Γ} turns out to be a vector bundle, since ξ_Γ has maximal rank everywhere. As usual, we write $V(\Gamma)$ for the set of vertices of Γ and $E(\Gamma)$ for the set of edges of Γ. We also denote by $H(\Gamma)$ the set of those half-edges of Γ which are not legs.

To describe N_{ξ_Γ}, we begin with a single curve $\widetilde{C}$ in $\overline{\mathcal{M}}_\Gamma$ and let $(C; \{x_p\}_{p\in P})$ be the corresponding curve in $\overline{\mathcal{M}}_{g,P}$. Thus, $\widetilde{C}$ is an L-pointed nodal curve, where $L = L(\Gamma)$ is the set of half-edges of Γ, and C is obtained from it by identifying the points labeled by ℓ and ℓ' whenever $e = \{\ell, \ell'\}$ is an edge of Γ; we denote by y_e the resulting node. Obviously, $\widetilde{C}$ is the partial normalization of C at the y_e. We set $D = \sum x_p$ and write $\widetilde{D}$ for the inverse image of D in $\widetilde{C}$ and R for the divisor in $\widetilde{C}$ consisting of the points which map to nodes of the form y_e. Consider the exact sequence (3.18) in Chapter XI, which we rewrite

$$\begin{aligned} 0 \to \operatorname{Ext}^1(\Omega^1_{\widetilde{C}}, \mathcal{O}_{\widetilde{C}}(-\widetilde{D} - R)) \to \operatorname{Ext}^1(\Omega^1_C, \mathcal{O}_C(-D)) \to \\ \bigoplus_{e\in E(\Gamma)} \operatorname{Ext}^1(\Omega^1_{C,y_e}, \mathcal{O}_{C,y_e}) \to 0\,. \end{aligned}$$

As we explained there, the terms of this exact sequence have the following interpretation. The one on the left is the tangent space to $\overline{\mathcal{M}}_\Gamma$ at $\widetilde{C}$, the middle one is the tangent space to $\overline{\mathcal{M}}_{g,P}$ at $(C; \{x_p\}_{p\in P})$, and the homomorphism connecting them is the differential of ξ_Γ at $\widetilde{C}$. Therefore, the term on the right is canonically isomorphic to the normal space to ξ_Γ at $\widetilde{C}$. On the other hand, formula (3.8) in Chapter XI shows that $\operatorname{Ext}^1(\Omega^1_{C,y_e}, \mathcal{O}_{C,y_e})$ is canonically isomorphic to the tensor product of the tangent spaces to $\widetilde{C}$ at the two points mapping to y_e. Hence, the normal space to ξ_Γ at $\widetilde{C}$ is the direct sum of these lines as e runs through the edges of Γ (cf. formula (3.19) in Chapter XI).

These considerations have a straightforward extension to families. As we know, an object F in $\overline{\mathcal{M}}_\Gamma$ is the datum of a family $X_v \to S$ of stable L_v-pointed curves of genus g_v for each vertex v of Γ. We let X be the disjoint union of the X_v. For each $\ell \in L$, we denote by σ_ℓ the corresponding section of $X \to S$. The clutching construction described in Section 7 of Chapter X yields a family $X' \to S$ of stable P-pointed genus g curves. Informally, X' is obtained from X by identifying the section σ_ℓ with the section $\sigma_{\ell'}$ whenever $\{\ell, \ell'\}$ is an edge of Γ. Then the normal bundle N_{ξ_Γ} at F is

$$N_{\xi_\Gamma,F} = \bigoplus_{\{\ell,\ell'\}\in E(\Gamma)} \sigma_\ell^* T_{X/S} \otimes \sigma_{\ell'}^* T_{X/S} = \bigoplus_{\{\ell,\ell'\}\in E(\Gamma)} \mathcal{L}_{\ell,F}^\vee \otimes \mathcal{L}_{\ell',F}^\vee ,$$

where $\mathcal{L}_\ell^\vee$ is the dual of the ℓth point bundle. This can be more precisely expressed as follows. The stack $\overline{\mathcal{M}}_\Gamma$ is the product

$$\overline{\mathcal{M}}_\Gamma = \prod_{v\in V(\Gamma)} \overline{\mathcal{M}}_{g_v,L_v} .$$

We denote by η_v the projection from $\overline{\mathcal{M}}_\Gamma$ to $\overline{\mathcal{M}}_{g_v,L_v}$. Moreover, for each half-edge ℓ, we denote by $v(\ell)$ the vertex of Γ it hangs from. Then

$$N_{\xi_\Gamma} = \bigoplus_{\{\ell,\ell'\}\in E(\Gamma)} \eta_{v(\ell)}^* \mathcal{L}_\ell^\vee \otimes \eta_{v(\ell')}^* \mathcal{L}_{\ell'}^\vee . \tag{3.2}$$

In the same vein, we can compute the "excess intersection bundles" for the intersections of two boundary strata. For this, we go back to Proposition (10.24) in Chapter XII. We have the cartesian diagram

$$\begin{array}{ccc} \overline{\mathcal{M}}_{\Gamma\Gamma'} = \coprod_{\Lambda\in\mathcal{G}_{\Gamma\Gamma'}} \overline{\mathcal{M}}_\Lambda & \xrightarrow{\coprod \xi_{\Lambda\Gamma}} & \overline{\mathcal{M}}_\Gamma \\ \big\downarrow & & \big\downarrow \xi_\Gamma \\ \overline{\mathcal{M}}_{\Gamma'} & \xrightarrow{\xi_{\Gamma'}} & \overline{\mathcal{M}}_{g,P} \end{array} \tag{3.3}$$

Then, following [274], Section 4.3, the excess bundle is given by

$$F_{\Gamma\Gamma'} = \bigoplus_{\Lambda\in\mathcal{G}_{\Gamma\Gamma'}} F_{\Lambda\Gamma\Gamma'}, \tag{3.4}$$

where $F_{\Lambda\Gamma\Gamma'}$ is the bundle on $\overline{\mathcal{M}}_\Lambda$ defined by

$$F_{\Lambda\Gamma\Gamma'} = \xi_{\Gamma'}^*(N_{\xi_{\Gamma'}})/N_{\xi_{\Lambda\Gamma}} . \tag{3.5}$$

The normal bundle to $\xi_{\Lambda\Gamma}$ can be computed in complete analogy with the computation (3.2). By definition of $\mathcal{G}_{\Gamma\Gamma'}$ we have two contractions $c\colon |\Lambda| \to |\Gamma|$ and $c' : |\Lambda| \to |\Gamma'|$ and a decomposition

$$E(|\Lambda|) = c^{-1}(E(|\Gamma|)) \cup c'^{-1}(E(|\Gamma'|)) \tag{3.6}$$

(cf. (10.23) in Chapter XII). The normal bundle to $\xi_{\Lambda\Gamma}$ is then given by

$$N_{\xi_{\Lambda\Gamma}} = \bigoplus_{\{\ell,\ell'\}\in E(|\Lambda|)\smallsetminus c^{-1}(E(|\Gamma|))} \eta_{v(\ell)}^* \mathcal{L}_\ell^\vee \otimes \eta_{v(\ell')}^* \mathcal{L}_{\ell'}^\vee . \tag{3.7}$$

It follows that

$$F_{\Lambda\Gamma\Gamma'} = \bigoplus_{\{\ell,\ell'\}\in c^{-1}(E(|\Gamma|))\cap c'^{-1}(E(|\Gamma'|))} \eta_{v(\ell)}^* \mathcal{L}_\ell^\vee \otimes \eta_{v(\ell')}^* \mathcal{L}_{\ell'}^\vee . \tag{3.8}$$

A corollary is a formula for the class of the determinant of N_{ξ_Γ}:

$$c_1(\wedge^{\max} N_{\xi_\Gamma}) = -\sum_{\ell\in H(\Gamma)} \eta_{v(\ell)}^* \psi_\ell . \tag{3.9}$$

A further important corollary is a formula for the pullback via a clutching morphism of the class δ of the boundary of moduli.

Proposition (3.10). *Let $H(\Gamma)$ be the set of those half-edges of Γ which are not legs. Then*

$$\xi_\Gamma^*\delta = -\sum_{\ell\in H(\Gamma)} \eta_{v(\ell)}^*\psi_\ell + \sum_{v\in V(\Gamma)} \eta_v^*\delta.$$

Here we shall give only a sketch of a possible proof. A full proof, along somewhat different lines, will be found in the next section (cf. Lemma (4.22)). We work in the analytic category. Set $n = |P|$ and let $e_1, \dots, e_k$ be the edges of Γ. Cover the boundary $\mathcal{D}$ of $\overline{\mathcal{M}}_{g,P}$ with a family $\{U_\alpha\}$ of bases of Kuranishi families $\pi_\alpha : \mathcal{C}_\alpha \to U_\alpha$. We may assume that each U_α is a polycylinder in $\mathbb{C}^{3g-3+n}$ with coordinates $t_{\alpha,1}, \dots, t_{\alpha,3g-3+n}$ and that $\mathcal{D}$ is defined in U_α by the equation $t_{\alpha,1}\cdots t_{\alpha,h} = 0$ for some $h \geq k$. We may also assume that the single node of the fiber of π_α at a general point of $t_{\alpha,i} = 0$ comes by clutching along the edge e_i of Γ for $i = 1, \dots, k$. Let V_α be the locus $t_{\alpha,1}\cdots t_{\alpha,k} = 0$ in U_α and consider the restricted family $\mathcal{C}_{\alpha|V_\alpha} \to V_\alpha$. This family carries k sections, labelled by the edges of Γ, entirely consisting of nodes in the fibers. Blowing up these sections yields a family $\widetilde{\pi}_\alpha : \widetilde{\mathcal{C}}_\alpha \to V_\alpha$, which is clearly a Kuranishi family "in $\overline{\mathcal{M}}_\Gamma$". As α varies, the V_α cover all of $\overline{\mathcal{M}}_\Gamma$. For each vertex $v \in V(\Gamma)$, we set $\mathcal{D}_v = \eta_v^*(\mathcal{D})$, where, as usual, $\eta_v : \overline{\mathcal{M}}_\Gamma \to \overline{\mathcal{M}}_{g_v,L_v}$ is the projection. We also write $\mathcal{D}' = \sum \mathcal{D}_v$.

If we set

$$f_{\alpha,1} = \prod_{i=1}^{k} t_{\alpha,i}\ , \qquad f_{\alpha,2} = \prod_{i=k+1}^{h} t_{\alpha,i}\ ,$$

then $\frac{1}{f_{\alpha,1}f_{\alpha,2}}$ is a local generator for the line bundle $\mathcal{O}_{\overline{\mathcal{M}}_{g,P}}(\mathcal{D})$. On the other hand, $f_{\alpha,2}$ is a local equation for $\mathcal{D}'$ in V_α, while $\frac{1}{f_{\alpha,1}}$ is a local generator for $\wedge^k N_{\xi_\Gamma}$, always on V_α. Proposition (3.10) now follows from (3.9).

4. The determinant of the cohomology and some applications.

Let F be a vector bundle on a scheme, or analytic space, X. Loosely speaking, the determinant of F is just $\wedge^{\max}F$. There is a problem with this definition, however, which suggests that we exercise more care. Suppose in fact that we are given an exact sequence of vector bundles

$$\mathcal{E}:\ 0 \to E \xrightarrow{j} F \xrightarrow{p} G \to 0\,. \tag{4.1}$$

We may set up an isomorphism

$$\varphi_\mathcal{E} : \wedge^{\max}E \otimes \wedge^{\max}G \to \wedge^{\max}F$$

by the following prescription. Let $e_1, \dots, e_n$ be sections of E, and $g_1, \dots, g_m$ sections of G, where n and m are the ranks of E and G. Write $g_i = p(f_i)$ for some section f_i of F. Then

$$\varphi_{\mathcal{E}}(e_1 \wedge \cdots \wedge e_n \otimes g_1 \wedge \cdots \wedge g_m) = j(e_1) \wedge \cdots \wedge j(e_n) \wedge f_1 \wedge \cdots \wedge f_m.$$

Clearly, this does not depend on the choice of lifting of $g_1, \dots, g_m$ to F. Now suppose that F is the direct sum of E and G and that j and p are the standard inclusion and projection. Inverting the roles of E and G, we get another exact sequence $0 \to G \to F \to E \to 0$, which gives rise to another isomorphism

$$\wedge^{\max} G \otimes \wedge^{\max} E \to \wedge^{\max} F.$$

Then the diagram

$$\begin{array}{ccc} \wedge^{\max} E \otimes \wedge^{\max} G & \searrow & \\ \downarrow & & \wedge^{\max} F \\ \wedge^{\max} G \otimes \wedge^{\max} E & \nearrow & \end{array}$$

commutes only up to sign, since $e_1 \wedge \cdots \wedge e_n \wedge g_1 \wedge \cdots \wedge g_m = -g_1 \wedge \cdots \wedge g_m \wedge e_1 \wedge \cdots \wedge e_n$ when the ranks of E and G are both odd.

A cure for this problem is to make the determinant "remember" the rank of F, or at least its parity. So one defines a *$\mathbb{Z}/2$-graded line bundle* (or graded line bundle for short) to be a pair (L, r), where L is an ordinary line bundle, and r is the assignment of an integer modulo 2 for each connected component of X, and the *determinant* of F to be the $\mathbb{Z}/2$-graded line bundle

$$\det F = (\wedge^{\max} F, \operatorname{rank} F)\,,$$

where the rank is to be taken modulo 2. To any isomorphism of vector bundles $f : A \to B$ there is attached a natural isomorphism $\det f : \det A \to \det B$.

A graded line bundle (L, a) will be said to be *even* if a is even and *odd* otherwise. The tensor product of two graded line bundles $A = (L, a)$ and $B = (M, b)$ is $A \otimes B = (L \otimes M, a + b)$, and we define the isomorphism

$$\tau = \tau_{A,B} : A \otimes B \to B \otimes A \tag{4.2}$$

by setting

$$\tau(\ell \otimes m) = (-1)^{ab} m \otimes \ell\,.$$

Even line bundles are therefore the same thing as ordinary line bundles. In the sequel, whenever we identify $A \otimes B$ with $B \otimes A$, we do so via the isomorphism τ. Notice that, if C is another graded line bundle, the diagram

$$\begin{array}{ccc} A\otimes B\otimes C & \xrightarrow{\tau_{A\otimes B,C}} & C\otimes A\otimes B \\ & {\scriptstyle \mathrm{id}\otimes\tau_{B,C}}\searrow \qquad \nearrow{\scriptstyle \tau_{A,C}\otimes\mathrm{id}} & \\ & A\otimes C\otimes B & \end{array}$$

commutes. This means that, given a tensor product of several graded line bundles and a second line bundle obtained from the first by shuffling the factors, there is a canonical isomorphism between the two which does not depend on the way the shuffling is performed, but only on the end result.

For any exact sequence as in (4.1), $\varphi_{\mathcal{E}} : \det E \otimes \det G \to \det F$ is an isomorphism of graded line bundles; when $F = E \oplus G$, the two exact sequences

$$\mathcal{E}:\ 0 \to E \to E\oplus G \to G \to 0,$$
$$\mathcal{E}':\ 0 \to G \to E\oplus G \to E \to 0$$

give rise to the commutative diagram

$$\begin{array}{ccc} \det E\otimes\det G & \overset{\varphi_{\mathcal{E}}}{\searrow} & \\ {\scriptstyle\tau}\downarrow & & \det(E\oplus G) \\ \det G\otimes\det E & \underset{\varphi_{\mathcal{E}'}}{\nearrow} & \end{array}$$

The unit graded line bundle, usually written $\mathbb{1}_X$, is $(\mathcal{O}_X, 0)$. Clearly, for any graded line bundle A, there are canonical isomorphisms beween $A\otimes\mathbb{1}$ and A, and between $\mathbb{1}\otimes A$ and A. There is also a notion of inverse of a graded line bundle $A = (L, a)$ on X. This is just $A^{-1} = (\mathcal{H}om(L, \mathcal{O}_X), a)$, which becomes a right inverse of A via the pairing

(4.3)
$$\begin{aligned} A\otimes A^{-1} &\xrightarrow{\sim} \mathbb{1}_X \\ \alpha\otimes\varphi &\mapsto \varphi(\alpha) \end{aligned}$$

and a left inverse of A via the composition of the above pairing with $\tau : A^{-1}\otimes A \to A\otimes A^{-1}$. If B is another graded line bundle, there is a canonical isomorphism between $B^{-1}\otimes A^{-1}$ and $(A\otimes B)^{-1}$, given by $\varphi\otimes\psi \mapsto \chi$ where, for sections α of A and β of B,

$$\chi(\alpha\otimes\beta) = \varphi(\beta)\psi(\alpha)\,.$$

The diagram

$$\begin{array}{ccc} B^{-1}\otimes A^{-1} & \longrightarrow & (A\otimes B)^{-1} \\ {\scriptstyle\tau_{A^{-1},B^{-1}}}\downarrow & & \downarrow{\scriptstyle\tau^{\vee}_{B,A}} \\ A^{-1}\otimes B^{-1} & \longrightarrow & (B\otimes A)^{-1} \end{array}$$

commutes, where the superscript $\vee$ stands for adjoint.

The notion of determinant extends to finite complexes of vector bundles on X. One defines the determinant of a complex $F^\bullet$ to be

$$\det(F^\bullet) = \cdots \otimes (\det F^q)^{(-1)^q} \otimes (\det F^{q-1})^{(-1)^{q-1}} \otimes \cdots .$$

Given a short exact sequence of complexes

$$\mathcal{E}: \; 0 \to E^\bullet \to F^\bullet \to G^\bullet \to 0,$$

there is the isomorphism

$$\varphi_{\mathcal{E}} : \det E^\bullet \otimes \det G^\bullet \to \det F^\bullet \tag{4.4}$$

defined by

$$\begin{aligned} &\cdots \otimes (\det E^q)^{(-1)^q} \otimes (\det E^{q-1})^{(-1)^{q-1}} \otimes \cdots \\ &\otimes \cdots \otimes (\det G^q)^{(-1)^q} \otimes (\det G^{q-1})^{(-1)^{q-1}} \otimes \cdots \\ &\cong \cdots \otimes (\det E^q)^{(-1)^q} \otimes (\det G^q)^{(-1)^q} \\ &\qquad \otimes (\det E^{q-1})^{(-1)^{q-1}} \otimes (\det G^{q-1})^{(-1)^{q-1}} \otimes \cdots \\ &\cong \cdots \otimes (\det F^q)^{(-1)^q} \otimes (\det F^{q-1})^{(-1)^{q-1}} \otimes \cdots = \det(F^\bullet) \end{aligned}$$

Clearly, if we view a vector bundle as a complex concentrated in degree zero, the notion of determinant and the isomorphism $\varphi_{\mathcal{E}}$ for complexes are direct generalizations of the analogous notions for vector bundles.

The determinant construction enjoys several formal properties, which we shall mostly state without proof, referring the reader to [425] for a complete treatment. We notice however that the proofs of many of these properties are rather straightforward but tedious verifications whose main difficulty is keeping the signs straight. First of all, the following is clear:

PROPERTY (4.5) (DETERMINANT 1). *The determinant and the isomorphisms $\varphi_{\mathcal{E}}$ are functorial in the base space X.*

Next, suppose that we are given a commutative diagram of complexes with exact rows and columns,

$$\begin{array}{ccccccccc} & & 0 & & 0 & & 0 & & \\ & & \downarrow & & \downarrow & & \downarrow & & \\ 0 & \longrightarrow & A^\bullet & \longrightarrow & B^\bullet & \longrightarrow & C^\bullet & \longrightarrow & 0 \\ & & \downarrow & & \downarrow & & \downarrow & & \\ 0 & \longrightarrow & A'^\bullet & \longrightarrow & B'^\bullet & \longrightarrow & C'^\bullet & \longrightarrow & 0 \\ & & \downarrow & & \downarrow & & \downarrow & & \\ 0 & \longrightarrow & A''^\bullet & \longrightarrow & B''^\bullet & \longrightarrow & C''^\bullet & \longrightarrow & 0 \\ & & \downarrow & & \downarrow & & \downarrow & & \\ & & 0 & & 0 & & 0 & & \end{array}$$

Denote by $\mathcal{E}_1$, $\mathcal{E}_2$, $\mathcal{E}_3$ (resp., $\mathcal{R}_1$, $\mathcal{R}_2$, $\mathcal{R}_3$) the first, second, and third columns (resp., rows) of this diagram. Then:

PROPERTY (4.6) (DETERMINANT 2). *The composite isomorphism*

$$\det A^\bullet \otimes \det C^\bullet \otimes \det A''^\bullet \otimes \det C''^\bullet \xrightarrow{\varphi_{\mathcal{R}_1}\otimes\varphi_{\mathcal{R}_3}} \det B^\bullet \otimes \det B''^\bullet \xrightarrow{\varphi_{\mathcal{E}_2}} \det B'^\bullet$$

is equal to the composite isomorphism

$$\begin{aligned} &\det A^\bullet \otimes \det C^\bullet \otimes \det A''^\bullet \otimes \det C''^\bullet \\ &\xrightarrow{\mathrm{id}\otimes\tau_{C^\bullet,A''^\bullet}\otimes\mathrm{id}} \det A^\bullet \otimes \det A''^\bullet \otimes \det C^\bullet \otimes \det C''^\bullet \cdots \\ &\xrightarrow{\varphi_{\mathcal{E}_1}\otimes\varphi_{\mathcal{E}_3}} \det A'^\bullet \otimes \det C'^\bullet \xrightarrow{\varphi_{\mathcal{R}_2}} \det B'^\bullet . \end{aligned}$$

The reader is encouraged to check that, given two complexes $A^\bullet$, $B^\bullet$, and letting $\mathcal{E}$, $\mathcal{E}'$ be the exact sequences $0 \to A^\bullet \to A^\bullet \oplus B^\bullet \to B^\bullet \to 0$ and $0 \to B^\bullet \to A^\bullet \oplus B^\bullet \to A^\bullet \to 0$, a special instance of property (4.6) yields the commutativity of

(4.7)
$$\begin{array}{ccc} \det A^\bullet \otimes \det B^\bullet & \xrightarrow{\varphi_{\mathcal{E}}} & \\ \tau\downarrow & & \det(A^\bullet \oplus B^\bullet) \\ \det B^\bullet \otimes \det A^\bullet & \xrightarrow{\varphi_{\mathcal{E}'}} & \end{array}$$

A fundamental and surprising property of determinants is the following.

LEMMA (4.8). *Let $A^\bullet$ be a finite acyclic complex of vector bundles on X. There is a canonical isomorphism*

$$\det(A^\bullet) \cong \mathbb{1}_X .$$

The proof is by induction on the length of $A^\bullet$. When $A^\bullet$ is the zero complex, its determinant is canonically isomorphic to $\mathbb{1}_X$ by definition. Next, suppose that $A^\bullet$ has length two, i.e., that it is of the form

$$\cdots \to 0 \to A^i \xrightarrow{u} A^{i+1} \to 0 \to \cdots ,$$

where u is an isomorphism. Then $\det(A^\bullet)$ equals $\det(A^{i+1}) \otimes \det(A^i)^{-1}$ or $\det(A^{i+1})^{-1} \otimes \det(A^i)$, depending on the parity of i. These two tensor products are canonically isomorphic to $\mathbb{1}_X$. For the first, the isomorphism is

$$\det(A^{i+1}) \otimes \det(A^i)^{-1} \xrightarrow{\det(u^{-1})\otimes\mathrm{id}} \det(A^i) \otimes \det(A^i)^{-1} \xrightarrow{\sim} \mathbb{1}_X ,$$

where the isomorphism on the right is the standard pairing (4.3). One proceeds similarly for the other tensor product. When the length of

$$\cdots \to A^{i-1} \xrightarrow{u_{i-1}} A^i \xrightarrow{u_i} A^{i+1} \to \cdots$$

is greater than two, we may break up $A^\bullet$ into shorter complexes as follows. Fix an index i. Then $K = \ker u_i$ is a vector bundle, and we get two acyclic complexes $B^\bullet$ and $C^\bullet$ by setting $B^h = A^h$ for $h < i$, $B^i = K$, $B^h = 0$ for $h > i$, and $C^\bullet = A^\bullet / B^\bullet$. If i is such that A^{i-1} and A^{i+1} are nonzero, $B^\bullet$ and $C^\bullet$ have shorter length than $A^\bullet$, and hence we may inductively assume that their determinants are canonically isomorphic to $\mathbb{1}_X$. We then define the canonical isomorphism from $\det(A^\bullet)$ to $\mathbb{1}_X$ as the composition

$$\det(A^\bullet) \xrightarrow{\varphi_{\mathcal{E}}^{-1}} \det(B^\bullet) \otimes \det(C^\bullet) \to \mathbb{1}_X \otimes \mathbb{1}_X \to \mathbb{1}_X ,$$

where $\mathcal{E}$ is the exact sequence $0 \to B^\bullet \to A^\bullet \to C^\bullet \to 0$. It follows from property (4.6) of determinants that this isomorphism does not depend on the choice of the index i.

The canonical isomorphism between the determinant of an acyclic complex and the unit line bundle is just a special case of a more general object. In fact, any quasi-isomorphism of finite complexes of vector bundles gives rise to an isomorphism between the respective determinants.

LEMMA (4.9) (DETERMINANT 3). *Let $f : A^\bullet \to B^\bullet$ be a quasi-isomorphism of finite complexes of vector bundles. There is an isomorphism*

$$\det f : \det(A^\bullet) \xrightarrow{\sim} \det(B^\bullet)$$

which depends only on the homotopy class of f. This makes det *into a functor*

$$\left\{ \begin{array}{c} \text{finite complexes of vector bundles} \\ \text{and quasi-isomorphisms} \end{array} \right\} \longrightarrow \left\{ \begin{array}{c} \text{graded line bundles} \\ \text{and isomorphisms} \end{array} \right\} .$$

Furthermore, for any commutative diagram

$$\begin{array}{ccccccccc} 0 & \longrightarrow & A_1^\bullet & \longrightarrow & A^\bullet & \longrightarrow & A_2^\bullet & \longrightarrow & 0 \\ & & \downarrow{\scriptstyle f_1} & & \downarrow{\scriptstyle f} & & \downarrow{\scriptstyle f_2} & & \\ 0 & \longrightarrow & B_1^\bullet & \longrightarrow & B^\bullet & \longrightarrow & B_2^\bullet & \longrightarrow & 0 \end{array}$$

with exact rows $\mathcal{E}$, $\mathcal{E}'$ such that f, f_1, f_2 are quasi-isomorphisms, the diagram

$$\begin{array}{ccc} \det(A_1^\bullet) \otimes \det(A_2^\bullet) & \xrightarrow{\varphi_{\mathcal{E}}} & \det(A^\bullet) \\ {\scriptstyle \det f_1 \otimes \det f_2} \Big\downarrow & & \Big\downarrow {\scriptstyle \det f} \\ \det(B_1^\bullet) \otimes \det(B_2^\bullet) & \xrightarrow{\varphi_{\mathcal{E}'}} & \det(B^\bullet) \end{array}$$

commutes. Finally, if we regard an isomorphism $g : A \to B$ of vector bundles as a quasi-isomorphism f of complexes concentrated in degree zero, $\det f$ is equal to $\det g$.

We shall just sketch an outline of the construction of $\det f$. Assume first that f is injective and that $C^\bullet = B^\bullet/A^\bullet$ is a complex of vector bundles. Then $C^\bullet$ is acyclic, and hence its determinant is canonically isomorphic to $1\!\!1_X$. The exact sequence

$$\mathcal{E}: \quad 0 \to A^\bullet \xrightarrow{f} B^\bullet \to C^\bullet \to 0 \tag{4.10}$$

gives rise to an isomorphism $\varphi_{\mathcal{E}} : \det(A^\bullet) \otimes \det(C^\bullet) \to \det(B^\bullet)$, and we define $\det f$ to be the composition

$$\det(A^\bullet) \xrightarrow{\sim} \det(A^\bullet) \otimes 1\!\!1_X \xrightarrow{\sim} \det(A^\bullet) \otimes \det(C^\bullet) \xrightarrow{\varphi_{\mathcal{E}}} \det(B^\bullet)\,.$$

One can give a similar definition of $\det f$ when f is onto. In this case, in fact, the kernel $K^\bullet$ of f is an acyclic complex of vector bundles, and from the exact sequence

$$\mathcal{E}': \quad 0 \to K^\bullet \to A^\bullet \xrightarrow{f} B^\bullet \to 0$$

one gets isomorphisms

$$\det(B^\bullet) \xrightarrow{\sim} 1\!\!1_X \otimes \det(B^\bullet) \xrightarrow{\sim} \det(K^\bullet) \otimes \det(B^\bullet) \xrightarrow{\varphi_{\mathcal{E}'}} \det(A^\bullet)\,,$$

and one defines $\det f$ to be the inverse of their composition. It is easy to show that, when the sequence (4.10) is split, and $g : B^\bullet \to A^\bullet$ is a left inverse of f, the determinants of f and g, as defined above, are inverse to each other.

Now we turn to the general case. We let $Z^\bullet$ be the mapping cylinder of f. Thus,

$$Z^i = A^i \oplus B^i \oplus A^{i+1}\,,$$

and the differential $d : Z^i \to Z^{i+1}$ is

$$d(a, b, a') = (da - a', db + f(a'), -da')\,.$$

Then

$$\begin{array}{cc} \alpha : A^\bullet \to Z^\bullet & \qquad \beta : B^\bullet \to Z^\bullet \\ \alpha(a) = (a, 0, 0) & \qquad \beta(b) = (0, b, 0) \end{array}$$

are injective quasi-isomorphisms whose cokernels are complexes of vector bundles, and we set $\det f = (\det\beta)^{-1} \circ \det\alpha$. Equivalently, we may notice that β has a left inverse $\gamma : Z^\bullet \to B^\bullet$ given by $\gamma(a, b, a') = f(a) + b$ with the property that $\gamma\alpha = f$, and we may set $\det f = \det\gamma \circ \det\alpha$. It is not difficult to show that the determinant as defined here is equal, in the special cases envisioned above, to the determinant as previously defined.

Finally, let $g : A^\bullet \to B^\bullet$ be another quasi-isomorphism and denote by $W^\bullet$ its mapping cylinder and by $\alpha' : A^\bullet \to W^\bullet$ and $\beta' : B^\bullet \to W^\bullet$

the analogues of α and β. Suppose that g is homotopic to f, i.e., that there is $h : A^\bullet \to B^\bullet[-1]$ such that

$$f - g = dh + hd\,.$$

Setting $u(a, b, a') = (a, b + h(a'), a')$ defines an isomorphism $u : Z^\bullet \to W^\bullet$, and the diagram

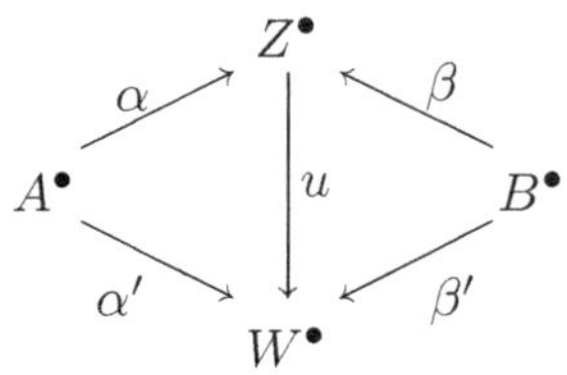

commutes. This shows that

$$(\det \beta')^{-1} \det \alpha' = (\det \beta')^{-1} \det u \det \alpha = (\det \beta)^{-1} \det \alpha\,,$$

completing the outline of the construction of $\det f$.

PROPERTY (4.11) (DETERMINANT 4). *Let there be given exact sequences*

$$\begin{aligned} \mathcal{E} : &\qquad 0 \to A^\bullet \xrightarrow{\alpha} B^\bullet \to 0 \to 0, \\ \mathcal{E}' : &\qquad 0 \to 0 \to B^\bullet \xrightarrow{\beta} C^\bullet \to 0. \end{aligned}$$

Then the isomorphism $\det A^\bullet \to \det B^\bullet$ *obtained by composing* $a \mapsto a \otimes 1$ *with* $\varphi_{\mathcal{E}}$ *is equal to* $\det \alpha$, *and the isomorphism* $\det B^\bullet \to \det C^\bullet$ *obtained by composing* $b \mapsto 1 \otimes b$ *with* $\varphi_{\mathcal{E}'}$ *is equal to* $\det \beta$.

It can be shown that in fact properties (4.5), (4.6), (4.9), and (4.11) (i.e., Determinant 1–4) characterize the determinant functor and the multiplication isomorphism φ.

When $\mathcal{F}$ is a coherent sheaf on a complete scheme X, we may define the determinant of the cohomology of $\mathcal{F}$ to be the line

$$\cdots \otimes (\det H^{q+1}(X, \mathcal{F}))^{(-1)^{q+1}} \otimes (\det H^q(X, \mathcal{F}))^{(-1)^q} \otimes \cdots\,.$$

More generally, suppose that we are given a proper morphism $f : X \to S$ and a coherent sheaf $\mathcal{F}$ on X. One would then be tempted to call determinant of the cohomology the line bundle

$$\cdots \otimes (\det R^{q+1} f_* \mathcal{F})^{(-1)^{q+1}} \otimes (\det R^q f_* \mathcal{F})^{(-1)^q} \otimes \cdots\,. \tag{4.12}$$

We have already encountered an instance of this construction, namely the Hodge line bundle for a family $\pi : X \to S$ of stable curves. This was defined as the determinant of $\pi_* \omega_\pi$. In fact, the higher direct image $R^1 \pi_* \omega_\pi$ is canonically trivial, so the Hodge line bundle turns out to be the determinant of the cohomology of the relative dualizing sheaf ω_π. Notice that, from this point of view, it lives in degree $g-1$, and not g.

The problem with (4.12) is that, for the time being, it only makes sense when all the direct images $R^q f_* \mathcal{F}$ happen to be locally free. The key to a solution is essentially Lemma (4.9), stating that a quasi-isomorphism of complexes of vector bundles induces a canonical isomorphism of the respective determinants. Roughly speaking, one may then define, locally on S, the determinant of the cohomology of $\mathcal{F}$ to be the determinant of a finite complex of vector bundles which calculates the direct images of $\mathcal{F}$, when such a complex exists. Here we will not go into the general theory of the determinant of the cohomology, which is rather delicate, and refer to [425] for a complete treatment, or to Chapter VI of [643] for a lucid summary. What we will present is a simplified treatment, mostly restricted to the morphisms and sheaves that will be of interest to us. Admittedly, our constructions will be rather ad hoc but will save us a considerable amount of work. First we need an algebraic lemma.

LEMMA (4.13). *Let $\pi : X \to S$ be a flat morphism of schemes, and let $\mathcal{F}$ be a coherent sheaf on X, flat over S. Denote by Z the set of points of X where $\mathcal{F}$ is not locally free. Then Z does not contain any component of any fiber of π.*

To see why this is true, recall that, on each sufficiently small open set U, $\mathcal{F}$ can be presented as the cokernel of a homomorphism $\alpha : \mathcal{O}_U^m \to \mathcal{O}_U^n$. Then $\mathcal{F}$ is not locally free precisely at those points where the rank of α drops. Thus, if an entire component of the fiber of π over a point $s \in S$ were contained in Z, the rank of α would drop along the whole component. This could already be detected on a sufficiently thick infinitesimal neighborhood of s. We are thus reduced to proving the lemma under the additional hypothesis that S is the spectrum of an artinian local ring A. Denote by $\mathfrak{m}$ the maximal ideal of A, let $U = \operatorname{Spec} B$ be an affine open subset of X, and denote by M the B-module which corresponds to the restriction of $\mathcal{F}$ to U. We must show that U cannot be contained in Z. The central fiber of $U \to S$, that is, $U_s = \operatorname{Spec} B/\mathfrak{m}B$, is reduced; shrinking U, we may even suppose that it is smooth. Let x be a point of U where the number of generators of $\mathcal{F}_x$ as an $\mathcal{O}_{X,x}$-module is minimum possible. Replacing U with a suitable neighborhood of x, we may then suppose that M is generated, as a B-module, by elements $m_1, \dots, m_n$ which give a minimal system of generators of $\mathcal{F}_y$ for every $y \in U$. This system of generators corresponds to a presentation

$$(4.14) \qquad 0 \to R \to B^n \to M \to 0\,,$$

and its minimality at every point of U means that $(B/P)^n \cong M/PM$ for every prime ideal P of B. This implies that all the entries of all the elements of R belong to the nilpotent radical of B, that is, to $\mathfrak{m}B$.

Now tensor (4.14) with $A/\mathfrak{m}$; as M is A-flat, we get an exact sequence. However, since $R \subset (\mathfrak{m}B)^n$, $R/\mathfrak{m}R$ maps to zero in $(B/\mathfrak{m}B)^n$. This means that $R = \mathfrak{m}R$. Multiplying both sides repeatedly by $\mathfrak{m}$ shows that $R = \mathfrak{m}^h R$ for every h. As $\mathfrak{m}$ is nilpotent, we conclude that $R = 0$ and hence that M is a free B-module. This concludes the proof.

Now let $\pi : X \to S$ be a family of nodal curves, and let $\mathcal{F}$ be a coherent sheaf on X, flat over S. We would like to define the determinant of the cohomology of $\mathcal{F}$; this will be a line bundle on S, which we will denote by $\mathrm{d}_\pi(\mathcal{F})$. The construction will be done locally on S, and we will then show that the resulting line bundles patch together canonically. The flatness assumption is actually unnecessary but simplifies the construction. We may cover S with open sets U such that there is an effective Cartier divisor D in $\pi^{-1}(U)$ which meets all the irreducible components of every fiber and does not contain any of them; in particular, D is relatively ample. We may thus suppose, replacing D with a multiple if necessary, that $R^1\pi_*\mathcal{F}(D) = 0$. By the previous lemma, we may also suppose that $\mathcal{F}$ is locally free at every point of D. For the purposes of the construction, a divisor with these properties will be said to be admissible. Then $\mathcal{F}$ is a subsheaf of $\mathcal{F}(D)$, and we set $\mathcal{F}(D)_{|D} = \mathcal{F}(D)/\mathcal{F}$. Moreover, $\pi_*\mathcal{F}(D)$ and $\pi_*\mathcal{F}(D)_{|D}$ are locally free, and there is an exact sequence

$$(4.15) \qquad 0 \to \pi_*\mathcal{F} \to \pi_*\mathcal{F}(D) \to \pi_*\mathcal{F}(D)_{|D} \to R^1\pi_*\mathcal{F} \to 0\,.$$

In other words, the two-stage complex of vector bundles

$$(E_D^0 \to E_D^1) \quad = \quad (\pi_*\mathcal{F}(D) \to \pi_*\mathcal{F}(D)_{|D})$$

calculates the direct image cohomology of $\mathcal{F}$. The idea is to define $\mathrm{d}_\pi(\mathcal{F})$ to be, locally, just $\det(E_D^\bullet)$. This requires checking independence on D. Let D' be another admissible divisor. Suppose first that $D' \geq D$. In this case there is an inclusion of complexes $j_{D'D} : E_D^\bullet \to E_{D'}^\bullet$ which is a quasi-isomorphism and hence an isomorphism $\det j_{D'D} : \det(E_D^\bullet) \to \det(E_{D'}^\bullet)$. We know from the properties of determinants that, when D'' is another admissible divisor such that $D'' \geq D'$, then

$$(4.16) \qquad \det(j_{D''D'} j_{D'D}) = \det j_{D''D'} \det j_{D'D}\,.$$

Next, we define a canonical isomorphism $\rho_{D'D} : E_D^\bullet \to E_{D'}^\bullet$, where D and D' are any two admissible divisors. Pick an admissible divisor D_1 such that $D_1 \geq D$ and $D_1 \geq D'$. We then let $\rho_{D'D}$ be the composition of $\det j_{D_1D}$ and of the inverse of $\det j_{D_1D'}$. Notice that, when $D' \geq D$, we may choose $D_1 = D'$, and then $\rho_{D'D}$ is just $\det j_{D'D}$. Of course, we must show that the choice of D_1 is irrelevant. This is immediate. If D_2 is another admissible divisor such that $D_2 \geq D$ and $D_2 \geq D'$, pick an admissible divisor D_3 such that $D_3 \geq D_1$ and $D_3 \geq D_2$. That

$(\det j_{D_1D'})^{-1}\det j_{D_1D} = (\det j_{D_3D'})^{-1}\det j_{D_3D} = (\det j_{D_2D'})^{-1}\det j_{D_2D}$ then follows at once from (4.16). It also follows immediately from (4.16) that, given three admissible divisors D, D' and D'', one has that

$$\rho_{D''D} = \rho_{D''D'}\rho_{D'D} .$$

A consequence of this formula is that the various $\det(E_D^\bullet)$ patch together canonically via the isomorphisms $\rho_{D'D}$ to yield a graded line bundle $\mathrm{d}_\pi(\mathcal{F})$ on S (also written simply $\mathrm{d}(\mathcal{F})$ when no confusion can arise). This is the *determinant of the cohomology* of $\mathcal{F}$ (relative to π).

One can similarly define the determinant of the (hyper)cohomology of a complex. Let $\mathcal{F}^\bullet$ be a finite complex of coherent sheaves on X. Let U be a sufficiently small open subset of S, and let D be a divisor in $\pi^{-1}(U)$ which is admissible for $\mathcal{F}^\bullet$ in the sense that it is admissible for each one of the $\mathcal{F}^i$ and that $\mathcal{F}^i \to \mathcal{F}^{i+1}$ is a morphism of vector bundles at each point of D for each i. Then we set

$$E_D^{i,0} = \pi_*(\mathcal{F}^i(D)) , \quad E_D^{i,1} = \pi_*(\mathcal{F}^i(D)_{|D}) ,$$

we regard the double complex $E_D^{\bullet\bullet}$ as a single complex graded by total degree, and we locally define $\mathrm{d}_\pi(\mathcal{F}^\bullet)$ to be its determinant.

The exact sequence (4.15) splits into two exact sequences

$$0 \to \pi_*\mathcal{F} \to E_D^0 \to Q \to 0 , \quad 0 \to Q \to E_D^1 \to R^1\pi_*\mathcal{F} \to 0 .$$

When $\pi_*\mathcal{F}$ and $R^1\pi_*\mathcal{F}$ are locally free, these are exact sequences of vector bundles, whence isomorphisms

$$\det(\pi_*\mathcal{F}) \otimes \det Q \xrightarrow{\sim} \det(E_D^0),$$
$$\det Q \otimes \det(R^1\pi_*\mathcal{F}) \xrightarrow{\sim} \det(E_D^1).$$

We then have a chain of isomorphisms

$$\begin{aligned}
&\det(E_D^1)^{-1} \otimes \det(E_D^0) \to \det(R^1\pi_*\mathcal{F})^{-1} \otimes \det(Q)^{-1} \otimes \det(\pi_*\mathcal{F}) \otimes \det(Q) \\
&\xrightarrow{\mathrm{id}\otimes\mathrm{id}\otimes\tau} \cdots\cdots \to \det(R^1\pi_*\mathcal{F})^{-1} \otimes \det(Q)^{-1} \otimes \det(Q) \otimes \det(\pi_*\mathcal{F}) \\
&\to \det(R^1\pi_*\mathcal{F})^{-1} \otimes \det(\pi_*\mathcal{F}) ,
\end{aligned}$$

showing that $\mathrm{d}_\pi(\mathcal{F}) \cong \det(R^1\pi_*\mathcal{F})^{-1} \otimes \det(\pi_*\mathcal{F})$ when $\pi_*\mathcal{F}$ and $R^1\pi_*\mathcal{F}$ are locally free, in agreement with our tentative definition (4.12). This justifies, in particular, our earlier claim that the Hodge line bundle is just $\mathrm{d}_\pi(\omega_\pi)$. A similar argument shows, more generally, that, for a complex $\mathcal{F}^\bullet$ whose hypercohomology sheaves are locally free,

$$\mathrm{d}_\pi(\mathcal{F}^\bullet) = \cdots \otimes \det(\mathbb{R}^{i+1}\pi_*(\mathcal{F}^\bullet))^{(-1)^{i+1}} \otimes \det(\mathbb{R}^i\pi_*(\mathcal{F}^\bullet))^{(-1)^i} \otimes \cdots .$$

Notice that the construction of the determinant of the cohomology is clearly compatible with base change. In particular, if s is a point of S and we set $X_s = \pi^{-1}(s)$ and $\mathcal{F}^i_s = \mathcal{F}^i \otimes \mathcal{O}_{X_s}$, then there is a natural isomorphism

$$\mathrm{d}(\mathcal{F}^\bullet) \otimes k(s) \cong \bigotimes (\det \mathbb{H}^q(X_s, \mathcal{F}^\bullet_s))^{(-1)^q} .$$

Next, let

$$0 \to \mathcal{E}^\bullet \to \mathcal{F}^\bullet \to \mathcal{G}^\bullet \to 0$$

be an exact sequence of finite complexes of coherent sheaves on X, all flat over S. We shall construct a canonical isomorphism of graded line bundles

$$\varphi : \mathrm{d}_\pi(\mathcal{E}^\bullet) \otimes \mathrm{d}_\pi(\mathcal{G}^\bullet) \xrightarrow{\sim} \mathrm{d}_\pi(\mathcal{F}^\bullet) . \tag{4.17}$$

We proceed locally. Let then U be an open subset of S, and let D be a Cartier divisor on $\pi^{-1}(U)$ which is admissible for $\mathcal{E}^\bullet$, $\mathcal{F}^\bullet$, and $\mathcal{G}^\bullet$. Set

$$\begin{aligned} E^{i,0}_D &= \pi_*(\mathcal{E}^i(D)) , & F^{i,0}_D &= \pi_*(\mathcal{F}^i(D)) , & G^{i,0}_D &= \pi_*(\mathcal{G}^i(D)) , \\ E^{i,1}_D &= \pi_*(\mathcal{E}^i(D)_{|D}) & F^{i,1}_D &= \pi_*(\mathcal{F}^i(D)_{|D}) & G^{i,1}_D &= \pi_*(\mathcal{G}^i(D)_{|D}) . \end{aligned}$$

Then the sequence

$$0 \to E^{\bullet\bullet}_D \to F^{\bullet\bullet}_D \to G^{\bullet\bullet}_D \to 0$$

is exact, whence a canonical isomorphism (4.17), at least on U. The various local representations we obtain by varying U patch together by canonicity.

Next, suppose that we are given a commutative diagram, with exact rows and columns, of complexes of coherent sheaves on X,

$$\begin{array}{ccccccccc}
 & & 0 & & 0 & & 0 & & \\
 & & \downarrow & & \downarrow & & \downarrow & & \\
0 & \longrightarrow & \mathcal{A}^\bullet & \longrightarrow & \mathcal{B}^\bullet & \longrightarrow & \mathcal{C}^\bullet & \longrightarrow & 0 \\
 & & \downarrow & & \downarrow & & \downarrow & & \\
0 & \longrightarrow & \mathcal{A}'^\bullet & \longrightarrow & \mathcal{B}'^\bullet & \longrightarrow & \mathcal{C}'^\bullet & \longrightarrow & 0 \\
 & & \downarrow & & \downarrow & & \downarrow & & \\
0 & \longrightarrow & \mathcal{A}''^\bullet & \longrightarrow & \mathcal{B}''^\bullet & \longrightarrow & \mathcal{C}''^\bullet & \longrightarrow & 0 \\
 & & \downarrow & & \downarrow & & \downarrow & & \\
 & & 0 & & 0 & & 0 & &
\end{array}$$

To each row, or column, of this diagram there is associated a "multiplication" isomorphism φ as in (4.17). An almost immediate, but very useful, consequence of property (4.6) of determinants is then the following.

LEMMA (4.18). *The composite isomorphism*

$$(\mathrm{d}_\pi(\mathcal{A}^\bullet)\otimes \mathrm{d}_\pi(\mathcal{C}^\bullet))\otimes(\mathrm{d}_\pi(\mathcal{A}''^\bullet)\otimes \mathrm{d}_\pi(\mathcal{C}''^\bullet)) \xrightarrow{\varphi\otimes\varphi} \mathrm{d}_\pi(\mathcal{B}^\bullet)\otimes \mathrm{d}_\pi(\mathcal{B}''^\bullet) \xrightarrow{\varphi} \mathrm{d}_\pi(\mathcal{B}'^\bullet)$$

is equal to the composite isomorphism

$$\begin{aligned} &\mathrm{d}_\pi(\mathcal{A}^\bullet)\otimes(\mathrm{d}_\pi(\mathcal{C}^\bullet)\otimes \mathrm{d}_\pi(\mathcal{A}''^\bullet))\otimes \mathrm{d}_\pi(\mathcal{C}''^\bullet)\\ &\quad\xrightarrow{\mathrm{id}\otimes\tau\otimes\mathrm{id}} (\mathrm{d}_\pi(\mathcal{A}^\bullet)\otimes \mathrm{d}_\pi(\mathcal{A}''^\bullet))\otimes(\mathrm{d}_\pi(\mathcal{C}^\bullet)\otimes \mathrm{d}_\pi(\mathcal{C}''^\bullet))\cdots\\ &\quad\xrightarrow{\varphi\otimes\varphi} \mathrm{d}_\pi(\mathcal{A}'^\bullet)\otimes \mathrm{d}_\pi(\mathcal{C}'^\bullet)\xrightarrow{\varphi} \mathrm{d}_\pi(\mathcal{B}'^\bullet)\,. \end{aligned}$$

As we said, the proof is straightforward. The statement is local on S. Let then D be a divisor in $\pi^{-1}(U)$, where U is a sufficiently small open subset of S, which is admissible for all the complexes involved. Set $A_D^{i,0}=\pi_*(\mathcal{A}^i(D))$, $A_D^{i,1}=\pi_*(\mathcal{A}^i(D)_{|D})$, and define in a similar way $B_D^{\bullet\bullet}$, $C_D^{\bullet\bullet}$, and so on. Then the diagram

$$\begin{array}{ccccccccc} & & 0 & & 0 & & 0 & & \\ & & \downarrow & & \downarrow & & \downarrow & & \\ 0 & \longrightarrow & A_D^{\bullet\bullet} & \longrightarrow & B_D^{\bullet\bullet} & \longrightarrow & C_D^{\bullet\bullet} & \longrightarrow & 0\\ & & \downarrow & & \downarrow & & \downarrow & & \\ 0 & \longrightarrow & A'^{\bullet\bullet}_D & \longrightarrow & B'^{\bullet\bullet}_D & \longrightarrow & C'^{\bullet\bullet}_D & \longrightarrow & 0\\ & & \downarrow & & \downarrow & & \downarrow & & \\ 0 & \longrightarrow & A''^{\bullet\bullet}_D & \longrightarrow & B''^{\bullet\bullet}_D & \longrightarrow & C''^{\bullet\bullet}_D & \longrightarrow & 0\\ & & \downarrow & & \downarrow & & \downarrow & & \\ & & 0 & & 0 & & 0 & & \end{array}$$

has exact rows and columns, and the claim follows by applying to it property (4.6) of determinants.

We finally discuss how the determinant of the cohomology construction interacts with relative duality. For simplicity, we limit ourselves to the case of a single vector bundle $\mathcal{F}$. We want to show that there is a canonical isomorphism

$$\mathrm{d}_\pi(\omega_\pi\otimes\mathcal{F}^\vee)\cong \mathrm{d}_\pi(\mathcal{F})\,.$$

This shows, in particular, that the Hodge line bundle can also be viewed as $\mathrm{d}_\pi(\mathcal{O}_X)$. To prove the formula, as usual, we proceed locally on S. Let then D be an admissible divisor for $\mathcal{F}$, and E and admissible divisor for $\omega_\pi\otimes\mathcal{F}^\vee$, both over an open subset U of S. We also assume, as we may, that D and E are disjoint. Consider the commutative diagram of exact sheaf sequences

$$
\begin{array}{ccccccccc}
 & & & & 0 & & 0 & & \\
 & & & & \downarrow & & \downarrow & & \\
 & & 0 & \longrightarrow & \omega_\pi \otimes \mathcal{F}^\vee(-D) & \longrightarrow & \omega_\pi \otimes \mathcal{F}^\vee & \longrightarrow \omega_\pi \otimes \mathcal{F}^\vee_{|D} \longrightarrow 0 & \\
 & & & & \downarrow & & \downarrow & & \\
 & & & & \omega_\pi \otimes \mathcal{F}^\vee(E) & = & \omega_\pi \otimes \mathcal{F}^\vee(E) & & \\
 & & & & \downarrow & & \downarrow & & \\
0 & \longrightarrow & \omega_\pi \otimes \mathcal{F}^\vee(E)_{|D} & \longrightarrow & \omega_\pi \otimes \mathcal{F}^\vee(E)_{|D+E} & \longrightarrow & \omega_\pi \otimes \mathcal{F}^\vee(E)_{|E} & \longrightarrow 0 & \\
 & & & & \downarrow & & \downarrow & & \\
 & & & & 0 & & 0 & &
\end{array}
$$

Notice that the bottom exact sequence is split. Taking direct images, we get another diagram with exact rows and columns,

$$
\begin{array}{ccccccccc}
 & & & & 0 & & 0 & & \\
 & & & & \downarrow & & \downarrow & & \\
 & & & & \pi_*(\omega_\pi \otimes \mathcal{F}^\vee(E)) & = & \pi_*(\omega_\pi \otimes \mathcal{F}^\vee(E)) & & \\
 & & & & \downarrow & & \downarrow & & \\
 & 0 \longrightarrow & \pi_*(\omega_\pi \otimes \mathcal{F}^\vee(E)_{|D}) & \rightarrow & \pi_*(\omega_\pi \otimes \mathcal{F}^\vee(E)_{|D+E}) & \rightarrow & \pi_*(\omega_\pi \otimes \mathcal{F}^\vee(E)_{|E}) & \rightarrow 0 & \\
 & & \| & & \downarrow & & \downarrow & & \\
0 \rightarrow \pi_*(\omega_\pi \otimes \mathcal{F}^\vee) & \rightarrow & \pi_*(\omega_\pi \otimes \mathcal{F}^\vee_{|D}) & \longrightarrow & R^1\pi_*(\omega_\pi \otimes \mathcal{F}^\vee(-D)) & \longrightarrow & R^1\pi_*(\omega_\pi \otimes \mathcal{F}^\vee) & \longrightarrow 0 & \\
 & & & & \downarrow & & \downarrow & & \\
 & & & & 0 & & 0 & &
\end{array}
$$

The complex which calculates $\mathrm{d}(\omega_\pi \otimes \mathcal{F}^\vee)$ is

$$A^\bullet = (\pi_*(\omega_\pi \otimes \mathcal{F}^\vee(E)) \to \pi_*(\omega_\pi \otimes \mathcal{F}^\vee(E)_{|E}))\,.$$

Setting

$$
\begin{aligned}
B^\bullet &= (\pi_*(\omega_\pi \otimes \mathcal{F}^\vee(E)) \to \pi_*(\omega_\pi \otimes \mathcal{F}^\vee(E)_{|D+E}))\,,\\
C^\bullet &= (0 \to \pi_*(\omega_\pi \otimes \mathcal{F}^\vee(E)_{|D}))\,,
\end{aligned}
$$

we see from the diagram that

$$
\begin{aligned}
\det(B^\bullet) \cong \det(C^\bullet) \otimes \det(A^\bullet) &\cong \det(\pi_*(\omega_\pi \otimes \mathcal{F}^\vee(E)_{|D}))^{-1} \otimes \det(A^\bullet)\,,\\
\det(B^\bullet) &\cong \det(R^1\pi_*(\omega_\pi \otimes \mathcal{F}^\vee(-D)))^{-1}\,,
\end{aligned}
$$

so that

$$
\begin{aligned}
\mathrm{d}_\pi(\omega_\pi \otimes \mathcal{F}^\vee)) \cong \det(A^\bullet) &\cong \det(\pi_*(\omega_\pi \otimes \mathcal{F}^\vee(E)_{|D})) \otimes \det(B^\bullet)\\
&\cong \det(\pi_*(\omega_\pi \otimes \mathcal{F}^\vee_{|D})) \otimes \det(B^\bullet)\\
&\cong \det(\pi_*(\omega_\pi \otimes \mathcal{F}^\vee_{|D})) \otimes \det(R^1\pi_*(\omega_\pi \otimes \mathcal{F}^\vee(-D)))^{-1}\,.
\end{aligned}
$$

This gives the desired isomorphism between $d_\pi(\mathcal{F})$ and $d(\omega_\pi \otimes \mathcal{F}^\vee)$, at least locally on S; in fact, up to a shift by 1 in degrees, the dual of the complex which calculates $d_\pi(\mathcal{F})$ is

$$\pi_*(\omega_\pi \otimes \mathcal{F}^\vee_{|D}) \to R^1\pi_*(\omega_\pi \otimes \mathcal{F}^\vee(-D)).$$

We leave to the reader checking that the isomorphism we have defined does not depend on the choice of D and E, and hence globalizes.

EXAMPLE (4.19) (THE BOUNDARY OF MODULI AS A DETERMINANT). The determinant of the cohomology construction is well suited to producing line bundles on moduli. We already encountered a first instance of this use when we constructed the Hodge line bundle. Here we would like to show that the line bundle $\mathcal{O}(\mathcal{D})$, where $\mathcal{D}$ is the boundary of moduli, can also be interpreted as a determinant of the cohomology. Let

$$\pi : X \to S$$

be a family of connected nodal curves of genus g. Consider the complex $\Omega^1_\pi \to \omega_\pi$, where the term on the left lives in degree -1, and the one on the right in degree 0, and set

$$L_\pi = d_\pi(\Omega^1_\pi \to \omega_\pi).$$

Of course, for this to make sense, we must verify that our assumptions for defining the determinant of the cohomology are met; in particular, we must check that Ω^1_π is S-flat. This can be done in a number of ways; here is a possible argument. First of all, we may assume that S is smooth, since any family of nodal curves can be locally obtained by pullback from a family of nodal curves over a smooth base (cf. Remark (6.10), Chapter XI). Shrinking S, we may assume that there exists an effective divisor D in X which cuts an ample divisor D_s on each fiber $X_s = \pi^{-1}(s)$ and does not contain nodes of the fibers. By part ii) of Proposition (3.5) in Chapter IX, it suffices to show that the Euler–Poincaré characteristic of $\Omega^1_\pi(nD) \otimes k(s)$ is independent of $s \in S$ for all integers n. On the other hand, $\Omega^1_\pi \otimes k(s) = \Omega^1_{X_s}$ sits in an exact sequence

$$0 \to \mathcal{K} \to \Omega^1_{X_s} \to \omega_{X_s} \to \mathcal{Q} \to 0$$

with $\mathcal{K}$ and $\mathcal{Q}$ concentrated at the nodes of X_s, where they have complex one-dimensional stalks (cf. sequence (2.20) in Chapter X and the discussion that follows it). Twisting by nD_s and taking Euler–Poincaré characteristics, we find that

$$\chi(\Omega^1_\pi(nD) \otimes k(s)) = \chi(\omega_{X_s}(nD_s)) = 2g - 2 + nd,$$

where d is the degree of D over S. This proves our claim.

The complex $A^\bullet$, where $A^0 = \omega_\pi$ and $A^i = 0$ for $i \neq 0$, is a subcomplex of $\Omega^1_\pi \to \omega_\pi$, and the quotient is the complex $B^\bullet$, where $B^{-1} = \Omega^1_\pi$ and $B^i = 0$ for $i \neq -1$. Thus,

$$L_\pi = \mathrm{d}_\pi(\omega_\pi)\,\mathrm{d}_\pi(\Omega^1_\pi)^{-1}.$$

The graded line bundle L_π is even, since $\mathrm{d}_\pi(\Omega^1_\pi)$ and $\mathrm{d}_\pi(\omega_\pi)$ have the same parity. Moreover, it has a canonical section. At points s such that X_s is smooth, this section is the isomorphism $\mathrm{d}(\Omega^1_{X_s}) \xrightarrow{\sim} \mathrm{d}(\omega_{X_s})$ induced by the isomorphism $\Omega^1_{X_s} \xrightarrow{\sim} \omega_{X_s}$. Elsewhere, we have to be more careful. We proceed locally on S, which we will hence assume to be as small as needed. Let D be a divisor in X which is admissible for both Ω^1_π and ω_π. The natural sheaf homomorphism $\rho_\pi : \Omega^1_\pi \to \omega_\pi$ induces homomorphisms

$$\begin{aligned} \rho^0_\pi &: \pi_*(\Omega^1_\pi(D)) \to \pi_*(\omega_\pi(D)),\\ \rho^1_\pi &: \pi_*(\Omega^1_\pi(D) \otimes \mathcal{O}_D) \to \pi_*(\omega_\pi(D) \otimes \mathcal{O}_D). \end{aligned}$$

The second of these is always an isomorphism, whence another isomorphism

$$\wedge^{\max}\rho^1_\pi : \wedge^{\max}\pi_*(\Omega^1_\pi(D) \otimes \mathcal{O}_D) \xrightarrow{\sim} \wedge^{\max}\pi_*(\omega_\pi(D) \otimes \mathcal{O}_D)$$

and a homomorphism

$$\wedge^{\max}\rho^0_\pi : \wedge^{\max}\pi_*(\Omega^1_\pi(D)) \to \wedge^{\max}\pi_*(\omega_\pi(D)).$$

Putting the two together, we get a homomorphism

$$(\wedge^{\max}\rho^1_\pi)^\vee \otimes \wedge^{\max}\rho^0_\pi : \mathrm{d}_\pi(\Omega^1_\pi) \to \mathrm{d}_\pi(\omega_\pi),$$

where $\vee$ stands for adjoint. It can be checked, along the lines followed in establishing the independence on D of the definition of determinant of the cohomology, that this homomorphism does not depend on the choice of D; details are left to the reader. We thus get a canonical section of $\mathrm{d}_\pi(\omega_\pi)\,\mathrm{d}_\pi(\Omega^1_\pi)^{-1}$, which we may indicate by $\det \rho_\pi$.

The various L_π define, in particular, a line bundle L on $\overline{\mathcal{M}}_{g,n}$, and the various $\det \rho_\pi$ a canonical section $\det \rho$ of L. Since ρ_π is an isomorphism on smooth fibers of π, this section does not vanish anywhere in $\mathcal{M}_{g,n}$. The upshot is that, denoting by $\mathcal{D}_1, \mathcal{D}_2, \dots$ the components of $\mathcal{D}$, $\det \rho$ gives an isomorphism between L and $\mathcal{O}(\sum n_i \mathcal{D}_i)$, where the n_i are nonnegative integers. We are going to show that in fact all the n_i are equal to 1, so that

$$L \cong \mathcal{O}(\mathcal{D}). \tag{4.20}$$

To do this, we shall examine in detail the case where S is a disk centered at s, X is smooth, and all the fibers of π are smooth except for X_s, which has a single node p. The curve X_s belongs to one of the components of $\mathcal{D}$, say to $\mathcal{D}_i$, and n_i is just the order of vanishing of $\det \rho_\pi$ at s. We shall now calculate this order. Let t ba a coordinate on S centered at s and pick analytic local coordinates x, y on X centered at p such that, in these coordinates, $\pi^*(t) = xy$. Let D be an admissible divisor in X, as in the construction of L_π. Look at the exact sequence

$$(4.21) \qquad 0 \to \mathcal{K} \to \Omega^1_{X_s}(D_s) \to \omega_{X_s}(D_s) \to \mathcal{Q} \to 0\,.$$

Here, $\mathcal{K}$ and $\mathcal{Q}$ are both concentrated at p, and their stalks at p are one-dimensional. The one of $\mathcal{K}$ is generated by the class of ydx, and the one of $\mathcal{Q}$ by the class of any differential having a pole at p. By the choice of D, $H^1(X_s, \Omega^1_{X_s}(D_s)) = H^1(X_s, \Omega^1_{X_s}(D_s)/\mathcal{K})$ vanishes, and hence the sequence (4.21) gives rise to the exact sequence

$$0 \to H^0(X_s, \mathcal{K}) \to H^0(X_s, \Omega^1_{X_s}(D_s)) \to H^0(X_s, \omega_{X_s}(D_s)) \to H^0(X_s, \mathcal{Q}) \to 0\,.$$

We may thus choose a frame $\varphi_1, \varphi_2, \ldots$ for $\pi_*(\Omega^1_\pi(D))$ in such a way that φ_1 reduces to the class of ydx in $H^0(X_s, \Omega^1_{X_s}(D_s))$, while the restrictions to X_s of the sections $\rho_\pi(\varphi_2), \rho_\pi(\varphi_3), \ldots$ are independent elements of $H^0(X_s, \omega_{X_s}(D_s))$. Notice that, by construction, these last differentials are regular along the two branches of the node. On the other hand, up to higher-order terms, $\rho_\pi(\varphi_1)$ is just

$$\rho_\pi(ydx) = xy\frac{dx}{x} = t\frac{dx}{x}\,.$$

This shows that $\rho_\pi(\varphi_1) = t\psi$, where ψ is a section of $\omega_\pi(D)$ on X which has poles at the node. Thus, $\psi, \rho_\pi(\varphi_2), \rho_\pi(\varphi_3), \ldots$ is a frame for $\pi_*(\omega_\pi(D))$, and $\det \rho_\pi$ vanishes to order 1 at s. This proves (4.20).

What we have just shown makes it possible to define the line bundle $\mathcal{O}(\mathcal{D})_\pi$ without stability assumptions on the family of nodal curves $\pi : X \to S$, simply by setting it equal to L_π; the resulting line bundles are clearly well behaved under base change. Likewise, it makes perfect sense to speak of the Chern class δ_π for any family of nodal curves as above. Given a family F of n-pointed nodal curves consisting of a family $\pi : X \to S$ plus n disjoint sections, one can also define a line bundle $\mathcal{O}(\mathcal{D}_i)_F$ for any component $\mathcal{D}_i$ of the boundary of $\overline{\mathcal{M}}_{g,n}$ and consider its Chern class $(\delta_i)_F$.

The interpretation of $\mathcal{O}(\mathcal{D})$ as a determinant leads to a proof of Proposition (3.10) which is different from the one sketched in the previous section. Let Γ be a connected P-pointed genus g graph. As usual, we denote by $V(\Gamma)$ the set of vertices of Γ. We also write $H(\Gamma)$ for the set of the half-edges of Γ which are not legs. Suppose that for each v, we are given a family $\pi_v : X_v \to S$ of connected nodal L_v-pointed genus g_v curves. If $\ell \in L_v$, we denote by σ_ℓ the corresponding section of π_v

and by $\mathcal{L}_\ell$ the corresponding point bundle on S (cf. (2.3)). We also set $D_\ell = \sigma_\ell(S)$. Let X be obtained from the disjoint union of the X_v by identifying D_h and $D_{h'}$ for each edge $\{h, h'\}$ of Γ, and let $\pi : X \to S$ be the obvious morphism. Then $\pi : X \to S$ is naturally a family of P-pointed connected nodal curves of genus g.

LEMMA (4.22). *There is a natural isomorphism*

$$\mathcal{O}(\mathcal{D})_\pi \cong \Big(\bigotimes_{v \in V(\Gamma)} \mathcal{O}(\mathcal{D})_{\pi_v} \Big) \otimes \Big(\bigotimes_{h \in H(\Gamma)} \mathcal{L}_h^{-1} \Big) .$$

Taking Chern classes of the two sides of the above formula, we get (3.10). Notice however that there are no stability or semistability assumptions in (4.22); this will turn out to be quite useful in the sequel.

Now we prove the lemma. We write $\widetilde{X}$ for the disjoint union of the X_v, $\widetilde{\pi}$ for its projection to S, and ν for the quotient morphism $\widetilde{X} \to X$. As by now customary, we denote by $E(\Gamma)$ the set of edges of Γ. If $e = \{h, h'\}$ is an edge of Γ, we set $\Sigma_e = \nu(D_h) = \nu(D_{h'})$. There are exact sequences

$$(4.23) \quad 0 \to \omega_{\widetilde{\pi}} \to \omega_{\widetilde{\pi}}\Big(\sum_{\ell \in H(\Gamma)} D_\ell \Big) \xrightarrow{Res} \bigoplus_{\{h,h'\} \in E(\Gamma)} (\mathcal{O}_{D_h} \oplus \mathcal{O}_{D_{h'}}) \to 0 ,$$

$$(4.24) \qquad 0 \to \omega_\pi \to \nu_*\Big(\omega_{\widetilde{\pi}}\Big(\sum_{\ell \in H(\Gamma)} D_\ell \Big)\Big) \xrightarrow{R} \bigoplus_{e \in E(\Gamma)} \mathcal{O}_{\Sigma_e} \to 0 ,$$

where Res stands for residue along the D_ℓ, and R is obtained by adding the residues along D_h and $D_{h'}$ for each edge $e = \{h, h'\}$. Taking determinants of the cohomology in the above exact sequences produces isomorphisms

$$(4.25) \qquad \mathrm{d}_{\widetilde{\pi}}(\omega_{\widetilde{\pi}}) \cong \mathrm{d}_\pi(\nu_*(\omega_{\widetilde{\pi}})) \cong \mathrm{d}_\pi(\nu_*(\omega_{\widetilde{\pi}}(\textstyle\sum D_\ell))) ,$$

$$(4.26) \qquad \mathrm{d}_\pi(\omega_\pi) \cong \mathrm{d}_\pi(\nu_*(\omega_{\widetilde{\pi}}(\textstyle\sum D_\ell))) ,$$

combining which gives

$$(4.27) \qquad \mathrm{d}_\pi(\omega_\pi) \cong \mathrm{d}_{\widetilde{\pi}}(\omega_{\widetilde{\pi}}) \cong \bigotimes_{v \in V(\Gamma)} \mathrm{d}_{\pi_v}(\omega_{\pi_v}) .$$

All these isomorphisms suffer from sign ambiguities. In fact, (4.25) involves the isomorphism

$$\mathrm{d}_{\widetilde{\pi}}(\textstyle\bigoplus(\mathcal{O}_{D_h} \oplus \mathcal{O}_{D_{h'}})) \cong \mathbb{1}_S ,$$

which depends on the choice of an ordering on each edge $\{h, h'\}$ and changes sign if one of these orderings is reversed. Likewise, (4.26) involves the isomorphism

$$\mathrm{d}_\pi(\textstyle\bigoplus \mathcal{O}_{\Sigma_e}) \cong \mathbb{1}_S ,$$

which depends on the choice of an ordering of the edges; when these are permuted, the isomorphism changes by the sign of the permutation. All these sign ambiguities, however, will be offset by other sign ambiguities, as we shall see below.

There is another exact sheaf sequence relating Kähler differentials on X and $\widetilde{X}$. This is

$$0 \to \mathcal{K} \to \Omega^1_\pi \xrightarrow{a} \nu_*(\Omega^1_{\tilde{\pi}}) \to 0\,,$$

where a is the pullback of forms. The kernel of a is a direct sum $\mathcal{K} = \bigoplus_{e\in E(\Gamma)} \mathcal{K}_e$, where $\mathcal{K}_e$ is a line bundle on Σ_e. Taking the determinants, we find the isomorphism

$$\mathrm{d}_\pi(\Omega^1_\pi) \cong \Big(\bigotimes_{e\in E(\Gamma)} \pi_*\mathcal{K}_e\Big) \otimes \mathrm{d}_{\tilde{\pi}}(\Omega^1_{\tilde{\pi}})\,. \tag{4.28}$$

We claim that

$$\pi_*\mathcal{K}_{\{h,h'\}} \cong \mathcal{L}_h \otimes \mathcal{L}_{h'}\,. \tag{4.29}$$

Putting this together with (4.27) and (4.28) yields the lemma.

The isomorphism (4.29) can be described as follows. Set $e = \{h, h'\}$ and let x and y be local coordinates on the two branches of X meeting at Σ_e, chosen so as to vanish on Σ_e. Then $\mathcal{K}_e$ is locally generated by the class of ydx. Say x is a coordinate on the branch indexed by h, and y a coordinate on the one indexed by h'. The isomorphism (4.29) associates to the class of ydx the section $\sigma_h^*(dx)\otimes\sigma_{h'}^*(dy)$ of $\mathcal{L}_h\otimes\mathcal{L}_{h'}$. We leave to the reader the task of checking that this is a good definition. We just observe that the isomorphism (4.29) depends on the choice of an ordering in each edge $\{h, h'\}$ and changes by a sign if one of these orderings is reversed, since the class of ydx is equal to the one of $-xdy$. In addition, the isomorphism between $\mathrm{d}_\pi(\mathcal{K})$ and $\bigotimes \pi_*\mathcal{K}_e$ depends on the choice of an ordering of $E(\Gamma)$, and a permutation of the edges changes it by the sign of the permutation itself. The sign ambiguities thus introduced neatly compensate those present in (4.27).

REMARK (4.30). In the course of the proof of (4.22) we also proved an important "additivity" property of the divisor class λ. Let Γ be a stable P-pointed graph of genus g. Recall (cf. (10.2) in Chapter XII) that $\overline{\mathcal{M}}_\Gamma$ is a product $\overline{\mathcal{M}}_\Gamma = \prod \overline{\mathcal{M}}_{g_v,L_v}$. For each $v \in V(\Gamma)$, let η_v be the projection from $\overline{\mathcal{M}}_\Gamma$ to $\overline{\mathcal{M}}_{g_v,L_v}$. Then formula (4.27) can be rewritten as

$$\xi_\Gamma^*\lambda = \sum_{v\in V(\Gamma)} \eta_v^*\lambda\,. \tag{4.31}$$

5. The Deligne pairing.

Let $\pi : X \to S$ be a family of nodal curves, and let L and M be two line bundles on X. One can consider the first Chern classes of L and M in the intersection ring of X, intersect them, and push forward the result to a cycle class on S. This is a codimension 1 class, that is, the class of a divisor. One can actually do better. The formalism of the Deligne pairing that we are now going to introduce produces, out of L and M, a line bundle on S whose Chern class is the above push-forward. The Deligne pairing enjoys a number of remarkable properties; for instance, it provides a concrete version, "without denominators," of the Riemann–Roch theorem for line bundles and for the map π.

Although the Deligne pairing is closely tied to the determinant of the cohomology, we will first define it in a way that does not involve determinants. It is only later that we shall make the connection with these. We shall use Weil reciprocity, which we discussed in Section 2 of Appendix B of Chapter VI, and whose statement we briefly recall. Let C be a smooth complete curve, and $D = \sum_{p\in C} n_p p$ a divisor on C. If f is a rational function on C whose divisor (f) is disjoint from D, we set

$$f(D) = \prod_{p\in C} f(p)^{n_p} .$$

The Weil reciprocity law says that, when f and g are rational functions on C with disjoint divisors, then

$$f((g)) = g((f)) . \tag{5.1}$$

A proof can be found, for instance, in the above-mentioned appendix or in [318]. Notice that C need not be connected; all we must ask is that f and g be nonzero on every component of C. Weil reciprocity holds also for nodal curves in the following sense. Let C be a possibly disconnected nodal curve, and let f and g be rational functions on C which do not vanish identically on any irreducible component of C and are regular and nonzero at all the nodes. Then, if the divisors of f and g are disjoint, Weil's identity (5.1) is valid. The proof simply amounts to noticing that what must be proved reduces to the Weil reciprocity formula for the pullbacks of f and g to the normalization of C. There is also a version "with parameters" of Weil's reciprocity, that is, a version for families of nodal curves. By this we mean the following. Let $\pi : X \to S$ be a family, algebraic or analytic, of nodal curves. If D is an effective relative Cartier divisor not containing nodes of fibers, $\pi_*\mathcal{O}_D$ is locally free, and there is a *norm map*

$$Norm_{D/S} : \pi_*\mathcal{O}_D \to \mathcal{O}_S \tag{5.2}$$

carrying a function h to the determinant of the multiplication by h, viewed as an endomorphism of the $\mathcal{O}_S$-module $\pi_*\mathcal{O}_D$. The norm map clearly commmutes with the multiplication of functions and, in particular, induces a sheaf homomorphism

$$Norm_{D/S} : \pi_* \mathcal{O}_D^\times \to \mathcal{O}_S^\times \, .$$

When f is a rational function on X whose divisor is disjoint from D, we set $f(D) = Norm_{D/S}(f)$; notice that this gives back the usual definition when S is a point. More generally, if D is a relative divisor, we can write it as $D_1 - D_2$, where D_1 and D_2 are effective, and we set $f(D) = Norm_{D_1/S}(f) Norm_{D_2/S}(f)^{-1}$, always under the assumption that the divisor of f is disjoint from D. This does not depend on the decomposition of D into a difference of effective divisors, since $f(E_1 + E_2) = f(E_1)f(E_2)$ when E_1 and E_2 are effective. Now let f and g be two meromorphic functions on X whose restrictions to every fiber satisfy the assumptions of Weil reciprocity. In other words, f and g should not vanish identically on any component of any fiber, should be regular and nonzero at all the nodes, and their divisors should be disjoint. Then (5.1) holds as an equality of functions on S. When S is reduced, this follows automatically from Weil reciprocity for a single curve. Otherwise, we notice that, by Remark (6.10) in Chapter XI, $X \to S$ can be embedded, locally on S in the analytic topology, in a family $\mathcal{X} \to \mathcal{S}$ such that $\mathcal{S}$ is smooth, and that f and g extend to meromorphic functions on all of $\mathcal{X}$ if $\mathcal{S}$ is small enough. Since $\mathcal{S}$ is reduced, Weil reciprocity holds for $\mathcal{X} \to \mathcal{S}$ and hence, by pullback, for $X \to S$.

If L is a line bundle, and ℓ is a meromorphic section of L, we denote by (ℓ) the divisor of ℓ, that is, the divisor of zeros of ℓ minus the divisor of poles of ℓ. Now consider two line bundles L and M on a nodal curve C and look at the vector space V freely generated by all pairs (ℓ, m), where ℓ is a rational section of L, and m is a rational section of M, satisfying conditions similar to those of Weil reciprocity, namely:

i) ℓ and m are nonzero on every component of C, and regular and nonzero at all the nodes of C;
ii) the divisors (ℓ) and (m) are disjoint.

Then consider the quotient of V modulo the equivalence relation generated by

$$\begin{aligned} (f\ell, m) &\sim f((m))(\ell, m)\,, \\ (\ell, gm) &\sim g((\ell))(\ell, m)\,, \end{aligned} \tag{5.3}$$

where f and g are rational functions on C. This space is the *Deligne pairing* of L and M and is denoted by $\langle L, M \rangle$. The class of the pair (ℓ, m) is denoted by $\langle \ell, m \rangle$. Using Weil reciprocity, we show that $\langle L, M \rangle$ is a line. We shall refer to the first kind of equivalence in (5.3) as an L-move and to the second as an M-move. It is immediate that the composition of two L-moves is an L-move, and similarly for M-moves. Also, given two pairs (ℓ, m) and (ℓ', m'), one can pass from one to a multiple of the other in at most three moves as follows. Pick a rational section μ of M whose divisor is disjoint from (ℓ) and (ℓ'), and write $\mu = gm$, $m' = g'\mu$, $\ell' = f\ell$, where f, g, and g' are rational functions. Then

$$(\ell', m') \sim g'((\ell'))(\ell', \mu) \sim g'((\ell'))f((\mu))(\ell, \mu) \sim g'((\ell'))f((\mu))g((\ell))(\ell, m).$$

It follows that the dimension of $\langle L, M\rangle$ is at most 1. Obviously, the roles of L and M can be reversed. It remains to show that a pair (ℓ, m) cannot be equivalent to a strict multiple of itself. This is a consequence of Weil reciprocity. Suppose in fact that we have a chain of moves taking a pair (ℓ, m) to a multiple of itself (a cycle, we will say). To show that the constant of proportionality must in fact be 1, we argue by induction on the length of the cycle. By one of the previous remarks we may suppose that, in the cycle, L-moves and M-moves alternate; for the same reason, we may suppose that the number of moves is even. The initial cases of the induction are the ones where the number of moves is 4 or 6 (2 moves means that they are both equalities, so there is nothing to prove). We will do the 4-moves case; the 6-moves case is similar and left to the reader. Suppose then that we have four pairs (ℓ, m), (ℓ', m), (ℓ, m'), and (ℓ', m'). To go from, say, (ℓ', m') to a multiple of (ℓ, m), we have two choices of path. We can go to a multiple of (ℓ, m') via an L-move and follow with an M-move, or go to a multiple of (ℓ', m) via an M-move and then follow with an L-move. We must show that we get the same result. In fact, writing $\ell' = f\ell$, $m' = gm$, where f and g are rational functions, we have

$$(\ell', m') \sim f((m'))(\ell, m') \sim f((m'))g((\ell))(\ell, m) = f((g))f((m))g((\ell))(\ell, m),$$
$$(\ell', m') \sim g((\ell'))(\ell', m) \sim g((\ell'))f((m))(\ell, m) = g((f))f((m))g((\ell))(\ell, m).$$

That the end results are equal follows from Weil reciprocity. Now we do the induction step. Suppose that we have a cycle of $n \geq 8$ moves. We short-circuit the cycle after three moves, going back to the initial point with three moves (the red links in the picture below, which illustrates the case $n = 8$, the original cycle consisting of the black links, and the labels indicating whether the corresponding links represent L-moves or M-moves). This breaks the cycle into two. There seems to be no gain, since we are left with a cycle of 6 moves, for which the result is known, and a cycle of n moves. Notice however that in the latter cycle there are two pairs of adjacent L-moves, or M-moves, each of which can be consolidated into a single move (the green links in Figure 1 below), so that we get in fact a cycle of length $n-2$, and the induction hypothesis applies.

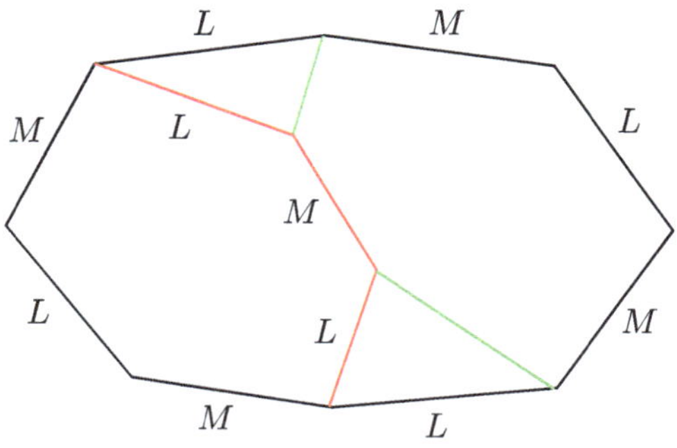

Figure 1. A cycle of 8 moves broken up in two cycles of 6.

The construction we have just completed can be carried out in essentially the same way when we are given a family $\pi : X \to S$ of nodal curves and L and M are line bundles on X. We proceed as follows. First, for any $s \in S$, we perform on the germ of the family at s the same construction we carried out for a single curve. This produces a rank 1 free $\mathcal{O}_{S,s}$-module $\langle L, M\rangle_s$. We now define a coherent sheaf on S by describing its sections. Given an open set U in S, a collection $\{u_s \in \langle L, M\rangle_s : s \in U\}$ is a section over U if and only if, for every $s \in U$, there are a neighborhood U' and rational sections ℓ of L and m of M over $\pi^{-1}(U')$ such that $u_t = \langle \ell, m\rangle$ for every t in U'. The sheaf thus constructed is a line bundle on S, called the *Deligne pairing* of L and M (relative to π) and denoted by $\langle L, M\rangle_\pi$. We will usually drop the subscript unless there is danger of confusion.

The Deligne pairing is clearly compatible with base change. Its main property is bilinearity, which means the following. Given line bundles L_1, L_2, M_1, and M_2 on X, there are well-defined isomorphisms

$$\begin{aligned} &\langle L_1, M\rangle_\pi \otimes \langle L_2, M\rangle_\pi \xrightarrow{\sim} \langle L_1 \otimes L_2, M\rangle_\pi\,, \\ &\langle L, M_1\rangle_\pi \otimes \langle L, M_2\rangle_\pi \xrightarrow{\sim} \langle L, M_1 \otimes M_2\rangle_\pi\,, \end{aligned} \tag{5.4}$$

given by

$$\begin{aligned} &\langle \ell_1, m\rangle \otimes \langle \ell_2, m\rangle \mapsto \langle \ell_1 \otimes \ell_2, m\rangle\,, \\ &\langle \ell, m_1\rangle \otimes \langle \ell, m_2\rangle \mapsto \langle \ell, m_1 \otimes m_2\rangle\,, \end{aligned}$$

plus isomorphisms

$$\begin{aligned} &\langle L, \mathcal{O}_X\rangle_\pi \xrightarrow{\sim} \mathcal{O}_S\,, \\ &\langle \mathcal{O}_X, M\rangle_\pi \xrightarrow{\sim} \mathcal{O}_S\,, \end{aligned} \tag{5.5}$$

given by

$$\langle \ell, 1\rangle \mapsto 1\,,$$
$$\langle 1, m\rangle \mapsto 1\,,$$

for any line bundles L and M.

In the rest of this section, when tensoring line bundles, to make the notation lighter, we will often drop the tensor product sign. The following are self-evident.

PROPOSITION (5.6). *Let $\pi : X \to S$ be a family of nodal curves, and let L_1, L_2, L_3, and M be line bundles on X. Then:*

i) *the composite isomorphisms*

$$\langle L_1, M\rangle_\pi \langle L_2, M\rangle_\pi \langle L_3, M\rangle_\pi \to \langle L_1L_2, M\rangle_\pi \langle L_3, M\rangle_\pi \to \langle L_1L_2L_3, M\rangle_\pi\,,$$
$$\langle L_1, M\rangle_\pi \langle L_2, M\rangle_\pi \langle L_3, M\rangle_\pi \to \langle L_1, M\rangle_\pi \langle L_2L_3, M\rangle_\pi \to \langle L_1L_2L_3, M\rangle_\pi$$

are equal.

ii) *the composite isomorphisms*

$$\langle L_1, \mathcal{O}_X\rangle_\pi \langle L_2, \mathcal{O}_X\rangle_\pi \to \langle L_1L_2, \mathcal{O}_X\rangle_\pi \to \mathcal{O}_S\,,$$
$$\langle L_1, \mathcal{O}_X\rangle_\pi \langle L_2, \mathcal{O}_X\rangle_\pi \to \langle L_1, \mathcal{O}_X\rangle_\pi \to \mathcal{O}_S\,,$$
$$\langle L_1, \mathcal{O}_X\rangle_\pi \langle L_2, \mathcal{O}_X\rangle_\pi \to \mathcal{O}_S\mathcal{O}_S \to \mathcal{O}_S$$

are equal.

Entirely analogous properties hold with respect to the second entry in the Deligne pairing.

Given two line bundles L and M on X, there is a canonical symmetry isomorphism

$$\langle L, M\rangle_\pi \xrightarrow{\tau} \langle M, L\rangle_\pi$$
$$\langle \ell, m\rangle \mapsto \langle m, \ell\rangle$$

which is clearly compatible with base change and with the operations (5.4) and (5.5).

PROPOSITION (5.7). *When $L = M$, the symmetry isomorphism τ is the multiplication by $(-1)^{\deg L}$.*

To prove this assertion notice, first of all, that the square of τ is the identity, and hence τ must be the multiplication by ± 1. To decide what the sign is, it suffices to do the case where S is a point, by continuity, and we may also suppose that X is a smooth curve, since the singular case follows from this by degeneration. As τ is clearly invariant under deformations of L, we may suppose that $L = \mathcal{O}(dp)$, where p is a point of X. Now set $\widetilde{\mathcal{X}} = X \times \mathbb{C}$ and let $\widetilde{\mathcal{L}}$ be the pullback of L to $\widetilde{\mathcal{X}}$. Blow up the point p in the fiber of $\widetilde{\mathcal{X}} \to \mathbb{C}$ above the origin, denote by $\mathcal{X}$ the resulting surface, by P the exceptional divisor, and by $\eta : \mathcal{X} \to \widetilde{\mathcal{X}}$ the blowdown map. Set $\mathcal{L} = \eta^*(\widetilde{\mathcal{L}})(-dP)$. Now $\mathcal{X} \to \mathbb{C}$ is a family of curves, each with a line bundle on. Except for the central one, all fibers are just X, and the line bundle just L. The central fiber instead is singular and consists of X with a $\mathbb{P}^1$ glued at p; as for the line bundle, it is trivial on X and equal to $\mathcal{O}(d)$ on $\mathbb{P}^1$. We are thus reduced to proving the theorem on $\mathbb{P}^1$ for the line bundle $\mathcal{O}(d)$ and, in fact, by the bilinearity of the Deligne pairing, for $\mathcal{O}(1)$. This is a simple explicit calculation. Let t_0, t_1 be homogeneous coordinates on $\mathbb{P}^1$, viewed as sections of $\mathcal{O}(1)$. Then

$$\begin{aligned}\langle t_1, t_0\rangle &= \left\langle \frac{t_1}{t_1 - t_0}(t_1 - t_0), t_0 \right\rangle = \frac{t_1}{t_1 - t_0}\Big|_{t_0=0} \langle t_1 - t_0, t_0\rangle = \langle t_1 - t_0, t_0\rangle \\ &= \left\langle t_1 - t_0, \frac{t_0}{t_1} t_1 \right\rangle = \langle t_1 - t_0, t_1\rangle = \left\langle \frac{t_1 - t_0}{t_0} t_0, t_1 \right\rangle = -\langle t_0, t_1\rangle\,,\end{aligned}$$

as claimed. This concludes the proof of (5.7).

What really gives the Deligne pairing its punch is that it can be expressed in terms of determinants of the cohomology.

THEOREM (5.8). *Let $\pi : X \to S$ be a family of nodal curves, and let L, M be line bundles on X. There is a canonical isomorphism*

$$\langle L, M\rangle_\pi \xrightarrow{\sim} \mathrm{d}_\pi(L \otimes M) \otimes \mathrm{d}_\pi(L)^{-1} \otimes \mathrm{d}_\pi(M)^{-1} \otimes \mathrm{d}_\pi(\mathcal{O}_X) \tag{5.9}$$

compatible with base change.

Proof. We will construct the required isomorphism locally on S (actually, between the stalks at every point of S of the two sheaves involved) and show that the local expressions we get patch together. Let then s be a point of S, and let U be a neighborhood of s. For the purposes of this proof, we shall call a divisor D on $\pi^{-1}(U)$ admissible if it is a relative divisor and does not pass through nodes of fibers. A rational section of a line bundle will be said to be admissible if its divisor is; this is a natural condition from the point of view of Weil reciprocity and just amounts to asking that the section in question be nonzero on all components of all fibers of π and that it be regular and nonzero at every node of every fiber. Suppose first that there exist on $\pi^{-1}(U)$ admissible regular sections ℓ of L and m of M. Suppose moreover that their divisors $Z = (\ell)$ and $W = (m)$ are disjoint. Under these assumptions, there is a commutative diagram of coherent sheaves on $\pi^{-1}(U)$ with exact rows and columns,

$$
\begin{array}{ccccccccc}
 & & 0 & & 0 & & 0 & & \\
 & & \downarrow & & \downarrow & & \downarrow & & \\
0 & \longrightarrow & \mathcal{O} & \xrightarrow{\times \ell} & L & \longrightarrow & L_{|Z} & \longrightarrow & 0 \\
 & & \downarrow{\scriptstyle \times m} & & \downarrow{\scriptstyle \times m} & & \downarrow{\scriptstyle \times m} & & \\
0 & \longrightarrow & M & \xrightarrow[\times \ell]{} & LM & \longrightarrow & LM_{|Z} & \longrightarrow & 0 \\
 & & \downarrow & & \downarrow & & \downarrow & & \\
0 & \longrightarrow & M_{|W} & \xrightarrow[\times \ell]{} & LM_{|W} & \longrightarrow & 0 & \longrightarrow & 0 \\
 & & \downarrow & & \downarrow & & \downarrow & & \\
 & & 0 & & 0 & & 0 & &
\end{array}
\tag{5.10}
$$

where $L_{|Z}$ stands for $L \otimes \mathcal{O}_Z$, and so on. This yields in particular the isomorphism

$$\alpha = \det(\times m) : \mathrm{d}(L_{|Z}) \to \mathrm{d}(LM_{|Z})\,.$$

Out of this isomorphism we shall produce an element

$$
\begin{aligned}
\Psi(\ell, m) &\in \operatorname{Hom}(\mathrm{d}(L)\,\mathrm{d}(M), \mathrm{d}(\mathcal{O})\,\mathrm{d}(LM)) \\
&\cong \mathrm{d}(LM) \otimes \mathrm{d}(L)^{-1} \otimes \mathrm{d}(M)^{-1} \otimes \mathrm{d}(\mathcal{O})\,.
\end{aligned}
$$

Here is the recipe. Pick generators a for $\mathrm{d}(\mathcal{O})$, a' for $\mathrm{d}(L)$, and b for $\mathrm{d}(M)$. Let a'' and b' be the generators of $\mathrm{d}(L_{|Z})$ and $\mathrm{d}(LM)$ such that, under the multiplication isomorphisms associated to the rows of (5.10),

$$a \otimes a'' \mapsto a'\,, \qquad b \otimes \alpha(a'') \mapsto b'\,.$$

We then define $\Psi(\ell, m)$ to be

$$a' \otimes b \mapsto a \otimes b'\,.$$

Let m' be another section of M with divisor disjoint from the divisor of ℓ, and ℓ' another section of L with divisor disjoint from the divisor of m. Write $m' = fm$, $\ell' = h\ell$, where f and h are meromorphic functions. Then $\det(\times m') : \mathrm{d}(L_{|Z}) \to \mathrm{d}(LM_{|Z})$ is clearly equal to $f(Z)\det(\times m)$. This proves the first of the following key identities:

$$\Psi(\ell, fm) = f((\ell))\Psi(\ell, m)\,, \qquad \Psi(h\ell, m) = h((m))\Psi(\ell, m)\,. \tag{5.11}$$

The second identity would also be proved if the roles of L and M, which enter asymmetrically in the definition of Ψ, could be reversed. Let us see that this is indeed the case, up to a harmless sign. We define the generator $\overline{b'}$ of $\mathrm{d}(LM)$ by the requirement that

$$a' \otimes \beta(c) \mapsto \overline{b'}$$

under the multiplication isomorphism associated to the middle column of (5.10), and the isomorphism

$$\Phi(\ell, m) \in \operatorname{Hom}(\mathrm{d}(L)\,\mathrm{d}(M), \mathrm{d}(\mathcal{O})\,\mathrm{d}(LM))$$

to be

$$a' \otimes b \mapsto a \otimes \overline{b'}\,.$$

The generators b' and $\overline{b'}$ are obtained by successive multiplications

$$(a \otimes c) \otimes (a'' \otimes 1) \mapsto b \otimes \alpha(a'') \mapsto b'\,,$$
$$(a \otimes a'') \otimes (c \otimes 1) \mapsto a' \otimes \beta(c) \mapsto \overline{b'}\,,$$

and it follows from Lemma (4.18) that $\overline{b'} = (-1)^{\deg(L)\deg(M)} b'$. Thus,

$$\Phi(\ell, m) = (-1)^{\deg(L)\deg(M)} \Psi(\ell, m)\,, \tag{5.12}$$

which in particular shows that the second equality in (5.11) holds. The identities (5.11) say that $\Psi(\ell, m)$ depends only on $\langle \ell, m \rangle$. This defines the isomorphism (5.9) in the case at hand.

We now turn to the general case. Let ℓ and m be admissible meromorphic sections of L and M over $\pi^{-1}(U)$ with disjoint divisors. Pick disjoint admissible effective divisors D and E such that ℓ and m are regular sections of $L(D)$ and $M(E)$, respectively, and that, moreover, D is disjoint from the divisor of m, and E from the one of ℓ. For good measure, we choose D and E to have even degree. This is not strictly necessary but will later spare us some sign headaches. Let Z and W be the divisors of zeros of ℓ and m as sections of $L(D)$ and $M(E)$. The divisors of ℓ and m as meromorphic sections of L and M are then $Z - D$ and $W - E$. By the special case treated above we may associate to ℓ and m an isomorphism

$$\overline{\Psi}(\ell, m) \in \operatorname{Hom}(\mathrm{d}(L(D))\,\mathrm{d}(M(E)), \mathrm{d}(\mathcal{O})\,\mathrm{d}(LM(D+E)))$$

such that

$$\overline{\Psi}(\ell, fm) = f(Z)\overline{\Psi}(\ell, m)\,, \qquad \overline{\Psi}(h\ell, m) = h(W)\overline{\Psi}(\ell, m)\,, \tag{5.13}$$

whenever f and h are meromorphic functions such that fm and $h\ell$ are regular sections of $M(E)$ and $L(D)$ with divisors disjoint from Z and W, respectively. The inclusions $L \to L(D)$, $M \to M(E)$ and $LM \to LM(D+E)$ give isomorphisms

$$\begin{aligned} &\mathrm{d}(L) \otimes \mathrm{d}(L(D)_{|D}) \xrightarrow{\sim} \mathrm{d}(L(D))\,, \\ &\mathrm{d}(M) \otimes \mathrm{d}(M(E)_{|E}) \xrightarrow{\sim} \mathrm{d}(M(E))\,, \\ &\mathrm{d}(LM) \otimes \mathrm{d}(LM(D+E)_{|D+E}) \xrightarrow{\sim} \mathrm{d}(LM(D+E))\,. \end{aligned} \tag{5.14}$$

On the other hand, since D and E are disjoint,

$$LM(D+E)_{|D+E} \cong LM(D)_{|D} \oplus LM(E)_{|E}\,,$$

whence an isomorphism

$$\mathrm{d}(LM(D+E)_{|D+E}) \cong \mathrm{d}(LM(D)_{|D})\,\mathrm{d}(LM(E)_{|E})\,.$$

Putting this together with the (5.14) shows that

$$\begin{aligned} &\mathcal{H}om(\mathrm{d}(L(D))\,\mathrm{d}(M(E)), \mathrm{d}(\mathcal{O})\,\mathrm{d}(LM(D+E))) \cong \\ &\qquad \mathcal{H}om(\mathrm{d}(L)\,\mathrm{d}(M), \mathrm{d}(\mathcal{O})\,\mathrm{d}(LM)))\otimes \\ &\qquad \mathcal{H}om(\mathrm{d}(L(D)_{|D})\,\mathrm{d}(M(E)_{|E}), \mathrm{d}(LM(D)_{|D})\,\mathrm{d}(LM(E)_{|E}))\,. \end{aligned} \tag{5.15}$$

As ℓ is regular and nonzero along E, and m regular and nonzero along D, there is an isomorphism

$$\begin{gathered} L(D)_{|D} \oplus M(E)_{|E} \xrightarrow{\sim} LM(D)_{|D} \oplus LM(E)_{|E} \\ (a,b) \mapsto (ma, \ell b) \end{gathered}$$

and hence an isomorphism of determinants

$$\Xi(\ell,m) : \mathrm{d}(LM(D)_{|D})\,\mathrm{d}(LM(E)_{|E}) \xrightarrow{\sim} \mathrm{d}(L(D)_{|D})\,\mathrm{d}(M(E)_{|E})\,.$$

Notice that, when f and h are meromorphic functions such that f is regular and nonzero along D, and h is regular and nonzero along E,

$$\Xi(\ell, fm) = f(D)^{-1}\Xi(\ell,m)\,, \qquad \Xi(h\ell, m) = h(E)^{-1}\Xi(\ell,m) \tag{5.16}$$

Keeping in mind the isomorphism (5.15), we now define

$$\Psi(\ell,m) = \overline{\Psi}(\ell,m) \otimes \Xi(\ell,m) \in \mathrm{Hom}(\mathrm{d}(L)\,\mathrm{d}(M), \mathrm{d}(\mathcal{O})\,\mathrm{d}(LM)))\,.$$

Taken together, (5.13) and (5.16) prove that (5.11) holds in general, where of course (ℓ) and (m) stand for the divisors of ℓ and m as meromorphic sections of L and M. This makes it possible to define the isomorphism (5.9) as the one which sends $\langle L, M\rangle$ to $\Psi(\ell,m)$, except for one detail: we must show that the end result does not depend on the choice of D and E. This is quite straightforward. By (5.12), it suffices to show independence from D. Let D' be another choice of divisor. Suppose first that $D' \geq D$ and set $F = D' - D$. Then, if Z' is the divisor of zeros of ℓ as a section of $L(D')$, $Z' = Z + F$. It follows immediately from the definitions that, if we replace D with D', $\overline{\Psi}(\ell,m)$ gets multiplied by $m(F)$, while $\Xi(\ell,m)$ gets multiplied by the inverse of the same quantity. Notice that there are no sign problems, since the divisors D, D', and F all have even degrees, and hence $\mathrm{d}(L(D)_{|D})$, and so on, are all even graded line bundles. Thus, $\Psi(\ell, m)$ does not change if we replace D with D'. In general, we may reach the same conclusion by choosing a D'' such that $D'' \geq D$, $D'' \geq D'$, and reducing to the previous case. Q.E.D.

The circle of ideas involved in the proof of (5.8), coupled with bilinearity, has several interesting consequences. Let D be any relative divisor not passing through nodes of fibers of $\pi : X \to S$. The sheaves $\pi_*(\mathcal{O}_D)$ and $\pi_*(M_{|D})$ are both locally free of rank equal to the degree of D over S. We may then define a line bundle $Norm_{D/S}(M_{|D})$ on S by setting

$$\begin{aligned} Norm_{D/S}(M_{|D}) &= \mathcal{H}om(\det(\pi_*(\mathcal{O}_D)), \det(\pi_*(M_{|D}))) \\ &\cong \det(\pi_*(\mathcal{O}_D))^{-1} \otimes \det(\pi_*(M_{|D})) \,. \end{aligned}$$

We may also define a norm map

$$Norm_{D/S} : \pi_*(M_{|D}) \to Norm_{D/S}(M_{|D})$$

as follows. If h is a section of $\pi_*(M_{|D})$, we let

$$\times h : \pi_*(\mathcal{O}_D) \to \pi_*(M_{|D})$$

be the multiplication by h, and we set

$$Norm_{D/S}(h) = \det(\times h) \,.$$

Then there is a canonical isomorphism

$$\begin{aligned} \langle \mathcal{O}_X(D), M \rangle &\xrightarrow{\sim} Norm_{D/S}(M_{|D}) \\ \langle 1, m \rangle &\mapsto Norm_{D/S}(m_{|D}) \end{aligned} \tag{5.17}$$

where m stands for a rational section of M whose divisor does not meet D and nodes in the fibers.

A special instance of the above occurs when $\pi : X \to S$ has a section $\sigma : S \to X$ not passing through nodes of the fibers, and we let D be the image of σ. Then, for any line bundle M on X, $\sigma^* M$ is canonically isomorphic to $Norm_{D/S}(M_{|D})$, and hence there is a canonical isomorphism

$$\langle \mathcal{O}_X(D), M \rangle \xrightarrow{\sim} \sigma^* M \,. \tag{5.18}$$

Taking $M = \omega_\pi(D)$ in (5.18) and observing that $\sigma^*(\omega_\pi(D))$ is canonically trivial by adjunction, one gets, using bilinearity, a canonical adjunction isomorphism

$$\langle \mathcal{O}_X(D), \omega_\pi \rangle \xrightarrow{\sim} \langle \mathcal{O}_X(D), \mathcal{O}_X(D) \rangle^{-1} \,. \tag{5.19}$$

Another consequence of (5.8) is that

$$c_1(\langle L, M\rangle_\pi) = \pi_*(c_1(L)\cdot c_1(M)) \tag{5.20}$$

in $A^\bullet(S)$, the Fulton–MacPherson operational intersection ring of S (cf. Chapter 17 of [275]). When $L = \mathcal{O}(D)$ with D a relative divisor, the identity follows easily from (5.17). In fact, let j be the inclusion of D in X and set $\alpha = \pi\circ j$. Then $D\cdot c_1(M) = j_*c_1(M_{|D})$ in the cycle ring of X, and hence $\pi_*(D\cdot c_1(M))$ is equal to $\alpha_*c_1(M_{|D}) = c_1(\wedge^{max}\alpha_*(M_{|D})) = c_1(Norm_{D/S}(M_{|D}))$. We now show how the general case can be reduced to this special one. Denoting by a and b the left and right sides of (5.20), what must be proved is that

$$a[W] = b[W]$$

in $A_\bullet(W)$ for any morphism $W\to S$ from a reduced and irreducible W to S. There is a proper birational morphism $\nu:\widehat{W}\to W$ such that $\widehat{W}$ is quasiprojective and normal. By compatibility with proper push-forwards, $a[W] = a\nu_*[\widehat{W}] = \nu_*(a[\widehat{W}])$, and similarly for $b[W]$; hence, we may suppose that W is quasiprojective and normal. By the naturality of the Deligne pairing we are then reduced to proving that $a[S] = b[S]$ under the additional assumption that S is reduced, irreducible, quasiprojective, and normal. Let $\widehat{X}$ be the normalization of X and denote by α the normalization map and by $\widehat{\pi}$ its composition with π. Then $\widehat{\pi}:\widehat{X}\to S$ is a family of nodal curves whose general fiber is smooth. It follows immediately from the definition of the Deligne pairing that $\langle\alpha^*L,\alpha^*M\rangle_{\widehat{\pi}} = \langle L, M\rangle_\pi$. Moreover, there is a string of equalities

$$\begin{aligned}\pi_*(c_1(L)c_1(M))[S] &= \pi_*(c_1(L)c_1(M)[X]) = \pi_*(c_1(L)c_1(M)\alpha_*[\widehat{X}])\\ &= \pi_*\alpha_*(\alpha^*(c_1(L)c_1(M))[\widehat{X}]) = \widehat{\pi}_*(c_1(\alpha^*L)c_1(\alpha^*M)[\widehat{X}])\\ &= \widehat{\pi}_*(c_1(\alpha^*L)c_1(\alpha^*M))[S]\,,\end{aligned}$$

so it suffices to prove that $c_1(\langle\alpha^*L,\alpha^*M\rangle)[S] = \widehat{\pi}_*(c_1(\alpha^*L)c_1(\alpha^*M))[S]$. In conclusion, it is sufficient to prove that $a[S] = b[S]$ under the assumption that S is reduced, irreducible, quasiprojective, and normal, and that the general fiber of $\pi: X\to S$ is smooth.

We may write $L = AB^{-1}$, where A and B are restrictions of very ample line bundles on a completion of X; by bilinearity, we may thus suppose that L is the restriction to X of a very ample line bundle on a completion of X. Let D be a general divisor in $|L|$, denote by T the locus of those points $s\in S$ such that D contains a component of X_s or passes through a node of X_s, and set $U = S\smallsetminus T$. By the generality of D, T has codimension at least 2 in S; it follows in particular that $j^*: A_{n-1}(S)\to A_{n-1}(U)$ is an isomorphism, where j stands for the inclusion of U in S, and n for the dimension of S. Thus, it suffices to show that $j^*(a[S]) = j^*(b[S])$, i.e., that $a[U] = b[U]$ in $A_{n-1}(U)$. As $D \cap \pi^{-1}(U)$ is a relative divisor for $\pi^{-1}(U) \to U$, this follows from the special case of our claim proved above.

Using the Deligne pairing, one can construct new classes in the Picard group of $\overline{\mathcal{M}}_{g,n}$. Consider a family F of stable n-pointed genus g curves, consisting of a family of curves $\pi : X \to S$ plus sections σ_i, $i = 1, \dots, n$, corresponding to divisors $D_i = \sigma_i(S)$. Set, for simplicity, $\widehat{\omega}_\pi = \omega_\pi(\sum D_i)$. Then $\langle \widehat{\omega}_\pi, \widehat{\omega}_\pi \rangle_\pi$ is a well-defined line bundle on S. Since the Deligne pairing is well behaved under base change, this defines a line bundle on $\overline{\mathcal{M}}_{g,n}$. The class of this line bundle in $\mathrm{Pic}(\overline{\mathcal{M}}_{g,n})$ is usually denoted by κ_1. We may somewhat improperly write

$$\kappa_1 = c_1(\langle \widehat{\omega}, \widehat{\omega} \rangle) \,. \tag{5.21}$$

We will see in Section 3 of Chapter XVII that this is a special case of a more general construction which yields a codimension a class κ_a in the cycle group of $\overline{\mathcal{M}}_{g,n}$ for any nonnegative integer a.

Of course, one can replace $\widehat{\omega}_\pi$ in one or both entries of the Deligne pairing with other "natural" line bundles and thus obtain other classes in the Picard group of moduli. However, these can all be expressed in terms of known classes using the bilinearity of the Deligne pairing and the adjunction formulas (5.18) and (5.19). For instance, taking $\omega_\pi = \widehat{\omega}_\pi(-\sum D_j)$ as one of the entries and $\mathcal{O}(D_i)$ as the other, produces the class ψ_i. More generally, taking $\widehat{\omega}_\pi^h(\sum a_i D_i)$ and $\widehat{\omega}_\pi^\ell(\sum b_i D_i)$ as entries, one gets a line bundle whose class is

$$h\ell\kappa_1 - \sum a_i b_i \psi_i \,. \tag{5.22}$$

In particular, if one replaces the logarithmic relative dualizing sheaf $\widehat{\omega}_\pi$ with the ordinary relative dualizing sheaf ω_π, one gets a divisor class $\widetilde{\kappa}_1$ which is related to κ_1 by

$$\widetilde{\kappa}_1 = \kappa_1 - \psi \,. \tag{5.23}$$

The class κ_1 satisfies an "additivity" property analogous to the one enjoyed by the class λ and expressed by formula (4.31). Using the same notation as in Remark (4.30), the additivity property of κ_1 can be written as follows:

$$\xi_\Gamma^* \kappa_1 = \sum_{v \in V(\Gamma)} \eta_v^* \kappa_1 \,. \tag{5.24}$$

In proving this formula, we adopt the notation used in the proof of Lemma (4.22). What must be shown is that

$$\langle \widehat{\omega}_\pi, \widehat{\omega}_\pi \rangle_\pi \cong \langle \widehat{\omega}_{\tilde{\pi}}, \widehat{\omega}_{\tilde{\pi}} \rangle_{\tilde{\pi}} \,. \tag{5.25}$$

A slight modification of (4.24) gives an exact sequence

$$0 \to \widehat{\omega}_\pi \to \nu_*(\widehat{\omega}_{\tilde{\pi}}) \xrightarrow{R} \bigoplus_{e \in E(\Gamma)} \mathcal{O}_{\Sigma_e} \to 0 . \tag{5.26}$$

Taking determinants yields

$$\mathrm{d}_\pi(\widehat{\omega}_\pi) \cong \mathrm{d}_{\tilde{\pi}}(\widehat{\omega}_{\tilde{\pi}}) . \tag{5.27}$$

Now look at the exact sequence

$$0 \to \mathcal{O}_X \to \nu_*(\mathcal{O}_{\widetilde{X}}) \xrightarrow{a} \bigoplus_{e \in E(\Gamma)} \mathcal{O}_{\Sigma_e} \to 0 , \tag{5.28}$$

where a stands for the difference of the restrictions to D_h and $D_{h'}$ for each edge $e = \{h, h'\}$. Of course, this requires choosing an ordering on each e. The sequence (5.26) can be viewed as being obtained from (5.28) by tensoring with $\widehat{\omega}_\pi$ and trivializing $\widehat{\omega}_\pi$ on each Σ_e, $e = \{h, h'\}$ by taking the residue along D_h or along $D_{h'}$, depending on the ordering chosen on e. We may also tensor (5.28) with $\widehat{\omega}_\pi^2$ to get

$$0 \to \widehat{\omega}_\pi^2 \to \nu_*(\widehat{\omega}_{\tilde{\pi}}^2) \xrightarrow{b} \bigoplus_{e \in E(\Gamma)} \mathcal{O}_{\Sigma_e} \to 0 , \tag{5.29}$$

where we have used again the fact that $\widehat{\omega}_\pi$, and hence its square, is canonically trivial on each D_h. In this case, again, the homomorphism b depends on the choice of an ordering on each edge of Γ. Taking determinants in (5.28) and (5.29), we get

$$\mathrm{d}_\pi(\mathcal{O}_X) \cong \mathrm{d}_{\tilde{\pi}}(\mathcal{O}_{\tilde{X}}) , \quad \mathrm{d}_\pi(\widehat{\omega}_\pi^2) \cong \mathrm{d}_{\tilde{\pi}}(\widehat{\omega}_{\tilde{\pi}}^2) . \tag{5.30}$$

Combining these equalities with (5.27) gives

$$\begin{aligned} \langle \widehat{\omega}_\pi, \widehat{\omega}_\pi \rangle_\pi &= \mathrm{d}_\pi(\widehat{\omega}_\pi^2)\, \mathrm{d}_\pi(\widehat{\omega}_\pi)^{-2}\, \mathrm{d}_\pi(\mathcal{O}_X) \\ &\cong \mathrm{d}_{\tilde{\pi}}(\widehat{\omega}_{\tilde{\pi}}^2)\, \mathrm{d}_{\tilde{\pi}}(\widehat{\omega}_{\tilde{\pi}})^{-2}\, \mathrm{d}_{\tilde{\pi}}(\mathcal{O}_{\tilde{X}}) = \langle \widehat{\omega}_{\tilde{\pi}}, \widehat{\omega}_{\tilde{\pi}} \rangle_{\tilde{\pi}} , \end{aligned}$$

as desired. Notice that the sign ambiguities in the identifications (5.30) introduced by the choice of an ordering on each edge of Γ cancel each other, and hence the isomorphism (5.25) is independent of these choices.

One of the most remarkable consequences of (5.8) is the following concrete Riemann–Roch formula.

THEOREM (5.31) (RIEMANN–ROCH). *Let $\pi : X \to S$ be a family of nodal curves, and let L be a line bundle on X. There is a canonical isomorphism of line bundles, compatible with base change:*

$$\mathrm{d}_\pi(L)^2 \xrightarrow{\sim} \langle L, L \otimes \omega_\pi^{-1}\rangle_\pi \otimes \mathrm{d}_\pi(\mathcal{O}_X)^2$$

To see this, just write $\langle L, L\omega_\pi^{-1}\rangle \cong \langle L, L^{-1}\omega_\pi\rangle^{-1}$ by bilinearity and then use (5.8) to get an isomorphism

$$\langle L, L^{-1}\omega_\pi\rangle^{-1} \cong \mathrm{d}(LL^{-1}\omega_\pi)^{-1}\,\mathrm{d}(L)\,\mathrm{d}(L^{-1}\omega_\pi)\,\mathrm{d}(\mathcal{O}_X)^{-1} \cong \mathrm{d}(L)^2\,\mathrm{d}(\mathcal{O}_X)^{-2},$$

where, of course, we have used that there are canonical isomorphisms $\mathrm{d}(\omega_\pi L^{-1}) \cong \mathrm{d}(L)$ and $\mathrm{d}(\omega_\pi) \cong \mathrm{d}(\mathcal{O}_X)$.

Naturally, the reason for calling (5.31) a Riemann–Roch theorem is that, taking first Chern classes in the rational intersection ring of S, dividing by 2, and using (5.20), one deduces from it the standard Riemann–Roch formula

$$c_1(R\pi_!L) = \pi_*\left(\frac{(c_1(L)\cdot c_1(L))}{2} - \frac{(c_1(L)\cdot c_1(\omega_\pi))}{2}\right) + c_1(R\pi_!\mathcal{O}_X).$$

However, (5.31) says much more as it gives a *canonical* isomorphism between the two line bundles involved, while standard Riemann–Roch only implies the existence of *some* isomorphism between some *undetermined powers* of the same line bundles. On the other hand, contrary to standard Riemann–Roch, (5.31) gives no information on the term $c_1(R\pi_!\mathcal{O}_X)$.

6. The Picard group of moduli space.

In this section we shall study a bit more in depth the Picard group of the moduli stack $\overline{\mathcal{M}}_{g,n}$, which has been introduced in Section 2. As shown by Theorem (5.6) in Chapter XII, $\overline{\mathcal{M}}_{g,n}$ is the quotient stack $[\mathrm{H}_{\nu,g,n}/PGL(r+1)]$, where $\mathrm{H}_{\nu,g,n}$ is the Hilbert scheme of stable n-pointed genus g curves embedded in $\mathbb{P}^r$ by the νth power of the log-canonical sheaf for some fixed $\nu \geq 3$ and $r = (2\nu-1)(g-1)+\nu n-1$. Thus, Corollary (2.20) applies and shows that $\mathrm{Pic}(\overline{\mathcal{M}}_{g,n})$ and $\mathrm{Pic}(\mathrm{H}_{\nu,g,n}, PGL(r+1))$ are isomorphic. Actually, even more is true.

PROPOSITION (6.1). *Let $\mathrm{H}_{\nu,g,n}$ be the Hilbert scheme parameterizing stable n-pointed genus g curves ν-log-canonically embedded in $\mathbb{P}^r$, $\nu \geq 3$, $r = (2\nu-1)(g-1)+\nu n-1$, and let $\mathrm{H}'_{\nu,g,n}$ be the open subset of $\mathrm{H}_{\nu,g,n}$ parameterizing smooth curves. Then:*

i) $\mathrm{Pic}(\overline{\mathcal{M}}_{g,n}) \to \mathrm{Pic}(\mathrm{H}_{\nu,g,n}, PGL(r+1))$ *and* $\mathrm{Pic}(\mathcal{M}_{g,n}) \to \mathrm{Pic}(\mathrm{H}'_{\nu,g,n}, PGL(r+1))$ *are group isomorphisms;*
ii) $\mathrm{Pic}(\mathrm{H}_{\nu,g,n}, PGL(r+1)) \to \mathrm{Pic}(\mathrm{H}_{\nu,g,n})^{PGL(r+1)}$ *and* $\mathrm{Pic}(\mathrm{H}'_{\nu,g,n}, PGL(r+1)) \to \mathrm{Pic}(\mathrm{H}'_{\nu,g,n})^{PGL(r+1)}$ *are group isomorphisms.*

As we have observed, i) follows immediately from (2.20). Concerning ii), we will prove just injectivity, which is the only part of the statement that we will use, referring to [558] for a proof of surjectivity. In the proof, we shall write H for $\mathrm{H}_{\nu,g,n}$ or $\mathrm{H}'_{\nu,g,n}$, and G for $PGL(r+1)$. We must show that a line bundle on H has at most one G-linearization or, equivalently, that the only G-linearization of $\mathcal{O}_H$ is the trivial one. We shall use the following elementary result.

LEMMA (6.2). *Let V, W be connected, reduced, and quasi-projective schemes. Then any $f \in H^0(V \times W, \mathcal{O}^\times_{V\times W})$ is of the form $f = hk$, where $h \in H^0(V, \mathcal{O}^\times_V)$ and $k \in H^0(W, \mathcal{O}^\times_W)$.*

To give a proof, we may argue as follows. Choose projective completions $\overline{V}$ of V and $\overline{W}$ of W. Up to replacing V and W with dense open subsets, we may suppose that $V = \overline{V} \smallsetminus L$ and $W = \overline{W} \smallsetminus M$, where L and M are hyperplane sections in some projective embedding. Now let h be the restriction of f to $V \cong V \times \{w\}$ for a general $w \in W$, and k the ratio $k'/f(v,w)$, where k' is the restriction of f to $W \cong \{v\} \times W$ for a general $v \in V$. The rational function f/hk is regular on all of $\overline{V} \times \overline{W}$ and hence constant. Since its value at (v,w) is 1, the lemma follows.

COROLLARY (6.3). *Let Γ be a connected algebraic group, and $\varphi : \Gamma \to \mathbb{C}^\times$ a morphism of schemes such that $\varphi(1_\Gamma) = 1$. Then φ is a group homomorphism.*

The proof is straightforward. For $\gamma, \gamma' \in \Gamma$, set $f(\gamma,\gamma') = \varphi(\gamma)\varphi(\gamma')/\varphi(\gamma\gamma')$. Then, by the lemma, $f(\gamma,\gamma') = h(\gamma)k(\gamma')$ for suitable nowhere vanishing functions h, k. Since $f(\gamma, 1_\Gamma) = 1$, h must be constant, and for the same reason k is also constant. Thus f is constant and in fact identically equal to 1, since $f(1_\Gamma, 1_\Gamma) = 1$. This proves our claim.

COROLLARY (6.4). $H^0(PGL(m), \mathcal{O}^\times_{PGL(m)}) = \mathbb{C}^\times$.

This corollary follows from the previous one by recalling that there are no nontrivial homomorphisms from $PGL(m)$ to $\mathbb{C}^\times$.

We now return to the proof of (6.1). Denote by p the projection of $G \times H$ to the second factor and by $\alpha : G \times H \to H$ the action morphism. Let $\rho : p^*\mathcal{O}_H \to \alpha^*\mathcal{O}_H$ be a G-linearization; this maps the pullback via p of the unit section of $\mathcal{O}_H$ to f times the pullback via α of the same unit section, where f is a nowhere vanishing regular function on $G \times H$. Lemma (6.2) and Corollary (6.4) say that f is the pullback via p of a function on H. On the other hand, the cocycle condition for ρ implies that f is identically equal to 1 on $\{1_G\} \times H$. Hence, f is identically equal to 1, that is, ρ is the trivial G-linearization. This concludes the proof.

REMARK (6.5). Part ii) of Proposition (6.1) will be particularly useful when proving equalities between classes in the Picard groups of moduli stacks. In fact, what its injectivity part asserts is essentially that an equality between such classes holds if and only if it holds for the universal family over the Hilbert scheme $\mathrm{H}_{\nu,g,n}$ (or $\mathrm{H}'_{\nu,g,n}$). Checking the latter will usually be far easier than checking equality directly over the moduli stack.

LEMMA (6.6). *The pullback homomorphisms* $\mathrm{Pic}(\overline{M}_{g,n}) \to \mathrm{Pic}(\overline{\mathcal{M}}_{g,n})$ *and* $\mathrm{Pic}(M_{g,n}) \to \mathrm{Pic}(\mathcal{M}_{g,n})$ *are injective with torsion cokernel. More precisely, there is a positive integer* k *such that* $k\,\mathrm{Pic}(\overline{\mathcal{M}}_{g,n}) \subset \mathrm{Pic}(\overline{M}_{g,n})$ *and* $k\,\mathrm{Pic}(\mathcal{M}_{g,n}) \subset \mathrm{Pic}(M_{g,n})$.

We write M for $M_{g,n}$ or $\overline{M}_{g,n}$, and $\mathcal{M}$ for $\mathcal{M}_{g,n}$ or $\overline{\mathcal{M}}_{g,n}$; the proof is the same in both cases. We know that M can be covered by open sets U_i of the form $U_i = B_i/G_i$, where B_i is the base of a Kuranishi family $\pi_i : X_i \to B_i$, and G_i is the automorphism group of the central fiber. Moreover, one can assume that the automorphism group of each fiber of π_i is a subgroup of G_i. Let L be a line bundle on M whose pullback to $\mathcal{M}$ is trivial and hence has a nowhere vanishing global section. The latter gives, in particular, a G_i-invariant section of the pullback of L to B_i and hence a nowhere vanishing section of L over U_i, for each i. These sections agree on overlaps of the U_i since they come from a global section over $\mathcal{M}$ and thus provide a trivialization of L. This proves that $\mathrm{Pic}(M)$ injects in $\mathrm{Pic}(\mathcal{M})$.

Start instead with a line bundle L on $\mathcal{M}$. We wish to show that there is a positive integer k, independent of L, such that L^k descends to a line bundle on M. We claim that we can take as k the product of the orders of the G_i. Set $X = \coprod X_i$, $B = \coprod B_i$, and let $\pi : X \to B$ be given locally by the π_i. As we know, we can view L as a line bundle on B, together with descent data relative to $B \to \mathcal{M}$. We must show that, if we raise L to the kth power, we get descent data also relative to $B \to M$. Now, the descent data for L provide identifications between fibers of L above points of B mapping to the same point of M. The only itch is that there may be ambiguities in these identifications, due to the presence of automorphisms of the fibers. We will be done if we show that raising L to the kth power gets rid of ambiguities. Let b be a point of B, and L_b the fiber of L at b. The automorphism group of the fiber $X_b = \pi^{-1}(b)$ acts linearly on L_b; since the latter is one-dimensional, the action is by multiplication by k_bth roots of unity, where k_b is the order of $\mathrm{Aut}(X_b)$. Since k_b divides k, the induced action on L_b^k is trivial. This shows that L^k descends to a line bundle on M and finishes the proof.

In the next section we shall need the following sharper result about the structure of the Picard groups of moduli stacks of stable curves. The proof will be given in Section 7 of Chapter XV, when we will have Teichmüller theory at our disposal.

PROPOSITION (6.7). $\mathrm{Pic}(\overline{\mathcal{M}}_{g,n})$ *is a free abelian group of finite rank, and* $\mathrm{Pic}(\overline{M}_{g,n})$ *is a subgroup of finite index.*

7. Mumford's formula.

Our goal in this section is to prove some fundamental relations between tautological classes. These are due to Mumford, to whom we owe the remarkable insight that Grothendieck's Riemann–Roch theorem is ideally suited to producing tautological relations. In particular, the formulas we are going to prove are reflections, in the context of moduli spaces of curves, of Noether's formula for the Euler–Poincaré characteristic of the structure sheaf of an algebraic surface.

Let $f : X \to H$ be a proper morphism of smooth schemes, and let $\mathcal{F}$ be a coherent sheaf on X. Grothendieck's Riemann–Roch theorem states that

$$\mathrm{ch}(f_!\mathcal{F}) \cdot \mathrm{td}(H) = f_*(\mathrm{ch}(\mathcal{F}) \cdot \mathrm{td}(X))$$

in $A^\bullet(H)$. Recall that ch( ) stands for the Chern character, while $\mathrm{td}(Y)$ stands for the Todd class $\mathrm{td}(T_Y)$ of the tangent sheaf of Y. Recall also that $\mathrm{td}(\mathcal{E})$ is a polynomial $T(c_1, c_2, \dots)$ in the Chern classes $c_i = c_i(\mathcal{E})$, where T is defined by

$$T(\sigma_1, \sigma_2, \dots) = F(x_1, x_2, x_3, \dots) = \frac{-x_1}{e^{-x_1}-1} \cdot \frac{-x_2}{e^{-x_2}-1} \cdot \frac{-x_3}{e^{-x_3}-1} \cdots ,$$
$$\sigma_i = i\text{th symmetric function of } x_1, x_2, \dots .$$

We set

$$\mathrm{td}^\vee(\mathcal{E}) = T^\vee(c_1, c_2, \dots),$$
$$T^\vee(\sigma_1, \sigma_2, \dots) = F(-x_1, -x_2, -x_3, \dots) = T(-\sigma_1, \sigma_2, -\sigma_3, \dots) .$$

It follows in particular that, if $\mathcal{E}$ is a vector bundle, then $\mathrm{td}^\vee(\mathcal{E}) = \mathrm{td}(\mathcal{E}^\vee)$. Now suppose that $f : X \to H$ is a family of nodal curves. From the exact sequence

$$0 \to f^*\Omega^1_H \to \Omega^1_X \to \Omega^1_{X/H} \to 0$$

and the multiplicativity of the Todd genus we get

$$\mathrm{td}(X) = \mathrm{td}^\vee(\Omega^1_X) = \mathrm{td}^\vee(\Omega^1_{X/H}) \cdot f^*(\mathrm{td}^\vee(\Omega^1_H)) .$$

Using the push–pull formula, the Riemann–Roch formula can thus be rewritten as

$$\mathrm{ch}(f_!\mathcal{F}) = f_*(\mathrm{ch}(\mathcal{F}) \cdot \mathrm{td}^\vee(\Omega^1_{X/H})) . \tag{7.1}$$

Recall that the low-degree terms of $\mathrm{td}^\vee(\mathcal{E})$ and $\mathrm{ch}(\mathcal{E})$ are

$$\mathrm{td}^\vee(\mathcal{E}) = 1 + \frac{c_1(\mathcal{E})}{2} + \frac{c_1(\mathcal{E})^2 + c_2(\mathcal{E})}{12} + \cdots ,$$
$$\mathrm{ch}(\mathcal{E}) = \mathrm{rank}(\mathcal{E}) + c_1(\mathcal{E}) + \frac{c_1(\mathcal{E})^2 - 2c_2(\mathcal{E})}{2} + \cdots .$$

In general, the degree i component of $\mathrm{ch}(\mathcal{E})$, which we indicate with $\mathrm{ch}_i(\mathcal{E})$, is of the form

$$\mathrm{ch}_i(\mathcal{E}) = (-1)^{i-1}\frac{c_i(\mathcal{E})}{(i-1)!} + P_i(c_1(\mathcal{E}),\ldots,c_{i-1}(\mathcal{E}))\,, \tag{7.2}$$

where P_i is a graded-homogeneous rational polynomial of degree i. We shall apply formula (7.1) to the situation in which $f : X \to H$, together with sections $\tau_1,\ldots,\tau_n$, is a family of stable n-pointed genus g curves. Equating terms of degree 1 in both sides of (7.1) and taking into account the explicit expression for the Chern character and the Todd genus, we get

(7.3)
$$c_1(f_!\mathcal{F}) = f_*\left(\frac{c_1(\mathcal{F})^2}{2} - c_2(\mathcal{F}) - \frac{c_1(\mathcal{F})\cdot c_1(\Omega^1_f)}{2} + \frac{c_1(\Omega^1_f)^2 + c_2(\Omega^1_f)}{12}\right).$$

Our next goal is to compute the first and second Chern classes of Ω^1_f. For this, denote by Σ the locus in X swept by the nodes in the fibers and by $\mathcal{I}$ its ideal sheaf. The exact sequence (2.20) specializes to

$$0 \to \Omega^1_f \to \omega_f \to \omega_f \otimes \mathcal{O}_\Sigma \to 0 \tag{7.4}$$

since X is smooth and the image of Ω^1_f in ω_f is just $\mathcal{I}\omega_f$. To calculate $c_1(\omega_f \otimes \mathcal{O}_\Sigma)$ and $c_2(\omega_f \otimes \mathcal{O}_\Sigma)$, we use again the Grothendieck Riemann–Roch formula, applied to the following situation. Let $j : Y \to Z$ be a codimension r closed immersion of smooth schemes, and let $\mathcal{G}$ be a coherent sheaf on Y. Then the Riemann–Roch formula gives

$$\mathrm{ch}(j_*\mathcal{G}) = \mathrm{ch}(j_!\mathcal{G}) = j_*(\mathrm{ch}(\mathcal{G})\cdot \mathrm{td}(Y))\cdot \mathrm{td}(Z)^{-1}\,.$$

For codimension reasons, the right-hand side only lives in degrees r or greater; on the other hand, the degree r term equals $\mathrm{rank}(\mathcal{G})[Y]$, where $[Y] = j_*(\mathbf{1}_Y)$ denotes the fundamental class of Y. Using (7.2), it follows that

$$c_i(j_*\mathcal{G}) = 0 \quad \text{when } i < r\,,$$
$$c_r(j_*\mathcal{G}) = (-1)^{r-1}(r-1)!\,\mathrm{rank}(\mathcal{G})[Y]\,.$$

We apply this mechanism to the inclusion of Σ in X and to the sheaf $\omega_f \otimes \mathcal{O}_\Sigma$. Via Whitney's formula for (7.4), we get

$$c_1(\Omega^1_f) = c_1(\omega_f)\,,$$
$$c_2(\Omega^1_f) = [\Sigma]\,.$$

We are now ready to apply (7.3) to various specific choices of the sheaf $\mathcal{F}$. We start with the case in which $\mathcal{F} = \omega_f^\nu$. Using the expressions we just derived for the first and second Chern classes of Ω^1_f and noticing that $c_2(\mathcal{F}) = 0$, we obtain

$$c_1(f_!(\omega_f^\nu)) = f_*\left(\frac{\nu^2 c_1(\omega_f)^2}{2} - \frac{\nu c_1(\omega_f)\cdot c_1(\omega_f)}{2} + \frac{c_1(\omega_f)^2 + [\Sigma]}{12}\right).$$

We next observe that $f_*[\Sigma] = \delta$ and use (5.22) and (5.20) to calculate the remaining terms. The result is

$$\lambda(\nu) = \frac{\nu^2 - \nu}{2}(\kappa_1 - \psi) + \frac{\kappa_1 - \psi + \delta}{12}, \tag{7.5}$$

where of course it is implicitly understood that the tautological classes on both sides of the equality are evaluated on the family at hand.

Now suppose that H is the Hilbert scheme $\mathrm{H}_{\nu,g,n}$ and $f : X \to H$, together with the sections τ_i, is the universal family over it. Formula (7.5) is an identity in $A^1(H)\otimes\mathbb{Q} = \mathrm{Pic}(H)\otimes\mathbb{Q}$. Thus (7.5) is valid in $\mathrm{Pic}(H)$ modulo torsion. On the other hand, its sides both come from $\mathrm{Pic}(\overline{\mathcal{M}}_{g,n})$, which is a subgroup of $\mathrm{Pic}(H)$ by (6.1). Since $\mathrm{Pic}(\overline{\mathcal{M}}_{g,n})$ has no torsion by (6.7), we conclude that (7.5) holds in $\mathrm{Pic}(\overline{\mathcal{M}}_{g,n})$. We summarize what has been proved in the following statement.

THEOREM (7.6). *For any choice of nonnegative integers g and n such that $2g - 2 + n > 0$,*

$$\kappa_1 = 12\lambda + \psi - \delta \tag{7.7}$$

in $\mathrm{Pic}(\overline{\mathcal{M}}_{g,n})$. Moreover, for any integer ν,

$$\lambda(\nu) = (6\nu^2 - 6\nu + 1)\lambda + \binom{n}{2}(\psi - \delta). \tag{7.8}$$

Formula (7.7) is just (7.5) for $\nu = 1$, and (7.8) is obtained by plugging into (7.5) the value of κ_1 given by (7.7).

REMARK (7.9). Recall that, according to formula (5.23), κ_1 can be expressed as the sum of ψ and of the modified kappa class $\widetilde{\kappa}_1$, and that the latter is just the class in $\mathrm{Pic}(\overline{\mathcal{M}}_{g,n})$ of the line bundle given, for any family $\pi : X \to S$ of stable n-pointed genus g curves, by the Deligne pairing $\langle\omega_\pi, \omega_\pi\rangle_\pi$. We thus get the following equivalent formulation of (7.7) which is sometimes useful:

$$\widetilde{\kappa}_1 = 12\lambda - \delta\,. \tag{7.10}$$

EXAMPLE (7.11). It is instructive to see what formula (7.7) means when $g = n = 1$. Let $f : X \to S$ be a family of nodal curves of genus 1 with a section $\sigma : S \to X$ which makes it a family of stable (or possibly semistable) 1-pointed curves. First of all, notice that ω_f is trivial on the fibers of f and hence is isomorphic to $f^*\sigma^*\omega_f$. It follows that $f_*\omega_f \cong \sigma^*\omega_f$, and hence that $\lambda = \psi$ on $\overline{\mathcal{M}}_{1,1}$. Another consequence of the triviality of ω_f on fibers is that $\langle \omega_f, \omega_f \rangle = \mathbb{1}_S$, and therefore

$$\begin{aligned}\kappa_1 &= c_1\langle \omega_f(D), \omega_f(D)\rangle = 2c_1\langle \omega_f, \mathcal{O}(D)\rangle + c_1\langle \mathcal{O}(D), \mathcal{O}(D)\rangle \\ &= -c_1\langle \mathcal{O}(D), \mathcal{O}(D)\rangle = \psi\,,\end{aligned}$$

since $\langle \omega_f(D), \mathcal{O}(D)\rangle = \mathbb{1}_S$ by adjunction, and hence $c_1\langle \mathcal{O}(D), \mathcal{O}(D)\rangle = -c_1\langle \omega_f, \mathcal{O}(D)\rangle = -\psi$. Summing up,

$$\kappa_1 = \psi = \lambda \quad \text{on } \overline{\mathcal{M}}_{1,1}\,. \tag{7.12}$$

A consequence is that formula (7.7) reduces to

$$12\lambda = \delta \quad \text{on } \overline{\mathcal{M}}_{1,1}\,. \tag{7.13}$$

Now let X be a smooth elliptic surface, and let $f : X \to S$ be the elliptic fibration. We assume that all the fibers of f are reduced and nodal and that f is relatively minimal in the sense that its fibers do not contain exceptional curves of the first kind. We also assume that f admits a section σ, and we set $D = \sigma(S)$. The fibration f, together with the section σ, is a family of semistable 1-pointed curves of genus 1. If we further assume that it is not locally a product, its moduli map $S \to \overline{M}_{1,1}$ is a ramified covering.

It is easy to relate the invariants appearing in (7.7) to the basic invariants of the surface X. The Picard group of $\overline{M}_{1,1}$ is infinite cyclic, and hence any relation among tautological classes is equivalent to the corresponding relation between the degrees of their pullbacks to S. In the calculations that follow, we shall improperly write $\kappa_1, \lambda, \psi, \delta$ for the pullbacks of these same classes to S. Observe that

$$c_1^2(X) = (\omega_X)^{\cdot 2} = (\omega_f)^{\cdot 2} + (f^*\omega_S)^{\cdot 2} + 2(\omega_f \cdot f^*\omega_S) = 0\,.$$

On the other hand, indicating with F a general fiber of f, the topological Euler characteristic of X is

$$\chi(X) = \chi(S)\chi(F) + \text{number of nodes in the fibers of } f = \deg\delta\,,$$

since in our case $\chi(F) = 0$. Noether's formula for X hence gives

$$\chi(\mathcal{O}_X) = \frac{c_1^2(X) + \chi(X)}{12} = \frac{\deg\delta}{12}\,. \tag{7.14}$$

Formula (7.7) follows from (7.12) and (7.14) if we can show that $\chi(\mathcal{O}_X) = \deg\lambda$. This is not difficult. In fact, by relative duality,

$$\deg\lambda = \deg(f_*\omega_f) = -\deg(R^1 f_*\mathcal{O}_X)\,,$$

while

$$\chi(\mathcal{O}_X) = \chi(f_*\mathcal{O}_X) - \chi(R^1 f_*\mathcal{O}_X) = \chi(\mathcal{O}_S) - \chi(R^1 f_*\mathcal{O}_X) = -\deg(R^1 f_*\mathcal{O}_X)$$

by the Riemann–Roch theorem on S. Summing up, we can say that formula (7.7) is essentially equivalent to Noether's formula for the surface X.

The same kind of argument that led to Theorem (7.6) can be used to derive other tautological relations. For instance, consider the sheaf $\mathcal{F} = \Omega^1_f \otimes \omega_f(D)$, where $f : X \to H$ is the universal family over $H = \mathrm{H}_{\nu,g,n}$, and D is the divisor of sections. Recall that, for a stable n-pointed curve $(C; p_1, \ldots, p_n)$, $H^0(C, \Omega^1_C \otimes \omega_C(\sum p_i))$ is the dual of $\mathrm{Ext}^1_{\mathcal{O}_C}(\Omega^1_C, \mathcal{O}_C(-\sum p_i))$, which in turn is the tangent space to the base of the Kuranishi family of $(C; p_1, \ldots, p_n)$ at its central point. Likewise, $H^1(C, \Omega^1_C \otimes \omega_C(\sum p_i))$ is dual to $\mathrm{Hom}_{\mathcal{O}_C}(\Omega^1_C, \mathcal{O}_C(-\sum p_i))$, which vanishes by Lemma (5.8) in Chapter XI. Therefore, in our case, $f_!\mathcal{F} = f_*\mathcal{F}$, and the PGL-invariant class

$$K_{\overline{\mathcal{M}}_{g,n}} = c_1(f_!\mathcal{F}) = c_1(f_*(\Omega^1_f \otimes \omega_f(D)))$$

can be interpreted as the canonical class of the moduli stack $\overline{\mathcal{M}}_{g,n}$. In order to apply formula (7.3), we need to compute the first and second Chern classes of $\Omega^1_f \otimes \omega_f(D)$. This can be done by the same procedure we used for Ω^1_f, and the result is:

$$c_1(\Omega^1_f \otimes \omega_f(D)) = c_1(\omega^2_f(D)) ,$$
$$c_2(\Omega^1_f \otimes \omega_f(D)) = [\Sigma] .$$

Plugging back into (7.3), we get the following result.

THEOREM (7.15).

$$K_{\overline{\mathcal{M}}_{g,n}} = 13\lambda + \psi - 2\delta .$$

Instead of asking for the canonical class of the *stack* $\overline{\mathcal{M}}_{g,n}$, one might ask for the canonical class of the *space* $\overline{M}_{g,n}$; the answer turns out to be slightly different.

COROLLARY (7.16) [353]. *For $g \geq 1$ and $g + n \geq 4$,*

$$K_{\overline{M}_{g,n}} = 13\lambda + \psi - 2\delta - \delta_{1,\emptyset} .$$

The reason for the discrepancy and for the limitations is the following. The morphism $f : \overline{\mathcal{M}}_{g,n} \to \overline{M}_{g,n}$ is ramified precisely at those points which correspond to curves with nontrivial automorphism group. Let $\Sigma \subset \overline{M}_{g,n}$ be the locus parameterizing these curves. The limitations in the statement insure first of all that Σ is not all of $\overline{M}_{g,n}$. On the other hand, Σ always contains the boundary divisor $\Delta_{1,\emptyset}$ since a curve with an unpointed elliptic tail has at least one nontrivial automorphism, i.e., the symmetry of the tail about the point of attachment to the rest of the curve. If the curve is general in $\Delta_{1,\emptyset}$, this is the only nontrivial automorphism it has. It follows that f is simply ramified along $\Delta_{1,\emptyset}$; in particular, the class of $f^*(\Delta_{1,\emptyset})$ is $2\delta_{1,\emptyset}$. A general point of any other boundary component corresponds to an automorphism-free curve. On the other hand, we know (cf. Proposition (2.5) in Chapter XII) that, under our assumptions, $\Sigma \cap M_{g,n}$ has codimension two or more. The corollary then follows from Theorem (7.15) and the Riemann–Hurwitz formula.

EXAMPLE (7.17). Similar ideas can be applied to calculate the canonical class of $\overline{M}_{g,n}$ in cases not covered by (7.16). We exemplify this for $\overline{M}_{1,1}$. Of course, we know the answer, since $\overline{M}_{1,1}$ is just $\mathbb{P}^1$, but it is instructive to arrive at it via an argument analogous to the one that gives (7.16). Every stable 1-pointed elliptic curve has an involution ι, the symmetry about the marked point. There are just two curves which have additional automorphisms. These are the quotient of $\mathbb{C}$ by the lattice generated by the fourth roots of unity, whose automorphism group is cyclic of order 4, and the quotient of $\mathbb{C}$ by the lattice generated by the sixth roots of unity, whose automorphism group is cyclic of order 6. We denote the corresponding points of $\overline{M}_{1,1}$ by p_4 and p_6.

Now consider the natural morphism $\pi : \overline{\mathcal{M}}_{1,1} \to \overline{M}_{1,1}$. Since the involution ι acts trivially on the bases of Kuranishi families, this morphism is ramified precisely at p_4 and p_6, with ramification indices 1 and 2. Thus, the divisor $\pi^*(p_4)$ is twice an effective divisor Q_4, and $\pi^*(p_6)$ is three times an effective divisor Q_6. Moreover, the Riemann–Hurwitz formula gives

$$(7.18)\qquad K_{\overline{\mathcal{M}}_{1,1}} = \pi^* K_{\overline{M}_{1,1}} + [Q_4] + 2[Q_6] = \pi^*\left(K_{\overline{M}_{1,1}} + \frac{1}{2}[p_4] + \frac{2}{3}[p_6]\right),$$

where $[x]$ stands for the class of x in the Picard group. Theorem (7.15), coupled with formula (7.7), gives

$$K_{\overline{\mathcal{M}}_{1,1}} = -\frac{10}{12}\delta\,.$$

Writing $K_{\overline{M}_{1,1}} = h\delta$, where h is a rational number to be determined, this formula, together with (7.18) and the observation that $[p_4] = [p_6] = \delta$, gives

$$h = -\frac{10}{12} - \frac{1}{2} - \frac{2}{3} = -2\,,$$

in accordance with the fact that the canonical bundle of $\mathbb{P}^1$ is $\mathcal{O}(-2)$.

8. The Picard group of the hyperelliptic locus.

As a warm-up to a finer analysis of moduli spaces, we shall analyze $\overline{\mathcal{H}}_g$, the closed substack of $\overline{\mathcal{M}}_g$ parameterizing stable hyperelliptic curves of genus $g > 1$, and compute its rational Picard group.

First of all, we have to convince ourselves that we are really dealing with a closed substack. In essence, what has to be shown is that, for any (algebraic) family $\overline{f} : \overline{X} \to \overline{S}$ of stable curves such that the induced morphism $\overline{S} \to \overline{\mathcal{M}}_g$ is étale, that is, for any family which is everywhere a Kuranishi family, the locus in $\overline{S}$ parameterizing hyperelliptic fibers of $\overline{f}$ is a closed subscheme. There are several ways of seeing this. Here is one. Let S be the open subscheme of $\overline{S}$ parameterizing smooth curves, and let $f : X \to S$ be the restriction of $\overline{f} : \overline{X} \to \overline{S}$ to S. Set $\mathcal{C} = X \times_S X$ and denote by $\pi : \mathcal{C} \to X$ the projection to the second factor and by $\Delta \subset \mathcal{C}$ the diagonal. Let x be a point of X and set $s = f(x)$. Then x is a hyperelliptic Weierstraß point on $X_s = f^{-1}(s)$ if and only if $h^1(X_s, \omega_{X_s}(-2x)) = h^0(X_s, \mathcal{O}_{X_s}(2x)) = 2$, that is, if and only if the surjective homomorphism

$$H^1(X_s, \omega_{X_s}(-2x)) \to H^1(X_s, \omega_{X_s}) \tag{8.1}$$

has a nonzero (actually, one-dimensional) kernel. On the other hand, if $\mathcal{F}$ is any coherent sheaf on $\mathcal{C}$, then $H^i(\pi^{-1}(x), \mathcal{F} \otimes \mathcal{O}_{\pi^{-1}(x)}) = 0$ for any $i > 1$ and any $x \in X$, since the fibers of π are one-dimensional. It is then a standard corollary of the basic theory of base change in cohomology that, if $\mathcal{F}$ is $\mathcal{O}_X$-flat, then

$$R^1\pi_*\mathcal{F} \otimes k(x) \to H^1(\pi^{-1}(x), \mathcal{F} \otimes \mathcal{O}_{\pi^{-1}(x)})$$

is an isomorphism for any x. Applying this to $\mathcal{F} = \omega_\pi(-2\Delta)$ and to $\mathcal{F} = \omega_\pi$, the homomorphism (8.1) can thus be identified with

$$R^1\pi_*(\omega_\pi(-2\Delta)) \otimes k(x) \to R^1\pi_*\omega_\pi \otimes k(x)\,.$$

It follows that the locus $W \subset X$ of those points x which are hyperelliptic Weierstraß points in $f^{-1}(f(x))$ is precisely the support of the kernel of

$$R^1\pi_*(\omega_\pi(-2\Delta)) \to R^1\pi_*\omega_\pi \cong \mathcal{O}_X$$

and hence is a closed subscheme of X. Thus, the locus H of those $s \in S$ such that X_s is hyperelliptic, which is clearly equal to $f(W)$, is a closed subscheme of S.

It is a consequence of Lemma (6.14) in Chapter XI that the locus in $\overline{S}$ parameterizing hyperelliptic fibers of $\overline{f} : \overline{X} \to \overline{S}$ is just the closure of H in $\overline{S}$ and hence is a closed subscheme of the latter. This shows that $\overline{\mathcal{H}}_g$ is indeed a closed substack of $\overline{\mathcal{M}}_g$. It follows from Lemma (6.15) in Chapter XI that $\overline{\mathcal{H}}_g$ is smooth of dimension $2g-1$. We shall indicate with $\mathcal{H}_g$ the open substack of $\overline{\mathcal{H}}_g$ parameterizing smooth hyperelliptic curves, and with $\overline{H}_g$ and H_g the coarse moduli counterparts of $\overline{\mathcal{H}}_g$ and $\mathcal{H}_g$.

There is a natural surjective morphism $M_{0,2g+2} \to H_g$ which associates to each $(2g+2)$-pointed genus 0 curve $(\mathbb{P}^1; p_1, \dots, p_{2g+2})$ the double covering C of $\mathbb{P}^1$ (unique up to isomorphism) branched at $p_1 + \dots + p_{2g+2}$; it follows, in particular, that H_g, and consequently $\overline{H}_g$, are irreducible. Since the isomorphism class of C does not depend on the numbering of the p_i, the morphism $M_{0,2g+2} \to H_g$ factors through $M_{0,2g+2}/\mathfrak{S}_{2g+2}$, where $\mathfrak{S}_{2g+2}$ is the symmetric group on $2g+2$ letters; actually, $M_{0,2g+2}/\mathfrak{S}_{2g+2} \to H_g$ is an isomorphism.

Proposition (8.2). $\mathrm{Pic}(\mathcal{H}_g)\otimes\mathbb{Q} = \mathrm{Pic}(H_g)\otimes\mathbb{Q} = 0$. *More precisely, there is a positive integer k such that L^k is trivial for any line bundle L on $\mathcal{H}_g$ or on H_g.*

Proof. Let L be a line bundle on $\mathcal{H}_g$. There is a positive integer h such that L^h is the pullback of a line bundle Q on H_g via $\mathcal{H}_g \to H_g$. The proof of this fact is exactly like the one of the analogous statement for $\mathcal{M}_{g,n}$ (cf. Lemma (6.6)) and will not be repeated here. Now let $\alpha : M_{0,2g+2} \to H_g$ be the natural morphism. The pullback α^*Q is a line bundle with a $\mathfrak{S}_{2g+2}$ action. We claim that the Picard group of $M_{0,2g+2}$ is trivial. Assuming this for the moment, a consequence is that α^*Q has a nowhere vanishing section s. Then

$$\bigotimes_{\sigma\in\mathfrak{S}_{2g+2}} \sigma^* s$$

is a $\mathfrak{S}_{2g+2}$-invariant section of $\alpha^*Q^{(2g+2)!}$ and hence descends to a nowhere vanishing section of $Q^{(2g+2)!}$. This proves the proposition with $k = (2g+2)!h$.

It remains to show that $\mathrm{Pic}(M_{0,2g+2})$ vanishes. A point of $M_{0,2g+2}$ is a $(2g+2)$-tuple $(p_1, \dots, p_{2g+2})$ of distinct points of $\mathbb{P}^1$, modulo automorphisms of the projective line. We can get rid of automorphisms by normalizing things so that $p_{2g} = 0$, $p_{2g+1} = 1$, and $p_{2g+2} = \infty$. We may then identify $M_{0,2g+2}$ with the open subset of $\mathbb{C}^{2g-1}$ consisting of the points $(z_1, \dots, z_{2g-1})$ such that $z_i \neq 0, 1$ for all i and $z_i \neq z_j$ for $i \neq j$. Our claim is thus a special case of the well-known fact that the complement in $\mathbb{P}^r$ of a bunch of hyperplanes has trivial Picard group. For completeness, here is a proof. Let $P_1, \dots, P_\ell$ be hyperplanes in $\mathbb{P}^r$, and let L be a line bundle on $U = \mathbb{P}^r \smallsetminus \bigcup P_i$. Then L extends as a coherent sheaf $\mathcal{L}$ to all of $\mathbb{P}^r$, and for some large n, the twist $\mathcal{L}(n)$ has a section which is nonzero on U. Let D be the divisor of the restriction of this section to U. Then the closure of D in $\mathbb{P}^r$ is linearly equivalent to dP_1 for some d, and hence L is isomorphic to the restriction to U of $\mathcal{O}_{\mathbb{P}^r}(d-n) \cong \mathcal{O}_{\mathbb{P}^r}((d-n)P_1)$, which is clearly trivial on the complement of P_1 and hence on U.

Q.E.D.

We shall now describe the boundary $\partial\mathcal{H}_g = \overline{\mathcal{H}}_g \smallsetminus \mathcal{H}_g$ of the moduli stack of hyperelliptic curves. The first remark is that it is a divisor, since it is cut out on $\overline{\mathcal{H}}_g$ by the boundary of $\overline{\mathcal{M}}_g$, which is itself a divisor. As we know, the components of $\partial\mathcal{M}_g$ are the divisor $\mathcal{D}_{irr}$ and the divisors $\mathcal{D}_{i,\emptyset}$ with $1 \leq i \leq \lfloor g/2 \rfloor$. The components of the boundary of $\partial\mathcal{H}_g$ are hence the components of the intersections of these with $\overline{\mathcal{H}}_g$. Recalling the classification of nodes of hyperelliptic curves described in Section 3 of Chapter X, we see that a hyperelliptic curve belongs to $\mathcal{D}_{i,\emptyset}$ if and only if it has a node of type δ_i. We write $\mathcal{D}_i$ to indicate the divisor in

$\overline{\mathcal{H}}_g$ consisting of curves with a node of type δ_i. The curves in $\mathcal{D}_i$ are obtained from hyperelliptic curves C_1 and C_2 of genera i and $g-i$ by identifying a Weierstraß point on C_1 with a Weierstraß point on C_2. It follows from the irreducibility of $\overline{H}_i$ and $\overline{H}_{g-i}$ that $\mathcal{D}_i$ is irreducible. To be more exact, this only covers the case $i \geq 2$. When $i=1$, we must take as C_1 a curve in $\overline{M}_{1,1}$ and attach it to C_2 at the marked point. When in addition $g-i=1$, C_2 must also be a curve in $\overline{M}_{1,1}$, attached to C_1 at the marked point. In any case, the irreducibility of $\mathcal{D}_i$ is guaranteed by the irreducibility of $\overline{M}_{1,1}$. It follows from part ii) of Lemma (6.15) in Chapter XI that the intersection of $\mathcal{D}_{i,\emptyset}$ with $\overline{\mathcal{H}}_g$ is transverse. In other words, $\mathcal{D}_{i,\emptyset}$ cuts on $\overline{\mathcal{H}}_g$ exactly $\mathcal{D}_i$ (and not a multiple). We write δ_i to indicate the class of $\mathcal{D}_i$ in $\mathrm{Pic}(\overline{\mathcal{H}}_g)$.

The situation for $\mathcal{D}_{irr}$ is completely different. We know, always from Section 3 of Chapter X, that nonseparating nodes come in different types, and we shall see that each of these corresponds to a different component of $\partial\mathcal{H}_g$. We let $\mathcal{E}_0$ be the locus of hyperelliptic curves with a node of type η_0, and $\mathcal{E}_i$, $i>0$, the locus of hyperelliptic curves with a pair of nodes of type η_i. These are clearly all divisors, since the type of a node is stable under deformation, and they are all irreducible. For instance, when $g>2$, a general curve in $\mathcal{E}_0$ is obtained from a smooth hyperelliptic curve C of genus $g-1$ by identifying two points which are conjugate under the hyperelliptic involution γ. The irreducibility of $\mathcal{E}_0$ then follows from the one of $\overline{H}_{g-1}$ and the one of $C/\gamma \cong \mathbb{P}^1$. As an added check, notice that dimensions match; from the above description the one of $\mathcal{E}_0$ turns out to be $2(g-1)-1+1=2g-2$, as expected. Genus 2 is somewhat special, in that a curve in $\mathcal{E}_0$ is obtained from a curve in $\overline{M}_{1,2}$ by identifying the two marked points. Irreducibility and dimension count go through as before. We now turn to $\mathcal{E}_i$ for $i>1$, leaving to the reader the task of dealing with the case $i=1$ following the lines of what we have done for $\mathcal{E}_0$ and $\mathcal{D}_i$. A general curve in $\mathcal{E}_i$ is obtained from a smooth hyperelliptic curve C_1 of genus i, a smooth hyperelliptic curve C_2 of genus $g-i-1$, a pair $\{p_1, p_1'\}$ of points on C_1 conjugate under the hyperelliptic involution γ_1, and a pair $\{p_2, p_2'\}$ of points on C_2 conjugate under the hyperelliptic involution γ_2 by identifying p_1 with p_2 and p_1' with p_2'. The irreducibility of $\mathcal{E}_i$ follows from the ones of $\overline{H}_i$, $\overline{H}_{g-i-1}$, $C_1/\gamma_1 \cong \mathbb{P}^1$, and $C_2/\gamma_2 \cong \mathbb{P}^1$. As for the dimension of $\mathcal{E}_i$, this turns out to be $2i-1+2(g-i-1)-1+1+1=2g-2$, as expected. We shall denote by η_i the class of $\mathcal{E}_i$ in $\mathrm{Pic}(\overline{\mathcal{H}}_g)$ for $i \geq 0$.

The divisor $\mathcal{D}_{irr}$ cuts $\overline{\mathcal{H}}_g$ transversely along $\mathcal{E}_0$. Instead, since nodes of type η_i, $i>0$, come in pairs, at a general point of $\mathcal{E}_i$, the divisor $\mathcal{D}_{irr}$ has two local components. It follows again from part ii) of Lemma (6.15) in Chapter XI that each one of these local components cuts $\overline{\mathcal{H}}_g$ transversely. The upshot is that the divisor $\mathcal{D}_{irr}$ cuts on $\overline{\mathcal{H}}_g$ the reducible divisor $\mathcal{E}_0 + 2\sum_{i>0}\mathcal{E}_i$. In terms of divisor classes, this says that

$$\delta_{irr} = \eta_0 + 2\sum_{i>0}\eta_i \qquad \text{on } \mathcal{H}_g. \tag{8.3}$$

As we observed, the rational Picard group of $\overline{\mathcal{H}}_g$ is generated by the classes of the components of the boundary, that is, by the δ_i and by the η_i. In particular, other natural classes such as λ must be expressible on $\overline{\mathcal{H}}_g$ as rational linear combinations of these boundary classes. The main result we wish to prove in this section is the following.

THEOREM (8.4). *For any $g \geq 2$, $\mathrm{Pic}(\overline{\mathcal{H}}_g) \otimes \mathbb{Q}$ is a vector space of dimension g over $\mathbb{Q}$, freely generated by the classes $\eta_0, \dots, \eta_{\lfloor (g-1)/2 \rfloor}$ and $\delta_1, \dots, \delta_{\lfloor g/2 \rfloor}$. Moreover, the following relation holds in $\mathrm{Pic}(\overline{\mathcal{H}}_g) \otimes \mathbb{Q}$:*

$$(8g+4)\lambda = g\eta_0 + 2\sum_{j=1}^{\lfloor \frac{g-1}{2} \rfloor}(j+1)(g-j)\eta_j + 4\sum_{j=1}^{\lfloor \frac{g}{2} \rfloor} j(g-j)\delta_j\,. \tag{8.5}$$

The rest of the section will be devoted to the proof of this theorem. We will use the following notational convention. Suppose that $\mathcal{L}$ is a line bundle on a stack $\mathcal{M}$, and let ℓ be its class in $\mathrm{Pic}(\mathcal{M})$. If F is a section of $\mathcal{M}$ over a complete curve S, $\mathcal{L}$ induces a line bundle $\mathcal{L}_F$ on S. We shall write $\deg_F \ell$ to indicate the degree of $\mathcal{L}_F$ and refer to this number as the degree of ℓ over F.

The strategy of the proof of (8.4) is very simple. To each family F of genus g hyperelliptic stable curves over a curve, we may associate the vector $\deg_F$ whose entries are the degrees $\deg_F \eta_i$, $i = 0, \dots, \lfloor (g-1)/2 \rfloor$, and $\deg_F \delta_i$, $i = 1, \dots, \lfloor g/2 \rfloor$. We shall produce g families $F_1, \dots, F_g$ such that the vectors $\deg_{F_1}, \dots, \deg_{F_g}$ are independent. This of course will prove the independence of the η_i and of the δ_i and hence, since we already know that these classes generate $\mathrm{Pic}(\overline{\mathcal{H}}_g) \otimes \mathbb{Q}$, the first part of the theorem. At the same time, we will show that the degrees of the two sides of (8.5) over F_i are equal, i.e., that

$$\begin{aligned}(8g+4)\deg_{F_i}\lambda = g\deg_{F_i}\eta_0 &+ 2\sum_{j=1}^{\lfloor \frac{g-1}{2} \rfloor}(j+1)(g-j)\deg_{F_i}\eta_j \\ &+ 4\sum_{j=1}^{\lfloor \frac{g}{2} \rfloor} j(g-j)\deg_{F_i}\delta_j\end{aligned} \tag{8.6}$$

for $i = 1, \dots, g$, thus completing the proof of the theorem.

First look at a stable $(2g+2)$-pointed curve of genus zero $(\Gamma; p_1, \dots, p_{2g+2})$. The partial normalization of Γ at a node q consists of two connected components, one containing α_1, the other α_2 marked points, with $\alpha_1 + \alpha_2 = 2g + 2$. We define the *index* of q to be $\alpha = \min\{\alpha_1, \alpha_2\}$. Thus $2 \leq \alpha \leq g + 1$. There exists a double covering $\beta : C \to \Gamma$ ramified at the p_i and at the nodes of odd index, as can easily be shown by induction on the number of components of Γ. It is also immediate to prove, using the Riemann–Hurwitz formula and induction on the number of components of Γ, that the genus of C is equal to g. The curve C is in general not stable, but just semistable. In fact, suppose that q is a node of Γ of index 2. Then one of the components of the partial normalization of Γ at q, call it Γ', is a smooth rational curve containing exactly two of the marked points. Thus, $C' = \beta^{-1}(\Gamma')$ is a double covering of Γ' branched at two points and hence is also smooth rational; moreover, C' meets the remainder of C at two points. We will call a component like C' a *bridge* of C. The stable model of Γ is obtained by contracting each bridge to a point and is clearly hyperelliptic. Each of the points originating from the contraction of a bridge is clearly a node of type η_0. Just as clearly, when q is a node of Γ of odd index $2j + 1$, the single point of C above it is a node of type δ_j. Likewise, there are two points above a node of Γ of even index $2j + 2 > 2$, and they constitute a pair of nodes of type η_j.

Now consider a family of stable $(2g + 2)$-pointed genus zero curves over a smooth curve S, consisting of a family $\pi : R \to S$ of nodal curves plus $2g + 2$ sections, which we view as divisors $D_1, \dots, D_{2g+2}$ in R. Suppose that $(\Gamma; p_1, \dots, p_{2g+2})$ is a fiber of $(\pi : R \to S; D_1, \dots, D_{2g+2})$. Then, locally near a node q of Γ, R is of the form $xy = t^h$, where t is a local coordinate on S; we will then say that h is the *multiplicity* of q. Let F be the stable model of a family of the form $f : X \to S$, where f is the composition of π and of a double covering $\mu : X \to R$ ramified along $D = \sum D_i$ and at the nodes of odd index in the fibers, for some family $(\pi : R \to S; D_1, \dots, D_{2g+2})$ as above. In other words, F is $f' : X' \to S$, where X' is obtained by contracting to points all the bridges in the fibers of f.

In practice, we will find it more convenient to work with $f : X \to S$ rather than with the family F. In particular, all the invariants of F can be readily calculated on $f : X \to S$. Observe first of all that, if q is a node of index 2 and multiplicity h in a fiber of π, then X' is of the form $xy = t^{2h}$ at the point resulting from the contraction of the corresponding bridge. Thus, the contribution to $\deg_F \eta_0$ of a node of index 2 and multiplicity h in a fiber of π is equal to $2h$. The contribution of a node q of even index $2j + 2 > 2$ to $\deg_F \eta_j$ is instead equal to the multiplicity of q. As for a node q of odd index $2j + 1$, since $\mu : X \to R$ is ramified at q, such a node is forced to have even multiplicity. If R is of the form $x'y' = t^{2h}$ at q, where t is a coordinate on S, then X is locally of the form $xy = t^h$, and μ is given by $x' = x^2$ and $y' = y^2$. Summing up, the contribution to $\deg_F \delta_j$ of a node of odd index $2j + 1$ is equal to one half of the multiplicity. We may summarize these observations as follows.

Denote by ε_j the number of nodes of index $2j+1$ on $\pi : R \to S$ and by ν_j the number of nodes of index $2j+2$, all counted according to their multiplicity. Then

$$\begin{aligned} &\deg_F \eta_0 = 2\nu_0\,, && \\ &\deg_F \eta_j = \nu_j\,, && j>0\,, \\ &\deg_F \delta_j = \tfrac{1}{2}\varepsilon_j\,, && j>0\,. \end{aligned} \tag{8.7}$$

To calculate $\deg_F \lambda$, we shall use relation (7.10), which implies that $(\omega_{f'} \cdot \omega_{f'})$ is equal to $12 \deg_F \lambda - \deg_F \delta$. Since ω_f is equal to the pullback of $\omega_{f'}$ to X by Theorem (6.7), part a), of Chapter X, this identity can be rewritten in the form

$$\begin{aligned} 12 \deg_F \lambda &= (\omega_f \cdot \omega_f) + \deg_F \delta \\ &= (\omega_f \cdot \omega_f) + \deg_F \eta_0 + 2\sum_{j>0} \deg_F \eta_j + \sum \deg_F \delta_j\,. \end{aligned} \tag{8.8}$$

Now $\mu^* D = 2D'$, where $D' = \sum D'_i$ is the ramification divisor of μ in X, and the Riemann–Hurwitz formula gives

$$\omega_f = \mu^* \omega_\pi(D')\,.$$

On the other hand, as D is a union of disjoint sections,

$$(\omega_\pi \cdot D) = -(D \cdot D)\,.$$

Thus,

$$\begin{aligned} (\omega_f \cdot \omega_f) &= \frac{1}{4}(\mu^*(\omega_\pi^2(D)) \cdot \mu^*(\omega_\pi^2(D))) \\ &= \frac{1}{2}(\omega_\pi^2(D) \cdot \omega_\pi^2(D)) \\ &= 2(\omega_\pi \cdot \omega_\pi) + 2(\omega_\pi \cdot D) + \frac{1}{2}(D \cdot D) \\ &= 2(\omega_\pi \cdot \omega_\pi) - \frac{3}{2}(D \cdot D)\,. \end{aligned} \tag{8.9}$$

To calculate $(\omega_\pi \cdot \omega_\pi)$, we may again use formula (7.10). Since $R \to S$ is a family of curves of genus zero, λ is trivial on it, and hence, using (8.7), we get

$$\begin{aligned} 2(\omega_\pi \cdot \omega_\pi) &= -2\sum \nu_j - 2\sum \varepsilon_j \\ &= -\deg_F \eta_0 - 2\sum_{j>0} \deg_F \eta_j - 4\sum \deg_F \delta_j\,. \end{aligned}$$

Combining this formula with (8.8) and (8.9) finally gives

$$8\lambda = -(D \cdot D) - 2\sum \deg_F \delta_j\,. \tag{8.10}$$

All the families F_i will be special instances of the family F defined above for specific choices of a family $(\pi : R \to S; D_1, \dots, D_{2g+2})$ of stable $(2g+2)$-pointed curves of genus zero. We start with F_1. In $\mathbb{P}^1 \times \mathbb{P}^1$ consider general divisors $B_1, B_2 \in |\mathcal{O}(1,1)|$ and general divisors $B_3, \dots, B_{2g+2} \in |\mathcal{O}(1,0)|$, and set $B = \sum B_i$. Then $\mathcal{O}(B) = \mathcal{O}(2g+2, 2)$ is a square, and the only singularities of B are nodes, in the number of $4g+2$. Let $\sigma : R \to \mathbb{P}^1 \times \mathbb{P}^1$ be the blow-up at the nodes of B and denote by $E_1, \dots, E_{4g+2}$ the exceptional curves, by D_i the proper transform of B_i, and by $\pi : R \to \mathbb{P}^1$ the composition of σ with the projection of $\mathbb{P}^1 \times \mathbb{P}^1$ to the second factor. Then $\pi : R \to \mathbb{P}^1$, together with the D_i, is a family of stable $(2g+2)$-pointed curves of genus zero. Moreover, if we set $D = \sum D_i$, then $\mathcal{O}(D) \cong \mathcal{O}(\sigma^* B - 2 \sum E_i)$ is a square, and hence we can form the two-sheeted covering $\mu : X \to R$ branched along D. The family F_1 is the stable model of $f : X \to \mathbb{P}^1$, where f is the composition of μ with π.

The invariants of F_1 are readily calculated. The nodes of the fibers are all of type η_0, and $\deg_{F_1} \eta_0$ is twice the number of nodes of B. Thus,

$$(8.11) \qquad \deg_{F_1} \eta_0 = 8g + 4\,, \qquad \deg_{F_1} \eta_j = \deg_{F_1} \delta_j = 0 \text{ for } j > 0\,.$$

On the other hand, $(D \cdot D)$ is equal to minus twice the number of nodes plus 4, to account for the self-intersections of B_1 and B_2, which are both equal to 2. Its value is thus equal to $-8g$, and hence $\deg_{F_1} \lambda = g$ by formula (8.10). This proves that (8.6) is valid for the family F_1.

The construction of F_{2h+1} for $h > 0$ is similar. Fix a point $p \in \mathbb{P}^1 \times \mathbb{P}^1$ and pick divisors $B_1, \dots, B_{2h+2} \in |\mathcal{O}(1,1)|$, general among those which pass through p, and general divisors $B_{2h+3}, \dots, B_{2g+2} \in |\mathcal{O}(1,0)|$. Set $B = \sum B_i$. The singularities of B are an ordinary $(2h+2)$-fold point at p and $(h+1)(4g-2h+1)$ nodes. Let $\sigma : R \to \mathbb{P}^1 \times \mathbb{P}^1$ be the blow-up at p and at the nodes, let $D = \sum D_i$ be the proper transform of B, and let E and $E_1, \dots, E_{(h+1)(4g-2h+1)}$ be the exceptional curves mapping to p and to the nodes. Then $\mathcal{O}(D) = \sigma^*(\mathcal{O}(2g+2, 2h+2))(-(2h+2)E - 2\sum E_i)$ is a square, and hence there is a two-sheeted covering $\mu : X \to R$ branched at D. The stable model of $f : X \to \mathbb{P}^1$, where f is the composition of μ, σ and the projection of $\mathbb{P}^1 \times \mathbb{P}^1$ to the second factor, is our family F_{2h+1}. It is clear that

$$(8.12) \qquad \begin{cases} \deg_{F_{2h+1}} \eta_0 = 2(h+1)(4g-2h+1)\,, \\ \deg_{F_{2h+1}} \eta_h = 1\,, \\ \deg_{F_{2h+1}} \delta_j = \deg_{F_{2h+1}} \eta_i = 0 \text{ for every } j \text{ and for } i \neq 0, h\,. \end{cases}$$

In view of (8.10), to compute $\deg_{F_{2h+1}} \lambda$, we must know $(D \cdot D)$. A straightforward computation gives

$$(D \cdot D) = -(2h+2)(4g-2h)\,.$$

Thus,

$$\begin{aligned}(8g+4)\deg_{F_{2h+1}}\lambda &= (h+1)(8g^2-4gh+4g-2h)\\ &= g\deg_{F_{2h+1}}\eta_0 + 2(h+1)(g-h)\deg_{F_{2h+1}}\eta_h\,,\end{aligned}$$

showing that (8.6) holds for F_{2h+1}.

The construction of F_{2h}, $h>0$, is not unlike the one of F_{2h+1}, but with a little twist. We start, as usual, with a divisor $B=\sum B_i$ in $\mathbb{P}^1\times\mathbb{P}^1$, where $B_1,\dots,B_{2h+1}$ are general among the members of $|\mathcal{O}(1,1)|$ passing through some fixed point p, and $B_{2h+2},\dots,B_{2g+2}$ are general divisors in $|\mathcal{O}(1,0)|$. We then blow up at the singular points of B, and let R' be the resulting surface, and $\sigma' : R'\to\mathbb{P}^1\times\mathbb{P}^1$ the blow-down map. We also denote by $H,H_1,\dots,H_{(2h+1)(2g-h+1)}$ the exceptional curves of the blow-up, where H contracts to p, and the H_i to the $(2h+1)(2g-h+1)$ nodes of B. We let p_1 be the image of p under the projection of $\mathbb{P}^1\times\mathbb{P}^1$ to the second factor and choose another point $p_2\in\mathbb{P}^1$ which is not the projection of any singular point of B. The next step is to base change via the double covering $\alpha:\mathbb{P}^1\to\mathbb{P}^1$ branched at p_1 and p_2; we call q_1 and q_2 the points of $\mathbb{P}^1$ such that $\alpha(q_1)=p_1$, $\alpha(q_2)=p_2$, and we let $\pi:R\to\mathbb{P}^1$ be the pullback of $R'\to\mathbb{P}^1$ under α. We also let D_i be the pullback via $R\to R'$ of the proper transform of B_i in R' and call $E,E_1,\dots,E_{(2h+1)(2g-h+1)}$ the inverse images of $H,H_1,\dots,H_{(2h+1)(2g-h+1)}$. We let $\sigma:R\to\mathbb{P}^1\times\mathbb{P}^1$ be the morphism obtained by base change from σ'. The fiber $\pi^{-1}(q_1)$ is the union of two smooth rational curves, one of which is E, attached at a single point q. By construction, q is a node of index $2h+1$ and multiplicity 2. Blowing it up produces a smooth surface $\widehat{R}$ fibered over $\mathbb{P}^1$ via a map $\widehat{\pi}$. We call $\widehat{E},\dots,\widehat{D}_1,\dots$ the pullbacks of $E,\dots,D_1,\dots$. The fiber $\widehat{\pi}^{-1}(q_1)$ is a chain of three smooth rational curves; one of the end components of the chain is $\widehat{E}$, and we call the middle one Γ. It is immediate to check that

$$\mathcal{O}(\textstyle\sum\widehat{D}_i+\Gamma)\cong\sigma^*(\mathcal{O}(2g+2,4h+2))(-2\sum\widehat{E}_i-(4h+2)\widehat{E}-2h\Gamma)\,.$$

In particular, $\mathcal{O}(\sum\widehat{D}_i+\Gamma)$ is a square, and hence there is a double covering $\widehat{\mu}:\widehat{X}\to\widehat{R}$ branched along Γ and the $\widehat{D}_i$. The divisor $\widehat{\mu}^*\Gamma$ is of the form $2\Gamma'$, where Γ' is exceptional of the first kind. Blowing down Γ' gives a surface X and a double covering $\mu:X\to R$ branched along the D_i and at q. We let $f:X\to\mathbb{P}^1$ be the composition of μ and π and let F_{2h} be its stable model. It is clear from the construction that

$$\begin{aligned}&\deg_{F_{2h}}\eta_0 = 4(2h+1)(2g-h+1)\,,\\ (8.13)\qquad &\deg_{F_{2h}}\delta_h = 1\,,\\ &\deg_{F_{2h}}\delta_j = \deg_{F_{2h}}\eta_i = 0 \ \text{ for every } j\neq h \text{ and every } i>0\,.\end{aligned}$$

A straightforward computation gives

$$(D\cdot D) = -2(2h+1)(4g-2h+1)\,,$$

and plugging these values into (8.10) shows that

$$(8g+4)\deg_{F_{2h}}\lambda = g\deg_{F_{2h}}\eta_0 + 4h(g-h)\deg_{F_{2h}}\delta_h\,.$$

To complete the proof of (8.4), one just needs to observe that it follows from formulas (8.11), (8.12), and (8.13) that the matrix whose entries are the degrees of the classes η_j, $j=0,\dots,\lfloor\frac{g-1}{2}\rfloor$, and δ_j, $j=1,\dots,\lfloor\frac{g}{2}\rfloor$, on the families $F_1,\dots,F_g$, is nonsingular, and hence the η_j and δ_j are independent.

REMARK (8.14). The integral Picard groups of $\mathcal{H}_g$ and $\overline{\mathcal{H}}_g$ are also known. The group $\mathrm{Pic}(\mathcal{H}_g)$ is finite cyclic of order $8g+4$ if g is odd and of order $4g+2$ if g is even [36]. On the other hand, $\mathrm{Pic}(\overline{\mathcal{H}}_g)$ is a free abelian group on g generators [147]; it must be observed, however, that the boundary classes $\eta_0,\dots,\eta_{\lfloor\frac{g-1}{2}\rfloor},\delta_1,\dots,\delta_{\lfloor\frac{g}{2}\rfloor}$ generate a proper subgroup of $\mathrm{Pic}(\overline{\mathcal{H}}_g)$.

A consequence of the absence of torsion in the Picard group of $\overline{\mathcal{H}}_g$ is that relation (8.5) is valid also over $\mathbb{Z}$. In a certain sense, this relation generalizes identity (7.13) which holds in genus 1. More precisely, it generalizes to all genera Mumford's identity

$$10\lambda = \delta_{irr} + 2\delta_1 \tag{8.15}$$

for $\mathrm{Pic}(\overline{\mathcal{M}}_2)$ [556].

9. Bibliographical notes and further reading.

For the various notions of quasi-coherent sheaf on a Deligne–Mumford stack, we refer to the general works on stacks mentioned in the bibliographical notes to Chapter XII.

The basic sources for the tautological line bundles over moduli are Mumford [555] and especially Knudsen [427]. The formulas for the pullback of the boundary classes via the clutching morphisms and for the self-intersections of the boundary classes, discussed in Sections 3 and 4, have been discovered, in various guises, by several people. For instance, they can be found in [427]. Our treatment of the excess intersection bundles for the intersections of boundary strata is inspired by the one by Graber and Pandharipande in Appendix A of [307].

The algebraic theory of the determinant of the cohomology is developed by Knudsen and Mumford [425], based in part on ideas of Grothendieck. A succinct and clear summary of the theory, with few proofs, is contained in Chapter 6 of Soulé's book [643]. The expression of the class of the boundary of moduli as a determinant is given by Knudsen in [427].

The Deligne pairing is introduced and studied by Deligne in [164] without resorting to Weil reciprocity, but rather deducing and "explaining" the latter via the pairing. The more elementary approach followed here is also due to Deligne [166].

Proposition (6.1) concerning the Picard group of $\overline{\mathcal{M}}_{g,n}$ is proved in Mumford [555]. The proof is based in part on results contained in [558] (specifically, on Proposition 1.4), which in turn depend on results of Rosenlicht [607] (Lemma (6.2) in the present chapter and its corollaries). Proposition (6.7), asserting in particular the absence of torsion in the Picard group of $\overline{\mathcal{M}}_{g,n}$, is also due to Mumford [555].

The fundamental formula (7.7) and its companion (7.8) are due to Mumford [555], while the expression for the canonical class of the moduli space $\overline{M}_{g,n}$ given by (7.16) first appears in Harris and Mumford [353].

Mumford's formula (7.15) has been rediscovered by string theorists in connection with Polyakov's integral (cf., for example, Polyakov [598], Beilinson and Manin [58], Manin [499], Catenacci–Cornalba–Martellini–Reina [111], and Biswas–Nag–Sullivan [75]).

The striking similarity between formula (7.8) and the coefficients of the central charge in the central extensions of the Virasoro algebra has been studied by a number of authors (cf., for example, Beilinson and Schechtman [60], Beilinson–Manin–Schechtman [59], Kontsevich [443], Kawamoto–Namikawa–Tsuchiya–Yamada [404], and Arbarello–De Concini–Kac–Procesi [33]).

The structure of the rational Picard group of the completed hyperelliptic locus $\overline{H}_g$ is studied in Cornalba and Harris [141], where (8.4) is proved. Arsie and Vistoli [36] show that the Picard group of $\mathcal{H}_g$ is cyclic of order $8g+4$ for odd g and cyclic of order $4g+2$ for even g, and Gorchinskiy and Viviani [301] exhibit eplicit geometric generators for these groups. That the Picard group of $\overline{\mathcal{H}}_g$ is a free abelian group of rank g is proved in [147].

Note: The exercises for this section will be found in the section of exercises for Chapter XVII

Chapter XIV.

Projectivity of the moduli space of stable curves

1. Introduction.

This chapter is entirely devoted to a proof of the projectivity of the moduli spaces of stable curves. The proof that we shall give here uses a minimum of geometric invariant theory, which is presented in the first section. In a purely GIT approach, when $g \geq 2$ and $k \geq 3$, one views $\overline{M}_g$ as the quotient H/PSL, where $H = \mathrm{H}_{k,g}$ is the Hilbert scheme of *stable* curves embedded in $\mathbb{P}^r$ by the k-canonical series, and $PSL = PSL(r+1, \mathbb{C})$ where $r = (2k-1)(g-1)-1$. Then one is asked to prove that the Hilbert point corresponding to a k-canonical stable curve C is stable in the GIT sense. Choosing a sufficiently large integer h, such a Hilbert point may be identified with the homomorphism

$$\varphi : \mathrm{Sym}^h H^0(C, \omega_C^k) \longrightarrow H^0(C, \omega_C^{kh}),$$

or equivalently, setting $N = \dim H^0(C, \omega_C^{kh})$ and identifying $\wedge^N H^0(C, \omega_C^{kh})$ with $\mathbb{C}$, we may view this Hilbert point as the point

$$[\wedge^N \varphi] \in H \subset \mathbb{P}V,$$

where V stands for the Nth exterior power of the hth symmetric power of the standard representation of $SL(r+1, \mathbb{C})$, and the action of $PSL(r+1, \mathbb{C})$ on H is induced by the action of $SL(r+1, \mathbb{C})$ on V. For a general k-canonical stable curve C, proving stability of the corresponding Hilbert point turns out to be quite demanding. Things are somewhat simpler, though far from trivial, for smooth curves. After introducing the basic tools of geometric invariant theory in Section 2, in Section 3 we prove that the Hilbert point of a *smooth* k-canonical curve is stable in the GIT sense. The role of this result in our proof of the projectivity of $\overline{M}_g$ is rather indirect; for example, we do not really use the fact that, among other things, it implies the quasi-projectivity of M_g. The way we proceed is the following. As we know from Section 9 of Chapter XII, there is a family of stable curves $\pi : \mathcal{X} \to Z$, parameterized by a complete scheme Z, whose corresponding moduli morphism $\eta : Z \to \overline{M}_g$ is finite. On Z, the Mumford class $\kappa_1 = \pi_*(c_1(\omega_\pi)^2)$ is well defined, and we use Seshadri's criterion to show that this class is ample, thus proving the projectivity of Z and hence that of $\overline{M}_g$. Seshadri's criterion is a precise requirement

E. Arbarello et al., *Geometry of Algebraic Curves*,
Grundlehren der mathematischen Wissenschaften 268,
DOI 10.1007/978-3-540-69392-5_6,

of positivity for the restriction of κ_1 to complete curves $S \subset Z$. To fulfill this requirement, we proceed as follows. Let $f : X \to S$ be the restriction of the family π to S. First suppose that the general fiber X_s of f is smooth, look at the k-canonical embedding $X_s \to \mathbb{P}^r$, and denote by $[\wedge^N \varphi_s] \in \mathbb{P}V$ the varying Hilbert point of $X_s \subset \mathbb{P}^r$ for s in S. This is where the GIT stability of k-canonical embedded *smooth* curves comes into play. Fix $s \in S$. To say that $[\wedge^N \varphi_s] \in \mathbb{P}V$ is GIT-stable means that there is a homogeneous polynomial P in the variables of V such that $P([\wedge^N \varphi_s]) = 0$, while $P([\wedge^N \varphi_{s'}]) \neq 0$ for a general point $s' \in S$. Now the collection of "vectors" $\Psi = \{P([\wedge^N \varphi_s])\}_{s \in S}$ looks like a (nonzero) section of some line bundle on S. The fundamental remark is that, if the degree of the homogeneous polynomial P is of the form $m(r+1)$, then Ψ is precisely a section of the mth power of the line bundle

$$\mathcal{L}_h = \det(f_*(\omega_f^{kh}))^{r+1} \otimes \det(f_*\omega_f^k))^{-hN}.$$

Thus $\mathcal{L}_h$ has a positivity property, and using the Riemann–Roch theorem and a few more elementary ingredients, this implies the positivity property for κ_1 required by Seshadri's criterion. In fact, all these results hold with obvious modifications in the case of stable pointed curves. Once one realizes this, a clear path appears to get rid of the hypothesis of smoothness of the general fiber of f. This simply consists, possibly after a finite base change, in normalizing the family X and resoning separately on each connected component of this normalization, which one considers, in the usual way, as a family of stable pointed curves with smooth general fiber.

It is interesting to notice that, as a byproduct, we shall obtain a number of inequalities involving the numerical invariants of families of stable curves, which, besides being essential for the proof, are quite remarkable and useful in themselves. Out of them we get the ampleness of $\kappa_1 + a\lambda$ for $a > -1$, the ampleness of ω_f when the general fiber of f is smooth and the family f not isotrivial, the nefness of λ, and the one of the point-bundle classes ψ_i. In addition, using the inequalities we just mentioned, it is possible to give a partial description of the ample cone of moduli space, as explained in the last section. Another application, also given in the last section, is a description of the Satake compactification of the moduli space of smooth curves of given genus.

2. A little invariant theory.

Let g be an integer greater than 1, k an integer greater than or equal to 3, and set $r = (2k-1)(g-1)-1$. As we have seen in Chapter XII, the moduli space of stable curves of genus g is the quotient H/PSL, where $H = \mathrm{H}_{k,g}$ is the subscheme of the Hilbert scheme parameterizing *stable* curves embedded in $\mathbb{P}^r$ by the k-canonical series, and $PSL = PSL(r+1, \mathbb{C})$. Concretely, for any sufficiently large fixed integer m, the point of H corresponding to a curve $C \subset \mathbb{P}^r$ is represented by the mth Hilbert point of C, that is, by

$$\varphi_m : H^0(\mathbb{P}^r, \mathcal{O}(m)) \longrightarrow H^0(C, \mathcal{O}_C(m))$$

or, taking exterior powers, by

$$\wedge^n \varphi_m : \wedge^n H^0(\mathbb{P}^r, \mathcal{O}(m)) \longrightarrow \wedge^n H^0(C, \mathcal{O}_C(m)),$$

where $n = h^0(C, \mathcal{O}_C(m))$. Since we are working up to homothety, it is permissible to identify $\wedge^n H^0(C, \mathcal{O}_C(m))$ to $\mathbb{C}$ and hence to think of $\wedge^n \varphi_m$ as a linear functional on $\wedge^n H^0(\mathbb{P}^r, \mathcal{O}(m))$. The standard action of $SL(r+1, \mathbb{C})$ on $\mathbb{C}^{r+1}$ induces a dual action on $H^0(\mathbb{P}^r, \mathcal{O}(1))$ and hence linear actions on

$$\wedge^n H^0(\mathbb{P}^r, \mathcal{O}(m)) = \wedge^n \operatorname{Sym}^m H^0(\mathbb{P}^r, \mathcal{O}(1))$$

and on its dual. Thus, if V stands for the nth exterior power of the mth symmetric power of the standard representation of $SL(r+1, \mathbb{C})$, H can be viewed as a subscheme of the projective space $\mathbb{P}V$; what is more important, the action of $PSL(r+1, \mathbb{C})$ on H is induced by the action of $SL(r+1, \mathbb{C})$ on V.

Our main goal in this chapter is to show that $\overline{M}_g$ is projective. If we wiew $\overline{M}_g$ as the quotient H/PSL, the most direct approach to the problem is to try to show that, if N is a large enough integer, then there are sufficiently many $SL(r+1, \mathbb{C})$-invariant homogeneous polynomials of degree N in the homogeneous coordinate ring $\oplus_{i \geq 0} \operatorname{Sym}^i V^\vee$ of $\mathbb{P}V$ to separate the different orbits of $PSL(r+1, \mathbb{C})$ in H, since these polynomials will then yield a projective embedding of $\overline{M}_g = H/PSL$. Although this is not the path that we will follow in the proof of projectivity, it will serve as a motivation for the definitions and results below.

Let then G be a complex linear algebraic group, that is, an affine subgroup of $GL(U)$ for some finite-dimensional complex vector space U, and let V be a complex finite-dimensional representation of G which is rational in the sense that the homomorphism $G \to GL(V)$ is a morphism of affine varieties. In what follows we shall assume that G is linearly reductive, that is, that any finite-dimensional representation of G is completely reducible or, in other words, can be written as a direct sum of irreducible representations. In practice, the only case we shall need is the one where $G = SL(n, \mathbb{C})$; some results, although stated in general, will be proved only for this particular group. If v is a nonzero point of V, we denote by $[v]$ the corresponding point of $\mathbb{P}V$. If x is a point of V or of $\mathbb{P}V$, we write $O(x)$ to denote its orbit under the action of G. We shall say that a nonzero point v of V is *semistable* if $\overline{O(v)}$ does not contain 0; we shall say that v is *stable* if the following two conditions are met:

i) the stabilizer G_v of v is finite;
ii) the orbit $O(v)$ is closed.

Equivalently, v is stable if its orbit is closed and of dimension equal to the dimension of G. The relevance of the notion of stability with respect to the problem of separating orbits is expressed by the following simple result. In the statement, $\operatorname{Sym} V^\vee$ stands for the polynomial algebra $\oplus_{i\geq 0} \operatorname{Sym}^i V^\vee$.

LEMMA (2.1). *Let x be a nonzero point of V and suppose that there is a homogeneous G-invariant polynomial $P \in \operatorname{Sym} V^\vee$ of positive degree such that $P(x) \neq 0$. Then x is semistable. Conversely, there is a positive integer m such that:*

i) *If x is any semistable point of V, there is a homogeneous G-invariant polynomial $P \in \operatorname{Sym} V^\vee$ of degree m such that $P(x) \neq 0$;*
ii) *If x is a stable point of V, then, for any nonzero $y \in V$ such that $[y] \notin O[x]$, there is a homogeneous G-invariant polynomial P of degree m such that $P(x) \neq 0$ but $P(y) = 0$.*

The first statement is obvious. For the proof of the remaining ones, we need some generalities on Reynolds operators. If W is a finite-dimensional representation of G, it can be written as $W = W^G \oplus W'$, where W^G consists of the invariants under the action of G, and W' is a direct sum of nontrivial irreducible representations. The Reynolds operator $E_W : W \to W$ is the projection onto W^G induced by this splitting. It is important to notice that, if $\varphi : W \to T$ is a morphism of finite-dimensional representations, then $E_T\varphi = \varphi E_W$; in fact, by Schur's lemma, $\varphi(W^G) \subset T^G$ and $\varphi(W') \subset T'$. This remark makes it possible to define Reynolds operators also for those infinite-dimensional representations which are unions of finite-dimensional ones. If W is one such representation, and w belongs to W, one picks a finite-dimensional subrepresentation T containing w and sets $E_W(w) = E_T(w)$; the compatibility of the Reynolds operator with morphisms of finite-dimensional representations implies that this does not depend on the choice of T. It is clear that Reynolds operators are also compatible with morphisms of representations which are unions of finite-dimensional ones. Notice finally that the property of being the union of finite-dimensional representations is inherited by subrepresentations.

Returning to the proof of the lemma, all the above considerations apply to the representation of G in the polynomial algebra $S = \operatorname{Sym} V^\vee$, since S is the union of the finite-dimensional subrepresentations $S_{\leq k} = \oplus_{i\leq k} \operatorname{Sym}^i V^\vee$. Now suppose that $x \in V$ is semistable, let $I \subset S$ be the ideal of $\overline{O(x)}$, and J the ideal consisting of all the elements of S without constant term. Clearly, I and J are G-invariant. Moreover, since $0 \notin \overline{O(x)}$, we can find $u \in I$ and $v \in J$ such that $u + v = 1$. Applying the Reynolds operator yields that $E_S u + E_S v = 1$. Since $E_S u \in I$, there must be a homogeneous component P of $E_S v$ that does not vanish on $\overline{O(x)}$ and hence, by G-invariance, at x.

The characterization of semistability we have just proved shows that the locus of semistable points is the complement in V of the closed subvariety defined by the ideal H generated by all the G-invariant homogeneous polynomials of positive degree. If $P_1, \ldots, P_\nu$ are homogeneous G-invariant generators for H, let μ be the product of the degrees of the P_i. If x is a semistable point of V, there is an i such that $P_i(x) \neq 0$; hence, there is a G-invariant homogeneous polynomial of degree μ, for example, $P_i^{\mu/\deg P_i}$, that does not vanish at x.

We now prove the last part of the lemma. If $x \in V$ is stable and $[y]$ does not belong to the orbit of $[x]$, then $O(\lambda y) \cap O(x) = \emptyset$ for any complex number λ. We claim that the same is true if we replace $O(\lambda y)$ with $\overline{O(\lambda y)}$. In fact, the complement of $O(\lambda y)$ in $\overline{O(\lambda y)}$ is a finite union of orbits whose dimension is strictly less than the dimension of $O(\lambda y)$. If we had that $\overline{O(\lambda y)} \cap O(x) \neq \emptyset$, then $O(x)$ would be contained in $\overline{O(\lambda y)} \setminus O(\lambda y)$ and hence would not be an orbit of maximal dimension, against the assumption that x be stable. Now we let I be the ideal of $O(x)$, J the ideal of the cone over $\overline{O(y)}$, and proceed as in the characterization of semistability. We write $1 = u + v = E_S u + E_S v$, where $u \in I$ and $v \in J$, and notice that, since $E_S u$ vanishes on $O(x)$, one of the homogeneous components of $E_S v$ does not vanish at x. This component is the polynomial P we are looking for; in fact, it vanishes at y since $E_S v$ does and J is a homogeneous ideal.

It remains to show that P can be chosen so that its degree does not depend on x and y. We define B to be the subset of $V \times V$ consisting of all couples (α, β) such that α is stable, β is semistable, and the orbit of $[\alpha]$ does not contain $[\beta]$. We also define A to be the subset of $V \times V$ consisting of all couples (α, β) such that β is semistable and there is a homogeneous polynomial vanishing at β but not at α. We have shown that $B \subset A$. Now we claim that A is Zariski-open in $V \times V$. In fact, let (x, y) be a point of A; pick G-invariant homogeneous polynomials R and Q such that R vanishes at y but not at x and that Q does not vanish at y. We may assume that R and Q have the same degree. If η is a point of V such that $Q(\eta) \neq 0$, set $R_\eta(\xi) = R(\xi) - \frac{R(\eta)}{Q(\eta)} Q(\xi)$; this is a polynomial in ξ of the same degree as R and such that $R_\eta(\eta) = 0$. Moreover, if (ξ, η) belongs to the Zariski-open subset $U \subset V \times V$ consisting of all couples (α, β) such that $R(\alpha)Q(\beta) - R(\beta)Q(\alpha) \neq 0$ and $Q(\beta) \neq 0$, then $R_\eta(\xi) \neq 0$. Thus U is contained in A. Since (x, y) clearly belongs to U, this shows that A is Zariski-open. By compactness, we may then find finitely many integers m_i and Zariski-open sets U_i covering A, and hence also B, with the property that, if (x, y) belongs to U_i, then there is a G-invariant homogeneous polynomial of degree m_i vanishing at y but not at x. Now suppose that x is stable and that y is not semistable. Then any G-invariant homogeneous polynomial Q of positive degree vanishes at y. On the other hand, we have seen that there is a Q of degree μ which does not vanish at x. Putting everything together, we conclude that, for any stable x and any y such that $[y]$ does not belong to the orbit of $[x]$, there is a G-invariant homogeneous polynomial of degree $m = \mu \prod m_i$ which vanishes at y but not at x. This concludes the proof of the lemma.

In practical cases, proving stability is usually far from easy. The main tool for doing it is the so-called Hilbert–Mumford numerical criterion. In order to state it, we recall that a one-parameter subgroup of a linear algebraic group G is a nontrivial homomorphism of algebraic groups $\lambda : \mathbb{C}^\times \to G$.

PROPOSITION (2.2) (HILBERT–MUMFORD CRITERION OF STABILITY). *Let G be a linearly reductive linear algebraic group, and let $\rho : G \to GL(V)$ be a finite-dimensional rational representation of G. A nonzero point $x \in V$ is stable if and only if, for any one-parameter subgroup $\lambda : \mathbb{C}^\times \to G$, $\rho\lambda(t)x$ diverges as $t \to 0$, while x is semistable if and only if, for any λ, $\rho\lambda(t)x$ does not go to zero as $t \to 0$.*

Notice that, since for any λ the map $t \mapsto \lambda(t^{-1})$ is also a one-parameter subgroup, in the statement of the proposition one can replace the words "as $t \to 0$" with "as $t \to \infty$." We also recall that any rational representation $\sigma : \mathbb{C}^\times \to GL(V)$ is diagonalizable, that is, there are integers $r_1, \dots, r_m$ such that, relative to a suitable basis for V, one can write

$$\sigma(t) = \begin{pmatrix} t^{r_1} & \cdot & \cdot & 0 \\ \cdot & \cdot & & \cdot \\ \cdot & & \cdot & \cdot \\ 0 & \cdot & \cdot & t^{r_m} \end{pmatrix}.$$

This can be seen as follows. Since the circle group $C = \{t \in \mathbb{C} \mid |t| = 1\}$ is compact, there is a C-invariant hermitian inner product $\langle\ ,\ \rangle$ on V. This means that

$$\langle \sigma(\exp(\sqrt{-1}\vartheta))v, \sigma(\exp(\sqrt{-1}\vartheta))v \rangle = \langle v, v \rangle \tag{2.3}$$

for any $v \in V$ and real ϑ. If we set

$$A = \frac{d}{dz}\sigma(\exp z)_{|z=0},$$

then $\sigma(\exp z) = \exp(zA)$, since both sides are solutions of the differential equation

$$\frac{dX}{dz} = AX$$

and obviously agree for $z = 0$. On the other hand, differentiating (2.3) with respect to ϑ and taking into account that one can also write

$$A = \frac{1}{\sqrt{-1}}\frac{d}{d\vartheta}\sigma(\exp(\sqrt{-1}\vartheta))_{|\vartheta=0},$$

we find that A is selfadjoint and hence diagonalizable. Thus, with respect to a suitable basis of V,

$$\sigma(\exp z) = \begin{pmatrix} \exp(r_1 z) & \cdot & \cdot & 0 \\ \cdot & & \cdot & \cdot \\ \cdot & & \cdot & \cdot \\ 0 & \cdot & \cdot & \exp(r_m z) \end{pmatrix},$$

and the r_i must be integers since the left-hand side does not change when z is replaced with $z + 2\pi\sqrt{-1}$.

Now we may go back to the proof of the proposition. That $\rho(\lambda(t))x$ behaves as stated, for any one-parameter subgroup λ, when x is stable or semistable, is not so interesting and is easy to prove; since we shall not need it, we leave it as an exercise for the reader. The converse is the really interesting part of the proposition; we shall prove it for $G = SL(n, \mathbb{C})$, which is the only case we need. For a complete proof, we refer to [558].

Assume first that $\rho(\lambda(t))x$ diverges, as $t \to 0$, for any one-parameter subgroup λ. To prove that x is stable, it suffices to show that the map $\xi : G \to V$ defined by $g \mapsto \rho(g)x$ is proper. In fact this, in addition to showing that the image of G, that is, the orbit of x, is closed also implies that all the fibers of ξ are complete. This applies in particular to the fiber over x, which is nothing but the stabilizer of x; since G is affine, G_x must then be finite. By the valuative criterion of properness, to show that ξ is proper, we must prove that, for any meromorphic map φ from the punctured disc Δ^* to G which does not extend across the puncture, also the map $t \mapsto \rho(\varphi(t))x$ does not extend. Now $\varphi(t)$ is an $n \times n$ matrix $(a_{ij}(t))$ with meromorphic entries and determinant identically equal to one; to say that it does not extend across the origin means that at least one of the entries is not holomorphic for $t = 0$. Permuting rows and columns, one may assume that the order of pole at 0 of a_{11} is greater than or equal to the order of pole of any other entry. Adding to the rows of $\varphi(t)$ the first row times suitable holomorphic functions, we can make a_{i1} zero for $i > 1$, and the same can be done with the columns. In other words, one can find matrices $A'(t)$ and $B'(t)$ of determinant equal to 1, holomorphic at 0, and such that

$$A'(t)\varphi(t)B'(t) = \begin{pmatrix} a_{11}(t) & 0 \\ 0 & \varphi_1(t) \end{pmatrix},$$

where φ_1 is an $(n-1) \times (n-1)$ matrix. We can repeat with φ_1 what has been done with φ. Continuing in this way, one finds holomorphic matrices $A(t)$ and $B(t)$ of determinant 1 such that $A(t)\varphi(t)B(t)$ is the diagonal matrix

$$\begin{pmatrix} t^{\alpha_1} f_1(t) & \cdot & \cdot & \cdot & 0 \\ \cdot & \cdot & & & \cdot \\ \cdot & & \cdot & & \cdot \\ \cdot & & & \cdot & \cdot \\ 0 & \cdot & \cdot & \cdot & t^{\alpha_n} f_n(t) \end{pmatrix},$$

where f_i is holomorphic at the origin and $f_i(0) \neq 0$ for each i. Since $\det(\varphi(t)) = 1$, we find that $\sum \alpha_i = 0$ and that $\prod f_i(t) \equiv 1$. Thus,

$$\varphi(t) = A(t)^{-1}\mu(t)C(t)\,, \tag{2.4}$$

where

$$C(t) = \begin{pmatrix} f_1(t) & \cdot & \cdot & 0 \\ \cdot & & \cdot & \cdot \\ \cdot & & \cdot & \cdot \\ 0 & \cdot & \cdot & f_n(t) \end{pmatrix} B(t)^{-1}$$

is holomorphic with determinant 1, and μ is the one-parameter subgroup

$$\mu(t) = \begin{pmatrix} t^{\alpha_1} & \cdot & \cdot & 0 \\ \cdot & & \cdot & \cdot \\ \cdot & & \cdot & \cdot \\ 0 & \cdot & \cdot & t^{\alpha_n} \end{pmatrix} .$$

That μ is nontrivial, that is, that the α_i are not all zero, follows from the fact that φ does not extend for $t = 0$.

Now suppose that $\rho(\varphi(t))x$ converges to some $y \in V$ as $t \to 0$. Then $\rho(\mu(t)C(t))x$ converges to $A(0)y$. We can diagonalize $\rho(\mu(t))$, that is, we can find a basis $v_1, \dots, v_m$ of V such that $\rho(\mu(t))v_i = t^{r_i}v_i$ for some integers r_i. Write $\rho(C(t))x = \sum b_i(t)v_i$. Since $\rho(\mu(t)C(t))x$ converges as $t \to 0$, we must have that $b_i(0) = 0$ whenever $r_i < 0$. Thus, $\rho(\mu(t)C(0))x$ and hence also $\rho(C(0)^{-1}\mu(t)C(0))x$ converge as t goes to zero. As $\lambda(t) = C(0)^{-1}\mu(t)C(0)$ is a one-parameter subgroup, this is a contradiction.

We next suppose that, for any one-parameter subgroup λ, $\rho(\lambda(t))x$ does not go to zero as $t \to 0$ and show that this implies semistability for x. The argument is very much the same as for the characterization of stability, so we only sketch it. We argue by contradiction and hence suppose that 0 belongs to the closure of the orbit of x. One can then find a meromorphic map $\varphi : \Delta^* \to G$ such that $\rho(\varphi(t))x$ goes to zero as $t \to 0$. The decomposition (2.4) still holds, so $\rho(\mu(t)C(t))x$ goes to zero as $t \to 0$. Using the same notation as in the characterization of stability, this implies that $b_i(0) = 0$ whenever $r_i \leq 0$. One concludes that, for the one-parameter subgroup $\lambda(t) = C(0)^{-1}\mu(t)C(0)$, $\lambda(t)x$ goes to zero as $t \to 0$, against the hypothesis.

3. The invariant-theoretic stability of linearly stable smooth curves.

In this section our main goal is to prove that the Hilbert points of smooth curves of positive genus embedded by their k-canonical series are stable for $k \geq 3$. In order to do this, it will be convenient to work with a slightly more general setup. Let then C be a smooth curve of genus $g \geq 1$, let L be a very ample line bundle on C, let $r + 1$ be the dimension of $H^0(C,L)$, and d the degree of L. Choosing a basis for $H^0(C,L)$ determines an embedding of C in $\mathbb{P}^r$. The mth Hilbert point of $C \subset \mathbb{P}^r$ is then the homomorphism

$$H^0(\mathbb{P}^r, \mathcal{O}_{\mathbb{P}^r}(m)) = \operatorname{Sym}^m H^0(C,L) \xrightarrow{\varphi_m} H^0(C,L^m) = H^0(C,\mathcal{O}_C(m))$$

or, taking exterior powers, the homomorphism

$$\begin{aligned}\wedge^n H^0(\mathbb{P}^r, \mathcal{O}_{\mathbb{P}^r}(m)) = \wedge^n \operatorname{Sym}^m H^0(C,L) &\xrightarrow{\wedge^n \varphi_m} \wedge^n H^0(C,L^m) \\ &= \wedge^n H^0(C,\mathcal{O}_C(m)),\end{aligned}$$

where n stands for the dimension of $H^0(C,L^m)$. As we have explained in the previous section, the homomorphism $\wedge^n\varphi_m$ can be viewed as a point of $\wedge^n \operatorname{Sym}^m V$, where V is the standard representation of $G = SL(r+1,\mathbb{C})$. We wish to give geometric conditions under which $\wedge^n\varphi_m$ is stable. We begin by translating the Hilbert–Mumford criterion of stability into one which is better adapted to our situation. Pick a basis $X_0,\dots,X_r$ of $H^0(C,L)$ and integers $\rho_0,\dots,\rho_r$, not all zero, such that $\sum\rho_i = 0$. The weight of a monomial $\prod X_i^{h_i}$ (with respect to $X_0,\dots,X_r$ and $\rho_0,\dots,\rho_r$) is defined to be $\sum h_i\rho_i$, while the weight of an element of $\operatorname{Sym}^m H^0(C,L)$ is the maximum of the weights of the monomials that compose it. The weight of an element $s \in H^0(C,L^m)$ is the minimum of the weights of the elements of $\operatorname{Sym}^m H^0(C,L)$ which are mapped to s by φ_m. Finally, the weight of a basis of $H^0(C,L^m)$ is the sum of the weights of its elements.

LEMMA (3.1). *The mth Hilbert point of $C \subset \mathbb{P}^r$ is stable if and only if, for any choice of a basis $X_0,\dots,X_r$ of $H^0(C,L)$ and of integers $\rho_0,\dots,\rho_r$, not all zero, such that $\sum\rho_i = 0$, $H^0(C,L^m)$ has a basis of negative weight.*

To prove this, pick a one-parameter subgroup $\gamma(t)$ of G. The action of γ on $H^0(C,L)$ is diagonalizable, that is, there is a basis $X_0,\dots,X_r$ of $H^0(C,L)$ such that $\gamma(t)X_i = t^{\rho_i}X_i$, where $\rho_0,\dots,\rho_r$ are integers such that $\sum\rho_i = 0$. If $v_1,\dots,v_n$ are elements of $\operatorname{Sym}^m H^0(C,L)$, we write

$$v_i = \sum_j P_{ij},$$

where the P_{ij} are monomials, and denote by ρ_{ij} the weight of P_{ij}. Thus,

$$\begin{aligned}\gamma(t^{-1})(\wedge^n\varphi_m)(v_1\wedge\cdots\wedge v_n) &= \varphi_m(\gamma(t)v_1)\wedge\cdots\wedge\varphi_m(\gamma(t)v_n) \\ &= \sum_{j_1,\dots,j_n} t^{(\sum_i \rho_{ij_i})}\varphi_m(P_{1j_1})\wedge\cdots\wedge\varphi_m(P_{nj_n}).\end{aligned}$$

To say that $\wedge^n\varphi_m$ is stable means that we may choose $v_1,\dots,v_n$ so that

$$\gamma(t^{-1})(\wedge^n\varphi_m)(v_1\wedge\cdots\wedge v_n)$$

diverges as $t \to 0$. In other words, there is a choice of $j_1, \ldots, j_n$ such that $\sum_i \rho_{ij_i} < 0$ and that

$$\varphi_m(P_{1j_1}) \wedge \cdots \wedge \varphi_m(P_{nj_n}) \neq 0 .$$

Thus, $\varphi_m(P_{1j_1}), \ldots, \varphi_m(P_{nj_n})$ is a basis of $H^0(C, L^m)$ of negative weight.

Suppose, conversely, that $\varphi_m(v_1), \ldots, \varphi_m(v_n)$ is a basis of $H^0(C, L^m)$ of negative weight and that each v_i is chosen in such a way that it has the same weight as $\varphi_m(v_i)$. Then, for any choice of $j_1, \ldots, j_n$,

$$\sum_i \rho_{ij_i} \leq \sum_i \max_j \rho_{ij} < 0 ,$$

so that $\gamma(t^{-1})(\wedge^n \varphi_m)(v_1 \wedge \cdots \wedge v_n)$ diverges as $t \to 0$. This shows that $\wedge^n \varphi_m$ is stable.

We now go back to the problem of finding geometric conditions which ensure that the Hilbert points of $C \subset \mathbb{P}^r$ are stable. The simplest one is probably linear stability, which we now define. If $C \subset \mathbb{P}^r$ is a curve which is not contained in any hyperplane, the *reduced degree* of C is

$$\frac{\deg C}{r} .$$

One says that C is *linearly stable* if, for any projection Γ of C from a linear subspace of $\mathbb{P}^r$, one has

$$\text{reduced degree of } C < \text{reduced degree of } \Gamma .$$

Let us return to our original situation where C is a smooth curve of genus $g \geq 1$ embedded in $\mathbb{P}^r$ by the complete linear system $|L|$. Then C is linearly stable if and only if, for any invertible proper subsheaf M of L such that $h^0(C, M) \neq \{0\}$, one has

$$\frac{\deg L}{h^0(C, L) - 1} < \frac{\deg M}{h^0(C, M) - 1} .$$

It follows from Clifford's theorem that, when the degree of L is greater than $2g$, then C is automatically linearly stable. In fact, let M be an invertible proper subsheaf of L such that $h^0(C, M) \neq \{0\}$. If $\deg M \geq 2g - 1$, then

$$\frac{\deg L}{h^0(C, L) - 1} = \frac{\deg L}{\deg L - g} < \frac{\deg M}{\deg M - g} = \frac{\deg M}{h^0(C, M) - 1} .$$

If, on the other hand, $\deg M \leq 2g - 2$, then Clifford's theorem implies that

$$\frac{\deg M}{h^0(C, M) - 1} \geq 2 ,$$

while

$$\frac{\deg L}{h^0(C,L)-1} = \frac{\deg L}{\deg L - g} \le \frac{2g+1}{g+1} < 2\,.$$

It follows in particular that C is linearly stable when $L = \omega_C^3$ and $g \ge 2$, when $L = \omega_C^2$ and $g > 2$, while the same kind of argument shows that C is linearly stable when $L = \omega_C$ and C is not elliptic or hyperelliptic.

The main result we shall prove is the following.

THEOREM (3.2). *Let g, d, and r be positive integers. Then there are infinitely many positive integers m with the following property. Let C be any smooth curve of genus g, and let L be any very ample line bundle of degree d on C with $h^0(C,L) = r+1$. Embed C in $\mathbb{P}^r$ via $|L|$. Then, if $C \subset \mathbb{P}^r$ is linearly stable, the mth Hilbert point of $C \subset \mathbb{P}^r$ is stable.*

The proof uses the criterion provided by Lemma (3.1). Thus, given a basis $X_0, \dots, X_r$ of $H^0(C,L)$ and integers $\rho_0 \le \dots \le \rho_r$, not all zero, such that $\sum \rho_i = 0$, we must find a basis of negative weight for $H^0(C, L^m)$ for infinitely many values of m. We denote by L_i the subsheaf of L generated by $X_0, \dots, X_i$, and by d_i its degree. Let p and N be positive integers, to be chosen later, and let

$$0 = h_0 < \dots < h_l = r$$

be a finite sequence of integers, also to be chosen later. Denote by $W_{j,k}$ the image of the map

$$\begin{aligned}\operatorname{Sym}^{N(p-k)} H^0(C, L_{h_j}) \otimes \operatorname{Sym}^{Nk} H^0(C, L_{h_{j+1}}) \otimes \operatorname{Sym}^{N} H^0(C, L)\\ \longrightarrow H^0(C, L^{N(p+1)})\,.\end{aligned}$$

These subspaces clearly provide a filtration

$$\begin{aligned} W_{0,0} \subset W_{0,1} \subset \dots \subset W_{0,p-1} \subset W_{1,0} \subset \cdots \\ \cdots \subset W_{l-1,p-1} \subset W_{l,0} \subset H^0(C, L^{N(p+1)})\,,\end{aligned}$$

and the weight of an element of $W_{j,k}$ does not exceed

$$q_{j,k} = N(p-k)\rho_{h_j} + Nk\rho_{h_{j+1}} + N\rho_r\,.$$

If we take into account that, since L is very ample, $W_{l,0}$ is equal to $H^0(C, L^{N(p+1)})$ if N is large enough (this is also a special case of Lemma (3.3) below) and denote by $w_{j,k}$ the dimension of $W_{j,k}$, it follows in particular that, for large N, $H^0(C, L^{N(p+1)})$ has a basis of weight not exceeding

$$\begin{aligned} q_{0,0}w_{0,0} &+ q_{0,1}(w_{0,1} - w_{0,0}) + \dots + q_{l,0}(w_{l,0} - w_{l-1,p-1}) \\ &= w_{0,0}(q_{0,0} - q_{0,1}) + \dots + w_{l-1,p-1}(q_{l-1,p-1} - q_{l,0}) \\ &\quad + N(p+1)\rho_r h^0(C, L^{N(p+1)})\,.\end{aligned}$$

Therefore, to estimate the weight of a basis of $H^0(C, L^{N(p+1)})$ from above, we need a good lower estimate for the $w_{j,k}$. The bound we need is provided by the simple result below; in the statement, if V is a nonzero subspace of $H^0(C,L)$, L_V stands for the subsheaf of L generated by V.

LEMMA (3.3). *Let C be a smooth curve of genus $g \geq 1$, L a very ample line bundle on C of degree d, and p a positive integer. Then there is an integer N_0, depending only on g, d, and p, such that for any $N > N_0$, any integer k such that $0 \leq k \leq p$, and any nonzero subspaces U and V of $H^0(C, L)$, the dimension of the image of*

$$\mathrm{Sym}^{N(p-k)} U \otimes \mathrm{Sym}^{Nk} V \otimes \mathrm{Sym}^N H^0(C, L) \longrightarrow H^0(C, L^{N(p+1)})$$

is at least

$$N(p-k)\deg(L_U) + Nk\deg(L_V).$$

To prove the lemma, it suffices to show that

(3.4)
$$\mathrm{Sym}^{N(p-k)} U \otimes \mathrm{Sym}^{Nk} V \otimes \mathrm{Sym}^N H^0(C, L) \longrightarrow H^0(C, L_U^{N(p-k)} \otimes L_V^{Nk} \otimes L^N)$$

is onto for large N, since $H^0(C, L_U^{N(p-k)} \otimes L_V^{Nk} \otimes L^N)$ is a subspace of $H^0(C, L^{N(p+1)})$ of dimension

$$N(p-k)\deg(L_U) + Nk\deg(L_V) + Nd + 1 - g \geq N(p-k)\deg(L_U) + Nk\deg(L_V)$$

for $Nd > 2g - 2$. Denote by Z the image of

$$\mathrm{Sym}^{p-k} U \otimes \mathrm{Sym}^k V \otimes H^0(C, L) \longrightarrow H^0(C, L_U^{p-k} \otimes L_V^k \otimes L).$$

Since the images of $\mathrm{Sym}^{p-k} U$ and $\mathrm{Sym}^k V$ in $H^0(C, L_U^{p-k})$ and $H^0(C, L_V^k)$ generate L_U^{p-k} and L_V^k, respectively, and L is very ample, the linear system $|Z|$ is a base-point-free very ample linear subsystem of $|L_U^{p-k} \otimes L_V^k \otimes L|$. Thus, the homomorphism

$$\mathrm{Sym}^N Z \to H^0(C, L_U^{N(p-k)} \otimes L_V^{Nk} \otimes L^N) \tag{3.5}$$

is onto for large N. Since (3.4) is the composition of (3.5) and of the surjective homomorphisms

$$\begin{aligned} &\mathrm{Sym}^{N(p-k)} U \otimes \mathrm{Sym}^{Nk} V \otimes \mathrm{Sym}^N H^0(C, L) \\ &\quad \to \mathrm{Sym}^N(\mathrm{Sym}^{p-k} U \otimes \mathrm{Sym}^k V \otimes H^0(C, L)), \\ &\mathrm{Sym}^N(\mathrm{Sym}^{p-k} U \otimes \mathrm{Sym}^k V \otimes H^0(C, L)) \to \mathrm{Sym}^N Z, \end{aligned}$$

it is surjective for large N, too.

The proof of Lemma (3.3) is complete, except for the fact that it remains to show that we can choose an N_0 that does not depend on U, V, C, L, or k, but only on g, d, and p. We may argue as follows. As we have seen, the validity of the estimate provided by Lemma (3.3) is implied, when $Nd > 2g - 2$, by the surjectivity of (3.5). If we consider

C as embedded in projective space by the very ample linear system $|Z|$, the surjectivity of (3.5) amounts to the surjectivity of

$$H^0(\mathbb{P}^M, \mathcal{O}_{\mathbb{P}^M}(N)) \to H^0(C, \mathcal{O}_C(N)),$$

where M is the dimension of $|Z|$, or, equivalently, to the vanishing of $H^1(\mathbb{P}^M, \mathcal{I}(N))$, where $\mathcal{I}$ is the ideal sheaf of C. Since, given g, d, and p, there are only finitely many possibilities for M and for the degree h of $|Z|$, that is, h does not exceed $(p+1)d$ and M does not exceed $\max((p+1)d/2, (p+1)d-g)$, that $H^1(\mathbb{P}^M, \mathcal{I}(N))$ vanishes for N beyond a uniform N_0 is implied by Lemma (4.1) of Chapter IX.

We now return to the proof of Theorem (3.2). It has been shown that, for large enough N, $H^0(C, L^{N(p+1)})$ has a basis of weight not exceeding

$$\begin{aligned} q_{0,0}w_{0,0} + q_{0,1}(w_{0,1} - w_{0,0}) + \cdots + q_{l,0}(w_{l,0} - w_{l-1,p-1}) \\ = w_{0,0}(q_{0,0} - q_{0,1}) + \cdots + w_{l-1,p-1}(q_{l-1,p-1} - q_{l,0}) \\ + N(p+1)\rho_r h^0(C, L^{N(p+1)}), \end{aligned}$$

where

$$q_{j,k} = N(p-k)\rho_{h_j} + Nk\rho_{h_{j+1}} + N\rho_r,$$

and $w_{j,k}$ is the dimension of the image of

$$\begin{aligned} &\mathrm{Sym}^{N(p-k)} H^0(C, L_{h_j}) \otimes \mathrm{Sym}^{Nk} H^0(C, L_{h_{j+1}}) \otimes \mathrm{Sym}^N H^0(C, L) \\ &\quad \longrightarrow H^0(C, L^{N(p+1)}). \end{aligned}$$

On the other hand, Lemma (3.3) shows that, for large N,

$$w_{j,k} \geq N(p-k)d_{h_j} + Nkd_{h_{j+1}}.$$

Combining everything, we find that, for large N, there is a basis of $H^0(C, L^{N(p+1)})$ of weight not greater than

$$\begin{aligned} &-\sum_{j=0}^{l-1}\sum_{k=0}^{p-1} N^2\left((p-k)d_{h_j} + kd_{h_{j+1}}\right)\left(\rho_{h_{j+1}} - \rho_{h_j}\right) \\ &\quad + N(p+1)\rho_r(N(p+1)d + 1 - g) \\ &= -N^2\sum_{j=0}^{l-1}\left(\frac{p^2+p}{2}d_{h_j} + \frac{p^2-p}{2}d_{h_{j+1}}\right)\left(\rho_{h_{j+1}} - \rho_{h_j}\right) + N^2(p+1)^2\rho_r d \\ &\quad + N(p+1)\rho_r(1-g) \\ &= -N^2p^2\sum_{j=0}^{l-1}\frac{d_{h_j} + d_{h_{j+1}}}{2}\left(\rho_{h_{j+1}} - \rho_{h_j}\right) \\ &\quad + N^2p\sum_{j=0}^{l-1}\frac{d_{h_{j+1}} - d_{h_j}}{2}\left(\rho_{h_{j+1}} - \rho_{h_j}\right) \\ &\quad + N^2(p+1)^2\rho_r d + N(p+1)\rho_r(1-g). \end{aligned}$$

What has to be shown is that this quantity, which we denote by w, is negative for an appropriate choice of p and of $h_0, \dots, h_l$. We have not yet used the assumption that $C \subset \mathbb{P}^r$ be linearly stable. This assumption implies, in particular, that $d_h > hd/r$ for $0 \le h < r$. Thus, if m_h stands for the minimum integer larger than hd/r and if we set

$$\varepsilon = \min\{m_h - hd/r \mid h = 0, \dots, r-1\},$$

we find that $\varepsilon > 0$ and that

$$d_h \ge \frac{hd}{r} + \varepsilon$$

for $0 \le h < r$. As a consequence,

$$\sum_{j=0}^{l-1} \frac{d_{h_j} + d_{h_{j+1}}}{2} \left(\rho_{h_{j+1}} - \rho_{h_j}\right) \ge \frac{d}{r} \sum_{j=0}^{l-1} \frac{h_j + h_{j+1}}{2} \left(\rho_{h_{j+1}} - \rho_{h_j}\right) + \varepsilon\left(\rho_r - \rho_0\right),$$

so that

$$\begin{aligned} w \le & -N^2 p^2 \frac{d}{r} \sum_{j=0}^{l-1} \frac{h_j + h_{j+1}}{2} \left(\rho_{h_{j+1}} - \rho_{h_j}\right) - N^2 p^2 \varepsilon \left(\rho_r - \rho_0\right) \\ & + N^2 pd(\rho_r - \rho_0) + N^2 (p+1)^2 \rho_r d\,. \end{aligned} \tag{3.6}$$

The time has come to choose p and $h_0, \dots, h_l$. The correct choice is provided by the following result.

LEMMA (3.7). *Let $\rho_0 \le \rho_1 \le \dots \le \rho_r$ be real numbers. Then*

$$\max_{l,h} \left(\sum_{j=0}^{l-1} \frac{h_j + h_{j+1}}{2} \left(\rho_{h_{j+1}} - \rho_{h_j}\right) \right) \ge r\rho_r - \frac{r}{r+1} \sum_i \rho_i\,, \tag{3.8}$$

where l varies among all integers between 1 and r, and h among all sequences of integers $0 = h_0 < h_1 < \dots < h_l = r$.

Before proving the lemma, we use it to conclude the proof of Theorem (3.2). Pick a sequence $0 = h_0 < h_1 < \dots < h_l = r$ for which the maximum in the left-hand side of (3.8) is attained. In our situation, $\sum \rho_i = 0$ and $\rho_0 < 0$, so that (3.8) and (3.6) yield

$$\begin{aligned} w &\le -N^2 p^2 d\rho_r - N^2 p^2 \varepsilon \left(\rho_r - \rho_0\right) + N^2 pd(\rho_r - \rho_0) + N^2 (p+1)^2 \rho_r d \\ &= -N^2 p(p\varepsilon - d)(\rho_r - \rho_0) + 2N^2 pd\rho_r + N^2 d\rho_r \\ &\le -N^2 p(p\varepsilon - d)(\rho_r - \rho_0) + 2N^2 pd(\rho_r - \rho_0) + N^2 d(\rho_r - \rho_0) \\ &= -N^2 (\rho_r - \rho_0)(p(p\varepsilon - d) - 2pd - d)\,. \end{aligned}$$

We choose a p so large that $p(p\varepsilon - d) - 2pd - d > 0$. Since $H^0(C, L^{N(p+1)})$ has a basis of weight at most w, it then has a basis of negative weight. This shows that the mth Hilbert point of $C \subset \mathbb{P}^r$ is stable for $m = N(p+1)$ and N sufficiently large.

To conclude the proof of Theorem (3.2), it remains to prove Lemma (3.7). The left-hand side of (3.8) is just the area of the convex hull A of the points $(0, \rho_r)$, $(0, \rho_0)$, $(1, \rho_1)$, ..., (r, ρ_r) in the plane (Figure 1 shows an example with $r = 3$).

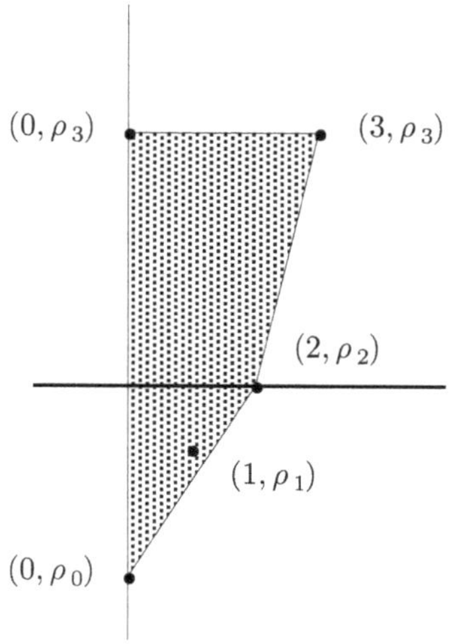

Figure 1.

Decreasing the ρ_i with $0 < i < r$ so as to bring the points (i, ρ_i) down onto the boundary of A obviously does not change the left-hand side of (3.8), while it increases the right-hand side. Therefore we may limit ourselves to proving (3.8) in the special case where all the points (i, ρ_i) lie on the lower boundary of A. In this case the area of A is just

$$\sum_{j=0}^{r-1} \frac{2j+1}{2} (\rho_{j+1} - \rho_j) .$$

But this quantity equals

$$\sum_{j=1}^{r} j(\rho_j - \rho_{j-1}) - \frac{1}{2}(\rho_r - \rho_0) = r\rho_r + \frac{1}{2}(\rho_r + \rho_0) - \sum_i \rho_i .$$

On the other hand,

$$\begin{aligned} r\rho_r + \frac{1}{2}(\rho_r + \rho_0) - \sum_i \rho_i &= r\rho_r + \frac{1}{2}(\rho_r + \rho_0) - \frac{1}{r+1}\sum_i \rho_i - \frac{r}{r+1}\sum_i \rho_i \\ &\geq r\rho_r - \frac{r}{r+1}\sum_i \rho_i \end{aligned}$$

since A is convex. This finishes the proof of Lemma (3.7) and hence of Theorem (3.2).

4. Numerical inequalities for families of stable curves.

The relative dualizing sheaf of a family of stable curves enjoys some remarkable positivity properties. We have already encountered a first instance of this in Chapter XI, Section 9, where we showed that the direct images of the relative dualizing sheaf and of its powers are semi-positive. In this section and in the next one, we shall present several other positivity results and some of their consequences. The proof of semi-positivity was based on curvature consideration. Here the methods will be mostly algebraic, and the key ingredient will be the invariant-theoretic stability of smooth linearly stable curves proved in the previous section. Stability enters the picture via the following elementary result.

LEMMA (4.1). *Let $f : X \to S$ be a family of nodal curves over a reduced and irreducible complete curve, let L be an invertible sheaf on X, and let h be a positive integer. For any $s \in S$, set $X_s = f^{-1}(s)$ and $L_s = L \otimes \mathcal{O}_{X_s}$. Suppose that the following conditions are satisfied:*

i) *$h^0(X_s, L_s)$ is independent of s;*
ii) *$h^1(X_s, L_s^h) = 0$ for any $s \in S$;*
iii) *for general $s \in S$, the complete linear system $|L_s|$ on X_s is base-point-free and very ample;*
iv) *for general $s \in S$, the hth Hilbert point of X_s, embedded in projective space via $|L_s|$, is semistable.*

Then, if we denote by ℓ the dimension of $H^0(X_s, L_s)$, by N the dimension of $H^0(X_s, L_s^h)$, and set

$$\mathcal{L}_h = \det(f_*(L^h))^{\ell} \otimes \det(f_*L)^{-hN} ,$$

there is a positive integer m such that the line bundle $\mathcal{L}_h^m$ has a nonzero section. If we substitute "stable" for "semistable" in iv) and suppose in addition that, as s varies, the hth Hilbert points of the X_s do not all belong to the same $SL(\ell)$-orbit, then there is a positive integer m such that for any $s' \in S$, there is a nonzero section of $\mathcal{L}_h^m$ vanishing at s'. Finally, the integer m can be chosen to depend only on h, ℓ, and N, and not on the particular family under consideration.

We begin by proving the first statement. We shall use the characterization of semistability provided by (2.1). Look at the homomorphism

$$\varphi : \wedge^N \operatorname{Sym}^h f_*L \to \det f_*L^h ;$$

choosing a basis for $H^0(X_s, L_s)$ gives an embedding of X_s in $\mathbb{P}^{\ell-1}$, and the fiber of φ over s, which we denote by φ_s, can be identifed with the hth Hilbert point of $X_s \subset \mathbb{P}^{\ell-1}$. Thus, there is an $SL(\ell, \mathbb{C})$-invariant homogeneous polynomial P which does not vanish at φ_s; we may assume that the degree of P is of the form $m\ell$, where m is an integer that depends only on h, ℓ, and N. The idea is to somehow "evaluate" P at φ; the result of this operation will be the section we wish to construct. To make sense out of this, it is convenient to work in a slightly more general setting.

Consider then a vector bundle E of rank ℓ and a line bundle H on S. Given a complex holomorphic representation

$$\rho : GL(\ell, \mathbb{C}) \to GL(V),$$

we can construct a new vector bundle E_ρ with typical fiber V and structure group $GL(\ell, \mathbb{C})$ by the following prescription. If $\{g_{\alpha\beta}\}$ is a system of transition functions for E with respect to an open cover $\{U_\alpha\}$, then a system of transition functions for E_ρ with respect to the same cover is $\{\rho(g_{\alpha\beta})\}$. Suppose moreover that we are given a bundle homomorphism

$$\psi : E_\rho \to H$$

and a $GL(\ell, \mathbb{C})$-invariant subspace $W \subset \operatorname{Sym}^k V$, and let $\sigma : GL(\ell, \mathbb{C}) \to GL(W)$ be the representation obtained by restriction from $\operatorname{Sym}^k \rho$. Thus, there are an inclusion of vector bundles

$$E_\sigma \hookrightarrow \operatorname{Sym}^k E_\rho$$

and, by composition with $\operatorname{Sym}^k \psi$, a homomorphism

$$\Psi : E_\sigma \to H^k.$$

The section $P(\varphi)$ we wish to construct will be a special instance of Ψ. We take $E = f_* L$, $H = \det f_*(L^h)$, and choose as ρ the Nth exterior power of the hth symmetric power of the standard representation of $GL(\ell, \mathbb{C})$. With these choices, $E_\rho = \wedge^N \operatorname{Sym}^h f_* L$, and we may take as ψ the homomorphism φ. As for W, our choice for it is the subrepresentation of $\operatorname{Sym}^{m\ell} V$ generated by P. This is a line; in fact, since P is $SL(\ell, \mathbb{C})$-invariant and of degree $m\ell$, an element $g \in GL(\ell, \mathbb{C})$ acts on it by multiplication by $\det g^{hNm}$. It follows that, in our concrete case, E_σ is the line bundle $\det(f_*(L))^{hNm}$, so that $P(\varphi) = \Psi$ is a section of $\det(f_*(L^h))^{m\ell} \otimes \det(f_* L)^{-hNm}$; since, by construction, it does not vanish at s, this proves the first part of the lemma.

To prove the second part, we distinguish two cases. If φ vanishes at s', so does $P(\varphi)$. Otherwise, for general s the hth Hilbert point of X_s is distinct from the one of $X_{s'}$. By (2.1) we may then choose P in such a way that it vanishes at s' but not at s. Thus $P(\varphi)$ vanishes at s' but is not identically zero. This finishes the proof of the lemma.

We next show how the preceding result can be used, in conjunction with the Grothendieck Riemann–Roch theorem, to produce numerical inequalities.

PROPOSITION (4.2). *Let $f : X \to S$ be a family of nodal curves over a reduced and irreducible complete curve, and let L be an invertible sheaf on X. For any $s \in S$, set $X_s = f^{-1}(s)$ and $L_s = L \otimes \mathcal{O}_{X_s}$. Suppose that*

i) *$h^0(X_s, L_s)$ is independent of s;*
ii) *for general $s \in S$, the complete linear system $|L_s|$ on X_s is base-point-free and very ample.*

Suppose moreover that there are infinitely many positive integers h for which

iii) *$h^1(X_s, L_s^h) = 0$ for every $s \in S$;*
iv) *for general $s \in S$, the hth Hilbert point of X_s, embedded in projective space via $|L_s|$, is semistable.*

Denote by d the relative degree of L and by ℓ the rank of f_*L. Then

$$\ell(L \cdot L) \geq 2d \deg(f_*L)\,.$$

We use the same notation as in Lemma (4.1). The lemma implies that the line bundle $\mathcal{L}_h = \det(f_*(L^h))^\ell \otimes \det(f_*L)^{-hN}$ has nonnegative degree for arbitrarily large values of h. We may evaluate this degree by means of the Grothendieck Riemann–Roch theorem, which yields

$$\deg f_*(L^h) = \frac{(L^h \cdot L^h)}{2} - \frac{(L^h \cdot \omega_f)}{2} + \deg_f \lambda\,,$$

since $R^1 f_*(L^h) = 0$ by iii). The Riemann–Roch theorem also says that

$$N = hd + 1 - g\,,$$

where g stands for the genus of the fibers of f. Putting everything together, we find that the degree of $\mathcal{L}_h$ is

$$(4.3)\quad \deg \mathcal{L}_h = \ell \frac{h^2}{2}(L \cdot L) - \ell \frac{h}{2}(L \cdot \omega_f) + \ell \deg_f \lambda - h(hd + 1 - g) \deg f_*L\,.$$

This is a polynomial in h; as it is nonnegative for arbitrarily large values of h, the coefficient of its leading term, that is,

$$\frac{\ell}{2}(L \cdot L) - d \deg(f_*L)\,,$$

must be nonnegative. This concludes the proof of (4.2).

When $R^1 f_*L = 0$, the conclusion of (4.2) can be put into a somewhat different form. In this case, in fact, the Riemann–Roch theorem gives

$$\deg f_*L = \frac{(L \cdot L)}{2} - \frac{(L \cdot \omega_f)}{2} + \deg_f \lambda\,, \quad d - \ell = g - 1\,,$$

so the inequality in (4.2) amounts to

$$(4.4) \qquad (1-g)(L\cdot L) + d(L\cdot \omega_f) - 2d \deg_f \lambda \geq 0\,.$$

Proposition (4.2), when coupled with the stability criterion (3.2), has surprisingly strong consequences. Consider, for example, a family $f : X \to S$ of semistable curves of genus g over a reduced and irreducible complete curve. Suppose that the general fiber of f is smooth. Set $L = \omega_f^k$, where k is a positive integer. It follows from (3.2) and the remarks immediately preceding it that all the assumptions of (4.2) are satisfied by f and L as soon as $k \geq 3$ when $g > 1$, or as soon as $k \geq 2$ when $g > 2$, or even for $k = 1$ if the general fiber of f is not elliptic or hyperelliptic.

Let us restrict, for the time being, to the case where $k > 1$. Since $R^1 f_* L = 0$, we may use (4.4), which, in the case at hand, translates into

$$(g-1)\big(k^2(\omega_f\cdot\omega_f) - 4k\deg_f \lambda\big) \geq 0\,.$$

Taking into account that $(\omega_f \cdot \omega_f) = 12\deg_f \lambda - \deg_f \delta$ (cf. Theorem (7.6) in Chapter XIII), we obtain the inequality

$$\left(12 - \frac{4}{k}\right)\deg_f \lambda - \deg_f \delta \geq 0$$

for $k > 2$ and $g > 1$ or for $k = 2$ and $g > 2$. In particular, we find that $10\deg_f \lambda - \deg_f \delta \geq 0$ for any $g > 2$ and that $11\deg_f \lambda - \deg_f \delta \geq 0$ for $g \geq 2$. The following result, however, shows that one can do much better.

THEOREM (4.5). *Let $f : X \to S$ be a family of semistable curves of genus g over a reduced and irreducible complete curve. Suppose that the general fiber of f is smooth. Then*

$$\left(8 + \frac{4}{g}\right)\deg_f \lambda - \deg_f \delta \geq 0\,.$$

To prove this, we consider two distinct cases. When $g > 1$ and the general fiber of f is not hyperelliptic, we apply (4.2) with $L = \omega_f$. We get

$$0 \leq g(\omega_f\cdot\omega_f) - (4g-4)\deg_f \lambda = (8g+4)\deg_f \lambda - g\deg_f \delta\,,$$

as desired. If instead the general fiber of f is hyperelliptic or elliptic, we may resort to formula (8.5) in Chapter XIII, coupled with formula (8.3) in the same chapter. In fact, $(i+1)(g-i) \geq g$ for all i between 0 and $g-1$, and similarly $4j(g-j) > g$ for all j such that $1 \leq j \leq g-1$. This finishes the proof of Theorem (4.5).

It should be observed that the inequality provided by (4.5) is optimal in the sense that the constant $(8+4/g)$ appearing in it cannot be replaced by any smaller number. In fact Theorem (8.4) in Chapter XIII shows that $\deg_f \delta$ equals $(8+4/g)\deg_f \lambda$ for families of hyperelliptic curves whose degenerate members have only singularities of type η_0 (cf. Section 3 of Chapter X). Families of this type with $\deg_f \delta$ different from zero do exist; an example is provided by the family F_1 constructed during the proof of Theorem (8.4) in Chapter XIII (cf. in particular formula (8.11) in the same chapter). Another example is given by a double covering of $Q = \mathbb{P}^1 \times \mathbb{P}^1$ branched along a general curve of type $(2g+2, 2m)$, $m \geq 1$, fibered over the second factor of Q.

On the other hand, better inequalities can be obtained if one is willing to attribute different "weights" to singular fibers of different kinds. For instance, Moriwaki has proved that, under the assumptions of Theorem (4.5),

$$(8g+4)\deg_f \lambda \geq g \deg_f \delta_{irr} + 4\sum_{j=1}^{\lfloor \frac{g}{2} \rfloor} j(g-j)\deg_f \delta_j\,. \tag{4.6}$$

This inequality should be confronted with identity (8.4) in Chapter XIII, which shows that in fact (4.6) is an equality when $f : X \to S$ is a family of hyperelliptic curves without singular fibers of type η_i for $i > 0$.

Another remark is that one cannot substitute the word "nodal" for "semistable" in the statement of (4.5). In fact, given a family of nodal curves over a curve, blowing up a smooth point of a fiber leaves the degree of λ unchanged but increases the degree of δ. Repeating this process we may construct families with the degree of λ fixed but with the degree of δ as large as we wish.

A first consequence of (4.5) is the following.

Corollary (4.7). *Let $f : X \to S$ be a family of nodal curves over a reduced and irreducible complete curve. If the general fiber of f is smooth, then* $\deg_f \lambda \geq 0$.

To prove this, note first that $f_*\omega_f = 0$ when $g = 0$, so that $\deg_f \lambda = 0$ in this case. In higher genus, since $\deg_f \lambda$ does not change under blow-up of smooth points in the fibers, we may assume that all the fibers of f are semistable. But then, since the general fiber of f is smooth, $\deg_f \delta \geq 0$, and the conclusion follows from (4.5).

There are several other inequalities that can be deduced from (4.5) or proved by similar methods. The first one we shall prove is a strengthening of (4.7). To state it, we introduce and briefly illustrate the notion of isotriviality. Consider a family F of n-pointed nodal curves over the base S. When S is irreducible, we say that F is *isotrivial* if a general point of S has a neighborhood, in the *analytic* topology, over which F is a product family; in general, we shall say that F is isotrivial when it is isotrivial over every irreducible component of S. Isotrivial families of stable curves are very close to being product families. Suppose in fact that F is a family of stable n-pointed curves, consisting of a family $f : X \to S$ of nodal curves plus sections $D_1, \dots, D_n$. Assume moreover that S is connected. The isotriviality of F implies that the natural map $S \to \overline{M}_{g,n}$ is constant in the neighborhood of a point of every component of S and hence constant. Thus, $f : X \to S$ is, analytically, a locally trivial bundle, and one can find local trivializations for it such that the D_i are horizontal sections. Let C be a typical fiber of f, let $q_1, \dots, q_n$ be the marked points on it, and let

$$T = \mathbf{Isom}_S((X; D_1, \dots, D_n), (C \times S; \{q_1\} \times S, \dots, \{q_n\} \times S))$$

be the scheme over S parameterizing isomorphisms, as n-pointed curves, between fibers of $Y \to S$ and of the product family $C \times S \to S$. Clearly, T is a finite unramified covering of S of degree $|\mathrm{Aut}((C; q_1, \dots, q_n))|$. Set $X' = X \times_S T$, let $f' : X' \to T$ be the projection to the second factor, and denote by D'_i the pullback of D_i to X'; by construction, the family of stable n-pointed curves consisting of $f' : X' \to T$, together with the sections $D'_1, \dots, D'_n$, is isomorphic to the product family $(C \times T \to T; \{q_1\} \times T, \dots, \{q_n\} \times T)$. We are now ready to state our next result.

PROPOSITION (4.8). *Let $f : X \to S$ be a nonisotrivial family of nodal curves over a reduced and irreducible complete curve. Suppose that the general fiber of f is smooth. Then*

$$\deg_f \lambda > 0 .$$

There is something to prove only in genus $g \geq 1$. The degree of λ does not change if we blow down a smooth rational component of a singular fiber touching the rest of the fiber at a single point. In the proof we may thus work under the additional hypothesis that $f : X \to S$ is a family of semistable curves. If the family contains singular fibers, then $\deg_f \delta > 0$, so we are done by (4.5). We are thus reduced to proving the proposition when all the fibers of f are smooth. The argument we shall give here is based on curvature considerations; an entirely algebraic argument will be found in the next section, right after the proof of (5.7). It was shown in Chapter XI that the Hodge bundle $f_*(\omega_f)$ has nonnegative curvature everywhere on S (cf. formula (9.7)). It follows that $\det f_*(\omega_f)$ also has non-negative curvature everywhere (so that in particular we get another proof of (4.7)). To conclude, it suffices to show that $\det f_*(\omega_f)$ has strictly positive curvature somewhere. If we denote by R the curvature form of $f_*(\omega_f)$, the curvature form of $\det f_*(\omega_f)$ is the trace of R. All we must show is that R is not identically zero, since we know it is nonnegative. The nonisotriviality assumption implies that, for general $s \in S$, the Kodaira–Spencer map

$$\rho : T_s(S) \to H^1(C, T_C)$$

is injective, where $C = f^{-1}(s)$. It then follows from the local Torelli theorem, or from the local Torelli theorem for hyperelliptic curves when all the fibers of f are hyperelliptic, that the same is true for the composition of ρ with the map

$$H^1(C, T_C) \to \mathrm{Hom}(H^0(C, \omega_C), H^1(C, \mathcal{O}_C)) .$$

For any tangent vector $v \in T_s(S)$, we denote by τ_v its image under this composed map; we have seen that τ_v is not zero if v is not. Now it has been proved in Section 9 of Chapter XI (cf. formulas (9.6) and (9.8)) that, for any abelian differential $\varphi \in H^0(C, \omega_C)$ and any $v \in T_s(S)$,

$$\iota_{\bar{v}} \iota_v \langle R\varphi, \varphi \rangle = -\langle \tau_v(\varphi), \tau_v(\varphi) \rangle ,$$

where ι_w stands for inner product with w. If v is not zero, $\tau_v(\varphi)$ is not zero for some φ, which shows that R does not vanish at s. This concludes the proof of the proposition.

A natural question that may be asked is in what cases is the inequality of (4.5) actually an equality. The answer is that equality occurs only in three situations: either the family $f : X \to S$ is isotrivial, or it is a family of elliptic curves, or it is a family of hyperelliptic curves whose degenerate members have only singularities of type η_0. For families whose general fiber is elliptic or hyperelliptic, this follows at once from formulas (8.5) and (8.3) in Chapter XIII, since $(i + 1)(g - i)$ is strictly larger than g for i between 1 and $\lfloor (g - 1/2) \rfloor$, while $4j(g - j)$ is strictly larger than g for j between 1 and $\lfloor (g/2) \rfloor$. If the general fiber of f is not elliptic or hyperelliptic, it is necessary to go back to the proof of (4.2) and, more precisely, to formula (4.3), applied to $L = \omega_f$. As the calculations leading to (4.5) imply, to say that $(8g + 4) \deg_f \lambda = g \deg_f \delta$ amounts to saying that the degree-two term of the right-hand side of (4.3), viewed as a polynomial in h, vanishes. It follows that the coefficient of the degree-one term, that is,

$$-\frac{g}{2}(\omega_f \cdot \omega_f) + (g - 1) \deg_f \lambda = -(5g + 1) \deg_f \lambda + \frac{g}{2} \deg_f \delta ,$$

must be nonnegative. This means that

$$g \deg_f \delta \geq (10g + 2) \deg_f \lambda ,$$

which, coupled with (4.5), gives $0 \geq (2g - 2) \deg_f \lambda$. In view of (4.8), this is possible only if $f : X \to S$ is isotrivial or $g \leq 1$.

It is instructive to notice that neither (4.5) nor (4.8) remain true if the assumption that the general fiber of f be smooth is dropped, as is shown by the following two examples.

EXAMPLE (4.9). Pick a nonisotrivial family of stable 1-pointed elliptic curves consisting of a family of curves $\varphi : Y \to S$ together with a section Γ. Then pick a smooth curve C of genus $g-1>0$ and a point q on it. Let $f: X \to S$ be the family of stable curves obtained from $\varphi : Y \to S$ and the constant family $C \times S \to S$ by identifying Γ with the constant section $\{q\} \times S$. We then have

$$\begin{aligned}\deg_f \lambda &= \deg_\varphi \lambda\,,\\ \deg_f \delta &= \deg_\varphi \delta + (\Gamma \cdot \Gamma)\,.\end{aligned}$$

On the other hand, we know that $12\deg_\varphi \lambda - \deg_\varphi \delta = 0$. As for $(\Gamma \cdot \Gamma) = -(\omega_\varphi \cdot \Gamma)$, it can be calculated by noticing that ω_φ is trivial on the fibers of φ and hence equal to $\varphi^*(\varphi_*\omega_\varphi)$, so that

$$(\Gamma \cdot \Gamma) = -\deg_\varphi \lambda\,.$$

Putting everything together, we conclude that

$$11\deg_f \lambda = \deg_f \delta\,,$$

which is incompatible with the inequality in (4.5) since $g \geq 2$ and $\deg_f \lambda > 0$.

EXAMPLE (4.10). Pick two smooth curves S and C of genera $i > 1$ and $g - i > 0$, and a point q on C. Let $f : X \to S$ be the family of stable curves obtained from the constant families $S \times S \to S$ and $C \times S \to S$ by identifying the diagonal section Δ of $S \times S \to S$ with the constant section $\{q\} \times S$ of $C \times S \to S$. Since an abelian differential on any fiber of f can be obtained simply by choosing an abelian differential on S and one on C, we conclude that $\deg_f \lambda = 0$. On the other hand, $f : X \to S$ is not isotrivial. This can be seen, for instance, by noticing that

$$\deg_f \delta = (\Delta \cdot \Delta) = 2 - 2i < 0\,.$$

A useful consequence of (4.5) and (4.8) is the following result.

COROLLARY (4.11). *Let $f : X \to S$ be a family of semistable curves such that S is a reduced and irreducible complete curve and the general fiber of f is smooth. Then $(\omega_f \cdot \omega_f) \geq 0$. Moreover, if $g > 1$ and $f : X \to S$ is not isotrivial, then $(\omega_f \cdot \omega_f) > 0$.*

To prove this, notice that, using the identity $(\omega_f \cdot \omega_f) = 12\deg_f \lambda - \deg_f \delta$, the inequality given by (4.5) can be written under the form

$$(\omega_f \cdot \omega_f) \geq 4\frac{g-1}{g}\deg_f \lambda\,. \tag{4.12}$$

The right-hand side of this inequality is nonnegative by (4.7), and if $g > 1$ and $f : X \to S$ is not isotrivial, it is strictly positive by (4.8).

Our next inequality has to do with self-intersections of sections of families of nodal curves.

PROPOSITION (4.13). *Let $f : X \to S$ be a family of nodal curves of genus $g > 0$ over a reduced and irreducible complete curve, and let $\Gamma \subset X$ be a section of f not passing through any of the singular points of the fibers. Suppose that the general fiber of f is smooth. Then*

$$(\Gamma \cdot \Gamma) \leq 0 .$$

Moreover, if $(\Gamma \cdot \Gamma) = 0$, then the family $f : X \to S$, together with the section Γ, is an isotrivial family of 1-pointed nodal curves.

If we perform a base change from S to its normalization, the self-intersection of Γ does not change. Therefore, in the proof we may assume that S is smooth. Suppose first that all the fibers of f are semistable. If $g = 1$, notice that, as we observed in Example (4.9),

$$(\Gamma \cdot \Gamma) = -(\omega_f \cdot \Gamma) = -\deg_f \lambda .$$

Thus $(\Gamma \cdot \Gamma) \leq 0$ by (4.7), while (4.8) implies that $f : X \to S$ is isotrivial when $(\Gamma \cdot \Gamma) = 0$. As we shall see in a moment, the same conclusions can be reached in genus $g > 1$ using the algebraic index theorem for the surface X. We briefly recall what this theorem says. Given a normal projective surface V, we denote by $\mathrm{Num}(V)$ the group of Cartier divisors on V modulo numerical equivalence.

THEOREM (4.14) (ALGEBRAIC INDEX THEOREM). *Let V be a normal projective surface. Then the intersection pairing on $\mathrm{Num}(V) \otimes \mathbb{R}$ is nondegenerate with signature $(1, n-1)$, where n is the rank of $\mathrm{Num}(V)$.*

Returning to the surface X, denote by F a fiber of f. Since $(F \cdot F) = 0$, the intersection form is not negative definite on the subgroup of $\mathrm{Num}(X)$ generated by ω_f, $\mathcal{O}(\Gamma)$, and $\mathcal{O}(F)$. The index theorem then implies that

$$\begin{aligned} 0 \leq \det &\begin{pmatrix} (\omega_f \cdot \omega_f) & (\omega_f \cdot \Gamma) & (\omega_f \cdot F) \\ (\Gamma \cdot \omega_f) & (\Gamma \cdot \Gamma) & (\Gamma \cdot F) \\ (F \cdot \omega_f) & (F \cdot \Gamma) & (F \cdot F) \end{pmatrix} \\ = \det &\begin{pmatrix} (\omega_f \cdot \omega_f) & -(\Gamma \cdot \Gamma) & 2g-2 \\ -(\Gamma \cdot \Gamma) & (\Gamma \cdot \Gamma) & 1 \\ 2g-2 & 1 & 0 \end{pmatrix} . \end{aligned}$$

Expanding this out yields the inequality

$$-2g(2g-2)(\Gamma \cdot \Gamma) \geq (\omega_f \cdot \omega_f) , \tag{4.15}$$

which, since $g > 1$, shows that $(\Gamma \cdot \Gamma) \leq 0$. Now suppose that $(\Gamma \cdot \Gamma) = 0$. Then (4.15) implies that

$$(\omega_f \cdot \omega_f) = 0 ,$$

which in turn implies, by (4.11), that $f : X \to S$ is isotrivial.

What we have shown so far is that $(\Gamma \cdot \Gamma) \leq 0$ and that $f : X \to S$ is isotrivial if $(\Gamma \cdot \Gamma) = 0$, under the additional hypothesis that all the fibers of f be semistable. Now we wish to show that the same conclusions hold even if this extra assumption is removed. The self-intersection of Γ does not change if we blow up the singular points of X. We may thus assume that X is smooth. Now let E be a smooth rational component of a fiber of f which meets the rest of the fiber at a single point. Then E is an exceptional curve of the first kind on X. Denote by X' the surface obtained by blowing it down and by Γ' the image of Γ in X'. The map f factors through a morphism $f' : X' \to S$, and Γ' is a section of f'. The self-intersection $(\Gamma' \cdot \Gamma')$ equals $(\Gamma \cdot \Gamma) + 1$ if Γ meets E and $(\Gamma \cdot \Gamma)$ if it does not. Hence, to prove our claims for $f : X \to S$ and Γ, it suffices to prove them for $f' : X' \to S$ and Γ'. Thus, by repeated blow-downs, the general case can be reduced to the case where all the fibers of f are semistable, which we already settled.

At this point we have shown, in every case, that $(\Gamma \cdot \Gamma) \leq 0$ and that $f : X \to S$ is isotrivial if $(\Gamma \cdot \Gamma) = 0$. All that remains to be done is to show that in this last case $f : X \to S$ is isotrivial as a family of curves 1-pointed by Γ and not only as a family of unpointed curves. As above, we may assume that every fiber of f is semistable. Let then $h : Y \to S$ be the stable model of $f : X \to S$, and D the image of Γ in Y (or, when $g = 1$, let $(h : Y \to S, D)$ be the stable model of the family of 1-pointed curves $(f : X \to S, \Gamma)$). The isotriviality of $h : Y \to S$ implies that, for $g = 1$, the natural map $S \to M_{1,1}$ is constant, so $(h : Y \to S, D)$ is also isotrivial. This settles the case $g = 1$. For $g > 1$, notice first that D is a section of h and that $(D \cdot D) \geq (\Gamma \cdot \Gamma)$, so that $(D \cdot D) = 0$. Next recall that, as we observed when discussing the notion of isotriviality, there is a finite unramified covering $T \to S$ such that, setting $Y' = Y \times_S T$ and letting $h' : Y' \to T$ be the projection to the second factor, $Y' \to T$ is isomorphic to a product family $C \times T \to T$. Denoting by D' the pullback of D to Y', we have

$$(\omega_{h'} \cdot D') = -(D' \cdot D') = 0\,.$$

On the other hand, if π denotes the projection of $C \times T$ to the first factor, $\omega_{h'} = \pi^*(\omega_C)$, so

$$(\omega_{h'} \cdot D') = d(2g-2)\,,$$

where d is the degree of the map $D' \to C$. Since $g > 1$, d must be zero. The conclusion is that $\pi(D')$ is a point and hence that D is a horizontal section of $h : Y \to S$. Proposition (4.13) is now fully proved.

It is perhaps worth noticing that the first part of (4.13) can also be deduced from the semipositivity property of the relative dualizing sheaf (Theorem (9.16) of Chapter XI). It suffices to notice that the homomorphism

$$f_*(\omega_f) \to f_*(\omega_{f|\Gamma})$$

is generically onto, since the dualizing sheaf of a smooth curve of positive genus is generated by its sections, and hence the semipositivity of $f_*(\omega_f)$ implies that

$$0 \leq \deg f_*(\omega_{f|\Gamma}) = (\omega_f \cdot \Gamma) = -(\Gamma \cdot \Gamma)\,.$$

An interesting consequence of (4.13) which anticipates, in a special situation, the ampleness results of the next section, is the following result, due to Arakelov.

COROLLARY (4.16). *Let $f : X \to S$ be a nonisotrivial family of stable curves of genus $g > 1$ such that S is a reduced and irreducible complete curve and a general fiber of f is smooth. Then ω_f is ample.*

The proof is based on Nakai's criterion of ampleness (cf. [355]). Since $(\omega_f \cdot \omega_f) > 0$ by (4.11), all we must show is that $(\omega_f \cdot C)$ is strictly positive for any reduced and irreducible curve $C \subset X$. If C is a component of a fiber of f, then $(\omega_f \cdot C) > 0$ since the dualizing sheaf of a stable curve is ample. Suppose instead that C is not a component of a fiber of f. By a finite number of blow-ups of singular points of fibers we can construct a new family $f' : X' \to S$ of semistable curves such that the proper transform C' of C does not pass through singular points of fibers of f'. Denoting by π the natural map from X' to X, we have that $\pi^*(\omega_f) = \omega_{f'}$, so $(\omega_f \cdot C) = (\omega_{f'} \cdot C')$. Set $X'' = X' \times_S C'$ and let $f'' : X'' \to C'$ be the projection to the second factor. Then $(\omega_{f'} \cdot C') = (\omega_{f''} \cdot C'')$, where C'' stands for the image of the diagonal of $C' \times_S C'$ in X''. On the other hand, $f'' : X'' \to C'$ is not isotrivial, and C'' is a section of f'' not passing through the singular points of the fibers, and so, by (4.13),

$$(\omega_f \cdot C) = (\omega_{f''} \cdot C'') = -(C'' \cdot C'') > 0\,.$$

This finishes the proof of (4.16).

REMARK (4.17). It is clear that (4.13) cannot be valid for families of curves of genus zero. In fact, if L is a line bundle of positive degree d on a curve S, the section of $\mathbb{P}(\mathcal{O} \oplus L)$ consisting of all the points of the form $[\alpha : 0]$ has self-intersection d. On the other hand, it is not difficult to show that, if C_1 and C_2 are disjoint sections of a family of nodal genus zero curves, then

$$(C_1 \cdot C_1) + (C_2 \cdot C_2) \leq 0\,.$$

In the next section we shall prove a generalization of (4.13) (Theorem (5.12)) which will cover—and hopefully clarify—the genus zero case as well.

5. Projectivity of moduli spaces.

The main result of this section is that the moduli spaces of stable curves are projective. More exactly, we shall prove the following.

THEOREM (5.1). *Let g and n be nonnegative integers such that $2g-2+n>0$. Then the class κ_1 is ample on $\overline{M}_{g,n}$.*

We will also be looking into the ampleness of other classes. For instance, we may ask for which values of a is $\kappa_1 + a\lambda$ ample. In genus zero λ vanishes, so the answer is already provided by (5.1). In positive genus we will be able to settle the question by proving the following result.

THEOREM (5.2). *Let g and n be integers such that $g \geq 1$, $n \geq 0$, $2g-2+n>0$, and let a be a rational number. Then the class $\kappa_1 + a\lambda$ is ample on $\overline{M}_{g,n}$ if, and only if, $a > -1$.*

Since, as we know,

$$\kappa_1 = 12\lambda - \delta + \psi\,,$$

a statement which is equivalent to the combination of (5.1) and (5.2) is that $b\lambda - \delta + \psi$ is ample if and only if $b > 11$ for $g \geq 1$ and that $-\delta + \psi$ is ample for $g = 0$.

We begin by outlining the strategy of the proof of (5.1) and (5.2). In Section 9 of Chapter XII we showed that there are complete schemes $\mathcal{X}$ and Z and morphisms

$$\begin{aligned} &\pi : \mathcal{X} \to Z\,, \\ &\sigma_i : Z \to \mathcal{X}\,, \quad i = 1, \dots, n\,, \\ &\eta : Z \to \overline{M}_{g,n} \end{aligned}$$

such that:

i) η is finite;
ii) $\pi : \mathcal{X} \to Z$ is a family of nodal curves, the σ_i are sections of π, and $(\pi : \mathcal{X} \to Z; \sigma_1, \dots, \sigma_n)$ is a family of stable n-pointed curves of genus g;
iii) for any $z \in Z$, the isomorphism class of $(\pi^{-1}(z); \sigma_1(z), \dots, \sigma_n(z))$ is $\eta(z)$.

Since η is a finite morphism, $\kappa_1 + a\lambda$ is ample if and only if $\eta^*(\kappa_1 + a\lambda)$ is ample. As a test for ampleness on Z, we shall use Seshadri's criterion (Proposition (9.11) in Chapter XI). As we know, this says that $\eta^*(\kappa_1 + a\lambda)$ is ample if, and only if, there is a positive constant k such that, for any reduced and irreducible complete curve $S \subset Z$,

$$(\eta^*(\kappa_1 + a\lambda) \cdot S) \geq k\, m(S)\,, \tag{5.3}$$

where $m(S)$ stands for the maximum among the multiplicities of points of S. Notice that *ii)* and *iii)* imply that, if S is as above and G is the family of n-pointed genus g curves on S obtained from $(\pi : \mathcal{X} \to Z, \sigma_1, \ldots, \sigma_n)$ by pullback, then

$$(\eta^*(\kappa_1 + a\lambda) \cdot S) = \deg_G \kappa_1 + a \deg_G \lambda .$$

Thus, Theorems (5.1) and (5.2) will be proved if we can prove the following four statements.

LEMMA (5.4). *Suppose $g \geq 1$. Then there is a nonisotrivial family F of stable n-pointed curves of genus g over a reduced and irreducible complete curve such that*

$$\deg_F \kappa_1 - \deg_F \lambda = 0 .$$

LEMMA (5.5). *For any family F of semistable n-pointed curves of genus g over a reduced and irreducible complete curve,*

$$\deg_F \kappa_1 - \deg_F \lambda \geq 0 .$$

LEMMA (5.6). *For any family F of nodal n-pointed curves of genus g over a reduced and irreducible complete curve,*

$$\deg_F \lambda \geq 0 .$$

LEMMA (5.7). *There is a positive constant α such that, for any nonisotrivial family F of stable n-pointed curves of genus g over a reduced and irreducible complete curve S,*

$$\deg_F \kappa_1 \geq \alpha\, m(S) .$$

In fact, (5.4) says that $\kappa_1 - \lambda$ is not ample for $g \geq 1$, (5.5) says that $\kappa_1 - \lambda$ is nef, (5.6) implies that λ is nef, and, in view of Seshadri's criterion, (5.7) says that κ_1 is ample. But now Seshadri's criterion also implies that, given an ample class γ_1 and a nef class γ_2 on a projective scheme, any linear combination $a_1\gamma_1 + a_2\gamma_2$ with $a_1 > 0$ and $a_2 \geq 0$ is ample. Thus (5.5), (5.6), and (5.7) imply that $\kappa_1 + a\lambda$ is ample for $a > -1$. On the other hand, if $\kappa_1 + a\lambda$ were ample for some $a \leq -1$, this same argument would prove the ampleness of $\kappa_1 - \lambda$, against (5.4).

Lemmas (5.4), (5.5), (5.6), and (5.7) will be proved using the same methods and some of the results of the previous section. We begin with (5.4). The proof is provided by a slight variant of Example (4.9). Choose a nonisotrivial family G of stable 1-pointed elliptic curves over a complete curve S consisting of a family of curves $\varphi : Y \to S$, together with a section $\Gamma \subset Y$. As we observed in example (4.9), for this family, we have

$$12 \deg_G \lambda - \deg_G \delta = 0 ,$$
$$\deg_G \psi = \deg_G \lambda ,$$

so that

$$\deg_G \kappa_1 - \deg_G \lambda = 12 \deg_G \lambda - \deg_G \delta = 0 .$$

This settles the case $g = 1$, $n = 1$. In the other cases, choose a smooth $(n+1)$-pointed genus $g-1$ curve $(C; q_0, \dots, q_n)$ and let G' be the constant family of smooth $(n+1)$-pointed genus $g-1$ curves consisting of $C \times S \to S$ together with the sections $\{q_i\} \times S$, $i = 0, \dots, n$. Then construct a family F of stable n-pointed curves of genus g out of G and G' by identifying Γ with $\{q_0\} \times S$. The family F is not isotrivial because G is not. On the other hand, by the additivity of λ and κ_1 (cf. identities (4.31) and (5.24) in Chapter XIII) we conclude that

$$\begin{aligned} \deg_F \kappa_1 - \deg_F \lambda &= \deg_G \kappa_1 + \deg_{G'} \kappa_1 - \deg_G \lambda - \deg_{G'} \lambda \\ &= \deg_G \kappa_1 - \deg_G \lambda \\ &= 0 . \end{aligned}$$

We next take on Lemma (5.5). If we replace F with a new family F' obtained from it via a finite base change of degree d, we have

$$\deg_{F'} \kappa_1 = d \deg_F \kappa_1 ; \quad \deg_{F'} \lambda = d \deg_F \lambda .$$

Therefore, to prove that the degree of $\kappa_1 - \lambda$ is nonnegative on F, it suffices to show that it is nonnegative on F'. We now claim that it suffices to prove (5.5) for those families whose general fiber is smooth. Suppose in fact that the general fiber of F is singular or, put otherwise, that the locus of the singular points in the fibers of F contains one-dimensional components. Denote by Σ the union of all such components, let X be the partial normalization along Σ of the total space of F, and let $f : X \to S$ be the projection. After a finite unramified base change, we may suppose that the inverse image of Σ in X is a disjoint union of sections. Thus, we may assume that F is obtained from the family

$$f : X \to S ,$$

plus sections $\Gamma_1, \dots, \Gamma_{n+2m}$ not passing through singularities of the fibers of f, by identifying Γ_{n+2i-1} to Γ_{n+2i} for $i = 1, \dots, m$ and keeping $\Gamma_1, \dots, \Gamma_n$ as sections of marked points. Notice that the general fiber of f is smooth but possibly disconnected. Let $X_1, \dots, X_h$ be the connected components of X and denote by f_i the restriction of f to X_i. From the fact that F is a family of *stable* n-pointed curves it follows that $f_i : X_i \to S$, together with the n_i sections Γ_j lying on it, is a family of *stable* n_i-pointed curves (cf. Remark (3.3) in Chapter X), which we shall denote by F_i. It then follows from the additivity of κ_1 and λ that

$$\deg_F \kappa_1 - \deg_F \lambda = \sum_i (\deg_{F_i} \kappa_1 - \deg_{F_i} \lambda) .$$

So, if (5.5) is known to hold for families whose general fiber is smooth, such as the F_i, it holds for F.

Next we shall prove (5.5) for a family $F = (f : X \to S, \Gamma_1, \dots, \Gamma_n)$ whose general fiber is smooth. We shall apply (4.2) to the line bundle $L = \omega_f^k(k \sum \Gamma_i)$, where k is a positive integer such that the relative degree of L is strictly larger than $2g$, so that all the hypotheses of (4.2) are satisfied by (3.2) and the remarks immediately preceding it. Notice that $R^1 f_* L = 0$. In fact, if C is any fiber of f and D is the divisor cut out on it by $\sum \Gamma_i$, using duality, we get that

$$h^1(C, \omega_C^k(kD)) = h^0(C, \omega_C^{-k+1}(-kD)) = 0\,,$$

since under our assumptions $\omega_C^{-k+1}(-kD)$ has nonpositive degree on every component of C and strictly negative degree on at least one of them. It follows that we can apply (4.2) under the form (4.4). For our choice of L, we have

$$\begin{aligned}(L \cdot L) &= k^2(\omega_f(\textstyle\sum\Gamma_i) \cdot \omega_f(\sum\Gamma_i)) = k^2 \kappa_1\,,\\ (L \cdot \omega_f) &= k(\omega_f(\textstyle\sum\Gamma_i) \cdot \omega_f) = k\kappa_1\,,\\ d &= k(2g - 2 + n)\,,\end{aligned}$$

so that (4.4) translates into

$$k(g - 1 + n)\deg_F \kappa_1 - (4g - 4 + 2n)\deg_F \lambda \geq 0\,. \tag{5.8}$$

Since $\deg_f \lambda \geq 0$, formula (5.8) gives the result, provided that we can find a k such that the relative degree of L is strictly larger than $2g$, that is, such that

$$k(2g - 2 + n) > 2g$$

and, simultaneously, such that

$$k(g - 1 + n) \leq 4g - 4 + 2n\,.$$

It is immediate to check that these two conditions are satisfied by $k = 1$ for $n \geq 3$, by $k = 2$ for $n = 2$ or $n = 1$, $g > 1$, and by $k = 3$ for $n = 0$. The only situation not covered by this argument is the one where $g = 1$, $n = 1$. In this case, however, we know that $\deg_F \kappa_1 - \deg_F \lambda = 0$, as we observed, for instance, while proving (5.4). This finishes the proof of (5.5).

It should be remarked that, unlike the one given by (4.5), the inequality provided by (5.5) is certainly not sharp for families whose general fiber is smooth, although (5.4) asserts that it is sharp for arbitrary families. For instance, the proof of (5.5) gives that, for $n \geq 3$,

$$(g - 1 + n)\deg_F \kappa_1 - (4g - 4 + 2n)\deg_F \lambda \geq 0\,,$$

which is better than (5.5) unless $g = 0$.

The proof of (5.6) has essentially already been given. The same argument as in the proof of (5.5) reduces us to the case where the general fiber of F is smooth. In this special case, observe that it suffices to know that the result holds for families of unpointed curves; that this is the case is the content of (4.7).

Our next task is to tackle (5.7). The proof will more or less retrace the steps of the proof of (5.5), with a few additional twists here and there. We begin by reducing to the case where the general fiber of F is smooth. The argument is the same as in the proof of (5.5), except for the fact that two extra precautions are necessary. First of all, we have to make sure that we can get by with a finite unramified base change of *bounded* degree d. In fact, under such a base change, the degree of κ_1 gets multiplied by d, but $m(S)$ remains unchanged, so knowing that the inequality in (5.7) is valid after base change yields the inequality

$$\deg_F \kappa_1 \geq \frac{\alpha}{d} m(S)$$

for the original family. This is fine, provided that d is not allowed to escape to infinity, since we are looking for an estimate which is *uniform*, that is, one which does not depend on the particular family under consideration, but only on g and n. Now, in the argument, the only place where a base change is needed is in reducing to a family which can be obtained from a family of curves with smooth general fiber by identifying sections in pairs. This can certainly be achieved via a base change whose degree does not exceed $b!$, where b is twice the number of singular points in a general fiber of F. All we have to do, then, is notice that, for any stable n-pointed curve C of genus g, the number of singular points of C is not larger than the dimension of $\overline{M}_{g,n}$, that is, not larger than $3g-3+n$.

A second precaution to be taken in reducing to proving (5.7) for families whose general fiber is smooth is to make sure that, starting with families of n-pointed genus g curves, there are only finitely many possibilities for the genus γ and the number ν of marked points of the fibers of the families produced by the reduction. This is easy. The genus γ cannot exceed g, while ν cannot be larger than n plus twice the number of singular points of a general fiber of the original family, so that $\nu \leq n + 2(3g-3+n)$.

We now come to the main point, that is, proving (5.7) for families whose general fiber is smooth. We shall use the following variant of (4.2).

LEMMA (5.9). *Let $f : X \to S$ be a family of nodal curves of genus g over a reduced and irreducible complete curve, and let L be an invertible sheaf on X. Let d be the relative degree of L. For any $s \in S$, set $X_s = f^{-1}(s)$ and $L_s = L \otimes \mathcal{O}_{X_s}$. Suppose that*

i) $h^0(X_s, L_s)$ *is independent of* s;
ii) *for general* $s \in S$, *the complete linear system* $|L_s|$ *on* X_s *is base-point-free and very ample.*

Suppose moreover that, denoting by ℓ *the rank of* f_*L, *there is a positive integer* h *such that*

iii) $h^1(X_s, L_s^h) = 0$ *for any* $s \in S$;
iv) *for general* $s \in S$, *the* h*th Hilbert point of* X_s, *embedded in projective space via* $|L_s|$, *is stable;*
v) *the* h*th Hilbert points of the* X_s *do not all belong to the same* $SL(\ell)$ *orbit.*

Then there is a positive constant c *such that*

$$\ell\frac{h^2}{2}(L\cdot L) - \ell\frac{h}{2}(L\cdot\omega_f) + \ell\deg_f\lambda - h(hd+1-g)\deg f_*L \geq c\,m(S)\,.$$

Moreover, c *can be chosen to depend only on* h, g, ℓ, *and* d, *and not on the particular family under consideration.*

In the proof we shall use the same notation as in (4.1), (4.2), and their proofs. Let s' be a point of maximum multiplicity in S. Lemma (4.1) implies that, for some $\mu > 0$ depending only on h, r, and $N = hd+1-g$, the μth power of the line bundle $\mathcal{L}_h = \det(f_*(L^h))^\ell \otimes \det(f_*L)^{-hN}$ has a nonzero section vanishing at s'. This gives

$$\mu\deg\mathcal{L}_h \geq m(S)\,.$$

On the other hand, the degree of $\mathcal{L}_h$ is given by (4.3), so we conclude that

$$\ell\frac{h^2}{2}(L\cdot L) - \ell\frac{h}{2}(L\cdot\omega_f) + \ell\deg_f\lambda - h(hd+1-g)\deg f_*L \geq \frac{1}{\mu}m(S)\,,$$

as desired.

Now let F be a family of stable n-pointed genus g curves as in the statement of (5.7), consisting of a family $f : X \to S$ of nodal curves plus sections $\Gamma_1, \dots, \Gamma_n$. Suppose that the general fiber of f is smooth. We wish to apply Lemma (5.9) to $f : X \to S$ and to the line bundle $L = \omega_f^k(k\sum\Gamma_i)$ for some large, but fixed, k. Fix a large h. Lemma (5.9) is applicable if the hth Hilbert points of the X_s do not all belong to the same orbit. Since for large h—how large depending only on g, n, and k—the isomorphism class of the couple (X_s, L_s) is completely determined by the hth Hilbert point of X_s, embedded by $|L_s|$, this is certainly the case unless $f : X \to S$ is isotrivial and the divisors cut out on its smooth fibers by $k \sum \Gamma_i$ all belong to the same linear system.

Since $L = \omega_f^k(k \sum \Gamma_i)$ and k is large, $R^1 f_* L = 0$, so the Riemann–Roch theorem gives

$$\deg f_* L = \frac{(L \cdot L)}{2} - \frac{(L \cdot \omega_f)}{2} + \deg_f \lambda\,,$$
$$d = k(2g - 2 + n)\,.$$

On the other hand, as we observed during the proof of (5.5),

$$(L \cdot L) = k^2 \deg_F \kappa_1\,,$$
$$(L \cdot \omega_f) = k \deg_F \kappa_1\,,$$

so Lemma (5.9) yields an inequality of the form

$$\frac{h^2}{2}[k^2(g-1+n)\deg_F \kappa_1 - k(4g-4+2n)\deg_F \lambda] + hl_1 + l_2 \geq c\,m(S)\,,$$

where l_1 and l_2 are linear combinations of $\deg_F \kappa_1$ and $\deg_F \lambda$ with integral coefficients not depending on h. For large h, this gives an inequality of the form

$$\deg_F \kappa_1 - a \deg_F \lambda \geq \alpha\, m(S)\,,$$

where α and a are positive constants. By (4.7), this implies that

$$\deg_F \kappa_1 \geq \alpha\, m(S)\,,$$

thus proving Lemma (5.7) for the families we are considering.

To fully prove (5.7), it remains to deal with the case where $f : X \to S$ is isotrivial with smooth general fiber and the divisors cut out on smooth fibers by $k \sum \Gamma_i$ all belong to the same linear system. We begin by noticing that $n \geq 2$. This is obvious for $g = 0$. For $g = 1$, if we had that $n = 1$, the family F would be isotrivial since $f : X \to S$ is. For $g \geq 2$, if n were equal to 1, the section Γ_1 could not be horizontal, and hence $k\Gamma_1$ could not cut linearly equivalent divisors on the fibers of f.

Suppose $2g-2+n > 2$. If $g \geq 2$, possibly after renumbering $\Gamma_1, \dots, \Gamma_n$, we may assume that the family of 1-pointed curves $(f : X \to S; \Gamma_n)$ is not isotrivial. If $g = 0$, in view of Remark (4.17), we may also assume that $(\Gamma_i \cdot \Gamma_i) \leq 0$ for $i = 2, \dots, n$ (in fact, as we shall see later, $(\Gamma_i \cdot \Gamma_i) < 0$ for every i). In every case we may assume that $F'' = (f : X \to S; \Gamma_1, \dots, \Gamma_{n-1})$ is a nonisotrivial family of $(n-1)$-pointed curves. Notice that the fibers of F'' are all semistable. Let $F' = (f' : X' \to S; \Gamma'_1, \dots, \Gamma'_{n-1})$ be the stable model of F'', and let

$\pi : X \to X'$ be the contraction map. Then part a) of Proposition (6.7) in Chapter X says that

$$\pi^*(\omega_{f'}(\textstyle\sum\Gamma'_i)) = \omega_f(\textstyle\sum_{i<n}\Gamma_i)\,,$$

so

$$\deg_{F''}\kappa_1 = (\omega_f(\textstyle\sum_{i<n}\Gamma_i)\cdot\omega_f(\textstyle\sum_{i<n}\Gamma_i)) = (\omega_{f'}(\textstyle\sum\Gamma'_i)\cdot\omega_{f'}(\textstyle\sum\Gamma'_i)) = \deg_{F'}\kappa_1\,.$$

Since $\deg_F\kappa_1 = \deg_{F''}\kappa_1 - (\Gamma_n\cdot\Gamma_n)$ and, by our choices for $g=0$ and by (4.13) otherwise, $(\Gamma_n\cdot\Gamma_n)\le 0$, we will be done if we know that (5.7) holds in genus g with $n-1$ marked points.

For $g\ge 2$, there is a further shortcut. Since, by our choices, Γ_n is not horizontal, $k\Gamma_n$ does not cut linearly equivalent divisors on all the smooth fibers of f. Thus, $k\sum\Gamma'_i$ does not cut linearly equivalent divisors on the smooth fibers of f', and so (5.7) has already been checked for F'.

If $g<2$, instead, iterating the same procedure reduces us, inductively, to the case of a family F such that $f : X\to S$ is isotrivial and $g=1$, $n=2$ or $g=0$, $n=4$. We let F', π, and so on, be as above. We also denote by Γ'_n the image of Γ_n in X'. We know that, after a finite unramified base change, the family F' becomes trivial; the degree of the base change in question does not exceed the order of the automorphism group of a typical fiber of F'. Since the order of the automorphism group of a 1-pointed stable curve of genus 1 is bounded, we may and will assume that F' is a product family. We may then view Γ'_n as the graph of a nonconstant morphism from S to the typical fiber of f'; we denote by d the degree of this morphism. Clearly,

$$d \ge m(S)\,. \tag{5.10}$$

We may write $\pi^*(\Gamma'_n) = \Gamma_n + E$, where E is an effective divisor which π contracts to a bunch of points. But then

$$\begin{aligned}
\deg_F\kappa_1 &= \deg_{F'}\kappa_1 - (\Gamma_n\cdot\Gamma_n)\\
&= -(\Gamma_n\cdot\Gamma_n)\\
&= (\omega_f(\textstyle\sum_{i<n}\Gamma_i)\cdot\Gamma_n)\\
&= (\pi^*(\omega_{f'}(\textstyle\sum\Gamma'_i))\cdot\pi^*(\Gamma'_n) - E)\\
&= (\omega_{f'}(\textstyle\sum\Gamma'_i)\cdot\Gamma'_n)\\
&= (2g-2+n-1)d\\
&= d\,.
\end{aligned}$$

This, combined with (5.10), finishes the proof of (5.7).

We pause for a moment to settle a point left open in the previous section. The proof of Proposition (4.8) we gave there is analytic in nature. Here we wish to sketch an algebraic proof based on (5.9). Let then $f : X \to S$ be a nonisotrivial family of nodal curves with smooth general fiber over a reduced and irreducible complete curve. We apply (5.9) to $L = \omega_f^k$ for large k. Arguing exactly as in the proof of (5.7), we get an inequality of the form

$$(12 - a) \deg_f \lambda - \deg_f \delta = \deg_f \kappa_1 - a \deg_f \lambda > 0 .$$

Since $\deg_f \lambda \geq 0$ by (4.7), while $\deg_f \delta \geq 0$ by the smoothness of the general fiber of f, this proves that $\deg_f \lambda > 0$.

Now that the question of the ampleness of κ_1, and hence of the projectivity of $\overline{M}_{g,n}$, has been settled, we may ask which other line bundles on moduli space are ample or nef. We shall not give a complete answer to this question but will be content with partial results. One such is given by (5.2). Another one is the following.

PROPOSITION (5.11). *The class λ is nef on $\overline{M}_{g,n}$ for any g and n. It is ample if and only if $g = 1$, $n = 1$ or $g = 0$, $n = 3$.*

The first statement is nothing but (5.6). As for the second, we know that, for $g = n = 1$, the class $\kappa_1 - \lambda$ vanishes, so $\lambda = \kappa_1$ is ample. In genus zero, λ vanishes, so it is not ample unless $\overline{M}_{0,n}$ is zero-dimensional, that is, unless $n = 3$. For all remaining values of g and n, we shall construct a nonisotrivial family $F_{g,n}$ of stable n-pointed genus g curves over a complete curve for which $\deg_{F_{g,n}} \lambda = 0$. Suppose first that $g > 1$, $n > 0$ or $g = 1$, $n > 1$. Pick any smooth curve S of genus g and distinct points $q_1, \dots, q_{n-1}$ on it. Let X be the surface obtained by blowing up $S \times S$ at the points (q_i, q_i), and let $f : X \to S$ be the composition of the contraction map $X \to S \times S$ with the projection of $S \times S$ to its second factor. For $i = 1, \dots, n-1$, let Γ_i be the proper transform of $\{q_i\} \times S$; let Γ_n be the proper transform of the diagonal of $S \times S$. The family $F_{g,n} = (f : X \to S; \Gamma_1, \dots, \Gamma_n)$ has all the required properties. It remains to construct $F_{g,0}$ for $g > 1$; this is done simply by identifying the two canonical sections of $F_{g-1,2}$. This finishes the proof.

Our next result is a generalization of Proposition (4.13).

THEOREM (5.12). *Let $F = (f : X \to S; \Gamma_1, \dots, \Gamma_n)$ be a family of n-pointed nodal curves of genus g over a reduced and irreducible complete curve. Suppose that $2g - 2 + n > 0$ and that the general fiber of f is smooth. Then*

$$(\Gamma_i \cdot \Gamma_i) \leq 0$$

for any i. Moreover, if $(\Gamma_i \cdot \Gamma_i) = 0$ for some i, then F is isotrivial.

We begin the proof by reducing to the case where F is a family of stable n-pointed curves. Arguing as in the proof of (4.13), we may assume that all the fibers of F are semistable. Let $F' = (f' : X' \to S; \Gamma'_1, \dots, \Gamma'_n)$ be the stable model of F, and let $\pi : X \to X'$ be the contraction map. Then $\pi^*(\Gamma'_i) = \Gamma_i + E$, where E is a linear combination with nonnegative coefficients of components of singular fibers of f, so

$$(\Gamma_i + E \cdot E) = 0 .$$

Thus,

$$\begin{aligned}(\Gamma'_i \cdot \Gamma'_i) &= (\Gamma_i + E \cdot \Gamma_i + E) = (\Gamma_i \cdot \Gamma_i) + (\Gamma_i \cdot E) + (\Gamma_i + E \cdot E)\\ &= (\Gamma_i \cdot \Gamma_i) + (\Gamma_i \cdot E) \geq (\Gamma_i \cdot \Gamma_i) .\end{aligned}$$

If (5.12) holds for F', it follows that $(\Gamma_i \cdot \Gamma_i) \leq 0$ for any i; if, on the other hand, $(\Gamma_i \cdot \Gamma_i) = 0$, then $(\Gamma'_i \cdot \Gamma'_i)$ must also vanish, so F' is isotrivial, and hence F is isotrivial, too.

We may therefore assume that F is a family of stable n-pointed curves. We shall use the algebraic index theorem for the surface X. Denote by C a fiber of f. Since $(C \cdot C) = 0$, the intersection form is not negative definite on the subgroup of $\mathrm{Num}(X)$ generated by $\omega_f(\sum \Gamma_j)$, $\mathcal{O}(\Gamma_i)$, and $\mathcal{O}(C)$. The index theorem then implies that

$$\begin{aligned} 0 &\leq \det \begin{pmatrix} (\omega_f(\sum\Gamma_j) \cdot \omega_f(\sum\Gamma_j)) & (\Gamma_i \cdot \omega_f(\sum\Gamma_j)) & (C \cdot \omega_f(\sum\Gamma_j)) \\ (\omega_f(\sum\Gamma_j) \cdot \Gamma_i) & (\Gamma_i \cdot \Gamma_i) & (C \cdot \Gamma_i) \\ (\omega_f(\sum\Gamma_j) \cdot C) & (\Gamma_i \cdot C) & (C \cdot C) \end{pmatrix} \\ &= \det \begin{pmatrix} \deg_F \kappa_1 & 0 & 2g-2+n \\ 0 & (\Gamma_i \cdot \Gamma_i) & 1 \\ 2g-2+n & 1 & 0 \end{pmatrix} . \end{aligned}$$

Expanding the determinant yields the inequality

$$\deg_F \kappa_1 + (2g-2+n)^2 (\Gamma_i \cdot \Gamma_i) \leq 0 . \tag{5.13}$$

Since $\deg_F \kappa_1 \geq 0$ by (5.7) (or by (5.5) and (5.6)), we conclude that $(\Gamma_i \cdot \Gamma_i) \leq 0$. If $(\Gamma_i \cdot \Gamma_i) = 0$, formula (5.13) implies that $\deg_F \kappa_1 = 0$, so F is isotrivial by (5.7). This concludes the proof of (5.12).

The result we have just proved leads almost immediately to another ampleness (or, rather, nefness) result for the moduli spaces of curves.

COROLLARY (5.14). *The class ψ_i is nef on $\overline{M}_{g,n}$ for any g and any n such that $2g-2+n > 0$ and for any i such that $1 \leq i \leq n$.*

Let $F = (f : X \to S; \Gamma_1, \ldots, \Gamma_n)$ be a family of stable n-pointed curves of genus g over a complete curve. We must show that $\deg_F \psi_i \geq 0$, that is, that $(\Gamma_i \cdot \Gamma_i) \leq 0$. If the general fiber of f is smooth, this follows from (5.12). Otherwise, modulo a finite base change, we may assume that there are $F_1, \ldots, F_h$, where F_j is a family of stable n_j-pointed curves over S with smooth general fiber, such that F is obtained from the F_j by identifying sections in pairs. But then Γ_i comes from one of the canonical sections of one of the F_j, and the result follows again from (5.12).

REMARK (5.15). For $n = 1$, Corollary (5.14) reduces to the statement, due to Arakelov and Mumford, that the relative dualizing sheaf of $\overline{\mathcal{C}}_g \to \overline{\mathcal{M}}_g$ is nef.

Combining (5.2) and (5.14) shows that $\kappa_1 + a\lambda + \sum b_i\psi_i$ is ample on $\overline{M}_{g,n}$ whenever $a > -1$ and $b_i \geq 0$ for every i. Of course, it would be nice to know which linear combinations of the form

$$\kappa_1 + a_{irr}\delta_{irr} + \sum a_{j,h}\delta_{j,h} + \sum b_i\psi_i$$

are ample. However, the answer to this question does not seem to be known. A much simpler special case of this problem, whose answer apparently is also unknown, is the following. Fix integers $g > 1$, $n > 0$, and a rational number $a > 11$. We know that $a\lambda - \delta + \psi$ is ample on $\overline{M}_{g,n}$. On the other hand, $a\lambda - \delta$ is not ample. In fact, for the family $F_{g,n}$ constructed in the proof of (5.11), which is not isotrivial, we have that $\deg_{F_{g,n}} \lambda = 0$, $\deg_{F_{g,n}} \delta = n - 1 \geq 0$, so that $\deg_{F_{g,n}}(a\lambda - \delta) \leq 0$. It follows that there is some ε, with $0 \leq \varepsilon < 1$, such that $a\lambda - \delta + \varepsilon\psi$ is nef but not ample on $\overline{M}_{g,n}$. The problem is to determine ε as a function of g, n, and a.

We have observed that the class λ is nef but ample on $\overline{M}_g$. Let L be the Hodge line bundle on $\overline{\mathcal{M}}_g$, whose Chern class is λ. Since $\kappa_1 = 12\lambda - \delta$ is ample and δ is the class of the boundary of moduli space, which is an effective divisor, a power of L is generated by its sections over M_g and separates points of M_g. It can be shown that, in fact, a suitable power of L is generated by its sections over all of $\overline{M}_g$. We shall not prove this result, but we shall briefly discuss some of its consequences. Pick an integer k such that L^k is generated by its sections and separates the points of M_g. Let $\varphi : \overline{M}_g \to \mathbb{P}^N$ be the map into projective space defined by $|L^k|$. We can assume, by taking k large enough, that $\Sigma_g = \varphi(\overline{M}_g)$ is a normal variety. Since φ is a birational map from $\overline{M}_g$ to Σ_g, its fibers are connected, by Zariski's main theorem. Let C_1 and C_2 be two stable curves of genus g. Denote by N_i the normalization of C_i *minus its rational components*. We claim that the two points $[C_1]$ and $[C_2]$ of $\overline{M}_g$ map to the same point of Σ_g if and only if N_1 and N_2 are isomorphic. Let us prove the "only if" part first. Since the fibers of φ are connected, if two points of $\overline{M}_g$ are mapped to the same point of Σ_g, there is a connected curve in $\overline{M}_g$ containing them and mapping to a point. Therefore it suffices to prove that N_1 and N_2 are isomorphic when C_1 and C_2 appear as fibers of a family $f : X \to S$ of stable curves such that S is a smooth connected complete curve and $\deg_f \lambda = 0$. Arguing as in the proof of (5.5), we may assume, possibly after a base change, that $f : X \to S$ can be constructed out of families $f_i : X_i \to S$, $i = 1, \dots, h$, of nodal curves with smooth general fiber by gluing along sections not meeting singular points of the fibers. For every i, we have that $\deg_{f_i} \lambda = 0$, by the additivity of λ, so (4.8) implies that the families $f_i : X_i \to S$ are all isotrivial. On the other hand, for any fiber C of f, the components of C, except possibly some of the rational ones, are precisely the typical fibers of the f_i. This implies that N_1 and N_2 are in fact isomorphic.

Now let us prove the converse. We will say that two stable curves of genus g are λ-equivalent if their classes are mapped to the same point of Σ_g by φ; notice that any two fibers of a family $f : X \to S$ of stable curves over a complete connected curve such that $\deg_f \lambda = 0$ are λ-equivalent. We wish to show that, if N_1 and N_2 are isomorphic, then C_1 and C_2 are λ-equivalent. Clearly, we have to worry only about the case where C_1 and C_2 are singular. Normalize C_1 at a singular point x; the result is either a stable 2-pointed curve $(M; p, q)$ of genus $g-1$ or a pair of stable 1-pointed curves $(M_1; p)$, $(M_2; q)$ of genera i and $g-i$. Suppose that we are in the first case. The stabilization construction in section 8 of Chapter X gives, starting from the product family $M \times M \to M$, a family $(f' : X' \to M; D'_1, D'_2, D'_3)$ of stable 3-pointed curves of genus $g-1$, where X is a suitable modification of $M \times M$ and the sections D'_1, D'_2, and D'_3 are the proper transforms of the diagonal and of $\{p\} \times M$ and $\{q\} \times M$, respectively. Let $(f : X \to M; D_1, D_2)$ be the stable model of $(f' : X' \to M; D'_1, D'_2)$, and let $h : Y \to M$ be the family of stable curves of genus g obtained from it by identifying the sections D_1 and D_2. By the additivity property of λ we have that $\deg_h \lambda = \deg_f \lambda = 0$, so any two fibers of h are λ-equivalent; moreover, the fiber of h at q is just C_1. The conclusion is, roughly speaking, that q can be moved to any position we wish on M without changing the λ-equivalence class of C_1. Of course, this has to be taken with a grain of salt. If q happens to be moved to a node y of M, this is replaced with a smooth rational curve E joining the two branches of y, and q ends up on E. Conversely, it may happen that moving q off the component E where it originally lay renders the resulting 2-pointed curve semistable but not stable; in this case, E has to be contracted. Clearly, all the above considerations apply not only to q, but also to p. When the normalization of C_1 at x is not connected, the same argument shows that, so far as the λ-equivalence class of C_1 is concerned, the positions of p and q on M_1 and M_2 are immaterial. Now pick any irreducible component of C_1; by what has just been observed, all the nodes of C_1 can be moved to this component without changing λ-equivalence class. In other words, we can find a curve C_1' in the λ-equivalence class of C_1 which is either irreducible and rational, or else consists of a nonrational irreducible component Γ_1 plus smooth components, each of them attached to Γ_1 at a single point; in addition, in the second case we can set things up so that the normalization of Γ_1 is any component of N_1 that we wish.

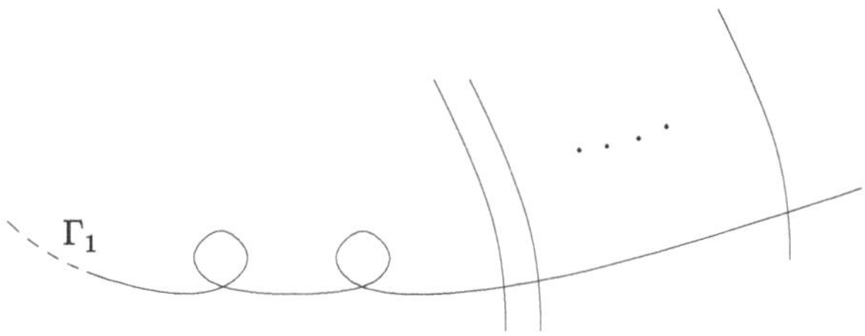

Figure 2.

Clearly, the same procedure can be applied to C_2; we denote by C_2' the resulting curve and by Γ_2 a component containing all the nodes of C_2'. We may certainly assume that Γ_1 and Γ_2 have the same normalization; it follows in particular that any component of C_1' different from Γ_1 is isomorphic to a component of C_2' different from Γ_2, and conversely. But then it is clear, by the principle that nodes can be "moved around" without effect on the λ-equivalence class of the curve involved, that C_1' and C_2' are λ-equivalent. This finishes the proof that the classes of C_1 and C_2 are mapped to the same point by φ if and only if N_1 and N_2 are isomorphic.

The variety Σ_g is called the *Satake compactification* of M_g. This is justified by the fact that φ maps M_g isomorphically onto $\varphi(M_g)$. The *boundary* of the Satake compactification is $\partial\Sigma_g = \Sigma_g \setminus \varphi(M_g)$. The description of the fibers of φ we have just given shows that those fibers that lie above points of the boundary of Σ_g all have positive dimension. Thus, the boundary of the Satake compactification has codimension greater than one; in fact, a closer look shows that all of its components have codimension three, except for $\varphi(\Delta_1)$, which has codimension two. This has interesting consequences. One of them is that any holomorphic function on M_g is constant, since it extends to Σ_g by Hartogs' theorem. Another important, and perhaps somewhat unexpected, consequence is that there exist, inside M_g, plenty of complete subvarieties of positive dimension.

6. Bibliographical notes and further reading.

The classic reference for geometric invariant theory is of course Mumford's book [558]. Other useful references are, for instance, the books by Dolgachev [183], Fogarty [265], and Mukai [541]. For the classical roots of the subject, the reader may wish to consult Hilbert's invariant theory papers [363] and Weyl's book [682].

The GIT stability of linearly stable curves was discovered by Mumford [558] (see also [555]), who showed stability for the *Chow points* of these curves, rather than for Hilbert points. The proof presented here is adapted from Gieseker's version [294], which gives stability for Hilbert points.

The central result of Section 4, namely Proposition (4.2), is taken from Cornalba and Harris [141], where a more general statement dealing with fibrations with higher-dimensional fibers is proved. The result depends on a rather strong Hilbert stability assumption. This has been slightly weakened by Bost [79], who reproved the result under the assumption of Chow stability of the general fiber.

Inequality (4.5) usually goes under the name of *slope inequality*, and is independently due to Cornalba and Harris [141] and Xiao Gang [688]; the proof presented here is essentially the one of [141]. Actually, Xiao proves the inequality, under the form (4.12), not just for semistable fibrations, but for general *relatively minimal fibrations*, that is, for fibrations $f : X \to S$ of a surface over a curve whose fibers do not contain exceptional curves of the first kind. As observed in Stoppino [647], the method of proof of [141] works without changes in this more general situation, provided that the fibers of f are not all hyperelliptic, while a slight generalization of the method covers also the hyperelliptic case. It must be observed that the slope inequality for hyperelliptic fibrations was already known to Horikawa and Persson.

It is natural to expect that families of curves with some special properties might satisfy improved versions of the slope inequality. This is indeed the case, as shown for families of curves with fixed gonality by Konno [441,442] and Barja and Stoppino [50], or for families of curves which are coverings of curves of lower genus by Barja [48], Barja and Zucconi [51], and Cornalba and Stoppino [150].

Moriwaki proves his inequality (4.6) in [527] (see also Moriwaki [526]). A simpler and more direct proof has been given by Yamaki [689], who also obtained more refined inequalities involving finer invariants for the configurations of nodes appearing in the singular fibers.

Corollary (4.16) is due to Arakelov [17]; as observed by Mumford (see, for instance, [556], Theorem 4.1), it implies the nefness of the relative dualizing sheaf of $\overline{\mathcal{C}}_g \to \overline{\mathcal{M}}_g$.

The classical geometers of the Italian school implicitly assumed the moduli spaces of smooth curves to be (quasi-projective) algebraic varieties. The first proof of the quasi-projectivity of M_g is due to Baily [42]. Two more proofs were given by Mumford in [558]. The first proof of the projectivity of $\overline{M}_{g,n}$ is due to Knudsen and Mumford [427]. Almost simultaneously, an alternative proof of projectivity via GIT was given by Mumford [555]. Another proof is given by Kollár [439], as an exemplification of a general projectivity criterion for complete moduli spaces. The proof of projectivity presented here is taken from [144]. Quasi-projectivity of moduli spaces is extensively studied by Viehweg [670].

The structure of the ample cone of $\overline{M}_{g,n}$ is not known. It has been conjectured by Fulton that the ample line bundles are precisely those that have positive intersection with all the one-dimensional strata in the stratification of the boundary of $\overline{M}_{g,n}$ by graph type (the so-called *topological stratification*). It has been shown by Gibney, Keel, and Morrison [291] that to prove the conjecture, it suffices to prove it in genus zero. The conjecture is known to be true only for very low values of g and n (cf. [291] and Farkas and Gibney [255]).

In the same order of ideas, another related open problem is the one of determining the effective cone of $\overline{M}_{g,n}$. If one defines the *slope* of a divisor of the form $a\lambda - b\delta$, where a and b are positive, to be the ratio a/b, an explicit conjectural lower bound for the slope was proposed by Harris and Morrison. The conjecture was however shown to be (narrowly) false by Farkas and Popa [257]. Other counterexamples are due to Khosla [415] and Farkas [250].

The moduli space $\overline{M}_g$ is known to be of general type for large g (at the time of this writing, for $g = 22$ and $g \geq 24$; cf. Harris and Mumford [353], Eisenbud and Harris [204], and Farkas [251]); a more detailed discussion of the Kodaira dimension of moduli spaces of curves will be found in the bibliographical notes to Chapter XXI. Thus, $\overline{M}_g$ has a canonical model for large g by [73]. Hassett has outlined a program to try and understand this model by first studying the log-canonical models $\operatorname{Proj}\left(\oplus_{n\geq 0}\Gamma(K_{\overline{\mathcal{M}}_g}(\alpha\mathcal{D})^n)\right)$, where $\mathcal{D}$ is the boundary of $\overline{\mathcal{M}}_g$, and $\alpha \in (0,1]$ is rational. In view of Theorem (7.15) in Chapter XIII, the log-canonical rings in question are just the rings of sections of powers of $a\lambda - \delta$ for $13/2 < a \leq 13$, and of course, for $a > 11$, the log-canonical model one gets is just $\overline{M}_g$. The first steps in the log-canonical program have been taken by Hassett and Hyeon in [359], where they study the case $10 < a \leq 11$, and in [360], where they study what happens for a near 10.

The expository papers [254] by Farkas and [531,532], by Morrison deal extensively with questions of ampleness and effectivity for divisors in $\overline{M}_{g,n}$, and with the Mori theory of moduli spaces of curves.

The Satake compactification was first described in Satake [613].

Chapter XV.
The Teichmüller point of view

1. Introduction.

In the present chapter we shall construct the Teichmüller space $\mathcal{T}_{g,n}$ for n-pointed genus g curves and show that it is homeomorphic to a ball in $\mathbb{C}^{3g-3+n}$. We shall discuss in greater details the Teichmüller modular group $\Gamma_{g,n}$ that was introduced in Chapter X, show that it acts on $\mathcal{T}_{g,n}$ properly discontinuously and with finite stabilizers, and that the quotient $\mathcal{T}_{g,n}/\Gamma_{g,n}$ coincides with the moduli space $M_{g,n}$.

Traditionally, to construct the Teichmüller space, one first introduces a topology in $\mathcal{T}_{g,n}$, either via the Teichmüller metric or else via the Fricke embedding, and *then* one introduces on $\mathcal{T}_{g,n}$ a complex structure and constructs a universal family. Here, we already have at our disposal Kuranishi families, and since the Teichmüller space locally looks like the base of a Kuranishi family, we only need to patch these local data together. As a result, the Teichmüller space is constructed right away as a complex manifold and is automatically equipped with a universal family obtained by gluing together Kuranishi families. This approach provides a number of shortcuts with respect to the traditional one. For instance, the fact that the mapping class group acts properly discontinuously can be proved in quite a direct way.

In the first section we give the basic definitions concerning Teichmüller structures and the mapping class group. Assume that $2g-2+n>0$. A Teichmüller structure on an n-pointed genus g curve $(C, x_1, \dots, x_n)$ is the datum of the isotopy class $[f]$ of an orientation-preserving homeomorphism

$$f : (C, x_1, \dots, x_n) \longrightarrow (\Sigma, p_1, \dots, p_n),$$

where $(\Sigma, p_1, \dots, p_n)$ is a fixed oriented n-pointed genus g topological surface. An isomorphism between curves with Teichmüller structure $(C, x_1, \dots, x_n, [f])$ and $(C', x'_1, \dots, x'_n, [f'])$ is an isomorphim of n-pointed curves

$$\varphi : (C, x_1, \dots, x_n) \longrightarrow (C', x'_1, \dots, x'_n)$$

such that $[f'\varphi] = [f]$. As a set, the Teichmüller space $\mathcal{T}_{g,n}$ is the set of isomorphism classes of n-pointed genus g curves with Teichmüller structure. A Teichmüller structure $[f]$ rigidifies the pointed curve $(C, x_1, \dots, x_n)$ in the sense that

$$\operatorname{Aut}(C, x_1, \dots, x_n, [f]) = 1. \tag{1.1}$$

E. Arbarello et al., *Geometry of Algebraic Curves*,
Grundlehren der mathematischen Wissenschaften 268,
DOI 10.1007/978-3-540-69392-5_7,

The notion of family of pointed curves with a Teichmüller structure is the intuitive one.

Now let $(C, x_1, \dots, x_n, [f])$ be a pointed curve with Teichmüller structure, let $\pi : \mathcal{C} \to (B, b_0)$ be a Kuranishi family for $(C, x_1, \dots, x_n)$, and let φ be the identification of the pointed curve $(C, x_1, \dots, x_n)$ with the central fiber of π. If B is a sufficiently small, we can find a C^∞ trivialization $F : \mathcal{C} \to \Sigma \times B$ such that $[F_{b_0}\varphi] = [f]$. Now all the fibers of π are equipped with a Teichmüller structure, and, in a sense that it is not difficult to make precise, the pair $(\pi, [F])$ is a Kuranishi family for the curve with Teichmüller structure $(C, x_1, \dots, x_n, [f])$. Using the universal property of these Kuranishi families and the fact that curves with a Teichmüller structure are rigid, it is possible to patch all these families into a single one. By construction the parameter space for this family coincides with the set $\mathcal{T}_{g,n}$. In this way we introduce on $\mathcal{T}_{g,n}$ the structure of a $(3g - 3 + n)$-dimensional manifold, and we construct over it a universal family of n-pointed genus g curves with Teichmüller structure.

It is important to observe that, by construction, the Teichmüller space satisfies a universal property with respect to families of curves with Teichmüller structure even when these families are merely continuous.

The mapping class group $\Gamma_{g,n}$ is the group of all isotopy classes of orientation-preserving homeomorphism of $(\Sigma, p_1, \dots, p_n)$ into itself. This group acts naturally on $\mathcal{T}_{g,n}$. The action of an element $[\gamma]$ on a Teichmüller structure $[f]$ is the obvious one: $[\gamma] \cdot [f] = [\gamma \circ f]$. Essentially by construction, $\Gamma_{g,n}$ acts on $\mathcal{T}_{g,n}$ as a properly discontinuous group of holomorphic transformations, giving the equality of analytic spaces

$$\mathcal{T}_{g,n}/\Gamma_{g,n} = M_{g,n} \,.$$

The basic example of Teichmüller space is the one of elliptic curves. The Teichmüller space $\mathcal{T}_{1,1}$ is just the upper half-plane $\mathbb{H}$, and the mapping class group $\Gamma_{1,1}$ is just $SL_2(\mathbb{Z})$. There are two features in this example that generalize to arbitrary Teichmüller spaces. The first one concerns the identification between $\Gamma_{1,1}$ and $SL_2(\mathbb{Z})$ which says, in particular, that a symplectic automorphism of the first homology group of an elliptic curve E is always induced by a homeomorphism of $(E, 0)$ into itself. In Section 3 we give a sketch of the proof of the Dehn–Nielsen realization theorem which generalizes this identification to higher genera. We look, for simplicity, at the unpointed case. Let Π be the fundamental group of the reference genus g surface Σ (as we shall mod out by inner automorphisms of Π, no reference to a base point is needed). There is a natural homomorphism

$$\begin{aligned} \Gamma_g &\to \mathrm{Out}(\Pi) = \mathrm{Aut}(\Pi)/\{\text{Inner automorphisms}\} \\ f &\mapsto [f_*] \end{aligned}$$

The Dehn–Nielsen realization theorem states that this establishes an isomorphism between Γ_g and the subgroup $\mathrm{Out}^+(\Pi)$ of $\mathrm{Out}(\Pi)$ consisting of those automorphisms of Π which "preserve orientation" in the sense that they induce a symplectic automorphism of $\Pi/[\Pi,\Pi]$. Using the fact that Σ is a $K(\Pi,1)$, the Dehn–Nielsen realization theorem can be easily reduced to the following statement, of which we give a complete proof: *let $f : X \to Y$ be a continuous map of compact connected orientable surfaces of the same genus. If the degree of f is ± 1, then f is homotopic to a homeomorphism.*

Another important feature of the Teichmüller space $\mathcal{T}_{1,1} = \mathbb{H}$ is that it is contractible. This too generalizes to higher genus. Indeed, a basic result, originally due to Fricke, asserts that $\mathcal{T}_{g,n}$ is homeomorphic to the unit ball in $\mathbb{C}^{3g-3+n}$. An immediate corollary is that $M_{g,n}$ is irreducible. The proof of Fricke's theorem that we shall present occupies Sections 4, 5, and 6, and is essentially due to Teichmüller. Let us briefly explain the main ideas that go into it. Fix a compact n-pointed connected Riemann surface S of genus g. Consider the $(3g-3+n)$-dimensional space $H^0(S, K_S^2(\sum p_i))$ of those quadratic differentials on S which are holomorphic except possibly for simple poles at the marked points $p_1, \dots, p_n$. For $\omega \in H^0(S, K_S^2(\sum p_i))$, set

$$\|\omega\| = \int_S dA_\omega$$

Here, if $\omega = f(z)dz^2$ is a local expression for ω in terms of a local coordinate z, the volume form dA_ω is defined by

$$dA_\omega = \frac{\sqrt{-1}}{2}|f| dz \wedge d\bar{z} = |f| dx \wedge dy\,,$$

where x and y stand for the real and imaginary parts of z. Look at the unit ball

$$B(K_S^2(\textstyle\sum p_i)) = \{\omega \in H^0(S, K_S^2(\sum p_i)) : \|\omega\| < 1\}$$

in $H^0(S, K_S^2(\sum p_i))$. Fricke's theorem is proved by exhibiting a specific homeomorphism

$$\Phi : B(K_S^2(\textstyle\sum p_i)) \to \mathcal{T}_{g,n}\,. \tag{1.2}$$

To define Φ, let $\omega \in B(K_S^2(\sum p_i))$ and set $k = \|\omega\| < 1$. We want to define a new Riemann surface $(S_\omega, q_1, \dots, q_n)$, together with an orientation-preserving homeomorphism

$$f_\omega : (S_\omega, q_1, \dots, q_n) \to (S, p_1, \dots, p_n)\,, \tag{1.3}$$

and then set

$$\Phi(\omega) = [S_\omega, q_1, \dots, q_n; [f_\omega]] \in \mathcal{T}_{g,n}\,.$$

The underlying topological manifold of S_ω will be the same as the one of S, and f_ω will be the identity map, so that in particular $q_i = p_i$ for all i. Let Z be the set of zeroes and poles of ω. We first introduce the new complex structure on $S \smallsetminus Z$. Around each point $p \in S \smallsetminus Z$ the quadratic differential ω defines a set of distinguished coordinates, any two of which differ at most by a sign and the addition of a constant. Such a coordinate z is simply defined by

$$z = \int \sqrt{\omega}\,. \tag{1.4}$$

Now define a new coordinate patch around p by performing the real linear transformation

$$z' = \frac{z + k\bar{z}}{1 - k}. \tag{1.5}$$

Clearly, a different choice of z changes z' by no more than a sign and the addition of a constant. Thus, these coordinate patches define a new holomorphic structure on $S \smallsetminus Z$. It can then be shown, quite easily, that this structure extends to one on all of S.

Using the invariance of domain, to prove that Φ is a homeomorphism, it is enough to prove that Φ is:

i) continuous,
ii) injective,
iii) closed,

and that

iv) $\mathcal{T}_{g,n}$ is connected.

Observe that by construction $(S_\omega, p_1, \dots, p_n)$ and f_ω depend continuously on ω. Therefore, we have a continuous family of curves with Teichmüller structure parameterized by $B(K_S^2(\sum p_i))$. By the universal property we mentioned above, we deduce that Φ is continuous.

The injectivity of Φ is the heart of the proof of Fricke's theorem. One must show that if $[f_\omega] = [f_{\omega'}]$, then $\omega = \omega'$. The homeomorphism f_ω is a particular case of what is called an admissible quasi-diffomorphism. One such is a homeomorphism $F : S' \to S$ between two surfaces which is differentiable outside a finite set of points Z and has bounded dilatation. To explain what the term dilatation means, we look at a local expression $w = w(z)$ for F away from Z. Suppose first that $w(z) = Kx + \sqrt{-1}y$ is an affine map. Then the dilatation is simply the constant K. If one writes $K = (1+k)/(1-k)$, then $k = w_{\bar{z}}/w_z$. For a general F, one looks at all the local expressions for F and takes the sup k_F of all the quantities $|w_{\bar{z}}/w_z|$. Then the dilatation of F is given by $K[F] = (1 + k_F)/(1 - k_F)$.

Teichmüller's Uniqueness Theorem states that if $f : (S_\omega, p_1, \dots, p_n) \to (S, p_1, \dots, p_n)$ is an admissible quasi-diffeomorphism with $[f] = [f_\omega]$, then $K[f] \geq K[f_\omega]$, and that equality holds if and only if $f = f_\omega$. The proof of Teichmüller's Uniqueness Theorem is given in Section 6 and is based on the geometry associated to a quadratic differential which we discuss in Section 5. The fact that both the injectivity and the closedness of Φ are a direct consequence of Teichmüller's Uniqueness Theorem is explained in Section 4. In the same section we deduce the connectedness of $\mathcal{T}_{g,n}$ from the basic existence theorem for the Beltrami equation.

In Section 7 we use the Teichmüller space and the fact that the modular group is generated by Dehn twists to prove several important results about the moduli stack $\overline{\mathcal{M}}_{g,n}$. The first one is that it is simply connected, the second is that its Picard group $\operatorname{Pic}(\overline{\mathcal{M}}_{g,n})$ is a free abelian group of finite rank, and that $\operatorname{Pic}(\overline{M}_{g,n})$ is a subgroup of finite index. Another consequence of this kind of arguments is the vanishing of the first integral cohomology group of $\overline{M}_{g,n}$.

In the last section we discuss a bordification of Teichmüller space. This is a space $\widehat{\mathcal{T}}_{g,n}$ containing $\mathcal{T}_{g,n}$ as an open subset, having the property that the action of $\Gamma_{g,n}$ on $\mathcal{T}_{g,n}$ extends to $\widehat{\mathcal{T}}_{g,n}$ in such a way that the quotient $\widehat{\mathcal{T}}_{g,n}/\Gamma_{g,n}$ coincides with $\overline{M}_{g,n}$. This bordification, which is closely related to the Fenchel–Nielsen coordinates, has an independent interest. From our point of view it will be one of the tools we will use in presenting Kontsevich's proof of the Witten conjecture. An essential prerequisite to understand the construction of $\widehat{\mathcal{T}}_{g,n}$ is the theory of oriented blow-ups discussed in Section 9 of Chapter X. To give a first idea of what $\widehat{\mathcal{T}}_{g,n}$ looks like, we consider the case of $\widehat{\mathcal{T}}_{1,1}$. It turns out that

$$\widehat{\mathcal{T}}_{1,1} = \mathbb{H} \cup \mathbb{P}^1(\mathbb{Q}) .$$

Here $\mathbb{H} = \mathcal{T}_{1,1}$ is the upper half-plane, and the points of $\mathbb{P}^1(\mathbb{Q}) \subset \partial\mathbb{H}$ are in a one-to-one correspondence with the free homotopy classes of simple closed curves on a reference surface of genus 1. These are the vanishing cycles, shrinking which we go to the boundary of $\mathcal{T}_{1,1}$ in all possible ways.

2. Teichmüller space and the mapping class group.

Fix, once and for all, an oriented genus g topological surface Σ and a finite subset $P \subset \Sigma$. We assume that $2g - 2 + |P| > 0$. Consider a P-pointed genus g curve $(C, \mathbf{x})$, where, as usual, $\mathbf{x} : P \to C$ is an injective map, and we set $x_p = \mathbf{x}(p)$ for every p. A *Teichmüller structure* on $(C, \mathbf{x})$ is the datum of the isotopy class $[f]$ of an orientation-preserving homeomorphism

$$f : (C, \mathbf{x}) \longrightarrow (\Sigma, P) ,$$

where the allowable isotopies are those which map x_p to p, for each p and *for every value of the parameter*. An isomorphism between curves with Teichmüller structure $(C, \mathbf{x}, [f])$ and $(C', \mathbf{x}', [f'])$ is an isomorphim of n-pointed curves

$$\varphi : (C, \mathbf{x}) \longrightarrow (C', \mathbf{x}')$$

such that $[f'\varphi] = [f]$. The set of isomorphism classes of P-pointed genus g curves with Teichmüller structure has a natural topology and complex structure which we are presently going to describe. The resulting space is called the *Teichmüller space of* (Σ, P) and is denoted by the symbol $\mathcal{T}_{\Sigma,P}$ or, more simply, by $\mathcal{T}_{g,P}$. The fact that this shorthand notation is not ambiguous will be clear later in this section. In this chapter we will normally assume that the points of P are numbered, and we will use the notation $\mathcal{T}_{(\Sigma,p_1,\dots,p_n)}$ as an alternative to $\mathcal{T}_{\Sigma,P}$, where $n = |P|$ and $\{p_1, \dots, p_n\} = P$. In this case, we shall normally write $\mathcal{T}_{g,n}$ for $\mathcal{T}_{(\Sigma,p_1,\dots,p_n)}$.

The point in $\mathcal{T}_{g,n}$ associated to the n-pointed curve $(C, x_1, \dots, x_n)$ and to the isotopy class $[f]$ will be denoted by the symbol $[C, x_1, \dots, x_n, [f]]$. We shall sometimes write $\mathcal{T}_g$ instead of $\mathcal{T}_{g,0}$.

It is important to remark that the Teichmüller structure $[f]$ rigidifies the pointed curve $(C, x_1, \dots, x_n)$ in the sense that

$$\operatorname{Aut}(C, x_1, \dots, x_n, [f]) = \{1\}. \tag{2.1}$$

As we shall see, it is exactly this rigidity property that makes the Teichmüller space a smooth manifold naturally equipped with a universal family. To prove (2.1), observe that in genus zero the number of marked point is at least three, and hence $(C, x_1, \dots, x_n)$ is already rigid, while, in general, an automorphism γ of $(C, x_1, \dots, x_n)$ such that $[f\gamma] = [f]$ must be homotopically trivial and, in particular, must induce the identity on integral cohomology. We may then appeal to the following elementary result.

LEMMA (2.2). *Let C be a smooth curve of genus $g \geq 1$. Let γ be an automorphism of C. Suppose that $\gamma_* : H_1(C, \mathbb{Z}) \to H_1(C, \mathbb{Z})$ is the identity and, in case $g = 1$, that γ fixes a point of C. Then γ is the identity.*

Proof. The assumptions immediately imply that γ induces the identity at the level of holomorphic forms. It follows in particular that γ acts trivially on the Jacobian $J(C)$. Another consequence is that γ commutes with the canonical map; thus, if the genus of C is greater than one, γ is either the identity or the hyperelliptic involution. In any case, γ has a fixed point q. But then the action of γ commutes with the Abel–Jacobi map $C \to J(C)$ based at q. The conclusion follows. Q.E.D.

To describe the complex structure of Teichmüller space, it is convenient to extend the notion of Teichmüller structure from individual curves to families. Consider then a family (holomorphic, differentiable, or even simply continuous) of n-pointed genus g curves, consisting of a fiber space $\pi : \mathcal{C} \to B$ plus sections $\sigma_i : B \to \mathcal{C}$, $i = 1, \dots, n$, and for each $b \in B$, set $C_b = \pi^{-1}(b)$. A *Teichmüller structure* on such a family is the datum of a Teichmüller structure $[f_b]$ on each fiber $(C_b, \sigma_1(b), \dots, \sigma_n(b))$, satisfying the following local triviality condition. It must be possible to cover B with open sets U over each of which there is a topological trivialization $(F, \pi) : \pi^{-1}(U) \to \Sigma \times U$, sending each section σ_i to the section $\{p_i\} \times U$, such that, for each $b \in U$, $[F_b] = [f_b]$, where F_b stands for the restriction of F to C_b. In order to show that this makes good sense, we must prove that this notion of local triviality is independent of the topological trivializations we choose. Let

$$(F, \pi) : \pi^{-1}(U) \to \Sigma \times U\,,$$
$$(G, \pi) : \pi^{-1}(U) \to \Sigma \times U$$

be two topological trivializations over the same connected open subset of B and suppose that $[F_b] = [G_b]$ at a point $b \in U$. Let b' be another point of U, and let $\gamma : [0, 1] \to U$ be a path joining b' to b. Then, as t varies between 0 and 1, the homeomorphisms

$$G_{\gamma(t)} \circ F_{\gamma(t)}^{-1} \circ F_{b'} : (C_{b'}, \sigma_1(b'), \dots, \sigma_n(b')) \to (\Sigma, p_1, \dots, p_n)$$

trace out an isotopy between $G_{b'}$ and $G_b \circ F_b^{-1} \circ F_{b'}$, which in turn is isotopic to $F_{b'}$. Thus, $[F_{b'}] = [G_{b'}]$ for all $b' \in U$.

What we have just shown implies, among other things, that, given two Teichmüller structures on the same family of curves, the locus of those points in the base of the family where the two Teichmüller structures agree is both open and closed. In particular, when the base is connected, the two Teichmüller structures are the same as soon as they agree at one point.

The notion of Teichmüller structure on a family of curves has been defined in terms of local topological trivializations. At the end of Section 4 we shall actually see that families of curves admitting a Teichmüller structure are topologically trivial and that any Teichmüller structure on such a family is induced by a global trivialization.

Let $(C, x_1, \dots, x_n, [f])$ be an n-pointed genus g curve with Teichmüller structure, and let

$$\begin{array}{ccc} C \overset{\varphi}{\cong} C_{u_0} \hookrightarrow \mathcal{C} \\ \pi \Big\downarrow \Big\uparrow \sigma_i, i = 1, \dots, n \\ u_0 \in U \end{array} \tag{2.3}$$

be a Kuranishi family for $(C, x_1, \dots, x_n)$, where φ is an isomorphism between the n-pointed curves $(C, x_1, \dots, x_n)$ and $(C_{u_0}, \sigma_1(u_0), \dots, \sigma_n(u_0))$.

Possibly after shrinking U, such a family admits a topological trivialization $(F, \pi) : \mathcal{C} \to \Sigma \times U$ such that $F_{u_0} \circ \varphi = f$ and hence can be endowed with a unique Teichmüller structure extending the one on $(C, x_1, \ldots, x_n)$. We thus get what we shall refer to as a *Kuranishi family for the curve with Teichmüller structure* $(C, x_1, \ldots, x_n, [f])$. The name is justified by the fact that such a family enjoys, with respect to deformations of curves with Teichmüller structure, a universal property exactly analogous to the one of standard Kuranishi families. In fact, one immediately checks that any holomorphic family of n-pointed genus g curves with Teichmüller structure over a base S having $(C, x_1, \ldots, x_n, [f])$ as fiber at $s_0 \in S$ arises, possibly after shrinking S, by pullback of the Kuranishi family for $(C, x_1, \ldots, x_n, [f])$ via a unique morphism $(S, s_0) \to (U, u_0)$. To do so, it suffices to use the universal property of ordinary Kuranishi families and the uniqueness of the Teichmüller structure extending the one on the fiber at s_0. It should be observed that, when U is small enough, the family we just constructed is Kuranishi at every point of U, as follows from the analogous property of standard Kuranishi families.

After these preliminaries we are ready to describe the topology and the complex structure on $\mathcal{T}_{g,n}$. The construction is entirely parallel to the one that gives the complex structure on the moduli space $M_{g,n}$, only simpler due to the rigidity of curves with Teichmüller structure.

Let $y = [C, x_1, \ldots, x_n, [f]]$ be a point of $\mathcal{T}_{g,n}$. Choose a Kuranishi family for $(C, x_1, \ldots, x_n, [f])$. As we explained, this can be constructed by putting on the family (2.3), where U is chosen to be connected and "sufficiently small," the unique Teichmüller structure extending $[f]$; for each $u \in U$, we denote by $[F_u]$ the Teichmüller structure on $(C_u, \sigma_1(u), \ldots, \sigma_n(u))$. By Proposition (6.5) we may suppose that the action of $G = \mathrm{Aut}(C, x_1, \ldots, x_n)$ on $(C, x_1, \ldots, x_n)$ extends to equivariant actions on $(\mathcal{C}, \sigma_1, \ldots, \sigma_n)$ and on U and that any isomorphisms between fibers of π is the restriction of the action of an element of G on $\mathcal{C}$; as a consequence, U/G injects in $M_{g,n}$. We claim that the natural map

$$\begin{aligned} \alpha : U &\longrightarrow \mathcal{T}_{g,n} \\ u &\mapsto [C_u, \sigma_1(u), \ldots, \sigma_n(u), [F_u]] \end{aligned} \tag{2.4}$$

is injective. We shall refer to such an α as a *standard coordinate patch* for $\mathcal{T}_{g,n}$ around the point y. The injectivity of α is an almost immediate consequence of the rigidity property of $(C, x_1 \ldots, x_n, [f])$. In fact, suppose that $\alpha(u) = \alpha(u')$ for two different points u and u' of U. This means that there must be an isomorphism $\psi : (C_u, \sigma_1(u), \ldots, \sigma_n(u)) \to (C_{u'}, \sigma_1(u'), \ldots, \sigma_n(u'))$ such that $[F_u] = [F_{u'} \circ \psi]$. By our assumptions, ψ is induced by an element $\rho \in G$. In other words, there is a commutative diagram

$$\begin{array}{ccc} \mathcal{C} & \xrightarrow{\tilde{\xi}} & \mathcal{C} \\ {\scriptstyle\pi}\downarrow & & \downarrow{\scriptstyle\pi} \\ U & \xrightarrow{\xi} & U \end{array} \tag{2.5}$$

where $\tilde{\xi}$ and ξ are automorphisms such that $\xi(u_0) = u_0$, $\xi(u) = u'$, $\tilde{\xi}$ carries sections orderly into sections, $\varphi^{-1}\tilde{\xi}_{u_0}\varphi = \rho$, and $\tilde{\xi}_u = \psi$. Here, for any $v \in U$, we write $\tilde{\xi}_v$ to indicate the isomorphism from $(C_v, \sigma_1(v), \dots, \sigma_n(v))$ to $(C_{\xi(v)}, \sigma_1(\xi(v)), \dots, \sigma_n(\xi(v)))$ induced by $\tilde{\xi}$. Now set $G_v = F_{\xi(v)} \circ \tilde{\xi}_v$. The isotopy classes $[G_v]$ define a new Teichmüller structure on the family (2.3). Since $[G_u] = [F_{u'} \circ \psi] = [F_u]$, we deduce that $[G_v] = [F_v]$ for every $v \in U$. In particular, for $v = u_0$, we get that $[f \circ \rho] = [f]$ and hence that ρ is isotopic to the identity and, by the rigidity property, that $\rho = 1$, proving our claim.

That the patches we have just described define a complex structure on $\mathcal{T}_{g,n}$ is a formal consequence of the universal property of Kuranishi families. In fact, if $\alpha : U \to \mathcal{T}_{g,n}$ and $\beta : V \to \mathcal{T}_{g,n}$ are two standard patches whose codomains have a point z in common, then the restriction to a neighborhood W of $\beta^{-1}(z)$ of the family of curves with Teichmüller structure over V is the pullback of the family over U via a holomorphic map $\varphi : W \to U$. On the other hand, φ clearly agrees with the restriction to W of $\alpha^{-1} \circ \beta$. Hence $\alpha^{-1} \circ \beta$ is holomorphic where defined. This completes the construction of a (possibly non-Hausdorff, for the moment) complex structure on $\mathcal{T}_{g,n}$. Later in this section we shall show that $\mathcal{T}_{g,n}$ is actually a Hausdorff topological space.

Disregarding the Teichmüller structures yields a map

$$m : \mathcal{T}_{g,n} \to M_{g,n} \,. \tag{2.6}$$

If $\alpha : U \to \mathcal{T}_{g,n}$ is a standard coordinate patch around $y = [C, x_1, \dots, x_n, [f]]$ as described above, and G is the automorphism group of $(C, x_1, \dots, x_n)$, then m can be locally identified with the quotient map $U \to U/G$. Therefore m is holomorphic.

We now wish to show that implicit in our construction of $\mathcal{T}_{g,n}$ is also the construction of a *universal family of pointed curves with Teichmüller structure*

$$\eta : \mathcal{X}_{g,n} \longrightarrow \mathcal{T}_{g,n} \,. \tag{2.7}$$

Indeed, over each standard coordinate patch $\alpha : U \to \mathcal{T}_{g,n}$ with $\alpha(u_0) = [C, x_1, \dots, x_n, [f]]$, we have a Kuranishi family of curves with Teichmüller structure, and two such families patch together via an isomorphism over U. This isomorphism is *unique*, since $(C, x_1, \dots, x_n, [f])$

is rigid. Via these patching data, one constructs the family (2.7) equipped with n canonical sections and a Teichmüller structure. Let us now describe its universal property. Any holomorphic family of n-pointed genus g curves with Teichmüller structure over a base T is canonically isomorphic to the pullback of the family $\mathcal{X}_{g,n} \to \mathcal{T}_{g,n}$ via a unique morphism $T \to \mathcal{T}_{g,n}$, as follows from the universal property of Kuranishi families. In other words, $\mathcal{T}_{g,n}$ represents the functor

$$T \to \left\{ \begin{matrix} \text{holomorphic families of } n\text{-pointed genus } g \\ \text{curves with Teichmüller structure over } T \end{matrix} \right\}$$

from analytic spaces to sets.

We stop briefly for two important remarks concerning the universal property of Teichmüller space. The first of these is that, as follows from the universal property, any isomorphism $(\Sigma, p_1, \dots, p_n) \to (R, q_1, \dots, q_n)$ of n-pointed genus g oriented topological surfaces induces a canonical isomorphism between the Teichmüller spaces $\mathcal{T}_{(\Sigma, p_1, \dots, p_n)}$ and $\mathcal{T}_{(R, q_1, \dots, q_n)}$, thus justifying the notation $\mathcal{T}_{g,n}$.

The second remark is that, since in our construction the universal property of the Teichmüller space is a reflection of the one of Kuranishi families and since, as we remarked in Section 7 of Chapter XI, the universal property of Kuranishi families also works in the context of C^m families of curves, the same is true for the universal property of Teichmüller space. More precisely, the notion of Teichmüller structure carries over, with obvious changes, to the context of C^m families of curves, and a direct consequence of Proposition (7.1) of Chapter XI is the following:

PROPOSITION (2.8). *Let $\alpha : X \to S$ be a C^m family of genus g curves with Teichmüller structure. Suppose $g \geq 2$. Let $\eta : \mathcal{X}_{g,n} \to \mathcal{T}_{g,n}$ be the universal family of pointed curves with Teichmüller structure. Let $f : S \to \mathcal{T}_{g,n}$ be the map which associates to each point of S the isomorphism class of the corresponding fiber, and let $F : X \to \mathcal{X}_{g,n}$ be the map whose restriction to the fiber X_s is the unique isomorphism of pointed curves with Teichmüller structure from X_s to $\eta^{-1}(f(s))$. Then the pair (f, F) is a morphism of C^m families of curves. In particular, f is of class C^m.*

We are going to use this proposition in the next sections.

The mapping class group

Recall from Section 9 of Chapter X that the mapping class group, also called Teichmüller modular group and denoted by $\Gamma_{\Sigma,P}$, or simply by $\Gamma_{g,P}$, is the group of all isotopy classes of orientation-preserving homeomorphism of (Σ, P) into itself. In accordance with our previous conventions, when the points of $P = \{p_1, \ldots, p_n\}$ are numbered, we usually write $\Gamma_{(\Sigma,p_1,\ldots,p_n)}$, or simply $\Gamma_{g,n}$, to denote the mapping class group. We also write Γ_g for $\Gamma_{g,0}$. The mapping class group acts naturally on $\mathcal{T}_{g,n} = \mathcal{T}_{(\Sigma,p_1,\ldots,p_n)}$, the action of an element $[\gamma]$ being given by

$$(2.9) \qquad [\gamma] \cdot [C, x_1, \ldots, x_n, [f]] = [C, x_1, \ldots, x_n, [\gamma \circ f]] .$$

The elements of $\Gamma_{g,n}$ act on $\mathcal{T}_{g,n}$ as holomorphic automorphisms. This is essentially a tautology. In fact, let $[\gamma]$ be an element of $\Gamma_{g,n}$, and let $\alpha : U \to \mathcal{T}_{g,n}$ be a standard coordinate patch obtained from a Kuranishi family (2.3) endowed with a Teichmüller structure $\{F_u\}_{u \in U}$. Changing the Teichmüller structure to $\{\gamma \circ F_u\}_{u \in U}$ yields a new coordinate patch $\beta : U \to \mathcal{T}_{g,n}$. Since $[\gamma]$ acts by replacing each Teichmüller structure $[f]$ with $[\gamma \circ f]$ it is obvious that, for any $u \in U$,

$$\beta^{-1}([\gamma] \cdot \alpha(u)) = u .$$

This shows that $[\gamma]$ acts holomorphically.

The map (2.6) can be identified with the quotient map from $\mathcal{T}_{g,n}$ to $\mathcal{T}_{g,n}/\Gamma_{g,n}$. In fact, it is clear that, if $y \in \mathcal{T}_{g,n}$ and $[\gamma] \in \Gamma_{g,n}$, then y and $[\gamma] \cdot y$ map to the same point of $M_{g,n}$. Conversely, to say that $m(y) = m(y')$, where $y = [C, x_1, \ldots, x_n, [f]]$ and $y' = [C', x'_1, \ldots, x'_n, [f']]$, means that there is an isomorphism $\varphi : (C, x_1, \ldots, x_n) \to (C', x'_1, \ldots, x'_n)$. But then $y' = [f' \varphi f^{-1}] \cdot y$. This shows that $M_{g,n} = \mathcal{T}_{g,n}/\Gamma_{g,n}$ set-theoretically. On the other hand, the map (2.6) is open by construction, so that $M_{g,n}$ and $\mathcal{T}_{g,n}/\Gamma_{g,n}$ coincide also as topological spaces.

It is also the case that, if y is as above, then the stabilizer of y can be identified with the automorphism group G of the n-pointed curve $(C, x_1, \ldots, x_n)$. To prove this, start with an automorphism ρ of $(C, x_1, \ldots, x_n)$ and set

$$\gamma_\rho = f \rho f^{-1} .$$

The map $\rho \mapsto [\gamma_\rho]$ is a homomorphism from G to $\Gamma_{g,n}$ and lands inside the stabilizer of y. In fact, $f\rho = \gamma_\rho f$, that is, ρ is an isomorphism between the curves with Teichmüller structure $(C, x_1, \ldots, x_n, [f])$ and $(C, x_1, \ldots, x_n, [\gamma_\rho f])$, whose classes in $\mathcal{T}_{g,n}$ are y and $[\gamma_\rho] \cdot y$, respectively. Conversely, let $[\gamma]$ be any element of the stabilizer of y. Then there is an element ψ of G such that $[f\psi] = [\gamma f]$. This translates into $[\gamma] = [\gamma_\psi]$, which shows that every element of the stabilizer of y is of the form $[\gamma_\rho]$ for some automorphism ρ of $(C, x_1, \ldots, x_n)$. It remains to show that such a ρ is unique. But this is clear since ρ is homotopic to $f^{-1}\gamma f$, and automorphisms of $(C, x_1, \ldots, x_n)$ are unique within their homotopy class.

We can be even more precise. Let $\alpha : U \to \mathcal{T}_{g,n}$ be a standard coordinate patch centered at y. Recall that the group G acts on U. We claim that this action and the one of G on $\mathcal{T}_{g,n}$ correspond to each other under α. The proof is quite similar to the one of the injectivity of α, of which we keep the notation. The patch α arises from a Kuranishi family (2.3) endowed with a Teichmüller structure $\{F_u\}_{u\in U}$. If ρ is an element of G, then there is a diagram (2.5), where $\tilde{\xi}$ and ξ are automorphisms such that $\xi(u_0) = u_0$ and $\tilde{\xi}_{u_0}\varphi = \varphi\rho$. The action of ρ on U is then given by $\rho{\cdot}u = \xi(u)$. We must show that $\alpha(\xi(u)) = [\gamma]{\cdot}\alpha(u)$, where $\gamma = \gamma_\rho = f\rho f^{-1}$. We define two new Teichmüller structures $\{G_u\}_{u\in U}$ and $\{H_u\}_{u\in U}$ on (2.3) by setting $G_u = F_{\xi(u)}\tilde{\xi}_u$ and $H_u = \gamma F_u$. It is a straightforward check to verify that $G_{u_0} = H_{u_0}$. It follows that $[G_u] = [H_u]$ for all u. Since $\tilde{\xi}_u$ is an isomorphism between the curves with Teichmüller structure $(C_u, \sigma_1(u), \ldots, \sigma_n(u), [G_u])$ and $(C_{\xi(u)}, \sigma_1(\xi(u)), \ldots, \sigma_n(\xi(u)), [F_{\xi(u)}])$, and since

$$[\gamma]\cdot[C_u, \sigma_1(u), \ldots, \sigma_n(u), [F_u]] = [C_u, \sigma_1(u), \ldots, \sigma_n(u), [H_u]]\,,$$

we conclude that

$$\begin{aligned}[\gamma_\rho]\cdot\alpha(u) &= [C_u, \sigma_1(u), \ldots, \sigma_n(u), [H_u]]\\ &= [C_u, \sigma_1(u), \ldots, \sigma_n(u), [G_u]] = \alpha(\xi(u))\,,\end{aligned}$$

as desired.

We end this section by showing that $\mathcal{T}_{g,n}$ is Hausdorff and, at the same time, that the action of $\Gamma_{g,n}$ on $\mathcal{T}_{g,n}$ is properly discontinuous. First, it is convenient to summarize some of the properties of the action of $\Gamma_{g,n}$ that we have discovered so far. Let y be a point of $\mathcal{T}_{g,n}$, and let G be the stabilizer of y. Then the following hold.

i) $M_{g,n} = \mathcal{T}_{g,n}/\Gamma_{g,n}$ is Hausdorff
ii) G is a finite group
iii) y has arbitrarily small neighborhoods V with the following properties:

iii.a) V is G-stable
iii.b) V/G injects into $M_{g,n}$
iii.c) the stabilizer of z is contained in G for any $z \in V$

In fact, we can take as V the images $\alpha(U)$ of standard patches $\alpha : U \to \mathcal{T}_{g,n}$. The only one of the above properties that may not be clear is perhaps *iii.c)*. To prove it, recall that the action of G on V corresponds to the action on the base U and total space $\mathcal{C}$ of the Kuranishi family (2.3) of the automorphism group of the central fiber and that every isomorphism between fibers of (2.3) is induced by an automorphism of the central fiber. In particular, this is true of the automorphisms of the fiber above a point $u \in U$. Since the group of these automorphisms corresponds to the stabilizer of $y = \alpha(u)$, property *iii.c)* follows.

The Hausdorffness of $\mathcal{T}_{g,n}$ and the proper discontinuity of the action of $\Gamma_{g,n}$ are formal consequences of properties *i)–iii)*. To prove that $\Gamma_{g,n}$ acts properly discontinuously on $\mathcal{T}_{g,n}$, we must first find, for any pair of points y and y' not belonging to the same $\Gamma_{g,n}$-orbit, neighborhoods W and W' such that $\gamma \cdot W \cap W' = \emptyset$ for all γ in $\Gamma_{g,n}$. Clearly, it suffices to take as W and W' the preimages of two disjoint neighborhoods of $m(y)$ and $m(y')$, which exist since $M_{g,n}$ is Hausdorff. To finish proving that the action of $\Gamma_{g,n}$ is properly discontinuous, we must find, for any $y \in \mathcal{T}_{g,n}$, a neighborhood V such that all $\gamma \in \Gamma_{g,n}$ such that $\gamma \cdot V \cap V \neq \emptyset$ belong to G, the stabilizer of y. We claim that a neighborhood satisfying properties *iii.a)–iii.c)* above will do. Suppose in fact that the intersection of $\gamma \cdot V$ and V is not empty, and let z be a point of it. We may then write $z = \gamma \cdot z'$, where z' is another point of V. By *iii.b)* there is $\rho \in G$ such that $z' = \rho \cdot z$, so that $z = \gamma\rho \cdot z$ and $\gamma\rho$ belongs to the stabilizer of z. Since this is contained in G, we conclude that γ belongs to G, as desired. This also shows that $\mathcal{T}_{g,n}$ is Hausdorff. In fact, we already know that points y and y' of $\mathcal{T}_{g,n}$ belonging to different orbits of $\Gamma_{g,n}$ have disjoint neighborhoods. Suppose instead that $y' = \gamma \cdot y$ for some $\gamma \in \Gamma_{g,n}$, and let V be a neighborhood of y satisfying *iii.a)–iii.c)*. Thus $\gamma \cdot V$ is a neighborhood of y'. On the other hand, if $y' \neq y$, that is, if γ does not belong to the stabilizer of y, then V and $\gamma \cdot V$ do not intersect, as we just proved.

3. A little surface topology.

This section collects a number of topological results which will be useful to better understand the nature of the Teichmüller space and of the mapping class group. For most of them no proof will be given here; instead, we will provide references to the relevant sources.

The first result that we would like to discuss has to do with the very definition of Teichmüller space. It is known that, if X and Y are surfaces without boundary, compact or not, two homeomorphisms $f_1, f_2 : X \to Y$ are homotopic if and only if they are isotopic. A proof can be found in [216] (cf. in particular Theorems 6.4 and A4). An immediate consequence is that a Teichmüller structure on an unpointed curve C can also be defined as the *homotopy* class of a homeomorphism from C to a reference surface Σ. For an n-pointed curve $(C, x_1, \dots, x_n)$, instead, a Teichmüller structure can be alternatively defined as the equivalence class of a homeomorphism from $(C, x_1, \dots, x_n)$ to a reference surface $(\Sigma, p_1, \dots, p_n)$ modulo the following relation: two homeomorphisms are considered equivalent if the induced homeomorphisms from $C \smallsetminus \{x_1, \dots, x_n\}$ to $\Sigma \smallsetminus \{p_1, \dots, p_n\}$ are homotopic. In order to convince ourselves that this is indeed the case, all we need is the following remark, whose easy proof is left to the reader. Let

$$f_1, f_2 : (C, x_1, \dots, x_n) \to (\Sigma, p_1, \dots, p_n)$$

be homeomorphisms; then any isotopy between the homeomorphisms from $C \setminus \{x_1, \dots, x_n\}$ to $\Sigma \setminus \{p_1, \dots, p_n\}$ they induce can be extended to an isotopy between f_1 and f_2 relative to $\{x_1, \dots, x_n\}$. Similarly, the mapping class group $\Gamma_{g,n}$ can also be defined as the group of homeomorphisms of $(\Sigma, p_1, \dots, p_n)$ modulo the equivalence relation defined above; in particular, Γ_g is the group of homeomorphisms of Σ modulo homotopy.

Another variant of the definitions of Teichmüller space and mapping class group, entirely equivalent to the original one, can be obtained by fixing a differentiable structure on the reference surface $(\Sigma, p_1, \dots, p_n)$ and replacing the words "homeomorphism" and "isotopy" with "diffeomorphism" and "differentiable isotopy" throughout. What this amounts to saying is that every class in $\Gamma_{g,n}$ contains a diffeomorphism and that two diffeomorphisms which are isotopic relative to $p_1, \dots, p_n$ are also differentiably isotopic; we shall comment on this later in this section, after we have discussed Dehn twists and Dehn's theorem. Thus, the Teichmüller space $\mathcal{T}_{g,n}$ is the set of isomorphism classes of objects $(C, x_1, \dots, x_n, \varphi)$, where $(C, x_1, \dots, x_n)$ is an n-pointed curve of genus g, and φ is the differentiable isotopy class of an oriented diffeomorphism from $(C, x_1, \dots, x_n)$ to $(\Sigma, p_1, \dots, p_n)$. Similarly,

$$\Gamma_{g,n} = \mathrm{Diff}_+(\Sigma, p_1, \dots, p_n) / \mathrm{Diff}_0(\Sigma, p_1, \dots, p_n)\,,$$

where $\mathrm{Diff}_+(\Sigma, p_1, \dots, p_n)$ stands for the group of orientation-preserving diffeomorphisms of $(\Sigma, p_1, \dots, p_n)$, and $\mathrm{Diff}_0(\Sigma, p_1, \dots, p_n)$ for its identity component, which is nothing but the group of diffeomorphisms of $(\Sigma, p_1, \dots, p_n)$ which are differentiably isotopic to the identity.

The second result we would like to mention is the so-called *Dehn–Nielsen theorem*. This is a highly nontrivial result but can be very simply stated. Recall that the group $\mathrm{Out}(G)$ of *outer automorphisms* of a group G is the quotient of $\mathrm{Aut}(G)$ by the subgroup of inner automorphisms. The Dehn–Nielsen theorem then says that every outer automorphism of the fundamental group of a compact orientable surface Σ is induced by a homeomorphism of Σ to itself. When the genus of Σ is zero, there is nothing to prove; therefore, in the following discussion, we shall implicitly assume that the genus g of Σ is strictly positive.

To put the theorem into perspective, and for future use, it is useful at this point to introduce the notion of exterior homomorphism of groups. Let G and H be two groups; the set of *exterior homomorphisms* from G to H, written

$$\mathrm{Hom}^{ext}(G, H)\,,$$

is the quotient of $\mathrm{Hom}(G, H)$ modulo the action of H given by inner automorphisms of the target. It must be stressed that this definition departs somewhat from common usage, in that normally the set of exterior homomorphisms is defined to be the quotient of $\mathrm{Hom}(G, H)$

modulo inner automorphisms of G. Ours is thus a coarser equivalence; however, when the homomorphisms involved are surjective, as will most often be the case, there is no difference between the two notions. In general, an advantage of our definition is that it makes sense to compose exterior homomorphism. The set of exterior isomorphisms from G to H will be denoted $\mathrm{Iso}^{ext}(G,H)$; in the special case where $H=G$, this is nothing but $\mathrm{Out}(G)$.

In proving the Dehn–Nielsen theorem we shall follow Seifert [621]. A standard result in algebraic topology (cf., for instance, [644], Chapter 8, Theorem 11) says that, if X and Y are connected spaces, and Y happens to be a $K(\pi,1)$, then the set of homotopy classes of maps from X to Y is in one-to-one correspondence with the set of exterior homomorphisms $\mathrm{Hom}^{ext}(\pi_1(X),\pi_1(Y))$ via the map $h\mapsto h_*$. Here, as we will almost always do in the sequel when dealing with exterior homomorphisms of fundamental groups, we have omitted mention of the base points; this causes no trouble since we are working modulo inner automorphisms. Clearly, this result comes close to proving the Dehn–Nielsen theorem, since the universal covering of Σ is homeomorphic to a disk, and hence Σ is a $K(\pi,1)$. What needs to be shown in order to complete the proof is that, if $f:\Sigma\to\Sigma$ induces an automorphism of the fundamental group of Σ, then f is homotopic to a homeomorphism of Σ to itself. It follows from the topological result quoted above that the inverse automorphism is induced by another map $f':\Sigma\to\Sigma$ and that $f'f$ is homotopic to the identity. Therefore the degree of f and f' is ±1, and hence the Dehn–Nielsen theorem is a consequence of the following result.

PROPOSITION (3.1). *Let $f:X\to Y$ be a continuous map of compact connected orientable surfaces of the same genus. If the degree of f is ±1, then f is homotopic to a homeomorphism.*

The results is obvious in genus zero, since maps from the sphere to itself are classified, up to homotopy, by their degree.

From now on we assume that the genus of X is positive. We shall find it useful to adopt the following terminology. A cell decomposition of a surface will be called a *triangle decomposition* if the closure of every 2-cell is cellularly homeomorphic to a triangle. We next put a riemannian metric on Y, cover Y with geodesically convex geodesic balls $U_1,\dots,U_n$, and choose a triangulation of X so fine that the image under f of the closure of every 2-symplex is contained in one of the U_i. We define a new map f' from X to Y as follows. Let $v_1,\dots,v_N$ be the vertices of the triangulation. For each i, pick a point q_i of Y close to $f(v_i)$ in such a way that the q_i are all distinct and no three of them lie on the same geodesic. We can choose the q_i to be as close as we wish to the $f(v_i)$; in particular, we may assume that, if v_j, v_h, and v_k are the vertices of a 2-symplex whose image is contained in U_i, then q_j, q_h, and q_k are also contained in U_i. The map f' is defined by mapping each vertex v_j to the corresponding q_j, each edge v_jv_h to the minimal geodesic arc joining q_j to q_h, and finally by extending the map to the interior of each 2-symplex $v_jv_hv_k$ so that the symplex itself maps homeomorphically onto the geodesic triangle with vertices q_j, q_h, and q_k. Clearly, if x is any point of X, $f(x)$ and $f'(x)$ lie in the same U_i and hence are joined by a unique minimal geodesic arc. This arc depends continuously on its endpoints and hence on x; thus f and f' are homotopic. Notice that the fibers of f' are all zero-dimensional. Moreover, the images of the edges of the triangulation subdivide Y in polygonal regions, which can be further subdivided in (geodesic) triangles. We thus get a triangulation of Y; clearly, this lifts to a triangle decomposition of X. The map f' is now what Seifert calls a folded covering. Formally, a *folded covering* is a cellular map of compact orientable surfaces endowed with a triangle decomposition such that every cell maps homeomorphically onto a cell.

By what we have said, we may assume that f is a folded covering. The next, and crucial, step is to get rid of the *folds*, that is, the pairs of adjacent 2-cells of X mapping to the same cell of Y. These come in three different types, illustrated in Figure 1 below.

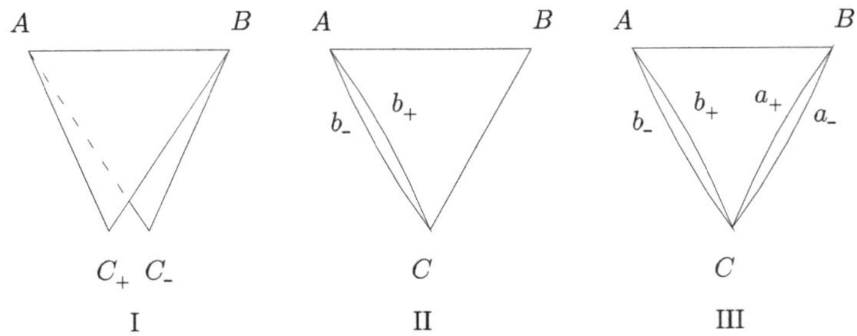

Figure 1.

The two flaps of a type I fold have in common only one edge, those of a type II fold have two edges in common, and those of a type III fold share one edge and the opposite vertex. The flaps of a fold cannot have all three edges in common, for in this case X would be a sphere.

The procedure for eliminating a fold is basically the same in all three cases. For folds of type I of II, we excise the fold along its boundary (AC_+BC_-A for type I and b_+b_- for type II), thus obtaining a surface with boundary E'; then we sew together the edges of the boundary by identifying AC_+ with AC_- and BC_+ with BC_- for type I and b_+ with b_- for type II. The result is a new surface X', endowed with a triangle decomposition, with two fewer 2-cells than X but homeomorphic to X. Moreover, the composition of the inclusion of E' in X with f descends to a folded covering $f' : X' \to Y$. In fact, we shall exhibit a homeomorphism

h between X' and X such that $f \circ h$ is homotopic to f'. Before we do so, however, let us turn to folds of type III. In the presence of such a fold, we cut X along a_+a_-; this produces a surface with boundary E. Then we identify a_+ to a_- on both sides of the cut, thereby obtaining a new folded covering $\widehat{X} \to Y$ of degree ± 1. The Euler characteristic of $\widehat{X}$ is $\chi(X)+2$, and hence $\widehat{X}$ cannot be connected since, as is well known, a compact connected orientable surface cannot map with nonzero degree to another surface of strictly larger genus. In fact, let $k : Z \to W$ be a map of compact connected orientable and oriented surfaces. For any nonzero class $\alpha \in H^1(W, \mathbb{Z})$, there is another class β such that $\int \alpha \cup \beta \neq 0$. On the other hand, if $\deg(k) \neq 0$, then

$$\int_Z k^*(\alpha) \cup k^*(\beta) = \deg(k) \int_W \alpha \cup \beta \neq 0 ,$$

showing that $k^* : H^1(W, \mathbb{Z}) \to H^1(Z, \mathbb{Z})$ is injective. The surface $\widehat{X}$ must then consist of two connected components X' and X'' whose genera add up to the genus of X. For the same reason as above, one of the two components, say X', must have the same genus as X, so X'' is a sphere and maps to Y with degree zero. We denote by E' and D the inverse images of X' and X'' in E; clearly, D is a disk, and X'' is obtained by identifying two complementary halves of its boundary. As was the case for folds of types I and II, $f' : X' \to Y$ is a folded covering of degree ± 1, and the triangle decomposition of X' has strictly fewer 2-cells than the one of X. In this case also, we shall produce a homeomorphism $h : X' \to X$ and a homotopy between f' and $f \circ h$.

The construction of this homotopy and of the homeomorphism h is the same for all three types of folds. To begin with, notice that, if we disregard the triangle decompositions, for all types of fold the construction of X' consists in punching out the interior of a disk D and identifying one half of the boundary with the complementary half. More precisely, we can choose "coordinates" such that the disk in question is the unit disk centered at the origin of the (x, y)-plane and the sewing process identifies each boundary point (x, y) with $(x, -y)$, while f maps these two points to the same point of Y. The restriction of f to D lifts to a map φ from D to the universal covering of Y. Since the latter is a disk, φ is homotopic, relative to ∂D, to a map which is constant on all segments $x =$ constant. Pushing this homotopy down to Y shows that f is homotopic to the map η which agrees with f outside of D and is defined on D by

$$\eta(x, y) = f(x, \sqrt{1 - x^2}) .$$

We next let $h_t : E' \to X$, $0 \leq t \leq 1$, be the "eyelids" homotopy illustrated in Figure 2 below.

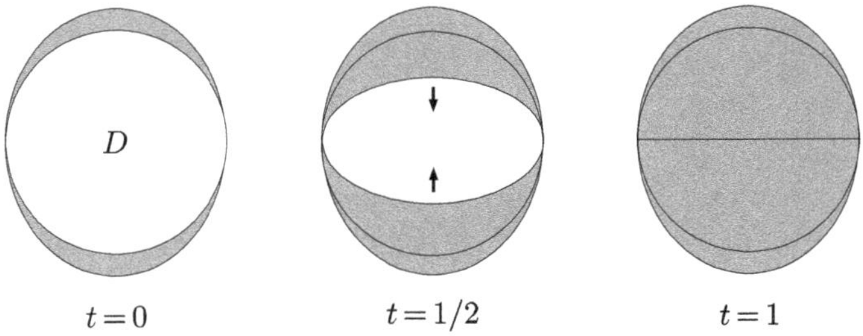

Figure 2.

More exactly, the homotopy can be described as follows. Pick some positive number $a < 1$. Then h_t is defined to be the identity outside the ellipse with equation $x^2 + a^2y^2 = 1$, while if (x, y) is a point of the ellipse and $a \leq s \leq 1$, one sets

$$h_t(x, sy) = \left(x, \frac{a(s-1)}{1-a} ty + sy \right).$$

Clearly, h_0 is the inclusion of E' in X, while h_1 drops to a homeomorphism $h : X' \to X$. For every t, the composition $\eta \circ h_t$ also descends to a map $F_t : X' \to Y$. This provides a homotopy between $f' = F_0$ and $\eta \circ h = F_1$. Recalling that η and f are homotopic, we then get a homotopy between f' and $f \circ h$.

Summing up, we have constructed a new map $f' \circ h^{-1} : X \to Y$, homotopic to f, which is a folded covering relative to a new triangle decomposition of X having strictly fewer 2-cells than the original one. The process can be repeated but has to stop after finitely many iterations. The end product is a folded covering $f'' : X \to Y$, homotopic to f and without folds. We claim that f'' is a homeomorphism. In fact, adjacent 2-cells of X map to the corresponding cells of Y with coherent orientation. By connectedness, then, either all cells of X map to the corresponding cells of Y with degree 1, or else they all map to them with degree -1. In either case each 2-cell of Y is the image of a unique 2-cell of X, since f'' has degree ± 1; thus f'' is a homeomorphims. This completes the proof of the Dehn–Nielsen theorem.

A corollary of the Dehn–Nielsen theorem is a new description of the mapping class group Γ_g and of the Teichmüller space $\mathcal{T}_g$. Let Σ be a reference genus g topological surface and denote by Π the fundamental group of Σ. As we have seen, the mapping class group may be viewed as a subgroup of $\mathrm{Out}(\Pi)$. We claim that this subgroup can be described in a purely algebraic way. Recall that Π is generated by elements α_i, β_i, $i = 1, \dots, g$, subject to the only relation

$$[\alpha_1, \beta_1] \cdots [\alpha_1, \beta_1] = 1,$$

where $[\alpha, \beta]$ stands for the commutator of α and β. The group $\Pi/[\Pi, \Pi] = H_1(\Sigma, \mathbb{Z})$ has a symplectic structure given by the intersection pairing and is isomorphic to $\mathbb{Z}^{2g}$ with the standard symplectic structure. In fact, one can set things up so that, denoting by a_i and b_i the images of α_i and β_i in $H_1(\Sigma, \mathbb{Z})$, the intersection matrix of the basis $a_1, \ldots, a_g, b_1, \ldots, b_g$ is the standard symplectic matrix

$$Q = \begin{pmatrix} 0 & I_g \\ -I_g & 0 \end{pmatrix}.$$

Then the mapping class group Γ_g is the subgroup $\mathrm{Out}^+(\Pi)$ of $\mathrm{Out}(\Pi)$ consisting of those automorphisms of Π which "preserve orientation" in the sense that they induce a symplectic automorphism of $\Pi/[\Pi, \Pi]$. This is an index two subgroup; in fact, it follows from the Dehn–Nielsen theorem that an automorphism of $\Pi/[\Pi, \Pi]$ induced by one of Π is always symplectic up to sign. Corresponding to this description of the mapping class group there is one of the Teichmüller space $\mathcal{T}_g$ as the set of isomorphism classes of data (C, φ), where C is a curve of genus g, and $\varphi \in \mathrm{Iso}^{ext}(\pi_1(C), \Pi)$ is an exterior isomorphism inducing a symplectic isomorphism $H_1(C, \mathbb{Z}) \to \Pi/[\Pi, \Pi]$.

There is an analogue of the Dehn–Nielsen theorem also in the n-pointed case (see, for instance, [692], Theorem V.9), which reads as follows. Let $(\Sigma, p_1, \ldots, p_n)$ be an n-pointed genus g surface, and let $\Pi = \Pi_{g,n}$ be the fundamental group of $\Sigma \smallsetminus \{p_1, \ldots, p_n\}$. Recall that Π is generated by elements α_i, β_i, $i = 1, \ldots, g$, and γ_j, $j = 1, \ldots, n$, subject to the only relation

$$[\alpha_1, \beta_1] \cdots [\alpha_1, \beta_1]\, \gamma_1 \cdots \gamma_n = 1\,.$$

Here γ_j is the class of a small positively oriented loop around p_j. Then an automorphism σ of Π comes from an orientation-preserving (resp., orientation-reversing) homeomorphism of $(\Sigma, p_1, \ldots, p_n)$ to itself if and only if there is a permutation ρ of $1, \ldots, n$ such that σ maps each γ_j to a conjugate of $\gamma_{\rho(j)}$ (resp., a conjugate of $\gamma_{\rho(j)}^{-1}$).

As in the unpointed case, a corollary is a description of the mapping class group $\Gamma_{g,n}$ as the subgroup of $\mathrm{Out}(\Pi)$ consisting of those automorphisms which leave the conjugacy class of each one of the γ_j fixed. Similarly, one can describe the Teichmüller space $\mathcal{T}_{g,n}$ as the set of isomorphism classes of data $(C, x_1, \ldots, x_n, \varphi)$, where $(C, x_1, \ldots, x_n)$ is an n-pointed curve of genus g, and $\varphi \in \mathrm{Iso}^{ext}(\pi_1(C \smallsetminus \{x_1, \ldots, x_n\}), \Pi_{g,n})$ is an exterior isomorphism which sends the class of a small positively oriented loop around x_j to a conjugate of γ_j for each j.

An immediate consequence of the fact that $\Gamma_{g,n}$ can be realized as a subgroup of $\mathrm{Out}(\Pi_{g,n})$ is that it is countable, since $\Pi_{g,n}$ is finitely generated.

The next result that we would like to discuss, originally due to Dehn, asserts that the mapping class group $\Gamma_{g,n}$ is generated by the isotopy classes of Dehn twists, which we introduced in Section 9 of Chapter X. Once this result is established, it is not too difficult to exhibit a finite set of Dehn twists generating the entire mapping class group. A set of generators for $\Gamma_{g,n}$ is provided by the Dehn twists around the colored cycles in Figure 3 below.

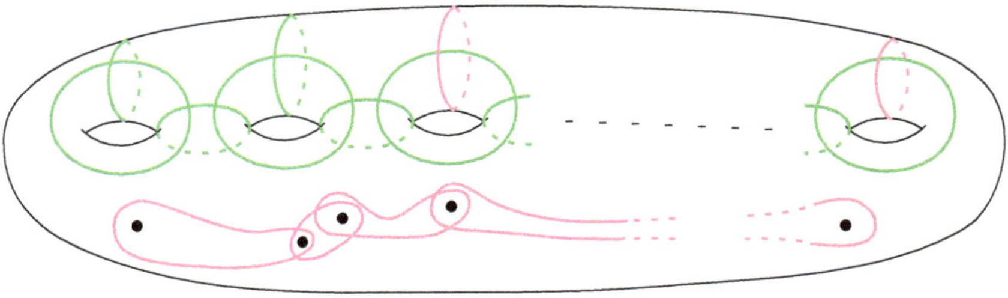

Figure 3.

For $n = 0$, this is due to Lickorish; a proof of Dehn's and Lickorish's theorems can be found in [74]. It has been shown by Humphries that the minimum number of Dehn twists needed to generate Γ_g is $2g + 1$ and that a set of generators is provided by the twists around the $2g + 1$ green cycles in Figure 3. Finally, it has been shown by Gervais that $\Gamma_{g,n}$ actually admits a finite *presentation*.

Dehn's theorem shows that, when Σ has a differentiable structure, every class in $\Gamma_{g,n}$ contains a diffeomorphism. In fact, we can put on Σ a piecewise linear structure by choosing a C^∞ triangulation of Σ. It is then known (cf., for instance, [216], Theorem A1) that any simple closed curve on a surface is ambient isotopic to a piecewise linear one. On the other hand, it is very easy to exhibit an ambient isotopy carrying a piecewise linear simple close curve to a smooth one. Thus, the mapping class group is generated by the Dehn twists associated to smooth simple closed curves, which, as we have seen, are diffeomorphisms. As we have mentioned earlier, it is also the case that isotopic diffeomorphisms are also differentiably isotopic; this, however, will not be needed in the sequel.

The final result we shall discuss concerns the relationship between the mapping class group Γ_g and the symplectic group $Sp_{2g}(\mathbb{Z})$. Let $a_1, \dots, a_g, b_1, \dots, b_g$ be a system of generators for $H_1(\Sigma, \mathbb{Z})$ whose intersection matrix is the standard symplectic matrix Q. We can then identify $Sp(H_1(\Sigma, \mathbb{Z}))$ with $Sp_{2g}(\mathbb{Z})$, and we have a homomorphism

$$\begin{aligned} \chi : \Gamma_g &\longrightarrow Sp_{2g}(\mathbb{Z}) \\ h &\mapsto h_* \end{aligned} \tag{3.2}$$

which we discussed in Section 9 of Chapter X. By the Dehn–Nielsen realization theorem, χ is surjective. Let us denote by $\tilde{\Gamma}_g$ its kernel, which is usually called the *Torelli group*. Set

$$\tilde{M}_g = \mathcal{T}_g / \tilde{\Gamma}_g .$$

In view of the surjectivity of χ, the geometrical interpretation of this moduli space is clear; it classifies isomorphism classes of pairs (C,ρ), where C is a smooth genus g curve, and ρ is a symplectic isomorphism between $H_1(C,\mathbb{Z})$, equipped with its intersection pairing, and the standard symplectic module $(\mathbb{Z}^{2g},Q)$. Exactly the same proof used to show the rigidity of a curve with Teichmüller structure yields the rigidity of the pair (C,ρ). Thus,

$$\mathrm{Aut}(C,\rho)=1. \tag{3.3}$$

As a consequence, the group $\tilde{\Gamma}_g$ acts freely on $\mathcal{T}_g$, so that $\tilde{M}_g$ is again a smooth manifold carrying a universal family of curves over it. Of course, $\tilde{M}_g$ is acted on by $Sp_{2g}(\mathbb{Z})$:

$$\begin{aligned} Sp_{2g}(\mathbb{Z})\times\tilde{M}_g &\to \tilde{M}_g \\ \sigma[C,\rho] &\mapsto [C,\sigma\rho]\,. \end{aligned}$$

Taking the quotient under this action simply forgets the symplectic isomorphism, so that

$$\tilde{M}_g/Sp_{2g}(\mathbb{Z})=M_g\,.$$

At this point it is natural to link the space $\tilde{M}_g$ with the Torelli map. Indeed, the moduli space $\tilde{M}_g$ is the natural domain of the Torelli map

$$\tau:\tilde{M}_g\to\mathbb{H}_g\,.$$

We recall how this is defined. Consider the universal family $\pi:\mathcal{C}\to\tilde{M}_g$. Each fiber C_y over a point y in $\tilde{M}_g$ is equipped with a symplectic isomorphism ρ_y such that $y=[C_y,\rho_y]$. This endows $H_1(C_y,\mathbb{Z})$ with a distinguished symplectic basis $a_1(y),\dots,a_g(y),b_1(y),\dots,b_g(y)$ and therefore with a unique basis $\omega_1(y),\dots,\omega_g(y)$ of holomorphic differential forms such that

$$\int_{a_i(y)}\omega_j(y)=\delta_{i,j}\,.$$

The Torelli map is defined by associating to y the matrix of b-periods of the curve C_y

$$\tau(y)=\left(\int_{b_i(y)}\omega_j(y)\right).$$

The image of this map is what we called the *Jacobian locus* $\mathcal{J}_g\subset\mathbb{H}_g$. Torelli's theorem (cf. Vol. I, Chapter VI, Section 3) tells us that, set-theoretically, $\tilde{M}_g=\mathcal{J}_g$. But while $\tilde{M}_g$ is a smooth variety, $\mathcal{J}_g$ is definitely not. In fact, the local Torelli theorem (Chapter XI, Section 8) or, equivalently, Nœther's theorem, tells us that the singularities of τ and therefore those of $\mathcal{J}_g$ arise exactly along the hyperelliptic locus.

4. Quadratic differentials and Teichmüller deformations.

In the next few sections we will prove the following remarkable result, originally due to Fricke.

THEOREM (4.1). *For any pair of nonnegative integers g and n such that $2g-2+n>0$, the Teichmüller space $\mathcal{T}_{g,n}$ is homeomorphic to the unit ball in $\mathbb{C}^{3g-3+n}$.*

An immediate consequence of this result is the irreducibility of $M_{g,n}$.

COROLLARY (4.2). *$M_{g,n}$ is irreducible for all g and all n such that $2g-2+n>0$.*

The proof of Theorem (4.1) we shall present is based on a fundamental result by Teichmüller which exhibits a *canonical* representative for each isotopy class of orientation-preserving homeomorphisms $f : (C, x_1, \dots, x_n) \to (S, p_1, \dots, p_n)$. This canonical representative is called a Teichmüller map and has the following two remarkable properties:

1) it is a diffeomorphism outside a finite number of points,
2) outside these points it can be locally described, *in a canonical way*, as a *real affine* transformation.

We fix a compact n-pointed connected Riemann surface S of genus g, where $2g-2+n>0$, and we denote by $\mathcal{T}_{g,n}$ the Teichmüller space of $(S, p_1, \dots, p_n)$. Let ω be a quadratic differential on S. If $\omega = f(z)dz^2$ is a local expression for it in terms of a local coordinate z, we define the (singular) volume form associated to ω to be

$$dA_\omega = \frac{\sqrt{-1}}{2}|f|dz \wedge d\bar{z} = |f|dx \wedge dy\,,$$

where x and y stand for the real and imaginary parts of z. One immediately checks that this definition is independent of the choice of local coordinate. Now consider the $(3g-3+n)$-dimensional space $H^0(S, K_S^2(\sum p_i))$ of those quadratic differentials on S which are holomorphic except possibly for simple poles at the marked points $p_1, \dots, p_n$. Notice that the form dA_ω is of class L^1, so that a norm can be introduced on this space by setting

$$\|\omega\| = \int_S dA_\omega$$

for any ω in $H^0(S, K_S^2(\sum p_i))$. Look at the unit ball in $H^0(S, K_S^2(\sum p_i))$,

$$B(K_S^2(\textstyle\sum p_i)) = \{\omega \in H^0(S, K_S^2(\sum p_i)) : \|\omega\| < 1\}\,.$$

We are going to define a map

$$\Phi : B(K_S^2(\textstyle\sum p_i)) \to \mathcal{T}_{g,n} \tag{4.3}$$

and prove the following more precise version of (4.1).

THEOREM (4.4). *Φ is a homeomorphism.*

The construction of Φ is straightforward and quite beautiful. Starting from the reference Riemann surface $(S, p_1, \dots, p_n)$ and a quadratic differential ω on it with norm $\|\omega\| = k < 1$, we want to define a new Riemann surface $(S_\omega, q_1, \dots, q_n)$ together with an orientation-preserving homeomorphism

$$f_\omega : (S_\omega, q_1, \dots, q_n) \to (S, p_1, \dots, p_n)\,. \tag{4.5}$$

We will refer to $(S_\omega, q_1, \dots, q_n)$ and f_ω as the *Teichmüller deformation* of S and the *Teichmüller map* associated to ω. The underlying topological manifold of S_ω will be the same as the one of S, and f_ω will be the identity map, so that in particular $q_i = p_i$ for all i. However, it is important to realize that, for $k \neq 0$, not only the complex structures of S and S_ω will be different, but their differentiable structures as well. In other words, f_ω will not be a diffeomorphism, although of course there will be some other map from $(S_\omega, q_1, \dots, q_n)$ to $(S, p_1, \dots, p_n)$ which is. We shall collectively refer to the zeroes and poles of ω as the *singularities* of ω. Let then Z be the set of all such singularities. We first introduce the new complex structure on $S \setminus Z$. The construction is based on the elementary but fundamental remark that around each point $p \in S \setminus Z$ the quadratic differential ω defines a set of distinguished coordinates, any two of which differ at most by a sign and the addition of a constant. Such a coordinate z is simply defined by the requirement that

$$\omega = (dz)^2 \quad \text{near } p\,.$$

In other terms,

$$z = \int \sqrt{\omega}\,. \tag{4.6}$$

The ambiguity intrinsic in this definition could be decreased by further imposing that z vanish at p, thus fixing it up to sign; in general, however, we will find it more convenient not to do so. We will call the coordinate z an *ω-coordinate* around the point p. Now define a new coordinate patch around p by performing the real linear transformation

$$z' = \frac{z + k\bar{z}}{1 - k}\,. \tag{4.7}$$

Clearly, a different choice of z changes z' by no more than a sign and the addition of a constant. Thus, these coordinate patches define a new holomorphic structure on $S \setminus Z$.

We will now show that this structure extends to one on all of S by constructing explicit coordinate patches around each point of Z; clearly, such an extension will be unique. Let then p be a singularity of ω, let

z be a local coordinate on S centered at p, and let $f(z)dz^2$ be a local expression for ω. We begin with the simpler case where f vanishes to even order $2n$ at p, so that it has a single-valued holomorphic square root $g(z)$ near p. We may then write $g(z)dz = dh$, where $h(z)$ is holomorphic and vanishes to order $n+1$ at p. Thus, replacing z with an $(n+1)$st root of h, we may suppose that $\omega = d(z^{n+1})^2$. It follows that z^{n+1} is an ω-coordinate at all points of S sufficiently close to p, but different from it, and hence, by virtue of (4.7), that

$$\xi(z) = \frac{z^{n+1} + k\bar{z}^{n+1}}{1-k}$$

is a local coordinate for the new complex structure on $S \setminus Z$ at all these points. Since $k < 1$, the function $1 + k\frac{\bar{z}^{n+1}}{z^{n+1}}$ takes its values in the half-plane of complex numbers with positive real part, where a single-valued determination of the $(n+1)$st root function exists. Set

$$\eta(z) = z\left(\frac{1 + k\frac{\bar{z}^{n+1}}{z^{n+1}}}{1-k}\right)^{\frac{1}{n+1}}$$

for $z \neq 0$ and $\eta(0) = 0$. The function η is continuous at p since $1+k\frac{\bar{z}^{n+1}}{z^{n+1}}$ is bounded. It is C^∞ away from p but not differentiable at p if $k \neq 0$. We will now show that η is a homeomorphism between an open neighborhood of p and an open neighborhood of the origin in the complex plane, and we will take it as our new coordinate around p. This is compatible with the new complex structure on $S \setminus Z$ since $\eta^{n+1} = \xi$. To prove our claim, notice first that η is injective. Suppose in fact that $\eta(a) = \eta(b)$. Then $\xi(a) = \xi(b)$, and hence $a^{n+1} = b^{n+1}$. It follows that

$$1 + k\frac{\bar{a}^{n+1}}{a^{n+1}} = 1 + k\frac{\bar{b}^{n+1}}{b^{n+1}}$$

and hence, by the definition of η, that $a = b$. To finish the proof of the claim, we must show that η is an open map. This follows immediately from the "invariance of domain" theorem or can be proved by the following direct argument. Since η is an immersion at all points different from p, it is enough to show that the image via η of any neighborhood of p contains a disc centered at the origin of $\mathbb{C}$. Observe that η, as a function of two real variables, is homogeneous of degree one. Since it is injective, the induced map from the S^1 of real lines through the origin to itself is also injective and hence onto. As $|\eta(z)| \geq |z|$ for all z, the claim follows.

We next deal with the case where f vanishes at p to odd order m or has a simple pole at p, in which case we set $m = -1$. Denote by q the map $\zeta \mapsto \zeta^2 = z$. The pulled-back differential $q^*\omega$ vanishes to order

$2m+2$ at the origin and hence is of the form $(dh)^2$, where $h(\zeta)$ is a suitable holomorphic function. In fact, a closer examination shows that h is of the form $\zeta^{m+2}u(z)$, where u does not vanish at the origin. Thus, replacing ζ with $\zeta v(\zeta^2)$ and z with $zv(z)^2$, where v is an $(m+2)$-nd root of u, we may suppose that $q^*\omega = d(\zeta^{m+2})^2$. This means that ζ^{m+2} is an ω-coordinate at all points of S close to p, but different from it. Thus,

$$\xi = \frac{\zeta^{m+2} + k\bar{\zeta}^{m+2}}{1-k}$$

is a local coordinate for the new complex structure on $S \setminus Z$ at these points. We then take as new coordinate at p the function $\eta(z)$ whose value is

$$\eta(z) = z\left(\frac{1 + k\frac{|z|^{m+2}}{z^{m+2}}}{1-k}\right)^{\frac{2}{m+2}} = \zeta^2\left(\frac{1 + k\frac{\bar{\zeta}^{m+2}}{\zeta^{m+2}}}{1-k}\right)^{\frac{2}{m+2}}$$

for $z \neq 0$ and zero for $z = 0$. The proof that this extends the new complex structure on $S \setminus Z$ is just as in the case where f vanishes to even order at p, once we observe that $\eta^{m+2} = \xi^2$. This finishes the construction of the complex structure of S_ω.

It is interesting to notice that there is a *canonically defined* quadratic differential ω' on S_ω. Away from the singularities of ω, this is simply the differential whose local expression is $(dz')^2$, where z' is as defined in (4.7). It is a trivial check, that we leave to the reader, to verify that ω' extends to a meromorphic differential on all of S_ω. In fact, one easily sees that the singularities of ω' are exactly those of ω, in the sense that the zeroes and poles of ω', together with their multiplicities, are the same as those of ω.

Now that the construction of S_ω is complete, we may define the map $\Phi : B(K_S^2(\sum p_i)) \to \mathcal{T}_{g,n}$ by setting

$$\Phi(\omega) = [S_\omega, p_1, \dots, p_n, [f_\omega]], \tag{4.8}$$

where $f_\omega : (S_\omega, p_1, \dots, p_n) \to (S, p_1, \dots, p_n)$ is the set-theoretic identity. As we already mentioned, the homeomorphism f_ω is usually called the Teichmüller map associated to ω. The price one is paying for the simplicity of the local representation (4.7) of the complex structure of S_ω is the failure of f_ω to be differentiable at the zeroes of ω.

It is crucial to observe that our construction of $(S_\omega, p_1, \dots, p_n)$ and f_ω depends continuously on $\omega \in B(K_S^2(\sum p_i))$. In fact, if we denote by $\mathcal{S}$ the disjoint union of all the S_ω and by f the map from $\mathcal{S}$ to $S \times B(K_S^2(\sum p_i))$ which sends $x \in S_\omega$ to $(f_\omega(x), \omega)$, what we just did was to put on $\mathcal{S} \to B(K_S^2(\sum p_i))$, together with the constant sections

$\{p_i\} \times B(K_S^2(\sum p_i))$, $i = 1, \ldots, n$, a structure of continuous family of compact n-pointed genus g Riemann surfaces for which f is a topological trivialization. The map f thus endows the family $\mathcal{S} \to B(K_S^2(\sum p_i))$, together with its distinguished sections, with a Teichmüller structure. By the universal property of the Teichmüller space (cf. Proposition (2.8)), the map Φ is therefore *continuous*. Notice that the continuity of Φ is a direct consequence of our constructing the Teichmüller space by glueing bases of Kuranishi families.

Our plan is to show how the proof of Theorem (4.4) directly reduces to the proof of two fundamental results, namely Teichmüller's uniqueness theorem and the theorem of existence of solutions of the Beltrami equation.

In the proof of Theorem (4.4), one of the central notions is that of Beltrami differential. On a connected Riemann surface S, we consider measurable sections of $T_S \otimes \overline{K}_S$, where T_S and K_S are, respectively, the complex tangent and cotangent bundles to S. These are vector-valued differentials on S which are locally of the form

$$\mu = \nu \frac{\partial}{\partial z} \otimes d\overline{z},$$

where z is a local coordinate, and ν is a measurable function. It makes sense to define a measurable function $|\mu|$ by setting it locally equal to $|\nu|$, since the latter is clearly independent of the choice of coordinate z. A *Beltrami differential* on the Riemann surface S is an L^∞ section of $T_S \otimes \overline{K}_S$ whose norm $\|\mu\| = \sup_S |\mu|$ is strictly less than 1. We shall really need only a particular kind of Beltrami differentials, namely those which are C^∞ everywhere, except at a finite number of points. These differentials will be called *admissible*. To a Beltrami differential one associates a perturbed version of the $\overline{\partial}$ operator on S by setting

$$\overline{\partial}_\mu = \overline{\partial} - \mu\,,$$

where μ acts on a function f as

$$\mu(f) = \nu \frac{\partial f}{\partial z} d\overline{z}\,.$$

The corresponding *Beltrami equation* is

$$\overline{\partial}_\mu u = 0\,,$$

that is,

$$u_{\overline{z}} = \nu(z) u_z\,. \tag{4.9}$$

The basic existence theorem for the Beltrami equation asserts that it has local solutions, in an appropriate generalized sense, which are homeomorphisms to open subsets of the complex plane. In the bibliographical note we will give references for the various versions of this result. Formally, the existence theorem we need is the following.

THEOREM (4.10). *Let $\nu(z, t_1, \ldots, t_n)$ be a C^∞ function on a neighborhood of the origin in $\mathbb{C} \times \mathbb{R}^n$. Suppose that $|\nu(z,t)| < 1$ for all values of z and $t = (t_1, \ldots, t_n)$. Then there exists a C^∞ function $w(z,t)$, also defined on a neighborhood of the origin, such that*

$$w_{\overline{z}} = \nu(z,t) w_z \,, \quad w_z(0,0) \neq 0 \,.$$

We shall also need the following uniqueness result.

LEMMA (4.11). *Let $\nu(z)$ be a function on a neighborhood of the origin in the complex plane which is C^∞ except at a finite set Z. Suppose that $|\nu(z)| < 1$ for all z. Let u be a homeomorphism from a neighborhood of the origin to an open subset of $\mathbb{C}$ which solves the Beltrami equation (4.9) away from Z. Let f be a bounded function on a neighborhood of the origin which is once differentiable away from Z. Then f satisfies (4.9) away from Z if and only if it is a holomorphic function of u.*

Proof. Suppose first that Z is empty and let w be the solution of the Beltrami equation provided by Theorem (4.10). A simple chain rule computation gives

$$f_{\overline{z}} - \nu f_z = (1 - |\nu|^2) \cdot f_{\overline{w}} \cdot \overline{w_z} \,.$$

Thus, f is a holomorphic function of w near the origin if and only if it is a solution of the Beltrami equation. We may apply this to the function u. Moreover, since u has an inverse, we also get that w is a holomorphic function of u. This proves the lemma when Z is empty. In the general case the above argument shows that f solves the Beltrami equation away from Z if and only if it is a holomorphic function of u there. The holomorphicity of f at points of Z follows from the Riemann extension theorem. Q.E.D.

Theorem (4.10), in its parameterless version, that is, for $n = 0$, and Lemma (4.11) say, in particular, that a C^∞ Beltrami differential μ on a Riemann surface S defines on S a new complex structure whose holomorphic functions are the solutions of the corresponding Beltrami equation. The reader should be warned that $\overline{\partial}_\mu$ is *not* the $\overline{\partial}$ operator of this complex structure, but just its component of type $(0,1)$ (with respect to the original structure). A more precise, though cumbersome, notation for $\overline{\partial}_\mu$ could thus be $\overline{\partial}_\mu^{(0,1)}$.

In the bibliographical notes we will give references for the proof of Theorem (4.10). One cheap way to prove it is to appeal to Theorem (7.7) of Chapter XI. Here is how the argument goes. Since the problem is of a local nature, we may alter ν outside a neighborhood of the origin. Hence we may assume that ν is defined and C^∞ on $\mathbb{C} \times U$, where U is a neighborhood of the origin in $\mathbb{R}^n$, and that there is a positive r

such that ν vanishes for $|z| > r$. Thus, we may view $\mu = \nu\,\partial/\partial z \otimes d\overline{z}$ as a family $\{\mu_t\}_{t\in U}$ of Beltrami differentials on $\mathbb{P}^1$, vanishing outside the disk of radius r centered at 0. Let H be the hyperplane bundle on $\mathbb{P}^1$; its smooth sections can be viewed as C^∞ functions f on $\mathbb{C}$ such that f/z extends in a C^∞ way across ∞. Since $\overline{\partial}_{\mu_t}$ is the standard $\overline{\partial}$ operator in a neighborhood of ∞, for any smooth section u of H, $\overline{\partial}_{\mu_t}u$ is a smooth H-valued $(0,1)$-form. Let ϑ_{μ_t} be the formal adjoint of $\overline{\partial}_{\mu_t}$ with respect to (say) the Fubini–Study metric. Then the differential operator $L_t = \vartheta_{\mu_t}\overline{\partial}_{\mu_t}$ is self-adjoint and strongly elliptic. If u is a section of H such that $L_t u = 0$, then $(\overline{\partial}_{\mu_t}u, \overline{\partial}_{\mu_t}u) = (u, L_t u) = 0$, and hence $\overline{\partial}_{\mu_t}u = 0$. In other words, the L_t-harmonic sections of H are just those sections which are holomorphic with respect to the complex structure defined by μ_t. The space of these sections has dimension 2 for any t, by the Riemann–Roch theorem, so that Theorem (7.7) of Chapter XI applies. Pick a μ_0-holomorphic section u which vanishes simply at 0 and set $w_t = F_t u$, where F_t is the harmonic projector associated to L_t. Then $\overline{\partial}_{\mu_t}w_t = 0$, and w_t depends differentiably on t, by Theorem (7.7) of Chapter XI. In other words, $w(z,t) = w_t(z)$ has all the required properties.

Admissible Beltrami differentials originate, in particular, from the so-called admissible quasi-diffeomorphisms. An orientation-preserving homeomorphism $F : S \to S'$ between two compact Riemann surfaces which is a diffeomorphism outside a finite set $Z \subset S$ is called a *quasi-diffeomorphism*. Pick holomorphic coordinates z and w around $p \in S \smallsetminus Z$ and $F(p) \in S'$, respectively. The condition that F be an orientation-preserving diffeomorphism on $S \smallsetminus Z$ tells us that, on $S \smallsetminus Z$, the Jacobian determinant of F is positive. Since, in local coordinates, the Jacobian is

$$|w_z|^2 - |w_{\overline{z}}|^2\,,$$

the local function

$$\nu(z) = \frac{w_{\overline{z}}}{w_z}$$

is C^∞ away from Z and of absolute value less than 1. It is a straightforward application of the chain rule to check that setting

$$\mu_F = \nu\frac{\partial}{\partial z} \otimes d\overline{z}$$

gives a well-defined section of $T_S \otimes \overline{K}_S$ which is C^∞ away from Z. The quasi-diffeomorphism F is said to be *admissible* if $\|\mu_F\| < 1$, i.e., if μ_F is a Beltrami differential. By Lemma (4.11), the complex structure of S' can be completely described in terms of the one of S and of the differential μ_F; a bounded function f on an open subset of S' is holomorphic if and only if $u = f \circ F$ is a solution of the Beltrami equation $\overline{\partial}_{\mu_F}u = 0$ away from Z.

LEMMA (4.12). *Let $F : S \to S'$ be a quasi-diffeomorphism. Then*

$$\|\mu_F\| = \|\mu_{F^{-1}}\|.$$

If $F' : S \to S''$ is another quasi-diffeomorphism, then $\mu_{F'} = \mu_F$ if and only if $F' \circ F^{-1} : S' \to S''$ is holomorphic.

The second assertion of the lemma follows immediately from Lemma (4.11), while another elementary chain rule computation shows that $|\mu_F(p)| = |\mu_{F^{-1}}(F(p))|$ for any $p \in S$, thus proving the first assertion.

It is convenient to introduce the concept of *dilatation* for a quasi-diffeomorphism $F : S \to S'$. This is simply defined to be

$$K[F] = \frac{1 + \|\mu_F\|}{1 - \|\mu_F\|}.$$

It follows from Lemma (4.12) that a quasi-diffeomorphism and its inverse have the same dilatation. It is also clear that F is admissible if and only if $K[F] < \infty$. The dilatation of the Teichmüller map f_ω associated to a quadratic differential $\omega \in B(K_S^2(\sum p_i))$ on S, as defined in (4.5), is just

$$K[f_\omega] = \frac{1+k}{1-k}, \quad \text{where} \quad k = \|\omega\|.$$

Also observe that if one sets $z = x + \sqrt{-1}y$ in (4.7), then

$$z' = Kx + \sqrt{-1}y, \quad \text{where} \quad K = \frac{1+k}{1-k}, \tag{4.13}$$

which justifies the word "dilatation."

We now turn to Theorem (4.4). Let S be a reference n-pointed, genus g Riemann surface with $2g - 2 + n > 0$. As we have announced, we shall rely on a fundamental result of Teichmüller, the so-called Teichmüller uniqueness theorem. This remarkable theorem asserts that in each isotopy class of orientation-preserving homeomorphisms $f : (C, x_1, \dots x_n) \to (S, p_1, \dots p_n)$, there is a *canonical* representative which is a Teichmüller map, that is, a map of the form f_ω as defined in (4.5).

The uniqueness theorem of Teichmüller asserts that, among all admissible quasi-diffeomorphisms isotopic to it, the Teichmüller map $f_\omega : (S_\omega, p_1, \dots, p_n) \to (S, p_1, \dots, p_n)$ is one with *minimal dilatation* and that it is uniquely characterized by this property.

THEOREM (4.14) (TEICHMÜLLER'S UNIQUENESS THEOREM). *Let $(S, p_1, \dots, p_n)$ be an n-pointed genus g Riemann surface with $2g-2+n > 0$, and let ω be a meromorphic quadratic differential on S which is holomorphic everywhere except possibly for simple poles at $p_1, \dots, p_n$. Assume that $\|\omega\| < 1$. Let $f : (S_\omega, p_1, \dots, p_n) \to (S, p_1, \dots, p_n)$ be an admissible quasi-diffeomorphism which is isotopic to f_ω (i.e., isotopic to the identity) relative to $p_1, \dots, p_n$. Then*

$$K[f] \geq K[f_\omega] = \frac{1 + \|\omega\|}{1 - \|\omega\|},$$

and equality holds if and only if $f = f_\omega$.

We will prove this theorem in Section 6. Assuming Teichmüller's uniqueness theorem, we are now going to prove that the map $\Phi : B(K_S^2(\sum p_i)) \to \mathcal{T}_{g,n}$ defined by (4.8) is *injective*. Assume that $\Phi(\omega_1) = \Phi(\omega_2)$. This means that there are an isomorphism $\varphi : (S_{\omega_1}, p_1, \dots, p_n) \to (S_{\omega_2}, p_1, \dots, p_n)$ and an isotopy $f_{\omega_1} \sim f_{\omega_2} \circ \varphi$ relative to $p_1, \dots, p_n$. Set $k_i = \|\omega_i\|$, $i = 1, 2$. Since φ is holomorphic, the first part of Teichmüller's uniqueness theorem, together with Lemma (4.12), tells us that

$$\frac{1+k_2}{1-k_2} = K[f_{\omega_2}] = K[f_{\omega_2} \circ \varphi] \geq K[f_{\omega_1}] = \frac{1+k_1}{1-k_1}\,.$$

Applying the same argument to the isotopy $f_{\omega_2} \sim f_{\omega_1} \circ \varphi^{-1}$, we obtain $k_1 = k_2$, and the second part of Teichmüller's uniqueness theorem then implies that

$$f_{\omega_1} = f_{\omega_2} \circ \varphi\,.$$

Let z_i be an ω_i-coordinate on S for $i = 1, 2$. Applying Lemma (4.12) to the composition $\varphi^{-1} \circ f_{\omega_2}^{-1} = f_{\omega_1}^{-1}$, we get

$$k_1 \frac{\partial}{\partial z_1} \otimes d\bar{z}_1 = \mu_{f_{\omega_1}^{-1}} = \mu_{f_{\omega_2}^{-1}} = k_2 \frac{\partial}{\partial z_2} \otimes d\bar{z}_2\,.$$

Since $k_1 = k_2$, this shows that

$$\overline{\frac{\partial z_2}{\partial z_1}} = \frac{\partial z_2}{\partial z_1}\,.$$

But z_2 is a holomorphic function of z_1. This means that $z_2 = c \cdot z_1$ for some real constant c. Hence $\omega_2 = (dz_2)^2 = c^2 \cdot (dz_1)^2 = c^2 \cdot \omega_1$. Since $\|\omega_1\| = k_1 = k_2 = \|\omega_2\|$, we get $c^2 = 1$, proving the injectivity of Φ.

We are now going to conclude the proof of Theorem (4.4) and hence of Theorem (4.1). The first step is the following.

PROPOSITION (4.15). *The Teichmüller map Φ is closed.*

Proof. Let $\{\omega_n\}$ be a sequence in $B(K_S^2(\sum p_i))$. Suppose that the sequence $y_n = \Phi(\omega_n)$ converges to $y \in \mathcal{T}_{g,n}$. We must prove that a subsequence of $\{\omega_n\}$ converges in $B(K_S^2(\sum p_i))$. Let $y = [C, x_1, \dots, x_n, [f]]$, where f is a diffeomorphism. As a neighborhood B of y in $\mathcal{T}_{g,n}$ we may take the basis of a Kuranishi family for $(C, x_1, \dots, x_n)$. This consists of a family of curves $\pi : \mathcal{C} \to B$ together with disjoint sections $\sigma_1, \dots, \sigma_n$, plus an identification between $(C, x_1, \dots, x_n)$ and $(\pi^{-1}(y), \sigma_1(y), \dots, \sigma_n(y))$. We can assume that there is a C^∞ trivialization $(F, \pi) : \mathcal{C} \to S \times B$ carrying each section σ_i to the constant section $\{p_i\} \times B$ and such that $F_{|\pi^{-1}(y)} = f$. We may further assume that $\{y_n\} \subset B$. Write $f_n = F_{|\pi^{-1}(y_n)}$. Set $\mu = \mu_f$ and $\mu_n = \mu_{f_n}$. Since f and f_n are C^∞, the Beltrami differentials μ and μ_n are also C^∞; since F is C^∞, $\{\mu_n\}$

converges uniformly to μ. Using Teichmüller's uniqueness theorem, we get

$$\begin{aligned}\frac{1+\|\mu\|}{1-\|\mu\|} &= \lim_{n\to\infty}\left(\frac{1+\|\mu_n\|}{1-\|\mu_n\|}\right) = \lim_{n\to\infty} K[f_n] \\ &\geq \lim_{n\to\infty} K[f_{\omega_n}] = \lim_{n\to\infty}\left(\frac{1+\|\omega_n\|}{1-\|\omega_n\|}\right).\end{aligned}$$

This means that, given a constant c such that $\|\mu\| < c < 1$, we have $\|\omega_n\| < c$ for large enough n. Hence a subsequence of $\{\omega_n\}$ converges.

We now come to the last ingredient of the proof of Theorem (4.4).

PROPOSITION (4.16). *The Teichmüller space $\mathcal{T}_{g,n}$ is connected.*

We can immediately see that this implies Theorem (4.4). In fact, Proposition (4.15), in addition to showing that $\Phi(B(K_S^2(\sum p_i)))$ is closed in $\mathcal{T}_{g,n}$, also shows that Φ gives a homeomorphism between $B(K_S^2(\sum p_i))$ and $\Phi(B(K_S^2(\sum p_i)))$. But then the "invariance of domain" theorem says that $\Phi(B(K_S^2(\sum p_i)))$ is open in $\mathcal{T}_{g,n}$, since $B(K_S^2(\sum p_i))$ and $\mathcal{T}_{g,n}$ are differentiable manifolds of dimension $6g-6+2n$.

We first prove Proposition (4.16) in the case $n=0$. An immediate consequence of Theorem (4.10) and Lemma (4.11) is the following.

LEMMA (4.17). *Let μ_t be a family of smooth Beltrami differentials on S, where t varies in an interval $I \subset \mathbb{R}$. Suppose that μ_t depends smoothly on t in the sense that it is C^∞ on $S \times I$ Then there are a differentiable family $\xi : Y \to I$ of Riemann surfaces and a differentiable trivialization $S \times I \to Y$ such that the Beltrami differential μ_{F_t} associated to F_t is μ_t.*

To prove Proposition (4.16), denote by x_0 the base point $[S,[1]]$ of $\mathcal{T}_S = \mathcal{T}_g$, let $x = [C,[f]]$ be another element of $\mathcal{T}_g$, where $f : C \to S$ is a diffeomorphism, and set $\mu_t = t\mu_{f^{-1}}$. By Lemma (4.17), there is a differentiable family of curves with Teichmüller structure over an interval I whose fiber at t is $[Y_t,[F_t^{-1}]]$. This comes from a differentiable map $\gamma : I \to \mathcal{T}_g$ by the universal property of Teichmüller space. To prove connectedness, we just have to show that $\gamma(0) = x_0$ and $\gamma(1) = x$; in other words, that $[Y_0,[F_0^{-1}]] = [S,[\mathrm{id}]]$ and $[Y_1,[F_1^{-1}]] = [C,[f]]$. Since, by construction, $\mu_{F_0^{-1}} = 0$ and $\mu_{F_1^{-1}} = \mu_{f^{-1}}$, this follows from the second part of Lemma (4.12). This proves (4.16) for $n = 0$.

To show that $\mathcal{T}_{g,n}$ is always connected, we argue by induction on n. The initial step has just been completed. For the induction step, look at the map

$$\eta : \mathcal{T}_{g,n} \to \mathcal{T}_{g,n-1} \tag{4.18}$$

given by $[C,x_1,\ldots,x_n,[f]] \mapsto [C,x_1,\ldots,x_{n-1},[f]]$, where of course in the left-hand side $[f]$ stands for the isotopy class of f relative to $x_1,\ldots,x_n$, while in the right-hand side it stands for the isotopy class of f relative to $x_1, \ldots, x_{n-1}$. Clearly, η is onto, while the local description of $\mathcal{T}_{g,n}$ as the base of a Kuranishi family implies that η is holomorphic and that its differential has maximum rank everywhere. To show that $\mathcal{T}_{g,n}$ is connected, assuming that this is known for $\mathcal{T}_{g,n-1}$, it suffices to show that the fibers of η are connected. This is easy. Consider two points of the same fiber of η. These are of the form $[C, x_1, \ldots, x_n, [f]]$ and $[C, x_1, \ldots, x_{n-1}, y_n, [f']]$, where f and f' are diffeomorphisms from C to S which are isotopic relative to $x_1, \ldots, x_{n-1}$. Pick an isotopy $\{F_t\}$ such that $F_0 = f$, $F_1 = f'$, and $F_t(x_i) = p_i$ for all t and for $i = 1, \ldots, n-1$. Then $\{[C, x_1, \ldots, x_{n-1}, F_t^{-1}(p_n), [F_t]]\}_{t \in [0,1]}$ is a continuous family of n-pointed curves with Teichmüller structure contained in the fiber $\eta^{-1}[C, x_1, \ldots, x_{n-1}, [f]]$ and joining $[C, x_1, \ldots, x_n, [f]]$ to $[C, x_1, \ldots, x_{n-1}, y_n, [f']]$. This concludes the proof of (4.16) and hence of Theorem (4.4).

We close by recording an amusing consequence of (4.4). The morphism (4.18) fits into a diagram

$$\begin{array}{ccc} \mathcal{X}_{g,n} & \xrightarrow{\ \xi\ } & \mathcal{X}_{g,n-1} \\ \big\downarrow{\scriptstyle \pi} & & \big\downarrow{\scriptstyle \pi'} \\ \mathcal{T}_{g,n} & \xrightarrow{\ \eta\ } & \mathcal{T}_{g,n-1} \end{array}$$

where $\pi : \mathcal{X}_{g,n} \to \mathcal{T}_{g,n}$ and $\pi' : \mathcal{X}_{g,n-1} \to \mathcal{T}_{g,n-1}$ are the universal families over the respective Teichmüller spaces. Denote by $\sigma_1, \ldots, \sigma_n$ and $\sigma'_1, \ldots, \sigma'_{n-1}$ the distinguished sections of π and π'. We know that, for any n, the universal family over $\mathcal{T}_{g,n}$ is topologically trivial, meaning that it is homeomorphic, as a fiber space over $\mathcal{T}_{g,n}$, to the product family $\mathcal{T}_{g,n} \times (S, p_1, \ldots, p_n)$. Composing σ_n with ξ yields a morphism $\vartheta : \mathcal{T}_{g,n} \to \mathcal{X}_{g,n-1}$ which can be informally described by

$$[C, x_1, \ldots, x_n, [f]] \mapsto x_n \in C \smallsetminus \{x_1, \ldots, x_{n-1}\}.$$

This morphism maps $\mathcal{T}_{g,n}$ onto the complement, inside $\mathcal{X}_{g,n-1}$, of the distinguished sections, which we denote by U. Clearly, when viewed as a morphism from $\mathcal{T}_{g,n}$ to U, ϑ is a topological covering. Furthermore, by what has been observed, U is topologically just $\mathcal{T}_{g,n-1} \times (S \smallsetminus \{p_1, \ldots, p_{n-1}\})$. Therefore $\mathcal{T}_{g,n}$ is the universal covering of U, and, since $\pi' \circ \vartheta = \eta$, *the fiber of (4.18) above* $[C, x_1, \ldots, x_{n-1}, [f]]$ *is the universal covering of* $C \smallsetminus \{x_1, \ldots, x_{n-1}\}$.

5. The geometry associated to a quadratic differential.

In this section we shall study some elementary aspects of the geometry attached to the (singular) metric defined on a Riemann surface by a nonzero holomorphic quadratic differential.

Consider a Riemann surface S (not necessarily compact) and a nonzero holomorphic quadratic differential $\omega \in H^0(S, K_S^2)$. Let Z be the set of zeroes of ω. On $S_0 = S \smallsetminus Z$ we introduce a hermitian metric, the so-called *ω-metric*. This is defined to be the metric with local expression

$$|f| dz\, d\bar{z}\,,$$

where $f dz^2$ is a local expression for ω. Notice that the volume form associated to the ω-metric is nothing but the form dA_ω we already encountered in Section 4; the latter is also equal to minus the imaginary part of the ω-metric. Notice also that the ω-metric is flat, since in ω-coordinates, that is, coordinates ζ such that $\omega = (d\zeta)^2$, its local expression is just $d\zeta\, d\bar{\zeta}$. In other words, $S \smallsetminus Z$, equipped with the ω-metric, looks locally like the euclidean plane. The length of a path $\gamma : [0,1] \to S_0$ with respect to the ω-metric will be called the *ω-length* of γ and denoted $l_\omega(\gamma)$. Since Z is a set of isolated points and since ω is regular at points of Z, it makes perfect sense to talk about the ω-length of a path in S and not only in S_0. The area of S in the ω-metric is

$$A_\omega(S) = \|\omega\| = \int_S dA_\omega\,.$$

Geodesics for the ω-metric will be called *ω-geodesics* when it is important to keep track of the differential ω; when no confusion is likely, we shall often refer to them simply as geodesics. Clearly, a curve is an ω-geodesic if and only if, at each of its points, it is a straight line in ω-coordinates. A more intrinsic way to express the condition that a curve α be an ω-geodesic is to say that

$$\arg(\omega)_{|\alpha} = \text{constant},$$

where, by definition, we set $\arg(\omega)_{|\alpha} = \arg\left(f(\alpha(t)) \cdot \alpha'(t)^2\right)$ if $f(z)dz^2$ is a local expression for ω. An ω-geodesic is said to be *horizontal* (resp., *vertical*) if

$$\arg(\omega)_{|\alpha} = 0\,, \qquad \left(\text{resp., } \arg(\omega)_{|\alpha} = \pi\right). \tag{5.1}$$

Hence, in terms of an ω-coordinate $\zeta = \xi + \sqrt{-1}\eta$, the horizontal (resp. vertical) geodesics are the curves $\eta = $ constant (resp., $\xi = $ constant). Clearly, a geodesic arc α in S_0 *locally minimizes distances* in the sense that, given any pair of points q and r on it which are sufficiently close to each other, the portion of α between q and r is an arc of minimal ω-length among all arcs joining q to r, and it is the unique one having this property. In fact, for arcs entirely contained in S_0, the local minimizing property characterizes geodesics. Since it makes sense to talk about the length of an arc in S (and not only in S_0), we will define an *ω-geodesic in*

S to be a path in S having the property of locally minimizing distances. Geodesics passing through a zero of ω will be called *singular.* We will show that, on a compact Riemann surface of genus $g > 1$, any two points can be joined by an ω-geodesic and that such a geodesic is unique within its homotopy class. Before proving this, we will study geodesics in the neighborhood of a zero of ω. Let p be such a zero, and let n be its order. Near p we can write, in a suitable coordinate ζ,

$$\omega = \left(d\zeta^{\frac{n+2}{2}}\right)^2 .$$

The multivalued map $w = \zeta^{\frac{n+2}{2}}$ can be described as the composition of the two maps $\zeta \mapsto z = \zeta^{n+2}$ and $z \mapsto w = \sqrt{z}$. The horizontal and vertical geodesics in the w-plane are depicted in Figure 4*a)*. Going back to the z-plane, we get the pattern of horizontal and vertical geodesics illustrated in Figure 4*b)*.

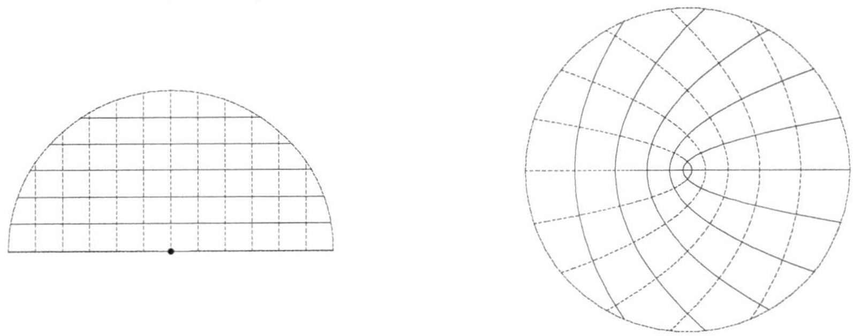

a) *b)*

Figure 4.

The horizontal and vertical geodesics in the ζ-plane are then as depicted in Figure 5 for low values of n.

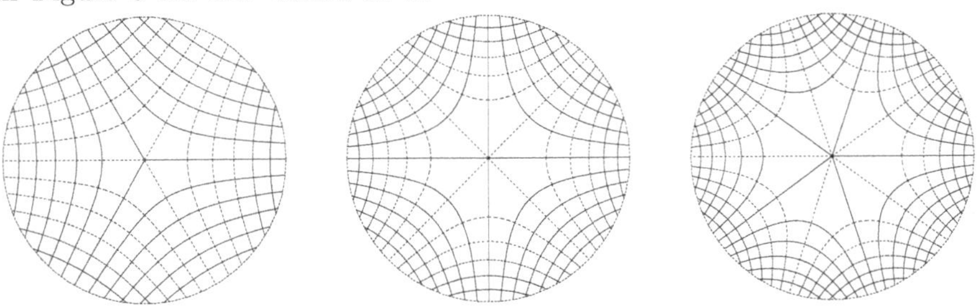

$n = 1$ $n = 2$ $n = 3$

Figure 5.

Notice that, at each zero of order n of ω, there are exactly $n + 2$ horizontal and $n + 2$ vertical singular ω-geodesic rays. Incidentally, we may also observe that, if ω is allowed to have poles, then the horizontal and vertical geodesics near a simple pole are as in Figure 4*b*).

For any real value of ϑ, the map $\zeta \mapsto w = \zeta^{\frac{n+2}{2}}$ defines biholomorphisms between the half-plane

$$\vartheta - \frac{\pi}{2} < \arg(w) < \vartheta + \frac{\pi}{2} \tag{5.2}$$

in the w-plane and each one of the sectors

$$\frac{2}{n+2}\vartheta - \frac{1-2j}{n+2}\pi < \arg(\zeta) < \frac{2}{n+2}\vartheta + \frac{1+2j}{n+2}\pi\,, \qquad j = 0, \ldots, n+1\,, \tag{5.3}$$

in the ζ-plane.

This has two important consequences. The first is that two points in the ζ-plane whose arguments differ by less than $\frac{2\pi}{n+2}$ can be joined by a unique geodesic arc not passing through the origin. In fact, for a suitable choice of ϑ, they lie in one of the sectors (5.3), and the corresponding points in the half-plane (5.2) are joined by a unique geodesic (a straight line). A corollary is that two rays in the ζ-plane forming an angle of less than $\frac{2\pi}{n+2}$ do not constitute a geodesic. In fact, any point of one ray is joined to any point of the other by a geodesic arc not passing through the origin; this is strictly shorter than the path going from the first point to the origin along one ray and then from the origin to the second point along the other ray, as is obvious if one compares the corresponding paths in the half-plane (5.2) (see Figure 6).

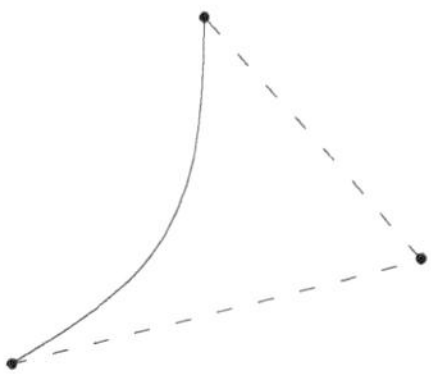

In the ζ-plane

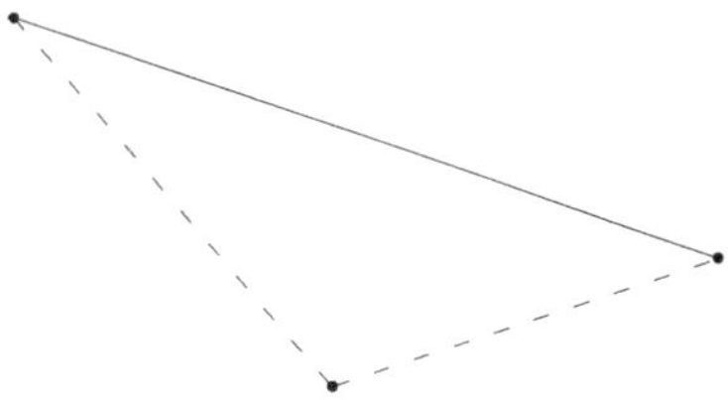

In the w-plane

Figure 6.

The second consequence is that any nonsingular geodesic is confined to the interior of a sector of angular width $\frac{2\pi}{n+2}$. Thus, pairs of rays forming angles not smaller than $\frac{2\pi}{n+2}$ constitute a geodesic.

These facts, taken together, yield a complete description of the ω-geodesics in the ζ-plane. These are either nonsingular, that is, connected components of preimages under the map w of straight lines not passing through the origin, or pairs of rays forming angles not smaller than $\frac{2\pi}{n+2}$. Furthermore, any two points in the ζ-plane are joined by a unique geodesic arc. The local analysis we have just carried out implies that, in several important respects, the ω-metric and the ω-geodesics on S

behave just like an ordinary Riemannian metric and its geodesics. For instance, if, given any two points p and q of S, we define $\rho_\omega(p,q)$ to be the infimum of all the ω-lengths of paths joining p and q, then ρ_ω is a distance function on S, and the topology it defines coincides with the manifold topology of S. Equally important, every point $p \in S$ has arbitrarily small open neighborhoods U which are *convex*, meaning that any two points of U are joined by one and only one geodesic arc entirely contained in U. At points of S_0 this is simply a consequence of the fact that S_0 looks locally like the Euclidean plane, while at points of Z it follows immediately from the local description of geodesics near a zero of ω. We can be even more precise. The *ω-geodesic balls*, or *geodesic balls* for short, are defined to be the balls for the ρ_ω distance on S. It is then clear that, for any point p of S, any sufficiently small geodesic ball centered at p is convex and biholomorphic to a disk.

We are now ready to prove the following theorem.

THEOREM (5.4). *Let S be a compact Riemann surface of genus $g > 1$. Let ω be a nonzero holomorphic quadratic differential on S. Then in each homotopy class with fixed endpoints of paths on S, there exists a unique ω-geodesic.*

First a few comments on the statement and the strategy of proof. It is a classical result in Riemannian geometry that the conclusion holds for geodesics on negatively curved Riemannian manifolds of dimension two. The analogy is further strengthened by the observation that, although the ω-metric is flat away from the zeroes of ω, it has in some sense infinitely negative curvature at these; for instance, this is the conclusion one is forced to if one *defines* curvature via the Gauss–Bonnet identity. Likewise, the proof we shall give is just a slight adaptation of one of the classical proofs for the Riemannian case. In particular, uniqueness will be shown by a Gauss–Bonnet-type argument, using the residue theorem in the form of the argument principle.

Now for the proof. We begin by showing existence, via a compactness argument. Let $p, q \in S$, and let l be the infimum of the ω-lengths of all arcs joining p and q and belonging to a given homotopy class $[\alpha]$. Let $\{\alpha^i\}$, $i = 1, 2, \ldots$, be a sequence of arcs such that $\alpha^i \in [\alpha]$ and $\lim_{i\to\infty} l_\omega(\alpha_i) = l$. Denote by $\mathcal{U}$ a covering of S with geodesic balls so small as to be convex and biholomorphic to a disk. Let $M > 0$ be the Lebesgue constant of $\mathcal{U}$. Clearly, any two points of S whose distance is less than M are joined by a minimal geodesic, and any geodesic arc in S whose length is less than M is necessarily minimal. It is thus possible to subdivide each arc α^i into n subarcs of length less than $M/2$, where n does not depend on i. Denote by $p_1^i, \ldots, p_n^i = q$ the end points of these subarcs and set $p_0^i = p$. Possibly replacing $\{\alpha^i\}$ with a subsequence, we may assume that, for each s, the sequence $\{p_s^i\}$ converges to a point $p_s \in S$. It follows that p_s and p_{s+1} can be joined by a unique minimal geodesic arc β_s, of length at most $M/2$.

Let β be the union of the arcs $\beta_0, \dots, \beta_{n-1}$. Clearly $l_\omega(\beta) \leq l$, and moreover $\beta \in [\alpha]$. In fact, if i is large enough, then for each s, the arc β_s and the portion of α^i between p_s and p_{s+1} are both contained in the geodesic ball of radius M centered at p_s and hence in an element U_s of $\mathcal{U}$. Then a homotopy between β and α^i can be constructed by putting together a homotopy, inside U_0, between β_0 and the portion of α^i between p_0 and p_1, a homotopy, inside U_1, between β_1 and the portion of α^i between p_1 and p_2, and so on. Therefore β minimizes the ω-length of paths from p to q belonging to $[\alpha]$. But this implies that β satisfies the local mimizing property, showing that it is an ω-geodesic.

As we announced, we will deduce uniqueness from a Gauss–Bonnet-type result. In order to state this, consider an open region P in a connected Riemann surface X and let $\tilde{\omega}$ be a holomorphic quadratic differential on X. We assume that P is homeomorphic to a disk, that the closure of P is homeomorphic to a closed disk, and that the boundary of P is the union of finitely many $\tilde{\omega}$-geodesic arcs whose interiors do not contain zeroes of $\tilde{\omega}$. Let $q_1, \dots, q_s$ be the points of ∂P where two of these arcs meet, let μ_j be the multiplicity of q_j as a zero of $\tilde{\omega}$, and denote by ϑ_j the interior angle formed by the subarcs of ∂P adjoining q_j for $j = 1, \dots, s$. By compactness, P contains a finite number of zeroes of $\tilde{\omega}$; let them be $p_1, \dots, p_r$, and let $\nu_i > 0$ be the multiplicity of p_i as a zero of $\tilde{\omega}$. The situation is illustrated in Figure 7.

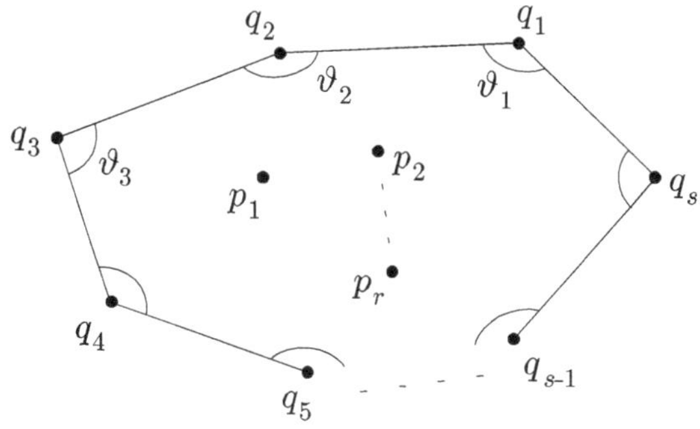

Figure 7.

Set $\nu = \sum_{i=1}^r \nu_i$. We claim that the following Gauss–Bonnet formula holds:

$$(\nu + 2)2\pi = \sum_{j=1}^{s} (2\pi - (\mu_j + 2)\vartheta_j). \tag{5.5}$$

To prove the claim, first observe that the uniformization theorem implies that a neighborhood U of the closure of P can be realized as an open subset of $\mathbb{C}$. If z is a global coordinate on U, we may write $\tilde{\omega} = F(z)dz^2$

on all U where F is holomorphic. The argument principle (i.e., the residue theorem) tells us that

$$\int_{\partial P} d(\arg F) = 2\pi\nu + \sum_{j=1}^{s} \mu_j \vartheta_j \,.$$

Since ∂P is a union of $\tilde{\omega}$-geodesics, we know that $\arg(F dz^2)$ is constant between consecutive q_j. In the above equality we can then substitute $d(\arg F)$ with $-2d(\arg dz)$ and obtain

$$2\pi\nu + \sum_{j=1}^{s} \mu_j \vartheta_j = -2 \int_{\partial P} d(\arg dz) = -2(2\pi - \sum_{j=1}^{s} (\pi - \vartheta_j)),$$

which yields (5.5).

We shall now use (5.5) to prove the uniqueness part of (5.4), arguing by contradiction. Let α and β be distinct ω-geodesics with the same initial point and the same end point and suppose that they are homotopic. Let $\pi : X \to S$ be the universal covering of S and set $\tilde{\omega} = \pi^*(\omega)$. Denote by $\tilde{\alpha}$ and $\tilde{\beta}$ liftings of α and β to X with the same initial point. Since α and β are homotopic, the end points of $\tilde{\alpha}$ and $\tilde{\beta}$ also coincide. We may choose two distinct points p and q, belonging both to $\tilde{\alpha}$ and to $\tilde{\beta}$, such that the portions of $\tilde{\alpha}$ and $\tilde{\beta}$ between p and q do not have other points in common. These two portions of $\tilde{\alpha}$ and $\tilde{\beta}$ then bound a simply connected region $P \subset X$ whose boundary is the union of finitely many smooth $\tilde{\omega}$-geodesic arcs. In applying formula (5.5) to the region P we may set things up so that $q_1 = p$, $q_s = q$. By the local characterization of singular geodesics, we know that

$$\vartheta_j \geq \frac{2\pi}{\mu_j + 2}, \quad j = 2, \ldots, s-1.$$

Plugging these inequalities into (5.5), we find that

$$4\pi \leq (\nu+2)2\pi = \sum_{j=1}^{s} (2\pi - (\mu_j+2)\vartheta_j) \leq 2\pi - (\mu_1+2)\vartheta_1 + 2\pi - (\mu_s+2)\vartheta_s < 4\pi,$$

thus reaching a contradiction. This finishes the proof of uniqueness and hence of the theorem.

The following corollary of Theorem (5.4) will be used in the proof of Teichmüller's uniqueness theorem.

COROLLARY (5.6). *Let S be a compact Riemann surface of genus $g > 1$. Let ω be a nonzero quadratic differential on S, and let $f : S \to S$ be a homeomorphism homotopic to the identity. Then there exists a nonnegative real number M such that*

$$l_\omega(f(\alpha)) > l_\omega(\alpha) - 2M$$

for every ω-geodesic path α.

To prove this, pick a homotopy $F : S \times [0,1] \to S$ between the identity map and f, so that $F(p,0) = p$ and $F(p,1) = f(p)$. Let α_p be the path defined by $\alpha_p(t) = F(p,t)$. By Theorem (5.4), there exists a unique ω-geodesic $\overline{\alpha}_p$ from p to $f(p)$ in the homotopy class of α_p. Set $M = \sup\{l_\omega(\overline{\alpha}_p) : p \in S\}$. By compactness M is a nonnegative real number. Let α be a path in S, and let p and q be its endpoints. There are homotopies with fixed endpoints between α and $\alpha_p f_*(\alpha)\alpha_q^{-1}$ and between $\alpha_p f_*(\alpha)\alpha_q^{-1}$ and $\overline{\alpha}_p f_*(\alpha)\overline{\alpha}_q^{-1}$. If α is an ω-geodesic, then

$$l_\omega(\alpha) \le l_\omega(\overline{\alpha}_p) + l_\omega(f(\alpha)) + l_\omega(\overline{\alpha}_q) \le l_\omega(f(\alpha)) + 2M\,.$$

We end this section with a remark about ω-coordinates. Due to the fact that these are determined only up to sign (and translation), there is in general no coherent way to put an orientation on all horizontal or vertical ω-geodesics in $S \setminus Z$. This is apparent in the vicinity of a zero of odd order of ω (see Figure 5). The problem does not occur when ω is the square of a holomorphic differential φ. One may then take as canonical coordinate ζ one for which $\varphi = d\zeta$. This coordinate is uniquely determined up to translation so that, writing $\zeta = \xi + \sqrt{-1}\eta$, the vector fields $\frac{\partial}{\partial \xi}$ and $\frac{\partial}{\partial \eta}$ are intrinsically defined. We shall refer to these as the *horizontal* and *vertical vector fields*, respectively; they are clearly defined on $S \setminus Z$, where they provide a coherent orientation on all the horizontal and vertical geodesics. In the general case, the best way around the absence of a such an orientation is to pass to a two-sheeted ramified covering $\pi : \tilde{S} \to S$ on which $\pi^*(\omega)$ admits a square root. Clearly π ramifies over the zeroes of ω of odd multiplicity and the horizontal and vertical $\pi^*(\omega)$-geodesics project onto the horizontal and vertical ω-geodesics. As explained above, the choice of a square root of $\pi^*(\omega)$ determines a canonical orientation on horizontal and vertical $\pi^*(\omega)$-geodesics.

6. The proof of Teichmüller's uniqueness theorem.

Let S and S' be two compact Riemann surfaces of genus $g > 1$. Let ω (resp., ω') be a nonzero quadratic differential on S (resp., on S'). We want to study an admissible quasi-diffeomorphism $f : S \to S'$ in terms of the metrics induced by ω and ω'. Set $Z = \{\text{zeroes of } \omega\} \cup \{\text{singularities of } f\}$. The basic invariant of f is the *stretching function* λ_f, which is the positive C^∞-function on $S \setminus Z$ defined in the following way. Let ζ (resp. ζ') be an ω-coordinate (resp., an ω'-coordinate) around p (resp., around $f(p)$). By abuse of language, write $\zeta' = f(\zeta)$ to indicate the local expression of f in these coordinates and set

$$\lambda_f = |f_\zeta + f_{\overline{\zeta}}|\,.$$

This is a well-defined function, since it does not change if ζ and ζ' are altered by the addition of constants or by changes of sign. Writing

$\zeta = \xi + \sqrt{-1}\eta$, we also have $\lambda_f = |f_\xi|$, so that λ_f represents the stretching realized by f in the ξ direction. We will reduce the study of the stretching function to two particular cases. The first is the one of the inverse of a Teichmüller map

$$f_\omega^{-1} : S \to S_\omega ,$$

where on S we take the ω-metric and on S_ω the metric attached to the canonical quadratic differential ω' constructed in Section 4. We then get

$$\zeta' = f_\omega^{-1}(\zeta) = \frac{\zeta + k\bar{\zeta}}{1-k} \quad \text{and} \quad \lambda_{f_\omega^{-1}} = \frac{1+k}{1-k} = K[f_\omega^{-1}] = K[f_\omega] .$$

The second case is the one in which $S = S'$, $\omega = \omega'$, and f is homotopic to the identity. In this case we have the following lemma.

LEMMA (6.1). *Let S be a compact Riemann surface of genus $g > 1$, and let ω be a nonzero holomorphic quadratic differential on S. Let $f : S \to S$ be an admissible quasi-diffeomorphism homotopic to the identity. Then*

$$\frac{\int_S \lambda_f dA_\omega}{\int_S dA_\omega} \geq 1 . \tag{6.2}$$

In other words, the average stretching of f with respect to the ω-metric is at least 1.

Proof. If ω is not already the square of a holomorphic differential, there is a two-sheeted covering $\pi : \tilde{S} \to S$ such that this is true for $\pi^*(\omega)$. One can define a "stretching function" $\lambda_{f\circ\pi}$ just as for f, and $\lambda_{f\circ\pi} = \pi^*(\lambda_f)$. Clearly,

$$\int_{\tilde{S}} \lambda_{f\circ\pi} dA_{\pi^*(\omega)} = 2 \int_S \lambda_f dA_\omega , \qquad \int_{\tilde{S}} dA_{\pi^*(\omega)} = 2 \int_S dA_\omega ,$$

so in this case the inequality to be proved is equivalent to

$$\frac{\int_{\tilde{S}} \lambda_{f\circ\pi} dA_{\pi^*(\omega)}}{\int_{\tilde{S}} dA_{\pi^*(\omega)}} \geq 1 . \tag{6.3}$$

We are thus reduced to proving (6.2) under the additional hypothesis that ω is the square of a holomorphic differential. As we observed at the end of the preceding section, the choice of a square root φ of ω determines a horizontal vector field away from the set Z of zeroes of ω. We let $(t, p) \mapsto \sigma_t(p)$, where t is a real number and $p \in S \smallsetminus Z$, be the flow generated by the horizontal vector field. For every t, the map σ_t is an isometry where defined. The paths $t \mapsto \sigma_t(p)$, that is, the integral curves of the horizontal vector field, are horizontal ω-geodesics. Their images are called the *horizontal trajectories* of ω. Fix an arbitrary positive number τ. One can then easily construct a subset $T \subset S$ which is the union of a finite number of horizontal segments with vertices at singular point of f or at zeroes of ω, having the property that, for $p \notin T$, $\sigma_t(p)$ is defined for all values of $t \in [-\tau, \tau]$. Since T has measure zero and σ_t is an isometry, we have

$$\begin{aligned}\int_S \lambda_f dA_\omega &= \frac{1}{2\tau}\int_{-\tau}^{\tau}\left(\int_S \lambda_f dA_\omega\right) dt = \frac{1}{2\tau}\int_{-\tau}^{\tau}\left(\int_{\sigma_t(S\smallsetminus T)} \lambda_f dA_\omega\right) dt \\ &= \frac{1}{2\tau}\int_{-\tau}^{\tau}\left(\int_{S\smallsetminus T} \sigma_t^*(\lambda_f dA_\omega)\right) dt = \frac{1}{2\tau}\int_{-\tau}^{\tau}\left(\int_{S\smallsetminus T} \sigma_t^*(\lambda_f) dA_\omega\right) dt \\ &= \int_{S\smallsetminus T}\left(\frac{1}{2\tau}\int_{-\tau}^{\tau}\sigma_t^*(\lambda_f)dt\right) dA_\omega = \int_S \left(\frac{l_\omega(f(\alpha_{\tau,p}))}{l_\omega(\alpha_{\tau,p})}\right) dA_\omega\,,\end{aligned}$$

where $\alpha_{\tau,p}$ is the restriction of the path $t \mapsto \sigma_t(p)$ to the interval $[-\tau, \tau]$. Using (5.6), we get

$$\frac{\int_S \lambda_f dA_\omega}{\int_S dA_\omega} \geq 1 - \frac{M}{\tau}.$$

The lemma is proved by letting τ tend to infinity.

We are now in a position to prove Teichmüller's uniqueness theorem (4.14). We first deal with the unpointed case. Let $f : S_\omega \to S$ be an admissible diffeomorphism homotopic to f_ω. Write $h = f^{-1}$, set $u = h \circ f_\omega : S_\omega \to S_\omega$, and apply Lemma (6.1) to u by taking on S_ω the canonical quadratic differential ω' constructed in Section 4. Using the Schwarz inequality, we get

$$\int_{S_\omega} \lambda_u^2 dA_{\omega'} \geq \frac{\left(\int_{S_\omega} \lambda_u dA_{\omega'}\right)^2}{\int_{S_\omega} dA_{\omega'}} \geq \int_{S_\omega} dA_{\omega'}\,.$$

Since $h = u \circ f_\omega^{-1}$, using ω-coordinates, we have

$$h(\zeta) = u\left(\frac{\zeta + k\bar{\zeta}}{1-k}\right),$$

so that

$$\lambda_h = K \cdot (\lambda_u \circ f_\omega^{-1})\,, \tag{6.4}$$

where $K = K[f_\omega] = K[f_\omega^{-1}] = \frac{1+k}{1-k}$. Also observe that K coincides with the Jacobian $J_{f_\omega^{-1}}$ of f_ω^{-1} and that $f_\omega^*(dA_\omega) = K^{-1} dA_{\omega'}$. We then have

$$\int_{S_\omega} dA_{\omega'} \leq \int_{S_\omega} \lambda_u^2 dA_{\omega'} = \int_S (\lambda_u \circ f_\omega^{-1})^2 J_{f_\omega^{-1}} dA_\omega = \int_S \lambda_h^2 K^{-1} dA_\omega\,.$$

In order to prove that $K[f] \geq K$, it suffices to show that $\lambda_h^2 \leq K[f] \cdot J_h$, because then

$$\begin{aligned} \int_{S_\omega} dA_{\omega'} &\leq \int_S \lambda_h^2 K^{-1} dA_\omega \\ &\leq K[f] \cdot K^{-1} \int_S J_h dA_\omega = K[f] \cdot K^{-1} \int_{S_\omega} dA_{\omega'} \,. \end{aligned} \tag{6.5}$$

But now $K[f] = K[h] = \frac{1+\|\mu_h\|}{1-\|\mu_h\|}$, while, in ω-coordinates, $\mu_h = \frac{h_{\bar\zeta}}{h_\zeta} \frac{\partial}{\partial\zeta} \otimes d\bar\zeta$. Hence,

$$K[f] \cdot J_h \geq \frac{|h_\zeta| + |h_{\bar\zeta}|}{|h_\zeta| - |h_{\bar\zeta}|} \cdot (|h_\zeta|^2 - |h_{\bar\zeta}|^2) \geq |h_\zeta + h_{\bar\zeta}|^2 = \lambda_h^2 \,. \tag{6.6}$$

The inequality $K[f] \geq K$ is thus established. If equality holds, looking at (6.5), we infer that $\lambda_h^2 = K[f] \cdot J_h$. But then (6.6) implies that

$$\begin{gathered} |h_\zeta + h_{\bar\zeta}| = |h_\zeta| + |h_{\bar\zeta}| \,, \\ \frac{1+k}{1-k} = K = \frac{1 + |h_{\bar\zeta}/h_\zeta|}{1 - |h_{\bar\zeta}/h_\zeta|} \,. \end{gathered}$$

The first of these equalities tells us that $\arg(h_{\bar\zeta}) = \arg(h_\zeta)$, while the second tells us that $|h_{\bar\zeta}/h_\zeta| = k$. Hence $h_{\bar\zeta} = kh_\zeta$, that is, h satisfies the Beltrami equation whose solutions are the holomorphic functions on S_ω. This means that $u = h \circ f_\omega$ is holomorphic. Since it is also homotopic to the identity, and $h = f^{-1}$, we conclude that $f = f_\omega$.

To prove (4.14) in general, we reduce to the unpointed case by means of a trick. If n is even, we let $\pi : \overline{S} \to S$ be a double covering branched at the p_i. If n is odd, we let $\pi : \overline{S} \to S$ be the composition of π'' and π', where $\pi' : S' \to S$ is a double covering branched at $p_1, \dots, p_{n-1}$, and $\pi'' : \overline{S} \to S'$ is a double covering branched at the two points of S' mapping to p_n. Notice that in any case $\overline{\omega} = \pi^*(\omega)$ is holomorphic and that π, viewed as a map from $\overline{S}_{\overline{\omega}}$ to S_ω, is holomorphic. Furthermore, the genus of $\overline{S}$ is strictly larger than 1 except when $g = 0$ and $n = 3, 4$. Leaving these two cases aside for the moment, we argue as follows. By assumption, there is a continuous family f_t of homeomorphisms from S_ω to S such that $f_0 = f_\omega$, $f_1 = f$, and that $f_t(p_i) = p_i$ for every t and every i. The isotopy $\{f_t\}$ lifts to an isotopy $\{\overline{f}_t : \overline{S}_{\overline{\omega}} \to \overline{S}\}$ with $\overline{f}_0 = f_{\overline{\omega}}$. We set $\overline{f} = \overline{f}_1$; clearly, $\overline{f}$ is a quasi-diffeomorphism. The unpointed case of the theorem then says that $K[f_{\overline{\omega}}] \leq K[\overline{f}]$ and that equality holds only when $f_{\overline{\omega}} = \overline{f}$. On the other hand, since π is holomorphic, both as a map from $\overline{S}$ to S and as a map from $\overline{S}_{\overline{\omega}}$ to S_ω, the dilatation of $\overline{f}$ is the same as the one of f, and the dilatation of $f_{\overline{\omega}}$ is the same as the one of f_ω. The conclusion of the theorem follows.

To treat the cases where $g = 0$ and n equals 3 or 4, observe that the isotopy $\{\overline{f}_t\}$ above leaves the points $\{q_1, \ldots, q_m\}$ of the preimage of $\{p_1, \ldots, p_n\}$ fixed. Thus, the construction of π and the argument above can be repeated for $(\overline{S}, q_1, \ldots, q_m)$. This yields the result since for $n = 4$, the genus of $\overline{S}$ is 1, and m is 4, while for $n = 3$, the genus of $\overline{S}$ is 0, and m is 6. The proof of Teichmüller's uniqueness theorem is now complete.

7. Simple connectedness of the moduli stack of stable curves.

This short section is devoted to proving that the moduli stacks of stable curves are simply connected, in a sense that will be made explicit below. Before we embark on formal definitions and proofs, let us give a heuristic argument that explains what is going on. For the sake of this argument, we shall pretend that $\overline{M}_{g,n}$ is smooth and that the action of the Teichmüller modular group $\Gamma_{g,n}$ on $\mathcal{T}_{g,n}$ is free; thus the fundamental group of $M_{g,n}$ is $\Gamma_{g,n}$. Since $\partial M_{g,n}$ has real codimension two in $\overline{M}_{g,n}$, the fundamental group of $M_{g,n}$ surjects onto the one of $\overline{M}_{g,n}$. On the other hand, we know that $\Gamma_{g,n}$ is generated by Dehn twists and that, by Picard–Lefschetz theory (see Section 9 of Chapter X), any Dehn twist corresponds to a small loop around $\partial M_{g,n}$. Clearly, in $\overline{M}_{g,n}$, any such loop can be filled in, showing that the fundamental group of $\overline{M}_{g,n}$ is trivial.

This line of reasoning can be made rigorous by introducing the concept of orbifold fundamental group. The orbifold fundamental group of $\mathcal{M}_{g,n}$ turns out to be $\Gamma_{g,n}$, and the rest of the argument can be adapted to show that the orbifold fundamental group of $\overline{\mathcal{M}}_{g,n}$ is trivial. Since this would carry us too far, we will follow a different strategy, due to Mumford [555].

We express the moduli stack of stable n-pointed genus g curves as a quotient $\overline{\mathcal{M}}_{g,n} = [H/G]$, where $H = \mathrm{H}_{\nu,g,n}$ is the Hilbert scheme of stable, n-pointed, ν-log-canonical curves of genus g with $\nu \geq 3$, and $G = PGL(r+1)$, where $r = (2\nu - 1)(g-1) + \nu n - 1$. By *topological covering of* $\overline{\mathcal{M}}_{g,n}$ we mean a G-equivariant topological covering of H. In particular, an unramified covering of $\overline{\mathcal{M}}_{g,n}$ "is" a topological covering. We will prove the following result.

THEOREM (7.1). *Every connected topological covering of $\overline{\mathcal{M}}_{g,n}$ is trivial.*

Proof. We write Δ to indicate the locus in H parameterizing singular curves. Let $H' \to H$ be a G-equivariant topological covering. Let $\pi : \mathcal{X} \to \mathcal{T}_{g,n}$ be the universal curve over Teichmüller space, and let D denote the divisor of its canonical sections. We let $\widetilde{H} \to \mathcal{T}_{g,n}$ be the bundle of projective frames in $\pi_*(\omega_\pi^\nu(\nu D))$. The group G acts freely on $\widetilde{H}$, and $\mathcal{T}_{g,n}$ is the quotient of $\widetilde{H}$ modulo G. The Teichmüller modular group $\Gamma = \Gamma_{g,n}$ also acts freely on $\widetilde{H}$, and the quotient is just $H \smallsetminus \Delta$. Let Δ' be the inverse image of Δ in H', and let $\widetilde{H}'$ be the fiber product of $\widetilde{H}$ and $H' \smallsetminus \Delta'$ over $H \smallsetminus \Delta$. Clearly, the group G acts freely also on $\widetilde{H}'$, and the maps $\widetilde{H}' \to \widetilde{H}$ and $\widetilde{H}' \to H' \smallsetminus \Delta'$ are G-equivariant. We then let $\mathcal{T}'_{g,n}$ be the quotient of $\widetilde{H}'$ modulo G. We have a commutative diagram

$$\begin{array}{ccccccc}
\mathcal{T}'_{g,n} & \longleftarrow & \widetilde{H}' & \longrightarrow & H' \smallsetminus \Delta' & \hookrightarrow & H' \\
\downarrow & & \downarrow & \overset{\alpha}{\nearrow} & \downarrow \gamma & & \downarrow \\
\mathcal{T}_{g,n} & \longleftarrow & \widetilde{H} & \xrightarrow{\ \beta\ } & H \smallsetminus \Delta & \hookrightarrow & H
\end{array}$$

where the arrow α is defined as follows. Since G acts freely on both ends of $\widetilde{H}' \to \widetilde{H}$, the map $\mathcal{T}'_{g,n} \to \mathcal{T}_{g,n}$ is a topological cover which splits completely since $\mathcal{T}_{g,n}$ is contractible. But then the covering $\widetilde{H}' \to \widetilde{H}$ also splits completely, and α is obtained by composing a section of this covering with the projection to $H' \smallsetminus \Delta'$.

Fix a base point h in $H \smallsetminus \Delta$, a base point $h'' \in \widetilde{H}$ such that $\beta(h'') = h$, and set $h' = \alpha(h'')$. We have to show that $\pi_1(H, h)/\pi_1(H', h')$ is trivial. Since Δ has real codimension two in H, $\pi_1(H, h) = \pi_1(H \smallsetminus \Delta, h)$ and $\pi_1(H', h') = \pi_1(H' \smallsetminus \Delta', h')$. But then

$$\pi_1(H, h)/\pi_1(H', h') = \pi_1(H \smallsetminus \Delta, h)/\pi_1(H' \smallsetminus \Delta', h')$$

is a quotient of $\pi_1(H \smallsetminus \Delta, h)/\pi(\widetilde{H}, h'') = \Gamma$. In conclusion, Γ surjects onto $\pi_1(H)/\pi_1(H')$. On the other hand, as we know, the mapping class group Γ is generated by Dehn twists, corresponding to elements in $\pi_1(H \smallsetminus \Delta)$ which are represented by small simple loops going around components of Δ. Each such loop is contractible in H, and hence $\pi_1(H)/\pi_1(H')$ must be trivial.

Q.E.D.

COROLLARY (7.2). $\mathrm{Pic}(\overline{\mathcal{M}}_{g,n})$ *has no torsion.*

To prove the corollary, first recall from Chapter XIII that $\mathrm{Pic}(\overline{\mathcal{M}}_{g,n})$ is the group of isomorphism classes of G-equivariant line bundles on H. We now argue by contradiction. Suppose that L is a torsion element in $\mathrm{Pic}(\overline{\mathcal{M}}_{g,n})$. We may assume that L is of prime order p. Thus, there is a G-equivariant isomorphism between L^p and $\mathcal{O}_H$. Consider the pth power map from L to $L^p \cong \mathcal{O}_H$ and denote by H' the inverse image of the unit section. Then H' is a G-equivariant connected cyclic unramified covering of H of degree p. This contradiction establishes the corollary.

We may now finish the proof of Proposition (6.7) in Chapter XIII. As the reader will recall, this asserts that $\mathrm{Pic}(\overline{\mathcal{M}}_{g,n})$ is a free abelian group of finite rank and that $\mathrm{Pic}(\overline{M}_{g,n})$ is a subgroup of finite index. The result we just proved, together with Lemma (6.6) in Chapter XIII, implies that $\mathrm{Pic}(\overline{M}_{g,n})$ is also torsion-free. Now look at the short exact sequence

$$0 \to H^1(\overline{M}_{g,n}, \mathbb{Z}) \to H^1(\overline{M}_{g,n}, \mathcal{O}_{\overline{M}_{g,n}}) \to \mathrm{Pic}(\overline{M}_{g,n}) \to H^2(\overline{M}_{g,n}, \mathbb{Z})$$

coming from the exponential sheaf sequence on $\overline{M}_{g,n}$. As we know, $\overline{M}_{g,n}$ is a projective variety and a (smooth) orbifold, and hence, as we observed in Section 4 of Chapter XII, its complex cohomology looks just like the one of a smooth projective variety; in particular, it has a Hodge decomposition. It follows that $\mathrm{Pic}^0(\overline{M}_{g,n}) = H^1(\overline{M}_{g,n}, \mathcal{O}_{\overline{M}_{g,n}})/H^1(\overline{M}_{g,n}, \mathbb{Z})$ is a complex torus. But then $\mathrm{Pic}^0(\overline{M}_{g,n})$ must vanish, since $\mathrm{Pic}(\overline{M}_{g,n})$ has no torsion, and hence $\mathrm{Pic}(\overline{M}_{g,n})$ is a subgroup of $H^2(\overline{M}_{g,n}, \mathbb{Z})$. Since the latter group is finitely generated, the same is true for $\mathrm{Pic}(\overline{M}_{g,n})$. This, in conjunction with the fact that $\mathrm{Pic}(\overline{\mathcal{M}}_{g,n})/\mathrm{Pic}(\overline{M}_{g,n})$ is a torsion group, immediately implies that $\mathrm{Pic}(\overline{\mathcal{M}}_{g,n})$ has finite rank and that $\mathrm{Pic}(\overline{M}_{g,n})$ has finite index in it, concluding the proof of Proposition (6.7) in Chapter XIII.

Here are a couple of other useful topological consequences of the results proved in this section.

COROLLARY (7.3). $H^1(\overline{M}_{g,n}, \mathbb{Z}) = H^1(\overline{M}_{g,n}, \mathcal{O}_{\overline{M}_{g,n}}) = \{0\}$.

Further results on the integral and rational cohomology of moduli spaces will be discussed in Chapter XIX.

8. Going to the boundary of Teichmüller space.

In this book we will not offer a systematic treatment of the various bordifications of Teichmüller spaces. We will content ourselves with constructing a space $\widehat{\mathcal{T}}_{S,P}$ containing $\mathcal{T}_{S,P}$ as an open set having the property that the action of $\Gamma_{S,P}$ on $\mathcal{T}_{S,P}$ extends to $\widehat{\mathcal{T}}_{S,P}$ in such a way that the quotient $\widehat{\mathcal{T}}_{S,P}/\Gamma_{S,P}$ coincides with $\overline{M}_{g,P}$. Aside from its independent interest, the reason for introducing this space is that it will serve well our purposes during the proof of Witten's conjecture on the intersection theory of $\overline{M}_{g,P}$. We shall see in Chapter XVIII that the space $\widehat{\mathcal{T}}_{S,P}$, although not particularly good-looking, is closely related to a very natural simplicial complex. As we mentioned in the introduction, the first nontrivial case of bordification is the one of $\mathcal{T}_{1,1}$, where $\widehat{\mathcal{T}}_{1,1} = \mathbb{H} \cup \mathbb{P}^1(\mathbb{Q})$.

One way to go to the boundary of Teichmüller spaces is via the so-called Fenchel–Nielsen coordinates. In the present section we will follow a different route. However, it could be useful for the reader to keep the idea of Fenchel–Nielsen coordinates in the background. So, we briefly explain what this idea is. For simplicity, we treat only the case of unpointed curves (of genus $g > 1$). On the reference surface S we fix a *pants decomposition.* By this we mean a system of $3g-3$ disjoint, simple closed curves $L_1, \dots, L_{3g-3}$ such that $S \setminus \{L_1, \dots, L_{3g-3}\}$ is the union of $2g-1$ pairs of pants (see Figure 8).

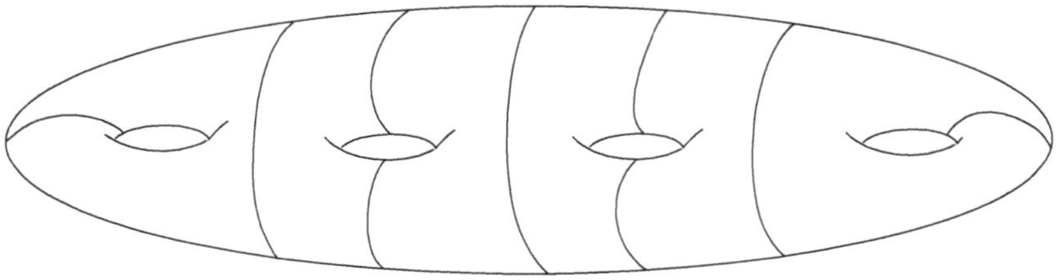

Figure 8.

By using a little hyperbolic geometry, one sees that a point $[C, f]$ of the Teichmüller space $\mathcal{T}_S$ is completely determined by the following procedure. First assign a length ρ_i to each curve L_i. In this way, each of the $2g-1$ pairs of pants has a completely determined complex structure. To get a complex structure on S and therefore a point $[C, f] \in \mathcal{T}_S$, one needs to give, for each $i = 1, \ldots, 3g-3$, a "rotation angle" θ_i which says how to glue the various boundaries of the pants in order to obtain a Riemann surface C. One can then check that the "polar coordinates" $(\rho_1 \ldots, \rho_{3g-3}, \theta_1 \ldots, \theta_{3g-3})$ are in fact global coordinates on $\mathcal{T}_S$, providing a homeomorphism

$$(\rho, \theta) : \mathcal{T}_S \longrightarrow \mathbb{R}_+^{3g-3} \times \mathbb{R}^{3g-3}.$$

The global coordinates (ρ, θ) are the Fenchel–Nielsen coordinates. When one is acting on $\mathcal{T}_S$ with powers of a Dehn twist based on L_i, one is adding multiples of $2\pi\sqrt{-1}$ to the angle coordinate θ_i. The Fenchel–Nielsen coordinates turn out to be real analytic but are impractical from the point of view of the complex structure of $\mathcal{T}_S$. Of course, for each pants decomposition of the surface S, one gets a different system of Fenchel–Nielsen coordinates. The idea of going to the boundary of $\mathcal{T}_S$ using Fenchel–Nielsen coordinates is very simple. One just lets some, or all, of the ρ-coordinates go to zero. In this way, one describes those stable curves where some, or all, of the cycles $L_1, \ldots, L_{3g-3}$ are pinched to a point (a node of the resulting curve). Of course, since the pants decomposition has been fixed beforehand, we cannot hope to get in this way all stable curves. So, one does this construction starting from all possible pants decompositions. At the end, one obtains a certain number of copies

$$U(L_1, \ldots, L_\delta) = (\overline{\mathbb{R}_+})^{3g-3} \times \mathbb{R}^{3g-3},$$

one for each pants decomposition, that must be glued together along the common open set $\mathcal{T}_S$. With care, this can be done, and the result is a manifold with corners. However, one problem still remains. Once we decide to send to zero one of the ρ-coordinates, say ρ_1, what happens to the corresponding θ-coordinate θ_1? Once ρ_1 is set to be equal to 0, the information given by θ_1 is useless. So, instead of gluing the copies of $(\overline{\mathbb{R}_+})^{3g-3} \times \mathbb{R}^{3g-3}$, one should really glue copies of

$$(\overline{\mathbb{R}_+})^{3g-3} \times \mathbb{R}^{3g-3}/\sim, \tag{8.1}$$

where

$$(\rho_1 \ldots, \rho_{3g-3}, \theta_1 \ldots, \theta_{3g-3}) \sim (\rho'_1 \ldots, \rho'_{3g-3}, \theta'_1 \ldots, \theta'_{3g-3})$$

if and only if $\rho_i = \rho'_i$ for all $i = 1, \ldots, 3g-3$, and $\theta_i = \theta'_i$ if $\rho_i \neq 0$. This equivalence relation provides a serious trauma to the uneventful topology of $(\overline{\mathbb{R}_+})^{3g-3} \times \mathbb{R}^{3g-3}$. The resulting quotient space is no longer first countable, nor it is locally compact. To cure these pathologies, one takes a coarser topology on the quotient set (8.1), namely the product topology on

$$\frac{\overline{\mathbb{R}}_+ \times \mathbb{R}}{\{0\} \times \mathbb{R}} \times \cdots \times \frac{\overline{\mathbb{R}}_+ \times \mathbb{R}}{\{0\} \times \mathbb{R}},$$

where, in each factor, a neighborhood basis of the point $[0,0] \in (\overline{\mathbb{R}}_+ \times \mathbb{R})/(\{0\} \times \mathbb{R})$ is given by the subsets

$$\{[x,y] : x < \epsilon\}, \quad \epsilon > 0.$$

Figure 9 shows an open set A in the quotient topology and an open set B in the coarser topology

Figure 9.

In this new topology, $U(L_1, \ldots, L_\delta)/\sim$ is still not locally compact, but it is first countable and Hausdorff. We will see that all these features appear in the topology of $\widehat{\mathcal{T}}_{S,P}$.

As we announced, we will not use Fenchel–Nielsen coordinates to construct the space $\widehat{\mathcal{T}}_{S,P}$. Instead, our construction of the space $\widehat{\mathcal{T}}_{S,P}$ will parallel the construction of the Teichmüller space $\mathcal{T}_{S,P}$. In constructing $\mathcal{T}_{S,P}$, we patched together bases of Kuranishi families having smooth curves as central fibers. In the construction of $\widehat{\mathcal{T}}_{S,P}$ the building blocks are *suitable real blow-ups* of bases of Kuranishi families having stable curves as central fibers.

For the convenience of the reader, we quickly go over the theory of real blow-ups, which we already developed in Section 9 of Chapter X.

One starts with a smooth complex manifold M and a divisor with normal crossing $D \subset M$. The real oriented blow-up of M along D is a manifold with corners denoted with the symbol $\mathrm{Bl}_D(M)$. As one immediately realizes, it suffices to give the definition of oriented blow-up in the *local case,* that is, where $M = \mathbb{C}^N$, and D is the union of $k \leq N$ coordinate hyperplanes in $\mathbb{C}^N$. Typically, we have

$$D = \{(z_1, \ldots, z_N) \in \mathbb{C}^N : z_1 \cdots z_k = 0\}.$$

Let S^1 denote the unit circle in the complex plane, which we alternatively view as $\mathbb{R}/2\pi\mathbb{Z}$. Then *the real oriented blow-up of* $\mathbb{C}^N$ *along* D is nothing but the closure in $\mathbb{C}^N \times (S^1)^k$ of the locus

$$\left\{(z_1,\dots,z_N,\zeta_1,\dots,\zeta_k) \in \mathbb{C}^N \times (S^1)^k : \frac{z_1}{|z_1|} = \zeta_1, \dots, \frac{z_k}{|z_k|} = \zeta_k\right\}.$$

The blow-up map is the natural projection

$$\begin{aligned} \tau : \mathrm{Bl}_D(\mathbb{C}^N) &\longrightarrow \mathbb{C}^N \\ (z_1,\dots,z_N,\zeta_1,\dots,\zeta_k) &\mapsto (z_1,\dots,z_N). \end{aligned}$$

The fiber structure of this map is clear. To describe it, look at the locally closed strata

$$D_I = \left\{(z_1,\dots,z_N) \in \mathbb{C}^N : z_{i_1}\cdots z_{i_s} = 0, \quad z_{j_1}\cdots z_{j_{k-s}} \neq 0\right\}, \tag{8.2}$$

where

$$I = \{i_1,\dots,i_s\} \subset \{1,\dots,k\}, \qquad \{j_1,\dots,j_{k-s}\} = \{1,\dots,k\} \smallsetminus I.$$

We have $D_\emptyset = \mathbb{C}^N \smallsetminus D$ and $D_{\{1,\dots,k\}} = D_1 \cap \cdots \cap D_k$. In general, the stratum D_I is a proper open subset of the linear subspace $D_{i_1} \cap \cdots \cap D_{i_s}$. Denote by $N_{D_i/\mathbb{C}^N}$ the normal bundle of D_i in $\mathbb{C}^N$ and by $\mathbb{S}(N_{D_i/\mathbb{C}^N})$ the associated S^1-bundle.[4] Then, over D_I, the map τ is nothing but the s-dimensional torus bundle

$$\tau : \left(\mathbb{S}(N_{D_{i_1}/\mathbb{C}^N}) \times \cdots \times \mathbb{S}(N_{D_{i_s}/\mathbb{C}^N})\right)_{|D_I} = \tau^{-1}(D_I) \longrightarrow D_I.$$

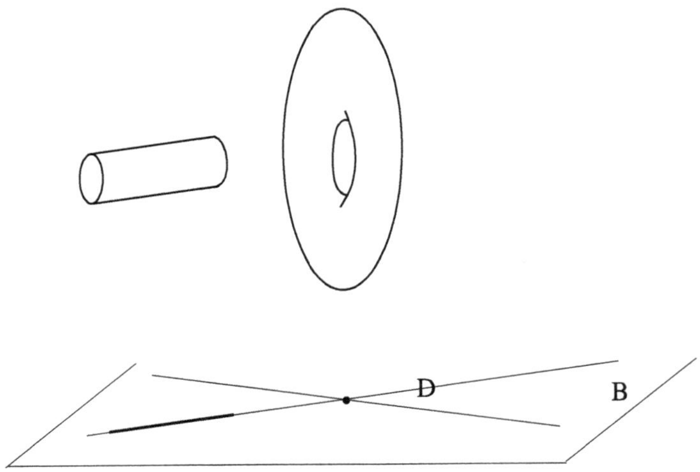

Figure 10.

[4]By way of notation, given a real vector space V, we denote by $\mathbb{S}(V)$ the *sphere of rays through the origin of* V. Thus,

$$\mathbb{S}(V) = (V \smallsetminus \{0\})/\sim \qquad \text{where} \quad v \sim u \iff u = \lambda v, \qquad \text{with} \quad \lambda > 0.$$

Next we briefly recall the real oriented blow-up associated to a stable curve. Given a stable P-pointed curve $(C, \{x_p\}_{p\in P})$ of genus g, we set

$$\mathrm{Sing}(C) = \{y_1, \dots, y_\delta\}, \qquad \mathrm{Reg}(C) = C \smallsetminus \mathrm{Sing}(C).$$

We denote by $\nu : N \to C$ the normalization map, and we set

$$\nu^{-1}(y_i) = \{r_i, s_i\}, \; i = 1, \dots, \delta, \qquad R = \{r_1, s_1, \dots, r_\delta, s_\delta\}.$$

Consider the real oriented blow-up of N along R,

$$\tau : \mathrm{Bl}_R(N) \longrightarrow N$$

We set

$$(8.3) \qquad \check{C} = \mathrm{Bl}_R(N), \qquad \sigma = \nu\tau : \check{C} \longrightarrow C.$$

By definition, the fiber of τ over a point $r \in R$ is $\mathbb{S}(T_r(N))$, which is a copy of S^1. Hence $\check{C}$ is a (possibly disconnected) Riemann surface with boundary (see Figure 11). We may construct a P-pointed oriented topological surface Σ of genus g by giving, for each point $y \in \mathrm{Sing}(C)$, an identification between the two boundary components $\tau^{-1}(r) = \mathbb{S}(T_r(N))$ and $\tau^{-1}(s) = \mathbb{S}(T_s(N))$, where $\{r, s\} = \nu^{-1}(y)$. Of course, in order to get an oriented surface, this identification should carry the orientation of $\tau^{-1}(r)$ into the opposite orientation of $\tau^{-1}(s)$. By construction, the surface Σ comes equipped with a system of simple closed curves $L = \{L_1, \dots, L_\delta\}$. We denote by

$$k : \check{C} \to \Sigma$$

the quotient map and by

$$(8.4) \qquad f : \Sigma \longrightarrow C$$

the natural contraction map defined by the property that $fk = \nu\tau$.

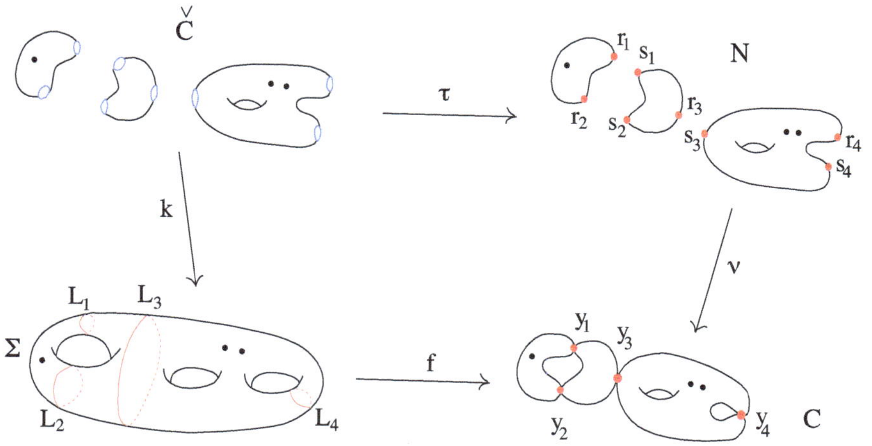

Figure 11.

Both $\mathbb{S}(T_r(N))$ and $\mathbb{S}(T_s(N))$ have a natural orientation. A way to identify them to obtain the oriented surface Σ is to give a nonzero vector (up to a positive proportionality factor) in $\mathrm{Hom}_{\mathbb{C}}(\overline{T}_r(N), T_s(N))$ or, equivalently, a vector on the circle $\mathbb{S}\,(T_r(N) \otimes T_r(N))$.

The δ-dimensional real torus

$$\mathbb{T}(C) = \prod_{y \in \mathrm{Sing}(C)} \mathbb{S}\,(T_r(N) \otimes_{\mathbb{C}} T_s(N))\,, \quad \nu^{-1}(y) = \{r, s\}\,, \tag{8.5}$$

is called the *torus of rays of* C. For each point v belonging to the torus of rays $\mathbb{T}(C)$, we have a way of constructing a surface Σ and maps $k : \check{C} \to \Sigma$ and $f : \Sigma \to C$ as above.

We start our discussion of the bordification $\widehat{\mathcal{T}}_{S,P}$ of $\mathcal{T}_{S,P}$ by describing it set-theoretically. First, we introduce the concept of marking for a P-pointed genus g stable curve $(C, \mathbf{x})$, where $\mathbf{x} = \{x_p\}_{p \in P}$. A *Teichmüller marking,* or simply *a marking for* $(C, \mathbf{x})$, is a continuous surjective map

$$f : (S, P) \to (C, \mathbf{x})$$

of P-pointed topological spaces (meaning that $f(p) = x_p$ for each $p \in P$) such that:

i) the preimage of the nodes of C under f is a set L consisting of δ disjoint, smooth, closed curves $L_1, \dots, L_\delta$;
ii) $f : S \smallsetminus L \to \mathrm{Reg}(C)$ is a homeomorphism;
iii) if $\check{S}_L$ denotes the compact surface with boundary obtained by cutting S along L, then f lifts to a homeomorphism $\check{f} : \check{S}_L \to \check{C}$.

We say that $(C, \mathbf{x}, f)$ is *equivalent* to $(C', \mathbf{x}', f')$, and write $(C, \mathbf{x}, f) \sim (C', \mathbf{x}', f')$, if there exists an isomorphism $\varphi : (C, \mathbf{x}) \to (C', \mathbf{x}')$ such that the diagram

$$\begin{array}{ccc} (S,P) & \xrightarrow{\ f\ } & (C,\mathbf{x}) \\ & \searrow^{f'} & \downarrow{\scriptstyle\varphi} \\ & & (C',\mathbf{x}') \end{array} \tag{8.6}$$

commutes, up to a homotopy relative to P such that, at every given time, the lifting property iii) above holds. We denote by $[C, \mathbf{x}, f]$ the equivalence class of a marked stable curve $(C, \mathbf{x}, f)$. Notice that this definition of marking is consistent with the one used for smooth curves, except for the direction of the arrow, which here goes from the reference surface (S, P) to the pointed curve $(C, \mathbf{x})$. This involves only minor adjustments. One of them regards the action of the mapping class group.

The mapping class group $\Gamma_{S,P}$ acts on $\widehat{\mathcal{T}}_{S,P}$ in the obvious way: if $[C, \mathbf{x}, k] \in \widehat{\mathcal{T}}_{S,P}$ and $[\gamma] \in \Gamma_{S,P}$, then

$$[\gamma] \cdot [C, \mathbf{x}, k] = [C, \mathbf{x}, k\gamma^{-1}].$$

Observe that, if $L = (L_1, \dots L_\delta)$ is the system of curves in S contracting to the nodes of C, then the subgroup $\Gamma(L) \subset \Gamma_{S,P}$ generated by Dehn twists along the curves L_i fixes the point $[C, \mathbf{x}, k]$. The projection map

$$\eta : \widehat{\mathcal{T}}_{S,P} \longrightarrow \overline{M}_{g,P}$$

sending $[C, \mathbf{x}, f]$ to $[C, \mathbf{x}]$ provides a set-theoretical identification

$$\overline{\eta} : \widehat{\mathcal{T}}_{S,P} / \Gamma_{S,P} = \overline{M}_{g,P}\,. \tag{8.7}$$

Perhaps, the only thing which is not immediately clear is why $\overline{\eta}$ is indeed an injection; the proof also explains the reason for the somewhat strange condition iii) in the definition of marking. To prove this point, it is enough to show that any pair of markings $k, k' : (S, P) \to (C, \mathbf{x})$ are in the same $\Gamma_{S,P}$-orbit. By assumption we have a homeomorphism

$$\check{k}'^{-1}\check{k} : \left(\check{S}_L, P\right) \to \left(\check{S}_{L'}, P\right),$$

where L and L' are two systems of smooth, closed, and disjoint curves in S. If the map $h = \check{k}'^{-1}\check{k}$ descends to a homeomorphism $S \to S$, we are done. In general, however, this does not happen because of the following simple reason. Let L_i^+ and L_i^- be the two connected components of the preimage of L_i in $\check{S}_L$, and let L'^+_i, L'^-_i be their analogues for L'_i. We may assume that $h(L_i^+) = L'^+_i$. There is no reason why h should be compatible with the identifications $a_i : L_i^+ \xrightarrow{\simeq} L_i^-$ and $b_i : L'^+_i \xrightarrow{\simeq} L'^-_i$. However, the maps ha_i and b_ih are isotopic for each i, and it is clear that all these isotopies can be realized as b_ih_t, where $h_t : \left(\check{S}_L, P\right) \to \left(\check{S}_{L'}, P\right)$ is an ambient isotopy such that h_ta_i is independent of t for all i. Now h_1 drops down to a well-defined homeomorphism $m : (S, P) \to (S, P)$, and the isotopy h_t induces an isotopy between $k : (S, P) \to (C, \mathbf{x})$ and $k'm$.

Of course, $\mathcal{T}_{S,P}$ is a subset of $\widehat{\mathcal{T}}_{S,P}$; when the topology of $\widehat{\mathcal{T}}_{S,P}$ will be available, we will see that it is indeed an open subset. We denote by $\partial\mathcal{T}_{S,P}$ the boundary of $\mathcal{T}_{S,P}$ in $\mathcal{T}_{S,P}$, i.e., the set $\widehat{\mathcal{T}}_{S,P} \smallsetminus \overline{\mathcal{T}}_{S,P}$. We can already anticipate that along the boundary $\partial\mathcal{T}_{S,P}$ the topology of $\widehat{\mathcal{T}}_{S,P}$ will not be locally compact. This can be already understood, heuristically, from the following consideration. Fix a point $z = [C, \mathbf{x}, k] \in \partial\mathcal{T}_{S,P}$. Look at the system L of curves contracting to the nodes of C via the marking.

The infinite group $\Gamma(L)$, on one hand, acts on the open set $\mathcal{T}_{S,P}$ with finite stabilizers and, on the other, *stabilizes the point* z. In particular, any open neighborhood U of $z \in \partial\mathcal{T}_{S,P}$ will have the property that $U \cap \mathcal{T}_{S,P}$ posseses infinitely many connected components.

As we remarked in Section 9 of Chapter X, the blow-up operations (8.3) and (8.4) that we performed on a single curve can be done simultaneously on all singular fibers of a (transverse) family of nodal curves. Here we will only be concerned with Kuranishi families (which are certainly transverse). The reader should go back to Proposition (9.16) and especially to Remark (9.17) in that chapter. Let us summarize what we proved there, having in mind what we presently need. We start with a Kuranishi family

$$\pi : X \to B$$

of P-pointed genus g curves. We let $D \subset B$ be the divisor parameterizing the singular fibers of π, and we look at the real oriented blow-up

$$\sigma : \mathrm{Bl}_D(B) \to B\,.$$

We let $\varphi : Y \to \mathrm{Bl}_D(B)$ be the pulled-back family. We set $E = \sigma^{-1}(D)$ so that φ concides with π over $\mathrm{Bl}_D(B) \smallsetminus E = B \smallsetminus D$. Then one can construct a diagram

$$\begin{array}{ccc} Z & \xrightarrow{\ \lambda\ } & Y \\ & \searrow^{\psi} & \downarrow{\scriptstyle\varphi} \\ & & \mathrm{Bl}_D(B) \end{array} \tag{8.8}$$

where $\psi : Z \to \mathrm{Bl}_D(B)$ is a real analytic family of compact orientable P-pointed genus g differentiable surfaces which agrees with φ away from E. Moreover, for each point $t \in \mathrm{Bl}_D(B)$, let

$$\lambda_t : Z_t \to Y_t$$

be the map induced by λ on the fibers over t. Suppose that Y_t has δ double points. Then their preimages form a set L_t consisting of δ smooth, closed, disjoint curves $L_{1,t}, \dots, L_{\delta,t}$, and $\lambda_t : Z_t \smallsetminus L_t \to \mathrm{Reg}(Y_t)$ is a homeomorphism of P-pointed surfaces. Over each point $t \in \mathrm{Bl}_D(B)$ the map $\lambda_t : Z_t \to Y_t$ is a marking of the stable P-pointed curve $(Y_t, \mathbf{y}_t)$. To get a fixed differentiable surface S into the picture instead of the varying differentiable surface Z_t, we pull back everything to the universal covering of $\mathrm{Bl}_D(B)$, which we denote by

$$\rho : \widetilde{B} \to \mathrm{Bl}_D(B)\,. \tag{8.9}$$

We then obtain the diagram

$$
\begin{array}{ccc}
\widetilde{Z} & \xrightarrow{\widetilde{\lambda}} & \widetilde{Y} \\
& \underset{\widetilde{\psi}}{\searrow} & \downarrow \widetilde{\varphi} \\
& & \widetilde{B}
\end{array}
\tag{8.10}
$$

Since the space $\mathrm{Bl}_D(B)$ contracts to the central fiber of σ, which is a torus $\mathbb{T}$ (see formula (9.13) in Chapter X), the universal covering $\widetilde{B}$ is contractible. Therefore, it is possible to choose a C^∞ trivialization

$$\tau : S \times \widetilde{B} \xrightarrow{\cong} \widetilde{Z} \tag{8.11}$$

and a diagram

$$
\begin{array}{ccc}
S \times \widetilde{B} & \xrightarrow{k} & \widetilde{Y} \\
& \searrow & \downarrow \widetilde{\varphi} \\
& & \widetilde{B}
\end{array}
\tag{8.12}
$$

where now

$$k_t : (S, P) \to (\widetilde{Y}_t, \mathbf{y}_t) \tag{8.13}$$

is a Teichmüller marking. We need to take a closer look at the composition

$$\beta = \rho\sigma : \widetilde{B} \to \mathrm{Bl}_D(B) \to B\,.$$

Choose $t \in \beta^{-1}(0)$. Look at the system of simple closed curves (the vanishing cycles) $L_t = \{L_{1,t}, \dots, L_{\delta,t}\} \subset S$ defined by the marking k_t of Y_t. The set $[L_t]$ consisting of the homotopy classes of the $L_{i,t}$ is independent of $t \in \beta^{-1}(0)$. Since we are tacitly considering homotopy classes, we will write L instead of L_t. Each stratum D_I of B corresponds to a set of vanishing cycles $\{L_{i_1}, \dots, L_{i_s}\}$, $i_j \in I$. We denote by $\Gamma(L_I)$ the abelian subgroup of $\Gamma_{S,P}$ generated by the Dehn twists along $L_{i_1}, \dots, L_{i_s}$, and we set $\Gamma(L) = \Gamma(L_{\{1,\dots\delta\}})$. We have the identifications

$$\mathbb{Z}^\delta = \pi_1((C)) = \pi_1(\mathrm{Bl}_D(B)) = \Gamma(L)\,.$$

The action of $\Gamma(L)$ on $\widetilde{B}$ has the following property. Given $t \in \widetilde{B}$ and $\gamma \in \Gamma(L)$, we have

$$[h_{\gamma t}] = [h_t \gamma^{-1}], \tag{8.14}$$

From this it is an exercise to check that, if $\beta(t) \in D_I$, then

$$\gamma t = t \quad \Longleftrightarrow \quad \gamma \in \Gamma(L_I)\,. \tag{8.15}$$

Fix a point $z \in B$ belonging to the stratum D_I. Set $s = |I|$ and let $\widetilde{\mathbb{T}}(C_z) \cong \mathbb{R}^s$ denote the universal cover of the s-dimensional torus $\mathbb{T}(C_z)$. Then each connected component of $\beta^{-1}(z)$ is isomorphic to a copy of $\widetilde{\mathbb{T}}(C_z)$. From the Fenchel–Nielsen point of view, one should imagine that the coordinates of $\widetilde{\mathbb{T}}(C_z)$ are the "useless" angle-coordinates θ_i, $i \in I$, corresponding to those radius-coordinates ρ_i which are identically zero on $\beta^{-1}(D_I)$. From $\widetilde{B}$ we construct a new set $\widehat{B}$ by contracting to a point each of the affine spaces appearing in the fibers of β. Let us do this in formulae in the case in which B is two-dimensional with coordinates z_1, z_2, and $D = \{z_1 z_2 = 0\}$. In this case the universal cover of $\mathrm{Bl}_D(B)$ is $\mathbb{R}^2_+ \times \mathbb{R}^2$, and $\beta : \widetilde{B} \to B$ is

$$(\rho_1, \rho_2, \vartheta_1, \vartheta_2) \mapsto (\rho_1 e^{\sqrt{-1}\vartheta_1}, \rho_2 e^{\sqrt{-1}\vartheta_2}).$$

Then

$$\beta^{-1}(0) = \widetilde{\mathbb{T}}(C) = \{(0, 0, \vartheta_1, \vartheta_2) : \vartheta_1, \vartheta_2 \in \mathbb{R}\}$$

gets contracted to a point in $\widehat{B}$. On the other hand, if $z = (0, z_2)$ with $z_2 = \rho e^{\sqrt{-1}\vartheta} \neq 0$, then

$$\beta^{-1}(z) = \widetilde{\mathbb{T}}(C_z) = \{(0, \rho, \vartheta_1, \vartheta_2) : \vartheta_1 \in \mathbb{R}, \vartheta_2 \equiv \vartheta \mod 2\pi\sqrt{-1}\} = \coprod_{k \in \mathbb{Z}} A_k\,,$$

where

$$A_k = \{(0, \rho, \vartheta_1, \vartheta + 2k\pi\sqrt{-1}) : \vartheta_1 \in \mathbb{R}\}\,,$$

and each A_k gets contracted to a single point of $\widehat{B}$ (see Figure 12).

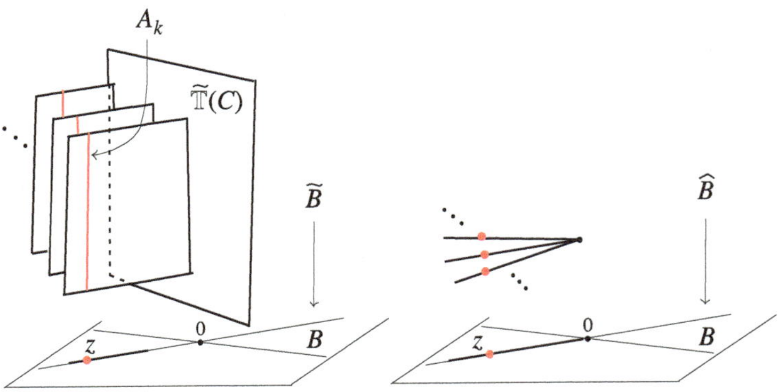

Figure 12.

We let $\mu : \widetilde{B} \to \widehat{B}$ denote the contraction map. By definition, the map $\beta : \widetilde{B} \to B$ drops to a map $\widehat{\beta} : \widehat{B} \to B$. We now describe the topology of $\widehat{B}$. We consider on $\widetilde{B}$ a topology τ which is weaker than the universal cover topology. In this topology, a set is open if and only if it is a connected component, in the universal cover topology, of the preimage under $\beta : \widetilde{B} \to B$ of an open set of B. We then put on $\widehat{B}$ the quotient topology. In Figure 13 we illustrate the case in which B is a one-dimensional disc; in the picture, the open sets are shaded.

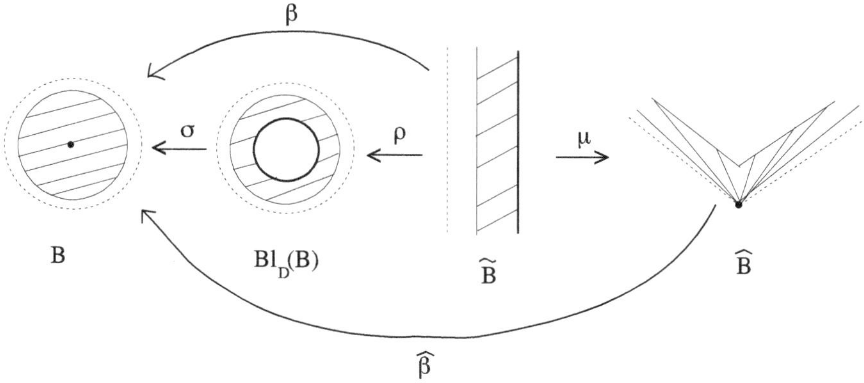

Figure 13.

The space $(\widetilde{B}, \tau)$ is still first countable, but it is neither contractible nor Hausdorff, and the blow-down map $\sigma : (\widetilde{B}, \tau) \to \mathrm{Bl}_D(B)$ is no longer continuous. The non-Hausdorffness of $(\widetilde{B}, \tau))$ is controllable, in the sense that pairs of points that cannot be separated necessarily belong to the same affine space appearing in the fibers of β. On the other hand, the maps $\beta : (\widetilde{B}, \tau) \to B$ and $\widehat{\beta} : \widehat{B} \to B$ are both continuous and open, and $\widehat{B}$ is Hausdorff, first countable, contractible but not locally compact. From the construction and from (8.15) it follows that $\Gamma(L)$ acts on $\widehat{B}$ and that

$$B = \widehat{B}/\Gamma(L)\,. \tag{8.16}$$

Moreover, for a point $z \in B$ belonging to the stratum D_I, the points in the fiber $\widehat{\beta}^{-1}(z)$ are in one-to-one correspondence with elements in $\Gamma(L)/\Gamma(L_I)$. Let $\hat{\pi} : \widehat{\mathcal{C}} \to \widehat{B}$ be the pull-back via $\widehat{\beta}$ of the Kuranishi family over B. Then diagram (8.12) drops down to a diagram

$$\begin{array}{ccccc} S \times \widehat{B} & \xrightarrow{h} & \widehat{\mathcal{C}} & \longrightarrow & \mathcal{C} \\ & \searrow & \downarrow{\scriptstyle \hat{\pi}} & & \downarrow{\scriptstyle \pi} \\ & & \widehat{B} & \xrightarrow{\widehat{\beta}} & B \end{array} \tag{8.17}$$

For each $t \in \widehat{B}$ with $\widehat{\beta}(t) = z$, set

$$h_t = h_{|S \times \{t\}} : S = S \times \{t\} \longrightarrow \widehat{C}_t = C_z\,.$$

For $t \in B$, set $x_i(t) = h_t(p_i) \in \widehat{C}_t$, $i = 1, \dots, n$. We may define a set-theoretical map (depending on the trivialization (8.11))

$$F = F_{\widehat{B}, \tau} : \widehat{B} \to \widehat{T}_{S,P} \tag{8.18}$$

by setting

$$F(t) = [\widehat{C}_t, x_1(t), \dots, x_n(t); h_t]\,.$$

LEMMA (8.19). *The map F is injective.*

Proof. First of all, we can always assume that two fibers of the original Kuranishi family are isomorphic only through an automorphism σ of the central fiber (which extends to an automorphism, again denoted by σ, of the entire family). But then the same assumption can be made on the family $\widehat{\mathcal{C}} \to \widehat{B}$, which is the pullback of the original Kuranishi family via $\widehat{\beta}$. We now proceed by contradiction. Were $F_{\widehat{B},\tau}$ not injective, there should be distinct points $t, t' \in \widehat{B}$ such that $[h_{t'}] = [\varphi h_t]$, for some isomorphism $\varphi : \widehat{C}_t \to \widehat{C}_{t'}$. In particular, we would have

$$(8.20)\qquad (h_{t'})_* = \varphi_*(h_t)_* : \mathbb{Z}^{2g} = H_1(S, \mathbb{Z}) \longrightarrow H_1(C_{t'}, \mathbb{Z})\,.$$

By the assumption we just made, the isomorphism φ is induced by an automorphism σ of the family $\widehat{\mathcal{C}} \to \widehat{B}$, so that

$$(8.21)\qquad \varphi = \sigma_t : \widehat{C}_t \longrightarrow \widehat{C}_{t'}\,.$$

If j_t denotes the inclusion of $\widehat{C}_t$ in $\widehat{\mathcal{C}}$, followed by the retraction onto $\widehat{C}_0 = C$, we have

$$(8.22)\qquad (j_{t'}\varphi)_* = (j_{t'}\sigma_t)_* = (\sigma j_t)_* : H_1(\widehat{C}_t, \mathbb{Z})) \to H_1(C, \mathbb{Z}))\,.$$

Now look at the homomorphism

$$(j_t h_t)_* : \mathbb{Z}^{2g} = H_1(S, \mathbb{Z}) \longrightarrow H_1(C, \mathbb{Z})\,.$$

This is a continuous family of surjections. It is therefore constant in t. From (8.20), (8.21), and (8.22) we get

$$\sigma_*(j_t h_t)_* = (j_{t'}\varphi)_*(h_t)_* = (j_{t'} h_{t'})_* = (j_t h_t)_*$$

Since $(j_t h_t)_*$ is surjective, this is possible only if σ, and therefore φ, is the identity. In particular, $\widehat{\beta}(t) = \widehat{\beta}(t')$ and $[h_{t'}] = [h_t]$. By (8.15) and (8.16), this is possible only if $t = t'$.

Q.E.D.

Clearly, varying the point $[C, x_1, \dots, x_n]$ in $\overline{M}_{g,n}$ and the trivialization τ, the images of the $F_{\widehat{B},\tau}$ cover $\widehat{\mathcal{T}}_{S,P}$. Moreover, the sets $F_{\widehat{B},\tau}(\widehat{B})$ form the basis for a topology on $\widehat{\mathcal{T}}_{S,P}$. We give $\widehat{\mathcal{T}}_{S,P}$ this topology.

From the description of this topology, it follows that $\mathcal{T}_{S,P}$ is open in $\widehat{\mathcal{T}}_{S,P}$ and that the map $\beta : \widehat{\mathcal{T}}_{S,P} \to \overline{M}_{g,P}$ is a continuous, open surjection. In particular, $\widehat{\mathcal{T}}_{S,P}/\Gamma_{S,P}$ is homeomorphic to $\overline{M}_{g,P}$. To make contact between the patches $F_{\widehat{B},\tau} : \widehat{B} \to \widehat{\mathcal{T}}_{S,P}$ and Fenchel–Nielsen coordinates, we make the following observation. Once the trivialization τ is chosen, we have a choice of a set $L = \{L_1, \dots, L_\delta\}$ of vanishing cycles automatically made for us. Now we can complete L to a pants decomposition $\mathcal{P}_L = \{L_1, \dots, L_\delta, L_{\delta+1}, \dots, L_{3g-3}\}$ of the surface S, and we can do this in various ways. For each of these ways, we have a set of Fenchel–Nielsen coordinates, and then we have to identify the various quotients (8.1). Now the patches $F_{\widehat{B},\tau}$ take care of all these identifications at once.

9. Bibliographical notes and further reading.

That homotopic homeomorphisms of compact surfaces are isotopic was first proved by Baer [40,41]. Epstein [216] reproved the result, extending it also to noncompact surfaces, possibly with boundary. The first proof of the Dehn–Nielsen theorem, based on hyperbolic geometry, was published in [566] by Nielsen, who however attributed the result to Dehn. Other proofs were later published by various authors; in this chapter we have followed the purely topological proof by Seifert [621]. A very clear and readable exposition of Nielsen's proof, integrated with substantial simplifications introduced by Nielsen himself in an unpublished manuscript, can be found in [646]. This paper also contains an interesting account of the early history of the theorem. A proof of the Dehn–Nielsen theorem for punctured surfaces is contained in [692], where a proof of Baer's theorem is also to be found. Dehn's theorem on the generation of the mapping class group by Dehn twists first appeared in [159] and was later rediscovered by Lickorish [477,478]. Lickorish's proof, with improvements, is reproduced in Birman's book [74]. Humphries' result on the minimum number of Dehn twists needed to generate Γ_g is proved in [377]. Gervais' finite presentation for the mapping class group is constructed in [283].

Regarding Theorem (4.10), the basic existence theorem for the Beltrami equation asserts that it has local solutions, in an appropriate generalized sense, which are homeomorphisms to open subsets of the complex plane. We do not need the full strength of this result, which is due to Morrey [528], but just the fact that the same conclusion holds under the stronger hypothesis that μ is C^∞. This is due to Korn [447] and Lichtenstein [476], who more generally deal with the case where μ satisfies a Hölder condition; a simplified proof of their result was given by Bers [64] and Chern [121] and can be found also in Chapter IV, Section 8, of [151].

Dependence on parameters in Theorem (4.10) is often not considered in the literature; a notable exception is [5], to which we might refer for a proof.

In the literature there are many proofs of Teichmüller's uniqueness theorem. We refer, for example, to [1], [278], or [472] and the bibliography therein.

Standard references for the classical theory of the Teichmüller space are Abikoff [1], Ahlfors [4], Bers [63], Gardiner [278], Grothendieck [325], Hubbard [372], Imayoshi-Taniguchi [385], Nag [560], and Tromba [664].

Our treatement of the bordification of $\mathcal{T}_{S,P}$ is an expansion of the point of view taken by Looijenga in [488]. A thorough treatement of real blow-ups can be found in the paper [373] by Hubbard, Papadopol, and Veselov.

10. Exercises.

A. Generators for the symplectic group

Let Π be the abstract fundamental group of a genus g surface. Put on $\Pi/[\Pi,\Pi] = \mathbb{Z}^{2g}$ and $\mathbb{Z}^{2g}/m\mathbb{Z}^{2g}$ the standard symplectic structures. We denote by $\mathrm{Aut}^+(\Pi)$ the group of automorphisms of Π such that the induced automorphism of $\Pi/[\Pi,\Pi]$ is symplectic. We then have homomorphisms

$$\mathrm{Aut}^+(\Pi) \to Sp_{2g}(R)\,, \tag{10.1}$$

where R denotes either $\mathbb{Z}$ or $\mathbb{Z}/m\mathbb{Z}$. The present series of exercises shows that these homomorphisms are onto.

The *elementary symplectic row operations* on a $2g \times 2g$ symplectic matrix M are the following.

i) For $1 \leq h \leq g$, replacing the h-th row with minus the $(g+h)$th, and the $(g+h)$th with the hth.
ii) For $1 \leq h,k \leq g$, interchanging the hth and kth rows, and the $(g+h)$th and $(g+k)$th rows.
iii) For $1 \leq h \leq 2g$, replacing the hth row with the sum of the hth and $(g+h)$th rows when $h \leq g$, and with the sum of the hth and $(h-g)$th rows when $h > g$.
iv) For $1 \leq h,k \leq g$, replacing the kth row with the sum of the hth and kth rows, and the $(g+h)$th row with the difference between the $(g+h)$th and $(g+k)$th.
v) For $1 \leq h \leq g$, multiplying the hth row by a unit a, and the $(g+h)$th by a^{-1}.

Elementary symplectic column operations are defined in a similar way.

A-1. Show that each of the elementary row (column) operations can be performed by multiplying M on the left (resp., right) by a suitable symplectic matrix. Denote these matrices by E_h, $T_{h,k}$, D_h, $S_{h,k}$, and $U_h(a)$.

A-2. Show that each symplectic matrix M can be reduced to the identity matrix by a sequence of elementary symplectic row and column operations.

A-3. Show that $Sp_{2g}(R)$ is generated by the matrices E_h, $T_{h,k}$, D_h, and $S_{h,k}$.

A-4. Show that (10.1) is onto by exibiting, for each of the matrices E_h, $T_{h,k}$, D_h, $S_{h,k}$, an automorphism of Π mapping to it.

B. **The local description of the bordification $\widehat{\mathcal{T}}_{S,P}$**

C-1. Following the construction of $\widehat{\mathcal{T}}_{S,P}$, construct in all details $\widehat{\mathcal{T}}_{1,1}$

C-2. Prove in detail the following assertions made in the text.

a) $\widehat{B}$ is Hausdorff, first countable, and contractible.
b) Assertion (8.15).
c) $\widehat{\beta} : \widehat{B} \to B$ is continuous and open.
d) Assertion (8.16).
e) For a point $z \in B$ belonging to the stratum D_I, the points in the fiber $\widehat{\beta}^{-1}(z)$ are in one-to-one correspondence with elements in $\Gamma(L)/\Gamma(L_I)$.
f) By varying the point $[C, x_1, \dots, x_n]$ in $\overline{M}_{g,n}$ and the trivialization τ, the sets $F_{\widehat{B},\tau}(B)$ cover $\widehat{\mathcal{T}}_{S,P}$ and form a basis for a topology on $\widehat{\mathcal{T}}_{S,P}$.

Chapter XVI.
Smooth Galois covers of moduli spaces

1. Introduction.

As we anticipated, moduli spaces of stable pointed curves can be represented as quotients of smooth varieties by finite groups. This is a very important fact, since varieties of this kind, even singular, have a naturally defined intersection theory. In the present chapter we shall describe this quotient representation, starting from the case of smooth curves where the constructions are considerably more transparent from a geometrical point of view.

In the first section, our objective is to construct a smooth variety X of which M_g is a quotient by a *finite* group K. We can already say that to make that construction, it will suffice to find a finite-index normal subgroup Λ of the mapping class group Γ_g such that Λ acts freely on $\mathcal{T}_g$, and then set

$$X = \mathcal{T}_g/\Lambda\,, \qquad K = \Gamma_g/\Lambda\,,$$

so that

$$M_g = X/K\,.$$

It is natural to look for subgroups $\Lambda \subset \Gamma_g$ a geometrical meaning.

Let then C be a smooth curve of genus g. As we explained in Section 3 of the preceding chapter, a Teichmüller structure on a curve C can be thought of as an exterior isomorphism $\alpha : \pi_1(C, x) \xrightarrow{\cong} \Pi$, where Π is the homotopy group of the reference surface S. Now let G be a group. A *Teichmüller structure of level* G on C is a surjective exterior homomorphism from $\pi_1(C, x)$ to G, that is, the equivalence class, modulo conjugation, of a surjective homomorphism

$$\alpha : \pi_1(C, x) \twoheadrightarrow G\,.$$

There is natural notion of equivalence of these structures, and the set of these equivalence classes is called the *moduli space of genus g curves with Teichmüller structure of level G*. Following word by word the construction of moduli space or of Teichmüller space, one can put on this set the structure of an analytic space which is denoted by ${}_GM_g$. Clearly, if the group G coincides with Π, the space ${}_GM_g$ coincides with the Teichmüller space.

E. Arbarello et al., *Geometry of Algebraic Curves*,
Grundlehren der mathematischen Wissenschaften 268,
DOI 10.1007/978-3-540-69392-5_8,

There is a little subtlety in the definition of ${}_GM_g$ which is worth mentioning right away. Unlike Teichmüller space, the spaces ${}_GM_g$ are not connected but are the disjoint unions of isomorphic components. The way to isolate a component is to fix an exterior surjective homomorphism

$$\psi : \Pi \twoheadrightarrow G$$

and insist that the G-structure $\alpha : \pi_1(C, x) \to G$ factors through ψ. The equivalence classes of G-structures factoring through ψ form a connected component of ${}_GM_g$, which is denoted by $M_g[\psi]$. The mapping class group Γ_g acts on Π, and it turns out that

$${}_GM_g = \bigcup_{\psi \mod \Gamma_g} M_g[\psi] .$$

A direct way to describe the connected component $M_g[\psi]$ is to look at the subgroup Λ_ψ consisting of all the classes $[\gamma] \in \Gamma_g$ such that $\psi[\gamma] = \psi$. Then

$$M_g[\psi] = \mathcal{T}_g / \Lambda_\psi .$$

There are cases where a single connected component $M_g[\psi]$ seems to be the natural object to be considered. On the other hand, we will see that, in some instance, especially in the compactification process, considering the whole of ${}_GM_g$, presents some definite advantage.

In many of the applications we will require that $\psi : \Pi \to G$ is a strongly characteristic quotient, meaning that ψ is surjective and that it is unique up to composition with automorphisms of G. One of the advantages of this assumption is that under the natural action of $\mathrm{Out}(G)$ on ${}_GM_g$, we have a set-theoretic equality $M_g = {}_GM_g / \mathrm{Out}(G)$.

As an example of strongly characteristic quotient of Π, one could take $G = H_1(S, \mathbb{Z})$ or $G = H_1(S, \mathbb{Z}/m\mathbb{Z})$. For a homomorphism ψ, to be strongly characteristic implies, in particular, that its kernel is carried to itself by every automorphism of Π. This yields a homomorphism

$$\Gamma_g = \mathrm{Out}^+(\Pi) \to \mathrm{Out}(G)$$

whose kernel is exacly Λ_ψ. If we denote by $\Gamma_g[\psi]$ the quotient Γ_g/Λ_ψ, we then have a Galois covering

$$M_g[\psi] \to M_g[\psi]/\Gamma_g[\psi] = M_g .$$

We are interested in the case in which $\Gamma_g[\psi]$ is finite and $M_g[\psi]$ is smooth. Subgroups of Γ_g for which this is the case abound. The first requirement is of course that G be a finite group, so that Λ_ψ has finite index in Γ_g. Now look at a point $z = [C, \alpha] \in M_g[\psi]$, where $\alpha : \pi_1(C) \to G$ factors through ψ. As usual, the local structure of $M_g[\psi]$ at z is of the form B/H, where B is the basis of a Kuranishi family for C, and H is the automorphism group of the pair (C,α). By definition, these automorphisms are the automorphisms of C inducing the identity on G. Thus, to get a smooth Galois cover of M_g, we must look for a ψ for which Λ_ψ has finite index in Γ_g, and H is trivial. The first example that comes to mind is to take as G the first homology group of the reference surface Σ with coefficients in $\mathbb{Z}/m\mathbb{Z}$ and look at the strongly characteristic quotient

$$\chi : \Pi \to H_1(\Sigma, \mathbb{Z}/m\mathbb{Z}) .$$

It is an elementary theorem by Serre that any automorphism of a curve of genus greater than 1 inducing the identity on the first homology group with coefficients in $\mathbb{Z}/m\mathbb{Z}$ is in fact the identity, as soon as $m \geq 3$. The space $M_g[\chi]$ is usually denoted by $M_g[m]$ and is called the moduli space of curves with level m structure. Similarly, one denotes the corresponding group Λ_χ with the symbol Λ_m. The conclusion is that, for $m \geq 3$, these spaces are smooth finite Galois covers of M_g. These spaces are fine moduli spaces in the sense that they represent an appropriate moduli functor and are equipped with a universal family of curves with level m strucure. As one may imagine, the same conclusions are valid for all moduli spaces $M_g[\psi]$ for which $\Lambda_\psi \subset \Lambda_m$ for some $m \geq 3$ or, as one says, for all the spaces $M_g[\psi]$ dominating $M_g[m]$ for some $m \geq 3$.

Of course, it is possible to rephrase the same results in terms of G-structures. In this case one can say that if G is a finite, strongly characteristic quotient of Π such that a quotient of G is isomorphic to $(\mathbb{Z}/m\mathbb{Z})^{2g}$ for some $m \geq 3$, then ${}_GM_g$ is smooth, and

$$M_g = {}_GM_g / \operatorname{Out}(G) . \tag{1.1}$$

In the second section, in order to prapare the study of Galois covers of the compactified space $\overline{M}_g$, we generalize Serre's result to the case of stable curves by proving that the only automorphism of a stable curve inducing the identity on the first homology group with coefficients in $\mathbb{Z}/m\mathbb{Z}$ with $m \geq 3$ is the identity.

In the third section we start with a smooth Galois cover $M_g[\psi] \to M_g$ and try to compactify it. There is a standard procedure to do that. One defines $\overline{M}_g[\psi]$ as the normal closure of $\overline{M}_g$ in the field of rational functions of $M_g[\psi]$. By normality, $\Gamma_g[\psi]$ acts on $\overline{M}_g[\psi]$, and

$$\overline{M}_g[\psi]/\Gamma_g[\psi] = \overline{M}_g .$$

Furthermore, $\overline{M}_g[\psi]$ is the parameter space for a family of genus g stable curves whose moduli map is the quotient $\overline{M}_g[\psi] \to \overline{M}_g$. The problem is to find a group G for which $\overline{M}_g[\psi]$ is smooth. The disadvantage of this approach is that there is no direct control on the boundary points of $\overline{M}_g[\psi]$. From an operative point of view, we will see that, in order to decide whether a boundary point is smooth, one has to *go around it* and use Picard–Lefschetz theory. The reason for this indirect approach is that $\overline{M}_g[\psi]$ no longer presents itself as the coarse moduli space of an algebraic stack, nor does it (coarsely) represent a functor, at least at first sight. Notwithstanding these difficulties, it turns out that it is possible to choose a finite group G for which $\overline{M}_g[\psi]$ is smooth.

The local picture for the map $\mu : \overline{M}_g[\psi] \to \overline{M}_g$ is as follows. Let z be a point in $\overline{M}_g[\psi]$ and write $\mu(z) = [C] \in M_g$, where C is a stable curve. Let B be the base of a Kuranishi family for C. Then there are a neighborhood V of z in $\overline{M}_g[\psi]$ and a ramified Galois cover $\alpha : V \to B$ with the following property. Let B^* be the locus in B parameterizing smooth curves in the given Kuranishi family and set $V^* = V \cap M_g[\psi]$. Then $\alpha : V^* \to B^*$ is a Galois cover, and there is an exact sequence

$$1 \to \pi_1(V^*, v) \xrightarrow{\alpha_*} \pi_1(B^*, b) = \mathbb{Z}^\delta \xrightarrow{PL} H \subset \Gamma_g[\psi] ,$$

where PL is the Picard–Lefschetz transformation and δ the is the number of nodes of C. This already indicates that the Picard–Lefschetz transformation governs the geometry of V^* and therefore of V, and in particular determines whether V is smooth or not. We exemplify this by showing that $\overline{M}_g[m]$ is never smooth.

In the fourth section we change our point of view. As usual, we start with smooth curves. Let G be a finite group, and C a smooth genus g curve. A G-cover of C is simply a principal G-bundle $P \to C$. It is straightforward to see that G-covers admit a moduli space and, more generally, form a stack. One also easily sees that there is a bijection between isomorphism classes of G-covers and isomorphism classes of curves with level G structure. But things are more delicate when considering isomorphism of G-covers versus isomorphism of curves with level G structure. In any event, when the center of G is trivial, one can prove that the stack of G-covers of smooth genus g curves is isomorphic to the stack of smooth genus g curves with G-structure.

While it is far from clear how to generalize the notion of G-structure from smooth curves to stable ones, this generalization comes quite naturally for G-covers. An admissible G-cover consists of a stable curve C, a morphism $f : P \to C$, where P is a connected nodal curve, and an action of G on P, compatible with f, such that:

i) every node of P maps to a node of C,
ii) away from the nodes f is a G-bundle,
iii) if p is a node of P and $\mathrm{Stab}_G(p)$ stands for the stabilizer of p in G, there is a local equation $\xi\eta = 0$ for P at p such that any element of $\mathrm{Stab}_G(p)$ acts as $(\xi, \eta) \mapsto (\zeta\xi, \zeta^{-1}\eta)$, where ζ is an rth root of unity.

The integer r in iii) is called the index of p. It easily follows that all the nodes of P lying above a same node q of C have the same index, so that one often says that r is the index of q. The entire Section 5 is devoted to the theory of admissible G-covers. Using techniques that, at this stage, should be completely familiar to the reader, we construct Kuranishi families for admissible G-covers, compute their tangent spaces, and prove an analogue of the stable reduction theorem for G-covers. Once this is done, one can imitate the construction of the analytic structure on $\overline{M}_g$ and construct a normal, compact analytic space $\mathrm{Adm}_g(G)$ which is a coarse *moduli space for admissible G-covers of genus g stable curves.* It is worth anticipating the following picture. Suppose that C has δ nodes $q_1, \dots, q_\delta$ and that $P \to C$ is an admissible G-cover. Let B and $\widetilde{B}$ be the bases of the Kuranishi families for C and for $P \to C$, respectively. Then one can always assume that B and $\widetilde{B}$ are polycylinders centered at the origin in the variables $t_1, \dots, t_{3g-3}$ and $z_1, \dots, z_{3g-3}$, respectively, and that the classifying map from $\widetilde{B}$ to B is

$$\begin{aligned} &\sigma : \widetilde{B} \to B \\ &t_j = z_j^{n_j}\,, \ j \leq \delta\,, \qquad t_j = z_j\,, \ j > \delta\,, \end{aligned} \tag{1.2}$$

where n_j is the index of the node q_j of C.

By construction, a basis for the analytic structure of $\mathrm{Adm}_g(G)$ is given by sets of the form $\widetilde{B}/H$, where H is the automorphism group of $P \to C$ as a G-cover. It is not difficult to see that the space $\mathrm{Adm}_g(G)$ is a coarse moduli space for an analytic stack (indeed a Deligne–Mumford stack) $\mathcal{A}dm_g(G)$, which is called *the stack of admissible G-covers of stable curves of genus g,* and that we have natural projections

$$\mathrm{Adm}_g(G) \to \overline{M}_g\,, \qquad \mathcal{A}dm_g(G) \to \overline{\mathcal{M}}_g\,.$$

We may now rephrase a remark we made earlier and say that if the center of G is trivial, the open substack of $\mathcal{A}dm_g(G)$ parameterizing G-covers of smooth curves coincides with the stack of curves with level G structure ${}_G M_g$. When the center of G is trivial, it is therefore natural to set

$$ {}_G\overline{M}_g = \mathrm{Adm}_g(G)\,, \qquad {}_G\overline{\mathcal{M}}_g = \mathcal{A}dm_g(G)\,. \tag{1.3}$$

Our next task is to find a group G for which ${}_G\overline{\mathcal{M}}_g$ coincides with ${}_G\overline{M}_g$. For this, we must find a group G having the property that any admissible G-cover of any stable genus g curve is automorphism-free. Once this is done, one can conclude that ${}_G\overline{M}_g$ is a smooth moduli space and that $\overline{M}_g$ is a quotient of it by the finite group $\mathrm{Out}(G)$.

In Section 6 we study automorphism groups of admissible G-covers. Starting with an admissible G-cover $\pi : P \to C$, there are various automorphism groups that are attached to this cover. The group $\mathrm{Aut}(P\to C)$ is the group of pairs (f,α) where $f: P\to P$ and $\alpha: C\to C$ are automorphisms such that $\alpha\pi = \pi f$. One can also look at the subgroup $\mathrm{Aut}(P/C)\subset \mathrm{Aut}(P\to C)$ formed by pairs of type (f,id_C). Finally, in both groups, we may look at the G-invariants subgroups $\mathrm{Aut}(P\to C)^G\subset \mathrm{Aut}(P\to C)$ and $\mathrm{Aut}(P/C)^G\subset \mathrm{Aut}(P/C)$.

By definition, the automorphism group of an admissible G-cover $f: P\to C$ is the group $\mathrm{Aut}(P\to C)^G$. In the entire section we work under the following assumptions:

$$
(1.4)\quad \begin{cases} \text{i)} & G \text{ is a strongly characteristic quotient of } \Pi;\\ \text{ii)} & \text{the center } Z(G) \text{ of } G \text{ is trivial;}\\ \text{iii)} & \text{a quotient of } G \text{ is isomorphic to } (\mathbb{Z}/m\mathbb{Z})^{2g} \text{ for some } m\geq 3. \end{cases}
$$

In this setting one can immediately prove that

$$\mathrm{Aut}(P\to C)^G = \mathrm{Aut}(P/C)^G\,.$$

Next, starting from a Kuranishi family $\pi:\mathcal{C}\to B$ for C, one looks at a Kuranishi family $\mathcal{P}\to\mathcal{Z}\to\widetilde{B}$ for $P\to C$ and gets diagrams

$$
(1.5)\qquad \begin{array}{ccc} \mathcal{P} & \longrightarrow & \mathcal{C}\\ \big\downarrow & & \big\downarrow{\scriptstyle\pi}\\ \widetilde{B} & \xrightarrow{\ \sigma\ } & B \end{array} \qquad\qquad \begin{array}{ccc} \mathcal{P}^* & \longrightarrow & \mathcal{C}^*\\ \big\downarrow & & \big\downarrow{\scriptstyle\pi}\\ \widetilde{B}^* & \xrightarrow{\ \sigma\ } & B^* \end{array}
$$

the second of which is obtained from the first by considering only smooth fibers. With some work one sees that

$$\mathrm{Aut}(P\to C)^G = \mathrm{Aut}(P/C)^G = \mathrm{Aut}(\mathcal{P}^*/\mathcal{C}^*)^G$$

and that there is an exact split sequence

$$
(1.6)\qquad 1\to G=\mathrm{Aut}(P_a/C_a)\xrightarrow{\alpha}\mathrm{Aut}(\mathcal{P}^*/\mathcal{C}^*)\xrightarrow{\beta}\mathrm{Aut}(\widetilde{B}^*/B^*)\to 1\,,
$$

where $a\in B^*$, $C_a=\pi^{-1}(a)$, and P_a is any fiber above a point of $\sigma^{-1}(a)$. The next observation is that this exact sequence can be identified with the exact sequence

$$
(1.7)\qquad 1\to \pi_1(C_a)/\pi_1(P_a)\xrightarrow{\alpha}\pi_1(\mathcal{C}^*)/\pi_1(\mathcal{P}^*)\xrightarrow{\beta}\pi_1(B^*)/\pi_1(\widetilde{B}^*)\to 1\,.
$$

Set $M=\pi_1(B^*)/\pi_1(\widetilde{B}^*)$. We have

$$M=\pi_1(B^*)/\pi_1(\widetilde{B}^*)=\bigoplus_{i=1}^{\delta}\mathbb{Z}/n_i\mathbb{Z}\,,$$

where we assume that C has δ nodes of indices $n_1, \ldots, n_\delta$. Thus $\mathrm{Aut}(\mathcal{P}^*/\mathcal{C}^*) \cong G \ltimes M$. Let $\tau : M \to \mathrm{Aut}(G)$ be the homomorphism defining this semidirect product. One then checks that the induced homomorphism $\overline{\tau} : M \to \mathrm{Out}(G)$ is induced by the Picard–Lefschetz transformation

$$PL : \pi_1(B^*, b) \to \Gamma_g \to \mathrm{Out}(G)\,. \tag{1.8}$$

Our task is to analyze $\mathrm{Aut}(\mathcal{P}^*/\mathcal{C}^*)^G$ and find conditions that make this group trivial. Now for an element $(h, m) \in G \ltimes M$ to be G-invariant, we must have

$$\tau(m)(g) = h^{-1}gh \quad \forall g \in G\,, \tag{1.9}$$

meaning that $\overline{\tau}(m)$ is the identity of $\mathrm{Out}(G)$. Thus, a condition for the automorphism group of $P \to C$ to be trivial is that M injects in $\mathrm{Out}(G)$ or, equivalently, that

$$\ker(PL) \subset n_1\mathbb{Z} \oplus \cdots \oplus n_\delta\mathbb{Z}\,.$$

It turns out that the reverse of this inclusion is trivial. The remaining part of this section, Section 7, and the technical Section 8 are devoted to the explicit description of a specific group $G_{\{p,q\}}$ with the following properties:

i) the order of any element of $G_{\{p,q\}}$ divides pq, where p and q are distinct primes,
ii) for every stable genus g curve, the Picard–Lefschetz transformation (1.8) has the property that

$$\ker PL = pq\mathbb{Z} \oplus \cdots \oplus pq\mathbb{Z}\,.$$

Once this is done, we set $G = G_{\{p,q\}}$ and the construction of a smooth moduli space ${}_G\overline{M}_g$ with

$$\overline{M}_g = {}_G\overline{M}_g / \mathrm{Out}(G)$$

is completed.

In the last section we generalize the preceding construction to the case of stable n-pointed curves. Let $(C; p_1, \ldots, p_n)$ be an n-pointed stable curve of genus g, and let G be a finite group. We define an admissible G-cover of $(C; p_1, \ldots, p_n)$ to be an admissible G-cover $P \to C$ plus the choice of a point $q_i \in P$ mapping to p_i for each i. It is then easy to define a stack ${}_G\overline{\mathcal{M}}_{g,n}$ and a space ${}_G\overline{M}_{g,n}$ and to verify that these coincide if $G = G_{\{p,q\}}$. In particular, in this case, ${}_G\overline{M}_{g,n}$ is smooth. One can then find a natural action of the group

$$\widetilde{H} = G^n \ltimes \mathrm{Aut}(G)$$

on ${}_G\overline{M}_{g,n}$ and verify that a specific quotient H of $\widetilde{H}$ acts effectively on ${}_G\overline{M}_{g,n}$ and that

$$\overline{M}_{g,n} = {}_G\overline{M}_{g,n}/H\,.$$

Finally, we observe that the fiber of the natural projection $\pi_{n+1} : {}_G\overline{M}_{g,n+1} \to {}_G\overline{M}_{g,n}$ at a point corresponding to an admissible G-cover $P \to C$ is the curve P.

2. Level structures on smooth curves.

Our main goal in this section is to show that the moduli spaces of smooth curves can be represented as quotients of smooth varieties modulo the action of finite groups. In this section and in the next few ones, we will only deal with the moduli space M_g, leaving for a later section the discussion of the case of pointed curves.

Let C be a smooth curve of genus g, and let G be a group. A *Teichmüller structure of level* G on C is a surjective exterior homomorphism from $\pi_1(C,x)$ to G, that is, the equivalence class, modulo conjugation, of a surjective homomorphism $\pi_1(C,x) \to G$. Notice that the choice of the base point x is irrelevant; in fact, if x' is another base point, the groups $\pi_1(C,x)$ and $\pi_1(C,x')$ can be identified, canonically up to inner automorphisms. An isomorphism between two curves C, C' with level G structure is an isomorphism $f : C \to C'$ such that, if the level structures are given by homomorphisms $\alpha : \pi_1(C,x) \to G$ and $\alpha' : \pi_1(C', f(x)) \to G$, then $\alpha = \beta \circ \alpha' \circ f_*$ for some inner automorphism β of G. The isomorphism class of a curve C with level G structure α will be denoted with $[C;\alpha]$, and the set of isomorphism classes of genus g curves with level G structure with ${}_GM_g$.

As in the case of Teichmüller structures (see Section 2 of Chapter XV), one can define the notion of family of curves with level G structure and use it to introduce a complex structure on ${}_GM_g$; quite naturally, the resulting space will be called the *moduli space of genus* g *curves with Teichmüller structure of level* G. The construction of the complex structure on ${}_GM_g$ is entirely similar to the one of the complex structures on M_g and $\mathcal{T}_g$, and will not be repeated here. Suffice it to say that, locally near a point x corresponding to a curve C with level G structure $\alpha : \pi_1(C) \to G$, the space ${}_GM_g$ looks like B/H, where B is the base of a suitable Kuranishi family for C, and H is the subgroup of $\mathrm{Aut}(C)$ consisting of those automorphisms which respect the level structure. In particular, the forgetful map ${}_GM_g \to M_g$ can be locally identified with $B/H \to B/\mathrm{Aut}(C)$ and hence is a morphism of analytic spaces. Another consequence of the local description of ${}_GM_g$ as B/H is that $H = \{1\}$ is a sufficient condition for ${}_GM_g$ to be smooth at x. In other words, if $\mathrm{Aut}(C,\alpha) = \{1\}$, then ${}_GM_g$ is smooth at the point corresponding to (C,α). When G is a finite group, the morphism ${}_GM_g \to M_g$ is a finite ramified covering, since the fundamental group of a compact Riemann surface is finitely generated, and hence ${}_G\mathcal{M}_g$ is an algebraic variety, by the generalized Riemann extension theorem.

The space ${}_GM_g$ has an "analytic stack" counterpart ${}_G\mathcal{M}_g$, whose objects are the families of smooth genus g curves with level G structure. We will not insist too much on this point of view, since we will be primarily interested in the cases in which ${}_G\mathcal{M}_g$ coincides with (the stack attached to) ${}_GM_g$, that is, the cases in which all genus g curves with level G structure are automorphism-free; examples of such a situation are not difficult to find, as we shall see below. Here we just notice that ${}_G\mathcal{M}_g$ is clearly smooth and that it is not difficult to show that, when G is a finite group, it comes from an algebraic stack which is in fact a Deligne–Mumford stack.

It is useful to describe in some detail the relations intercurring between ${}_GM_g$ and the Teichmüller space $\mathcal{T}_g$. Fix a reference genus g surface Σ and denote by Π the fundamental group $\pi_1(\Sigma)$. Choose a specific surjective exterior homomorphism $\psi : \Pi \to G$. Then to each point $[C, [f]]$ of the Teichmüller space $\mathcal{T}_g = \mathcal{T}_\Sigma$ we can associate a surjective exterior homomorphism $\alpha : \pi_1(C) \to G$ by composing $f_* : \pi_1(C) \to \Pi$ with ψ. We thus get a map

$$t_\psi : \mathcal{T}_g \to {}_GM_g \,,$$

which is readily seen to be an open morphism of analytic spaces. Notice however that t_ψ is not necessarily onto. To see why this is the case, observe first that the mapping class group $\Gamma_g = \Gamma_\Sigma$ acts on the right, via composition, on the set of surjective exterior homomorphisms from Π to G; thus, if $\psi : \Pi \to G$ is a surjective exterior homomorphism and $[\xi] \in \Gamma_g$, then $\psi[\xi] = \psi \circ \xi_*$. We then claim that, if $\psi' : \Pi \to G$ is another surjective exterior homomorphism, then $t_\psi(\mathcal{T}_g)$ and $t_{\psi'}(\mathcal{T}_g)$ have a nonempty intersection if and only if ψ and ψ' belong to the same Γ_g-orbit, in which case $t_\psi(\mathcal{T}_g) = t_{\psi'}(\mathcal{T}_g)$; thus,

$$(2.1) \qquad {}_GM_g = \coprod_{\psi \bmod \Gamma_g} t_\psi(\mathcal{T}_g) \,.$$

Suppose in fact that ψ and ψ' belong to the same orbit and, more precisely, that $\psi' = \psi[\xi]$, where $[\xi]$ is an element of Γ_g. Then, for any $x \in \mathcal{T}_g$, we have that

$$t_{\psi'}(x) = t_{\psi[\xi]}(x) = t_\psi([\xi](x)) \,.$$

This shows that $t_{\psi'}(\mathcal{T}_g) = t_\psi(\mathcal{T}_g)$. Conversely, suppose that $t_\psi(x) = t_{\psi'}(x')$, where $x = [C, [f]]$ and $x' = [C', [f']]$. This means that there is an isomorphism $\beta : C' \to C$ such that the equality of exterior homomorphisms

$$\psi' \circ f'_* = \psi \circ f_* \circ \beta_*$$

holds. Set $\xi = f \circ \beta \circ f'^{-1}$. Then $[\xi]$ is an element of Γ_g, and $\psi' = \psi[\xi]$, showing that ψ and ψ' belong to the same Γ_g-orbit. If the group G is finite, ${}_GM_g$ has finitely many connected components, since it is an algebraic variety, as we have already observed. We set

$$M_g[\psi] = t_\psi(\mathcal{T}_g)$$

and denote by $\mathcal{M}_g[\psi]$ the stack counterpart of $M_g[\psi]$. The space $M_g[\psi]$ is called the moduli space of genus g curves with ψ-structure. We also let Λ_ψ be the subgroup of Γ_g consisting of all those classes $[\xi]$ such that $\psi[\xi] = \psi$. Clearly, $t_\psi(x) = t_\psi(y)$ if and only if there is a $[\xi] \in \Lambda_\psi$ such that $[\xi](x) = y$. Thus,

$$M_g[\psi] = \mathcal{T}_g/\Lambda_\psi \,.$$

Clearly, when G is a finite group, Λ_ψ is a subgroup of finite index, and $M_g[\psi]$ is an algebraic variety.

If Λ_ψ happens to be a normal subgroup of Γ_g, then $M_g[\psi] \to M_g$ is a Galois covering, and

$$M_g = M_g[\psi]/\Gamma_g[\psi] \,,$$

where $\Gamma_g[\psi]$ denotes the quotient group Γ_g/Λ_ψ. A sufficient condition for the normality of Λ_ψ is that the kernel of ψ be a *characteristic* subgroup of Π, meaning that it is carried to itself by every automorphism of Π. In fact, if the kernel of ψ is characteristic, every automorphism γ of Π passes to the quotient and induces an automorphism $\overline{\gamma}$ of G; this yields a homomorphism

$$\Gamma_g = \mathrm{Out}^+(\Pi) \to \mathrm{Out}(G) \tag{2.2}$$

whose kernel is just Λ_ψ. In this case the action of Γ_g on $M_g[\psi]$ is easy to describe: if C is a genus g curve endowed with a level structure $\alpha : \pi_1(C) \to G$, an element $\gamma \in \Gamma_g$ acts by sending $[C;\alpha]$ to $[C;\overline{\gamma}\alpha]$. If the kernel of $\psi : \Pi \twoheadrightarrow G$ is characteristic, we shall sometimes say that ψ is a *characteristic (exterior) homomorphism* and that G is a *characteristic quotient* of Π. The groups G that we shall use to exhibit the moduli spaces of genus g curves as quotients of smooth schemes modulo the actions of finite groups will actually enjoy the more stringent property of being strongly characteristic quotients of Π in the following sense. Given groups H and G, we shall say that G is a *strongly characteristic quotient* of H if there is a surjective homomorphism $H \to G$, and this homomorphism is unique up to composition with automorphisms of G; equivalently, if there is one and only one subgroup K of H such that H/K is isomorphic to G. In this case, we shall also say that K is a *strongly characteristic subgroup* of H.

It is evident that, if K is a strongly characteristic subgroup of H, then K is also characteristic. The converse, however, is not true, as the following simple example shows. Set $H = \mathfrak{S}_3 \times C_2$, where $\mathfrak{S}_3$ is the symmetric group on three elements, and C_2 is cyclic of order two, and let K be the subgroup of H generated by C_2 and by a 3-cycle. The subgroup K is characteristic, since C_2 is the center of H and K/C_2 is characteristic in $H/C_2 \cong \mathfrak{S}_3$. On the other hand, H/K is isomorphic to $H/\mathfrak{S}_3$.

Strongly characteristic quotients abound. Perhaps the simplest significant example is the following.

Example (2.3). Let H be a finitely generated group, and let m be a positive integer. We denote by $H^{(2),m}$ the normal subgroup of H generated by commutators and mth powers and set $G = H/H^{(2),m}$. We claim that G is a strongly characteristic quotient of H. In fact, let $H \to G$ be a surjective homomorphism, and let K be its kernel. Since G is abelian, K contains the commutator subgroup, and since the order of any element of G is a divisor of m, K contains all mth powers. It follows that $K \supset H^{(2),m}$. But then K must coincide with $H^{(2),m}$ since G is a finite group.

By construction, $M_g[\psi]$ parameterizes genus g curves C endowed with a level G structure $\alpha : \pi_1(C) \to G$ which admits a factorization $\alpha = \psi \circ \beta$, where $\beta \in \mathrm{Iso}^{ext}(\pi_1(C), \Pi)$ is an orientation-preserving exterior isomorphism. We may call such a level structure a *level ψ structure.* Now suppose that, for every point of $M_g[\psi]$, the corresponding curve with level G structure has no nontrivial automorphisms. We know that in this case $M_g[\psi]$ is smooth; moreover, by patching together Kuranishi families, it is possible to construct a universal family

$$\eta : \mathcal{X}[\psi] \to M_g[\psi] \tag{2.4}$$

of smooth genus g curves with level ψ structure. In other words, $\mathcal{M}_g[\psi]$ coincides with $M_g[\psi]$. By construction, the moduli map of this family is the canonical projection

$$\mu = \mu_\psi : M_g[\psi] \to M_g \,.$$

This discussion suggests us a way of achieving what is our main goal in this section, that is, to express M_g as the quotient of a smooth variety by a finite group. It suffices to find a finite group G and a surjective homomorphism $\psi : \Pi \to G$ such that every genus g curve with level ψ structure is automorphism-free.

The first idea that comes to mind actually works. Let m be a positive integer, and let

$$\chi : \Pi \to H_1(\Sigma, \mathbb{Z}/m\mathbb{Z}) \tag{2.5}$$

be the homomorphism obtained by composing $\Pi \to \Pi/[\Pi,\Pi] = H_1(\Sigma,\mathbb{Z})$ with the "reduction modulo m" homomorphism $H_1(\Sigma,\mathbb{Z}) \to H_1(\Sigma,\mathbb{Z}/m\mathbb{Z})$.

A level χ structure on a genus g curve C is also called a *Jacobi structure of level* m or, more briefly, a *level* m *structure*. An equivalent, and more usual, definition is that such a structure is an isomorphism

$$H_1(C, \mathbb{Z}/m\mathbb{Z}) \to H_1(\Sigma, \mathbb{Z}/m\mathbb{Z}) \tag{2.6}$$

respecting the symplectic structures given by the intersection pairings. To see why the two definitions are equivalent, notice first that, if χ is as in (2.5), then a level χ structure certainly induces a symplectic isomorphism (2.6). Conversely, let such an isomorphism α be given, and let β be another symplectic isomorphism induced by some orientation-preserving homeomorphism from C to Σ, so that in particular β comes from a level χ structure on C. Since $\alpha \circ \beta^{-1}$ is a symplectic automorphism of $H_1(\Sigma, \mathbb{Z}/m\mathbb{Z})$, to show that α also comes from a level χ structure, it suffices to show that the natural homomorphism

$$\Gamma_\Sigma \to Sp(H_1(\Sigma, \mathbb{Z}/m\mathbb{Z})) \tag{2.7}$$

from Γ_Σ to the group of symplectic automorphisms of $H_1(\Sigma, \mathbb{Z}/m\mathbb{Z})$ is surjective. Since $H_1(\Sigma, \mathbb{Z})$ and $H_1(\Sigma, \mathbb{Z}/m\mathbb{Z})$ can be identified with $\mathbb{Z}^{2g}$ and $(\mathbb{Z}/m\mathbb{Z})^{2g}$, endowed with the standard symplectic structures, this follows from Exercises A-1–A-4 in Chapter XV. This argument also shows that $\Gamma_g[\chi]$ coincides with the symplectic group $Sp(H_1(\Sigma, \mathbb{Z}/m\mathbb{Z}))$.

It is a consequence of Example (2.3) that $H_1(\Sigma, \mathbb{Z}/m\mathbb{Z})$ is a strongly characteristic quotient of Π. In fact, the kernel of (2.5) is $\Pi^{(2),m}$, the subgroup of Π generated by all commutators and all mth powers. We shall write $M_g[m]$, $\Lambda_{[m]}$, and $\Gamma_g[m]$ to indicate $M_g[\chi]$, Λ_χ, and $\Gamma_g[\chi]$, respectively. As we have seen,

$$\Gamma_g[m] = Sp_{2g}(\mathbb{Z}/m\mathbb{Z}) .$$

Moreover, $M_g[m]$, the space of genus g curves with level m structure, parameterizes isomorphism classes of pairs (C, ρ) where

$$\rho : H_1(C, \mathbb{Z}/m\mathbb{Z}) \to (\mathbb{Z}/m\mathbb{Z})^{2g}$$

is a symplectic isomorphism. We will show that $M_g[m]$ is smooth when $m \geq 3$. For this, it suffices to show that, if $m \geq 3$, then the pairs (C, ρ) are rigid, i.e., that

$$\mathrm{Aut}(C, \rho) = 1.$$

In other words, it is enough to prove the following statement.

PROPOSITION (2.8). *Let C be a smooth curve of genus $g \geq 1$, and let $m \geq 2$ be an integer. Let φ be an automorphism of C and denote by $\overline{\varphi}_*$ the induced automorphism of $H_1(C, \mathbb{Z}/m\mathbb{Z})$. Suppose that $\overline{\varphi}_* = 1$ and, in case $g = 1$, that φ fixes a point of C. Then either $\varphi = 1$, or $m = 2$ and φ has order two.*

Proof. To begin with, by Lemma (2.2) in Chapter XV, it is sufficient to show that the conclusion holds for the induced homomorphism $\varphi_* : H_1(C,\mathbb{Z}) \to H_1(C,\mathbb{Z})$ in integral homology. Since the assumptions imply that φ has finite order, the proposition is thus a consequence of the following elementary algebraic result.

LEMMA (2.9). *Let ρ be an automorphism of $\mathbb{Z}^N$, and let $m \geq 2$ be an integer. Suppose that ρ has finite order and that the automorphism of $(\mathbb{Z}/m\mathbb{Z})^N$ obtained from ρ by reduction modulo m is the identity. Then either $\rho = 1$, or $\rho^2 = 1$ and $m = 2$.*

Proof. Suppose that ρ is not the identity. Let n be the order of ρ, and let p be a prime factor of n. By assumption, we can write $\rho^{n/p} = 1 + m\alpha$, where $\alpha \neq 0$. Then

$$1 = (1+m\alpha)^p = 1 + pm\alpha + \binom{p}{2} m^2\alpha^2 + \cdots + m^p\alpha^p,$$

so that

$$pm\alpha = -\binom{p}{2} m^2\alpha^2 - \cdots - m^p\alpha^p.$$

Since $\alpha \neq 0$, there exists a surjective homomorphism $F : \mathrm{Im}(\alpha) \to d\mathbb{Z}$ for some $d \neq 0$. Pick a v such that $F(\alpha(v)) = d$ and set $dn_i = F(\alpha^i(v))$ for $i \geq 2$. We then have that

$$pm = -\binom{p}{2} m^2 n_2 - \binom{p}{3} m^3 n_3 - \cdots - m^p n_p\,.$$

Since the right-hand side is divisible by m^2, we must have $p = m$. If $m \geq 3$, the right-hand side is divisible by p^3, which is absurd. Hence $m = p = 2$.

We can say at this point that either ρ is the identity, or else $m = 2$ and the order of ρ is a power of two; in particular, the lemma is fully proved except when $m = 2$. We shall show that, if the order of ρ is assumed to be strictly greater than 2, one reaches a contradiction. Replacing ρ with one of its powers, one can assume that $\rho^4 = 1$ but $\rho^2 \neq 1$. Write $\rho = 1 + 2\beta$, where β is an endomorphism of $\mathbb{Z}^N$. Then $\rho^2 = 1 + 4\beta + 4\beta^2$, and hence the reduction modulo 4 of ρ^2 is the identity. The case $m = 4$ of the lemma then shows that ρ^2 is the identity, a contradiction. This completes the proof of (2.9) and hence also the one of (2.8).

REMARK (2.10). It should be mentioned, for completeness, that the endomorphisms ρ of $\mathbb{Z}^N$ whose square is the identity and whose reduction modulo 2 is also the identity are very easy to describe: they are just those endomorphisms whose matrix, relative to a suitable basis of $\mathbb{Z}^N$, is diagonal with diagonal entries all equal to ± 1. In fact, if we write $\rho = 1 + 2\alpha$, to say that $\rho^2 = 1$ means that $\alpha + \alpha^2 = 0$. This immediately implies that $v + \alpha(v)$ is ρ-invariant and $\alpha(v)$ anti-invariant for any $v \in \mathbb{Z}^N$. Thus, denoting by V_0 and V_1 the images of $1 + \alpha$ and α, we find that $\mathbb{Z}^N = V_0 \oplus V_1$, that V_0 and V_1 are ρ-invariant, and that ρ acts as the identity on V_0 and as minus the identity on V_1.

We say that the level structure associated to a surjective homomorphism $\psi : \Pi \to G$ *dominates* the one associated to another surjective homomorphism $\varphi : \Pi \to H$ if φ factors into the composition of ψ with some homomorphism $G \to H$. Obviously, if every genus g curve with level φ structure is automorphism-free, then the same is true for every curve with level ψ structure. We can therefore summarize the results obtained so far in the following theorem.

THEOREM (2.11). *Let $\psi : \Pi \to G$ be a surjective exterior homomorphism defining a level structure which dominates a level m structure for some $m \geq 3$. Suppose that G is a finite group. Then $M_g[\psi]$ is a smooth finite ramified covering of M_g. Moreover, $\mathcal{M}_g[\psi] = M_g[\psi]$; in other words, over $M_g[\psi]$ there exists a universal family $\eta : \mathcal{X}[\psi] \to M_g[\psi]$ of smooth genus g curves with level ψ structure. If ψ is characteristic, then $M_g[\psi]$ is a finite Galois covering of M_g with Galois group $\Gamma_g[\psi]$; the stabilizer of a point of $M_g[\psi]$ under the action of $\Gamma_g[\psi]$ coincides with the automorphism group of the corresponding curve.*

We shall now state and prove a variant of (2.11) which is often more practical to use. As usual, we denote by Π the fundamental group of the reference genus g surface Σ. Let G be a finite strongly characteristic quotient of Π. The group of exterior automorphisms of G acts on ${}_GM_g$ via composition, and this action corresponds, via (2.2), to the one of Γ_g on the individual components $M_g[\psi]$ of ${}_GM_g$. Let $[C;\alpha]$ and $[C';\alpha']$ be two points of ${}_GM_g$ and suppose that there exists an isomorphism $f : C \to C'$. Since G is a strongly characteristic quotient of Π and hence of $\pi_1(C)$, there exists an automorphism β of G which makes the diagram

$$\begin{array}{ccc} \pi_1(C) & \xrightarrow{\ \alpha\ } & G \\ {\scriptstyle f_*}\downarrow & & \downarrow{\scriptstyle \beta} \\ \pi_1(C') & \xrightarrow{\ \alpha'\ } & G \end{array}$$

commutative. This shows in particular that, as a set, M_g is the quotient of ${}_GM_g$ modulo $\mathrm{Out}(G)$.

THEOREM (2.12). *Let G be a finite strongly characteristic quotient of Π. Suppose that some quotient of G is isomorphic to $(\mathbb{Z}/m\mathbb{Z})^{2g}$ for some $m \geq 3$. Then*

i) ${}_G\mathcal{M}_g = {}_GM_g$;

ii) $M_g = {}_G M_g / \operatorname{Out}(G)$;
iii) $\mathcal{M}_g = [{}_G M_g / \operatorname{Out}(G)]$.

The assumptions say in particular that any level G structure on a smooth genus g curve dominates a level m structure; thus i) holds. That ii) is true set-theoretically has been observed in the discussion preceding the statement; the validity of ii) then follows from the normality of M_g. It remains to prove iii). We shall just sketch the outline of a proof, leaving it to the reader to formalize the argument and fill in the details. Start with an object in $\mathcal{M}_g$, that is, with a family $f : X \to S$ of smooth genus g curves. Consider the set P of pairs (s, α), where $s \in S$, and α is a surjective exterior homomorphism from $\pi_1(X_s)$ to G, with the obvious scheme structure. The group $\operatorname{Out}(G)$ acts on P by composition and makes $P \to S$ into a principal $\operatorname{Out}(G)$-bundle. Moreover, P maps $\operatorname{Out}(G)$-equivariantly to ${}_G M_g$ by sending (s, α) to $[X_s; \alpha]$. We have thus produced a section of $[{}_G M_g / \operatorname{Out}(G)]$ over S.

Conversely, start with a principal $\operatorname{Out}(G)$-bundle $Q \to S$ plus an $\operatorname{Out}(G)$-equivariant map $Q \to {}_G M_g$. The group $\operatorname{Out}(G)$ acts equivariantly on the universal family $X \to {}_G M_g$. Thus this action pulls back to an equivariant action on $Y = X \times_{{}_G M_g} Q \to Q$, which is clearly free. Taking quotients yields a family $Y/\operatorname{Out}(G) \to Q/\operatorname{Out}(G) = S$ of smooth genus g curves.

We leave it to the reader to check that the two constructions we have described are inverse to each other and give an isomorphism between $\mathcal{M}_g$ and $[{}_G M_g / \operatorname{Out}(G)]$.

In our construction of smooth Galois covers of M_g the key ingredient has been Proposition (2.8). It is clear that, if we want to construct smooth Galois covers of $\overline{M}_g$, we shall need an analogous statement concerning stable curves. In the next section we shall address this preliminary question.

3. Automorphisms of stable curves.

The main purpose of this section is to prove the following generalization of Proposition (2.8).

PROPOSITION (3.1). *Let $m \geq 3$ be an integer. Let X be a stable curve of genus $g \geq 2$. Let φ be an automorphism of X, and let $\overline{\varphi}_*$ be the induced automorphism on $H_1(X, \mathbb{Z}/m\mathbb{Z})$. Suppose that $\overline{\varphi}_* = 1$. Then $\varphi = 1$.*

Let $\{x_i\}_{i \in I}$ be the set of double points of X. Let

$$X = \bigcup_{\alpha \in A} X_\alpha$$

be the decomposition of X into irreducible components. Denote by Σ_α the normalization of X_α and let g_α be its genus. Let Γ be the dual graph of X. Recall that the set V of vertices and the set E of edges of Γ are in one-to-one correspondence, respectively, with the set of irreducible components and with the set of nodes of X. We denote by v_α the node corresponding to the component X_α and by t_i the edge corresponding to the node x_i. We denote by $|\Gamma|$ the one-dimensional CW-complex associated to Γ.

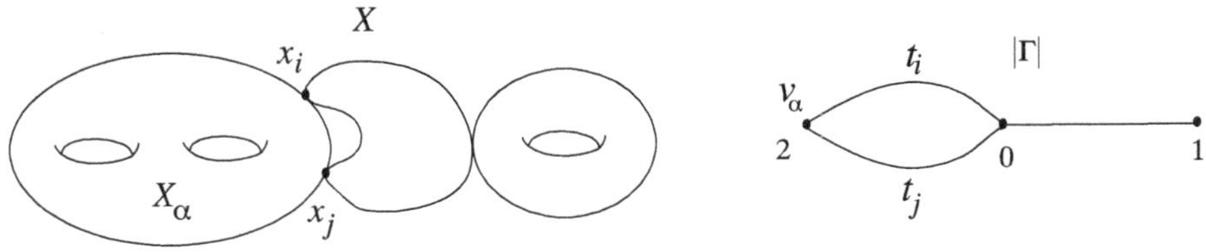

Figure 1.

Consider next the CW-complex Σ' obtained from X by substituting an edge t'_i to each double point x_i. Of course, Σ' is homotopically equivalent to X, and we define a continuous map q from Σ' to $|\Gamma|$ sending t'_i to t_i and contracting each Σ_α to v_α:

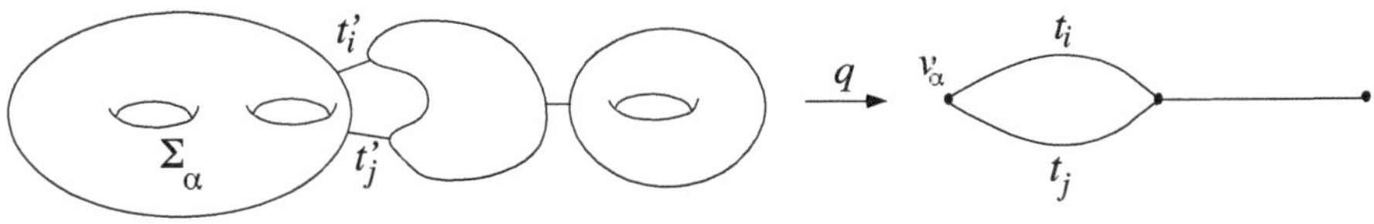

Figure 2.

For any commutative ring with unit R, we then have the exact sequence

$$0 \to \bigoplus_{\alpha \in A} H_1(\Sigma_\alpha, R) \to H_1(X, R) \xrightarrow{q_*} H_1(|\Gamma|, R) \to 0\,. \tag{3.2}$$

To prove this, we compare the long exact homology sequences of the pairs $(\Sigma', \cup\Sigma_\alpha)$ and $(|\Gamma|, \cup\{v_\alpha\})$. The central portions of the these reduce to

$$\begin{array}{ccccccccc} 0 & \longrightarrow & H_1(\cup\Sigma_\alpha) & \longrightarrow & H_1(\Sigma') & \longrightarrow & H_1(\Sigma', \cup\Sigma_\alpha) & \longrightarrow & H_0(\cup\Sigma_\alpha) \\ & & \downarrow & & \downarrow & & \| & & \| \\ & & 0 & \longrightarrow & H_1(|\Gamma|) & \longrightarrow & H_1(|\Gamma|, \cup\{v_\alpha\}) & \longrightarrow & H_0(\cup\{v_\alpha\}) \end{array}$$

and a trivial diagram chase proves (3.2). Taking $R = \mathbb{Z}$, it follows from (3.2) that $H_1(X, \mathbb{Z})$ has no torsion, since both $\bigoplus_\alpha H_1(\Sigma_\alpha, \mathbb{Z})$ and $H_1(|\Gamma|, \mathbb{Z})$ are torsion-free.

When $R = \mathbb{Z}/m\mathbb{Z}$, the exact sequence (3.2) specializes to

$$0 \to \bigoplus_{\alpha \in A} H_1(\Sigma_\alpha, \mathbb{Z}/m\mathbb{Z}) \to H_1(X, \mathbb{Z}/m\mathbb{Z}) \xrightarrow{q_*} H_1(|\Gamma|, \mathbb{Z}/m\mathbb{Z}) \to 0. \tag{3.3}$$

The fact that φ acts trivially on the middle term of this sequence tells us that it also does on the two extremes. As a first consequence, we may infer that $\varphi(X_\alpha) = X_\alpha$, and therefore $\varphi(v_\alpha) = v_\alpha$, as long as $g_\alpha \geq 1$. The following result implies that, in fact, all vertices and edges (and even half-edges) of Γ are left fixed by φ.

LEMMA (3.4). *Let $m \geq 3$ be an integer. Let Γ be the graph of a stable curve. Let ξ be an automorphism of Γ fixing all univalent and bivalent vertices and such that ξ_* is the identity on $H^1(|\Gamma|, \mathbb{Z}/m\mathbb{Z})$. Then ξ is the identity.*

Proof. We prove the lemma by induction on the number $|V|+|E|$, taking as a basis for the induction the trivial case $|V| = 1$, $|E| = 0$. The induction step is as follows. First, we can assume that there is no edge in Γ with just one vertex v. If not, this edge forms a loop c in Γ which represents a nonzero element in $H_1(|\Gamma|, \mathbb{Z}/m\mathbb{Z})$. Since ξ_* is the identity on $H_1(|\Gamma|, \mathbb{Z}/m\mathbb{Z})$, we have that $\xi(c) = c$ and $\xi(v) = v$. Removing c yields a subgraph Γ' which is carried to itself by ξ. The restriction of ξ to Γ' satisfies the assumptions of the lemma. Hence, by induction hypothesis, ξ is the identity on Γ' and therefore on Γ. We shall now argue by contradiction, assuming that ξ is not the identity. First of all, we will show that one can assume that there is no fixed vertex $v \in \Gamma$. There are various cases to be considered. Suppose first that a fixed vertex v is univalent, as in Figure 3 below.

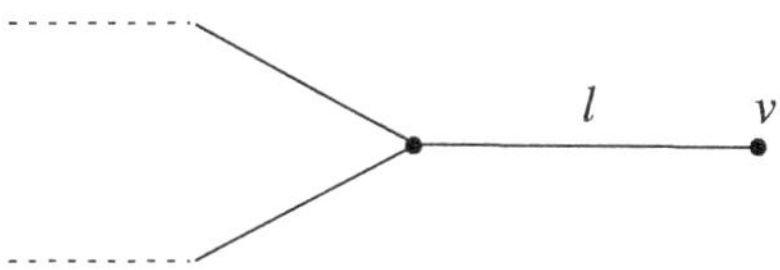

Figure 3.

Then, removing l and v, we may proceed by induction as above. In the remaining cases the vertex v must appear as in one of the following pictures, where ξ either fixes or exchanges l and l':

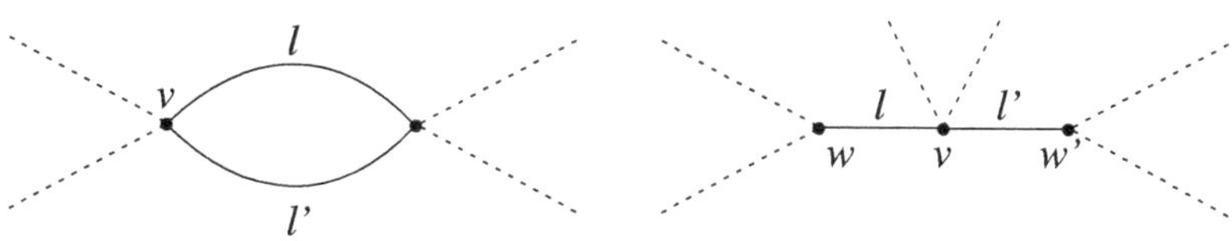

Figure 4.

In the first case, since ξ_* is the identity, and $m \geq 3$, ξ cannot exchange l and l' . Then, contracting l and l', we conclude, by induction, that ξ is the identity. In the second case, we contract again l and l' to obtain a graph Γ' on which ξ induces the identity mapping. Now, $\xi(l) = l'$ only if w and w' are univalent, which we already excluded.

Hence, ξ is the identity on Γ as well. We can also assume that there is no edge carried to itself by ξ, otherwise we could contract it and, again, proceed by induction. Finally, we can assume that there is no simple open, connected chain $\delta \subset \Gamma$ such that $\xi(\delta) = \delta$, because otherwise there would be either a fixed vertex or a fixed edge.

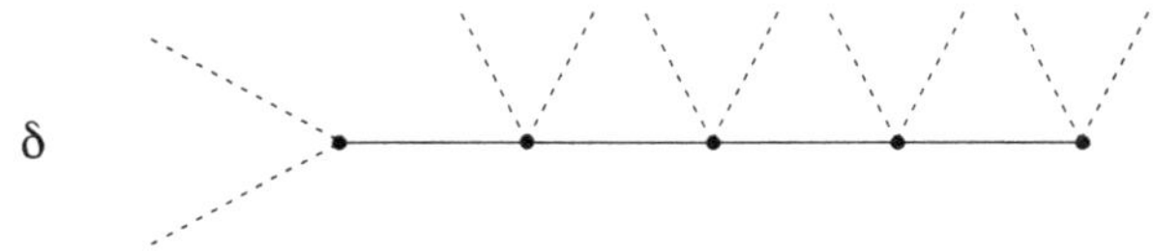

Figure 5.

Since we can assume that all of its vertices are at least trivalent, the graph Γ must contain at least two distinct simple loops c and c' with a common vertex v. By hypothesis, ξ carries c into c and c' into c'. If $\delta = c \cap c'$, then $\xi(\delta) = \delta$. Clearly, δ is a nonempty disjoint union of vertices and connected chains. Contracting each connected component of δ to a point, we see, by induction, that ξ carries each connected component of δ into itself. But we already excluded this. Q.E.D.

We return to the proof of Proposition (3.1). We now know that each component of X is carried to itself by φ. Applying Proposition (2.8), we conclude that, when $g_\alpha \geq 1$, the restriction of φ to X_α is the identity. To deal with those components X_α such that $g_\alpha = 0$, it suffices to notice that, since φ acts trivially on the half-edges of Γ, it induces an automorphism of Σ_α fixing the marked points. As there are at least three of these, this automorphism must be the identity. In conclusion, φ restricts to the identity on all components of X. This concludes the proof of (3.1).

4. Compactifying moduli of curves with level structure; a first attempt.

The notion of level G structure does not easily generalize to the case of singular stable curves. In order to construct a satisfactory compactification of $M_g[\psi]$, we will have to adopt a different point of view. We shall do this in a later section; in the present one we describe an indirect approach to the compactification problem, which has the drawback of not corresponding to any obvious moduli problem. As a consequence, the nature of this compactification at the boundary will have to be inferred by indirect means. This is where the Picard–Lefschetz theory will come into play.

Consider a moduli space $M_g[\psi]$, where ψ is a level structure dominating a level m structure with $m \geq 3$. In particular, $M_g[\psi]$ is smooth and admits a universal family $\mathcal{X}[\psi] \to M_g[\psi]$ of genus g curves with level ψ structure. We define a compactification $\overline{M}_g[\psi]$ of $M_g[\psi]$ as the normalization of $\overline{M}_g$ in the field of rational functions of $M_g[\psi]$. We let $\mu : M_g[\psi] \to M_g$ denote the natural projection and

$$\overline{\mu} : \overline{M}_g[\psi] \to \overline{M}_g$$

the normalization map. Since μ is finite and $M_g[\psi]$ is normal, we can identify $M_g[\psi]$ with the open subset $\mu^{-1}(M_g)$ of $\overline{M}_g[\psi]$.

We will prove that there exists a family of genus g stable curves

$$\overline{\eta} : \overline{\mathcal{X}}[\psi] \to \overline{M}_g[\psi] \tag{4.1}$$

whose restriction to $M_g[\psi]$ is the universal family $\eta : \mathcal{X}[\psi] \to M_g[\psi]$ and whose moduli map is $\overline{\mu}$. To do this, it will suffice to extend $\mathcal{X}[\psi] \to M_g[\psi]$ locally near any point of $\partial M_g[\psi] = \overline{M}_g[\psi] \setminus M_g[\psi]$, since any two local extensions are canonically isomorphic, by the uniqueness part of stable reduction (cf. Chapter X, Section 5.).

A key tool will be the result that follows. Let C_0 be a stable genus g curve, and

$$\pi : \mathcal{C} \to (B, b_0) \tag{4.2}$$

a standard Kuranishi family for it; thus, $\mathrm{Aut}(C_0)$ acts on B, and any isomorphism between fibers of π comes from an automorphism of C_0. We may suppose that B is an open set in $\mathbb{C}^{3g-3}$, that b_0 is the origin, and that the locus in B parameterizing singular curves is a union of coordinate hyperplanes; we let B^* be the complement of this locus and set $\mathcal{C}^* = \pi^{-1}(B^*)$.

LEMMA (4.3). *Let $\alpha : U \to B^*$ be a connected unramified covering and suppose that $\mathcal{Y} = \mathcal{C}^* \times_{B^*} U \to U$ is endowed with a level m structure with $m \geq 3$. If two fibers $\mathcal{Y}_u$ and $\mathcal{Y}_{u'}$ of $\mathcal{Y} \to U$ are isomorphic as curves with level m structure, then $\alpha(u) = \alpha(u')$, and the isomorphism between them covers the identity on $\mathcal{C}_{\alpha(u)}$.*

Proof. Given an automorphism $\sigma \in \mathrm{Aut}(C_0)$, we denote by $f_\sigma : B \to B$ and $F_\sigma : \mathcal{C} \to \mathcal{C}$, the isomorphisms induced by σ. We now invoke Lemma (9.19) of Chapter X telling us that C_0 is a deformation retract of $\mathcal{C}$. We make the identification

$$H_1(\mathcal{C}, \mathbb{Z}/m\mathbb{Z}) = H_1(C_0, \mathbb{Z}/m\mathbb{Z}) = (\mathbb{Z}/m\mathbb{Z})^{2g-\delta'}, \quad \delta' \leq \delta,$$

where δ is the number of nodes of C_0, and δ' the number of nonseparating ones. The inclusion $j_b : \mathcal{C}_b \hookrightarrow \mathcal{C}$ induces a homomorphism

$$j_{b*} : H_1(\mathcal{C}_b, \mathbb{Z}/m\mathbb{Z}) \longrightarrow H_1(\mathcal{C}, \mathbb{Z}/m\mathbb{Z}) = H_1(C_0, \mathbb{Z}/m\mathbb{Z}) = (\mathbb{Z}/m\mathbb{Z})^{2g-\delta'}$$

which depends continuously on b for $b \in B^*$ and has the property that, for any automorphism σ and points b and $b' = f_\sigma(b)$, the diagram

$$(4.4) \qquad \begin{array}{ccc} H_1(\mathcal{C}_b, \mathbb{Z}/m\mathbb{Z}) & \xrightarrow{F_{\sigma *}} & H_1(\mathcal{C}_{b'}, \mathbb{Z}/m\mathbb{Z}) \\ \downarrow j_{b*} & & \downarrow j_{b'*} \\ H_1(C_0, \mathbb{Z}/m\mathbb{Z}) & \xrightarrow{\sigma_*} & H_1(C_0, \mathbb{Z}/m\mathbb{Z}) \end{array}$$

commutes.

Suppose that there is an isomorphism of curves with level m structure from $\mathcal{Y}_u$ to $\mathcal{Y}_{u'}$. If we forget about the level structures, this is just an isomorphism $\mathcal{C}_{\alpha(u)} \to \mathcal{C}_{\alpha(u')}$. As such, it comes from an automorphism σ of C_0 and gives rise to a diagram (4.4), with $b = \alpha(u)$, $b' = \alpha(u')$. The level m structure which exists on $\mathcal{Y} \to U$ by assumption consists of isomorphisms

$$\lambda_v : (\mathbb{Z}/m\mathbb{Z})^{2g} \to H_1(\mathcal{Y}_v, \mathbb{Z}/m\mathbb{Z}) = H_1(\mathcal{C}_{\alpha(v)}, \mathbb{Z}/m\mathbb{Z}), \quad v \in U,$$

depending continuously on v. In particular, the composition

$$(\mathbb{Z}/m\mathbb{Z})^{2g} \xrightarrow{\lambda_v} H_1(\mathcal{C}_{\alpha(v)}, \mathbb{Z}/m\mathbb{Z}) \xrightarrow{(j_{\alpha(v)})_*} (\mathbb{Z}/m\mathbb{Z})^{2g-\delta'}$$

is independent of v. By construction, the restriction of F_σ to $\mathcal{C}_{\alpha(u)}$, viewed as an isomorphism onto $\mathcal{C}_{\alpha(u')}$, is compatible with the level structures λ_u and $\lambda_{u'}$; this just means that $\lambda_{u'} = \lambda_u \circ (F_\sigma)_*$. Combining this with diagram (4.4) shows that

$$\sigma_* \circ (j_{\alpha(u)})_* \circ \lambda_u = (j_{\alpha(u')})_* \circ \lambda_{u'} .$$

On the other hand, as we just observed, $(j_{\alpha(u)})_* \circ \lambda_u = (j_{\alpha(u')})_* \circ \lambda_{u'}$. Since $(j_{\alpha(u)})_* \circ \lambda_u$ is onto, we conclude that σ_* is the identity. By Lemma (3.1), σ must also be the identity. Q.E.D.

Let us go back to the problem of constructing a universal family on $M_g[\psi]$. As we have noticed, it suffices to solve this problem locally along the boundary. Let x be a point of $\partial M_g[\psi]$, and let C_0 be a stable genus g curve representing the point $\overline{\mu}(x) \in \overline{M}_g$. Consider the Kuranishi family (4.2) for C_0 and the moduli map $\nu : B \to \overline{M}_g$. As a neighborhood of x in $\overline{M}_g[\psi]$ we choose the connected component of $\overline{\mu}^{-1}\nu(B)$ containing x, and we call it V. We also set

$$V^* = V \cap M_g[\psi]$$

and we look at the restriction to V^* of the universal family $\mathcal{X}[\psi] \to M_g[\psi]$:

$$(4.5) \qquad \mathcal{X}[\psi]_{|V^*} \to V^* .$$

We will show that there are a morphism $\alpha : V^* \to B$ and a cartesian diagram

$$
\begin{array}{ccc}
\mathcal{X}[\psi]_{|V^*} & \xrightarrow{\ A\ } & \mathcal{C} \\
\big\downarrow & & \big\downarrow \\
V^* & \xrightarrow{\ \alpha\ } & B
\end{array}
\tag{4.6}
$$

The Riemann extension theorem will then imply that α extends to a morphism $\overline{\alpha} : V \to B$, and via this morphism one can extend the family (4.5) to all of V.

To find α and A, we proceed as follows. Let W be the set of all pairs consisting of a point $b \in B^*$ and of a level ψ structure on $\mathcal{C}_b$:

$$
W = \{(b, \lambda) : b \in B^*, \ \ \lambda : \pi_1(\mathcal{C}_b) \to G\},
$$

where of course λ is assumed to factor through $\psi : \Pi \to G$. It is easy to see that W carries a natural analytic structure which makes it an unramified covering of B^*. Indeed, W is nothing but the Hilbert scheme of maps

$$
W = \mathbf{Isom}_{M_g[\psi] \times_{M_g} B^*}(p^*\mathcal{X}[\psi], q^*(\mathcal{C}_{|B^*}),
\tag{4.7}
$$

where p and q be the projections of $M_g[\psi] \times_{M_g} B^*$ to its two factors; in simple terms, W is the set of triples (y, b, ϕ), where $y \in M_g[\psi]$, $b \in B^*$, and $\phi : \mathcal{C}_b \to X_y$ is an isomorphism. Here X_y stands for the fiber of $\mathcal{X}[\psi] \to M_g[\psi]$ at y. One passes from one description to the other as follows. Given a point (y, b, ϕ) of (4.7), we get a level ψ structure on $\mathcal{C}_b$ by composing $\phi_* : \pi_1(\mathcal{C}_b) \to \pi_1(X_y)$ with the canonical level ψ structure on X_y. Conversely, given a level ψ structure $\pi_1(\mathcal{C}_b) \to G$, there are a point $y \in M_g[\psi]$ and an isomorphism $\mathcal{C}_b \to X_y$ compatible with the respective level ψ structures; the isomorphism is unique, since we are dealing with level structures dominating a level m structure with $m \geq 3$. We have a commutative diagram

$$
\begin{array}{ccc}
W & \xrightarrow{\ \vartheta\ } & B^* \\
{\scriptstyle\tau}\big\downarrow & & \big\downarrow{\scriptstyle\nu} \\
M_g[\psi] & \xrightarrow{\ \mu\ } & M_g
\end{array}
$$

We denote by $\rho : \mathcal{Y} \to W$ the pullback, via ϑ, of the Kuranishi family (4.2). By construction, this is canonically isomorphic to the pullback of $\mathcal{X}[\psi] \to M_g[\psi]$ via τ. As such, it is a family of curves with level ψ structure.

We claim that on each connected component of W the morphism τ is injective. Since we are working with level structures dominating level m structures with $m \geq 3$, this readily follows from Lemma (4.3). In fact, the lemma says that, if $\tau(b, \lambda) = \tau(b', \lambda')$, then $b = b'$, and the isomorphism between $\mathcal{C}_b$ and $\mathcal{C}_{b'}$ taking λ into λ' is the identity, so that $\lambda = \lambda'$.

It is clear that $\tau(W) = \mu^{-1}\nu(B) \cap M_g[\psi]$ and hence that one component of W maps isomorphically to V^* via τ. Composing the inverse of this isomorphism with ϑ yields the desired morphism $\alpha : V^* \to B$. We summarize the results obtained so far in the following:

THEOREM (4.8). *Let ψ be a level structure dominating a level m structure with $m \geq 3$. Consider the natural projection $\mu : M_g[\psi] \to M_g \subset \overline{M}_g$. Let $\overline{M}_g[\psi]$ be the normalization of $\overline{M}_g$ in the field of rational function of $M_g[\psi]$. Then there exists a family of genus g stable curves $\overline{\eta} : \overline{\mathcal{X}}[\psi] \to \overline{M}_g[\psi]$ extending the universal family $\eta : \mathcal{X}[\psi] \to M_g[\psi]$ of smooth, genus g curves with level ψ structure, whose moduli map is the projection $\overline{M}_g[\psi] \to \overline{M}_g$. Furthermore, when Λ_ψ is a normal subgroup of $\Gamma_g[\psi]$, in particular when ψ is characteristic, the action of $\Gamma_g[\psi]$ extends to $\overline{M}_g[\psi]$, and $\overline{M}_g$ is the quotient of $\overline{M}_g[\psi]$ under this action.*

The only part of the theorem that has not been fully proved so far is the last statement. Using the fact that the action of $\Gamma_g[\psi]$ on $M_g[\psi]$ is compatible with the projection $\mu : M_g[\psi] \to M_g$, we see that, by the universal property of the normalization, this action extends to $\overline{M}_g[\psi]$. Furthermore, since $\overline{M}_g[\psi]/\Gamma_g[\psi]$ and $\overline{M}_g$ are both normal and have the same field of rational functions,

(4.9) $$\overline{M}_g[\psi]/\Gamma_g[\psi] = \overline{M}_g\,.$$

In proving Theorem (4.8) we actually produced a tool to characterize the local nature of $\overline{M}_g[\psi]$ around points of its boundary $\partial M_g[\psi] = \overline{M}_g[\psi] \smallsetminus M_g[\psi]$. Here is where the Picard–Lefschetz transformation turns out to be very useful.

The local monodromy representation

Let us look at the local picture in a neighborhood of a boundary point $x \in \partial M_g[\psi]$, in the case where Λ_ψ is a normal subgroup of Γ_g. Going back to the notation used in the proof of (4.8), we have the following diagram

$$\begin{array}{ccc} W & \xrightarrow{\quad\vartheta\quad} & B^* \subset B \\ {\scriptstyle\tau}\big\downarrow & & \big\downarrow{\scriptstyle\nu} \\ V \subset \overline{M}_g[\psi] & \xrightarrow{\quad\mu\quad} & \overline{M}_g \end{array}$$

The morphism μ is a ramified Galois covering with groups $\Gamma_g[\psi]$. We also know that there is a connected component W' of W mapping isomorphically to $V^* = V \cap M_g[\psi]$, via τ. Recall that $\alpha = \vartheta \circ (\tau_{|W'})^{-1}$: $V^* \to B^*$ is a Galois covering with group

$$H \subset \{\gamma \in \Gamma_g[\psi] : \gamma(V^*) = V^*\}.$$

Of course, the extension $\overline{\alpha} : V \to B$ is a (ramified) Galois covering with group H. We claim that H is the image of the Picard–Lefschetz representation, or equivalently that there is an exact sequence

$$1 \to \pi_1(V^*, v) \xrightarrow{\alpha_*} \pi_1(B^*, b) \xrightarrow{PL} H \subset \Gamma_g[\psi].$$

In fact, let $\rho : I = [0,1] \to B^*$ be a closed loop based at a point b and consider $\mathcal{C}_b$ as the reference surface on which Γ_g acts. Let $v = (b, \lambda) \in V^*$ be a point over b, so that λ is ψ-structure on $\mathcal{C}_b$. We may think of λ as the class of a homomorphism $\lambda : \pi_1(\mathcal{C}_b) \to G$. Trivialize $\mathcal{C} \to B$ over I:

$$F : \rho^*\mathcal{C} \to \mathcal{C}_b \times I.$$

The lifting of ρ to V^* with initial point v is given by

$$\widetilde{\rho}(t) = (\rho(t), \lambda \circ (F_t)_*).$$

This shows that $\pi_1(V^*, v)$ is inded the kernel of the Picard–Lefschetz representation $PL : \pi_1(B^*, b) \to \Gamma_g[\psi]$.

The group H is called the *local monodromy group*. It is useful to notice that to calculate the Picard–Lefschetz representation, we do not actually need to use all the special properties of B employed in the proof of (4.3) and (4.8); any convex neighborhood of the origin in B will do.

We claim that the Picard–Lefschetz representation completely determines the covering $\overline{\alpha} : V \to B$. In fact, let $\overline{\beta} : U \to B$ be another covering with U connected and normal, set $U^* = \overline{\beta}^{-1}(B^*)$, $\beta = \overline{\beta}_{|U^*}$, and pick a base point $u \in U^*$ such that $\beta(u) = b$. Suppose that $\beta : U^* \to B^*$ is an unramified Galois covering and that the image of $\pi_1(U^*, u)$ in $\pi_1(B^*, b)$ coincides with the one of $\pi_1(V^*, v)$. This assumption implies that there is an isomorphism $\xi : U^* \to V^*$ sending u to v and such that $\alpha \circ \xi = \beta$. By the Riemann extension theorem, ξ extends to an isomorphism $\overline{\xi} : U \to V$ satisfying $\overline{\alpha} \circ \overline{\xi} = \overline{\beta}$.

REMARK (4.10). This observation provides us with a practical recipe for constructing a neighborhood of a boundary point $x \in \partial\overline{M}_g[\psi]$. Consider the Kuranishi family (4.2) for the stable curve C_0 corresponding to x and let b_0 be a point of B^*. Suppose we are given the following data.

a) A connected, normal analytic variety U.
b) A finite, surjective morphism $\beta : U \to B$ such that:
c) $U^* = \beta^{-1}(B^*) \to B^*$ is a topological covering corresponding to the subgroup $\ker(PL) \subset \pi_1(B^*, b_0)$, where PL stands for the level ψ Picard–Lefschetz representation.

We can then conclude that U is isomorphic to a neighborhood of x in $\overline{M}_g[\psi]$.

It follows, in particular, that the smoothness of $\overline{M}_g[\psi]$ at a boundary point can be decided, in principle, just by examining the corresponding Picard–Lefschetz representation. We will immediately use this observation to show that $\overline{M}_g[m]$ is *never* smooth if $g > 2$. Consider a stable curve C_0 of genus g with two smooth irreducible components, C_1 and C_2, meeting at exactly two points q_1 and q_2, the nodes of C_0. Consider also a Kuranishi family for C_0 as in (4.2) and let C be a general fiber of it. This is illustrated in Figure 6, where γ_1 and γ_2 are the vanishing cycles corresponding to q_1 and q_2.

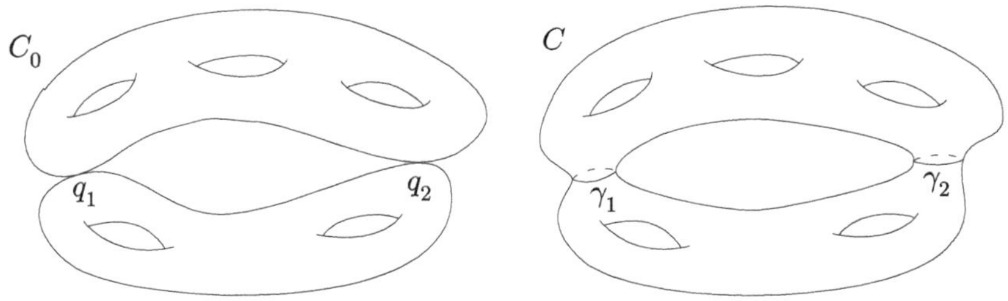

Figure 6.

The singular fibers of the Kuranishi family are parameterized by the locus $\{z_1 z_2 = 0\}$. We denote by B^* the complement of this locus in B, and we choose a base point $b \in B^*$. Then

$$\pi_1(B^*, b) = \mathbb{Z}\lambda_1 \oplus \mathbb{Z}\lambda_2 ,$$

where λ_i is a loop around the locus $z_i = 0$. The local monodromy group H is the image of $\pi_1(B^*, b)$ under the Picard–Lefschetz representation

$$PL : \mathbb{Z}\lambda_1 \oplus \mathbb{Z}\lambda_2 \to Sp(H_1(C, \mathbb{Z}/m\mathbb{Z})) .$$

Since γ_1 is homologous to $\pm\gamma_2$, $PL(\lambda_1) = PL(\lambda_2)$. We then have

$$PL(h\lambda_1 + k\lambda_2)(x) = x + (h + k)(x \cdot \gamma_1)\gamma_1 .$$

It follows that $h\lambda_1 + k\lambda_2$ belongs to the kernel of PL if and only if $h + k \equiv 0 \mod m$.

Now denote by U the analytic subvariety of $B \times \mathbb{C}$ defined by the equation $z_1 z_2 = \zeta^m$, where ζ is the standard coordinate in $\mathbb{C}$. The variety U is normal, and the projection $\beta : U \to B$ is a degree m covering ramified along $z_1 z_2 = 0$. We set $U^* = \beta^{-1}(B^*)$ and choose a point $u \in U$ lying above b. We will show that the image of $\pi_1(U^*, u)$ in $\pi_1(B^*, b)$ coincides with the kernel of the homomorphism PL. As we have observed, this will imply that a neighborhood of the point $[C_0]$ in $\overline{M}_g[m]$ is analytically isomorphic to U. Since the origin is a singular point of U, the conclusion will be that $\overline{M}_g[m]$ is singular at $[C_0]$.

The kernel K of PL has index m in $\pi_1(B^*, b)$, and the same is true for $\pi_1(U^*, u)$, since β has degree m. Hence, to show that the image of $\pi_1(U^*, u)$ is equal to K, it will suffice to show that it contains K. But this is clear. In fact, if we write $u = (\underline{z}_1, \ldots, \underline{z}_{3g-3}, \underline{\zeta})$, and h and k are integers whose sum is of the form rm for some integer r, then the path

$$[0,1] \ni t \mapsto (\underline{z}_1 e^{2\pi\sqrt{-1}ht}, \underline{z}_2 e^{2\pi\sqrt{-1}kt}, \underline{z}_3, \ldots, \underline{z}_{3g-3}, \underline{\zeta} e^{2\pi\sqrt{-1}rt})$$

is a closed loop in U^* which maps to $h\lambda_1 + k\lambda_2$.

EXERCISE (4.11). Let ψ be a level structure dominating a level m structure with $m \geq 3$. Let γ_1 and γ_2 be two nonseparating cycles, not homotopic to each other, on the reference genus g surface Σ. Suppose that there exists a diffeomorphism f of Σ such that $[f] \in \Lambda_\psi$ and $f(\gamma_1) = \gamma_2$. Show that $\overline{M}_g[\psi]$ is singular.

5. Admissible G-covers.

G-covers versus level G structures.

It is useful to link the notion of level G structure with the closely allied but more geometric notion of G-cover. Let G be a finite group, and let C be a smooth curve of genus $g > 1$. A *G-cover* of C is an unramified Galois covering $P \to C$ plus an isomorphism between G and $\mathrm{Aut}(P/C)$. Equivalently, a G-cover is a principal G-bundle over C. An isomorphism between two G-covers $\alpha : P \to C$ and $\beta : P' \to C'$ is a commutative diagram

$$\begin{array}{ccc} P & \xrightarrow{\ f\ } & P' \\ {\scriptstyle\alpha}\downarrow & & \downarrow{\scriptstyle\beta} \\ C & \xrightarrow{\ \varphi\ } & C' \end{array}$$

where f and φ are isomorphisms, and f is compatible with the given identifications of G with $\mathrm{Aut}(P/C)$ and $\mathrm{Aut}(P'/C')$. Hence, G-covers and their isomorphisms form a *groupoid*, that is, a category all of whose morphisms are isomorphisms. Curves with a level G structure also form a groupoid, and there is a functor F which associates to each G-cover a curve with level G structure. In fact, any choice of base point $y \in P$ determines a surjective homomorphism $\pi_1(C, x) \to \mathrm{Aut}(P/C) = G$, where $x = \alpha(y)$. Clearly, changing base point merely changes this homomorphism by a conjugation. It is important to notice that, conversely, any conjugation by an element of $\pi_1(C, x)$ can be realized by a change of the base point y inside $\alpha^{-1}(x)$.

The functor F determines a *bijection* between isomorphism classes of G-covers and isomorphism classes of curves with level G structure. In fact, given a surjective homomorphism $\sigma : \pi_1(C, x) \to G$, there is a Galois covering $\alpha : P \to C$ such that the kernel of σ is $\alpha_*(\pi_1(P, y))$, where y is any point of $\alpha^{-1}(x)$, and the group of deck transformations can be identified with $\pi_1(C,x)/\pi_1(P,y) \cong G$. Now let $\beta : Q \to C$ be another G-cover such that the corresponding level G structure is the same as the one given by σ. This means that the homomorphism $\tau : \pi_1(C,x) \to G = \mathrm{Aut}(Q/C)$ is the composition of σ with conjugation by an element of G. By the discussion above, choosing a suitable base point $y' \in Q$ lying above x, we may suppose that in fact $\sigma = \tau$. In particular, the images of $\pi_1(P,y)$ and $\pi_1(Q,y')$ in $\pi_1(C,x)$ are equal. Therefore, there is a unique isomorphism $\varphi : P \to Q$ such that $\beta\varphi = \alpha$ and $\varphi(y) = y'$. Since $\sigma = \tau$, the induced identification $\mathrm{Aut}(P/C) = \mathrm{Aut}(Q/C)$ is compatible with the identification of the two sides with G.

It is important to notice that, while an isomorphism of G-covers determines an isomorphism between the corresponding curves with level G structure, this correspondence, although surjective, is not a bijection. This is best seen by looking at automorphisms. We have the following commutative diagram of group homomorphisms:

$$\begin{array}{ccc}
\{\text{Automorphisms of } G\text{-cover } P \to C\} & \xrightarrow{\ \Psi\ } & \mathrm{Aut}(C) \\
\Big\downarrow F & \overset{\Phi}{\nearrow} & \\
\{\text{Automorphisms of } (C \text{ plus level } G \text{ structure})\} & &
\end{array}$$

While Φ is always an injection, in general Ψ is not. In fact, if $\alpha : P \to C$ is a G-cover, a deck transformation φ is an automorphism of the G-cover if and only if $h\varphi(p) = \varphi(hp)$ for any $h \in G$ and any $p \in P$. Since φ corresponds to the action of a specific element $k \in G$, this amounts to saying that $hkp = khp$ for any $h \in G$ and any $p \in P$; as P is a principal G-bundle, this is equivalent to $hk = kh$ for any $h \in G$. Thus, $\ker(\Psi) = Z(G)$, where $Z(G)$ is the center of G.

We can therefore conclude that, when the center of G is trivial, the moduli space ${}_GM_g$ of curves with level G structure can be thought of as the moduli space of G-covers of genus g curves.

Limits of G-covers.

As we already pointed out, the notion of level G structure does not easily generalize to the case of singular curves, and this is the main reason why we have taken a somewhat indirect approach to compactifying moduli of curves with level structure. By contrast, there is a natural notion of limit for G-covers when the underlying curve degenerates to a stable one. This is the notion of admissible G-cover. Let $\mathcal{C} \to S$ be a family of connected nodal curves. A *family of degree d admissible covers of $\mathcal{C}$ over S* is a finite morphism $\mathcal{P} \to \mathcal{C}$ such that

(1) $\mathcal{P} \to S$ is a family of connected nodal curves;

(2) every node of a fiber of $\mathcal{P} \to S$ maps to a node of the corresponding fiber of $\mathcal{C} \to S$;
(3) away from the nodes, $\mathcal{P} \to \mathcal{C}$ is étale of constant degree d;
(4) the local picture of $\mathcal{P} \to \mathcal{C} \to S$ near a node is

$$\begin{aligned} \{(\xi, \eta, s) : \xi\eta = f(s)\} \to \{(x, y, s) : xy = f(s)^r\} \to S \\ (\xi, \eta, s) \mapsto (\xi^r, \eta^r, s) \mapsto s \end{aligned} \tag{5.1}$$

for some integer $r > 0$.

The integer r appearing in (5.1) will be called the ramification index or simply the *index* of the cover at the node.

Consider a family of admissible covers as above. Suppose that, in addition, we are given an action of a finite group G on $\mathcal{P}$ which is compatible with $\mathcal{P} \to \mathcal{C}$. We say that these data define a *family of admissible G-covers of $\mathcal{C}$ over S* if

(1) $\mathcal{P} \to \mathcal{C}$ is a principal G-bundle away from the nodes;
(2) If p is a node of a fiber of $\mathcal{P} \to S$, and $\mathrm{Stab}_G(p)$ stands for the stabilizer of p in G, then any element of $\mathrm{Stab}_G(p)$ acts, in the representation (5.1), as $(\xi, \eta, s) \mapsto (\zeta\xi, \zeta^{-1}\eta, s)$, where ζ is an rth root of unity.

There is an obvious notion of morphism of families of admissible G-covers. If $\mathcal{P} \to \mathcal{C} \to S$ and $\mathcal{Q} \to \mathcal{Z} \to T$ are two such families, a morphism from the first to the second is a commutative diagram

$$\begin{array}{ccc} \mathcal{P} & \longrightarrow & \mathcal{Q} \\ \downarrow & & \downarrow \\ \mathcal{C} & \longrightarrow & \mathcal{Z} \\ \downarrow & & \downarrow \\ S & \longrightarrow & T \end{array}$$

such that both squares are cartesian and $\mathcal{P} \to \mathcal{Q}$ commutes with the action of G.

Let C be a nodal curve, let $P \to C$ be an admissible G-cover, and let x be a node of C. Then the index of the cover at any node p of P mapping to x equals the order of the stabilizer of p and depends only on x. Therefore, in the sequel, we shall often refer to this integer as the index of the G-cover at x.

Notice that, for any admissible G-cover $P \to C$, the curve C is stable if and only if P is. In fact, denote by P' and C' the smooth, possibly noncompact, curves obtained from P and C by removing all nodes. Then P' is a topological covering of C'. To say that C is stable is equivalent to saying that every connected component of C' has negative Euler–Poincaré characteristic, and similarly for P. Now let C_1 be a component of C', and P_1 a component of P' mapping to it. Our claim follows from the fact that the Euler–Poincaré characteristic of P_1 equals the Euler–Poincaré characteristic of C_1 times the degree of the map $P_1 \to C_1$.

In the sequel, when we speak of admissible covers, we shall implicitly assume, without mentioning it, that the curves involved are stable. There will be very few occasions when we shall have to deal with nonstable curves, and we will warn the reader when this occurs.

We will construct a moduli space for admissible G-covers using the same approach that we used to construct the moduli space of stable curves, by gluing together Kuranishi families. The compactness of the resulting space will then be a consequence of a stable reduction theorem which we are now going to state and prove. Let $\pi : \mathcal{C} \to \Delta$ be a family of stable curves of genus g over the disc $\Delta = \{z \in \mathbb{C} : |z| < 1\}$, with all fibers smooth except possibly the central one, and set $\Delta^* = \Delta \smallsetminus \{0\}$, $\mathcal{C}^* = \pi^{-1}(\Delta^*)$. Let $\mathcal{P}^* \to \mathcal{C}^* \to \Delta^*$ be a family of G-covers.

PROPOSITION (5.2). *Possibly after a base change of the form $z \mapsto z^\nu$, the family $\mathcal{P}^* \to \mathcal{C}^* \to \Delta^*$ extends to a family of admissible G-covers over Δ.*

Proof. Up to base change in Δ, we may assume that $\mathcal{P}^* \to \Delta^*$ extends to a family $\mathcal{P} \to \Delta$ of stable curves; we shall denote by P the central fiber, and by C the central fiber of $\mathcal{C} \to \Delta$. By the basic results on extending isomorphisms in families of stable curves, the action of G on $\mathcal{P}^*$ extends to an action on $\mathcal{P}$; we write $\mathcal{Q}$ to denote the quotient $\mathcal{P}/G$ and notice that it equals $\mathcal{C}$ away from the central fiber. The surface $\mathcal{Q}$ is normal, and hence its singular points are isolated. Since $\mathcal{P}^* \to \mathcal{C}^*$ is unramified, the purity of the branch locus theorem implies that the ramification locus of $\mathcal{P} \to \mathcal{Q}$ consists of singular points of P and, possibly, of components of P. The latter, however, never occurs. In fact, if a component E of P belonged to the ramification locus, this would mean that there is an element h of G which is not the identity but acts as such on E. This is impossible since h would then have to act nontrivially in the normal direction to E and therefore would have to interchange the fibers of $\mathcal{P} \to \mathcal{Q}$, which is not the case. In conclusion, the isotropy group $\mathrm{Stab}_G(p)$ of a point p of $\mathcal{P}$ is trivial unless p is a singular point of P. To analyze the way $\mathrm{Stab}_G(p)$ acts on a neighborhood of p, we shall use the following lemma.

LEMMA (5.3). *Let f be a holomorphic function on a neighborhood of the origin in $\{(x, y, z) \in \mathbb{C}^3 : xy = z^m\}$. Suppose that f vanishes only on $\{x = 0\}$ and that it equals x times a unit at a general point of this locus. Then $f = ux$ on a neighborhood of the origin, where u is a nowhere vanishing function.*

This is obvious when $m = 1$. In general, we blow up repeatedly $X = \{xy = z^m\}$ until we get a smooth surface Y, in which the singular point of X has been replaced by a chain $E_1, \ldots, E_{m-1}$ of smooth rational curves with self-intersection -2. We can number these curves so that E_i meets E_{i-1} and E_{i+1} for $i = 1, \ldots, m-1$, where we write E_0 and E_m to denote the proper transforms of the loci $\{x = 0\}$ and $\{y = 0\}$. If we view x and f as functions on Y, the divisors they define have zero intersection with E_i for $i = 1, \ldots, m-1$. Since x and f both vanish to order m along E_0 and do not vanish along E_m, the only possibility is that $(f) = mE_0 + (m-1)E_1 + \cdots + E_{m-1}$, and similarly for x. Thus $f = ux$, where u is a unit on Y. Since u is constant on $E_1 + \cdots + E_{m-1}$, it descends to a holomorphic function on X, and the result follows.

Let h be an element of $\mathrm{Stab}_G(p)$ which does not interchange the two branches of P at p (we shall later see that this is always the case). If we represent a neighborhood of p as $\{xy = z^m\}$, then, applying Lemma (5.3) to h^*x and h^*y, we find that the action of h is of the form

$$h^*x = (a + \text{higher-order terms})x\,, \quad h^*y = (b + \text{higher-order terms})y\,,$$

where a and b are nonzero constants. Taking into account that $h^*x\, h^*y = (h^*z)^m = z^m$, we find that $b = a^{-1}$. It follows that h acts as multiplication by a^{-1} on the tangent space to $\{y = 0\}$ at p and as multiplication by a on the tangent space to $\{x = 0\}$ at p. The constant a is a kth root of unity, where k is the order of h. Since h has finite order, its action on the branch $\{y = 0\}$ (say) is linear, in suitable coordinates. This implies that, if $a = 1$, then h acts as the identity on the component of P containing $\{y = 0\}$ and hence on all of P.

We now show that an element h of $\mathrm{Stab}_G(p)$ cannot interchange the two branches of P at p. If it did, we would have that

$$h^*x = uy = (a + \text{higher order terms})y\,,$$
$$h^*y = u^{-1}x = (a^{-1} + \text{higher order terms})x\,.$$

Thus $h^{*2}x = (1 + \text{higher-order terms})x$. As we just observed, this means that h^2 is the identity. As a consequence, $x = h^{*2}x = u^{-1}h^*(u)x$, and hence $h^*u = u$. Let v be a square root of u; then $h^*v = v$, since the origin is left fixed by h and v does not vanish there. Set $x' = v^{-1}x$ and $y' = vy$. It is clear that $x'y' = z^m$ and that h^* interchanges x' and y'. Therefore, changing "coordinates," we may assume that $h^*x = y$ and $h^*y = x$. But then the points where $x = y$ are fixed under the action of h. This, however, contradicts the fact that $\mathcal{P}^* \to \mathcal{C}^*$ is unramified.

The upshot of the above discussion is that $\mathrm{Stab}_G(p)$ injects into the automorphism group of the tangent space to (say) the branch $\{y = 0\}$ at the origin and hence is cyclic. Let h be a generator of $\mathrm{Stab}_G(p)$, and let k be its order. We know that $h^*x = x(\zeta + \text{higher-order terms})$, where ζ is a primitive kth root of unity. Now write

$$x' = \frac{1}{k}\sum_{i=0}^{k-1}\frac{1}{\zeta^i}h^{*i}(x)\,.$$

Clearly, $h^*(x') = \zeta x'$, and moreover $x' = ux$, where u is a unit. We then set $y' = u^{-1}y$; it follows from $x'y' = z^m$ that $h^*(y') = \zeta^{-1}y'$. The conclusion is that we may assume that the action of any element of $\mathrm{Stab}_G(p)$ on $\{xy = z^m\}$ is of the form

$$(x, y, z) \mapsto (\eta x, \eta^{-1} y, z) \tag{5.4}$$

for some kth root of unity η.

The description of the action of $\mathrm{Stab}_G(p)$ implies that, near p, the morphism $\mathcal{P} \to \mathcal{Q}$ can be identified with

$$(x, y, z) \mapsto (\xi, \eta, z) = (x^k, y^k, z),$$

so that $\mathcal{Q}$ can be locally identified with $\{\xi\eta = z^{mk}\}$. This means that $\mathcal{P} \to \mathcal{Q}$ is a family of admissible covers. In fact, the explicit form (5.4) of the action of the stabilizers of points says that it is a family of admissible G-covers. To conclude the proof of (5.2), it remains to show that $\mathcal{Q}$ coincides with $\mathcal{C}$. Since $\mathcal{Q}$ and $\mathcal{C}$ agree everywhere, except possibly on the central fiber, it suffices to show that $\mathcal{Q} \to \Delta$ is a family of stable curves, that is, that Q is stable. But this is clear, since we observed earlier that, for any admissible G-cover $P' \to Q'$, the curve Q' is stable if and only if P' is. This finishes the proof of (5.2).

The moduli space of admissible G-covers.

We are now going to construct Kuranishi families for admissible G-covers and compute their tangent spaces. Fix an admissible G-cover $f : P \to C$. Let g be the genus, and $q_1, \dots, q_\delta$ the nodes of C, and let $\tilde{g}$ be the genus of P. Assume that there are m_j nodes of P in the preimage of q_j under f. Each of these nodes has index $n_j = |G|/m_j$. Fix a Kuranishi family

$$\mathcal{Y} \to (U, u_0), \qquad P = Y_{u_0}, \tag{5.5}$$

for P where, as usual, Y_u stands for the fiber above u. The automorphism group of P acts on U, and, as usual, we may assume that

$$\mathrm{Stab}(u) = \mathrm{Aut}(Y_u) \subset \mathrm{Aut}(P)$$

for each $u \in U$. We may suppose that U is an open subset of $\mathbb{C}^{3\tilde{g}-3}$, that u_0 is the origin, and that $\mathrm{Aut}(P)$ acts on U via linear transformations. The group G is identified with a subgroup of $\mathrm{Aut}(P/C) \subset \mathrm{Aut}(P)$. We let $V = U^G$ be the fixed point subvariety of the action of G on U; by construction, it is a linear subvariety, and its tangent space at u_0 is T^G_{U,u_0}. It follows that, for every $u \in V$, the automorphism group of Y_u contains G. We set $\mathcal{P} = \mathcal{Y}_{|V}$ and $\mathcal{Z} = \mathcal{P}/G$. We will show that

$$\mathcal{P} \to \mathcal{Z} \to V \tag{5.6}$$

is a family of admissible G-covers and, indeed, a Kuranishi family for $f : P \to C$. The key point is proving the first of these assertions, from which the second readily follows. Indeed, let

$$\mathcal{Q} \to \mathcal{X} \to (S, s_0) \tag{5.7}$$

be a deformation of $f : P \to C$. By the universal property of (5.5), possibly after shrinking S, there is a unique cartesian diagram

$$\begin{array}{ccc} \mathcal{Q} & \xrightarrow{\Phi} & \mathcal{Y} \\ \downarrow & & \downarrow \\ (S, s_0) & \xrightarrow{\varphi} & (U, u_0) \end{array} \tag{5.8}$$

inducing the identity on P. The group G acts on all four corners of the diagram (the action on S is the trivial one), and, again by universality, all the morphisms appearing in the diagram are compatible with these actions. In particular, since G acts trivially on S, φ factors through $V = U^G$. Moreover, as $\mathcal{X}$ is the quotient of $\mathcal{Q}$ by G, the diagram gives rise to a morphism of families of admissible G-covers

$$\begin{array}{ccc} \mathcal{Q} & \longrightarrow & \mathcal{P} \\ \downarrow & & \downarrow \\ \mathcal{X} & \longrightarrow & \mathcal{Z} \\ \downarrow & & \downarrow \\ (S, s_0) & \longrightarrow & (V, u_0) \end{array}$$

It remains to show that (5.6) is a family of admissible G-covers. This could be done directly using the same method we employed to prove Proposition (5.2), but we prefer to reach the same result by giving an explicit elementary construction of a Kuranishi family for $P \to C$, which we will then show to coincide with (5.6).

Before we do it, we analyze the tangent space to V at u_0. As we know, the tangent space to U at u_0 is the middle term of

$$0 \to H^1(P, \mathcal{H}om(\Omega^1_P, \mathcal{O}_P)) \to \operatorname{Ext}^1(\Omega^1_P, \mathcal{O}_P) \to H^0(P, \mathcal{E}xt^1(\Omega^1_P, \mathcal{O}_P)) \to 0\,. \tag{5.9}$$

There are natural compatible actions of G on all the terms of the sequence, so that, taking invariants, we get another exact sequence

$$0 \to H^1(P, \mathcal{H}om(\Omega^1_P, \mathcal{O}_P))^G \to \operatorname{Ext}^1(\Omega^1_P, \mathcal{O}_P)^G \to H^0(P, \mathcal{E}xt^1(\Omega^1_P, \mathcal{O}_P))^G \to 0\,, \tag{5.10}$$

whose middle term is the tangent space to V at u_0. We shall study the other two terms of (5.10) separately, beginning with the one on the left. Let $\alpha : N \to C$ be the normalization map of C. We denote by

D the divisor consisting of all the points of N mapping to nodes of C. Similarly, let $\beta : \widetilde{N} \to P$ be the normalization of P, and $\widetilde{D}$ the divisor of all the points of $\widetilde{N}$ mapping to nodes of P; moreover, let $\widetilde{f} : \widetilde{N} \to N$ be the morphism induced by f. Recall that $\mathcal{H}om(\Omega^1_C, \mathcal{O}_C) \cong T_N(-D)$ and $\mathcal{H}om(\Omega^1_P, \mathcal{O}_P) \cong T_{\widetilde{N}}(-\widetilde{D})$. A local coordinate calculation immediately shows that $\widetilde{f}^*(T_N(-D)) = T_{\widetilde{N}}(-\widetilde{D})$. It follows that

$$
\begin{aligned}
H^1(\mathcal{H}om(\Omega^1_P, \mathcal{O}_P))^G &= H^1(T_{\widetilde{N}}(-\widetilde{D}))^G \\
&= H^1(T_N(-D)) = H^1(\mathcal{H}om(\Omega^1_C, \mathcal{O}_C)) .
\end{aligned} \tag{5.11}
$$

This can be interpreted, in geometrical terms, as asserting that the G-invariant first-order deformations of P which do not smooth the nodes can be also viewed as the first-order deformations of C which do not smooth the nodes, or as the first-order deformations of $(N; D)$, or as the G-invariant first-order deformations of $(\widetilde{N}; \widetilde{D})$.

We now turn to $H^0(P, \mathcal{E}xt^1(\Omega^1_P, \mathcal{O}_P))^G$. As we know,

$$
H^0(P, \mathcal{E}xt^1(\Omega^1_P, \mathcal{O}_P)) = \bigoplus_{p \in P_{\mathrm{sing}}} \mathrm{Ext}^1_{\mathcal{O}_{P,p}}(\Omega^1_{P,p}, \mathcal{O}_{P,p}) .
$$

As G acts trivially on C, it is thus sufficient to concentrate our attention on a single node q of C and on the nodes $p_1, \dots, p_m$ of P mapping to it. Let γ be an element of G which belongs to the stabilizer of p_i. Since $P \to C$ is an admissible G-cover, in suitable coordinates, γ acts on one branch of the node as multiplication by a root of unity ζ, and on the other as multiplication by ζ^{-1}. It then follows, using, for instance, formula (3.7) in Chapter XI, that γ acts as the identity on $\mathrm{Ext}^1(\Omega^1_{P,p_i}, \mathcal{O}_{P,p_i})$. Since G acts transitively on $\{p_1, \dots, p_m\}$, this implies that

$$
\Big(\bigoplus_{i=1}^{m} \mathrm{Ext}^1_{\mathcal{O}_{P,p_i}}(\Omega^1_{P,p_i}, \mathcal{O}_{P,p_i}) \Big)^G = \mathrm{Ext}^1_{\mathcal{O}_{P,p_j}}(\Omega^1_{P,p_j}, \mathcal{O}_{P,p_j})
$$

for any j. Thus, the dimension of $H^0(P, \mathcal{E}xt^1(\Omega^1_P, \mathcal{O}_P))^G$ equals δ, the number of nodes of C. Combining this with (5.11), we find in particular that

$$
\dim \big(\mathrm{Ext}^1_{\mathcal{O}_P}(\Omega^1_P, \mathcal{O}_P)^G \big) = \dim \big(\mathrm{Ext}^1_{\mathcal{O}_C}(\Omega^1_C, \mathcal{O}_C) \big) = 3g - 3 .
$$

We now return to the problem of showing that (5.6) is a family of admissible G-covers. As we announced, we shall rely on an alternate and very explicit construction of a Kuranishi family for $P \to C$, patterned after the construction of Kuranishi families for nodal curves given in Section 3 of Chapter XI (cf. in particular the proof of Theorem (3.17)). We begin by choosing $h = 3g - 3 - \delta$ general points $p_1, \dots, p_h$ on N, not belonging to D. As we explained in Section 2 of Chapter XI, performing independent Schiffer variations at each one of the p_i yields a Kuranishi family

$$\mathcal{N} \to (B', b'_0), \quad \mathcal{D}$$

for the 2δ-pointed curve $(N; D)$. Since $\widetilde{f} : \widetilde{N} \to N$ is unramified at the p_i, each one of these Schiffer variations can be pulled back to all the sheets of $\widetilde{f}$. The result is a deformation

$$\widetilde{\mathcal{N}} \to (B', b'_0), \quad \widetilde{\mathcal{D}} \tag{5.12}$$

of $(\widetilde{N}; \widetilde{D})$. The action of G on $\widetilde{N}$ extends to an action on $(\widetilde{\mathcal{N}}, \widetilde{\mathcal{D}})$, compatible with (5.12). In fact, away from the p_i, $\widetilde{\mathcal{N}} \to B'$ is a product family, while G acts simply transitively on the points of $\widetilde{N}$ mapping to p_i for each i and consequently, by construction, on the corresponding Schiffer variations. Recall that the individual sections in $\mathcal{D}$ and $\widetilde{\mathcal{D}}$ come in pairs, corresponding to the nodes of C and P. Identifying the two sections in each pair produces deformations

$$\mathcal{X}' \to (B', b'_0), \qquad \mathcal{Q}' \to (B', b'_0) \tag{5.13}$$

of C and P, respectively. The Kodaira–Spencer homomorphism identifies T_{B', b'_0} to $H^1(C, \mathcal{H}om(\Omega^1_C, \mathcal{O}_C))$. Moreover, G acts on $\mathcal{Q}'$, and the quotient modulo this action is $\mathcal{X}'$.

The next step is to smooth the nodes in the fibers of $\mathcal{Q}' \to (B', b'_0)$, in a way that is compatible with the action of G. To keep the notation simple, we shall do this when C has a single node, that is, when $\delta = 1$; the general case can be done by repeating at every node the procedure we are going to outline. We denote by q the node of C, and by $p_1, \dots, p_m$ the nodes of P. The index of $P \to C$ at q is $n = |G|/m$. Observe that, by construction, the two families (5.13) come with trivializations near the nodes and that, in the case of the second family, these trivializations are compatible with the action of G. This means the following. The locus of nodes in the fibers of $\mathcal{X}' \to (B', b'_0)$ is a section Σ, while the locus of nodes in the fibers of $\mathcal{Q}' \to (B', b'_0)$ consists of m sections $\Sigma_1, \dots, \Sigma_m$ with Σ_i passing through p_i. Then a neighborhood of Σ_i is of the form

$$W_i = B' \times \{(\xi, \eta) : \xi\eta = 0, |\xi| < \varepsilon, |\eta| < \varepsilon\}$$

with the action of any element γ of G fixing p_i given by

$$(b, \xi, 0) \mapsto (b, \zeta\xi, 0), \quad (b, 0, \eta) \mapsto (b, 0, \zeta^{-1}\eta), \tag{5.14}$$

where ζ is a suitable nth root of unity. Similarly, a neighborhood of Σ is of the form

$$W = B' \times \{(x, y) : xy = 0, |x| < \varepsilon, |y| < \varepsilon\},$$

and the quotient morphism $\mathcal{Q}' \to \mathcal{X}'$ is given by

$$(b, \xi, 0) \mapsto (b, \xi^n, 0), \quad (b, 0, \eta) \mapsto (b, 0, \eta^n).$$

The smoothing procedure described in the proof of Theorem (3.17) in Chapter XI (cf. in particular formulas (3.15)) makes it possible to replace a neighborhood of $\xi = \eta = 0$ in $W_i \times \{z \in \mathbb{C} : |z| < \varepsilon\}$ with a neighborhood of $\xi = \eta = 0$ in

$$W_i' = B' \times \{(\xi, \eta) : \xi\eta = z\}$$

for $i = 1, \ldots, m$. We denote by $\mathcal{Q} \to \widetilde{B}$ the resulting family, where $\widetilde{B} = B' \times \{z \in \mathbb{C} : |z| < \varepsilon\}$. The smoothing procedure is compatible with the action of G. If γ is an element of G belonging to the stabilizer of p_i, acting on W_i as in (5.14), it acts on W_i' by

$$(5.15). \qquad (b, \xi, \eta, z) \mapsto (b, \zeta\xi, \zeta^{-1}\eta, z).$$

If we set $\widetilde{\mathcal{X}} = \mathcal{Q}/G$, then

$$(5.16) \qquad \mathcal{Q} \to \widetilde{\mathcal{X}} \to \widetilde{B}$$

is a family of admissible G-covers. In fact, by the explicit form of the action, the quotient of W_i' modulo the stabilizer of p_i is

$$(5.17) \qquad W_i'/\operatorname{Stab}_G(p_i) = B' \times \{(x, y) : xy = z^n\},$$

and the quotient map is

$$(b, \xi, \eta, z) \mapsto (b, \xi^n, \eta^n, z).$$

The fiber of (5.16) at $\widetilde{b}_0 = (b_0', 0)$ is just our original G-cover $P \to C$.

We claim that $\mathcal{Q} \to \widetilde{B}$ can be identified with $\mathcal{P} \to V$, after suitably shrinking the bases, and that the identification can be chosen to be compatible with the G-actions. This is straightforward; look at diagram (5.8) with $(S, s_0) = (\widetilde{B}, \widetilde{b}_0)$. Since $T_{\widetilde{B}, \widetilde{b}_0} = \operatorname{Ext}^1(\Omega^1_P, \mathcal{O}_P)^G$ by construction, in this case $d\varphi$ sends $T_{\widetilde{B}, \widetilde{b}_0}$ isomorphically to T_{V, u_0}. Thus, after suitably shrinking $\widetilde{B}$ and U, φ maps $\widetilde{B}$ isomorphically to V, and Φ maps $\mathcal{Q}$ isomorphically to $\mathcal{P}$; moreover, as we observed, Φ is compatible with the action of G. This proves our claim; it follows that $\mathcal{P} \to \mathcal{Z} = \mathcal{P}/G \to V$ is a family of admissible G-covers, because $\mathcal{Q} \to \widetilde{\mathcal{X}} = \mathcal{Q}/G \to \widetilde{B}$ is.

As we have mentioned, the construction we have just completed carries over, without difficulties, to the case in which C has any number

δ of nodes. Let $\pi : \mathcal{C} \to (B, b_0)$ be a Kuranishi family for C. By its universal property there is a cartesian diagram

$$\begin{array}{ccc} \mathcal{Z} & \xrightarrow{\quad\rho\quad} & \mathcal{C} \\ \tilde{\pi}\downarrow & & \downarrow\pi \\ (V, u_0) = (\widetilde{B}, \widetilde{b}_0) & \xrightarrow{\quad\sigma\quad} & (B, b_0) \end{array} \tag{5.18}$$

inducing the identity on C. Because of the explicit construction of $\mathcal{P} \to \mathcal{Z} \to V$ and in particular because of formula (5.17), we can arrange things so that B and $\widetilde{B}$ are polycylinders centered at the origin in the variables $t_1, \dots, t_{3g-3}$ and $z_1, \dots, z_{3g-3}$, respectively, that b_0 and $\widetilde{b}_0$ are their respective origins, and that the classifying map from $\widetilde{B}$ to B is

$$\begin{aligned} &\sigma : \widetilde{B} \to B \\ &t_j = z_j^{n_j}\,, \ j \leq \delta\,, \qquad t_j = z_j\,, \ j > \delta\,. \end{aligned} \tag{5.19}$$

Here, n_j stands for the index of $P \to C$ at the node q_j of C.

Remark (5.20). This last observation shows, in particular, that for the family (5.6) the n_jth power of the Dehn twist associated to the node q_j of C lifts to the product of the Dehn twists associated to the nodes of P lying above q_j.

Having constructed Kuranishi families for admissible G-covers, we may replicate, essentially word by word, the steps that led to the analytic structure on $\overline{M}_g$ and construct in this by now familiar way a normal, compact analytic space $\mathrm{Adm}_g(G)$ which is a coarse *moduli space for admissible G-covers of genus g stable curves.* By construction, a basis for the analytic structure of $\mathrm{Adm}_g(G)$ is given by sets of the type $\widetilde{B}/H$, where $\widetilde{B}$ is the base of a Kuranishi family for a G-cover $P \to C$, and H is the automorphism group of $P \to C$ as a G-cover. It is also evident, from our definitions, that the space $\mathrm{Adm}_g(G)$ is a coarse moduli space for an analytic stack $\mathcal{A}dm_g(G)$, which is called *the stack of admissible G-covers of stable curves of genus g,* and that we have natural projections

$$\mathrm{Adm}_g(G) \to \overline{M}_g\,, \qquad \mathcal{A}dm_g(G) \to \overline{\mathcal{M}}_g\,.$$

It would not be too hard to check that, in general, $\mathcal{A}dm_g(G)$ is an algebraic stack, and indeed a Deligne–Mumford stack, but we will need this only in the cases we are really interested in, namely the ones in which $\mathcal{A}dm_g(G)$ is actually equal to $\mathrm{Adm}_g(G)$.

Suppose that the group G has trivial center. As we remarked at the beginnig of the section, in this case the notions of G-cover of a smooth genus g curve and the one of smooth genus g curve with level G structure coincide. In particular the open substack of $\mathcal{A}dm_g(G)$ parameterizing G-covers of smooth curves coincides with the stack of curves with level G structure ${}_G\mathcal{M}_g$. When the center of G is trivial, it is therefore natural to set

$$(5.21) \qquad {}_G\overline{M}_g = \mathrm{Adm}_g(G)\,, \qquad {}_G\overline{\mathcal{M}}_g = \mathcal{A}dm_g(G)\,.$$

Our next task is to find a group G for which ${}_G\overline{\mathcal{M}}_g$ coincides with ${}_G\overline{M}_g$. For this, we must find a group G having the property that any admissible G-cover of any stable genus g curve is automorphism-free.

6. Automorphisms of admissible covers.

As we have explained, we need to find a group G such that the automorphism group of any admissible G-cover is trivial. In this section we shall construct such a group; the proof that it has the required property will be carried out in this section and in the following two. Specifically, in this section we shall establish sufficient conditions for the triviality of the automorphism group of any admissible G-cover. Some of these conditions are of algebraic nature and can be verified fairly easily for the group we shall construct. There is however an additional condition, of a more geometric nature, which is considerably more delicate, and whose verification will occupy all of the following two sections.

We begin by establishing some notation. Let $\pi : X \to Y$ be a morphism of schemes. We shall denote by $\mathrm{Aut}(X\to Y)$ the group of pairs (f, α) where $f : X \to X$ and $\alpha : Y \to Y$ are automorphisms such that $\alpha\pi = \pi f$. We denote by $\mathrm{Aut}(X/Y) \subset \mathrm{Aut}(X\to Y)$ the subgroup formed by pairs of type (f, id_Y). Thus, there is an exact sequence

$$(6.1) \qquad 0 \to \mathrm{Aut}(X/Y) \to \mathrm{Aut}(X\to Y) \to \mathrm{Aut}(Y)\,.$$

Suppose that a group G acts on X via automorphisms of X over Y. We then let $\mathrm{Aut}(X\to Y)^G \subset \mathrm{Aut}(X\to Y)$ and $\mathrm{Aut}(X/Y)^G \subset \mathrm{Aut}(X/Y)$ denote the subgroups of G-invariant elements. In explicit terms, the elements of $\mathrm{Aut}(X \to Y)^G$ are pairs $(f, \alpha) \in \mathrm{Aut}(X \to Y)$ such that $gf(x) = f(gx)$ for every $g \in G$. In analogy with (6.1), we have an exact sequence

$$(6.2) \qquad 0 \to \mathrm{Aut}(X/Y)^G \to \mathrm{Aut}(X\to Y)^G \to \mathrm{Aut}(Y)\,.$$

With this notation, the automorphism group of an admissible G-cover $f : P \to C$ is nothing but the group $\mathrm{Aut}(P\to C)^G$. Let us observe that there are injective homomorphisms

$$\phi : G \to \mathrm{Aut}(P/C)\,, \qquad \phi_{\mathrm{inv}} : Z(G) \to \mathrm{Aut}(P/C)^G$$

and that, while nothing can be said, in general, about the surjectivity of ϕ and ϕ_{inv}, both ϕ and ϕ_{inv} are in fact isomorphisms when $f : P \to C$ is a smooth G-cover.

Let Π be the homotopy group of a fixed genus g reference surface, and let $P \to C$ be an admissible G-cover, where G is a finite group, and C has genus g. Our first goal is to prove a criterion for the vanishing of $\mathrm{Aut}(P{\to}C)^G$ under the assumption that G has the following properties:

$$
(6.3)\quad \begin{cases} \text{i)} & G \text{ is a strongly characteristic quotient of } \Pi; \\ \text{ii)} & \text{the center } Z(G) \text{ of } G \text{ is trivial}; \\ \text{iii)} & \text{a quotient of } G \text{ is isomorphic to } (\mathbb{Z}/m\mathbb{Z})^{2g} \text{ for some } m \geq 3. \end{cases}
$$

For the time being, let G be any finite group, not necessarily satisfying i), ii), and iii). Assume that C has δ nodes $p_1, \ldots, p_\delta$ and that the index of G at p_i is n_i, $i = 1, \ldots, \delta$. Let $\pi : \mathcal{C} \to (B, b_0)$ be a Kuranishi family for C. We may suppose that B is a polycylinder centered at the origin in the variables $t_1, \ldots, t_{3g-3}$ and that $b_0 = 0$. Consider the base change diagram (5.18) and the Kuranishi family (5.6) for $P \to C$. We have a diagram

$$
(6.4)\qquad \begin{array}{ccccc} \mathcal{P} & \xrightarrow{\lambda} & \mathcal{Z} & \xrightarrow{\rho} & \mathcal{C} \\ & \searrow^{\gamma} & \downarrow{\scriptstyle\widetilde{\pi}} & & \downarrow{\scriptstyle\pi} \\ & & \widetilde{B} & \xrightarrow{\sigma} & B \end{array}
$$

where the square is cartesian, $C = \pi^{-1}(0) = \widetilde{\pi}^{-1}(0)$, $P = \gamma^{-1}(0)$, and λ extends $P \to C$. Recall that σ may be assumed to be given by (5.19). We set

$$
B^* = B \smallsetminus \{\textstyle\prod_{i=1}^{\delta} t_i = 0\}, \quad \widetilde{B}^* = \sigma^{-1}(B^*), \quad \mathcal{C}^* = \pi^{-1}(B^*),
$$
$$
\mathcal{Z}^* = \rho^{-1}(\mathcal{C}^*), \quad \mathcal{P}^* = \gamma^{-1}(\widetilde{B}^*),
$$

and obtain an induced diagram

$$
(6.5)\qquad \begin{array}{ccccc} \mathcal{P}^* & \xrightarrow{\lambda} & \mathcal{Z}^* & \xrightarrow{\rho} & \mathcal{C}^* \\ & \searrow^{\gamma} & \downarrow{\scriptstyle\widetilde{\pi}} & & \downarrow{\scriptstyle\pi} \\ & & \widetilde{B}^* & \xrightarrow{\sigma} & B^* \end{array}
$$

Our first observation is the following.

LEMMA (6.6). *Let $P \to C$ be an admissible G-cover. Suppose that* i) *and* iii) *in (6.3) hold. Then*

$$
\mathrm{Aut}(P{\to}C)^G = \mathrm{Aut}(P/C)^G .
$$

Proof. By assumption, $\widetilde{\pi} : \mathcal{Z}^* \to \widetilde{B}^*$ can be thought of as a family of smooth genus g curves with level m structure. Let $\beta \in \mathrm{Aut}(P{\to}C)^G$. By the universal property of the Kuranishi family $\mathcal{P} \to \mathcal{Z} \to \widetilde{B}$, the element β acts as an automorphism on $\mathcal{Z} \to \widetilde{B}$ and hence on $\mathcal{Z}^* \to \widetilde{B}^*$, viewed as a family of curves with level m structure. By applying Lemma (4.3) to the case in which $\alpha = \sigma$ and $\mathcal{Y} = \mathcal{Z}^*$ we deduce that β lies over the identity automorphism of $\mathcal{C} \to B$. Since

$$\mathrm{Aut}(\mathcal{C} \to B) = \mathrm{Aut}(C)$$

by the universal property of the Kuranishi family $\mathcal{C} \to B$, the image of β under the natural projection $\mathrm{Aut}(P \to C)^G \to \mathrm{Aut}(C)$ is the identity. Q.E.D.

Look at the families (6.4) and (6.5). Using Lemma (6.6), the universal property of the Kuranishi family for $P \to C$, and the lemma on extension of isomorphisms of families of stable curves (Corollary (5.4) in Chapter X), we obtain the following identifications:

$$\mathrm{Aut}(P \to C)^G = \mathrm{Aut}(P/C)^G = \mathrm{Aut}(\mathcal{P}/\mathcal{C})^G = \mathrm{Aut}(\mathcal{P}^*/\mathcal{C}^*)^G . \tag{6.7}$$

Thus, the problem of finding a group G such that $\mathrm{Aut}(P \to C)^G$ is trivial translates into the one of finding a G such that $\mathrm{Aut}(\mathcal{P}^*/\mathcal{C}^*)^G = \{1\}$. We start by studying $\mathrm{Aut}(\mathcal{P}^*/\mathcal{C}^*)$. Choose $b \in B^*$ and let $a \in \widetilde{B}^*$ be a point lying above it. Set $C_a = \widetilde{\pi}^{-1}(a)$ and $P_a = \gamma^{-1}(a)$. First of all, there is an injective homomorphism $\alpha : \mathrm{Aut}(P_a/C_a) \to \mathrm{Aut}(\mathcal{P}^*/\mathcal{C}^*)$, obtained by composing the identifications $\mathrm{Aut}(P_a/C_a) = G = \mathrm{Aut}(\mathcal{P}^*/\mathcal{Z}^*)$ with $\mathrm{Aut}(\mathcal{P}^*/\mathcal{Z}^*) \to \mathrm{Aut}(\mathcal{P}^*/\mathcal{C}^*)$. Next observe that any automorphism of $\mathcal{P}^*$ inducing the identity on $\mathcal{C}^*$ determines an automorphism of $\mathcal{Z}^*$ inducing the identity on $\mathcal{C}^*$ and hence an automorphism of $\widetilde{B}^*$ inducing the identity on B^*. This defines a homomorphism $\beta : \mathrm{Aut}(\mathcal{P}^*/\mathcal{C}^*) \to \mathrm{Aut}(\widetilde{B}^*/B^*)$. Our next result makes essential use of assumption i), that is, of the fact that G is a strongly characteristic quotient of Π.

LEMMA (6.8). *If* i) *in* (6.3) *holds, the sequence*

$$1 \to \mathrm{Aut}(P_a/C_a) \xrightarrow{\alpha} \mathrm{Aut}(\mathcal{P}^*/\mathcal{C}^*) \xrightarrow{\beta} \mathrm{Aut}(\widetilde{B}^*/B^*) \to 1 \tag{6.9}$$

is split exact.

Proof. That the image of α is contained in the kernel of β is clear. To prove the converse, recall that $\mathrm{Aut}(P_a/C_a) = \mathrm{Aut}(\mathcal{P}^*/\mathcal{Z}^*)$ and observe that any automorphism of $\mathcal{Z}^*$ inducing the identity on $\widetilde{B}^*$ and on $\mathcal{C}^*$ must be the identity. Proving the surjectivity of β requires a little more work. First of all, any element of $\mathrm{Aut}(\widetilde{B}^*/B^*) = \mathrm{Aut}(\widetilde{B}/B)$ pulls back to an automorphism of $\mathcal{Z} = \mathcal{C} \times_B \widetilde{B}$ lying above the identity of $\mathcal{C}$. This defines an isomomorphism $j : \mathrm{Aut}(\widetilde{B}^*/B^*) \to \mathrm{Aut}(\mathcal{Z}^*/\mathcal{C}^*)$. Next, choose a section D of $\pi : \mathcal{C} \to B$; of course, D does not meet any node in the fibers of π. Clearly, $\widetilde{D} = D \times_B \widetilde{B}$ is a section of $\mathcal{Z} \to \widetilde{B}$, and its inverse image via λ is a disjoint union of $|G|$ sections of $\mathcal{P} \to \widetilde{B}$; let E be one of these. Let m be an element of $\mathrm{Aut}(\widetilde{B}^*/B^*)$. Clearly, $\widetilde{D}$ is invariant under $j(m) \in \mathrm{Aut}(\mathcal{Z}^*/\mathcal{C}^*)$. We must find a lifting $k(m) : \mathcal{P}^* \to \mathcal{P}^*$ of $j(m)$. We shall define $k(m)$ fiber by fiber. Let x be a point of $\widetilde{B}^*$; we denote by z_x (resp., p_x) the unique point of $\widetilde{D}$ (resp., E) above x, and by Z_x the fiber of $\mathcal{Z} \to \widetilde{B}$ at x. The automorphism $j(m)$ sends Z_x isomorphically to $Z_{m(x)}$, and z_x to $z_{m(x)}$. If we identify Z_x and $Z_{m(x)}$ to $C_{\sigma(x)}$ via ρ, this isomorphism corresponds to the identity on $C_{\sigma(x)}$ by construction. The theory of topological coverings then says that $Z_x \to Z_{m(x)}$ lifts uniquely to an isomorphism $P_x \to P_{m(x)}$ sending p_x to $p_{m(x)}$ if and only if

$$\rho_*\lambda_*(\pi_1(P_x, p_x)) = \rho_*\lambda_*(\pi_1(P_{m(x)}, p_{m(x)})) \subset \pi_1(C_{\sigma(x)}, \rho(z_x)).$$

That this is true follows at once from the fact that $\rho_*\lambda_*(\pi_1(P_x, p_x))$ is a strongly characteristic subgroup of $\pi_1(C_{\sigma(x)}, \rho(z_x))$ for any x, by assumption i).

The fiberwise liftings $P_x \to P_{m(x)}$ we have constructed patch together, by uniqueness and continuity, to yield the desired lifting $k(m) : \mathcal{P}^* \to \mathcal{P}^*$ of $j(m) : \mathcal{Z}^* \to \mathcal{Z}^*$. By construction, it is clear that $k : \mathrm{Aut}(\widetilde{B}^*/B^*) \to \mathrm{Aut}(\mathcal{P}^*/\mathcal{C}^*)$ is a group homomorphism.

Q.E.D.

We finally come to the key lemma regarding the automorphism group of an admissible G-cover. At this stage, we will make full use of assumptions i), ii), and iii) in (6.3). To set up, we fix a point $b \in B^*$ and consider the Picard–Lefschetz transformation

(6.10) $$PL : \pi_1(B^*, b) \to \Gamma_g \to \mathrm{Out}(G).$$

We identify $\pi_1(B^*, b)$ with $\mathbb{Z}^\delta = \mathbb{Z}\gamma_1 \oplus \cdots \oplus \mathbb{Z}\gamma_\delta$, where γ_i is the class of a simple loop around $t_i = 0$.

LEMMA (6.11). *Let $P \to C$ be an admissible G-cover of a genus g stable curve C. Assume that C has δ nodes $p_1, \ldots, p_\delta$ and that the index of $P \to C$ at p_i is n_i, $i = 1, \ldots, \delta$. Suppose that (6.3) hold and that*

(6.12) $$\ker PL = n_1\mathbb{Z} \oplus \cdots \oplus n_\delta\mathbb{Z}.$$

Then $\mathrm{Aut}(P \to C)^G = \{1\}$.

Proof. Set

$$M = \mathrm{Aut}(\widetilde{B}^*/B^*) = \bigoplus_{i=1}^{\delta} \mathbb{Z}/n_i\mathbb{Z}.$$

Our assumptions imply that M injects into $\mathrm{Out}(G)$. Rewrite the sequence (6.9) as

$$1 \to G \xrightarrow{\alpha} \mathrm{Aut}(\mathcal{P}^*/\mathcal{C}^*) \xrightarrow{\beta} M \to 1\,. \tag{6.13}$$

By (6.7), we must show that $\mathrm{Aut}(\mathcal{P}^*/\mathcal{C}^*)^G = \{1\}$. In diagram (6.5), both σ and ρ are Galois covers with Galois group M, while λ is a Galois cover with Galois group G. By Lemma (6.8), the order of $\mathrm{Aut}(\mathcal{P}^*/\mathcal{C}^*)$ is $\deg(\rho)\deg(\lambda)$, and hence $\rho\lambda$ is a Galois cover as well. It follows that the sequence (6.9) can also be rewritten as

$$1 \to \pi_1(C_a)/\pi_1(P_a) \xrightarrow{\alpha} \pi_1(\mathcal{C}^*)/\pi_1(\mathcal{P}^*) \xrightarrow{\beta} \pi_1(B^*)/\pi_1(\widetilde{B}^*) \to 1\,. \tag{6.14}$$

As we know, this exact sequence exhibits the middle term as a semidirect product. Indeed, the action of the abelian group $M = \pi_1(B^*)/\pi_1(\widetilde{B}^*)$ on G is induced by the action of $\pi_1(B^*)$ on $\pi_1(C_a)$, by virtue of the fact that $\pi_1(P_a)$ is a characteristic subgroup of $\pi_1(C_a)$. At the level of outer automorphisms of $\pi_1(C_a)$, this is nothing but the Picard–Lefschetz transformation. We denote by $\tau : M \to \mathrm{Aut}(G)$ the homomorphism exhibiting $\mathrm{Aut}(\mathcal{P}^*/\mathcal{C}^*)$ as the semidirect product of G and M and by $\overline{\tau} : M \to \mathrm{Out}(G)$ the induced (Picard–Lefschetz) homomorphism. The group G acts on the sequence (6.13), the action on the right-hand term being trivial. Passing to the invariants, we have the exact sequence

$$1 \to Z(G) \xrightarrow{\alpha} \mathrm{Aut}(\mathcal{P}^*/\mathcal{C}^*)^G \xrightarrow{\beta} M\,. \tag{6.15}$$

The condition for an element $(h, m) \in G \ltimes M = \mathrm{Aut}(\mathcal{P}^*/\mathcal{C}^*)$ to be G-invariant is that

$$(g, 1)^{-1}(h, m)(g, 1) = (h, m) \qquad \forall g \in G\,.$$

This translates into

$$\tau(m)(g) = h^{-1}gh \qquad \forall g \in G\,, \tag{6.16}$$

meaning that $\overline{\tau}(m)$ is the identity of $\mathrm{Out}(G)$. As we observed, M injects in $\mathrm{Out}(G)$ by assumption. Thus m must be zero. But then (6.16) forces h to be in the center of G, which, by assumption, is trivial. In conclusion, $\mathrm{Aut}(\mathcal{P}^*/\mathcal{C}^*)^G$ must be the trivial group.

Q.E.D.

REMARK (6.17). Let C be as in the statement of Lemma (6.11), let G be a finite group, and let $P \to C$ be an admissible G-cover. For $i = 1, \dots, \delta$, denote by n_i the index of $P \to C$ at p_i. Then the inclusion

$$\ker PL \supset n_1\mathbb{Z} \oplus \cdots \oplus n_\delta\mathbb{Z}\,.$$

always holds. In fact, let $\gamma_1, \ldots, \gamma_\delta$ be the standard generators for $\pi_1(B^*, b)$. As in the proof of the preceding lemma, consider the G-cover $P_a \to C_a = C_b$. Going around γ_i changes this G-cover by the automorphism $PL(\gamma_i) \in \mathrm{Out}(G)$. On the other hand, since over $\widetilde{B}$ a family of G-covers is defined, when one goes around any loop in $\widetilde{B}^*$, the G-cover $P_a \to C_a$ goes back to itself. Since the inverse image in $\widetilde{B}^*$ of $n_i\gamma_i$ is a closed loop, it follows that $PL(n_i\gamma_i) = 1$ for any i, which is exactly what we wanted to show.

The group

We are now going to construct a finite group G satisfying (6.3) and (6.12). First some simple preliminaries about strongly characteristic quotients.

LEMMA (6.18). a) *Let L, H, K be groups and assume that there are surjective homomorphisms $\beta : L \to H$ and $\alpha : H \to K$. If K is a strongly characteristic quotient of L, then it is also a strongly characteristic quotient of H.*
b) *Let L be a group, H a strongly characteristic subgroup of L, and K a strongly characteristic subgroup of H. Then K is a strongly characteristic subgroup of L.*

We first prove a). Let $\alpha' : H \to K$ be a surjective homomorphism. Then the kernels of $\alpha\beta$ and $\alpha'\beta$ are equal, since K is a strongly characteristic quotient of L. Thus, $\ker(\alpha) = \beta(\ker(\alpha\beta))$ is equal to $\ker(\alpha') = \beta(\ker(\alpha'\beta))$.

We now come to b). Let $\alpha : L \to L/K$ be a surjective homomorphism. The kernel of the composition of α with the natural homomorphism $L/K \to L/H$ is H, since the latter is a strongly characteristic subgroup of L. Thus $K' = \ker(\alpha) \subset H$. We claim that there exists a commutative diagram

$$\begin{array}{ccccccccc} 1 & \longrightarrow & H/K' & \longrightarrow & L/K' & \longrightarrow & L/H & \longrightarrow & 1 \\ & & u\downarrow & & v\downarrow & & w\downarrow & & \\ 1 & \longrightarrow & H/K & \longrightarrow & L/K & \longrightarrow & L/H & \longrightarrow & 1 \end{array} \tag{6.19}$$

where the vertical arrows are isomorphisms. If we can prove the claim, $K = K'$ since K is strongly characteristic in H, and we are done. The isomorphism v comes from α by passage to the quotient, and the isomorphism w exists since L/H is a characteristic quotient of L/K' by part a) of the lemma. This shows that we can complete the diagram with an isomorphism u, concluding the proof. Q.E.D.

We now begin the construction of a finite group satisfying (6.3) and (6.12). As usual, given an integer $m \geq 2$ and a group G, we write $G^{(2),m}$ to denote the subgroup of G generated by commutators and mth powers.

We denote by Σ_m a Galois covering of Σ having $H_1(\Sigma, \mathbb{Z}/m\mathbb{Z}) = \Pi/\Pi^{(2),m}$ as Galois group. We shall normally denote the group $H_1(\Sigma, \mathbb{Z}/m\mathbb{Z})$ by G_m and write its operation multiplicatively.

Now let p and q be distinct odd primes. Denote by $\Sigma_{[\frac{q}{p}]}$ a Galois covering of Σ_p with Galois group $H_1(\Sigma_p, \mathbb{Z}/q\mathbb{Z})$, and denote this group by H; the operation in H will normally be written additively. We also denote by $G_{[\frac{q}{p}]}$ the Galois group of the composite covering $\Sigma_{[\frac{q}{p}]} \to \Sigma$. The group $G_{[\frac{q}{p}]}$ is nothing but $\Pi/{\Pi'}^{(2),q}$, where $\Pi' = \Pi^{(2),p}$, and hence is the middle term of the exact sequence

$$1 \to \Pi'/{\Pi'}^{(2),q} \to \Pi/{\Pi'}^{(2),q} \to \Pi/\Pi^{(2),p} \to 1 \,. \tag{6.20}$$

It follows from part b) of Lemma (6.18) that $G_{[\frac{q}{p}]}$ is a strongly characteristic quotient of Π, since ${\Pi'}^{(2),q}$ is a strongly characteristic subgroup of $\Pi' = \Pi^{(2),p}$, and $\Pi^{(2),p}$ is a strongly characteristic subgroup of Π, by Example (2.3). The sequence (6.20) can be identified with

$$0 \to H = H_1(\Sigma_p, \mathbb{Z}/q\mathbb{Z}) \to G_{[\frac{q}{p}]} \to H_1(\Sigma, \mathbb{Z}/p\mathbb{Z}) = G_p \to 1 \,. \tag{6.21}$$

As p and q are relatively prime, by Sylow's theorem this exact sequence exhibits $G_{[\frac{q}{p}]}$ as a semidirect product of H and G_p. If we set-theoretically identify $G_{[\frac{q}{p}]}$ with $H \times G_p$, the operation in it is given by

$$(h,k)(h',k') = (h + k_*(h'), kk') \,, \tag{6.22}$$

where $k \in G_p$ is viewed as a deck transformation of $\Sigma_p \to \Sigma$, and k_* stands for the pushforward in homology with $\mathbb{Z}/q\mathbb{Z}$-coefficients. In other words, the homomorphism from $G_{[\frac{q}{p}]}$ to $\mathrm{Aut}(H)$ induced by conjugation drops down to the homomorphism $G_p \to \mathrm{Aut}(H)$ which associates to each $k \in G_p$ the pushforward k_*.

Since p and q are relatively prime, H is a completely reducible representation of G_p. We write H^{inv} to denote the invariants, and H^n to denote the sum of all nontrivial irreducible subrepresentations of H, so that

$$H = H^{inv} \oplus H^n \,.$$

Any element $h \in H$ can be written uniquely as a sum of elements $h^{inv} \in H^{inv}$ and $h^n \in H^n$. An explicit expression for h^{inv} is

$$h^{inv} = \frac{1}{p^{2g}} \sum_{k \in G_p} k_*(h) \,.$$

Let us prove a number of useful properties of the groups we have just introduced.

1) *G_p acts faithfully on H.*

Let k be an element of G_p and interpret it as a deck transformation of $\Sigma_p \to \Sigma$. Put an auxiliary complex structure on Σ and the induced complex structure on Σ_p. Then k acts analytically. If k acts trivially on $H = H_1(\Sigma_p, \mathbb{Z}/q\mathbb{Z})$, then k is the identity element of G_p by Proposition (2.8).

2) *The center of $G_{[{q \atop p}]}$ is H^{inv}.*

Suppose that $(h,k)(h',k') = (h',k')(h,k)$ for every (h',k') in $G_{[{q \atop p}]}$. Taking $k' = 1$ and using (6.22), we get $h + k_*(h') = h' + h$, so that $k_*(h') = h'$ for any h'. By 1), this implies that $k = 1$. Now the above commutation relation reduces to $h+h' = k'_*(h)+h'$, so that $h = k'_*(h)$ for every $k' \in G_p$, showing that h is invariant.

3) *H^{inv} is canonically isomorphic to $G_q \cong H_1(\Sigma, \mathbb{Z}/q\mathbb{Z})$.*

Since p and q are relatively prime, we have

$$H^{inv} = H_1(\Sigma_p, \mathbb{Z}/q\mathbb{Z})^{G_p} = H_1(\Sigma_p/G_p, \mathbb{Z}/q\mathbb{Z}) = H_1(\Sigma, \mathbb{Z}/q\mathbb{Z}) \cong G_q\,.$$

Now set

$$G^n_{[{q \atop p}]} = G_{[{q \atop p}]}/H^{inv}\,.$$

4) *The center of $G^n_{[{q \atop p}]}$ is trivial.*

It suffices to notice that $G_{[{q \atop p}]}$ is the direct product of $G^n_{[{q \atop p}]}$ and H^{inv}.

The diagram

$$\begin{array}{ccccc}
G_{[{q \atop p}]} & \longrightarrow & G_{[{q \atop p}]}/H^n = H^{inv} \times G_p = G_q \times G_p = G_{pq} & \longrightarrow & G_q \\
\downarrow & & \downarrow & & \downarrow \\
G^n_{[{q \atop p}]} = G_{[{q \atop p}]}/H^{inv} & \longrightarrow & G_{[{q \atop p}]}/H = G_p & \longrightarrow & \{1\}
\end{array}$$

and its analogue with the roles of p and q interchanged can be combined and completed to a commutative diagram

$$\begin{array}{ccccc}
G_{\{p,q\}} & \longrightarrow & G_{[{p \atop q}]} & \longrightarrow & G^n_{[{p \atop q}]} \\
\downarrow & & \downarrow & & \downarrow \\
G_{[{q \atop p}]} & \longrightarrow & G_{pq} & \longrightarrow & G_q \\
\downarrow & & \downarrow & & \downarrow \\
G^n_{[{q \atop p}]} & \longrightarrow & G_p & \longrightarrow & \{1\}
\end{array}$$

where $G_{\{p,q\}} = G_{\left[{q \atop p}\right]} \times_{G_{pq}} G_{\left[{p \atop q}\right]}$. Since the degrees of the two arrows in the top left corner are relatively prime, any homomorphism $\Pi \to G_{\{p,q\}}$ obtained by combining surjective homomorphisms $\Pi \to G_{\left[{q \atop p}\right]}$ and $\Pi \to G_{\left[{p \atop q}\right]}$ is onto. On the other hand, if $\alpha : \Pi \to G_{\{p,q\}}$ is any surjective homomorphism, the kernel of α is the intersection of the kernels of the composite homomorphisms $\alpha_1 : \Pi \to G_{\left[{q \atop p}\right]}$ and $\alpha_2 : \Pi \to G_{\left[{p \atop q}\right]}$; as such, it does not depend on α since $G_{\left[{q \atop p}\right]}$ and $G_{\left[{p \atop q}\right]}$ are strongly characteristic quotients of Π. Thus, $G_{\{p,q\}}$ is a strongly characteristic quotient of Π as well. Moreover, since $G_{\{p,q\}}$ is isomorphic to $G^n_{\left[{q \atop p}\right]} \times G^n_{\left[{p \atop q}\right]}$, its center is trivial.

Summing up, the group $G = G_{\{p,q\}}$ we have constructed satisfies (6.3). In the next two sections we shall show that it also enjoys the crucial property (6.12) for any admissible G-cover of any stable genus g curve, so that, as Lemma (6.11) implies, the automorphism group of any such cover is trivial.

7. Smooth covers of $\overline{M}_g$.

Following Looijenga [487], Brylinski [92], and Abramovich, Corti and Vistoli [2], we are now going to prove the central result of this chapter. As usual, we denote by Π the fundamental group of a reference topological surface Σ of genus g.

THEOREM (7.1). *Let g be an integer ≥ 2. There exists a finite strongly characteristic quotient G of Π such that ${}_G\overline{\mathcal{M}}_g = {}_G\overline{M}_g$. Moreover,*

$$\overline{M}_g = {}_G\overline{M}_g / \operatorname{Out}(G)\,;$$

thus, the Deligne–Mumford compactification $\overline{M}_g$ is the quotient of a smooth complete variety by a finite group. In particular, given an exterior homomorphism $\psi : \Pi \to G$, the variety $\overline{M}_g[\psi]$ is smooth, and the Deligne–Mumford compactification $\overline{M}_g = \overline{M}_g[\psi]/\Gamma_g[\psi]$ is the quotient of a smooth complete irreducible variety by a finite group.

Let $G = G_{\{p,q\}}$ be the group constructed in the previous section. The key statement in the theorem is the first one, and the remaining ones follow readily from it. The second statement, in particular, is a consequence of the fact that any two admissible G-covers of a genus g stable curve differ by an automorphism of G, since G is a strongly characteristic quotient of Π. To prove the first part of the theorem, it suffices to prove the following result.

LEMMA (7.2). *Let $G = G_{\{p,q\}}$, where p and q are distinct odd primes. Let C be a stable curve of genus g, and let $P \to C$ be an admissible G-cover. Then $\operatorname{Aut}(P \to C)^G = \{1\}$.*

Since G satisfies (6.3), in view of Lemma (6.11), to prove this lemma, it is enough to show that G also satisfies (6.12). Let $p_1, \ldots, p_\delta$ be the nodes of C, and let n_i be the index of p_i, $i = 1, \ldots, \delta$. Recall that n_i is just the order of a generator of the stabilizer in G of any point of P lying above p_i. It is immediate to check, by inspection, that the order of any element of G divides pq. In particular, n_i divides pq. Consider a Kuranishi family $\pi : \mathcal{C} \to B$ for C. As usual, we may assume that

$$B = \{(t_1, \ldots, t_{3g-3}) \in \mathbb{C}^{3g-3} : |t_i| < 1, i = 1, \ldots, 3g-3\}$$

and that the fiber over a general point of $t_i = 0$ has only one node, which tends to q_i as one approaches the central fiber. We let $\widetilde{\pi} : \widetilde{\mathcal{C}} \to \widetilde{B}$ be the family of stable curves obtained from $\pi : \mathcal{C} \to B$ via the base change

(7.3)
$$\begin{gathered} \widetilde{B} \longrightarrow B \\ (z_1, \ldots, z_{3g-3}) \mapsto (z_1^{n_1}, \ldots, z_\delta^{n_\delta}, z_{\delta+1}, \ldots, z_{3g-3}) . \end{gathered}$$

Finally we let

(7.4)
$$\mathcal{P} \to \widetilde{\mathcal{C}} \to \widetilde{B}$$

be a Kuranishi family for $P \to C$. Denote by Θ the dual graph of C, and let $E(\Theta)$ be the set of its edges, so that the cardinality of $E(\Theta)$ is δ. It is clear that $\pi_1(B^*, b)$ can be identified with the free abelian (multiplicative) group generated by the edges of Θ:

(7.5)
$$\pi_1(B^*, b) = \prod_{e \in E(\Theta)} \langle e \rangle \cong \mathbb{Z}^\delta ,$$

where $\langle e \rangle$ stands for the infinite cyclic group generated by e. Now consider the Picard–Lefschetz homomorphism

(7.6)
$$PL : \pi_1(B^*, b) \to \Gamma_g \to \mathrm{Out}(G)$$

From Lemma (6.11) and Remark (6.17) it follows that we must prove that

(7.7)
$$\ker(PL) \subset \prod_{e \in E(\Theta)} \langle e^{pq} \rangle .$$

Let

(7.8)
$$u = \prod_{e \in E(\Theta)} e^{u_e}$$

be an element of $\pi_1(B^*, b)$. We must show that u_e is divisible by pq for every e when u belongs to $\ker(PL)$. A consequence of the assumption that u maps to the identity element of $\mathrm{Out}(G)$ is that it also maps to the identity of $\mathrm{Out}(G_{[\frac{q}{p}]})$. We will prove that this implies that u_e is divisible by q for every e. Reversing the roles of p and q, this will also show that it is divisible by p. As p and q are relatively prime, we will then be done.

REMARK (7.9). The proof shows in particular that, when $G = G_{\{p,q\}}$ and $P \to C$ is an admissible G-cover, the index of any node of C is exactly equal to pq.

We now go back to the proof of (7.7). We need some preliminaries. Let L be a finitely generated free abelian group (a *lattice* for short). A sublattice $L' \subset L$ is said to be *primitive* if L/L' has no torsion or, equivalently, if L is the direct sum of L' and of another sublattice. Now let C_b be the fiber of the Kuranishi family $\mathcal{C} \to B$ at the base point b. As we observed, it follows from (3.2) that $H_1(C,\mathbb{Z})$ has no torsion, and then (9.24) in Chapter X shows that the image of $H^1(|\Theta|,\mathbb{Z})$ in $H_1(C_b,\mathbb{Z})$ is a primitive sublattice. We recall that this image is just the subgroup of $H_1(C_b,\mathbb{Z})$ generated by the homology classes of the vanishing cycles. From now on we will routinely identify $H^1(|\Theta|,\mathbb{Z})$ and this subgroup. We shall write $[e]$ to denote the homology class of the vanishing cycle attached to the node corresponding to the edge e of Θ with some fixed orientation. For the rest of this section, when dealing with integral homology and cohomology, we shall also omit any mention of the group of coefficients.

Let K be the kernel of the natural homomorphism $G \to G_p$ and set $Q = P/K$. Then $Q \to C$ is an admissible G_p-cover. We let $\mathcal{Q} \to \mathcal{C}' \to B'$ be a Kuranishi family for this cover, where B' is a polycylinder in the coordinates $\zeta_1, \dots, \zeta_{3g-3}$, and $\mathcal{C}' \to B'$ is the pullback of $\mathcal{C} \to B$ via a base change of the form $t_i = \zeta_i^{m_i}$ with $m_i = 1$ for $i > \delta$. As we know, the integers m_j, $j \leq \delta$, are just the indices of $Q \to C$ at the nodes of C. Although this is not strictly necessary, we begin by computing these indices. We change notation slightly and write m_e to indicate the index of $Q \to C$ at the node corresponding to the edge e of Θ.

LEMMA (7.10). *The index m_e equals p if e is a nondisconnecting edge of Θ and 1 if e is disconnecting.*

Proof. We begin by observing that the index is obviously invariant under specialization, and hence it suffices to prove the lemma in the case where C has a single node. Since G_p is a direct product of cyclic groups of order p, the index m_e must divide p. Suppose that e is nondisconnecting. Then $[e] \in H^1(|\Theta|)$ is not zero; hence it follows from (9.24) that the rank of $H_1(C)$ equals $2g-1$. On the other hand, if m_e were equal to 1, P would be a topological Galois cover of C with Galois group G_p. This is impossible, since the corresponding surjective exterior homomorphism $\pi_1(C) \to G_p$ would have to factor through $H_1(C,\mathbb{Z}/p\mathbb{Z})$, whose order is p^{2g-1} and hence strictly smaller than the one of G_p. Now suppose that e is disconnecting. Recall that m_e can also be interpreted as follows. Let $b' \in B'$ be a point mapping to b, and let $Q_{b'}$ be the fiber of $\mathcal{Q} \to B'$ at b'. Then m_e is the degree of the covering $\tilde{\gamma} \to \gamma$, where γ is a vanishing cycle on C_b, and $\tilde{\gamma}$ a vanishing cycle in $Q_{b'}$ lying above it. View γ as a closed loop; lifting this loop to $Q_{b'}$ yields a closed loop, since γ is homologically trivial and $Q_{b'} \to C_b$ is an abelian cover. Thus $m_e = 1$.

Q.E.D.

We denote by Ξ the graph of Q. If e is an edge of Θ and f is an edge of Ξ lying above it, we indicate this fact by writing f/e. This is the same as saying that the node corresponding to f maps to the one associated to e. We write g' to indicate the genus of Q which, by the Riemann–Hurwitz formula, equals $p^{2g}(g-1)+1$. We also write $E_0(\Theta)$ to indicate the set of disconnecting edges of Θ, and $E_1(\Theta)$ to indicate the set of nondisconnecting ones. Denote by $\pi_1(B^*, b)^p$ the subgroup of pth powers in $\pi_1(B^*, b)$. Then Remark (5.20) and Lemma (7.10) show that the restriction $\pi_1(B^*, b)^p \to \Gamma_g = \Gamma_{C_b}$ of the Picard–Lefschetz homomorphism lifts to a homomorphism

$$\pi_1(B^*, b)^p \to \Gamma_{g'} = \Gamma_{Q_{b'}},$$

where b' is a point of B' mapping to b, and $Q_{b'}$ the corresponding fiber of $\mathcal{Q} \to B'$. If $e \in E_0(\Theta)$, this homomorphism sends e^p to the product of the pth powers of the Dehn twists associated to the edges of Ξ lying over e. If instead $e \in E_1(\Theta)$, then e^p gets sent to the product of the Dehn twists associated to the edges of Ξ lying over e. Thus, if u is as in (7.8), then u^p acts on $H_1(Q_{b'})$ via

(7.11) $$x \mapsto T_u(x) = x + \sum_{\substack{f/e \\ e\in E_0(\Theta)}} pu_e(x\cdot[f])[f] + \sum_{\substack{f/e \\ e\in E_1(\Theta)}} u_e(x\cdot[f])[f].$$

Now suppose that u belongs to $\ker(PL)$ and hence maps to the identity element of $\mathrm{Out}(G_{[\frac{q}{p}]})$. Recalling the exact sequence (6.21), this implies in particular that u acts on $H = H_1(Q_{b'}, \mathbb{Z}/q\mathbb{Z})$ as conjugation by an element of G_p. It follows that u^p acts on $H_1(Q_{b'}, \mathbb{Z}/q\mathbb{Z})$ as the identity. On the other hand, this action is described by the modulo q reduction of (7.11). To better analyze what this means, it is convenient to identify the groups $H_1(Q_{b'}) \otimes H_1(Q_{b'})$ and $\mathrm{Hom}(H_1(Q_{b'}), H_1(Q_{b'}))$ via the homomorphism $x\otimes y \mapsto (-\cdot x)y$; this is an isomorphism since the intersection pairing is nondegenerate and unimodular. We may then view the endomorphism $Z_u = T_u - \mathrm{id}$ as an element of $H_1(Q_{b'}) \otimes H_1(Q_{b'})$, and the assertion that u^p acts on $H_1(Q_{b'}, \mathbb{Z}/q\mathbb{Z})$ as the identity translates into

(7.12)
$$Z_u = \sum_{\substack{f/e \\ e\in E_0(\Theta)}} pu_e\,[f]\otimes[f] + \sum_{\substack{f/e \\ e\in E_1(\Theta)}} u_e\,[f]\otimes[f] \;\in\; q(H_1(Q_{b'})\otimes H_1(Q_{b'})).$$

It is apparent from its expression that, in fact, Z_u belongs to the subgroup of the symmetric square $\mathrm{Sym}^2 H^1(|\Xi|) \subset \mathrm{Sym}^2 H_1(Q_{b'})$ generated by the tensors of the form $[f]\otimes[f]$, where f runs through all the edges of Ξ. Notice that $\mathrm{Sym}^2 H_1(Q_{b'})$ is a primitive sublattice of $H_1(Q_{b'})\otimes H_1(Q_{b'})$ and that $\mathrm{Sym}^2 H^1(|\Xi|)$ is a primitive sublattice of $\mathrm{Sym}^2 H_1(Q_{b'})$, since $H^1(|\Xi|)$ is primitive in $H_1(Q_{b'})$.

To conclude, we invoke the following crucial result.

LEMMA (7.13). *The tensors of the form $[f]\otimes[f]$, where f runs through all the edges of Ξ, are mutually distinct, linearly independent, and generate a primitive sublattice of* $\operatorname{Sym}^2 H^1(|\Xi|)$

Most of the next section will be occupied by the proof of this lemma. Assuming its validity, we may argue as follows. Denote by L the subgroup of $H_1(Q_{b'})\otimes H_1(Q_{b'})$ generated by the elements $[f]\otimes[f]$. A consequence of (7.13) and of the remarks immediately preceding it is that L is a primitive sublattice of $H_1(Q_{b'})\otimes H_1(Q_{b'})$. But then (7.12) implies that, in fact, Z_u belongs to qL. Since the elements of the form $[f]\otimes[f]$ are independent, this means that all the coefficients in (7.12) are divisible by q. As p and q are relatively prime, we conclude that q divides u_e for every $e\in E(\Theta)$, as desired. Interchanging the roles of p and q, we find that p also divides u_e for all e. Since p and q are relatively prime, the conclusion is that all the u_e are divisible by pq. Modulo proving (7.13), this finishes the proof of (7.7) and hence of (7.2) and of Theorem (7.1).

REMARK (7.14). The reader will notice a significant difference between the statement of Theorem (7.1) and its counterpart in the open case, namely Theorem (2.12). In fact, while the latter asserts that $\mathcal{M}_g$ is the stack quotient of ${}_GM_g$ modulo a finite group, no corresponding statement is to be found in (7.1). The reason is that the analogue of what is asserted by (2.12) is simply not true for $\overline{\mathcal{M}}_g$. In fact, if $\overline{\mathcal{M}}_g$ were the stack quotient of a smooth scheme X modulo the action of a finite group, then X would be an unramified covering of $\overline{\mathcal{M}}_g$. On the other hand, we observed in Section 7 of Chapter XV that $\overline{\mathcal{M}}_g$ is simply connected, that is, does not have connected unramified coverings other than itself. Therefore, $\overline{\mathcal{M}}_g$ would have to be isomorphic to a component of X, and we know very well that it is not a scheme!

What is true instead is that the stack quotient $[{}_G\overline{M}_g/\operatorname{Out}(G)]$ is a ramified covering of $\overline{\mathcal{M}}_g$. It is one of the quirks allowed by the concept of stack that this covering is in fact an isomorphism over the dense open substack $\mathcal{M}_g$. Ramification is confined to the boundary of $\overline{\mathcal{M}}_g$. If $P\to C$ is an admissible G-cover, the covering is local-analytically just the natural morphism $\widetilde{B}\to B$, where B and $\widetilde{B}$ are bases of Kuranishi families for C and for $P\to C$, respectively. As such, in suitable coordinates it is given by (5.19), where $n_j=pq$ for every j, as observed in Remark (7.9).

Since we are at it, let us prove a connectedness property of the graph Ξ that will be used in a crucial way in the proof of Lemma (7.13) to be presented in the next section. First we need a definition. Let Ω be the dual graph of a nodal curve, and let e and f be edges of Ω. We say that $\{e,f\}$ is a *disconnecting pair of edges* if the graph obtained from Ω by removing e and f is not connected; notice that we are not requiring e and f to be distinct.

LEMMA (7.15). *Ξ does not contain disconnecting pairs of edges.*

Proof. Recall from the proof of (7.1) that Ξ is the dual graph of Q, where $Q \to C$ is an admissible G_p-cover of a stable genus g curve. Let $\tilde{e}$ and $\tilde{f}$ be edges of Ξ and denote by e and f their respective images in Θ, the dual graph of C. We wish to show that the pair $\{\tilde{e}, \tilde{f}\}$ does not disconnect Ξ. The first step is to observe that this property, if valid, is preserved under specialization, and hence it is sufficient to verify it under the further assumption that e and f are the only edges of Θ.

We shall need the following remark. Consider the Kuranishi family $\mathcal{Q} \to \mathcal{C}' \to B'$ for the admissible G_p-cover $Q \to C$ and let $Q_{b'} \to C_b$ be a general fiber; here b' is a general point of B', and b its image in B. Of course, $Q_{b'}$ and C_b are smooth, and $Q_{b'} \to C_b$ is an ordinary G_p-cover. We denote by γ_e and γ_f the vanishing cycles in C_b associated to the nodes of C corresponding to the edges e and f. The inverse images of γ_e and γ_f in $Q_{b'}$ are disjoint unions of circles, which are just the vanishing cycles associated to the singularities of P. The vertices of Θ can be identified with the connected components of $C_b \smallsetminus \gamma_e \smallsetminus \gamma_f$, and the edges with γ_e and γ_f. Similarly, vertices and edges of Ξ can be identified, respectively, with the connected components of the complement in P of the vanishing cycles and with the vanishing cycles themselves. These identifications are clearly compatible with the action of G_p. If v is a vertex of Θ, we shall denote by H_v the stabilizer of a vertex of Ξ mapping to v; as G_p is abelian, this is independent of the particular vertex we choose. It is easy to describe the group H_v concretely. If C_b^v stands for the connected component of $C_b \smallsetminus \gamma_e \smallsetminus \gamma_f$ corresponding to v, then H_v is just the image of $H_1(C_b^v, \mathbb{Z}/p\mathbb{Z})$ in $H_1(C_b, \mathbb{Z}/p\mathbb{Z}) = G_p$. It is also easy to describe the stabilizer of an edge of Ξ, using Lemma (7.10). This result says that the stabilizer of $\tilde{e}$ has order p if e is a disconnecting edge of Θ and order 1 if e is nondisconnecting.

We now proceed with the proof of (7.15). Up to interchanging the roles of e and f, there are exactly six possibilities for Θ, which are illustrated in Figure 7.

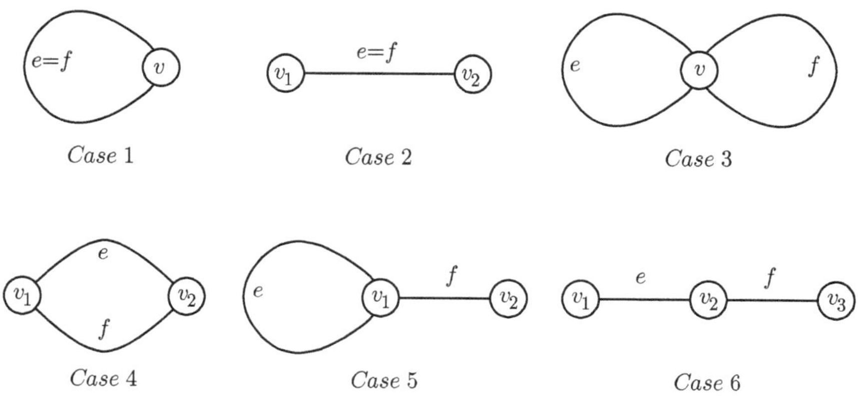

Figure 7.

The basic cases are 1 and 2, which will be treated separately. The other cases are in some sense derived ones and will be handled together. We shall write V_i to indicate the set of all the vertices of Ξ mapping to v_i, and g_i to denote the genus of the component of C corresponding to v_i.

Case 1. Although in (7.15) we are implicitly assuming that Θ has genus at least 2, we shall carry out our analysis also for $g = 1$. The group H_v has index p in G_p, and hence Ξ has p vertices. The number of edges of Ξ is p^{2g-1}, as follows for instance from (7.10). Let w_0 and w_1 be distinct vertices of Ξ which are connected by an edge. As we observed, this edge has an order p stabilizer, hence its orbit under the action of H_v consists of (at least) p^{2g-2} edges, all connecting w_0 and w_1. Let γ be an element of G_p such that $\gamma(w_0) = w_1$ and set $w_i = \gamma^i(w_0)$ for $i = 1, \dots, p$; in particular, $w_p = w_0$. For each i, applying γ^i transforms the p^{2g-2} edges connecting w_0 and w_1 into p^{2g-2} edges connecting w_i and w_{i+1}. This accounts for all the edges of Ξ. Thus, for $p > 2$, Ξ consists of p vertices $w_0 = w_p, w_1, \dots, w_{p-1}$ and of p^{2g-1} edges, p^{2g-2} of them connecting w_i and w_{i+1} for $i = 0, \dots, p-1$. If instead $p = 2$, then Ξ has two vertices, connected by 2^{2g-1} edges. It follows that the removal of an edge does not disconnect Ξ. If $g \geq 2$, two vertices of Ξ, if connected by an edge, are in fact connected by strictly more than two edges. Thus, removing two edges cannot disconnect Ξ.

Case 2. Clearly, $g = g_1 + g_2$. The groups H_{v_1} and H_{v_2} have in common just the identity element and have orders p^{2g_1} and p^{2g_2}, respectively. Thus, the cardinality of V_1 is p^{2g_2}, and the one of V_2 is p^{2g_1}. Notice that H_{v_1} fixes every element of V_1 but acts transitively on V_2, and symmetrically for H_{v_2}. Since there is at least one edge of Ξ joining a vertex in V_1 and a vertex in V_2, this implies that every element of V_1 is connected by an edge to every element of V_2; moreover, since the intersection of H_{v_1} and H_{v_2} is reduced to the identity, such an edge is unique. Since $g_1, g_2 \geq 1$, p^{2g_1} and p^{2g_2} are both strictly larger than 2, and we conclude that Ξ does not contain disconnecting pairs of edges.

Cases 3–6. In each of these cases the graph Ξ is a specialization of the one described in case 1 or of the one described in case 2. We let $\tilde{e}$ and $\tilde{f}$ be fixed edges of Ξ mapping to e and f, respectively. If we remove from Ξ all the edges lying over e and, in case 6, also V_1, we are left with a graph consisting of n connected components, where $n = p$ in cases 3, 4, and 5, and $n = p^{2g_1}$ in case 6. In cases 4, 5, and 6 each of these components is as described in case 2, while in case 3 it is as described in case 1. It follows that the component containing $\tilde{f}$ remains connected even after removal of this edge. On the other hand, if we collapse each component to a point, we obtain from Ξ a graph as the one described in case 1 or in case 2, which therefore is not disconnected by $\tilde{e}$. Thus removing $\tilde{e}$ and $\tilde{f}$ from Ξ leaves us with a connected graph.

Q.E.D.

8. Totally unimodular lattices.

Let L be a lattice. As we recalled in the preceding section, a sublattice L' of L is said to be *primitive* if there is a direct sum decomposition $L = L' \oplus L''$ for some sublattice L''. If A is the matrix expressing a system of generators for L' in terms of a basis for L, then L' is primitive if and only if a maximal rank square submatrix of A has determinant equal to ± 1. Given a subset E of L, we let $\langle E \rangle$ denote the sublattice of L generated by E.

DEFINITION (8.1). *A finite subset E of a lattice L is said to be unimodular if $\langle E \rangle$ is a primitive sublattice of L and if, for every subset $F \subset E$, one has*

$$\langle F \rangle = \langle E \rangle \iff \langle F \rangle \otimes \mathbb{Q} = \langle E \rangle \otimes \mathbb{Q} .$$

A consequence of the preceding definition is that any subset of a unimodular subset $E \subset L$ generates a primitive sublattice of L. In fact, if $G \subset E$, we may choose elements $e_1, \dots, e_n$ of E which form a basis of $\langle E \rangle \otimes \mathbb{Q}$ modulo $\langle G \rangle \otimes \mathbb{Q}$. Thus, $\langle G \cup \{e_1, \dots, e_n\} \rangle \otimes \mathbb{Q}$ equals $\langle E \rangle \otimes \mathbb{Q}$, and hence $\langle E \rangle = \langle G \cup \{e_1, \dots, e_n\} \rangle = \langle G \rangle \oplus \langle \{e_1, \dots, e_n\} \rangle$.

It follows that a finite subset E of L is unimodular if and only if every subset of E generates a primitive sublattice of L.

The aim of this section is to prove the following result.

THEOREM (8.2). *Let E be a unimodular subset of a lattice L with the property that $E \cap (-E) = \emptyset$. Then the elements of $\mathrm{Sym}^2(L)$ of the form $e \otimes e$, $e \in E$, are mutually distinct, linearly independent, and generate a primitive sublattice of $\mathrm{Sym}^2(L)$.*

Notice that it suffices to prove the theorem under the additional assumption that E generates L. In fact, if this is not the case, we may write $L = \langle E \rangle \oplus L'$ for some sublattice L', and hence $\mathrm{Sym}^2(\langle E \rangle)$ is a direct summand of $\mathrm{Sym}^2(L)$.

It is convenient to express Theorem (8.2) in terms of matrices. Suppose that L is of rank n and let $E = \{v_1, \dots, v_m\}$. Choose a basis for L; expressing the vectors $v_1, \dots, v_m$ in terms of this basis, we may view them as columns of an $n \times m$ integral matrix A. The condition that E be unimodular is then equivalent to the condition that, for any set of colums of A, the submatrix of A formed by these columns contain a maximal-rank square submatrix of determinant ± 1. A matrix with this property is said to be *unimodular.*

In the assumptions of the preceding theorem, the further requirement is made that $E \cap (-E) = \emptyset$. This translates into the condition that no column of A be zero or proportional to another column or, as one says, that A be *reduced.* Theorem (8.2) or, rather, the special case of the theorem when E is also assumed to generate L, may now be rephrased as follows.

THEOREM (8.3). *Let $m \geq n$ be positive integers, and let $A = (a_{ij})$ be an integral, maximal-rank, unimodular, reduced $n \times m$ matrix. Then the quadratic polynomials*

$$\left(\sum_{i=1}^{n} a_{ij} x_i\right)^2, \qquad j = 1, \ldots, m,$$

in the indeterminates $x_1, \ldots, x_n$ are linearly independent and generate a primitive sublattice of the lattice

$$\bigoplus_{i=1,\ldots,n} \mathbb{Z} \cdot x_i^2 \oplus \left(\bigoplus_{1 \leq i < j \leq n} \mathbb{Z} \cdot 2x_i x_j\right).$$

Proof. Since (8.3) is a direct translation of (8.2), the validity of its assumptions and of its conclusion are insensitive to the accidents of the translation. Thus they are not affected if we shuffle the columns of A or if we change basis in L, that is, if we multiply A on the left by a square integral matrix of determinant ± 1. Using operations of this sort, we may assume that A is in the *normal form*

$$A = (I, N), \tag{8.4}$$

where I is the $n \times n$ identity matrix. One advantage of the normal form is that a unimodular matrix A as in (8.4) is in fact *totally unimodular*, meaning that all its square submatrices have determinants equal to 1, -1, or 0.

We now proceed by induction on n, the case $n = 1$ being trivial. Delete the first row and the first column of A. The resulting matrix A' is totally unimodular, of maximal rank, but perhaps nonreduced. Since A is in normal form and reduced, we see that A' has no zero columns. Notice that the conclusion of the theorem is unchanged if we multiply any number of columns of A by -1. We will do so when needed. Suppose that a column K of A' appears twice. The reducedness of A forces K to appear at most three times, corresponding to the following three possible columns of A:

$$\begin{pmatrix} 0 \\ K \end{pmatrix}, \quad \begin{pmatrix} 1 \\ K \end{pmatrix}, \quad \begin{pmatrix} -1 \\ K \end{pmatrix}.$$

Actually, the last two columns cannot appear simultaneously in the matrix A, for otherwise A would contain one of the following two nonunimodular submatrices

$$\begin{pmatrix} 1 & -1 \\ 1 & 1 \end{pmatrix}, \quad \begin{pmatrix} 1 & -1 \\ -1 & -1 \end{pmatrix}.$$

Denote by P the submatrix of A composed by the columns whose corresponding subcolumns in A' appear exactly twice. From what we just remarked, if a column K appears twice in A', the column $\begin{pmatrix} 0 \\ K \end{pmatrix}$ must appear in A. Thus, changing the signs of entire columns of A and reordering the columns, we can put P in the form

$$P = \begin{pmatrix} 0 & \dots & 0 & 1 & \dots & 1 \\ & P' & & & P' & \end{pmatrix}.$$

Now we choose a new basis, consisting of the first column of A, that is, $\begin{pmatrix} 1 \\ 0 \\ \vdots \end{pmatrix}$, a maximal independent set of columns of $\begin{pmatrix} 0 & \dots & 0 \\ & P' & \end{pmatrix}$, plus other columns of A as needed. We rewrite A in terms of this new basis and permute the columns of the resulting matrix so as to put it in normal form. This puts P' in the form

$$P' = \begin{pmatrix} I & Q \\ \mathbf{0} & \mathbf{0} \end{pmatrix},$$

where I and $\mathbf{0}$ denote an identity matrix and a zero matrix of the appropriate size. We want to show that Q is empty. Suppose it is not. Since A is reduced, P is too, so each column of Q must have at least two nonzero entries. Thus, there must be a submatrix M of P of the form

$$M = \begin{pmatrix} 0 & 0 & 0 & 1 & 1 & 1 \\ 1 & 0 & \varepsilon & 1 & 0 & \varepsilon \\ 0 & 1 & \eta & 0 & 1 & \eta \end{pmatrix},$$

where $\varepsilon = \pm 1$ and $\eta = \pm 1$. If M_{ijk} is the submatrix made up of the ith, jth, and kth columns of M, then one checks that $M_{345} = -\varepsilon - \eta$, $M_{156} = -\eta + 1$, $M_{246} = \varepsilon - 1$, and these cannot be simultaneously equal to 0, 1, or -1. Hence Q is empty.

This means that, after a suitable permutation of its columns, the matrix A takes the form

$$\begin{pmatrix} & 1 & \dots & 1 & 0 & \dots & 0 & 1 & \dots & 1 \\ I_n & & & & & & & & & \\ & & P' & & & B & & & C & \end{pmatrix},$$

where I_n is the $n \times n$ identity matrix, and where

$$P' = \begin{pmatrix} I \\ \mathbf{0} \end{pmatrix}.$$

Let p, q, and r be the numbers of columns of P', B, and C, respectively, so that $n + p + q + r = m$. Set

$$B = (b_{ij})_{i=2,\dots,n,\, j=1,\dots,q}\,, \quad C = (c_{ij})_{i=2,\dots,n,\, j=1,\dots,r}\,.$$

The matrix

$$(I_{n-1} \quad B \quad C)$$

is totally unimodular, reduced, and of maximal rank. By induction, the polynomials

$$\begin{cases} x_j^2 \,, & j = 2, \ldots, n \,, \\ \left(\sum_{i=2}^{n} b_{ij} x_i \right)^2 \,, & j = 1, \ldots, q \,, \\ \left(\sum_{i=2}^{n} c_{ij} x_i \right)^2 \,, & j = 1, \ldots, r \,, \end{cases}$$

are linearly independent and span a primitive sublattice of

$$\bigoplus_{i=2,\ldots,n} \mathbb{Z} \cdot x_i^2 \oplus \left(\bigoplus_{2 \leq i < j \leq n} \mathbb{Z} \cdot 2x_i x_j \right) .$$

We must show that the polynomials

$$\begin{cases} x_j^2 \,, & j = 1, \ldots, n \,, \\ (x_1 + x_j)^2 \,, & j = 2, \ldots, p \,, \\ \left(\sum_{i=2}^{n} b_{ij} x_i \right)^2 \,, & j = 1, \ldots, q \,, \\ \left(x_1 + \sum_{i=2}^{n} c_{ij} x_i \right)^2 \,, & j = 1, \ldots, r \,, \end{cases}$$

are linearly independent and span a primitive sublattice of

$$\bigoplus_{i=1,\ldots,n} \mathbb{Z} \cdot x_i^2 \oplus \left(\bigoplus_{1 \leq i < j \leq n} \mathbb{Z} \cdot 2x_i x_j \right) .$$

The above system of polynomials is equivalent to the system

$$\begin{cases} x_j^2 \,, & j = 1, \ldots, n \,, \\ 2x_1 x_j \,, & j = 2, \ldots, p \,, \\ \left(\sum_{i=2}^{n} b_{ij} x_i \right)^2 \,, & j = 1, \ldots, q \,, \\ \sum_{i=p+1}^{n} 2c_{ij} x_1 x_i + \left(\sum_{i=2}^{n} c_{ij} x_i \right)^2 \,, & j = 1, \ldots, r \,. \end{cases}$$

We are then reduced to the following statement. Let

$$L_1 = \bigoplus_{i=1}^{s} \mathbb{Z} \cdot e_i \quad \subset \quad L_2 = \bigoplus_{i=1}^{s+t} \mathbb{Z} \cdot e_i$$

be two lattices. Suppose that

$$(8.5) \qquad v_1, \dots, v_h, w_1, \dots, w_r \in L_1$$

are independent and span a primitive sublattice. Let $p < t$, and let $u_1, \dots, u_r$ be elements of $\langle e_{s+p+1}, \dots, e_{s+t} \rangle$. Then

$$(8.6) \qquad v_1, \dots, v_h, w_1 + u_1, \dots, w_r + u_r, e_{s+1}, \dots, e_{s+p} \in L_2$$

are independent and span a primitive sublattice. Looking at the matrices of (8.5) and (8.6) with respect to the basis $e_1, \dots, e_{s+t}$, the result follows at once (in our case, v_j corresponds to $(\sum_{i=2}^{n} b_{ij} x_i)^2$, w_j to $(\sum_{i=2}^{n} c_{ij} x_i)^2$, u_j to $\sum_{i=p+1}^{n} 2c_{ij} x_1 x_i$, and $e_{s+1}, \dots, e_{s+p}$ to $x_1^2, 2x_1x_2, \dots, 2x_1x_p$).

Q.E.D.

We are now going to prove Lemma (7.13). We need an auxiliary result and some terminology. As in the previous section, when dealing with integral homology and cohomology, no mention will be made of the coefficient ring. Let Ω be the dual graph of a stable curve and fix, once and for all, an orientation on each of its edges. The sets $E(\Omega)$ and $V(\Omega)$ of edges and vertices of Ω can be regarded as bases of $C_1(|\Omega|)$ and $C_0(|\Omega|)$, respectively. If e is an edge, by abuse of notation the same symbol e will be used also to indicate the element of the dual basis of $C^1(|\Omega|)$ corresponding to it. The same convention will be used for vertices. The class of the cocycle e in $H^1(|\Omega|)$ will be denoted by $[e]$. To make contact with the notation employed in the previous section, suppose that Ω is the graph of the central fiber X_0 of a family of curves whose general fiber is smooth. Then, if we denote by X a smooth fiber close to X_0, the inclusion $H^1(|\Omega|) \hookrightarrow H_1(X)$ we discussed in Section 9 of Chapter X sends $[e]$ to the class of the vanishing cycle corresponding to e. Thus, the present use of the notation $[e]$ is consistent with the use of the same notation in the previous section.

LEMMA (8.7). *Let Ω be the dual graph of a stable curve. The set $E = \{[e] : e \in E(\Omega)\} \subset H^1(|\Omega|)$ is unimodular. If, in addition, Ω contains no disconnecting pairs of edges, then $E \cap (-E) = \emptyset$.*

Proof. Set $p = \operatorname{rank} H^1(|\Omega|)$. Choose edges $e_1, \dots, e_p$ such that $\langle [e_1], \dots, [e_p] \rangle \otimes \mathbb{Q} = H^1(|\Omega|, \mathbb{Q})$. We must show that

$$(8.8) \qquad \langle [e_1], \dots, [e_p] \rangle = H^1(|\Omega|)\,.$$

Let $e_1, \ldots, e_p, f_1, \ldots, f_q$ be an enumeration of the edges of Ω. We view the edges as a basis for $C^1(\Omega)$. To prove (8.8), it suffices to show that, for $j = 1, \ldots, q$,

$$f_j \in \langle e_1, \ldots, e_p \rangle + \delta C^0(\Omega)\,.$$

Let Ω' be the graph obtained from Ω by removing the edges $e_1, \ldots, e_p$. Denote by δ' the coboundary in $C^\bullet(\Omega')$. Clearly, Ω' is acyclic. In fact, the support of any cycle $c \in H_1(|\Omega|') \subset H_1(|\Omega|)$ is disjoint from the e_i, so that $[e_i](c) = 0$ for $i = 1, \ldots, p$, showing that $c = 0$. It follows that Ω' is a disjoint union of trees and that $H^1(|\Omega|') = 0$. We conclude that $f_j \in \delta' C^0(\Omega')$ for $j = 1, \ldots, q$. But, given a vertex v of Ω', we have

$$\delta v = \sum_s \pm f_{i_s} + \sum_t \pm e_{i_t} = \delta' v + \sum_t \pm e_{i_t}\,,$$

so that $\delta' C^0(\Omega') \subset \langle e_1, \ldots, e_p \rangle + \delta C^0(\Omega)$.

The first claim in the statement of the lemma is proved. Now let e be an edge of Ω, and let Ω' be the graph obtained from Ω by removing e. Looking at the exact cohomology sequence of the pair $(|\Omega|, |\Omega'|)$, we see that $[e] \in H^1(|\Omega|)$ is zero if and only if e disconnects Ω. Similarly, two distinct nondisconnecting edges e and f of Ω constitute a disconnecting pair if and only if $[e] = \pm[f]$, as one sees by examining the exact sequence of the pair $(|\Omega|, |\Omega''|)$, where Ω'' stands for the graph obtained from Ω by removing e and f. These two observations, combined, prove the second claim in the statement of the lemma.

Q.E.D.

Lemma (7.13) follows immediately by combining (8.2), (8.7), and (7.15). Theorem (7.1), and in particular the existence of a smooth Galois cover of $\overline{M}_g$, is now completely proved.

9. Smooth covers of $\overline{M}_{g,n}$.

Our aim in this section is to extend the results of the previous ones to the moduli spaces of stable n-pointed curves. As we shall see, most of the work has already been done. Let $(C; p_1, \ldots, p_n)$ be an n-pointed nodal curve of genus g, and let G be a finite group. We define an *admissible G-cover of* $(C; p_1, \ldots, p_n)$ to be an admissible G-cover $P \to C$ plus the choice of a point $q_i \in P$ mapping to p_i for each i; here we are not asking that C and P be stable. Notice that $(P; q_1, \ldots, q_n)$ is not necessarily stable even when $(C; p_1, \ldots, p_n)$ is. Of course, one can also speak of families of admissible G-covers of a family $\mathcal{C} \to S$, $\sigma_i : S \to \mathcal{C}$, $i = 1, \ldots, n$, of n-pointed stable curves. One such is just a family $\mathcal{P} \to \mathcal{C} \to S$ of admissible G-covers of $\mathcal{C} \to S$, viewed as a family of unpointed curves, plus a choice of liftings of the sections σ_i to sections of $\mathcal{P} \to S$. There is also an obvious extension to the n-pointed case of the notion of morphism of families of admissible G-covers.

Kuranishi families of admissible G-covers of stable n-pointed genus g curves can be constructed exactly as in the unpointed case. The same goes for the corresponding moduli stacks and spaces; these will be denoted ${}_G\overline{\mathcal{M}}_{g,n}$ and ${}_G\overline{M}_{g,n}$, respectively. If $(P; q_1, \dots, q_n) \to (C; p_1, \dots, p_n)$ is an admissible G-cover, with $(C; p_1, \dots, p_n)$ stable, the space T classifying its first-order deformations coincides with $\mathrm{Ext}^1(\Omega^1_P, \mathcal{O}_P(-\widetilde{D}))^G$ and sits in the exact sequence

$$0 \to H^1(C, \mathcal{H}om(\Omega^1_C, \mathcal{O}_C(-D))) \to T \to \bigoplus_{x \in \mathrm{Sing}(C)} \mathrm{Ext}^1_{\mathcal{O}_{P,\overline{x}}}(\Omega^1_{P,\overline{x}}, \mathcal{O}_{P,\overline{x}}) \to 0\,,$$

where $D = \sum p_i$, $\widetilde{D} = \sum Gq_i$, and $\overline{x}$ is an arbitrarily chosen node of P lying above the node x of C. We leave to the reader the easy task of filling in the details.

From now on we let $G = G_{\{p,q\}}$ be the group constructed in Section 6, where p and q are distinct odd primes. We wish to show that admissible G-covers $(P; q_1, \dots, q_n) \to (C; p_1, \dots, p_n)$, with $(C; p_1, \dots, p_n)$ stable of genus g, are automorphism-free.

Lemma (9.1). *Set $G = G_{\{p,q\}}$, where p and q are distinct odd primes, and let $(C; p_1, \dots, p_n)$ be a stable n-pointed curve of genus $g > 1$. Then any admissible G-cover of $(C; p_1, \dots, p_n)$ has no nontrivial automorphisms.*

Proof. For $n = 0$, this is just Lemma (7.2). Let $f : (P; q_1, \dots, q_n) \to (C; p_1, \dots, p_n)$ be an admissible G-cover, and let α be an automorphism of it. Then α sends each component of P lying above a pointed component of C to itself. In fact, if E is one such component, and a marked point p_i is contained in its image in C, then E contains γq_i for some $\gamma \in G$, and $\alpha\gamma q_i = \gamma\alpha q_i = \gamma q_i$.

Now let E' be a component of P, and let E be its image in C. Suppose that, if we ignore the marked points, the presence of E makes C nonstable. This occurs if and only if E is smooth rational and is joined to the remaining components of C either at a single point a or else at two points a and b. In either case, we denote by C' the curve obtained from C by collapsing E to a point and by P' the curve obtained from P by collapsing each component of $f^{-1}(E)$ to a point. We wish to show that $P' \to C'$ is an admissible G-cover. In the first case, $E \smallsetminus \{a\} \cong \mathbb{C}$ is simply connected. Since $E' \smallsetminus f^{-1}(a)$ is a connected topological covering of $E \smallsetminus \{a\}$, E' must map isomorphically to E. Thus the inverse image of E consists of $|G|$ copies of E, each attached to the rest of P at a single point, and each one of these curves destabilizes P. Furthermore, f is unramified at the points of attachment. All this implies that $P' \to C'$ is an admissible G-cover.

Now we deal with the second case, namely the one of a component E attached to the rest of C at two points a and b. In this case $E \smallsetminus \{a, b\} \cong \mathbb{C}^\times$, and hence $E' \to E$ is a cyclic covering of some degree ν, totally ramified at two points a' and b' lying above a and b, respectively. In other words, this covering can be identified with the covering $\mathbb{P}^1 \to \mathbb{P}^1$ given by $z \to z^\nu$, the points a and a' with the origin, and the points b and b' with the point at infinity. Any element γ of the stabilizer of a' in G also stabilizes b', and it follows from what we just said that if γ acts as multiplication by a νth root of unity ζ on the tangent space to E' at a', then it acts as multiplication by ζ^{-1} on the tangent space to E' at b'. Now denote by P_1 and P_2 the components of P meeting E' at a' and b', respectively. By the definition of admissible G-cover, γ must act as multiplication by ζ^{-1} on the tangent space to P_1 at a' and as multiplication by ζ on the tangent space to P_2 at b'. Again it follows that $P' \to C'$ is an admissible G-cover. Iterating this argument shows that, if $\widehat{P}$ denotes the stable model of P, and $\widehat{C}$ the stable model of C, all viewed as unpointed curves, then $\widehat{P} \to \widehat{C}$ is an admissible G-cover.

Any automorphism α of the admissible G-cover $(P; q_1, \dots, q_n) \to (C; p_1, \dots, p_n)$ drops down to an automorphism α' of $P' \to C'$. Suppose that α' is the identity. If E and E' are as above, α sends both E and E' to themselves and fixes at least three points on each, namely the marked points and the points of attachment to the rest of C and P. Since E and E' are smooth rational, α must act as the identity on them. We conclude that α must be the identity. This observation makes it possible to show, inductively on the number of destabilizing components of C (viewed as an unpointed curve), that if α is not the identity, it must induce a nontrivial automorphism of $\widehat{P} \to \widehat{C}$. On the other hand, we know that no such automorphism exists by Lemma (7.2). The conclusion is that α must be the identity. Q.E.D.

We can say at this point that the stack ${}_G\overline{\mathcal{M}}_{g,n}$ and the space ${}_G\overline{M}_{g,n}$ coincide for $G = G_{\{p,q\}}$; in particular, ${}_G\overline{M}_{g,n}$ is smooth. We are going to see that, in addition, $\overline{M}_{g,n}$ is the quotient of ${}_G\overline{M}_{g,n}$ modulo the action of a finite group H. To describe H, consider the natural action of $\mathrm{Aut}(G)$ on the n-fold product $G \times \cdots \times G$ and form the corresponding semidirect product

$$\widetilde{H} = G^n \ltimes \mathrm{Aut}(G).$$

The group $\widetilde{H}$ acts on ${}_G\overline{M}_{g,n}$ as follows. Let m be a point of ${}_G\overline{M}_{g,n}$ corresponding to an admissible covering $P \to C$, an injective homomorphism $\alpha : G \to \mathrm{Aut}(P/C)$, and points $q_1, \dots, q_n$ in P, and let $h = (\gamma_1, \dots, \gamma_n, \varphi)$ be an element of $\widetilde{H}$, where $\gamma_1, \dots, \gamma_n$ belong to G, and φ to $\mathrm{Aut}(G)$. Then

$$h \cdot m = \left[P \to C\,,\ \alpha \circ \varphi^{-1} : G \hookrightarrow \mathrm{Aut}(P/C)\,,\ \alpha\varphi^{-1}(\gamma_1)q_1, \dots, \alpha\varphi^{-1}(\gamma_n)q_n\right]$$

We leave it to the reader to check that this is indeed an action. There are elements $h \in \widetilde{H}$ which do not act effectively on ${}_G\overline{M}_{g,n}$, that is, are such that $(P \to C, \alpha \circ \varphi^{-1}, \alpha\varphi^{-1}(\gamma_1)q_1, \dots, \alpha\varphi^{-1}(\gamma_n)q_n)$ is isomorphic to

$(P \to C, \alpha, q_1, \ldots, q_n)$ for any choice of the latter. These form a normal subgroup K of $\widetilde{H}$, and we set

$$H = {}_G H_{g,n} = \widetilde{H}/K\,.$$

The elements of K can be explicitly described; they are precisely those elements $(\gamma, \ldots, \gamma, \varphi)$, $\gamma \in G$, such that φ is the inner automorphism $\ell \mapsto \gamma^{-1}\ell\gamma$. In particular, K is isomorphic to G. Notice also that, in the unpointed case, $\widetilde{H} = \mathrm{Aut}(G)$, and K is the subgroup of inner automorphism. From this point of view, then H is a direct generalization of $\mathrm{Out}(G)$.

Summing up, what we have shown so far proves the following statement in genus $g \geq 2$.

THEOREM (9.2). *Let g and n be nonnegative integers such that $2g - 2 + n > 0$. Let Π be the fundamental group of a reference genus g Riemann surface. There exists a finite strongly characteristic quotient G of Π such that ${}_G\overline{\mathcal{M}}_{g,n} = {}_G\overline{M}_{g,n}$. Moreover,*

$$\overline{M}_{g,n} = {}_G\overline{M}_{g,n}/{}_GH_{g,n}\,,$$

where ${}_GH_{g,n}$ is the quotient of the semidirect product of G^n and $\mathrm{Aut}(G)$, the latter acting on the former in the natural way, modulo the subgroup generated by elements of the form $(\gamma, \ldots, \gamma, \varphi)$, where $\gamma \in G$, and $\varphi \in \mathrm{Aut}(G)$ is the inner automorphism $\ell \mapsto \gamma^{-1}\ell\gamma$. Thus, the Deligne–Mumford compactification $\overline{M}_{g,n}$ is the quotient of a smooth complete variety by a finite group. Finally,

$$\mathcal{M}_{g,n} = [{}_GM_{g,n}/{}_GH_{g,n}]\,.$$

To finish the proof of the theorem, we must deal with genera 0 and 1. In genus zero there is nothing to prove, as $M_{0,n}$ and $\overline{M}_{0,n}$ are smooth and coincide with their stack counterparts. We therefore assume that $g = 1$. We fix some odd prime p and choose as group G the direct sum of two copies of $\mathbb{Z}/p\mathbb{Z}$. We may view G as $H^1(\Sigma, \mathbb{Z}/p\mathbb{Z})$, where Σ is a reference Riemann surface of genus 1. Notice that G is a strongly characteristic quotient of the fundamental group of Σ. We wish to show that any admissible G-cover of a stable n-pointed genus 1 curve has no nontrivial automorphisms. We begin with the case $n = 1$. Let $(C; x)$ be a stable 1-pointed genus 1 curve, and $f : (P; y) \to (C; x)$ an admissible G-cover. Suppose first that C is smooth. Any G-cover automorphism of $(P; y) \to (C; x)$ induces the identity on C, by Proposition (2.8), and hence consists simply of a deck transformation. However, as this has to leave y fixed, it must be trivial. In conclusion, $(P; y) \to (C; x)$ has no nontrivial automorphisms.

We now wish to show that the same is true when C is singular. In this case C can be obtained from a smooth rational curve E via identification of two points a and b. Let z be the affine coordinate on $E \setminus \{b\}$ such that $z(a) = 0$, $z(x) = 1$ (and of course $z(b) = \infty$). It is easy to give an explicit description of $P \to C$ and of the action of G, up to isomorphism. For each $i \in \mathbb{Z}/p\mathbb{Z}$, let E_i, a_i, b_i, z_i be copies of E, a, b, z, let $\eta_i : E_i \to E_{i+1}$ be the natural identification, and let $\xi_i : E_i \to E$ be the pth power map $z = z_i^p$. We also let $f_i : E_i \to C$ be the composition of ξ_i and of the quotient map $E \to C$. Then P is the quotient of the disjoint union of the E_i modulo identification of b_i with a_{i+1} for each i, and the map $f : P \to C$ is the unique map whose restriction to E_i is f_i for each i. The action of G on P is as follows. There are generators γ_1 and γ_2 of G such that γ_1 acts by sending each E_i to E_{i+1} via η_i, while γ_2 acts as multiplication by ζ on each E_i, where ζ is a nontrivial pth root of unity; in other words, $z_i(\gamma_2(e)) = \zeta z_i(e)$ for each $e \in E_i$. As for the marked point y, it is the point $z_i = \zeta^j$ for some i and j.

Let α be an automorphism of the admissible G-cover $(P; y) \to (C; x)$. If γ is an element of G, then $\alpha(\gamma y) = \gamma\alpha(y) = \gamma y$. Thus, the p^2 points γy, $\gamma \in G$, are all fixed points of α. In particular, on each E_i there are at least p smooth points which are left fixed by α. A consequence is that each component of P is mapped to itself by α. Since these components are all smooth rational, and they contain at least $p \geq 3$ fixed points, the restriction of α to each E_i is the identity. Thus, α is the identity.

The case $n > 1$ can be reduced to the case $n = 1$ exactly as we reduced the case $n > 0$ to the case $n = 0$ in genus $g > 1$. The proof of Theorem (9.2) is now complete.

REMARK (9.3). When G is as in the statement of Theorem (9.2), it is easy to describe the projection morphism $\pi_{n+1} : {}_G\overline{M}_{g,n+1} \to {}_G\overline{M}_{g,n}$. On ${}_G\overline{M}_{g,n}$ there is a universal family of admissible G-covers

$$\mathcal{P} \to \mathcal{C} \to {}_G\overline{M}_{g,n} .$$

Then ${}_G\overline{M}_{g,n+1}$ can be identified with $\mathcal{P}$, and the projection morphism π_{n+1} with $\mathcal{P} \to {}_G\overline{M}_{g,n}$. In particular, the fiber of π_{n+1} at a point corresponding to an admissible G-cover $P \to C$ is just P.

REMARK (9.4). Let p and q be distinct odd primes, let $(C; p_1, \dots, p_n)$ be a stable n-pointed curve of genus g, and let $(P; q_1, \dots, q_n) \to (C; p_1, \dots, p_n)$ be an admissible G-cover, where $G = G_{\{p,q\}}$ if $g \geq 2$ and $G = \mathbb{Z}/p\mathbb{Z} \oplus \mathbb{Z}/p\mathbb{Z}$ if $g = 1$. If $g \geq 2$, every node of C has index pq, as can be easily proved by reducing to the unpointed case (cf. Remark (7.9)), retracing the steps of the proof of Lemma (9.1). One similarly shows that, when $g = 1$, every node of C has index p.

This allows us to accurately describe the structure of the morphism ${}_G\overline{M}_{g,n}/H \to \overline{\mathcal{M}}_{g,n}$ near the boundary. Let $P \to C$ be an admissible G-cover, where C is a stable n-pointed curve of genus g with δ nodes. Consider a Kuranishi family for C, parameterized by a polycylinder B in the coordinates $t_1, \dots, t_{3g-3+n}$; we may assume that C corresponds to the origin and that the locus in B parameterizing singular curves is $\{t_1 \cdots t_\delta = 0\}$. Likewise, let $\widetilde{B}$ be the base of a Kuranishi family for $P \to C$. Again, we may assume that $\widetilde{B}$ is a polycylinder in coordinates $z_1, \dots, z_{3g-3+n}$, that $P \to C$ corresponds to the origin, and that the locus in $\widetilde{B}$ parameterizing singular covers is $\{z_1 \cdots z_\delta = 0\}$. What we just observed about the indices of the singular points of C then shows that we may set things up so that the natural morphism $\widetilde{B} \to B$ is of the form

$$(z_1, \dots, z_{3g-3+n}) \mapsto (z_1^k, \dots, z_\delta^k, z_{\delta+1}, \dots, z_{3g-3+n}),$$

where $k = pq$ for $g \geq 2$, and $k = p$ for $g = 1$. Thus, while ${}_G\overline{M}_{g,n}/H \to \overline{\mathcal{M}}_{g,n}$ is an isomorphism away from $\partial\mathcal{M}_{g,n}$, locally along the boundary it looks in some sense like a cyclic covering of degree pq (or degree p in genus 1) branched along $\partial\mathcal{M}_{g,n}$.

Remark (9.5). We shall study the diagram

$$\begin{array}{ccc} \overline{M}_{g,n+2} & \xrightarrow{\pi_2} & \overline{M}_{g,n+1} \\ {\scriptstyle\pi_1}\downarrow & & \downarrow{\scriptstyle\pi} \\ \overline{M}_{g,n+1} & \xrightarrow{\pi} & \overline{M}_{g,n} \end{array} \tag{9.6}$$

where all the maps are projections, π_1 and π corresponding to omitting the last marked point, and π_2 to omitting the next to last one, using the explicit representation of the spaces involved as quotients of smooth varieties by finite groups. To this end, we write

$$\overline{M}_{g,n} = M/H\,, \qquad \overline{M}_{g,n+1} = M_1/H_1\,, \qquad \overline{M}_{g,n+2} = M_2/H_2\,,$$

where

$$\begin{aligned} M &= {}_G\overline{M}_{g,n}\,, & M_1 &= {}_G\overline{M}_{g,n+1}\,, & M_2 &= {}_G\overline{M}_{g,n+2}\,, \\ H &= {}_GH_{g,n}\,, & H_1 &= {}_GH_{g,n+1}\,, & H_2 &= {}_GH_{g,n+2}\,. \end{aligned}$$

Diagram (9.6) corresponds to a commutative diagram

$$\begin{array}{ccc} H_2 & \longrightarrow & H_1 \\ \downarrow & & \downarrow{\scriptstyle\rho} \\ H_1 & \xrightarrow{\rho} & H \end{array}$$

of surjective group homomorphisms and to a commutative diagram

$$\begin{array}{ccc} M_2 & \xrightarrow{\beta} & M_1 \\ {\scriptstyle\alpha}\downarrow & & \downarrow{\scriptstyle\gamma} \\ M_1 & \xrightarrow{\gamma} & M \end{array} \tag{9.7}$$

of equivariant morphisms. From these data one gets a morphisms

$$\begin{array}{ccc} M_2/H_2 \longrightarrow (M_1 \times_M M_1)/(H_1 \times_H H_1) & \xrightarrow{\nu} & M_1/H_1 \times_{M/H} M_1/H_1 \\ \| & & \| \\ \overline{M}_{g,n+2} & \longrightarrow & \overline{M}_{g,n+1} \times_{\overline{M}_{g,n}} \overline{M}_{g,n+1} \end{array}$$

The middle space $(M_1 \times_M M_1)/(H_1 \times_H H_1)$ is normal; in fact, $M_1 \times_M M_1$ is normal since its singularities are all conical ones of the form $x_1 y_1 = x_2 y_2$. On the other hand, we claim that, when $(g,n) \neq (2,0),(1,1)$, the morphism ν is generically one-to-one, and hence $(M_1 \times_M M_1)/(H_1 \times_H H_1)$ is the normalization of $\overline{M}_{g,n+1} \times_{\overline{M}_{g,n}} \overline{M}_{g,n+1}$. This is immediate. Let (p,q), (p',q') be points of $M_1 \times_M M_1$ which map to the same point of $M_1/H_1 \times_{M/H} M_1/H_1$. Thus, $\gamma(p) = \gamma(q)$, $\gamma(p') = \gamma(q')$, and there are $h,k \in H_1$ such that $p' = hp$, $q' = kq$. Then

$$\rho(h)\gamma(p) = \gamma(p') = \gamma(q') = \rho(k)\gamma(q)$$

by equivariance. In other words, $\rho(h^{-1}k)$ belongs to the stabilizer of $\gamma(p) = \gamma(q)$. Since a general n-pointed curve of genus g is automorphism-free, for general p, the stabilizer of $\gamma(p)$ is trivial. This means that $(h,k) \in H_1 \times_H H_1$. Thus, (p,q) and (p',q') represent the same point of $(M_1 \times_M M_1)/(H_1 \times_H H_1)$. This proves out claim. We leave it to the reader to figure out what happens when (g,n) equals $(2,0)$ of $(1,1)$.

10. Bibliographical notes and further reading.

Moduli spaces of curves and abelian varieties with level structures are studied by Mumford in [558]. The notion of a genus g curve with a Teichmüller structure of level G has been first introduced by Deligne and Mumford in [167]. In [92], Brylinski describes, via monodromy, the boundary of the compactification of a moduli space of curves with level structures. In [487], Looijenga introduces the notion of a Prym level structure on a smooth curve and proves that the compactification X of the moduli space of curves with Prym level structure is smooth and that $\overline{M}_g = X/G$, with G a finite group. Pikaart and de Jong in [594] find new smooth compactifications of moduli of curves with level structure, and Boggi and Pikaart [76] treat the case of pointed curves, thus exhibiting $\overline{M}_{g,n}$ as a quotient of a smooth variety by a finite group. In our treatement we follow Abramovich, Corti, and Vistoli [2]. As we explain in the text, the advantage of their approach is that admissible G-covers give a modular interpretation of the smooth variety X of which $\overline{M}_{g,n}$ is a quotient by a finite group.

11. Exercises.

1. Prove that Poincaré duality holds for $M_{g,n}$ and $\overline{M}_{g,n}$. Specifically, if X is either $M_{g,n}$ or $\overline{M}_{g,n}$, Y is a compact subvariety of X and

$Z \subset Y$ is closed, then there is an isomorphism

$$H_{2d-k}(X \smallsetminus Z, X \smallsetminus Y; \mathbb{Q}) \cong H^k(Y, Z; \mathbb{Q}), \tag{11.1}$$

where $d = \dim_{\mathbb{C}} X = 3g - 3 + n$.

2. Let d be as above. Let $Y = \partial M_{g,n}$ and let

$$Z = \{[(C, x_1, \ldots, x_n)] \in Y : C \text{ has at least two nodes}\}.$$

Show Lefschetz duality (cf. [644], Chapter 6, Section 6, Theorem 19)

$$H_{d-s-1}(Y \smallsetminus Z) = H^s(Y, Z). \tag{11.2}$$

Chapter XVII. Cycles in the moduli spaces of stable curves

1. Introduction.

In 1982, at the beginning of Part II of [556], Mumford writes:

Whenever a variety or a topological space is defined by some universal property, one expects that by virtue of its defining property, it possesses certain cohomolgy classes called tautological classes. The standard example is a Grasmannian, e.g., the Grassmannian Grass of k-planes in $\mathbb{C}^n$. By its very definition there is a universal bundle E on Grass of rank k, and this induces Chern classes $c_l(E)$, $1 \leq l \leq k$, in both the cohomology ring of Grass and the Chow ring of Grass. These two rings are, in fact, isomorphic and generated as rings by $\{c_l(E)\}$. Moreover one gets tautological relations from the fact that E is a sub-bundle of the trivial bundle $\mathbb{C}^n \times Grass$. This gives an exact sequence

$$0 \to E \to \mathcal{O}^n \to F \to 0\,, \qquad F \text{ a bundle of rank } n-k,$$

hence

$$(1 + c_1(E) + \cdots + c_k(E))_l^{-1} = 0\,, \quad l > n-k\,.$$

As is well known, these are a complete set of relations for the cohomology and Chow rings of Grass.

We shall begin a program of the same sort for the Chow ring (or cohomology ring) of $\overline{\mathcal{M}}_g$. Our purpose is merely to identify a natural set of tautological classes and some tautological relations. To what extent these lead to a presentation of either ring is totally unclear at the moment.

Although already initiated in [555], the first study of the tautological ring of moduli spaces is contained in this seminal, highly original paper. There the definition of the κ classes is given, and relations are discovered between the κ classes and the Hodge classes and among the κ classes themselves. This is done via the flatness of the Gauss–Manin connection and via a nonstandard (at the time) use of the Grothendieck–Riemann–Roch theorem. We present this theory in Sections 3, 4, and 5. In these sections we also study the behavior of the κ classes and of the ψ classes under the various operations of clutching and projection as well as the intersection theory of the various boundary strata of $\overline{\mathcal{M}}_{g,P}$.

E. Arbarello et al., *Geometry of Algebraic Curves*,
Grundlehren der mathematischen Wissenschaften 268,
DOI 10.1007/978-3-540-69392-5_9,

The foundational Part I of the paper by Mumford mentioned above, dealing with the correct definition of the Chow ring of $\overline{\mathcal{M}}_{g,P}$, is made easier for us by the work done in Chapter XVI. We present this foundational material in Section 2.

Section 6 is mostly expository. There we describe a set of conjectures due to Faber, Getzler, Looijenga, and Pandharipande about the Gorenstein nature of the tautological ring of moduli spaces. There are only two results in this direction that are proved in this book. The first, due to Looijenga, is the vanishing on $\overline{\mathcal{M}}_g$ of every polynomial in the κ classes of degree greater that $g-2$. A corollary of this vanishing theorem is an important result, originally due to Diaz, asserting that the maximal dimension of a complete subvariety of M_g does not exceed $g-2$. The second, due to Faber, is the nonvanishing of the class κ_{g-2} on M_g. Looijenga's vanishing theorem will be proved in Chapter XXI, and Faber's nonvanishing theorem will be proved in Chapter XX.

In the final Section 7 we present Keel's description of the Chow ring of $\overline{\mathcal{M}}_{0,P}$, and we offer a direct computation of $A^1(\overline{\mathcal{M}}_{0,P})$.

2. Algebraic cycles on quotients by finite groups.

In this section we introduce the cycle rings of the moduli stacks of curves. The cleanest approach would be to define cycle rings for general Deligne–Mumford stacks and work with these systematically. However, we shall follow a more down-to-earth approach, which is made possible by the fact that, as we saw in Chapter XVI, the moduli spaces of curves are all quotients of smooth varieties by finite group actions. For the approach via stacks, we refer to [671] or [298].

Of course, the difficulty in defining an intersection product on the cycle groups in question is that moduli spaces are singular. The singularities, however, are mild enough, as they are just quotient ones. The first one to overcome the difficulties intrinsic in the singular nature of moduli spaces was Mumford in [556]. One of his main tools was to exhibit the moduli spaces of curves as global quotients of Cohen–Macaulay varieties by finite groups. The subsequent discovery by Looijenga [487] that one can in fact use *smooth* varieties instead of Cohen–Macaulay ones, which we presented in Chapter XVI, greatly simplifies Mumford's approach. While Mumford has to construct an ad hoc adaptation of Fulton–MacPherson's operational theory, one can instead use standard cycle theory straight out of the box, that is, as presented in Fulton's book [275].

In the following treatment, all cycle groups will be with $\mathbb{Q}$-coefficients; for simplicity, this will not be reflected in the notation. Our exposition will be punctuated by systematic references to specific statements from Fulton's book [275]. We start by considering, quite generally, a reduced and irreducible quasiprojective variety X, acted on by a finite group G; we do not assume that the action of G on X is faithful, i.e., that

$\rho : G \to \mathrm{Aut}(X)$ is injective, and we denote by G' and G'' the image and the kernel of ρ, respectively. We also set

$$\delta = \delta(X, G) = |G''|\,. \tag{2.1}$$

Let $q : X \to X/G$ be the quotient morphism. Observe that X is flat over X/G, and hence there exists a well-defined flat pullback map $q^* : A_\bullet(X/G) \to A_\bullet(X)$. The image $q^*(A_\bullet(X/G))$ is the group of G-invariant classes $A_\bullet(X)^G$, and the homomorphism $q_*q^* : A_\bullet(X/G) \to A_\bullet(X/G)$ is just multiplication by $|G'|$ (cf. Example 1.7.6 on page 20 of [275]); in particular, q^* identifies $A_\bullet(X/G)$ to $A_\bullet(X)^G$.

Now suppose that X is smooth. One may then define an intersection product in $A_\bullet(X/G)$ by setting

$$x \cdot y = \frac{1}{|G'|} q_*(q^*(x) \cdot q^*(y))$$

(cf. [275], Example 8.3.12). One immediately checks that this definition makes $A_\bullet(X/G)$ into a commutative ring with unit $\mathbf{cl}(X/G)$, the fundamental class[5] of X/G, and that $q^* : A_\bullet(X/G) \to A_\bullet(X)$ is a ring homomorphism. Notice that, since $q^*q_*y = \sum_{g \in G'} g_*y$, the push–pull formula

$$q_*(y \cdot q^*(x)) = q_*(y) \cdot x$$

is valid.

At first sight, the multiplicative structure on the cycle ring of X/G seems to depend on a specific representation of the latter as quotient of a smooth variety by a finite group. This, however, is not the case. Recall in fact from [275], Chapter 17, that one can define bivariant cycle groups $A^\bullet(S \to T)$ for a morphism $S \to T$. These include as special cases both the ordinary cycle groups $A_\bullet(V)$ of a general variety V, as well as "cohomological" cycle groups $A^\bullet(V)$. In fact, $A_k(V) = A^{-k}(V \to \text{point})$, and one sets $A^k(V) = A^k(V \xrightarrow{\text{id}} V)$. In addition, there are "cup" intersection products $A^h(V) \times A^k(V) \to A^{h+k}(V)$ and "cap" products $A^h(V) \times A_k(V) \xrightarrow{\cap} A_{k-h}(V)$. The intersection product makes $A^\bullet(V)$ into an associative and commutative ring with unit, and for any morphism $f : V \to W$, there are pullback homomorphisms $f^* : A^h(W) \to A^h(V)$ respecting the ring structure. Moreover, if f is proper, the push–pull formula

$$f_*(f^*\alpha \cap \beta) = \alpha \cap f_*(\beta)$$

is valid. The crucial fact (cf. [275], Example 17.4.10) is that, when $V = X/G$ with X smooth of dimension d and G a finite group, the cap

[5]We would prefer to adopt a more standard notation for the fundamental class of a variety V, like, for instance, $[V]$. Unfortunately, when applied to a quotient, this conflicts with the standard notation for stack quotient.

product homomorphisms $\cap\mathbf{cl}(X/G) : A^h(X/G) \to A_{d-h}(X/G)$ give a *ring isomorphism*

$$A^\bullet(X/G) \xrightarrow{\cap\mathbf{cl}(X/G)} A_\bullet(X/G)\,, \tag{2.2}$$

which can be viewed as an analogue of Poincaré duality. It follows in particular that the product structure in $A_\bullet(X/G)$ does not depend on the quotient representation used. This also allows us to freely move between a homological notation and a cohomological one for cycle rings of quotients of smooth varieties by finite groups, and shows that, for morphisms between such varieties, there are good notions of pushforward and pullback satisfying all the usual properties. Under the isomorphism (2.2), the identity element $1_{X/G}$ corresponds to the fundamental class $\mathbf{cl}(X/G)$.

We may now introduce the cycle groups of the stack quotient $[X/G]$. We let $\pi : X \to [X/G]$ be the quotient morphism, and $p : [X/G] \to X/G$ the natural morphism, so that $q = p \circ \pi$. We set

$$A_\bullet([X/G]) = A_\bullet(X/G)$$

as rings. We would like to define pullback and pushforward maps

$$p^* : A_\bullet(X/G) \to A_\bullet([X/G])\,, \qquad p_* : A_\bullet([X/G]) \to A_\bullet(X/G)$$

so that the usual properties of pullback and pushforward are satisfied; in particular, p^* must be a ring homomorphism. We are then forced to set

$$p^* = \text{identity}\,, \qquad p_* = \text{multiplication by } \frac{1}{\delta}\,,$$

where δ is as defined in (2.1). The fact that the pushforward p_* is not necessarily the identity is the main reason why we distinguish between the rings $A_\bullet([X/G])$ and $A_\bullet(X/G)$, although these are canonically isomorphic via p^*. The formula for p_* expresses that the "degree" of p is $1/\delta$. This is in accordance with the intuition that the degree of π should be $|G|$ and that

$$\deg q = |G'| = \frac{1}{\delta}|G| = \deg p \deg \pi\,.$$

One naturally sets $\pi_* = p_*^{-1} q_*$ and $\pi^* = q^* p^{*-1}$. It is useful to notice that δ is the minimum of the orders of the automorphism groups of objects in $[X/G](\mathbb{C})$ and hence is an intrinsic invariant, independent of the particular presentation of $[X/G]$ as a quotient. In fact, up to isomorphism, such an object is a G-equivariant morphism $\chi : G \to X$, and it is immediate to check that its automorphisms are precisely the morphisms $G \to G$ of the form $g \mapsto g\gamma$, where γ belongs to the stabilizer of $x = \chi(1)$. On the other hand, there is a nonempty Zariski-open subset of X whose points all have stabilizer equal to G''.

It is straightforward to check that π^* is a ring homomorphism and that the push–pull formula is valid both for π and for p. In particular, the product in $A_\bullet([X/G])$ can be written

$$x \cdot y = \frac{1}{|G|}\pi_*(\pi^*(x) \cdot \pi^*(y)) . \tag{2.3}$$

The multiplicative identity in $A_\bullet([X/G])$ is the fundamental class $\mathbf{cl}([X/G]) = p^*\mathbf{cl}(X/G)$. More generally, let V be an integral subvariety of X/G and set $Y = q^{-1}(V)_{red}$. Thus Y is a reduced, G-invariant subvariety of X, and hence determines an integral substack $[Y/G]$ of $[X/G]$. This substack has a fundamental class in $A_\bullet([X/G])$, which is defined as

$$\mathbf{cl}([Y/G]) = \frac{\delta(X,G)}{\delta(Y,G)}p^*[V] .$$

The pushforward formula

$$p_*\mathbf{cl}([Y/G]) = \frac{1}{\delta(Y,G)}[V] \tag{2.4}$$

holds. In particular,

$$p_*\mathbf{cl}([X/G]) = \frac{1}{\delta}\mathbf{cl}(X/G) . \tag{2.5}$$

Since $\mathbf{cl}([X/G]) = p^*\mathbf{cl}(X/G)$ by definition, the exact analogue of (2.2) holds also in the stack context.

At first sight, there seem to be no really compelling reasons for distinguishing between $A^\bullet([X/G])$ and $A^\bullet(X/G)$, unless δ is different from 1. On the other hand, even when $\delta = 1$, there are definitely more fundamental classes of substacks than there are fundamental classes of subschemes, as follows from (2.4). In other words, if $[X/G]$ is a stack and X/G the corresponding space, and we write $A^\bullet([X/G])_\mathbb{Z}$ (resp., $A^\bullet(X/G)_\mathbb{Z}$) for the subgroup of $A^\bullet([X/G])$ generated by fundamental classes of substacks (resp., the subgroup of $A^\bullet(X/G)$ generated by fundamental classes of subschemes), then the inclusion $A^\bullet(X/G)_\mathbb{Z} \subset A^\bullet([X/G])_\mathbb{Z}$ is in general strict.

Concerning Chern classes of vector bundles on a stack of the form $[X/G]$, recall that such a vector bundle can be viewed as a G-equivariant bundle E on X, meaning that E is a vector bundle on X plus a lifting of the G-action on X. The Chern classes of E are naturally G-invariant elements of $A^\bullet(X)$ and therefore give well-defined Chern classes $c_i(E) \in A^i([X/G])$.

We now apply what we have learned to define cycle rings of moduli stacks of curves. Unfortunately, as we know (cf. Remark (7.14) in Chapter XVI), these are in general not quotients of smooth varieties modulo finite groups. However, this is true for moduli *spaces*. Let then $M = X/G$ be a moduli space of curves, and $\mathcal{M}$ the corresponding moduli stack. We then have natural morphisms

$$[X/G] \xrightarrow{\alpha} \mathcal{M} \xrightarrow{\beta} X/G = M$$

whose composition, with our previous notation, is p. We set

$$A^\bullet(\mathcal{M}) = A^\bullet(X/G) = A^\bullet([X/G])$$

and naturally let both α^* and β^* be the identity maps. It remains to define fundamental classes of substacks of $\mathcal{M}$. Let Y be an integral substack of $\mathcal{M}$. Then $\alpha^{-1}(Y)$ is supported on an integral substack Y' of $[X/G]$, and its fundamental class is an integral multiple $k[Y']$ of $[Y']$. One then defines $[Y]$ by setting

$$\alpha^*[Y] = k[Y'] .$$

Loosely speaking, $[Y]$ is k times $[Y']$. Going back to the situation described in Remark (9.4) of Chapter XVI, we see that, when $\mathcal{M} = \overline{\mathcal{M}}_{g,n}$, two cases can occur. Either Y is not contained in $\partial\overline{\mathcal{M}}_{g,n}$, in which case $k = 1$, or Y lies entirely in the boundary and k is the product of two fixed distinct odd primes or a fixed odd prime for $g = 1$.

As for $A^\bullet(X/G)_\mathbb{Z} \subset A^\bullet([X/G])_\mathbb{Z}$, also the inclusion $A^\bullet(M)_\mathbb{Z} \subset A^\bullet(\mathcal{M})_\mathbb{Z}$ is in general strict. For instance, the class of the substack $\overline{\mathcal{H}}_g$ of the moduli stack $\overline{\mathcal{M}}_g$ parameterizing hyperelliptic curves does not belong to $A^\bullet(\overline{M}_g)_\mathbb{Z}$, while its double does, and is the class of the hyperelliptic locus $\overline{H}_g \subset \overline{M}_g$.

3. Tautological classes on moduli spaces of curves.

As we saw in Section 10 of Chapter XII, the basic morphisms between moduli stacks of curves are the projection maps

$$\pi_x : \overline{\mathcal{M}}_{g,P\cup\{x\}} \longrightarrow \overline{\mathcal{M}}_{g,P} ,$$

the clutching maps

$$\xi_\Gamma : \overline{\mathcal{M}}_\Gamma \to \overline{\mathcal{M}}_{g,P} ,$$

and the sections

$$\sigma_p : \overline{\mathcal{M}}_{g,P} \to \overline{\mathcal{C}}_{g,P} = \overline{\mathcal{M}}_{g,P\cup\{x\}} , \qquad p \in P .$$

As we explained in Chapter XII, the morphism π_x can be identified with the projection

$$\pi : \overline{\mathcal{C}}_{g,P} \to \overline{\mathcal{M}}_{g,P}$$

to moduli space from the universal curve. Given a stable graph Γ, we define a cycle class $\delta_\Gamma \in A^\ell(\overline{\mathcal{M}}_{g,P})$, where ℓ is the number of edges of Γ, by the prescription

$$\delta_\Gamma = \frac{1}{|\operatorname{Aut}(\Gamma)|}\xi_{\Gamma*}(1_{\overline{\mathcal{M}}_\Gamma}). \tag{3.1}$$

The class δ_Γ is nothing but the fundamental class of the substack $\mathcal{D}_\Gamma$ parameterizing those curves whose graph is Γ or a specialization of Γ, corresponding to the stratum Δ_Γ in moduli space. We shall refer to classes of this kind as *boundary classes.* We already encountered the codimension one boundary classes in Section 2 of Chapter XIII. Recall that these are δ_{irr}, which corresponds to the graph in Figure 1,

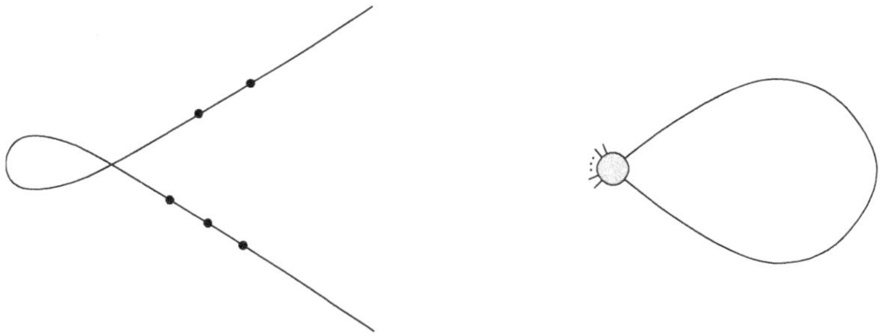

Figure 1. A general curve in Δ_{irr} and its graph

and the classes $\delta_{\mathcal{P}}$, where $\mathcal{P}$ is a "stable bipartition" of (g, P). In practice, given a bipartition $\mathcal{P} = \{(h, A), (k, B)\}$, where of course $g = h + k$ and $P = A \coprod B$, we shall write $\delta_{h,A}$, or equivalently $\delta_{k,B}$, instead of $\delta_{\mathcal{P}}$. Here are the corresponding graph and a sketch of the general curve in $\Delta_{h,A}$.

Figure 2. A general curve in $\Delta_{h,A}$ and its graph

We shall also write $\xi_{h,A}$, or $\xi_{k,B}$, instead of $\xi_{\mathcal{P}}$. In what follows, when Γ is a graph with a single edge, we shall use the symbol $\mathcal{D}_\Gamma$ to indicate not only the substack, but also the corresponding stack divisor. We recall that the section $\sigma_p : \overline{\mathcal{M}}_{g,P} \to \overline{\mathcal{C}}_{g,P} = \overline{\mathcal{M}}_{g,P\cup\{x\}}$ can be identified with the map $\xi_{0,\{p,x\}} : \overline{\mathcal{M}}_{g,P} \simeq \overline{\mathcal{M}}_{0,\{p,x,t\}} \times \overline{\mathcal{M}}_{g,(P\cup\{t\})\smallsetminus\{p\}} \to \overline{\mathcal{M}}_{g,P\cup\{x\}}$. In particular, the image of σ_p is just the divisor $\mathcal{D}_{0,\{p,x\}}$, which will often be denoted simply by D_p. We set

$$D = \sum_{p \in P} D_p .$$

One can define other natural classes by means of the morphisms σ_p and π. The class $\psi_p \in A^1(\overline{\mathcal{M}}_{g,P})$ of the point bundle $\mathcal{L}_p$, which we already encountered in Section 2 of Chapter XIII, is

$$\psi_p = \sigma_p^* c_1(\omega_\pi) , \tag{3.2}$$

where ω_π is the relative dualizing sheaf of $\pi : \overline{\mathcal{C}}_{g,P} \to \overline{\mathcal{M}}_{g,P}$. Now let K_π be the first Chern class of the relative log-canonical sheaf $\omega_\pi(D)$. Generalizing the definition of κ_1 in Chapter XIII, we set

$$\kappa_a = \pi_*(K_\pi^{a+1}) \in A^a(\overline{\mathcal{M}}_{g,P}) , \quad a = 0, 1, \dots . \tag{3.3}$$

In particular, $\kappa_0 = 2g - 2 + |P|$. The classes κ_a and ψ_p are usually called *Mumford–Morita–Miller classes.* A variant of these are the classes

$$\widetilde{\kappa}_a = \pi_*(\widetilde{K}_\pi^{a+1}) ,$$

where $\widetilde{K}_\pi = c_1(\omega_\pi)$. The $\widetilde{\kappa}_a$ are related to the κ_a by the formula

$$\widetilde{\kappa}_a = \kappa_a - \sum \psi_p^a . \tag{3.4}$$

The proof of this formula is straightforward and is based on the observation that taking residues along D_p gives a canonical trivialization of $\sigma_p^*(K_\pi)$, so that $K_\pi \cdot D_p = 0$ in $A^\bullet(\overline{\mathcal{C}}_{g,P})$, while $\sigma_p^*(D_p) = -\sigma_p^*(\widetilde{K}_\pi)$. One then has

$$\begin{aligned} \widetilde{\kappa}_a &= \pi_*((K_\pi - D)^{a+1}) = \pi_*(K_\pi^{a+1} + (-1)^{a+1} \textstyle\sum D_p^{a+1}) \\ &= \kappa_a + (-1)^{a+1} \textstyle\sum \pi_*(D_p^{a+1}) . \end{aligned}$$

On the other hand,

$$\pi_*(D_p^{a+1}) = \sigma_p^*(D_p^a) = (-1)^a \sigma_p^*(\widetilde{K}_\pi)^a = (-1)^a \psi_p^a . \tag{3.5}$$

Additional natural, or tautological, classes are the *Hodge classes,* which generalize the codimension one Hodge class λ first encountered in Chapter XIII. We look at the Hodge bundle $\pi_*\omega_\pi$ for the universal curve $\pi : \overline{\mathcal{C}}_{g,P} \to \overline{\mathcal{M}}_{g,P}$, and we set

$$\lambda_i = c_i(\pi_*\omega_\pi) = c_i(\pi_!\omega_\pi) . \tag{3.6}$$

More generally, for any $\nu > 0$, we set

$$\lambda_i(\nu) = c_i(\pi_! \omega_\pi^\nu)\,. \tag{3.7}$$

The use of $\pi_!$ is necessitated by the fact that the higher direct images of ω_π^ν need not be trivial, as is instead the case for $\nu = 1$; thus, the first Chern class of $\pi_! \omega_\pi^\nu$ and the one of $\pi_* \omega_\pi^\nu$ are in general different when $\nu \neq 1$. Of course, $\lambda_i(1) = \lambda_i$, $\lambda_1(\nu) = \lambda(\nu)$, and $\lambda_1 = \lambda$.

4. Tautological relations and the tautological ring.

The tautological classes and maps defined in the previous section satisfy a number of intriguing relations. Consider the diagram

$$\begin{array}{ccc} \overline{\mathcal{M}}_{g,P\cup\{x,y\}} & \xrightarrow{\pi_x} & \overline{\mathcal{M}}_{g,P\cup\{y\}} \\ {\scriptstyle \pi_y}\downarrow & & \downarrow{\scriptstyle \pi_y} \\ \overline{\mathcal{M}}_{g,P\cup\{x\}} & \xrightarrow{\pi_x} & \overline{\mathcal{M}}_{g,P} \end{array} \tag{4.1}$$

which, in view of the identification between the universal curve $\overline{\mathcal{C}}_{g,P}$ and the moduli space $\overline{\mathcal{M}}_{g,P\cup\{y\}}$, we may also reinterpret as

$$\begin{array}{ccc} \overline{\mathcal{C}}_{g,P\cup\{x\}} & \xrightarrow{\rho} & \overline{\mathcal{C}}_{g,P} \\ {\scriptstyle \alpha}\downarrow & & \downarrow{\scriptstyle \beta} \\ \overline{\mathcal{M}}_{g,P\cup\{x\}} & \xrightarrow{\pi} & \overline{\mathcal{M}}_{g,P} \end{array} \tag{4.2}$$

We shall write τ_t, $t \in P \cup \{x\}$ (resp., τ'_p, $p \in P$), for the canonical sections of α (resp., of β), and D_t, $t \in P \cup \{x\}$ (resp., D'_p, $p \in P$), for the corresponding divisors. We also set

$$D = \sum_{t \in P\cup\{x\}} D_t\,, \quad D' = \sum_{p \in P} D'_p\,.$$

Recall from part a) of Proposition (6.7) in Chapter X that

$$\rho^*(\omega_\beta(D')) = \omega_\alpha(D - D_x)\,. \tag{4.3}$$

Noticing that D_x, viewed as a divisor in $\overline{\mathcal{M}}_{g,P\cup\{x,y\}}$, is just $\mathcal{D}_{0,\{x,y\}}$, this implies the identity

$$\rho^*(K_\beta) = K_\alpha - \delta_{0,\{x,y\}} \tag{4.4}$$

in $A^\bullet(\overline{\mathcal{M}}_{g,P\cup\{x,y\}})$. On the other hand, the pullback of D'_p is $D_p + E_p$, where E_p is the divisor in $\overline{\mathcal{C}}_{g,P\cup\{x\}}$ defined as follows. Look at fibers of

α containing a smooth rational component meeting the rest of the fiber at only one point, i.e., a "rational tail." Then E_p is the divisor swept out by the rational tails containing only two marked points, namely the ones labeled by p and x. If we set $E = \sum E_p$, recalling (4.3), we get

$$\rho^*(D') = D - D_x + E\,, \quad \rho^*(\omega_\beta) = \omega_\alpha(-E)\,. \tag{4.5}$$

One of our goals in this section is to describe pullbacks and pushforwards of tautological classes via the natural maps between moduli spaces. We begin by noticing that (4.5) immediately implies that

$$\pi_x^* \psi_p = \psi_p - \delta_{0,\{p,x\}} \tag{4.6}$$

in $A^\bullet(\overline{\mathcal{M}}_{g,P\cup\{x\}})$, for $\tau_p^*(E) = \mathcal{D}_{0,\{p,x\}}$. On the other hand, the relative dualizing sheaf ω_α is trivial on E_p. This implies that $\psi_p \cdot \delta_{0,\{p,x\}} = 0$. Since x plays no special role in this identity, this shows more generally that

$$\psi_p \cdot \delta_{0,\{p,p'\}} = 0 \quad \text{in } A^\bullet(\overline{\mathcal{M}}_{g,P}) \tag{4.7}$$

for any $p, p' \in P$, $p \neq p'$. An easy induction, based on (4.6) and (4.7), gives

$$\psi_p^a = \pi^* \psi_p^a + \pi^* \psi_p^{a-1} \cdot \delta_{0,\{p,x\}} \tag{4.8}$$

for all $a > 0$. We now come to two important identities, which usually go under the names of *string equation* and *(generalized) dilaton equation.*

PROPOSITION (4.9). *Let $\pi = \pi_x$ be the projection*

$$\pi_x : \overline{\mathcal{M}}_{g,P\cup\{x\}} \to \overline{\mathcal{M}}_{g,P}\,,$$

and let a_t, $t \in P \cup \{x\}$, be nonnegative integers. Then:

(String Equation) $\displaystyle \pi_*\Big(\prod_{p\in P} \psi_p^{a_p}\Big) = \sum_{a_p>0} \psi_p^{a_p-1} \prod_{q\neq p} \psi_q^{a_q}\,;$

(Dilaton Equation) $\displaystyle \pi_*\Big(\psi_x^{a_x+1} \prod_{p\in P} \psi_p^{a_p}\Big) = \kappa_{a_x} \prod_{p\in P} \psi_p^{a_p}\,.$

The standard dilaton equation is the special case of the generalized one in which $a_x = 0$. Since we observed that $\kappa_0 = 2g - 2 + |P|$, this can be written as

$$\pi_*\Big(\psi_x \prod_{p\in P} \psi_p^{a_p}\Big) = (2g - 2 + |P|) \prod_{p\in P} \psi_p^{a_p}\,. \tag{4.10}$$

Actually, the names "string equation" and "dilaton equation" are usually reserved for the equalities between intersection numbers that the two identities in (4.9) give in the top-dimensional case, that is, for $\sum a_p = 3g - 3 + |P| + 1$ and for $\sum a_p = 3g - 3 + |P|$ (and $a_x = 0$), respectively.

Coming to the proof of (4.9), we first tackle the string equation. We express each power $\psi_p^{a_p}$ such that $a_p > 0$ using formula (4.8) and notice that $\delta_{0,\{p,x\}} \cdot \delta_{0,\{p',x\}} = 0$ for $p \neq p'$ since the corresponding divisors do not meet. This shows that

$$\prod_{p\in P} \psi_p^{a_p} = \sum_{a_p>0} \delta_{0,\{p,x\}} \cdot \pi^*\Big(\psi_p^{a_p-1} \prod_{q\neq p} \psi_q^{a_q}\Big) + \pi^*\Big(\prod_{p\in P} \psi_p^{a_p}\Big).$$

Notice that $\pi_*\pi^* = 0$ since $\pi_*\pi^* z = \pi_*(\pi^* z \cdot 1) = z \cdot \pi_* 1$ by the push–pull formula, and $\pi_* 1 = 0$ by dimension reasons. Hence, applying π_* to both sides gets rid of the last summand on the right. On the other hand, the push–pull formula shows that

$$\pi_*\Big(\delta_{0,\{p,x\}} \cdot \pi^*\Big(\psi_p^{a_p-1} \prod_{q\neq p} \psi_q^{a_q}\Big)\Big) = \pi_*(\delta_{0,\{p,x\}}) \cdot \psi_p^{a_p-1} \prod_{q\neq p} \psi_q^{a_q} = \psi_p^{a_p-1} \prod_{q\neq p} \psi_q^{a_q}.$$

In fact, $\pi_*(\delta_{0,\{p,x\}}) = 1$ since $\mathcal{D}_{0,\{p,x\}}$ is a section of π. This proves the string equation.

The proof of the dilaton equation is a bit more delicate. First of all, recall that $\delta_{0,\{p,x\}} \cdot \psi_x = 0$ by (4.7). Expressing each power $\psi_p^{a_p}$ such that $a_p > 0$ via formula (4.8) and using this observation, the left-hand side of the dilaton equation reduces to

$$\pi_*\Big(\psi_x^{a_x+1} \prod_{p\in P} \pi^*(\psi_p^{a_p})\Big) = \pi_*(\psi_x^{a_x+1}) \prod_{p\in P} \psi_p^{a_p}.$$

To prove the full dilaton equation, it thus suffices to prove the special case where $a_p = 0$ for all $p \in P$. In other words, we must show that

$$\pi_*(\psi_x^a) = \kappa_{a-1} \tag{4.11}$$

when $a > 0$. In proving this formula, it will be convenient to look at our basic diagram in the form (4.1); however, we shall keep using the names α, β, and so on for the maps in the diagram. Identity (4.11) is obtained by pushing forward via π the more fundamental equality

$$\psi_x^a = K_\pi^a, \tag{4.12}$$

which we now prove. Recall first that, as we already observed, taking residues along $D_x = \mathcal{D}_{0,\{x,y\}} \subset \overline{\mathcal{M}}_{g,P\cup\{x,y\}}$ gives a canonical trivialization of the log-canonical sheaf $\omega_\alpha(D)$ along this divisor; it follows that

$$K_\alpha \cdot \delta_{0,\{x,y\}} = 0. \tag{4.13}$$

Another useful elementary observation is that, for any cycle class γ on $\overline{\mathcal{M}}_{g,P\cup\{x,y\}}$,

$$\tau_x^*\gamma = \alpha_*(\gamma \cdot \delta_{0,\{x,y\}}) . \tag{4.14}$$

Since the sections τ_t are disjoint, we can rewrite ψ_x as

$$\psi_x = \tau_x^*(K_\alpha - \delta_{0,\{x,y\}}) , \tag{4.15}$$

and hence

$$\psi_x^a = \alpha_*((K_\alpha - \delta_{0,\{x,y\}})^a \cdot \delta_{0,\{x,y\}}) . \tag{4.16}$$

Now,

$$(K_\alpha - \delta_{0,\{x,y\}})^a \cdot \delta_{0,\{x,y\}} = (-1)^a {\delta_{0,\{x,y\}}}^{a+1} = (K_\rho - \delta_{0,\{x,y\}})^a \cdot \delta_{0,\{x,y\}} ,$$

where the second equality is obtained from the first reversing the roles of α and ρ. On the other hand, reversing the roles of α and ρ in formula (4.4) gives

$$\alpha^*(K_\pi) = K_\rho - \delta_{0,\{x,y\}} .$$

Putting everything together, we find that

$$\begin{aligned}\psi_x^a &= \alpha_*((K_\rho - \delta_{0,\{x,y\}})^a \cdot \delta_{0,\{x,y\}}) = \alpha_*(\alpha^*(K_\pi^a) \cdot \delta_{0,\{x,y\}}) \\ &= K_\pi^a \cdot \alpha_*(\delta_{0,\{x,y\}}) = K_\pi^a ,\end{aligned}$$

as desired. This completes the proof of (4.9).

Another useful property of the classes κ is that, on $\overline{\mathcal{M}}_{g,P\cup\{x\}}$, one has

$$\pi_x^*(\kappa_a) = \kappa_a - \psi_x^a . \tag{4.17}$$

An immediate consequence of (4.17) is

$$\pi_{x*}(\kappa_a) = \kappa_{a-1} . \tag{4.18}$$

To prove (4.17), we use the somewhat unusual commutation relation expressed by the following key result.

LEMMA (4.19). *Consider diagram (4.2). Then* $\pi^*\beta_* = \alpha_*\rho^*$.

Before proving the lemma, observe that (4.17) immediately follows from it by the string of equalities

$$\begin{aligned}\pi^*\kappa_a &= \pi^*\beta_* K_\beta^{a+1} = \alpha_*\rho^* K_\beta^{a+1} = \alpha_*((K_\alpha - \delta_{0,\{x,y\}})^{a+1}) \\ &= \alpha_*(K_\alpha^{a+1} - ((K_\alpha - \delta_{0,\{x,y\}})^a \cdot \delta_{0,\{x,y\}}) \\ &= \kappa_a - \alpha_*((K_\alpha - \delta_{0,\{x,y\}})^a \cdot \delta_{0,\{x,y\}}) \\ &= \kappa_a - \psi_x^a ,\end{aligned}$$

where we have also used formulas (4.4), (4.13), and (4.16).

Proof of Lemma (4.19). We shall consider indifferently diagram (4.2) or its incarnation (4.1). Also, we may as well work with their analogues for moduli spaces, since the rational Chow rings of moduli stacks are just the Chow rings of the underlying moduli spaces. In any case we shall retain the names α, β, ρ, and π for the morphisms. The diagram to consider can thus be written as follows:

(4.20)
$$\begin{array}{ccc} \overline{M}_{g,n+2} & \xrightarrow{\ \rho\ } & \overline{M}_{g,n+1} \\ {\scriptstyle\alpha}\big\downarrow & & \big\downarrow{\scriptstyle\beta} \\ \overline{M}_{g,n+1} & \xrightarrow{\ \pi\ } & \overline{M}_{g,n} \end{array}$$

We shall use the explicit representation of moduli spaces as quotients of smooth varieties by finite groups described in Chapter XVI. More specifically, as in Remark (9.5) in that chapter, we write

$$\overline{M}_{g,n} = M/H , \qquad \overline{M}_{g,n+1} = M_1/H_1 , \qquad \overline{M}_{g,n+2} = M_2/H_2 ,$$

where

$$\begin{aligned} M &= {}_G\overline{M}_{g,n} , & M_1 &= {}_G\overline{M}_{g,n+1} , & M_2 &= {}_G\overline{M}_{g,n+2} , \\ H &= {}_GH_{g,n} , & H_1 &= {}_GH_{g,n+1} , & H_2 &= {}_GH_{g,n+2} . \end{aligned}$$

Consider the diagram

(4.21)
$$\begin{array}{ccc} M_2 & \xrightarrow{\ \rho\ } & M_1 \\ {\scriptstyle\alpha}\big\downarrow & & \big\downarrow{\scriptstyle\beta} \\ M_1 & \xrightarrow{\ \pi\ } & M \end{array}$$

and recall that there are natural surjective homomorphisms $H_2 \to H_1 \to H$ making all the morphisms in (4.21) equivariant. Since the Chow ring of a quotient of a smooth variety by a finite group K is just the subring of K-invariants, to prove Lemma (4.19), it suffices to prove its analogue for diagram (4.21). This diagram can be expanded to

(4.22)

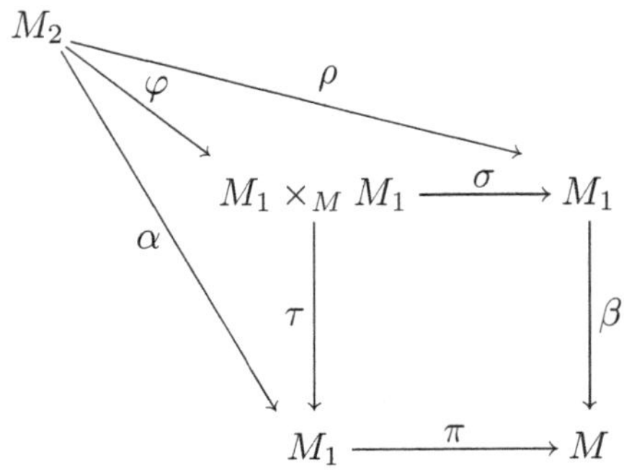

By Example 17.4.1 in [275],

$$\pi^* \beta_* = \tau_* \sigma^* .$$

In fact, the four morphisms involved in this identity are all proper l.c.i. morphisms of codimension -1, so in particular the excess normal bundle does not actually intervene in the formulas. To conclude, we have to show that

$$\varphi_* \varphi^* = \mathrm{id} . \tag{4.23}$$

If φ were a blow-up, this would be just part b) of Proposition 6.7 of [275]. As it turns out, φ is not quite a blow-up but can be described as follows. We can view $\beta : M_1 \to M$ as a family of n-pointed nodal curves; these curves are admissible G-covers of stable, n-pointed, genus g curves and may not be stable themselves. Base change via π gives the morphism $\tau : M_1 \times_M M_1 \to M_1$, which can be viewed as a family of n-pointed nodal curves with an extra section δ corresponding to the diagonal. It follows from Remark (9.3) in Chapter XVI that applying to these data the stabilization procedure described in Section 8 of Chapter X produces $\alpha : M_2 \to M_1$. Let us recall that the stabilization procedure consists in performing a modification of $M_1 \times_M M_1$ along two disjoint loci. The first of these is the locus Z_1 where the extra section δ meets one of the n preexisting ones; here the modification is just the blow-up along Z_1. The second locus Z_2 is the one where δ hits nodes in the fibers of τ; the modification procedure along Z_2 is described analytically by formula (8.5) in Chapter X. We denote by W the total transform of Z_2. Now let $\chi : Q \to M_1 \times_M M_1$ be the blow-up along $Z_1 \cup Z_2$. We claim that $\chi = \varphi \circ \psi$, where $\psi : Q \to M_2$ is the blow-up along W. There is something to prove only along Z_2. Locally along Z_2, the variety M_2 is described by the equations

$$xy = t = \xi\eta , \qquad \lambda\xi = \mu x , \qquad \lambda y = \mu\eta ,$$

where λ, μ are homogeneous coordinates. Locally along W, the variety Q is described by the equations

$$\begin{array}{llll} xy = \xi\eta\,, & ab = cd\,, & ay = bx\,, & a\xi = cx\,, \\ a\eta = dx\,, & b\xi = cy\,, & b\eta = dy\,, & c\eta = d\xi\,, \end{array}$$

where a, b, c, d are homogeneous coordinates, and the morphism ψ is defined by

$$\begin{array}{llll} \lambda = d\,, & \mu = b\,, & \text{when} & (d,b) \neq (0,0)\,, \\ \lambda = a\,, & \mu = c\,, & \text{when} & (a,c) \neq (0,0)\,. \end{array}$$

The reader can directly check that ψ, as defined above, is a blow-up. Alternatively, one can invoke a classical result by Fujiki and Nakano [564,272] stating that a sufficient condition for a morphism $f : X \to Y$ of smooth varieties to be a blow-up is that there exists a proper codimension k smooth subvariety $Z \subset Y$ such that the fibers of f at points of Z are $(k-1)$-dimensional projective spaces, while the remaining fibers are single points.

At this point it is immediate to check that (4.23) holds. In fact, as both χ and ψ are blow-ups, $\chi_*\chi^* = \mathrm{id}$ and $\psi_*\psi^* = \mathrm{id}$ by part b) of Proposition 6.7 of [275], and hence,

$$\varphi_*\varphi^* = \varphi_*\psi_*\psi^*\varphi^* = \chi_*\chi^* = \mathrm{id}\;.$$

Q.E.D.

Further remarkable identities can be obtained by repeated applications of the dilaton equation. To explain how this goes, it is convenient to state the results in the case where the marked points are indexed by sets of the form $\{1, \ldots, n\}$, rather than by general ones. Using (4.17) and the push–pull formula, and repeatedly applying the dilaton equation to the projections

$$\pi_n : \overline{\mathcal{M}}_{g,n} \to \overline{\mathcal{M}}_{g,n-1}\,, \quad \pi_{n-1} : \overline{\mathcal{M}}_{g,n-1} \to \overline{\mathcal{M}}_{g,n-2}\,, \quad \ldots,$$

we get

$$\begin{aligned} &(\pi_{n-1}\pi_n)_*(\psi_1^{a_1} \cdots \psi_{n-2}^{a_{n-2}}\psi_{n-1}^{a_{n-1}+1}\psi_n^{a_n+1}) \\ &\qquad = \pi_{n-1*}(\psi_1^{a_1} \cdots \psi_{n-2}^{a_{n-2}}\psi_{n-1}^{a_{n-1}+1}\kappa_{a_n}) \\ &\qquad = \pi_{n-1*}(\psi_1^{a_1} \cdots \psi_{n-1}^{a_{n-1}+1}(\pi_{n-1}{}^*(\kappa_{a_n}) + \psi_{n-1}^{a_n})) \\ &\qquad = \psi_1^{a_1} \cdots \psi_{n-2}^{a_{n-2}}(\kappa_{a_{n-1}}\kappa_{a_n} + \kappa_{a_{n-1}+a_n})\,, \end{aligned} \tag{4.24}$$

$$\begin{aligned} &(\pi_{n-2}\pi_{n-1}\pi_n)_*(\psi_1^{a_1} \cdots \psi_{n-3}^{a_{n-3}}\psi_{n-2}^{a_{n-2}+1}\psi_{n-1}^{a_{n-1}+1}\psi_n^{a_n+1}) \\ &\qquad = (\psi_1^{a_1} \cdots \psi_{n-3}^{a_{n-3}})(\kappa_{a_{n-2}}\kappa_{a_{n-1}}\kappa_{a_n} + \kappa_{a_{n-2}}\kappa_{a_{n-1}+a_n} \\ &\qquad\quad + \kappa_{a_{n-1}}\kappa_{a_{n-2}+a_n} + \kappa_{a_n}\kappa_{a_{n-2}+a_{n-1}} + 2\kappa_{a_{n-2}+a_{n-1}+a_n})\,, \end{aligned}$$

and so on. In general, one finds the formulas

$$(\pi_{k+1}\cdots\pi_n)_*(\psi_1^{a_1}\cdots\psi_k^{a_k}\psi_{k+1}^{a_{k+1}+1}\cdots\psi_n^{a_n+1}) = \psi_1^{a_1}\cdots\psi_k^{a_k}R_{a_{k+1}\dots a_n}\,, \tag{4.25}$$

where $R_{a_{k+1}\dots a_n}$ is a polynomial in the κ classes. A compact expression for $R_{b_1\dots b_l}$, due to Faber, is the following (Exercise C-3):

$$R_{b_1\dots b_l} = \sum_{\sigma\in\mathfrak{S}_l}\kappa_\sigma\,, \tag{4.26}$$

where $\mathfrak{S}_l$ stands for the symmetric group on l letters, and κ_σ is defined as follows. Write the permutation σ as a product of $\nu(\sigma)$ disjoint cycles, including 1-cycles: $\sigma = \gamma_1\dots\gamma_{\nu(\sigma)}$, where we think of $\mathfrak{S}_l$ as acting on the l-tuple $(b_1,\dots,b_l)$. Denote by $|\gamma|$ the sum of the elements of a cycle γ. Then

$$\kappa_\sigma = \kappa_{|\gamma_1|}\kappa_{|\gamma_2|}\cdots\kappa_{|\gamma_{\nu(\sigma)}|}\,. \tag{4.27}$$

Formula (4.25), together with the string equation, implies the remarkable fact that the intersection numbers of the classes ψ_i and κ_i on a fixed $\overline{\mathcal{M}}_{g,n}$ is completely determined by the intersection theory of the ψ_i alone on all the $\overline{\mathcal{M}}_{g,\nu}$ with $\nu\ge n$, and conversely. A special case of this is Witten's remark [684] that knowing the intersection numbers of the κ's on $\overline{\mathcal{M}}_{g,0}$ is equivalent to knowing the intersection numbers of the ψ's on all the $\overline{\mathcal{M}}_{g,n}$.

When thinking of the advantages coming from considering *all the spaces* $\overline{M}_{g,n}$ *at once, for all* g *and* n, the following thought of Grothendieck [332] comes to mind.

Parmi ces multiplicités modulaires, ce sont celles de Mumford–Deligne pour les courbes algébriques "stables" de genre g, à ν points marquès, que je note $\widehat{M}_{g,\nu}$ (compactification de la multiplicité "ouverte" $M_{g,\nu}$ correspondant aux courbes lisses), qui depuis quelques deux ou trois années ont exercé sur moi une fascination particulière, plus forte peut-être qu'aucun autre objet mathématique à ce jour. A vrai dire, il s'agit plutôt du système de toutes les multiplicités $M_{g,\nu}$ pour g, ν variables, liées entre elles par un certain nombre d'opérations fondamentales (telles les opérations de "bouchage de trous" i.e. de "gommage" de points marqués, celles de "recollement" et les opérations inverses), qui sont le reflet en géométrie algébrique absolue de caractéristique zéro (pour le moment) d'opérations géométriques familières du point de vue de la "chirurgie" topologique ou conforme des surfaces.[6]

---

[6] *Among these modular multiplicities, it is those of Mumford–Deligne for the "stable" algebraic curves of genus g, with ν marked points, which I denote*

We will see the idea of considering *all the spaces* $\overline{M}_{g,n}$ *at once* working in its full force in Chapter XX, when presenting Witten's conjecture and Kontsevich's proof of it.

As we have said, one of our goals in this section is to describe how the tautological classes pull back under the projection and clutching maps defined in Chapter XII. Much of the work has already been done. For convenience, we collect the results we have already proved, and a few new ones, in a series of lemmas. We begin with the projection $\pi_x : \overline{\mathcal{M}}_{g,P\cup\{x\}} \to \overline{\mathcal{M}}_{g,P}$.

LEMMA (4.28). i) $\pi_x^*(\kappa_a) = \kappa_a - \psi_x^a$;
 ii) $\pi_x^*(\psi_p) = \psi_p - \delta_{0,\{p,x\}}$ *for any* $p \in P$;
 iii) $\pi_x^*(\delta_\Gamma) = \sum_{v\in V} \delta_{\Gamma_v}$, *where* Γ_v *is the* $P \cup \{x\}$*-marked graph obtained from* Γ *by letting* $P_v \cup \{x\}$ *be the index set for the vertex* v.
 In particular:
 iv) $\pi_x^*(\delta_{irr}) = \delta_{irr}$;
 v) $\pi_x^*(\delta_{a,A}) = \delta_{a,A} + \delta_{a,A\cup\{x\}}$.

Part i) of the lemma is just formula (4.17), and part ii) is (4.6), while iii), iv) and v) are clear.

We next turn to the clutching morphisms studied in Section 10 of Chapter XII. Consider a clutching map $\xi_\Gamma : \overline{\mathcal{M}}_\Gamma \to \overline{\mathcal{M}}_{g,P}$. Recall that

$$\overline{\mathcal{M}}_\Gamma = \prod_{v\in V(\Gamma)} \overline{\mathcal{M}}_{g_v,L_v}\,, \tag{4.29}$$

where $V(\Gamma)$ stands for the set of vertices of the graph Γ, and L_v for the set of the half-edges issuing from v. We let $H(\Gamma)$ be the set of those half-edges of Γ which are not legs, and we set $H_v = L_v \cap H(\Gamma)$. We denote by η_v the projection from $\overline{\mathcal{M}}_\Gamma$ to the factor $\overline{\mathcal{M}}_{g_v,L_v}$. The first pull-back formula regards the ψ-classes and is trivial:

$$\xi_\Gamma^*(\psi_p) = \psi_p \tag{4.30}$$

The second result regards the Mumford classes κ_a, and is a generalization to higher a of formula (5.24) in Chapter XIII.

---

by $\widehat{M}_{g,\nu}$ *(compactification of the "open" multiplicity* $M_{g,\nu}$ *corresponding to nonsingular curves), which for the last two or three years have exercised a particular fascination over me, perhaps even stronger than any other mathematical object to this day. Indeed, it is more the system of* <u>*all*</u> *the multiplicities* $M_{g,\nu}$ *for variable* g *and* ν*, linked together by a certain number of fundamental operations (such as the operation of "plugging holes", i.e., "erasing" marked points, and of "glueing" and the inverse operations) which are the reflection, in absolute algebraic geometry in characteristic zero (for the moment) of geometric operations familiar from the point of view of the topological or conformal "surgery" of surfaces.*

LEMMA (4.31).

$$\xi_\Gamma^*(\kappa_a) = \sum_{v\in V(\Gamma)} \eta_v^*(\kappa_a)\,.$$

To prove the lemma, consider the diagram

$$\begin{array}{ccc} \mathcal{Y} & \xrightarrow{\ \varphi\ } & \overline{\mathcal{C}}_{g,P} \\ {\scriptstyle \pi'}\big\downarrow & & \big\downarrow{\scriptstyle \pi} \\ \overline{\mathcal{M}}_\Gamma & \xrightarrow{\ \xi_\Gamma\ } & \overline{\mathcal{M}}_{g,P} \end{array}$$

where

$$\mathcal{Y} = \coprod_{v\in V(\Gamma)} \left(\overline{\mathcal{C}}_{g_v,L_v} \times \prod_{\substack{u\in V(\Gamma)\\ u\neq v}} \overline{\mathcal{M}}_{g_u,L_u} \right),$$

and φ is defined by gluing along sections in the manner prescribed by Γ. We also let π'_v and φ_v be the restrictions of π' and φ to the vth component of $\mathcal{Y}$. The morphism π'_v is endowed with canonical sections S_ℓ, $\ell\in L_v$. Now the point is that, by the very definition of dualizing sheaf, $\varphi_v^*(K) = c_1(\omega_{\pi'_v}(\sum S_\ell))$. The lemma follows.

The third pullback formula deals with the boundary strata themselves. From Section 2 of Chapter XIII we recall the formulae (3.4) and (3.8) for the excess intersection bundles. Given strata $\overline{\mathcal{M}}_\Gamma$ and $\overline{\mathcal{M}}'_\Gamma$, the fiber product $\overline{\mathcal{M}}_{\Gamma\Gamma'} = \overline{\mathcal{M}}_\Gamma \times_{\overline{\mathcal{M}}_{g,P}} \overline{\mathcal{M}}'_\Gamma$ decomposes as the disjoint union of strata $\overline{\mathcal{M}}_\Lambda$, where Λ runs over a set $\mathcal{G}_{\Gamma\Gamma'}$ of representatives of isomorphism classes of P-marked, genus g dual graphs equipped with two contractions $c: |\Lambda| \to |\Gamma|$ and $c': |\Lambda| \to |\Gamma'|$ such that

$$E(|\Lambda|) = c^{-1}E(|\Gamma|) \cup c'^{-1}E(|\Gamma'|)\,. \tag{4.32}$$

Moreover, for each component $\overline{\mathcal{M}}_\Lambda$ of $\overline{\mathcal{M}}_{\Gamma\Gamma'}$, the excess intersection bundle is given by

$$F_{\Lambda\Gamma\Gamma'} = \bigoplus_{\{\ell,\ell'\}\in c^{-1}E(|\Gamma|)\cap c'^{-1}E(|\Gamma'|)} \eta_{v(\ell)}^*\mathcal{L}_\ell^\vee \otimes \eta_{v(\ell')}^*\mathcal{L}_{\ell'}^\vee\,. \tag{4.33}$$

Denoting by $\xi_{\Lambda\Gamma}$ the clutching map from $\overline{\mathcal{M}}_\Lambda$ to $\overline{\mathcal{M}}_\Gamma$, we get
(4.34)

$$(\xi_\Gamma)^*(\delta_{\Gamma'}) = \sum_{\Lambda\in\mathcal{G}_{\Gamma\Gamma'}} (\xi_{\Lambda\Gamma})_* \left(\prod_{\{\ell,\ell'\}\in c^{-1}E(|\Gamma|)\cap c'^{-1}E(|\Gamma'|)} (-\eta_{v(\ell)}^*(\psi_\ell) - \eta_{v(\ell')}^*(\psi_{\ell'})) \right).$$

For future reference, we are going to make this formula explicit for the pullback of boundary divisors to boundary divisors. In this case the formula is quite simple and could also be dealt with directly. We have to consider two types of clutching morphisms. First of all, we have the morphism

$$\xi_{irr} : \overline{\mathcal{M}}_{g-1,P\cup\{q,r\}} \to \overline{\mathcal{M}}_{g,P}$$

associated to the graph with one vertex and one edge (cf. Figure 4 in Section 10 of Chapter XII). From formula (4.34) we get the following:

LEMMA (4.35). i) $\xi_{irr}^*(\delta_{irr}) = \delta_{irr} - \psi_q - \psi_r + \sum_{q\in B, r\notin B} \delta_{b,B}$;

ii) $\xi_{irr}^*(\delta_{a,A}) = \begin{cases} \delta_{a,A} & \text{if } g = 2a,\ A = P = \emptyset, \\ \delta_{a,A} + \delta_{a-1,A\cup\{q,r\}} & \text{otherwise.} \end{cases}$

The second codimension one clutching morphism is

$$\xi_{h,H} : \overline{\mathcal{M}}_{h,H\cup\{q\}} \times \overline{\mathcal{M}}_{g-h,(P\smallsetminus H)\cup\{r\}} \to \overline{\mathcal{M}}_{g,P} ,$$

which is associated to the a graph $\Gamma_{h,H}$ with two vertices and one edge (cf. figure 4 in Section 10 of Chapter XII). We set $k = g-h$, $K = P \smallsetminus H$. We denote by η_1 and η_2 the projections of $\overline{\mathcal{M}}_{h,H\cup\{q\}} \times \overline{\mathcal{M}}_{k,(K)\cup\{r\}}$ to the first and second factors. If B is a subset of P, we set $B^c = P \smallsetminus B$, and we have:

LEMMA (4.36). i) $\xi_{h,H}^*(\delta_{irr}) = \eta_1^*(\delta_{irr}) + \eta_2^*(\delta_{irr})$.

ii) *$\xi_{h,H}^*(\delta_{b,B})$ is of the form $\eta_1^*(\zeta_1) + \eta_2^*(\zeta_2)$, where ζ_1 is a cycle class on $\overline{\mathcal{M}}_{h,H\cup\{q\}}$, and ζ_2 is a cycle class on $\overline{\mathcal{M}}_{k,K\cup\{r\}}$. The class ζ_1 is as follows.*

a) *If $H = P$, then*

$$\zeta_1 = \begin{cases} \delta_{2h-g,P\cup\{q\}} - \psi_q & \text{if } (b,B) = (h,P) \text{ or } (b,B) = (k,\emptyset), \\ \delta_{b,B} + \delta_{b+h-g,B\cup\{q\}} & \text{otherwise.} \end{cases}$$

b) *If $H \neq P$, then*

$$\zeta_1 = \begin{cases} -\psi_q & \text{if } (b,B) = (h,H) \text{ or } (b,B) = (k,K), \\ \delta_{b,B} & \text{if } B \subset H \text{ and } (b,B) \neq (h,H), \\ \delta_{b+h-g,(B\setminus K)\cup\{q\}} & \text{if } B \supset K \text{ and } (b,B) \neq (k,K), \\ 0 & \text{otherwise.} \end{cases}$$

The class ζ_2 is described by the analogues of a) *and* b), *where h is replaced with k, H with K, and q with r.*

Sometimes it is convenient to have pullback formulas not only for the ξ maps, but also for the simpler map

(4.37) $$\vartheta : \overline{\mathcal{M}}_{a,A\cup\{q\}} \to \overline{\mathcal{M}}_{g,P} ,$$

which associates to any $A \cup \{q\}$-pointed genus a curve the P-pointed genus g curve obtained by gluing to it a *fixed* $A^c \cup \{r\}$-pointed genus $g-a$ curve C via identification of q and r.

LEMMA (4.38). *The following pullback formulas hold:*

i) $\vartheta^*(\kappa_1) = \kappa_1$;

ii) $\vartheta^*(\psi_p) = \begin{cases} \psi_p & \text{if } p \in A, \\ 0 & \text{if } p \in A^c; \end{cases}$

iii) $\vartheta^*(\delta_{irr}) = \delta_{irr}$.

Suppose $A = P$. *Then*

$$\text{iv) } \vartheta^*(\delta_{b,B}) = \begin{cases} \delta_{2a-g,P\cup\{q\}} - \psi_q & \text{if } (b,B) = (a,P)\,, \\ & \text{or } (b,B) = (g-a,\emptyset), \\ \delta_{b,B} + \delta_{b+a-g,B\cup\{q\}} & \text{otherwise.} \end{cases}$$

Suppose $A \neq P$. *Then*

$$\text{iv') } \vartheta^*(\delta_{b,B}) = \begin{cases} -\psi_q & \text{if } (b,B) = (a,A) \text{ or} \\ & (b,B) = (g-a,A^c), \\ \delta_{b,B} & \text{if } B \subset A \text{ and } (b,B) \neq (a,A), \\ \delta_{b+a-g,(B\smallsetminus A^c)\cup\{q\}} & \text{if } B \supset A^c \text{ and} \\ & (b,B) \neq (g-a,A^c), \\ 0 & \text{otherwise.} \end{cases}$$

The considerations developed so far bring naturally into the picture the so-called *system of tautological rings* $\{R^\bullet(\overline{\mathcal{M}}_{g,n})\}_{g,n}$. This is defined to be the smallest set of $\mathbb{Q}$-subalgebras

$$R^\bullet(\overline{\mathcal{M}}_{g,n}) \subset A^\bullet(\overline{\mathcal{M}}_{g,n})\,,$$

which is closed under the following homomorphisms:

i)
$$(\pi_x)_* : A^\bullet(\overline{\mathcal{M}}_{g,P\cup\{x\}}) \longrightarrow A^\bullet(\overline{\mathcal{M}}_{g,P})$$

for all g and P.

ii)
$$(\xi_{irr})_* : A^\bullet(\overline{\mathcal{M}}_{g-1,P\cup\{q,r\}}) \longrightarrow A^\bullet(\overline{\mathcal{M}}_{g,P})\,,$$
$$(\xi_{h,H})_* : A^\bullet(\overline{\mathcal{M}}_{h,H\cup\{q\}}) \otimes A^\bullet(\overline{\mathcal{M}}_{g-h,(P\smallsetminus H)\cup\{r\}}) \longrightarrow A^\bullet(\overline{\mathcal{M}}_{g,P})$$

for all g and P, and for all $h \leq g$ and $H \subset P$.

We are now in a position to describe a set of additive generators for $R^\bullet(\overline{\mathcal{M}}_{g,P})$. Let Γ be a P-marked, genus g dual graph, and consider the stack $\overline{\mathcal{M}}_\Gamma$ defined by (4.29) and, for every $v \in V$, denote by F_v an arbitrary monomial in the κ classes and the ψ classes of $\overline{\mathcal{M}}_{g_v,L_v}$.

PROPOSITION (4.39). *A set of additive generators for $R^\bullet(\overline{\mathcal{M}}_{g,P})$ is given by the classes of the form $(\xi_\Gamma)_*(\prod_{v\in V} F_v)$.*

Proof. First of all, the ψ-classes belong to $R^\bullet(\overline{\mathcal{M}}_{g,P})$ because of (3.5). It then follows from (4.11) that also the κ-classes belong to $R^\bullet(\overline{\mathcal{M}}_{g,P})$. Since the maps ξ_Γ can be thought of as compositions of the codimension 1 clutching maps ξ_{irr} and $\xi_{h,H}$, also the classes $(\xi_\Gamma)_*(\prod_{v\in V} F_v)$ all belong to $R^\bullet(\overline{\mathcal{M}}_{g,P})$, by the definition of the tautological system of rings. To show that the span of the classes $(\xi_\Gamma)_*(\prod_{v\in V} F_v)$ is closed under multiplication, we proceed as follows. Look at a product $(\xi_\Gamma)_*(\prod_{v\in V} F_v)\cdot(\xi_{\Gamma'})_*(\prod_{v'\in V'} F'_{v'})$. Using formulae (4.30), Lemma (4.31), and push–pull, we are reduced to proving that $(\xi_\Gamma)_*(1)\cdot(\xi_{\Gamma'})_*(1)$ lies in the span of the classes $(\xi_\Gamma)_*(\prod_{v\in V} F_v)$. But, using push–pull again, we immediately see that this is a rational multiple of $(\xi_\Gamma)_*(\xi_\Gamma)^*(\delta_{\Gamma'})$, and we are done by formula (4.34). To prove that the classes $(\xi_\Gamma)_*(\prod_{v\in V} F_v)$ actually span $R^\bullet(\overline{\mathcal{M}}_{g,P})$, it now suffices to show that the system of rings they define is closed under the homomorphisms $(\pi_x)_*$, $(\xi_{irr})_*$, and $(\xi_{h,H})_*$. Closure under the last two is obvious. To prove closure under $(\pi_x)_*$, we first observe that for a suitable Γ', we have $(\pi_x)_*\xi_{\Gamma *}=\xi_{\Gamma' *}(\pi_x)_*$, and then we conclude using (4.11), (4.18), and the push–pull formula. Q.E.D.

The tautological ring $R^\bullet(\mathcal{M}_{g,P})$ of the moduli space of smooth curves is simply defined as the ring generated by the ψ-classes and the κ-classes. We will further discuss tautological rings in Section 6.

5. Mumford's relations for the Hodge classes.

In Chapter XIII we showed, among other things, that the Hodge class λ is a linear combination of κ_1, ψ, and δ (cf. Theorem (7.6)). This can be easily generalized to the higher λ_a, using the same methods, by means of the Grothendieck Riemann–Roch theorem. To keep things simple, we begin our computations in $A^\bullet(\mathcal{M}_{g,n})$, that is, modulo boundary classes.

Consider the projection $\pi:\mathcal{C}=\mathcal{C}_{g,n}\to\mathcal{M}_{g,n}$. By definition, the Hodge bundle is $\mathbb{E}=\pi_*\omega_\pi$, and the Hodge classes λ_i are given by $\lambda_i=c_i(\mathbb{E})$. The Grothendieck Riemann–Roch formula says in our case that

$$\mathrm{ch}(\pi_!\omega_\pi)=\pi_*(\mathrm{ch}(\omega_\pi)\cdot\mathrm{td}^\vee(\omega_\pi))\,, \tag{5.1}$$

since $\Omega^1_C=\omega_C$ for a smooth curve C. Recall that the kth Bernoulli number B_k is defined by the identity of formal power series

$$\frac{x}{e^x-1}=\sum_{k=0}^{\infty}B_k\frac{x^k}{k!}$$

and that $B_1 = -1/2$, while $B_k = 0$ for k odd and greater than 1. We set

$$\widetilde{B}_k = (-1)^k B_{2k}\,, \tag{5.2}$$

so that $\widetilde{B}_k > 0$. We have

$$\frac{x}{e^x - 1} = 1 - \frac{1}{2}x + \sum_{k=1}^{\infty}(-1)^{k-1}\widetilde{B}_k \frac{x^{2k}}{2k!}\,. \tag{5.3}$$

Furthermore, the identity

$$\widetilde{B}_k = \frac{2(2k!)}{(2\pi)^{2k}} \cdot \zeta(2k) \tag{5.4}$$

holds, where ζ is the Riemann zeta function. Finally, for a vector bundle E, recall the formula

$$\sum_{j=0}^{\infty} c_j(E)t^j = \exp\left(\sum_{s=1}^{\infty}(-1)^{s-1}(s-1)!\,\mathrm{ch}_s(E)t^s\right), \tag{5.5}$$

relating the Chern character with the Chern classes, and also recall that

$$\mathrm{ch}_s(E) = \frac{(-1)^{s-1}c_s(E)}{(s-1)!} + \text{polynomial in lower Chern classes.} \tag{5.6}$$

By definition,

$$\mathrm{ch}(\widetilde{K}) = e^{\widetilde{K}}\,, \qquad \mathrm{td}^\vee(\omega_\pi) = \frac{\widetilde{K}}{e^{\widetilde{K}} - 1}\,,$$

where we write $\widetilde{K}$ for $\widetilde{K}_\pi = c_1(\omega_\pi)$. Since $R^1\pi_*\omega_\pi$ is trivial, we have

$$\begin{aligned}
\mathrm{ch}(\pi_*\omega_\pi) &= \mathrm{ch}(R^1\pi_*\omega_\pi) + \pi_*\left(\frac{\widetilde{K}}{1 - e^{-\widetilde{K}}}\right) = 1 + \pi_*\left(\frac{-\widetilde{K}}{e^{-\widetilde{K}} - 1}\right) \\
&= 1 + \pi_*\left(1 + \frac{1}{2}\widetilde{K} + \sum_{i=1}^{\infty}\frac{(-1)^{i-1}\widetilde{B}_i}{(2i)!}\widetilde{K}^{2i}\right) \\
&= 1 + \frac{2g-2}{2} + \pi_*\left(\sum_{i=1}^{\infty}\frac{(-1)^{i-1}\widetilde{B}_i}{(2i)!}\widetilde{K}^{2i}\right) \\
&= g + \sum_{i=1}^{\infty}\frac{(-1)^{i-1}\widetilde{B}_i}{(2i)!}\widetilde{\kappa}_{2i-1}\,.
\end{aligned} \tag{5.7}$$

From (5.5) we get

$$\sum_{j=1}^{\infty}\lambda_j t^j = \exp\left(\sum_{i=1}^{\infty}\frac{(-1)^{i-1}\widetilde{B}_i}{2i(2i-1)}\cdot\widetilde{\kappa}_{2i-1}t^{2i-1}\right). \tag{5.8}$$

The classes λ_j can thus be expressed in terms of the classes $\widetilde{\kappa}_{2i-1}$ and hence, by (3.4), in terms of the classes ψ_p and κ_i, with i odd. As a consequence, they all belong to the tautological ring $R^\bullet(\mathcal{M}_{g,P})$. For example,

$$\lambda_1 = \frac{1}{12}\widetilde{\kappa}_1\,, \quad \lambda_2 = \frac{1}{2}\lambda_1^2 = \frac{1}{288}\widetilde{\kappa}_1^2\,, \quad \lambda_3 = \frac{1}{6}\left(\frac{\widetilde{\kappa}_1}{13}\right)^3 - \frac{1}{360}\widetilde{\kappa}_3\,.$$

The first formula is, of course, Mumford's formula (Theorem (7.6) in Chapter XIII), modulo the boundary term.

We are now going to carry out the same computation for $\pi : \overline{\mathcal{C}}_{g,P} \to \overline{\mathcal{M}}_{g,P}$. From formula (5.7) we see that dealing with pointed curves is actually uninfluential. Thus, we will first perform our computation in $\overline{\mathcal{M}}_g$ and will then make the appropriate changes to get a formula in $\overline{\mathcal{M}}_{g,P}$. For simplicity, we set $\overline{\mathcal{C}} = \overline{\mathcal{C}}_g$. When dealing with stable curves, formula (5.1) must be replaced by

$$\text{(5.9)} \qquad \operatorname{ch}(\pi_!\omega_\pi) = \pi_*(\operatorname{ch}(\omega_\pi)\cdot \operatorname{td}^\vee(\Omega^1_\pi))\,.$$

Since $R^1\pi_*\omega_\pi$ is trivial, we have

$$\text{(5.10)} \qquad \operatorname{ch}(\pi_*\omega_\pi) = 1 + \pi_*(e^{\widetilde{K}}\cdot \operatorname{td}^\vee(\Omega^1_\pi))\,.$$

The sheaf homomorphism $\rho : \Omega^1_\pi \to \omega_\pi$ is injective, and hence $\mathcal{I} = \Omega^1_\pi \otimes \omega_\pi^{-1}$ is a sheaf of ideals in $\mathcal{O}_{\overline{\mathcal{C}}}$. Moreover, ρ is not an isomorphism precisely at the nodes of the fibers of $\overline{\mathcal{C}} \to \overline{\mathcal{M}}_g$. We can in fact be more precise. As we know, analytically, the "local picture" of $\overline{\mathcal{C}} \to \overline{\mathcal{M}}_g$ is just

$$(x, y, t_2, t_3, \dots, t_{3g-3}) \longmapsto (t_1 = xy, t_2, t_3, \dots, t_{3g-3})\,,$$

and ρ is locally described, in terms of the local generators dx and dy for Ω^1_π and dx/x for ω_π, by

$$\rho(dx) = x\frac{dx}{x}\,, \qquad \rho(dy) = -y\frac{dx}{x}\,.$$

Thus, locally, $\mathcal{I} = (x, y)$. This implies that $\mathcal{I}$ is the ideal sheaf of a *smooth* codimension two closed substack of $\overline{\mathcal{C}}$, which we denote by Σ. The homomorphism ρ fits in an exact sequence

$$0 \to \Omega^1_\pi \xrightarrow{\rho} \omega_\pi \to \omega_\pi \otimes \mathcal{O}_\Sigma \to 0$$

which is the exact analogue, in the present context, of (7.4) in Chapter XIII. From this we deduce that

$$\text{(5.11)} \qquad \operatorname{td}^\vee(\Omega^1_\pi) = \operatorname{td}^\vee(\omega_\pi)\operatorname{td}^\vee(\mathcal{O}_\Sigma)^{-1}\,.$$

Before studying the inclusion $\Sigma \subset \overline{\mathcal{C}}$ and the Todd class of $\mathcal{O}_\Sigma$ in the world of stacks, let us look at what happens with schemes. Suppose that we have an inclusion $j : Z \hookrightarrow X$ of a smooth codimension two subvariety in a smooth variety. Then

$$\mathrm{td}^\vee(\mathcal{O}_Z)^{-1} = 1 + Q(\mathrm{ch}_1(\mathcal{O}_Z), \mathrm{ch}_2(\mathcal{O}_Z), \dots),$$

where Q is a polynomial. Using the Grothendieck Riemann–Roch theorem for j, one sees that $\mathrm{ch}_1(\mathcal{O}_Z)$ is the pushforward via j_* of a polynomial P in the first and second Chern classes of the normal bundle $N_{Z/X}$. To find this polynomial, by the splitting principle, it suffices to consider the case in which $Z = D_1 \cdot D_2$. From the exact sequence

$$0 \to \mathcal{O}_X(-D_1 - D_2) \to \mathcal{O}_X(-D_1) \oplus \mathcal{O}_X(-D_2) \to \mathcal{O}_X \to \mathcal{O}_Z \to 0$$

we get

$$\begin{aligned}\mathrm{td}^\vee(\mathcal{O}_Z) &= \mathrm{td}^\vee(\mathcal{O}_X(-D_1))^{-1} \cdot \mathrm{td}^\vee(\mathcal{O}_X(-D_2))^{-1} \cdot \mathrm{td}^\vee(\mathcal{O}_Z(-D_1 - D_2)) \\ &= \left(\frac{-D_1}{e^{-D_1} - 1}\right)^{-1} \cdot \left(\frac{-D_2}{e^{-D_2} - 1}\right)^{-1} \cdot \left(\frac{-D_1 - D_2}{e^{-D_1 - D_2} - 1}\right).\end{aligned}$$

Thus,

$$\begin{aligned}\mathrm{td}^\vee(\mathcal{O}_Z)^{-1} - 1 &= \left(\frac{-D_1}{e^{-D_1} - 1}\right) \cdot \left(\frac{-D_2}{e^{-D_2} - 1}\right) \cdot \left(\frac{e^{-D_1 - D_2} - 1}{-D_1 - D_2}\right) \\ &= \frac{1}{D_1 + D_2}\left(D_1\left(\frac{D_2}{1 - e^{-D_2}} - 1\right) + D_2\left(\frac{D_1}{1 - e^{-D_1}} - 1\right)\right) \\ &= \frac{D_1 \cdot D_2}{D_1 + D_2} \sum_{i=1}^{\infty} \frac{(-1)^{i-1}\widetilde{B}_i}{(2i)!}(D_1^{2i-1} + D_2^{2i-1}) \\ &= j_*\left(\sum_{i=1}^{\infty} \frac{(-1)^{i-1}\widetilde{B}_i}{(2i)!}\left(\frac{D_1^{2i-1} + D_2^{2i-1}}{D_1 + D_2}\right)\right) \\ &= j_*\left(\sum_{i=1}^{\infty} \frac{(-1)^{i-1}\widetilde{B}_i}{(2i)!}\left(D_1^{2i-2} - D_1^{2i-3}D_2 + \dots + D_2^{2i-2}\right)\right) \\ &= j_* P(D_1 + D_2, D_1 \cdot D_2) \\ &= j_* P(c_1(N_{Z/X}), c_2(N_{Z/X})).\end{aligned}$$

We now return to $\Sigma \subset \overline{\mathcal{C}}$. We shall show that the components of Σ can be precisely described in terms of clutching morphisms. For this, we view $\overline{\mathcal{C}}$ as $\overline{\mathcal{M}}_{g,\{p\}}$. Consider the following diagrams

$$
\begin{array}{ccc}
\overline{\mathcal{M}}_{g-1,\{x,y\}} \times \overline{\mathcal{M}}_{0,\{a,b,p\}} & \longrightarrow & \overline{\mathcal{M}}_{g,\{p\}} = \overline{\mathcal{C}} \\
\cong\big\downarrow & \overset{\overline{\xi}_0}{\nearrow} & \big\downarrow \pi \\
\overline{\mathcal{M}}_{g-1,\{x,y\}} & \xrightarrow{\ \xi_{irr}\ } & \overline{\mathcal{M}}_g
\end{array}
$$

$$
\begin{array}{ccc}
\overline{\mathcal{M}}_{h,\{x\}} \times \overline{\mathcal{M}}_{g-h,\{y\}} \times \overline{\mathcal{M}}_{0,\{a,b,p\}} & \longrightarrow & \overline{\mathcal{M}}_{g,\{p\}} = \overline{\mathcal{C}} \\
\cong\big\downarrow & \overset{\overline{\xi}_h}{\nearrow} & \big\downarrow \pi \\
\overline{\mathcal{M}}_{h,\{x\}} \times \overline{\mathcal{M}}_{g-h,\{y\}} & \xrightarrow{\ \xi_h\ } & \overline{\mathcal{M}}_g
\end{array}
\qquad 1 \le h \le g-1\,,
$$

in which the top arrows are the clutching morphisms that identify x with a and y with b. The stack Σ is the disjoint union of components, each of which is the image of one of the $\overline{\xi}_h$. Moreover, the $\overline{\xi}_h$ are isomorphisms onto the corresponding component of Σ, with two exceptions, $\overline{\xi}_0$ and $\overline{\xi}_{g/2}$ (the latter, of course, only for even g), which are both unramified double covers of stacks. In fact, we know from Theorem (10.11) in Chapter XII that a clutching map along a graph Γ is the composition of the quotient by $\mathrm{Aut}(\Gamma)$ and a normalization map. In all the cases at hand, the normalization map is the identity since Σ is smooth. As for $\mathrm{Aut}(\Gamma)$, it is trivial except for $h = 0$ or $h = g/2$, when it has order two. Finally, for each $h \neq 0, g/2$, the image of $\overline{\xi}_h$ is obviously equal to the one of $\overline{\xi}_{g-h}$ and to the image of no other $\overline{\xi}_j$. All in all, we can conclude that, setting

$$
X_0 = \overline{\mathcal{M}}_{g-1,\{x,y\}}\,, \qquad X_h = \overline{\mathcal{M}}_{h,\{x\}} \times \overline{\mathcal{M}}_{g-h,\{y\}}\,, \qquad X = \coprod_{h=0}^{g-1} X_h\,,
$$

the obvious morphism

$$
\overline{\xi} : X \to \Sigma \subset \overline{\mathcal{C}}
$$

is an unramified degree two covering of stacks.

Now we must compute the normal bundle $N_{\Sigma/\overline{\mathcal{C}}}$. By what we have just observed,

$$
\overline{\xi}^* N_{\Sigma/\overline{\mathcal{C}}} = N_{\overline{\xi}}\,,
$$

where $N_{\overline{\xi}}$ is the normal bundle to the morphism $\overline{\xi}$. On the other hand, we know how to compute normal bundles of clutching maps and hence of $\overline{\xi}$. For $h > 0$, we let η_x be the projection of $X_h = \overline{\mathcal{M}}_{h,\{x\}} \times \overline{\mathcal{M}}_{g-h,\{y\}}$ to the first factor, and η_y the projection to the second factor. For $h = 0$, instead, we let $\eta_x = \eta_y$ be the identity. Then formula (3.2) in Chapter XIII immediately implies that

$$
N_{\overline{\xi}_h} = \eta_x^* \mathcal{L}_x \oplus \eta_y^* \mathcal{L}_y\,.
$$

Set

$$\widetilde{\psi}_x = c_1(\eta_x^* \mathcal{L}_x)\,, \qquad \widetilde{\psi}_y = c_1(\eta_y^* \mathcal{L}_y)\,.$$

Since

$$\begin{aligned} 2P(c_1(N_{\Sigma/\overline{C}}), c_2(N_{\Sigma/\overline{C}})) &= \overline{\xi}_* \overline{\xi}^* P(c_1(N_{\Sigma/\overline{C}}), c_2(N_{\Sigma/\overline{C}})) \\ &= \overline{\xi}_* P(c_1(N_{\overline{\xi}}), c_2(N_{\overline{\xi}}))\,, \end{aligned}$$

we conclude that
(5.12)

$$\mathrm{td}^\vee(\mathcal{O}_\Sigma)^{-1} = 1 + \frac{1}{2}\overline{\xi}_* \left(\sum_{i=1}^{\infty} \frac{(-1)^{i-1}\widetilde{B}_i}{(2i)!} \left(\widetilde{\psi}_x^{2i-2} - \widetilde{\psi}_x^{2i-3}\widetilde{\psi}_y + \cdots + \widetilde{\psi}_y^{2i-2} \right) \right).$$

We next observe that $\overline{\xi}_h^* \omega_\pi$ is trivial. In proving this, we write ξ_0 for ξ_{irr}. Consider the diagram

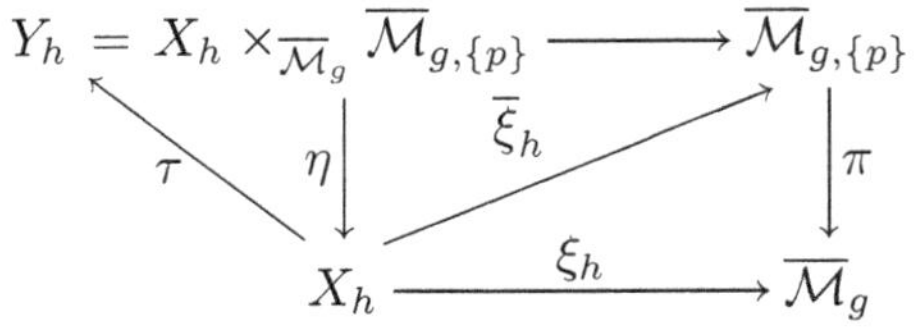

where the section τ is the pullback of $\overline{\xi}_h$ and consists entirely of singular points. We have to show that $\tau^*\omega_\eta = \overline{\xi}_h^*\omega_\pi$ is trivial. Let $\widehat{Y}_h$ be the partial normalization of Y_h along τ. The section τ "splits" into two sections $\tau_1, \tau_2 : X_h \to \widehat{Y}_h$. Taking residues along either one of these trivializes $\tau^*\omega_\eta$.

Let us now go back to (5.10) and (5.11). We have

$$\begin{aligned} \mathrm{ch}(\pi_*\omega_\pi) &= 1 + \pi_*(e^{\widetilde{K}} \cdot \mathrm{td}^\vee(\Omega_\pi^1)) \\ &= 1 + \pi_*(e^{\widetilde{K}} \cdot \mathrm{td}^\vee(\omega_\pi) \cdot \mathrm{td}^\vee(\mathcal{O}_\Sigma)^{-1}) \\ &= 1 + \pi_* \left(e^{\widetilde{K}} \cdot \frac{\widetilde{K}}{e^{\widetilde{K}} - 1} + \left(\mathrm{td}^\vee(\mathcal{O}_\Sigma)^{-1} - 1 \right) \right). \end{aligned}$$

The last equality comes from (5.12) and the fact that $\overline{\xi}_h^*\omega_\pi$ is trivial, so that $\widetilde{K} \cdot \overline{\xi}_{h*}(\alpha) = 0$ for every α. Putting everything together, we finally get
(5.13)

$$\begin{aligned} \mathrm{ch}(\pi_*\omega_\pi) =& g + \sum_{i=1}^{\infty} \frac{(-1)^{i-1}\widetilde{B}_i}{(2i)!} \left(\widetilde{\kappa}_{2i-1} + \frac{1}{2}\xi_* \left(\widetilde{\psi}_x^{2i-2} - \widetilde{\psi}_x^{2i-3}\widetilde{\psi}_y + \cdots + \widetilde{\psi}_y^{2i-2} \right) \right) \\ =& g + \sum_{i=1}^{\infty} \frac{B_{2i}}{(2i)!} \left(\widetilde{\kappa}_{2i-1} + \frac{1}{2}\xi_* \left(\widetilde{\psi}_x^{2i-2} - \widetilde{\psi}_x^{2i-3}\widetilde{\psi}_y + \cdots + \widetilde{\psi}_y^{2i-2} \right) \right), \end{aligned}$$

where $\xi = \pi \circ \overline{\xi}$. In $\overline{\mathcal{M}}_{g,P}$ the formula is exactly the same, with the definitions of X, $\overline{\xi}$, and ξ being obvious generalizations of those given in the unpointed case.

We end this section by proving yet another result of Mumford, namely the following:

PROPOSITION (5.14). *The tautological ring $R^\bullet(\mathcal{M}_{g,P})$ is generated by the classes $\kappa_1, \ldots, \kappa_{g-2}, \psi_1, \ldots, \psi_n$, where $n = |P|$.*

Again, as the classes ψ_i play no role, we will restrict our attention to $\mathcal{M}_g$. On the universal curve $\pi : \mathcal{C} \to \mathcal{M}_g$ we may consider the evaluation map $\pi^*\mathbb{E} \to \omega_\pi$, where $\mathbb{E}$ is the Hodge bundle. This is a surjective map of locally free sheaves, hence its kernel is a locally free sheaf of rank $g-1$. We then get

$$c_j(\pi^*\mathbb{E} - \omega_\pi) = 0\,, \qquad j \geq g\,.$$

On the other hand,

$$c(\pi^*\mathbb{E} - \omega_\pi) = \pi^*(1 + \lambda_1 + \cdots + \lambda_g)(1 + K_\pi)^{-1}\,.$$

Thus, for $j \geq g$, we get

$$K_\pi^j - \pi^*(\lambda_1)K_\pi^{j-2} + \cdots + (-1)^g \pi^*(\lambda_g)K_\pi^{j-g}.$$

Pushing down this relation to $\mathcal{M}_g$, we obtain

$$\begin{aligned} \kappa_{j-1} - \lambda_1\kappa_{j-2} + \cdots + (-1)^g\lambda_g\kappa_{j-g-1} &= 0 \qquad \text{if} \quad j \geq g+2\,,\\ \kappa_g - \lambda_1\kappa_{g-1} + \cdots + (-1)^g\lambda_g(2g-2) &= 0 \qquad \text{if} \quad j = g+1\,,\\ \kappa_{g-1} - \lambda_1\kappa_{g-2} + \cdots + (-1)^{g-1}\lambda_{g-1}(2g-2) &= 0 \qquad \text{if} \quad j = g\,. \end{aligned} \tag{5.15}$$

Formula (5.8) tells us that $\lambda_1, \ldots, \lambda_g$ are polynomials in the odd κ_i for $i \leq g$. The first of the above relations tells us, inductively, that κ_j is a polynomial in $\kappa_1, \kappa_2, \ldots, \kappa_g$ when $j \geq g+1$. It remains to show that κ_g and κ_{g-1} are polynomials in the lower κ_i. Before doing this, we prove the following result.

PROPOSITION (5.16). *Let $\pi : \mathcal{C} \to \mathcal{M}_g$ be the universal curve and consider the Hodge bundle $\mathbb{E} = \pi_*\omega_\pi$ and its dual $\mathbb{E}^\vee$. Then, $c(\mathbb{E})c(\mathbb{E}^\vee) = 1$, or equivalently $\mathrm{ch}(\mathbb{E}) + \mathrm{ch}(\mathbb{E}^\vee) = 0$, or, again equivalently, $\mathrm{ch}_{2k}(\mathbb{E}) = 0$ for $k \geq 1$.*

Proof. Consider the relative de Rham complex

$$\omega_\pi^\bullet : \quad 0 \to \mathcal{O}_{\mathcal{C}} \xrightarrow{d} \omega_\pi \to 0\,,$$

where d is the differential along the fibers of π. Its hypercohomology sequence is

$$\cdots \to \pi_*\mathcal{O}_\mathcal{C} \xrightarrow{d} \pi_*\omega_\pi \to \mathbb{R}^1\pi_*\omega_\pi^\bullet \to R^1\pi_*\mathcal{O}_\mathcal{C} \xrightarrow{\alpha} R^1\pi_*\omega_\pi \to \cdots, \tag{5.17}$$

where we are using the identifications $\mathbb{R}^q\pi_*(0 \to \omega_\pi) = R^{q-1}\pi_*\omega_\pi$. The homomorphism d in (5.17) is clearly zero, and α is also zero since its Serre dual $\pi_*\mathcal{O}_\mathcal{C} \to \pi_*\omega_\pi$ is just d. The sequence (5.17) thus reduces to

$$0 \to \pi_*\omega_\pi \to \mathbb{R}^1\pi_*\omega_\pi^\bullet \to R^1\pi_*\mathcal{O}_\mathcal{C} \to 0,$$

which, by the Serre duality, can be rewritten as

$$0 \to \mathbb{E} \to \mathbb{R}^1\pi_*\omega_\pi^\bullet \to \mathbb{E}^\vee \to 0. \tag{5.18}$$

On the other hand, the vector bundle in the middle of this sequence is isomorphic to $R^1\pi_*\mathbb{C} \otimes \mathcal{O}_{\mathcal{M}_g}$. As $R^1\pi_*\mathbb{C}$ is a local system, $c(\mathbb{R}^1\pi_*\omega_\pi^\bullet) = c(\mathbb{E})c(\mathbb{E}^\vee) = 1$.

Q.E.D.

End of the proof of Proposition (5.14). We must prove that κ_g and κ_{g-1} are polynomials in the lower κ_i. From Proposition (5.16) we know that $\mathrm{ch}_{2k}(\mathbb{E}) = 0$, so that, by (5.6), the even λ_i can be expressed as polynomials in the lower odd λ_i. Using this fact and the second and third relations in (5.15), we must prove our assertion for κ_g only when g is odd and for κ_{g-1} only when g is even. From (5.6) and (5.7) we get

$$\frac{(-1)^{i-1}\widetilde{B}_i}{(2i)!}\kappa_{2i-1} = \mathrm{ch}_{2i-1}(\mathbb{E}) = \frac{\lambda_{2i-1}}{(2i-2)!} + \text{polynomial in lower } \lambda_j.$$

Using this and the second relation in (5.15), for $g = 2i$, we get the following system:

$$\begin{cases} (-1)^{i-1}\widetilde{B}_i\kappa_{g-1} = g(g-1)\lambda_{g-1} + \text{ lower terms,} \\ \kappa_{g-1} = 2(g-1)\lambda_{g-1} + \text{ lower terms.} \end{cases}$$

For $g = 2i - 1$, using the third relations in (5.15), we get instead:

$$\begin{cases} (-1)^{i-1}\widetilde{B}_i\kappa_g = g(g+1)\lambda_g + \text{ lower terms,} \\ \kappa_g = (2g-2)\lambda_g + \text{ lower terms.} \end{cases}$$

In order to express κ_{g-1} and κ_g in terms of lower κ_i, it therefore suffices to show that

$$\widetilde{B}_i \neq (-1)^{i-1}i, \qquad \widetilde{B}_i \neq \frac{(-1)^{i-1}i(2i-1)}{2(i-1)}.$$

This is done by inspection if $i \leq 10$. In the other cases one uses (5.4) to get

$$\widetilde{B}_i > 2\left(\frac{i}{\pi e}\right)^{2i} > 2i \quad \text{if } i \geq 11.$$

Q.E.D.

REMARK (5.19). As Mumford observes in Section 5 of [556], although the Gauss–Manin connection does not extend regularly to $\mathbb{R}^1\overline{\pi}_*\omega^\bullet_{\overline{\pi}}$, where $\overline{\pi} : \overline{\mathcal{C}} \to \overline{\mathcal{M}}_g$, it has regular singularities with a nilpotent polar part. This is enough to conclude that one can extend Proposition (5.16) to $\overline{\mathcal{M}}_g$. This result will be used only in the proof of Faber's nonvanishing Theorem (7.1) in Chapter XX.

Let us briefly digress on Proposition (5.16) and on Mumford's observation. We shall discuss this from two perspectives. One is to infer it directly from a more general result in the literature. The other is to deduce it as a very special case of a result, also in the literature, in Hodge theory. We have included the latter because it connects to other parts of this book and also because it may be of independent interest.

We shall do the second of these first. For this, we assume given a compact, complex manifold $\overline{S}$, and a reduced normal crossing divisor D in $\overline{S}$. Letting $S = \overline{S} \smallsetminus D$ be the complement of D, we assume given over S a polarized variation of Hodge structure of weight n:

$$(\mathcal{H}_{\mathbb{Z}}, \mathcal{F}^p, Q, \nabla)$$

(cf. Chapter 4 in [155]). Here, $\mathcal{H}_{\mathbb{Z}}$ is a local system, and

$$\nabla : \mathcal{H} \to \Omega^1 \otimes_{\mathcal{O}_S} \mathcal{H}$$

is an integrable connection on $\mathcal{H} := \mathcal{H}_{\mathbb{Z}} \otimes \mathcal{O}_S$ with $\nabla(\mathcal{H}_{\mathbb{Z}}) = 0$. The $\mathcal{F}^p$ are holomorphic sub-bundles of $\mathcal{H}$, and

$$Q : \mathcal{H}_{\mathbb{Z}} \otimes \mathcal{H}_{\mathbb{Z}} \to \mathbb{Z}$$

is a ∇-horizontal, non-degenerate form with $Q(e, f) = (-1)^n Q(f, e)$ and such that, at each point $s \in S$, the data

$$(H_s, F^p_s, Q_s)$$

give a polarized Hodge structure of weight n. Here H_s, F^p_s denote the fibers at s of $\mathcal{H}, \mathcal{F}^p$. We also have the usual transversality condition

$$\nabla : \mathcal{F}^p \to \mathcal{F}^{p-1} \otimes \Omega^1_S, \tag{5.20}$$

which is nontrivial only when $n \geqq 2$.

It is known that the local monodromies around the branches of D are quasi-unipotent, and we shall assume that they are unipotent. Then there are the canonical Deligne extensions $\mathcal{H}_e, \mathcal{F}^p_e$ of $\mathcal{H}, \mathcal{F}^p$ to $\overline{S}$ such that the connection ∇ extends to

$$\nabla_e : \mathcal{H}_e \to \mathcal{H}_e \otimes \Omega^1_{\overline{S}}(\log D),$$

i.e., the extended connection has regular singular points (cf. II, 5.3 and 5.4 in [160]). We then have the following:

THEOREM (5.21). *The Chern classes of the canonical extension satisfy*

$$c_k(\mathcal{H}_e) = 0, \qquad k \geqq 1.$$

This theorem follows from a much stronger result due to Cattani–Kaplan–Schmid (cf. 5.23 in [155]). To state this, we recall that the Hodge bundles

$$\mathcal{H}^{p,q} = \mathcal{F}^p / \mathcal{F}^{p+1}$$

have canonical Hermitian metrics constructed from the polarizing forms. We denote by

$$c_k(\Theta_{\mathcal{H}^{p,q}})$$

the associated Chern forms. Denoting by $\mathcal{H}_e^{p,q} = \mathcal{F}_e^p / \mathcal{F}_e^{p+1}$ the canonically extended Hodge bundles, the result of [155] is:

The forms $c_k(\Theta_{\mathcal{H}^{p,q}})$ define closed, (k,k) currents on $\overline{S}$, and the resulting cohomology classes $[c_k(\Theta_{\mathcal{H}^{p,q}})] \in H^{2k}(\overline{S})$ give the Chern classes $c_k(\mathcal{H}_e^{p,q})$ of the canonically extended Hodge bundles.

In particular, for the total Chern class, we have

$$c(\mathcal{H}_c) = \prod_{p=n}^{0} [c(\Theta_{\mathcal{H}^{p,q}})]. \tag{5.22}$$

Next, the curvature forms of the Hodge bundles have been computed [155], and from this it follows that *at the level of forms on S*

$$\prod_{p=n}^{0} c(\Theta_{\mathcal{H}^{p,q}}) = 1. \tag{5.23}$$

There is a subtlety here in that (5.23) does *not* hold for the Chern forms for the universal Hodge bundles over the period domain (loc. cit.). It is only by imposing the differential constraint (5.20) that we obtain (5.23) at the form level.

To conclude Theorem (5.21), we need to know that the cup-product in cohomology on the RHS may be computed by first multiplying on S the forms $c(\Theta_{\mathcal{H}^{p,q}})$, and secondly so multiplying these forms which are singular along D results in a well-defined, closed current whose cohomology class is given by (5.22). We remark that this is a very special circumstance, in that one cannot usually multiply distributions. That it is possible in the present situation follows from the calculations of the curvature in [155] (cf. the proof of (5.22)). The case needed for (5.16) may also be worked out using the curvature calculations in Section 9 of Chapter XI.

To apply the general result (5.21) to the situation at hand, for the universal curve, we have

$$\begin{array}{ccc} \mathcal{C}_g & \subset & \overline{\mathcal{C}}_g \\ \downarrow & & \downarrow \\ \mathcal{M}_g & \subset & \overline{\mathcal{M}}_g \end{array}$$

and the essential points are:

(i) the local monodromies around the branches of $\overline{\mathcal{M}}_g \setminus \mathcal{M}_g$ are unipotent;

(ii) we have

$$\pi_*(\omega_{\overline{\mathcal{C}}_g/\overline{\mathcal{M}}_g}) = \mathcal{H}_e^{1,0}. \tag{5.24}$$

The first of these is a consequence of the Picard–Lefschetz formula (cf. Section 9 of Chapter X). The second is the theorem on regular singular points (Théorème 7.9 in [160]). The point is that there are a priori two extensions of the Hodge bundle $\mathcal{H}^{1,0} = \pi_*(\omega_{\mathcal{C}_g/\mathcal{M}_g})$ to a bundle on $\overline{\mathcal{M}}_g$, and the theorem that the Gauss–Manin connection has regular singular points is exactly that these two extensions coincide. This concludes our discussion of the Hodge-theoretic approach to Proposition (5.16).

The other approach deals with the situation where $S, \overline{S}$, and D are as above, and we are given a holomorphic vector bundle $\mathcal{H}_e \to \overline{S}$ together with an integrable connection

$$\nabla_e \mathcal{H}_e \to \mathcal{H}_e \otimes \Omega^1_{\overline{S}}(\log D)$$

having regular singular points along D. The situation need not arise from a variation of Hodge structure. Write $D = \cup_{i=1}^r D_i$, where the D_i are irreducible, smooth, and meet transversely. We do not assume that the local monodromies $T_i = \exp \Gamma_i$ are unipotent or, equivalently, that Γ_i is nilpotent. If we denote by

$$\Gamma_i = \mathrm{Res}_{D_i}(\nabla_e)$$

the residue of ∇_e along D_i, then

$$\Gamma_i \in \mathrm{Hom}_{\mathcal{O}_{\overline{S}}}(\mathcal{H}_e, \mathcal{H}_e\big|_{D_i}).$$

In case the Γ_i are nilpotent, $(\mathcal{H}_e, \nabla_e)$ is Deligne's canonical extension to $\overline{S}$ of $(\mathcal{H}, \nabla)$ on S. In general, there is a technical assumption on the eigenvalues of the T_i that will be satisfied when the T_i are not unipotent. Denoting by $[D_i] \in H^2(\overline{S})$ the fundamental class of D_i and by $N_p(\mathcal{H}_e)$ the pth Newton class of $\mathcal{H}_e$ (see below), the result of Appendix B in [218] is the following:

THEOREM (5.25). *With the above notation,*

$$N_p(\mathcal{H}_e) = (-1)^p \sum_{\alpha_1+\cdots+\alpha_r=p} \binom{p}{\alpha} \mathrm{Tr}(\Gamma_1^{\alpha 1} \circ \cdots \circ \Gamma_r^{\alpha_r})[D_1]^{\alpha_1} \cdots [D_r].$$

Here, we recall that the $N_p(\mathcal{H}_e)$ determine the $c_k(\mathcal{H}_e)$ and vice versa. As before, (5.21) is a trivial consequence of the general result (5.25), since in that case the Γ_i are commuting nilpotent transformations.

As the proof (5.25) given in [218] is short and elegant, we shall summarize it here. The first step is to recall the definition of the *Atiyah class* $\alpha(\mathcal{F}) \in H^1(\Omega^1_X(\mathrm{Hom}(\mathcal{F},\mathcal{F}))$ associated to a vector bundle $\mathcal{F} \to X$ over a complex manifold X (cf. [39]). If f_{ij} are the transition functions relative to trivializations of $\mathcal{F}$ over open sets $U_i \subset X$ giving a covering of X, the $\mathrm{Hom}(\mathcal{F},\mathcal{F})$-valued 1-forms $f_{ij}^{-1} \cdot df_{ij}$ give a Čech representative of $\alpha(\mathcal{F})$. A Dolbeault representative is given by choosing a C^∞ connection of type $(1,0)$ with connection matrix θ_i, relative to the given trivialization of $\mathcal{F}$, in U_i. Then the $(1,1)$ part Θ of the curvature, given in U_i by $\Theta_i = \overline{\partial}\theta_i$, represents $\alpha(\mathcal{F})$ in Dolbeault cohomology.

If X is a compact Kähler manifold and if $\mathcal{F}$ admits a flat, or even a holomorphic, connection, then $\alpha(\mathcal{F}) = 0$. For X compact Kähler, the Newton classes $N_p(\mathcal{F})$ are represented by formally taking the sum over the pth powers of the roots of the characteristic polynomial

$$\det(\lambda I - \alpha(\mathcal{F})) \in \sum_p H^p(\Omega^p_X)\lambda^p$$

of $\alpha(\mathcal{F})$.

In the case at hand, we take $X = \overline{S}$ and $\mathcal{F} = \mathcal{H}_e$. Since ∇_e has regular singular points, there is an exact sequence

(5.26)

$$0 \to \Omega^1_{\overline{S}}(\mathrm{Hom}(\mathcal{H}_e,\mathcal{H}_e)) \to \Omega^1_{\overline{S}}(\mathrm{Hom}(\mathcal{H}_e,\mathcal{H}_e)(\log D)) \overset{\mathrm{Res}}{\to} \mathrm{Hom}_{\mathcal{O}_D}(\mathcal{H}_e,\mathcal{H}_e) \to 0.$$

The basic observation (cf. B.2 in [218]) is:

In the exact cohomology sequence of (5.26), $\mathrm{Res}_D(\nabla_e) \in H^0(\mathrm{Hom}_{\mathcal{O}_D}(\mathcal{H}_e,\mathcal{H}_e))$ has coboundary $\delta(\mathrm{Res}_D(\nabla_e)) = -\alpha(\mathcal{H}_e)$.

We may intuitively think of

$$\mathrm{Res}_D(\nabla_e) = \sum_i \Gamma_i \delta_{D_i},$$

and when this is worked out precisely, which is essentially a local issue in a punctured polycylinder, (5.25) is a consequence.

To conclude Mumford's observation, one must proceed as above and apply the theorem on regular singular points of the Gauss–Manin connection to conclude that $\pi_*(\omega_{\overline{\mathcal{C}}_g/\overline{\mathcal{M}}_g})$ coincides with Deligne's canonical extension of $\pi_*(\omega_{\mathcal{C}_g/\mathcal{M}_g})$.

6. Further considerations on cycles on moduli spaces.

The present section is largely expository in nature. We illustrate a number of properties of the tautological classes, some of which are still conjectural at the time of this writing, and some whose proofs involve concepts which go well beyond the scope of this volume. In the bibliographical notes we will give some indication of the progress that has been made in this general area up to the time of publication of this book.

The first set of properties and conjectures were brought to light by Faber in [231]. To state them, let us recall that a finite-dimensional graded $\mathbb{Q}$-algebra

$$R^\bullet = \bigoplus_{\nu \geq 0} R^\nu$$

is a *Gorenstein graded algebra with socle in degree* D if there exists an isomorphism

$$\varphi : R^D \to \mathbb{Q}$$

for which the bilinear pairing induced, via φ, by the multiplication

$$R^s \times R^{D-s} \to R^D \xrightarrow{\sim} \mathbb{Q}$$

is nondegenerate. For example, the cohomology ring of a compact D-(real)dimensional manifold is Gorenstein with socle in degree D, and the same is true for the Chow ring of a compact, D-(complex)dimensional algebraic manifold. The Gorenstein conjecture for $\mathcal{M}_{g,n}$ consists in the following three statements.

- The vanishing statement: $R^i(\mathcal{M}_{g,n}) = 0$ for $i > g - 2 + n$.
- The socle statement: $R^{g-2+n}(\mathcal{M}_{g,n}) \cong \mathbb{Q}$.
- The perfect duality statement: there is a perfect pairing

$$R^s(\mathcal{M}_{g,n}) \times R^{g-2+n-s}(\mathcal{M}_{g,n}) \to R^{g-2+n}(\mathcal{M}_{g,n}) .$$

Similarly, regarding $\overline{\mathcal{M}}_{g,n}$, we have the following three statements, the first one of which is obvious.

- The vanishing statement: $R^i(\overline{\mathcal{M}}_{g,n}) = 0$ for $i > 3g - 3 + n$.
- The socle statement: $R^{3g-3+n}(\overline{\mathcal{M}}_{g,n}) \cong \mathbb{Q}$.
- The perfect duality statement: there is a perfect pairing

$$R^s(\overline{\mathcal{M}}_{g,n}) \times R^{3g-3+n-s}(\overline{\mathcal{M}}_{g,n}) \to R^{3g-3+n}(\overline{\mathcal{M}}_{g,n}) .$$

In the bibliographical notes we will give the references for the parts of these conjectures that are known to hold at the time of writing this book. These are the vanishing and socle statements for $\mathcal{M}_{g,n}$ [489], [231], [237] and the socle statement for $\overline{\mathcal{M}}_{g,n}$ [308], [310], [239]. The perfect duality statements have been checked only in low genus. To a greater or lesser extent, many of these proofs involve the theory of the moduli space of stable maps, which we do not cover in this book.

Let us look at the Gorenstein conjecture for $\mathcal{M}_g$. It says that $R^\bullet(\mathcal{M}_g)$ behaves like the (p,p)-part of the cohomology ring of a $(g-2)$-dimensional complex projective manifold. Thus, the entire structure of the ring is determined by the top intersections of the κ-classes, i.e., by the expressions of the cycles $\prod \kappa_{a_i}$, with $\sum a_i = g-2$, in terms of some fixed generator of $R^{g-2}(\mathcal{M}_g)$. These intersection cycles, in turn, are completely determined by the intersection cycles $\pi_*(\psi_1^{d_1+1}\cdots\psi_n^{d_n+1})$ appearing in formula (4.25), where $\pi : \overline{\mathcal{M}}_{g,n} \to \overline{\mathcal{M}}_g$ is the projection (cf. Exercise C-4). In [489], Looijenga shows that the dimension of $R^{g-2}(\mathcal{M}_g)$ is at most 1. In Chapter XX we will prove the following fundamental result, originally due to Faber [231,237], which, coupled with Looijenga's result, exhibits a generator of $R^{g-2}(\mathcal{M}_g)$.

THEOREM (6.1). $\kappa_{g-2} \neq 0$.

Moreover, the following remarkable formula is obtained in [305] for $n=3$ and in [481] for the general case:

(6.2)
$$\pi_*(\psi_1^{d_1+1}\cdots\psi_n^{d_n+1}) = \frac{(3g-3+n)!(2g-1)!!}{(2g-1)!\prod_{j=1}^{n}(2d_j+1)!!}\,\kappa_{g-2} \qquad \text{if } \sum d_j = g-2\,.$$

It follows from the preceding considerations and from Theorem (6.1) that, if the perfect duality statement for $\mathcal{M}_g$ holds, then formula (6.2) completely determines the multiplicative structure of $R^\bullet(\mathcal{M}_g)$.

In Section 4 of Chapter XXI we will prove the following vanishing statement due to Looijenga [489].

THEOREM (6.3). $R^i(\mathcal{M}_g) = 0$ *for* $i > g-2$.

A corollary of this theorem is a remarkable result previously proved by Diaz [169].

THEOREM (6.4). *There is no complete subvariety of* M_g *of dimension greater that* $g-2$.

Proof. Let $X \subset M_g$ be a d-dimensional complete subvariety. Since κ_1 is ample on $\overline{M}_g$, the intersection number $(\kappa_1^d \cdot 1_X)$ is positive. But then, by Theorem (6.3), we must have $d \leq g-2$.

Q.E.D.

There are more refined versions of the Gorenstein conjectures concerning the moduli spaces of stable pointed curves of compact type $\mathcal{M}_{g,n}^{ct}$ and the moduli spaces of stable pointed curves with rational tails $\mathcal{M}_{g,n}^{rt}$. For these, we refer the interested reader to [233].

Another conjecture by Faber is that the ample class κ_1 may satisfy a *Hard Lefschetz* property in $R^\bullet(\mathcal{M}_g)$ in the sense that multiplication by $(\kappa_1)^{g-2-2i}$ should map $R^i(\mathcal{M}_g)$ isomorphically onto $R^{g-2-i}(\mathcal{M}_g)$ for $0 \leq i \leq (g-2)/2$.

7. The Chow ring of $\overline{M}_{0,P}$.

In this section we illustrate a result of Keel describing the Chow ring of the moduli space of pointed curves of genus 0 (cf. [408]). For brevity, when writing the boundary divisors of $\overline{M}_{0,P}$, we will drop the reference to the genus ($g = 0$), and we will write Δ_S instead of $\Delta_{0,S}$. Similarly, for the divisor classes we will write δ_S instead of $\delta_{0,S}$. In what follows it will be convenient to assume that $P = \{1, 2, \ldots, n\}$. Keel's theorem is the following.

THEOREM (7.1). *The Chow ring $A^*(\overline{M}_{0,P})$ is generated by the classes δ_S with $S \subset P$ and $|S| \geq 2$, $|S^c| \geq 2$. The relations among these generators are generated by the following ones:*

1) $\delta_S = \delta_{S^c}$;
2) *For any quadruple of distinct elements $i, j, k, l \in P$,*

$$\sum_{i,j\in S;\, k,l\notin S} \delta_S = \sum_{i,k\in S;\, j,l\notin S} \delta_S = \sum_{i,l\in S;\, j,k\notin S} \delta_S;$$

3) $\delta_S\delta_T = 0$, *unless* $S \subset T$, $S \supset T$, $S \subset T^c$ *or* $S \supset T^c$.

Moreover, $A^*(\overline{M}_{0,P}) = H^*(\overline{M}_{0,P}; \mathbb{Z})$.

We will not give a proof of this theorem. We will limit ourselves to a number of comments, and we will only give the complete computation of the first Chow group $A^1(\overline{M}_{0,P})$, showing that it coincides with $H^2(\overline{M}_{0,P}; \mathbb{Z})$. From now on we shall only use integer cohomology, and we will drop any mention of the coefficients in the notation.

The relations 1), 2), and 3) can be easily established. Relations 1) are obvious. Relations 3) follow immediately from the observation that Δ_S and Δ_T are disjoint except in the cases indicated. To get the relations in 2), look at the morphism

$$\pi_{i,j,k,l} : \overline{M}_{0,P} \longrightarrow \overline{M}_{0,\{i,j,k,l\}}$$

defined by forgetting all the points in P with the exception of the ones labelled by i, j, k, l and stabilizing the resulting curve. Look at the divisor classes $\delta_{\{i,j\}}$ and $\delta_{\{i,k\}}$ on $\overline{M}_{0,\{i,j,k,l\}}$. The pullbacks of $\delta_{\{i,j\}}$ and $\delta_{\{i,k\}}$ via $\pi_{i,j,k,l}$ are given, respectively, by

$$\sum_{i,j\in S;\, k,l\notin S} \delta_S \quad \text{and} \quad \sum_{i,k\in S;\, j,l\notin S} \delta_S\,.$$

The fact that $\delta_{\{i,j\}} = \delta_{\{i,k\}} \in A^1(\overline{M}_{0,\{i,j,k,l\}}) = A^1(\mathbb{P}^1)$ gives, by pullback, the first relation in 2). The second is obtained similarly.

In view of 1), from now on we will only consider subsets $S \subset P$ with the property

$$|S \cap \{1, 2, 3\}| \leq 1\,. \tag{7.2}$$

Let us now explain what is the main idea in Keel's presentation of the Chow ring $A^*(\overline{M}_{0,P})$. The moduli space $\overline{M}_{0,P}$ is a smooth variety which is birationally equivalent to the product $(\mathbb{P}^1)^{n-3}$ of $n-3$ copies of $\mathbb{P}^1$, where $n = |P|$. Indeed, the open subset $M_{0,P}$ is an affine variety consisting of $\mathbb{C}^{n-3}$ with some hyperplanes removed:

$$M_{0,P} = \mathbb{C}^{n-3} \smallsetminus (\{z_i = 0\} \cup \{ z_i = 1\} \cup \{z_i = z_j \,, i, j = 1, \dots, n-3 \,, i \neq j\}).$$

Keel expresses this birational equivalence explicitly, as a series of blow-ups, starting from $(\mathbb{P}^1)^{n-3}$. Each of these blow-ups is centered along a *smooth* codimension 2 subvariety which is carefully described. The reason for doing this is that the following general lemma on Chow rings of blow-ups holds.

LEMMA (7.3). *Let Z be a smooth, codimension d subvariety of a smooth variety X. Let $i : Z \to X$ be the inclusion, and N the normal bundle of Z in X. Assume that $i^* : A^*(X) \to A^*(Z)$ is surjective and denote by $\widetilde{X}$ the blow-up of X along Z. Then*

$$A^*(\widetilde{X}) \cong \frac{A^*(X)[T]}{(P(T), \; T \cdot Ker(i^*))} \,,$$

where $P(T) \in A^*(X)[T]$ is such that $P(0) = [Z]$ and $i^*P(T) = T^d + c_1(N)T^{d-1} + \cdots + c_d(N)$. Moreover, if $A^*(X) = H^*(X)$ and $A^*(Z) = H^*(Z)$, then $A^*(\widetilde{X}) = H^*(\widetilde{X})$.

The procedure to explicitly express $\overline{M}_{0,P}$ as a series of blow-ups, starting from $(\mathbb{P}^1)^{n-3}$, is inductive on n. The basic step in this procedure is to look at the morphism

$$\eta = (\pi, \pi_{1,2,3,n+1}) : \overline{M}_{0,P\cup\{n+1\}} \longrightarrow \overline{M}_{0,P} \times \overline{M}_{0,\{1,2,3,n+1\}} = \overline{M}_{0,P} \times \mathbb{P}^1$$

(where π forgets the point $n+1$, and $\pi_{1,2,3,n+1}$ forgets all but the points $1, 2, 3, n+1$) and describe it as a series of blow-ups along smooth codimension 2 centers. It is clear that η is a contraction whose exceptional divisor is

$$\sum_{S \subset P} \Delta_{S\cup\{n+1\}} \,,$$

by (7.2). To express this contraction as a sequence of blow-ups along smooth centers, one proceeds as follows. Consider the diagram

$$\begin{array}{ccc} \overline{M}_{0,P\cup\{n+1\}} & \xrightarrow{\;\eta = \eta^{(1)}\;} & \overline{M}_{0,P} \times \overline{M}_{0,\{1,2,3,n+1\}} = B^{(1)} \\ & \nwarrow^{\sigma_i} & \uparrow \sigma_i^{(1)} \\ & & \overline{M}_{0,P} \end{array}$$

where, for $i = 1, \ldots, n$, σ_i is the natural section, $\eta^{(1)}$ and $B^{(1)}$ are defined by the respective equalities, and $\sigma_i^{(1)} = \eta^{(1)}\sigma_i$. Thus $\sigma_i^{(1)}$ is a section of the projection $pr_1 : B^{(1)} \to \overline{M}_{0,P}$. We can embed a boundary divisor Δ_S of $\overline{M}_{0,P}$ as a smooth codimension 2 subvariety of $B^{(1)}$ via $\sigma_i^{(1)}$. If we do this starting from $i \in S$, the result is independent of i. We set

$$D_S^{(1)} = \sigma_i^{(1)}(\Delta_S) \subset B^{(1)}, \quad i \in S .$$

We have:

$$\begin{array}{ccc}
\overline{M}_{0,P\cup\{n+1\}} & \xrightarrow{\eta^{(1)}} & B^{(1)} = \overline{M}_{0,P} \times \overline{M}_{0,\{1,2,3,n+1\}} \supset D_S^{(1)} \\
& {\scriptstyle \pi} \searrow & \big\downarrow {\scriptstyle pr_1} \\
& & \overline{M}_{0,P} \supset \Delta_S
\end{array}$$

Let $\pi^{(1)} : B^{(2)} \to B^{(1)}$ be the blow-up along the union of the $D_S^{(1)}$ with $|S^c| = 2$. Since the $D_S^{(1)}$ with $|S^c| = 2$ are disjoint, the center of this blow-up is smooth. One can easily check that the map $\eta^{(1)}$ factors through $\pi^{(1)}$, so that one can define sections $\sigma_i^{(2)} : \overline{M}_{0,P} \to B^{(2)}$ of $pr_1 \circ \pi^{(1)}$. As before, one sets

$$D_S^{(2)} = \sigma_i^{(1)}(\Delta_S) \subset B^{(2)}, \quad i \in S ,$$

and one gets

$$\begin{array}{ccc}
\overline{M}_{0,P\cup\{n+1\}} & \xrightarrow{\eta^{(2)}} & B^{(2)} \supset D_S^{(2)} \\
& {\scriptstyle \pi} \searrow & \big\downarrow {\scriptstyle pr_1 \circ \pi^{(1)}} \\
& & \overline{M}_{0,P} \supset \Delta_S
\end{array}$$

One checks that the codimension 2 subvarieties $D_S^{(2)}$ with $|S^c| = 3$ are disjoint. Then one lets $B^{(3)}$ be the blow-up of $B^{(2)}$ along the union of these subvarieties. It is now clear how to proceed. Once the factorization $\eta^{(k-1)} = \pi^{(k-1)}\eta^{(k)}$ is established, one can define sections $\sigma_i^{(k)} : \overline{M}_{0,P} \to B^{(k)}$ and get the diagram

$$\begin{array}{ccc}
\overline{M}_{0,P\cup\{n+1\}} & \xrightarrow{\eta^{(k)}} & B^{(k)} \supset D_S^{(k)} \\
& {\scriptstyle \pi} \searrow & \big\downarrow \\
& & \overline{M}_{0,P} \supset \Delta_S
\end{array}$$

where the $D_S^{(k)}$ with $|S^c| = k+1$, are smooth, disjoint, and of codimension 2 in $B^{(k)}$. Then one looks at the blow-up $\pi^{(k)} : B^{(k+1)} \to B^{(k)}$ with center the union of these subvarieties. What becomes progressively more involved to prove is that $\eta^{(k)}$ factors through $\pi^{(k)}$. Once this is done, there is only one more thing to check, namely that, for each k, the map $\eta^{(k)}$ only contracts divisors $\Delta_{S\cup\{n+1\}}$ with $|S^c| \geq k+1$, so that the map $\eta^{(n-2)} : \overline{M}_{0,P\cup\{n+1\}} \to B^{(n-2)}$ must be an isomorphism.

Finally, observe that the equality $A^*(\overline{M}_{0,P}) = H^*(\overline{M}_{0,P})$ follows from the lemma in view of the inductive procedure and of the fact that the varieties $D_S^{(k)}$ are themselves isomorphic to products of moduli spaces of stable pointed curves of genus 0.

In the exercises, the reader is aked to perform Keel's construction in the first nontrivial case, namely the one of $\overline{M}_{0,5}$.

We now concentrate our attention on the first Chow group $A^1(\overline{M}_{0,P})$. Let us first observe that

$$A^1(\overline{M}_{0,P}) = \mathrm{Pic}(\overline{M}_{0,P}) = H^2(\overline{M}_{0,P}, \mathbb{Z})\,.$$

This follows directly from the fact that $\overline{M}_{0,P}$ is birationally equivalent to $(\mathbb{P}^1)^{n-3}$, so that $h^{0,1}(\overline{M}_{0,P}) = h^{0,2}(\overline{M}_{0,P}) = 0$. To show that $\mathrm{Pic}(\overline{M}_{0,P})$ is generated by boundary divisors is also easy. Let L be a line bundle on $\overline{M}_{0,P}$. Its restriction to the affine open set $M_{0,P}$ is trivial. Therefore there is a meromorphic section s of L which is nowhere vanishing when restricted to $M_{0,P}$. This means that the divisor of s is of the form $(s) = \sum n_i\Delta_i$, where the Δ_i's are boundary divisors, so that $L = \mathcal{O}(-\sum n_i\Delta_i)$. We can then conclude that $H^2(\overline{M}_{0,P}, \mathbb{Z}) = A^1(\overline{M}_{0,P})$ is generated by the classes δ_S with $S \subset P$ such that $|S| \geq 2$, $|S^c| \geq 2$ with the following relations:

1) $\delta_S = \delta_{S^c}$;
2) For any quadruple of distinct elements $i, j, k, l \in P$,

$$\sum_{i,j\in S;\, k,l\notin S} \delta_S = \sum_{i,k\in S;\, j,l\notin S} \delta_S = \sum_{i,l\in S;\, j,k\notin S} \delta_S.$$

Observe that, in genus zero, the ψ classes are linear combinations of boundary classes. More precisely, we have the following result.

LEMMA (7.4). *For any choice of distinct elements $x, y, z \in P$, the following relation holds in $A^1(\overline{M}_{0,P})$:*

$$\psi_z = \sum_{\substack{z\in A\\ x,y\notin A}} \delta_A\,.$$

This follows immediately from (4.6) and part v) of Lemma (4.28) by induction on $|P|$.

Keel's result for $H^2(\overline{M}_{0,P}, \mathbb{Z}) = A^1(\overline{M}_{0,P})$ is the following.

Proposition (7.5). *Let $P = \{1, 2, \dots, n\}$. Assume that $n \geq 4$. Fix distinct elements $i, j, k \in P$. Then $H^2(\overline{M}_{0,P}, \mathbb{Z})$ is freely generated by the classes*

$$(7.6) \qquad \delta_S \quad \text{with} \quad i \in S \quad \text{and} \quad 2 \leq |S| \leq n-3 \quad \text{or} \quad S = \{j,k\}\,.$$

In particular,

$$\operatorname{rank}\left(H^2(\overline{M}_{0,P}, \mathbb{Z})\right) = \operatorname{rank}\left(A^1(\overline{M}_{0,P})\right) = 2^{n-1} - \binom{n-1}{2} - 1\,.$$

Proof. Let B be the subgroup of $H^2(\overline{M}_{0,P}, \mathbb{Z})$ generated by the elements in (7.6). The only boundary divisors not appearing in (7.6) are the divisors $\delta_{\{s,t\}}$, where $s \neq i$, $t \neq i$, and $(s,t) \neq (j,k)$. We write relation 2) for the quadruple i, k, s, t as

$$\delta_{\{i,k\}} + \delta_{\{s,t\}} + \sum_{\substack{i,k \in S;\, s,t \notin S \\ 3 \leq |S| \leq n-3}} \delta_S = \delta_{\{i,t\}} + \delta_{\{s,k\}} + \sum_{\substack{i,t \in S;\, k,s \notin S \\ 3 \leq |S| \leq n-3}} \delta_S\,.$$

This shows that $\delta_{s,t} \equiv \delta_{s,k}$ modulo B. Now start from the 4-tuple i, j, k, s and get $\delta_{s,k} \equiv 0$ modulo B.

Denote by $\mathcal{B}^P_{i,j,k}$ the elements in (7.6) We must now prove that the elements of $\mathcal{B}^P_{i,j,k}$ are linearly independent. We proceed by induction on $n = |P|$. The case $n = 4$ is trivial. We assume that $n \geq 5$. Suppose that there is the relation

$$(7.7) \qquad \nu \delta_{j,k} + \sum_{i \in S,\, |S| \leq n-3} \nu_S \delta_S = 0\,.$$

Fix $a \in P \smallsetminus \{i\}$. We are going to pullback this relation to $\overline{M}_{0,P \cup \{x\} \smallsetminus \{i,a\}}$ via the map ϑ defined in (4.37), where now $A = P \smallsetminus \{i,a\}$ and $g = 0$. Let us first assume that $a \neq j$ and $a \neq k$. Recalling iv) and iv') in Lemma (4.38), the pullback of the left-hand side of (7.7) via ϑ is given by

$$(7.8) \qquad \sum_{S \supset \{i,a\},\, |S| \leq n-3} \nu_S \delta_{S \cup \{x\} \smallsetminus \{i,a\}} - \nu_{\{i,a\}} \psi_x + \nu \delta_{\{j,k\}}\,.$$

We now use the expression of ψ_x given in Lemma (7.4), where we are considering ψ_x as an element of $H^2(\overline{M}_{0,P \cup \{x\} \smallsetminus \{i,a\}})$. We get

$$(7.9) \qquad \psi_x = \delta_{\{j,k\}} + \sum_{\substack{x \in T \subset P \cup \{x\} \smallsetminus \{i,a\} \\ j,k \in T^c,\, |T| \leq n-4}} \delta_T\,.$$

Relation (7.8) becomes

$$\sum_{\substack{S \supset \{i,a\},\, |S| \leq n-3 \\ j \text{ or } k \in S}} \nu_S \delta_{S \cup \{x\} \smallsetminus \{i,a\}}$$
$$+ \sum_{\substack{S \supset \{i,a\},\, |S| \leq n-3 \\ j,k \in S^c}} (\nu_S - \nu_{\{i,a\}}) \delta_{S \cup \{x\} \smallsetminus \{i,a\}} + (\nu - \nu_{\{i,a\}}) \delta_{\{j,k\}}\,.$$

We can now apply the induction hypothesis to $\mathcal{B}_{x,j,k}^{P\cup\{x\}\smallsetminus\{i,a\}}$. Since a is arbitrary, as long as $a \neq j$ and $a \neq k$, we deduce that

$$\begin{aligned} \nu_{\{i,a\}} &= \nu \quad \text{for } a \neq j,\ a \neq k, \\ \nu_S &= 0 \quad \text{for } j \text{ or } k \in S,\ i \in S,\ S \neq \{i,j\},\ S \neq \{i,k\}, \text{ and } |S| \leq n-3, \\ \nu_S &= \nu \quad \text{for } j,k \in S^c,\ i \in S, \text{ and } |S| \leq n-3. \end{aligned}$$

Using again the general expression (7.4) for ψ_i in terms of the boundary divisors, the original relation (7.7) can then be written as

$$\begin{aligned} \nu_{\{i,j\}}\delta_{\{i,j\}} + \nu_{\{i,k\}}\delta_{\{i,k\}} + \nu\Big(\sum_{\substack{2\leq|S|\leq n-3 \\ i\in S;\, j,k\in S^c}} \delta_S\Big) + \nu\delta_{\{j,k\}} \\ = \nu_{\{i,j\}}\delta_{\{i,j\}} + \nu_{\{i,k\}}\delta_{\{i,k\}} + \nu\psi_i = 0\,. \end{aligned}$$

Pulling back this relation to $\overline{M}_{0,P\cup\{x\}\smallsetminus\{j,k\}}$, we get $\nu = 0$. Let $l \neq k$. Pulling back the resulting relation to $\overline{M}_{0,\{i,j,l,x\}}$, we get $\nu_{\{i,j\}} = 0$. But then $\nu_{\{i,k\}} = 0$ as well.

Q.E.D.

8. Bibliographical notes and further reading.

As we mentioned in the introduction, Mumford laid the foundations of the intersection theory on moduli spaces of curves in [556]. At the time of Mumford's writing, Looijenga's theorem exhibiting $\overline{M}_{g,P}$ as the quotient of a smooth variety by a finite group was not yet available. Now that it is, these foundations are greatly simplified. Taking another viewpoint, one may invoke the intersection theory on Deligne–Mumford stacks as developed by Vistoli in [671] and proceed from there. The classes $\widetilde{\kappa}_i$ where introduced by Mumford in [556] and later, from a slightly different point of view, by Morita [518] [519]. Morita [519] and Miller [512] independently proved that the monomials of degree less than or equal to $2N$ in the $\widetilde{\kappa}_i$'s are linearly independent in $H^*(M_g;\mathbb{Q})$, as long as $g \geq 3N$. In his original definition, given a family of stable pointed curves $\pi : \mathcal{C} \to S$, Mumford uses the symbol κ_i instead of $\widetilde{\kappa}_i$ to denote the class $\pi_*(c_1(\omega_\pi)^{i+1})$. As proposed in [27], we reserve the symbol κ_i for the class $\pi_*(c_1(\omega_\pi(D)^{i+1})$. The reason is that, with this modification, the classes κ_i enjoy the properties (4.11), (4.17), (4.18) and satisfy Lemma (4.18).

Formula (4.34) for the intersection of boundary strata was freely used, at least at the divisorial level, before being written down explicitly by Graber and Pandharipande in [307]. In the same paper Proposition (4.39) is proved.

In the bibliographical notes to Chapter XXI we will point the reader to the important work done in the direction of expressing geometrically defined cycles in $\overline{M}_{g,n}$ as, for example, the ones coming from Brill–Noether Theory, in terms of tautological classes.

Computations of Chow rings of moduli spaces of curves in low genus are given by Mumford [556], Faber [227], [228], and Izadi [393]. An excellent source for characteristic classes is Morita's monograph [522].

Based on low genus computations, Faber, in 1993, made his conjecture on the Gorenstein nature of the tautological ring of $\mathcal{M}_g$. These conjectures appeared in [231] together with the basic ideas for the proof of the socle statement. The proof in the text (Section 7 in Chapter XX) is taken from [237]. An independent proof is given by Graber and Vakil [308], [310], see also Bryan and Pandharipande [91], Bertram, Cavalieri, and Todorov [66], and Cavalieri [113]. The vanishing statement of this conjecture, also proved in our text (Section 4 in Chapter XXI), is proved by Looijenga in [489], where it is also proved that $\dim R^{g-2}(\mathcal{M}_g) \leq 1$. The first vanishing statement appears in Getzler's paper [285] where the vanishing is conjectured in codimension $> g$. This conjecture was proved in cohomology by Ionel in [386]. The fact that $\kappa_1, \dots, \kappa_{g-2}$ generate the tautological ring of $\mathcal{M}_g$ was proved by Mumford in [556]. Faber in [231] conjectured that $\kappa_1, \dots, \kappa_{[g/3]}$ generate $R^\bullet(\mathcal{M}_g)$. This was proved by Morita in cohomology [523] and by Ionel [387] in the Chow ring. The Gorenstein conjecture for $\overline{\mathcal{M}}_{g,P}$ appears as a question in [335] and as a formal conjecture in [584]. The socle part of this conjecture is proved by Graber and Vakil [308], [310] and by Faber and Pandharipande [239]. The remarkable formula (6.2) is proved by Goulden, Jackson, and Vakil [305] for $n = 3$ and by Liu and Xu for the general case. Excellent references for the tautological ring and the various conjectures are by Faber [231], [233], Hain and Looijenga [335], Vakil [666], Graber and Vakil [310], Faber and Pandharipande [239], and Graber and Pandharipande [307].

We shall give references for Diaz's result and related questions in the bibliography of the last chapter.

9. Exercises.

A. Degrees on curves of various line bundles on moduli I.

In each of the following, we are going to specify a one-parameter family $\pi : S \to B$ of curves. The content of the problem will then be to carry out the following:

a. Either verify that all members of the family are stable, or determine the stable limits (and the base change needed to arrive at the stable reduction);
b. Identify the total space S of the family;
c. Identify, as an element of the Picard group $\operatorname{Pic}(S)$, the relative dualizing sheaf ω of the family;
d. Calculate the self-intersection $\kappa = \omega^2$;

e. Describe the direct image $\pi_*\omega$ and determine in particular its first Chern class λ;
f. Calculate the degree d_a of the pullback $\phi^*(\delta_a)$ to B of the boundary divisor δ_a, where $\phi : B \to \overline{M}_g$ is the moduli map; and finally
g. Verify the Mumford relation $12\lambda = \kappa + \delta$ for the family.

In those problems where one carries out a stable reduction in order to arrive at the family in question, since one of the members of the given family is unstable, we will take the required base changes to be branched over the point in question, plus one general other point.

A-1. Let C and $D \subset \mathbb{P}^2$ be general plane quartic curves, and $\{C_t\}$ the pencil of curves they span (that is, the zero loci of the polynomials $t_0F + t_1G$, where F and G are the defining equations of C and D).

A-2. Let C and $D \subset \mathbb{P}^2$ be a general pair of plane quartic curves simply tangent to one another at a point p, and $\{C_t\}$ the pencil of curves they span.

A-3. Let $X \subset \mathbb{P}^3$ be a general quartic surface, and $\{C_b = X \cap H_t\}$ a general pencil of plane sections of X.

A-4. Let $t_1, \ldots, t_7$ be distinct nonzero scalars and take $\{C_t\}$ to be the family of hyperelliptic curves

$$y^2 = (x - t_1) \cdot \ldots \cdot (x - t_7)(x - t^2).$$

(Why are we putting a t^2 in the last factor instead of t?)

A-5. Let C and $D \subset \mathbb{P}^1 \times \mathbb{P}^1$ be general curves of type $(2,3)$, and $\{C_t\}$ the pencil of curves they span.

A-6. Let $p \in \mathbb{P}^2$ be any point, C and D general plane quartic curves double at p, and $\{C_t\}$ the normalizations of the curves in the pencil they span.

A-7. Let C be a general plane quartic with a cusp, D a general quartic, and $\{C_t\}$ the pencil of curves they span.

A-8. Let C be a general plane quartic with a tacnode, D a general quartic, and $\{C_t\}$ the pencil of curves they span.

A-9. Let C be a double conic, D a general quartic, and $\{C_t\}$ the pencil of curves they span.

B. Degrees on curves of various line bundles on moduli II.

In the preceding batch, we considered a number of one-parameter families of stable curves, each of which was generically smooth; for each, we asked to calculate the degrees of the associated divisor classes λ,

δ, and κ. We are now going to do the same for the following one-parameter families of singular stable curves. In each of the following cases, accordingly, the problem is to determine the degrees of the divisor classes λ, δ, and κ on the specified one-parameter family of stable curves.

B-1. Let C be a fixed curve of genus $g-1$, and $p \in C$ a fixed point. Consider the family of stable curves of genus g obtained by identifying p with another point $q \in C$ (recall that we identified in class the stable limit of these curves as $q \to p$).

B-2. Let C be a fixed curve of genus $g-1$, and $p \in C$ a fixed point; let $\{E_t \subset \mathbb{P}^2\}$ be a general pencil of plane cubic curves, and q a base point of the pencil. Consider the family of stable curves of genus g obtained by identifying $p \in C$ with $q \in E_t$.

B-3. Let C be a fixed curve of genus $g-2$, and $p, q \in C$ two fixed points; let $\{E_t\}$ be a general pencil of plane cubic curves, and r and s two base points of the pencil. Consider the family of stable curves of genus g obtained by identifying $p \in C$ with $r \in E_t$ and $q \in C$ with $s \in E_t$.

B-4. Let C and D be fixed curves of genera α and $g-\alpha$, respectively, and let $p \in C$ be a fixed point. Consider the family of stable curves of genus g obtained by identifying $p \in C$ with a point $q \in D$.

B-5. Let $p \in \mathbb{P}^2$ be any point, and $\{C_t \subset \mathbb{P}^2\}$ a general pencil of plane quartic curves double at p. Show that for an open subset of $t \in \mathbb{P}^1$, the curve C_t is a stable curve of genus 3, and consider the family of stable curves of genus 3 obtained by completing this family (recall that we identified in class the stable limits of this family).

B-6. Let $Q \cong \mathbb{P}^1 \times \mathbb{P}^1$ be a smooth quadric in $\mathbb{P}^3$; let $L \subset Q$ be a line, and let $\{C_t \subset Q\}$ be a general pencil of curves of type $(2,2)$ on Q. Show that for an open subset of $t \in \mathbb{P}^1$, the curve obtained from C_t by identifying the two points of intersection of L with C_t is stable, and consider the family of stable curves of genus 2 obtained by completing this family. Verify the relation $10\lambda = \delta_0 + 2\delta_1$.

B-7. Do the preceding problem with $\{C_t \subset Q\}$ a general pencil of curves of type $(2,g)$ on Q, and L a line of the ruling meeting the curves C_t twice. Again, verify the relation $10\lambda = \delta_0 + 2\delta_1$.

B-8. Let $\{C_t \subset \mathbb{P}^2\}$ be a general pencil of cubic curves, and p and q base points of the pencil. Consider the family of stable curves of genus 2 obtained from the C_t by identifying p and q.

B-9. Do the preceding problem with $\{C_t \subset \mathbb{P}^2\}$ a general pencil of quartic curves to arrive at a family of stable curves of genus 4.

C. Miscellaneous exercises

C-1. Show that $\overline{M}_{0,5}$ can be obtained from $\mathbb{P}^1 \times \mathbb{P}^1$ by blowing up the three points $(0,0)$, $(1,1)$, and (∞,∞). As such, $\overline{M}_{0,5}$ is isomorphic to the Del Pezzo surface F_5. Recognize that the 10 exceptional curves of F_5 correspond to the 10 boundary divisors of $\overline{M}_{0,5}$. Compare Keel's presentation of $A^*(F_5)$ with the one obtained by exhibiting F_5 as the blow-up of $\mathbb{P}^2$ in four points.

C-2. Let $f : X \to H$ be the universal family over $H = \mathrm{H}_{\nu,g,n}$, and let $D_1, \dots, D_n$ be its canonical sections. Let $\nu, a_1, \dots, a_n$ be integers. Prove an analogue of formula (7.5), Chapter XIII for the class in $\mathrm{Pic}(\overline{\mathcal{M}}_{g,n})$ given by $c_1(f_!(\omega_f^\nu(\sum a_i D_i)))$.

C-3. Prove Faber's formula (4.26).

C-4. Using Lemma (4.28) i), show that the cycles $\prod \kappa_{a_i}$ can be expressed in terms of the cycles $\pi_*(\psi_1^{d_1+1} \cdots \psi_n^{d_n+1})$, where $\pi : \overline{\mathcal{M}}_{g,n} \to \overline{\mathcal{M}}_{g,n}$ is the projection.

C-5. Give an example for which $\pi_! \omega_\pi^\nu \neq \pi_* \omega_\pi^\nu$.

C-6. Show that on $\mathcal{M}_g$ the κ_i's are polynomials in the $\lambda_i(2)$'s and, viceversa, the $\lambda_i(2)$'s are polynomials in the κ_i's.

Chapter XVIII.
Cellular decomposition of moduli spaces

1. Introduction.

In this chapter we will introduce a number of simplicial complexes associated to a pointed oriented surface (S, P). These simplicial complexes are used to define $\Gamma_{S,P}$-equivariant cell decompositions of the Teichmüller space $\mathcal{T}_{S,P}$ and of its bordification $\widehat{\mathcal{T}}_{S,P}$. These decompositions descend to orbicell-decompositions of the moduli space $M_{g,P}$ and of suitable compactifications of it. In Section 2 we will introduce the arc complex $A(S,P)$ and the subspace $|A^0(S,P)| \subset |A(S,P)|$ defined by proper simplices.

To explain what these complexes are, let us limit ourselves to the case of 1-pointed Riemann surfaces. We fix a compact oriented surface S of genus g and a point $p \in S$, and we denote by $\mathcal{A}$ the set of isotopy classes, relative to p, of loops in S based at p which are not homotopically trivial. The *arc complex* $A = A(S,p)$ is the simplicial complex whose k-simplices are given by $(k+1)$-tuples $a = ([\alpha_0], \dots, [\alpha_k])$ of distinct classes in $\mathcal{A}$ which are representable by a $(k+1)$-tuple $(\alpha_0, \dots, \alpha_k)$ of loops intersecting only at p. The geometric realization of A is denoted by $|A|$. A simplex $([\alpha_0], \dots, [\alpha_k]) \in A$ is said to be *proper* if $S \smallsetminus \cup_{i=0}^k \alpha_i$ is a disjoint union of discs. The improper simplices form a subcomplex of A denoted by A_∞. We set

$$A^0 = A \smallsetminus A_\infty \quad \text{and} \quad |A^0| = |A| \smallsetminus |A_\infty| \,.$$

The mapping class group $\Gamma_{g,1} = \Gamma_{S,p}$ acts on A in the obvious way. For $a = ([\alpha_0], \dots, [\alpha_k]) \in A$ and $[\gamma] \in \Gamma_{g,1}$, we define

$$[\gamma] \cdot a = ([\gamma\alpha_0], \dots, [\gamma\alpha_k]) \,. \tag{1.1}$$

Let $\mathcal{T}_{g,1} = \mathcal{T}_{S,p}$ be the Teichmüller space of 1-pointed genus g Riemann surfaces. The fundamental result is the following theorem, originally due to Mumford (see the bibliograhical notes at the end of this chapter).

THEOREM (1.2). *There is a $\Gamma_{g,1}$-equivariant homeomorphism*

$$\Psi\colon \quad \mathcal{T}_{g,1} \to |A^0| \,.$$

E. Arbarello et al., *Geometry of Algebraic Curves*,
Grundlehren der mathematischen Wissenschaften 268,
DOI 10.1007/978-3-540-69392-5_10,

As an example, let us describe the complex A^0 in the case of one-pointed curves of genus one. The set $\mathcal{A}$ is the subset of the fundamental group $\pi_1(S,p)$ consisting of those nonzero elements which are represented by loops with no self-intersections. Since $\pi_1(S,p) \cong \mathbb{Z}^2$, we have

$$\mathcal{A} = \{[\alpha] = (m,n) \in \mathbb{Z}^2 \smallsetminus \{(0,0)\} : \gcd(m,n) = 1\} \cong \mathbb{P}^1(\mathbb{Q}) .$$

Given two vertices of A, say $[\alpha] = (m,n)$ and $[\beta] = (m',n')$, the condition for the pair $([\alpha],[\beta])$ to be a 1-simplex is that α and β intersect exactly in p, which translates into the condition $mn' - m'n = \pm 1$. With the same reasoning, one sees that the only two 2-simplices that are incident to $([\alpha],[\beta])$ are $([\alpha],[\beta],[\alpha]+[\beta])$ and $([\alpha],[\beta],[\alpha]-[\beta])$, and clearly there are no 3-simplices. It is also evident that A_∞ coincides with the set of vertices of A. One way to picture this cell complex is to make use of the upper half-plane and its hyperbolic geometry.

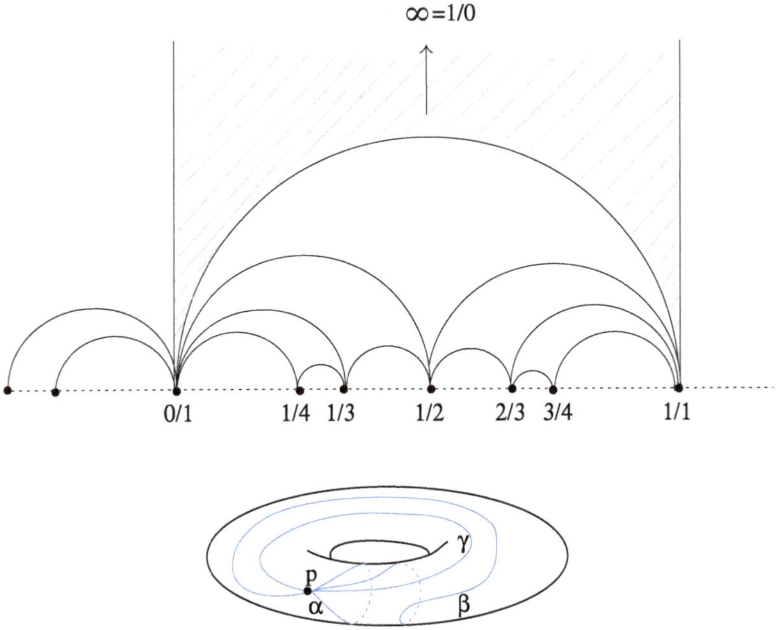

Figure 1.

In Figure 1, the shaded region is the top-dimensional cell $([\alpha],[\beta],[\gamma])$, where $\alpha = (0,1)$, $\beta = (1,1)$, and $\gamma = (1,0)$. Since the Teichmüller space of 1-pointed curves of genus one can be identified with the upper half-plane $\mathbb{H}$, the picture above shows how plausible is the existence of the isomorphism (1.2), at least in this particular case. Actually this picture tells us even more, namely that

$$|A| = \mathbb{H} \cup \mathbb{P}^1(\mathbb{Q}) .$$

Recalling from the introduction to Chapter XV and from exercise C-1 in the same chapter that also the bordification $\widehat{\mathcal{T}}_{1,1}$ of $\mathcal{T}_{1,1}$ can be identified with $\mathbb{H} \cup \mathbb{P}^1(\mathbb{Q})$, the above picture makes plausible also the following result.

THEOREM (1.3). *There is a $\Gamma_{g,1}$-equivariant continuous extension of Ψ*

$$\widehat{\Psi}\colon \quad \widehat{\mathcal{T}}_{g,1} \to |A|\,.$$

In Section 3 we describe the space $|A_0(S,P)|$ in terms of ribbon graphs. A ribbon graph is just the Poincaré dual of an arc system. In the following picture, the arc system is the blue one, and its dual is the red ribbon graph.

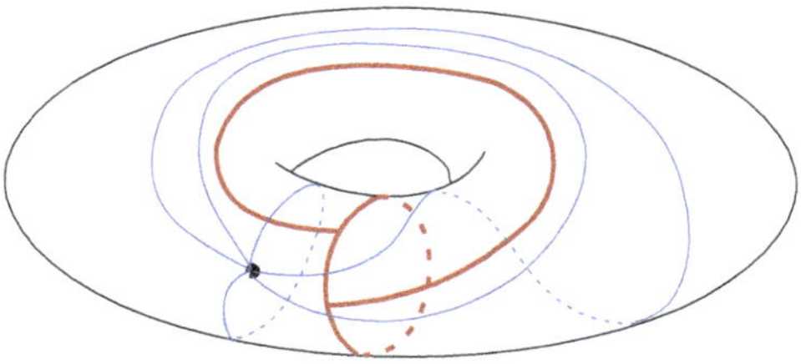

Figure 2.

It turns out that, in proving Theorem (1.2), it is very natural to use the language of ribbon graphs. Let us give an idea of how ribbon graphs enter the picture when defining the map Ψ. Let (C,p) be a 1-pointed smooth curve of genus g. The uniformization theorem for Riemann surfaces provides the surface $C \smallsetminus \{p\}$ with a canonical hyperbolic metric, the Poincaré metric. In this metric the point p appears as a *cusp* (see Figure 3). This cusp has infinite distance from the points in $C \smallsetminus \{p\}$. The set of horocycles around p is a canonical family of simple closed curves in $C \smallsetminus \{p\}$, contracting to p. Choose a small constant c and a horocycle η of length c. Given a general point $q \in C \smallsetminus \{p\}$, there is a *unique* shortest geodesic joining q to η. On the other hand, it turns out that the locus of points q in C having the property that there exist two or more shortest geodesics from q to η is a connected graph $G \subset C$ which is called *the hyperbolic spine of* (C,p). The fundamental property of this graph is that $C \smallsetminus G$ is a disc Δ.

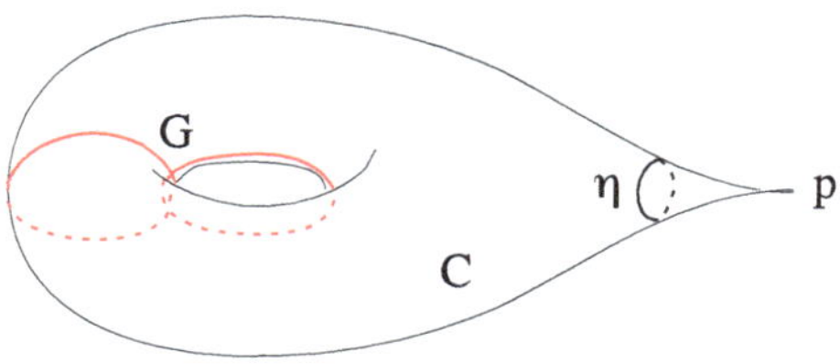

Figure 3.

Suppose that G has $k+1$ edges. Then the Poincaré dual of G is a graph consisting of $k+1$ loops $\alpha_0, \dots, \alpha_k$ based at p, which in fact give rise to a simplex $a = (\alpha_0, \dots, \alpha_k) \in A^0$. To get a point in $|a|$, we need a $(k+1)$-tuple of numbers $\mu_0, \dots, \mu_k$ with $\sum \mu_i = 1$. Now each loop α_i corresponds, by duality, to an edge e_i of G. Draw the geodesics from the vertices of e_i to p. These geodesics cut out on η two arcs. Using elementary hyperbolic geometry, one can see that these two arcs have the same length a_i. This is the coefficient one assigns to the loop α_i. We finally choose the constant c in such a way as to get $\sum \mu_i = 1$.

In order to prove Theorem (1.3), or rather its generalization to the case of n-pointed curves, we introduce the notion of stable ribbon graph, and we connect this notion to the arc complex $A(S, P)$. This is done in Section 9. As we already remarked, the main tools in proving Theorem (1.2), Theorem (1.3), and their generalizations to the case of n-pointed curves are the uniformization theorem and the hyperbolic geometry of the Siegel upper half-plane. These are reviewed in Sections 5 and 6.

In Section 7, using the hyperbolic spine, we define a map

$$\Psi : \mathcal{T}_{S,P} \times \Delta_P \to |A^0(S, P)|,$$

where Δ_P is the standard $(|P|-1)$-dimensional simplex. In Section 8 we prove the bijectivity of Ψ. In Section 10, we define an extension

$$\widehat{\Psi} : \widehat{\mathcal{T}}_{S,P} \times \Delta_P \to |A(S, P)|$$

of Ψ, and in Section 11 we prove its continuity. While the continuity of Ψ is quite straightforward, the proof of the continuity of $\widehat{\Psi}$ is rather involved. However, the extension $\widehat{\Psi}$ is only used in Section 7 of the next chapter in order to give a combinatorial expression for the first Chern classes of the point-bundles $\mathcal{L}_p$ over $\overline{M}_{g,P}$. This combinatorial expression is one of the essential tools in Kontsevich's proof of the Witten conjecture, as we will see in Chaper XX. Upon first approach, the reader shoud follow the proof of Theorem (2.7) considering only the case of $\mathcal{T}_{S,P}$. In Section 4 we explain the idea behind the proof of this result (which is formally stated in Theorem (2.7)). As a corollary of this theorem, we get a homeomorphism

$$M_{g,P} \times \Delta_P \cong |A_0(S, P)|/\Gamma_{S,P}.$$

In the final section we define a suitable compactification $\overline{M}'_{g,P}$ of $M_{g,P}$, and, using the map $\widehat{\Psi}$, we extend to it the orbicellular decomposition of $M_{g,P}$.

2. The arc system complex.

Throughout, we fix a compact oriented surface S and a finite, *nonempty* subset $P \subset S$. We let g be the genus of S and set $n = |P|$. We denote by $\mathcal{A}$ the set of isotopy classes, relative to P, of arcs and loops in S intersecting P exactly at their endpoints, and where the loops are additionally required to be homotopically nontrivial in the complement of P minus their base-point.

DEFINITION (2.1). *The arc complex $A(S,P)$ is the simplicial complex whose k-simplices are given by $(k+1)$-tuples $a = ([\alpha_0], \dots, [\alpha_k])$ of distinct classes in $\mathcal{A}$ which are representable by a $(k+1)$-tuple $(\alpha_0, \dots, \alpha_k)$ of arcs and loops intersecting only in P. The geometric realization of $A(S,P)$ is denoted by $|A(S,P)|$.*

We denote by Δ_P the simplex having the points of P as vertices. In other words, Δ_P consists of the formal linear combinations

$$\sum_{p \in P} r_p p \,,$$

where the r_p are nonnegative real numbers such that $\sum r_p = 1$. Any numbering of the points of P provides an identification of Δ_P with the standard n-simplex $\{(r_1, \dots, r_n) \in \mathbb{R}^n_{\geq 0} : \sum r_i = 1\}$. We define a simplicial map

$$\lambda : |A(S,P)| \to \Delta_P \tag{2.2}$$

by sending each vertex $[\gamma]$ represented by an arc (resp., a loop) to the midpoint of the 1-simplex in Δ_P spanned by its endpoints (resp., the vertex of Δ_P represented by its base point). The map λ is then defined by simplicial extension. The explicit expression of λ is the following. Let $a = ([\alpha_0], \dots, [\alpha_k])$ be a simplex in $|A(S,P)|$. Let $l \in |a|$, and let

$$l = (l_{\alpha_0}, \dots, l_{\alpha_k})$$

be the expression of l in barycentric coordinates. Then

$$\lambda(l) = (r_p)_{p \in P} \,, \qquad r_p = \frac{1}{2} \sum_{\alpha_j \ni p} \epsilon_{\alpha_j} l_{\alpha_j} \,, \tag{2.3}$$

where $\epsilon_{\alpha_j} = 2$ if α_j is a loop at p and $\epsilon_{\alpha_j} = 1$ otherwise. The number r_p is equal to zero exactly when none of the curves $\alpha_0, \dots, \alpha_k$ passes through p.

DEFINITION (2.4). *A simplex $([\alpha_0], \dots, [\alpha_k])$ is said to be proper if $S \setminus \cup_{i=0}^k \alpha_i$ is the disoint union of discs each containing at most one of the points $p_1, \dots, p_n$.*

The improper simplices form a subcomplex of $A(S,P)$ denoted by $A_\infty(S,P)$. We set

$$A^0(S,P) = A(S,P) \smallsetminus A_\infty(S,P) \quad \text{and} \quad |A^0(S,P)| = |A(S,P)| \smallsetminus |A_\infty(S,P)| .$$

In Figure 4 we give an example of an improper (left) and of a proper (right) 2-simplex in the case $g = 1$, $n = 2$.

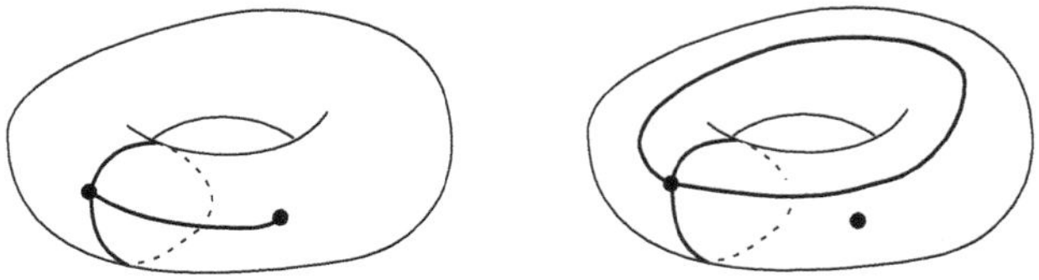

Figure 4.

The mapping class group $\Gamma_{S,P}$ acts on $A(S,P)$ in the obvious way. For $a = ([\alpha_0], \dots, [\alpha_k]) \in A$ and $[\gamma] \in \Gamma_{S,P}$, we define

$$[\gamma] \cdot a = ([\gamma\alpha_0], \dots, [\gamma\alpha_k]) . \tag{2.5}$$

Moreover,

$$\lambda([\gamma] \cdot a) = \lambda(a) . \tag{2.6}$$

Let $\widehat{\mathcal{T}}_{S,P}$ be the partial compactification of the Teichmüller space $\mathcal{T}_{S,P}$ given in Section 8 of Chapter XV. The main theorem of this chapter will be the following.

THEOREM (2.7). *There exists a $\Gamma_{S,P}$-equivariant homeomorphism*

$$\Psi : \mathcal{T}_{S,P} \times \Delta_P \longrightarrow |A^0(S,P)| . \tag{2.8}$$

This homeomorphism extends to a $\Gamma_{S,P}$-equivariant, continuous, surjective map

$$\widehat{\Psi} : \widehat{\mathcal{T}}_{S,P} \times \Delta_P \longrightarrow |A(S,P)| \tag{2.9}$$

such that the composition of $\widehat{\Psi}$ with the the map λ coincides with the projection of $\widehat{\mathcal{T}}_{S,P} \times \Delta_P$ on its second factor. The map $\widehat{\Psi}$ induces a continuous, open, surjective map

$$\overline{\Phi} : \overline{M}_{g,P} \times \Delta_P \longrightarrow |A(S,P)|/\Gamma_{S,P} \tag{2.10}$$

and a homeomorphism

$$\Phi : M_{g,P} \times \Delta_P \longrightarrow |A^0(S,P)|/\Gamma_{S,P} . \tag{2.11}$$

Upon first approach, the reader should follow the proof of Theorem (2.7) only for the case of Ψ and $\mathcal{T}_{S,P}$, leaving aside stable curves and stable ribbon graphs. The existence and continuity of the extension $\widehat{\Psi}$ is more complex and will be needed only in the next chapter in order to give a combinatorial expression for the first Chern classes of the point bundles $\mathcal{L}_p$ over $\overline{M}_{g,P}$.

From the case $g = 1$, $n = 1$, which we illustrated in the introduction (see Figure 1), it is already evident that the complex $A(S,P)$ is not locally finite. This feature will be reflected in the topology of $|A(S,P)|$. Let us spend a few words about this topology.

Given a finite-dimensional simplicial complex K, we denote by K^n the set of n-dimensional simplices of K. There are two natural topologies on the geometric realization $|K|$ of K. The first one is the *coherent topology.* Its defining property is that a subset $U \subset |K|$ is open in this topology if and only if $U \cap |s|$ is open in $|s|$ for each simplex s of K. The second is the *metric topology.* To describe it, first identify points $p \in |K|$ with functions $p : K^0 \to [0,1]$ having the following two properties:

$$s_p = \{v \in K^0 : p(v) \neq 0\} \quad \text{is a simplex of} \quad K\,,$$

$$\sum_{v \in K^0} p(v) = 1\,.$$

Of course, s_p is the least-dimensional simplex s such that $p \in |s|$, and we can think of $(p(v))_{v \in K}$ as the barycentric coordinates of p. One can then introduce a metric in $|K|$ by setting

$$d(p,q) = \sum_{v \in K} |p(v) - q(v)|\,.$$

REMARK (2.12). Observe that $d(p,q) = 2$ if and only if p and q lie in the interior of simplices with disjoint closures.

The topology induced by this metric is the metric topology. It is obviously Hausdorff and coarser than the coherent topology. It is also clear that, if $|K|$ is not locally finite, then the metric topology, unlike the coherent topology, is not locally compact. Provisionally, we will denote by $|K|_d$ the set $|K|$ equipped with the metric topology. Suppose now that K is acted on by a group Γ. We can then form the complex K/Γ by letting $(K/\Gamma)^n = K^n/\Gamma$. We will assume that any element $\gamma \in \Gamma$ fixing a simplex a of K also fixes each vertex of a. If this is not the case, we can always pass to the second baricentric subdivison of K, where the above property holds. In this way, we may identify $|K/\Gamma|$ with $|K|/\Gamma$. Moreover, we have the following result.

LEMMA (2.13). *The spaces $|K/\Gamma|_d$ and $|K|_d/\Gamma$ are homeomorphic (via the above identification). In particular, $|K|_d/\Gamma$ is Hausdorff.*

Proof. First of all, observe that, given a simplicial complex L and its first barycentric subdivision L', there is a canonical homeomorphism $|L|_d \cong |L'|_d$. Since K'/Γ can be identified with $(K/\Gamma)'$, it suffices to prove the lemma up to barycentric subdivisions of K. To prove the lemma, we must show that, given $[p]$ and $[q]$ in $|K/\Gamma|_d$, we have

$$d_{K/\Gamma}([p],[q]) = \min_{p' \in [p], q' \in [q]} d_K(p', q') . \tag{2.14}$$

By using (2.12) for both K and K/Γ, we may as well assume that $d_{K/\Gamma}([p],[q]) < 2$. If $[p]$ and $[q]$ belong to the interior of the same simplex of $|K/\Gamma|$, then (2.14) is clear. We may then assume that $[p]$ and $[q]$ lie in the interior of distinct simplices of $|K/\Gamma|$, whose closures have nonempty intersection. Let σ and τ be liftings to K of these two simplices. A moment of reflection will convince the reader that (2.14) could be easily proved if one could assume that there exists a simplex τ' in the orbit $\Gamma\tau$ with the property that $\sigma \cap \Gamma\tau = \sigma \cap \tau'$. On the other hand, by passing to the second barycentric subdivision of K, one can always make this assumption. Q.E.D.

From now on we will drop the subscript d to denote the metric topology. The symbol $|K|$ will always denote the geometric realization of K equipped with the metric topology.

3. Ribbon graphs.

As in the previous section, we fix a compact topological surface S of genus g and a finite subset $P \subset S$; we set $n = |P|$.

It is very useful to describe the simplices of $|A(S,P)|$ in terms of ribbon graphs. In this section we give this description, restricting our attention to the simplices contained in $|A^0(S,P)|$.

DEFINITION (3.1). *A ribbon graph is a one-dimensional cell complex G without isolated vertices such that, for each vertex v of G, a cyclic order is given on the set of half-edges which are incident to v. The number of half-edges which are incident to v is called the valency of v.*

A more abstract but useful definition is the following.

DEFINITION (3.2). *A ribbon graph G is a triple $(X(G), \sigma_0, \sigma_1)$, where $X(G)$ is a nonempty set, σ_0 a permutation of $X(G)$, and σ_1 a fixed point-free involution of $X(G)$. The set of orbits of σ_1 is denoted by $X_1(G)$, and its elements are called the edges of G. The set of orbits of σ_0 is denoted by $X_0(G)$, and its elements are called the vertices of G.*

It is clear that, given a ribbon graph G as in the first definition, we can take $X(G)$ to be the set of its half-edges. In this case, σ_1 exchanges the two halves of each edge, while σ_0 sends each half-edge h to the one following it in the cyclic order around the vertex v which is incident to

h. Another way of thinking of $X(G)$ is to view its elements as the edges of G together with an orientation. When taking this attitude, we write elements of $X(G)$ with an arrow,

$$\vec{e} \in X(G),$$

and we call them *oriented edges*. We leave it as an easy exercise to prove that definitions (3.1) and (3.2) are equivalent.

All ribbon graphs considered from now on will implicitly be assumed to be connected. The cyclic order of half-edges around the vertices of a ribbon graph G allows one to draw a "ribbon" around G. To do this, one embeds each edge separately in a strip, and then one joins the various strips following the cyclic order at each vertex, as illustrated in Figure 5.

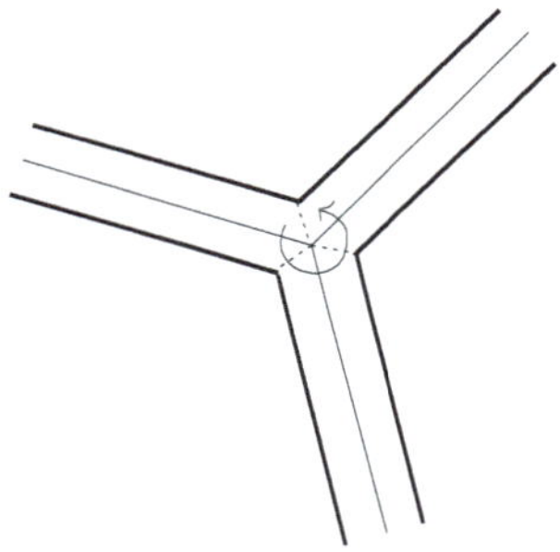

Figure 5.

The resulting object is an orientable surface with boundary $S(G)$, in which the graph G is embedded. Sometimes we will say that the surface $S(G)$ is *the ribbon along* G. Suppose that $S(G)$ has ν boundary components. Attaching a disc to each boundary component of $S(G)$, we obtain a compact orientable topological surface $\overline{S}$ together with ν points $p_1, \ldots, p_\nu$ (the centers of the discs). Thus, the ribbon graph G is embedded in $S \smallsetminus \{p_1, \ldots, p_\nu\}$, and $\overline{S} \smallsetminus G$ is the union of ν discs.

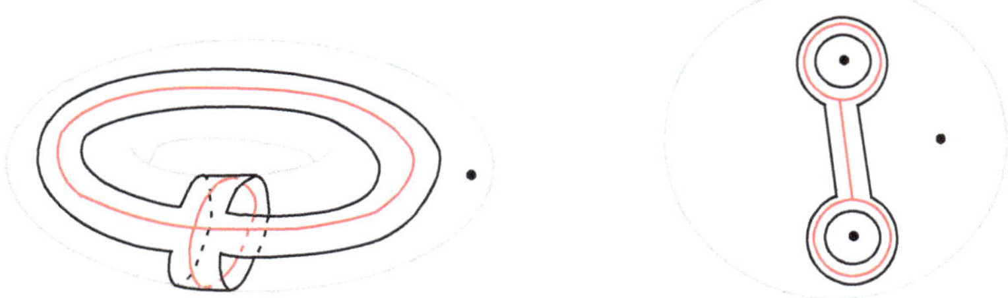

Figure 6. Ribbons of genus 1 with 1 boundary component (left) and of genus 0 with 3 boundary components (right).

Definition (3.3). *Given a ribbon graph* G, *one says that* G *has genus* γ *and* ν *boundary components if* $S(G)$ *is a genus* γ *surface with* ν *boundary components.*

The boundary components of the surface $S(G)$ can be described combinatorially as follows. Consider the permutation $\sigma_2 = \sigma_0^{-1}\sigma_1$. Then the set $X_2(G)$ of orbits of $X(G)$ under the action of σ_2 is in one-to-one correspondence with the connected components of the boundary of $S(G)$, i.e., with the "holes" of $S(G)$. In fact, as illustrated in Figure 7, σ_1 sends a half-edge $h = h(e)$, based at a vertex v, to the other half $h'(e)$ of the same edge e. On the other hand, σ_0^{-1} sends $h'(e)$ to the half-edge $\sigma_2(h)$ preceding $h'(e)$ in the cyclic order determined by the vertex v' at which $h'(e)$ is based, so that the orbit of h under σ_2 determines a boundary component of $S(G)$. The converse assertion is also easy to establish.

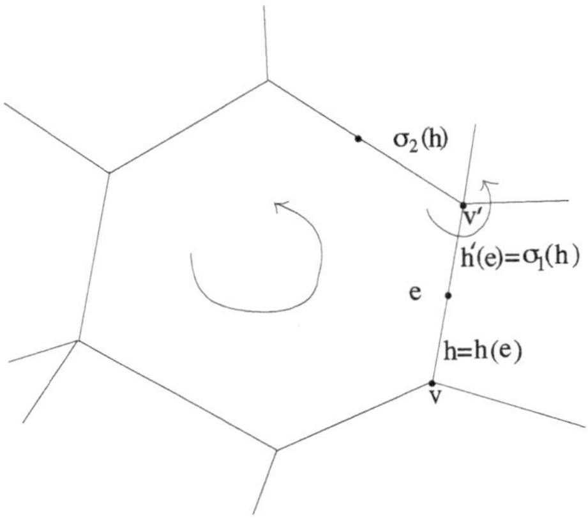

Figure 7.

Given an oriented edge $\vec{e} \in X(G)$, we will denote by

$$[\vec{e}]_0\,,\ [\vec{e}]_1\,,\ [\vec{e}]_2 \tag{3.4}$$

the orbits of $\vec{e}$ under the action of σ_0, σ_1, and σ_2, respectively. Hence, $[\vec{e}]_0$ is a vertex, $[\vec{e}]_1$ is an edge, and $[\vec{e}]_2$ is a boundary component of G. Regarding edges, sometimes, instead of writing $[\vec{e}]_1$, we will simply write e.

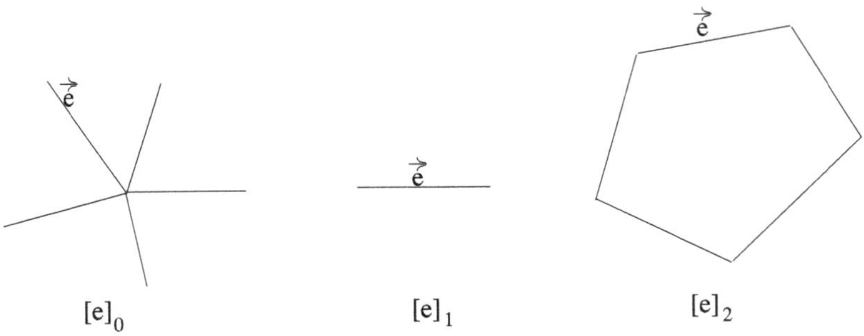

Figure 8.

An important notion concerning ribbon graphs is the one of duality.

DEFINITION (3.5). *Given a ribbon graph $G = (X(G), \sigma_0, \sigma_1)$, the dual of G is the ribbon graph $G^* = (X(G), \sigma_2^{-1}, \sigma_1)$.*

From the definition it is clear that the vertices and the boundary components of G^* are in a one-to-one correspondence with the boundary components and the vertices of G, respectively, while an edge joining two vertices of G^* corresponds to the edge of G separating the two corresponding boundary components of G. Hence, we have canonical identifications:

$$X_0(G^*) = X_2(G) \quad X_1(G^*) = X_1(G) \quad X_2(G^*) = X_0(G)\,.$$

When G is viewed as embedded in the compact surface $\overline{S}$, the graph G^* is nothing but the Poincaré dual of G.

DEFINITION (3.6). *Let T be a finite set, and G a ribbon graph. A T-marking of G is an injection*

$$\mathbf{x} : T \hookrightarrow X_0(G) \cup X_2(G)$$

whose range contains $X_2(G)$, so that all of the boundary components and some of the vertices of G are marked. Moreover, all vertices of valency one or two should be marked.

DEFINITION (3.7). *A unital metric m on a ribbon graph G is the assignment of a positive real number l_e for each edge $e \in X_1(G)$ in such a way that $\sum_{e \in X_1(G)} l_e = 1$.*

Given a T-marked ribbon graph with unital metric $(G, \mathbf{x}, m)$, for $t \in T$, we set

$$\begin{array}{ll} r_t = 0 & \text{if } t \text{ labels a vertex,} \\ r_t = \frac{1}{2}\sum_{\vec{e} \in [t]} l_e & \text{otherwise,} \end{array}$$

where $[t]$ denotes the boundary component of G labelled by t. We refer to $\{r_t\}_{t \in T}$ as the T-tuple of *half-perimeters* of G. Observe that, by definition of (unital) metric, we have that $\sum_{t \in T} r_t = 1$.

DEFINITION (3.8). *Given two T-marked ribbon graphs $(G, \mathbf{x})$ and $(G', \mathbf{x}')$, an isomorphism between the two is a bijection $\varphi : X(G) \to X(G')$ such that $\varphi \circ \sigma_{i,G} = \sigma_{i,G'} \circ \varphi$ for $i = 0, 1, 2$ and such that $\mathbf{x}' = \overline{\varphi} \circ \mathbf{x}$, where $\overline{\varphi} : X_0(G) \cup X_2(G) \to X_0(G) \cup X_2(G)$ is the map induced by φ. When the two marked ribbon graphs in question have unital metric, an isomorphism between them respecting the metrics is called an isometry.*

Let $(G, \mathbf{x}, m)$ be a P-marked ribbon graph of genus g with unital metric. We let P' be the subset of P indexing boundary components of G. Formally,

$$P' = \mathbf{x}^{-1}(X_2(G))\,. \tag{3.9}$$

Let $f : |G| \to S \smallsetminus P'$ be an embedding such that $S \smallsetminus P'$ deformation retracts onto $f(|G|)$. Then each connected component of $S \smallsetminus f(|G|)$ contains exactly one point $p \in P'$; let U_p be the component containing p. Then $U_p \smallsetminus \{p\}$ retracts onto a boundary component of $f(|G|)$. In the following, when speaking of embeddings of P-marked ribbon graphs, we shall always assume that they are is *compatible with the P-marking*, in the sense that $U_p \smallsetminus \{p\}$ retracts onto the boundary component labeled by p for every $p \in P'$. We write $[f]$ to indicate the isotopy class of f. The 4-tuple $(G, \mathbf{x}, m, [f])$ and another 4-tuple $(G', \mathbf{x}', m', [f'])$ are said to be isomorphic if there is an isometry between the corresponding marked graphs with unital metric that takes the isotopy class $[f]$ into $[f']$.

We will now define a natural map

(3.10)
$$q : |A^0(S,P)| \longrightarrow \left\{ \begin{array}{c} P\text{-marked genus } g \text{ embedded} \\ \text{ribbon graphs with unital metric} \end{array} \right\} \Big/ \text{isomorphism}$$
$$= \{(G, \mathbf{x}, m, [f])\} \Big/ \text{isomorphism}\,.$$

As we shall see, this map simply consists in passing from a proper arc system to its Poincaré dual. More precisely, start with a *proper* simplex $a = ([\alpha_0], \ldots, [\alpha_k]) \in A(S,P)$. We can take representatives $\alpha_0, \ldots, \alpha_k$ which are simple arcs and loops intersecting only in the points of P. We set
$$G^a = \cup_{i=0}^k \alpha_i\,.$$
We can view G^a as a (connected) ribbon graph embedded in S. For this graph, the permutation σ_0 is the counterclockwise rotation of edges issuing from a point $p \in P$, while σ_1 is the orientation-reversing operator. By Poincaré duality also the dual graph
$$G_a = (G^a)^*$$
is embedded in the surface S, and we let $f : G_a \to S$ be this embedding.

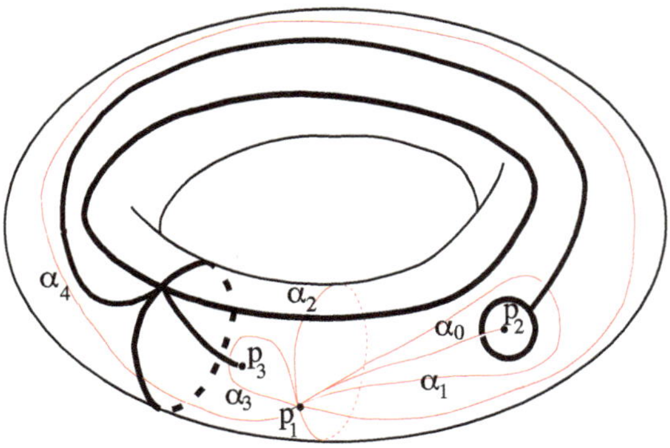

Figure 9.

In Figure 9, (S,P) is a 3-pointed genus 1 surface, $a = ([\alpha_0],[\alpha_1],[\alpha_2],[\alpha_3]) \in A^0(S,P)$ is a proper 3-simplex, and the marked ribbon graph G_a is the one drawn with a thick line. In G_a there are two marked boundary components and one marked vertex.

The ribbon graph G_a is called the *ribbon graph associated to the proper simplex a*. Of course, one has that $(G_a)^* = G^a$. Since a is a proper simplex relative to a pointed surface (S,P), we have a canonical marking $\mathbf{y} : P \to X_2(G^a) \cup X_0(G^a)$. Then also the associated ribbon graph G_a inherits a canonical P-marking $\mathbf{x} : P \to X_0(G_a) \cup X_2(G_a)$.

Finally observe that giving a point l in the interior $|a|^\circ$ of a simplex $|a| \subset |A^0|$ is tantamount to equipping the marked graph G^a (and therefore, dually, the graph G_a) with a unital metric which we denote by m_l. Now the map (3.10) is defined by

$$|a|^0 \ni l \mapsto q(l) = [(G_a, \mathbf{x}, m_l, [f])] . \tag{3.11}$$

We leave the proof of the following theorem as an exercise for the reader.

THEOREM (3.12). *The map*

$$q : |A^0(S,P)| \longrightarrow \left\{ \begin{matrix} P\text{-marked genus } g \text{ embedded} \\ \text{ribbon graphs with unital metric} \end{matrix} \right\} \Big/ \text{isomorphism}$$

defined by (3.11) is a bijection. Moreover, suppose that $l \in |a|^0$ and $q(l) = [(G_a, \mathbf{x}, m_l, [f])]$. Then the point $\lambda(l) \in \Delta_P$ is given by the P-tuple of half-perimeters of the metrized ribbon graph $(G_a, \mathbf{x}, m_l)$. Finally, if Γ_a denotes the stabilizer in $\Gamma_{S,P}$ of a simplex a, we have

$$\Gamma_a = \mathrm{Aut}((G_a, \mathbf{x})) . \tag{3.13}$$

Once the isomorphism (2.8) in Theorem (2.7) will be proved, the bijection q will make it possible to interpret the resulting cellular decomposition in terms of ribbon graphs; each cell will correspond to a P-marked, embedded ribbon graph $(G, \mathbf{x}, [f])$, up to isomorphism. Points in this cell will correspond to unital metrics on $(G, \mathbf{x})$.

Sometimes, but not always, thinking of $|A^0(S,P)|$ in terms of ribbon graphs can be very useful. Indeed, if one wants to have a clear idea of how the various cells of $|A^0(S,P)|$ are assembled together, the picture offered by ribbon graphs is preferable. In fact, the motion from one cell to an adjacent one can handily be described in terms of *Feynman moves*. The typical Feynman move is illustrated in Figure 10: the edge e is first contracted and then expanded as indicated.

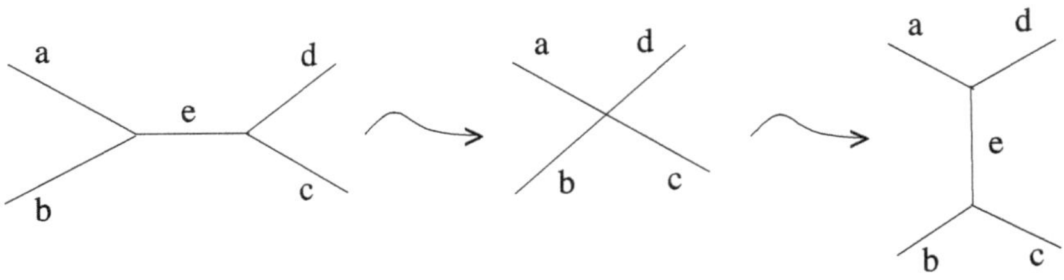

Figure 10.

To give an example, let us look at the case in which S is the Riemann sphere and P consists of three points. In the picture below we draw three simplices in $A(S,P)$,

$$a = ([\alpha],[\beta],[\gamma]), \quad b = ([\alpha],[\beta]), \quad c = ([\alpha],[\beta],[\delta]).$$

Clearly, these are proper simplices, i.e., they belong to $A^0(S,P)$, and of course b is a common face of a and c. But from the picture it is not immediately clear how to pass from $|a|$ to $|c|$ by crossing $|b|$.

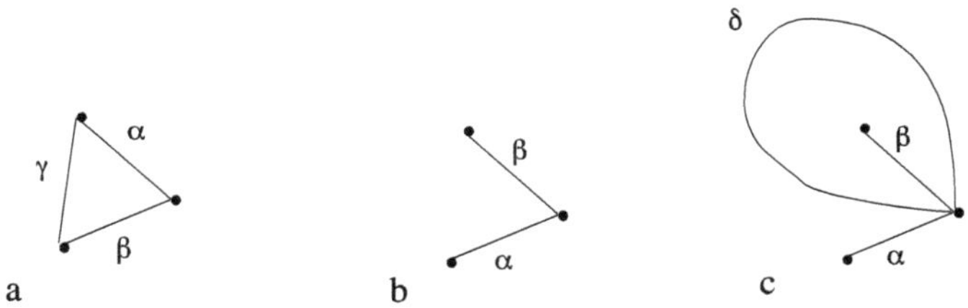

Figure 11.

On the other hand, thinking dually at the corresponding ribbon graphs G_a, G_b, and G_c, this passage immediately reveals itself as a simple Feynman move.

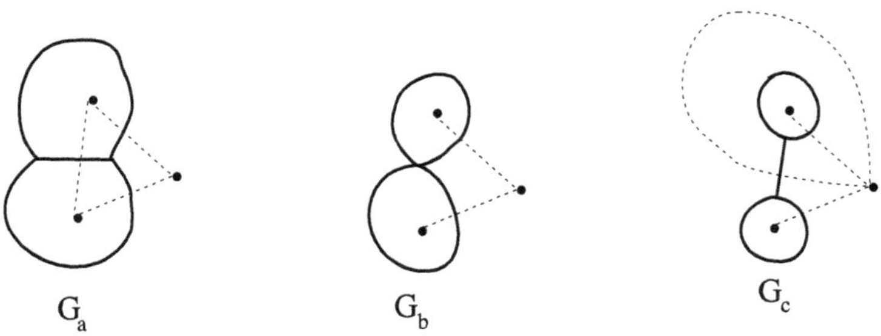

Figure 12.

Nevertheless, there is an advantage in working with arc complexes instead of ribbon graphs. This advantage is clear when one is considering the entire complex $A(S,P)$ and not just $A^0(S,P)$. As we shall see in the next section, the description of $|A(S,P)|$ in terms of ribbon graphs is more cumbersome, makes the description of the $\Gamma_{S,P}$-action less natural, and, to a certain extent, hides the simplicial nature of $A(S,P)$.

4. The idea behind the cellular decomposition of $M_{g,n}$.

Let us now give a brief sketch of how Theorem (2.7) is proved. Actually, we will limit ourselves to giving an idea of how the map (2.8) is defined. Let $(C, \mathbf{x})$ be a P-pointed smooth curve of genus g. Here, as usual, $\mathbf{x} : P \to C$ is an injective map. Write $n = |P|$, let $p_1, \dots, p_n$ be the elements of P, and set $x_i = \mathbf{x}(p_i)$, for $i = 1, \dots, n$. As we shall explain in the next section, the uniformization theorem for Riemann surfaces provides the surface $C \setminus \mathbf{x}(P)$ with a canonical hyperbolic metric, the Poincaré metric. In this metric the points $x_1, \dots, x_n$ appear as cusps. These cusps have infinite distance from the points in $C \setminus \mathbf{x}(P)$. For each cusp x_i, the set of horocycles around x_i is a canonical family of simple closed curves in $C \setminus \mathbf{x}(P)$, contracting to x_i. Given a point $r = (r_1, \dots, r_n) \in \Delta_P$, it is possible to choose a small constant c and a set $\eta_1, \dots, \eta_n$ of nonintersecting horocycles, one for each cusp, such that the length of η_i is cr_i (the constant c turns out to be uninfluential). For the sake of simplicity, let us assume that all the r_i are positive. Set $\eta = \eta_1 \cup \dots \cup \eta_n$. In general, given a point $q \in C \setminus \mathbf{x}(P)$, there is a *unique* shortest geodesic joining q to η. The main theorem we will prove in the next sections is that the locus of points q in C having the property that there are two or more shortest geodesics from q to η is in fact a ribbon graph $G \subset C$, which is called the hyperbolic spine of $(C, \mathbf{x}, r)$. This ribbon graph has the property that $C \setminus G$ is a union of n discs $D_1, \dots, D_n$, centered at the points x_i. Hence, the P-marking on C induces a P-marking on (the boundary components of) G. Furthermore, we can define a metric in G in the following way. Look at an edge e of G. This edge is on the boundary of one or two of the discs $D_1, \dots, D_n$. Suppose that e is in the boundary of D_i and D_j (and it may happen that $i = j$). Draw the geodesics from the vertices of e to x_i and x_j (see Figures 13 and 14). These geodesics cut out two arcs on η_i and η_j. Using elementary hyperbolic geometry, we will see that these two arcs have the same length $a(e)$. This is the length we assign to the edge e. In this way we equip G with a (unital) metric. In conclusion, the spine G is a P-marked metrized ribbon graph, embedded in C compatibly with the marking.

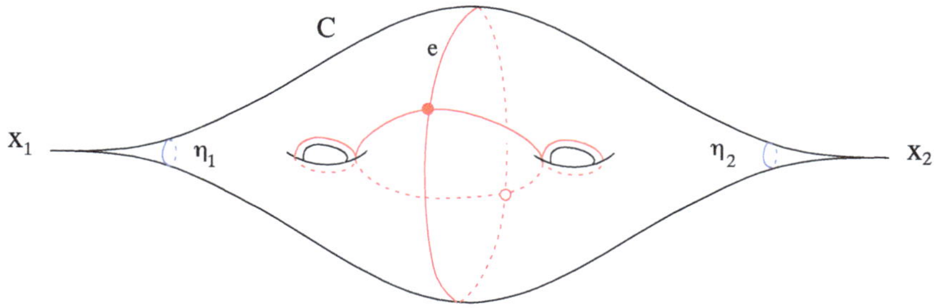

Figure 13.

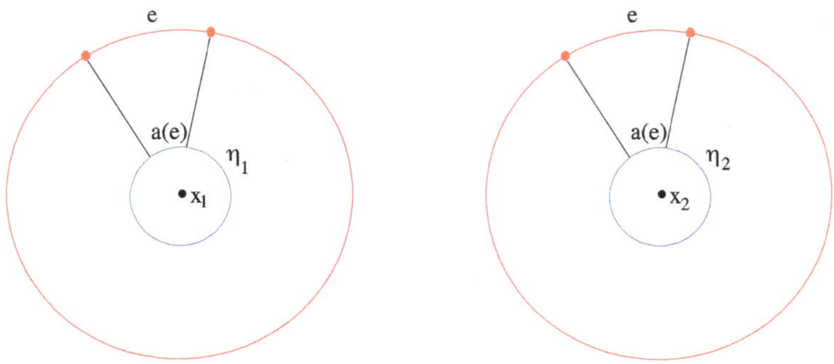

Figure 14.

Now suppose that a marking $f : (S, P) \to (C, \mathbf{x})$ is given. Then $f^{-1}(G) \subset S$ is a P-marked metrized ribbon graph embedded in S. Therefore, by (3.12), it determines a point of $|A^0(S, P)|$. By definition, this point is the image of $([C, \mathbf{x}, f], r) \in \mathcal{T}_{S,P} \times \Delta_P$ under the map Ψ. When some of the r_i are equal to zero, then the corresponding discs D_i shrink to their centers x_i, which then fall inside the graph G. As we shall see, the hyperbolic spine construction uses in an essential way the fact that the r_i are not all equal to zero.

We need to recall some basic facts about uniformization and hyperbolic geometry. The following two sections are devoted to this.

5. Uniformization.

As is well known, the *uniformization theorem* states that a simply connected Riemann surface is analytically equivalent to one of the following: the Riemann sphere $\widehat{\mathbb{C}}$, the complex plane $\mathbb{C}$, or the upper half-plane $\mathbb{H} = \{z \in \mathbb{C} : \operatorname{Im} z > 0\}$. Riemann surfaces whose universal cover is $\mathbb{C}$ are called *parabolic,* while those having $\mathbb{H}$ as universal cover are called *hyperbolic.*

DEFINITION (5.1). *Let C be Riemann surface. Let S denote one of the three surfaces $\widehat{\mathbb{C}}$, $\mathbb{C}$, $\mathbb{H}$. An analytic (universal) cover $f : S \to C$ is called a uniformization map.*

Uniformization depends continuously on parameters. In other words, the following result holds.

PROPOSITION (5.2). *Let $f : \tilde{\mathcal{C}} \to T$ be a continuous family of simply connected Riemann surfaces parameterized by a connected topological space T. Then all the fibers $\tilde{C}_t$, $t \in T$, are isomorphic to a fixed Riemann surface S, where S coincides with one of the three surfaces $\widehat{\mathbb{C}}$, $\mathbb{C}$, $\mathbb{H}$. Moreover, there is a homeomorphism $F : S \times T \to \tilde{\mathcal{C}}$ such that $pr_T = fF$ and $F_t : S \times \{t\} \to \tilde{C}_t$ is an analytic isomorphism.*

This result is well known. We will content ourself with giving a brief sketch of its proof. One way to prove the uniformization theorem is to deduce it from the Hodge decomposition theorem. Let us recall the statement of this theorem. Given a Riemann surface C, we consider the de Rham complex $(\mathcal{A}^\bullet(C), d)$. The complex structure on C induces a star operator $* : \mathcal{A}^p(C) \to \mathcal{A}^{2-p}(C)$ with the property that $*^2 = (-1)^p$. In terms of this operator one defines the adjoint $\delta = - * d *$ of d and the Laplace operator $\Delta = \delta d + d\delta : \mathcal{A}^1(C) \to \mathcal{A}^1(C)$. For $\omega, \varphi \in \mathcal{A}^1(C)$, set

$$\langle \omega, \varphi \rangle_S = \int_S \omega \wedge *\overline{\varphi}, \quad ||\omega||_S^2 = \langle \omega, \omega \rangle_S .$$

Also set

$$\begin{aligned} &\mathcal{A}^1(C)_{L^2} = \{\omega \in \mathcal{A}^1(C) \, : \, ||\omega||_S < \infty\}, \\ &\mathcal{H}^1(C) = \{\omega \in \mathcal{A}^1(C)_{L^2} \, : \, \Delta\omega = 0\}, \\ &E(C) = \{\omega \in \mathcal{A}^1(C) \, : \omega = df \, , \, f \in \mathcal{A}^0(C)\}, \\ &E'(C) = \{\omega \in \mathcal{A}^1(C) \, : \omega = \delta\mu \, , \, \mu \in \mathcal{A}^2(C)\}. \end{aligned}$$

The space $\mathcal{H}^1(C)$ is the space of *harmonic 1-forms on C*. Under the bilinear form $\langle \ , \ \rangle$, the space $\mathcal{A}^1(C)_{L^2}$ becomes a pre-Hilbert space, and the Hodge decomposition theorem states that there is an orthogonal decomposition

$$\mathcal{A}^1(C)_{L^2} = \mathcal{H}^1(C) \oplus E(C) \oplus E'(C).$$

Given a simply connected Riemann surface C, one may prove the uniformization theorem in three steps. The first one is the following.

LEMMA (5.3). *Let C be a simply conected Riemann surface, and let p be a point in C. Then there exists a meromorphic function f on C having only a simple pole at p, whose real part u satisfies the following Dirichlet conditions. If U is a sufficiently small neighborhood of p, then*

$$\begin{aligned} &||du||^2_{C-\overline{U}} < \infty, \\ &\langle du, dh \rangle = 0 \quad \forall h \in \mathcal{A}^0(C) \text{ with } h \equiv 0 \text{ near } p \text{ and } ||dh|| < \infty. \end{aligned} \tag{5.4}$$

The second step is the following.

LEMMA (5.5). *The function f constructed in the preceding theorem yields an injective analytic map $f : C \to \mathbb{C}$.*

The third step is, of course, the Riemann mapping theorem. The map f is the uniformizing map.

As we shall presently see, the validity of Proposition (5.2), that is, the continuous dependence on parameters of the uniformization, really rests on the proof of Lemma (5.3), so we will give a quick proof of this lemma, and we refer the reader to [645, page 220] for the proof of Lemma (5.5).

Proof of Lemma (5.3). Fix a point $p \in C$. As a first step, we show that there exists a unique differential ω having the following properties.

First of all, it is harmonic and exact in C except at p, where it has a singularity of the form dz/z^2, for some holomorphic coordinate z defined in a neighborhood U of p. Secondly, it satisfies the Dirichlet conditions (5.4) whith ω in place of du. To construct ω, let $\lambda(z)$ be a C^∞ function with support in U and identically equal to 1 near p. Set $\psi = d(\frac{\lambda(z)}{z})$. Then $\psi - \sqrt{-1} * \psi$ belongs to $\mathcal{A}^1(C)_{L^2}$. By the Hodge decomposition theorem we can write $\psi - \sqrt{-1} * \psi = \varphi + df + \delta\mu$, where φ is harmonic. It is now an easy matter to show that $\omega = \psi - df$ has all the required properties. Next, set $du = (\omega + \overline{\omega})/2$. Then u is a real harmonic function which is regular on C except at p where it has a singularity of the form $\mathrm{Re}(1/z)$. Moreover, du satisfies the Dirichlet conditions (5.4). Finally, we look at the differential $du + \sqrt{-1} * du$. If we can show that this differential form is exact, then any integral f of it has all the required properties. Thus it remains to show that $*du$ is exact. Let γ be a simple closed curve in C. We must show that $\int_\gamma *du = 0$. As C is simply connected, $C \setminus \gamma$ consists of two connected components which we call A and B. Assume that $p \in B$. Let T be a tubular neighborhood of γ, not containing p, and choose a C^∞ function h on C which is identically equal to 1 in the closure of A and identically equal to 0 in $B \setminus B \cap T$. Then

$$0 = \langle du, dh \rangle = \int_C dh \wedge *du = \int_{T\cap B} dh \wedge *du = \int_{\partial(T\cap B)} h * du - \int_{T\cap B} h\, d * du\,.$$

As u is harmonic in $C \setminus p$, we have $d * du = 0$. On the other hand, $\int_{\partial(T\cap B)} h * du = \int_\gamma *du$.

Q.E.D.

From the proof just presented we see that the fact that the uniformization map $f : C \to \mathbb{C}$ depends continuously on parameters really reduces to the fact that the Hodge decomposition does, and this is a well-known fact.

We will use the following corollary of the preceding proposition.

PROPOSITION (5.6). *Let $\pi : \mathcal{C} \to T$ be a continuous family of Riemann surfaces, parameterized by a contractible space T. Then all the fibers C_t, $t \in T$, of π have the same universal cover S, where S coincides with one of the three surfaces $\widehat{\mathbb{C}}$, $\mathbb{C}$, $\mathbb{H}$. Moreover there is a continuous map $G : S \times T \to \mathcal{C}$ such that $pr_T = \pi G$ and $G_t : S \times \{t\} \to C_t$ is a uniformization map.*

Proof. Let $\alpha : \tilde{\mathcal{C}} \to \mathcal{C}$ be a universal cover. Consider the map $f = \pi\alpha : \tilde{\mathcal{C}} \to T$. Since T is contractible, the fibers $\tilde{C}_t$ of this map are simply connected, so that $\alpha_t : \tilde{C}_t \to C_t$ is a universal cover and, as such, inherits a complex structure. These complex structures make $f : \tilde{\mathcal{C}} \to T$ into a continuous family of simply connected Riemann surfaces to which we can apply the preceding proposition. Q.E.D.

6. Hyperbolic geometry.

As we know, a Riemann surface whose universal cover is isomorphic to the upper half-plane $\mathbb{H}$ is said to be hyperbolic. From the uniformization theorem it follows that the only nonhyperbolic Riemann surfaces are $\widehat{\mathbb{C}}$, $\mathbb{C}$, $\mathbb{C}^* = \mathbb{C}\smallsetminus\{0\}$, and the complex tori. We also recall that a *Riemann surface of finite type* (g, n, m) is a Riemann surface obtained from a compact Riemann surface by removing n points and m disjoint closed sets, each homeomorphic to a closed disc. From now on we will concentrate our attention on hyperbolic Riemann surfaces, and in particular we will study the hyperbolic geometry of hyperbolic Riemann surfaces of finite type. In this section we collect, without proof, a number of basic results in the hyperbolic geometry of Riemann surfaces. These results are standard and can be found in several textbooks and articles, a list of which can be found in the bibliographical notes to this chapter.

Fuchsian groups.

The automorphism group of $\mathbb{H}$ is readily described. It is the group of *Möbius transformations with real coefficients:*

$$\begin{aligned}\mathrm{Aut}(\mathbb{H}) &= \left\{ z \mapsto \frac{az+b}{cz+d} \ : \ a,b,c,d \in \mathbb{R},\ ad-bc=1 \right\} \\ &\cong PSL(2,\mathbb{R}) = SL(2,\mathbb{R})/\{\pm I\}\,.\end{aligned}$$

Thus, a hyperbolic Riemann surface C is of the form

$$C = \mathbb{H}/\Gamma\,,$$

where Γ is a discrete subgroup of $PSL(2,\mathbb{R})$, i.e., a *Fuchsian group.* Since $\mathbb{H} \to C$ is a topological cover, we may assume that Γ acts on $\mathbb{H}$ without fixed points. Every element in $PSL(2,\mathbb{R})$, viewed as an automorphism of $\widehat{\mathbb{C}}$, has two, possibly coincident, fixed points which, when not lying on the extended real line $\widehat{\mathbb{R}} = \mathbb{R} \cup \{\infty\}$, come in pairs of complex conjugate points. An element T of $PSL(2,\mathbb{R})$ is said to be *elliptic* if $\mathrm{Tr}(T)^2 < 4$, *parabolic* if $\mathrm{Tr}(T)^2 = 4$, and *hyperbolic* if $\mathrm{Tr}(T)^2 > 4$. Elliptic elements have two complex conjugate fixed points, hyperbolic ones two fixed points on $\widehat{\mathbb{R}}$, and parabolic ones a single fixed point on $\widehat{\mathbb{R}}$. It follows that the Fuchsian group Γ contains only hyperbolic and parabolic elements. One can easily check that a hyperbolic element is conjugate in $PSL(2,\mathbb{R})$ to a *dilatation* $z \mapsto \lambda z$ for some $\lambda > 0$, $\lambda \neq 1$, while a parabolic element is conjugate to a *translation* $z \mapsto z+b$ for some $b \in \mathbb{R}$, $b \neq 0$.

The Poincaré metric.

The upper half-plane $\mathbb{H}$ can be equipped with the Poincaré metric given by

$$ds^2 = \frac{dx^2+dy^2}{y^2}\,.$$

In this metric, the area element is given by

$$dA = \frac{dx \wedge dy}{y^2}.$$

It is easy to check that the elements of $PSL(2, \mathbb{R})$ are orientation-preserving isometries for the Poincaré metric. It should also be noted that the group of *all isometries* of $\mathbb{H}$ is generated by $PSL(2, \mathbb{R})$ and by the transformation $z \mapsto -\bar{z}$. Now let C be a hyperbolic Riemann surface, and $\pi : \mathbb{H} \to C$ a uniformization map. Via π, the Poincaré metric on $\mathbb{H}$ induces a metric on C. Since any two uniformization maps for C differ by an element of $PSL(2, \mathbb{R})$, this metric is well defined, and it is called *the Poincaré metric of* C.

Recall that, on a Riemann surface C, a *conformal metric* is given, in a local coordinate $z = x + iy$, by

$$\lambda(z)^2(dx^2 + dy^2),$$

where $\lambda(z)$ is a C^∞ function on C. The curvature of a conformal metric is given by

$$K = -\Delta \log \lambda.$$

It follows that, on a Riemann surface C, the Poincaré metric is a conformal metric of constant curvature $K = -1$. The following easy lemma characterizes the Poincaré metric.

LEMMA (6.1). *The Poincaré metric on a hyperbolic Riemann surface is the unique metric which is conformal, complete, and with constant curvature* $K = -1$.

Geodesics on surfaces of finite type.

The geodesics for the Poincaré metric on $\mathbb{H}$ are the half-circles and the half-lines perpendicular to the real axis $\mathbb{R}$, so that any pair of points of $\mathbb{H}$ is joined by a geodesic arc. These geodesics satisfy the usual length-minimizing property in the sense that if γ is an arc of geodesic joining two points of $\mathbb{H}$, the length of γ is the infimum of the lenghts of all rectifiable arcs joining the two points in question. The Gauss–Bonnet theorem for the Poincaré metric implies that the area of a geodesic triangle only depends on the interior angles and says more precisely that

$$\int_V dA = \pi - \alpha - \beta - \gamma \tag{6.2}$$

for any geodesic triangle V with vertices in $\mathbb{H} \cup \widehat{\mathbb{R}}$ and interior angles α, β, γ.

We are now going to study geodesics on a hyperbolic Riemann surface C, especially in the case in which C is of finite type. Let then $C = \mathbb{H}/\Gamma$ be a hyperbolic surface. To better understand the geometry of C, it is perhaps useful to introduce the *Dirichlet fundamental region for* Γ. This region is constructed in the following way. Fix a point $p \in \mathbb{H}$. For each element $T \in \Gamma \smallsetminus \{1\}$, look at the geodesic joining p to $T(p)$. Consider first the geodesic segment $[p, T(p)]$ and then the geodesic γ passing through the midpoint of this segment and perpendicular to it.

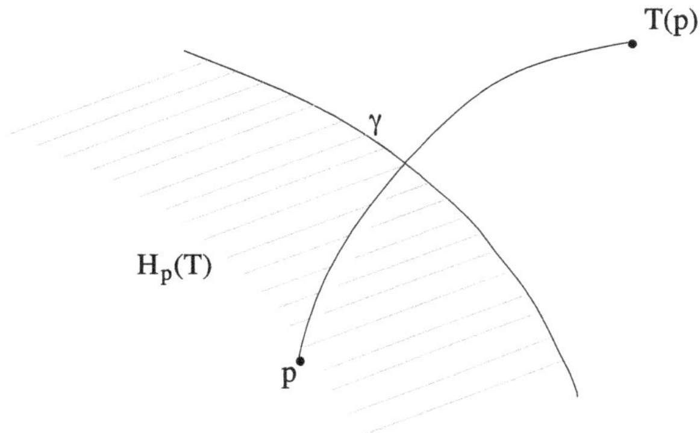

Figure 15.

The geodesic γ divides $\mathbb{H}$ in two regions, one of which contains the point p. We call $H_p(T)$ the closure of this region, and we set

$$F = \bigcap_{T \in \Gamma \smallsetminus \{1\}} H_p(T) .$$

The region F is a fundamental region for Γ in the sense that

$$\bigcup_{T \in \Gamma} T(F) = \mathbb{H} \qquad \text{and} \qquad \overset{o}{F} \cap T(\overset{o}{F}) = \emptyset \quad \text{for} \quad T \in \Gamma \smallsetminus \{1\} .$$

The sides of F are geodesic segments and come in pairs. To obtain the surface C, pairs of sides of F are identified by elements of Γ. The elements of Γ performing these identifications form a system of generators for Γ.

The picture of a fundamental region is particularly simple when the Riemann surface C is of finite type. Let us assume that C has n punctures and m boundary components, which we denote by $\delta_1, \dots, \delta_m$. In this case the fundamental region has a finite number of sides. If we take the closure $\overline{F}$ of F in $\mathbb{H} \cup \widehat{\mathbb{R}}$, then we see in $\widehat{\mathbb{R}}$ a finite number of vertices of $\overline{F}$, each corresponding to a puncture of C. Such a vertex $v \in \widehat{\mathbb{R}}$ is called an *improper vertex* of F, and it is the fixed point of a parabolic element $T_v \in \Gamma$. The two sides of F concurring in v are paired by T_v. Every parabolic element in Γ is a power of some element of the form T_v, where $v \in \widehat{\mathbb{R}}$ is a vertex of *some* fundamental region for Γ.

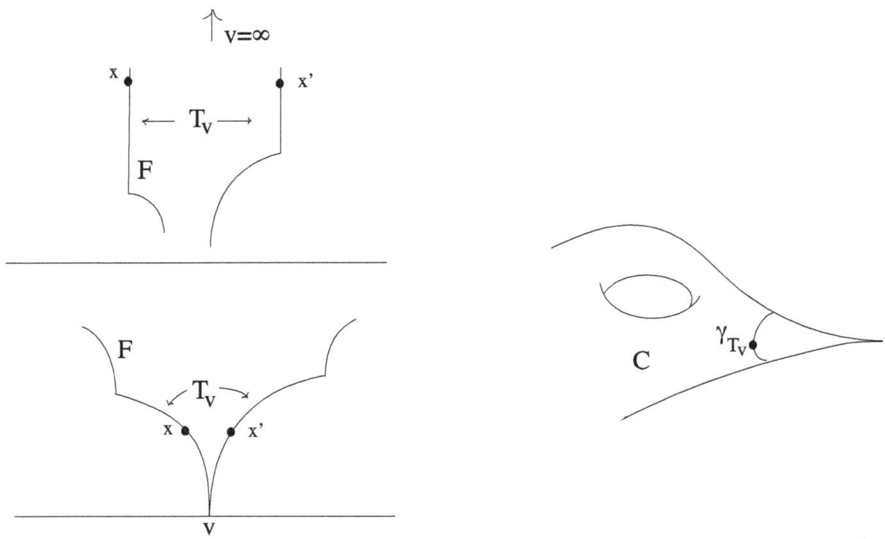

Figure 16.

In the context of hyperbolic geometry the punctures of C are also called the *cusps of* C. Apart from the improper vertices, the intersection of $\overline{F}$ with $\widehat{\mathbb{R}}$ also consists of a finite number of intervals corresponding to the boundary components of C. These are called the *improper sides of* F. The two sides of F abutting at the vertices of an improper side are paired by a *hyperbolic* transformation $T \in \Gamma$.

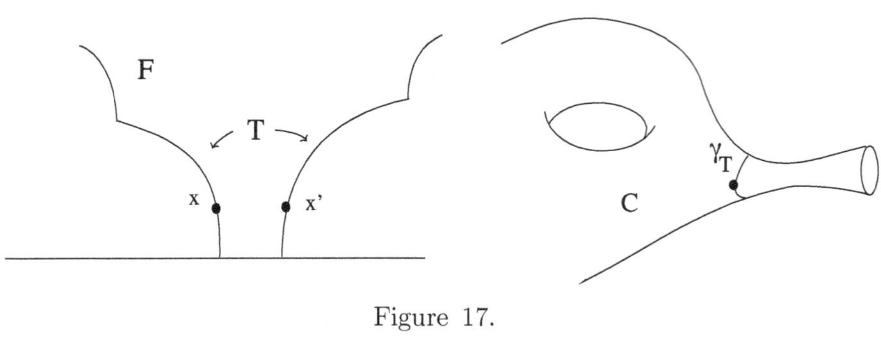

Figure 17.

We introduce the following notation, as illustrated in Figures 16 and 17. Let $T = T_v$ (resp., T) be a parabolic (resp., hyperbolic) element of Γ pairing two sides of F concurring in an improper vertex v (resp., pairing two improper sides of F). Let x and x' be points on these sides which are paired by T. We let γ_T denote the loop in C which is the image of a simple arc in F joining x and x', and we denote by $[[\gamma_T]]$ the free homotopy class of γ_T. We will show that, when T is hyperbolic, $[[\gamma_T]]$ contains a (unique) geodesic representative, while it contains no geodesic representative when T is parabolic. We need the following definition.

DEFINITION (6.3). *Let $T \in PSL(2, \mathbb{R})$ be a hyperbolic element. The geodesic in $\mathbb{H}$ joining the two fixed points of T is called the axis of T and is denoted by α_T.*

By definition, if z_1, $z_2 \in \widehat{\mathbb{R}}$ are the two fixed points of T, then α_T is the semicircle in $\mathbb{H}$ joining z_1 and z_2 and orthogonal to the real axis.

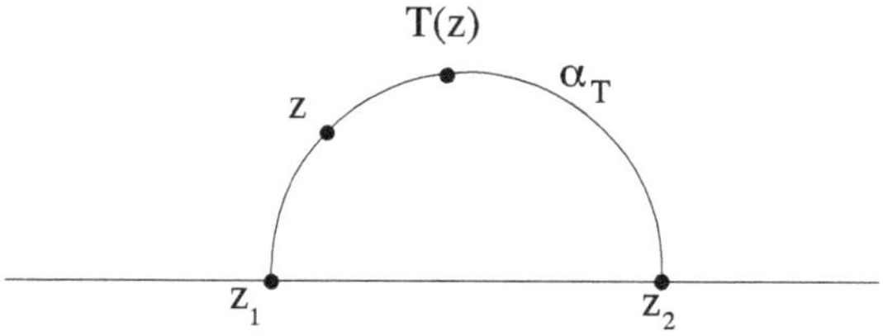

Figure 18.

Observe that T leaves α_T invariant. If z is a point of α_T, then also $T(z)$ is, and, under the projection $\mathbb{H} \to C$, the arc of α_T joining z to $T(z)$ projects to the closed *simple* geodesic

$$\beta_T = \alpha_T / < T > ,$$

which is a representative of the free homotopy class of γ_T. This construction also shows that this is the unique simple closed geodesic representative (it may well happen that there are representatives that consist of a geodesic loop that crosses itself). It is not difficult to show that the length of the geodesic β_T in the Poincaré metric of C is given by the formula

$$4 \cosh^2 \left(\frac{l(\beta_T)}{2} \right) = \mathrm{Tr}^2(T) .$$

Since both sides of this equation are invariant under conjugation of T, it suffices to prove this formula when $T(z) = \lambda z$, which is straightforward. It is interesting to look at this formula in the case in which T is the hyperbolic transformation pairing the two sides that are adjacent to an improper side. In the picture below we show the geodesics $\alpha_T \subset \mathbb{H}$ and $\beta_T \subset C$. If, in a family, the boundary component δ tends to a puncture, then the hyperbolic transformation T tends to a parabolic one, and the length of β_T tends to zero.

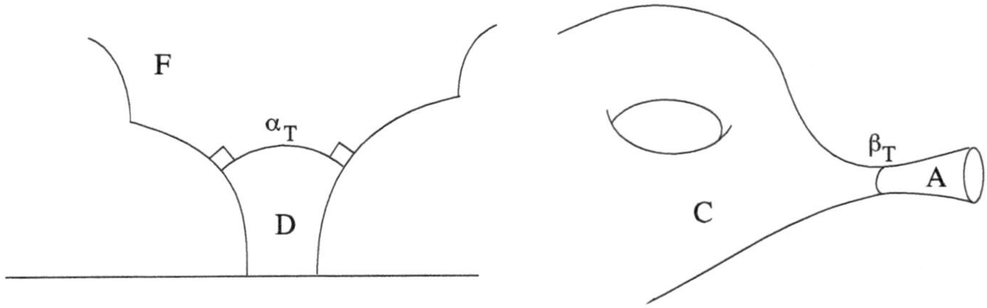

Figure 19.

This leads us to analyzing the case of a parabolic transformation. Let us consider a parabolic transformation T corresponding to a cusp of C, as we explained above. Look at a fundamental region F for the group Γ. Up to conjugation, we can always assume that $T(z) = z + a$, so that the improper vertex is ∞. Any horizontal line $y = c$ is T-invariant, so that any segment $\{0 \leq x \leq a,\ y = c\}$ with $c > 0$ and sufficiently large projects onto a simple closed curve in C belonging to the free homotopy class of γ_T. These curves are called *horocycles.* The region $\{0 \leq x \leq a,\ y \geq c\} \subset \mathbb{H}$ is called a *horocycle region.*

DEFINITION (6.4). *A horocycle region in $\mathbb{H}$ is said to be standard if it has unit (hyperbolic) area. If T is a parabolic element of Γ corresponding to a cusp, we denote by $H(T) \subset \mathbb{H}$ a standard horocycle region.*

DEFINITION (6.5). *A horocycle region on a hyperbolic Riemann surface C is a portion of C which is isometric to $\{z : \mathrm{Im}(z) \geq 1\}/ < z \mapsto z + a >$ for some $a > 0$. Its boundary is called a horocycle.*

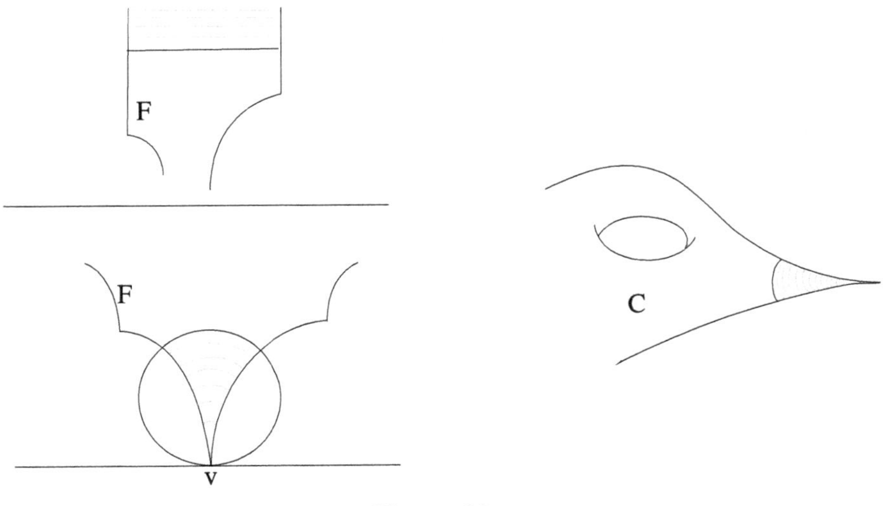

Figure 20.

In the preceding picture the two shaded areas on the left are horocycle regions with finite and infinite vertices in $\mathbb{H}$, while the shaded area on the right is the corresponding horocycle region in the Riemann surface $C = \mathbb{H}/\Gamma$. Of course, a horocycle is not a geodesic. Indeed, by the above description, when T is parabolic, there is no geodesic representative of the free homotopy class of γ_T. The following results holds.

LEMMA (6.6). *Under the projection $\mathbb{H} \to C = \mathbb{H}/\Gamma$, the surface $H(T)/ < T >$ injects in C. Its image is called a standard horocycle region for C. The standard horocycle regions of C are disjoint.*

In what follows we will also need the following elementary result.

LEMMA (6.7). *Let σ_1 and σ_2 be two disjoint horocycles in $\mathbb{H}$. Then the locus of points in $\mathbb{H}$ which are equidistant from σ_1 and σ_2 is a geodesic.*

Proof. It is an exercise to prove that, under the action of $PSL(2,\mathbb{R})$, we can assume that σ_1 and σ_2 have the same euclidean radius. But in this case the result is trivial.

Q.E.D.

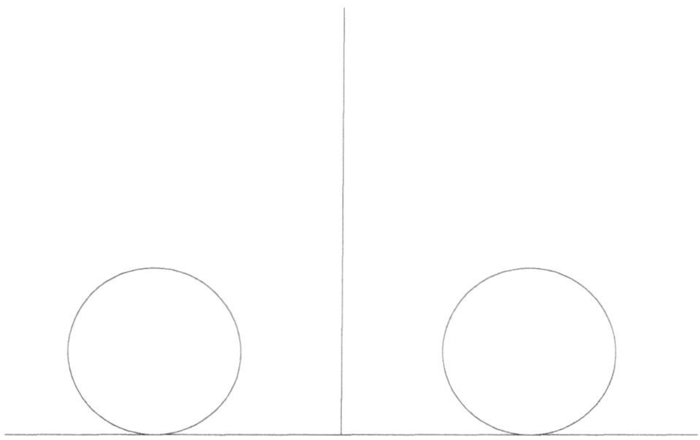

Figure 21.

The intrinsic metric for surfaces of finite type.

Let $C = \mathbb{H}/\Gamma$ be a hyperbolic Riemann surface of finite type. Let $\delta_1, \dots, \delta_m$ be its boundary components. Consider the double C^d of C, which is a punctured surface of finite type without boundary components. The *intrinsic metric on C* is, by definition, the metric on C induced by the Poincaré metric on C^d via the canonical injection $C \subset C^d$. Notice that the boundary curves of C are geodesics of finite length in the intrinsic metric. In fact, if δ is such a boundary curve, let α be the unique geodesic in C^d which is freely homotopic to δ. If J is the canonical involution of C^d, also $J(\alpha)$ is a geodesic, freely homotopic to δ. But then $\alpha = J(\alpha) = \delta$.

We are now going to construct a subsurface $C^k \subset C$, which is called the Nielsen kernel of C, and a surface C^e, plus a canonical immersion $C \subset C^e$, which is called the Nielsen extension of C. Let us start with the *Nielsen kernel.* Look at a fundamental region $F \subset \mathbb{H}$ for Γ. The boundary components of C correspond to intervals $I_1, \dots I_m \subset \widehat{\mathbb{R}}$ with

$$\bigcup_{j=1}^{m} I_j = \overline{F} \cap \widehat{\mathbb{R}}, \quad I_j \cap I_k = \emptyset, \quad j \neq k\,.$$

Let us fix our attention on one of these intervals $I = I_j$. Then there is a hyperbolic element T pairing the two sides of F abutting at the endpoints of I. Look, as in Figure 19, at the region $D \subset F$ bounded by these two sides, by the interval I, and by the axis α_T of T. Look at the simple closed geodesic β_T which is the image of α_T in C. The annular subregion

$$A = D/\langle T \rangle \subset C$$

is called the *funnel adjacent to* β_T (see Figure 19). For each interval I_j, consider the corresponding funnel A_j, $j = 1, \dots, m$. By construction, these funnels are disjoint, and the Nielsen kernel of C is defined to be the Riemannn surface

$$C^k = C \smallsetminus \{A_1 \cup \cdots \cup A_m\}\,.$$

It is an exercise to prove that the modulus of the annular region A_j is given by

$$M(A_j) = \frac{\pi^2}{l_j}\,,$$

where l_j is the hyperbolic length of β_{T_j}. We next define the *Nielsen extension of* C. For each of the m boundary curves δ of C, denote by $l(\delta)$ its length in the intrinsic metric of C. Set

$$\rho_\delta = e^{\frac{\pi^2}{l_\delta}}\,.$$

Consider the annulus $A_\delta = \{z \in \mathbb{C} : 1 < |z| < \rho_\delta\}$ and the first quadrant Q in $\mathbb{H}$. Equip A_δ with the Poincaré metric via the map $f : Q \to A_\delta$ defined by

$$f(z) = e^{-2\pi\sqrt{-1}\frac{\log z}{l_\delta}}\,.$$

For this metric, the circle $|z| = \rho_\delta$ is a geodesic of length l_δ. Attach A_δ, equipped with this metric to C, equipped with the intrinsic metric, via an isometry between δ and $|z| = \rho_\delta$. One thus obtains a surface C^e endowed with a Riemannian metric. This is the Nielsen extension of C. We have the following elementary result.

Lemma (6.8). *Let C be Riemann surface of finite type. Then*

a) *the restriction of the Poincaré metric on C to its Nielsen kernel C^k coincides with the intrinsic metric of C^k;*
b) *the Nielsen extension C^e is the unique surface whose Nielsen kernel is C.*

In particular, we have

$$(C^e)^k = C\,, \qquad (C^k)^e = C\,.$$

Combining a) and b), we also have that

c) *the restriction of the Poincaré metric on C^e to its kernel C coincides with the intrinsic metric of C.*

Continuously varying metrics.

Lemma (6.9). *Let $\pi : \mathcal{C} \to T$ be a continuous family of hyperbolic Riemann surfaces of finite type, parameterized by a locally contractible space T. Then both the Poincaré and the intrinsic metric on the fibers of π vary continuously.*

Proof. The problem is local. We may therefore assume that T is contractible. By Proposition (5.6), there is a continuous map $G : \mathbb{H} \times T \to \mathcal{C}$ such that $pr_T = \pi G$ and $G_t : \mathbb{H} \times \{t\} \to C_t$ is a uniformization map. For each $t \in T$, the Poincaré metric on the fiber $C_t = \pi^{-1}(t)$ is the one induced by the Poincaré metric on $\mathbb{H}$ via G_t. It follows that the Poincaré metric on C_t varies continuously with $t \in T$. This result implies the analogous one on the intrinsic metric. In fact, by the construction and the uniqueness of the Nielsen extension, also the family of Riemannn surfaces $\pi^e : \mathcal{C}^e \to T$, whose fiber $(\pi^e)^{-1}(t)$ is the Nielsen extension $(C_t)^e$ of C_t, is a continuous family. Therefore, the Poincaré metric on fibers of this family varies continuously. As a consequence of point c) of the previous lemma, the intrinsic metric on fibers of π also varies continuously.

Q.E.D.

The Collar Lemma.

Let C be a hyperbolic surface. Let γ be a simple closed geodesic. The subdomain

$$K(\gamma, \eta) = \{x \in C : d(x, \gamma) < \eta\}$$

is called a *collar of γ with width η* if it is an annular neighborhood of γ.

Lemma (6.10) (The Collar Lemma). *There exist a universal (small) constant L and a real function λ defined in the interval $[0, L]$ and strictly decreasing from ∞ to 0 having the following property. Given* *any complete hyperbolic surface C and any simple closed geodesic α of length $l(\alpha) < L$, the set*

$$K(\alpha) = \{x \in C : d(x, \alpha) < \lambda(l(\alpha))\}$$

is isometric to a collar about α bounded by two curves of length L and constant geodesic curvarture. Moreover, any two such collars are disjoint and are also disjoint from the standard horocycle regions of C.

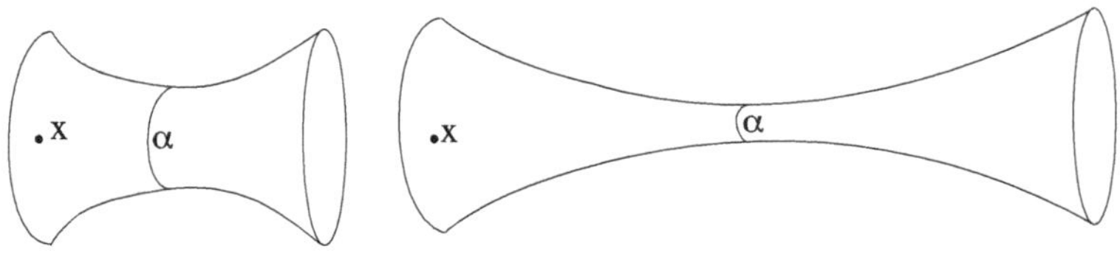

Figure 22.

7. The hyperbolic spine and the definition of Ψ.

We fix, once and for all, a smooth stable P-pointed genus g curve $(C; \mathbf{x})$. As usual, we set $x_p = \mathbf{x}(p)$ for every $p \in P$. We set $X = C \smallsetminus \mathbf{x}(P)$. We also fix a uniformization map $u : \mathbb{H} \to X$, and we equip X with the Poincaré metric. We shall refer to the x_p as the cusps of X. Let Γ be the Fuchsian group corresponding to u, so that

$$\mathbb{H}/\Gamma = X .$$

Let a function

$$r : P \to [0, 1]$$

be given and set $r_p = r(p)$. The number r_p is called the *weight of x_p*. Assume that

$$\sum_{p \in P}^{n} r_p = 1 . \tag{7.1}$$

Set

$$P^+ = \{p \in P : r_p \neq 0\}; \quad \mathbf{x}^+ = \{x_p : p \in P^+\}; \quad \mathbf{x}^0 = \{x_p : p \notin P^+\} .$$

For each $p \in P^+$, choose a horocycle η_p around the cusp x_p of length

$$l(\eta_p) = c r_p , \quad p \in P^+ , \tag{7.2}$$

where c is a (small) positive constant, to be chosen later. A first requirement, which is fulfilled if c is small, is that the horocycle regions bounded by the η_p are disjoint. Denote by H_p the region bonded by η_p and set $H = \cup\{H_p : p \in P^+\}$. When we want to stress the dependence of H on the particular choice of the function r, we write $H(r)$ instead of H. We set

$$G^0 = \{x \in X : \exists \text{ at least 2 shortest geodesics from } x \text{ to } H\},$$
$$V = \{x \in X : \exists \text{ at least 3 shortest geodesics from } x \text{ to } H\}.$$

For a point $x \in X$, we denote by $d(x, H)$ the distance between x and H. The notion of shortest geodesic from x to H makes sense also when x is a cusp belonging to $\mathbf{x}^0$ and therefore at infinite distance from H. Such a shortest geodesic is, by definition, a geodesic γ from x to H having the property that, for every $y \in \gamma \cap X$, the segment of γ between y and H is a shortest geodesic. In what follows we will often consider the closure of G^0 in C. We set

$$G = \text{closure of } G^0 \text{ in } C.$$

THEOREM (7.3). *The subset $G \subset C$ is a connected (ribbon) graph with finitely many vertices and finitely many edges. The edges of G are (the closure of) geodesic segments for the Poincaré metric of X. The set of vertices of G is the set $V \cup \mathbf{x}^0$. The vertices in V all have valency greater than or equal to 3. The vertices in $\mathbf{x}^0$ may be bivalent or univalent. Finally,*

$$C \setminus G = \coprod_{p \in P^+} \Delta_p, \tag{7.4}$$

where Δ_p is topologically a disc with center $x_p \in \mathbf{x}^+$. Moreover, the graph G is independent of c, provided that c is sufficiently small.

Proof. From Lemma (6.7) it follows that G is a union of geodesic segments. For $x \in C$, denote by $\nu(x)$ the number of shortest geodesic arcs from x to H (for $x \in V$, this will turn out to be the valency of x as a vertex of G). Let us show that $\nu(x)$ is a locally finite function. Of course, we may restrict to the case where $x \notin H$. Denote by $d = d(x, H)$ the distance between x and H. This distance could be infinite if $x \in \mathbf{x}^0$. Choose a horocycle $\eta = \eta_p$ and a shortest geodesic segment γ from x to η.

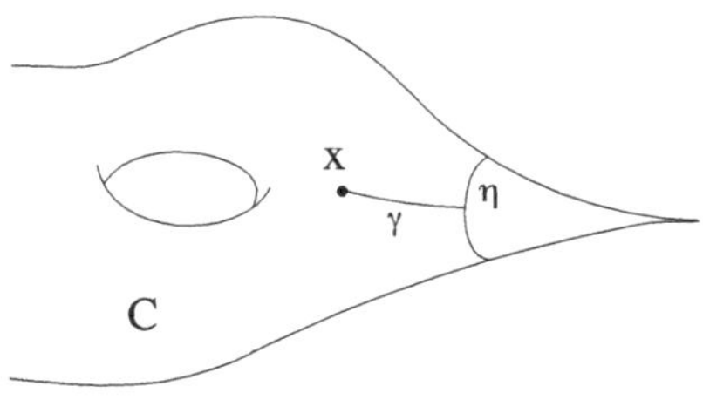

Figure 23.

We may lift this picture from X to $\mathbb{H}$. The inverse image of $H = H_1 \cup \cdots \cup H_s$ under $\pi : \mathbb{H} \to X$ is a countable union of disjoint horodiscs. We can assume that one of the horodiscs in the preimage of the horocycle region bounded by η is the half-plane $D_0 = \{z \in \mathbb{H} : \operatorname{Im} z \geq b\}$ (see Figure 24). Set

$$\pi^{-1}(H) = D_0 \cup D_1 \cup D_2 \cup \ldots$$

We may also assume that a lift of x is a point h with $\operatorname{Re}(h) = 0$. Since $x \notin H$, $\operatorname{Im}(h) < b$, and a lifting σ_0 of γ is given by the vertical segment joining h to ib (see Figure 24). Let us first assume that d is finite. The locus of points in $\mathbb{H}$ having hyperbolic distance not greater than d from h is a euclidean disc B with center on the imaginary axis. By construction, this disc is either disjoint from or tangent to the discs of $\pi^{-1}(H)$. But since the discs $D_0, D_1, D_2, \ldots$ are disjoint, only finitely many of them can be tangent to ∂B. Therefore, $\nu(x)$ is finite, and we also see that it is an upper semicontinuous function of x. Set $\nu = \nu(x)$. The ν shortest geodesic segments from x to H are the bijective images under π of the ν geodesic segments $\sigma_0, \sigma_1, \ldots, \sigma_{\nu-1}$ from h to the points of tangency between ∂B and the D_i.

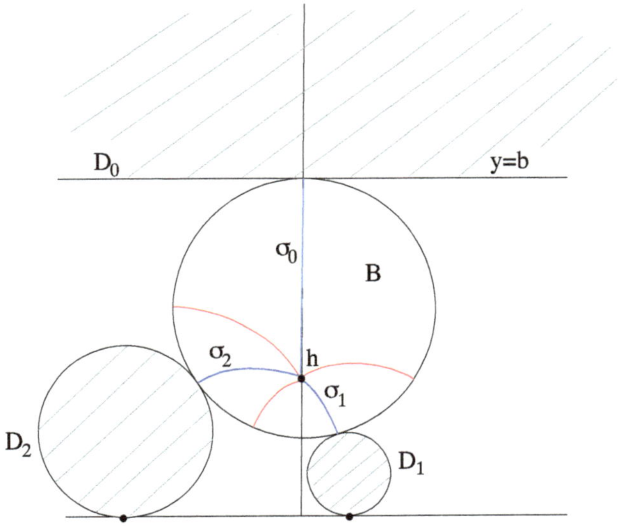

Figure 24.

The disc B is divided into ν sectors by the σ_i. Each sector is bounded by two geodesic segments σ_i and σ_j going from h to horodiscs D_i and D_j, respectively. This sector is divided into two parts by the geodesic segment which is the locus of points in B having the same distance from the horodiscs D_i and D_j (these are the red segments in Figure 24). In a small neighborhood of h, this configuration of geodesics maps homeomorphically under π onto a small neighborhood of x in G. Now let us show that $\nu = \nu(x)$ is finite also when $x \in \mathbf{x}^0$. Choose a horocycle ε around x which is disjoint from H. Let γ be a shortest geodesic from x to H and assume that the other end of γ lies on η_p for some p. The geodesic γ must then be perpendicular to both ε and η_p. Let v be the point of intersection of γ and ε. Arguing by contradiction, suppose that $\nu(x)$ is not finite. Then, by compactness, there is a sequence $\{v_k\}_{k \in \mathbb{N}}$ of points in ε converging to v, where each v_k is the intersection point of ε and a shortest geodesic γ_k from x to H. For large k, the endpoint of γ_k must necessarily lie in some η_q. But then there are two geodesic segments that are perpendicular to both horocycles ε and η_q, which is absurd.

At this stage, we know that G is a graph in C with a finite number of geodesic edges and a finite number of vertices, that the vertices of G^0 in X are exactly the points of V, and that each point x of V is a $\nu(x)$-valent vertex of G. We are now going to prove that $G^0 \cup \mathbf{x}^0$ is a deformation retract of $C \smallsetminus \mathbf{x}^+$. Since this last set is connected, we then get, as a consequence, that $G = G^0 \cup \mathbf{x}^0$, so that the points of $\mathbf{x}^0$ are (possibly univalent or bivalent) vertices of G. Let $x \in C \smallsetminus H$ and suppose that $x \notin G^0 \cup \mathbf{x}^0$. Then there is exactly one horocycle η_p which is nearest to x. Let γ be the shortest geodesic in X joining x and η_p. Now flow along γ away from η_p. Follow the deformed point on γ. Two cases may occur:

1) The deformed point x eventually lands in G^0.
2) It does not. Then η_p continues to be the nearest horocycle to x, while its distance from x continues to increase. Hence, x must converge to some point $x_q \in \mathbf{x}^0$.

In this way we obtain the desired retraction in the complement of H.

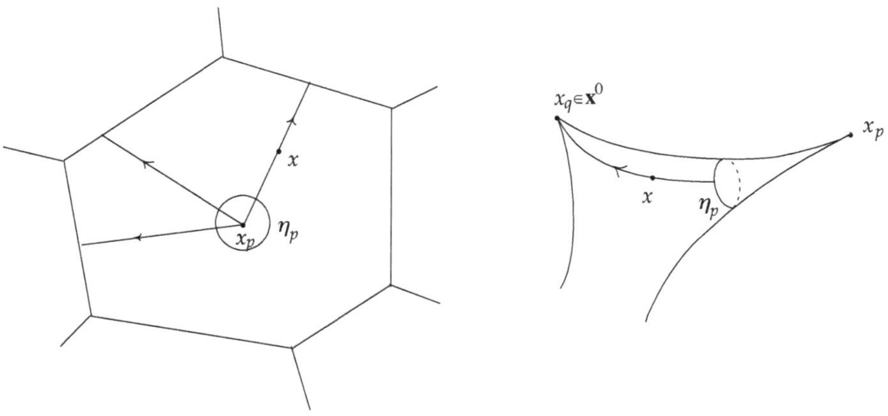

Figure 25. The retraction flow in cases 1) and 2).

If instead $x \in H_p \smallsetminus \{x_p\}$, it lies on a unique geodesic orthogonal to η_p. Then flow along this geodesic away from x_p until outside H_p and then continue as in the preceding case. This completes the definition of the retraction to $G^0 \cup \mathbf{x}^0$.

Now $C \setminus G$ is a union of components each containing exactly one point of $\mathbf{x}^+$, and $G \cap \mathbf{x}^+ = \emptyset$. Each retraction line contains exactly one point of $\mathbf{x}^+$. It follows that each component of $C \setminus G$ is topologically a disc with center a point of $\mathbf{x}^+$. From the construction it also follows that, as long as the constant c is small, the graph G does not change. Q.E.D.

DEFINITION (7.5). *The ribbon graph $G \subset C$ constructed in the preceding theorem is called the hyperbolic spine of $(C, \mathbf{x}, r)$.*

REMARK (7.6). It is important to remark that, in order to construct the spine G, we needed that $\mathbf{x}^+ \neq \emptyset$, i.e., that $r_p > 0$ for at least one p.

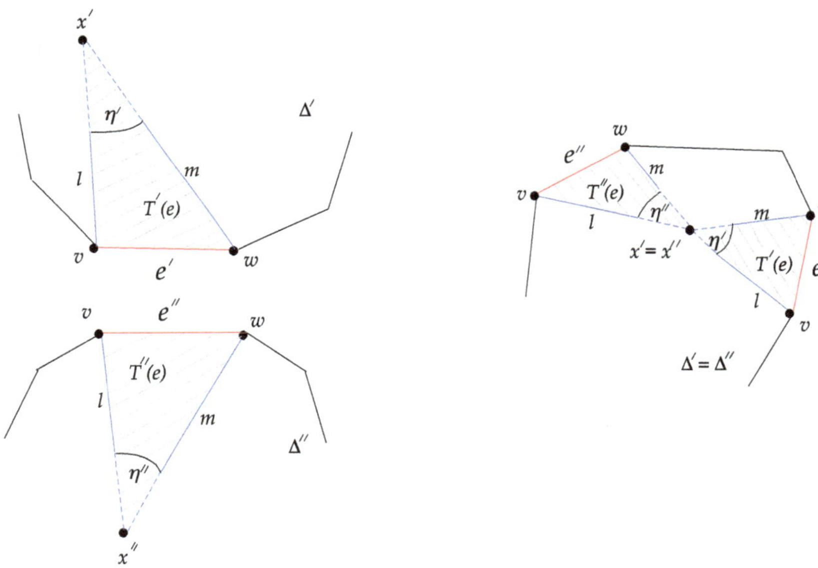

Figure 26.

From the construction of the spine G we can deduce a canonical decomposition of the curve C into a finite number of hyperbolic triangles. These triangles come naturally in pairs. Fix an edge e of G and consider the decomposition of $C \setminus G$ given in (7.4). Obviously, the edge e appears exactly twice in this decomposition, either as a pair of edges e' and e'' of two distinct discs Δ' and Δ'', or as a pair of distinct edges e' and e'' in the same disc $\Delta' = \Delta''$ (see Figure 26). Let $x', x'' \in \mathbf{x}^+$ be the "centers" of Δ' and Δ'', respectively. Join x' to the endpoints of e' with two geodesics in Δ'. Of course, these two geodesic may coincide if e' happens to be a closed loop. We denote by $T'(e)$ the region bounded by these three geodesic segments. We define $T''(e)$ similarly. Denote by v

and w the two, possibly coincident, vertices of e and by η' and η'' the two horocycles around x' and x'', respectively. Set

$$l = d(v, \eta') = d(v, \eta''), \quad m = d(w, \eta') = d(w, \eta'').$$

It is useful to lift this picture to $\mathbb{H}$, as we implicitly did in the course of the proof of the preceding theorem (see Figure 24), because there the hyperbolic geometry is more transparent. To simplify the notation, we will denote an object (edge, horocycle, etc.) and its lifting to $\mathbb{H}$ with the same symbol. We are looking at the two hyperbolic triangles $T'(e)$ and $T''(e)$ in $\mathbb{H}$. A first picture is the following. Up to a non-Euclidean transformation, we can assume that the two horocycles have equal radii and are based at two points x' and x'' of the real axis. Therefore, the edge e lies in the vertical line passing through the midpoint of the segment $[x', x'']$.

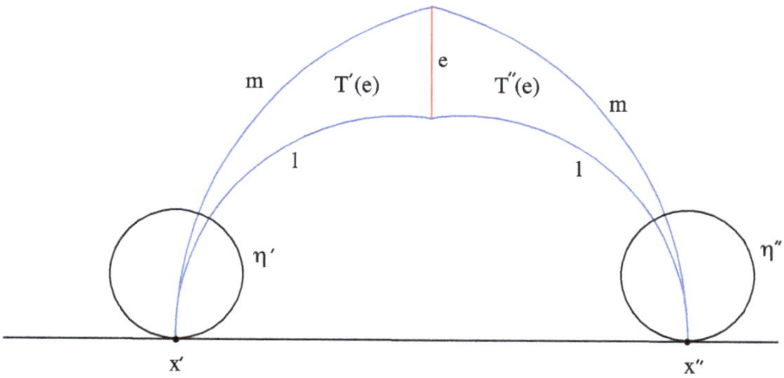

Figure 27.

From this picture we see that the length of the portion of the horocycle η' contained in $T(e')$ is equal to the length of the portion of the horocycle η'' contained in $T(e'')$. We denote this length by $a(e)$. Thus,

$$a(e) = l(\eta' \cap T(e')) = l(\eta'' \cap T(e'')).$$

Another way to lift the two triangles $T(e')$ and $T(e'')$ is by letting one of the two cusps, say x', be the point at infinity in $\mathbb{H}$. Look at the two triangular region defined above (see Figure 27). We may also assume that a lift of the horocycle η' around x' is the horizontal line $L = \{y = 1\}$. A lifting of the portion of the horocycle η' contained in $T(e')$ is a segment of length $a(e)$ on L. Starting from L and going down, towards the real axis, draw two vertical segments of lengths l and m, respectively. Then join the two endpoints of these segments with a geodesic. This is a lifting of e to $\mathbb{H}$, and we denote it again with the letter e. Finally, perform a hyperbolic reflection of the picture obtained so far about the edge e. This is the hyperbolic model of the "triangles" adjacent to the edge e that we will most often use.

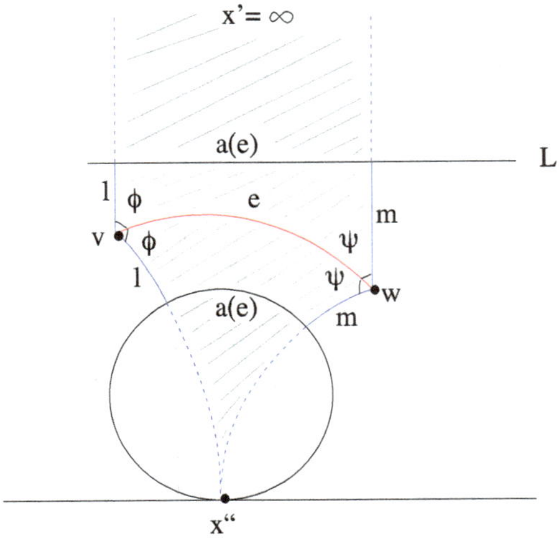

Figure 28. The shaded area is the building block $B(a(e), l, m)$.

We will denote with the symbol $B(a(e), l, m)$ the union of the two hyperbolic triangles which is the shaded area in the above picture. It only depends on the three lengths $a(e)$, l, and m. Since among the vertices of G are also the cusps belonging to $\mathbf{x}^0$, it may well happen that one or both vertices of e are in $\mathbf{x}^0$. In these cases the region $B(a(e), l, m)$ degenerates to the ones depicted in Figure 29, as l or m, or both, become infinite.

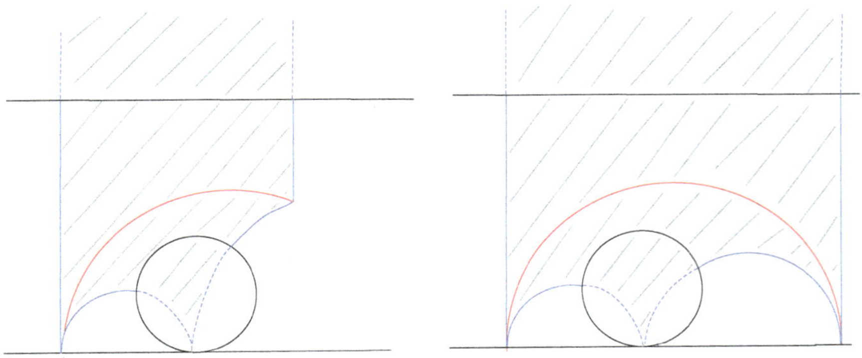

Figure 29. Degenerate cases for the building blocks $B(a(e), l, m)$.

Following the notation introduced in Definition (3.2), we consider the sets $X_0(G)$ and $X_2(G)$ of vertices and boundary components of G. We have

$$X_0(G) = V \cup \mathbf{x}^0 , \qquad X_2(G) = \mathbf{x}^+ .$$

Also, recalling (7.2) and (7.1), we have

$$\sum_{e \in X_1(G)} 2a(e) = \sum_{p \in P^+} l(\eta_p) = \sum_{p \in P^+} cr_p = c , \qquad cr_p = \sum_{e \in x_p} a(e) , \quad p \in P^+ ,$$

where, in the last sum, x_p is viewed as a point of $X_2(G)$, i.e., as the collection of edges around x_p. We set

$$\mu(e) = \frac{2a(e)}{c}\,. \tag{7.7}$$

It follows that

(7.8)

$$\mu : X_1(G) \to (0,1]\,, \qquad \sum_{e\in X_1(G)} \mu(e) = 1\,, \qquad \frac{1}{2}\sum_{e\in x_p} \mu(e) = r_p\,, \quad p \in P^+\,.$$

We are now in a position to define the map (2.8)

$$\Psi : \mathcal{T}_{S,P} \times \Delta_P \longrightarrow |A^0(S,P)|$$

in the statement of Theorem (2.7). Start with a point $([C,\mathbf{x},f],r) \in \mathcal{T}_{S,P} \times \Delta_P$, where $(C,\mathbf{x})$ is a P-pointed genus g Riemann surface, $f : S \to C$ is a diffeomeorphism such that $x_p = f(p)$ for every $p \in P$ and $r = \{r_p\}_{p\in P} \in \Delta_P$. Following Theorem (7.3), we consider the spine $G \subset C$ attached to the data $(C,\mathbf{x},r)$. Look at the graph $f^{-1}(G) \subset S$ (by abuse of notation, we often identify G and $f^{-1}(G)$). By duality, i.e., via the bijection (3.10), this ribbon graph determines a simplex $\sigma \in A^0(S,P)$. The vertices of this simplex are the arcs in S which are dual of the edges of G:

$$\sigma = (e^*)_{e\in X_1(G)}\,.$$

We then define

$$\Psi\left([C,\mathbf{x},f],r\right) = \sum_{e\in X_1(G)} \mu(e)e^* \in |\sigma| \subset |A^0(S,P)|\,. \tag{7.9}$$

One property of Ψ which immediately follows from its definition is that it is $\Gamma_{S,P}$-equivariant. Our next task is to prove that Ψ is bijective.

8. The equivariant cellular decomposition of Teichmüller space.

This section is devoted to the proof of the bijectivity of the map Ψ defined in (7.9).

PROPOSITION (8.1). *The map Ψ is bijective.*

We begin the proof by showing that Ψ is surjective. Start from a point z belonging to the interior of a simplex $|\sigma| \subset |A^0(S,P)|$ and let $G = G_\sigma \subset S$ be the ribbon graph which is the dual of σ. We may then write

$$z = \sum_{e\in X_1(G)} \mu(e)e^* \in \mathring{|\sigma|} \subset |A^0(S,P)| \quad \text{with} \quad \sum_{e\in X_1(G)} \mu(e) = 1\,. \tag{8.2}$$

Recalling definition (3.6), we set $P^0 = X_0(G) \cap P$ and $P^+ = P \smallsetminus P^0$. The graph G induces a topological triangulation of S, obtained by joining the points of P^+ to the vertices of G. For each edge e of G, set $a(e) = c\mu(e)/2$. Assign to each vertex v a number $l_v \in \mathbb{R} \cup \{\infty\}$. Then construct, as in Figures 28 and 29, the building blocks $B(a(e), l_v, l_w)$, where v and w are the vertices of e. Each one of these building blocks carries a natural hyperbolic structure. Assemble these blocks, following the rules dictated by the topological triangulation induced by G. In fact, the role of G will be exactly this: it tells us how to match the various building blocks. Since we are gluing blocks along geodesics we get, as a result, a hyperbolic structure on $S \smallsetminus P$ except, perhaps, at the *finite vertices of* G. Here we recall that

$$X_0(G) = V \cup P^0 .$$

We also recall that the vertices of V are called the finite vertices. To obtain a hyperbolic structure, we must verify that, at each finite vertex of G, the angles of the various triangles in the building blocks add up to 2π. More precisely, we must prove that there exists a choice of the quantities $\{l_v\}_{v \in X_0(G)}$ for which these angles do add up to 2π. It will turn out that this choice is in fact unique. Set $k = |V|$ and denote by $\mathbf{l} = \{l_v\}_{v \in V}$ an element of $\mathbb{R}^V_+$. Define the function

$$F : \mathbb{R}^V_+ \to \mathbb{R}^V$$

by setting

$$F(\mathbf{l}) = \{F_v(\mathbf{l})\}_{v \in V} ,$$

where $F_v(\mathbf{l})$ is the sum of the angles of the blocks $B(a(e), l_v, l_w)$ having v as a vertex. Looking at Figure 28, the angle contribution coming from the block $B(a(e), l_v, l_w)$ is either 2φ or 4φ, depending on whether the vertices v and w represent distinct or coincident vertices of G. As we just said, our goal is to prove that

$$(2\pi, \ldots, 2\pi) \in \operatorname{image}(F) . \tag{8.3}$$

To study the function F, it is essential to express it as a sum of contributions coming from the single blocks. As the horocyclic lengths $a(e)$ are given, we write $B(l, m) = B(a(e), l, m)$, where $l = l_v$ and $m = l_w$. Consider the angles φ and ψ indicated in Figure 28. Both φ and ψ are functions $\varphi(l, m)$ and $\psi(l, m)$ of l and m. Clearly, we have $\psi(l, m) = \varphi(m, l)$. Setting

$$\xi(l, m) = \varphi(l, m) + \varphi(m, l) ,$$

we get, by the Gauss–Bonnet theorem, that

$$\xi(m, l) = \pi - A(l, m) ,$$

where $A(l,m)$ is the area of either one of the two triangles forming $B(l,m)$. Clearly, $\varphi(l,m)$ is a real C^∞ function defined in $\mathbb{R}^2_+$. Let us see that it can be extended to a C^∞ function defined in $\mathbb{R}^2$. For this, we use the fact that, in our construction, we are free to choose arbitrarily the (small) positive constant c. To be precise, we should really write $\varphi(l,m,c)$ instead of $\varphi(l,m)$. Now go back to Figure 28 and move the line L from level $y=1$ to level $y=K$ for some $K>1$. Then all the horocyclic lengths get divided by K, and we have

$$\varphi(l,m,c)=\varphi(l+\log K, m+\log K, c/K)\,.$$

This simple remark allows us to extend the map φ to all of $\mathbb{R}^2$. The function φ has a number of properties whose (easy) verification is left to the reader:

a) φ is strictly decreasing in l and stricly increasing in m;
b) ξ is strictly decreasing in l and m;
c) $\lim\limits_{l,m\to-\infty}\xi=\pi$;
d) As l tends to $-\infty$ and m stays bounded from below, φ tends to π.

Finally, we notice that φ can be continuously extended when l, m, or both take on the value ∞ and that (see Figure 29)

e) $\varphi(\infty,m)=0$.

To stress the fact that each function φ comes from an edge e of G, we will write φ_e instead of φ. We have

$$F_v(1)=2\sum_{e\in[v]}\varphi_e(l_v,l_w)\,,$$

where $[v]$ is the set of edges incident to the vertex $v\in V$, and where $w\in X_0(G)$ is the other vertex of e. It could be that $v=w$; in this case the term $\varphi_e(l_v,l_w)$ appears twice in the above sum. Since the functions φ_e have been extended to $\mathbb{R}^2$, we obtain a C^∞ extension of F to $\mathbb{R}^V$. We are now going to describe the image of F in $\mathbb{R}^V$. Given a subset $W\subset V$, we define E_W as the subset of those edges of G which are incident to some vertex of W. We set $E_W=E^1_W\cup E^2_W$, where the edges of E^2_W have both ends in W, while those of E^1_W have only one. We consider the open convex set Z of $\mathbb{R}^V$ defined by

$$Z=\{\{x_v\}_{v\in V}\in\mathbb{R}^V : x_v>0\ \ \forall v\in V\,,\ \sum_{v\in W}x_v<2\pi|E_W|\ \ \forall W\subset V\}\,.$$

We want to prove the following lemma.

LEMMA (8.4). *The map F is a homeomorphism between $\mathbb{R}^V$ and Z.*

Proof. We already noticed that F is continuous. We claim that, to prove the lemma, is suffices to prove the following points:

A) image$(F) \subset Z$;
B) F is injective;
C) F is closed.

Indeed, the invariance of domain theorem, the continuity of F, and point B) imply that F is open. The convexity and hence the connectivity of Z, together with point C), tells us that image$(F) = Z$. We next prove A), B), and C). For the proof of A), first observe that, by definition, $F_v(\mathbf{l}) > 0$ for every $v \in V$. Also,

$$\begin{aligned}\sum_{v\in W} F_v(\mathbf{l}) &= 2\sum_{v\in W}\sum_{e\in[v]} \varphi_e(l_v, l_w)\\ &= \sum_{v\in W,\, e\in E^1_W,} 2\varphi_e(l_v,l_w) + \sum_{v\in W,\, e\in E^2_W,} 2\xi_e(l_v,l_w) < 2\pi|E^1_W| + 2\pi|E^2_W|.\end{aligned}$$

This proves A). For the proof of B), suppose that $\mathbf{l}$ and $\mathbf{l}'$ are two distinct points of $\mathbb{R}^V$. Set $W = \{v \in V : l'_v > l_v\}$. We may assume that $W \neq \emptyset$. Define $\mathbf{l}''$ by setting $l''_v = \max(l_v, l'_v)$. From a) and b) above it follows that

$$\sum_{v\in V} F_v(\mathbf{l}'') < \sum_{v\in V} F_v(\mathbf{l})\,.$$

As a consequence, $F_v(\mathbf{l}'') < F_v(\mathbf{l})$ for some $v \in V$. Using a) we see that v must belong to W, so that, using the fact that $l''_v = l'_v$ and a) again, we have

$$F_v(\mathbf{l}') \le F_v(\mathbf{l}'') < F_v(\mathbf{l})\,.$$

This shows that $F_v(\mathbf{l}) \neq F_v(\mathbf{l}')$, proving B). Instead of proving C), we will prove the stronger statement that F is proper. Let $\{\mathbf{l}^i\}_{i=1,2,\dots}$ be a sequence in $\mathbb{R}^V$ that escapes to infinity, i.e., that tends to the point at infinity in the one-point compactification of $\mathbb{R}^V$. Arguing by contradiction, suppose that its image $\{F(\mathbf{l}^i)\}_{i=1,2,\dots}$ does not escape to infinity in Z; then, passing to a subsequence, we may assume that that $F_v(\mathbf{l}^i)$ is bounded away from 0 for every v and that $\sum_{v\in W} F_v(\mathbf{l}^i)$ is bounded away from $2\pi|E_W|$ for every $W \subset V$. Passing to a further subsequence, we may assume that there is a decomposition $V = V^- \cup V^+ \cup V^0$, where l^i_v tends to $-\infty$ if $v \in V^-$, to $+\infty$ if $v \in V^+$, and to a finite limit if $v \in V^0$. Notice that $V^- \cup V^+$ cannot be empty since $\{\mathbf{l}^i\}$ is assumed to escape to infinity. If $v \in V^+$, we get that $F_v(\mathbf{l}^i)$ tends to 0 by e), so that $F(\mathbf{l}^i)$ tends to the boundary of Z, a contradiction. Therefore V^+ must be empty. Set $W = V^-$ and get

$$\sum_{v\in W} F_v(\mathbf{l}^i) = \sum_{v\in W,\, e\in E^1_W,} 2\varphi_e(l^i_v, l^i_w) + \sum_{v\in W,\, e\in E^2_W,} 2\xi_e(l^i_v, l^i_w)\,.$$

By d), the first sum on the right-hand side tends to $2\pi|E^1_W|$, while the second tends to $2\pi|E^2_W|$, by virtue of c). Again, $F(\mathbf{l}^i)$ tends to the boundary of Z, a contradiction. Thus F is proper.

Q.E.D.

Going back to the proof of Proposition (8.1), we may now prove (8.3). We must show that $(2\pi,\dots,2\pi)\in Z$. Equivalently, for each $W\subset V$, we must show that $2\pi|W|<2\pi|E_W|$. The vertices in V all have valency greater than or equal to 3. Hence, $2|E^2_W|+|E^1_W|\geq 3|W|$, and therefore $2|E_W|\geq 3|W|$, proving (8.3). We are now at the following stage. We started from a the point

$$z\in|A(S,P)|$$

as in (8.2). Giving such a point is like giving an embedded, metrized, marked ribbon graph $G_\sigma\subset S$. Out of this datum we constructed the various building blocks $B(a(e),l,m)$. By pasting together these building blocks, we constructed a P-pointed genus g curve $(C,\mathbf{x})$, where the underlying topological surface of C is S, and $\mathbf{x}:P\to C$ is just the inclusion $P\subset S$. We also constructed a connected ribbon graph $G\subset C$, isomorphic to the dual of σ, and a hyperbolic metric on $X=C\smallsetminus\mathbf{x}(P)$. By construction, this hyperbolic metric is complete and with curvature $K=-1$, so that it coincides with the Poincaré metric. Also, the various horocycle regions in the building blocks of C fit together to form horocycle regions H_p, $p\in P^+$, where $\partial H_p=\eta_p$ is a horocycle of length $r_p=\frac{1}{2}\sum_{e\in x_p}\mu(e)$ (see (7.8)). Furthermore, the graph G is contained in the spine of $(C,\mathbf{x},r)$. Let us provisionally denote by G' this spine. We want to show that $G'=G$. By construction, $G\subset G'$, $\mathbf{x}^0\subset G$, and $\mathbf{x}^0\subset G'$. Moreover, both $C\smallsetminus G$ and $C\smallsetminus G'$ are unions of discs centered at the points of $\mathbf{x}^+$. The only way in which G and G' could possibly differ is by a union of trees. Each one of these trees would have to terminate in one or more univalent vertices belonging to G' and not to G. But we know that the univalent vertices of both G and G' belong to $\mathbf{x}^0\subset G\cap G'$.

As a final step, we let $f:S\to C$ be the identity map. We then get

$$\Psi([C,\mathbf{x},f],r))=z$$

This proves the surjectivity of Ψ. It is now an exercise, which is left to the reader, to prove that, in this way, we actually defined an inverse of Ψ. This concludes the proof of Proposition (8.1).

Using Lemma (6.9), it is not difficult to show that Ψ is a continuous map. In order to show that it is a homeomorphism, it would then remain to show that it is closed. We will follow a different route. We will first extend Ψ to a continuous surjection $\widehat{\Psi}:\widehat{\mathcal{T}}_{S,P}\times\Delta_P\to|A(S,P)|$ and then deduce from this that Ψ is closed.

If one tries to extend the map Ψ to $\widehat{\mathcal{T}}_{S,P} \times \Delta_P$, the first rough idea that comes to mind is this: take a point $([C, \mathbf{x}, f], r) \in \widehat{\mathcal{T}}_{S,P} \times \Delta_P$ and perform the hyperbolic spine construction on each irreducible component of the stable curve C. This idea works, with one catch, namely that the spine construction can be performed only on those components containing at least one marked point x_p with $r_p > 0$. What we need is a notion of stable ribbon graph taking this particular feature into account.

9. Stable ribbon graphs.

In this section we will generalize the notion of ribbon graph in order to extend the map q of Theorem (3.12) to the whole of $|A(S,P)|$, where, as usual, S is a fixed topological surface of genus g, and P is a fixed nonempty finite subset of S. We first need some definitions.

We shall use the following notational conventions. Given a P-pointed dual graph γ, for any vertex v of γ, we indicate with P_v the subset of P indexing legs emanating from v, and by Q_v the set of halves of edges of γ emanating from v.

Definition (9.1). *A stable P-marked ribbon graph of genus g consists of data*

$$(\gamma, V^+, \{(G_v, \mathbf{z}_v)\}_{v \in V^+}) , \tag{9.2}$$

where

1) *γ is the dual graph of a stable P-pointed curve of genus g.*
2) *V^+ is a nonempty subset of the set V of vertices of γ such that:*
   i) *if $v \in V^+$, then $P_v \neq \emptyset$.*
   ii) *if v and w are in $V^0 = V \setminus V^+$, there is no edge in γ connecting v and w.*
3) *For each vertex $v \in V^+$, $(G_v, \mathbf{z}_v)$ is a $(P_v \cup Q_v)$-marked genus g_v ribbon graph such that $\mathbf{z}_v(Q_v) \subset X_0(G_v)$. In other words, we require the set Q_v to mark only vertices of G_v, and not boundary components.*

We will normally write $\mathbf{x}_v$ and $\mathbf{y}_v$ to indicate the restrictions of $\mathbf{z}_v$ to P_v and Q_v, respectively, and use the somewhat improper but suggestive notation $\mathbf{x}_v \cup \mathbf{y}_v$ to indicate $\mathbf{x}_v$. Often, as for ordinary ribbon graphs, we use the shorthand notation $(G, \mathbf{z})$ to denote the data (9.2).

The vertices, the edges, and the boundary components of a stable ribbon graph are, by definition, those of the various ribbon graphs G_v, for $v \in V^+$.

Definition (9.3). *A unital metric on a stable, P-marked ribbon graph (9.2) is the datum of a (not necessarily unital) metric on each ribbon graph $(G_v, \mathbf{x}_v \cup \mathbf{y}_v)$ having the property that the sum of the lengths of all the edges of all the G_v is equal to 1, and such that, moreover, the half-perimeter r_q is equal to zero for every $q \in Q_v$.*

We shall use the following terminology. A topological space $\overline{S}$, together with an injective mapping $\mathbf{x} : P \to S$, is said to be a *nodal P-pointed topological surface of genus g* if it is the underlying topological space of a stable P-pointed genus g curve $(C, \mathbf{x})$. The nodes of $\overline{S}$ and the irreducible components of $\overline{S}$ are the ones of C. The normalization of $\overline{S}$ is the underlying topological space of the normalization of C. We will say that $(S, \mathbf{x})$ is *stable* if $(C, \mathbf{x})$ is.

Given a finite set $L = \{L_1, \ldots, L_\delta\}$ of disjoint simple closed curves in S which do not intersect P, we let S_L be the topological space obtained by contracting each of the λ_i to a point, and let

$$\phi_L : S \to S_L \tag{9.4}$$

be the contraction map. Then S_L is naturally a P-pointed nodal topological surface.

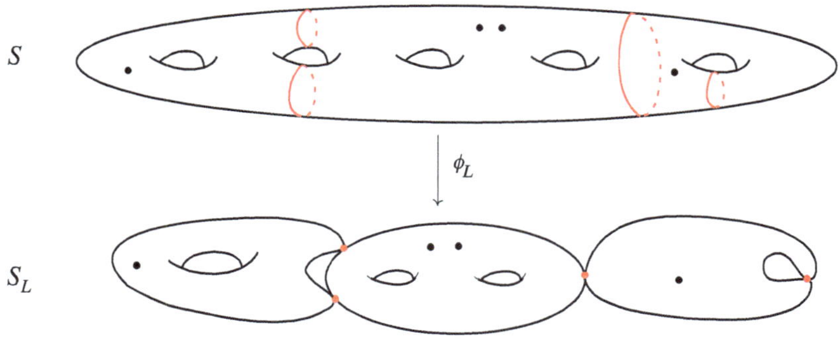

Figure 30.

Suppose that S_L is stable of genus g, and let $\gamma = \gamma_L$ be its dual graph. For each vertex v of γ, let $S_v = S_{L,v}$ be the corresponding component of the normalization of S_L.

Consider a stable P-marked ribbon graph

$$(G, \mathbf{x}) = \left(\gamma, V^+, \{(G_v, \mathbf{x}_v \cup \mathbf{y}_v)\}_{v \in V^+}\right).$$

An embedding of $(G, \mathbf{x})$ in the normalization $\widetilde{S}_L$ of S_L is a collection $f = \{f_v\}_{v \in V^+}$, where f_v is an embedding

$$f_v : (G_v, \mathbf{x}_v \cup \mathbf{y}_v) \hookrightarrow (S_v, P_v \cup Q_v)$$

such that $f_v(G_v)$ is a deformation retract of $S_v \smallsetminus P'_v$. We denote by $[f]$ the collection of isotopy classes $\{[f_v]\}_{v \in V^+}$.

The components S_v of $\widetilde{S}_L$ corresponding to vertices $v \in V^+$ are called the *positive components* of S_L, and the ones corresponding to vertices $v \in V^0$ are called the *ghost components* of S_L.

The notions of isometry between stable, P-marked ribbon graphs with unital metrics and the one of isomorphism between stable, embedded, P-marked ribbon graphs with unital metrics are the obvious ones.

Figure 31 illustrates the case of a stable 4-marked embedded ribbon graph of genus 5. The grey surface is the ghost component and corresponds to the unique (grey) vertex of V^0. The vertices of V^+ are the white ones. In the course of the proof of the theorem that follows it will be clear why ghost components are called as such.

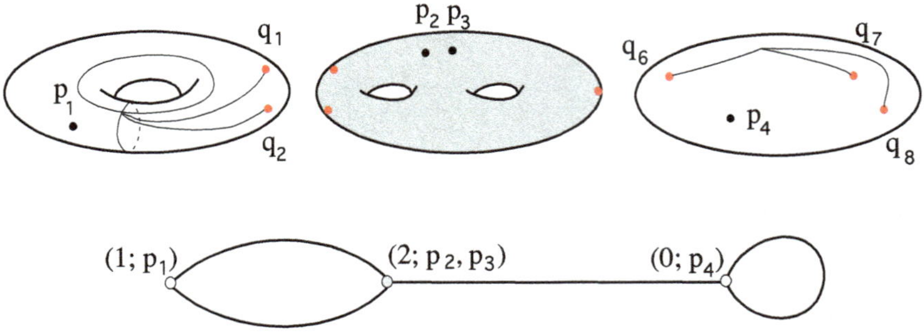

Figure 31.

The following theorem expresses points of $|A(S,P)|$ in terms of isomorphism classes of stable, marked, embedded ribbon graphs with unital metric. In proving this theorem we will understand the meaning behind the definition of ghost component. In most applications a ghost component is just a component of S not containing any point of P.

THEOREM (9.5). *The bijection q of Theorem (3.12) extends to a bijection*

$$\overline{q} : |A(S,P)| \longleftrightarrow \left\{ \begin{array}{c} \text{stable } P\text{-marked genus } g \text{ embedded} \\ \text{ribbon graphs with unital metric} \end{array} \right\} \Big/ \text{isomorphism}\,.$$

In this bijection, as in Theorem (3.12), given a simplex $a \in A(S,P)$ and a point $l \in |a|^0$, the point $\overline{q}(l)$ is the isomorphisom class of an embedded stable P-pointed genus g ribbon graph $(G_a, \mathbf{x}, m_l, [f])$ with unital metric m_l. The point $\lambda(l) \in \Delta_P$ is given by the $|P|$-tuple of half-perimetrs of the metrized stable ribbon graph $(G_a, \mathbf{x}, m_l)$. The embedded stable graph $(G_a, \mathbf{x}, [f])$ only depends on a. The stabilizer Γ_a of a in $\Gamma_{S,P}$ possesses a normal subgroup Γ'_a acting as the identity on $|a|$ and such that

$$\Gamma_a/\Gamma'_a = \mathrm{Aut}((G_a, \mathbf{x}))\,. \tag{9.6}$$

Proof. Let

$$a = ([\alpha_0], \ldots, [\alpha_k]) \in A(S,P)\,, \qquad l \in |a|^0\,. \tag{9.7}$$

Let $\Sigma_1, \ldots \Sigma_s$ be the connected components of $S \setminus \bigcup_{i=0}^{k} \alpha_i$ which are either isomorphic to a disc containing more than one point of P, or not isomorphic to a disc. If no such component exists, it means that $a \in A^0(S,P)$, and then we set $\overline{q}(l) = q(l)$. Otherwise, choose in the interior of each Σ_j a surface F_j such that ∂F_j is a disjoint union of circles and such that F_j is a deformation retract of Σ_j; if F_j happens to be a cylinder not containing any of the points of P, then take as F_j one of its meridians (look at F_2 in the first row of Figure 32). By construction we have

$$\bigcup_{i=0}^{k} \alpha_i \subset S \smallsetminus \{\bigcup_{i=1}^{s} F_j\}. \tag{9.8}$$

In the Figure 32, S is a genus 5 surface, $|P| = 4$, the α_i are the blue curves, and we have two components of $S \smallsetminus \{\alpha_0 \cup \cdots \cup \alpha_6\}$ which are not homeomorphic to a disk containing at most one point of P.

Next collapse to a point each connected component of the boundaries of the F_j to obtain a stable P-pointed genus g topological surface S_γ with dual graph γ. Let V be the set of vertices of γ and define $V^0 \subset V$ to be the subset of those vertices v such that S_v does not contain any of the (preimages of) the arcs α_i. These S_v are the ghost components. In the case of Figure 32, there is only one of them, namely the shaded one, and the corresponding vertex in γ is also shaded.

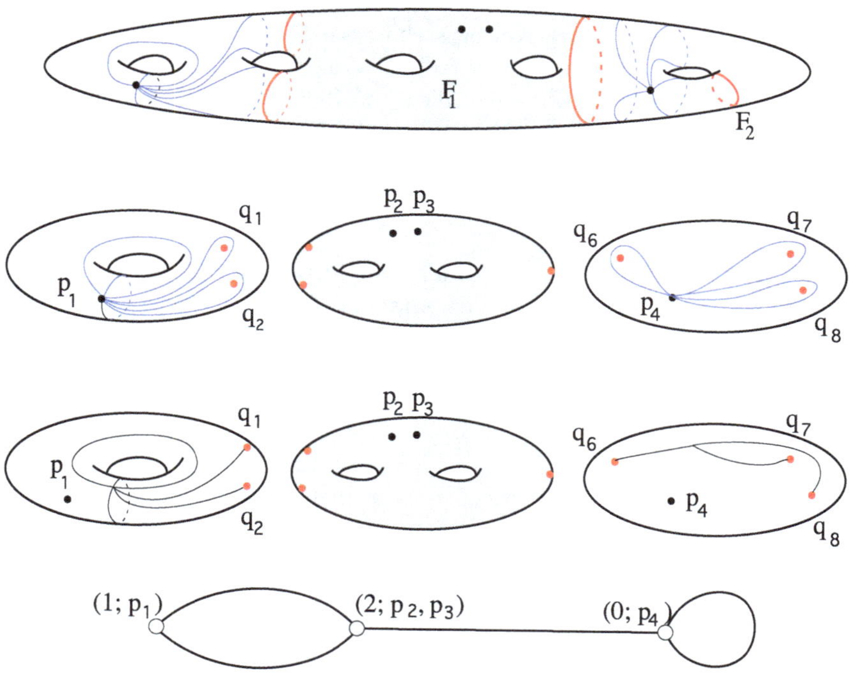

Figure 32.

Set $V^+ = V \smallsetminus V^0$. By construction, V^+ satisfies conditions i) and ii) of Definition (9.1). Because of (9.8), the arc system a defines an arc system a_v in S_v for each $v \in V^+$ (see the second row from the top in Figure 32). By definition, a_v is a *proper* simplex belonging to $A^0(S_v, P_v \cup Q_v)$. Here, as usual, $P_v = P \cap S_v$, and Q_v stands for the set of preimages in S_v of the nodes of S_γ. Moreover, the point $l \in |a|^0$ determines, up to an obvious renormalization of barycentric coordinates, a point $l_v \in |a_v|^0 \subset A^0(S_v, P_v \cup Q_v)$ for each $v \in V^+$. Observe that, by construction, no point $q \in Q_v$ belongs to a_v. Therefore the qth coordinate of $\lambda_v(l_v) \in \Delta_{P_v \cup Q_v}$ is equal to zero. By Theorem (3.12), i.e., by duality, the point

$$q(l_v) = (G_v, \mathbf{x}_v \cup \mathbf{y}_v, m_{l_v}, [f_v])$$

is an embedded metrized $(P_v \cup Q_v)$-marked ribbon graph of genus g_v. (Look at the the third row from the top in Figure 32). By construction, the data $(\gamma, V^+, \{(G_v, \mathbf{x}_v \cup \mathbf{y}_v, m_{l_v}, [f_v])\}_{v \in V^+})$ satisfy all the conditions of Definitions (9.1) and (9.3) and thus define an embedded stable P-marked genus g ribbon graph with unital metric, which we denote by $(G_a, \mathbf{x}, m_l, [f])$. We then set

$$\overline{q}(l) = (G_a, \mathbf{x}, m_l, [f])\,.$$

It is now an easy matter to check that $\overline{q}$ is indeed a bijection. The inverse of $\overline{q}$ is obtained by performing in reverse the steps depicted in Figure 32. One last remark concerns the assertion regarding the stabilizer Γ_a of a in $\Gamma_{S,P}$. We will limit ourselves to the description of its normal subgroup Γ'_a. The resulting identification (9.6) will follow immediately from this description. Going back to (9.8), Γ'_a is the subgroup of $\Gamma_{S,P}$ whose elements are represented by diffeomorphisms of (S, P) which are isotopic to the identity on $S \setminus \{\bigcup_{i=0}^{s} F_j\}$. Q.E.D.

10. Extending the cellular decomposition to a partial compactification of Teichmüller space.

In this section we define a surjective $\Gamma_{S,P}$-equivariant extension

$$\widehat{\Psi} : \widehat{\mathcal{T}}_{S,P} \times \Delta_P \longrightarrow |A(S,P)| \tag{10.1}$$

of the bijection $\Psi : \mathcal{T}_{S,P} \times \Delta_P \to |A^0(S,P)|$. In the next section we will proceed to prove that this extension is a continuous map.

The set-theoretical definition of $\widehat{\Psi}$ is as follows. Let $([C, \mathbf{x}, f], r)$ be a point in $\widehat{\mathcal{T}}_{S,P} \times \Delta_P$. This means that $(C, \mathbf{x})$ is a stable P-pointed genus g curve, $f : (S,P) \to (C, \mathbf{x})$ is a marking, and $r : P \to [0,1]$ is such that $\sum_{p \in P} r_p = 1$. Set

$$P^+ = \{p \in P : r_p > 0\}\,, \quad P^0 = \{p \in P : r_p = 0\}\,.$$

Denote by $\mathbf{x}^+$ (resp., $\mathbf{x}^0$) the restriction of $\mathbf{x}$ to P^+ (resp., P^0). By definition of marking, we have a collection $\{L_1, \dots, L_\delta\}$ of disjoint simple closed smooth curves in S such that each L_i contracts under f to a node of C. Moreover, if $L = L_1 \cup \dots \cup L_\delta$, the map f restricts to a homeomorphism $S \smallsetminus L \to C \smallsetminus \mathrm{Sing}(C)$.

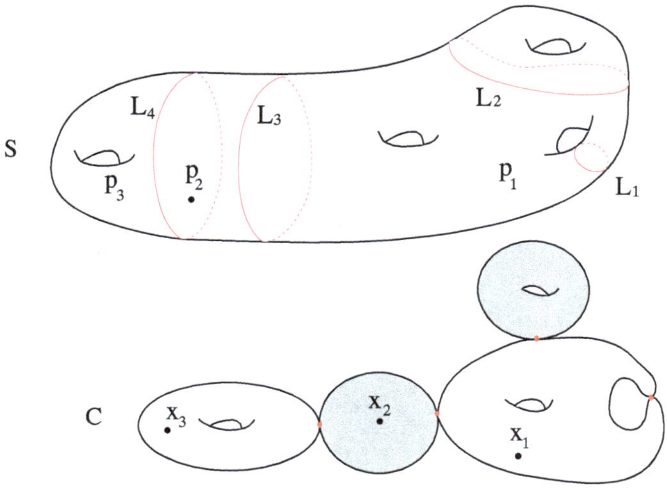

Figure 33. Here $r_2 = 0, r_1 > 0, r_3 > 0$, and the ghost components are the grey ones.

Let γ be the dual graph of C. Recalling the notation of the previous section and the map (9.4), we denote by

$$f_L : S_L \longrightarrow C$$

the map through which f naturally factors. Let V be the set of vertices of γ. For each $v \in V$, let C_v be the corresponding component of the normalization of C, let

$$f_v : S_v \longrightarrow C_v$$

be the map induced by f_L, and let

$$\mathbf{x}_v : P_v \to C_v \,, \quad \mathbf{y}_v : Q_v \to C_v \,,$$

be the restrictions of f to P_v and of f_v to Q_V. Set

$$P_v^+ = P^+ \cap P_v \,, \quad P_v^0 = P^0 \cap P_v \,,$$
$$V^+ = \{v \in V : P_v^+ \neq \emptyset\} \,, \quad V^0 = V \smallsetminus V^+ \,.$$

If $v \in V^+$, we say that C_v is a positive component of C. If $v \in V^0$, we say that C_v is a ghost component of C. For each $v \in V^+$, we define

$$r^v : P_v \cup Q_v \longrightarrow [0,1]$$

by setting

$$r_t^v = \begin{cases} \rho_v r_t & \text{if } t \in P_v, \\ 0 & \text{if } t \in Q_v, \end{cases}$$

where the constant ρ_v is chosen so that $\sum_{p \in P_v} r_p^v = 1$. In view of Proposition (8.1), we may consider the bijective maps

$$\Psi_v : \mathcal{T}_{S_v, P_v \cup Q_v} \times \Delta_{P_v \cup Q_v} \longrightarrow |A^0(S_v, P_v \cup Q_v)| .$$

Via Theorem (3.12), we may interpret the image under Ψ_v of the point

$$([C_v, \mathbf{x}_v \cup \mathbf{y}_v, f_v], r^v) \in \mathcal{T}_{S_v, P_v \cup Q_v} \times \Delta_{P_v \cup Q_v}$$

as an embedded metrized $(P_v \cup Q_v)$-marked genus g_v ribbon graph. In fact, this is exactly how the map Ψ_v was constructed; the ribbon graph in question is nothing but the hyperbolic spine of $(C_v, \mathbf{x}_v \cup \mathbf{y}_v, r^v)$. We denote this embedded metrized spine by $(G_v, \mathbf{x}_v \cup \mathbf{y}_v, m_v)$. The set of data

$$(\gamma, V^+, \{(G_v, \mathbf{x}_v \cup \mathbf{y}_v, m_v, [f_v])\}_{v \in V^+})$$

is an embedded stable P-marked metrized ribbon graph of genus g. Under the bijection $\overline{q}$ of Theorem (9.5), it corresponds to a point $l \in |a|^0 \subset A(S, P)$. We then set

$$\widehat{\Psi}(([C, \mathbf{x}, f], r)) = l .$$

This defines the $\Gamma_{S,P}$-equivariant surjection (10.1) extending Ψ.

One can be slightly more explicit about the definition of Ψ. By duality, i.e., via the map q of Theorem (3.12), the collection of metrized, embedded ribbon graphs $(G_v, \mathbf{x}_v \cup \mathbf{y}_v, m_v, [f_v])$ determines a collection of simplices $a_v \subset A(S_v, P_v \cup Q_v)$ together with points $l_v \in |a_v|^0$. Since the points in Q_v have half-perimeters equal to zero, none of these points lies in one of the simple closed curves forming the simplex a_v. But then these curves can be seen as curves in S forming the arc system $a \subset A(S, P)$ (look again at Figure 32, from the bottom up). Denote by E the set of all the edges of the graphs G_v as v varies in V^+. Then we can set

$$a = (e^*)_{e \in E} .$$

Also, following (7.7) and (7.8), we can define, for each $v \in V^+$, the corresponding function μ_v. These functions, after a suitable rescaling, define a function

$$(10.2) \qquad \mu : E \to (0, 1] , \qquad \sum_{e \in E} \mu(e) = 1 , \qquad \frac{1}{2} \sum_{e \in [x_p]} \mu(e) = r_p , \quad p \in P .$$

We may now set, exactly as in (7.9),

$$\widehat{\Psi}\left(([C, \mathbf{x}, f], r)\right) = \sum_{e \in E} \mu(e) e^* \in |a|^0 \subset |A(S,P)| \,. \tag{10.3}$$

Our next task is to prove that $\widehat{\Psi}$ is continuous.

11. The continuity of $\widehat{\Psi}$.

The aim of this section is to prove the following theorem.

THEOREM (11.1). *The map* $\widehat{\Psi} : \widehat{\mathcal{T}}_{S,P} \times \Delta_P \to |A(S,P)|$ *is continuous.*

As we already pointed out, the continuity of $\widehat{\Psi}$ in $\mathcal{T}_{S,P} \times \Delta_P$ is a quite straigthforward consequence of Lemma (6.9). The difficulty really comes from the boundary points in $\widehat{\mathcal{T}}_{S,P}$.

Before proving the theorem, we would like to go back to the construction of $\widehat{\mathcal{T}}_{S,P}$ and the description of its topology. Start with a point $[C, \mathbf{x}, f] \in \widehat{\mathcal{T}}_{S,P}$. Look at a Kuranishi family for $(C, \mathbf{x})$ and denote it by $\pi : \mathcal{C} \to B$. As in Section 8 of Chapter XV, out of this Kuranishi family, one can construct a continuous family of stable, pointed curves

$$\widehat{\pi} : \widehat{\mathcal{C}} \longrightarrow \widehat{B} \tag{11.2}$$

with central fiber isomorphic to $(C, \mathbf{x})$ and parameterized by a contractible topological space $\widehat{B}$, which plays, in this context, the role of a Kuranishi family for $[C, \mathbf{x}, f]$. The fiber of $\widehat{\pi}$ over a point $t \in \widehat{B}$ is an n-pointed stable curve which we denote by $(C_t, \mathbf{x}_t)$. The family $\widehat{\pi}$ comes equipped with a continuous map

$$F : S \times \widehat{B} \longrightarrow \widehat{\mathcal{C}} \tag{11.3}$$

having the following properties. First of all, $pr_{\widehat{B}} = \widehat{\pi} F$. Secondly, there is a set of disjoint smooth, simple, closed curves $L = L_1 \cup \cdots \cup L_\delta$ contained in S such that, for each $t \in \widehat{B}$, the set $L_t = F_t^{-1}(\mathrm{Sing}(C_t))$ is a, possibly empty, union of connected components of L, while $L_{t_0} = L$. Here

$$F_t : S \longrightarrow C_t \tag{11.4}$$

denotes the restriction of F to $S \cong S \times \{t\}$. Moreover,

$$F_t : (S \smallsetminus L_t) \times \{t\} \longrightarrow C_t \smallsetminus \mathrm{Sing}(C_t) \tag{11.5}$$

is a homeomorphism. Finally, for the central fiber over t_0, we have

$$(C_{t_0}, \mathbf{x}_{t_0}, F_{t_0}) \cong (C, \mathbf{x}, f) \,. \tag{11.6}$$

As we explained in Section 8 of Chapter XV, the space $\widehat{B}$ can be identified with a neighborhood of $[C, \mathbf{x}, f]$ in $\widehat{\mathcal{T}}_{S,P}$. Theorem (11.1) is then reduced to the following lemma.

LEMMA (11.7). *The map*

$$
\begin{aligned}
\widehat{\Psi} : \widehat{B} \times \Delta_P &\longrightarrow |A(S,P)|\,, \\
(t,r) &\mapsto \widehat{\Psi}(([C_t,\mathbf{x}_t,F_t],r))\,,
\end{aligned}
\tag{11.8}
$$

is continuous at the point (t_0, r^0) *for any* $r^0 \in \Delta_P$.

Proof. Denote by $\widehat{f} : \widehat{B} \to B$ the natural projection and recall the stratification $\{D_I\}$, $I \subset \{1,\ldots,\delta\}$ defined in Section 8 of Chapter XV. Set

$$\widehat{D}_I = \widehat{f}^{-1}(D_I)\,.$$

Since $\widehat{B}$ is first countable, to test the continuity of $\widehat{\Psi}$ at (t_0, r^0), we may consider a sequence

$$\{(t_n, r^n)\}_{n\in\mathbb{N}} \subset \widehat{B} \times \Delta_P \quad \text{with} \quad \lim_{n\to\infty}(t_n, r^n) = (t_0, r^0)\,,$$

and we may also assume that

$$\{(t_n, r^n)\}_{n\in\mathbb{N}} \subset \widehat{D}_I$$

for some fixed index I. We are therefore reduced to proving that, for each I, the map

$$\widehat{\Psi} : \left(\widehat{D}_I \cup \{t_0\}\right) \times \Delta_P \longrightarrow |A(S,P)|$$

is sequentially continuous at (t_0, r^0) for every $r^0 \in \Delta_P$. One of the advantages is that the fibers of $\widehat{\pi}$ over $\widehat{D}_I$ have fixed topological type. In conclusion, we may replace the family (11.2) with a family

$$\widehat{\pi} : \widehat{\mathcal{C}} \longrightarrow \widehat{D}$$

parameterized by some locally contractible space $\widehat{D}$ (the space $\widehat{D}_I \cup \{t_0\}$ is one such), and there exists a continuous map F as in (11.3) (with $\widehat{D}$ replacing $\widehat{B}$) satisying the properties illustrated in (11.4) and (11.5). We may also assume that (11.6) holds for some point t_0 in $\widehat{D}$ and that one of the following holds:

i) C_t is smooth for every $t \in \widehat{D} \smallsetminus \{t_0\}$;
ii) $F_t^{-1}(\mathrm{Sing}(C_t)) = L_1 \cup \cdots \cup L_\nu\,, \quad \nu \le \delta$, for every $t \in \widehat{D} \smallsetminus \{t_0\}$

We will now prove the continuity at (t_0, r^0) in case i). We will see from our argument that case ii) can be reduced to case i) by simultaneously normalizing the fibers of the family $\widehat{\pi} : \mathcal{C} \to \widehat{D}$.

We denote by $y_1, \ldots, y_\delta$ the nodes of the central fiber C_{t_0}. Write $n = |P|$, $P = \{p_1, \ldots, p_n\}$, where the p_i are numbered so that $r_i^0 = r_{p_i}^0 > 0$ for $i = 1, \ldots, s$ and $r_i^0 = 0$ for $i = s+1, \ldots, n$. We may then restrict our attention to the open face $U = \{r \in \Delta_P : r_i > 0\,,\ i = 1, \ldots, s\}$. For each $t \in \widehat{D} \smallsetminus \{t_0\}$, we equip C_t with the Poincaré metric. As we pointed out in Lemma (6.9), the Poincaré metric varies continuously with t. For each $t \in \widehat{D} \smallsetminus \{t_0\}$ and $r \in U$, we denote by $H_t^1(r), \ldots H_t^s(r)$ the horocycle regions in C_t corresponding to r. Here $H_t^i(r)$ is the horocycle region around $x_i(t)$, $i = 1, \ldots s$. We also consider standard horocycle regions (see (6.4)) $H_t^1, \ldots, H_t^n$, one for each point $x_i(t) = \mathbf{x}_t(p_i)$, and we may assume that $H_t^i(r) \subset H_t^i$ for $i = 1, \ldots, s$. For each $t \in \widehat{D} \smallsetminus \{t_0\}$ and for each $i = 1, \ldots, \delta$, we denote by $\gamma_t^i \subset C_t$ the unique geodesic representative in the homotopy class $[F_t(L_i)]$. As t tends to t_0, the geodesic γ_t^i tends to the point y_i. There is no loss in generality in assuming that $F_t(L_i) = \gamma_t^i$. In fact, it is possible to change F in order to obtain this property, without changing the homotopy class of the maps $F_t : (S, P) \to (C_t, \mathbf{x}_t)$. We are now going to use the Collar Lemma (6.10). For each $t \in B \smallsetminus \{t_0\}$ and for each $i = 1, \ldots, \delta$, we choose a collar $A_t^i \subset C_t$ around γ_t^i having the following properties:

1) the A_t^i are mutually disjoint;
2) the A_t^i are disjoint from the standard horocycle regions $H_t^1, \ldots, H_t^n$;
3) as t tends to t_0, the collar A_t^i shrinks to the point y_i.

This setup is depicted in Figure 34.

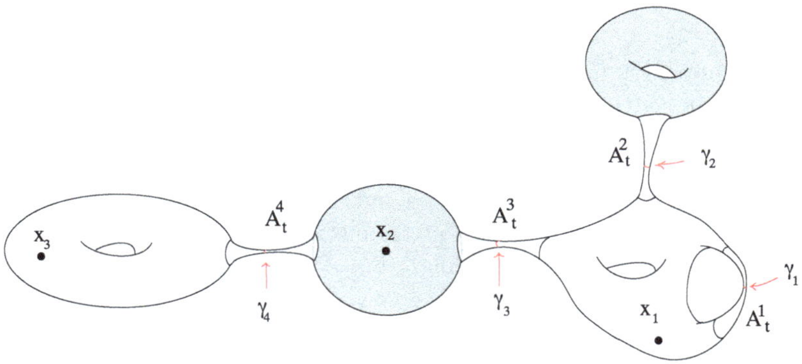

Figure 34.

We set

$$A_t = A_t^1 \cup \cdots \cup A_t^\delta\,.$$

Let C' be the normalization of a positive component of $(C_{t_0}, \mathbf{x}_{t_0}, r^0)$. Denote by $\zeta_1, \ldots, \zeta_\nu$ the points of C' mapping, via the normalization map, to the nodes of C_{t_0}. Denote by C'_t the connected component of $C_t \smallsetminus \{\gamma_t^1, \ldots, \gamma_t^\delta\}$ which tends to $C' \smallsetminus \{\zeta_1, \ldots, \zeta_\nu\}$ as t tends to t_0. In particular, as t tends to t_0, the set $A_t \cap C'_t$ tends to the set $\{\zeta_1, \dots, \zeta_\nu\}$, which is viewed as a set of punctures of C'. The family of stable curves $\widehat{\pi} : \widehat{\mathcal{C}} \to \widehat{D}$ induces a family

$$\alpha : \mathcal{C}' \to \widehat{D}$$

of Riemann surfaces of finite type with $\alpha^{-1}(t) = C'_t$ and $\alpha^{-1}(t_0) = C' \smallsetminus \{\zeta_1, \dots, \zeta_\nu\}$ (see Figure 35 below). We equip $C' \smallsetminus ((\mathbf{x}(P) \cap C') \cup \{\zeta_1, \dots, \zeta_\nu\})$ with the Poincaré metric and $C'_t \smallsetminus (\mathbf{x}_t(P) \cap C'_t)$ with the metric induced by the Poincaré metric of $C_t \smallsetminus \mathbf{x}_t(P)$ for every $t \in \widehat{D}$.

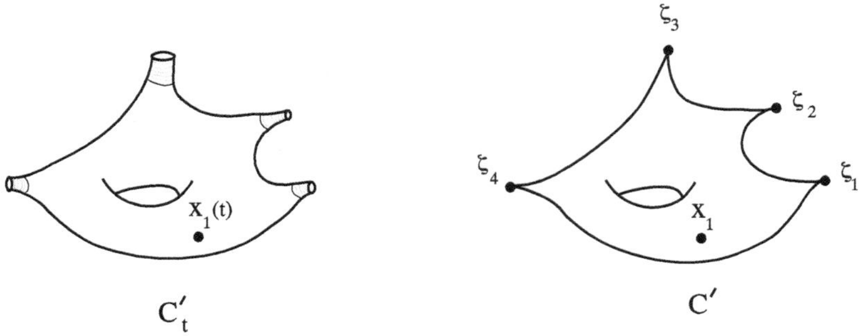

Figure 35. A general fiber and the central fiber of the family α relative to the positive component on the RHS of Figure 34. The shaded region is $A_t \cap C'_t$.

It is important to remark that, in this way, we obtain a continuously varying metric on the fibers of α. To prove this assertion, consider the Nielsen extension C'^e_t of C'_t for every $t \in \widehat{D}$ and notice that C', as a pointed curve, coincides with its Nielsen extension. Also notice that, since the curves γ^i_t along which we cut C_t are geodesics, the Poincaré metric on C'^e_t coincides with the given metric on C'_t. Our remark is then a consequence of Lemmas (6.8) and (6.9).

We will use the following terminology. A point $z \in C'_t \subset C_t$ with $t \neq t_0$ is said to be an *intruder* if one of the horocycle regions nearest to z lies in a component of $C_t \smallsetminus \{\gamma^1_t, \dots, \gamma^\delta_t\}$ which is *different from* C'_t. In Figure 36 the point z_n is an intruder in C'_{t_n}.

Claim (11.9). *By changing the A^i_t without altering properties 1), 2), and 3), one can make the following further assumption:*

4) *the intruders in C'_t are all contained in $A_t \cap C'_t$.*

In particular, as t tends to t_0, the set of intruders in C'_t tends to the set $\{\zeta_1, \dots, \zeta_\nu\}$, viewed as punctures of C'.

Proof. It suffices to show that, if $\{t_n\}_{n\in\mathbb{N}}$ is a sequence in $\widehat{D}$ tending to t_0 and if $z_n \in C_{t_n}$ is an intruder and $z = \lim_{n\to\infty} z_n \in C'$, then $z \in \{\zeta_1,\dots,\zeta_\nu\}$. Let $\mathbf{x}^+(P)\cap C' = \{x'_1,\dots,x'_k\}$; since C' is (the normalization of) a positive component of C, this set is nonempty. Let $x'_i(t) \in \mathbf{x}_t^+(P)\cap C'_t$ be a section of α with $x'_i(t_0) = x'_i$ for $i = 1,\dots,k$. Denote by d_n the minimum distance in C'_{t_n} between p_n and the points $x'_i(t_n)$ for $i = 1,\dots,k$. We claim that $\lim_{n\to\infty} d_n = \infty$. Suppose in fact that H_n is a horocycle region in C_{t_n} lying in a component of $C_{t_n} \smallsetminus \{\gamma^1_{t_n},\dots,\gamma^\delta_{t_n}\}$ which is different from C'_{t_n} and has shortest distance d'_n from p_n. From the collar lemma it follows that d'_n tends to infinity, since a geodesic joining z_n to H_n must pass through a collar K around one of the shrinking γ_i.

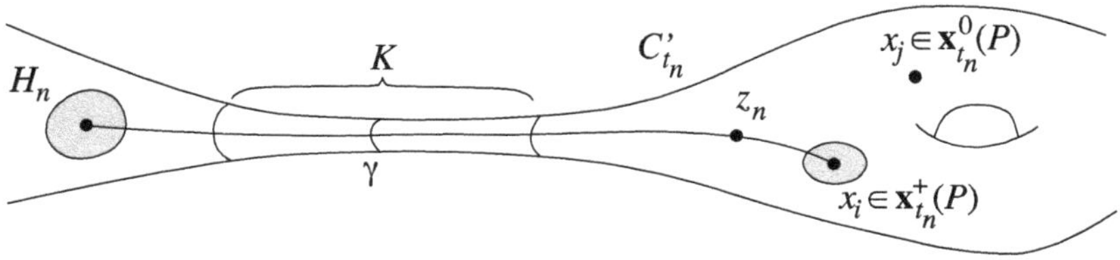

Figure 36.

Since d'_n tends to infinity, the same must be true for d_n. A priori, again as a direct consequence of the collar lemma, only two possibilities may occur. The first one is that the limit point z is one of the points $\zeta_1,\dots,\zeta_\nu$, which is what we want. The second is that $z \in \mathbf{x}^0(P)\cap C'$. We will show that this cannot happen. Let σ_n be the shortest geodesic joining z_n with H_n and look at the intersection $\sigma'_n = \sigma_n \cap C'_{t_n}$. The limit of these geodesic is a curve σ joining $z \in \mathbf{x}^0(P)\cap C'$ to one of the points in $\{\zeta_1,\dots,\zeta_\nu\}$ (in fact, exactly the point to which the shortest geodesic of the collar K shrinks). But this is not possible because, on one hand, the points in σ'_n are all intruders, so that

$$\sigma = \lim_{n\to\infty} \sigma_n \subset \mathbf{x}^0(P)\cap C' \cup \{\zeta_1,\dots,\zeta_\nu\},$$

and, on the other, σ is a nonconstant path. Q.E.D. (for the Claim).

In the final part of the proof of Lemma (11.7), given a finite set T, it will be convenient to view a T-marking of a Riemann surface or algebraic curve E, somewhat improperly, as a *subset* of E. With this in mind, let G' be the spine for $(C',\mathbf{x}',r')$, where

$$\mathbf{x}' = (\mathbf{x}\cap C')\cup\{\zeta_1,\dots,\zeta_\nu\},\quad (\mathbf{x}')^0 = (\mathbf{x}^0\cap C')\cup\{\zeta_1,\dots,\zeta_\nu\},$$
$$(\mathbf{x}')^+ = \mathbf{x}^+\cap C',\quad r'_{|(\mathbf{x}')^+} = r^0_{|\mathbf{x}^+\cap C'}.$$

We now set

$$C''_t = C'_t \smallsetminus A_t \cap C'_t.$$

For $t \in \widehat{B} \smallsetminus \{t_0\}$ and $r \in \Delta_P$, we let $G_t(r)$ denote the spine of $(C_t, \mathbf{x}_t, r)$. Set

$$G''_t(r) = G_t(r) \cap C''_t\,.$$

Consider the triple $(C''_t, \mathbf{x}''_t, r'')$ where

$$\mathbf{x}''_t = (\mathbf{x}_t \cap C''_t)\,, \quad (\mathbf{x}''_t)^0 = \mathbf{x}^0_t \cap C''_t\,, \quad (\mathbf{x}''_t)^+ = \mathbf{x}^+_t \cap C''_t\,, \quad r''_{|(\mathbf{x}''_t)^+} = r_{|\mathbf{x}^+_t \cap C''_t}\,.$$

Set $(\mathbf{x}''_t)^+ = \{x''_1(t), \dots, x''_k(t)\}$ and denote by $H''_i(t, r'')$ a horocycle region around $x''_i(t)$ having boundary of length $r''(x''_i(t))$. Set $H''(t,r) = H''_1(t,r'') \cup \dots \cup H''_k(t,r'')$. The basic remark, following directly from the claim (11.9), is that $G''_t(r)$ is a spine for $(C''_t, \mathbf{x}''_t, r'')$. We mean by this that the graph $G''_t(r)$ is the locus of points of $p \in C''_t$ such that there are at least two shortest geodesics from p to $H''(t,r)$. Exactly in the same way we may characterize the graph G'. Consider the family of Riemann surfaces of finite type $\beta : \mathcal{C}'' \to \widehat{B}$ having the property that

$$\beta^{-1}(t) = C''_t\,, \qquad \beta^{-1}(t_0) = C'\,.$$

Since the induced metric on the fibers of β varies continuously with t, and since the horocycle regions vary continuously with t and r, it follows that the graph $G''_t(r)$ tends to G' as (t,r) tends to (t_0, r^0). Using again our claim (11.9), we see that that, as (t,r) tends to (t_0, r^0), the intersection of the spine $G_t(r)$ of $(C_t, \mathbf{x}_t, r)$ with the fiber C'_t tends to the spine G' of $(C', \mathbf{x}', r')$.

We claim that this finishes the proof of Lemma (11.7). Going back to the definition (10.3) of $\widehat{\Psi}$, recall that the arc system σ appearing in it is defined as the union of arc systems σ_i and that each one of these is the dual of a corresponding spine of the normalization of a positive component of the central fiber $C_{t_0} = C$. It follows that, if $\sigma_t(r)$ is the dual of $G_t(r)$, then, as (t,r) tends to (t_0, r^0), the arc system $\sigma_t(r)$ tends to σ. It is now an easy matter to check that also the coordinates μ vary continuously with t and r. Q.E.D.

Let us return to the proof of Theorem (2.7). At this stage, we know that the map

$$\widehat{\Psi} : \widehat{\mathcal{T}}_{S,P} \times \Delta_P \longrightarrow |A(S,P)|$$

is continuous. We also know that its restriction

$$\Psi : \mathcal{T}_{S,P} \times \Delta_P \longrightarrow |A^0(S,P)|$$

is bijective. This last fact and the way in which $\widehat{\Psi}$ is defined also shows that $\widehat{\Psi}$ is surjective. Let us now look at the quotient map

$$\overline{\Phi} : \overline{M}_{g,P} \times \Delta_P \longrightarrow |A(S,P)|/\Gamma_{S,P}\,. \tag{11.10}$$

We know that $\overline{\Phi}$ is a continuous surjection. Since $\overline{M}_{g,P}$ is compact and $|A(S,P)|/\Gamma_{S,P}$ is Hausdorff (see Lemma (2.13)), it is also closed. Finally, the bijectivity of Ψ implies that the map

$$\Phi : M_{g,P} \times \Delta_P \longrightarrow |A^0(S,P)|/\Gamma_{S,P} \tag{11.11}$$

is also bijective (we know that it is continuous). Since $\overline{\Phi}$ is closed and the preimage of $|A^0(S,P)|/\Gamma_{S,P}$ under $\overline{\Phi}$ is $M_{g,P} \times \Delta_P$, the map Φ is also closed. So it is a homeomorphism as claimed. If the action of $\Gamma_{S,P}$ on $\mathcal{T}_{S,P} \times \Delta_P$ and on $|A^0(S,P)|$ were free, we could immediately conclude that Ψ is also a homeomorphism at the level of Teichmüller space. The only thing we know is that these actions have finite stabilizers. This is good enough. The proof of Theorem (2.7) is now complete.

12. Odds and ends.

In most applications it will be important to restrict the attention to pointed stable curves $(C, x_1, \dots, x_n)$ having *positive weights* $(r_1, \dots, r_n)$ at the points $(x_1, \dots, x_n)$. Relative to these weights, the ghost components of C are those components not containing any of the points $(x_1, \dots, x_n)$.

Let us look at the map

$$\lambda : |A(S,P)| \longrightarrow \Delta_P$$

and its restriction to $|A^0(S,P)|$. We would like to describe the preimage under λ of the interior $\overset{\circ}{\Delta}_P$ of Δ_P. To this end, consider the subcomplex $A'_\infty(S,P)$ of the arc system complex $A(S,P)$ whose k-simplices are given by $(k+1)$-tuples $a = ([\alpha_0], \dots, [\alpha_k])$ having the property that there is a connected component of $S \smallsetminus \bigcup_{i=0}^{k} \alpha_i$ containing a point of P. Also, we denote by $A''_\infty(S,P)$ the subcomplex of $A(S,P)$ whose k-simplices are given by $(k+1)$-tuples $a = ([\alpha_0], \dots, [\alpha_k])$ having the property that there is a connected component of $S \smallsetminus \bigcup_{i=0}^{k} \alpha_i$ which either contains a point of P or is not a disc. The subcomplex $A'_\infty(S,P)$ is contained in $A''_\infty(S,P)$, and both subcomplexes are $\Gamma_{S,P}$-invariant. We set

$$\begin{aligned} A'(S,P) &= A(S,P) \smallsetminus A'_\infty(S,P)\,, \qquad & |A'(S,P)| &= |A(S,P)| \smallsetminus |A'_\infty(S,P)|\,, \\ A'^0(S,P) &= A(S,P) \smallsetminus A''_\infty(S,P)\,, \qquad & |A'^0(S,P)| &= |A(S,P)| \smallsetminus |A'_\infty(S,P)|\,. \end{aligned}$$

The simplices of $A'(S,P)$ are arc systems $a = ([\alpha_0], \dots, [\alpha_k])$ such that no connected component of $S \smallsetminus \bigcup_{i=0}^{k} \alpha_i$ contains a point of P. The simplices of $A'^0(S,P)$ have the further property that every connected component of $S \smallsetminus \bigcup_{i=0}^{k} \alpha_i$ is a disc (not containig points of P). Both $|A'^0(S,P)|$

and $|A'(S,P)|$ can be readily described in terms of ribbon graphs. The points of $|A'^0(S,P)|$ (resp., $|A'(S,P)|$) correspond to embedded P-marked genus g metrized ribbon graphs (resp., stable ribbon graphs) all of whose half-perimeters are positive. In particular, these graphs have exactly $|P|$ boundary components. We have

$$|A'^0(S,P)| \subset |A'(S,P)|\,.$$

It is clear that

$$\lambda^{-1}(\mathring{\Delta}_P) = |A'(S,P)|\,, \qquad \lambda^{-1}(\mathring{\Delta}_P) \cap |A^0(S,P)| = |A'^0(S,P)|\,.$$

We next consider the continuous $\Gamma_{S,P}$-invariant surjection

$$\widehat{\Psi} : \widehat{\mathcal{T}}_{S,P} \times \mathring{\Delta}_P \longrightarrow |A'(S,P)|$$

and its restriction

$$\Psi : \mathcal{T}_{S,P} \times \mathring{\Delta}_P \longrightarrow |A'^0(S,P)|\,,$$

which is a $\Gamma_{S,P}$-invariant homeomorphism. As in (11.10) and (11.11), these maps induce a continuous surjection

$$\overline{\Phi} : \overline{M}_{g,P} \times \mathring{\Delta}_P \to |A'(S,P)|/\Gamma_{S,P} \tag{12.1}$$

and a homeomorphism

$$\Phi : M_{g,P} \times \mathring{\Delta}_P \to |A'^0(S,P)|/\Gamma_{S,P}\,. \tag{12.2}$$

We would like to give an explicit description of a topological quotient $\overline{M}'_{g,P}$ of $\overline{M}_{g,P}$ having the property that $\overline{\Phi}$ drops to a *homeomorphism*

$$\overline{\Phi}' : \overline{M}'_{g,P} \times \mathring{\Delta}_P \xrightarrow{\cong} |A'(S,P)|/\Gamma_{S,P}\,. \tag{12.3}$$

Let us go back to the construction of the map $\widehat{\Psi}$ and hence of the map $\overline{\Phi}$. We see that the ghost components play no role in the definition of $\overline{\Phi}$ and in fact contribute to its not being injective. It is therefore natural to construct a new space where the ghost components disappear. This space is defined as a quotient of $\overline{M}_{g,P}$ modulo an equivalence relation, as we will presently explain. Any stable P-pointed curve $(C, \mathbf{x})$ of genus g can be canonically decomposed as the union of two curves $C = C^+ \cup C^0$, where C^+ is the union of all the components of C containing marked points, and C^0 is the union of those containing no marked points. In the present context, these last are the ghost components of $(C, \mathbf{x})$. Let $\xi_1, \dots, \xi_u$ be the points that C^+ has in common with C^0. We say that $[(C, \mathbf{x})]$ is equivalent to $[(C',\mathbf{x}')]$ if there is a family of nodal curves $\{C_s^0\}_{s\in S}$ over a connected base S, together with sections of smooth points $\tau_1,\dots,\tau_u$, with the property that $(C,\mathbf{x})$ (resp., $(C',\mathbf{x}')$) can be obtained from C^+ and C_s^0 (resp., $C_{s'}^0$) by identifying ξ_i with $\tau_i(s)$ (resp., with $\tau_i(s')$) for $i=1,\dots,u$.

Another way to describe the equivalence class of $[(C,\mathbf{x})]$ is the following. Keep C^+ fixed and smooth out all the singularities of C that are not on C^+, if any. The resulting curve $(C',\mathbf{x}')$ is of type $C'=C^+\cup C_1\cup\dots\cup C_k$, where C_i is a smooth curve of genus g_i attached to C^+ at ν_i of its points. Moreover, $C_i\cap C_j=\emptyset$ when $i\neq j$. Then the equivalence class of $[(C,\mathbf{x})]$ is a copy of $(\overline{M}_{g_1,\nu_1}\times\dots\times\overline{M}_{g_k,\nu_k})/G$ for some finite group G. In Figure 37 we show the passage from C to C'. The ghost components are the grey ones, $k=2$, $g_1=1$, $g_2=3$, $\nu_1=1$, and $\nu_2=2$.

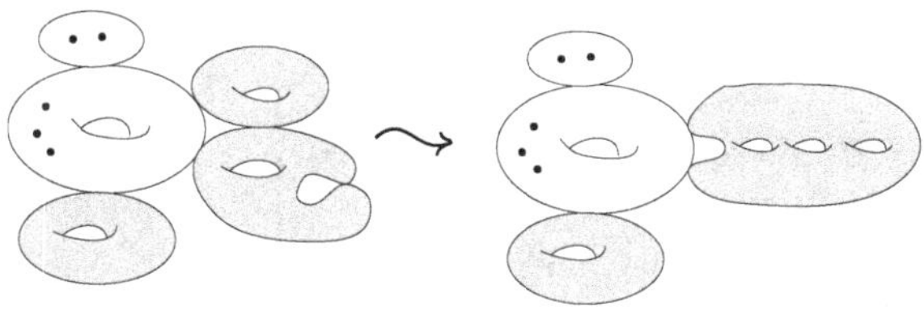

Figure 37.

It is easy to verify that the one we just described is an equivalence relation. By construction, the map $\overline{\Phi}$, which is a continuous open surjection, drops to a homeomorphism (12.3). We let

$$Q:\overline{M}_{g,P}\to\overline{M}'_{g,P} \tag{12.4}$$

denote the natural projection.

We need one last variation of the maps Ψ and $\widehat{\Psi}$. When looking at stable curves $(C,\mathbf{x},r)$ with positive weights, it is also useful to relax the condition that the sum of the weights be equal to 1. Consider the map

$$\rho:\mathbb{R}_+^P\longrightarrow\Delta_P$$

defined by $\rho(r)=r/(\sum r_i)$. We define the space $\overline{\mathcal{T}}_{g,P}^{comb}$ via the cartesian diagram

$$\begin{array}{ccc}\overline{\mathcal{T}}_{g,P}^{comb} & \longrightarrow & |A(S,P)| \\ \big\downarrow & & \big\downarrow\lambda \\ \mathbb{R}_+^P & \xrightarrow{\ \sigma\ } & \Delta_P\end{array} \tag{12.5}$$

By abuse of notation we denote by $\lambda : \overline{\mathcal{T}}^{comb}_{g,P} \to \mathbb{R}^P_+$ the natural projection. The points of $\overline{\mathcal{T}}^{comb}_{g,P}$ can be interpreted as isomorphism classes of embedded stable P-marked genus g metrized ribbon graphs whose all half-perimeters are positive, but where the metric is no longer unital. Starting from $|A'^0(S,P)|$, we define in a similar way the space $\mathcal{T}^{comb}_{g,P}$. We also set

$$M^{comb}_{g,P} = \mathcal{T}^{comb}_{g,P}/\Gamma_{S,P}\,, \qquad \overline{M}^{comb}_{g,P} = \overline{\mathcal{T}}^{comb}_{g,P}/\Gamma_{S,P}\,.$$

We then have the following identifications:

(12.6)
$$\overline{\mathcal{T}}^{comb}_{g,P} = \left\{ \begin{array}{c} \text{stable } P\text{-marked genus } g \\ \text{embedded ribbon graphs with} \\ \text{metric and positive perimeters} \end{array} \right\} \Big/ \text{isomorphism}\,,$$

(12.7)
$$\mathcal{T}^{comb}_{g,P} = \left\{ \begin{array}{c} P\text{-marked genus } g \text{ embedded} \\ \text{ribbon graphs with metric} \\ \text{and positive perimeters} \end{array} \right\} \Big/ \text{isomorphism}\,,$$

(12.8)
$$\overline{M}^{comb}_{g,P} = \left\{ \begin{array}{c} \text{stable } P\text{-marked genus } g \\ \text{ribbon graphs with metric} \\ \text{and positive perimeters} \end{array} \right\} \Big/ \text{isomorphism}\,,$$

(12.9)
$$M^{comb}_{g,P} = \left\{ \begin{array}{c} P\text{-marked genus } g \text{ ribbon graphs} \\ \text{with metric and positive perimeters} \end{array} \right\} \Big/ \text{isomorphism}\,.$$

We have homeomorphisms

(12.10)
$$\overline{H} : \overline{M}'_{g,P} \times \mathbb{R}^P_+ \xrightarrow{\simeq} \overline{M}^{comb}_{g,P}$$

and

(12.11)
$$H : M_{g,P} \times \mathbb{R}^P_+ \xrightarrow{\simeq} M^{comb}_{g,P}\,.$$

For $r \in \mathbb{R}^P_+$, we set

$$\overline{M}^{comb}_{g,P}(r) = \lambda^{-1}(r)\,, \qquad M^{comb}_{g,P}(r) = \lambda^{-1}(r) \cap M^{comb}_{g,P}\,.$$

These spaces have an obvious description in terms of (stable) ribbon graphs with *fixed perimeters*. The homeomorphisms (12.10) and (12.11) induce homeomorphisms

(12.12)
$$\overline{h} : \overline{M}'_{g,P} \xrightarrow{\cong} \overline{M}^{comb}_{g,P}(r)\,, \qquad h : M_{g,P} \xrightarrow{\cong} M^{comb}_{g,P}(r)\,.$$

Finally, we define the map

$$f : \overline{M}_{g,P} \longrightarrow \overline{M}^{comb}_{g,P}(r) \tag{12.13}$$

as the composition of the contraction $Q : \overline{M}_{g,P} \to \overline{M}'_{g,P}$ given in (12.4) and of the homeomorphism $\overline{h}$. By construction, the restriction of f to $M_{g,P}$ induces the homeomorphism h.

13. Bibliographical notes and further reading.

The idea of using ribbon graphs to give a cellular decomposition of the moduli space $M_{g,n}$ with $n \geq 1$ goes back to Mumford and independently to Thurston (unpublished). It first appeared in Harer's paper [340] (see also [339]). Mumford's idea was to use the theory of Jenkins–Strebel quadratic differentials (as developed in Strebel's book [648]; see also [31] for a treatment aimed at the case of pointed curves). An alternative approach, also leading to a cell decomposition of moduli spaces, consists in using hyperbolic geometry. This approach is due to Bowditch and Epstein [80] and Penner [589]. Both the Strebel differentials method and the hyperbolic geometry method lead quite directly to a $\Gamma_{g,n}$-invariant cellular decomposition of $\mathcal{T}_{g,n}$. The relation between the cellular decompositions obtained via these two procedures is studied by Mondello [516]. In our treatment we follow the hyperbolic geometry method of Bowditch and Epstein [80]. The reason is that hyperbolic geometry appears to be more suited to treating the extension of the cellular decomposition of $\mathcal{T}_{g,n}$ to its bordification $\widehat{\mathcal{T}}_{g,n}$. A study of the extension to $\widehat{\mathcal{T}}_{g,n}$ of the cellular decomposition via Strebel differentials can be found in [488].

A proof of the uniformization theorem along the lines sketched in our text can be found in Springer's book [645]. As far as the hyperbolic geometry of Riemann surfaces is concerned, standard references are Ford [268], Ratcliffe [602], Jost [399], Imayoshi and Taniguchi [385], and Benedetti and Petronio [61]. A proof of the Collar Lemma can be found in Keen [411], Halpern [336], Matelski [502], and Matsuzaki [504]. The properties of the Nielsen extension of a Riemann surface are studied in Bers [65].

Chapter XIX.

First consequences of the cellular decomposition

1. Introduction.

In this chapter we will illustrate some of the consequences of the (orbi)cellular decomposition of $M_{g,P}$ and $\overline{M}'_{g,P}$ described in the previous chapter. The fundamental result in that chapter is the existence of a $\Gamma_{S,P}$-equivariant homeomorphism

$$\Psi : \mathcal{T}_{S,P} \times \Delta_P \longrightarrow |A^0(S,P)| \,. \tag{1.1}$$

In Section 2 we present the first consequence of the existence of such a homeomorphism, namely the vanishing Theorem (2.2) stating that the rational homology of $M_{g,P}$ vanishes in sufficiently high degree. To prove the vanishing theorem, we will show that there is a natural $\Gamma_{S,P}$-equivariant retraction of $|A^0(S,P)|$ into a lower-dimensional complex. This retraction can be well understood in the case of one-pointed genus one curves. Figure 1 in the introduction to Chapter XVIII gives a description of the complex $|A^0(S,P)|$ when $g(S) = |P| = 1$. Figure 1 below shows how this two-dimensional complex retracts to a one-dimensional spine.

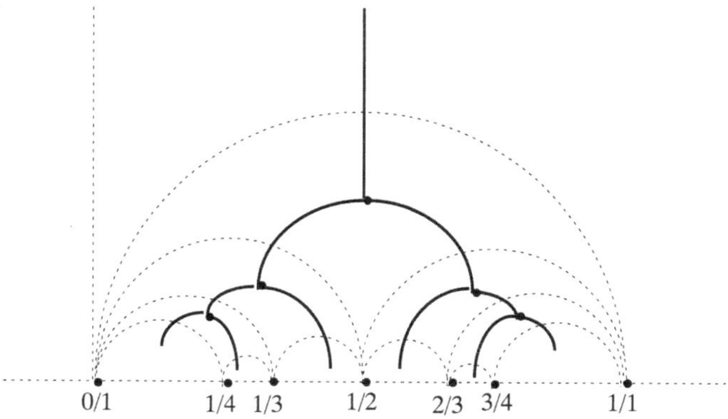

Figure 1.

There is another case in which a vanishing theorem can be easily established, namely the case of n-pointed genus zero curves. Fixing three of the n points as 0, 1, and ∞, we realize $M_{0,n}$ as $\mathbb{C}^{n-3}$ with coordinates $z_1, \ldots, z_{n-3}$ minus the hyperplanes $z_i = 0$, $z_i = 1$, and $z_i = z_j$ for $i \neq j$.

E. Arbarello et al., *Geometry of Algebraic Curves*,
Grundlehren der mathematischen Wissenschaften 268,
DOI 10.1007/978-3-540-69392-5_11, © Springer-Verlag Berlin Heidelberg 2011

As such, $M_{0,n}$ is an affine variety, and thus its homology vanishes in degrees strictly greater than $n-3$. It would be tempting to think that, also when the genus is positive, the vanishing statement of Theorem (2.2) is the consequence of an algebro-geometric property of $M_{g,n}$, as in the case of genus 0. Indeed, the above vanishing theorem would be an immediate consequence of the Mayer–Vietoris sequence if one could establish the following conjecture.

CONJECTURE (1.2) (Looijenga). *Let g and n be nonnegative integers such that $2g-2+n>0$. Then $M_{g,n}$ is the union of g affine subsets if $n \neq 0$, and is the union of $g-1$ affine subsets if $n=0$.*

In the *Esquisse d'un programme* [332] Grothendieck observed:

> *Il est vrai que les multiplicités modulaires ne sont pas affines (sauf pour des petites valeurs de g), mais il suffirait qu'une telle $M_{g,\nu}$ de dimension N (ou plutôt un revêtement fini convenable) soit réunion de $N-2$ ouverts affines, donc que $M_{g,\nu}$ ne soit pas "trop proche d'une variété compacte".*[7]

In Section 3 we show how the vanishing Theorem (2.2) can be used to compare the cohomology of $\overline{M}_{g,n}$ with the cohomology of its boundary strata. For this, we use the exact sequence of *rational* cohomology with compact support of the pair $(\overline{M}_{g,n}, \partial M_{g,n})$, and we deduce that the restriction map

$$H^k(\overline{M}_{g,n}) \to H^k(\partial M_{g,n})$$

is injective for $k \leq d(g,n)$ and is an isomorphism for $k < d(g,n)$, where $d(g,n)$ is the constant defined in (2.6). Let $D_1, D_2, \dots, D_N$ be the irreducible components of the boundary $\partial M_{g,n}$. As we showed in Chapter XII (cf. diagram (10.7)), each of these components is the image of a map $\mu_i : X_i \to \overline{M}_{g,n}$, where X_i can be of two different kinds. Either $X_i = \overline{M}_{g-1,n+2}$, and μ_i is obtained by identifying the last two marked points of each $(n+2)$-pointed curve of genus $g-1$, or $X_i = \overline{M}_{a,h+1} \times \overline{M}_{b,k+1}$, where $a+b=g$, $h+k=n$, $2a-2+h \geq 0$, $2b-2+k \geq 0$, and μ_i is obtained by identifying the $(h+1)$st point of an $(h+1)$-pointed genus a curve with the $(k+1)$-st point of a $(k+1)$-pointed genus b curve. We then consider the composition

$$\sigma : \coprod_{i=1}^{N} X_i \to \partial M_{g,n} \to \overline{M}_{g,n} \, .$$

[7] *It is true that modular multiplicities are not affine (except for small values of g), but it would be enough if such an $M_{g,\nu}$ of dimension N (or rather, a suitable finite covering) were a union of $N-2$ affine open sets, making $M_{g,\nu}$ "not too close to being a compact variety".*

Using a little bit of Hodge theory, we show that the composition map σ induces an injective homomorphism

$$H^k(\overline{M}_{g,n}) \to \bigoplus_{i=1}^{N} H^k(X_i)$$

for $k \leq d(g,n)$. This result prepares the ground for various induction arguments. We give an example of such an argument, by inductively proving the vanishing of $H^1(\overline{M}_{g,n}, \mathbb{Q})$, for all g and n. Indeed it is also possible to prove that the third and fifth rational cohomology groups of $\overline{M}_{g,n}$ vanish. On the other hand, not all odd cohomology of $\overline{M}_{g,n}$ vanishes, and, at the end of the section, we show that $H^{11,0}(\overline{M}_{1,11}) \neq 0$.

In Section 4, using the knowledge of $H^2(M_{0,P}, \mathbb{Q})$ (cf. Proposition (7.5) in Chapter XVII), we inductively compute $H^2(\overline{M}_{g,n}, \mathbb{Q})$, showing that it is generated by tautological classes. The precise statement is given in Theorem (4.1). The inductive procedure relies heavily on the pullback formulae for tautological classes which we proved in Lemmas (4.35), (4.36), and (4.38) of Chapter XVII. For the rest, it consists of rather straightforward linear algebra arguments. The only slight difficulties occur in the initial cases of the induction, and they are due to the fact that, in genus less than or equal to two, the tautological classes satisfy a number of nontrivial relations.

In Section 5 we turn our attention to the rational cohomology of moduli spaces of smooth curves. After a brief overview on what is known about the stable cohomology of $M_{g,n}$ at the time of this writing, we compute $H^1(M_{g,n}, \mathbb{Q})$ and $H^2(M_{g,n}, \mathbb{Q})$ by adapting to the orbifolds $\overline{M}_{g,n}$ and $M_{g,n}$ the Deligne–Gysin spectral sequence relating the cohomology of a smooth variety $\overline{X}$ with the cohomology of $X = \overline{X} \smallsetminus D$, where D is a divisor with normal crossings.

The last three sections have a more subsidiary character. The topics treated there will be of central use in the next chapter, during the proof of the main theorem concerning the intersection of tautological classes in $\overline{M}_{g,n}$.

In Section 6 we describe in more detail the orbicell decomposition of $M_{g,n}$. In particular, we explain how automorphism groups of curves act on the various cells of the decomposition.

In Section 7 we give a combinatorial expression for the cohomology classes $\psi_p \in H^2(\overline{M}_{g,P}, \mathbb{Q})$. The main result is the following. Let us look at the continuous map

$$f : \overline{M}_{g,P} \longrightarrow \overline{M}_{g,P}^{comb}(r)$$

defined in the last section of the previous chapter (formula (12.13)). Its restriction to $M_{g,P}$ is a homeomorphism onto $M_{g,P}^{comb}(r)$. Consider on $\overline{M}_{g,P}$ the tautological line bundle $\mathcal{L}_p$. By definition $\psi_p = c_1(\mathcal{L}_p)$. We define a line bundle $\mathcal{L}_p^{comb}$ on $\overline{M}_{g,P}^{comb}(r)$ such that $f^*(\mathcal{L}_p^{comb}) = \mathcal{L}_p$. We then define a piecewise-linear form $\omega_p \in H^2_{PL}(\overline{M}_{g,P}^{comb}(r))$ such that

$$[\omega_p] = c_1(\mathcal{L}_p^{comb}).$$

The definition of ω_p is as follows. Let a be a cell of $\overline{M}_{g,P}^{comb}(r)$, and let G be the corresponding ribbon graph. Let r_p be the perimeter of the boundary component of G marked by p, let $\{e_1, \ldots, e_\nu\}$ be the edges (with possible repetitons) of this boundary component, and denote by l_e the lenght of e for any edge e. Then

$$(1.3) \qquad (\omega_p)_{|a} = \sum_{1 \le s < t \le \nu - 1} d\left(\frac{l_{e_s}}{r_p}\right) \wedge d\left(\frac{l_{e_t}}{r_p}\right).$$

To prove this equality, we make use of the homeomorphism (12.10) in Chapter XVIII and therefore of the *full simplicial complex* $A(S, P)$ (and not just of $A^0(S, P)$). A consequence of this computation is the following. Suppose that one wants to compute the intersection number

$$\int_{\overline{M}_{g,P}} \prod_{p \in P} \psi_p^{d_p},$$

where

$$\sum_{p \in P} d_p = 3g - 3 + |P|.$$

Then one can reduce the above computation to a combinatorial one, since

$$(1.4) \qquad \int_{\overline{M}_{g,P}} \prod_{p \in P} \psi_p^{d_p} = \int_{\overline{M}_{g,P}^{comb}(r)} \prod_{p \in P} \omega_p^{d_p} = \int_{M_{g,P}^{comb}(r)} \prod_{p \in P} \omega_p^{d_p}.$$

In the final section we give a combinatorial expression of a volume form on $M_{g,n}$ expressed in terms of the cellular decomposition.

As is clear from what we have said, the main objects of study in this chapter are the homology and cohomology groups of moduli spaces. It may be reassuring to remark right away that, as far as *rational* (co)homology is concerned, there is no difference between the one of the *space* $\overline{M}_{g,n}$ and the one of the *orbifold* (or stack) $\overline{\mathcal{M}}_{g,n}$. The same applies to the open moduli $M_{g,n}$ and $\mathcal{M}_{g,n}$. Therefore, in the notation for rational homology and cohomology of moduli, we will go back and forth between spaces and orbifolds without further warning.

2. The vanishing theorems for the rational homology of $M_{g,P}$.

In this section we will prove a first important theorem about the vanishing of the homology of $M_{g,n}$ in high degree. As we shall see, for positive genus, this result is a direct consequence of the cellular decomposition described in the preceding section. We will use the following notation:

$$(2.1) \qquad e(g,n) = \begin{cases} n-3 & \text{if } g=0\,, \\ 4g-5 & \text{if } g>0,\ n=0\,, \\ 4g-4+n & \text{if } g>0,\ n>0\,. \end{cases}$$

THEOREM (2.2) (Harer). *$H_k(M_{g,n},\mathbb{Q}) = 0$ for $k > e(g,n)$.*

Proof. Let us start with the case $g=0$. Fixing three of the n points as 0, 1, and ∞, we can realize $M_{0,n}$ as $\mathbb{C}^{n-3}$ with coordinates $z_1,\dots,z_{n-3}$ minus the hyperplanes $z_i=0$, $z_i=1$, and $z_i=z_j$ for $i\neq j$. As such, $M_{0,n}$ is seen to be an affine variety, so that its homology vanishes in degrees strictly greater than $\dim M_{0,n} = n-3$.

From now on we assume $g\geq 1$. We are first going to prove the theorem in the case $n=1$. We will later show how to reduce the other cases to this one. We fix a reference 1-pointed genus g Riemann surface $(S,\{p\})$ with $p\in S$. In this case the simplex $\Delta_{\{p\}}$ appearing in (1.1) is a zero-dimensional cell, and from Theorem (2.7) in Chapter XVIII we have a $\Gamma_{g,1}$-equivariant homeomorphism

$$\Psi : \mathcal{T}_{g,1} \longrightarrow |A^0(S,\{p\})|\,.$$

We set $A^0 = A^0(S,\{p\})$ and recall that, by definition, $|A^0| = |A| \smallsetminus |A_\infty|$. We first need the following general lemma.

LEMMA (2.3). *Let A be a finite-dimensional simplicial complex. Let $B\subset A$ be a subcomplex. Set $C = A\smallsetminus B$. Let A^1 and B^1 be the first baricentric subdivisions of A and B, respectively. Let D be the subcomplex of A^1 whose vertices are barycenters of simplices of C. In particular, D has dimension $\mu-\nu$, where μ (resp., ν) is the maximal (resp., minimal) dimension of a simplex belonging to C, and $|D|\subset|C| = |A|\smallsetminus|B|$. Then there is a deformation retraction of $|C|$ onto $|D|$. Moreover, if A is acted on by a group G and if the subcomplex B is preserved by this action, then the above deformation can be assumed to be G-equivariant.*

Proof. By definition, the vertices of a k-simplex of A^1 are barycenters $b_0,\dots,b_k$ of a strictly increasing sequence $a_0 < a_1 < \cdots < a_k$ of simplices of A. Such a simplex is in D if and only if a_0 is a simplex of C. Let $p\in|C| = |A^1|\smallsetminus|B^1|$. Write $p = \sum_{i=0}^k \lambda_i b_i$ and let s be the minimum index such that a_s belongs to C, so that $s\leq k$. Setting

$$p' = \left(\sum_{i=s}^{k}\lambda_i\right)^{-1}\cdot\sum_{i=s}^{k}\lambda_i b_i \quad\text{and}\quad H(p,t) = (1-t)p + tp'$$

gives the desired retraction. In the presence of a G-action this retraction is clearly G-equivariant. Q.E.D.

To apply this lemma to our situation, let us estimate the maximum dimension μ and minimum dimension ν of simplices belonging to A^0. Let then $a = (\alpha_0, \dots, \alpha_k)$ be a simplex in A^0, and let N be the number of connected components of $S \setminus \cup_{i=0}^{k} \alpha_i$. These components are discs, so that $2 - 2g = N - k$. In particular, $\nu \geq 2g - 1$. Now suppose that a has maximal dimension. Then each of the above connected components must be bounded by 3 among the arcs $\alpha_0, \dots, \alpha_k$. It follows that $N = \frac{2}{3}(k+1)$. Therefore, $k = 6g - 4$ and $\mu - \nu \leq 4g - 3$.

COROLLARY (2.4). *The complex $|A^0(S, \{p\})|$ can be $\Gamma_{g,1}$-equivariantly deformed to a complex of dimension less than or equal to $4g - 3$.*

COROLLARY (2.5). *$H_k(M_{g,1}[\varphi], L) = 0$ for any level structure φ, for any local system of $\mathbb{Q}$-vector spaces L, and for any $k > 4g - 3$.*

In particular, the theorem is proved in case $n = 1$. In the introduction we illustrated the retraction of $|A(S, \{p\})|$ to a one-dimensional spine in the case $g = 1$, $n = 1$ (see Figure 1).

We are now going to treat the cases where $n \neq 1$. Let $(\Sigma; x_1, \dots, x_n)$ be a reference n-pointed genus g Riemann surface satisfying the stability condition $2 - 2g + n < 0$. We consider on $M_{g,n}$ the level structure given by the surjective homomorphism

$$\varphi_n : \pi_1(\Sigma \setminus \{x_1, \dots, x_n\}) \to H_1(\Sigma; \mathbb{Z}/3\mathbb{Z}).$$

For brevity, we write $\varphi = \varphi_n$. It follows from Theorem (2.11) of Chapter XVI that the moduli space $M_{g,n}[\varphi]$ is smooth. Since $M_{g,n}$ is a quotient of $M_{g,n}[\varphi]$ by the finite group $\Gamma_{g,n}[\varphi]$, the vanishing of $H_k(M_{g,n}[\varphi], \mathbb{Q})$ implies the vanishing of $H_k(M_{g,n}, \mathbb{Q})$. It is then sufficient to prove our vanishing statement for the homology of $M_{g,n}[\varphi]$. Look at the universal family $\pi : \mathcal{C} \to M_{g,n}[\varphi]$. Let $D \subset \mathcal{C}$ be the divisor which is the sum of the images of the n sections of π. By definition, $\mathcal{C} \setminus D$ is isomorphic to $M_{g,n+1}[\varphi]$. It follows that the morphism $\eta : M_{g,n+1}[\varphi] \to M_{g,n}[\varphi]$ is a topologically locally trivial fibration with fiber F homeomorphic to $\Sigma \setminus \{x_1, \dots, x_n\}$. Let L be a local system of $\mathbb{Q}$-vector spaces on $M_{g,n+1}[\varphi]$. The E^2-term of the Leray spectral sequence with coefficients in L of the fibration η is given by

$$E^2_{p,q} = H_p(M_{g,n}[\varphi]; \mathcal{H}_q(\eta; L)),$$

where $\mathcal{H}_q(\eta; L)$ is the local system of the qth homology groups of the fibers of η with coefficients in L. When $n > 0$, the homological dimension of the fiber F is equal to 1. It follows that if $H_k(M_{g,n}[\varphi]; L')$ vanishes for $k > k_0$ and for any local system L', then $H_k(M_{g,n+1}[\varphi]; L)$ vanishes for $k > k_0 + 1$. In view of (2.5), this shows, inductively on n, that $H_k(M_{g,n}[\varphi]; L)$ vanishes for any local system of $\mathbb{Q}$-vector spaces L and any $k > 4g - 4 + n$ whenever $n \geq 1$. In particular, this proves Theorem

(2.2) for $n \geq 1$. To treat the case $n = 0$ it is convenient, although not strictly necessary, to switch to cohomology. We have then to show that $H^p(M_g[\varphi]; \mathbb{Q})$ vanishes for $p > 4g - 5$. Look at the cohomology Leray spectral sequence for $\eta: M_{g,1}[\varphi] \to M_g[\varphi]$, whose E_2 term is

$$E_2^{p,q} = H^p(M_g[\varphi]; R^q\eta_*\mathbb{Q}) .$$

In this case, the fibers of η are compact Riemann surfaces and carry a canonical orientation; this gives a canonical section ω, and hence a trivialization, of the rank 1 local system $R^2\eta_*\mathbb{Q}$. Cupping with ω, viewed as an element of $E_2^{0,2} = H^0(M_g[\varphi]; R^2\eta_*\mathbb{Q})$, gives homomorphisms $E_2^{p,q} \to E_2^{p,q+2}$, compatible with differentials. In particular, $E_2^{p,0} \to E_2^{p,2}$ is an isomorphism. Thus, $d_2: E_2^{p,2} \to E_2^{p+2,1}$ vanishes, since $d_2: E_2^{p,0} \to E_2^{p+2,-1}$ obviously does, and $d_2: E_2^{p-2,3} \to E_2^{p,2}$ also vanishes since $E_2^{p,q} = 0$ for $q > 2$. The analogues of these statements hold for all E_r with $r \geq 2$; the upshot is that $E_2^{p,2} = E_\infty^{p,2}$. This shows that, if $H^p(M_g[\varphi]; \mathbb{Q}) \cong E_2^{p,2}$ does not vanish, neither does $H^{p+2}(M_{g,1}[\varphi]; \mathbb{Q})$. In conjunction with (2.5), this proves our claim.

As we remarked in Chapter XII, Section 4, Poincaré duality holds for $M_{g,n}$, so that we can express Theorem (2.2) in terms of cohomology with compact support. We set

$$(2.6) \qquad d(g,n) = \begin{cases} n-4 & \text{if } g = 0 , \\ 2g-2 & \text{if } n = 0 , \\ 2g-3+n & \text{if } g \neq 0,\ n \neq 0 . \end{cases}$$

Dualizing Theorem (2.2), we then get the following:

THEOREM (2.7). *$H_c^k(M_{g,n}; \mathbb{Q}) = 0$ for $k \leq d(g,n)$.*

3. Comparing the cohomology of $\overline{M}_{g,n}$ to the one of its boundary strata.

One way to effectively use the vanishing theorems (2.2) and (2.7) is to look at the exact sequence of *rational* cohomology with compact support

$$\cdots \to H_c^k(M_{g,n}) \to H^k(\overline{M}_{g,n}) \to H^k(\partial M_{g,n}) \to H_c^{k+1}(M_{g,n}) \to \cdots$$

Using (2.7), it follows that:

PROPOSITION (3.1). *The restriction map*

$$H^k(\overline{M}_{g,n}) \to H^k(\partial M_{g,n})$$

is injective for $k \leq d(g,n)$, and is an isomorphism for $k < d(g,n)$.

Now let us consider the irreducible components $D_1, D_2, \dots, D_N$ of the boundary $\partial M_{g,n}$. As we showed in Chapter XII (cf. diagram (10.7)), each of these components is the image of a map $\mu_i : X_i \to \overline{M}_{g,n}$ where X_i can be of two different kinds. Either $X_i = \overline{M}_{g-1,n+2}$, and μ_i is obtained by identifying the last two marked points of each $(n+2)$-pointed curve of genus $g-1$, or $X_i = \overline{M}_{a,h+1} \times \overline{M}_{b,k+1}$, where $a+b=g$, $h+k=n$, $2a-2+h \geq 0$, $2b-2+k \geq 0$, and μ_i is obtained by identifying the $(h+1)$st point of an $(h+1)$-pointed genus a curve with the $(k+1)$st point of a $(k+1)$-pointed genus b curve. Let us then consider the composition

$$\sigma : \coprod_{i=1}^{N} X_i \to \partial M_{g,n} \hookrightarrow \overline{M}_{g,n}\,.$$

We will presently prove a stronger version of Proposition (3.1) stating that, in degrees not greater than $d(g,n)$, also the composition σ induces an injection in cohomology. This result is somewhat surprising, especially when the degree k is less than $d(g,n)$. In this case, it implies that the kth degree cohomology of $\partial M_{g,n}$ injects in the kth degree cohomology of the disjoint union of the X_i, showing that the intricate geometry arising from the way in which the various strata of $\partial M_{g,n}$ intersect each other does not contribute to the cohomology, at least in relatively low degree.

THEOREM (3.2). *The map σ induces an injective homomorphism*

$$H^k(\overline{M}_{g,n}) \hookrightarrow \bigoplus_{i=1}^{N} H^k(X_i)$$

for $k \leq d(g,n)$.

Proof. As we proved in Section 9 of Chapter XVI, each of the X_i is the quotient of a smooth complete variety Z_i by a finite group G_i. It will then suffice to prove the injectivity of the map

$$H^k(\overline{M}_{g,n}) \to \bigoplus_{i=1}^{N} H^k(Z_i)$$

in the given range. For this, we use the following result in Hodge theory due to Deligne (cf. [165], Proposition (8.2.5)).

THEOREM (3.3). *Let Y be a complete variety. If $u : X \to Y$ is a proper surjective morphism and X is smooth, then the weight k quotient of $H^k(Y,\mathbb{Q})$ is the image of $H^k(Y,\mathbb{Q})$ in $H^k(X,\mathbb{Q})$.*

We use this result by taking as Y the boundary $\partial M_{g,n}$ and as X the disjoint union of the Z_i. In particular, $H^k(\partial M_{g,n})/W_{k-1}(H^k(\partial M_{g,n}))$ injects in $H^k(X)$. By the previous proposition we know that, in the given range, the map

$$\rho : H^k(\overline{M}_{g,n}) \to H^k(\partial M_{g,n})$$

is injective. It remains to show that the image of ρ does not intersect $W_{k-1}(H^k(\partial M_{g,n}))$. But this is evident because ρ is a morphism of Hodge structures and hence is strictly compatible with filtrations. Hence,

$$\rho(H^k(\overline{M}_{g,n})) \cap W_{k-1}(H^k(\partial M_{g,n})) = \rho(W_{k-1}(H^k(\overline{M}_{g,n}))) = \rho(\{0\}) = 0\,,$$

since $H^k(\overline{M}_{g,n})$ is of pure weight k. The proof of Theorem (3.2) is now complete. Q.E.D.

As a first elementary application of Theorem (3.2), we shall reprove the vanishing of the first homology of $\overline{M}_{g,n}$ (cf. Corollary (7.3) in Chapter XV).

COROLLARY (3.4). *$H^1(\overline{M}_{g,n}, \mathbb{Q}) = 0$ for all g and n such that $2g-2+n > 0$.*

Proof. The theorem is true for $\overline{M}_{0,3} = \{pt\}$ and $\overline{M}_{1,1} = \mathbb{P}^1$. Except in these two cases, $1 \leq d(g,n)$, so that the homomorphism $H^1(\overline{M}_{g,n}) \to \overset{N}{\underset{i=1}{\oplus}} H^1(X_i)$ is always injective. Now X_i is either $\overline{M}_{g-1,n+2}$ or $\overline{M}_{a,h+1} \times \overline{M}_{b,k+1}$, where $a+b=g$, $h+k=n$. Therefore, by the Künneth formula, $H^1(\overline{M}_{g,n})$ injects in a direct sum of first cohomology groups of moduli spaces $\overline{M}_{p,\nu}$ where either $p<g$, or $p=g$ and $\nu<n$. The result follows by double induction on g and n. Q.E.D.

It turns out that also the third and the fifth rational cohomology groups of $\overline{M}_{g,n}$ vanish. The difficulty in these two cases is that, to make the induction work, one has to deal separately with a number of initial cases of increasing complexity. In a series of exercises the reader will be asked to treat some of these cases (see Exercises A) and B)). Of course, one may ask what happens in general with the odd-degree cohomology of $\overline{M}_{g,n}$. The first case in which a nonvanishing result is known is the following, regarding the degree 11 cohomology.

THEOREM (3.5) (Deligne). $H^{11,0}(\overline{\mathcal{M}}_{1,11}) \neq 0$.

Proof. Set $\mathcal{M} = \overline{\mathcal{M}}_{1,1}$ and consider the universal family (in the orbifold, or in the stack sense) $\pi : \mathcal{C} \to \mathcal{M}$. Take the 10-fold fiber product of $\mathcal{C}$ over $\mathcal{M}$ and call it X:

$$X = \mathcal{C} \times_{\mathcal{M}} \mathcal{C} \times_{\mathcal{M}} \cdots \times_{\mathcal{M}}\,.$$

Denote by $p : X \to \mathcal{M}$ the canonical projection and by $p_i : X \to \mathcal{C}$ the projection to the ith factor, so that $\pi \circ p_i = p$. From Section 10 of Chapter XII we know that $\overline{\mathcal{M}}_{1,11}$ is a modification of X. It then suffices to show that $H^{11,0}(X)$ does not vanish. Consider the exact sequence

$$0 \to p^*\Omega^1_{\mathcal{M}} \to \Omega^1_X \to \Omega^1_{X/\mathcal{M}} \to 0\,.$$

To produce a nonzero section of $\bigwedge^{11}\Omega^1_X$, it is enough to produce one of $p^*\Omega^1_{\mathcal{M}}\otimes\bigwedge^{10}\Omega^1_{X/\mathcal{M}}$. On the other hand,

$$\Omega^1_{X/\mathcal{M}} = \bigoplus_{i=1}^{10} p_i^*(\omega_\pi)\,,$$

so that

$$p^*\Omega^1_{\mathcal{M}}\otimes\bigwedge^{10}\Omega^1_{X/\mathcal{M}} \cong p^*\Omega^1_{\mathcal{M}}\otimes\left(\bigotimes_{i=1}^{10} p_i^*(\omega_\pi)\right).$$

In Section 7 of Chapter XIII we proved that, if $\mathbb{E}$ is the Hodge bundle on $\mathcal{M}=\overline{\mathcal{M}}_{1,1}$, then

$$\omega_\pi = \pi^*(\mathbb{E})\,.$$

In the same chapter we also showed that $\Omega^1_{\mathcal{M}}$, the canonical bundle of $\overline{\mathcal{M}}_{1,1}$, is given by the formula

$$\Omega^1_{\mathcal{M}} \cong \mathbb{E}^{13}\otimes\mathcal{L}(-2\infty),$$

where $\mathcal{L}$ and ∞ are, respectively, the point bundle on $\overline{\mathcal{M}}_{1,1}$ and the boundary point of $\overline{\mathcal{M}}_{1,1}$ (cf. Theorem (7.15)), and furthermore (formulas (7.12) and (7.13)) that

$$\mathcal{L}\cong\mathbb{E}\,,\qquad \mathbb{E}^{12}\cong\mathcal{O}(\infty)\,.$$

Putting everything together, we get

$$p^*\Omega^1_{\mathcal{M}}\otimes\bigwedge^{10}\omega_\pi \cong p^*\left(\Omega^1_{\mathcal{M}}\otimes\mathbb{E}^{10}\right)\cong p^*\left(\mathbb{E}^{24}(-2\infty)\right)\cong p^*\mathcal{O}_{\mathcal{M}}\,.$$

It follows that $p^*\Omega^1_{\mathcal{M}}\otimes\bigwedge^{10}\Omega^1_{X/\mathcal{M}}$ has a nontrivial section, as announced.

Q.E.D.

4. The second rational cohomology group of $\overline{M}_{g,n}$.

In this section we shall outline the proof of one of the main results in this chapter, which completely describes the second-degree rational cohomology of $\overline{M}_{g,n}$. As usual, we will use both the notation $\overline{M}_{g,n}$ and $\overline{M}_{g,P}$, where $|P|=n$. When writing cohomology groups, we will no longer mention the ring of coefficients; this ring will always be the field $\mathbb{Q}$. We shall freely use the notation and results of Chapter XVII regarding the degree two tautological classes κ_1, $\psi_1,\dots,\psi_n$ and the boundary classes $\delta_{irr},\delta_{a,A}$. In what follows, for any integer a, we write δ_a to denote the sum of all classes $\delta_{a,A}$; notice that, in case $g=2a$, the summand $\delta_{a,A}=\delta_{a,A^c}$ occurs only once, and not twice, in this sum.

THEOREM (4.1). *For any g and n such that $2g-2+n>0$, $H^2(\overline{M}_{g,n})$ is generated by the classes $\kappa_1, \psi_1, \dots, \psi_n$, δ_{irr} and the classes $\delta_{a,A}$ such that $0 \le a \le g$, $2a-2+|A| \ge 0$, and $2(g-a)-2+|A^c| \ge 0$. The relations among these classes are as follows.*

a) *If $g>2$, all relations are generated by those of the form*

$$\delta_{a,A} = \delta_{g-a,A^c}\,. \tag{4.2}$$

b) *If $g=2$, all relations are generated by the (4.2) plus the following one:*

$$5\kappa_1 = 5\psi + \delta_{irr} - 5\delta_0 + 7\delta_1\,. \tag{4.3}$$

c) *If $g=1$, all relations are generated by the (4.2) plus the following ones:*

$$\kappa_1 = \psi - \delta_0\,, \tag{4.4}$$

$$12\psi_p = \delta_{irr} + 12\sum_{\substack{S\ni p\\ |S|\ge 2}} \delta_{0,S}\,. \tag{4.5}$$

d) *If $g=0$, all relations are generated by the (4.2) and by the following ones:*

$$\sum_{\substack{A\ni p,q\\ A\not\ni r,s}} \delta_{0,A} = \sum_{\substack{A\ni p,r\\ A\not\ni q,s}} \delta_{0,A} = \sum_{\substack{A\ni p,s\\ A\not\ni q,r}} \delta_{0,A}\,, \tag{4.6}$$

where p,q,r,s is an arbitrary 4-tuple of distinct elements in P, and

$$\begin{aligned} \kappa_1 &= \sum_{A\not\ni x,y} (|A|-1)\delta_{0,A}\,,\\ \psi_z &= \sum_{\substack{A\ni z\\ A\not\ni x,y}} \delta_{0,A}\,,\\ \delta_{irr} &= 0\,. \end{aligned} \tag{4.7}$$

All the statements in Theorem (4.1) concerning the *relations* among the tautological classes appear as exercises at the end of this chapter (see Exercises C)). The part of Theorem (4.1) stating that $H^2(\overline{M}_{g,n})$ is generated by tautological classes and boundary classes is proved by induction. In this section we will limit ourselves to proving the induction step, assuming the initial cases. The latter consist of a number of results regarding curves of genus less than or equal to 2, which appear as exercises at the end of the chapter (see Exercises A) and D)).

In the proof of Theorem (4.1) we will make essential use of Theorem (3.2). It is therefore important to recall from Section 4 of Chapter XVII the restriction maps from $\overline{M}_{g,n}$ to its boundary components. The basic maps describing these restrictions are

$$\xi : \overline{M}_{g-1,P\cup\{q,r\}} \to \overline{M}_{g,P}\,,$$
$$\eta : \overline{M}_{a,A\cup\{s\}} \times \overline{M}_{g-a,A^c\cup\{t\}} \to \overline{M}_{g,P}\,,$$

where, for simplicity, we set $\xi = \xi_{irr}$ and $\eta = \xi_{a,A}$. For the convenience of the reader, we recall how they are defined. The image under ξ of a $P\cup\{q,r\}$-pointed genus $g-1$ curve is obtained by identifying the points labeled by q and r; likewise, the image under η of a pair consisting of an $A\cup\{s\}$-pointed genus a curve and an $A^c\cup\{t\}$-pointed genus $g-a$ curve is the P-pointed curve of genus g obtained by identifying the points labelled by s and t. Actually, instead of the maps η, we will consider the maps

$$\vartheta : \overline{M}_{a,A\cup\{s\}} \to \overline{M}_{g,P}$$

which associate to any $A\cup\{s\}$-pointed genus a curve the P-pointed genus g curve obtained by gluing to it a *fixed* $A^c\cup\{t\}$-pointed genus $g-a$ curve C via identification of s and t. The reason why, when dealing with second cohomology groups of $\overline{M}_{g,n}$, we consider the maps ϑ instead of the maps η is the following. By Corollary (3.4), the first cohomology of $\overline{M}_{\gamma,\nu}$ always vanishes. Therefore, by the Künneth formula,

(4.8)
$$H^2\left(\overline{M}_{a,A\cup\{q\}} \times \overline{M}_{g-a,A^c\cup\{r\}}\right) \cong H^2\left(\overline{M}_{a,A\cup\{q\}}\right) \oplus H^2\left(\overline{M}_{g-a,A^c\cup\{r\}}\right)\,.$$

Thus, knowing how the natural classes pull back under ϑ actually tells us how they pull back under η. It is important to stress that, although of course ϑ depends on the choice of C, any two choices give rise to homotopic maps so that, in cohomology, the pullback map ϑ^* is independent of the choice of C. In Lemmas (4.35), (4.36), and (4.38) of Chapter XVII we gave explicit formulas describing how the degree 2 tautological classes and boundary classes pull back under ξ, η, and ϑ.

In computing the second cohomology group of $\overline{M}_{g,n}$, what really makes the inductive procedure work is the following consequence of Theorem (3.2) and Corollary (3.4).

LEMMA (4.9). *The pullback map* $\xi^* : H^2(\overline{M}_{g,P}) \to H^2(\overline{M}_{g-1,P\cup\{q,r\}})$ *is injective for any* $g \geq 2$.

Proof. By Theorem (3.2) and by the remark we just made, it suffices to prove that, if x is any element of $H^2(\overline{M}_{g,P})$ such that $\xi^*(x) = 0$, then x pulls back to zero under any one of the maps

$$\vartheta^* : H^2\left(\overline{M}_{g,P}\right) \longrightarrow H^2\left(\overline{M}_{a,A\cup\{s\}}\right)\,,$$

where $0 \leq a \leq g$ and $A \subset P$, $2a - 1 + |A| > 0$. Look at the fixed $A^c \cup \{t\}$-pointed genus $g-a$ curve C which we attach to the varying $A \cup \{s\}$-pointed genus a curve, via identification of s and t, to obtain the map ϑ. If $g - a \neq 0$, we can assume that C has one node. We may therefore assume that C is obtained from an $A^c \cup \{t, r, q\}$-pointed genus $g-a-1$ curve C' by identifying q and r. Therefore, if we let

$$\vartheta' : \overline{M}_{a,A\cup\{s\}} \longrightarrow \overline{M}_{g-1,P\cup\{q,r\}}$$

be the map obtained by identifying the point s of an $A \cup \{s\}$-pointed genus a curve with the point t of the curve C', we have

$$\xi\vartheta' = \vartheta \,.$$

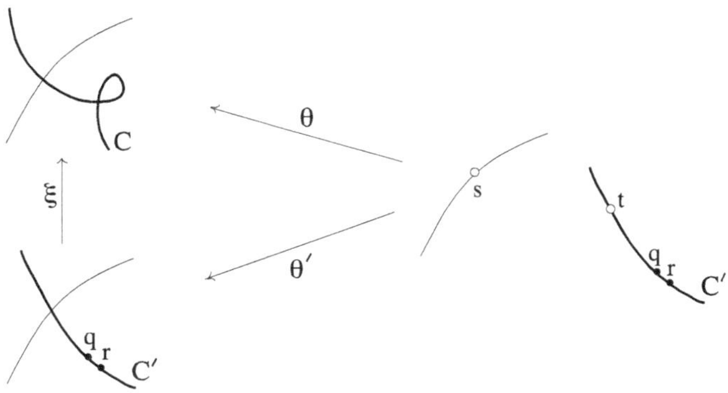

Figure 2.

Therefore, the condition $\xi^*(x) = 0$ implies that $\vartheta^*(x) = 0$ when $g - a \neq 0$. In particular, this proves the theorem for $n = 0$ and all $g \geq 2$. Starting from these cases, we proceed to prove the theorem by induction on n. But to prove the theorem for the pair (g, n), we need to show that the condition $\xi^*(x) = 0$ implies that $\vartheta^*(x) = 0$ *only* when $g - a = 0$, i.e., when $a = g$, the other cases having already been taken care of. To treat this remaining case, we look at the map

$$\vartheta_1 : \overline{M}_{g-1,A\cup\{s,q,r\}} \longrightarrow \overline{M}_{g-1,P\cup\{q,r\}}$$

obtained by attaching a fixed $A^c \cup \{t\}$-pointed curve C of genus 0 to a varying $A \cup \{s, q, r\}$-pointed curve of genus $g-1$, by identifying the point s with the point t. We denote by ξ_1 the map from $M_{g-1,A\cup\{s,q,r\}}$ to $M_{g,A\cup\{s\}}$ obtained by identifying q with r. We then have

$$\vartheta_1{}^*\xi^* = \xi_1{}^*\vartheta^* : H^2\left(\overline{M}_{g,P}\right) \longrightarrow H^2\left(\overline{M}_{g-1,A\cup\{s,q,r\}}\right) .$$

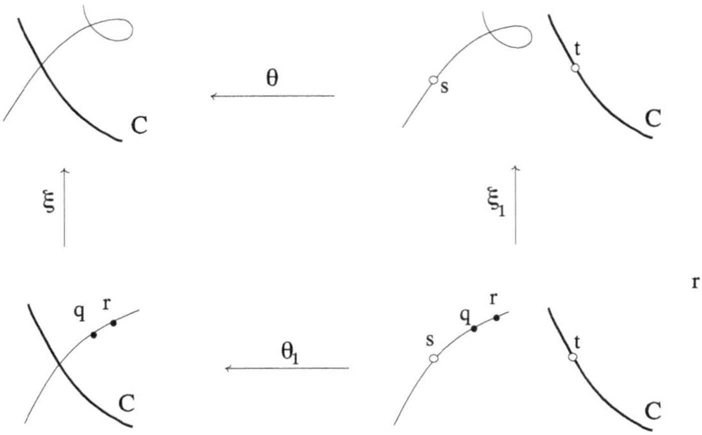

Figure 3.

By induction hypothesis, the map ξ_1^* is injective. Thus, again, the condition $\xi^*(x) = 0$ implies that $\vartheta^*(x) = 0$.

Q.E.D.

Proof of Theorem (4.1). As we already mentioned, the part of the theorem regarding the relations between the tautological classes and the boundary classes appears as exercises at the end of this chapter (see Exercises C)). We only have to prove that $H^2(\overline{M}_{g,n})$ is generated by tautological classes and boundary classes. Since this is true for $g = 0$ by Keel's theorem (see Proposition (7.5) of Chapter XVII), we can assume that $g > 0$.

We proceed by double induction on (g, n). The initial cases of the induction are those for which $2 > d(g, n)$ and $g > 0$, i.e., the cases $(1, 1)$ and $(1, 2)$. These two cases will be treated in Exercises A) at the end of this chapter. Now comes the inductive step.

Our strategy for the inductive step is quite simple. The idea is to use Theorem (3.2) or, when possible, the stronger result of Lemma (4.9). Suppose that we want to show that $H^2(\overline{M}_{g,n})$ is generated by tautological classes, assuming that the same is known to be true in genus less than g, or in genus g but with fewer than n marked points. Theorem (3.2) shows that $H^2(\overline{M}_{g,n})$ injects into the direct sum of the second cohomology groups of the X_i. By induction hypothesis, these are generated by tautological classes, all relations among which are known. Formulae (4.35), (4.36), and (4.38) of Chapter XVII completely describe the images of tautological classes under each map $H^2(\overline{M}_{g,n}) \to H^2(X_i)$. Thus, at least in principle, we can decide which classes in $\bigoplus H^2(X_i)$ come from tautological classes in $H^2(\overline{M}_{g,n})$. On the other hand, if α is any class in $H^2(\overline{M}_{g,n})$ and we denote by α_i its restriction to X_i, the α_i satisfy obvious compatibility relations on the "intersections" of the X_i. Let V be the subspace of $\bigoplus H^2(X_i)$ defined by these compatibility relations.

Since, by induction, the spaces $H^2(X_i)$ are generated by tautological classes, V can be completely described, at least in principle, using (4.35), (4.36), and (4.38) of Chapter XVII. We will show, in essence, that V coincides with the subspace generated by the images of the tautological classes of $H^2(\overline{M}_{g,n})$. By the injectivity of $H^2(\overline{M}_{g,n}) \to \bigoplus H^2(X_i)$, this will conclude the proof.

The inductive steps for $g = 1$ and $g = 2$ have to be treated separately. The reason is that, in these two cases, the presence of relations between the tautological classes makes the argument more cumbersome. But, more to the point, in these cases we cannot use the full strength of Lemma (4.9), which only holds for $g \geq 2$. The reader will be guided in the proof of these induction steps in a set of exercises at the end of this chapter (see Exercises D)). Here we give the induction step assuming $g \geq 3$. Because of this assumption, we can use Lemma (4.9) both for genus g and for genus $g-1$.

Let then $g \geq 3$ be an integer, and let P be a finite set. If P is not empty, let p be a fixed element of P. Let q, r be distinct and not belonging to P. Let $\xi : \overline{M}_{g-1,P\cup\{q,r\}} \to \overline{M}_{g,P}$ be the map that is obtained by identifying the points labeled by q and r. We wish to show that $H^2(\overline{M}_{g,P})$ is generated by tautological classes, assuming that the analogous statement is known to hold for $\overline{M}_{\gamma,\nu}$ whenever $\gamma < g$ or $\gamma = g$ and $\nu < |P|$. We will do this only for $P \neq \emptyset$, the argument for $P = \emptyset$ being entirely similar. Let y be any element of $H^2(\overline{M}_{g,P})$. The pullback $\xi^*(y)$ is invariant under the operation of interchanging q and r. Therefore, by the induction assumptions, it is a linear combination of κ_1, the ψ_i, $i \in P$, $\psi_q + \psi_r$, δ_{irr}, and the classes $\delta_{u,U}$, $\delta_{u,U\cup\{q,r\}}$, and $\delta_{u,U\cup\{q\}} + \delta_{u,U\cup\{r\}}$, where u is any integer between 0 and g, and U runs through all subsets of P containing p; when $g = 3$, we can even do without κ_1. Formulas (4.35) of Chapter XVII tell us that there is a linear combination z of tautological classes such that the pullback of $x = y - z$ is of the form

$$
\begin{aligned}
\xi^*(x) = f \cdot (\psi_q + \psi_r) &+ \sum_{\substack{p\in U\subset P \\ 0\leq u\leq g-2}} g_{u,U}\delta_{u,U\cup\{q,r\}} \\
&+ \sum_{\substack{p\in U\subset P \\ 0\leq u\leq g-1}} h_{u,U}\cdot(\delta_{u,U\cup\{q\}} + \delta_{u,U\cup\{r\}})
\end{aligned}
\tag{4.10}
$$

for suitable coefficients $f, g_{u,U}, h_{u,U}$. In case $g = 3$, we may even assume, using (4.3), that $f = 0$. We will show that, in fact, $\xi^*(x) = 0$. Lemma (4.9) will then tell us that x itself vanishes, proving that y is a linear combination of tautological classes, as desired. Suppose $s \notin P \cup \{q, r\}$ and let $\vartheta : \overline{M}_{g-1,P\cup\{s\}} \to \overline{M}_{g,P}$ be the map that is obtained by attaching a fixed elliptic tail at the point labeled by s. Look at the diagram

$$\begin{array}{ccc} \overline{M}_{g-1,P\cup\{s\}} & \xrightarrow{\ \vartheta\ } & \overline{M}_{g,P} \\ {\scriptstyle\varphi}\downarrow & \nearrow{\scriptstyle\xi} & \\ \overline{M}_{g-1,P\cup\{q,r\}} & & \end{array}$$

where φ attaches the point labeled by t of a sphere marked by $\{t,q,r\}$ to the point labeled by s of a variable curve in $\overline{M}_{g-1,P\cup\{s\}}$. This diagram is commutative up to homotopy. The identity $\varphi^*\xi^*(x) = \vartheta^*(x)$, together with formulas (4.10) and (4.35), (4.36), (4.38) of Chapter XVII, applied to φ, implies that

$$\vartheta^*(x) = \sum_{\substack{p\in U \\ 0\le u\le g-2}} g_{u,U}\delta_{u,U\cup\{s\}}\,. \tag{4.11}$$

Now consider the commutative diagram

$$\begin{CD} \overline{M}_{g-2,P\cup\{q,r,s\}} @>{\beta}>> \overline{M}_{g-1,P\cup\{q,r\}} \\ @V{\gamma}VV @VV{\xi}V \\ \overline{M}_{g-1,P\cup\{s\}} @>{\vartheta}>> \overline{M}_{g,P} \end{CD} \tag{4.12}$$

where γ and β are the analogues of ξ and ϑ, respectively. If we write down explicitly the identity $\gamma^*\vartheta^*(x) = \beta^*\xi^*(x)$ using formulas (4.10) and (4.11), plus formulas (4.35), (4.36), (4.38) of Chapter XVII, we get the relation

$$\begin{aligned} &\sum_{\substack{p\in U \\ 0\le u\le g-2}} g_{u,U}(\delta_{u,U\cup\{s\}} + \delta_{u-1,U\cup\{s,q,r\}}) \\ &\qquad = f\cdot(\psi_q+\psi_r) + \sum_{\substack{p\in U \\ 0\le u\le g-2}} g_{u,U}(\delta_{u,U\cup\{q,r\}} + \delta_{u-1,U\cup\{s,q,r\}}) \\ &\qquad\quad + \sum_{\substack{p\in U \\ 0\le u\le g-1}} h_{u,U}\cdot(\delta_{u,U\cup\{q\}} + \delta_{u,U\cup\{r\}} + \delta_{u-1,U\cup\{s,q\}} \\ &\qquad\quad + \delta_{u-1,U\cup\{s,r\}}) \end{aligned} \tag{4.13}$$

in $H^2(\overline{M}_{g-2,P\cup\{q,r,s\}})$. If $g\ge 4$, all the tautological classes appearing in (4.13) are independent, so $f = g_{u,U} = h_{u,U} = 0$ for all u and U. When $g=3$, we already know that $f=0$; since the boundary classes are independent in genus 1, we conclude that $g_{u,U} = h_{u,U} = 0$ for all u and U in this case as well. This shows that $\xi^*(x)=0$, as desired.

Q.E.D.

5. A quick overview of the stable rational cohomology of $M_{g,n}$ and the computation of $H^1(M_{g,n})$ and $H^2(M_{g,n})$.

A fundamental theorem of Harer, which we will state precisely below, asserts that for $g >> k$, there is an isomorphism between $H^k(M_g;\mathbb{Q})$ and $H^k(M_{g+1};\mathbb{Q})$. This theorem is the mapping class group counterpart of a theorem of Borel on the stable real cohomology of arithmetic groups [77,78]. As a consequence of this theorem, it makes good sense to talk about the stable rational cohomology of M_g. This stable cohomology has been the object of extensive research over an extended period of time.

The first result pointing in the direction of stabilization regards the low-degree cohomology which, if one may say so, stabilizes immediately, as the following theorem asserts.

THEOREM (5.1). *The following hold true:*

i) $H^1(M_{g,n},\mathbb{Q}) = 0$ *for any* $g \geq 1$ *and any* n *such that* $2g-2+n > 0$.
ii) $H^2(M_{g,n},\mathbb{Q})$ *is freely generated by* $\kappa_1, \psi_1, \ldots, \psi_n$ *for any* $g \geq 3$ *and any* n. $H^2(M_{2,n},\mathbb{Q})$ *is freely generated by* $\psi_1, \ldots, \psi_n$ *for any* n, *while* $H^2(M_{1,n},\mathbb{Q})$ *vanishes for all* n.

The result on H^1 is due to Mumford ([551], Theorem 1) for $n = 0$ and to Harer [337] for arbitrary n, while the one on H^2 is due to Harer [337]. Let us remark that both parts of the theorem can be considerably strengthened. As we said in Chapter XV, Section 7, the orbifold fundamental group of $\mathcal{M}_{g,n}$ is the Teichmüller modular group $\Gamma_{g,n}$. On the other hand, it is known that $\Gamma_{g,n}$ equals its commutator subgroup for $g > 2$. For $n = 0$, this is Theorem 1 of [599], while for arbitrary n, it is Lemma 1.1 of [337]. Thus, the first integral homology group of $\mathcal{M}_{g,n}$ vanishes for $g > 2$. Likewise, the main result of [337] actually computes the second integral homology of $\mathcal{M}_{g,n}$ for $g > 4$; this turns out to be free of rank $n+1$ for any n.

The fundamental stability result is Harer's stability theorem, first proved in [338] and then improved by Ivanov [391,392].

THEOREM (5.2) (Stability Theorem). *For* $k < \frac{1}{2}(g-1)$, *there are isomorphisms*

$$H^k(M_g;\mathbb{Q}) \cong H^k(M_{g+1};\mathbb{Q}) \cong H^k(M_{g+2};\mathbb{Q}) \cong \cdots .$$

In view of this theorem, one says that k *is in the stable range* if $k < \frac{1}{2}(g-1)$. In [342], Harer extended the range of validity of the theorem to $k < \frac{2}{3}(g-1)$.

Let us give a complex geometrical description of how one compares the rational cohomology of M_g with the one of M_{g+1}. Denote by S the stratum of the boundary of M_{g+1} whose points represent stable irreducible curves with exactly one node. This stratum is Zariski-open in Δ_{irr} and is an orbifold which can be viewed as the quotient of $M_{g,2}$

modulo the involution interchanging the labels of the two marked points. Thus, the natural projection $\pi : M_{g,2} \to M_g$ factors through a morphism $p : S \to M_g$. Now consider a tubular neighborhood T of S in $\overline{M}_{g+1}$ and denote by φ the embedding of $T \smallsetminus S$ in M_{g+1}. Finally, view $T \smallsetminus S$ as a fiber bundle over S and denote by $\sigma : T \smallsetminus S \to S$ the natural projection. The stability theorem asserts that the two homomorphisms

$$H^k(M_g;\mathbb{Q}) \xrightarrow{(p\sigma)^*} H^k(T \smallsetminus S;\mathbb{Q}) \xleftarrow{\varphi^*} H^k(M_{g+1};\mathbb{Q})$$

are both isomorphisms in the stable range.

Miller [512] and Morita [519] independently proved the following theorem.

THEOREM (5.3). *Let k be in the stable range. Then the natural homomorphism*

$$\mathbb{Q}[x_1, x_2, \dots] \to H^\bullet(M_g;\mathbb{Q})$$

sending x_i to κ_i, $i = 1, 2, \dots$, is injective up to degree $2k$.

The crowning result of this theory is the proof of Mumford's conjecture by Madsen and Weiss.

THEOREM (5.4) (Mumford's conjecture). *Let k be in the stable range. Then the natural homomorphism*

$$\mathbb{Q}[x_1, x_2, \dots] \to H^\bullet(M_g;\mathbb{Q})$$

sending x_i to κ_i, $i = 1, 2, \dots$, is an isomorphism up to degree $2k$.

The first proof of Mumford's conjecture is given in [495] (see also [494]). A second, different proof based on the work of Segal [616], Tillmann [662], and Madsen and Tillmann [493] is given by Galatius, Tillmann, Madsen, and Weiss in [277].

Finally, the case of pointed curves is taken care of by the following theorem of Looijenga [490].

THEOREM (5.5). *Let k be in the stable range. Then the natural homomorphism*

$$H^\bullet(M_g;\mathbb{Q})[y_1, y_2, \dots, y_n] \to H^\bullet(M_{g,n};\mathbb{Q})$$

sending y_i to ψ_i, $i = 1, 2, \dots, n$, is an isomorphism up to degree $2k$.

In conclusion, in the stable range, the cohomology ring of $M_{g,n}$ is isomorphic to a polynomial ring in the variables $y_1, \dots, y_n$ and $x_1, x_2, \dots$, where $\deg(x_j) = 2j$ for $j = 1, 2, \dots$ and $\deg(y_i) = 2$ for $i = 1, 2, \dots, n$.

In [593], Pikaart proves the following interesting result.

THEOREM (5.6). *Let k be in the stable range. Then the natural homomorphism*

$$H^k(\overline{M}_{g,n};\mathbb{Q}) \to H^k(M_{g,n};\mathbb{Q})$$

is surjective. In particular, when k is in the stable range, the mixed Hodge structure on $H^k(M_{g,n};\mathbb{Q})$ is pure of weight k.

Excellent survey articles with extensive bibliography on the cohomology of moduli spaces are Harer's [339], Hain and Looijenga's [335], and Morita's [521]. They all appeared before the proof of Mumford's conjecture.

Excellent surveys on the ideas surrounding the proof of Mumford's conjecture are Tilmann's ICM talk [663], Ekedahl [207], and Eliashberg, Galatius, and Mishachev [210]. A brief and lucid sketch of the ideas involved in the work of Tillmann, Madsen, and Weiss is contained in Kirwan's ICM talk [419].

The proofs of Harer's stability theorem, of Miller's and Morita's injectivity theorem, and, even more so, of Mumford's conjecture, require mathematical tools that go far beyond the reach of this book. On the other hand, the tools at our disposal allow us to compute the first and second rational cohomology groups of $M_{g,n}$ for $g > 0$ and $n \geq 0$, that is, to prove Theorem (5.1).

Before we embark on the actual proof, we need to recall Deligne's Gysin spectral sequence computing the cohomology of nonsingular varieties (see 3.2.4.1 in [162], or [322]). Let X be a complex manifold, and let D be a divisor with normal crossings in X. Set $V = X \setminus D$. The Gysin spectral sequence we are going to define computes the cohomology of V. Locally in X, the divisor D looks like a union of coordinate hyperplanes. Let $D^{[p]}$ denote the union of the points of multiplicity at least p in D, and let $\widetilde{D}^{[p]}$ be the normalization of $D^{[p]}$. We set

$$\widetilde{D}^{[0]} = D^{[0]} = X\,.$$

In concrete terms, a point y of $\widetilde{D}^{[p]}$ can be thought of as the datum of a point x in $D^{[p]}$ and of p local components of D through x. Let $E_p(y)$ denote the set of these components. The sets $E_p(y)$ form a local system on $\widetilde{D}^{[p]}$, which will be denoted by E_p. By definition, the set of orientations of $E_p(y)$ is the set of generators of $\wedge^p\mathbb{Z}^{E_p(y)}$. The local system of orientations of E_p defines on $\widetilde{D}^{[p]}$ a local system of rank one

$$\varepsilon^p = \wedge^p\mathbb{Q}^{E_p}\,.$$

There is a natural inclusion of $D^{[p]}$ in $D^{[p-1]}$, but, in general, there is no natural map from $\widetilde{D}^{[p]}$ to $\widetilde{D}^{[p-1]}$. What we have instead is a natural correspondence between the two. Let $\widetilde{D}^{[p-1,p]}$ be the space whose points are pairs (y, L), where $y = (x, E_p(y)) \in \widetilde{D}^{[p]}$ and $L \in E_p(y)$. We then have morphisms

$$
\begin{array}{ccc}
\widetilde{D}^{[p-1,p]} & \xrightarrow{\xi^{[p-1,p]}} & \widetilde{D}^{[p-1]} \\
\pi^{[p-1,p]} \big\downarrow & & \\
\widetilde{D}^{[p]} & &
\end{array}
\tag{5.7}
$$

$$\pi^{[p-1,p]}(y, L) = y\,; \qquad \xi^{[p-1,p]}(y, L) = (x, E_p(y) \smallsetminus \{L\})\,.$$

Furthermore, there is a natural isomorphism

$$(\pi^{[p-1,p]})^* \varepsilon^p \cong (\xi^{[p-1,p]})^* \varepsilon^{p-1}\,.$$

One can then define a generalized Gysin homomorphism

$$H^i(\widetilde{D}^{[p]}; \varepsilon^p) \xrightarrow{\pi^*} H^i(\widetilde{D}^{[p-1,p]}; \pi^* \varepsilon^p) \cong H^i(\widetilde{D}^{[p-1,p]}; \xi^* \varepsilon^{p-1}) \xrightarrow{\xi_*} H^{i+2}(\widetilde{D}^{[p-1]}; \varepsilon^{p-1}),$$

where, for brevity, we have written π and ξ for $\pi^{[p-1,p]}$ and $\xi^{[p-1,p]}$, respectively. We may now state Deligne's theorem.

THEOREM (5.8). *There is a spectral sequence, abutting at $H^*(V, \mathbb{Q})$ and with $E_2 = E_\infty$, whose E_1-term is given by*

$$E_1^{-p,q} = \begin{cases} H^{q-2p}(\widetilde{D}^{[p]}, \varepsilon^p) & \text{for } p > 0\,, \\ H^q(X, \mathbb{Q}) & \text{for } p = 0\,, \\ 0 & \text{for } p < 0\,. \end{cases}$$

Moreover, the differential

$$d_1 \colon H^{q-2p}(\widetilde{D}^{[p]}, \varepsilon^p) \to H^{q-2p+2}(\widetilde{D}^{[p-1]}, \varepsilon^{p-1})$$

is the Gysin homomorphism $(\xi^{[p-1,p]})_* (\pi^{[p-1,p]})^*$.

Let us now observe that the above theorem also holds in the case in which X is an orbifold and D is an orbifold divisors with normal crossings. Indeed, in view of the local nature of their proofs, one can easily see that Propositions 3.6 and 3.13 in [160] and Proposition 3.1.8 in [162] all hold in the orbifold situation. We can then apply the preceding theorem to the situation in which $X = \overline{\mathcal{M}}_{g,P}$ and $V = \mathcal{M}_{g,P}$. We set $\mathcal{D} = X \smallsetminus V = \partial \mathcal{M}_{g,P}$. Here we ask the reader to go back to Section 10 of Chapter XII, where we studied the stratification of $\mathcal{M}_{g,P}$ in terms of stable graphs. Let us choose, once and for all, a representative in each isomorphism class of stable P-pointed, genus g graphs with exactly p edges, and let $\mathcal{G}^p_{g,P}$ denote the (finite) set of these representatives. We then have

$$\mathcal{D}^{[p]} = \bigcup_{\Gamma \in \mathcal{G}^p_{g,P}} \mathcal{D}_\Gamma\,,$$

while the normalization $\widetilde{\mathcal{D}}^{[p]}$ of $\mathcal{D}^{[p]}$ is given by

$$\widetilde{\mathcal{D}}^{[p]} = \coprod_{\Gamma \in \mathcal{G}^p_{g,P}} \widetilde{\mathcal{D}}_\Gamma \,, \tag{5.9}$$

where, according to Proposition (10.11) of Chapter XII,

$$\widetilde{\mathcal{D}}_\Gamma = [\overline{\mathcal{M}}_\Gamma / \operatorname{Aut}(\Gamma)] \,.$$

We also have $\widetilde{\mathcal{D}}^{[0]} = \mathcal{D}^{[0]} = \overline{\mathcal{M}}_{g,P}$. Notice that diagram (5.7) decomposes as the "disjoint union" of diagrams

$$\begin{array}{ccc} \widetilde{\mathcal{D}}_{\Gamma',\Gamma} & \xrightarrow{\xi_{\Gamma',\Gamma}} & \widetilde{\mathcal{D}}_{\Gamma'} \\ {\scriptstyle \pi_{\Gamma',\Gamma}}\big\downarrow & & \\ \widetilde{\mathcal{D}}_\Gamma & & \end{array}$$

where (Γ', Γ) runs through all pairs of graphs with $\Gamma \in \mathcal{G}^p_{g,P}$, $\Gamma' \in \mathcal{G}^{p-1}_{g,P}$, and $\Gamma < \Gamma'$. We now adapt Deligne's theorem to the above situation, and we get the following result.

THEOREM (5.10). *There is a spectral sequence, abutting at $H^*(\mathcal{M}_{g,P}, \mathbb{Q})$ and with $E_2 = E_\infty$, whose E_1-term is given by*

$$E_1^{-p,q} = \begin{cases} \bigoplus_{\Gamma \in \mathcal{G}^p_{g,P}} H^{q-2p}(\widetilde{\mathcal{D}}_\Gamma, \varepsilon^p) & \text{for } p > 0, \\ H^q(\overline{\mathcal{M}}_{g,P}, \mathbb{Q}) & \text{for } p = 0, \\ 0 & \text{for } p > 0. \end{cases}$$

Moreover, the differential

$$d_1 \colon \bigoplus_{\Gamma \in \mathcal{G}^p_{g,P}} H^{q-2p}(\widetilde{\mathcal{D}}_\Gamma, \varepsilon^p) \to \bigoplus_{\Gamma' \in \mathcal{G}^{p-1}_{g,P}} H^{q-2p+2}(\widetilde{\mathcal{D}}_{\Gamma'}, \varepsilon^{p-1})$$

of this spectral sequence is the Gysin map

$$\sum_{\Gamma \in \mathcal{G}^p_{g,P},\, \Gamma' \in \mathcal{G}^{p-1}_{g,P},\, \Gamma < \Gamma'} (\xi_{\Gamma',\Gamma})_* (\pi_{\Gamma',\Gamma})^* \,.$$

We may now go back to the proof of Theorem (5.1). For simplicity, in the notation for cohomology groups, we omit any mention of the coefficients. We assume $\mathbb{Q}$-coefficients throughout.

Proof of Theorem (5.1). We denote by $\mathcal{D}_1, \dots, \mathcal{D}_N$ the components of the boundary of $\overline{\mathcal{M}}_{g,P}$, so that $\cup \mathcal{D}_i = \mathcal{D}^{[1]}$; thus, $\mathcal{D}_i = \mathcal{D}_\Gamma$ for some

$\Gamma \in \mathcal{G}^1_{g,P}$. Clearly, $\widetilde{\mathcal{D}}^{[1]} = \coprod \widetilde{\mathcal{D}}_i$, where $\widetilde{\mathcal{D}}_i$ stands for the normalization of $\mathcal{D}_i$.

We first compute $H^1(M_{g,P})$. The only nonzero terms in Deligne's spectral sequence that are relevant to the computation of $H^1(M_{g,P})$ are $E_1^{-1,2}$, $E_1^{0,2}$, $E_1^{0,1}$. On the other hand, $E_1^{0,1}$ equals $H^1(\overline{M}_{g,P})$ which, by Corollary (3.4), vanishes. It follows that

$$H^1(M_{g,P}) = \ker(d_1^{-1,2} : E_1^{-1,2} \to E_1^{0,2}) .$$

But $d_1^{-1,2}$ is the Gysin map

$$\bigoplus_{i=1}^{N} H^0(\widetilde{\mathcal{D}}_i) \longrightarrow H^2(\overline{M}_{g,P}) . \tag{5.11}$$

From Theorem (4.1) we know that the boundary classes in $H^2(\overline{M}_{g,P})$ are linearly independent as long as $g > 0$. This means that $d_1^{-1,2}$ is injective. This proves the vanishing of $H^1(M_{g,P})$ when $g > 0$. We next consider the second rational cohomology group of $M_{g,P}$. The only nonzero terms in Deligne's spectral sequence that are relevant to the computation of $H^2(M_{g,P})$ are $E_1^{-2,4}$, $E_1^{-1,4}$, $E_1^{-1,3}$, and $E_1^{0,2}$. Since

$$E_1^{-1,3} = \bigoplus_{i=1}^{N} H^1(\widetilde{\mathcal{D}}_i) ,$$

and since each $\widetilde{\mathcal{D}}_i$ is a quotient of a product of moduli spaces of stable pointed curves by finite groups, the term $E_1^{-1,3}$ vanishes by virtue of Corollary (3.4). We then have

$$H^2(M_{g,P}) = \ker(d_1^{-2,4}) \oplus \operatorname{coker}(d_1^{-1,2}) .$$

From Theorem (4.1) it follows that

$$\operatorname{coker}(d_1^{-1,2}) = \begin{cases} \mathbb{Q}\langle \kappa_1, \psi_1, \dots, \psi_n \rangle & \text{for } g > 2 , \\ \mathbb{Q}\langle \psi_1, \dots, \psi_n \rangle & \text{for } g = 2 , \\ 0 & \text{for } g = 1 . \end{cases}$$

It remains to show that $\ker(d_1^{-2,4}) = 0$. The homomorphism

$$d_1^{-2,4} : H^0(\widetilde{\mathcal{D}}^{[2]}, \varepsilon^2) \longrightarrow H^2(\widetilde{\mathcal{D}}^{[1]}, \varepsilon^1)$$

is the Gysin map

$$\bigoplus_{\Gamma \in \mathcal{G}^2_{g,P}} H^0(\widetilde{\mathcal{D}}_\Gamma, \varepsilon^2) \longrightarrow \bigoplus_{i=1}^{N} H^2(\widetilde{\mathcal{D}}_i) . \tag{5.12}$$

The graphs of stable P-pointed genus g curves with two double points are shown in Figure 4 below, where $C \cup D \cup B = A \cup B = P$ and $c + d + b = a + b = g$.

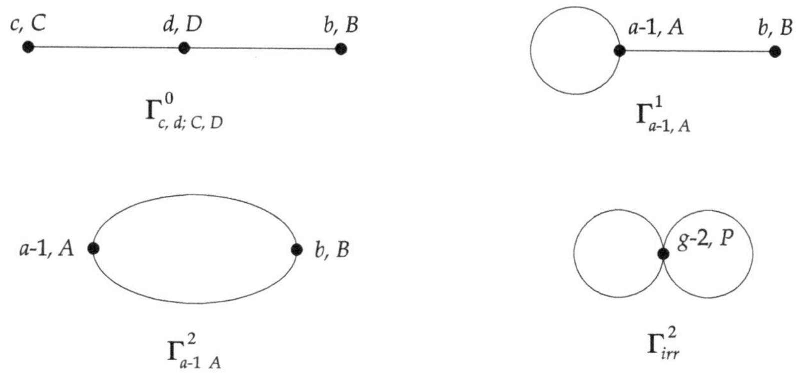

Figure 4.

We now consider an element $\delta \in H^0(\widetilde{\mathcal{D}}^{[2]}, \varepsilon^2)$. Observe that, when $\mathcal{D}_\Gamma$ is part of the self-intersection of one of the $\mathcal{D}_i$, we have $H^0(\widetilde{\mathcal{D}}_\Gamma; \varepsilon^2) = 0$ since ε^2 is not trivial. This rules out the graphs $\Gamma^0_{c,g-2c;\emptyset,P}$, $\Gamma^2_{a-1,A}$, Γ^2_{irr}. In all other cases, $H^0(\widetilde{\mathcal{D}}_\Gamma, \varepsilon^2) = H^0(\widetilde{\mathcal{D}}_\Gamma)$. We may then write

$$\delta = \sum \alpha^0_{c,d;C,D}\, 1_{\Gamma^0_{c,d;C,D}} + \sum \alpha^1_{a-1,A}\, 1_{\Gamma^1_{a-1,A}},$$

where 1_Γ is a generator of $H^0(\widetilde{\mathcal{D}}_\Gamma)$, and where $(d, D) \neq (g - 2c, P)$. Proceeding by contradiction, we assume that the image of δ under (5.12) is zero. Let $g = a + b$ and assume that $a > 0$. Fix a subset $A \subset P$, set $B = P \setminus A$, and assume that $(a, A) \neq (g, P)$. By the Künneth formula, one of the summands $H^2(\widetilde{\mathcal{D}}_i)$ in the right-hand side of (5.12) contains a summand of the form $H^2(\overline{\mathcal{M}}_{a,A\cup\{x\}}) \otimes H^0(\overline{\mathcal{M}}_{b,B\cup\{y\}})$. Let $\pi_{a,A}$ be the composition of the map (5.12) with the projection onto this summand. This homomorphism vanishes identically on a certain number of summands in the left-hand side of (5.12). Taking this into account, $\pi_{a,A}$ can be viewed as a homomorphism

$$\pi_{a,A}\colon \left(\bigoplus_{c+d=a,\ C\cup D=A} H^0\left(\widetilde{\mathcal{D}}_{\Gamma^0_{c,d;C,D}}\right)\right) \oplus H^0\left(\widetilde{\mathcal{D}}_{\Gamma^1_{a-1;A}}\right) \longrightarrow H^2(\overline{\mathcal{M}}_{a,A\cup\{x\}}).$$

The summands in the domain of $\pi_{a,A}$ play, with respect to $\overline{\mathcal{M}}_{a,A\cup\{x\}}$, the same role that the summands $H^0(\widetilde{\mathcal{D}}_1), \dots, H^0(\widetilde{\mathcal{D}}_N)$ play for $\overline{\mathcal{M}}_{g,P}$ in the Gysin homomorphism (5.11). Since $a > 0$, the homomorphism $\pi_{a,A}$ is injective for all pairs (a, A) with $a > 0$ and $(a, A) \neq (g, P)$. It follows that the coefficients $\alpha^0_{c,d;C,D}$ and $\alpha^1_{a-1;A}$ are all zero. Q.E.D.

6. A closer look at the orbicell decomposition of moduli spaces.

In the last section of the previous chapter we introduced the space $\mathcal{T}_{g,P}^{comb}$ (cf. (12.6)) and the homeomorphisms

$$\begin{aligned} \overline{H} : \overline{M}'_{g,P} \times \mathbb{R}_+^P &\xrightarrow{\simeq} \overline{M}_{g,P}^{comb}, \\ H : M_{g,P} \times \mathbb{R}_+^P &\xrightarrow{\simeq} M_{g,P}^{comb}, \\ \overline{h} : \overline{M}'_{g,P} &\xrightarrow{\simeq} \overline{M}_{g,P}^{comb}(r), \\ h : M_{g,P} &\xrightarrow{\simeq} M_{g,P}^{comb}(r) \end{aligned}$$

(cf. (12.10), (12.11), and (12.12) in the same chapter). Let us look more closely at the nature of the cellular decompositions of moduli spaces implicit in those homeomorphisms. Let us start with a point $[G, \mathbf{x}, m, [f]] \in \mathcal{T}_{g,P}^{comb}$, that is, with (an isomorphism class of) a P-marked, genus g, embedded ribbon graph with metric. There is a unique cell containing this point in its interior. We denote it by $a_{(G,\mathbf{x},[f])}$. Its dimension is equal to the number of edges of G, and one moves in this cell by varying the lengths of the edges of G or, which is the same, the metric m. Therefore,

$$\mathcal{T}_{g,P}^{comb} \supset |a_{(G,\mathbf{x},[f])}| \cong \mathbb{R}_+^{X_1(G)},$$

where, as usual, $X_1(G)$ denotes the set of edges of G. By varying $[f]$, the cells $a_{(G,\mathbf{x},[f])}$ can all be canonically identified. When no confusion is possible, we will drop the reference to f in the symbols. We may also observe that to a top-dimensional cell $a_{(G,\mathbf{x})}$ of $\mathcal{T}_{g,P}^{comb}$ there corresponds a ribbon graph G with maximum number of edges. It is an easy exercise (cf. Exercise F-1) to show that the maximum number of edges of a ribbon graph G of genus g with n boundary components is equal to $6g-6+3n$ and that, when this number is attained, all the vertices of G are *trivalent*. This confirms the fact that

$$\dim \mathcal{T}_{g,P}^{comb} = 6g - 6 + 3n. \tag{6.1}$$

Now look at the projection

$$\pi : \mathcal{T}_{S,P}^{comb} \to M_{g,P}^{comb} = \mathcal{T}_{S,P}^{comb}/\Gamma_{S,P}.$$

Given a cell a in $\mathcal{T}_{g,P}^{comb}$, we have

$$\pi(|a|) = |a|/\Gamma_a,$$

where Γ_a is the stabilizer of a in $\Gamma_{S,P}$. As we remarked in Theorem (3.12) of Chapter XVIII, if $(G_a, \mathbf{x}, [f])$ is the embedded ribbon graph corresponding to a, this stabilizer is nothing but the automorphism group of the P-marked ribbon graph $(G_a, \mathbf{x})$, so that

$$\pi(|a|) = |a| / \mathrm{Aut}(G_a, \mathbf{x}) .$$

The group $\mathrm{Aut}(G_a, \mathbf{x})$ can also be interpreted as the group of isometries of the P-marked ribbon graph with metric $(G_a, \mathbf{x}, m)$, where m is a metric for which all the edges have equal length. Let

$$([C, \mathbf{x}], r) \in M_{g,P} \times \mathbb{R}_+^P$$

be such that

$$H(([C, \mathbf{x}], r)) = [G_a, \mathbf{x}, m] \ \in \ |a| / \mathrm{Aut}(G_a, \mathbf{x}) \ \subset \ M_{S,P}^{comb} .$$

Then, thinking of how the hyperbolic spine construction is defined, we have the obvious identification

$$\mathrm{Aut}(G_a, \mathbf{x}) = \mathrm{Aut}(C, \mathbf{x}) . \tag{6.2}$$

The picture for $\overline{\mathcal{T}}_{g,P}^{comb}$ and $\overline{M}_{g,P}^{comb}$ is not substantially different, with only one catch. Let again π denote the projection from $\overline{\mathcal{T}}_{g,P}^{comb}$ to $\overline{M}_{g,P}^{comb}$. Let a be a cell in $\overline{\mathcal{T}}_{g,P}^{comb}$ corresponding to an embedded stable P-pointed genus g ribbon graph $(G_a, \mathbf{x}, [f])$. As we remarked in Theorem (9.5), the stabilizer of a in $\Gamma_{S,P}$ has a normal subgroup Γ'_a such that

$$\Gamma_a / \Gamma'_a = \mathrm{Aut}(G_a, \mathbf{x}) .$$

However, Γ'_a acts as the identity on $|a|$. Therefore, we have again

$$\pi(|a|) = |a| / \mathrm{Aut}(G_a, \mathbf{x}) .$$

As before, denote by m a metric on $(G_a, \mathbf{x})$ in which all edges have equal length and suppose that $([C, \mathbf{x}], r)$ is a point in $\overline{M}'_{g,P} \times \mathbb{R}_+^P$ such that

$$\overline{H}(([C, \mathbf{x}], r)) = [G_a, \mathbf{x}, m] \ \in \ |a| / \mathrm{Aut}(G_a, \mathbf{x}) \ \subset \ \overline{M}_{S,P}^{comb} .$$

Then

$$\mathrm{Aut}(G_a, \mathbf{x}) = \mathrm{Aut}(C, \mathbf{x}) / \mathrm{Aut}^0(C, \mathbf{x}) , \tag{6.3}$$

where $\mathrm{Aut}^0(C, \mathbf{x})$ is the subgroup of those automorphisms of C inducing the identity on its positive components, i.e., on the irreducible components containing points of $\mathbf{x}$.

Even the innocent-looking cases of $M_{0,3}$ and $M_{1,1}$ are not completely trivial from the point of view of ribbon graphs. Let us first look at $M_{0,3}^{comb} = \overline{M}_{0,3}^{comb} \cong \{pt\} \times \mathbb{R}_+^3$. In this case, the top-dimensional orbicells are the four three-dimensional orbicells corresponding to the $\{1,2,3\}$-marked ribbon graphs in Figure 5.

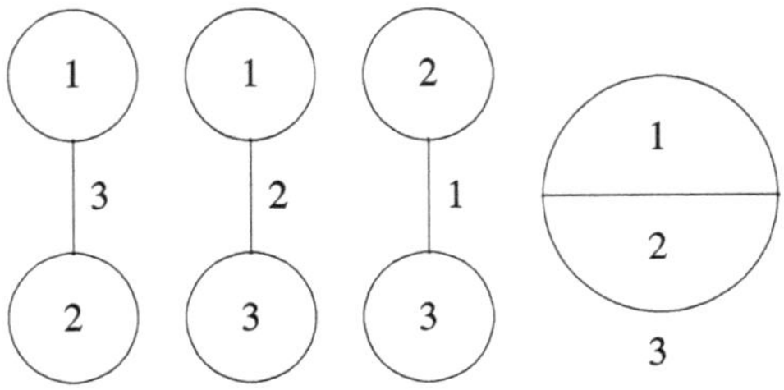

Figure 5.

Here, of course, we are listing only the isomorphism classes of $\{1,2,3\}$-marked ribbon graphs. The entire space $M_{0,3}^{comb}$ is the product of $\mathbb{R}^+$ times the space which is schematically depicted in Figure 6.

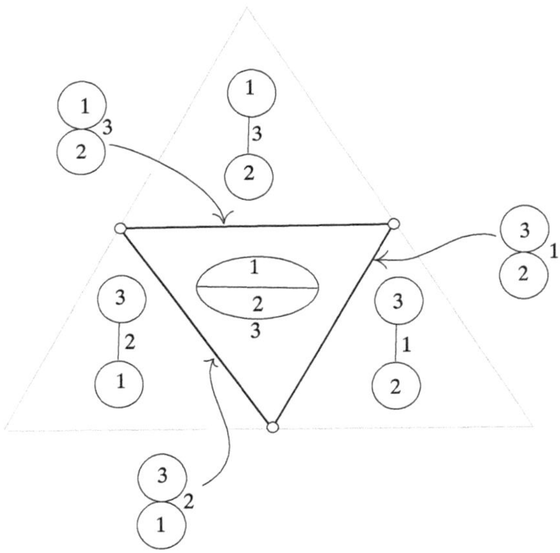

Figure 6.

In this space, each of the four 2-simplices and each of the three inner edges corresponds to a $\{1,2,3\}$-marked ribbon graph. One moves inside each open simplex by changing the metric on the corresponding graph, i.e., the lengths of the edges of the graph. Then one passes from one simplex to an adjacent one via a simple Feynman move.

The case of $M_{1,1}^{comb}$ is readily described. There is one top three-dimensional orbicell corresponding to the 1-marked ribbon graph (G_2, x) in Figure 7

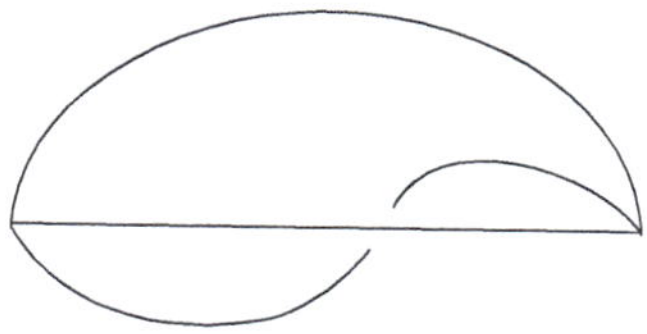

Figure 7.

and a two-dimensional orbicell corresponding to the 1-marked ribbon graph (G_1, x) in Figure 8.

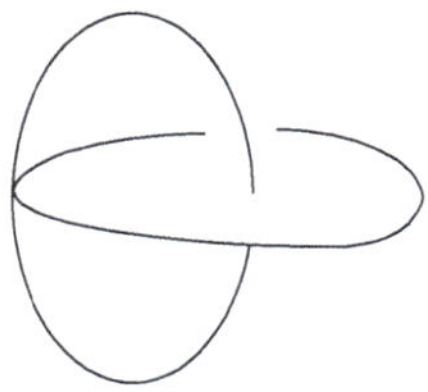

Figure 8.

Observe that

$$\operatorname{Aut}((G_2, x)) \cong \mathbb{Z}_6\,, \quad \text{and} \quad \operatorname{Aut}((G_1, x)) \cong \mathbb{Z}_4\,. \tag{6.4}$$

Therefore, the two orbicells of $M_{1,1}^{comb}$ look like

$$\mathbb{R}_+^3/\mathbb{Z}_6\,, \qquad \mathbb{R}_+^2/\mathbb{Z}_4\,.$$

Notice however that in both cases the automorphism group of the graph does not act faithfully. In fact, the order two element acts like the identity. We may now compute the virtual Euler–Poincaré characteristic of $M_{1,1}^{comb}(1)$, which is the Euler–Poincaré characteristic of the orbifold $\mathcal{M}_{1,1}$ (cf. Section 9 of Chapter XX). We get

$$\chi(M_{1,1}^{comb}(1)) = \chi(\mathcal{M}_{1,1}) = -\frac{1}{|\operatorname{Aut}((G_1, x))|} + \frac{1}{|\operatorname{Aut}((G_2, x))|} = -\frac{1}{12}\,. \tag{6.5}$$

The space $\overline{M}_{1,1}^{comb}$ has one more cell, namely the one-dimensional cell corresponding to the stable graph given by the data

$$(\gamma, G_v, \{x, y, z\})$$

illustrated in Figure 9 below.

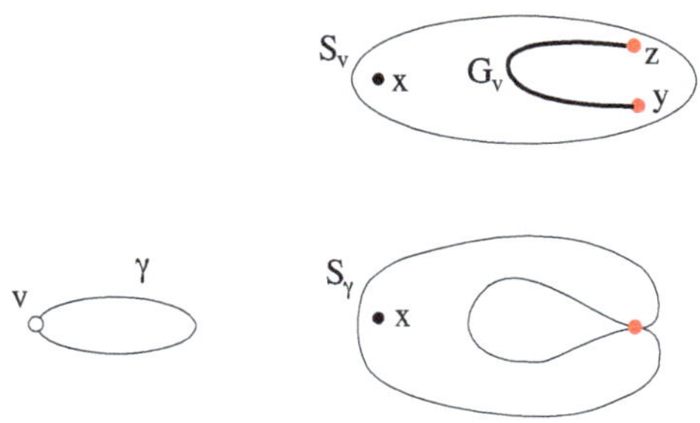

Figure 9.

Here S_γ is a nodal $\{x\}$-pointed elliptic curve, γ is its dual graph, and v is the only vertex of γ, while $(S_v, \{x, y, z\})$ is the three-pointed sphere which is the normalization of S_γ, and G_v is an arc in $S_v \setminus \{x\}$ joining y and z.

7. Combinatorial expression for the classes ψ_i.

Let us recall from (2.3) of Chapter XIII the definition of the point bundles $\mathcal{L}_p$, $p \in P$, on the moduli stack $\overline{\mathcal{M}}_{g,P}$. If $\eta : \mathcal{C} \to \overline{\mathcal{M}}_{g,P}$ is the universal curve, and σ_p the pth section, then

$$\mathcal{L}_p = \sigma_p^*(\omega_\eta),$$

where ω_η is the relative dualizing sheaf. As we observed in Chapter XIII, $\mathcal{L}_p$ does not descend to a line bundle on the moduli space $\overline{M}_{g,P}$, due to the fact that the automorphism groups of curves act nontrivially on the fibers of $\mathcal{L}_p$. Thus, in a sense, the "fiber" of $\mathcal{L}_p$ over a point $[(C, \mathbf{x})] \in \overline{M}_{g,P}$ is not the cotangent space to C at x_p, but rather its quotient by the action of the automorphism group of $(C, \mathbf{x})$. Another way of viewing $\mathcal{L}_p$ is to recall from Chapter XVI that $\overline{M}_{g,P}$ is the quotient of a complete smooth variety X by a finite group G and that we may arrange for X to be equipped with a universal family. We denote by $\widetilde{\mathcal{L}}_p$ the pullback of $\mathcal{L}_p$ to X. The action of G lifts equivariantly to $\widetilde{\mathcal{L}}_p$, and we may view $\mathcal{L}_p$ as being described by the datum of $\widetilde{\mathcal{L}}_p$ together with this action. From the finiteness of G it follows that, for sufficiently divisible N, the group G acts trivially on $\widetilde{\mathcal{L}}_p^{\otimes N}$. Therefore, $\widetilde{\mathcal{L}}_p^{\otimes N}$ drops to a bona fide line bundle on $\overline{M}_{g,P}$, which we denote by $\mathcal{L}_p^N$.

In what follows, we will always treat $\mathcal{L}_p$ as an ordinary line bundle on $\overline{M}_{g,P}$, leaving to the reader the task of checking that this always makes sense. For example, when writing the first Chern class of $\mathcal{L}_p$, we use the symbol $c_1(\mathcal{L}_p)$ with the meaning

$$c_1(\mathcal{L}_p) = \frac{1}{N} c_1(\mathcal{L}_p^N).$$

Recall that one normally writes

$$\psi_p = c_1(\mathcal{L}_p).$$

As we have seen in Section 4 of Chapter XVII, the intersection theory of the classes ψ_p completely determines the intersection theory of all tautological classes. In this section, using the isomorphism between $\overline{M}'_{g,P}$ and $\overline{M}^{comb}_{g,P}$, we will find a combinatorial formula for the classes ψ_p. In what follows we shall freely use the notation established in Section 3 of Chapter XVIII.

A way to compute Chern classes is by transgression. Look at the S^1-bundle

$$\overline{\pi} : S^1(\mathcal{L}_p) \to \overline{M}_{g,P}$$

associated to $\mathcal{L}_p$. In the Leray spectral sequence of $\overline{\pi}$, the transgression is the homomorphism

$$E_2^{01} = H^0(\overline{M}_{g,P}, R^1\overline{\pi}_*(\mathbb{Q})) \cong \mathbb{Q} \xrightarrow{d_2} E_2^{2,0} = H^2(\overline{M}_{g,P}, \mathbb{Q}),$$

and, by definition,

$$c_1(\mathcal{L}_p) = d_2(1). \tag{7.1}$$

Since there is an exact sequence

$$E_2^{01} \overset{d_2}{\to} E_2^{2,0} \overset{\overline{\pi}^*}{\to} E_\infty^{2,0} \subset H^2(S^1(\mathcal{L}_p), \mathbb{Q}),$$

it follows that a de Rham representative of ψ_p is the class of a closed 2-form $\omega \in \mathcal{E}^2(\overline{M}_{g,P})$ such that

$$\overline{\pi}^*(\omega) = d\varphi, \qquad \int_{S^1} \varphi = 1 \tag{7.2}$$

for some 1-form φ. Of course, the functoriality of the first Chern class follows from the functoriality of the Leray spectral sequence.

Now let us fix $r = \sum r_p \cdot p \in \Delta_P$, and let us look at the map

$$f : \overline{M}_{g,P} \longrightarrow \overline{M}_{g,P}^{comb}(r)$$

obtained by composing the contraction $Q : \overline{M}_{g,P} \to \overline{M}'_{g,P}$ with the homeomorphism $\overline{h} : \overline{M}'_{g,P} \to \overline{M}_{g,P}^{comb}(r)$ (cf. (12.13) in the last section of the previous chapter). By the way Q was constructed, the line bundle $\mathcal{L}_p$ restricts to a trivial line bundle on the fibers of Q and therefore drops to a well-defined line bundle $\mathcal{L}'_p$ on $\overline{M}'_{g,P}$ with $Q^*(\mathcal{L}'_p) = \mathcal{L}_p$. Let us denote by $\mathcal{L}_p^{comb}$ the pullback of $\mathcal{L}'_p$ to $\overline{M}_{g,P}^{comb}(r)$ via $(\overline{h})^{-1}$, so that

$$\overline{h}^*(\mathcal{L}_p^{comb}) = \mathcal{L}'_p, \qquad f^*(\mathcal{L}_p^{comb}) = \mathcal{L}_p.$$

We shift our attention to $\mathcal{L}_p^{comb}$, with the goal of giving a combinatorial expression for its first Chern class. Before doing this, we need to recall a few facts about piecewise linear forms (PL-forms for short).

Let K be a simplicial complex. By definition, $|K|$ is a union of simplices. One such simplex is the image of a continuous map

$$\sigma : \Delta_n = \left\{ (t_0, \ldots, t_n) \in \mathbb{R}^{n+1} : 0 \le t_i \le 1, \sum_{i=1}^{n} t_i = 1 \right\} \longrightarrow |K|,$$

which is injective in $\overset{\circ}{\Delta}_n$. A PL-form φ of degree ν on $|K|$ is a collection

$$\varphi = \{\varphi_\sigma\}_{\sigma \in K},$$

where

$$\varphi_\sigma = \sum \varphi_{i_1 \dots i_\nu} dt_{i_1} \wedge \dots \wedge dt_{i_\nu}, \quad 0 \le i_k \le \dim \sigma,$$

is a ν-form on the hyperplane $\sum t_i = 1$ in $\mathbb{R}^{\dim \sigma + 1}$, having as coefficients polynomials in the t_i with rational coefficients, and such that

$$(\varphi_\sigma)_{|\tau} = \varphi_\tau$$

whenever τ is a face of σ. One can define the complex of PL-forms on K. Its cohomology is denoted by $H^*_{PL}(K)$. Notice that a piecewise-linear ν-form does not define a ν–form on X. For instance, in $\mathbb{R}^2 = U \cup Z \cup V$ where $U = \{y > 0\}$, $Z = \{y = 0\}$, and $V = \{y < 0\}$, the datum $\{(dx + dy)_U, (dx)_U, (dx - dy)_U\}$ defines a piecewise-linear form. It is therefore a remarkable fact that the complex of piecewise-linear forms on X actually computes the rational cohomology of X, i.e., that

$$H^*_{PL}(K) \cong H^*(K, \mathbb{Q}).$$

In conclusion, to compute the first Chern class of $\mathcal{L}_p^{comb}$, one can use PL-forms on $\overline{M}_{g,P}^{comb}(r)$. There is only one small caveat. The space $\overline{M}_{g,P}^{comb}(r)$ is not exactly a simplicial complex, but rather an orbisimplicial complex. As we explained in the preceding section, this space is not a union of simplices but of quotients of simplices by the action of finite groups. We must accordingly adapt the notion of PL-form by requiring that φ_σ should be Γ_σ-invariant, where Γ_σ is the finite group acting on the simplex σ.

We will next define a piecewise-linear form representing the first Chern class of $\mathcal{L}_p^{comb}$. As we explained in the last section, an orbicell of $\overline{M}_{g,P}^{comb}(r)$ is of the form

$$|a|/\Gamma_a,$$

where a is an arc system corresponding to a stable P-pointed ribbon graph $(G_a, \mathbf{x})$ whose pth half-perimeter is equal to r_p. Moreover, $\Gamma_a = \mathrm{Aut}((G_a, \mathbf{x}))$. The barycentric coordinates relative to the cell $|a|$ are the lengths

$$\{l_e\}_{e \in X_1(G_a)}$$

of the edges of G_a. Let $[(C, \mathbf{x})] \in \overline{M}_{g,P}$ be a point such that

$$f[(C, \mathbf{x})] \in |a|/\Gamma_a \subset \overline{M}_{g,P}^{comb}(r). \tag{7.3}$$

The map f is obtained via the hyperbolic spine construction. Since we are interested in the line bundle $\mathcal{L}_p$, we look at the boundary component of G_a marked by p. For simplicity, we denote this boundary component with the same symbol x_p used to designate the point of C marked by p. Thus,

$$x_p \in X_2(G_a)\,.$$

We may consider x_p as a cyclically ordered set of oriented edges of G_a

$$x_p = (\vec{e}_1, \dots, \vec{e}_\nu)\,. \tag{7.4}$$

We denote by e_s the edge corresponding to $\vec{e}_s$ for $s = 1, \dots, \nu$. Thus, $\{e_1, \dots, e_\nu\}$ is a set of edges, with possible repetitions. A repetition happens exactly when the edge in question bounds, on both sides, the same boundary component of G_a. Set

$$(\omega_p)_a = \sum_{1 \le s < t \le \nu - 1} d\left(\frac{l_{e_s}}{r_p}\right) \wedge d\left(\frac{l_{e_t}}{r_p}\right)\,. \tag{7.5}$$

Observe that the sum extends only up to $\nu - 1$. It is important to realize that, since r_p has been fixed, we have that $dl_{e_1} + \cdots + dl_{e_\nu} = dr_p = 0$ and, more importantly, that the expression of $(\omega_p)_a$ depends only on the cyclic order of the edges around the pth marked boundary component, and not on the particular ordering (7.4). Since the stabilizer Γ_a acts on $\{e_1, \dots, e_\nu\}$ as a cyclic group, the piecewise-linear 2-form $(\omega_p)_a$ is well defined on the orbisimplex $|a|/\Gamma_a$. Since the "faces" of $|a|$ are obtained by setting some of the lengths l_{e_i} equal to zero, it is clear that (7.5) gives a well-defined, piecewise-linear 2-form

$$\omega_p = \{(\omega_p)_a\}_{|a| \subset \overline{M}^{comb}_{g,P}(r)}$$

on $\overline{M}^{comb}_{g,P}(r)$.

LEMMA (7.6). *For each $p \in P$ and for each $r \in \mathbb{R}^P_+$,*

$$[\omega_p] = c_1(\mathcal{L}_p^{comb}) \in H^2_{PL}(\overline{M}^{comb}_{g,P}(r))\,.$$

In particular,

$$[f^*(\omega_p)] = c_1(\mathcal{L}_p) \in H^2(\overline{M}_{g,P}, \mathbb{Q})\,.$$

Proof. Let us first describe the S^1-bundle

$$\pi : S^1(\mathcal{L}_p^{comb\vee}) \longrightarrow \overline{M}^{comb}_{g,P}(r)$$

associated to the dual of $\mathcal{L}_p^{comb}$. As above, fix an orbicell $c = |a|/\Gamma_a$ in $\overline{M}^{comb}_{g,P}(r)$. We wish to describe the restriction of $S^1(\mathcal{L}_p^{comb\vee})$ to a. Look at the stable graph G_a corresponding to a and denote by ∂_p its pth boundary component. A point in $(\mathcal{L}_p^{comb\vee})_{|a}$ is a pair (m,η), where m is a metric on G_a, that is to say, a point of a, and η is a point on ∂_p. Let $[(C,\mathbf{x})]$ and m be as in (7.3). The isomorphism

$$(h^{-1})^* : S^1(\mathcal{L}_p'^{\vee})_{|a} \xrightarrow{\cong} S^1(\mathcal{L}_p^{comb\vee})_{|a}$$

associates to a nonzero tangent vector $v \in T_{x_p}(C)$ the point $\eta \in \partial_p$ obtained as the intersection between ∂_p and the geodesic through x_p with tangent vector v. We can identify the fibers of π over points of a with $\mathbb{R}/r_p\mathbb{Z}$. We can then take as local coordinates on $S^1(\mathcal{L}_p^{comb\vee})_{|a}$ the lengths l_e of the edges of G_a together with the "angle coordinate" θ on $\mathbb{R}/r_p\mathbb{Z}$. On each fiber of π, a ν-tuple of cyclically ordered points $\overline{\theta}_1,\dots,\overline{\theta}_\nu$ is given, where $\overline{\theta}_i$ is the initial point of the oriented edge $\vec{e}_i$. Let $\theta_s \in [0,r_p)$ be the angle coordinate of $\overline{\theta}_s$. We have

$$l_{e_s} = \begin{cases} \theta_{s+1}-\theta_s & \text{if } s=1,\dots,\nu-1\,, \\ \theta_1-\theta_\nu+r_p & \text{if } s=\nu\,. \end{cases}$$

Figure 10 illustrates the situation we just described.

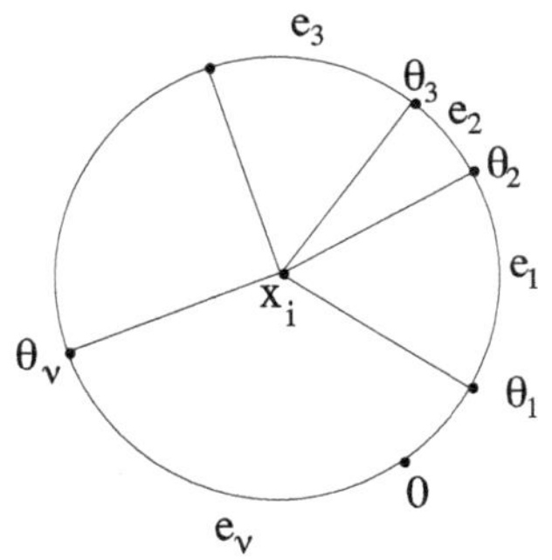

Figure 10.

Now we set

$$\varphi_{\pi^{-1}(a)} = \sum_{s=1}^{\nu} \left(\frac{l_{e_s}}{r_p}\right) d\left(\frac{\theta_s}{r_p}\right)$$

Then $\varphi = \{\varphi_{\pi^{-1}(a)}\}$ gives a well-defined piecewise-linear 1-form on $S^1(\mathcal{L}_p^{comb\vee})$. Moreover,

$$d\varphi_{\pi^{-1}(a)} = -\sum_{1\le s<t\le \nu-1} d\left(\frac{l_{e_s}}{r_p}\right)\wedge d\left(\frac{l_{e_t}}{r_p}\right) = -\pi^*((\omega_p)_a)\,.$$

Finally,

$$\int_{\text{fiber of }\pi} \varphi = \sum_{s=1}^{\nu} \frac{l_{e_s}}{r_p}\int_0^{r_p}\frac{d\theta_s}{r_p} = 1\,.$$

As we remarked in (7.2), $-[\omega_p]$ is the first Chern class of $\mathcal{L}_p^{comb\vee}$.

Q.E.D.

We stress again that the basic reason for going through all the trouble of compactifying the combinatorial moduli space $M_{g,P}^{comb}(r)$ is that this makes it possible to calculate intersection numbers of the ψ classes as *combinatorial* integrals over the *open* moduli. In fact, by Lemma (7.6), the following formula holds:

$$(7.7) \qquad \int_{\overline{M}_{g,P}} \prod_{p\in P} \psi_p^{d_p} = \int_{\overline{M}_{g,P}^{comb}(r)} \prod_{p\in P} \omega_p^{d_p} = \int_{M_{g,P}^{comb}(r)} \prod_{p\in P} \omega_p^{d_p} .$$

In one of the exercises at the end of this chapter we ask the reader to use the combinatorial expression for ψ_p to show that

$$(7.8) \qquad \int_{\overline{M}_{1,1}} \psi_p = \frac{1}{24} .$$

8. A volume computation.

In this section we will introduce two volume forms on $M_{g,P}^{comb}$ and discuss how they are related. We ask the reader to recall the definition of the space $M_{g,P}^{comb}(r)$ and of the piecewise linear 2-forms ω_p for $p \in P$.

The first piecewise linear 2-form we put on $M_{g,P}^{comb}(r)$ is

$$(8.1) \qquad \Omega = \sum_{p\in P} r_p^2 \omega_p .$$

Set $d = 3g-3+n$ and consider the top-dimensional form Ω^d. The crucial observation is the following.

PROPOSITION (8.2). *The form Ω^d is an orientation form on $M_{g,P}^{comb}(r)$.*

An immediate consequence of this proposition is that the form

$$(8.3) \qquad \Omega^d \wedge (dr_1 \wedge \cdots \wedge dr_n) ,$$

where $r = (r_1, \dots, r_n)$ varies in $\mathbb{R}_+^n$, is an orientation form on $M_{g,P}^{comb}$. Before proving the proposition, we need to establish the following lemma.

LEMMA (8.4). *On each top-dimensional orbicell of $M_{g,P}^{comb}(r)$, the form Ω is a nondegenerate (symplectic) form.*

Proof. Let a be a top-dimensional orbicell of $M_{g,P}^{comb}(r)$. Assume that $a = a_{(G,x)}$ for some P-marked ribbon graph (G, x). Since a is top-dimensional, G is a trivalent graph. Since the half-perimeters r_p are fixed, the cotangent bundle $T_a^\vee$ is the trivial bundle on a generated by the 1-forms dl_e, with $e \in X_1(G)$, subject to the relations

$$\sum_{\vec{e} \in x(p)} dl_e = 0 \qquad \text{for all } p \in P\,.$$

Dually, the tangent bundle T_a is a trivial bundle generated by the vector fields

$$\sum_{e \in X_1(G)} c_e \frac{\partial}{\partial l_e} \qquad \text{such that } \sum_{\vec{e} \in x(p)} c_e = 0\,.$$

We are now going to define a bundle homomorphism

$$B_a : T_a^\vee \to T_a$$

and prove that it is the inverse of $\Omega_a/4$, where Ω_a is the restriction of Ω to a. Let e be an edge of G. We want to define $B_a(dl_e)$. Let $\vec{e}$ be an orientation of e. The two end-points of e are $v = [\vec{e}]_0$ and $w = [\sigma_1(\vec{e})]_0$. Let $(\vec{e}, \vec{f}_1, \vec{f}_2)$ be the circular ordering of the half-edges at v, and likewise let $(\vec{e}, \vec{h}_1, \vec{h}_2)$ be the circular ordering of the half-edges at w. Define

$$B_a(dl_e) = -\frac{\partial}{\partial l_{f_1}} + \frac{\partial}{\partial l_{f_2}} - \frac{\partial}{\partial l_{h_1}} + \frac{\partial}{\partial l_{h_2}}\,. \tag{8.5}$$

It is obvious that B_a is well defined; in fact,

$$B_a\Big(\sum_{\vec{e} \in x(p)} dl_e\Big) = 0 \qquad \text{for all } p \in P\,. \tag{8.6}$$

Denote by W_+ and W_- the boundary components of G having e as one of their edges. Finally, denote by F (resp., H) the boundary component of G having f_1 and f_2 (resp., h_1 and h_2) in its boundary.

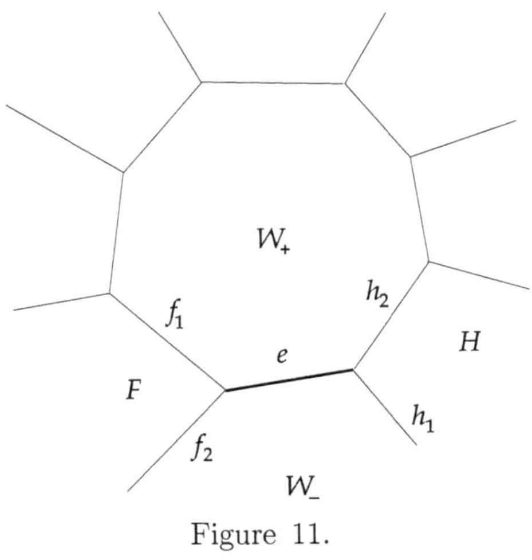

Figure 11.

Of course, it may well happen that W_+, W_-, F, and H are not all distinct, and the same might happen for f_1, f_2, h_1, h_2. As an exercise, the reader will see that this fact is immaterial for the computation that follows. We wish to show that

$$\Omega_a(B_a(dl_e)) = 4dl_e \,. \tag{8.7}$$

Using in an essential way the fact that the perimeters are constant, it is immediate to see that the only nonzero summands in this computation are:

$$r_F^2 \omega_F \left(\frac{\partial}{\partial l_{f_2}} - \frac{\partial}{\partial l_{f_1}} \right) = -dl_{f_1} - dl_{f_2} \,,$$
$$r_H^2 \omega_H \left(\frac{\partial}{\partial l_{h_2}} - \frac{\partial}{\partial l_{h_1}} \right) = -dl_{h_1} - dl_{h_2} \,,$$
$$r_{W_+}^2 \omega_{W_+} \left(\frac{\partial}{\partial l_{h_2}} - \frac{\partial}{\partial l_{f_1}} \right) = dl_{h_2} + dl_{f_1} + 2dl_e \,,$$
$$r_{W_-}^2 \omega_{W_-} \left(\frac{\partial}{\partial l_{f_2}} - \frac{\partial}{\partial l_{h_1}} \right) = dl_{f_2} + dl_{h_1} + 2dl_e \,.$$

This proves the lemma.

Q.E.D.

We now come to the proof of Proposition (8.2). Suppose that two top-dimensional orbicells a and a' of $M_{g,P}^{comb}(r)$ have a common face b. We must prove that Ω_a^d and $\Omega_{a'}^d$ induce on b opposite orientations. We are in the situation of a Feynman move:

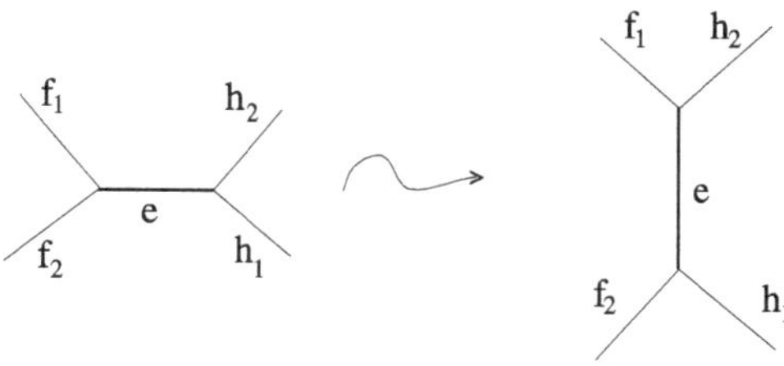

Figure 12.

Consider the exact sequence

$$0 \to T_b \to T_{a|_b} \to N_{b/a} \to 0 \,,$$

where $N_{b/a}$ is the normal bundle. The orientation induced by Ω_a^d on b is given by the image of Ω_a^d under the canonical isomorphism between $\bigwedge^d T_{a|_b}^\vee \otimes N_{b/a}$ and $\bigwedge^{d-1} T_b^\vee$. This isomorphism is given by contracting against $\frac{\partial}{\partial l_e}$. We then get

$$\iota_{\frac{\partial}{\partial l_e}}(\Omega_a^d) = (3g-3+n) \left(\iota_{\frac{\partial}{\partial l_e}}(\Omega_a) \wedge \Omega_a^{d-1} \right) .$$

We know that $\Omega_{a|b} = \Omega_{a'|b}$. On the other hand, looking at the picture above, we get

$$\iota_{\frac{\partial}{\partial l_e}}(\Omega_a) = \frac{1}{4}(-dl_{f_1} + dl_{f_2} - dl_{h_1} + dl_{h_2}) = -\iota_{\frac{\partial}{\partial l_e}}(\Omega_{a'}) .$$

Q.E.D.

From now on, using the PL structure, we will treat the bundles T_a, T_b, $T_{\mathbb{R}_+^n}$, and so on, as complex vector spaces. At this point we have a well-defined volume form

$$d\mu_{M_{g,P}^{comb}} = \left| \frac{\Omega^d}{d!} \wedge (dr_1 \wedge \cdots \wedge dr_n) \right|$$

on $M_{g,P}^{comb}$. Now fix a top-dimensional cell $a = a_G$ in $M_{g,P}^{comb}(r)$. We may also look at the top-dimensional cell b in $M_{g,P}^{comb}$ corresponding to the same ribbon graph G. While in a the perimeters of G are frozen, in b they are free to assume any value in $\mathbb{R}_+$. Set $N = 6g - 6 + 3n$, and let $e_1, \ldots, e_N$ be the edges of G. The coordinates $l_{e_1}, \ldots, l_{e_N}$ are the natural coordinates on b, and $|dl_{e_1} \wedge \cdots \wedge dl_{e_N}|$ is the natural volume form on b. The following remarkable theorem shows how simply these volumes forms are related with $d\mu_{M_{g,P}^{comb}}$.

THEOREM (8.8) (Kontsevich).

$$d\mu_{M_{g,P}^{comb}} = \left| \frac{\Omega_a^d}{d!} \wedge (dr_1 \wedge \cdots \wedge dr_n) \right| = 2^{2n+5g-5} |dl_{e_1} \wedge \cdots \wedge dl_{e_N}| .$$

Proof. First, we introduce a bit of notation. Let V be a vector space, and let $\mathbf{v} = (v_1, \ldots, v_r)$ be a system of vectors in V. We set $\bigwedge \mathbf{v} = v_1 \wedge \cdots \wedge v_r$. Also, if $\alpha : V \to W$ is a vector space homomorphism, we set $\alpha(\mathbf{v}) = (\alpha(v_1), \ldots, \alpha(v_r))$.

Let $\mathbf{h}$ be a basis of $T_a^\vee$, let $\mathbf{h}^\vee$ be the dual basis of T_a, and write

$$\frac{(\Omega_a)^d}{d!} = \rho_a \cdot \bigwedge \mathbf{h} ,$$

where ρ_a is a constant. Let A be the matrix of B_a in the given bases and write

$$\Omega_a = \sum c_{i,j} h_i \wedge h_j ,$$

where $C = (c_{i,j})$ is antisymmetric. Then, by the definition of Pfaffian,

$$\rho_a = \mathrm{Pfaff}(C) .$$

The fact that B_a is the inverse of $\Omega_a/4$, i.e., that A is the inverse of $C/4$, gives

$$\det A^{-1} = \left(\frac{\rho_a}{4^d}\right)^2 . \tag{8.9}$$

It will be useful to reinterpret the determinant of A via the formalism of determinants of complexes. View the homomorphism $B_a : T_a^\vee \to T_a$ as an acyclic complex $U^\bullet$ concentrated in degrees 0 and 1. Then we know, from the theory of determinants developed in Section 4 of Chapter XIII, that there is a canonical isomorphism $\det U^\bullet \cong \mathbb{C}$ and that

$$(\textstyle\bigwedge B_a(\mathbf{h}))^{-1} \otimes \bigwedge \mathbf{h} \mapsto 1 .$$

It follows that

$$(\textstyle\bigwedge \mathbf{h}^\vee)^{-1} \otimes \bigwedge \mathbf{h} \mapsto \det A .$$

Let G be the ribbon graph corresponding to a, and let b be the top-dimensional cell of $M_{g,P}^{comb}$ corresponding to the same ribbon graph. In a the perimeters of G are fixed, while in b they may assume any value in $\mathbb{R}_+$. The same formula (8.5) defining B_a can be used to define a homomorphism $B_b : T_b^\vee \to T_b$, which will no longer be an isomorphism. In fact, in view of (8.6), there is an exact sequence of vector spaces

$$V^\bullet : \quad 0 \to T_{\mathbb{R}_+^n}^\vee \to T_b^\vee \xrightarrow{B_b} T_b \to T_{\mathbb{R}_+^n} \to 0 , \tag{8.10}$$

which can be viewed as being obtained by joining the exact sequence

$$0 \to T_a \to T_b \to T_{\mathbb{R}_+^n} \to 0 \tag{8.11}$$

to its dual via the identification $B_a : T_a^\vee \xrightarrow{\simeq} T_a$. We shall regard $V^\bullet$ as a complex concentrated in degrees $-1, 0, 1, 2$.

Let $\mathbf{r}$ be the basis $(dr_1, \dots, dr_n)$ of $T_{\mathbb{R}_+^n}^\vee$, and let $\mathbf{r}^\vee$ be the dual basis of $T_{\mathbb{R}_+^n}$. Thus, $(\mathbf{h}, \mathbf{r})$ is a basis of $T_b^\vee$. It is clear that

$$\textstyle\bigwedge \mathbf{r}^\vee \otimes (\bigwedge(\mathbf{h}^\vee, \mathbf{r}^\vee))^{-1} \otimes \bigwedge(\mathbf{h}, \mathbf{r}) \otimes (\bigwedge \mathbf{r})^{-1} \mapsto \det A$$

under the isomorphism

$$\eta : \det V^\bullet \to \mathbb{C} .$$

This formula, together with (8.9), has the remarkable consequence that

$$\textstyle\bigwedge \mathbf{r}^\vee \otimes \dfrac{\Omega_a^d}{d!} \wedge (\bigwedge \mathbf{r}) \otimes \dfrac{\Omega_a^d}{d!} \wedge (\bigwedge \mathbf{r}) \otimes \bigwedge \mathbf{r}^\vee \mapsto 4^{2d} , \tag{8.12}$$

where we have used the isomorphisms $\det T_b^\vee \cong (\det T_b)^{-1}$ and $(\det T_{\mathbb{R}_+^n}^\vee)^{-1} \cong \det T_{\mathbb{R}_+^n}$.

Recall that our goal is to show that the constant c defined by

$$\frac{\Omega_a^d}{d!} \wedge \bigwedge \mathbf{r} = c \cdot \bigwedge \mathbf{k},$$

where $\mathbf{k}$ is the basis $(dl_{e_1}, \ldots, dl_{e_N})$, is equal to $2^{5g-5+2n}$. Formula (8.12) says that

$$\bigwedge \mathbf{r}^\vee \otimes \bigwedge \mathbf{k} \otimes \bigwedge \mathbf{k} \otimes \bigwedge \mathbf{r}^\vee \mapsto \frac{4^{2d}}{c^2}$$

under η. Keeping in mind that $d = 3g - 3 + n$, one must prove that

$$\eta(\bigwedge \mathbf{r}^\vee \otimes \bigwedge \mathbf{k} \otimes \bigwedge \mathbf{k} \otimes \bigwedge \mathbf{r}^\vee) = 2^{2g-2}.$$

Now observe that the complex $V^\bullet$ in formula (8.10) can be obtained by tensoring with $\mathbb{C}$ the *complex of abelian groups*

$$\begin{gathered} E^\bullet : \quad 0 \to E^{-1} \xrightarrow{\delta_{-1}} E^0 \xrightarrow{\delta_0} E^1 \xrightarrow{\delta_1} E^2 \to 0 \\ E^{-1} = \mathbb{Z}^{X_2(G)}, \quad E^0 = \mathbb{Z}^{X_1(G)}, \quad E^1 = \mathbb{Z}^{X_1(G)}, \quad E^2 = \mathbb{Z}^{X_2(G)}, \end{gathered}$$

where the differentials are defined as follows. Given $\vec{e} \in X(G)$, let $e_1, \ldots, e_k$ be the cyclic order of the sides of $[\vec{e}]_2$. If f_1, f_2, h_1, h_2 are as in Figure 11, with $e = [\vec{e}]_1$, $W_+ = [\vec{e}]_2$, and $W_- = [\sigma_1(\vec{e})]_2$, then

$$\delta_{-1}([\vec{e}]_2) = e_1 + \cdots + e_k, \quad \delta_0([\vec{e}]_1) = f_1 - f_2 + h_1 - h_2, \quad \delta_1([\vec{e}]_1) = [\vec{e}]_2 + [\sigma_1(\vec{e})]_2.$$

It is essential to observe also that, under the isomorphism $E^\bullet \otimes \mathbb{C} \cong V^\bullet$, the canonical bases of E^{-1}, E^0, E^1, E^2 correspond, respectively, to the bases $\mathbf{r}$ of V^{-1}, $\mathbf{k}$ of V^0, $\mathbf{k}^\vee$ of V^1, and $\mathbf{r}^\vee$ of V^2.

The situation we are studying is a particular case of the following more general setting. Let $A^\bullet$ be a finite complex of free abelian groups of finite rank. Suppose that the complex of $\mathbb{C}$-vector spaces $A^\bullet \otimes \mathbb{C}$ is exact; this is equivalent to saying that the cohomology of $A^\bullet$ is torsion. For each i, let $\mathbf{h}_i = (h_{i,1}, h_{i,2}, \ldots, h_{i,n_i})$ be a basis for A^i. Then

$$\bigwedge \mathbf{h} = \quad \cdots \otimes \bigwedge \mathbf{h}_i^{(-1)^i} \otimes \bigwedge \mathbf{h}_{i-1}^{(-1)^{i-1}} \otimes \cdots$$

is an element of $\det(A^\bullet \otimes \mathbb{C})$ which maps, via the canonical isomorphism

$$\det(A^\bullet \otimes \mathbb{C}) \xrightarrow{\simeq} \mathbb{C} \tag{8.13}$$

given by the theory of determinants, to a nonzero complex (in fact, rational) number t. Different bases of each group A^i are related by a unimodular transformation, and hence changing bases changes the number t at most by a sign. We define the *torsion of the complex* $A^\bullet$ to be

$$\tau(A^\bullet) = |t|^{-1}.$$

From this point of view, the constant $\eta(\bigwedge \mathbf{r}^\vee \otimes \bigwedge \mathbf{k} \otimes \bigwedge \mathbf{k} \otimes \bigwedge \mathbf{r}^\vee)$ that we must calculate is the reciprocal of the torsion of $E^\bullet$. Thus, we must show that

$$\tau(E^\bullet) = 2^{2-2g}\,. \tag{8.14}$$

Suppose that

$$\mathcal{E}: \quad 0 \to A^\bullet \xrightarrow{\alpha} B^\bullet \xrightarrow{\beta} C^\bullet \to 0$$

is an exact sequence of finite complexes of free abelian groups of finite rank with torsion cohomology. Then

$$\tau(B^\bullet) = \tau(A^\bullet)\tau(C^\bullet)\,. \tag{8.15}$$

In fact, we may find a basis $\mathbf{b}_i$ of B^i of the form $(\alpha(h_1), \dots, \alpha(h_n), k_1, \dots, k_m)$, where $\mathbf{a}_i = (h_1, \dots, h_n)$ is a basis of A^i, and $\mathbf{c}_i = (\beta(k_1), \dots, \beta(k_m))$ is a basis of C^i. Recall from Section 4 of Chapter XIII that there is a canonical isomorphism

$$\varphi_{\mathcal{E}} : \det A^\bullet \otimes \det C^\bullet \xrightarrow{\simeq} \det B^\bullet\,.$$

Then

$$\varphi_{\mathcal{E}}(\bigwedge \mathbf{a} \otimes \bigwedge \mathbf{c}) = \pm \bigwedge \mathbf{b}\,,$$

which proves (8.15).

We will use the following simple relation between the torsion of the complex $A^\bullet$ and the order of its cohomology groups.

LEMMA (8.16). *Let $A^\bullet$ be a finite complex of free abelian groups of finite rank, with torsion cohomology groups. Then*

$$\tau(A^\bullet) = \prod_i |H^i(A^\bullet)|^{(-1)^i}\,.$$

The proof is by induction on the length of the complex $A^\bullet$, starting from the case where this length does not exceed two. As usual, we write $B^i = B^i(A^\bullet)$ to indicate the coboundaries in A^i, and $Z^i = Z^i(A^\bullet)$ to indicate the cocycles. To do the initial case, start with a complex

$$\cdots \to 0 \to A^i \xrightarrow{\alpha} A^{i+1} \to 0 \to \cdots\,.$$

The group B^{i+1} has finite index in $A^{i+1} = Z^{i+1}$, so there are a basis $\mathbf{h} = (h_1, \dots, h_n)$ of A^{i+1} and integers $d_1, \dots, d_n$ such that $(d_1 h_1, \dots, d_n h_n)$ is a basis of B^{i+1}, and a basis $\mathbf{l} = (\ell_1, \dots, \ell_n)$ of A^i such that $\alpha(\ell_j) = d_j h_j$ for every j. Now,

$$\bigwedge \mathbf{l}^{(-1)^i} \otimes (d_1 h_1 \wedge \cdots \wedge d_n h_n)^{(-1)^{i+1}} = \bigwedge \mathbf{l}^{(-1)^i} \otimes \bigwedge \alpha(\mathbf{l})^{(-1)^{i+1}}$$

maps to ± 1 via (8.13) by definition. On the other hand,

$$\begin{aligned}\bigwedge \mathbf{l}^{(-1)^i} \otimes (d_1 h_1 \wedge \ldots \wedge d_n h_n)^{(-1)^{i+1}} &= (\textstyle\prod d_j)^{(-1)^{i+1}} \bigwedge \mathbf{l}^{(-1)^i} \otimes \bigwedge \mathbf{h}^{(-1)^{i+1}} \\ &= |H^{i+1}(A^\bullet)|^{(-1)^{i+1}} \bigwedge \mathbf{l}^{(-1)^i} \otimes \bigwedge \mathbf{h}^{(-1)^{i+1}} .\end{aligned}$$

Thus,

$$\tau(A^\bullet) = |H^{i+1}(A^\bullet)|^{(-1)^{i+1}} ,$$

as claimed. To do the induction step, we let q be such that $A^i = 0$ for $i < q$ but $A^q \neq 0$, and we view $A^\bullet$ as the middle term of an exact sequence

$$0 \to D^\bullet \to A^\bullet \to F^\bullet \to 0 ,$$

where

$$\begin{aligned} D^\bullet &= \quad \cdots \to 0 \to A^q \to Z^{q+1} \to 0 \to \cdots , \\ F^\bullet &= \quad \cdots \to 0 \to 0 \to B^{q+2} \to A^{q+2} \to \cdots . \end{aligned}$$

Since the claim is valid for $D^\bullet$ and $F^\bullet$ by induction hypothesis, it follows from (8.15) and from

$$\begin{array}{lll} H^i(D^\bullet) = H^i(A^\bullet) , & H^i(F^\bullet) = 0 & \text{for } i = q, q+1, q+2 , \\ H^i(D^\bullet) = 0 , & H^i(F^\bullet) = H^i(A^\bullet) & \text{otherwise} \end{array}$$

that it is valid for $A^\bullet$ as well. This completes the proof of (8.16).

In view of (8.16), formula (8.14) is an immediate consequence of the following result.

LEMMA (8.17). $H^{-1}(E^\bullet) \cong 0$, $H^0(E^\bullet) \cong H^2(E^\bullet) \cong \mathbb{Z}/2\mathbb{Z}$, $H^1(E^\bullet) \cong (\mathbb{Z}/2\mathbb{Z})^{2g}$.

To prove the lemma, we need an intermediary observation.

SUBLEMMA (8.18). *Every cycle z in $E^\bullet$ has the property that $2z$ is a boundary.*

Proof. It is obvious that δ_{-1} is injective. We next look at δ_0. Formula (8.7), suitably interpreted, says that there exists an operator $\tilde{\Omega} : \mathbb{Z}^{X_1(G)} \to \mathbb{Z}^{X_1(G)}$ such that

$$\tilde{\Omega}\delta_0(z) \equiv 4z \mod \operatorname{image}(\delta_{-1}) . \tag{8.19}$$

Next, choose an oriented edge $\vec{e}_B \in B$ for each boundary component B. For each oriented edge $\vec{e}$ in B, define a positive integer $v(\vec{e})$ in the following way. Write the cyclic order of the edges belonging to the boundary component $B = [\vec{e}]_2$ as $(e_1, \ldots, e_k)$, where $e_1 = e$. Then $e_B = e_i$

for some i, $1 \leq i \leq k$, and we set $v(\vec{e}) = i$. Given a chain $\sum c_e e \in \mathbb{Z}^{X_1(G)}$, set

$$k_e = c_e + \sum_{i=2}^{v(\vec{e})} c_{e_i} + \sum_{i=2}^{v(\sigma_1(\vec{e}))} c_{f_j} ,$$

where $(e, e_2, \ldots, e_k)$ and $(e, f_2, \ldots, f_h)$ are the cyclically ordered edges around the two (possibly coincident) boundary components $[\vec{e}]_2$ and $[\sigma_1(\vec{e})]_2$ having e as an edge. We claim that, if $\delta_1(\sum c_e e) = 0$, then $2(\sum c_e e) = \delta_0(\sum k_e e)$. This computation is an exercise. To explain it, drawing a picture helps. The coefficient of (say) a in the expression of $\delta_2(\sum k_e e)$ is given by

$$-k_{v_1} + k_{v_2} - k_{v_3} + k_{v_4} ,$$

where v_1, v_2, v_3, v_3 are as in Figure 13.

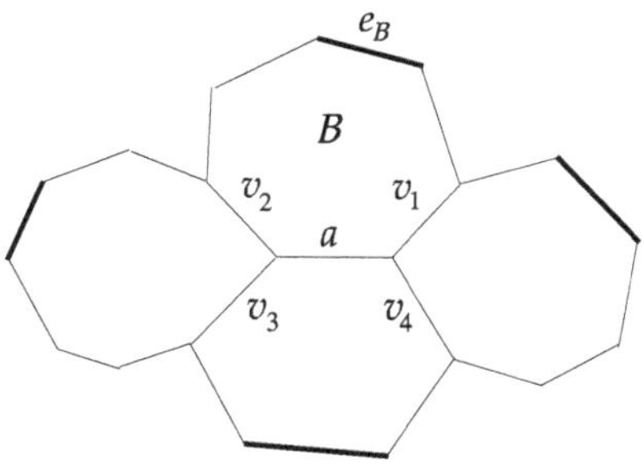

Figure 13.

In this picture we drew four distinct boundary components, but of course in general they are not necessarily distinct, and the same comment can be made about the edges v_1, v_2, v_3, v_3. But, as usual, one can easily see that being distinct or not plays no role in the computation. If the marked edges, like e_B, are as in the above figure, then one immediately computes $-k_{v_1} + k_{v_2} - k_{v_3} + k_{v_4}$ and checks that it equals $2c_a$. If one or more among the marked edges coincide with a, then in the computation one has to take into account the fact that, for each boundary component B,

$$\sum_{\vec{e} \in B} c_e = 0 ,$$

which is a direct consequence of the fact that $\delta_1(\sum c_e e) = 0$. Finally, let $B = [\vec{e}]_2 \in E^2$. Consider the vertex $x = [\vec{e}_B]_0$, i.e., the initial point of $\vec{e}_B$. Let e_0, e_1, e_B be the edges with vertex x. Then

$$2B = \delta_1(e_B - e_0 + e_1) .$$

The proof of the sublemma is complete.

Proof of Lemma (8.17). The immediate consequence of the sublemma is that $H^k(E^\bullet)$ is a $\mathbb{Z}/2\mathbb{Z}$-vector space. Set $h^i = \dim_{\mathbb{Z}/2\mathbb{Z}} H^i(E^\bullet)$ and $\nu^i = \dim_{\mathbb{Z}/2\mathbb{Z}} H^i(E^\bullet \otimes \mathbb{Z}/2\mathbb{Z})$. Then $\nu^i = h^i + h^{i+1}$ by the universal coefficient theorem. Since δ_{-1} is injective, $h^{-1} = 0$.

It is not difficult to check that the complex $E^\bullet \otimes \mathbb{Z}/2\mathbb{Z}$ is obtained by gluing the complex

$$(8.20) \quad E_1^\bullet : \quad 0 \to (\mathbb{Z}/2\mathbb{Z})^{X_2(G)} \xrightarrow{\partial_2} (\mathbb{Z}/2\mathbb{Z})^{X_1(G)} \xrightarrow{\partial_1} (\mathbb{Z}/2\mathbb{Z})^{X_0(G)} \to 0\,,$$

defined by the cell decomposition of the surface S associated to the ribbon graph G and computing the mod 2 homology of S, with its dual complex

$$(8.21) \quad E_2^\bullet : \quad 0 \to (\mathbb{Z}/2\mathbb{Z})^{X_0(G)} \xrightarrow{\partial_1^\vee} (\mathbb{Z}/2\mathbb{Z})^{X_1(G)} \xrightarrow{\partial_2^\vee} (\mathbb{Z}/2\mathbb{Z})^{X_2(G)} \to 0\,.$$

In fact, it is an easy exercise to verify that

$$(8.22) \qquad \delta_{-1} = \partial_2\,, \quad \delta_0 = \partial_1^\vee \partial_1\,, \quad \delta_1 = \partial_2^\vee\,.$$

It follows that

$$H^{-1}(E^\bullet \otimes \mathbb{Z}/2\mathbb{Z}) = H^{-1}(E_1^\bullet) = H_2(S;\mathbb{Z}/2\mathbb{Z}) = \mathbb{Z}/2\mathbb{Z}\,,$$
$$H^{2}(E^\bullet \otimes \mathbb{Z}/2\mathbb{Z}) = H^{2}(E_2^\bullet) = H^2(S;\mathbb{Z}/2\mathbb{Z}) = \mathbb{Z}/2\mathbb{Z}\,.$$

Now, the image of ∂_1 has codimension 1 in $\mathbb{Z}/2\mathbb{Z}^{X_0(G)}$, and hence $\ker \partial_1^\vee$ is one-dimensional. Thus,

$$\ker \partial_1^\vee = \Big\langle \partial_1\Big(\sum_{e \in X_1(G)} e\Big)\Big\rangle \subset \operatorname{image}(\partial_1)\,,$$

and hence $\operatorname{image}(\partial_1^\vee \partial_1)$ has codimension 1 in $\operatorname{image}(\partial_1^\vee)$, so that

$$\nu^1 = \dim_{\mathbb{Z}/2\mathbb{Z}} H^1(E_2^\bullet) + 1 = 2g + 1.$$

Combining this with $\nu^{-1} = \nu^2 = 1$ and $h^{-1} = 0$, we conclude, via the universal coefficient theorem, that $h^{-1} = 0$, $h^0 = h^2 = 1$, $h^1 = 2g$.

Q.E.D.

9. Bibliographical notes and further reading.

The vanishing of the high-degree cohomology of moduli spaces is due to Harer (see [339], the bibliography therein, and also Looijenga's paper [488]).

The first part of Theorem (5.1) is due to Mumford ([551], Theorem 1) for $n = 0$ and to Harer ([337], Lemma 1.1) for arbitrary n, while the second is due to Harer [337]. In [341] Harer computes the third rational homology group of $M_{g,n}$ and in a later unpublished work also computes the fourth rational homology group of moduli spaces. Using Harer's computations, Edidin [187] computes the codimension two homology of the moduli space of stable curves. The computations of the first and second rational cohomology of $\overline{M}_{g,n}$ and $M_{g,n}$ presented in the text are taken from [28] and [30].

Deligne's result on the nonvanishing $H^{11,0}(\overline{\mathcal{M}}_{1,1})$ first appeared in [163].

The Deligne spectral sequence computing the cohomology of the complement of a divisor with normal crossings is introduced in [162] (see also Griffiths and Schmid [322] and Morgan [321]).

The combinatorial expression for the first Chern class ψ_p of the point bundle $\mathcal{L}_p$ on $\overline{M}_{g,P}$ is one of the central ingredients in Kontsevich's proof of Witten's conjecture and appear in [444] (see also [486]). The volume computation in Section 7 is also due to Kontsevich and first appeared in [444].

The computation of the Picard group of $\overline{M}_{g,P}$ proposed in Exercises E-1 and E-2 is taken from [26]. For more refined results on the Picard group of $\overline{M}_{g,P}$, we refer to the papers of Mestrano [510,509] and Kouvidakis [448].

10. Exercises.

In Section 4 we presented the general lines of the proof of Theorem (4.1), omitting a certain number of details. The aim of the series of exercises A, B, C, D, E (taken from [28]) is to fill these gaps and to prove additional results on the low-cohomology groups of $\overline{M}_{g,n}$.

A. Low genus computations.

A-1. Show that $\chi(M_{0,n}) = (-1)^{n-3}(n-3)!$

A-2. Denote by $M'_{0,4}$ and $M'_{0,5}$ the quotients of $M_{0,4}$ and $M_{0,5}$, respectively, modulo the operation of interchanging the labeling of two of the marked points. Show that $\chi(M'_{0,4}) = 0$ and $\chi(M'_{0,5}) = 1$.

A-3. Look at $\pi : M_{1,2} \to M_{1,1}$ and prove that $\chi(M_{1,2}) = \chi(M_{1,1}) = 1$.

A-4. Suppose that X is a quasi-projective algebraic variety and let

$$X = \overline{X}_d \supset \overline{X}_{d-1} \supset \cdots \supset \overline{X}_1 \supset \overline{X}_0$$

be a filtration of X by closed subvarieties. Suppose that $X_i = \overline{X}_i \smallsetminus \overline{X}_{i-1}$ is of pure complex dimension i (or is empty) for every i. Show that, for the Euler characteristic with compact supports, one has $\chi_c(X) = \sum \chi_c(X_i)$. Show that if, in addition, the Poincaré duality holds for the X_i, and X is compact, then $\chi(X) = \sum \chi(X_i)$.

A-5. Look at $\pi : M_{1,3} \to M_{1,2}$ and prove that $\chi(M_{1,3}) = 0$. **Hint:** denote by X the locus in $M_{1,2}$ of all curves $(C; p_1, p_2)$ such that p_2 is a 2-torsion point with respect to the group law with origin in p_1. We also denote by x the point of $M_{1,2}$ corresponding to the covering C of $\mathbb{P}^1$ ramified at ∞ and at the third roots of unity, with p_1 lying above ∞ and p_2 above 0, and denote by U the complement of $X \cup \{x\}$. Then

$$\chi(U) = \chi(M_{1,2}) - \chi(X) - 1\,,$$
$$\chi(M_{1,3}) = \chi(\pi^{-1}(U)) + \chi(\pi^{-1}(X)) + \chi(\pi^{-1}(x))\,.$$

A-6. Use the the stratification by graph type to show that $\chi(\overline{M}_{1,2}) = 4$ and $\chi(\overline{M}_{1,3}) = 12$.

A-7. Show that $h^2(\overline{M}_{1,2}) = 2$ and $h^2(\overline{M}_{1,3}) = 5$ and, in both cases, find the generators for the corresponding cohomology group.

A-8. Using the preceding exercises, show that $h^3(\overline{M}_{1,3}) = 0$.

B. The third cohomology group of $\overline{M}_{g,n}$.

B-1. Let k be an odd integer, h a nonnegative integer and suppose $H^q(\overline{M}_{g,n}) = 0$ for all odd $q \leq k$, all $g \leq h$, and all n such that $q > d(g,n)$. Then $H^q(\overline{M}_{g,n}) = 0$ for all odd $q \leq k$, all $g \leq h$, and all n.

B-2. Prove that $H^3(\overline{M}_{g,n}) = 0$ for all g and n.

C. Relations among tautological classes in $H^2(\overline{M}_{g,P})$.

C-1. Show that, for any choice of distinct elements $x, y, z \in P$, the following relations hold in $H^2(\overline{M}_{0,P})$:

$$\psi_z = \sum_{\substack{A \ni z \\ A \not\ni x,y}} \delta_{0,A}\,, \tag{10.1}$$

$$\kappa_1 = \sum_{A \not\ni x,y} (|A| - 1)\delta_{0,A}\,. \tag{10.2}$$

These, together with the relation $\delta_{irr} = 0$ and relations (4.2) and (4.6), generate all relations in $H^2(\overline{M}_{0,P})$ among the natural classes κ_1, ψ_i, δ_{irr}, and $\delta_{0,A}$ with $|A| \geq 2$ and $|A^c| \geq 2$. (Hint: use induction on $|P|$ and Lemma (4.28) of Chapter XVII.)

C-2. Show that the classes $\delta_{0,S\cup\{q,r\}}$, where S runs through all subsets of P with at most $|P|-2$ elements, are independent in $H^2(\overline{M}_{0,P\cup\{q,r\}})$. (Hint: Let x, y be any two distinct points of P. Let $\vartheta : \overline{M}_{0,\{x,y,q,t\}} \to \overline{M}_{0,P\cup\{q,r\}}$ be the morphism consisting in attaching a fixed tail at the point labeled t and use Lemma (7.5) in Chapter XVII.)

C-3. Prove that the following relations hold in $H^2(\overline{M}_{1,P})$. For any $p \in P$:

$$\kappa_1 = \psi - \delta_0 \,, \tag{10.3}$$

$$12\psi_p = \delta_{irr} + 12 \sum_{\substack{S \ni p \\ |S| \geq 2}} \delta_{0,S} \,. \tag{10.4}$$

(Hint: use (7.13) in Chapter XIII and induction on $|P|$.)

C-4. Show that the homomorphism $\xi^* : H^2(\overline{M}_{1,P}) \to H^2(\overline{M}_{0,P\cup\{q,r\}})$ maps δ_{irr} to zero.

C-5. Show that the classes δ_{irr} and the classes $\delta_{1,S}$ are linearly independent. (Hint: use exercise 2C) and the map $\xi : \overline{M}_{0,P\cup\{q,r\}} \to \overline{M}_{1,P}$.)

C-6. Prove that the following relation holds in $H^2(\overline{M}_{2,P})$:

$$5\kappa_1 = 5\psi + \delta_{irr} - 5\delta_0 + 7\delta_1 \,. \tag{10.5}$$

(Hint: use (8.15) in Chapter XIII and induction on $|P|$.)

C-7. Prove that that the boundary classes and the classes ψ_p, $p \in P$, are independent in $H^2(\overline{M}_{2,P})$ modulo the trivial relations (4.2). (Hint: use the map $\xi : \overline{M}_{1,P\cup\{q,r\}} \to \overline{M}_{2,P}$.)

C-8. Let $g \geq 3$. When $P \neq \emptyset$, fix an element $p \in P$ and show that κ_1, the ψ_i, δ_{irr}, and the $\delta_{a,A}$ such that $p \in A$ are independent. When $P = \emptyset$, instead, show the independence of κ_1, the ψ_i, δ_{irr}, and the δ_a with $2a \leq g$. (Hint: use the map $\xi : \overline{M}_{g-1,P\cup\{q,r\}} \to \overline{M}_{g,P}$.)

D. The second cohomology group of $\overline{M}_{1,n}$ and $\overline{M}_{2,n}$.

D-1. Show that the kernel of $\xi^* : H^2(\overline{M}_{1,P}) \to H^2(\overline{M}_{0,P\cup\{x,y\}})$ is one-dimensional and is generated by δ_{irr}.

D-2. Show that $H^2(\overline{M}_{1,P})$ is generated by boundary classes. (Hint: proceed by induction on $n = |P|$. The initial cases are taken care of by Exercise A-7. Denote by $V = V_{1,P}$ the subspace of $H^2(\overline{M}_{1,P})$ generated by the elements $\delta_{1,S}$, where S runs through all subset of P with at most $n-2$ elements. To prove the claim, it suffices to show that the morphism ξ^* vanishes modulo V. Follow the steps below.

i) Consider the basis $\mathcal{B}$ constructed in Proposition (7.5) of Chapter XVII and relative to the three points $x, y, z \in P$. Let $\alpha \in H^2(\overline{M}_{1,P})$. Write $\xi^*\alpha$ in terms of the basis $\mathcal{B}$:

$$
\begin{aligned}
\xi^*\alpha = a_{\{x,y\}}\delta_{\{x,y\}} &+ \sum_{S\subset X, z\in S, |S|\geq 2, |X\smallsetminus S|\geq 1} a_S\delta_S \\
&+ \sum_{S\subset X, z\in S, |X\smallsetminus S|\geq 3} b_S\delta_{S\cup\{x,y\}} \\
&+ \sum_{S\subset X, z\in S, |X\smallsetminus S|\geq 2} c_S(\delta_{S\cup\{x\}} + \delta_{S\cup\{y\}}) .
\end{aligned}
\tag{10.6}
$$

ii) Consider a subset R of P such that $|P \smallsetminus R| \geq 2$ and look at the morphism

$$\vartheta_R : \overline{M}_{1,R\cup\{u\}} \to \overline{M}_{1,P}$$

defined by attaching a varying stable genus 1 $R\cup\{u\}$-pointed curve to a fixed stable genus zero $(P \smallsetminus R) \cup \{v\}$-pointed curve C_0 via identification of the points labeled u and v. Using maps of type ϑ_R, for various choices of R, show that by adding to α suitable combination of boundary classes on may assume that

$$\xi^*\alpha = \sum_{z\in S\subset X,\, |X\smallsetminus S|\geq 2} c_S(\delta_{S\cup\{x\}} + \delta_{S\cup\{y\}}) . \tag{10.7}$$

iii) Consider the square

$$
\begin{array}{ccc}
\overline{M}_{0,R\cup\{x,y,u\}} & \xrightarrow{\eta_R} & \overline{M}_{0,P\cup\{x,y\}} \\
\eta \big\downarrow & & \big\downarrow \xi \\
\overline{M}_{1,R\cup\{u\}} & \xrightarrow{\vartheta_R} & \overline{M}_{1,P}
\end{array}
$$

where η is the morphism obtained by identifying the points labeled x and y, while η_R is obtained by identifying the point labeled u on the varying curve in $\overline{M}_{0,R\cup\{x,y,u\}}$ with the point labeled v on the fixed curve C_0. Starting from the equality $\eta_R^*\xi^*\alpha = \eta^*\vartheta_R^*\alpha$, where $R \subset P$ and $|P \smallsetminus R| \geq 2$, prove, by descending induction on S, that $c_S = 0$.)

D-3. Show that, for any finite set P, the space $H^2(\overline{M}_{2,P})$ is generated by the classes ψ_q with $q \in P$ and by the boundary classes δ_{irr}, $\delta_{1,A}$, $\delta_{2,B}$, where A and B run through all subsets of P such that $|B^c| \geq 2$ and such that, if $P \neq \emptyset$, then A contains a preassigned point $p \in P$.

(Hint: Set $n = |P|$. Do the cases $n = 0, 1$ directly, then proceed by induction on n. Let $\alpha \in H^2(\overline{M}_{2,P})$. Follow the steps below.

i) For each subset $R \subset P$ such that R^c contains two or more points and $p \in R^c$, consider the map

$$\vartheta_R : \overline{M}_{2,R\cup\{z\}} \to \overline{M}_{2,P} .$$

Show that by adding to α a suitable linear combination of δ_{irr} and of the ψ_i one can assume that

$$\xi^*(\alpha) = \sum_{p\in S\subset P} a_S \delta_{1,S} + \sum_{S\subset P,\, |S^c|\geq 1} c_S(\delta_{1,S\cup\{x\}} + \delta_{1,S\cup\{y\}}) , \tag{10.8}$$

$$\vartheta_R^*(\alpha) = b_z^R \psi_z + \sum_{S\subset R,\, |S|\geq 2} c_S^R \delta_{0,S} + \sum_{S\subset R,\, |S|\geq 1} d_S^R \delta_{0,S\cup\{z\}} + \sum_{S\subset R} h_S^R \delta_{1,S} . \tag{10.9}$$

ii) Set $R = P \setminus \{p, q\}$ for some $q \in P$ with $q \neq p$ and look at the diagram

$$\begin{array}{ccc} \overline{M}_{1,R\cup\{x,y,z\}} & \xrightarrow{\ \nu\ } & \overline{M}_{1,P\cup\{x,y\}} \\ \mu \big\downarrow & & \big\downarrow \xi \\ \overline{M}_{2,R\cup\{z\}} & \xrightarrow{\ \vartheta_R\ } & \overline{M}_{2,P} \end{array}$$

Combine this diagram with diagram (4.12) to conclude that, possibly adding to α suitable linear combinations of tautological classes, one obtains that $\xi^*(\alpha) = 0$. Recall that ξ^* is injective.)

E. The Picard group of $\overline{\mathcal{M}}_{g,n}$.

In Proposition (6.7) of Chapter XIII we proved that $\mathrm{Pic}(\overline{\mathcal{M}}_{g,n})$ is a free abelian group of finite rank and that $\mathrm{Pic}(\overline{M}_{g,n})$ is a subgroup of finite index. The aim of these exercises, based on [26], is to show that for $g \geq 3$, the Picard group $\mathrm{Pic}(\overline{\mathcal{M}}_{g,n})$ is freely generated by the classes $\lambda, \psi_1, \dots, \psi_n$ and by the boundary classes δ_{irr}, and $\delta_{a,A}$ with $0 \leq a \leq g$, $2a - 2 + |A| \geq 0$, and $2(g-a) - 2 + |A^c| \geq 0$ (where, as usual, we identify $\delta_{a,A}$ with δ_{g-a,A^c}). We give a guide for the proof of this statement in the unpointed case $n = 0$. The reader will imitate this argument to prove the general case. We must then show that, for $g \geq 3$, the Picard group $\mathrm{Pic}(\overline{\mathcal{M}}_g)$ is freely generated by λ, δ_{irr}, $\delta_1, \dots, \delta_{[g/2]}$. From now on we set $\delta_0 = \delta_{irr}$. The strategy to prove this result is the following. Set $[g/2] = k$ and construct k families of stable genus g curves $G_1, \dots, G_k$

parameterized by irreducible curves. Set $G_i = (\pi_i : \mathcal{C}_i \to S_i)$. Consider the integral matrix

$$\eta(G_1, \dots, G_{k+2}) = \begin{pmatrix} \deg_{G_1} \lambda & \deg_{G_1} \delta_0 & \cdots & \deg_{G_1} \delta_k \\ \deg_{G_2} \lambda & \deg_{G_2} \delta_0 & \cdots & \deg_{G_2} \delta_k \\ \vdots & \vdots & & \vdots \\ \deg_{G_{k+2}} \lambda & \deg_{G_{k+2}} \delta_0 & \cdots & \deg_{G_{k+2}} \delta_k \end{pmatrix}.$$

Each element η in $\mathrm{Pic}(\overline{M}_g)$ can be written in the form $\eta = a\lambda + \sum b_i \delta_i$. Set $d_i = \deg_{G_i} \eta$. Then

$$\begin{pmatrix} d_1 \\ d_2 \\ \vdots \\ d_{k+2} \end{pmatrix} = \eta(G_1, \dots, G_{k+2}) \begin{pmatrix} a \\ b_0 \\ \vdots \\ b_k \end{pmatrix}.$$

If we can produce two different sets of families $G_1, \dots, G_{k+2}$ such that the corresponding values of η are relatively prime, we can conclude that a and the b_i are integers. We are now going to describe a number of families.

The family Λ_h, $2 \le h \le g$. Let Z be a smooth K3 surface of degree $2h-2$ in $\mathbb{P}^h$ or, when $h = 2$, a double covering of $\mathbb{P}^2$ ramified along a sextic. Consider on Z a Lefschetz pencil of hyperplane sections. Blow up the base locus of this pencil and obtain another surface Y. The curves of the pencil appear in Y as fibers of a map $\varphi : Y \to B = \mathbb{P}^1$ and the exceptional curves appear as sections $E_1, \dots, E_h$ of φ. Fix a genus $g-h$ curve Γ and a point γ on it. Construct a new surface X by joining the surface Y and the surface $\Gamma \times \mathbb{P}^1$ along E_1 and $\{\gamma\} \times \mathbb{P}^1$. We thus get a family $f : X \to \mathbb{P}^1 = B$. This is the family Λ_h.

The family F_h, $g \ge 3$, $2 \le 2h \le g-1$. Fix smooth curves C_1, C_2, and Γ of genera h, $g-h-1$, and 1, and points $x_1 \in C_1$, $x_2 \in C_2$, $\gamma \in \Gamma$. Consider the surfaces $Y_1 = C_1 \times \Gamma$, $Y_2 = (\Gamma \times \Gamma$ blown up at $(\gamma,\gamma))$, $Y_3 = C_2 \times \Gamma$, and set:

$A = \{x_1\} \times \Gamma$, $\quad B = \{x_2\} \times \Gamma$;

$E =$ exceptional divisor in the blow-up of $\Gamma \times \Gamma$ at (γ,γ);

$\Delta =$ proper transform of the diagonal in the blow-up of $\Gamma \times \Gamma$ at (γ,γ);

$S =$ proper transform of $[\gamma] \times \Gamma$.

We construct a surface X by identifying S with A and Δ with B. This surface comes naturally equipped with a projection $f : X \to \Gamma$. This is the family F_h.

The family F. Consider a general pencil of conics in the plane. Blow up $\mathbb{P}^2$ at the four base points of the pencil. Denote by $\psi : X \to \mathbb{P}^2$ the blow-up, by $E_1, \dots, E_4$ the exceptional divisors of ψ, and by $\varphi : X \to \mathbb{P}^1$ the resulting conic bundle. Let C be a fixed curve of genus $g-3$, and p_1, p_2, p_3, p_4 four points of C. Construct a surface Y by setting

$$Y = (X \textstyle\coprod (C \times \mathbb{P}^1))/(E_i \sim \{p_i\} \times \mathbb{P}^1 \,,\ i = 1, \dots, 4)\,.$$

This is the family F.

The family F'. Let $\psi : X \to \mathbb{P}^2$, $E_1, \dots, E_4$, and $\varphi : X \to \mathbb{P}^1$ be as above. Let C_1 be a smooth elliptic curve, and C_2 a smooth curve of genus $g-3$. Let p_1 be a point of C_1 and p_2, p_3, p_4 points of C_2. Set

$$Y = (X \textstyle\coprod (C_1 \times \mathbb{P}^1) \coprod (C_2 \times \mathbb{P}^1))/(E_i \sim \{p_i\} \times \mathbb{P}^1 \,,\ i = 1, \dots, 4)\,.$$

We thus get a family $f : Y \to \mathbb{P}^1$ of stable curves of genus g. This is the family F'.

Following our strategy, the proof that $\mathrm{Pic}(\overline{\mathcal{M}}_g)$ is freely generated by λ, $\delta_0, \delta_1, \dots, \delta_{[g/2]}$ is concluded by doing the following exercise.

E-1. i) If $g = 2m+1$, then

$$\eta(\Lambda_h, F, F_1, \dots, F_m) = (-1)^{m+1}(h+1)\,.$$

ii) If $g = 2m+2$, then

$$\eta(\Lambda_h, F, F', F_1, \dots, F_m) = (-1)^m(h+1)\,.$$

E-2. Compute $\mathrm{Pic}(\overline{\mathcal{M}}_{g,n})$ by imitating the computation of $\mathrm{Pic}(\overline{\mathcal{M}}_g)$.

F. **Ribbon graphs.**

F-1. Show that the maximum number of sides of a ribbon graph G of genus g with n boundary components is equal to $6g-6+3n$ and that, when this number is attained, all the vertices of G are trivalent.

F-2. Prove equality (6.2).

F-3. Compute $\chi(M_{0,3}^{comb})$.

F-4. Use the combinatorial expression for ψ_p to show that $\int_{\overline{M}_{1,1}} \psi_p = \frac{1}{24}$.

Chapter XX.

Intersection theory of tautological classes

1. Introduction.

In this chapter we present a circle of ideas introduced by Witten leading to a conjecture, bearing his name, regarding the intersection numbers of tautological classes on $\overline{M}_{g,n}$. As conjectured by Witten and first proved by Kontsevich, the generating series F of these numbers satisfies differential equations

$$L_n(e^F) = 0\,, \quad n \geq -1\,. \tag{1.1}$$

To be more specific, the intersection numbers at the center of Witten's conjecture are the numbers $< \tau_{d_1} \cdots \tau_{d_n} >$ defined by

$$< \tau_{d_1} \cdots \tau_{d_n} >= \int_{\overline{M}_{g,n}} \psi_1^{d_1} \cdots \psi_n^{d_n}\,, \tag{1.2}$$

where one sets $< \tau_{d_1} \cdots \tau_{d_n} >= 0$ if $d_1 + \cdots + d_n \neq 3g - 3 + n$. As we observed in Section 4 of Chapter XVII, it is remarkable that the knowledge of the numbers (1.2) *for all g and n* implies the knowledge of all the intersection numbers among all the tautological classes for all g and n. We recall that, by definition, the tautological classes are the boundary classes, δ_{irr} and $\delta_{a,A}$, the Mumford classes $\kappa_\nu \in H^{2\nu}(\overline{M}_{g,n}, \mathbb{Q})$, and the point bundle classes $\psi_i \in H^2(\overline{M}_{g,n}, \mathbb{Q})$, $i = 1, \ldots, n$. The generating series for intersection numbers (1.2) is the formal power series

$$F(t_0, t_1, \dots) = \sum_{d_1 \geq 0, \dots, d_n \geq 0} \frac{1}{n!} < \tau_{d_1} \cdots \tau_{d_n} > t_{d_1} \cdots t_{d_n}\,. \tag{1.3}$$

The differential operators L_n in (1.1) are defined by

$$\begin{aligned} L_n = &-(2n+3)!!\frac{\partial}{\partial t_{n+1}} + \sum_{i=0}^{\infty} \frac{(2n+2i+1)!!}{(2i-1)!!} t_i \frac{\partial}{\partial t_{i+n}} \\ &+ \frac{1}{2} \sum_{r+s+1=n} (2r+1)!!(2s+1)!! \frac{\partial^2}{\partial t_r \partial t_s}\,. \end{aligned}$$

E. Arbarello et al., *Geometry of Algebraic Curves*,
Grundlehren der mathematischen Wissenschaften 268,
DOI 10.1007/978-3-540-69392-5_12, © Springer-Verlag Berlin Heidelberg 2011

In Section 2 of this chapter we show how equations (1.1) allow one to recursively compute all the intersection numbers (1.2), starting from the first one,

$$< \tau_0\tau_0\tau_0 >= \int_{\overline{M}_{0,3}} 1 = 1\,.$$

In Section 3 we describe the connection between equations (1.1) and the KdV hierarchy. This is very important from a conceptual point of view in that it can be used as one of the tools to prove Witten's conjecture (1.1), and also from a more practical point of view if one is interested in the actual computation of the intersection numbers (1.2).

An important remark is that the operators L_n satisfy the Virasoro-type commutation relations

$$[L_m, L_n] = (m-n)L_{m+n} \quad \text{for} \quad n, m \geq -1\,. \tag{1.4}$$

For this reason, equations (1.1) are often referred to as Virasoro equations. The first two equations $L_{-1}(e^F) = 0$ and $L_0(e^F) = 0$ are called, respectively, the string equation and the dilaton equation. Looking at what these equations mean at the level of intersection numbers, one finds exactly the relations we already proved in Proposition (4.9) of Chapter XVII. In view of the commutation relations (1.4), in order to prove Witten's conjecture (1.1), it suffices to prove the single equation $L_2(e^F) = 0$. We will verify this relation in Section 6. To do so, we will have to transform the generating series F to apply effectively the operator L_2 to e^F.

In Section 4, we come to Kontsevich's idea of using the cellular decomposition of $M_{g,n}$ we described in Chapter XVIII. The first step is to give a combinatorial expression ω_i for the cohomology class ψ_i in terms of this cellular decomposition. This is what we did in Section 7 of Chapter XIX (cf. formula (7.5) in that chapter). Then

$$< \tau_{d_1}\cdots\tau_{d_n} >= \int_{\overline{M}_{g,n}} \psi_1^{d_1}\cdots\psi_n^{d_n} = \int_{M_{g,n}^{comb}(r)} \omega_1^{d_1}\cdots\omega_n^{d_n}\,, \tag{1.5}$$

and one can hope that the right-hand side of this equality can be expressed solely in terms of ribbon graphs. This is indeed possible, and after nontrivial combinatorial computations, also involving Theorem (8.8) in Chapter XIX, one can prove the following remarkable statement. Let $\Lambda_1, \ldots, \Lambda_N$ be variables in $\mathbb{R}_+^N$. Set $\Lambda = \mathrm{diag}(\Lambda_1, \ldots, \Lambda_N)$ and

$$t_i(\Lambda) = -(2i-1)!!\,\mathrm{Tr}\left(\Lambda^{-(2i+1)}\right) = -(2i-1)!!\left(\Lambda_1^{-(2i+1)} + \cdots + \Lambda_N^{-(2i+1)}\right).$$

Look at the generating series (1.3) in the new variables $\Lambda_1, \ldots, \Lambda_N$. Then

$$F(t_0(\Lambda), t_1(\Lambda), \ldots) = \sum_{G\in\overline{\mathcal{G}}^{3,c}} \frac{(\sqrt{-1}/2)^{|X_0(G)|}}{|\operatorname{Aut}(G)|} \prod_{e\in X_1(G)} \frac{2}{\Lambda_e + \Lambda_{e'}}\,, \tag{1.6}$$

where $\overline{\mathcal{G}}^{3,c}$ is the set of isomorphism classes of 3-valent, connected ribbon graphs with N-colored boundary components. It should be observed that the Newton polynomials $\operatorname{Tr}\Lambda^{-j}$ are independent functions of the Λ_i for i and j less than or equal to N. For this reason, we are interested in the above combinatorial expression especially when the number N of the variables $\Lambda_1, \dots \Lambda_N$ tends to infinity.

Expression (1.6), remarkable as it is, does not make it any easier to verify the equation $L_2(e^F) = 0$. The fundamental reason to express the generating series (1.3) as an infinite sum indexed by ribbon graphs, as in (1.6), is that these sums naturally appear in the asymptotic expansions of matrix integrals. We discuss matrix integrals and asymptotic expansions in Section 5. A typical example of asymptotic expansion is the following. On the space $\mathcal{H}_N$ of $N \times N$ hermitian matrices one considers the measure

$$d\mu_N = \frac{1}{c_N} e^{-\frac{1}{2}\operatorname{Tr}(X^2)} dX\,, \tag{1.7}$$

where

$$dX = \prod_i dX_{ii} \prod_{i<j} d\operatorname{Re}(X_{ij})\, d\operatorname{Im}(X_{ij})\,, \quad c_N = \int_{\mathcal{H}_N} e^{-\frac{1}{2}\operatorname{Tr}(X^2)} dX\,.$$

Then there is an asymptotic expansion for $t \to 0$,

$$\int_{\mathcal{H}_N} e^{t\frac{\operatorname{Tr} X^3}{3}} d\mu_N = \sum_{G \in \overline{\mathcal{G}}^3} \frac{1}{|\operatorname{Aut}(G)|} t^{|X_0(G)|}\,, \tag{1.8}$$

where $\overline{\mathcal{G}^3}$ denotes the isomorphism classes of N-colored trivalent ribbon graphs. Kontsevich devises a matrix model tailored on the right-hand side of (1.6). The model is based on the following Gaussian measure. Fix a real nonsingular diagonal matrix $\Lambda = \operatorname{diag}(\Lambda_1, \dots, \Lambda_N)$ and set

$$d\mu_{\Lambda,N} = \frac{1}{c_{\Lambda,N}} e^{-\frac{1}{2}\operatorname{Tr}\Lambda X^2} dX\,, \tag{1.9}$$

where the normalization factor $c_{\Lambda,N}$ is defined as

$$c_{\Lambda,N} = \int_{\mathcal{H}_N} e^{-\frac{1}{2}\operatorname{Tr}\Lambda X^2} dX = \prod_{i,j=1}^{N} (\Lambda_i + \Lambda_j)^{-\frac{1}{2}}\,. \tag{1.10}$$

In analogy with (1.8), one gets the asymptotic expansion for $\Lambda^{-1} \to 0$

$$\int_{\mathcal{H}_N} e^{\frac{\sqrt{-1}}{6}\operatorname{Tr} X^3} d\mu_{\Lambda,N} = \sum_{G \in \overline{\mathcal{G}}^3} \frac{(\sqrt{-1}/2)^{|X_0(G)|}}{|\operatorname{Aut}(G)|} \prod_{e \in X_1(G)} \frac{2}{\Lambda_e + \Lambda_{e'}}\,. \tag{1.11}$$

The only difference between the right-hand side of this asymptotic equation and the right-hand side of (1.6) is that here we are summing over all trivalent ribbon graphs, while there we confine ourselves to connected ones. But a standard trick to pass from a sum over connected graphs to a sum over possibly disconnected ones is to take an exponential. Combining (1.6) and (1.9), one gets the following asymptotic expression for the exponential of the generating function (1.3):

$$\exp(F(t_0(\Lambda), t_1(\Lambda), \dots)) = \int_{\mathcal{H}_N} e^{\frac{\sqrt{-1}}{6} \operatorname{Tr} X^3} d\mu_{\Lambda,N} \,.$$

At this stage it is possible to check the equation $L_2(\exp F) = 0$. This is done in Section 6, with the help of an algebraic theorem by Di Francesco, Itzykson, and Zuber stating that, given any differential operator

$$D = D\left(\frac{\partial}{\partial t_0}, \frac{\partial}{\partial t_1}, \frac{\partial}{\partial t_2}, \dots\right) \in \mathbb{C}\left[\frac{\partial}{\partial t_0}, \frac{\partial}{\partial t_1}, \frac{\partial}{\partial t_2}, \dots\right],$$

there exists a polynomial

$$P_D(\operatorname{Tr} X, \operatorname{Tr} X^3, \operatorname{Tr} X^5, \dots) \in \mathbb{C}[\operatorname{Tr} X, \operatorname{Tr} X^3, \operatorname{Tr} X^5, \dots]$$

such that

$$D \int_{\mathcal{H}_N} e^{\frac{\sqrt{-1}}{6} \operatorname{Tr} X^3} d\mu_{\Lambda,N} = \int_{\mathcal{H}_N} P_D e^{\frac{\sqrt{-1}}{6} \operatorname{Tr} X^3} d\mu_{\Lambda,N} \,.$$

This brings the proof of Witten's conjecture to a computational conclusion.

In outlining this proof of Witten's conjecture, we omitted to mention a delicate point regarding the combinatorial expression (1.5) of the intersection numbers $< \tau_{d_1} \cdots \tau_{d_n} >$. In order for the second equality in (1.5) to make sense, the following steps are necessary. One must first find a suitable orbicell compactification $\overline{M}_{g,n}^{comb}(r)$ of $M_{g,n}^{comb}(r)$. Then one should extend the piecewise linear 2-form ω_i, defined on $M_{g,n}^{comb}(r)$, to a piecewise linear 2-form, denoted with the same letter, defined on the compactification $\overline{M}_{g,n}^{comb}(r)$. Finally, one should extend the homeomorphism $h : M_{g,n} \to M_{g,n}^{comb}(r)$ to a continuous surjection $f : \overline{M}_{g,n} \to \overline{M}_{g,n}^{comb}(r)$ and verify that $f^*(\omega_i) = \psi_i$. It is precisely to be able to perform these constructions that, in Theorem (2.7) of Chapter XVIII, we went through the difficult process of finding a $\Gamma_{S,P}$-equivariant, continuous, surjective map

$$\widehat{\Psi} : \quad \widehat{\mathcal{T}}_{S,P} \times \Delta_P \longrightarrow |A(S,P)| \tag{1.12}$$

extending the more intuitive $\Gamma_{S,P}$-equivariant homeomorohism

$$\Psi : \quad \mathcal{T}_{S,P} \times \Delta_P \longrightarrow |A^0(S,P)| \,. \tag{1.13}$$

Section 7 contains a proof of Faber's nonvanishing theorem (Theorem (6.1) of Chapter XVII).

Sections 8 and 9 are devoted to the computation of the virtual Euler–Poincaré characteristic of the moduli spaces of smooth pointed curves. This beautiful computation was originally done by Harer and Zagier, but here we follow the shortcut suggested by Kontsevich.

Finally, Section 10 contains a brief introduction to Gromov–Witten invariants.

2. Witten's generating series.

In Section 3 of Chapter XVII we introduced a set of cycle classes in $H^*(\overline{M}_{g,n},\mathbb{Q})$. These classes are of two types. There are the boundary classes $\delta_{p,A}$ and the Mumford–Morita–Miller classes. This last set of classes is formed by the Mumford classes $\kappa_\nu \in H^{2\nu}(\overline{M}_{g,n},\mathbb{Q})$ and by the degree two classes $\psi_1,\dots,\psi_n \in H^2(\overline{M}_{g,n},\mathbb{Q})$. The basic problem in the intersection theory of moduli spaces is to compute the intersection numbers of these classes. This means computing

$$\int_{\overline{M}_{g,n}} \alpha\,,$$

where α is any top-degree monomial in the above classes, the top degree being the number $2d$, where $d = 3g-3+n$ is the dimension of $\overline{M}_{g,n}$. As we observed in Chapter XVII, due to the orbifold, or stack, nature of $\overline{M}_{g,n}$, the above intersection numbers are, in general, rational numbers. Among the monomials in the tautological classes, there are those involving only the classes ψ_i. We denote the corresponding intersection numbers with the symbol $<\tau_{d_1},\dots,\tau_{d_n}>$, where $d_i \geq 0$, $i=1,\dots,n$, and

$$(2.1)\qquad <\tau_{d_1}\cdots\tau_{d_n}>=\begin{cases}\int_{\overline{M}_{g,n}}\psi_1^{d_1}\cdots\psi_n^{d_n} & \text{if } d_1+\cdots+d_n = d\,,\\ 0 & \text{if } d_1+\cdots+d_n \neq d\,.\end{cases}$$

Sometimes, when we want to recall to which moduli space a given intersection number belongs, we write $<\tau_{d_1}\cdots\tau_{d_n}>_{g,n}$ instead of $<\tau_{d_1}\cdots\tau_{d_n}>$. In Section 4 of Chapter XVII we observed the very remarkable fact that the knowledge of the numbers (2.1) *for all g and n* implies the knowledge of all the intersection number among all the above cycle classes for all g and n.

We next organize the intersection numbers (2.1) in a formal power series in infinitely many variables

$$(2.2)\qquad F(t_0,t_1,\dots) = \sum_{\substack{d_1,\dots,d_n\\ d_i\geq 0}} \frac{1}{n!} <\tau_{d_1}\cdots\tau_{d_n}> t_{d_1}\cdots t_{d_n}\,.$$

This is the generating series for the intersection numbers. The exponential of this series is called the *partition function*

$$(2.3)\qquad Z(t_0,t_1,\dots) = \exp(F(t_0,t_1,\dots)).$$

Notice that, because of the definition of the symbol $< \tau_{d_1} \cdots \tau_{d_n} >$, the exponential Z makes good sense, since its coefficients involve only finite sums of monomials in the coefficients of F. In order to state the main theorem of this chapter, we must introduce the Virasoro operators L_n, $n \geq -1$. Recall that, for an odd positive integer k, the double factorial $k!!$ is by definition the product of all the odd positive integers up to k. The Virasoro operators are defined by the following formulae:

(2.4)
$$\begin{aligned}
L_{-1} &= -\frac{\partial}{\partial t_0} + \sum_{i=1}^{\infty} t_i \frac{\partial}{\partial t_{i-1}} + \frac{t_0^2}{2}, \\
L_0 &= -3 \cdot \frac{\partial}{\partial t_1} + \sum_{i=0}^{\infty} (2i+1) t_i \frac{\partial}{\partial t_i} + \frac{1}{8}, \\
L_1 &= -5 \cdot 3 \cdot \frac{\partial}{\partial t_2} + \sum_{i=0}^{\infty} (2i+1)(2i+3) t_i \frac{\partial}{\partial t_{i+1}} + \frac{1}{2} \frac{\partial^2}{\partial t_0^2}, \\
L_2 &= -7 \cdot 5 \cdot 3 \cdot \frac{\partial}{\partial t_3} + \sum_{i=0}^{\infty} (2i+1)(2i+3)(2i+5) t_i \frac{\partial}{\partial t_{i+2}} + 3 \frac{\partial^2}{\partial t_0 \partial t_1}, \\
&\cdots\cdots\cdots \\
L_k &= -(2k+3)!! \cdot \frac{\partial}{\partial t_{k+1}} + \sum_{i=0}^{\infty} \frac{(2k+2i+1)!!}{(2i-1)!!} t_i \frac{\partial}{\partial t_{i+k}}, \\
&\quad + \frac{1}{2} \sum_{r+s+1=k} (2r+1)!!(2s+1)!! \frac{\partial^2}{\partial t_r \partial t_s}.
\end{aligned}$$

The main theorem of this chapter, whose proof will be completed in Section 6, is the following.

THEOREM (2.5). *$L_n Z = 0$ for all $n \geq -1$.*

We now wish to digress briefly on how to use the above identities to compute intersection numbers. The first remark regards the derivatives of the function F.

LEMMA (2.6). *Let F be the generating series (2.2). Then*

$$\frac{\partial F}{\partial t_k} = \sum_{d_1,\ldots,d_n} \frac{1}{n!} < \tau_k \tau_{d_1} \cdots \tau_{d_n} > t_{d_1} \cdots t_{d_n}, \qquad k \geq 0,$$

$$\left(\sum_{d=\max(0,-k)}^{\infty} c_d t_d \frac{\partial}{\partial t_{d+k}} \right) F =$$

$$\sum_{\substack{d_1,\ldots,d_n \\ d_i+k \geq 0}} \frac{1}{n!} \sum_{i=1}^{n} c_{d_i} < \tau_{d_1} \cdots \tau_{d_i+k} \cdots \tau_{d_n} > t_{d_1} \cdots t_{d_n}, \quad k \in \mathbb{Z},$$

where the c_d are constants.

Proof. It is useful to write

$$<\tau_{d_1}\cdots\tau_{d_n}>=<\tau_0^{n_0}\tau_1^{n_1}\tau_2^{n_2}\cdots>,$$

where n_0 is the number of d_i that are equal to 0, n_1 is the number of d_i that are equal to 1, and so on. We can then write

$$F(t_0,t_1,t_2,\dots)=\sum_{n_0,n_1,n_2,\dots}<\tau_0^{n_0}\tau_1^{n_1}\tau_2^{n_2}\cdots>\left(\frac{t_0^{n_0}}{n_0!}\frac{t_1^{n_1}}{n_1!}\frac{t_2^{n_2}}{n_2!}\cdots\right),$$

where the sequences $\{n_0,n_1,n_2,\dots\}$ have only finitely many nonzero terms. It is now clear that taking the derivative with respect to t_k gives the first relation. For the second relation, suppose for simplicity that $k\geq 0$. We have

$$\left(\sum_{d=0}^{\infty}c_d t_d\frac{\partial}{\partial t_{d+k}}\right)F=$$

$$\left(\sum_{d=0}^{\infty}c_d t_d\frac{\partial}{\partial t_{d+k}}\right)\left(\sum_{e_1,\dots,e_{n-1}}\frac{1}{(n-1)!}<\tau_{e_1}\cdots\tau_{e_{n-1}}>t_{e_1}\cdots t_{e_{n-1}}\right)=$$

$$\sum_{d=0}^{\infty}\left(\sum_{e_1,\dots,e_{n-1}}\frac{c_d}{(n-1)!}<\tau_{d+k}\tau_{e_1}\cdots\tau_{e_{n-1}}>t_d t_{e_1}\cdots t_{e_{n-1}}\right)=$$

$$\sum_{d=0}^{\infty}\left(\sum_{e_1,\dots,e_{n-1}}\frac{c_d}{n!}\left[<\tau_{d+k}\tau_{e_1}\cdots\tau_{e_{n-1}}>t_d t_{e_1}\cdots t_{e_{n-1}}+\cdots\right.\right.$$

$$\left.\left.+<\tau_{e_1}\cdots\tau_{e_{n-1}}\tau_{d+k}>t_{e_1}\cdots t_{e_{n-1}}t_d\right]\right)=$$

$$\sum_{d_1,\dots,d_n}\sum_{i=1}^{n}\frac{c_{d_i}}{n!}<\tau_{d_1}\cdots\tau_{d_i+k}\cdots\tau_{d_n}>t_{d_1}\cdots t_{d_n}.$$

Q.E.D.

Using the above lemma, we may now write down the equations $L_n Z=0$ in terms of the coefficients of F. The equation $L_{-1}Z=0$, which is called the *string equation,* immediately translates into

$$<\tau_0\tau_{d_1}\cdots\tau_{d_n}>=\sum_{i \text{ s.t. } d_i-1\geq 0}<\tau_{d_1}\cdots\tau_{d_i-1}\cdots\tau_{d_n}>. \tag{2.7}$$

The equation $L_0 Z=0$, which is called the *dilaton equation,* also easily translates into

$$<\tau_1\tau_{d_1}\cdots\tau_{d_n}>=(2g-2+n)<\tau_{d_1}\cdots\tau_{d_n}>. \tag{2.8}$$

These equations are respectively the first and second equations in Proposition (4.9) of Chapter XVII which we already proved! Set $\underline{d} = d_1, \dots, d_n$ and $\tau_{\underline{d}} = \tau_{d_1} \cdots \tau_{d_n}$. Again using the lemma, the equation $L_k Z = 0$, for $k \geq 1$, translates into

(2.9)

$$\begin{aligned} &< \tau_{k+1}\tau_{\underline{d}} >_{g,n+1} \\ &\quad = \frac{1}{(2k+3)!!}\left[\sum_{j=1}^{n} \frac{(2k+2d_j+1)!!}{(2d_j-1)!!} < \tau_{d_1} \cdots \tau_{d_j+k} \cdots \tau_{d_n} >_{g,n} \right. \\ &\quad + \frac{1}{2} \sum_{r+s=k-1} (2r+1)!!(2s+1)!! < \tau_r\tau_s\tau_{\underline{d}} >_{g-1,n+2} \\ &\quad \left. + \frac{1}{2} \sum_{r+s=k-1} (2r+1)!!(2s+1)!! \sum_{I\subset\{1,\dots,n\}} < \tau_r\tau_{\underline{d}_I} >_{p,a+1} < \tau_s\tau_{\underline{d}_{I^c}} >_{q,b+1} \right] . \end{aligned}$$

Here, as usual, I^c denotes the complement of I in $\{1,\dots,n\}$, $a = |I|$, $b = n-a$, $p+q = g$.

Let us look more closely, at least from a formal point of view, at the shape of equation (2.9). The left-hand side of the equality can be viewed as the integral of a top-degree form on $\overline{M}_{g,n+1}$ and, of course, also as an integral over $\overline{M}_{g,n}$:

$$< \tau_{k+1}\tau_{\underline{d}} >= \int_{\overline{M}_{g,n+1}} \psi_1^{d_1} \cdots \psi_n^{d_n}\psi_{n+1}^{k+1} = \int_{\overline{M}_{g,n}} \pi_*\left(\psi_1^{d_1} \cdots \psi_1^{d_n}\psi_{n+1}^{k+1}\right) ,$$

where

$$k+1+d_1+\cdots+d_n = 3g-3+n .$$

On the right-hand side, the first group of terms can be viewed as integrals over $\overline{M}_{g,n}$, the second group of terms as integrals over $\overline{M}_{g-1,n+2}$, while each summand of the third group corresponds to an integral over $\overline{M}_{p,\mu+1} \times \overline{M}_{q,\nu+1}$, where $p+q = g$ and $\mu+\nu = n$. As such, the last two groups of terms can be viewed as contributions coming from the boundary divisors of $\overline{M}_{g,n}$. These terms correspond to the degree two part of the differential operators L_k, which exists only for $k \geq 1$. The main difference betweeen the equations $L_k Z = 0$ for $k \geq 1$ and the first two equations $L_{-1}Z = L_0 Z = 0$ is exactly the presence of these terms.

By definition, we have

$$F(t_0,0,0,\dots) = \log Z(t_0,0,0,\dots) = \frac{1}{6}t_0^3 ,$$

where the coefficient of t_0^3 is

$$< \tau_0\tau_0\tau_0 >= \int_{\overline{M}_{0,3}} 1 = 1 .$$

We have the following key lemma.

LEMMA (2.10). *A solution Z to the system $L_n Z = 0$, $n \geq -1$, is completely determined by the initial condition $\log Z(t_0, 0, 0, \dots) = \frac{1}{6} t_0^3$. Equivalently, assuming Theorem (2.5), the intersection numbers $< \tau_{d_1} \cdots \tau_{d_n} >$ can be recursively computed, using the identities $L_n Z = 0$, $n \geq -1$, starting from $< \tau_0 \tau_0 \tau_0 >= 1$.*

Before proving the lemma, let us compute, as an example, the intersection number

$$< \tau_1 >= \int_{M_{1,1}} \psi_1 \,.$$

From the string equation we get $< \tau_1 >=< \tau_0 \tau_2 >$. From $L_1 Z = 0$ we get

$$< \tau_2 \tau_0 >= \frac{1}{15} \left(3 < \tau_1 > + \frac{1}{2} < \tau_0 \tau_0 \tau_0 > \right) .$$

Thus $< \tau_1 >= 1/24$. The proof of Lemma (2.10) is an expansion of this computation.

Proof of Lemma (2.10). We proceed by induction on g. Assume that we know all intersection numbers $< \tau_{d_1} \cdots \tau_{d_\nu} >_{p,\nu}$ for all $p \leq g-1$ and for all ν. We must then show how to compute $< \tau_{d_1} \cdots \tau_{d_n} >_{g,n}$ for all n, using the identities $L_k Z = 0$, $k \geq -1$. We first compute $< \tau_k >_{g,1}$, so that $k = 3g - 2$. We proceed exactly as in the computation of $< \tau_1 >$. From the string equation we get $< \tau_0 \tau_{k+1} >_{g,2} =< \tau_k >_{g,1}$. Now we use that $L_k Z = 0$ and get

$$< \tau_k >_{g,1} =< \tau_{k+1} \tau_0 >_{g,2} = \tfrac{1}{2k+3} < \tau_k >_{g,1} + \left\{ \begin{array}{c} \text{linear combination} \\ \text{of terms known by} \\ \text{induction hypothesis} \end{array} \right\} .$$

Finally, (2.9) and the string equation tell us that, for a fixed g, once the intersection numbers $< \tau_{d_1} \cdots \tau_{d_\nu} >_{p,\nu}$ are known for all pairs (p, ν) with $p \leq g - 1$ and for all pairs (g, ν) with $\nu \leq n$, then they are also known for the pair $(g, n + 1)$. Q.E.D.

We next come to a very important remark, namely that the operators L_k satisfy the commutation relations

(2.11) $$[L_m, L_n] = (m - n) L_{m+n} \quad \text{for} \quad n, m \geq -1 \,.$$

The proof of these relations is a straightforward computation. One refers to this fundamental property by saying that the operators L_n with $n \geq -1$ form a "half" of a Virasoro algebra. Since the operators L_n satisfy the commutation relation (2.11), in order to prove Theorem (2.5), it suffices to show that $L_{-1} Z = L_0 Z = L_2 Z = 0$. So we need only to prove the equation $L_2 Z = 0$. At the time of writing this book, no direct algebro-geometrical proof of the identity $L_2 Z = 0$ is yet available. We will prove it using the cellular decomposition of $M_{g,n}$ and the properties of matrix integrals. Indeeed, in Section 4 we will transform the generating function (2.2) into an infinite sum indexed by ribbon graphs of all genera and with increasing number of boundary components. The exponential of this sum will then be interpreted as the asymptotic expansion of a matrix integral. This asymptotic expression for the partition function (2.3) will allow us to check that

$$L_2 Z = 0\,, \tag{2.12}$$

which, as we just noticed, implies Theorem (2.5).

3. Virasoro operators and the KdV hierarchy.

In this section we will establish the connection between the Virasoro equations $L_k Z = 0$ and the KdV hierarchy of differential equations. This connection is of great practical use in the explicit computation of the intersection numbers $< \tau_{d_1} \cdots \tau_{d_n} >$.

We shall introduce the KdV hierarchy in the the so-called Gelfand–Dikii form. In the bibliographical notes we will give references connecting the multiple ways of looking at the KdV hierarchy. A formal power series $U(t_0, t_1, t_2, \dots)$ is said to satisfy the KdV hierarchy in Gelfand–Dikii form if it satisfies the differential equations

$$\frac{\partial U}{\partial t_i} = \frac{\partial}{\partial t_0} R_i[U]\,, \quad i \geq 0\,, \tag{3.1}$$

where the $R_i[U]$ are differential polynomials (i.e., polynomials in U and in its derivatives with respect to t_0) defined recursively by

$$R_0 = U\,, \qquad \frac{\partial R_{n+1}}{\partial t_0} = \frac{1}{2n+1}\left(\frac{\partial U}{\partial t_0} + 2U\frac{\partial}{\partial t_0} + \frac{1}{4}\frac{\partial^3}{\partial t_0^3}\right) R_n\,.$$

The fact that there exist differential polynomials satisfying this recursion follows from the theory of the Riccati equation (cf. [178], Chapter 12). The first differential polynomials are

$$\begin{aligned} R_0[U] &= U\,, \\ R_1[U] &= \tfrac{3}{2}U^2 + \tfrac{1}{4}U''\,, \\ R_2[U] &= \tfrac{15}{18}U^3 + \tfrac{5}{12}UU'' + \tfrac{1}{48}U''''\,. \end{aligned}$$

The following result is due to Dijkgraaf, Verlinde, and Verlinde [181].

THEOREM (3.2). *Let $F = F(t_0, t_1, t_2, \dots)$ be a formal power series. Assume that $F(t_0, 0, 0, \dots) = \frac{1}{6}t_0^3$. Set $Z = e^F$ and $U = \partial^2 F/\partial t_0^2$. Then Z satisfies the Virasoro equations $L_k Z = 0$, $k \geq -1$, if and only* *if Z satisfies the string equation $L_{-1}Z = 0$ and U satisfies the KdV hierarchy (3.1).*

Before proving this theorem, we need to establish some notation. We set

$$\ll \tau_{d_1} \cdots \tau_{d_n} \gg = \frac{\partial}{\partial t_{d_1}} \frac{\partial}{\partial t_{d_2}} \cdots \frac{\partial}{\partial t_{d_n}} F . \tag{3.3}$$

This notation is suggested by the fact that

$$< \tau_{d_1} \cdots \tau_{d_n} > = \left(\frac{\partial}{\partial t_{d_1}} \frac{\partial}{\partial t_{d_2}} \cdots \frac{\partial}{\partial t_{d_n}} F \right)_{|t_i=0} , \tag{3.4}$$

as follows from Lemma (2.6). In particular, we have

$$\ll \tau_0 \tau_0 \tau_{d_1} \cdots \tau_{d_n} \gg = \frac{\partial}{\partial t_{d_1}} \frac{\partial}{\partial t_{d_2}} \cdots \frac{\partial}{\partial t_{d_n}} U . \tag{3.5}$$

We next prove the following analogue of Lemma (2.10).

LEMMA (3.6). *A formal power series $F = F(t_0, t_1, t_2, \ldots)$ satisfying the initial condition $F(t_0, 0, 0, \ldots) = \frac{1}{6} t_0^3$ and such that $U = \partial^2 F / \partial t_0^2$ satisfies the KdV hierarchy (3.1) and $Z = e^F$ satisfies the string equation $L_{-1} Z = 0$ is completely determined by these conditions.*

Proof. The fact that U is completely determined by the KdV hierarchy and the initial condition $U(t_0, 0, 0, \ldots) = t_0$ is obvious. Let us see how one can determine F or, equivalently, all the intersection numbers $< \tau_{d_1} \cdots \tau_{d_n} >$. First of all observe that, using the string equation, we have

$$< \tau_k > = < \tau_{k+2} \tau_0 \tau_0 > = \left(\frac{\partial U}{\partial t_{k+2}} \right)_{|t_0 = t_1 = \cdots = 0} . \tag{3.7}$$

Hence, the intersection numbers $< \tau_k >$ (i.e., the intersection numbers of $M_{g,1}$) are completely determined by the KdV and string equations. Notice that in (3.7) we necessarily have $k = 3g - 2$. To see how all the intersection numbers $J = < \tau_{d_1} \cdots \tau_{d_n} >$ are determined, we proceed by induction. First of all, using the string equation, we may assume that there is no τ_0 entry in the expression of J. Since $\sum d_i = 3g - 3 + n$ we have $d_i \leq 3g - 2$. By induction, we may assume that the KdV and string equations determine the intersection number J whenever all the d_i are positive and one of them, say d_1, is greater than or equal to N. The initial step of the descending induction corresponds to the case $N = 3g - 2$. We want to show how the KdV and the string equations determine the intersection number J when all the d_i are positive and one of them, say d_1, is greater than or equal to $N-1$. The intersection number

$$(3.8) \qquad <\tau_{d_1+2}\tau_{d_2}\cdots\tau_{d_n}\tau_0\tau_0>=\left(\frac{\partial}{\partial t_{d_1+2}}\frac{\partial}{\partial t_{d_2}}\cdots\frac{\partial}{\partial t_{d_n}}U\right)_{|t_i=0}$$

is clearly determined by the KdV and string equations. Now use the string equation twice and get

$$<\tau_{d_1+2}\tau_{d_2}\cdots\tau_{d_n}\tau_0\tau_0>=J+H\,,$$

where the term H is a sum of intersection numbers involving τ_{d_1+2} and τ_{d_1+1}. If none of these summands contains a τ_0 entry, we are done by induction. But this may not be the case since some of the d_i with $i\geq 2$ could be equal to either 1 or 2. So, there could be summands in H of these three types:

$$(3.9) \qquad \begin{array}{lll} a) & <\tau_{d_1+2}\tau_{h_2}\cdots\tau_{h_{n-2}}\tau_0\tau_0>\,, & h_i>0\,,\\ b) & <\tau_{d_1+2}\tau_{f_2}\cdots\tau_{f_{n-1}}\tau_0>\,, & f_i>0\,,\\ c) & <\tau_{d_1+1}\tau_{e_2}\cdots\tau_{e_{n-1}}\tau_0>\,, & e_i>0\,. \end{array}$$

Terms of type a) are known by induction, as shown in (3.8). The remaining two terms occur only if some of the d_i with $i\geq 2$ are equal to 1. Let then K be the number of d_i, with $i\geq 2$, which are equal to 1. If $K=0$, we just proved that H can be expressed in terms of known intersection numbers, and we are done. We proceed by induction on K. We may then assume that the KdV and the string equation determine all intersection numbers $J=<\tau_{d_1}\cdots\tau_{d_n}>$ in which all the d_i are positive, one of them, say d_1, is greater than or equal to $N-1$, and there are at most $K-1$ among the d_i, with $i\geq 2$, which are equal to 1. Consider an intersection number J in which K among the d_i, with $i\geq 2$, are equal to 1. We proceed as above. Look at the summands of H. Now also the terms of type c) are known by induction since, for them, there are at most $K-1$ among the e_i that are equal to 1. We apply the string equation to the terms of type b). Then, we get either terms known by induction or else terms that are again intersection numbers of type b) but involving curves with progressively fewer punctures. The process terminates when we get intersection numbers in $M_{g,1}$ which are known by the first step of the induction. Q.E.D.

Sketch of Proof of Theorem (3.2). By the uniqueness statements of Lemmas (2.10) and (3.6), it suffices to prove that, if F satisfies the KdV and string equation, then it also satisfies the Virasoro equations $L_nZ=0$. Setting $U=\partial^2F/\partial t_0^2$ in the KdV equation (3.1), we get

$$\frac{\partial^2}{\partial t_0^2}\frac{\partial F}{\partial t_{n+1}}=\frac{1}{2n+1}\left(\frac{\partial U}{\partial t_0}\frac{\partial}{\partial t_0}+2U\frac{\partial^2}{\partial t_0^2}+\frac{1}{4}\frac{\partial^4}{\partial t_0^4}\right)\frac{\partial F}{\partial t_n}\,.$$

It can be proved, with a rather lengthy computation, that the quantities $L_n Z/Z$ satisfy exactly the same type of equations, namely that

$$\frac{\partial^2}{\partial t_0^2}\left(\frac{L_{n+1}Z}{Z}\right) = \frac{1}{2n+1}\left(\frac{\partial U}{\partial t_0}\frac{\partial}{\partial t_0} + 2U\frac{\partial^2}{\partial t_0^2} + \frac{1}{4}\frac{\partial^4}{\partial t_0^4}\right)\left(\frac{L_n Z}{Z}\right).$$

A very clear proof of this fact is contained in Section 6 of [286]. By assumption, $L_{-1}Z = 0$. Suppose inductively that $L_n Z = 0$. Then

$$\frac{\partial^2}{\partial t_0^2}\left(\frac{L_{n+1}Z}{Z}\right) = 0\,.$$

Apply L_{-1} to this equality, use that $L_{-1}Z = [L_{-1}, L_{n+1}]Z = 0$, and get

$$\left(2t_0\frac{\partial}{\partial t_0} + 1\right)\left(\frac{L_{n+1}Z}{Z}\right) = 0\,.$$

This means that $L_{n+1}Z/Z$ is homogeneous of degree $1/2$, which is impossible for a nonzero ratio of power series. Thus $L_{n+1}Z = 0$. Q.E.D.

4. The combinatorial identity.

In Chapters XVIII and XIX we discussed the cellular decomposition of moduli spaces of pointed curves in terms of ribbon graphs. We start by reviewing the part of that discussion which is relevant to our present purposes. The main tool is the cellular decomposition of $M_{g,P}$ based on the hyperbolic spine of a P-pointed genus g Riemann surface. Recall from Section 12 of Chapter XVIII (formula (12.11)) the homeomorphism

$$H : M_{g,P} \times \mathbb{R}_+^P \longrightarrow M_{g,P}^{comb}\,.$$

The space $M_{g,P}^{comb}$ is an orbicell complex which can be described as follows. The orbicells of $M_{g,P}^{comb}$ are labeled by isomorphism classes of P-marked ribbon graphs of genus g. Here and in the sequel, P is a finite set $P = \{p_1, \dots, p_n\}$, and we will use interchangeably the symbols $M_{g,P}$ and $M_{g,n}$, and likewise for the symbols $M_{g,P}^{comb}$ and $M_{g,n}^{comb}$.

We recall that a P-marked ribbon graph G can be abstractly thought of as a 4-tuple $G = (X(G), \sigma_0, \sigma_1, \mathbf{x})$, where $X(G)$ is the set of oriented edges of G, σ_1 is an involution on $X(G)$ whose orbits form the set $X_1(G)$ of edges G, σ_0 is a permutation of $X(G)$ whose orbits form the set $X_0(G)$ of vertices of G, while the orbits of $\sigma_2 = {\sigma_0}^{-1}\sigma_1$ form the set $X_2(G)$ of boundary components of G, and finally the marking $\mathbf{x}$ is an injection $\mathbf{x} : P \to X_0(G) \cup X_2(G)$ whose domain contains $X_2(G)$. Here we will consider, without further notice, only ribbon graphs such that the domain of $\mathbf{x}$ is equal to $X_2(G)$. As usual, we will also write x_p for $\mathbf{x}(p)$.

Given a P-marked ribbon graph G of genus g, we let $|a(G)|^0$ denote the open cell consisting of all possible assignments of a positive length l_e to each edge e of G, i.e., of a metric on G. If $N = |X_1(G)|$ is the number of edges of G, then $|a(G)|^0 = \mathbb{R}_+^N$. An orbicell of $M_{g,P}^{comb}$ is of the form $|a(G)|^0/\operatorname{Aut}(G)$, where $\operatorname{Aut}(G)$ is the automorphism group of the P-marked graph G. Of course, some of the faces of $|a(G)|$ also give rise to orbicells of $M_{g,P}^{comb}$. These faces are obtained by letting some of the lenghts of the edges of G tend to zero. But this can be done only under the condition that the resulting graph is still a P-marked ribbon graph of genus g. In particular, for every $p \in P$, the pth perimeter should always be positive; thus,

$$r_p = \sum_{\vec{e} \in x(p)} l_e > 0\,.$$

The homeomorphism h assigns to a point $[(C, \mathbf{x}, r)] \in M_{g,P} \times \mathbb{R}_+^P$ the isomorphism class of the metrized ribbon graph $G \subset C$ which is the hyperbolic spine of $(C, \mathbf{x}, r))$ (cf. Definition (3.6) of Chapter XVIII). We also recall from Section 7 of Chapter XIX that we denote by $M_{g,P}^{comb}(r)$ the fiber over r of the perimeter map

$$\lambda : M_{g,P}^{comb} \longrightarrow \mathbb{R}_+^P$$

and that, for every $r \in \mathbb{R}_+^P$, we have a homeomorphism

$$h : M_{g,P} \longrightarrow M_{g,P}^{comb}(r)\,.$$

As we explained in Section 7 of Chapter XIX, going around the pth boundary cycle in each ribbon graph defines a line bundle $\mathcal{L}_p^{comb}$ on $M_{g,P}^{comb}(r)$ whose pullback via h is $\mathcal{L}_p$, so that

$$\psi_{p|M_{g,P}} = c_1(\mathcal{L}_p) = h^*\left(c_1(\mathcal{L}_p^{comb})\right)\,.$$

On the other hand, we defined a degree two PL-form ω_p on $M_{g,P}^{comb}(r)$ representing the first Chern class of $\mathcal{L}_p^{comb}$ and whose restriction to the open cell $|a(G)|^0$ is given by

$$(4.1) \qquad (\omega_p)_{|_{|a(G)|^0}} = \sum_{1 \le i < j \le k-1} d\left(\frac{l_{e_i}}{r_p}\right) \wedge d\left(\frac{l_{e_j}}{r_p}\right),$$

where $\{\vec{e}_1, \dots, \vec{e}_k\} = x_p$, meaning that $e_1, \dots, e_k$ are the edges around the pth boundary component of G. We now recall Lemma(7.6) of Chapter XIX. According to this lemma, the 2-form ω_p extends to a 2-form defined on a compactification $\overline{M}_{g,P}^{comb}(r)$ of $M_{g,P}^{comb}(r)$, which we denote with the same symbol ω_p. Moreover, the homeomorphism h extends to a continuous map $f : \overline{M}_{g,P} \to \overline{M}^{comb}_{g,P}(r)$, and $f^*[\omega_p] = \psi_p$. Writing ω_i for ω_{p_i} and ψ_i for ψ_{p_i}, we get

$$< \tau_{d_1} \cdots \tau_{d_n} >= \int\limits_{\overline{M}_{g,n}} \psi_1^{d_1} \cdots \psi_n^{d_n} = \int\limits_{\overline{M}^{comb}_{g,n}(r)} \omega_1^{d_1} \cdots \omega_n^{d_n} = \int\limits_{M^{comb}_{g,n}(r)} \omega_1^{d_1} \cdots \omega_n^{d_n} .$$

(We write precisely these equalities as, in the previous chapter, we went through the trouble of compactifying $M^{comb}_{g,n}$.) Also observe that in the last integral we may as well take as domain of integration the open dense set which is the union of all cells $|a(G)|^0/\operatorname{Aut}(G)$, with G a *trivalent ribbon graph.*

We now recall the volume form $\Omega^d/d!$ on $M^{comb}_{g,n}(r)$ which we studied in Section 8 of Chapter XIX. Here $d = 3g - 3 + n$, and

$$\Omega = \sum_{p \in P} r_p^2 \omega_p$$

is a PL-form of degree 2 which is nondegenerate symplectic on the open strata of $M^{comb}_{g,n}(r)$. We have

$$\begin{aligned} \operatorname{vol}\left(M^{comb}_{g,n}(r)\right) &= \int\limits_{M^{comb}_{g,n}(r)} \frac{\Omega^d}{d!} = \frac{1}{d!} \int\limits_{M^{comb}_{g,n}(r)} \left(r_1^2\omega_1 + \cdots + r_n^2\omega_n\right)^d \\ &= \sum_{d_1+\cdots+d_n=d} \prod_{i=1} \frac{r_i^{d_i}}{d_i!} < \tau_{d_1} \cdots \tau_{d_n} > . \end{aligned} \tag{4.2}$$

Applying the Laplace transform to (4.2) gives

$$\begin{aligned} &\int\limits_{\mathbb{R}^n_{\geq 0}} e^{-\sum \lambda_i r_i} \int\limits_{M^{comb}_{g,n}(r)} \frac{\Omega^d}{d!} dr_1 \cdots dr_n \\ &\qquad = \sum_{d_1+\cdots+d_n=d} < \tau_{d_1} \cdots \tau_{d_n} > \prod_{i=1}^n \frac{1}{d_i!} \int_0^\infty r_i^{2d_i} e^{-\lambda_i r_i} dr_i \\ &\qquad = \sum_{d_1+\cdots+d_n=d} < \tau_{d_1} \cdots \tau_{d_n} > \prod_{i=1}^n \frac{2d_i!}{d_i!} \lambda_i^{-2(d_i+1)} , \end{aligned}$$

where $\lambda_1, \dots, \lambda_n$ are now complex variables with positive real part. Let $|a(G)|^0/\operatorname{Aut}(G)$ be an open top-dimensional orbicell of $M^{comb}_{g,n}$, and let $e_1, \dots, e_K$ be the edges of G, where $K = 6g - 6 + 3n$. Then Theorem (8.8) of the preceding chapter says that

$$\frac{\Omega^d}{d!} dr_1 \cdots dr_n = 2^{2n+5g-5} dl_{e_1} \cdots dl_{e_K}$$

on $|a(G)|^0$. Combining this with the previous identities, we get

$$
\begin{aligned}
&\sum_{d_1+\cdots+d_n=d} <\tau_{d_1}\cdots\tau_{d_n}> \prod_{i=1}^{n} \frac{2d_i!}{d_i!}\lambda_i^{-(2d_i+1)} \\
&= \int_{\mathbb{R}^n_{\geq 0}} e^{-\sum \lambda_i r_i} \int_{M^{comb}_{g,n}(r)} \frac{\Omega^d}{d!} dr_1\cdots dr_n \\
&= \int_{M^{comb}_{g,n}} e^{-\sum \lambda_i r_i} \frac{\Omega^d}{d!} dr_1\cdots dr_n \\
&= \sum_{G\in\mathcal{G}^3_{g,n}} \frac{2^{2n+5g-5}}{|\operatorname{Aut}(G)|} \int_{|a(G)|} e^{-\sum\lambda_i r_i} dl_{e_1}\cdots dl_{e_K},
\end{aligned} \tag{4.3}
$$

where $\mathcal{G}^{3,c}_{g,n}$ stands for the set of isomorphism classes of connected 3-valent P-marked ribbon graphs. Now, for each edge $e=[\vec{e}]_1\in X_1(G)$, we have two, possibly coincident, boundary components of G having e as one of their edges. These are, by definition, $[\vec{e}]_2$ and $[\sigma_1(\vec{e})]_2$. Think of $\lambda=(\lambda_1,\dots,\lambda_n)$ as a map $\lambda: X_2(G)\to\mathbb{C}^n$. Set

$$\lambda_e+\lambda_{e'}=\lambda([\vec{e}]_2)+\lambda([\sigma_1(\vec{e})]_2)\,.$$

Writing $r_i=\sum_{\vec{e}\in x(p_i)} l_e$ for $i=1,\dots,n$, we get $\sum\lambda_i r_i=\sum_{e\in X_1(G)}(\lambda_e+\lambda_{e'})l_e$, so that

$$\int_{|a(G)|} e^{-\sum\lambda_i r_i} dl_{e_1}\cdots dl_{e_K} = \prod_{e\in X_1(G)} \int_0^\infty e^{-l_e(\lambda_e+\lambda_{e'})} dl_e = \prod_{e\in X_1(G)} \frac{1}{\lambda_e+\lambda_{e'}}\,.$$

Observing that $(2d_i-1)!!=2^{-d_i}(2d_i)!/(d_i)!$, formula (4.3) becomes

$$
\begin{aligned}
&\sum_{d_1+\cdots+d_n=d} <\tau_{d_1}\cdots\tau_{d_n}> \prod_{i=1}^{n} \frac{(2d_i-1)!!}{\lambda_i^{2d_i+1}} \\
&\qquad = \sum_{G\in\mathcal{G}^{3,c}_{g,n}} \frac{2^{-|X_0(G)|}}{|\operatorname{Aut}(G)|} \prod_{e\in X_1(G)} \frac{2}{\lambda_e+\lambda_{e'}}\,.
\end{aligned} \tag{4.4}
$$

This is the main combinatorial identity that we will need. To recast it in a different form, we consider N-colored ribbon graphs, where N is a positive integer. An N-coloring of (the boundary components of) a graph G is simply a map

$$X_2(G)\to\{1,\dots,N\}\,.$$

When G is P-marked, we may view such a coloring as a map from P to $\{1,\dots,N\}$, and we write $j(i)=j(p_i)=j_i$. The main combinatorial identity now has the following fundamental consequence.

THEOREM (4.5). *Let*

$$F(t_0, t_1, \dots) = \sum_{d_1,\dots,d_n} \frac{1}{n!} < \tau_{d_1} \cdots \tau_{d_n} > t_{d_1} \cdots t_{d_n} .$$

Let $\Lambda_1, \dots, \Lambda_N$ *be complex variables with positive real part. Set* $\Lambda = \operatorname{diag}(\Lambda_1, \dots, \Lambda_N)$ *and*

$$t_i(\Lambda) = -(2i-1)!! \operatorname{Tr} \Lambda^{-(2i+1)} = -(2i-1)!! \left(\Lambda_1^{-(2i+1)} + \cdots + \Lambda_N^{-(2i+1)} \right).$$

Then

$$F(t_0(\Lambda), t_1(\Lambda) \dots) = \sum_{G \in \overline{\mathcal{G}}^{3,c}} \frac{(\sqrt{-1}/2)^{|X_0(G)|}}{|\operatorname{Aut}(G)|} \prod_{e \in X_1(G)} \frac{2}{\Lambda_e + \Lambda_{e'}}, \tag{4.6}$$

where $\overline{\mathcal{G}}^{3,c}$ *is the set of isomorphism classes of trivalent, connected ribbon graphs with* N*-colored boundary components.*

Proof.

$$F(t_0(\Lambda), t_1(\Lambda), \dots) = \sum_{\substack{n \geq 1 \\ d_1,\dots,d_n \geq 0}} \frac{1}{n!} < \tau_{d_1} \cdots \tau_{d_n} > t_{d_1}(\Lambda) \cdots t_{d_n}(\Lambda)$$

$$= \sum_{\substack{n \geq 1 \\ d_1,\dots,d_n \geq 0}} \frac{(-1)^n}{n!} < \tau_{d_1} \cdots \tau_{d_n} > \sum_{1 \leq j_1,\dots,j_n \leq N} \prod_{i=1}^{n} \frac{(2d_i - 1)!!}{\Lambda_{j_i}^{2d_i+1}} .$$

On the other hand, writing down (4.4) in the coordinates $\Lambda_{j_1}, \dots, \Lambda_{j_n}$ for each possible coloring j and then summing over all colorings yields

$$\begin{aligned} &\sum_{d_1,\dots,d_n \geq 0} < \tau_{d_1} \cdots \tau_{d_n} > \sum_{1 \leq j_1,\dots,j_n \leq N} \prod_{i=1}^{n} \frac{(2d_i - 1)!!}{\Lambda_{j_i}^{2d_i+1}} \\ &\qquad = \sum_{G \in \widehat{\mathcal{G}}_{g,n}^{3,c}} \frac{2^{-|X_0(G)|}}{|\operatorname{Aut}(G)|} \prod_{e \in X_1(G)} \frac{2}{\Lambda_e + \Lambda_{e'}}, \end{aligned} \tag{4.7}$$

where $\widehat{\mathcal{G}}_{g,n}^{3,c}$ is the set of isomorphism classes of trivalent connected genus g ribbon graphs with n marked boundary components which are also N-colored, and where

$$\Lambda_e + \Lambda_{e'} = \Lambda_{j([\vec{e}]_2)} + \Lambda_{j([\sigma_1(\vec{e})]_2)} . \tag{4.8}$$

Now the theorem follows immediately from (4.7) by the following considerations. First of all, we reinterpret the factor $1/n!$. Let G be an element of $\widehat{\mathcal{G}}_{g,n}^{3,c}$, and Γ the corresponding element of $\overline{\mathcal{G}}^{3,c}$. The elements of $\mathrm{Aut}(\Gamma)$ permute the boundary components, giving a homomorphism $\sigma : \mathrm{Aut}(\Gamma) \to \mathfrak{S}_n$ whose kernel is $\mathrm{Aut}(G)$. Moreover, $|\mathfrak{S}_n/\sigma(\mathrm{Aut}(\Gamma))|$ is equal to the number m_Γ of nonisomorphic elements of $\widehat{\mathcal{G}}^{3,c}$ whose corresponding unmarked graph is Γ. Thus,

$$n! = m_\Gamma \cdot \frac{|\,\mathrm{Aut}(\Gamma)|}{|\,\mathrm{Aut}(G)|}\,.$$

Next, $2|X_1(G)| = 3|X_0(G)|$ since G is trivalent. Combining this with $|X_0(G)| - |X_1(G)| + |X_2(G)| = 2 - 2g$ shows that $2n = 2|X_2(G)| \equiv |X_0(G)|$ modulo 4, and hence $(-1)^n = (\sqrt{-1})^{|X_0(G)|}$.

Q.E.D.

It is clear that we are interested in the case in which the number N of the variables $\Lambda_1, \dots \Lambda_N$ tends to infinity. In fact, the Newton polynomials $\mathrm{Tr}\,\Lambda^{-j}$ are independent functions of the Λ_i for i and j less than or equal to N.

As we shall see in the next section, the combinatorial formula for $F(t_0(\Lambda), t_1(\Lambda), \dots)$ will allow us to express the partition function $Z(t_0(\Lambda), t_1(\Lambda), \dots) = \exp F(t_0(\Lambda), t_1(\Lambda), \dots)$ as the asymptotic expansion of a matrix integral, and this will be the key step in proving the main theorem (2.5) or, which is the same, in checking equation (2.12).

5. Feynman diagrams and matrix models.

Before introducing matrix models, it is instructive to start from the classical example of the Gaussian measure on the real line $\mathbb{R}$, whose density is

$$d\mu_{\mathbb{R}} = \frac{1}{\sqrt{2\pi}} e^{\frac{-x^2}{2}} dx\,. \tag{5.1}$$

This measure has the property that

$$\mu_{\mathbb{R}}(\mathbb{R}) = \int_{\mathbb{R}} d\mu_{\mathbb{R}} = 1\,. \tag{5.2}$$

Given any real- or complex-valued function f on $\mathbb{R}$ the *expectation value* or the *average* of f with respect to the above Gaussian measure is by definition

$$< f >= \int_{\mathbb{R}} f d\mu_{\mathbb{R}}\,.$$

For instance, the so-called *mean* of the Gaussian measure, i.e., the expectation value of x, is equal to zero, while its *variance*, i.e., the expectation value of x^2, is equal to 1. Hence, the normalization property (5.2) and the last two properties can be written as

$$< 1 >= 1\,, \quad < x >= 0\,, \quad < x^2 >= 1\,.$$

The notion of Gaussian measure naturally generalizes to $\mathbb{R}^d$. We fix a real positive symmetric $d \times d$ matrix A, and we consider the measure

$$d\mu = e^{-\frac{1}{2}(Ax,x)}dx\,,$$

where $(\ ,\ )$ denotes the euclidean inner product. We have

$$\int_{\mathbb{R}^d} d\mu = \frac{(2\pi)^{d/2}}{\sqrt{detA}}\,.$$

We then normalize the measure μ and define

$$d\mu_{\mathbb{R}^d} = \frac{\sqrt{detA}}{(2\pi)^{d/2}}d\mu\,.$$

As before, we denote by $< f >$ the expectation value of a function $f : \mathbb{R}^d \to \mathbb{C}$ with respect to the above Gaussian measure; thus,

$$< f >= \int_{\mathbb{R}^d} f d\mu_{\mathbb{R}^d}\,.$$

The basic tool in computing expectation values with respect to the Gaussian measure is Wick's lemma, which gives the expectation values of monomials in the variables $x_1, \ldots, x_d$.

WICK'S LEMMA (5.3). *The following relations hold:*

1. $< x_{\nu_1} x_{\nu_2} \cdots x_{\nu_n} >= 0$ *if n is odd,*
2. $< x_{\nu_1} x_{\nu_2} >= (A^{-1})_{\nu_1 \nu_2}\,,$
3. $< x_{\nu_1} x_{\nu_2} \cdots x_{\nu_{2n}} >= \sum_P < x_{\nu_{s_1}} x_{\nu_{s_2}} > \cdots < x_{\nu_{s_{2n-1}}} x_{\nu_{s_{2n}}} >,$

where P is the set of all distinct pairings, i.e., decompositions

$$\{1, \ldots, 2n\} = \{s_1, s_2\} \cup \cdots \cup \{s_{2n-1}, s_{2n}\}\,.$$

The numbers $< x_{\nu_1} x_{\nu_2} >= (A^{-1})_{\nu_1 \nu_2}$ are called the propagators of the measure $d\mu_{\mathbb{R}^d}$.

Proof. Set $x = (x_1, \ldots, x_d)$. Since A is a symmetric matrix, passing from x to $-x$ gives the first assertion. We next consider the expression $< e^{t(y,x)} >$, where t is a real parameter, and $y = (y_1, \ldots, y_d)$ is a nonzero vector. Substituting x with $t(A^{-1}y) - x$, one gets

$$< e^{t(y,x)} >= e^{\frac{t^2}{2}(A^{-1}y,y)}\,. \tag{5.4}$$

Comparing the coefficients of t^2 of both sides of this equation gives the second assertion. We finally prove the last claim. Look at the coefficient of t^{2n} in (5.4). From the right-hand side one gets

$$\frac{1}{(2n)!} \sum_{\nu_1, \ldots, \nu_{2n}} y_{\nu_1} y_{\nu_2} \cdots y_{\nu_{2n}} < x_{\nu_1} x_{\nu_2} \cdots x_{\nu_{2n}} >\,. \tag{5.5}$$

On the other hand, the coefficient of t^{2n} of the left-hand side of (5.4) is

$$\begin{aligned}(5.6)\qquad &\frac{1}{2^n n!}(A^{-1}y,y)^n \\ &= \frac{1}{2^n n!}\sum_{\nu_1,\dots,\nu_{2n}} y_{\nu_1}y_{\nu_2}\cdots y_{\nu_{2n}} < x_{\nu_1}x_{\nu_2} > \cdots < x_{\nu_{2n-1}}x_{\nu_{2n}} > .\end{aligned}$$

Observe that the expectation $< x_{\nu_1}x_{\nu_2}\cdots x_{\nu_{2n}} >$ is invariant with respect to the action of $\mathfrak{S}_{2n}$ permuting the factors of the product $x_{\nu_1}x_{\nu_2}\cdots x_{\nu_{2n}}$. It follows that the coefficient of $y_{\nu_1}y_{\nu_2}\cdots y_{\nu_{2n}}$ in (5.5) is equal to

$$\frac{1}{(2n)!}|\mathfrak{S}_{2n}/H|^{-1}\cdot < x_{\nu_1}x_{\nu_2}\cdots x_{\nu_{2n}} >= |H|^{-1}\cdot < x_{\nu_1}x_{\nu_2}\cdots x_{\nu_{2n}} >,$$

where H is the subgroup consisting of those elements $\sigma \in \mathfrak{S}_{2n}$ having the property that $\nu_{\sigma(i)} = \nu_i$ for every $i = 1,\dots,2n$. On the other hand, the coefficient of $y_{\nu_1}y_{\nu_2}\cdots y_{\nu_{2n}}$ in (5.6) is

$$\frac{1}{2^n n!}\sum_{[\tau]\in\mathfrak{S}_{2n}/H} < x_{\nu_{\tau(1)}}x_{\nu_{\tau(2)}} > \cdots < x_{\nu_{\tau(2n-1)}}x_{\nu_{\tau(2n)}} >,$$

where $[\tau]$ denotes the class of τ in $\mathfrak{S}_{2n}/H$. In conclusion, we have

$$< x_{\nu_1}x_{\nu_2}\cdots x_{\nu_{2n}} >= \frac{1}{2^n n!}\sum_{\tau\in\mathfrak{S}_{2n}} < x_{\nu_{\tau(1)}}x_{\nu_{\tau(2)}} > \cdots < x_{\nu_{\tau(2n-1)}}x_{\nu_{\tau(2n)}} > .$$

Observe finally that the group $\mathfrak{S}_{2n}$ acts transitively on the set P of pairings, with stabilizer isomorphic to $(\mathbb{Z}_2)^n \times \mathfrak{S}_n$. The third claim of the lemma follows from this. Q.E.D.

We now wish to compute, in a particular example, the expectation value of a function f which is not a monomial in the x_i. This will bring into the picture the concept of asymptotic expansion. The example we have in mind is the one of the function

$$(5.7)\qquad f = e^{\frac{t}{3!}x^3},$$

so that $d = 1$. But first we need a definition.

DEFINITION (5.8). *Given a holomorphic function f defined on a domain $D \subset \mathbb{C}$ with $0 \in \partial D$, one says that a formal power $\sum_{i=0}^{\infty} a_i t^i$ is an asymptotic expansion of f at 0, and one writes*

$$(5.9)\qquad f(t) \underset{t\to 0}{\asymp} \sum_{i=0}^{\infty} a_i t^i$$

if

$$\lim_{\substack{t\to 0\\ t\in D}} \frac{f(t) - \sum_{i=0}^{n} a_i t^i}{t^{n+1}} = a_{n+1}$$

for all $n \geq 0$.

It follows from the definition that if an asymptotic expansion exists, it is unique. Obviously, if $f(t)$ is holomorphic in a neighborhood of 0, then its Taylor series at 0 is the asymptotic expansion of f at 0. To give an example in the nonholomorphic case, we consider the function $e^{\frac{1}{t}}$. Setting $D = \{t \in \mathbb{C} : \mathrm{Re}(t) < 0\}$, one sees that

$$e^{\frac{1}{t}} \underset{t\to 0}{\asymp} 0 .$$

From the definition two important facts follow. If

$$f(t) \underset{t\to 0}{\asymp} \sum_{i=0}^{\infty} a_i t^i , \qquad g(t) \underset{t\to 0}{\asymp} \sum_{i=0}^{\infty} b_i t^i ,$$

then

$$f(t)+g(t) \underset{t\to 0}{\asymp} \sum_{i=0}^{\infty} a_i t^i + \sum_{i=0}^{\infty} b_i t^i , \qquad f(t)g(t) \underset{t\to 0}{\asymp} \left(\sum_{i=0}^{\infty} a_i t^i\right)\left(\sum_{i=0}^{\infty} b_i t^i\right).$$

It also follows that if

$$L = a_n(t)\frac{d^n}{dt^n} + \cdots + a_0(t) \tag{5.10}$$

is a differential operator with, say, polynomial coefficients, then

$$L(f) \underset{t\to 0}{\asymp} L\left(\sum_{i=0}^{\infty} a_i t^i\right).$$

A consequence that will be relevant for our purposes is the following.

LEMMA (5.11). *Let f, D, and $\sum_{i=0}^{\infty} a_i t^i$ be as in Definition (5.8), so that*

$$f(t) \underset{t\to 0}{\asymp} \sum_{i=0}^{\infty} a_i t^i .$$

Let L be a differential operator as in (5.10). Assume that $L(f) = 0$. Then $L\left(\sum_{i=0}^{\infty} a_i t^i\right) = 0$.

To give a very quick idea of how we will use this lemma, let us go back to our partition function $Z(t_\bullet(\Lambda)) = \exp F(t_\bullet(\Lambda))$. Via the theory of matrix models, we will be able to find a holomorphic function $H(\Lambda)$, defined in a domain containing ∞ in its boundary, and to show that

$$H(\Lambda) \underset{\Lambda^{-1}\to 0}{\asymp} Z(t_\bullet(\Lambda)) .$$

We will then proceed to prove directly that for the matrix model H, the identity $L_2(H) = 0$ holds.

Let us return to the expectation value of the function (5.7) for the standard Gaussian measure (5.1) on the real line. The elementary but basic remark is that, although the Taylor series in t for the exponential $e^{\frac{t}{3!}x^3}$ is not uniformly convergent on the real line, we can still interchange the summation and integral signs in the asymptotic sense:

(5.12)

$$< e^{\frac{t}{3!}x^3} > = \frac{1}{\sqrt{2\pi}} \int_{\mathbb{R}} e^{\frac{t}{3!}x^3} e^{-\frac{x^2}{2}} dx \underset{t \to 0}{\asymp} \sum_{i=0}^{\infty} \left(\frac{1}{\sqrt{2\pi}} \int_{\mathbb{R}} \frac{x^{3i}}{i!(3!)^i} e^{-\frac{x^2}{2}} dx \right) t^i$$
$$= \sum_{i=0}^{\infty} \frac{1}{i!(3!)^i} < x^{3i} > t^i .$$

The verification of this asymptotic equality is immediate. In our case the limit of the quotient (5.9) is

$$\lim_{t \to 0} \left(\frac{1}{\sqrt{2\pi}} \int_{\mathbb{R}} \left(\sum_{i=n+1}^{\infty} \frac{x^{3i}}{i!(3!)^i} t^{i-n+1} \right) e^{-\frac{x^2}{2}} dx \right) ,$$

and, as t tends to zero in a small sector having 0 as its vertex, the integral converges, and the above limit is $(1/(n+1)!(3!)^{n+1}) < x^{3(n+1)} >$.

We now use Wick's lemma, and we get

$$< e^{t\frac{x^3}{3!}} > \asymp \sum_{i=0}^{\infty} \frac{1}{i!(3!)^i} < x^{3i} > t^i = \sum_{i=0}^{\infty} \frac{|P_{3i}|}{i!(3!)^i} t^i ,$$

where P_{3i} is the set of all distinct pairings of $\{1, \dots, 3i\}$. The basic idea now is to express the coefficient $\frac{|P_{3i}|}{i!(3!)^i}$ in terms of Feynman diagram. Denote by $\mathcal{F}_{i,3}$ the set of isomorphism classes of graphs (the Feynman diagrams) with i vertices, all of which are trivalent. Here by graph we just mean a finite one-dimensional simplicial complex with no isolated vertices. We next associate to each element in P_{3i} an element of $\mathcal{F}_{i,3}$. Fix i points. At each of these points draw three half-edges originating from it and number the $3i$ half-edges so obtained in an arbitrary way. Each element of P_{3i} gives a way of joining the half-edges to get an element in $\mathcal{F}_{i,3}$.

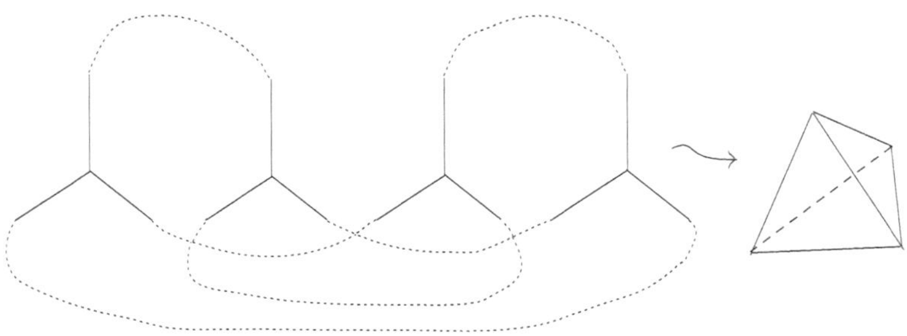

Figure 1. Case $i = 4$

Permuting the vertices, or permuting the half-edges at each vertex, does not change the isomorphism class of the graph constructed above. To be precise, we are looking at the action of the group $\mathfrak{S}_i \times (\mathfrak{S}_3)^i$ on P_{3i}. The number of orbits of this action is $|\mathcal{F}_{i,3}|$. Identify an element $\gamma \in P_{3i}$ with a graph Γ as we just explained. Via this identification, the stabilizer of γ under the action of $\mathfrak{S}_i \times (\mathfrak{S}_3)^i$ can be thought of as the automorphism group of Γ. Since $|\mathfrak{S}_i \times (\mathfrak{S}_3)^i| = i!(3!)^i$ (and we now see why we divided the variable t by 3!), we have

$$< x^{3i} >= \frac{|P_{3i}|}{i!(3!)^i} = \sum_{\Gamma \in \mathcal{F}_{i,3}} \frac{1}{|\operatorname{Aut}(\Gamma)|}. \tag{5.13}$$

In conclusion we have

$$< \exp(\tfrac{t}{3!}x^3) > \underset{t \to 0}{\asymp} \sum_{i=0}^{\infty} \sum_{\Gamma \in \mathcal{F}_{i,3}} \frac{1}{|\operatorname{Aut}(\Gamma)|} t^i . \tag{5.14}$$

The right-hand side of (5.14) bears some similarity to the combinatorial expression of the function $F(t_\bullet(\Lambda))$ given in Theorem (4.5). What we will do next is to express that combinatorial expression as an expectation value relative to some Gaussian measure. For this, we need to introduce matrix models.

In what follows we will have to use asymptotic expansions in more than one variable. Let f be a holomorphic function on a product S of sectors with vertex at the origin (a *polysector*) in $\mathbb{C}^n$. Consider a formal power series $\sum_I a_I z^I$, where $z = (z_1, \ldots, z_n)$ and $I = (i_1, \ldots, i_n)$ with the i_j nonnegative integers. We will say that $\sum_I a_I z^I$ is an *asymptotic expansion of* f *at* 0, and we write

$$f \underset{z \to 0}{\asymp} \sum_I a_I z^I$$

if, for every bounded polysector T whose closure is contained in $S \cup \{0\}$ and every $n \geq 0$, there is a constant C such that

$$\left| f(z) - \sum_{|I| \leq n} a_I z^I \right| \leq C \|z\|^{n+1}, \qquad z \in T .$$

Asymptotic expansions in several variables enjoy the analogues of all the properties of asymptotic expansions in one variable, with one notable exception, namely that in general one cannot differentiate. In fact, there are functions which have an asymptotic expansion but whose derivatives do not [362]. However, if f has the additional property that its derivatives are bounded on each bounded polysector T whose closure is contained in $S \cup \{0\}$, then differentiation is possible. In fact, for any differential operator L with constant coefficients,

$$Lf \underset{z\to 0}{\asymp} L\left(\sum_I a_I z^I\right). \tag{5.15}$$

All the functions we will consider will enjoy the extra condition, so that one can freely use (5.15).

Let us denote by $\mathcal{H}_N$ the space of $N \times N$ hermitian matrices. Set

$$dX = \prod_i dX_{ii} \prod_{i<j} d\,\mathrm{Re}(X_{ij})\, d\,\mathrm{Im}(X_{ij}).$$

The role played in the Gaussian measure on $\mathbb{R}^d$ by the inner product (Ax, y) is now going to be played by the inner product on $\mathcal{H}_N$ defined by

$$(X,Y)_N = \mathrm{Tr}(XY), \quad X, Y \in \mathcal{H}_N.$$

The corresponding Gaussian measure is given by

$$d\mu_N = \frac{1}{c_N} e^{-\frac{1}{2}\mathrm{Tr}(X^2)} dX, \tag{5.16}$$

where

$$c_N = \int_{\mathcal{H}_N} e^{-\frac{1}{2}\mathrm{Tr}(X^2)} dX.$$

The most important feature of this Gaussian measure is its invariance under the unitary group $U(N)$. Since, in general, one is interested in expectation values that can be expressed in terms of eigenvalues of matrices in $\mathcal{H}_N$, one considers integrals

$$< P >= \int_{\mathcal{H}_N} P(X) d\mu_N, \tag{5.17}$$

where P is a $U(N)$-invariant function in the sense that

$$P(gXg^{-1}) = P(X) \quad \text{for } g \in U(N).$$

Typically, one takes as P a function of the traces of powers of X. Let us give right away an example of such a computation. In fact, let us consider, in the context of matrix integrals, the example which is the direct analogue of (5.7), that is, let us compute

$$< e^{\frac{t}{3}\mathrm{Tr}\, X^3} > . \tag{5.18}$$

It will soon become clear why, in this case, we are dividing by 3 and not by 3!. Also in this case we have the asymptotic expansion

$$(5.19) \qquad < e^{\frac{t}{3}\operatorname{Tr} X^3} > \underset{t\to 0}{\asymp} \sum_{i=0}^{\infty} \frac{1}{i!3^i} < (\operatorname{Tr} X^3)^i > t^i \,.$$

We must then compute the terms $< (\operatorname{Tr} X^3)^i >$. It is at this point that ribbon graphs will appear. By Wick's lemma the propagators of the Gaussian measure $d\mu_N$ are given by

$$(5.20) \qquad < x_{ij}x_{kl} >= \delta_{il}\delta_{jk} \,.$$

To avoid unnecessary notational complications, let us look at the coefficient of t^2 in (5.19), which is already an interesting one. Using Wick's lemma, we have

$$\begin{aligned} < \operatorname{Tr}(X^3)^2 > &= \sum_{\substack{i_1,i_2,i_3=1\\ j_1,j_2,j_3=1}}^{N} < x_{i_1i_2}x_{i_2i_3}x_{i_3i_1}x_{j_1j_2}x_{j_2j_3}x_{j_3j_1} > \\ &= \sum < x_{i_1i_2}x_{i_2i_3} >< x_{i_3i_1}x_{j_1j_2} >< x_{j_2j_3}x_{j_3j_1} > \\ &\quad + \sum < x_{i_1i_2}x_{j_1j_2} >< x_{i_3i_1}x_{j_2j_3} >< x_{i_2i_3}x_{j_3j_1} > \\ &\quad + \sum < x_{i_1i_2}x_{j_1j_2} >< x_{i_3i_1}x_{j_3j_1} >< x_{i_2i_3}x_{j_2j_3} > + \cdots \,, \end{aligned}$$

where we have written only three of fifteen summations. Look, for example, at the last summation. For each summand, say $< x_{i_1i_2}x_{j_1j_2} >< x_{i_3i_1}x_{j_2j_3} >< x_{i_2i_3}x_{j_3j_1} >$, draw two vertices with three half ribbon edges stemming from each of them. Label the first triple of half ribbon edges with (i_1i_2), (i_2i_3), (i_3i_1), and the second with (j_1j_2), (j_2j_3), (j_3j_1) then join the half ribbon edge (st) with (kl) if and only if $< x_{st}x_{kl} >= 1$. Figure 2 shows the graph corrresponding to the term $< x_{i_1i_2}x_{i_2i_3} >< x_{i_3i_1}x_{j_1j_2} >< x_{j_2j_3}x_{j_3j_1} >$.

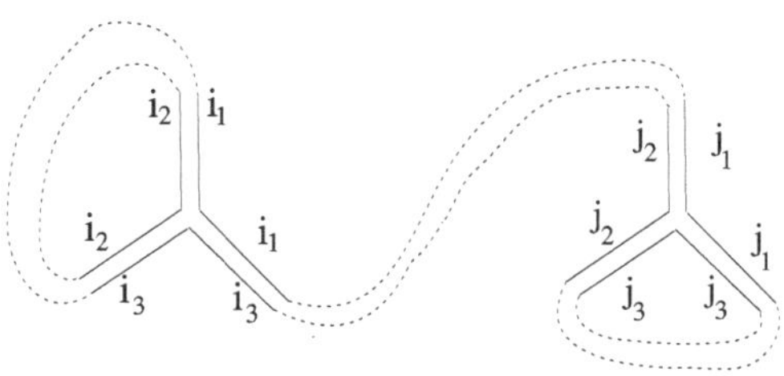

Figure 2.

We get the ribbon graphs in Figure 3, of which two are of genus 0, and one of genus 1.

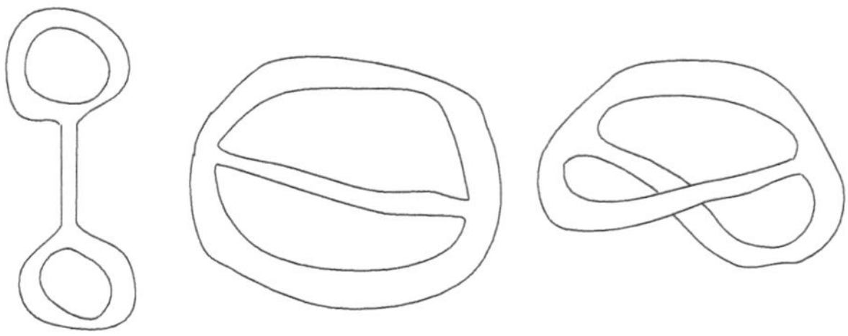

A B C

Figure 3.

The boundary components of these ribbon graphs are naturally colored with colors in the set $\{1, \ldots, N\}$. For instance, looking at Figure 2, we see that the three boundary components are colored with the three colors $(i_2,\ i_1 = j_1 = j_2 = i_3,\ j_3)$. In the sum above, nine summands give a graph of type A, three a graph of type B, and three a graph of type C. Since each of the above ribbon graphs is colored, we get $< \mathrm{Tr}(X^3)^2 >= 12N^3 + 3N$. The coefficient of t^2 in the asymptotic expansion of (5.18) is equal to $\frac{<\mathrm{Tr}(X^3)^2>}{2!3^2} = \frac{N^3}{2} + \frac{N^3}{6} + \frac{N}{6}$. Observe that the automorphism group of a graph of type A is cyclic of order 2 while the automorphism group of a graph of type B or C is $\mathfrak{S}_3$. So we can write

$$\frac{< \mathrm{Tr}(X^3)^2 >}{2!3^2} = \sum_{G \in \overline{\mathcal{G}}^3(2)} \frac{1}{|\mathrm{Aut}(G)|},$$

where $\overline{\mathcal{G}^3(2)}$ denotes the set of isomorphism classes of N-colored trivalent ribbon graphs with 2 vertices. For the general term $\frac{<\mathrm{Tr}(X^3)^i>}{i!3^i}$ of the asymptotic expansion (5.18), we proceed as we did for (5.13). The only difference is that, since we want to preserve the cyclic order of the edges around a given vertex, we use the cyclic group C_3 of order 3 instead of the symmetric group $\mathfrak{S}_3$ (this is why in (5.18) we divided by 3 and not by 3!). We then look at the action on P_{3i} of $\mathfrak{S}_i \times (C_3)^i$, and we get

$$(5.21)\quad < e^{\frac{t}{3}\mathrm{Tr}\, X^3} >= \sum_{i=0}^{\infty} \left(\sum_{G \in \overline{\mathcal{G}}^3(i)} \frac{1}{|\mathrm{Aut}(G)|} \right) t^i = \sum_{G \in \overline{\mathcal{G}}^3} \frac{1}{|\mathrm{Aut}(G)|} t^{|X_0(G)|},$$

where $\overline{\mathcal{G}^3}$ denotes the isomorphism classes of N-colored trivalent ribbon graphs, and $\mathcal{G}^3(i)$ denotes the isomorphism classes of N-colored trivalent ribbon graphs with i vertices.

In the sum above we are summing over all, *possibly disconnected,* N-colored trivalent ribbon graphs. Now there is a remarkable and elementary relation between a sum over connected ribbon graphs and a sum over possibly disconnected ribbon graphs, based on the observation that, if a ribbon graph G is the disjoint union of n_1 copies of G_1, n_2 copies of

G_2, ..., n_h copies of G_h, where $G_1, \ldots, G_h$ are nonisomorphic connected graphs, then

$$\operatorname{Aut}(G) = \prod_{i=1}^{h} (\mathfrak{S}_{n_i} \times \operatorname{Aut}(G_i)^{n_i}) ,$$

so that

$$\frac{1}{|\operatorname{Aut}(G)|} = \prod_{i=1}^{h} \frac{1}{n_i!} \frac{1}{|\operatorname{Aut}(G_i)|^{n_i}} .$$

The elementary relation we were alluding to, whose proof is left to the reader as an exercise, is the following:

$$(5.22) \qquad \sum_{G \in \overline{\mathcal{G}}^3} \frac{1}{|\operatorname{Aut}(G)|} t^{|X_0(G)|} = \exp\left(\sum_{G \in \overline{\mathcal{G}}^{3,c}} \frac{1}{|\operatorname{Aut}(G)|} t^{|X_0(G)|} \right) .$$

The asymptotic expansion (5.21) now looks closer to the expression of the series $F(t_\bullet(\Lambda)$ given in Theorem (4.5). To be able to express $F(t_\bullet(\Lambda))$ as an expectation value, Kontsevich introduces the following ad hoc matrix model. Let $\Lambda = \operatorname{diag}(\Lambda_1, \ldots, \Lambda_N)$ be a diagonal matrix whose entries are complex numbers with positive real part. Set

$$(5.23) \qquad d\mu_{\Lambda,N} = \frac{1}{c_{\Lambda,N}} e^{-\frac{1}{2} \operatorname{Tr} \Lambda X^2} dX ,$$

where the normalization factor $c_{\Lambda,N}$ is defined as

$$(5.24) \qquad c_{\Lambda,N} = \int_{\mathcal{H}_N} e^{-\frac{1}{2} \operatorname{Tr} \Lambda X^2} dX = \prod_{i,j=1}^{N} (\Lambda_i + \Lambda_j)^{-\frac{1}{2}} .$$

We also define expectation values

$$< P(X) >_{\Lambda,N} = \int_{\mathcal{H}_N} P(X) d\mu_{\Lambda,N} .$$

Now recall formula (4.6), which we rewrite here:

$$(5.25) \qquad F(t_0(\Lambda), t_1(\Lambda), \ldots) = \sum_{G \in \overline{\mathcal{G}}^{3,c}} \frac{(\sqrt{-1}/2)^{|X_0(G)|}}{|\operatorname{Aut}(G)|} \prod_{e \in X_1(G)} \frac{2}{\Lambda_e + \Lambda_{e'}} .$$

The reason why the measure $d\mu_{\Lambda,N}$ is a natural choice, with respect to this formula, is that its propagators are given by

$$(5.26) \qquad < x_{ij}, x_{kl} >_{\Lambda,N} = \delta_{il} \delta_{jk} \frac{2}{\Lambda_i + \Lambda_j} .$$

This is an easy computation which is left to the reader. Another remark is that we have an asymptotic expansion

$$(5.27)\qquad < e^{\frac{\sqrt{-1}}{6}\operatorname{Tr}X^3} >_{\Lambda,N} \underset{\Lambda^{-1}\to 0}{\asymp} \sum_{i=0}^{\infty} (\sqrt{-1}/2)^i \frac{1}{i!3^i} < (\operatorname{Tr}X^3)^i >_{\Lambda,N} .$$

This asymptotic expansion is proved in the same way that (5.12) and (5.19) were proved. In fact, one can reduce this computation to those by making the change of variables $X \mapsto \sqrt{\Lambda}X$. We now proceed exactly as in the proof of (5.21), and we get

$$(5.28)\quad < e^{\frac{\sqrt{-1}}{6}\operatorname{Tr}X^3} >_{\Lambda,N} \underset{\Lambda^{-1}\to 0}{\asymp} \sum_{G\in\overline{\mathcal{G}}^3} \frac{(\sqrt{-1}/2)^{|X_0(G)|}}{|\operatorname{Aut}(G)|} \prod_{e\in X_1(G)} \frac{2}{\Lambda_e+\Lambda_{e'}} .$$

The only difference between the right-hand side of this asymptotic equation and the right-hand side of (5.25) is the fact that here we are summing over all graphs, while there we confine ourselves to connected ones. But formula (5.22) tells us what to do, and we finally get

$$< e^{\frac{\sqrt{-1}}{6}\operatorname{Tr}X^3} >_{\Lambda,N} \underset{\Lambda^{-1}\to 0}{\asymp} \exp\left(F(t_0(\Lambda), t_1(\Lambda), \dots)\right) .$$

In conclusion, we proved the following:

THEOREM (5.29). *Let*

$$F(t_0,t_1,\dots) = \sum_{d_1,\dots d_n} \frac{1}{n!} < \tau_{d_1}\cdots\tau_{d_n} > t_{d_1}\cdots t_{d_n} .$$

Set $Z(t_0,t_1,\dots) = \exp F(t_0,t_1,\dots)$. *Let* $\Lambda_1,\dots,\Lambda_N$ *be complex variables with positive real part. Set* $\Lambda = \operatorname{diag}(\Lambda_1,\dots,\Lambda_N)$ *and*

$$t_i(\Lambda) = -(2i-1)!!\operatorname{Tr}\Lambda^{-(2i+1)} = -(2i-1)!!\left(\Lambda_1^{-(2i+1)} + \dots + \Lambda_N^{-(2i+1)}\right) .$$

Then

$$(5.30)\qquad \int_{\mathcal{H}_N} e^{\frac{\sqrt{-1}}{6}\operatorname{Tr}X^3} d\mu_{\Lambda,N} \underset{\Lambda^{-1}\to 0}{\asymp} Z(t_0(\Lambda), t_1(\Lambda), \dots),$$

where

$$d\mu_{\Lambda,N} = \frac{e^{-\frac{1}{2}\operatorname{Tr}\Lambda X^2}dX}{\int_{\mathcal{H}_N} e^{-\frac{1}{2}\operatorname{Tr}\Lambda X^2}dX} .$$

In the next section we will use this asymptotic equality to prove the identity $L_2Z = 0$. This will complete the proof of the main theorem (2.5).

6. Kontsevich's matrix model and the equation $L_2 Z = 0$.

To prove the identity $L_2 Z = 0$, we are going to use a theorem of Di Francesco, Itzykson, and Zuber which we are now going to state. The general problem that these authors address is the following. We view the expectation value

$$
(6.1)\qquad
\begin{aligned}
< e^{\frac{\sqrt{-1}}{6}\operatorname{Tr} X^3} >_{\Lambda,N} &= \int_{\mathcal{H}_N} e^{\frac{\sqrt{-1}}{6}\operatorname{Tr} X^3} d\mu_{\Lambda,N} \\
&= \frac{\int_{\mathcal{H}_N} e^{\frac{\sqrt{-1}}{6}\operatorname{Tr} X^3} e^{-\frac{1}{2}\operatorname{Tr}\Lambda X^2} dX}{\int_{\mathcal{H}_N} e^{-\frac{1}{2}\operatorname{Tr}\Lambda X^2} dX}
\end{aligned}
$$

in Kontsevich's matrix model as a function of the variables $t_0, t_1, \ldots$, where

$$t_i = -(2i-1)!! \operatorname{Tr} \Lambda^{-(2i+1)} .$$

Now consider any differential operator with constant coefficients

$$
(6.2)\qquad D \in \mathbb{C}\left[\frac{\partial}{\partial t_0}, \frac{\partial}{\partial t_1}, \ldots\right] .
$$

The problem is to express

$$D < e^{\frac{\sqrt{-1}}{6}\operatorname{Tr} X^3} >_{\Lambda,N}$$

as a matrix integral. The main theorem in [168] says that this is possible.

THEOREM (6.3). *There is a 1–1 correspondence*

$$
\begin{aligned}
\mathbb{C}\left[\frac{\partial}{\partial t_0}, \frac{\partial}{\partial t_1}, \frac{\partial}{\partial t_2}, \ldots\right] &\to \mathbb{C}[\operatorname{Tr} X, \operatorname{Tr} X^3, \operatorname{Tr} X^5, \ldots] \\
D\left(\tfrac{\partial}{\partial t_0}, \tfrac{\partial}{\partial t_1}, \tfrac{\partial}{\partial t_2}, \ldots\right) &\mapsto P_D(\operatorname{Tr} X, \operatorname{Tr} X^3, \operatorname{Tr} X^5, \ldots)
\end{aligned}
$$

such that

$$D < e^{\frac{\sqrt{-1}}{6}\operatorname{Tr} X^3} >_{\Lambda,N} = < P_D \cdot e^{\frac{\sqrt{-1}}{6}\operatorname{Tr} X^3} >_{\Lambda,N} .$$

The proof of this theorem gives also a constructive procedure for finding the polynomial P_D, once the differential operator D is given. For example, one gets the following formulae, where we set

$\ll f \gg_{\Lambda,N} = < f e^{\frac{\sqrt{-1}}{6}\operatorname{Tr} X^3} >_{\Lambda,N}.$

$$
\begin{aligned}
\frac{\partial}{\partial t_0} < e^{\frac{\sqrt{-1}}{6}\operatorname{Tr} X^3} >_{\Lambda,N} &= \ll \operatorname{Tr} X \gg_{\Lambda,N}\,, \\
\frac{\partial}{\partial t_1} < e^{\frac{\sqrt{-1}}{6}\operatorname{Tr} X^3} >_{\Lambda,N} &= \ll \tfrac{1}{12}\operatorname{Tr} X^3 + \tfrac{1}{24} \gg_{\Lambda,N}\,, \\
\frac{\partial}{\partial t_2} < e^{\frac{\sqrt{-1}}{6}\operatorname{Tr} X^3} >_{\Lambda,N} &= \ll \tfrac{1}{15\cdot 16}\operatorname{Tr} X^5 + \tfrac{1}{24}(\operatorname{Tr} X)^2 \gg_{\Lambda,N}\,, \\
\frac{\partial^2}{\partial t_0^2} < e^{\frac{\sqrt{-1}}{6}\operatorname{Tr} X^3} >_{\Lambda,N} &= \ll (\operatorname{Tr} X)^2 \gg_{\Lambda,N}\,, \\
\frac{\partial^2}{\partial t_0 \partial t_1} < e^{\frac{\sqrt{-1}}{6}\operatorname{Tr} X^3} >_{\Lambda,N} &= \ll \tfrac{1}{12}\operatorname{Tr} X^3 \operatorname{Tr} X + \tfrac{13}{24}\operatorname{Tr} X \gg_{\Lambda,N}\,, \\
\frac{\partial}{\partial t_3} < e^{\frac{\sqrt{-1}}{6}\operatorname{Tr} X^3} >_{\Lambda,N} &= \ll \tfrac{1}{5\cdot 3\cdot 7\cdot 64}\operatorname{Tr} X^7 + \tfrac{1}{5\cdot 3\cdot 16}\operatorname{Tr} X^3 \operatorname{Tr} X\,, \\
&\quad + \tfrac{1}{5\cdot 3\cdot 4}\operatorname{Tr} X \gg_{\Lambda,N}\,.
\end{aligned} \tag{6.4}
$$

The proof of Theorem (6.3) is completely algebraic. It uses the theory of Schur characters and a remarkable formula of Harish-Chandra for integrals over symmetric domains. Since the proof in [168] is very detailed and clear, there is no point in reproducing it here.

Now we can proceed to the proof of the identity $L_2 Z = 0$. To do this, we will closely follow [685]. Let us remark that, in principle, all the equations $L_n Z = 0$ of the Virasoro list in Theorem (2.5) could in fact be deduced from Theorem (6.3). We can also anticipate that for $L_{-1}Z = 0$, one needs only the first equation in (6.4), for $L_0 Z = 0$, only the second, for $L_2 Z = 0$, the first, the third, and the forth, and, finally for $L_3 Z = 0$, one needs all the equations in (6.4). To give an idea of these computations, we will first verify the string equation $L_{-1}Z = 0$ and then we will go directly to the equation $L_2 Z = 0$.

To simplify the notation, it will pay to make the following change of variables. We set

$$Y = -\sqrt{-1}X\,, \quad \Theta = -\Lambda\,, \quad \Theta = \operatorname{diag}(\theta_1, \dots, \theta_N)\,.$$

We also set

$$W = \prod_{i,j}(-\theta_i - \theta_j)^{-1/2}\,, \quad I = -\tfrac{1}{6}\operatorname{Tr} Y^3 + \tfrac{1}{2}\operatorname{Tr}\Theta Y^2\,.$$

Of course, we think of $Z = Z(t_\bullet)$ as a function of Θ by setting

$$t_i = (2i-1)!!\operatorname{Tr}\Theta^{-(2i+1)}\,. \tag{6.5}$$

Recalling (5.24), the function Z satisfies the (asymptotic) equality

$$W \cdot Z = \int_{\mathcal{H}_N^a} e^{-I} dY\,, \tag{6.6}$$

where $\mathcal{H}_N^a$ is the space of anti-hermitian matrices, and Θ is a strictly negative diagonal matrix. In particular, the integral on the right-hand side of (6.6) is strongly convergent and differentiation under the integral sign is allowed. Let us reprove the string equation $L_{-1}Z = 0$, that is,

$$\left(-\frac{\partial}{\partial t_0} + \sum_{i=1}^{\infty} t_i \frac{\partial}{\partial t_{i-1}} + \frac{t_0^2}{2}\right) Z = 0\,. \tag{6.7}$$

We act on both sides of (6.6) with the Euler-type operator

$$\operatorname{Tr} \Theta^{-1} \frac{\partial}{\partial \Theta} = \sum_{i=1}^{N} \theta_i^{-1} \frac{\partial}{\partial \theta_i}$$

We get

$$\operatorname{Tr} \Theta^{-1} \frac{\partial}{\partial \Theta} W = -\frac{1}{2} \left(\sum_{j=1}^{N} \theta_j^{-1}\right)^2 W = -\frac{t_0^2}{2} W\,, \tag{6.8}$$

Using (6.5) twice we also get

$$\operatorname{Tr} \Theta^{-1} \frac{\partial}{\partial \Theta} t_{i-1} = -t_i \quad \text{for } i \geq 1\,, \tag{6.9}$$

which implies

$$\operatorname{Tr} \Theta^{-1} \frac{\partial}{\partial \Theta} Z = -\sum_{i=1}^{\infty} t_i \frac{\partial}{\partial t_{i-1}} Z\,. \tag{6.10}$$

Summing up, we obtain

$$\operatorname{Tr} \Theta^{-1} \frac{\partial}{\partial \Theta} (W \cdot Z) = -W \left(\sum_{i=1}^{\infty} t_i \frac{\partial}{\partial t_{i-1}} + \frac{t_0^2}{2}\right) Z\,. \tag{6.11}$$

Looking at the other side of equation (6.6), in order to prove (6.5), it suffices to show that

$$\operatorname{Tr} \Theta^{-1} \frac{\partial}{\partial \Theta} \int_{\mathcal{H}_N^a} e^{-I} dY = -W \frac{\partial Z}{\partial t_0}\,. \tag{6.12}$$

In the proof and in the computations that follow in this section, we will use the following elementary facts, whose proof is left as an exercise. Let M be an $N \times N$ matrix. Then

$$\sum_{ij} M_{i,j} \frac{\partial}{\partial Y_{ij}} e^{\operatorname{Tr}(\Theta Y^n)} = n \operatorname{Tr}(M Y^{n-1}) e^{\operatorname{Tr}(\Theta Y^n)}\,.$$

Moreover, if M depends polynomially on Θ and Y,

$$\int_{\mathcal{H}_N^a} dY \frac{\partial}{\partial Y_{ij}} (M_{ij})\, e^{-I} dY = 0 \tag{6.13}$$

for any i and j. Returning to (6.12), using the above formulas and derivation under the integral sign, we get

$$\begin{aligned} \operatorname{Tr} \Theta^{-1} \frac{\partial}{\partial \Theta} \int_{\mathcal{H}_N^a} e^{-I} dY &= \int_{\mathcal{H}_N^a} (\tfrac{-1}{2}) \operatorname{Tr} \Theta^{-1} Y^2 e^{-I} dY \\ &= \int_{\mathcal{H}_N^a} \left(- \operatorname{Tr} \left(\Theta^{-1} \frac{\partial}{\partial Y} \right) - \operatorname{Tr} Y \right) e^{-I} dY \\ &= - \int_{\mathcal{H}_N^a} \operatorname{Tr} Y e^{-I} dY \,. \end{aligned} \tag{6.14}$$

But now the first identity in (6.4) gives (6.12). Q.E.D.

We now come to the equation $L_2 Z = 0$, that is, to

$$\left(-7 \cdot 5 \cdot 3 \cdot \frac{\partial}{\partial t_3} + \sum_{i=0}^{\infty} (2i+1)(2i+3)(2i+5) t_i \frac{\partial}{\partial t_{i+2}} + 3 \frac{\partial^2}{\partial t_0 \partial t_1} \right) Z = 0 \,. \tag{6.15}$$

To prove the validity of this identity, we will proceed exactly as in the proof of $L_{-1} Z = 0$. In fact, we will follow that proof step by step. The only difference is that the calculations are more complicated.

First of all, we prove a formula analogous to (6.11), but this time we act with the operator $\operatorname{Tr} \Theta^5 \frac{\partial}{\partial \Theta}$. The analogue of (6.8) is

$$\operatorname{Tr} \Theta^5 \frac{\partial}{\partial \Theta} W = -W \left(N \operatorname{Tr} \Theta^4 - \operatorname{Tr} \Theta \operatorname{Tr} \Theta^3 + \tfrac{1}{2} (\operatorname{Tr} \Theta^2)^2 \right) .$$

The analogue of (6.9) is given by the following identities:

$$\begin{aligned} \operatorname{Tr} \Theta^5 \frac{\partial}{\partial \Theta} t_0 &= - \operatorname{Tr} \Theta^3 \,, \\ \operatorname{Tr} \Theta^5 \frac{\partial}{\partial \Theta} t_1 &= -3 \operatorname{Tr} \Theta \,, \\ \operatorname{Tr} \Theta^5 \frac{\partial}{\partial \Theta} t_{i+2} &= -(2i+5)(2i+3)(2i+1) t_i \quad \text{for } i \geq 0 \,. \end{aligned}$$

Combining all these, we get the following analogue of (6.11):

$$\begin{aligned} &\operatorname{Tr} \Theta^5 \frac{\partial}{\partial \Theta} (W \cdot Z) \\ &\quad = W \left(\sum_{i=0}^{\infty} (2i+5)(2i+3)(2i+1) t_i \frac{\partial}{\partial t_{i+2}} - 3 \operatorname{Tr} \Theta \frac{\partial}{\partial t_1} \right. \\ &\qquad \left. - \operatorname{Tr} \Theta^3 \frac{\partial}{\partial t_0} - N \operatorname{Tr} \Theta^4 + \operatorname{Tr} \Theta \operatorname{Tr} \Theta^3 - \tfrac{1}{2} (\operatorname{Tr} \Theta^2)^2 \right) Z \,. \end{aligned} \tag{6.16}$$

One can already foresee that the extra terms appearing in the right-hand side of the above identity will have to be eliminated. Next we consider the analogue of formula (6.14). We get

$$\begin{aligned}\operatorname{Tr}\Theta^5\frac{\partial}{\partial\Theta}(WZ) &= \operatorname{Tr}\Theta^5\frac{\partial}{\partial\Theta}\int_{\mathcal{H}_N^a} e^{-I}dY \\ &= -\tfrac{1}{2}\int_{\mathcal{H}_N^a}\operatorname{Tr}(\Theta^5Y^2)e^{-I}dY\,.\end{aligned}\tag{6.17}$$

Now let us transform (6.15) using the formulae (6.4) to express $\frac{\partial}{\partial t_3}Z$ and $\frac{\partial^2}{\partial t_0\partial t_1}Z$ in terms of expectation values of polynomials in odd powers of traces. We then get that the formula we have to prove is

$$\begin{aligned}&W\left(\sum_{i=0}^{\infty}(2i+1)(2i+3)(2i+5)t_i\frac{\partial}{\partial t_{i+2}}\right)Z \\ &\quad=\int_{\mathcal{H}_N^a}\left(\tfrac{1}{64}\operatorname{Tr}Y^7+\tfrac{3}{16}\operatorname{Tr}Y\operatorname{Tr}Y^3+\tfrac{1}{8}\operatorname{Tr}Y^8\right)e^{-I}dY\,.\end{aligned}\tag{6.18}$$

Comparing (6.16) and (6.17), and using the expressions of $\frac{\partial}{\partial t_0}Z$ and $\frac{\partial}{\partial t_1}Z$ given in (6.4), we are reduced to proving the following relation:

$$\begin{aligned}0=\int_{\mathcal{H}_N^a}&\tfrac{1}{64}\operatorname{Tr}Y^7+\tfrac{3}{16}\operatorname{Tr}Y\operatorname{Tr}Y^3+\tfrac{1}{8}\operatorname{Tr}Y^8 \\ &-\tfrac{1}{2}\operatorname{Tr}\Theta^5Y^2+N\operatorname{Tr}\Theta^4-\operatorname{Tr}\Theta^3\operatorname{Tr}\Theta+\tfrac{1}{2}(\operatorname{Tr}\Theta^2)^2 \\ &+\operatorname{Tr}\Theta^3\operatorname{Tr}Y+\tfrac{1}{4}\operatorname{Tr}\Theta\operatorname{Tr}Y^3+\tfrac{1}{8}\operatorname{Tr}\Theta)e^{-I}dY\,.\end{aligned}\tag{6.19}$$

Denote by Qe^{-I} the integrand in (6.19). If $M(\Theta_i,Y_{ij})$ is an $N\times N$ matrix whose entries are polynomials in the entries of Θ and Y, we set

$$\frac{\partial}{\partial Y}M=\sum_{i,j}\frac{\partial}{\partial Y_{ij}}M_{ij}\,.$$

Recalling (6.13), our task is therefore to find a matrix M such that

$$\frac{\partial}{\partial Y}M(e^{-I})=Qe^{-I}\,.$$

An elementary computation shows that the following choice works!

$$\begin{aligned}M=\;&\Theta^4Y-\Theta^3Y\Theta+\tfrac{1}{2}\Theta^2Y\Theta^2+\Theta^3Y^2-\Theta^2Y\Theta Y+\tfrac{3}{2}\Theta Y\Theta^2Y \\ &-\tfrac{1}{4}\Theta Y\Theta Y\Theta-\Theta^2Y^2\Theta+\Theta^2Y^3-\Theta Y\Theta Y^2-\tfrac{3}{2}\Theta Y^2\Theta Y \\ &+\tfrac{9}{8}Y\Theta Y\Theta+Y\Theta^2Y^2\tfrac{1}{2}+\tfrac{3}{2}\Theta Y^4-\tfrac{11}{8}Y\Theta Y^3-\tfrac{1}{16}Y^2\Theta Y^2 \\ &+\tfrac{1}{32}Y^5-\tfrac{N}{8}\Theta Y+\tfrac{5N}{2}\Theta Y+\tfrac{3N}{2}\Theta^2-2\operatorname{Tr}\Theta^2-\tfrac{1}{16}\operatorname{Tr}Y^2 \\ &-\tfrac{3}{8}\operatorname{Tr}\Theta Y+\tfrac{1}{4}Y\operatorname{Tr}Y-\tfrac{5}{2}Y\operatorname{Tr}\Theta+\tfrac{1}{2}\Theta\operatorname{Tr}Y\,.\end{aligned}$$

This proves that $L_2 Z = 0$ and therefore also the main Theorem (2.5).

Q.E.D.

7. A nonvanishing theorem.

As an application of the intersection theory developed so far, we are going to present an interesting result due to Faber (cf. [230], [237], and [236]). We already discussed this result in Section 6 of Chapter XVII, in the context of the so-called Gorenstein conjectures (see also the bibliographical notes to that chapter). Faber's theorem is the following.

THEOREM (7.1). *$\kappa_{g-2} \neq 0$ as an element of $R^{g-2}(\mathcal{M}_g)$.*

The first step in the proof is a lemma regarding the product of the two Hodge classes λ_g and λ_{g-1}. Recall that, if $\pi : \mathcal{C} \to \mathcal{M}_g$ is the universal family, then the Hodge bundle is defined to be $\mathbb{E} = \mathbb{E}_g = \pi_*(\omega_\pi)$, and that $\lambda_i = c_i(\mathbb{E})$.

LEMMA (7.2). *The class $\lambda_g \lambda_{g-1}$ vanishes on the boundary $\partial\mathcal{M}_g = \overline{\mathcal{M}}_g \smallsetminus \mathcal{M}_g$.*

Proof. Consider the irreducible components $\mathcal{D}_{irr}$, $\mathcal{D}_h$, $h = 1, \ldots, [\frac{g}{2}]$, of the boundary $\partial\mathcal{M}_g$. We have finite morphisms

$$\xi_{irr} : \overline{\mathcal{M}}_{g-1,\{x,y\}} \to \mathcal{D}_{irr} \subset \overline{\mathcal{M}}_g\,, \qquad \xi_h : \overline{\mathcal{M}}_{h,x} \times \overline{\mathcal{M}}_{g-h,y} \to \mathcal{D}_h \subset \overline{\mathcal{M}}_g\,.$$

It then suffices to show that $\xi_{irr}^*(\lambda_g\lambda_{g-1}) = 0$ and that $\xi_h^*(\lambda_g\lambda_{g-1}) = 0$ for $h = 1, \ldots, [\frac{g}{2}]$. First of all, taking residues along the section labeled by x gives a surjective homomorphism

$$\xi_{irr}^*(\mathbb{E}) \to \mathcal{O}_{\overline{\mathcal{M}}_{g-1,\{x,y\}}} \to 0\,.$$

It follows that $\xi_{irr}^*\lambda_g = \xi_{irr}^* c_g(\mathbb{E})$ vanishes. Now look at $\mathcal{D}_h$. The pullback $\xi_h^*(\mathbb{E})$ decomposes into a direct sum

$$\xi_h^*(\mathbb{E}) = p_1^*(\mathbb{E}_h) \oplus p_2^*(\mathbb{E}_{g-h})\,,$$

where p_1 and p_2 are the projections from $\overline{\mathcal{M}}_{h,x} \times \overline{\mathcal{M}}_{g-h,y}$ to its two factors. Thus,

$$\begin{aligned} \xi_h^*\lambda_g &= p_1^*\lambda_h \cdot p_2^*\lambda_{g-h}\,, \\ \xi_h^*\lambda_{g-1} &= p_1^*\lambda_h \cdot p_2^*\lambda_{g-h-1} + p_1^*\lambda_{h-1} \cdot p_2^*\lambda_{g-h}\,. \end{aligned} \tag{7.3}$$

We now recall Remark (5.19) in Chapter XVII and use the relation $c(\mathbb{E})c(\mathbb{E}^\vee) = 1$ which, in genus h, we write as

$$(1 + \lambda_1 + \lambda_2 + \cdots + \lambda_h)(1 - \lambda_1 + \lambda_2 - \cdots + (-1)^h\lambda_h) = 1\,. \tag{7.4}$$

We see, in particular, that $\lambda_h^2 = 0$ in genus h and $\lambda_{g-h}^2 = 0$ in genus $g-h$. Looking at (7.3), we get the lemma.

Q.E.D

We now come to the proof of Theorem (7.1). We will show that

$$\int_{\overline{M}_g} \kappa_{g-2}\lambda_{g-1}\lambda_g = \frac{\widetilde{B}_g(g-1)!}{2^g(2g)!}, \tag{7.5}$$

where $\widetilde{B}_i$ is the modified ith Bernoulli number defined in (5.2) of Chapter XVII. Of course, this proves the theorem since $\widetilde{B}_g \neq 0$. First of all, we prove that

$$\lambda_g\lambda_{g-1} = (-1)^{g-1}(2g-1)!\,\mathrm{ch}_{2g-1}\mathbb{E}. \tag{7.6}$$

To see this, we go back to formula (5.5) in Chapter XVII, where we set $E = \mathbb{E}$, and take the logarithmic derivative of both sides with respect to t. Using (7.4), we find

$$\begin{aligned}\sum_{s=1}^{\infty}(-1)^{s-1}s!\,\mathrm{ch}_s(\mathbb{E})t^{s-1} &= \left(\sum_{j=0}^{\infty} j\lambda_j t^{j-1}\right)\left(\sum_{j=0}^{\infty}\lambda_j t^j\right)^{-1}\\ &= \left(\sum_{j=0}^{\infty} j\lambda_j t^{j-1}\right)\left(\sum_{j=0}^{\infty}\lambda_j (-t)^j\right).\end{aligned}$$

Equating terms degree $2g-2$ yields

$$(2g-1)!\,\mathrm{ch}_{2g-1}\mathbb{E} = (-1)^{g-1}g\lambda_{g-1}\lambda_g + (-1)^g(g-1)\lambda_{g-1}\lambda_g = (-1)^{g-1}\lambda_{g-1}\lambda_g,$$

proving (7.6). Using Mumford's formula (5.13) in Chapter XVII and (7.6), we see that (7.5) is equivalent to

(7.7)

$$\int_{\overline{M}_g} \widetilde{\kappa}_{g-2}\cdot\left(\widetilde{\kappa}_{2g-1} + \frac{1}{2}\sum_{h=0}^{g-1}\xi_{h*}\left(\widetilde{\psi}_x^{2g-2} - \widetilde{\psi}_x^{2g-3}\widetilde{\psi}_y + \cdots + \widetilde{\psi}_y^{2g-2}\right)\right) = \frac{g!}{2^{g-1}(2g)!}.$$

We would like to rewrite this equality in terms of Witten's *tau* symbols. First of all, using formula (4.25) in Chapter XVII, with $n = 2$ and $a_1 = g-1$, $a_2 = 2g$, we have

$$\int_{\overline{M}_g} \widetilde{\kappa}_{g-2}\widetilde{\kappa}_{2g-1} = <\tau_{g-1}\tau_{2g}> - <\tau_{3g-2}>. \tag{7.8}$$

To translate the remaining part of (7.7), that is, the contributions from $\mathcal{D}_0 = \mathcal{D}_{irr}$ and $\mathcal{D}_h$, $h = 1, \ldots, g-1$, in terms of Witten's *tau* symbols, requires more care. The end result of the computations we will carry out is stated in the following lemma.

LEMMA (7.9).

i) $\int_{\overline{M}_g} \widetilde{\kappa}_{g-2}\xi_{0*}\left(\widetilde{\psi}_x^{2g-2} - \widetilde{\psi}_x^{2g-3}\widetilde{\psi}_y + \cdots + \widetilde{\psi}_y^{2g-2}\right)$
$= \sum_{j=0}^{2g-2}(-1)^j < \tau_{2g-2-j}\tau_j\tau_{g-1} >;$

ii) $\int_{\overline{M}_g} \widetilde{\kappa}_{g-2}\xi_{h*}\left(\widetilde{\psi}_x^{2g-2} - \widetilde{\psi}_x^{2g-3}\widetilde{\psi}_y + \cdots + \widetilde{\psi}_y^{2g-2}\right)$
$= (-1)^{g-h} < \tau_{3h-g}\tau_{g-1} >< \tau_{3(g-h)-2} >$
$+(-1)^h < \tau_{3h-2} >< \tau_{2g-3h}\tau_{g-1} >$

for $h = 1, \dots, g-1$.

Proof. In our computations we will use the cartesian diagrams

$$
\text{(7.10)}\qquad
\begin{array}{ccc}
\overline{\mathcal{M}}_{g-1,\{p,x,y\}} & \xrightarrow{\ \xi_0'\ } & \overline{\mathcal{M}}_{g,\{p\}} \\
{\scriptstyle \pi'}\downarrow & & \downarrow{\scriptstyle \pi} \\
\overline{\mathcal{M}}_{g-1,\{x,y\}} & \xrightarrow{\ \xi_0\ } & \overline{\mathcal{M}}_g
\end{array}
$$

$$
\text{(7.11)}\qquad
\begin{array}{ccc}
(\overline{\mathcal{M}}_{h,\{p,x\}} \times \overline{\mathcal{M}}_{g-h,\{y\}})\coprod(\overline{\mathcal{M}}_{h,\{x\}} \times \overline{\mathcal{M}}_{g-h,\{p,y\}}) & \xrightarrow{\ \xi_h'\ } & \overline{\mathcal{M}}_{g,\{p\}} \\
{\scriptstyle \pi'}\downarrow & & \downarrow{\scriptstyle \pi} \\
\overline{\mathcal{M}}_{h,\{x\}} \times \overline{\mathcal{M}}_{g-h,\{y\}} & \xrightarrow{\ \xi_h\ } & \overline{\mathcal{M}}_g
\end{array}
$$

$$1 \le h \le g-1\,,$$

where ξ_0' and ξ_h' are the obvious clutching maps. We use diagram (7.10) to prove the first formula in the statement of the lemma and diagram (7.11) to prove the second one. Since the two computations are entirely similar, we will perform only the second one, leaving the first (which is slightly easier) to the reader. Set

$$A = \widetilde{\psi}_x^{2g-2} - \widetilde{\psi}_x^{2g-3}\widetilde{\psi}_y + \cdots + \widetilde{\psi}_y^{2g-2}\,.$$

The same argument as in the proof of (4.19) in Chapter XVII shows that $\pi^*\xi_{h*} = \xi_{h*}'\pi'^*$. Thus,

$$
\text{(7.12)}\qquad
\begin{aligned}
\widetilde{\kappa}_{g-2}\cdot\xi_{h*}A &= \pi_*\psi_p^{g-1}\cdot\xi_{h*}A \\
&= \pi_*(\psi_p^{g-1}\cdot\pi^*\xi_{h*}A) \\
&= \pi_*(\psi_p^{g-1}\cdot\xi_{h*}'\pi'^*A)\,.
\end{aligned}
$$

We write

$$\pi'^*A = \widetilde{\psi_x'}^{2g-2} - \widetilde{\psi_x'}^{2g-3}\widetilde{\psi_y'} + \cdots + \widetilde{\psi_y'}^{2g-2}\,,$$

where, on each of the two connected components of

$$Y = (\overline{\mathcal{M}}_{h,\{p,x\}} \times \overline{\mathcal{M}}_{g-h,\{y\}}) \coprod (\overline{\mathcal{M}}_{h,\{x\}} \times \overline{\mathcal{M}}_{g-h,\{p,y\}}),$$

the class $\widetilde{\psi'_x}$ (resp., $\widetilde{\psi'_y}$) is the pullback of the class ψ_x (resp., ψ_y) under the projection onto the first (resp., the second) factor. Similarly, on the first (resp., the second) connected components of Y, the class $\widetilde{\psi'_p} = \xi_h'^{\,*}\psi_p$ is equal to the pullback of ψ_p under the projection onto the first (resp., the second) factor. Using (7.12), the projection formula for ξ'_h, and, in the last step, Fubini's theorem, we get

$$\begin{aligned}
&\int_{\overline{M}_g} \widetilde{\kappa}_{g-2}\xi_{h*}\left(\widetilde{\psi}_x^{2g-2} - \widetilde{\psi}_x^{2g-3}\widetilde{\psi}_y + \cdots + \widetilde{\psi}_y^{2g-2}\right) \\
&= \int_{\overline{M}_g} \pi_*\xi'_{h*}\left(\widetilde{\psi'_p}^{g-1}\left(\widetilde{\psi'_x}^{2g-2} - \widetilde{\psi'_x}^{2g-3}\widetilde{\psi'_y} + \cdots + \widetilde{\psi'_y}^{2g-2}\right)\right) \\
&= \int_Y \widetilde{\psi'_p}^{g-1}\left(\widetilde{\psi'_x}^{2g-2} - \widetilde{\psi'_x}^{2g-3}\widetilde{\psi'_y} + \cdots + \widetilde{\psi'_y}^{2g-2}\right) \\
&= \int_{\overline{\mathcal{M}}_{h,\{p,x\}} \times \overline{\mathcal{M}}_{g-h,\{y\}}} \widetilde{\psi'_p}^{g-1}\left(\widetilde{\psi'_x}^{2g-2} - \widetilde{\psi'_x}^{2g-3}\widetilde{\psi'_y} + \cdots + \widetilde{\psi'_y}^{2g-2}\right) + \\
&\quad \int_{\overline{\mathcal{M}}_{h,\{x\}} \times \overline{\mathcal{M}}_{g-h,\{p,y\}}} \widetilde{\psi'_p}^{g-1}\left(\widetilde{\psi'_x}^{2g-2} - \widetilde{\psi'_x}^{2g-3}\widetilde{\psi'_y} + \cdots + \widetilde{\psi'_y}^{2g-2}\right) \\
&= (-1)^{g-h} < \tau_{3h-g}\tau_{g-1} >< \tau_{3(g-h)-2} > + (-1)^h < \tau_{3h-2} >< \tau_{2g-3h}\tau_{g-1} > .
\end{aligned}$$

Q.E.D.

We now use Exercise A-3, stating that

$$< \tau_{3j-2} > = \frac{1}{24^j j!}, \tag{7.13}$$

and we obtain

$$\begin{aligned}
&\frac{1}{2}\sum_h^{g-1} \widetilde{\kappa}_{g-2}\overline{\xi}_{h*}\left(\widetilde{\psi}_x^{2g-2} - \widetilde{\psi}_x^{2g-3}\widetilde{\psi}_y + \cdots + \widetilde{\psi}_y^{2g-2}\right) \\
&\qquad = \sum_{h=1}^{g-1} \frac{(-1)^{g-h}}{24^{g-h}(g-h)!} < \tau_{3h-2}\tau_{g-1} > .
\end{aligned} \tag{7.14}$$

From (7.8), from the first formula in Lemma (7.9), and from (7.14) we then get

$$\begin{aligned}
&\int_{\overline{M}_g} \widetilde{\kappa}_{g-2}\left(\widetilde{\kappa}_{2g-1} + \frac{1}{2}\sum_{h=0}^{g-1}\xi_{h*}\left(\widetilde{\psi}_x^{2g-2} - \widetilde{\psi}_x^{2g-3}\widetilde{\psi}_y + \cdots + \widetilde{\psi}_y^{2g-2}\right)\right) \\
&= < \tau_{g-1}\tau_{2g} > - < \tau_{3g-2} > + \frac{1}{2}\sum_{j=0}^{2g-2}(-1)^j < \tau_{2g-2-j}\tau_j\tau_{g-1} > \\
&\quad + \sum_{h=1}^{g-1}\frac{(-1)^{g-h}}{24^{g-h}(g-h)!} < \tau_{3h-2}\tau_{g-1} > .
\end{aligned}$$

Thus, in order to prove (7.7) and therefore Theorem (7.1), it suffices to prove the equalities:

$$\sum_{j=0}^{2g-2} (-1)^j < \tau_{2g-2-j}\tau_j\tau_{g-1} >= \frac{g!}{2^{g-2}(2g)!}, \tag{7.15}$$

$$\sum_{h=1}^{g} \frac{(-1)^{g-h}}{24^{g-h}(g-h)!} < \tau_{3h-g}\tau_{g-1} >= \frac{1}{24^g g!}. \tag{7.16}$$

These equalities are the object of Exercises A-4–A-8.

8. A brief review of equivariant cohomology and the virtual Euler–Poincaré characteristic.

The purpose of this section is to collect a series of results about group homology and cohomology providing the link between the homology of moduli spaces and the cohomology of mapping class groups. The detailed proofs of the result illustrated in this section can be found in [88].

In what follows we will use the following notation. Let Γ be a group. We shall consistently view any (left) Γ-module M (i.e., any left $\mathbb{Z}\Gamma$-module) as a Γ-bimodule by setting $mg = g^{-1}m$ for any $g \in \Gamma$ and $m \in M$. Given a Γ-module M, we set

$$M_\Gamma = \mathbb{Z} \otimes_{\mathbb{Z}\Gamma} M \cong M/R,$$

where $\mathbb{Z}$ is considered as a trivial Γ-module, and R is the submodule of M generated by the elements of type $gm - m$ with $m \in M$ and $g \in \Gamma$. Given Γ-modules M and N, sometimes we will write $M \otimes_\Gamma N$ to mean $M \otimes_{\mathbb{Z}\Gamma} N$. We have

$$M \otimes_\Gamma N = (M \otimes_{\mathbb{Z}} N)_\Gamma, \tag{8.1}$$

where the Γ-module structure on $M \otimes_{\mathbb{Z}} N$ is given by $g(m \otimes n) = gm \otimes gn$. If the group Γ acts *freely* on a CW complex Z, then we have

$$C_\bullet(Z/\Gamma) = C_\bullet(Z)_\Gamma. \tag{8.2}$$

Here and in the sequel, all the complexes are taken with integral coefficients. Now let Y be a $K(\Gamma, 1)$. Recall that this means that Y is a connected CW complex with $\pi_1(Y) = \Gamma$ whose universal cover $\widetilde{Y}$ is contractible; for instance, we can take $Y = B\Gamma$ and $\widetilde{Y} = E\Gamma$. The *homology of* Γ *with integral coefficients* is defined by

$$H_\bullet(\Gamma; \mathbb{Z}) = H_\bullet(Y; \mathbb{Z}).$$

Similarly, for the cohomology, we set

$$H^\bullet(\Gamma; \mathbb{Z}) = H^\bullet(Y; \mathbb{Z}).$$

More generally, suppose that Γ acts on a cell complex X. Then the *Γ-equivariant homology of X* is defined as follows. Consider the cell complex $\widetilde{Y} \times X$. The free action of Γ on $\widetilde{Y}$ and the action of Γ on X induce a free action of Γ on $\widetilde{Y} \times X$. We set

$$X_\Gamma = (\widetilde{Y} \times X)/\Gamma ,$$

and we define the Γ-equivariant homology of X by setting

$$H_\bullet^\Gamma(X;\mathbb{Z}) = H_\bullet(X_\Gamma;\mathbb{Z}) .$$

Similarly, we define the Γ-equivariant cohomology of X by setting

$$H_\Gamma^\bullet(X;\mathbb{Z}) = H^\bullet(X_\Gamma;\mathbb{Z}) .$$

When X reduces to a point $\{pt\}$, we have

$$H_\bullet^\Gamma(\{pt\};\mathbb{Z}) = H_\bullet(\Gamma;\mathbb{Z}) , \quad H_\Gamma^\bullet(\{pt\};\mathbb{Z}) = H^\bullet(\Gamma;\mathbb{Z}) .$$

Since Γ acts freely on $Z = \widetilde{Y} \times X$, using (8.1) and (8.2), we get

$$H_\bullet^\Gamma(X;\mathbb{Z}) = H_\bullet\big(C_\bullet(\widetilde{Y}) \otimes_\Gamma C_\bullet(X)\big) . \tag{8.3}$$

In particular,

$$H_\bullet(\Gamma;\mathbb{Z}) = H_\bullet\big(C_\bullet(\widetilde{Y})_\Gamma\big) . \tag{8.4}$$

Equality (8.3) exhibits the Γ-equivariant homology as the homology of the total complex of the double complex $C_\bullet(\widetilde{Y}) \otimes_\Gamma C_\bullet(X)$. As such, it is the abutment of two spectral sequences. The first one is nothing but the Leray spectral sequence of the fibration $\pi : X_\Gamma \to Y = \widetilde{Y}/\Gamma$ having X as fiber. The second one is more subtle and can be interpreted in terms of the projection $\sigma : X_\Gamma \to X/\Gamma$. This is not a fibration unless the action of Γ on X is free. The fiber of σ over an orbit $\Gamma x \in X/\Gamma$ is the quotient $\widetilde{Y}/\Gamma_x$, where Γ_x is the stabilizer of x in Γ. In particular, the fiber of σ over Γx is a $K(\Gamma_x, 1)$. We can then expect the homology of these stabilizers to appear in the spectral sequence. We will discuss this momentarily.

Looking at (8.3) and (8.4), it is natural to make the following definitions. Suppose that $C_\bullet$ is a complex of Γ-modules. Then we define the homology of Γ with coefficient in $C_\bullet$ as

$$H_\bullet(\Gamma, C_\bullet) = H_\bullet\big(C_\bullet(\widetilde{Y}) \otimes_\Gamma C_\bullet\big) .$$

With this notation $H_\bullet^\Gamma(X;\mathbb{Z}) = H_\bullet(\Gamma, C_\bullet(X))$. If $C_\bullet$ consists of a single module M concentrated in degree 0, we have the homology of Γ with coefficients in the Γ-module M:

$$H_\bullet(\Gamma, M) = H_\bullet\big(C_\bullet(\widetilde{Y}) \otimes_\Gamma M\big) .$$

Considering the ring $\mathbb{Z}$ as a trivial Γ-module, we get back (8.4). Observe that, according to the previous definition, we have

$$H_0(\Gamma, M) = M_\Gamma \,. \tag{8.5}$$

Now let us go back to the double complex $C_\bullet(\widetilde{Y}) \otimes_\Gamma C_\bullet(X)$ whose total complex computes $H_\bullet^\Gamma(X;\mathbb{Z})$. The E^1-term of the second spectral sequence is given by

$$E^1_{pq} = H_q\big(C_\bullet(\widetilde{Y}) \otimes_\Gamma C_p(X)\big) = H_q(\Gamma, C_p(X)) \,.$$

We will give an alternative expression for the cohomology of Γ with coefficient in $C_p(X)$. Let X_p be the set of p-cells of X so that

$$C_p(X) = \bigoplus_{\sigma \in X_p} \mathbb{Z}_\sigma \,.$$

Here $\mathbb{Z}_\sigma$ is a copy of $\mathbb{Z}$ in which the two generators correspond to the two possible orientations of σ. Let

$$\Gamma_\sigma = \{g \in \Gamma : g\sigma = \sigma\}$$

be the stabilizer of σ. Then $\mathbb{Z}_\sigma$ is a Γ_σ-module, where $g \in \Gamma_\sigma$ acts as $+1$ if g is orientation-preserving and as -1 if it is orientation-reversing. Let Σ_p be a set of representatives of X_p/Γ. The Γ-module structure of $C_p(X)$ can be described as follows:

$$C_p(X) = \bigoplus_{\sigma \in \Sigma_p} \left(\bigoplus_{g \in \Gamma/\Gamma_\sigma} g\mathbb{Z}_\sigma \right) = \bigoplus_{\sigma \in \Sigma_p} (\mathbb{Z}\Gamma \otimes_{\mathbb{Z}\Gamma_\sigma} \mathbb{Z}_\sigma) \,.$$

We claim that

$$H_q(\Gamma, \mathbb{Z}\Gamma \otimes_{\mathbb{Z}\Gamma_\sigma} \mathbb{Z}_\sigma) = H_q(\Gamma_\sigma, \mathbb{Z}_\sigma) \,.$$

In fact, since $\widetilde{Y}/\Gamma_\sigma$ is a $K(\Gamma_\sigma, 1)$, we have

$$\begin{aligned} H_q(\Gamma, \mathbb{Z}\Gamma \otimes_{\mathbb{Z}\Gamma_\sigma} \mathbb{Z}_\sigma) &= H_q\big(C_\bullet(\widetilde{Y}) \otimes_{\mathbb{Z}\Gamma} (\mathbb{Z}\Gamma \otimes_{\mathbb{Z}\Gamma_\sigma} \mathbb{Z}_\sigma)\big) \\ &= H_q\big(C_\bullet(\widetilde{Y}) \otimes_{\Gamma_\sigma} \mathbb{Z}_\sigma\big) = H_q(\Gamma_\sigma, \mathbb{Z}_\sigma) \,. \end{aligned}$$

We have proved the following proposition.

Proposition (8.6). *Let Γ be a group acting on a cell complex X. Then there is a spectral sequence abutting to the equivariant homology $H_\bullet^\Gamma(X, \mathbb{Z})$ whose E^1-term is given by*

$$E^1_{pq} = H_q(\Gamma, C_p(X)) = \bigoplus_{\sigma \in \Sigma_p} H_q(\Gamma_\sigma, \mathbb{Z}_\sigma) \,,$$

where Σ_p is a set of representatives for the p-cells of X/Γ.

A completely similar result holds in cohomology. Moreover, one can substitute the module of coefficients $\mathbb{Z}$, which is a trivial Γ-module, with any Γ-module M. In this case one computes the equivariant homology $H^{\Gamma}_{\bullet}(X, M) = H_{\bullet}(\Gamma, C_{\bullet}(X) \otimes M)$, while the E^1-term of the above spectral sequence is given by

$$E^1_{pq} = H_q(\Gamma, C_p(X) \otimes M) = \bigoplus_{\sigma \in \Sigma_p} H_q(\Gamma_\sigma, M_\sigma),$$

where $M_\sigma = \mathbb{Z}_\sigma \otimes M$. Two cases are particularly interesting. The first is where Γ acts freely on X and $M = \mathbb{Z}$. Then the spectral sequence degenerates at E^2. Recalling (8.5) and (8.2), we get

$$H^{\Gamma}_{\bullet}(X; \mathbb{Z}) = H_{\bullet}(C_{\bullet}(X)_{\Gamma}) = H_{\bullet}(X/\Gamma; \mathbb{Z}).$$

For the second case, we assume that the action of Γ on X admits only finite stabilizers. This implies that the groups Γ_σ are finite groups. In this case it is useful to take rational coefficients. This means that we take $M = \mathbb{Q}$ with trivial Γ-action. Now, it is a rather elementary result (see [88], III.10) that the homology groups of a finite group are torsion groups. It follows that, in this case too, the above spectral sequence degenerates, and one gets

$$H^{\Gamma}_{\bullet}(X; \mathbb{Q}) = H_{\bullet}(C_{\bullet}(X) \otimes_{\Gamma} \mathbb{Q}) = H_{\bullet}(X/\Gamma; \mathbb{Q}).$$

If, moreover, X is contractible, one gets

$$H_{\bullet}(\Gamma; \mathbb{Q}) = H_{\bullet}(X/\Gamma; \mathbb{Q}),$$

and, for the obvious reason, one says that X/Γ *is a rational* $K(\Gamma, 1)$. An example of this situation is provided by the mapping class group $\Gamma_{g,P}$ of a P-pointed genus g surface acting on the Teichmüller space $\mathcal{T}_{g,P}$. Since $\mathcal{T}_{g,P}$ is contractible and the action of $\Gamma_{S,P}$ has finite stabilizers, the moduli space $M_{g,P}$ is a rational $K(\Gamma_{g,P}, 1)$, so that

$$H_{\bullet}(\Gamma_{g,P}; \mathbb{Q}) = H_{\bullet}(M_{g,P}; \mathbb{Q}).$$

Completely analogous results hold in cohomology.

In order to avoid technical difficulties, from now on we will consider only groups Γ such that there exists a contractible complex X, on which Γ acts, having the following properties:

1) For each cell σ, the stabilizer Γ_σ is a finite group.
2) X/Γ has finitely many cells.
3) There exists a finite index subgroup $\Gamma' \subset \Gamma$ acting freely on X.

These properties imply that X/Γ is a rational $K(\Gamma,1)$, while X/Γ' is a $K(\Gamma',1)$. The mapping class group $\Gamma_{g,n}$ satisfies these properties. In fact, we can take as X the Teichmüller space $\mathcal{T}_{g,n}$ and as Γ' any level structure $\Gamma_{g,n}[m]$ with $m \geq 3$, (cf. Theorem (2.11) of Chapter XVI).

Let then X be a contractible complex satisfying conditions 1), 2), and 3). Let Σ be a set of representatives for the cells of X/Γ. We define the *virtual Euler–Poincaré characteristic of* Γ by setting

$$\chi_{virt}(\Gamma) = \sum_{\sigma\in\Sigma}(-1)^{\dim\sigma}\frac{1}{|\Gamma_\sigma|}\,. \tag{8.7}$$

We also say that $\chi_{virt}(\Gamma)$ is the *virtual Euler–Poincaré characteristic of* X/Γ, and we write

$$\chi_{virt}(X/\Gamma) = \chi_{virt}(\Gamma)\,. \tag{8.8}$$

Let us show that $\chi_{virt}(\Gamma)$ only depends on Γ. For this, it will suffice to prove the following three facts. The first is that, if Γ' acts freely on X, then

$$\chi_{virt}(\Gamma') = \sum_{i\geq 0}(-1)^i \operatorname{rank}(H_i(\Gamma';\mathbb{Z}))\,. \tag{8.9}$$

The second is that

$$\chi_{virt}(\Gamma) = \frac{\chi_{virt}(\Gamma')}{[\Gamma:\Gamma']}\,. \tag{8.10}$$

The third is that the expression on the right-hand side of (8.10) does not depend on the choice of the finite index subgroup Γ' acting freely on X. The first assertion is a direct consequence of the fact that Γ' acts freely on X, so that $H_\bullet(\Gamma';\mathbb{Z}) = H_\bullet(X/\Gamma';\mathbb{Z})$. For the second assertion, let Σ' be a set of representatives for the cells of X/Γ'. Then

$$[\Gamma:\Gamma']\cdot\chi_{virt}(\Gamma) = \sum_{\sigma\in\Sigma}(-1)^{\dim\sigma}\frac{[\Gamma:\Gamma']}{|\Gamma_\sigma|} = \sum_{\tau\in\Sigma'}(-1)^{\dim\tau} = \chi_{virt}(\Gamma')\,.$$

For the third assertion, let Γ'' be another finite index subgroup of Γ acting freely on X. Set $\Gamma''' = \Gamma'\cap\Gamma''$. Then Γ''' is a finite index subgroup of Γ' acting freely on X, and

$$\frac{\chi_{virt}(\Gamma')}{[\Gamma:\Gamma']} = \frac{\chi_{virt}(\Gamma''')/[\Gamma':\Gamma''']}{[\Gamma:\Gamma']} = \frac{\chi_{virt}(\Gamma''')}{[\Gamma:\Gamma''']}\,,$$

and similarly $\chi_{virt}(\Gamma'')/[\Gamma':\Gamma''] = \chi_{virt}(\Gamma''')/[\Gamma:\Gamma''']$.

We may now prove the following lemma.

LEMMA (8.11). *Set* $\chi_{g,n} = 2 - 2g - n$. *Then*

$$\chi_{virt}\left(\Gamma_{g,n+1}\right) = \chi_{virt}\left(\Gamma_{g,n}\right)\chi_{g,n}\,. \tag{8.12}$$

Proof. To prove this formula, we look at the natural projection

$$\eta : \mathcal{T}_{g,n+1} \longrightarrow \mathcal{T}_{g,n}\,.$$

This induces the projection

$$\overline{\eta} : \mathcal{T}_{g,n+1}/\Gamma_{g,n+1}[m] = M_{g,n+1}[m] \longrightarrow M_{g,n}[m] = \mathcal{T}_{g,n}/\Gamma_{g,n}[m]$$

between moduli spaces with level m structures. The map $\overline{\eta}$ is a fibration with fiber $S \smallsetminus \{x_1, \dots, x_n\}$, where S is a genus g Riemann surface, and $x_1, \dots, x_n$ are distinct points on it. It follows that

$$\chi_{virt}\left(\Gamma^{(m)}_{g,n+1}\right) = \chi\left(M_{g,n+1}[m]\right) = \chi\left(M_{g,n}[m]\right)\chi_{g,n} = \chi_{virt}\left(\Gamma^{(m)}_{g,n}\right)\chi_{g,n}\,.$$

Formula (8.12) follows immediately from (8.10) and from the fact that $[\Gamma_{g,n+1} : \Gamma_{g,n+1}[m]] = [\Gamma_{g,n} : \Gamma_{g,n}[m]]$. Indeed, the group $\Gamma_{g,n}[m]$ is the image of $\Gamma_{g,n+1}[m]$ under the natural surjection $\Gamma_{g,n+1} \to \Gamma_{g,n}$.

9. The virtual Euler–Poincaré characteristic of $M_{g,n}$.

In this section we compute the virtual Euler–Poincaré characteristic of $M_{g,n}$. Let (S, P) be a fixed P-pointed suface of genus g with $|P| = n$. Recall, from Section 12 of Chapter XVIII, the $\Gamma_{g,P}$-equivariant homeomorphism

$$\Psi : \mathcal{T}_{S,P} \times \mathring{\Delta}_P \longrightarrow |A'^0(S,P)|\,.$$

Also recall that the equivalence classes of simplices in $A'^0(S,P)/\Gamma_{S,P}$ are in one-to-one correspondence with the set

$$\mathcal{G}^n_g = \left\{ \begin{array}{c} \text{isomorphism classes of connected genus } g \text{ ribbon} \\ \text{graphs with } n \text{ marked boundary components} \end{array} \right\}.$$

LEMMA (9.1).

$$\chi_{virt}(M_{g,n}) = \sum_{G \in \mathcal{G}^n_g} \frac{(-1)^{V(G)}}{|\operatorname{Aut}(G)|}\,,$$

where $V(G)$ *is the number of vertices of* G.

The appearance of the cardinality of $\operatorname{Aut}(G)$ in the above expression stems from the fact that, in the one-to-one corrrespondence between equivalence classes of simplices in $A'^0(S,P)/\Gamma_{S,P}$ and elements of $\mathcal{G}^n_g$, the automorphism group of a ribbon graph G coincides with the stabilizer in $\Gamma_{S,P}$ of the corresponding simplex a_G. What, at first sight, looks odd is the sign $(-1)^{V(G)}$. In fact, the dimension of a_G is equal to the number $E(G)$ of edges of G and not to $V(G)$. Indeed, we have

$$V(G) - E(G) = 2 - 2g - n\,. \tag{9.2}$$

This discrepancy depends on the fact that $A'^0(S,P)$ is obtained by subtracting the subcomplex $A''_\infty(S,P)$ from the complex $A(S,P)$. As such, $A'^0(S,P)$ is not a simplicial complex, and one cannot directly use formula (8.7) to compute the virtual Euler–Poincaré characteristics of $M_{g,n}$. To understand what is going on, we go back to Lemma (2.3) of Chapter XIX.

We have a simplicial complex A, a subcomplex $B \subset A$, and we set $C = A \smallsetminus B$. We assume that A is acted on by a group Γ and that B is preserved by this action. Let A^1 and B^1 be the first barycentric subdivision of A and B, respectively. Let D be the subcomplex of A^1 whose vertices are barycenters of simplices of C. Then there is a G-equivariant retraction of $|C|$ onto $|D|$. We are now going to define a Γ-equivariant cell decomposition of $|D|$ from which one may compute the virtual Euler–Poincaré characteristic of $|D|/\Gamma$. If b (resp., b_0, b_1, etc.) is a vertex of D, we will denote by a (resp., a_0, a_1, etc.) the corresponding simplex of A. For each vertex b of D, we define the *star of* b as the subset of simplices of D given by

$$Star(b) = \{\langle b, b_1, \dots, b_k\rangle \in D_k : a < a_1 < \cdots < a_k\}\,.$$

We then set

$$|Star(b)| = \Bigg(\bigcup_{\sigma \in Star(b)} |\sigma|\Bigg)^{\circ}.$$

From the definition it follows that $|Star(b)|$ is a cell, that

$$\dim|Star(b)| = \dim|A| - \dim|a|\,,$$

and that we have a bona fide Γ-equivariant cell decomposition

$$|D| = \bigcup_{a \in C} |Star(b)|\,.$$

It is also clear that the stabilizer Γ_a of a under the action of Γ on C coincides with the stabilizer of $|Star(b)|$ under the action of Γ on $|D|$. Now suppose that $|C|$ and hence $|D|$ is contractible. Then

$$\chi_{virt}(\Gamma) = \chi_{virt}(|D|/\Gamma) = \sum_{[a]\in C/\Gamma} \frac{(-1)^{\operatorname{codim}_A(a)}}{|\Gamma_a|}\,.$$

Proof of Lemma (9.1). Apply the formula we just wrote to the case $A = A(S,P)$, $B = A''_\infty(S,P)$, $C = A'^0(S,P)$, and $\Gamma = \Gamma_{S,P}$ and notice that, by (9.2),

$$\operatorname{codim}_A(a_G) = 6g - 6 + 3n - E(G) \equiv V(G) \mod 2\,.$$

Q.E.D.

Set

$$\chi_{g,n} = \chi(S \setminus P) = 2 - 2g - n\,.$$

Following Kontsevich, we will compute the formal series

$$Y(t) = \sum_{\substack{g\geq 0,\, n>0\\ \chi_{g,n}<0}} \frac{\chi_{virt}(M_{g,n})}{n!} t^{\chi_{g,n}}$$

as the asymptotic expansion of an integral which we will then compute in a different way. The comparison will yield the result.

LEMMA (9.3).

$$\log\left(\sqrt{\frac{t}{2\pi}}\int_{\mathbb{R}} \exp\left(-t\sum_{j=2}^{\infty}\frac{x^j}{j}\right)dx\right) \underset{t^{-1}\to 0}{\asymp} Y(t)\,. \tag{9.4}$$

Before proving the lemma, we need to fix some notation. We will consider k-tuples of integers

$$\nu_\bullet = (\nu_1, \nu_2, \nu_3, \ldots, \nu_k)\,.$$

We will always assume that

$$\nu_1 = \nu_2 = 0 \qquad \text{and} \quad \nu_1 + 2\nu_2 + 3\nu_3 + \cdots + k\nu_k \equiv 0 \mod 2\,,$$

and for simplicity, we also assume that $\nu_s = 0$ for $s > k$. We set

$$E_{\nu_\bullet} = \frac{1}{2}(\nu_1 + 2\nu_2 + 3\nu_3 + \cdots + k\nu_k)\,, \qquad V_{\nu_\bullet} = \nu_1 + \cdots + \nu_k\,.$$

We also set

$$\mathcal{G}_{\nu_\bullet} = \left\{\begin{array}{l}\text{isomorphism classes of ribbon graphs}\\ \text{with } \nu_j \text{ vertices of valency } j = 1,2,3,\ldots\end{array}\right\}.$$

We will denote with the same symbol a ribbon graph and its isomorphism class. From the definitions it follows that $E_{\nu_\bullet}$ and $V_{\nu_\bullet}$ are, respectively, the number of edges and the number of vertices of a ribbon graph belonging to $\mathcal{G}_{\nu_\bullet}$. Next we set

$$\mathcal{G}_{\nu_\bullet}^{(n)} = \left\{ \begin{array}{l} \text{isomorphism classes of ribbon graphs with } n \text{ boundary} \\ \text{components and with } \nu_j \text{ vertices of valency } j = 1, 2, 3, \dots \end{array} \right\},$$

so that

$$\mathcal{G}_{\nu_\bullet} = \mathcal{G}_{\nu_\bullet}^{(1)} \cup \dots \cup \mathcal{G}_{\nu_\bullet}^{(E_{\nu_\bullet}+1)} .$$

We also set

$$\mathcal{G}_{\nu_\bullet}^{n} = \left\{ \begin{array}{c} \text{isomorphism classes of ribbon graphs with} \\ n \textit{ marked} \text{ boundary components and with} \\ \nu_j \text{ vertices of valency } j = 1, 2, 3, \dots \end{array} \right\}.$$

Notice that, while in the definition of $\mathcal{G}_g^n$ the ribbon graphs in question are supposed to be connected, no such restriction is made in the definitions of $\mathcal{G}_{\nu_\bullet}$, $\mathcal{G}_{\nu_\bullet}^{(n)}$, and $\mathcal{G}_{\nu_\bullet}^{n}$.

Proof of Lemma (9.3). The proof of this lemma is really a repetition of the argument used to prove formula (5.14). Again, we are using $d\mu = \frac{1}{\sqrt{2\pi}} \exp(-\frac{x^2}{2})dx$ as probability measure on $\mathbb{R}$. Changing variable from x to $y/\sqrt{t}$, we obtain

$$\begin{aligned} \sqrt{\frac{t}{2\pi}} \int_{\mathbb{R}} \exp\left(-t\sum_{j=2}^{\infty} \frac{x^j}{j}\right) dx &= \frac{1}{\sqrt{2\pi}} \int_{\mathbb{R}} \exp\left(-\sum_{j=3}^{\infty} \frac{y^j}{j} \frac{1}{(\sqrt{t})^{j-2}}\right) e^{-\frac{y^2}{2}} dy \\ &= \left\langle \exp\left(-\sum_{j=3}^{\infty} \frac{y^j}{j} \frac{1}{(\sqrt{t})^{j-2}}\right)\right\rangle \\ &\underset{t^{-1}\to 0}{\asymp} \sum_{\nu_\bullet} \prod_{i=1}^{k} \frac{1}{\nu_i!(i)^{\nu_i}} (-1)^{V_{\nu_\bullet}} < y^{2E_{\nu_\bullet}} > t^{V_{\nu_\bullet} - E_{\nu_\bullet}} \\ &= \sum_{\nu_\bullet} \prod_{i=1}^{k} \frac{1}{\nu_i!(i)^{\nu_i}} (-1)^{V_{\nu_\bullet}} \left|P_{2E_{\nu_\bullet}}\right| t^{V_{\nu_\bullet} - E_{\nu_\bullet}} , \end{aligned}$$

where P_X denotes the set of pairings of a set of cardinality X, and where, to get the last equality, we used Wick's lemma. The group

$$H = \prod_{i=1}^{k} \mathfrak{S}_{\nu_i} \times C_i^{\nu_i} .$$

operates on $P_{2E_{\nu_\bullet}}$. The set of orbits of this action coincides with $\mathcal{G}_{\nu_\bullet}$. Given an (unmarked) ribbon graph G, assign to it a marking in an arbitrary way, and provisionally denote by $\mathrm{Aut}^{un}(G)$ the automorphism group of the unmarked graph G and by $\operatorname{Aut}(G)$ the automorphism group of G considered as a marked graph. Then

$$\begin{aligned}\prod_{i=1}^{k}\frac{1}{\nu_i!(i)^{\nu_i}}|P_{2E\nu_\bullet}| &= \sum_{G\in\mathcal{G}_{\nu_\bullet}}\frac{1}{|\operatorname{Aut}^{un}(G)|}\\ &= \sum_{n=1}^{E_{\nu_\bullet}+1}\sum_{G\in\mathcal{G}^{(n)}_{\nu_\bullet}}\frac{1}{|\operatorname{Aut}^{un}(G)|}\\ &= \sum_{n=1}^{E_{\nu_\bullet}+1}\sum_{G\in\mathcal{G}^{n}_{\nu_\bullet}}\frac{1}{n!}\frac{1}{|\operatorname{Aut}(G)|}\,.\end{aligned}$$

It follows that

$$\log\left(\sqrt{\frac{t}{2\pi}}\int_{\mathbb{R}}\exp\left(-t\sum_{j=2}^{\infty}\frac{x^j}{j}\right)dx\right)$$

$$\underset{t^{-1}\to 0}{\asymp}\sum_{g\geq 0,n>0,\,\chi_{g,n}<0}\frac{1}{n!}\left(\sum_{G\in\mathcal{G}^n_g}(-1)^{V(G)}\frac{1}{|\operatorname{Aut}(G)|}\right)t^{\chi_{g,n}}\,,$$

where, as usual, taking the logarithm has the effect of changing a sum over graphs into a sum over connected graphs. Q.E.D.

Let us denote by $\mathcal{A}(f)$ the asymptotic expansion, as $t^{-1}\to 0$, of a function $f=f(t)$. From the proof of the above lemma one easily gets the following equality of formal power series:

$$\lim_{N\to\infty}\mathcal{A}\left(\log\left(\sqrt{\frac{t}{2\pi}}\int_{\mathbb{R}}\exp\left(-t\sum_{j=2}^{2N}\frac{x^j}{j}\right)dx\right)\right)=Y(t)\,. \tag{9.5}$$

We are now going to compute in a different way the left-hand side of (9.5).

LEMMA (9.6).

$$\lim_{N\to\infty}\mathcal{A}\left(\log\left(\sqrt{\frac{t}{2\pi}}\int_{\mathbb{R}}\exp\left(-t\sum_{j=2}^{2N}\frac{x^j}{j}\right)dx\right)\right)=\log\left(\frac{e^t}{t^t\sqrt{2\pi t}}\Gamma(t+1)\right)\,.$$

Proof. Write

$$\int_{-\infty}^{\infty}\exp\left(-t\sum_{j=2}^{2N}\frac{x^j}{j}\right)dx=\left(\int_{-\infty}^{-1}+\int_{-1}^{1}+\int_{1}^{\infty}\right)\exp\left(-t\sum_{j=2}^{2N}\frac{x^j}{j}\right)dx\,.$$

The integrand is a rapidly decreasing function of x in the intervals $(-\infty,-1]$ and $[1,\infty)$. Therefore, the asymptotic expansions of the first and third integrals, as $t^{-1}\to 0$, are the zero series. On the other hand, we have

$$\lim_{N\to\infty}\exp\left(-t\sum_{j=2}^{2N}\frac{x^j}{j}\right)=(1-x)^t e^{tx}$$

absolutely in the interval $[-1,1]$. We then have

$$\begin{aligned}\lim_{N\to\infty}\mathcal{A}\left(\int_{-\infty}^{\infty}\exp\left(-t\sum_{j=2}^{2N}\frac{x^j}{j}\right)dx\right)&=\mathcal{A}\left(\int_{-1}^{1}\exp\left(-t\sum_{j=2}^{2N}\frac{x^j}{j}\right)dx\right)\\&=\mathcal{A}\left(\int_{-1}^{1}(1-x)^t e^{tx}dx\right)\\&=\mathcal{A}\left(e^t\int_0^2 y^t e^{-ty}dy\right)\\&=\mathcal{A}\left(e^t\int_0^\infty y^t e^{-ty}dy\right)\\&=\mathcal{A}\left(\frac{e^t}{t^{t+1}}\int_0^\infty u^t e^u du\right)\\&=\mathcal{A}\left(\frac{e^t}{t^{t+1}}\Gamma(t+1)\right),\end{aligned}$$

where the fourth equality is a consequence of the fact that the asymptotic expansion of $\int_2^\infty y^t e^{-ty}dy$ for $t^{-1}\to 0$ is the zero series. Q.E.D.

Putting the above two lemmas together, we get

$$Y(t)=\mathcal{A}\left(\log\frac{\Gamma(t+1)}{2\pi}-t\log t+t-\frac{1}{2}\log t\right). \tag{9.7}$$

To complete our computation of the series $Y(t)$, we make the following remarks. First of all, from Lemma (8.11), we have

$$\chi_{virt}(M_{g,n+1})=\chi_{virt}(M_{g,n})\chi_{g,n}\,,\qquad \chi_{g,n+1}=\chi_{g,n}-1\,. \tag{9.8}$$

As a consequence,

$$\frac{\chi_{virt}(M_{g,n+1})}{(n+1)!}t^{\chi_{g,n+1}}=\frac{d}{dt}\left(\frac{1}{n+1}\frac{\chi_{virt}(M_{g,n})}{n!}t^{\chi_{g,n}}\right).$$

Set

$$A(t)=A_0(t)+A_1(t)\,,$$

where

$$A_0(t) = t\log(t) - t\,, \qquad A_1(t) = \sum_{g\geq 1} \chi(M_{g,1}) t^{\chi_{g,1}}\,.$$

We have

$$\begin{aligned} Y(t) &= \sum_{g\geq 1, n\geq 1} \frac{\chi(M_{g,n})}{n!} t^{\chi_{g,n}} + \sum_{n\geq 3} \frac{\chi(M_{0,n})}{n!} t^{\chi_{0,n}} \\ &= \sum_{\nu\geq 0} \frac{1}{(\nu+1)!} \frac{d^\nu}{dt^\nu} A_1(t) + \sum_{\nu\geq 3} \frac{1}{(\nu+1)!} \frac{d^\nu}{dt^\nu} A_0(t) \\ &= \sum_{\nu\geq 0} \frac{1}{(\nu+1)!} \frac{d^\nu}{dt^\nu} A(t) - A_0(t) - \frac{1}{2}\frac{d}{dt} A_0(t) \\ &= \int_t^{t+1} A(x)dx - t\log(t) + t - \frac{\log(t)}{2}\,. \end{aligned}$$

From (9.7) it follows that

$$\int_t^{t+1} A(x)dx = \mathcal{A}\left(\log\frac{\Gamma(t+1)}{\sqrt{2\pi}}\right)\,.$$

Since $B(x) = \frac{1}{2} - x + x\frac{d}{dx}\log\Gamma(x)$ satisfies the identity $\int_t^{t+1} B(x)dx = \log\frac{\Gamma(t+1)}{\sqrt{2\pi}}$, we get

$$A(t) = \mathcal{A}\left(\frac{1}{2} - t + t\frac{d}{dt}\log\Gamma(t)\right)\,.$$

Now, Stirling's asymptotic expansion of the logarithm of the Γ function is given by

$$\begin{aligned} \log\Gamma(t) - \frac{1}{2}\log(2\pi) - \left(t - \frac{1}{2}\right)\log(t) + t \underset{t^{-1}\to 0}{\asymp} &\sum_{\nu=1}^{\infty} \frac{B_{\nu+1}}{(\nu+1)\nu} t^{-\nu} \\ &= \sum_{\nu=1}^{\infty} \frac{\zeta(-\nu)}{-\nu} t^{-\nu}\,, \end{aligned}$$

where the B_ν are the Bernoulli numbers. It follows that

$$\sum_{g\geq 1} \chi(M_{g,1}) t^{1-2g} = \sum_{g\geq 1} \zeta(1-2g) t^{1-2g}\,.$$

Using (9.8), we then get the following theorem.

THEOREM (9.9).

$$\chi_{virt}(M_{0,n}) = (-1)^{n-3}(n-3)! \qquad \text{for } n \geq 3,$$
$$\chi_{virt}(M_{g,n}) = (-1)^n \frac{(2g+n-3)!}{(2g-2)!}\zeta(1-2g) \qquad \text{for } g \geq 1,\ 2g-2+n>0.$$

When $g = n = 1$, we get back the result anticipated in formula (6.5) of Chapter XIX.

10. A very quick tour of Gromov–Witten invariants.

The theory of Gromov–Witten invariants goes well beyond the scope of this volume and should constitute the subject of another book.

Here we would like to give the idea of what this theory is, in a very quick overview, and, for obvious reasons, we will keep the number of bibliographical references to a bare minimum.

After Gromov's work, enumerative geometry takes a sharp detour. The enumerative problems, per se, lose their center stage position, and their solution starts being viewed as a way of producing invariants of the ambient space. For instance, the fact that there are (as Chasles, de Jonquières, and Zeuthen knew) 640 rational nodal plane quartic through 11 general points in $\mathbb{P}^2$ is certainly a property regarding quartic curves, but, in a deeper sense, it is a characteristic property of the projective plane $\mathbb{P}^2$.

Fix a smooth projective variety V. This will be the ambient space we want to explore. The idea is to use curves as probes. More precisely, one defines the following, seemingly uncomputable, numbers. Fix a homology class β in $H_2(V,\mathbb{Z})$. The latter is a discrete abelian group, and an element of it should be viewed as a discrete invariant as, for instance, the degree of a plane curve. Next look at all genus g curves C and maps $f : C \to V$ such that $[f(C)] = \beta$ (i.e., curves of fixed "degree" and genus). Also fix a certain number of subvarieties $V_1, \dots, V_n \subset V$ and suppose that there is some way of saying that there is only a finite number N of curves C and maps f such that

$$[f(C)] = \beta, \qquad f(C) \text{ meets each of the } V_i. \tag{10.1}$$

This number N is a Gromov–Witten invariant of V. It is denoted with the symbol

$$< [V_1], \dots, [V_n] >^V_{\beta,g} .$$

In other words,

$$< [V_1], \dots, [V_n] >^V_{\beta,g} = \left| \left\{ \begin{array}{c} \text{pairs } ((C,p_1,\dots,p_n);f) \text{ with } C \text{ of} \\ \text{genus } g,\ p_i \in C,\ f : C \to V, \\ [f(C)] = \beta,\ f(p_i) \in V_i,\ i = 1,\dots,n \end{array} \right\} \Big/ \text{iso} \right| ,$$

where $((C,p_1,\dots,p_n),f)$ is isomorphic to $((C',p'_1,\dots,p'_n),f')$ if there is a bianalytic map $\varphi : C \to C'$ such that $f = f' \circ \varphi$ and $\varphi(p_i) = p'_i$, $i = 1,\dots,n$.

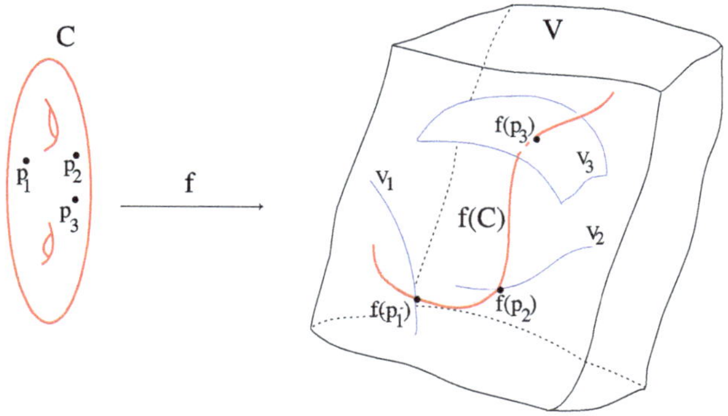

Figure 4.

Kontsevich introduced the appropriate *compact* moduli space to express these numbers as intersection numbers. It is the *moduli space of stable maps:*

$$\overline{M}_{g,n}(V,\beta) = \left\{ \begin{array}{c} \text{pairs } ((C,p_1,\ldots,p_n);f) \text{ with } (C,p_1,\ldots,p_n) \\ \text{an } n\text{-pointed nodal curve of genus } g, \\ f: C \to V \text{ a stable map, } [f(C)] = \beta. \end{array} \right\} \Big/ \text{iso}\,.$$

A *stable map* is simply an analytic map having the following restrictive property. If it happens to contract some of the irreducible components of the pointed curve C to a point, which it may, then that component (as pointed curve) should only have a finite number of automorphisms. There are natural *evaluation maps:*

$$\begin{aligned} \mathrm{ev}_i : \overline{M}_{g,n}(V,\beta) &\longrightarrow V \\ [(C,p_1,\ldots,p_n);f] &\mapsto f(p_i) \end{aligned} \tag{10.2}$$

and the Gromov–Witten invariants of V are easily expressed as intersection numbers on $\overline{M}_{\underline{g,n}}(V,\beta)$ by pulling back the Poincaré duals ω_{V_i} of the classes $[V_i]$ to $\overline{M}_{g,n}(V,\beta)$, via the various evaluation maps, and then intersecting them:

$$< [V_1],\ldots,[V_n] >^V_{\beta,g} = \int_{\overline{M}_{g,n}(V,\beta)} \mathrm{ev}_1{}^*(\omega_{V_1}) \wedge \cdots \wedge \mathrm{ev}_n^*(\omega_{V_n})\,. \tag{10.3}$$

(see, for example, [276] and [497]).

It is a rather extraordinary fact that the Gromov–Witten invariants define a deformation of the multiplicative structure of $H_\bullet(V)$. To explain this point, consider a basis $[U_0],\ldots,[U_N]$ of $H_\bullet(M)$ and the classical intersection matrix

$$g_{ij} = \begin{cases} [U_i]\cdot[U_j] = \displaystyle\int_V \omega_{U_i} \wedge \omega_{U_i} & \text{if } i+j = \dim(V), \\ 0 & \text{otherwise.} \end{cases} \tag{10.4}$$

This matrix captures the classical multiplicative structure of $H_\bullet(M)$. Now one forms the so-called *Gromov–Witten potential*

$$\Phi(x_0,\dots,x_N) = \sum_{g,n_0+\cdots+n_N\geq 3}\ \sum_{\beta\in H_2(V)} < [U_0]^{n_0}\cdots[U_N]^{n_N} >^V_{\beta,g} \frac{x_0^{n_0}}{n_0!}\cdots\frac{x_N^{n_N}}{n_N!}$$

and defines a new *quantic* product by setting

$$[U_i]*[U_j] = \sum_{kl} \frac{\partial^3\Phi}{\partial x_i\partial x_j\partial x_k} g^{kl}[U_l]\,.$$

It turns out that this is an associative product in $H_\bullet(V)\otimes\mathbb{Q}[[x_0,x_1,\dots,x_N]]$, and it can be easily checked that the constant term of $[U_i]*[U_j]$, as a series in the x_i, is the classical intersection product $[U_i]\cdot[U_j]$. Now, the associativity equation for the quantum product $*$ gives differential equations for the Gromov–Witten potential Φ, the so-called WDVV equations. These equations, in some cases, may be used to completely determine a certain number of GW-invariants.

It came as a surprise when Kontsevich [446], in 1993, was able to solve by this method the classical enumerative problem of determining the number N_d of degree d genus 0 plane nodal curves passing through $3d-1$ points in general position in $\mathbb{P}^2$. Indeed, he showed that these numbers can be computed inductively starting with Euclid's 1st axiom $N_1=1$. The fact that through 5 points in general position there passes a unique conic ($N_2=1$) was determined by Apollonius. Chasles proved that through 8 general points there pass 12 nodal cubics ($N_3=12$). As we already mentioned, he, de Jonquières, and Zeuthen proved that $N_4=620$. Schubert computed $N_5=87304$, but there was no hope at that time to see that, say, $N_8=13525751027392$. It turns out that, when $V=\mathbb{P}^2$, the quantum part (i.e., the nonconstant part) Φ_q of the Gromov–Witten potential Φ can be written in the form

$$\Phi_q(x,y) = \sum_{d=1}^{\infty} N_d \frac{y^{3d-1}}{(3d-1)!}e^{dx}\,. \tag{10.5}$$

There are only two variables because $H_\bullet(\mathbb{P}^2)$ is two-dimensional. Now the associativity equation for $*$ implies that

$$\frac{\partial^3\Phi_q}{\partial y^3} = \left(\frac{\partial^3\Phi_q}{\partial x^2\partial y}\right)^2 - \frac{\partial^3\Phi_q}{\partial x\partial y^2}\frac{\partial^3\Phi_q}{\partial y^3}\,, \tag{10.6}$$

and this gives the nontrivial recursive formula

$$N_d = \sum_{\substack{d_1+d_2=d\\ d_1,d_2>0}} N_{d_1}N_{d_2}\left(d_1^2d_2^2\binom{3d-4}{3d_1-2} - d_1^3d_2\binom{3d-4}{3d_1-1}\right)\,,\quad d>1\,,$$

computing N_d for all $d > 1$ from $N_1 = 1$.

In general, the computation of the Gromov–Witten invariants involves:

a) The solution of a highly nontrivial foundational problem. This is done by Ruan and Tian [609] in the symplectic setting and by Behrend and Fantechi [57] in the algebro-geometric one.
b) The idea of using the Atiyah–Bott localization theorem. This was first conceived by Kontsevich and then very efficiently put to work by Graber and Pandharipande [306].
c) The solution of a general conjecture by Eguchi, Hori, and Xiong [193]. This conjecture goes also under the name of Virasoro conjecture (see Getzler [286]) and is solved in some important cases by Liu and Tian [482], Okounkov and Pandharipande [574,575,576] and Givental [299].

We briefly comment on these three points only looking at the algebro-geometric approach. For the approach via symplectic geometry, we refer the reader to the beautiful survey by Li and Tian [473].

a) K. Behrend and B. Fantechi were able to construct a fundamental class for $\overline{M}_{g,n}(V,\beta)$, for any projective variety V. They called it the *virtual fundamental class.* This major achievement allowed the Gromov–Witten theory to start on very firm ground. After their work the symbol

$$\int_{\overline{M}_{g,n}(V,\beta)}$$

makes perfect sense. The fact that a fundamental class exists is a remarkable fact in view of the nature of $\overline{M}_{g,n}(V,\beta)$, which may be as badly behaved as a Hilbert scheme. What comes to help in defining the virtual fundamental class is the infinitesimal deformation theory for curves.

b) When a variety M possesses a torus action, then, using equivariant cohomology, Atiyah and Bott gave a way of reducing the computation of integrals of the type $\int_M \Phi$ to a sum of contribution of the type $\int_F (i^*(\Phi)/E(N_F))$, where F is a connected component of the fixed locus of the action, i is the inclusion, N_F is the normal bundle, and $E(N_F)$ is its Euler class. Now consider the case of Gromov–Witten invariants. The variety in question is $\overline{M}_{g,n}(V,\beta)$. If the target variety V admits a torus action, as happens when $V = \mathbb{P}^r$, then this action lifts to $\overline{M}_{g,n}(V,\beta)$. Graber and Pandharipande proved (1999) that the Atiyah–Bott localization formula works also in the context of the virtual fundamental class, so one may use localization when needed. The pleasant aspect of this idea (originally due to Kontsevich) is that, very often, the components of the fixed locus of a torus action on $\overline{M}_{g,n}(V,\beta)$ are simply copies of $\overline{M}_{p,\nu}$, i.e., moduli spaces of curves. This happens because fixed points of the action appear in correspondence to stable maps $C \to V$ that happen to collapse to a point an irreducible component Γ of C. If that component is ν-pointed and of genus p, the fixed locus picks up a component isomorphic to $\overline{M}_{p,\nu}$. In conclusion, in good cases, one may reduce the computation of GW-invariants to integrals over the much better behaved moduli spaces of curves, a substantial improvement.

c) To state the Virasoro conjecture, we need one more ingredient. Until now, we ralated the moduli spaces of stable maps $\overline{M}_{g,n}(V,\beta)$ with target variety V via the evaluation maps (2), and by means of these we defined the GW-invariants (10.3). But the moduli space $\overline{M}_{g,n}(V,\beta)$ is also closely related to the usual moduli space of stable curves via the obvious forgetful map,

$$\begin{gathered}\overline{M}_{g,n}(V,\beta) \longrightarrow \overline{M}_{g,n}\\ [C,p_1,\ldots,p_n;f] \mapsto [C,p_1,\ldots,p_n].\end{gathered}$$

Via this map, one can pull back to $\overline{M}_{g,n}(V,\beta)$ the classes $\psi_1,\ldots,\psi_n \in H^2(\overline{M}_{g,n})$. We call these pullbacks by the same name. We can then enrich the GW-invariants by mixing them with the ψ_i. We let $\{\gamma_a : a \in A\}$ be a basis of $H^\bullet(V)$. Write

$$\begin{aligned}&< \tau_{k_1}(\gamma_{a_1})\ldots\tau_{k_n}(\gamma_{a_n}) >^V_{g,\beta}\\ &\qquad = \int_{\overline{M}_{g,n}(V,\beta)} \psi_1^{k_1}\wedge\cdots\wedge\psi_n^{k_n}\wedge \mathrm{ev}_1^*(\gamma_{a_1})\wedge\cdots\wedge\mathrm{ev}_n^*(\gamma_{a_n})\,.\end{aligned}$$

It is important to notice that, when the target variety V reduces to a point $\{pt\}$, then the invariants $< \tau_{k_1}([pt])\ldots\tau_{k_n}([pt]) >^{\{pt\}}_{g,[pt]}$ reduce to Witten's invariants (1.2). Following Witten's procedure, one then introduces the Gromov–Witten potential

$$\begin{aligned}&F(\ldots t_{k_i a_i}\ldots)\\ &\quad = \sum_{n=0}^{\infty}\frac{1}{n!}\sum_{\substack{k_1,\ldots,k_n\\ a_1,\ldots,a_n}}\sum_{\beta\in H_2(V)} q^\beta < \tau_{k_1}(\gamma_{a_1})\ldots\tau_{k_n}(\gamma_{a_n}) >^V_{g,\beta} t_{k_1a_1}\ldots t_{k_na_n}\,.\end{aligned}$$

Eguchi, Hori, and Xiong define differential operators $\widetilde{L}_k$, $k = -1,0,1,2,\ldots$, of order less than or equal to 2 in the variables $t_{k_ia_i}$, satisfying the Virasoro conditions $[\widetilde{L}_m,\widetilde{L}_n] = (m-n)\widetilde{L}_{m+n}$, and conjecture that

$$\widetilde{L}_k(\exp(F)) = 0\,,\quad k = -1,0,1,2,\ldots\,. \tag{10.7}$$

This is the Virasoro conjecture. The operators $\widetilde{L}_k$ defined by Eguchi, Hori, and Xiong are a direct generalization of the ones introduced by Witten for the case $V = \{pt\}$, and the conjecture (10.7) reduces to Witten's conjecture when the target variety V is a single point.

At the time of writing this book, the Virasoro conjecture is open for a general projective variety.

11. Bibliographical notes and further reading.

There are several proofs of Witten's conjecture. In our text we followed Kontsevich's original paper [444] (see also Looijenga's Bourbaki talk [486]), but at the end of the proof we followed the shortcut proposed by Witten [685]. This shortcut has been analyzed in depth by Fiorenza and Murri [263].

We also departed from Kontsevich's paper as we are using hyperbolic geometry, instead of the theory of Jenkins–Strebel differentials, in order to define the cellular decomposition of moduli spaces, the reason being that the cellular decomposition obtained in this way seems more suitable to be extended to the boundary of moduli.

There are many sources for the graph enumeration techniques and for matrix integrals as, for instance, the papers of Bessis, Di Francesco, Itzykson, Zuber, Kontsevich, Mulase, and Zvonkin [388], [69], [390], [389], [168], [445], [545], [543], [693].

The cell decomposition of moduli spaces makes it possible to define a number of combinatorial cycles in rational homology. A comparison between combinatorial cycles and algebro-geometric cycles can be found in the papers of Penner, Mondello, Igusa, [591], [516], [384], [383], and in [27].

The first algebro-geometric proof of Witten's conjecture (and by this we mean one not using the cellular decomposition of moduli spaces in terms of ribbon graphs) is due to Okounkov and Pandharipande. The starting point of Okounkov–Pandharipande's proof is the fundamental work [209] by Ekedahl, Lando, Shapiro, and Vainshtein, where the authors provide a remarkable link between Hurwitz numbers and the intersection theory on the moduli space of curves. Let d be a positive integer, and let $\mu = (\mu_1, \dots, \mu_l)$ be a partition of $d = |\mu|$ into l parts. Let $g \geq 0$ be an integer. Denote by $H_{g,\mu}$ the Hurwitz number associated to these data, which may be defined as a weighted count of genus g degree d covers of $\mathbb{P}^1$ with ramification profile over $\infty \in \mathbb{P}^1$ given by μ and simply branched over $(2g-2+|\mu|+l)$ points. The formula for Hurwitz numbers proved in [209], the so-called ELSV formula, is

$$H_{g,\mu} = \frac{(2g-2+|\mu|+l)!}{\mathrm{Aut}(\mu)} \prod_{i=1}^{l} \frac{\mu_i^{\mu_i}}{\mu_i!} \int_{\overline{M}_{g,1}} \frac{\sum_{k=0}^{g}(-1)^k \lambda_k}{\prod_{i=1}^{l}(1-\mu_i\psi_i)} . \tag{11.1}$$

A considerable part of [572] is devoted to rederiving the above equality. This involves the localization-of-virtual-class techniques first introduced by Kontsevich and developed in great depth by Pandharipande and his collaborators. In his previous paper [570], while studying the generating function of Hurwitz numbers, Okounkov proposed to use formula (11.1) and the Toda equations to prove Witten's conjecture. This program, although not literally, was carried out in [572]. More precisely, the idea is that from this formula one can derive Kontsevich's fundamental combinatorial identity (4.6). Looking at the RHS of (11.1), Okounkov and Pandharipande make their first remarkable observation, namely that, via Laplace transform, the asymptotic behavior of Hurwitz numbers is governed by the LHS of (4.6). To get Kontsevich's identity, they proceed in two steps. By dissecting the surface C along paths joining ramification points, they first establish that another way to look at the asymptotics of Hurwitz numbers is to relate them with the asymptotics of random trees, and then they relate these to ribbon graphs, showing that the asymptotics of Hurwitz numbers are also governed by the RHS of 6. In this way Okounkov and Pandharipande are able to short-circuit the need for a combinatorial expression of the ψ_i which was a delicate point of Kontsevich's proof.

A different proof of the ELSV formula is given by Graber and Vakil [309], who refined an idea of Fantechi and Padharipande [245].

A shorter algebro-geometric proof of Witten's conjecture is given by Kazarian and Lando [407]. Also in this proof the starting point is the ELSV formula (or better its inverse) which expresses Hodge integrals in terms of Hurwitz numbers. Then the authors use a combinatorial trick to express the ψ-integrals in terms of Hodge integrals (and, therefore, in terms of Hurwitz numbers via the ELSV formula). This allows them to write the generating function of ψ-integrals in terms of generating functions of Hurwitz numbers. At this point they use a result of Okounkov [570], which says that the generating function of Hurwitz numbers satisfies the KP hierarchy, and this implies Witten's conjecture. The key idea of inverting the ELSV formula was already contained in the papers of Shadrin [638] and Zvonkine [694].

A completely different and rather astonishing proof of Witten's conjecture has been given by Mirzakhani [513]. In her paper she establishes a relationship between the Weil–Petersson symplectic form on the moduli space of hyperbolic Riemann surfaces with geodesic boundary components of given length, and the intersection numbers of the ψ-classes on moduli spaces of curves, and as a biproduct she obtains a result which is equivalent to Witten's conjecture (cf. also [546]).

A treatement of the link between matrix models and the KdV equation can be found in the book by Kaku [402]. This link was found in the late 1980s by a group of theoretical physicists, Kazakov [405], Brézin and Kazakov [86], Douglas [184], Douglas and Shenker [185], Gross and Migdal [323], in connection with the nonperturbative description of 2D gravity. The discrete model used in this description and the corresponding partition function led to the Toda lattice, and passing from a discrete to a continuous parameter the KdV equation made its appearance. In the same period Witten, Dijkgraaf, Verlinde, and Verlinde [684], [182], [181],

[179] discovered that the same partition function arises in 2D topological gravity, and in this setting the string equation and the KdV equation appear, so to speak, on equal footing by virtue of the Virasoro operators.

There are many sources for the general theory of the KdV and KP equations. Among them are the ones by Segal and Wilson [617], Dickey [178], Date, Kashiwara, Jimbo, and Miwa [158], Kac and Raina [401], and Mulase [544] (cf. also [21]).

The virtual Euler–Poincaré characteristic of $M_{g,n}$ was first computed by Harer and Zagier in their fundamental paper [344] and by Penner [589,590]. In [498], Manin gives a formula for the Poincaré polynomial of $\overline{\mathcal{M}}_{0,n}$. This formula was independently established by Getzler in [284], where the $\mathfrak{S}_n$-invariant Poincaré polynomial of $\overline{\mathcal{M}}_{0,n}$ is also computed. The Euler–Poincaré characteristic of $\overline{\mathcal{M}}_{g,n}$ has been computed by Bini and Harer in [72].

An important generalization of Witten's conjecture is the so-called r-spin Witten's conjecture. Building on work by Givental [300], this conjecture has been proved by Faber, Shadrin, and Zvonkine in [240] (see also Polishchuk [597]).

12. Exercises.

A. Virasoro equations and intersection numbers

We presented two ways of recursively computing intersection numbers. One is to use the Virasoro equations $L_n Z = 0$, and the other is to use the string equation coupled with the KdV hierarchy. The reader should compare these two methods in explicit cases. We propose, as exercises, the following computations. For this, we go back to the notation of (3.3), (3.4), and (3.5). We also set

$$F_g(t_0, t_1, \dots) = \sum_{\substack{d_1,\dots,d_n \; d_i \geq 0 \\ d_1+\cdots+d_n-n=3g-3}} \frac{1}{n!} < \tau_{d_1} \cdots \tau_{d_n} > t_{d_1} \cdots t_{d_n}\,, \tag{12.1}$$

so that $F = \sum_0^\infty F_g$. Furthermore, we set

$$\ll \tau_{d_1} \cdots \tau_{d_n} \gg_g = \frac{\partial}{\partial t_{d_1}} \frac{\partial}{\partial t_{d_2}} \cdots \frac{\partial}{\partial t_{d_n}} F_g\,. \tag{12.2}$$

A-1. Prove that

$$\begin{aligned} \ll \tau_k \tau_0 \tau_0 \gg = \frac{1}{2k+1} \Big(& \ll \tau_{k-1} \tau_0 \gg \ll \tau_0^3 \gg \\ & + 2 \ll \tau_{k-1} \tau_0^2 \gg \ll \tau_0^2 \gg + \frac{1}{4} \ll \tau_{k-1} \tau_0^4 \gg \Big) \end{aligned}$$

and

$$\ll \tau_k\tau_0\tau_0 \gg_g = \frac{1}{2k+1}\Bigg(\sum_{h=0}^{g} \ll \tau_{k-1}\tau_0 \gg_h \ll \tau_0^3 \gg_{g-h} + 2\sum_{h=0}^{g} \ll \tau_{k-1}\tau_0^2 \gg_h \ll \tau_0^2 \gg_{g-h} + \frac{1}{4} \ll \tau_{k-1}\tau_0^4 \gg_{g-1}\Bigg).$$

A-2. Prove that

$$< \tau_0^2\tau_3 >= \frac{1}{24}, \qquad < \tau_0^2\tau_2^2 > - < \tau_0^2\tau_3 >= \frac{1}{8},$$
$$< \tau_2\tau_3 >= \frac{29}{5760}, \qquad < \tau_2^3 >= \frac{7}{240}.$$

A-3. Prove that

$$< \tau_{3g-2} >_{g,1}= \frac{1}{(24)^g g!}.$$

In the next series of exercises the reader will prove the two formulas (7.15) and (7.16). The proof follows Faber [230], Faber and Pandharipande [237,236], and an observation by Dijkgraaf regarding the *three-point function* (cf. [237]). Set

$$\tau(z) = \sum_{n\geq 0} \tau_n z^n .$$

The general three-point function is defined as

$$E(x,y,z) =< \tau(x)\tau(y)\tau(z) >= \sum_{a,b,\geq 0} < \tau_a\tau_b\tau_c > x^a y^b z^c . \tag{12.3}$$

It can be shown that the KdV equation can be written in terms of the three-point function $E(x,y,z)$ as follows:

$$\begin{aligned}\left(2x\frac{\partial}{\partial x}+1\right)&\left((x+y+z)^2E(x,y,z)\right) = E(x,0,0)(y+z)^2E(0,y,z)\\ &+ xE(x,y,0)zE(0,0,z) + xE(x,0,z)yE(0,y,0) + x(x+y+x)E(x,y,z)\\ &+ 2xE(x,0,0)(y+z)E(0,y,z) + 2x(x+y)E(x,y,0)E(0,0,z)\\ &+ 2x(x+z)E(x,0,z)E(0,y,0) + \frac{1}{4}(x+y+z)^4xE(x,y,z) .\end{aligned}$$

In the exercises below we will not use the general three-point function but only two particular cases of it, namely the two-point function $E(0,w,z)$

and the special three-point function $E(w,z,-z)$, and we will derive the differential equations they satisfy.

A-4. Let $T_{\underline{d}} = \prod_{j=0}^{n} \tau_j^{d_j}$. Use Exercise A-1 to prove that

$$(2k+1) < \tau_k \tau_0^2 T_{\underline{d}} > = \sum_{0 \le a_j \le d_j} \left(\prod_{j=0}^{n} \binom{d_j}{a_j} \right) (< \tau_{k-1} \tau_0 T_{\underline{a}} >< \tau_0^3 T_{\underline{b}} > + 2 < \tau_{k-1} \tau_0^2 T_{\underline{a}} >< \tau_0^2 T_{\underline{b}} >) + \frac{1}{4} < \tau_{k-1} \tau_0^4 T_{\underline{d}} > ,$$

where $\underline{a} = \{a_1, \ldots, a_n\}$, and $T_{\underline{d}} = T_{\underline{a}} T_{\underline{b}}$.

A-5. Use the preceding exercise, with $k = a$ and $T_{\underline{d}} = \tau_b$, to prove that the two-point function

$$D(w,z) = E(0,w,z) =< \tau_0 \tau(w) \tau(z) >= \sum_{a,b \ge 0} < \tau_0 \tau_a \tau_b > w^a z^b$$

satisfies the differential equation

$$\left(2w\frac{\partial}{\partial w} + 1\right)((w+z)D(w,z)) = wD(w,z) + D(w,0)zD(0,z) + 2wD(w,0)D(0,z) + \frac{1}{4}(w+z)^3 wD(w,z) . \tag{12.4}$$

A-6. Show that the unique solution $D(w,z)$ of the differential equation (12.4) satisfying $D(w,0) = \exp(w^3/24)$ and $D(0,z) = \exp(z^3/24)$ is given by

$$D(w,z) = \exp\left(\frac{(w^3+z^3)}{24}\right) \sum_{k \ge 0} \frac{k!}{(2k+1)!} [\tfrac{1}{2} wz(w+z)]^k .$$

Show that all the terms of total degree $3g$ in $D(w,z)D(-w,0)$ have degree at least g in z, so that, in particular, the coefficient of $w^{2g+j} z^{g-j}$ in $D(w,z)D(-w,0)$ is zero. Deduce that, for all $j \ge 1$,

$$\sum_{h=0}^{g} \frac{(-1)^{g-h}}{24^{g-h}(g-h)!} < \tau_0 \tau_{3h-g+j} \tau_{g-j} >= 0 . \tag{12.5}$$

A-7. Use (12.5), the string equation, and recursion to prove formula (7.16).

A-8. a) Use Exercise A-4 with $k = a$ and $T_{\underline{d}} = \tau_b \tau_c$, together with the definition of $D(w,z)$ in Exercise A-5, to prove that the special three-point function

$$F(w,z) = E(w,z,-z) = \sum_{a,b,c \ge 0} < \tau_a \tau_b \tau_c > (-1)^c w^a z^{b+c}$$

satisfies the differential equation

(12.6)
$$4w^2F(w,z)+2w^3\frac{\partial F}{\partial w}(w,z)=w(2w+z)D(w,z)D(0,-z)$$
$$+w(2w-z)D(w,-z)D(0,z)+\frac{1}{4}w^5F(w,z)\,.$$

b) Show that the unique solution of this differential equation is given by

(12.7)
$$F(w,z)=\exp\left(\frac{w^3}{24}\right)\sum_{a,b\geq 0}w^{3a}(wz^2)^b\frac{(a+b)!}{2^{a+b-1}(2a+2b+2)!}\binom{a+b+1}{2a+1}.$$

c) Looking at the coefficient of $w^g z^{2g}$ in (12.7), prove formula (7.15).

A-9. Compute

$$\int_{\overline{M}_{1,2}}\xi_{irr*}(\kappa_1)\qquad \int_{\overline{M}_{1,2}}\xi_{1,\emptyset_*}(\kappa_1\times 1)\,,\qquad \int_{\overline{M}_{1,2}}\kappa_2\,,\qquad \int_{\overline{M}_{1,2}}\kappa_1^2\,.$$

B. **Asymptotic expansions and graphs**

B-1. Generalize (5.14) and prove that

$$\left\langle \exp\left(\sum_{s=1}^{2k}\frac{t_s x^s}{s!}\right)\right\rangle \underset{t\to 0}{\asymp} \sum_{\Gamma\in\mathcal{F}_k}\frac{1}{|\mathrm{Aut}(\Gamma)|}\prod_{s=1}^{2k}t_s^{v_s(\Gamma)}\,,$$

where $\mathcal{F}_k$ is the set of isomorphism classes of Feynman diagrams with vertices of valency not exceeding $2k$, and where $v_s(\Gamma)$ denotes the number of vertices of Γ with valency equal to s.

B-2. Recall from (5.16) the measure $d\mu_N$ on the space of $N\times N$ hermitian matrices and recall from (5.17) the meaning of the expectation value. Generalize (5.21) to prove the following asymptotic equality:

$$\left\langle \exp\left(\sum_{s=1}^{2k}\frac{t_s\operatorname{Tr}X^s}{s!}\right)\right\rangle \asymp \sum_{G\in\mathcal{G}_k}\frac{N^{|X_2(G)|}}{|\mathrm{Aut}(G)|}\prod_{s=3}^{2k}t_s^{v_s(G)}\,,$$

where $\mathcal{G}_k$ is the set of isomorphism classes of ribbon graphs with vertices of valency greater than 2 and not exceeding $2k$, and where $v_s(G)$ denotes the number of vertices of G with valency equal to s.

B-3. Prove equality (5.22).

B-4. Prove the second equality in (5.24).

B-5. Prove the equality in (5.26).

B-6. Using the change of variable $X \mapsto \sqrt{\Lambda}X$, prove the asymptotic equality (5.27).

C. Virtual Euler–Poincaré characteristic

Actually, (8.12) is a consequence of a more general result.

C-1. Given an exact sequence of groups

$$1 \longrightarrow \Gamma' \longrightarrow \Gamma \longrightarrow \Gamma'' \longrightarrow 1,$$

deduce from the Hochschild–Serre spectral sequence (cf. [88], VII.3) that

$$\chi_{virt}(\Gamma) = \chi_{virt}(\Gamma')\chi_{virt}(\Gamma'').$$

C-2. Prove the existence of an exact sequence

$$1 \longrightarrow \pi_1(S \smallsetminus \{x_1, \dots x_n\}) \longrightarrow \Gamma_{g,n+1} \longrightarrow \Gamma_{g,n} \longrightarrow 1.$$

Chapter XXI.
Brill–Noether theory on a moving curve

1. Introduction.

As we explained in the preface, when this volume was first conceived, it was planned that a major part of it would be devoted to Brill–Noether theory and its interactions with the moduli theory of curves. Over the years, the focus of the book changed, with the main theme becoming the geometry of the moduli spaces of curves. This left very little space for Brill–Noether theory, which in the meantime had developed at a rapid pace; it was thus impossible to give an up-to-date account of it. On the other hand, the alternative of leaving out Brill–Noether theory entirely seemed impractical, if for no other reason, because various foundational results announced in Volume 1, such as, for instance, Petri's statement, still had to be further discussed and proved.

We thus settled on a compromise solution. When Volume 1 was published, a draft version of parts of the intended second volume already existed. We decided to use this draft, with a few changes and additions, as the basis for the present chapter. As a consequence, the chapter has a bit of an archeological flavor, in content and in style of exposition, and does not completely reflect the present state of Brill–Noether theory. Rather, it gives a snapshot of what the theory looked like some twenty-five years ago. Nevertheless, we believe it may still be useful to the student of the subject.

The most fundamental question of the Brill–Noether theory is:

For which values of r and d does a
general curve of genus g possess a g^r_d?

In Theorem (2.3) of Chapter V we showed that *any* smooth curve of genus g possesses a g^r_d as long as the Brill–Noether number

$$\rho = g - (r+1)(g-d+r)$$

is nonnegative. If we think of the g^r_d as being given by an $(r+1)$-dimensional subspace W of the space of sections $H^0(C,L)$ of some degree d line bundle L on C, the nonnegativity of the Brill–Noether number says that the dimension of the codomain of the multiplication map

$$\mu_{0,W} : W \otimes H^0(C, \omega_C \otimes L^{-1}) \to H^0(C, \omega_C)$$

E. Arbarello et al., *Geometry of Algebraic Curves*,
Grundlehren der mathematischen Wissenschaften 268,
DOI 10.1007/978-3-540-69392-5_13,

is greater than or equal to the one of its domain; here and in the sequel we write μ_0 instead of $\mu_{0,W}$ when $W = H^0(C, L)$. The crudest statement about linear series on general curves states that, if a smooth curve C possesses a g^r_d with negative Brill–Noether number, then the curve is special in the sense of moduli. A more refined statement was implicitly assumed in a long forgotten paper by K. Petri (see (1.7) and the footnote on page 215 of Chapter V).

THEOREM (1.1) (PETRI'S STATEMENT, FIRST VERSION). *Let C be a general curve of genus g, and let L be a line bundle on C. Then the cup-product homomorphism μ_0 is injective.*

As we saw in Chapters IV and V, this statement is completely equivalent to the following two statements regarding the so called Brill–Noether varieties of linear series on a smooth curve C (see (1.6), Chapter V).

THEOREM (1.2) (PETRI'S STATEMENT, SECOND VERSION). *Let C be a general curve of genus g. Let d and $r \geq 0$ be integers. Then the variety $G^r_d(C)$ is smooth of dimension ρ.*

THEOREM (1.3) (PETRI'S STATEMENT, THIRD VERSION). *Let C be a general curve of genus g. Let d and r be integers such that $r \geq 0$ and $r > d - g$. Then the variety $W^r_d(C)$ has dimension ρ, and its singular locus is $W^{r+1}_d(C)$.*

Of course, in the above statements, having negative dimension means being empty. In this chapter, among other things, we prove the three equivalent statements above.

In Sections 2 and 3 we introduce the relative Picard variety $\mathcal{P}ic^d(p)$, parameterizing degree d line bundles on the fibers of a family $p : \mathcal{C} \to S$ of smooth pointed curves, together with the basic Brill–Noether varieties $\mathcal{C}^r_d(p)$, $\mathcal{W}^r_d(p)$, and $\mathcal{G}^r_d(p)$. In Section 4, using the language just introduced, we digress to prove a theorem by Looijenga stating that the tautological ring of M_g vanishes in degree greater than $g - 2$.

In Section 5 we go back to the Brill–Noether varieties and compute their Zariski tangent spaces in terms of the fundamental homomorphisms μ_0 and μ_1. From a closer analysis of the μ_1 homomorphism we deduce, in Section 6, the smoothness of $\mathcal{G}^1_d$, and, in preparation for the following section, we present Voisin's cohomological interpretation of the μ_1 homomorphism. In Section 7 we give Lazarsfeld's elegant proof of Theorem (1.1).

In Section 8 we introduce Horikawa's theory of deformations of mappings, and we use it to give yet another cohomological interpretation of the homomorphisms μ_0 and μ_1. We then apply Horikawa's theory to rigorously justify various naive moduli counts. In passing, we prove the theorem of de Franchis asserting the finiteness of the number of nonconstant morphisms from a fixed smooth curve to a (possibly variable) smooth curve of genus ≥ 2. The starting point of Horikawa's theory is the elementary observation that infinitesimal deformations of a morphism $\phi : C \to M$ from a curve to a manifold are classified by $H^0(C, N_\phi)$, where N_ϕ is the normal bundle to ϕ. In Section 9 we then interpret the torsion subsheaf $\mathcal{K}_\phi$ of N_ϕ and realize that, along a deformation whose Horikawa class lies in $H^0(C, \mathcal{K}_\phi)$, the complexity of the ramification of ϕ decreases. In Section 10 we apply this principle to prove that the dimension of the Severi variety $\Sigma_{d,g}$ of plane curves of given degree and genus (whose irreducibility has been proved by Joe Harris) is of dimension $3d + g - 1$ and that its general point corresponds to a nodal curve. We then prove that, for any pair of integers (d, r) with $d \geq r \geq 2$, there exist irreducible, nondegenerate nodal curves of degree d and genus g for every value of g between 0 and the Castelnuovo bound $\pi(d, r)$.

In Section 11 we introduce the Hurwitz scheme and prove its irreducibility. In Section 12, after revisiting some classical results by Castelnuovo and Segre on plane curves, we show that the most general curve possessing a g^1_d with negative Brill–Noether number does in fact possess a unique such g^1_d. Finally, in Section 13 we prove some unirationality results concerning loci of curves with small gonality and, in so doing, we rediscover the classical result asserting the unirationality of M_g for $g \leq 10$.

2. The relative Picard variety.

In Sections 2 and 3 of Chapter IV we constructed the Poincaré line bundles and the basic varieties of the Brill–Noether theory for a fixed smooth genus g curve. In this section and in the following one we are going to repeat the same construction for moving curves. Some of the steps in these relative constructions are the straightforward generalizations of their counterparts in the case of a fixed curve. We will therefore omit repetitive details. We suggest to the reader to go back and forth between this section and the corresponding section in Chapter IV. All our constructions can be performed both in the analytic and in the algebraic category, but we will be mostly concerned with the algebraic one. From now on, unless otherwise specified, we restrict our attention to smooth curves of genus $g > 1$.

Our first object of study is the relative Picard variety. Here is the basic existence theorem.

THEOREM (2.1). *Let d be an integer. Let $p : \mathcal{C} \to S$ be a family of smooth curves of genus $g > 1$ parameterized by a scheme S. Suppose that p admits a section. Then there exist a scheme over S*

$$\mathbf{Pic}^d(p) \to S$$

and a line bundle $\mathcal{L}_d = \mathcal{L}_d(p)$ over $\mathcal{C} \times_S \mathbf{Pic}^d(p)$ which restricts to a degree d line bundle on each fiber of p and which satisfies the following *universal property. For every morphism $f : T \to S$ and every line bundle $\mathcal{L}$ on $\mathcal{C} \times_S T$, restricting to a degree d line bundle on each fiber of $q : \mathcal{C} \times_S T \to T$, there exists a unique lifting $\varphi : T \to \mathbf{Pic}^d(p)$ of f such that*

$$\mathcal{L} = (\mathrm{id} \times \varphi)^* \mathcal{L}_d \otimes q^*(\mathcal{Q})$$

for some line bundle $\mathcal{Q}$ on T. The line bundle $\mathcal{L}_d$ is called a Poincaré bundle of degree d.

Before proving this theorem, we make a number of remarks. The first one is that when S is a point and $\mathcal{C}$ is a single curve C, then $\mathbf{Pic}^d(p) = \mathrm{Pic}^d(C)$, and the line bundle $\mathcal{L}_d$ has the properties of the Poincaré line bundle constructed in Lemma 2.2 of Chapter IV.

The second remark is that, if σ is a section of p and $\Sigma \subset \mathcal{C}$ is the corresponding divisor, then there is a unique Poincaré line bundle which restricts to the trivial bundle on $\Sigma \times_S \mathbf{Pic}^d(p) \cong \mathbf{Pic}^d(p)$. We will say that this particular Poincaré line bundle is *normalized with respect to σ*.

The third remark is that, because of the universal property, given a morphism of families of smooth, 1-pointed, genus g curves

$$\begin{array}{ccc} \mathcal{X} & \xrightarrow{F} & \mathcal{C} \\ {\scriptstyle q}\downarrow & & \downarrow{\scriptstyle p} \\ T & \xrightarrow{f} & S \end{array}$$

we have

$$\mathbf{Pic}^d(q) = T \times_S \mathbf{Pic}^d(p)\,.$$

In particular, the fiber over $s \in S$ of $\mathbf{Pic}^d(p)$ is $\mathrm{Pic}^d(C_s)$, where, as usual, we set $C_s = p^{-1}(s)$. Another consequence of this remark is that, in order to construct the pair $(\mathbf{Pic}^d(p), \mathcal{L}_d)$, with $\mathcal{L}_d$ normalized, it suffices to do so locally in S because, by the universal property, there is a canonical way to piece together the local constructions.

One futher remark is that one may rephrase Theorem (2.1) by saying that the S-scheme $\mathbf{Pic}^d(p)$ represents the *relative degree d Picard functor*

$$\mathcal{P}ic^d_{\mathcal{C}/S} : Sch/S \to Sets \tag{2.2}$$

defined by

$$\mathcal{P}ic^d_{\mathcal{C}/S}(T) = \mathrm{Pic}^d(\mathcal{C} \times_S T)/\,\mathrm{Pic}(T).$$

Here the quotient denotes the set $\mathrm{Pic}^d(\mathcal{C} \times_S T)$ modulo the equivalence relation that declares $[\mathcal{L}_1], [\mathcal{L}_2] \in \mathrm{Pic}^d(\mathcal{C} \times_S T)$ equivalent if and only if $\mathcal{L}_1 \otimes \mathcal{L}_2^{-1} \cong q^*(\mathcal{Q})$, where $q : \mathcal{C} \times_S T \to T$ and $\mathcal{Q} \in \mathrm{Pic}(T)$. In the literature a scheme representing this functor is often denoted with the symbol $\mathbf{Pic}^d_{\mathcal{C}/S}$, but here we find it more convenient to denote it with the symbol $\mathbf{Pic}^d(p)$. Of course, in the definition of the relative Picard functor, we could have started from a family $p : \mathcal{C} \to S$ of smooth curves of genus g without requiring the existence of a section. What turns out to be true is that the relative Picard functor is representable only when $p : \mathcal{C} \to S$ has a section. The theorem proves the positive part of this assertion.

The final remark is that it suffices to prove Theorem (2.1) for a single integer d. To see this, we make our first use of the existence of a section of $p : \mathcal{C} \to S$. Let σ be a section of p, and let $\Sigma \subset \mathcal{C}$ be the corresponding relative Cartier divisor. For every S-scheme T, we let Σ_T denote the pullback of Σ to $\mathcal{C} \times_S T$. Then, given integers d and e, we can define an isomorphism of functors

$$\mu : \mathcal{P}ic^e_{\mathcal{C}/S} \xrightarrow{\cong} \mathcal{P}ic^d_{\mathcal{C}/S} \tag{2.3}$$

by setting

$$\mu_T([\mathcal{L}]) = [\mathcal{L} \otimes \mathcal{O}((d-e)\Sigma_T)] .$$

It thus suffices to represent the functor $\mathcal{P}ic^d_{\mathcal{C}/S}$ for a single value of d.

Proof of Theorem (2.1). Let H be the Hilbert scheme of ν-log-canonical smooth 1-pointed genus g curves (cf. Section 5 in Chapter XI), and let

$$\begin{array}{c} \mathcal{C} \subset \mathbb{P}^N \times H \\ \varphi \Big\downarrow \quad\quad\quad \\ H \quad\quad\quad \end{array} \tag{2.4}$$

be the universal family. Given a family

$$p : \mathcal{C} \to S \tag{2.5}$$

of 1-pointed genus g curves, we let $\Sigma \subset \mathcal{C}$ denote the divisor defined by the section of p. Taking local frames for the locally free sheaf $p_*(\omega_p(\Sigma)^\nu)$, we may cover S by open sets U so that the family $p_U = p_{|U}$ is the pullback of the universal family (2.4) via a morphism $\eta : U \to H$. By our third remark above, in order to construct the pair $(\mathbf{Pic}^d(p), \mathcal{L}_d(p))$, it suffices to construct the pairs $(\mathbf{Pic}^d(p_U), \mathcal{L}_d(p_U))$, and therefore the pair $(\mathbf{Pic}^d(\varphi), \mathcal{L}_d(\varphi))$, for the universal family (2.4). In conclusion, since H is smooth, this reasoning tells us that we may prove the existence of the relative Picard scheme for a family (2.5) under the additional assumption that this is a family of ν-log canonically embedded, smooth, 1-pointed, genus g curves parameterized by a smooth (connected) variety S:

$$\begin{array}{c} \mathcal{C} \subset \mathbb{P}^N \times S \\ p \Big\downarrow \quad\quad\quad \\ S \quad\quad\quad \end{array} \tag{2.6}$$

Under these assumptions, we consider the relative symmetric product

$$\mathcal{C}_d \to S,$$

which may be viewed as the Hilbert scheme

$$\mathcal{C}_d = \mathrm{Hilb}^d_{\mathcal{C}/S}\,.$$

The *relative universal effective divisor of degree d*

$$\begin{array}{ccc} \mathcal{D} \subset \mathcal{C} \times \mathcal{C}_d & \xrightarrow{\ \pi\ } & \mathcal{C}_d \\ \big\downarrow & & \\ S & & \end{array} \tag{2.7}$$

is just the universal family over $\mathrm{Hilb}^d_{\mathcal{C}/S}$. The relative universal divisor of degree d satisfies the universal property which is the relative counterpart of the universal property enjoyed by the universal divisor of degree d over a fixed curve (see Lemma 2.1, Chapter IV). To define the scheme $\mathbf{Pic}^d(p)$, the idea is simply to take the quotient of $\mathcal{C}_d$ modulo linear equivalence.

We ask the reader to go back to the beginning of Section 3 in Chapter XII, where we introduced the notion of equivalence relation R on a scheme X, and where we explained what an effective quotient X/R is. The existence of such a quotient is a rare occurrence. A beautiful result of Grothendieck describes a situation in which effective quotients exist.

THEOREM (2.8). ("Scholie", Section XX of [329]) *Let $X \to S$ be a projective morphism. Let $R \underset{h}{\overset{f}{\rightrightarrows}} X$ be a schematic equivalence relation such that h is proper and flat. Then there exists an effective quotient $Y = X/R$. Moreover, Y is projective over S, and the quotient map $X \to Y$ is faithfully flat.*

We are going to assume this result; for a proof, we refer the reader to the lucid, self-contained account by Nitsure that can be found in [243], Section 5.6.3, Chapter 5. Look at the universal divisor (2.7). Suppose that $d >> 0$ so that $\pi_*\mathcal{O}(\mathcal{D})$ is locally free. Set

$$R = \mathbb{P}\pi_*\mathcal{O}(\mathcal{D}) \xrightarrow{h} \mathcal{C}_d\,.$$

Clearly, R is a smooth variety, projective over S, whose underlying set may be identified with the set

$$\{(D, D') \in \mathcal{C}_d \times_S \mathcal{C}_d \mid D \sim D'\}\,.$$

We may thus think of R as a smooth S-subvariety of $\mathcal{C}_d \times_S \mathcal{C}_d$ and assume that the morphism h is induced by the projection of $\mathcal{C}_d \times_S \mathcal{C}_d$ on one of its factors. Since the subvarieties S, $\mathcal{C}_d$, and R are all smooth, the fact that R is a set-theoretical equivalence relation makes R, automatically, into a schematic equivalence relation. As the morphism h is proper and flat, Grothendieck's theorem tells us that the quotient $\mathcal{C}_d/R$ exists. We set

$$\mathbf{Pic}^d(p) = \mathcal{C}_d/R\,,$$

and we let

$$u : \mathcal{C}_d \to \mathbf{Pic}^d(p) \tag{2.9}$$

denote the quotient morphism. We now construct the Poincaré line bundle. To construct $\mathcal{L}_d = \mathcal{L}_d(p)$, we fix a section σ of p and denote by $\Sigma \subset \mathcal{C}$ the relative divisor it defines. The morphism

$$\mathcal{C}_{d-1} = S \times_S \mathcal{C}_{d-1} \xrightarrow{\sigma\times\mathrm{id}} \mathcal{C} \times_S \mathcal{C}_{d-1} \to \mathcal{C}_d$$

defines a relative divisor

$$\mathcal{C}_\sigma \subset \mathcal{C}_d\,.$$

Exactly as in the case of a single curve (i.e., the case where S is a single point), one verifies that

$$\mathcal{L} = (\mathrm{id}_{\mathcal{C}} \times u)_* \mathcal{O}(\mathcal{D} - \pi^* \mathcal{C}_\sigma)$$

is a Poincaré line bundle on $\mathcal{C} \times_S \mathbf{Pic}^d(p)$ (see Lemma 2.2, Section 2, Chapter IV).

Q.E.D.

Of course, we could have constructed the relative Picard variety in the analytic category. In this category it is much easier to take quotients. For this reason, in the analytic construction of $\mathbf{Pic}^d(p)$ one can circumvent the use of Grothendieck's theorem (2.8). The most expedient way to proceed is the following. By the same reduction argument we used in the algebraic case, in the analytic case we can limit ourselves to constructing the relative Picard variety for a local Kuranishi family

$$\varphi : \mathcal{C} \to B \tag{2.10}$$

of smooth, genus g curves. Since the problem is local on S, we can furthermore assume that (2.10) admits an analytic section σ. We may also fix a frame for $\varphi_* \omega_\varphi$, where ω_φ is the sheaf of relative holomorphic differentials, that is, a basis $\omega_1(b), \dots, \omega_g(b)$ for $H^0(C_b, \omega_{C_b})$ varying holomorphically with b. Using the period map, described in Section 8 of Chapter XI, one then constructs the relative Jacobian

$$\mathcal{J}(\varphi) \to B\,.$$

We then have an analytic map

$$u : \mathcal{C}_d \to \mathcal{J}(\varphi) \tag{2.11}$$

defined in terms of the familiar abelian sums

$$u_b(D) = \left(\dots, \sum_{i=1}^{d} \int_{\sigma(b)}^{p_i} \omega_{\nu,b}\,, \dots \right),$$

where $D = p_1 + \dots + p_d$ and $u_b = u_{|C_b}$. When $d >> 0$, the morphism u identifies $\mathcal{J}(\varphi)$ with the quotient of $\mathcal{C}_d$ modulo linear equivalence. This is how we define $\mathbf{Pic}^d(\varphi)$ for $d >> 0$ in the analytic category. Then, using the section σ, and therefore the isomorphism (2.3), we define $\mathbf{Pic}^d(\varphi)$ for every $d \in \mathbb{Z}$. The construction of the relative Poincaré line bundle is exactly as in the algebraic case. In the analytic category we also have the exponential sequence

$$0 \to \mathbb{Z} \to \mathcal{O}_\mathcal{C} \to \mathcal{O}_\mathcal{C}^\times \to 0\,,$$

and we get the identification

$$\mathbf{Pic}^0(\varphi) = R^1\varphi_*\mathcal{O}_\mathcal{C}/R^1\varphi_*\mathbb{Z}\,.$$

Let us finally comment on the existence of the relative Picard variety for families $p : \mathcal{C} \to S$ that do not admit a section. We start with the analytic case. In this setting, using Kuranishi families, we may always assume that S is covered by small open sets where a section exists, and we may fix one such section for each open set U of this cover. One then constructs $\mathbf{Pic}^d(p_U)$ and uses the universal property to patch together these local pieces and define an analytic variety $\mathbf{Pic}^d(p)$. What cannot be constructed is a Poincaré line bundle, since a Poincaré line bundle which is normalized with respect to one section may not be normalized with respect to another one.

To see that $\mathbf{Pic}^d(p)$ exists in the algebraic category also when p has no sections, we shall use the following lemma.

LEMMA (2.12). *Let $Z \xrightarrow{\pi'} S' \xrightarrow{f} S$ be morphisms of schemes with f étale and finite. Denote by Z_{an}, π'_{an}, and so on, the underlying analytic spaces and morphisms. Suppose that there is a cartesian diagram of analytic spaces*

$$\begin{array}{ccc} Z_{an} & \xrightarrow{\widehat{F}} & \widehat{Y} \\ {\scriptstyle \pi'_{an}}\downarrow & & \downarrow{\scriptstyle \widehat{\pi}} \\ S'_{an} & \xrightarrow{f_{an}} & S_{an} \end{array} \tag{2.13}$$

Then there are a scheme Y *and morphisms of schemes* $Z \xrightarrow{F} Y \xrightarrow{\pi} S$ *such that*

$$\begin{array}{ccc} Z & \xrightarrow{F} & Y \\ {\scriptstyle \pi'}\downarrow & & \downarrow{\scriptstyle \pi} \\ S' & \xrightarrow{f} & S \end{array}$$

is cartesian and has (2.13) as underlying diagram of analytic spaces. Moreover, Y *is unique up to a unique isomorphism.*

Proof. The morphism $\widehat{F}$ is étale and finite, and $\widehat{Y}$ is a quotient of Z_{an} modulo the equivalence relation

$$\widehat{R} = Z_{an} \times_{\widehat{Y}} Z_{an} \subset Z_{an} \times Z_{an} .$$

Since $Z_{an} = S'_{an} \times_{S_{an}} \widehat{Y}$, we have

$$\widehat{R} = S'_{an} \times_{S_{an}} Z_{an} = R_{an} ,$$

where $R = S' \times_S Z = (S' \times_S S') \times_{S'} Z$ is an equivalence relation on Z. On the other hand,

$$p_1 : R = S' \times_S Z \to Z$$

is étale and finite since f is. From Theorem (2.8) it follows that there exists an effective quotient Z/R. We then set $Y = Z/R$ and let $F : X \to Y$ be the quotient map and $\pi : Y \to S$ the natural projection. The uniqueness of Y follows from the uniqueness of quotients.

Q.E.D.

We may now prove the existence of $\mathbf{Pic}^d(p)$ in the category of schemes. Start with an algebraic family $p : \mathcal{C} \to S$ of smooth curves. For any point of S, we may find a Zariski-open neighborhood U and a finite étale base change $f : U' \to U$ such that the pulled-back family $p' : \mathcal{C}' \to U'$ has a section. There is a natural projection map of analytic spaces

$$\mathbf{Pic}^d(p') \to \mathbf{Pic}^d(p_U) ,$$

where $p_U : \mathcal{C}_U \to U$ is the restriction of $p : \mathcal{C} \to S$ to U. It follows from Lemma (2.12) that $\mathbf{Pic}^d(p_U)$ can be given a scheme structure. The different $\mathbf{Pic}^d(p_U)$ patch together by the uniqueness part of the same lemma.

One may wonder what are the functorial properties of the scheme $\mathbf{Pic}^d(p)$ when $p : \mathcal{C} \to S$ has no sections. Does it represent a functor? The answer is positive but subtle. It represents the functor $(\mathcal{P}ic^d_{\mathcal{C}/S})_{\text{ét}}$ defined as follows. Endow the category of S-schemes with the étale topology and consider the contravariant functor $\mathcal{P}ic^d_{\mathcal{C}/S}$ as a presheaf in this topology. Then $(\mathcal{P}ic^d_{\mathcal{C}/S})_{\text{ét}}$ is the sheafification of $\mathcal{P}ic^d_{\mathcal{C}/S}$. For an exhaustive discussion of these matters, see Kleiman's Chapter 9 in [243].

3. Brill–Noether varieties on moving curves.

In this section we define the basic varieties of the Brill–Noether theory for moving curves. In simple terms, we will redo, with holomorphic dependence on parameters, the constructions we carried out in Section 3 of Chapter IV. Let

$$p : \mathcal{C} \to S \tag{3.1}$$

be a family of smooth curves of genus $g > 1$ parameterized by a scheme or analytic space S. As customary, we set $C_s = p^{-1}(s)$. The first Brill–Noether variety we will consider is denoted $\mathcal{C}^r_d$ and parameterizes effective divisors D of fixed degree d in the fibers of p such that $r(D) \geq r$ for some fixed r. Thus, set theoretically,

$$\operatorname{supp}(\mathcal{C}^r_d) = \{(s, D) : s \in S,\ D \in (C_s)_d \text{ such that } h^0(C_s, \mathcal{O}_{C_s}(D)) \geq r+1\}. \tag{3.2}$$

The second Brill–Noether variety, instead, parameterizes degree d line bundles L on the fibers of p with $h^0(L) \geq r+1$ and is denoted by $\mathcal{W}^r_d(p)$. Again set-theoretically,

$$\operatorname{supp}(\mathcal{W}^r_d(p)) = \{(s, D) : s \in S,\ L \in \operatorname{Pic}^d(C_s) \text{ such that } h^0(C_s, L) \geq r+1\}. \tag{3.3}$$

The mechanism which makes it possible to put a scheme structure on these varieties is the same in both cases, and we briefly review it next.

Let E be a coherent sheaf on a scheme or analytic space X and suppose that there is an exact sequence $\mathcal{O}^a_X \xrightarrow{\alpha} \mathcal{O}^b_X \to E \to 0$ (a *presentation* of E). The *hth Fitting ideal* of E is the ideal sheaf $I_h \subset \mathcal{O}_X$ generated by the minors of α of size $b-h+1$. The important fact about this ideal is that it depends only on E and not on the presentation (cf., for instance, [567]); it is also clear that I_h is functorial under base change. Since any coherent sheaf locally has a presentation, these properties imply that it makes sense to speak of the hth Fitting ideal of a coherent sheaf on X, even in the absence of a global presentation. The subvariety of X defined by the hth Fitting ideal of E is precisely the locus of those x such that $\dim(E \otimes k(x)) \geq h$.

Now let $f : Y \to X$ be a family of smooth curves of genus g, and L a line bundle of relative degree d on Y. Since L is flat over X, the basic theory of base change in cohomology (reviewed, for instance, at the beginning of Section 3 of Chapter IX) implies that there is, locally on X, a complex

$$\alpha : K^0 \to K^1 \tag{3.4}$$

of free sheaves which calculates, functorially with respect to base change, the direct images of L. In other words, there is an exact sequence

$$0 \to f_*L \to K^0 \xrightarrow{\alpha} K^1 \to R^1f_*L \to 0\,,$$

and the same is true after an arbitrary base change. We will denote the subscheme defined by the ith Fitting ideal of R^1f_*L with X^r. Its support is the locus of all $x \in X$ such that $R^1f_*L \otimes k(x) = H^1(Y_x, L_x)$ has dimension at least i or, by Riemann–Roch, the locus of all x such that $H^0(Y_x, L_x)$ has dimension at least $r+1$, where $r = d - g + i$.

Returning to the Brill–Noether varieties, we apply this construction to define $\mathcal{C}^r_d$ and $\mathcal{W}^r_d(p)$. For the first one, we look at the projection

$$\pi : \mathcal{C} \times_S \mathcal{C}_d \to \mathcal{C}_d$$

and at the line bundle $\mathcal{O}(\mathcal{D})$, where $\mathcal{D} \subset \mathcal{C} \times_S \mathcal{C}_d$ is the universal divisor (cf (2.7)). We then define

$$\mathcal{C}^r_d \subset \mathcal{C}_d$$

to be the S-subscheme defined by the $(g-d+r)$th Fitting ideal of $R^1\pi_*\mathcal{O}(\mathcal{D})$. By what we have just said, (3.2) holds.

To define $\mathcal{W}^r_d(p)$, we provisionally need to assume that p has a section, so that there exists a Poincaré line bundle $\mathcal{L}_d$ on $\mathcal{C} \times_S \mathbf{Pic}^d(p)$. We then define

$$\mathcal{W}^r_d(p) \subset \mathbf{Pic}^d(p)$$

to be the S-subscheme defined by the $(g-d+r)$th Fitting ideal of $R^1q_*\mathcal{L}_d$, where

$$q : \mathcal{C} \times_S \mathbf{Pic}^d(p) \to \mathbf{Pic}^d(p) \tag{3.5}$$

is the projection to the second factor. Again, (3.3) clearly holds. The crucial remark to be made here is that $\mathcal{W}^r_d(p)$ does not depend on the particular Poincaré bundle $\mathcal{L}_d$ and therefore, more importantly, does not depend on the choice of a section for p. Indeed, if $\mathcal{Q}$ is a line bundle on $\mathbf{Pic}^d(p)$, then, for every $i \geq 0$, the ith Fitting ideal of $R^1q_*\mathcal{L}_d$ coincides with the ith Fitting ideal of $R^1q_*(\mathcal{L}_d \otimes q^*\mathcal{Q})$. Recalling that the Fitting rank of a coherent sheaf $\mathcal{F}$ is defined to be the largest integer h such that the hth Fitting ideal of $\mathcal{F}$ vanishes, a consequence of the functorial properties of Fitting ideals is that the scheme $\mathcal{W}^r_d(p)$ represents the functor

$$\begin{aligned} &Sch/S \longrightarrow Sets \\ &T \mapsto \left\{[\mathcal{L}] \in \mathcal{P}ic^d_{\mathcal{C}/S}(T) : \text{Fitting-rank}(R^1{p_T}_*\mathcal{L}) \geq g-d+r\}\right\}, \end{aligned} \tag{3.6}$$

where $p_T : \mathcal{C} \times_S T \to T$ is the projection.

Exactly as in Proposition (3.4) of Chapter IV, the functorial properties of Fitting ideals, the ones of the universal divisor and the ones of the Poincaré line bundle imply the scheme-theoretical equality

$$u^{-1}(\mathcal{W}^r_d(p)) = \mathcal{C}^r_d\,, \tag{3.7}$$

where $u : \mathcal{C}_d \to \mathbf{Pic}^d(p)$ is the Abel–Jacobi map (2.9). We also have a relative counterpart of Lemma (2.3) of Chapter IV. Namely, there exist isomorphisms φ and ψ making the following diagram commute:

$$\begin{array}{ccc} T_{\mathcal{C}_d/S} & \xrightarrow{du} & u^* T_{\mathbf{Pic}^d(p)/S} \\ \downarrow{\scriptstyle\psi} & & \downarrow{\scriptstyle\varphi} \\ \pi_*\mathcal{O}_{\mathcal{D}}(\mathcal{D}) & \xrightarrow{\delta} & R^1\pi_*\mathcal{O}_{\mathcal{C}\times_S\mathcal{C}_d} \end{array} \tag{3.8}$$

In this diagram, $T_{\mathcal{C}_d/S}$ and $T_{\mathbf{Pic}^d(p)/S}$ denote the relative tangent sheaves, and the matrix of du is the moving Brill–Noether matrix. In other words, if $D = q_1 + \cdots + q_d$ and $\eta_1, \dots, \eta_g$ is a local frame for $p_*\omega_p$, then

$$(du)_{s,D} = \begin{pmatrix} \eta_{1,s}(q_1) & \dots & \eta_{g,s}(q_1) \\ \vdots & & \vdots \\ \eta_{1,s}(q_d) & \dots & \eta_{g,s}(q_d) \end{pmatrix},$$

where, as usual, the above matrix should be replaced by the one on page 159 of Chapter IV when the points q_i are not all distinct. On the other hand, the coboundary homomorphism δ is part of a locally free presentation of $R^1\pi_*\mathcal{O}_{\mathcal{C}\times_S\mathcal{C}_d}(\mathcal{D})$:

$$\pi_*\mathcal{O}_{\mathcal{D}}(\mathcal{D}) \xrightarrow{\delta} R^1\pi_*\mathcal{O}_{\mathcal{C}\times_S\mathcal{C}_d} \to R^1\pi_*\mathcal{O}_{\mathcal{C}\times_S\mathcal{C}_d}(\mathcal{D}) \to 0\,.$$

Thus, diagram (3.8) expresses the fact that the defining ideal of $\mathcal{C}^r_d$ in $\mathcal{C}_d$, which is by definition the $(g-d+r)$th Fitting ideal of $R^1\pi_*\mathcal{O}_{\mathcal{C}\times_S\mathcal{C}_d}(\mathcal{D})$, is generated by the $(g-d+r)\times(g-d+r)$ minors of the moving Brill–Noether matrix.

We now come to the third, and probably most important, Brill–Noether variety for moving curves. While $\mathcal{W}^r_d(p)$ parameterizes *complete* degree d linear series of dimension greater than or equal to r on the fibers of (3.1), the scheme $\mathcal{G}^r_d(p)$ we are going to construct parameterizes all g^r_d's on the fibers of (3.1).

To describe the construction, it is convenient to go back to the setup consisting of a family $f : Y \to X$ of smooth curves of genus g and a line bundle L of relative degree d on Y. Suppose first that a complex (3.4) calculating the direct images of L exists on all of X.

Let $\mathbb{G}(r+1, K^0) \xrightarrow{\beta} X$ be the Grassmannian bundle of $(r+1)$-planes in the fibers of K^0, and let $j : V \hookrightarrow \beta^* K^0$ be the universal subbundle on $\mathbb{G}(r+1, K^0)$. Consider the composite homomorphism

$$V \xrightarrow{j} \beta^* K^0 \xrightarrow{\beta^*(\alpha)} \beta^* K^1$$

and pick a global frame $e_1, \ldots, e_b$ for $\beta^* K^1$. For any local section v of V, we may write

$$\beta^*(\alpha) \circ j(v) = \sum a_i e_i$$

for suitable functions a_i. We let J be the ideal sheaf generated by all the a_i for all possible choices of local section v, and we let

$$\widetilde{X}^r \subset \mathbb{G}(r+1, K^0)$$

be the subscheme defined by J. We also denote by W the restriction of V to $\widetilde{X}^r$.

Now let $\gamma : T \to X$ be a morphism, denote by p_Y and p_T the projections of $Y \times_X T$ to its two factors, and suppose we are given a rank $r+1$ locally free subsheaf F of $p_{T*}(p_Y^* L)$ with the property that the homomorphism

$$F \otimes k(t) \to H^0(p_T^{-1}(t), p_Y^* L \otimes k(t)) = H^0(Y_{\gamma(t)}, L_{\gamma(t)}) \tag{3.9}$$

is injective for all $t \in T$. For brevity, we shall refer to such a datum as a *family of* g^r_d*'s in* f_*L *parameterized by* T. The prototypical such object is the bundle W on $\widetilde{X}^r$. In fact, by the definition of $\widetilde{X}^r$, W is a subsheaf of the kernel of the restriction to $\widetilde{X}^r$ of $\beta^*(\alpha) : \beta^* K^0 \to \beta^* K^1$. On the other hand, by the characteristic property of $K^0 \to K^1$, this kernel is just the pushforward to $\widetilde{X}^r$ of the pullback of L to $\widetilde{X}^r \times_X Y$.

Going back to $\gamma : T \to X$ and F, the latter is a subsheaf of $p_{T*}(p_Y^* L) = \ker(\gamma^* K^0 \to \gamma^* K^1)$, and the injectivity of (3.9) says that it is in fact a vector subbundle of K^0. By the universal property of the Grassmannian, F is the pullback of the universal subbundle V via a unique morphism $T \to \mathbb{G}(r+1, K^0)$. In fact, this morphism lands in $\widetilde{X}^r$, since F is a subsheaf of the kernel of $\gamma^* K^0 \to \gamma^* K^1$. Conversely, any morphism $T \to \widetilde{X}^r$ gives, by pullback of W, a family of g^r_d's in f_*L parameterized by T. This means that $\widetilde{X}^r$ represents the functor

$$\begin{aligned} &Sch/X \longrightarrow Sets \\ &T \mapsto \{\text{families of } g^r_d\text{'s in } f_*L \text{ parameterized by } T\}\,. \end{aligned} \tag{3.10}$$

In particular, $\widetilde{X}^r$ does not depend on the choice of the complex $K^\bullet$.

Before we show how to remove the assumption that there exists, globally on X, a complex (3.4) calculating the direct images of L, we pause for a side remark. If we fix trivializations for K^0 and K^1, the homomorphism α can be regarded as a morphism $\sigma : X \to M(a,b)$ to the variety of $a \times b$ matrices, where a and b are the ranks of K^0 and K^1. Then X^r is the pullback via σ of the generic determinantal variety $M_k(a,b)$ consisting of all matrices of rank at most k, with $k = b-r+d-g$ (cf. Section 2 of Chapter II). Similarly, $\widetilde{X}^r$ is the pullback, via the same morphism σ, of the canonical desingularization $\widetilde{M}_k(a,b)$ of $M_k(a,b)$, also discussed in Section 2 of Chapter II.

To prove that one can do away with the assumption that there exists a complex (3.4) on all of X, we may argue as follows. We know that X can be covered with open subsets on each of which a complex (3.4) exists. Thus, for each such open set U, we can construct $\widetilde{U}^r$, which represents the functor (3.10), of course with X replaced by U. On the other hand, the construction of $\widetilde{U}^r$ is clearly compatible with base change. In particular, if V is another open subset of X on which a complex (3.4) exists, $\widetilde{U \cap V}^r$ is the same as the pullback to $U \cap V$ of $\widetilde{U}^r$ or of $\widetilde{V}^r$. In other words, the various $\widetilde{U}^r$ patch together canonically by universality. The resulting scheme is again denoted by $\widetilde{X}^r$ and represents the functor (3.10). What we have proved can be summarized in the following statement.

LEMMA (3.11). *Let $f : Y \to X$ be a family of smooth curves of genus g, and L a line bundle of relative degree d on Y. Then there is a scheme $\widetilde{X}^r$ over X which represents the functor (3.10).*

We now have at our disposal the tools needed to construct $\mathcal{G}^r_d(p)$, where $p : \mathcal{C} \to S$ is the family (3.1). We assume that p has a section. The scheme $\mathcal{G}^r_d(p)$ is a special instance of the scheme $\widetilde{X}^r$ constructed above. We take

$$\begin{aligned} X &= \mathbf{Pic}^d(p)\,, \\ Y &= \mathcal{C} \times_S \mathbf{Pic}^d(p)\,, \\ f &= \text{the projection } \mathcal{C} \times_S \mathbf{Pic}^d(p) \to \mathbf{Pic}^d(p)\,, \\ L &= \text{a Poincaré line bundle } \mathcal{L}_d\,. \end{aligned}$$

The universal property (3.11) enjoyed by $\widetilde{X}^r$ implies that $\mathcal{G}^r_d(p)$ represents a well-defined geometric functor. To explain what this is, we make a formal definition.

DEFINITION (3.12). *Let $p : \mathcal{C} \to S$ be a family of smooth curves of genus g. A* family of g^r_d's on $p : \mathcal{C} \to S$ parameterized by an S-scheme *$f : T \to S$ is a pair $(\mathcal{L}, \mathcal{H})$, where $\mathcal{L}$ is a line bundle on $\mathcal{C}_T = \mathcal{C} \times_S T$ whose restriction to each fiber of $p_T : \mathcal{C}_T \to T$ has degree d, and $\mathcal{H}$ is a locally free subsheaf of $p_{T*}\mathcal{L}$ of rank $r+1$ such that, for each $t \in T$, the fiber homomorphism*

$$\mathcal{H} \otimes k(t) \to H^0(p_T^{-1}(t), \mathcal{L} \otimes k(t))$$

is injective. Two families $(\mathcal{L}, \mathcal{H})$ *and* $(\mathcal{L}', \mathcal{H}')$ *are said to be equivalent if there are a line bundle* $\mathcal{Q}$ *on* T *and an isomorphism* $\mathcal{L}' \xrightarrow{\simeq} \mathcal{L} \otimes p_T^* \mathcal{Q}$ *inducing an isomorphism* $\mathcal{H}' \xrightarrow{\simeq} \mathcal{H} \otimes \mathcal{Q}$.

The universal property of $\mathcal{G}_d^r(p)$ may now be expressed by the following result.

THEOREM (3.13). *Let* $p : \mathcal{C} \to S$ *be a family of smooth, curves of genus* $g > 1$. *Suppose that* p *admits a section. Then there exist an* S*-scheme* $\mathcal{G}_d^r(p)$ *representing the functor*

$$\begin{aligned} &Sch/S \longrightarrow Sets \\ &T \mapsto \left\{ \begin{array}{c} \text{equivalence classes of families of } g_d^r\text{'s} \\ \text{on } p : \mathcal{C} \to S \text{ parameterized by } T \end{array} \right\} \end{aligned} \tag{3.14}$$

and a morphism

$$\chi : \mathcal{G}_d^r(p) \to \mathbf{Pic}^d(p)$$

over S, *factoring through the inclusion* $\mathcal{W}_d^r(p) \subset \mathbf{Pic}^d(p)$. *The morphism* χ *corresponds to the forgetful morphism of functors which associates to a family* $(\mathcal{L}, \mathcal{H})$ *of* g_d^r*'s the line bundle* $\mathcal{L}$.

An obvious but important onbservation is that, when the family $p : \mathcal{C} \to S$ consists of a single curve C, i.e., when S is a point, the varieties $\mathcal{W}_d^r(p)$ and $\mathcal{G}_d^r(p)$ coincide with the varieties $W_d^r(C)$ and $G_d^r(C)$ defined in Section 3 of Chapter IV. It should also be noticed that the morphism

$$\chi : \mathcal{G}_d^r(p) \to \mathcal{W}_d^r(p)$$

is biregular off $\mathcal{W}_d^{r+1}(p)$ and that, over each point $w \in \mathcal{W}_d^{r+1}(p)$ corresponding to a degree d line bundle L on C, the fiber $\chi^{-1}(w)$ is the Grassmannian $\mathbb{G}(r+1, H^0(C, L))$.

Let us briefly comment on the existence of the varieties $\mathcal{W}_d^r(p)$ and $\mathcal{G}_d^r(p)$ when the family $p : \mathcal{C} \to S$ admits no section. Here we may repeat, word by word, the argument at the end of Section 2 where we showed how to define $\mathbf{Pic}^d(p)$ when p has no sections. To this end, we start from a family $p : \mathcal{C} \to S$ of smooth, genus g curves. Since local analytic section of p exist, we may define $\mathcal{W}_d^r(p)$ and $\mathcal{G}_d^r(p)$ as analytic spaces. After shrinking S, if necessary, we pass to a finite étale cover $f : S' \to S$ such that the pulled-back family $p' : \mathcal{C}' \to S'$ has a section. Thus, both $\mathcal{W}_d^r(p')$ and $\mathcal{G}_d^r(p')$ are algebraic. By universality, $\mathcal{W}_d^r(p') = \mathcal{W}_d^r(p) \times_S S'$ and $\mathcal{G}_d^r(p') = \mathcal{G}_d^r(p) \times_S S'$. We may then apply Lemma (2.12). Of course, when p has no sections, neither $\mathcal{W}_d^r(p)$ represents the functor (3.6), nor $\mathcal{G}_d^r(p)$ represents the functor (3.14).

Finally, let us introduce the *Brill–Noether subloci of* M_g. Let $p : \mathcal{C} \to S$ be a family of smooth, genus g curves. Look at the induced map $p_d : \mathcal{C}_d \to S$. We define the subscheme $S_d^r \subset S$ as the image of $\mathcal{C}_d^r$ via p_d:

$$S_d^r = p_d(\mathcal{C}_d^r)\,. \tag{3.15}$$

Attached to the family p is the moduli morphism

$$m : S \to M_g\,.$$

In the algebraic category we say that (3.1) is a *family of genus g curves with general moduli* if the moduli map is dominant, while in the analytic category a family of genus g curves with general moduli is one for which $m(S)$ is open in M_g. For example, the Kuranishi family (2.10) is a family with general moduli.

NOTATION (3.16). *For a family $p : \mathcal{C} \to S$ of genus $g > 1$ curves which is a Kuranishi family at every point of S, we often write $\mathbf{Pic}_g^d$, $\mathcal{W}_{g,d}^r$, and $\mathcal{G}_{g,d}^r$ or, even more simply, $\mathbf{Pic}^d$, $\mathcal{W}_d^r$, and $\mathcal{G}_d^r$ instead of $\mathbf{Pic}^d(p)$, $\mathcal{W}_d^r(p)$, and $\mathcal{G}_d^r(p)$.*

Consider a family of curves with general moduli and assume that its moduli map is surjective. We then define the subvarity $M_{g,d}^r \subset M_g$ by setting

$$M_{g,d}^r = m(S_d^r)\,. \tag{3.17}$$

The variety $M_{g,d}^r$ is the locus in M_g described by curves possessing a g_d^r. One of the basic tasks of the Brill–Noether theory for moving curves is to study the local nature of $\mathcal{W}_d^r$ and $\mathcal{G}_d^r$ and to compute the dimension of the irreducible components of $M_{g,d}^r$.

The following diagram summarizes the configuration of Brill–Noether varieties for a family $p : \mathcal{C} \to S$:

$$\begin{array}{ccccc}
\mathcal{C}_d^r \subset \mathcal{C}_d & \xrightarrow{\;u_d\;} & \mathbf{Pic}^d(p) \supset \mathcal{W}_d^r(p) & \xleftarrow{\;\chi\;} & \mathcal{G}_d^r(p) \\
{\scriptstyle \pi_d}\downarrow & \swarrow{\scriptstyle p_d} & & & \\
S_d^r \subset S & \xrightarrow{\;m\;} & M_g \supset M_{g,d}^r & &
\end{array} \tag{3.18}$$

An important sublocus of M_g is the one whose points represent curves satisfying Petri's condition, in the following sense.

DEFINITION (3.19). *A smooth genus g curve C is said to satisfy Petri's condition if for every line bundle L on C, the cup product map*

$$\mu_0 : H^0(C, L) \otimes H^0(C, \omega_C L^{-1}) \to H^0(C, \omega_C)$$

is injective.

PROPOSITION (3.20). *The locus of points in M_g parameterizing curves satisfying Petri's condition is open.*

Proof. Look at a Kuranishi family $p : \mathcal{C} \to B$ and consider the projection $q_d^r : \mathcal{G}_d^r(p) \to B$. Then by iii), Proposition (4.1), Chapter IV, we have that $\{b \in B : C_b$ satisfies Petri's condition$\} = \bigcup_{r,d} \{b \in B : (q_d^r)^{-1}(b)$ is smooth$\}$. Q.E.D.

Clearly, Petri's statement (1.1) says that a general curve satisfies Petri's condition. In view of Proposition (3.20), to prove (1.1), it suffices to exhibit *one* curve satisfying Petri's condition.

We now make some dimension-theoretic considerations preliminary to the computation of the tangent spaces to $\mathbf{Pic}^d$, $\mathcal{W}_d^r$, and $\mathcal{G}_d^r$. Recall the Brill–Noether number

$$\rho = g - (r+1)(g-d+r).$$

From Propositions (4.1) and (4.2) in Chapter IV one gets the following result.

PROPOSITION (3.21). *When $g \geq 2$ every component of $\mathcal{G}_d^r$ has dimension at least $3g-3+\rho$. Similarly, when $r \geq 0$ and $r \geq g-d$, every component of $\mathcal{W}_d^r$ has dimension at least $3g-3+\rho$, and every component of $\mathcal{C}_d^r$ has dimension at least $3g-3+\rho+r$.*

A number of remarks are in order.

REMARK (3.22). Note that Proposition (3.21) has meaning even when ρ is negative, since it still may happen that $\rho + 3g - 3$ is nonnegative. For example, $\mathcal{W}_2^1$ has always dimension $3g-3+(2-g) = 2g-1$. From the existence theorems proved in Chapter VII we already know that the fibers of $\mathcal{W}_d^r(p) \to B$, where $p : \mathcal{C} \to B$ is a Kuranishi family, have dimension at least ρ. On the other hand, there are certainly curves, e.g., the Castelnuovo extremal curves $C \subset \mathbb{P}^r$ of degree $d > 2r$ encountered in Section 2 of Chapter III, for which ρ is negative but $W_d^r(C)$ is nonempty, and consequently,

$$\dim W_d^r(C) > \rho.$$

There still remain the possibility that the whole variety $\mathcal{W}_d^r$ of special linear series has the *expected dimension* given by

$$\dim \mathcal{W}_d^r = 3g - 3 + \rho. \tag{3.23}$$

For example, look at the case of smooth, degree d plane curves. It is clear that

$$\begin{aligned}\dim \mathcal{W}_d^2 &= h^0(\mathbb{P}^2, \mathcal{O}(d)) - 1 - \dim \mathrm{Aut}(\mathbb{P}^2)\\ &= \frac{d(d+3)}{2} - 8\\ &= \frac{(d-1)(d-2)}{2} + 3d - 9\\ &= g + 3d - 9\\ &= 4g - 3 - 3(g - d + 2)\\ &= 3g - 3 + \rho\,,\end{aligned}$$

so that (3.23) holds in this case. Lest we become too optimistic, suppose that we go to the next nontrivial family of Castelnuovo extremal curves, which is the case of space curves of degree $d = 8$ of maximal genus. Thus $r = 3$ and $g = 9$. Moreover, by the lemma on page 119 of Chapter III, the general curve of this type is the complete intersection of a nonsingular quadric Q and a quartic. The left-hand side of (3.23) is given by

$$\begin{aligned}\dim \mathcal{W}_8^3 &= h^0(Q, \mathcal{O}(4)) - 1 - \dim \mathrm{Aut}(Q)\\ &= 24 - 6 = 18\,,\end{aligned}$$

while the right-hand side is

$$4g - 3 - 4(g - 8 + 3) = 17\,,$$

so (3.23) does not hold.

Remark (3.24). Proposition (3.21) does *not* assert that the varieties involved are nonempty when their expected dimension is nonnegative. In fact, this may be quite false. For instance, for $r = d = 2$, the number $3g - 3 + \rho$ is nonnegative as soon as $g \geq 3$, but $\mathcal{G}_d^r$ is empty because of Clifford's theorem.

Remark (3.25). Proposition (3.21) covers the genera greater than 1. When $g = 0$ or $g = 1$, $\mathcal{G}_d^r$ is smooth of dimension $g + \rho$ while, for $r \geq d - g$, $\mathcal{W}_d^r$ and $\mathcal{C}_d^r$ are smooth of dimensions $g + \rho$ and $g + \rho + r$.

4. Looijenga's vanishing theorem.

As announced in Section 6 of Chapter XVII, to which we refer for the notation, we are now going to prove a vanishing theorem for the tautological ring of $\mathcal{M}_g$, due to Looijenga [489]. The proof of this theorem involves the basic theory of g_d^1's on moving curves, which is now at our disposal. The statement we want to prove is the following.

Theorem (4.1). *$R^i(\mathcal{M}_g) = 0$ for $i > g - 2$.*

Before giving the proof of this result, we need to make some preliminary consideration. Instead of working with the stack $\mathcal{M}_g$, we express M_g as the quotient of a smooth variety M by a finite group G. We can also assume that there is a family of smooth genus g curves

$$\pi : \mathcal{C} \to M$$

whose moduli map is surjective. We can take as M the moduli space of genus g curves with an appropriate level structure. Points in $\mathcal{C}$ will be denoted as pairs (t, x) where $t \in M$ and $x \in C_t = \pi^{-1}(t)$. We also consider the n-fold product

$$\mathcal{C}^n = \mathcal{C} \times_M \cdots \times_M \mathcal{C}$$

and the n-fold symmetric product $\mathcal{C}_n = \mathcal{C}^n / \mathfrak{S}_n$. We denote by ω_i the relative dualizing sheaf of the projection $\pi_i : \mathcal{C}^n \to \mathcal{C}^{n-1}$ obtained by omitting the ith component and by $\widetilde{K}_i$ its first Chern class as an element of $A^1(\mathcal{C}^n)$. We will prove the following proposition

PROPOSITION (4.2). *Let $\pi : \mathcal{C} \to M$ be as above. Then any monomial of degree $d > g + n - 2$ in $\widetilde{K}_1, \dots, \widetilde{K}_n \in A^1(\mathcal{C}^n)$ vanishes.*

From this proposition Theorem (4.1) follows easily. In fact we have the cartesian diagrams

$$\begin{array}{ccc} \mathcal{C}^n & \xrightarrow{\ \alpha_i\ } & \mathcal{C} \\ {\scriptstyle \pi_i}\big\downarrow & & \big\downarrow{\scriptstyle \pi} \\ \mathcal{C}^{n-1} & \xrightarrow{\ \beta\ } & M \end{array} \qquad\qquad \begin{array}{ccc} \mathcal{C}^n & \xrightarrow{\ \pi_i\ } & \mathcal{C}^{n-1} \\ {\scriptstyle \pi_n}\big\downarrow & & \big\downarrow{\scriptstyle \pi'_{n-1}} \\ \mathcal{C}^{n-1} & \xrightarrow{\ \pi'_i\ } & \mathcal{C}^{n-2} \end{array} \tag{4.3}$$

where α_i is the projection from $\mathcal{C}^n$ to the ith factor. With an obvious notation we have

$$\alpha_i^*(\widetilde{K}) = \widetilde{K}_i , \qquad \beta^*(\widetilde{\kappa}_a) = \widetilde{\kappa}_a , \qquad \pi_n^*(\widetilde{K}'_i) = \widetilde{K}_i .$$

The first and third equalities follow from the cartesianness of diagrams (4.3). For the second, again by cartesianness and by Example 17.4.1 in [275] (cf. the proof of Lemma (4.19) in Chapter XVII), we have, for any $a \geq 0$,

$$\begin{aligned} \beta^* \widetilde{\kappa}_a &= \beta^* \pi_* \widetilde{K}^{a+1} \\ &= (\pi_i)_* \alpha_i^* (\widetilde{K}^{a+1}) \\ &= (\pi_i)_* (\widetilde{K_i}^{a+1}) \\ &= \widetilde{\kappa}_a . \end{aligned}$$

From this, using induction and the push–pull formula, it readily follows that, if $\rho = \beta\pi_i : \mathcal{C}^n \to M$ is the natural projection, then

$$\rho_* (\widetilde{K_1}^{a_1+1} \cdots \widetilde{K_n}^{a_n+1}) = \widetilde{\kappa}_{a_1} \cdots \widetilde{\kappa}_{a_n} .$$

It is now clear how Theorem (4.1) follows from Proposition (4.2). This proposition is proved by stratifying $\mathcal{C}^n$ and by using the following elementary lemma.

LEMMA (4.4). *Let X be a scheme, and let $X = V_0 \supset V_1 \supset \cdots \supset V_N$ be a stratification by closed subschemes. Let $L_1, \dots, L_N$ be line bundles on X such that $c_1(L_i)$ vanishes on $V_{i-1} \smallsetminus V_i$, $i = 1, \dots, N$. Then $c_1(L_1) \cdots c_1(L_N)$ is supported on V_N.*

From now on we fix an integer $n >> 0$ and an integer $d \geq g + n$. We will construct a stratification

$$(4.5) \qquad \mathcal{C}^n = V_0 \supset V_1 \supset \cdots \supset V_{d-1} = \emptyset$$

of $\mathcal{C}^n$ such that

$$(4.6) \qquad \widetilde{K}_{h|V_{i-1} \smallsetminus V_i} = 0\,, \quad i = 1, \dots, d-1\,, \ \ h = 1, \dots, n\,.$$

Proposition (4.2) will then be a consequence of the elementary lemma above.

In order to construct the above stratification, we introduce a number of Brill–Noether varieties. Consider the family of 1-pointed curves

$$(4.7) \qquad q : \mathcal{X} = \mathcal{C} \times_M \mathcal{C} \to \mathcal{C}\,.$$

We set $\mathbf{Pic}^i = \mathbf{Pic}^i(q)$, $\mathcal{G}^1_d = \mathcal{G}^1_d(q)$, and we have natural morphisms

$$\eta = \eta_i : \mathbf{Pic}^i \to \mathcal{C}\,, \qquad \zeta : \mathcal{G}^1_d \to \mathcal{C}\,.$$

We denote by $\mathbf{x}$ the image in $\mathbf{Pic}^1$ of the section of q, so that $\mathbf{x} \cdot \eta_1^{-1}((t,x)) = [x]$. We are going to construct subloci of $\mathcal{G}^1_d$ of $\mathcal{G}^1_d \times_M \mathcal{C}_d$ and of $\mathcal{G}^1_d \times_M \mathcal{C}_d \times \mathcal{C}^n$. The definition of the scheme structure of these subloci will be evident and left to the reader. A point in $\mathcal{G}^1_d$ is a triple (t, x, P), where t is a point in M, x a point on C_t, and P is a pencil (i.e., a g^1_d) on C_t. The first sublocus consists of those pencils P for which $dx \in P$.

$$\mathcal{G}^1_{d\star} = \{(t, x, P) \in \mathcal{G}^1_d : dx \in P\}\,.$$

This implies that, for $(t, x, P) \in \mathcal{G}^1_{d\star}$, the morphism $\varphi_P : C \to \mathbb{P}^1$ associated to P has degree less than or equal to d and is totally ramified at x. In particular, if $d \leq g$, the point x is a Weierstrass point of C. Next, we set

$$\mathcal{V} = \{(t, x, P, D) \in \mathcal{G}^1_{d\star} \times_M \mathcal{C}_d : D \in P\,, \ |\operatorname{supp}(D)| \leq d-1\}\,.$$

In this definition, D is a divisor in P containing a point of multiplicity greater than one, which is then either a base point of P or a ramification point of $\phi_P : C \to \mathbb{P}^1$. Set

$$(4.8) \qquad D = mx + D'\,, \quad \text{with} \quad x \notin \operatorname{supp}(D')\,.$$

If $D' \neq 0$, then ϕ_P is a $(d-m)$-sheeted ramified cover of $\mathbb{P}^1$ and may think that

$$\phi_P^{-1}(\infty) = (d-m)x\,, \quad \phi_P^{-1}(0) = D'\,.$$

In what follows it will be useful to order the points of the divisor D appearing in (4.8) or at least some of them. For this, we set

$$Z = \{(t, x, P, D, x_1, \dots, x_n) \in \mathcal{V} \times_M \mathcal{C}^n : x_i \in \operatorname{supp}(D)\,,\ i = 1, \dots, n\}\,.$$

As we anticipated, all these loci have a natural scheme structure whose definition is left to the reader. For example,

$$\mathcal{G}^1_{d\star} = \chi^{-1}([d\mathbf{x}])\,,$$

where $\chi : \mathcal{G}^1_d \to \mathbf{Pic}^d$ is the natural projection. Since $d >> g$, it is clear from its definition that $\mathcal{G}^1_{d\star}$ has the structure of a $(d-g-1)$-dimensional projective bundle over $\mathcal{C}$. In particular, $\mathcal{G}^1_{d\star}$ is irreducible of dimension $2g+d-3$. The natural projection from Z to $\mathcal{G}^1_{d\star}$ is finite-to-one, so that also Z has dimension equal to $2g+d-3$. We now stratify Z by setting

$$Z^k = \{(t, x, P, D, x_1, \dots, x_n) \in Z : |\operatorname{supp}(D) \smallsetminus \{x\}| \le d-k-1\}\,,$$

so that

$$Z = Z^0 \supset Z^1 \supset \cdots \supset Z^{d-1}\,.$$

The last stratum is somewhat special: for any point $(t, x, P, D, x_1, \dots, x_n) \notin Z^{d-1}$, the two divisors dx and D completely determine the pencil P, while, on the other hand, any point in Z^{d-1} is of the form $(t, x, P, dx, x, \dots, x,)$. In particular, under the natural (finite-to-one) projection $Z \to \mathcal{G}^1_{d\star}$, the stratum Z^{d-1} maps isomorphically onto $\mathcal{G}^1_{d\star}$, so that Z^{d-1} is an irreducible component of Z. Notice that each Z^k is closed and that, by Riemann's existence theorem, $Z^{k-1} \smallsetminus Z^k$ is nonempty.

The family (4.7) pulls back to a family $\gamma : \mathcal{Y} \to Z$, and we have commutative diagrams

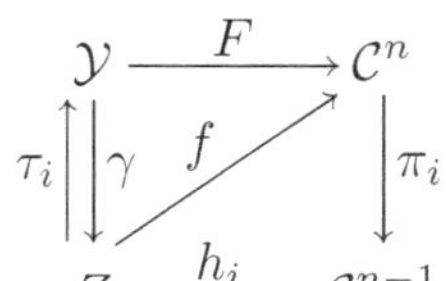

where

$$F((y, t, x, P, D, x_1, \dots, x_n)) = (x_1, \dots, x_n)\,,$$
$$h_i((t, x, P, D, x_1, \dots, x_n)) = (x_1, \dots, \widehat{x}_i, \dots, x_n)\,,$$

the section $\tau_i : Z \to \mathcal{Y}$ of γ is defined by

$$\tau_i(t, x, P, D, x_1, \dots, x_n) = (x_i, t, x, P, D, x_1, \dots, x_n)\,, \quad i = 1, \dots, n\,, \tag{4.9}$$

and $f : Z \to \mathcal{C}^n$ by

$$f((t, x, P, D, x_1, \dots, x_n)) = (x_1, \dots, x_n)\,.$$

In particular,

$$f^*\widetilde{K}_i = \tau_i^* c_1(\omega_\gamma)\,. \tag{4.10}$$

Next, we set

$$X^k = \overline{Z^k \smallsetminus Z^{d-1}}\,,$$

so that X_k is the union of all the irreducible components of Z^k that are distinct from Z^{d-1}. We have

a) $X^0 \supset \cdots \supset X^{d-1}$;
b) $X^0 \neq Z$;
c) $X^{d-1} = \emptyset$.

Finally, we set $V_i = f(X^i) \subset \mathcal{C}^n$ for $i = 0, \ldots, d-1$. We claim that in this way we obtain a stratification (4.5) satisfying (4.6), thus concluding the proof of Theorem (4.1). The first thing to show is that $V_0 = \mathcal{C}^n$ or, equivalently, that $f : X^0 \to \mathcal{C}^n$ is surjective. Given $x_1, \ldots, x_n$ in $C = C_t$, we must find a point $x \in C$, a divisor D in $|dx|$ with a point of multiplicity greater than one and such that $x_i \in \operatorname{supp}(D)$. For this, it suffices to find points $y_1, \ldots, y_{d-n-1}$ such that $dx \sim 2x_1 + x_2 + \ldots x_n + y_1 + \cdots + y_{d-n-1}$. This is possible since, for $d - n \geq g$, the morphism

$$\begin{gathered} C^{d-n} \to J(C) \\ (x, y_1, \ldots, y_{d-n-1}) \mapsto [-dx + 2x_1 + x_2 + \ldots x_n + y_1 + \cdots + y_{d-n-1}] \end{gathered}$$

is surjective by Jacobi's inversion theorem. Now we have a stratification (4.5). It is also clear that the map $f : X^0 \to \mathcal{C}^n$ is proper. We will next prove the following two lemmas, which are the heart of the entire argument.

LEMMA (4.11). *$f^*\widetilde{K}_{i|Z^{k-1}\smallsetminus Z^k} = 0$ for $k < d$ and $i = 1, \ldots, n$.*

LEMMA (4.12). *Set $U_{k-1} = X^{k-1} \cap f^{-1}(V_{k-1} \smallsetminus V_k) = X^{k-1} \smallsetminus (X^{k-1} \cap f^{-1}(V_k))$. Then U_{k-1} is contained in $Z^{k-1} \smallsetminus Z^k$.*

These two lemmas prove that the stratification (4.5) satisfies property (4.6). Indeed, since $U_{k-1} \subset Z^{k-1} \smallsetminus Z^k$, it follows from Lemma (4.12) that $f^*\widetilde{K}_i$ vanishes on U_{k-1}. But $f : U_{k-1} \to V_{k-1} \smallsetminus V_k$ is proper and onto, and thus $\widetilde{K}_i$ vanishes on $V_{k-1} \smallsetminus V_k$.

Proof of Lemma (4.11). Let W be a connected component of $Z^{k-1} \smallsetminus Z^k$. Look at the restriction to W of the family $\gamma : \mathcal{Y} \to Z$, which, by abuse of notation, we denote by $\gamma : \mathcal{Y} \to W$, and also consider the sections τ_i defined in (4.9). We have two relative divisors $\mathcal{D}$ and $\mathcal{D}_\infty$ on $\mathcal{Y}$. If D and D_∞ are the restrictions of $\mathcal{D}_0$ and $\mathcal{D}_\infty$ to the fiber of γ over a point $w = (t, x, P, D, x_1, \ldots, x_n) \in W$, then

$$D_\infty = dx\,, \qquad D = mx + D' \qquad \text{with } x \notin \operatorname{supp}(D'),\ |\operatorname{supp}(D')| = d - k\,.$$

The relevant remark is that, by the way W is defined, the integer m and the multiplicity ν_i of x_i in D are independent of $w \in W$ (of course, if $x_i = x$, then $\nu_i = m$). For every $w \in W$, there exists a finite morphism $\pi_t : C_t \to \mathbb{P}^1$ with $\pi_t^*(0) = D'$ and $\pi_t^*(\infty) = (d-m)x$. Setting $y_i = \pi_t(x_i)$, such a morphism determines isomorphisms

(4.13)
$$\mathbb{C} = \mathfrak{m}_{\mathbb{P}^1, y_i}/\mathfrak{m}^2_{\mathbb{P}^1, y_i} \xrightarrow{\simeq} \mathfrak{m}^{\nu_i}_{C_t, x_i}/\mathfrak{m}^{\nu_i+1}_{C_t, x_i} = T^\vee_{x_i}(C_t)^{\otimes \nu_i} = (\tau_i^* \omega_\gamma)^{\otimes \nu_i} \otimes k(w)\,.$$

Were the morphisms π_t uniquely defined, the isomorphism (4.13) would give a trivialization of $\tau_i^*(\omega_\gamma)^{\nu_i}$, thus proving Lemma (4.11), in view of (4.10). But the morphism π_t is defined only up to a multiplicative action of $\mathbb{C}^\times$. Its nonuniqueness can be cured as follows. Let R be the part of the ramification divisor of π_t lying over $\mathbb{P}^1 \smallsetminus \{0, \infty\}$. Write $(\pi_t)_*(R) = \sum_i r_i z_i$ and normalize π_t in such a way that $\prod_i z_i^{r_i} = 1$. At this stage, π_t is defined up to multiplication by an rth root of unity, where $r = \sum r_i$. We then get a canonical isomorphism $\mathbb{C} \xrightarrow{\simeq} T^\vee_{x_i}(C_t)^{\otimes r\nu_i} = (\tau_i^* \omega_\gamma)^{\otimes r\nu_i} \otimes k(w)$.

Q.E.D.

Proof of Lemma (4.12). It suffices to show that $f(X^{k-1} \cap Z^{d-1}) \subset f(X^k)$. Fix a point $z \in X^{k-1} \cap Z^{d-1}$ and a one-dimensional analytic arc in X^{k-1} passing through z and meeting Z^{d-1} only at z. Restricting the family of curves over Z to this arc, we get an analytic family of pointed curves over $\Delta = \{t \in \mathbb{C} : |t| < \varepsilon\}$. More precisely, we have an analytic family of triples $\{C_t, x_t, P_t\}_{t\in\Delta}$, where C_t is a smooth curve, $x_t \in C_t$, and P_t is a pencil on C_t containing dx_t. Moreover, there exists an analytic family of divisors $\{D_t\}_{t\in\Delta}$, $D_t \in P_t$, such that, for $t \neq 0$, we can write

$$D_t = mx_t + D'_t\,, \qquad \text{with } x_t \notin \operatorname{supp}(D'_t),\ |\operatorname{supp}(D'_t)| = d-k\,.$$

Writing $Q_t = P_t - mx_t$, $\nu = d - m$, and $r = k - m$, we are in the assumptions of the following lemma.

LEMMA (4.14). *Let ν be a positive integer, and let $\{C_t, x_t, Q_t\}_{t\in\Delta}$ be an analytic family of triples, where C_t is a smooth curve, $x_t \in C_t$, and Q_t is a pencil on C_t containing νx_t. Suppose that there exists an analytic family of divisors $\{D'_t\}_{t\in\Delta}$, $D'_t \in Q_t$, such that, for $t \neq 0$, $\operatorname{supp}(D'_t)$ is disjoint from x_t and has $\nu - r$ points, while $D'_0 = \nu x_0$. Then Q_0 can be written as $Q_0 = rx_0 + Q'$ for some pencil Q'.*

Assume this lemma. Then the pencil P_0 is of the form $(m+r)x_0 + Q' = kx_0 + Q'$ and therefore contains a divisor E different from dx_0, with $|\operatorname{supp}(E) \smallsetminus \{x\}| \leq d-k-1$, which means that $f(z) \in f(X^k)$. This concludes the proof of Lemma (4.12). It now remains to prove Lemma (4.14). For this, observe that if $r = 0$, there is nothing to prove, while for $r > 0$, that is, when D_t contains a ramification point of the pencil Q_t, Lemma (4.14) is an immediate consequence of the following result.

LEMMA (4.15). *Let ν and $\{C_t, x_t, Q_t\}_{t\in\Delta}$ be as in Lemma (4.14). Suppose that the pencil Q_t is base-point-free for $t \neq 0$. For $t \neq 0$, let R_t be the intersection with $C_t \smallsetminus \{x_t\}$ of the ramification divisor of the morphism associated to Q_t. Let R_0 be the limit of R_t as t tends to 0. The the multiplicity of x_0 in R_0 equals the multiplicity of x_0 as a fixed point of Q_0.*

Proof. Let (z,t) be local coordinates around $x_0 \in C_0$ in the total space of $\{C_t\}_{t\in\Delta}$, chosen so that the image of the section $t \mapsto x_t$ is $z = 0$. We may choose generators f_1 and f_2 of the pencil Q_t which, in terms of these coordinates, have local expressions

$$f_1 = z^d, \qquad f_2 = \sum_{i\neq d} a_i(t) z^i,$$

where f_2 is divisible neither by t nor by z. The multiplicity of x_0 as a base point of Q_0 is the least index h for which $a_h(0) \neq 0$ $(h < d)$. In the given chart the divisor R_t is the locus where $\frac{\partial}{\partial z}(f_2/f_1)$ vanishes. This is the divisor of $\sum_{i\neq d}(i-d)a_i(t)z^i$ for $t \neq 0$. This function is not divisible by either z or t, so that R_0 is the divisor of zeroes of $\sum_{i\neq d}(i-d)a_i(0)z^i$, proving that x_0 occurs with multiplicity h in R_0.

Q.E.D.

In the original paper [489] an additional important result is proved. Namely, it is shown that all classes in $R^{g-2}(\mathcal{M}_g)$ are proportional to the class $[\mathcal{H}_g]$ of the hyperelliptic locus. This, together with Theorem (7.1) of Chapter XX, shows that $[\mathcal{H}_g] \neq 0$ and that $R^{g-2}(\mathcal{M}_g) = \mathbb{Q} \cdot [\mathcal{H}_g]$.

5. The Zariski tangent spaces to the Brill–Noether varieties.

We start with a smooth curve C of genus $g > 1$ and a Kuranishi family

$$p : \mathcal{C} \to (B, b_0) \tag{5.1}$$

for $C \cong C_{b_0}$. We assume, in addition, that $\mathcal{C} \to B$ is a Kuranishi family for every one of its fibers. We set

$$\mathbf{Pic}^d = \mathbf{Pic}^d(p), \quad \mathcal{W}^r_d = \mathcal{W}^r_d(p), \quad \mathcal{G}^r_d = \mathcal{G}^r_d(p). \tag{5.2}$$

Let $\ell \in \mathbf{Pic}^d$ be a point corresponding to a degree d line bundle L on C. We wish to interpret in cohomological terms the exact sequence

$$0 \to T_\ell(\mathrm{Pic}^d(C)) \to T_\ell(\mathbf{Pic}^d) \to T_{b_0}(B) \to 0.$$

We know how to do this for the two extreme terms. In Section 2 of Chapter IV and in Section 2 of Chapter XI we showed that there are isomorphisms

$$T_\ell(\mathrm{Pic}^d(C)) \cong H^1(C, \mathcal{O}_C), \quad T_{b_0}(B) \cong H^1(C, T_C).$$

To interpret the middle term, we need some preparation. Set $S = \operatorname{Spec}\mathbb{C}[\varepsilon]$ and denote by s_0 the closed point of S. A *first-order deformation of the pair* (C, L) is, by definition, a 4-tuple $(\varphi, \mathcal{L}, \alpha, \beta)$ where

$$\varphi : \mathcal{X} \to S\,, \qquad \alpha : C \xrightarrow{\simeq} X_{s_0} \tag{5.3}$$

is a first-order deformation of C, $\mathcal{L}$ is a degree d line bundle on $\mathcal{X}$, and $\beta : L \to \alpha^*\mathcal{L}$ is an isomorphism. An isomorphism between two first-order deformations $(\varphi, \mathcal{L}, \alpha, \beta)$ and $(\varphi', \mathcal{L}', \alpha', \beta')$ of (C, L) is an isomorphism $\eta : \mathcal{X} \to \mathcal{X}'$ of deformations of C (see Chapter XI, Section 2), plus an isomorphism between $\mathcal{L}$ and $\eta^*\mathcal{L}'$ inducing the identity on L. From the universal property of the Poincaré line bundle described in Theorem (2.1) it follows that $T_\ell(\mathbf{Pic}^d)$ is in one-to-one correspondence with the set of isomorphism classes of first-order deformations of (C, L). To describe these, we proceed as follows. As in (2.3), Chapter XI, we imagine C as being given by transition data $\{U_\alpha, z_\alpha, f_{\alpha\beta}(z_\beta)\}$, and the first-order deformation (5.3) can be thought of as being given by gluing the $U_\alpha \times S$ via the identifications

$$z_\alpha = \tilde{f}_{\alpha\beta}(z_\beta, \varepsilon) = f_{\alpha\beta}(z_\beta) + \varepsilon b_{\alpha\beta}(z_\beta)\,, \tag{5.4}$$

whereas φ is given, on $U_\alpha \times S$, by the projection to S. The Kodaira–Spencer class $[\vartheta] \in H^1(C, T_C)$ of the first-order deformation φ is given by the cocycle

$$\vartheta = \{\vartheta_{\alpha\beta}\}\,, \qquad \vartheta_{\alpha\beta} = b_{\alpha\beta}\frac{\partial}{\partial z_\alpha}\,.$$

Next, we let $\{g_{\alpha\beta}\}$ be the transition functions for L relative to the cover $\mathcal{U} = \{U_\alpha\}$. We may think of $\mathcal{L}$ as being given, relative to the cover $\{U_\alpha \times S\}$, by transition functions

$$\tilde{g}_{\alpha\beta}(z_\beta, \varepsilon) = g_{\alpha\beta}(z_\beta) + \varepsilon a_{\alpha\beta}(z_\beta) \tag{5.5}$$

satisfying the usual cocycle rule, which, in this case, translates into

$$g_{\alpha\beta}g_{\beta\gamma} = g_{\alpha\gamma}\,, \tag{5.6}$$

$$a_{\alpha\gamma}g_{\alpha\gamma}^{-1} = \frac{\partial g_{\alpha\beta}}{\partial z_\beta}g_{\alpha\beta}^{-1}b_{\beta\gamma} + a_{\alpha\beta}g_{\alpha\beta}^{-1} + a_{\beta\gamma}g_{\beta\gamma}^{-1}\,. \tag{5.7}$$

Set

$$\begin{aligned} \phi &= \{\phi_{\alpha\beta}\} \in C^1(\mathcal{U}, \mathcal{O}_C)\,, & \phi_{\alpha\beta} &= a_{\alpha\beta}g_{\alpha\beta}^{-1}\,, \\ \eta &= \{\eta_{\alpha\beta}\} \in C^1(\mathcal{U}, \omega_C)\,, & \eta_{\alpha\beta} &= g_{\alpha\beta}^{-1}dg_{\alpha\beta}\,. \end{aligned} \tag{5.8}$$

Notice that η is actually a cocycle representing the first Chern class $c_1(L) \in H^1(C, \omega_C)$. Then (5.7) means that

$$\delta\phi + \eta \cdot \vartheta = 0\,, \tag{5.9}$$

where $\eta\cdot\vartheta \in C^2(\mathcal{U}, \mathcal{O}_C)$ is the cup product of η and ϑ. We are now going to construct a rank 2, locally free sheaf Σ_L on C whose first cohomology group parameterizes isomorphism classes of first-order deformations of (C, L). In fact, Σ_L will fit into the exact sequence

$$(5.10) \qquad 0 \to \mathcal{O}_C \to \Sigma_L \xrightarrow{\tau} T_C \to 0\,.$$

We define Σ_L to be the rank two locally free $\mathcal{O}_C$-module whose sections are the differential operators of order less than or equal to 1 acting on sections of L. To describe Σ_L locally and to construct (5.10), we choose, on each open set U_α, a trivialization $\chi_\alpha : L_{|U_\alpha} \to \mathcal{O}_{U_\alpha}$ so that $\chi_\alpha\chi_\beta^{-1}(1) = g_{\alpha\beta}$ and set $\xi_\alpha = \chi_\alpha^{-1}(1) \in \Gamma(U_\alpha, L)$, which implies

$$(5.11) \qquad \xi_\beta = g_{\alpha\beta}\xi_\alpha\,, \quad \text{in} \quad U_\alpha \cap U_\beta\,.$$

Then, on U_α, the sheaf Σ_L is free and generated by the two sections 1_α (the constant 1) and D_α, where

$$(5.12) \qquad D_\alpha(f\xi_\alpha) = \frac{\partial f}{\partial z_\alpha}\xi_\alpha\,.$$

The symbol map $\tau : \Sigma_L \to T_C$ in (5.10) sends D_α to $\frac{\partial}{\partial z_\alpha}$. Computing $D_\alpha(f\xi_\beta)$ via (5.11) gives

$$(5.13) \qquad D_\alpha = g_{\alpha\beta}^{-1}\frac{\partial g_{\alpha\beta}}{\partial z_\alpha}1_\beta + \frac{\partial z_\beta}{\partial z_\alpha}D_\beta\,.$$

This means that the extension class of (5.10) is precisely

$$\eta \in H^1(C, \omega_C) \cong H^1(C, \mathcal{H}om(T_C, \mathcal{O}_C))\,.$$

Thus, given d, all the Σ_L's for $L \in \operatorname{Pic}^d(C)$ are (noncanonically) isomorphic. Now (5.13), coupled with $\delta\vartheta = 0$ and (5.7), says that the cochain

$$(5.14) \qquad \sigma = \{\sigma_{\alpha\beta}\} \in C^1(\mathcal{U}, \Sigma_L)\,, \qquad \sigma_{\alpha\beta} = b_{\alpha\beta}D_\alpha - \phi_{\alpha\beta}1_\alpha$$

is in fact a cocycle. Its cohomology class $[\sigma]$ will be called the *Kodaira–Spencer class of the first-order deformation of* (C, L) and maps, via $H^1(C, \Sigma_L) \to H^1(C, T_C)$, to the Kodaira-Spencer class of the first-order deformation of C. In complete analogy with the case of deformation of curves, every 1-cocycle with coeficients in Σ_L defines a first-order deformation of (C, L), and cohomologous cycles give rise to isomorphic deformations. In other words, isomorphism classes of first-order deformations of (C, L) are in one-to-one correspondence with the elements of $H^1(C, \Sigma_L)$, that is, we get a bijection

$$H^1(C, \Sigma_L) \xrightarrow{\cong} T_\ell(\mathbf{Pic}^d)$$

which can be easily seen to be linear. In conclusion, we have proved the following result.

PROPOSITION (5.15). *Let $p : \mathcal{C} \to (B, b_0)$ be a Kuranishi family for a smooth curve C. Let ℓ be the point of $\mathbf{Pic}^d = \mathbf{Pic}^d(p)$ corresponding to a degree d line bundle L on C. Then the tangent space to $\mathbf{Pic}^d$ at ℓ is naturally identified with $H^1(C, \Sigma_L)$. Moreover, there is an isomorphism of exact sequences*

$$\begin{array}{ccccccccc} 0 & \longrightarrow & H^1(C,\mathcal{O}_C) & \longrightarrow & H^1(C,\Sigma_L) & \xrightarrow{\ \tau\ } & H^1(C,T_C) & \longrightarrow & 0 \\ & & \cong\uparrow & & \cong\uparrow & & \cong\uparrow & & \\ 0 & \longrightarrow & T_\ell(\mathrm{Pic}^d(C)) & \longrightarrow & T_\ell(\mathbf{Pic}^d) & \longrightarrow & T_{b_0}(B) & \longrightarrow & 0 \end{array}$$

where the top exact sequence is the cohomology exact sequence of (5.10), and the vertical arrows are Kodaira–Spencer maps.

We are now in a position to identify the Zariski tangent spaces to $\mathcal{W}^r_d$ and $\mathcal{G}^r_d$. Our description will be analogous to that of the Zariski tangent space to $W^r_d(C)$ as being given by

$$T_L(W^r_d(C)) = (\mathrm{Image}\,\mu_0)^\perp ,$$

where $h^0(C, L) = r + 1$, and

$$\mu_0 : H^0(C, L) \otimes H^0(C, \omega_C L^{-1}) \to H^0(C, \omega_C)$$

is the multiplication map (see Propositions (4.1) and (4.2), Chapter IV). The main philosophical difference is that derivatives of sections will appear in the formulas in an essential way. Actually, exactly as in Section 4 of Chapter IV, it is better to first determine the Zariski tangent spaces to points of $\mathcal{G}^r_d$ and then analyze the case of $\mathcal{W}^r_d$.

Fix a g^r_d on our curve C. This means that we have a degree d line bundle L on C and an $(r+1)$-dimensional subspace W of $H^0(C, L)$. We let w denote the point of $\mathcal{G}^r_d$ corresponding to these data. Look at the natural morphism

$$\chi : \mathcal{G}^r_d \to \mathbf{Pic}^d$$

and set $\ell = \chi(w)$. The fiber of χ over ℓ is the Grassmannian of $(r+1)$-dimensional subspaces of $H^0(C, L)$, so that we have the exact sequence

$$0 \longrightarrow \mathrm{Hom}(W, H^0(C,L)/W) \longrightarrow T_w(\mathcal{G}^r_d) \xrightarrow{d\chi} T_\ell(\mathbf{Pic}^d)\,. \tag{5.16}$$

From the universal properties of $\mathcal{G}^r_d$ and $\mathbf{Pic}^d$ it follows that the space $d\chi(T_w(\mathcal{G}^r_d))$ can be described as the set of isomorphism classes of first-order deformations

$$(\mathcal{X} \to S = \mathrm{Spec}\,\mathbb{C}[\varepsilon]\,,\ \mathcal{L}) \tag{5.17}$$

of (C, L) for which every section in W extends to a section of $\mathcal{L}$. Let the deformation (5.17) be given by patching data (5.4) and (5.5). Let s be a section in $W \subset H^0(C, L)$ given by holomorphic functions $s_\alpha(z_\alpha)$ satifying

(5.18) $$s_\alpha = g_{\alpha\beta} s_\beta \quad \text{in} \quad U_\alpha \cap U_\beta\,.$$

An extension $\widetilde{s}$ of s to a section of $\mathcal{L}$ corresponds to the datum of functions

(5.19) $$\widetilde{s}_\alpha(z_\alpha, \varepsilon) = s_\alpha(z_\alpha) + \varepsilon t_\alpha(z_\alpha)$$

satisfying

(5.20) $$\widetilde{s}_\alpha = \widetilde{g}_{\alpha\beta} \widetilde{s}_\beta\,,$$

which is equivalent to (5.6) plus

(5.21) $$t_\alpha = \phi_{\alpha\beta} s_\alpha - b_{\alpha\beta} \frac{\partial s_\alpha}{\partial z_\alpha} + g_{\alpha\beta} t_\alpha\,,$$

where $b_{\alpha\beta}$ and $\phi_{\alpha\beta}$ are as in (5.4) and (5.8). If we denote by t the cochain in $C^0(\mathcal{U}, L)$ whose local representatives are the t_α, then (5.21) reads

(5.22) $$\delta t = \sigma \cdot s\,,$$

where the dot stands for the cup product induced by the $\mathbb{C}$-linear pairing

$$\Sigma_L \otimes_{\mathbb{C}} L \to \mathcal{O}_C$$

(notice that this pairing is not $\mathcal{O}_C$-linear because differentiation is involved). Therefore, if s extend to a section of $\mathcal{L}$, (5.22) can be solved, i.e., $[\sigma] \cdot s = 0$ in $H^1(C, \Sigma_L)$. Conversely, if this happens, (5.20) defines an extension of s to a section of $\mathcal{L}$.

This discussion shows that the isomorphism classes of the first-order deformations $(\mathcal{X} \to S, \mathcal{L})$ of (C, L) such that every section of $W \subset H^0(C, L)$ extends to section of $\mathcal{L}$ correspond precisely to the elements of the subspace of $H^1(C, \Sigma_L)$ defined by

(5.23) $$\{\sigma \in H^1(C, \Sigma_L) : \sigma \cdot s = 0 \text{ for all } s \in W\}\,.$$

If we dualize the cup product map

$$H^1(C, \Sigma_L) \to \operatorname{Hom}(W, H^1(C, L))\,,$$

we get a homomorphism

$$\mu_W : W \otimes H^0(C, \omega_C \otimes L^{-1}) \to H^0(C, \omega_C \otimes \Sigma_L^\vee) \tag{5.24}$$

and a commutative diagram

$$\begin{array}{ccc} W \otimes H^0(C, \omega_C \otimes L^{-1}) & \xrightarrow{\quad \mu_{0,W} \quad} & H^0(C, \omega_C) \\ & \searrow^{\mu_W} & \uparrow \\ & & H^0(C, \omega_C \otimes \Sigma_L^\vee) \end{array} \tag{5.25}$$

where the vertical arrow is the dual of $H^1(C, \mathcal{O}_C) \to H^1(C, \Sigma_L)$. The subspace of $H^1(C, \Sigma_L)$ given by (5.23) can also be described as $(\operatorname{Image} \mu_W)^\perp = \ker \mu_W^\vee$. In conclusion, we have the following result.

PROPOSITION (5.26). *Let $W \subset H^0(C, L)$ be a g_d^r on a smooth curve C. Let w be the corresponding point in $\mathcal{G}_d^r$. Let $\chi : \mathcal{G}_d^r \to \mathbf{Pic}^d$ be the natural projection. Then*

$$d\chi(T_w(\mathcal{G}_d^r)) = (\operatorname{Image} \mu_W)^\perp = \ker \mu_W^\vee \,, \tag{5.27}$$

and there is an exact sequence

$$0 \longrightarrow \operatorname{Hom}(W, H^0(C, L)/W) \longrightarrow T_w(\mathcal{G}_d^r) \xrightarrow{d\chi} (\operatorname{Image} \mu_W)^\perp \longrightarrow 0 \,. \tag{5.28}$$

In particular,

$$\begin{aligned} \dim T_w(\mathcal{G}_d^r) &= (r+1)(\overline{r} - r) + 4g - 3 - (r+1)(g - d + \overline{r}) + \dim \ker(\mu_W) \\ &= 3g - 3 + \rho + \dim(\ker \mu_W) \,, \end{aligned}$$

where $h^0(C, L) = \overline{r} + 1$, and $\rho = g - (r+1)(g - d + r)$ is the Brill–Noether number.

When $W = H^0(C, L)$, we usually write μ_L, or simply μ, for μ_W. Since the morphism $\chi : \mathcal{G}_d^r \to \mathcal{W}_d^r$ is biregular off $\mathcal{W}_d^{r+1}$, we have the following result.

PROPOSITION (5.29). *Let L be a degree d line bundle on a smooth curve C and set $r + 1 = h^0(C, L)$. Let ℓ be the point in $\mathcal{W}_d^r$ corresponding to (C, L). Then*

$$T_\ell(\mathcal{W}_d^r) = (\operatorname{Image} \mu)^\perp = \ker \mu^\vee \,.$$

In particular,

$$\dim T_\ell(\mathcal{W}_d^r) = 3g - 3 + \rho + \dim(\ker \mu) \,,$$

where $\rho = g - (r+1)(g - d + r)$ is the Brill–Noether number.

Keeping the notation introduced above and recalling (3.21), we get the following:

COROLLARY (5.30). *The cup product map* μ_W *(resp.,* μ*) is injective if and only if* $\mathcal{G}^r_d$ *(resp.,* $\mathcal{W}^r_d$*) is smooth and of dimension*

$$3g - 3 + \rho = 4g - 3 - (r+1)(g-d+r)$$

at w *(resp., at* ℓ*).*

Recalling Definition (3.19), the preceding corollary, together with the commutativity of diagram (5.25), gives the following result.

COROLLARY (5.31). *If the curve* C *satisfies Petri's condition, then*

$$\dim_\ell \mathcal{W}^r_d = \dim_w \mathcal{G}^r_d = 3g - 3 + \rho$$

for every r *and* d *and for every* $\ell \in W^r_d(C) \smallsetminus W^{r+1}_d(C)$ *and every* $w \in G^r_d(C)$. *Moreover,* $\mathcal{G}^r_d$ *is smooth along* $G^r_d(C)$*, and the points of* $W^r_d(C)$ *which are singular in* $\mathcal{W}^r_d$ *are precisely those belonging to* $\mathcal{W}^{r+1}_d$.

6. The μ_1 homomorphism.

We now turn to a typical Kodaira–Spencer question: *Given a line bundle* L *on a smooth curve* C*, how can one interpret, in cohomological terms, the obstructions to extending* L*, together with a subspace* W *of its space of sections, along any deformation of* C*?*

In this section we will discuss first-order obstructions. It will turn out that these are completely described in terms of a homomorphism

$$\mu_{1,W} : \ker \mu_{0,W} \to H^0(C, \omega_C^2)\,,$$

where, as usual,

$$\mu_{0,W} : W \otimes H^0(C, \omega_C \otimes L^{-1}) \to H^0(C, \omega_C)$$

is the multiplication map. When $W = H^0(C, L)$, we write μ_L, $\mu_{0,L}$, $\mu_{1,L}$, or simply μ, μ_0, μ_1 when no confusion is likely, instead of μ_W, $\mu_{0,W}$, $\mu_{1,W}$. The map $\mu_{1,W}$ is defined to be the unique map which makes the following diagram commute:

$$\begin{array}{ccc}
W \otimes H^0(C, \omega_C \otimes L^{-1}) & \xrightarrow{\ \mu_{0,W}\ } & H^0(C, \omega_C) \\
\Vert & & \uparrow \\
W \otimes H^0(C, \omega_C \otimes L^{-1}) & \xrightarrow{\ \mu_W\ } & H^0(C, \omega_C \otimes \Sigma_L^\vee) \\
\uparrow & & \uparrow \\
\ker \mu_{0,W} & \xrightarrow{\ \mu_{1,W}\ } & H^0(C, \omega_C^2) \\
 & & \uparrow \\
 & & 0
\end{array} \tag{6.1}$$

where the exact sequence on the right is a piece of the cohomology exact sequence of the dual of (5.10) tensored with ω_C,

$$0 \to \omega_C^2 \to \omega_C \otimes \Sigma_L^\vee \to \omega_C \to 0\,.$$

We consider a Kuranishi family

$$p : \mathcal{C} \to (B, b_0)$$

for $C \cong C_{b_0}$ as in (5.1) and adopt the conventions (5.2). Write d for the degree of L, $r+1$ for the dimension of W, w for the point of $\mathcal{G}_d^r$ corresponding to W, and ℓ for its image in $\mathcal{W}_d^r$. As usual, we also write B_d^r for the image of $\mathcal{W}_d^r$ in B. Dualizing (6.1) and recalling (5.26), we get a commutative diagram

$$(6.2)\qquad \begin{array}{ccccc} d\chi(T_w(\mathcal{G}_d^r)) & \cong & \ker \mu_W^\vee & & \\ \cap & & \cap & & \\ T_\ell(\mathbf{Pic}^d) & \cong & H^1(C, \Sigma_L) & \xrightarrow{\mu_W^\vee} & \mathrm{Hom}(W, H^1(C, L)) \\ \Big\downarrow d\pi & & \Big\downarrow & & \Big\downarrow \\ T_{b_0}(B) & \cong & H^1(C, T_C) & \xrightarrow{\mu_{1,W}^\vee} & (\ker \mu_{0,W})^\vee \end{array}$$

where $\mathbf{Pic}^d = \mathbf{Pic}^d(p) \xrightarrow{\pi} B$ is the natural morphism. To say that a curve with general moduli possesses a g_d^r implies that the image of the morphism

$$\pi \circ \chi : \mathcal{G}_d^r \to B$$

is an open set, so that by Sard's theorem

$$d\pi : \ker \mu_W^\vee \to H^1(C, T_C)$$

must be surjective, and therefore $\mu_{1,W}^\vee$ must vanish. To summarize, we have proved the following lemma.

LEMMA (6.3). *Let C be a general curve of genus $g > 1$. Let L be any line bundle on C. Then*

$$\mu_{1,L} : \ker \mu_{0,L} \to H^0(C, \omega_C^2)$$

is zero (i.e., $\ker \mu_1 = \ker \mu_0$).

It is easy to give an explicit description of $\mu_{1,W}$. Consider an element $\sum s_i \otimes r_i$ of $\ker \mu_{0,W}$. Thus,

$$\sum s_i r_i = 0\,.$$

Think of the s_i (resp., r_i) as being given by holomorphic functions $s_{i\alpha}$ (resp., $r_{i\alpha}$) satisfying

$$s_{i\alpha} = g_{\alpha\beta} s_{i\beta}\,, \qquad \left(\text{resp., } r_{i\alpha} = g_{\alpha\beta}^{-1} \frac{dz_\beta}{dz_\alpha} r_{i\beta}\right).$$

Let

$$\sigma = \{\sigma_{\alpha\beta}\}\,, \qquad \sigma_{\alpha\beta} = \xi_{\alpha\beta} D_\alpha + \phi_{\alpha\beta} 1_\alpha$$

be a 1-cocycle with coefficients in Σ_L and $\widetilde{\sigma}$ its image in $Z^1(\mathcal{U}, T_C)$:

$$\widetilde{\sigma}_{\alpha\beta} = \xi_{\alpha\beta} \frac{\partial}{\partial z_\alpha}\,.$$

If we denote by $\langle\ ,\ \rangle$ the duality pairing, we have

$$\left\langle \mu_{1,W}\left(\sum s_i \otimes r_i\right),\ [\widetilde{\sigma}]\right\rangle = \left\langle \mu_W\left(\sum s_i \otimes r_i\right),\ [\sigma]\right\rangle = \sum \langle [\sigma] s_i, r_i\rangle\,.$$

The right-hand side of this equality is the integral over C of the cohomology class in $H^1(C, \omega_C)$ represented by the 1-cocycle

$$\phi_{\alpha\beta} \sum s_{i\alpha} r_{i\alpha} + \xi_{\alpha\beta} \sum \frac{ds_{i\alpha}}{dz_\alpha} r_{i\alpha} = \xi_{\alpha\beta} \sum \frac{ds_{i\alpha}}{dz_\alpha} r_{i\alpha} \tag{6.4}$$

(recall that $\sum s_i r_i = 0$ by assumption). Also notice that

$$\begin{aligned}\sum \frac{ds_{i\alpha}}{dz_\alpha} r_{i\alpha} &= \left(\frac{dz_\beta}{dz_\alpha}\right)^2 \sum \frac{ds_{i\beta}}{dz_\beta} r_{i\beta} + g_{\alpha\beta}^{-1} \frac{dz_\beta}{dz_\alpha} \frac{dg_{\alpha\beta}}{dz_\alpha} \sum s_{i\beta} r_{i\beta} \\ &= \left(\frac{dz_\beta}{dz_\alpha}\right)^2 \sum \frac{ds_{i\beta}}{dz_\beta} r_{i\beta}\,;\end{aligned} \tag{6.5}$$

hence $\{\sum (ds_{i\alpha}/dz_\alpha) r_{i\alpha}\}$ is a globally defined section of ω_C^2. It is then clear that the right-hand side of (6.4) is the cup product of $\widetilde{\sigma}$ with this section. Therefore,

$$\mu_{1,W}\left(\sum s_i r_i\right) = \left\{\sum \frac{ds_{i\alpha}}{dz_\alpha} r_{i\alpha}\right\} = \left\{-\sum \frac{dr_{i\alpha}}{dz_\alpha} s_{i\alpha}\right\}. \tag{6.6}$$

We end this section by giving the first applications of these considerations and of Corollary (5.30). We will study one-dimensional linear series.

PROPOSITION (6.7). *Let C be a smooth curve of genus $g > 1$. Let L be a line bundle on C, and let W be a two-dimensional subspace of $H^0(C, L)$, that is, an arbitrary g^1_d on C. Let Δ be the fixed divisor of the linear series $|W|$. Then*

$$\ker \mu_{0,W} \cong H^0(C, \omega_C L^{-2}(\Delta))\,.$$

If C is general, then $\ker \mu_{0,W} = 0$.

Proof. The isomorphism between $\ker \mu_{0,W}$ and $H^0(C, \omega_C L^{-2}(\Delta))$ is just the base-point-free pencil trick (cf. Chapter III, page 126). Let s_1, s_2 be a basis of W. Suppose that $\mu_0(s_1 \otimes r_1 + s_2 \otimes r_2) = 0$. If C is general, by Lemma (6.3) we must also have $\mu_1(s_1 \otimes r_1 + s_2 \otimes r_2) = 0$. Using the notation of the previous discussion, this means that

$$\begin{gathered} s_{1\alpha} r_{1\alpha} + s_{2\alpha} r_{2\alpha} = 0 \,. \\ \frac{ds_{1\alpha}}{dz_\alpha} r_{1\alpha} + \frac{ds_{2\alpha}}{dz_\alpha} r_{2\alpha} = 0 \,. \end{gathered}$$

If r_1, r_2 are not identically zero, then the determinant

$$s_{1\alpha} \frac{ds_{2\alpha}}{dz_\alpha} - s_{2\alpha} \frac{ds_{2\alpha}}{dz_\alpha}$$

must vansh identically. But this means that the meromorphic function s_2/s_1 is constant, a contradiction.

Q.E.D.

Despite its deceiving simplicity, the above result is a cornerstone in the study of special divisors. In fact, it has been discovered and rediscovered, from many different points of view, by a large number of authors, over an extended period of time (cf. the bibliographical notes at the end of this chapter). Probably the earliest and certainly one of the clearest formulations of this result was given by Francesco Severi [634]:

Sopra una curva di genere p (> 1) a moduli generali, la serie doppia di ogni serie lineare, almeno ∞^1, di ordine ν qualunque esistente sulla curva, è non speciale (e quindi è $\nu \geq \frac{p}{2} + 1$).[8]

This is essentially equivalent to Proposition (6.7), which is what we called the first version of Petri's statement for the case $r = 1$. It is interesting to notice that, in the case $r = 1$, Petri's statement gives the following stronger implication, which is actually false for general r.

PROPOSITION (6.8). *If $g > 1$, $d \geq 2$, and $d \leq g + 1$, then $\mathcal{G}^1_d$ is smooth, the singular locus of $\mathcal{W}^1_d$ is $\mathcal{W}^2_d$, and*

$$\dim \mathcal{G}^1_d = \dim \mathcal{W}^1_d = 3g - 3 + \rho = 2g + 2d - 5 \,.$$

Proof. First of all, notice that $\mathcal{G}^1_d$ is nonempty since every hyperelliptic curve has a g^1_d. Secondly, observe that, by Lemma (3.5) of Chapter IV,

[8] *On a curve of genus p (> 1) with general moduli, the double of any linear series of dimension ≥ 1 and arbitrary degree ν is nonspecial (and thus $\nu \geq \frac{p}{2} + 1$).*

no component of $\mathcal{W}^1_d$ is entirely contained in $\mathcal{W}^2_d$. By Corollary (5.30) it suffices to show that $\ker \mu_W = 0$ for any line bundle L on a smooth genus g curve C and any two-dimensional subspace $W \subset H^0(C, L)$. Now, $\ker \mu_W$ is contained in $\ker \mu_{1,W} \subset \ker \mu_{0,W}$. Thus, assuming that $\ker \mu_W \neq 0$, we can use the same argument as in the proof of Proposition (6.7) to reach a contradiction.

Q.E.D.

For example, for curves of genus 4 the projection $\mathcal{W}^1_3 \to B$ has one-dimensional fibers over the hyperelliptic locus and, outside this set, is a two-sheeted covering map, branched along the locus of curves lying on singular quadrics. In general, since the fiber of $\mathcal{G}^r_d$ (resp., $\mathcal{W}^r_d$) over a point $b \in B$ is $G^r_d(C_b)$ (resp., $W^r_d(C_b)$), the only thing that one can deduce from the second and third version of Petri's statement is that $\mathcal{G}^r_d$ (resp., $\mathcal{W}^r_d$) is smooth outside a subvariety that projects to a *proper* subvariety of B (resp., outside the union of $\mathcal{W}^{r+1}_d$ and a subvariety that projects to a *proper* subvariety of B). In general the analogue of Proposition (6.8) is false when $r \geq 2$. However, in Section 10 we will prove a result about the structure of $\mathcal{G}^2_d$ which is more precise than the one implied by Petri's statement in the case $r = 2$.

At this point we know that, when the dimension of $W^1_d(C)$ exceeds the expected one, the curve C has to be special in the sense of moduli. In Section 5 of Chapter IV, our discussion of Martens' and Mumford's theorems showed that it is possible to explicitly describe those curves C for which the dimension of $W^1_d(C)$ is close to its maximum possible value. Along these lines, a natural, more general question is the following. How special (in the sense of moduli) is a curve C carrying a g^1_d for which the Brill–Noether number ρ is negative? In terms of diagram (3.18) our question is: what is the dimension of $M^1_{g,d}$? Or, which is the same, what is the dimension of B^1_d? Again the answer is provided by Severi [634]:

La più generale curva di genere $p > 1$, che soddisfa alla condizione di contenere una g^1_ν (almeno) di ordine $\nu < \frac{p}{2} + 1$ (e perciò sia a moduli particolari) dipende precisamente da $3p - 3 - i$ moduli, ove i è l'indice di specialità della serie doppia della g^1_ν.[9]

This has been rediscovered by Farkas [260]. More precisely:

PROPOSITION (6.9) (SEVERI–FARKAS). *Let k be the minimum possible value of $h^1(C, L^2(-\Delta))$, where C belongs to $M^1_{g,d}$, L runs through all*

[9] *The most general curve of genus $p > 1$ having the property of possessing (at least) a g^1_ν of degree $\nu < \frac{p}{2} + 1$ (which makes the curve one with particular moduli) depends exactly on $3p - 3 - i$ moduli, where i is the index of speciality of the double of the linear series g^1_ν.*

degree d line bundles on C with $h^0(C,L) \geq 2$, and Δ stands for the base locus of $|L|$. If $d < g/2+1$ (i.e., if $\rho < 0$), then

$$\dim M^1_{g,d} = 3g - 3 - k\,.$$

In particular, $\dim M^1_{g,d} \leq 3g - 3 + \rho$.

This follows directly from Proposition (5.26) and Proposition (6.7). It is now natural to ask: when $d < g/2+1$, is it true that $\dim M^1_{g,d} = 3g - 3 + \rho$? To further justify this question, observe that, if $\rho < 0$, then it is natural to expect that:

a) *the most general curve C of genus g possessing a g^1_d will possess only a finite number of them;*
b) *at least one g^1_d on C will have no base points.*

If this where the case, the preceding proposition would show that $\dim M^1_{g,d} = 3g - 3 + \rho$. This agrees with the following naive count of moduli. Let $R = 2g + 2d - 2$ be the number of branch points of the morphism to $\mathbb{P}^1$ attached to a base-point-free g^1_d. Then we should have

$$\dim M^1_{g,d} = R - 3 = 3g - 3 + \rho\,.$$

In Section 8 we will present a very nice argument of Beniamino Segre which will enable us to prove the above statement. We will in fact proceed to show that

The most general curve C of genus g possessing a g^1_d with $d < g/2+1$ has in fact a unique g^1_d.

In Section 8 we will give a cohomological interpretation of the μ_1 homomorphism using the normal bundle sequence (see diagram (8.10)). Following Voisin [672], an equivalent interpretation can be given directly, and we will present it now as a preparation for the next section. Let L be a line bundle on a smooth curve C and assume that it is base-point-free. Consider the evaluation map

$$\mathrm{ev}_{C,L} : H^0(C,L) \otimes \mathcal{O}_C \to L$$

and let M_L be its kernel. Since L is base-point-free, M_L is locally free of rank r (the reader should prove this as an exercise), and we have an exact sequence of vector bundles on C

$$0 \to M_L \to H^0(C,L) \otimes \mathcal{O}_C \to L \to 0\,. \tag{6.10}$$

This sequence is simply the pullback to C, via the morphism $\phi_L : C \to \mathbb{P}^r = \mathbb{P}H^0(C,L)^\vee$, of the Euler sequence

$$0 \to \Omega^1_{\mathbb{P}^r} \to \mathcal{O}^{r+1}_{\mathbb{P}^r} \to \mathcal{O}_{\mathbb{P}^r}(1) \to 0\,.$$

The bundle M_L defined by the sequence (6.10) is called the *Lazarsfeld–Mukai bundle* [314,464,465,533], an object of paramount importance. Tensoring (6.10) with $\omega_C L^{-1}$, we get

$$0 \to M_L \otimes \omega_C L^{-1} \to H^0(C,L) \otimes \omega_C L^{-1} \to \omega_C \to 0\,. \tag{6.11}$$

Passing to cohomology, we see that

$$H^0(C, M_L \otimes \omega_C L^{-1}) = \ker \mu_0\,. \tag{6.12}$$

Tensoring the evaluation map with the derivation $d : \mathcal{O}_C \to \omega_C$, one obtains a homomorphism

$$\mathrm{ev}_{C,L} \otimes d : H^0(C,L) \otimes \mathcal{O}_C \to \omega_C L$$

whose restriction to M_L is $\mathcal{O}_C$-linear. Tensoring with $\omega_C L^{-1}$, this yields a homomorphism of $\mathcal{O}_C$ modules

$$s : M_L \otimes \omega_C L^{-1} \to \omega_C^2\,. \tag{6.13}$$

Using (6.12), one readily sees that the homomorphism induced by s on global sections is nothing but μ_1.

There is a completely straightforward generalization of the μ_1 map. Given two line bundles L and L' on a smooth curve C, let $\mu_{0,L,L'} : H^0(C,L) \otimes H^0(C,L') \to H^0(C,LL')$ be the multiplication map, and define

$$\mu_{1,L,L'} : \ker(\mu_{0,L,L'}) \to H^0(C, \omega_C LL') \tag{6.14}$$

by setting $\mu_{1,L,L'}(\sum_i s_i \otimes t_i) = \sum_i (s_i dt_i - t_i ds_i)$. We will call this the *generalized* μ_1 *map*. For more information on this map, see the bibliographical notes.

7. Lazarsfeld's proof of Petri's conjecture.

In this section we present an elegant proof of Petri's conjecture, first conceived by Lazarsfeld; we will follow a simplified argument, which we owe to Pareschi.

We recall, from Proposition (3.20), that the locus in M_g described by curves satisfying Petri's condition (3.19) is open, so that, in order to prove Petri's statement (1.1), it suffices to produce a single genus g curves that satisfies Petri's condition. It is also clear that, in Petri's statement (1.1), we can limit ourselves to complete, base-point-free linear series. Lazarsfeld's idea is to find this curve as a hyperplane section of a K3 surface. Of course, such a curve is by no means a general curve of genus g, at least when $g > 11$. Indeed, a simple moduli count shows that genus g hyperplane sections of K3 surfaces depend on at most $g + 19$ moduli.

Lazarsfeld's result is the following:

THEOREM (7.1). *Let X be a K3 surface with* $\mathrm{Pic}(X) = \mathbb{Z} \cdot [C_0]$, *where* C_0 *is a smooth irreducible curve of genus* g. *Then the general element in* $|C_0|$ *satisfies Petri's condition.*

Since, as is well known, for any $g \geq 2$, there are K3 surfaces X with $\mathrm{Pic}(X) = \mathbb{Z} \cdot [C_0]$ and C_0 smooth, irreducible and of genus g, the theorem above implies Petri's statement (1.1). As we shall see from its proof, Theorem (7.1) holds under a weaker assumption. In fact, it is not necessary to assume that $\mathrm{Pic}(X) = \mathbb{Z} \cdot [C_0]$. It is enough to assume that the linear series $|C_0|$ contains no reducible or nonreduced curve.

Proof of Theorem (7.1). Let C be a general element of $|C_0|$. We proceed by contradiction, assuming that on C there exists a complete, base-point-free linear series for which the μ_0 map is not injective. Let L be the line bundle on C corresponding to this linear series, and set $d = \deg L$ and $r + 1 = h^0(C, L)$. By assumption, the kernel of

$$\mu_0 : H^0(C, L) \otimes H^0(C, \omega_C L^{-1}) \to H^0(C, \omega_C)$$

is not zero. In particular, $h^0(C, \omega_C L^{-1}) \neq 0$. The linear system $|C|$ cuts out on C the canonical series. Let $U \subset |C| \cong \mathbb{P}^g$ be the nonempty Zariski-open subset consisting of the smooth divisors in $|C|$. Consider the tautological family

$$\text{(7.2)} \qquad \begin{array}{c} \mathcal{C} \subset X \times U \\ f \big\downarrow \qquad\quad \\ U \qquad\quad \end{array}$$

We assume that C is the fiber over the point $u \in U$. The characteristic map $T_u(U) \to H^0(C, N_{C/X})$ can be canonically identified with the map $H^0(X, \mathcal{O}_X(C))/H^0(X, \mathcal{O}_X) \to H^0(C, \mathcal{O}_C(C))$ coming from the cohomology exact sequence of

$$0 \to \mathcal{O}_X \to \mathcal{O}_X(C) \to \mathcal{O}_C(C) \to 0 .$$

Since X is a regular surface, it is an isomorphism, and the Kodaira–Spencer map of the family f can be identified with the coboundary map

$$\rho : T_u(U) \cong H^0(C, N_{C/X}) \to H^1(C, T_C) .$$

The natural projection

$$p : \mathcal{W}^r_d(f) \to U$$

is onto by assumption. Let $w \in W^r_d(C)$ be the point corresponding to the linear series $|L|$. Since C is a general element of U, we may assume that dp is surjective at w. From Proposition (5.29) and diagrams (6.1), (6.2) it follows that

$$\operatorname{Image}(\rho \circ dp) \subset \operatorname{Image}(\rho) \cap (\operatorname{Image} \mu_1)^{\perp},$$

where, as usual,

$$\mu_1 : \ker \mu_0 \to H^0(C, \omega_C^2).$$

In other words, we have

$$(7.3) \quad \operatorname{Image}(dp) \subset \rho^{-1}\left(\operatorname{Image}(\rho) \cap (\operatorname{Image} \mu_1)^{\perp}\right) = (\operatorname{Image}(\rho^{\vee} \circ \mu_1))^{\perp}.$$

Set

$$(7.4) \qquad \mu_{1,X} = \rho^{\vee} \circ \mu_1 : \ker \mu_0 \to H^1(C, \omega_C N_{C/X}^{\vee}).$$

Since dp is onto, it follows from (7.3) that $(\operatorname{Image} \mu_{1,X})^{\perp} = H^0(C, N_{C/X})$, implying that $\mu_{1,X} = 0$. To prove Theorem (7.1), it then suffices to prove the following result.

PROPOSITION (7.5). *Let X be a K3 surface with* $\operatorname{Pic}(X) = \mathbb{Z} \cdot [C_0]$, *where C_0 is a smooth irreducible curve of genus g. Then for any smooth curve $C \in |C_0|$ and any base-point-free line bundle L on C, the map $\mu_{1,X}$ is injective.*

Proof. The first step is to give a cohomological interpretation of the homomorphism $\mu_{1,X}$ analogous to the one given for μ_1 at the end of the preceding section. View the line bundle L as a coherent sheaf on X and denote by F_L the kernel of the evaluation map

$$(7.6) \qquad \operatorname{ev}_{X,L} : H^0(C, L) \otimes \mathcal{O}_X \to L.$$

Since L is base-point-free, $\operatorname{ev}_{X,L}$ is onto, hence F_L is locally free of rank $r+1$, and we have an exact sequence of sheaves on X

$$(7.7) \qquad 0 \to F_L \to H^0(C, L) \otimes \mathcal{O}_X \to L \to 0.$$

By Porteous' formula (formula (4.2) in Chapter II) or, more simply, by an elementary local calculation, it follows that $\det F_L = \mathcal{O}_X(-C)$. Now go back to the end of the preceding section, consider the vector bundle M_L on C, and notice that $\det M_L = L^{-1}$. Look at the natural surjection $F_{L|C} \to M_L \to 0$. Since X is a K3 surface, $\mathcal{O}_C(C) = \omega_C$, and hence

$$\det F_{L|C} \otimes \det M_L^{-1} = L \otimes \mathcal{O}_C(-C) = L\omega_C^{-1}.$$

Thus, we get an exact sequence

$$0 \to L\omega_C^{-1} \to F_{L|C} \to M_L \to 0,$$

which, after tensoring with $\omega_C L^{-1}$, becomes

$$(7.8) \qquad 0 \to \mathcal{O}_C \to F_L \otimes \omega_C L^{-1} \to M_L \otimes \omega_C L^{-1} \to 0.$$

Passing to cohomology, we get an exact sequence

$$0 \to H^0(C, \mathcal{O}_C) \to H^0(C, F_L \otimes \omega_C L^{-1}) \to H^0(C, M_L \otimes \omega_C L^{-1}) \xrightarrow{\delta} H^1(C, \mathcal{O}_C).$$

Now the proposition follows from the next two lemmas.

LEMMA (7.9). *The coboundary map δ coincides, up to multiplication by a nonzero scalar, with the homomorphism $\mu_{1,X}$.*

LEMMA (7.10). *Under the assumptions of Proposition (7.5), $h^0(C, F_L \otimes \omega_C L^{-1}) = 1$.*

Proof of Lemma (7.9). Exactly as we did at the end of the preceding section, to give a cohomological interpretation of $\mu_{1,X}$, we tensor the evaluation map (7.6) with the derivation $d : \mathcal{O}_X \to \Omega^1_X$, and we obtain a homomorphism

$$\mathrm{ev}_{X,L} \otimes d : H^0(C, L) \otimes \mathcal{O}_X \to L \otimes \Omega^1_X$$

whose restriction to F_L is $\mathcal{O}_X$-linear. Tensoring with $\omega_C L^{-1}$, we obtain a homomorphism of $\mathcal{O}_C$-modules

$$t : F_L \otimes \omega_C L^{-1} \to \Omega^1_X \otimes \omega_C \tag{7.11}$$

fitting in a commutative diagram

$$\begin{array}{ccccccccc} 0 & \longrightarrow & \mathcal{O}_C & \longrightarrow & F_L \otimes \omega_C L^{-1} & \longrightarrow & M_L \otimes \omega_C L^{-1} & \longrightarrow & 0 \\ & & \| & & \downarrow{\scriptstyle t} & & \downarrow{\scriptstyle s} & & \\ 0 & \longrightarrow & \mathcal{O}_C & \longrightarrow & \Omega^1_X \otimes \omega_C & \longrightarrow & \omega_C^2 & \longrightarrow & 0 \end{array}$$

where the top row is (7.8), s is as in (6.13), and the bottom row is obtained by tensoring with ω_C the dual of

$$0 \to T_C \to T_{X|C} \to N_{C/X} \to 0\,.$$

In particular, the coboundary map $H^0(C, \omega_C^2) \to H^1(C, \mathcal{O}_C)$, up to multiplication by a nonzero scalar, coincides with $\rho^\vee$. Recalling from the end of the preceding section that the homomorphism induced by s on global sections is μ_1, the lemma follows at once.

Q.E.D.

Proof of Lemma (7.10). First of all, we show that F_L has the following properties:

i) $h^0(X, F_L) = h^1(X, F_L) = 0$;
ii) $F_L^\vee$ is generated by its global sections away from a finite set of points;
iii) $h^0(X, F_L \otimes F_L^\vee) = h^0(C, F_L \otimes \omega_C L^{-1})$.

To prove i), it suffices to look at the cohomology sequence of (7.7) and to recall that $h^1(X, \mathcal{O}_X) = 0$. To prove ii), we look at the sequence (7.7) and apply $\mathcal{H}om(-, \mathcal{O}_X)$. Since $\mathcal{E}xt^1(L, \mathcal{O}_X) = \mathcal{E}xt^1(\mathcal{O}_C, \mathcal{O}_X) \otimes L^{-1} = \omega_C L^{-1}$, we get an exact sequence

$$0 \to H^0(C, L)^\vee \otimes \mathcal{O}_X \to F_L^\vee \to \omega_C L^{-1} \to 0\,. \tag{7.12}$$

Since $H^1(X, \mathcal{O}_X) = 0$, the sequence (7.12) remains exact if we pass to global sections. Since the first term of (7.12) is free and, as $h^0(C, \omega_C L^{-1}) \neq 0$, the last one is generated by its global sections away from a finite set of points, ii) follows. To prove iii), it suffices to tensor (7.12) with F_L and use i).

Lemma (7.10) is now reduced to proving that $h^0(X, F_L \otimes F_L^\vee) = 1$, in other words, that $F_L^\vee$ has no nontrivial endomorphisms. Suppose that there is a nontrivial endomorphism $\varphi : F_L^\vee \to F_L^\vee$. Let λ be an eigenvalue of $\varphi(x)$ for some $x \in X$. The determinant of the endomorphism $\psi = \varphi - \lambda \cdot 1$ vanishes at x, but since $\det \psi \in H^0(X, \det(F_L \otimes F_L^\vee)) = \mathbb{C}$, it vanishes identically on X. Thus, substituting φ with ψ, we can assume that φ is everywhere of less than maximal rank. We set $A = \operatorname{Image} \varphi$, $B = \operatorname{coker} \varphi$, and $B' = B/\mathcal{T}(B)$, where $\mathcal{T}(B)$ is the torsion subsheaf of B. By construction, both A and B' have positive ranks. Then

$$[C] = c_1(F_L^\vee) = c_1(A) + c_1(B') + c_1(\mathcal{T}(B)) \tag{7.13}$$

in $\operatorname{Pic}(X)$. Since, by assumption, $\operatorname{Pic}(X) \cong \mathbb{Z}$, and since $c_1(\mathcal{T}(B))$ is a nonnegative linear combination of the codimension one irreducible components of the support of $\mathcal{T}(B)$, to reach a contradiction, it suffices to show that the remaining two summands in (7.13) are represented by effective divisors. Now A and B' are torsion-free sheaves of positive rank and are both quotients of $F_L^\vee$. It follows from ii) that they are both generated by their global sections away from a finite set of points. Also, since $h^0(X, F_L) = 0$, neither A nor B' can be trivial. We are then reduced to the following general lemma.

LEMMA (7.14). *Let E be a nontrivial, torsion-free sheaf on a smooth projective surface X. Assume that E is generated by its global sections away from a finite set of points. Then $c_1(E)$ is represented by an effective divisor.*

Proof. Suppose first that E is locally free. Pick $\operatorname{rank}(E)$ sections which generate E at a general point of X. Their wedge product is a nonzero global section of the determinant of E which must vanish somewhere, since E is nontrivial. Hence, its zero locus is an effective divisor representing $c_1(E)$. In case E is not free, the canonical inclusion of E in its double dual $E^{\vee\vee}$ is an isomorphism outside a finite set of points (cf. [429], Chapter V, Section 5). Thus, $c_1(E) = c_1(E^{\vee\vee})$, and $E^{\vee\vee}$ is also generated by its global sections away from a finite set of points. As $E^{\vee\vee}$ is locally free, we are done.

Q.E.D.

This concludes the proof of Proposition (7.5) and therefore of Theorem (7.1).

8. The normal bundle and Horikawa's theory.

Our study of special divisors on a curve, and especially of the fundamental map μ_0, has centered around the intrinsic geometry of a moving curve and its moving Picard variety. We now take a different point of view which will enable us to interpret our cohomological approach and our intrinsic constructions in terms of the extrinsic geometry of the curve

We begin by recalling some elementary facts concerning the theory of deformations of mappings as developed by Horikawa [367,368,369,370]. Let C be a nonsingular curve, and let M be an n-dimensional complex manifold, which is not assumed to be compact. Let

$$\phi : C \to M \tag{8.1}$$

be a nonconstant analytic map. A *deformation* of ϕ parameterized by a pointed analytic space (S, s_0) is a diagram

$$\begin{array}{ccc} C \xrightarrow{\ \gamma\ } C_{s_0} = p^{-1}(s_0) \subset \mathcal{C} & \xrightarrow{\ \tilde{\phi}\ } & M \\ {\scriptstyle p}\big\downarrow & & \\ (S, s_0) & & \end{array} \tag{8.2}$$

where

$$p : \mathcal{C} \to (S, s_0)\,, \quad \gamma : C \xrightarrow{\simeq} C_{s_0} \tag{8.3}$$

is a deformation of C, and $\tilde{\phi}$ is a morphism such that

$$\tilde{\phi} \circ \gamma = \phi\,.$$

The notions of *morphism*, *isomorphism*, and *equivalence* between deformations of maps are the obvious extensions of the analogous notions for deformations of manifolds. A *first-order deformation of* ϕ is simply a deformation of ϕ parameterized by $\operatorname{Spec} \mathbb{C}[\varepsilon]$.

Let us go back to the morphism (8.1). We shall consistently use the symbol T to denote the sheaf $\phi^*(T_M)$. We define the *normal sheaf* N_ϕ *to the map* ϕ to be the cokernel of the injective sheaf homomorphism $d\phi : T_C \to T$. We then have an exact sequence of sheaves

$$0 \to T_C \xrightarrow{d\phi} T \to N_\phi \to 0\,. \tag{8.4}$$

Of course, when ϕ is an embedding, N_ϕ is just the normal bundle to C in M. Occasionally we shall write N for N_ϕ when no confusion is likely to occur.

Horikawa's theory sets up a one-to-one correspondence between the vector space $H^0(C, N_\phi)$ and the set of equivalence classes of first-order deformations of ϕ. The computations needed to associate to a first-order deformation of ϕ a section of N_ϕ closely parallel the various deformation-theoretic arguments that we have encountered in the previous chapters. We ask for a little patience on the part of the reader, while once again we go through the ritual of translating geometric situations into patching data and concocting cohomology classes out of these.

As usual, we imagine C as being given by transition data $\{U_\alpha, z_\alpha, f_{\alpha\beta}(z_\beta)\}$ where

$$\begin{cases} \mathcal{U} = \{U_\alpha\} & \text{is a finite cover of } C \text{ by coordinate disks,} \\ z_\alpha & \text{is a holomorphic coordinate in } U_\alpha, \\ z_\alpha = f_{\alpha\beta}(z_\beta) & \text{in } U_\alpha \cap U_\beta. \end{cases}$$

We may assume that, for every α,

$$\phi(U_\alpha) \subset V_\alpha \subset M\,,$$

where each V_α is a coordinate patch in M with coordinates $w_\alpha = (w_\alpha^1, \dots, w_\alpha^n)$. Set

$$w_\alpha = g_{\alpha\beta}(w_\beta) \quad \text{in} \quad V_\alpha \cap V_\beta$$

and let

$$w_\alpha = \psi_\alpha(z_\alpha)$$

be the expression of the map ϕ in terms of these local coordinates. The functions $f_{\alpha\beta}$, $g_{\alpha\beta}$, ψ_α obviously satisfy the compatibility conditions

$$\begin{cases} f_{\alpha\beta}(f_{\beta\gamma}(z_\gamma)) = f_{\alpha\gamma}(z_\gamma) \\ g_{\alpha\beta}(\psi_\beta(z_\beta)) = \psi_\alpha(f_{\alpha\beta}(z_\beta)) \end{cases} \quad \text{in } U_\alpha \cap U_\beta. \tag{8.5}$$

Now consider a first-order deformation (8.2) of ϕ. This is given by:

i) transition functions for $\mathcal{C}$,

$$z_\alpha = \tilde{f}_{\alpha\beta}(z_\beta, \varepsilon) = f_{\alpha\beta}(z_\beta) + \varepsilon b_{\alpha\beta}(z_\beta)\,;$$

ii) local expressions for $\tilde{\phi}$,

$$w_\alpha = \tilde{\psi}_\alpha(z_\alpha, \varepsilon) = \psi_\alpha(z_\alpha) + \varepsilon a_\alpha(z_\alpha)\,. \tag{8.6}$$

The cocycle $\left\{b_{\alpha\beta}\frac{\partial}{\partial z_\alpha}\right\}$ represents the Kodaira–Spencer class $\theta \in H^1(C, T_C)$ of the first-order deformation (8.3). The data i) and ii) must satisfy the compatibility conditions

$$\begin{cases} \tilde{f}_{\alpha\beta}(\tilde{f}_{\beta\gamma}(z_\gamma, \varepsilon), \varepsilon) = \tilde{f}_{\alpha\gamma}(z_\gamma, \varepsilon)\,, \\ g_{\alpha\beta}(\tilde{\psi}_\beta(z_\beta, \varepsilon)) = \tilde{\psi}_\alpha(\tilde{f}_{\alpha\beta}(z_\beta, \varepsilon), \varepsilon)\,. \end{cases}$$

Expanding these in power series in ε, it is easy to check that these conditions are equivalent to (8.5) together with

$$\sum_j \frac{\partial g_{\alpha\beta}}{\partial w_\beta^j} a_\beta^j = a_\alpha + \frac{\partial \psi_\alpha}{\partial z_\alpha} b_{\alpha\beta}\,. \tag{8.7}$$

This formula shows that the 0-cochain

$$\left\{\sum a_\alpha^i \frac{\partial}{\partial w_\alpha^i}\right\} \in C^0(\mathcal{U}, T)$$

maps to a *section* s of N_ϕ, which is called the *Horikawa class of the first-order deformation* (8.2). Of course, s depends on the choice of a local parameter ε. Once this is fixed, however, one can easily show that s does not depend on the other choices we have made and that in this way we get a one-to-one correspondence between $H^0(C, N_\phi)$ and the set of equivalence classes of first-order deformations of ϕ. Moreover, formula (8.7) also says that the coboundary homomorphism

$$\delta : H^0(C, N_\phi) \to H^1(C, T_C)$$

maps the Horikawa class s to the Kodaira–Spencer class θ.

More generally, given a family of smooth curves

$$p : \mathcal{C} \to S\,, \tag{8.8}$$

a point $s_0 \in S$, and a morphism

$$\psi : \mathcal{C} \to M\,,$$

we can define a *characteristic homomorphism*

$$T_{s_0}(S) \to H^0(C_{s_0}, N_{\psi_{s_0}})\,,$$

where ψ_{s_0} stands for the restriction of ψ to C_{s_0}, as follows. Given $v \in T_{s_0}(S)$ corresponding to a morphism $\sigma : \operatorname{Spec}\mathbb{C}[\varepsilon] \to S$, we map v to the Horikawa class of the pullback of (8.8) via σ. When $\psi_{s_0} : C_{s_0} \to M$ is an embedding, we clearly get back the classical characteristic homomorphism, as defined in Section 5 of Chapter IX.

Let C be a smooth curve of genus g. Let L be a line bundle on C of degree d, and W an $(r+1)$-dimensional subspace of $H^0(C, L)$. Assume that the g_d^r corresponding to W has no base points. Choose a basis $X_0, \dots, X_r$ of W and let

$$\phi : C \to \mathbb{P}^r \cong \mathbb{P}W^\vee$$

be the map corresponding to this choice of basis. As above, we set

$$T = \phi^*(T_{\mathbb{P}^r}).$$

We wish to show that it is possible to interpret the basic homomorphisms

$$\begin{gathered}\mu_{0,W} : W \otimes H^0(C, \omega_C L^{-1}) \to H^0(C, \omega_C),\\ \mu_{1,W} : \ker \mu_0 \to H^0(C, \omega_C^2)\end{gathered}$$

in terms of T and N_ϕ. In order to do so, we combine the exact sequences (5.10) and (8.4) with the pullback to C of the Euler sequence for $\mathbb{P}^r$

$$0 \to \mathcal{O}_{\mathbb{P}^r} \to \mathcal{O}_{\mathbb{P}^r}(1)^{\oplus(r+1)} \to T_{\mathbb{P}^r} \to 0.$$

This pulled-back sequence is the middle column of the diagram

$$\begin{array}{ccccccccc} & & 0 & & 0 & & & & \\ & & \downarrow & & \downarrow & & & & \\ & & \mathcal{O}_C & = & \mathcal{O}_C & & & & \\ & & \downarrow & & \downarrow{\scriptstyle\sigma} & & & & \\ 0 & \longrightarrow & \Sigma_L & \xrightarrow{\lambda_W} & L^{\oplus(r+1)} & \longrightarrow & N_\phi & \longrightarrow & 0 \\ & & \downarrow & & \downarrow{\scriptstyle\tau} & & \Vert & & \\ 0 & \longrightarrow & T_C & \xrightarrow{d\phi} & T & \longrightarrow & N_\phi & \longrightarrow & 0 \\ & & \downarrow & & \downarrow & & & & \\ & & 0 & & 0 & & & & \end{array} \tag{8.9}$$

where

$$\begin{aligned}\sigma(f) &= (fX_0, \dots, fX_r),\\ \tau(\ell_0, \dots, \ell_r) &= \sum \ell_i \frac{\partial}{\partial X_i},\end{aligned}$$

and λ_W is defined as follows. Let ∇ be a local section of Σ_L, that is, a differential operator of order at most one operating on sections of L. Set

$$\lambda_W(\nabla) = (\nabla X_0, \dots, \nabla X_r).$$

We claim that diagram (8.9) is commutative. If ∇ is of order zero, i.e., if it is the multiplication by a function f, then clearly

$$\lambda_W(\nabla) = \sigma(f).$$

For the general case, we may find open sets U_α covering C and a local coordinate z_α on each U_α such that

$$\nabla = a + bD_\alpha \quad \text{on } U_\alpha ,$$

where D_α is given by (5.12). We then have

$$\tau\lambda_W(\nabla) = \sum \nabla X_i \frac{\partial}{\partial X_i} = b \sum \frac{\partial X_i}{\partial z_\alpha} \frac{\partial}{\partial X_i}$$

by Euler's theorem. On the other hand,

$$b \sum \frac{\partial X_i}{\partial z_\alpha} \frac{\partial}{\partial X_i} = d\phi \left(b \frac{\partial}{\partial z_\alpha} \right) ,$$

proving the commutativity of (8.9). We may also notice that, for each smooth point $\phi(p)$ on the curve $\phi(C)$, the image

$$\lambda_W(\Sigma_L \otimes k(p))$$

defines a two-dimensional subspace of $\mathbb{C}^{r+1}$ containing the one-dimensional space corresponding to $\sigma(\mathcal{O}_{C,p})$.

The relevant part of the cohomology diagram of (8.9) can be conveniently displayed in the following manner:

(8.10)

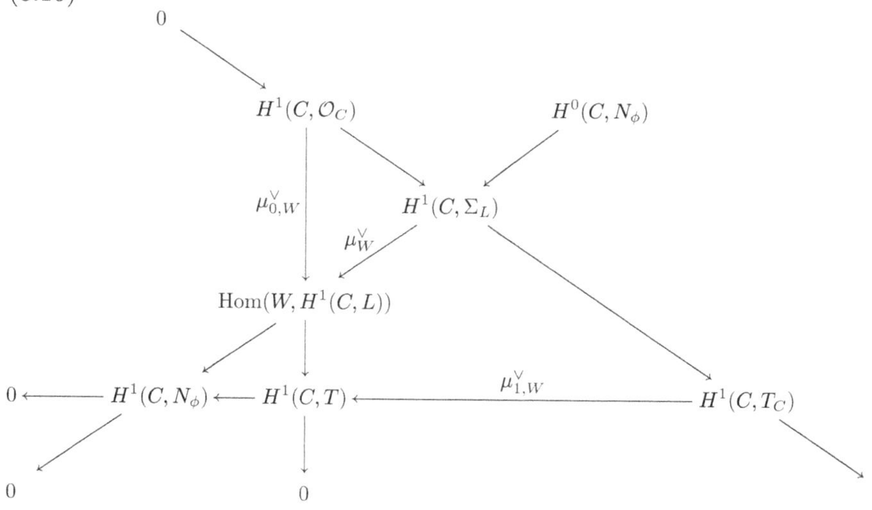

Here we have used the identifications

$$\text{(8.11)} \qquad \begin{aligned} \mathrm{Hom}(W, H^1(C, L)) &\cong H^1(C, L^{\oplus(r+1)}) , \\ H^1(C, T) &\cong \mathrm{coker}(\mu_{0,W}^\vee) . \end{aligned}$$

Dualizing diagram (8.9), we see that the second identification in (8.11) is nothing but the isomorphism (6.12). A straightforward consequence of diagram (8.10) is the following.

LEMMA (8.12). *There are natural identifications*

$$\ker \mu_{0,W} \cong H^1(C, \phi^*(T_{\mathbb{P}^r}))^\vee ,$$
$$\ker \mu_{1,W} \cong \ker \mu \cong H^1(C, N)^\vee .$$

Now notice that the first version (1.1) of Petri's statement easily follows from the analogous assertion concerning g^r_d's *without base points.* We are then in a position to give another, quite suggestive formulation of Petri's statement which brings to light its projective nature.

THEOREM (8.13) (PETRI'S STATEMENT - FOURTH VERSION). *Let C be a general curve of genus g, and let*

$$\phi : C \to \mathbb{P}^r, \qquad r \geq 1 ,$$

be a nondegenerate morphism. Then $H^1(C, N_\phi) = 0$.

Observe that, when $r = 1$, the normal sheaf N_ϕ is supported on the ramification divisor of ϕ, so that, obviously, $H^1(C, N_\phi) = 0$. This reproves Petri's statement for g^1_d's. Notice also that another immediate consequence of diagram (8.10) is that $H^1(C, N_\phi)$ vanishes when L is nonspecial.

As Horikawa [368,369] showed, there is an analogue of the Kuranishi family for deformations of maps. Let $\phi : C \to M$ be a nonconstant morphism from the smooth curve C to the complex manifold M. Then there exists a universal deformation

$$\begin{array}{ccc} C \xrightarrow{\ \cong\ } X_{u_0} = q^{-1}(u_0) \subset & \mathcal{X} & \xrightarrow{\ \Phi\ } M \\ & \Big\downarrow q & \\ & (U, u_0) & \end{array} \tag{8.14}$$

of ϕ, in the following sense. For any deformation (8.2) of $\phi : C \to M$ with S connected and "sufficiently small," there is a unique morphism

$$\begin{array}{ccc} \mathcal{C} & \xrightarrow{\ \alpha\ } & \mathcal{X} \\ \Big\downarrow & & \Big\downarrow \\ (S, s_0) & \longrightarrow & (U, u_0) \end{array}$$

of deformations of C such that $\Phi \circ \alpha = \tilde{\phi}$. Of course, such a universal deformation is essentially unique if we allow U to be shrunk as needed. Horikawa proved the following.

THEOREM (8.15). *Any nonconstant morphism $\phi : C \to M$ from a smooth curve to a complex manifold admits a universal deformation (8.14). Moreover,*

$$\dim(U') \geq h^0(C, N_\phi) - h^1(C, N_\phi) \tag{8.16}$$

for every component U' of U at u_0, and the characteristic map $T_{u_0}(U) \to H^0(C, N_\phi)$ is an isomorphism.

Instead of proving the theorem in its full generality, we shall limit ourselves to the case where M is a projective manifold. Suppose first that the genus g of C is at least 2. Let $\xi : \mathcal{Y} \to (B, b_0)$ be a Kuranishi family for C and consider the Hilbert scheme $H = \mathbf{Hom}_B(\mathcal{Y}, M \times B)$ parameterizing morphisms from fibers of ξ to M (cf. Section 7 of Chapter IX). We let u_0 be the point of H corresponding to the morphism ϕ and denote by U a small connected neighborhood of u_0 in H. We then set $\mathcal{X} = \mathcal{Y} \times_B U$. The inclusion morphism $U \to H$ corresponds to a U-morphism $\mathcal{X} \to M \times U$; composing with the projection to M yields a morphism $\Phi : \mathcal{X} \to M$ which, by construction, is a deformation of ϕ. To see that the deformation thus constructed satisfies the required universal property, we argue as follows. Consider a deformation (8.2) of $\phi : C \to M$. After suitably shrinking S there is a unique morphism

$$\begin{array}{ccc} \mathcal{C} & \longrightarrow & \mathcal{Y} \\ \downarrow & & \downarrow \\ (S, s_0) & \longrightarrow & (B, b_0) \end{array} \tag{8.17}$$

of deformations of C. The morphism $(\tilde{\phi}, p) : \mathcal{C} \to M \times S$ is an S-morphism from $\mathcal{C} \cong \mathcal{Y} \times_B S$ to $M \times S \cong (M \times B) \times_B S$ and as such is represented by a unique B-morphism $S \to H$ which, possibly after further shrinking S, factors through U.

The universal property we have just proved shows in particular that the characteristic homomorphism $T_{u_0}(U) \to H^0(C, N_\phi)$ is an isomorphism. It remains to prove the dimensionality statement (8.16). For this, we appeal to Theorem (8.20) in Chapter IX. The scheme $H = \mathbf{Hom}_B(\mathcal{Y}, M \times B)$ is an open subset of the Hilbert scheme $\mathrm{Hilb}_{(\mathcal{Y} \times M)/B}$, and the point u_0 attached to the morphism ϕ is just the graph Γ of ϕ, viewed as a subscheme of $\xi^{-1}(b_0) \times M \cong C \times M$. Since M is smooth, Γ is regularly embedded in $C \times M$. Theorem (8.20) of Chapter IX then says that every component of U at u_0 has dimension at least

$$\begin{aligned} &h^0(\Gamma, N_{\Gamma/(C \times M)}) - h^1(\Gamma, N_{\Gamma/(C \times M)}) + \dim(B) \\ &\quad = \chi(N_{\Gamma/(C \times M)}) + 3g - 3\,. \end{aligned} \tag{8.18}$$

To compute this lower bound, we need to describe the normal bundle $N_{\Gamma/(C \times M)}$. If we identify C with Γ via $j = (\mathrm{id}, \phi) : C \to C \times M$, this bundle is nothing but the normal bundle N_j to the morphism j and sits in an exact sequence

$$0 \to T_C \xrightarrow{dj} j^*T_{C\times M} \xrightarrow{\alpha} N_j \to 0\,.$$

Let $\beta : T_C \to j^*T_{C\times M} \cong T_C \oplus \phi^*T_M$ be the homomorphism $x \mapsto (x,0)$. It is readily seen that the composition $\alpha\circ\beta : T_C \to N_j$ is injective and that its cokernel is just N_ϕ. The lower bound (8.18) thus becomes

$$\chi(N_\phi) + \chi(T_C) + 3g - 3 = \chi(N_\phi) = h^0(C, N_\phi) - h^1(C, N_\phi)\,, \tag{8.19}$$

as claimed. This concludes the proof when $g \geq 2$.

We shall now sketch how the above argument can be modified when $g < 2$. Suppose $g = 1$. Pick a point $x \in C$ such that $\phi(x)$ is neither a singular point of $\phi(C)$ nor a critical value for ϕ. Then pick a smooth divisor $D \subset M$ through $\phi(x)$ which does not meet the singular locus of $\phi(C)$ and is transverse to $\phi(C)$. Consider a Kuranishi family for the 1-pointed curve $(C;x)$, consisting of a deformation $\xi : \mathcal{Y} \to (B, b_0)$ of C plus a section $\sigma : B \to \mathcal{Y}$ passing through x. Denote by $H \subset \mathbf{Hom}_B(\mathcal{Y}, M\times B)$ the Hilbert scheme parameterizing those morphisms from fibers of ξ to M which send the marked point to D. Then H has codimension 1 in (each component of) $\mathbf{Hom}_B(\mathcal{Y}, M\times B)$. As for $g \geq 2$, we let u_0 be the point of H corresponding to the morphism ϕ and denote by U a small connected neighborhood of u_0 in H. The family $\mathcal{X} \to (U, u_0)$ and the morphism Φ are defined as for $g \geq 2$. Now consider the deformation (8.2) of $\phi : C \to M$. If S is replaced with a suitably small connected analytic neighborhood of s_0, there is a unique section $\tau : S \to \mathcal{C}$ passing through x (or rather through $\gamma(x)$) such that $\tilde{\phi}(\tau(s)) \in D$ for all $s \in S$. This section makes $\mathcal{C} \to (S, s_0)$ into a deformation of the 1-pointed curve $(C;x)$. Hence, possibly after further shrinking S, there is a unique morphism (8.17) *of deformations of* 1*-pointed curves.* The remainder of the proof of the universal property of (8.14) is the same as for $g \geq 2$. The proof of the lower bound on the dimension of U is also essentially the same. It suffices to notice that, since H has codimension 1 in $\mathbf{Hom}_B(\mathcal{Y}, M\times B)$, the lower bound (8.18) is replaced by

$$h^0(\Gamma, N_{\Gamma/(C\times M)}) - h^1(\Gamma, N_{\Gamma/(C\times M)}) + \dim(B) - 1 = \chi(N_{\Gamma/(C\times M)})\,.$$

When $g = 0$, the argument is similar to the one for $g = 1$. We choose three points x_1, x_2, x_3 on C and three divisors D_1, D_2, D_3 through $\phi(x_1), \phi(x_2), \phi(x_3)$ with the same properties enjoyed by x and D in the genus 1 case, let H be the Hilbert schemes parameterizing morphisms from C to M mapping x_i to D_i for $i = 1, 2, 3$, and proceed exactly as above.

REMARK (8.20). When $M = \mathbb{P}^r$, it is possible to give an alternate proof of Theorem (8.15) without resorting to the Hilbert scheme and to Theorem (8.20) in Chapter IX, but using instead the universal property of $\mathcal{G}^r_d$ and the dimension estimate (3.21). We shall limit ourselves to a nondegenerate morphism $\phi : C \to \mathbb{P}^r$, where C is a smooth curve of genus $g \geq 2$.

Denote by L the pullback to C of the hyperplane bundle on $\mathbb{P}^r$, by d its degree, and set $W = H^0(\mathbb{P}^r, \mathcal{O}(1)) \subset H^0(C, L)$. Let $\xi : \mathcal{Y} \to (B, b_0)$ be a Kuranishi family for C. Denote by w the point in $\mathcal{G}^r_d(\xi)$ corresponding to W. As usual, we denote by χ the projection from $\mathcal{G}^r_d(\xi)$ to $\mathbf{Pic}^d(\xi)$ and by π the projection from $\mathbf{Pic}^d(\xi)$ to B. By Proposition (3.21), every component of $\mathcal{G}^r_d(\xi)$ at w has dimension at least $h^1(C, T_C) + \rho$, where $\rho = g - (r+1)(g-d+r)$ is the Brill–Noether number. Let V be a small neighborhood of w in $\mathcal{G}^r_d(\xi)$, let $\mathcal{L}$ be the pullback to $V \times_B \mathcal{Y}$ of a degree d Poincaré line bundle, and let $\eta : V \times_B \mathcal{Y} \to V$ be the natural projection. By definition of $\mathcal{G}^r_d(\xi)$, there is a locally free rank $r+1$ subsheaf $\mathcal{H}$ of $\eta_* \mathcal{L}$, which is the family of g^r_d's on V. Shrinking V if necessary, we may assume that $\mathcal{H}$ is free.

Now we let U be the bundle of projective frames of $\mathcal{H}$, so that $U \to V$ is a principal $PGL(r+1, \mathbb{C})$-bundle, and we denote by u_0 the point in U corresponding to the morphism ϕ. Consider the family of curves $\mathcal{X} = U \times_B \mathcal{Y} \to (U, u_0)$. The frames parameterized by U define a morphism $\Phi : \mathcal{X} \to \mathbb{P}^r$ which, by construction, is a deformation of ϕ. The universal property of this deformation follows immediately from the one of $\mathcal{G}^r_d(\xi)$. In turn, the universal property shows that the characteristic homomorphism $T_{u_0}(U) \to H^0(C, N_\phi)$ is an isomorphism. It remains to prove (8.16). This is a simple dimension count. If $\mathcal{G}'$ is any component of $\mathcal{G}^r_d(\xi)$, we denote by U' the corresponding component of U. Looking at diagram (8.9) and using Proposition (5.26), we get

$$\begin{aligned} h^0(C, N_\phi) - h^1(C, N_\phi) &= \chi(L^{\oplus(r+1)}) - \chi(\mathcal{O}_C) - \chi(T_C) \\ &= (r+1)(d-g+1) + g - 1 + h^1(C, T_C) \\ &= (r+1)^2 - 1 + \rho + h^1(C, T_C) \\ &\leq \dim \mathcal{G}' + (r+1)^2 - 1 = \dim U' . \end{aligned}$$

As a first application of the theory of deformations of mappings, we shall study the following basic question. Consider smooth curves C of genus g which admit a nonconstant morphism $\phi : C \to X$ of degree d onto some smooth curve X of genus $g' > 0$. The question we ask is: on how many moduli does such a C depend? The answer is best expressed in terms of the ramification divisor $R \subset C$ of ϕ and of its degree w, and goes under the name of *Riemann's moduli count.* Of course, w is related to the other invariants of ϕ by the Riemann–Hurwitz formula

$$2g - 2 = d(2g' - 2) + w . \tag{8.21}$$

Riemann's moduli count asserts essentially that the moduli of a curve C as above arise from two independent sources. One can vary the moduli of X, or one can freely move the w points of R. This gives a total of

$$3g' - 3 + w = 2g - 2 + (3 - 2d)(g' - 1) \tag{8.22}$$

moduli. The rest of the section will be devoted to justifying this heuristic answer. More precisely, we shall prove the following result.

THEOREM (8.23) (RIEMANN'S MODULI COUNT). *Let $d > 1$, $g > 1$, and $g' > 0$ be integers. Denote by $M_g(d, g') \subset M_g$ the locus of those smooth genus g curves which are degree d ramified coverings of smooth genus g' curves. Then $M_g(d, g')$ is nonempty if and only if $2g - 2 \geq d(2g' - 2)$; in this case it is a closed subscheme of M_g of pure dimension $3g' - 3 + w$, where $w = 2g - 2 - d(2g' - 2)$.*

If a covering of the type described in the statement of the theorem exists, then w is the degree of its ramification divisor because of the Riemann–Hurwitz formula (8.21) and hence is nonnegative. The converse is also standard and follows, for instance, from Riemann's existence theorem. Here is an alternate direct proof.

Since w is even, we may write $w = 2s$. There exists an unramified covering $f : Y \to X$ of degree d, since $H_1(X, \mathbb{Z}) \cong \mathbb{Z}^{2g'}$, and hence $\pi_1(X)$ contain (normal) subgroups of any integer index. Now pick distinct points $p_1, \ldots, p_s$ in X and pairs of distinct points $q_i, q_i' \in Y$, $i = 1, \ldots, s$, such that $f(q_i) = f(q_i') = p_i$ for every i. We also choose an open neighborhood U_i of p_i for each i; we perform the choice so that the U_i are disjoint and biholomorphic to disks. We also denote by V_i the connected component of $f^{-1}(U_i)$ containing q_i and by V_i' the one containing q_i'. Now we fix our attention on one specific i. We may assume that there is a local coordinate x centered at p_i such that U_i is the disk $\{|x| < 1\}$. We denote by y and y' the local coordinates on Y at q_i and q_i' obtained by composing x with f. Thus, $V_i = \{|y| < 1\}$, $V_i' = \{|y'| < 1\}$, and the map f is given, in these local coordinates, by $x = y$ and by $x = y'$. Choose a small positive ε and set

$$W = \{(z, t) \in \mathbb{C}^2 : z^2 + t^2 = \varepsilon^2, |z| < 1\}.$$

The morphism $h : W \to U_i$ defined by $x = z$ is a branched covering, with simple ramification precisely at the two points of W lying above $x = \pm\varepsilon$, i.e., at the points $t = 0$, $z = \pm\varepsilon$. When δ is a sufficiently small positive number, the region of W where $|z| > 1 - \delta$ is the disjoint union of two annuli A and A', both mapped biholomorphically to $\{x : 1 > |x| > 1 - \delta\}$ by h. Now we remove from Y the regions $\{|y| \leq 1 - \delta\}$ and $\{|y'| \leq 1 - \delta\}$, and we glue W to the resulting Riemann surface by identifying A to $\{y : 1 > |y| > 1 - \delta\}$ via $y = z$ and A' to $\{y' : 1 > |y'| > 1 - \delta\}$ via $y' = z$. The result is a complete Riemann surface Y' together with a branched covering $Y' \to X$ which agrees with h on W and with f elsewhere; the covering is simply ramified at two points. We may perform this operation simultaneously at all the p_i. What we obtain is a degree d branched covering $C \to X$ with $w = 2s$ simple ramification points. Thus, the genus of C is g by the Riemann–Hurwitz formula.

REMARK (8.24). The existence argument outlined above, with a slight modification, works also when $g' = 0$. One takes as Y the disjoint union of d copies of X, and then proceeds as before. The only issue is the connectedness of C. When $g' = 0$, the Riemann–Hurwitz identity immediately implies that $s \geq d - 1$, and hence s is just large enough to make it possible, by suitably choosing the q_i and q_i', to connect all the d sheets of Y.

In addition, the condition that $g > 1$ is clearly unnecessary. In fact, with a little more care, the argument proves the following. Let $g \geq 0$ and $d > 1$ be integers. Let X be a smooth curve of genus g'. Suppose that $w = 2g - 2 - d(2g' - 2) \geq 0$. Let $x_1, \dots, x_w$ be distinct points of X. Then there exist a smooth genus g curve C and a degree d covering $C \to X$ simply ramified at w points mapping to $x_1, \dots, x_w$.

Having disposed of the problem of existence, we next explain why $M_g(d, g')$ is a subscheme of M_g. First of all, fix an integer $n \geq 3$, set $S = M_g[n]$, and let

$$\alpha : \mathcal{C} \to S$$

be the universal family over S. Similarly, when $g' > 1$, set $T = M_{g'}[n]$, and let

$$\beta : \mathcal{X} \to T$$

be the universal family over T. When $g' = 1$, we take as $\mathcal{X} \to T$ the universal family of 1-pointed genus 1 curves over $M_{1,1}[n]$. Then we let

$$H_d \subset \mathbf{Hom}_{S \times T}(\mathcal{C} \times T, S \times \mathcal{X})$$

be the Hilbert schemes parameterizing coverings of degree d from fibers of α to fibers of β. A first observation is that H_d is of finite type over $\mathbb{C}$. To see this, let $\Gamma \subset C \times X$ be the graph of a degree d morphism $C \to X$, where C and X are smooth curves of genera g and g'. Let p_1 and p_2 be the projections from $C \times X$ to C and X, respectively, and let F_i be a fiber of p_i for $i = 1, 2$. Clearly,

$$(F_1 \cdot \Gamma) = 1\,, \quad (F_2 \cdot \Gamma) = d\,, \quad (\Gamma \cdot \Gamma) = d(2 - 2g')\,, \tag{8.25}$$

where the third identity follows from the second and from the fact that $\omega_{C \times X/C}(\Gamma)$ is trivial on Γ by adjunction. Now set $L = p_1^*\omega_C \otimes p_2^*\omega_X$ when $g' > 1$ and $L = p_1^*\omega_C \otimes p_2^*\mathcal{O}(x)$, where x is the marked point on X, for $g' = 1$. Then L^3 is a very ample line bundle on $C \times X$, and hence

Γ can be realized as a curve of degree n in $\mathbb{P}^r$, where r and n depend only on g, g', and d. This proves our claim.

We next show that H_d is proper over $S \times T$. By the properness of the Hilbert scheme, it suffices to show that, given smooth curves C and X of genera g and g', a divisor D in $C \times X$ which is the limit of points of H_d, i.e., of graphs of degree d morphisms from curves of genus g to curves of genus g', is in fact the graph of a morphism. Such a D must have the same numerical characters as a true graph. Keeping the notation introduced above, these are expressed by the identities (8.25). It follows from the first of these that D is of the form $\Gamma' + E_1 + \cdots + E_k$, where Γ' is the graph of a morphism, and the E_i are fibers of p_1, and hence that D is numerically equivalent to $\Gamma' + kF_1$. Thus $(F_2 \cdot \Gamma') = d - k$, and therefore,

$$(\Gamma' \cdot \Gamma') = (d-k)(2-2g')\,.$$

But then

$$d(2-2g') = (D \cdot D) = (d-k)(2-2g') + 2k = d(2-2g') + 2kg'\,,$$

which proves that $k = 0$ since $g' > 0$ by assumption. The conclusion is that $D = \Gamma'$ is a graph, proving our contention.

Since $M_g(d, g')$ is the image of H_d under the composition of the projection $H_d \to S \times T$, of the projection $S \times T \to S = M_g[n]$, and of the moduli map $M_g[n] \to M_g$, all of which are proper, it follows that $M_g(d, g')$ is a closed subscheme of M_g.

We finally turn to the heart of Theorem (8.23), which is the dimension statement. The dimension of H_d is readily calculated. By Theorem (8.15), the space of degree d morphisms from a variable smooth genus g curve to a *fixed* smooth curve of genus g' is smooth of dimension w. Thus,

$$\dim(H_d) = \begin{cases} 3g - 3 + w & \text{if } g' > 1\,, \\ w + 1 & \text{if } g' = 1\,. \end{cases} \tag{8.26}$$

This already suffices to conclude that the dimension of $M_g(d, g')$ does not exceed $3g' - 3 + w$, since in the $g' = 1$ case one of the dimensions of H_d is accounted for by the translations of the target elliptic curve. The key to the converse inequality is provided by the following classical result.

THEOREM (8.27) (DE FRANCHIS' THEOREM). *Let $g \geq 1$, $g' \geq 2$, and $d \geq 1$ be integers, and let C be a smooth curve of genus g. Then:*

i) *up to isomorphism, there are only finitely many genus g' curves X such that there exists a nonconstant morphism from C to X;*
ii) *for any smooth genus g' curve X, there are only finitely many nonconstant morphisms from C to X;*

iii) *up to isomorphism, there are only finitely many genus 1 curves X such that there exists a degree d morphism from C to X;*
iv) *for any smooth genus 1 curve X, up to composition with a translation of X, there are only finitely many degree d morphisms from C to X.*

Proof. The main difference between i) and iii), and between ii) and iv), is that, because of the Riemann–Hurwitz formula, the degree of a nonconstant morphism from a genus g curve to a curve of genus $g' \geq 2$ is bounded above by the ratio between $2g - 2$ and $2g' - 2$, while no such bound exists for morphisms to a curve of genus one. Thus, in proving statements i) and ii), there is no loss of generality in restricting to morphisms of fixed degree d. This will allow us to deal simultaneously with i) and iii), and with ii) and iv).

We begin by proving iii) and iv) when $g = 1$. Choose an origin e in C; this determines a group law on C. Then any X is isomorphic to the quotient of C modulo a subgroup of order d. Since there are finitely many such subgroups, iii) follows. Now choose an origin e' in X and suppose that we are given a degree d morphism $\phi : C \to X$. Composing ϕ with a translation, we may suppose that $\phi(e) = e'$ and hence that ϕ is a group homomorphism. Now, two such morphisms sharing the same kernel are obtained from one another by composing with a group automorphism of X. Since there are finitely many group automorphisms of X and there are finitely many possible kernels for ϕ, we get iv).

From now on, we assume that $g > 1$. As we observed during the second part of the proof of Theorem (8.23), the scheme H_d parameterizing degree d morphisms from smooth genus g curves (with a suitable level structure) to smooth genus g' or genus 1 curves (also with a suitable level structure) is of finite type over $\mathbb{C}$. Hence, to prove any one of statements i)–iv), it suffices to show that the morphisms in question are rigid or, in the case of iv), rigid up to translation. Let us start with statements ii) and iv). Let $\phi : C \to X$ be a nonconstant morphism of degree d. From the normal bundle sequence

$$0 \to T_C \to \phi^* T_X \to N_\phi \to 0$$

we get an exact sequence

$$0 \to H^0(C, \phi^* T_X) \to H^0(C, N_\phi) \to H^1(C, T_C) \,. \tag{8.28}$$

When $g' > 1$, T_X, and hence its pullback via ϕ, have negative degree, and hence $H^0(C, N_\phi)$ injects in $H^1(C, T_C)$. This means that one cannot deform ϕ without at the same time deforming C, which proves ii). When $g' = 1$, instead, T_X is trivial, so $H^0(C, N_\phi) \to H^1(C, T_C)$ has a one-dimensional kernel. This means that the dimension of the Hilbert scheme $\mathbf{Hom}^d(C, X)$ of degree d morphisms from C to X does not exceed 1.

On the other hand, composing ϕ with all possible translations in X yields a one-dimensional complete subscheme of $\mathbf{Hom}^d(C,X)$, and hence a component of $\mathbf{Hom}^d(C,X)$. Since $\mathbf{Hom}^d(C,X)$ is of finite type over $\mathbb{C}$, this proves iv).

To prove i) and iii), we follow a different approach. Let ϕ be as above, and let

$$Q = C \times_X C \subset C \times C$$

be the equivalence relation associated to it. Since the automorphism group of X is finite (which, incidentally, is also a special case of ii)), it will suffice to show that Q is rigid inside $C \times C$. We write $Q = \Delta + V$, where Δ is the diagonal in $C \times C$. For $1 \le i < j \le 4$, let

$$\pi_{i,j} : C \times C \times C \times C \to C \times C$$

be the projection to the ith and jth factors, and set $\Delta_{i,j} = \pi_{i,j}^{-1}(\Delta)$. Given two correspondences on C, that is, two divisors A and B in $C \times C$, we define their *composition* $A \circ B$ to be

$$A \circ B = (\pi_{1,4})_*((A \times B) \cdot \Delta_{2,3}) .$$

Now, the intersection multiplicity of A and B is equal to the intersection multiplicity of $A \times B$ with the diagonal in $(C \times C) \times (C \times C)$, that is, with the degree of the zero-cycle

$$(A \times B) \cdot \Delta_{1,3} \cdot \Delta_{2,4} .$$

On the other hand, when B is symmetric (with respect to the interchange of factors), this degree is clearly equal to

$$\deg((A \times B) \cdot \Delta_{1,4} \cdot \Delta_{2,3}) ,$$

i.e., by the projection formula, to the degree of $(A \circ B) \cdot \Delta$. In conclusion, *when B is symmetric,* the intersection multiplicity of A and B is

$$(A \cdot B) = (A \circ B \cdot \Delta) . \tag{8.29}$$

We shall apply these considerations to calculate the self-intersection of V. Since ϕ has degree d, it is clear that

$$Q \circ Q = dQ = d\Delta + dV .$$

On the other hand,

$$Q \circ Q = \Delta \circ \Delta + \Delta \circ V + V \circ \Delta + V \circ V = \Delta + 2V + V \circ V ,$$

whence

$$V \circ V = (d-1)\Delta + (d-2)V .$$

Thus,

$$(V \cdot V) = (V \circ V \cdot \Delta) = (d-1)(\Delta \cdot \Delta) + (d-2)(V \cdot \Delta)\,. \tag{8.30}$$

The self-intersection of the diagonal equals $2-2g$, and we claim that $(V \cdot \Delta)$ is equal to w, the degree of the ramification divisor of ϕ. To see this, let p be a point of C and choose coordinates x centered at p and t centered at $\phi(p)$ with respect to which ϕ is given by $t = x^m$. Choose as coordinates on $C \times C$ at (p,p) two copies x_1, x_2 of x. In these coordinates, a local equation for Q is $x_1^m = x_2^m$, and one for Δ is $x_1 = x_2$. Hence a local equation for V is

$$\sum_{h=0}^{m-1} x_1^h x_2^{m-h-1} = 0\,.$$

It follows that the intersection multiplicity of V and Δ at (p,p) is $m-1$, which proves our claim. Using the Riemann–Hurwitz formula, (8.30) can thus be rewritten as

$$(V \cdot V) = (d-1)(2-2g) + (d-2)r = 2-2g + d(d-2)(2-2g')\,.$$

This quantity is strictly negative when $g' \geq 1$ and $g > 1$, and hence there is a component D of V such that $(D \cdot V) < 0$. This component must then be rigid, in the following sense. Let $\phi_t : C \to X_t$, where t varies in a small disk centered at $0 \in \mathbb{C}$, be a family of degree d morphisms to smooth curves of genus g', with $X_0 = X$ and $\phi_0 = \phi$. Let Q_t be the equivalence relation on C determined by ϕ_t. Then D is a component of Q_t for every t. Let Q' be the equivalence relation generated by D. If $Q' = Q$, Q is rigid, and we are done. Otherwise, set $C' = C/Q'$. Then each ϕ_t factors as

$$\begin{array}{ccc} C & \xrightarrow{\phi_t} & X_t \\ & \searrow \quad \nearrow_{\phi'_t} & \\ & C' & \end{array}$$

where the degree of ϕ'_t is at least 2 but strictly less than d, and we can proceed by induction on d. The initial cases of the induction are those in which $g = 1$, those we did at the beginning of the proof, and those in which d is prime. In these latter cases, in fact, Q' is necessarily equal to Q.

Q.E.D.

It is now possible to complete the proof of Theorem (8.23). A consequence of de Franchis' theorem is that the fibers of the composite morphism

$$f : H_d \to S \times T \to S = M_g[n] \to M_g$$

have dimension zero when $g' > 1$, and dimension 1 when $g' = 1$. Combining this with (8.26) shows that the dimension of $M_g(d, g') = f(H_d)$ is $3g - 3 + w$, concluding the proof of Theorem (8.23).

REMARK (8.31). It is natural to ask what happens to Riemann's moduli count when $g' = 0$. It is easy to see that (8.23) is false in this case for a number of reasons. First of all, $M_g(d, g')$ is not closed. It is easy to construct degenerations of degree d morphisms from smooth curves to $\mathbb{P}^1$ which are not themselves morphisms of the same kind; in a sense, they may be regarded as morphisms of lower degree from smooth curves to $\mathbb{P}^1$, or alternatively as degree d morphisms from nodal curves to $\mathbb{P}^1$. More importantly, the dimension count fails completely for a very simple reason. When $g' = 0$, for any fixed g the total ramification index w grows linearly with d, so that the quantity $-3 + r = 3g' - 3 + w$ can be arbitrarily large while $3g - 3$ stays fixed.

Some dimensional information can be obtained from the exact sequence (8.28). A degree d morphism $\phi : C \to \mathbb{P}^1$ corresponds to a g^1_d on C, that is, to a vector subspace $W \subset H^0(C, \phi^*\mathcal{O}(1))$. Then (8.28), together with (8.12), implies that, at C, the genus g curves which are d-sheeted ramified coverings of $\mathbb{P}^1$ depend on at most $w - \dim(\ker(\mu_{0,W}))$ moduli.

Notice that, for $d > 1$, $g > 1$, and $g' \geq 1$, the moduli number $3g' - 3 + w$ given by Riemann's count is strictly less than $3g - 3$. In this case, in fact,

$$3g' - 3 + w = 2g - 2 + (3 - 2d)(g' - 1) \leq 2g - 2 < 3g - 3 .$$

An immediate consequence is the following very well-known and classical result concerning correspondences on curves with general moduli.

COROLLARY (8.32) (RIEMANN–HURWITZ). *Let C be a general curve of genus $g > 1$. Let*

$$\phi : C \to C'$$

be a nonconstant morphism of C onto a curve C'. Then either ϕ is birational, or else C' is rational.

This result and Petri's statement for g^1_d's (cf. Proposition (6.7)) may be combined to derive the following simple but very useful conclusion.

PROPOSITION (8.33). *Let C be a general curve of genus g. Let*

$$\phi : C \to \mathbb{P}^r, \quad r \geq 2 ,$$

be a nondegenerate morphism corresponding to a **special** *g^r_d. Then ϕ is not composed with an involution (i.e., ϕ gives a birational map of C onto its image).*

Proof. Set $\Gamma = \phi(C)$. Since C is general, we have just proved that either ϕ is not composed with an involution or else Γ is rational. We now show that the second case cannot occur. Suppose it does. Let n be the degree of Γ. Obviously, n is at least equal to 2. Let Δ be the inverse image of a point of Γ, and let D be a divisor of the g^r_d corresponding to ϕ. Since we are assuming that Γ is rational, we have

$$(8.34) \qquad \begin{cases} D \sim n\Delta\,, \\ h^0(C, \mathcal{O}(\Delta)) \geq 2\,. \end{cases}$$

Since, by assumption, $|D|$ is special, we also have

$$h^0(C, \omega_C(-n\Delta)) \neq 0\,, \quad n \geq 2\,.$$

This and (8.34) contradict Proposition (6.7).

Q.E.D.

The conclusion of the preceding proposition is no longer valid if we remove the assumption that ϕ correspond to a special g^r_d. However, suppose that the g^r_d is *complete* and nonspecial (the case of a special g^r_d has already been disposed of). Let m be the degree of ϕ, and let n be the degree of $\phi(C)$. Obviously, $n \geq r$, that is,

$$(8.35) \qquad m \leq \frac{d}{r} = \frac{g}{r} + 1\,,$$

where the second step follows from the Riemann–Roch theorem. Now assume that ϕ is composed with an involution. This means that Γ is rational and that C carries a g^1_m. But then (8.35) contradicts the Dimension Theorem of Brill–Noether theory (Theorem (1.5) in Chapter V) unless $r = 2$ and g is even. In this case $d = g + 2$. This last case can in fact occur. Suppose that C has even genus. Again by the Dimension Theorem, C can be realized as a branched covering of $\mathbb{P}^1$ of degree $(g+2)/2$. Let ϕ be the composition of this covering with the Veronese embedding of $\mathbb{P}^1$ in $\mathbb{P}^2$. The corresponding g^2_{g+2} is then nonspecial by Proposition (8.33).

9. Ramification.

It turns out that, when studying special linear series from the point of view of deformations of mappings, the presence of ramification plays a very special role and has, in fact, unexpected consequences. We shall explore these using the formalism of infinitesimal deformations, as developed in Section 8. There will be no change in notation. Let $\phi : C \to M$ be a nonconstant analytic map of a smooth curve C into an n-dimensional complex manifold M. The divisor R of zeroes of the differential of ϕ will be called the *ramification divisor of* ϕ. The sheaf homomorphism

$$d\phi : T_C \to T = \phi^* T_M$$

extends to a homomorphism

$$T_C(R) \to T$$

which has maximal rank everywhere. We then get a commutative diagram

$$\text{(9.1)}\qquad \begin{array}{ccccccccc} & & & & & & 0 & & \\ & & & & & & \downarrow & & \\ & & 0 & & & & \mathcal{K}_\phi & & \\ & & \downarrow & & & & \downarrow & & \\ 0 & \longrightarrow & T_C & \longrightarrow & T & \longrightarrow & N_\phi & \longrightarrow & 0 \\ & & \downarrow & & \| & & \downarrow & & \\ 0 & \longrightarrow & T_C(R) & \longrightarrow & T & \longrightarrow & N'_\phi & \longrightarrow & 0 \\ & & \downarrow & & & & \downarrow & & \\ & & \mathcal{Q}_\phi & & & & 0 & & \\ & & \downarrow & & & & & & \\ & & 0 & & & & & & \end{array}$$

where $N' = N'_\phi$ is a *locally free sheaf* of rank $n-1$, and $\mathcal{K} = \mathcal{K}_\phi$, $\mathcal{Q} = \mathcal{Q}_\phi$ are (noncanonically) isomorphic to the structure sheaf $\mathcal{O}_R$. This implies, in particular, that

$$\text{(9.2)}\qquad H^1(C, N) \cong H^1(C, N') .$$

We then see that the third version (8.13) of Petri's statement is really an assertion about N', and, in a sense, N' is better behaved than N.

We shall devote this section to the study of the skyscraper sheaf $\mathcal{K}$ and of the first-order deformations of ϕ corresponding to its sections. Write

$$R = \sum_{i=1}^{s} \nu_i p_i ,$$

where the p_i are distinct points. Choose a cover $\mathcal{U} = \{U_\alpha\}$ of C by coordinate open sets in such a way that each p_i lies in only one open set U_{α_i}, that the local coordinate z_{α_i} on U_{α_i} vanishes at p_i, and that $U_{\alpha_i} \cap U_{\alpha_j} = \emptyset$ if $i \neq j$. Then, clearly, every element of $H^0(C, \mathcal{K})$ comes from a cochain

$$\{a_\alpha\} \in C^0(\mathcal{U}, T)$$

of the form

$$(9.3)\qquad \begin{cases} a_{\alpha_i} = z_{\alpha_i}^{-\nu_i} c_i(z_{\alpha_i}) \dfrac{\partial \psi_{\alpha_i}}{\partial z_{\alpha_i}}\,, & i = 1, \dots, s\,, \\ a_\beta = 0\,, & \beta \neq \alpha_i\,,\ i = 1, \dots, s\,, \end{cases}$$

where the c_i are polynomials of degree at most $\nu_i - 1$. Recalling the basic formula (8.7), we see that the Kodaira–Spencer class of the section of $\mathcal{K}$ we just constructed is given by

$$\begin{cases} b_{\alpha_i \beta} = -z_{\alpha_i}^{-\nu_i} c_i(z_{\alpha_i})\,, & i = 1, \dots, s\,, \\ b_{\alpha\beta} = 0\,, & \alpha \neq \alpha_i \neq \beta\,. \end{cases}$$

As we know from Chapter XI, Section 2, in case $g > 0$, $c_i(z_{\alpha_i}) = z_{\alpha_i}^{\nu_i - 1}$, and $c_j = 0$ for $j \neq i$, this class is known as a *Schiffer variation* centered at the point p_i. Such a class is a nonzero element of $H^1(C, T_C)$ which corresponds to a first-order deformation "changing the complex structure of C only at the point p_i." More geometrically, the Schiffer variations centered at p_i describe the line in $H^1(C, T_C)$ which corresponds to the point p_i under the bicanonical map

$$\phi_{\omega^2} : C \to \mathbb{P}H^1(C, T_C).$$

In a similar way the points in $\mathbb{P}H^1(C, T_C)$ corresponding to the classes defined by

$$\begin{array}{ll} b_{\alpha_i \beta} = z^{-s}, & 1 \leq s \leq h+1, \\ b_{\alpha\beta} = 0\,, & \alpha \neq \alpha_i \neq \beta\,, \end{array}$$

span the hth osculating space to the bicanonical curve at $\phi_{\omega_C^2}(p_i)$. Let us look at the expressions (9.3) describing the elements of $H^0(C, \mathcal{K})$. The fact that $a_\beta = 0$ when β is not one of the α_i can be interpreted as follows. Let

$$\begin{array}{ccc} \mathcal{C} & \xrightarrow{\ \tilde{\phi}\ } & M \\ \Big\downarrow{\scriptstyle p} & & \\ \operatorname{Spec} \mathbb{C}[\varepsilon] & & \end{array}$$

be a first-order deformation of ϕ corresponding to a section of $\mathcal{K}$. Set $\Gamma = \phi(C)$. Then $\tilde{\phi}$ factors through the inclusion of Γ in M:

$$\begin{array}{ccc} \mathcal{C} & \xrightarrow{\ \tilde{\phi}\ } & M \\ & \searrow \quad \nearrow & \\ & \Gamma & \end{array}$$

To put it differently, *the first-order deformations of ϕ corresponding to sections of $\mathcal{K}$ leave the image curve fixed* (up to first-order, of course).

The following simple example illustrates this phenomenon, which is local in character. Let x, y be affine coordinates in $\mathbb{C}^2$. Let

$$\psi(z) = (z^2, z^3)\,, \qquad |z| < 1,$$

be the parametric equation of an ordinary cusp

$$y^2 = x^3. \tag{9.4}$$

Then $\mathcal{K}$ is supported at $z = 0$, and its sections are all multiples of

$$a(z) = \frac{1}{z}\frac{d\psi}{dz} = (2, 3z)\,.$$

The corresponding first-order deformation is

$$\tilde{\psi}(z, \varepsilon) = (z^2 + 2\varepsilon, z^3 + 3\varepsilon z)\,. \tag{9.5}$$

One checks that $\tilde{\psi}$ satisfies equation (9.4) modulo ε^2.

We can therefore say that in general *the presence of "cusps" in the image curve $\Gamma = \phi(C)$ implies that there exists, infinitesimally, more than one non-singular model of Γ.*

Looking again at the example of the ordinary cusp, if we interpret ε as a finite instead of an infinitesimal parameter in (9.5), we notice that for $\varepsilon \neq 0$, the curve

$$z \to \tilde{\psi}(z, \varepsilon)$$

has *a node and no cusps.* This simple remark can be generalized in a way that will turn out to be very useful for our purposes.

In order to explain this, we first have to analyze in more detail the local geometry of unibranch singularities (e.g., cuspidal points). Let Δ be the unit disc in $\mathbb{C}$, and let

$$\psi : \Delta \to \mathbb{C}^n, \qquad n > 1,$$

be a nonconstant holomorphic map such that

$$\psi(0) = 0\,.$$

We let $\mathcal{O}_1$ be the local ring of Δ at 0, $\mathcal{O}_2$ the local ring of $\mathbb{C}^n$ at 0, and set

$$\mathcal{O} = \psi^*(\mathcal{O}_2)\,,$$

where $\psi^* : \mathcal{O}_2 \to \mathcal{O}_1$ is the pullback homomorphism. We denote by $\mathfrak{m}$, $\mathfrak{m}_1$, $\mathfrak{m}_2$ the maximal ideals of $\mathcal{O}$, $\mathcal{O}_1$, $\mathcal{O}_2$, respectively. We choose a minimal set of generators for $\mathfrak{m}$ as follows. Let $k_1 \geq 1$ be the minimum among the integers h such that

$$\mathfrak{m} \cap \mathfrak{m}_1^h \not\subset \mathfrak{m}_1^{h+1},$$

and let g_1 be an element of $\mathfrak{m} \cap \mathfrak{m}_1^{k_1}$ which does not belong to $\mathfrak{m}_1^{k_1+1}$. If g_1 does not generate $\mathfrak{m}$, let $k_2 > k_1$ be the minimum among the integers h such that

$$\mathfrak{m} \cap \mathfrak{m}_1^h \not\subset \mathfrak{m}_1^{h+1} + \mathcal{O}g_1,$$

and let g_2 be an element of $\mathfrak{m} \cap \mathfrak{m}_1^{k_2}$ which does not belong to $\mathfrak{m}_1^{k_2+1} + \mathcal{O}g_1$. If g_1, g_2 do not generate $\mathfrak{m}$, we let $k_3 > k_2$ be the minimum among the integers h such that

$$\mathfrak{m} \cap \mathfrak{m}_1^h \not\subset \mathfrak{m}_1^{h+1} + \mathcal{O}g_1 + \mathcal{O}g_2,$$

and so on. This process ends after at most n steps, and we are left with a minimal set of generators $g_1, \ldots, g_\ell$ of $\mathfrak{m}$ such that

$$\begin{cases} \operatorname{ord}_0(g_i) = k_i\,, \\ k_1 < k_2 < \ldots < k_\ell\,, \end{cases} \tag{9.6}$$

and, moreover,

(9.7)

for every i, k_i is not a linear combination of $k_1, \ldots, k_{i-1}$ with nonnegative integral coefficients.

Clearly, $g_1, \ldots, g_\ell$ are uniquely determined up to multiplication by units of $\mathcal{O}$, hence, $k_1, \ldots, k_\ell$ are invariants of ψ. The number $k_1 - 1$ is called the *ramification index* of ψ at 0. The point $0 \in \Delta$ is called a ramification point of ψ if $k_1 > 1$. We will say that this ramification point is *nondegenerate* in case $\ell > 1$; we then call k_2 the *type* of the ramification point. We may choose a minimal set of generators $g'_1, \ldots, g'_n$ for m_2 in such a way that

$$\begin{aligned} &\psi^*(g'_i) = g_i\,, && i = 1, \ldots, \ell\,, \\ &\psi^*(g'_i) = 0\,, && i > \ell\,. \end{aligned}$$

This means that we may choose local coordinates $w_1, \ldots, w_n$ centered at 0 on $\mathbb{C}^n$ and a local coordinate z centered at 0 on Δ in such a way that, in these coordinates, ψ is given by

$$\begin{aligned} &w_1 = z^{k_i} + \text{ higher-order terms}, && i = 1, \ldots, \ell\,, \\ &w_i = 0\,, && i > \ell\,, \end{aligned}$$

where $k_1, \dots, k_\ell$ satisfy (9.6) and (9.7). Now set

$$\Delta^2 = \{(\xi, t) \in \mathbb{C}^2 : |\xi| < 1,\ |t| < 1\}$$

and let

$$\psi : \Delta^2 \to \mathbb{C}^n$$

be a holomorphic mapping. For each t, we set

$$\psi_t(\xi) = \psi(\xi, t).$$

We assume that, for each t, ψ_t is not constant and has only one ramification point. Clearly, the index of this ramification point is independent of t; let it be equal to $h - 1$. Assume, moreover, that for general t, the ramification point of ψ_t is nondegenerate. Let

$$V \subset \Delta^2$$

be the locus of the ramification points of ψ_t as t varies. Since V projects in 1–1 fashion onto $\{t \in \mathbb{C} : |t| < 1\}$, this projection is a biholomorphic isomorphism. Then, in suitable coordinates,

$$\begin{array}{ll} t,\ z = z(\xi, t) & \text{for } \Delta^2, \\ w_1, \dots, w_n & \text{for } \mathbb{C}^n, \end{array}$$

the locus V is defined by the equation $z = 0$, and we may write ψ, in local coordinates, as

$$w_i = p_i(z^h, t) + \gamma_i(t) z^k + [k+1], \qquad i = 1, \dots, n, \tag{9.8}$$

where

$$\left\{ \begin{array}{l} p_i(\zeta, t) \text{ is a polynomial in } \zeta \text{ of degree } m \geq 1; \\ h \text{ does not divide } k; \\ k > hm; \\ \text{the } \gamma_i \text{ are not all identically zero;} \\ [k+1] \text{ stands for terms of order higher than } k \text{ in } z. \end{array} \right. \tag{9.9}$$

Clearly, when $\gamma_i(t) \neq 0$ for at least one i, k is the type of the ramification point of ψ_t. On the other hand, when $\gamma_i(t) = 0$ for every i, the ramification point of ψ_t is degenerate, or else its type is larger than k. If we agree to say that the type of a degenerate ramification point is equal to $+\infty$, this shows that *the type of ψ_t is an upper semicontinuous function of t.*

We are now in a position to understand the phenomenon of "disappearance of cusps" that we discussed in the previous example.

Proposition (9.10). *Let $\mathcal{C} \xrightarrow{p} \Delta = \{t \in \mathbb{C} : |t| < 1\}$ be a smooth analytic family of curves, and let*

$$\phi : \mathcal{C} \to M$$

be an analytic morphism of $\mathcal{C}$ into a complex n-dimensional manifold M, $n \geq 2$. Set $C_t = p^{-1}(t)$, $\phi_t = \phi|_{C_t}$. Assume that

a) *ϕ_t is a birational map of C_t onto $\phi(C_t)$ for all t;*
b) *the number, the indices, and the types of the ramification points of ϕ_t are independent of t.*

Then the Horikawa class of $(\mathcal{C}, p, \phi)$ at $t = 0$ does not belong to $H^0(C_0, \mathcal{K}_{\phi_0})$, unless it is zero.

Proof. Let V be the locus traced out by the ramification points of the ψ_t. Assumption b) implies that V is a topological covering of Δ. We can then choose a coordinate cover $\{U_\alpha\}$ of $\mathcal{C}$ with coordinates (z_α, t) such that $V \cap U_\alpha = \{z_\alpha = 0\}$. We may also assume that $\phi(U_\alpha)$ is contained in a coordinate patch $W_\alpha \subset M$. Let $w_\alpha = (w_\alpha^1, \dots, w_\alpha^n)$ be local coordinates on W_α, and let

$$w_\alpha = \psi_\alpha(z_\alpha, t)$$

be the expression of ϕ in these coordinates. The Horikawa class of $(\mathcal{C}, p, \phi)$ at $t = 0$ is the section s of N_{ϕ_0} represented by

$$\left\{ \frac{\partial \psi_\alpha}{\partial t}(z_\alpha, 0) \right\}.$$

Recalling (9.3), to say that this class belongs to $H^0(C_0, \mathcal{K}_{\phi_0})$ means that there are meromorphic functions f_α such that

$$\frac{\partial \psi_\alpha}{\partial t}(z_\alpha, 0) = f_\alpha(z_\alpha) \frac{\partial \psi_\alpha}{\partial z_\alpha}(z_\alpha, 0).$$

To complete the proof, it is now sufficient to prove the following lemma.

Lemma (9.11). *Set $\Delta^2 = \{(z, t) \in \mathbb{C}^2 : |z| < 1, |t| < 1\}$ and let*

$$\psi : \Delta^2 \to \mathbb{C}^n, \qquad n > 1,$$

be an analytic map. Assume that, for each t, $\psi_t(z) = \psi(z, t)$ is injective, has only one ramification point, and that the index and type of this point are independent of t. Assume also that there exists a meromorphic function $f(z)$ such that

$$\frac{\partial \psi}{\partial t}(z, 0) = f(z) \frac{\partial \psi}{\partial z}(z, 0). \tag{9.12}$$

Then $f(z)$ is holomorphic.

Proof. The statement of the lemma is clearly invariant under an arbitrary change of coordinates in $\mathbb{C}^n$ and a change of coordinates of the type $z' = z'(z,t)$, $t' = t$ in Δ^2. Let $V \subset \Delta^2$ be the locus of the ramification points of the ψ_t. By assumption, V maps in a one-to-one fashion onto $\{t \in \mathbb{C} : |t| < 1\}$. Therefore, V is a smooth subvariety of Δ^2, and we may assume that its equation is $z = 0$. Let h be the ramification index of ψ_t. With suitable choices of coordinates in $\mathbb{C}^n$ and Δ^2, we may assume that

$$\psi = (\psi_1, \dots, \psi_n)$$

is given by

$$\psi_i(z,t) = p_i(z^h, t) + \gamma_i(t) z^k + [k+1], \qquad i = 1, \dots, n, \tag{9.13}$$

where conditions (9.9) are satisfied. We can also assume that

$$\psi_1(z,t) = \alpha(t) + z^h. \tag{9.14}$$

The fact that the type of the ramification point of ψ_t is constant implies that one of the γ_i, say γ_2, does not vanish at $t = 0$. Assumptions (9.12) and (9.14) give

$$\frac{\partial \alpha}{\partial t}(0) = h\, f(z) z^{h-1}.$$

Therefore $f(z) = c\, z^{1-h}$, where c is a constant. Using (9.12) and (9.13), we obtain

$$\begin{aligned} &\frac{\partial p_2}{\partial t}(z^h, 0) + \frac{\partial \gamma_2}{\partial t}(0) z^k + [k+1] \\ &\qquad = c\left(h \frac{\partial p_2}{\partial \zeta}(z^h, 0) + k\gamma_2(0) z^{k-h} + [k-h+1] \right). \end{aligned}$$

Since $\gamma_2(0) \neq 0$, we must have $c = 0$. Therefore, in this coordinate system, f is identically zero; in particular, it is holomorphic.

Q.E.D.

If we had to give a capsule statement of Proposition (9.10), it could be the following. *Suppose that we are given a deformation of*

$$\phi : C \to M,$$

parameterized by a disc $\Delta = \{t \in \mathbb{C} : |t| < 1\}$, *whose Horikawa class at* $t = 0$ *lies in* $H^0(C, \mathcal{K}_\phi)$; *then, along such a deformation, the complexity of the ramification of* ϕ *decreases.*

As a first application of these techniques, we wish to prove the following consequences of Petri's statement (1.3).

PROPOSITION (9.15). *Let C be a general curve of genus g. Let r and d be nonnegative integers such that $r \geq 2$ and*

$$r = d - g \quad \text{if } d \geq 2g - 1,$$

$$\left.\begin{array}{l} r \geq d - g \\ \rho = g - (r+1)(g - d + r) \geq 0 \end{array}\right\} \quad \text{if } d \leq 2g - 2.$$

Let L be a line bundle on C corresponding to a general point of $W_d^r(C)$. Then

i) *$|L|$ has no base points;*
ii) *If*

$$\phi : C \to \mathbb{P}^r$$

is the morphism corresponding to $|L|$, then ϕ is not composed with an involution and is a local immersion.

Proof. For $g = 0, 1$, the result is obviously true; therefore, from now on we assume that $g \geq 2$. By Petri's statement (1.3), every component of $W_d^r(C)$ has dimension ρ. On the other hand, again by the same result, the sublocus of $W_d^r(C)$ consisting of g_d^r's with base points has dimension at most equal to

$$\dim W_{d-1}^r(C) + 1 < \rho.$$

Therefore $|L|$ has no base points. In Proposition (8.33) we proved that ϕ is not composed with an involution, with the possible exception of the case in which g is even, $d = g + 2$, $r = 2$, and $|L|$ is composed with a $g^1_{(g+2)/2}$. But this case cannot occur because such an L does not correspond to a general point of $W_{g+2}^2(C)$. In fact, the third version of Petri's statement says that there is only a finite number of such L's, whereas the dimension of $W_{g+2}^2(C)$ is equal to g. We now proceed to show that ϕ is a local immersion. Suppose it is not and let s be a nonzero section of $\mathcal{K}_\phi$. Since C is general, $H^1(C, N_\phi)$ vanishes by Petri's statement in its fourth version (Theorem (8.13)). It then follows from Theorem (8.15) that any section of $H^0(C, N_\phi)$ can be integrated to a deformation of ϕ parameterized by a one-dimensional disk. Therefore, there exists a deformation $\widetilde{\phi}$ of ϕ, parameterized by a one-dimensional disk, whose Horikawa class at $t = 0$ is s. By Proposition (9.10) the number, the index, and the type of the ramification points of $\widetilde{\phi}_t$ cannot be constant in any neighborhood of 0. This contradicts the assumption that C and L are general.

Q.E.D.

By degenerating to a rational cuspidal curve, Eisenbud and Harris prove a much stronger theorem (Theorem (1.8) in [197]), showing that, under the same assumption of Proposition (9.15), when $r \geq 3$, the morphism ϕ is actually a *smooth embedding.* This is the theorem we stated in Section 1 of Chapter V.

Let us finally remark that, *in case* $r = 2$, it can be proved that, under the assumptions of Proposition (9.15), *the only singularities of* $\phi(C)$ *are nodes.* In fact, in Theorem (10.7) we shall prove a much stronger statement which will not involve the generality of C in the sense of moduli but just the generality of the pair (C, g^2_d) as a point of an irreducible component of $\mathcal{W}^2_d$.

In Proposition (9.15) we saw that, when both the curve C and the line bundle L are general, the ramification divisor of the morphism attached to L is equal to zero. The following proposition gives a bound for the degree of the ramification divisor for any morphism from a general curve to projective space.

PROPOSITION (9.16). *Let C be a general curve of genus g, and let*

$$\phi : C \to \mathbb{P}^r, \qquad r \geq 2,$$

be a nondegenerate morphism which is not composed with an involution. Let d be the degree of $\phi(C)$, and let R be the ramification divisor of ϕ. Then

$$\deg R \leq \rho,$$

where, as usual,

$$\rho = g - (r+1)(g - d + r).$$

In particular, when $r = 2$,

$$\deg R \leq 3d - 2g - 6. \tag{9.17}$$

Proof. We argue by contradiction, assuming that a general curve of genus g admits a map to projective space as in the statement whose ramification divisor has degree strictly bigger than ρ. By Theorem (8.15) there exist a family of smooth genus g curves $q : \mathcal{Y} \to U$ such that the image of the moduli map $U \to M_g$ is open and a morphism $\psi : \mathcal{Y} \to \mathbb{P}^r$ with the following properties. For $u \in U$, denote by Y_u the fiber $q^{-1}(u)$ and by ψ_u the restriction of ψ to Y_u; then, for every $u \in U$, the morphism ψ_u is nondegenerate and not composed with an involution, and the degree of its ramification divisor is strictly larger than ρ. In particular, if u is a general point of U, the coboundary map $\delta : H^0(C, N_{\psi_u}) \to H^1(Y_u, T_{Y_u})$ is onto.

We may also assume that, as u varies in U, the number, index, and type of the ramification points of ψ_u do not change. We then fix a general point u in U, set $C = Y_u$, $\phi = \psi_u$, and we denote by R the ramification divisor of ϕ. It follows from Proposition (9.10) that

$$\delta(H^0(C, \mathcal{K}_\phi)) = 0$$

and hence that the surjective map $H^0(C, N_\phi) \to H^1(C, T_C)$ factors through $H^0(C, N'_\phi)$, so that

$$h^0(C, N'_\phi) \geq 3g - 3 + (r+1)^2 - 1 . \tag{9.18}$$

On the other hand, N'_ϕ is of rank $r - 1$, and from (9.1) and (8.9) it follows that

$$\wedge^{r-1}(N'_\phi) = L^{r+1} \otimes \omega_C(-R) ,$$

where L is the pullback via ϕ of the hyperplane bundle on $\mathbb{P}^r$. Since C is general, we have $h^1(C, N_\phi) = h^1(C, N'_\phi) = 0$. Therefore, by the Riemann–Roch theorem for vector bundles,

$$h^0(C, N'_\phi) = (r+1)d + 2g - 2 - \deg R + (r-1)(1-g) .$$

Putting this equality together with (9.18), we get

$$\deg R \leq \rho .$$

Q.E.D.

10. Plane curves.

In this section we try to answer the question: what is the dimension of $\mathcal{G}^2_d$? Our discussion will parallel the treatment of $\mathcal{G}^1_d$ presented at the end of Section 6. The situation here is complicated by the fact that $M^2_{g,d}$ need not be irreducible, contrary, as we shall see in the next section, to what occurs for $M^1_{g,d}$.

THEOREM (10.1). *Let*

$$p : \mathcal{C} \to U$$

be a smooth family of curves of genus g parameterized by a smooth connected variety U. Suppose that

$$\psi : \mathcal{C} \to \mathbb{P}^2$$

is a morphism such that the restriction of ψ to each fiber of p is not composed with an involution and that, for general $u \in U$, the homomorphism

$$T_u(U) \to H^0(C, N_\phi)$$

is injective, where $C = p^{-1}(u)$, $\phi = \psi|_C$. Then

$$\dim U \leq 3d + g - 1,$$

where d is the degree of $\phi(C)$. In particular, if $g \geq 2$ and X is a component of $\mathcal{G}^2_d$ whose general point corresponds to a curve C of genus

g equipped with a basepoint-free g^2_d which is not composed with an involution, then

$$\dim X = 3g - 3 + \rho = 3d + g - 9.$$

Proof. If $\dim U \leq g$, there is nothing to prove; therefore, we assume that $\dim U \geq g+1$. The assumption that u is a general point of U implies, by Proposition (9.10), that

$$g + 1 \leq \dim U \leq h^0(C, N'_\phi)\,.$$

Therefore N'_ϕ is nonspecial. On the other hand, from (9.1) and (8.9) we get

$$N'_\phi = \omega_C L^3(-R)\,, \tag{10.2}$$

where R is the ramification divisor of ϕ, and L is the pullback via ϕ of the hyperplane bundle. Thus,

$$\dim U \leq h^0(C, N'_\phi) \leq 3d + g - 1\,.$$

As for the second statement, by Proposition (3.21) it suffices to prove that the dimension of X does not exceed $3d+g-9$. Let U be the bundle of projective frames for the universal g^2_d over X. Then there is a family of curves and morphisms

$$\begin{array}{ccc} \mathcal{X} & \longrightarrow & \mathbb{P}^2 \\ \big\downarrow & & \\ U & & \end{array}$$

satisfying all the assumptions of the first part of our theorem. Thus,

$$\dim X = \dim U - \dim(\mathrm{Aut}(\mathbb{P}^2)) \leq 3d + g - 9\,.$$

Q.E.D.

Remark (10.3). In the above theorem the restriction to g^2_d's which are not composed with an involution is a very natural one. In fact, let Y be a component of the variety of isomorphism classes of pairs (C, W) with C a smooth curve of genus g and W a g^2_d on C of the form $\phi^*(W')$, where $\phi : C \to C'$ is a degree n morphism to a curve of genus g', and W' is a base-point-free $g^2_{d'}$ on C' which is not composed with an involution. Of course, $d = nd'$. We denote by w the degree of the ramification divisor of ϕ. When $g' \geq 2$, it is clear from Theorem (10.1) that

$$\dim Y = 3d' + g' - 9 + w\,, \tag{10.4}$$

and the reader may easily verify that the same is true also for $g' = 0, 1$. Thus (10.4) is a lower bound for the dimension of the component of $\mathcal{G}^2_d$ containing Y and may well be strictly larger than the expected dimension

$3d+g-9$. In fact, using the Riemann–Hurwitz formula, it is immediate to check that this happens precisely when

$$w > (n-1)(6d'+2g'-2)$$

or equivalently when

$$g > (2n-1)(g'-1) + 3d'(n-1) + 1\,. \tag{10.5}$$

This provides plenty of examples of components of $\mathcal{G}^2_d$ of dimension larger than the expected one and hence of components whose general member is necessarily composed with an involution.

It is interesting to notice that the two kinds of components of $\mathcal{G}^2_d$, those whose general member is composed with an involution and those whose general member is not, may very well coexist for given g and d. In fact, the former exist if and only if g is not less that the right-hand side of (10.5) for some $g' \geq 0$ and some factorization $d = nd'$ with $n > 1$, while components of the second kind exist if and only if

$$g \leq \frac{(d-1)(d-2)}{2}\,, \tag{10.6}$$

as we shall see later in this section. On the other hand, the right-hand side of (10.6) can be larger than the one of (10.5) since, for fixed n and g', it grows quadratically in d', as opposed to linearly.

Theorem (10.1) is really a statement about continuous systems of plane curves. We recall that a plane curve of degree d can be thought of as a point in the projective space

$$\mathbb{P}^N\,, \quad N = \frac{d(d+3)}{2}\,.$$

A subvariety of this $\mathbb{P}^N$ is then called a *continuous system of plane curves of degree d*. Let us denote by

$$\Sigma_{d,g} \subset \mathbb{P}^N$$

the continuous system of all irreducible plane curves of degree d whose normalization has genus g.

THEOREM (10.7). *Let Σ be any irreducible component of $\Sigma_{d,g}$. Then*

$$\dim \Sigma = 3d + g - 1.$$

Moreover, a general point of Σ corresponds to a plane irreducible curve of degree d having

$$\delta = \frac{(d-1)(d-2)}{2} - g$$

nodes and no other singularity.

Proof. The dimensionality statement is essentially the second part of (10.1). To see it, we limit ourselves to the case $g \geq 2$, leaving to the reader the easy task of modifying the proof to make it work when $g = 0$ or $g = 1$. If U is a sufficiently small neighborhood of a general point of Σ, then by simultaneously normalizing the curves corresponding to the points of U, we get a family of smooth curves of genus g

$$p : \mathcal{C} \to U \subset \Sigma$$

and a morphism

$$\psi : \mathcal{C} \to \mathbb{P}^2$$

such that for every $x \in U$, $\psi(p^{-1}(x))$ is the plane curve corresponding to x. By associating to each $x \in U$ the curve $p^{-1}(x)$ and the linear series giving the morphism $\psi_x : p^{-1}(x) \to \mathbb{P}^2$, we get an open morphism

$$f : U \to \mathcal{G}_d^2$$

whith fiber dimension equal to $8 = \dim PGL(3)$. This, combined with the second part of (10.1) and the openness of f, gives

$$\dim(\Sigma) = 3d + g - 1 .$$

Having proved the dimensionality statement, we now prove the second part of (10.7). We first observe that, if

$$\phi : C \to \mathbb{P}^2$$

is the normalization morphism for the plane curve corresponding to a general point x of Σ, noticing that $\dim \Sigma \geq g + 1$ and arguing as in the proof of (10.1), we may conclude that $H^1(C, N_\phi)$ vanishes. By Theorem (8.15), this implies that the elements of $H^0(C, \mathcal{K}_\phi)$ are unobstructed. Using again the generality of x and Proposition (9.10), we conclude that $\mathcal{K}_\phi = 0$, proving that ϕ is an immersion and showing, at the same time, that N_ϕ is a line bundle of degree equal to $2g - 2 + 3d$. Set $\Gamma = \phi(C)$. Suppose that there are two points p_1 and p_2 of C such that $\phi(p_1) = \phi(p_2)$ and that the two branches of Γ in $\phi(p_1)$ corresponding to p_1 and p_2 have a contact of order $h > 1$. By our previous remark about the degree of N_ϕ, we have that

$$H^1(C, N_\phi(-p_1 - p_2)) = 0 .$$

Therefore, there is a section s of N_ϕ which vanishes at p_1 but not at p_2. Since $H^1(C, N_\phi) = 0$, the class s is unobstructed and hence corresponds to a one-parameter family of deformations of ϕ

$$\begin{array}{ccc} \mathcal{C} & \xrightarrow{\tilde{\phi}} & \mathbb{P}^2 \\ \downarrow{\scriptstyle p} & & \\ \Delta & & \end{array}$$

where

$$\Delta = \{t \in \mathbb{C} : |t| < \varepsilon\},$$

such that $p^{-1}(0) = C$, $\tilde{\phi}|_C = \phi$. Along this deformation the two branches of Γ corresponding to p_1 and p_2 are deformed, for any $t \neq 0$, into analytic arcs which do not have contacts of order h (or higher), in contradiction with the generality of x. To see this, choose local coordinates (t, z_1) and (t, z_2) on $\mathcal{C}$ centered at p_1 and p_2, respectively. Upon choosing suitable local coordinates centered at $\phi(p_1) \in \mathbb{P}^2$, $\tilde{\phi}$ can be represented, near p_1 and p_2, by $\mathbb{C}^2$-valued functions $f_1(t, z_1), f_2(t, z_2)$ such that

$$\begin{aligned}
&f_1(0,0) = f_2(0,0) = 0\,, && \\
(10.8)\qquad &\frac{\partial f_1}{\partial z_1}(t, z_1) \neq 0 && \forall\, t, z_1\,, \\
&\frac{\partial f_2}{\partial z_2}(t, z_2) \neq 0 && \forall\, t, z_2\,, \\
(10.9)\qquad &\frac{\partial f_1}{\partial t}(0,0) \wedge \frac{\partial f_1}{\partial z_2}(0,0) = 0\,, && \\
(10.10)\qquad &\frac{\partial f_2}{\partial t}(0,0) \wedge \frac{\partial f_2}{\partial z_2}(0,0) \neq 0\,. &&
\end{aligned}$$

From (10.10) it follows that, changing local coordinates in $\mathbb{P}^2$, we may assume that $f_2(t, z_2) = (t, z_2)$. Write $f_1(t, z_1) = (a(t, z_1), b(t, z_1))$. To say that the two branches of Γ under consideration have a contact of order h at $\phi(p_1)$ means that

$$\begin{aligned}
(10.11)\qquad &\frac{\partial^i a}{\partial z_1^i}(0,0) = 0\,, \qquad i = 0, \dots, h-1\,, \\
&\frac{\partial^h a}{\partial z_1^h}(0,0) \neq 0\,.
\end{aligned}$$

We now argue by contradiction, assuming that for every t, there is a point $(t, z(t))$ (necessarily unique) such that $z_1 \mapsto f_1(t, z_1)$ has a contact of order h with $z_2 \mapsto (t, z_2)$ at $f_1(t, z(t))$. In other words, assume that there is $z(t)$ such that

$$\begin{aligned}
&a(t, z(t)) = t\,, \\
&\frac{\partial^i a}{\partial z_1^i}(t, z(t)) = 0\,, \qquad i = 1, \dots, h-1\,.
\end{aligned}$$

Since we must have

$$\frac{\partial^h a}{\partial z_1^h}(t, z(t)) \neq 0 \qquad \text{for small } t\,,$$

by the implicit function theorem $z(t)$ is a holomorphic function of t. Differentiating the identity $a(t, z(t)) = t$ and using (10.11), we get that $\frac{\partial a}{\partial t}(0,0) = 1$. Since $\frac{\partial a}{\partial z_1}(0,0) = 0$, formula (10.9) implies that $\frac{\partial f_1}{\partial z_1}(0,0) = 0$, contradicting (10.8).

Having established that all multiple points of Γ are ordinary, assume that Γ has an n-fold point q with $n > 2$. Let $p_1, \dots, p_n$ be the points of C which map to q and notice that $d > n$. Therefore,

$$H^1(C, N_\phi(-p_1 - \cdots - p_n)) = 0\,,$$

and there is a section of N_ϕ which vanishes at $p_2, \dots, p_n$ but not at p_1. Since $H^1(C, N_\phi) = 0$, this class is not obstructed and corresponds to a family of deformations of ϕ along which the n-fold point q is transformed into finitely many points of strictly lower multiplicity. This, again, contradicts the generality of x.

Q.E.D.

The following theorem was established by Harris [349], proving a long-standing conjecture by F. Severi.

THEOREM (10.12). *$\Sigma_{d,g}$ is irreducible.*

We shall not deal with this result but will devote our attention to other, less difficult, aspects of the geometry of plane curves.

In Theorem (10.1) we computed the exact dimension of any component of $\mathcal{G}^2_d$ whose general point corresponds to a morphism of a genus g curve in $\mathbb{P}^2$ which is not composed with an involution. The problem arises to determine whether such a component exists, i.e., to decide for which values of g and d there exist irreducible plane curves of degree d and geometric genus g (here and in what follows, by geometric genus we mean the genus of the normalization). The answer to this question was found by Severi [633], who, more specifically, showed that:

THEOREM (10.13). *Given a positive integer d, for every value of g such that*

$$0 \le g \le \frac{(d-1)(d-2)}{2}\,,$$

there exists an irreducible plane curve of degree d and geometric genus g having at most nodes as singularities.

More generally, Tannenbaum [652] proves the following.

THEOREM (10.14) (SEVERI-TANNENBAUM). *Given any integer $r \ge 2$ and any integer d such that $d \ge r$, for every value of g such that*

$$0 \le g \le \pi(d, r)\,,$$

where $\pi(d,r)$ is the Castelnuovo bound, there exists an irreducible nondegenerate curve in $\mathbb{P}^r$ of degree d and geometric genus g having at most nodes as singularities.

For an extensive discussion of Castelnuovo's bound and Castelnuovo extremal curves, we refer to Chapter I. Severi's idea is to construct his curves by suitably smoothing the union of d general lines at a selected group of nodes. Tannenbaum follows a similar procedure on certain special kinds of smooth rational scrolls in $\mathbb{P}^r$. The basic tool is the following lemma.

LEMMA (10.15). *Let X be a smooth surface. Let C be a connected nodal curve in X. Assume that, for every component Γ of C, $(\omega_X \cdot \Gamma) < 0$. Let $p_1, \dots, p_h$ be the nodes of C. Then for every integer k, $0 \le k \le h$, there exists a flat family*

$$\begin{array}{ccc} \mathcal{C} & \hookrightarrow & X \times \Delta \\ \downarrow \pi & & \\ \Delta & & \end{array}$$

where $\Delta = \{t \in \mathbb{C} : |t| < \delta\}$, with the following properties:

i) *For any $t \neq 0$, $C_t = \pi^{-1}(t)$ has exactly $h-k$ nodes and no other singularities.*
ii) *$C_0 = C$.*
iii) *The family $\mathcal{C} \to \Delta$ is analytically locally trivial at any point of C different from $p_1, \dots, p_k$.*

The proof of this result, although conceptually quite simple, requires the introduction of a new tool, which can be thought of as an analogue, in the case of curves with nodes, of Horikawa's theory. This in turn is a classical subject which has been revived, among others, by Wahl [677]. Following a brief intuitive remark, we shall give a sketchy presentation of those results of Wahl which are relevant to our purpose.

The intuitive remark is simply that, if $f(x,y,t) = 0$ gives a family of plane curves C_t having nodes $p_1(t), \dots, p_n(t)$, then the curve given by $\frac{\partial f}{\partial t}(x,y,t) = 0$ passes through the points $p_\nu(t)$.

Formally, let X be a smooth surface, and let C be a reduced curve on X having, as its only singularities, nodes $p_1, \dots, p_h$. Let k be an integer such that $0 \le k \le h$. Set

$$\widehat{N}_k = \Big(\prod_{i=k+1}^{h} \mathfrak{m}_{p_i} \Big) (\mathcal{O}_X(C) \otimes \mathcal{O}_C),$$

where $\mathfrak{m}_{p_i}$ stands for the ideal sheaf of the point p_i in $\mathcal{O}_C$. Let $\widetilde{C}$ be the normalization of C, and let $\phi : \widetilde{C} \to X$ be the normalization morphism (of course, $\widetilde{C}$ may be disconnected). The connection between the sheaves $\widehat{N}_k$ and the normal sheaf N_ϕ is given by a natural isomorphism

$$\widehat{N}_0 \cong \phi_*(N_\phi) \tag{10.16}$$

or, equivalently, by an isomorphism

$$\phi^*(\widehat{N}_0) = \phi^*\mathcal{O}(C)\left(-\sum(q_i + r_i)\right) \cong N_\phi ,$$

where $\phi^{-1}(p_i) = \{q_i, r_i\}$. One of Wahl's result is the following.

PROPOSITION (10.17). *There is a natural bijection between* $H^0(C, \widehat{N}_k)$ *and the set of isomorphism classes of flat families*

$$\begin{array}{ccc} \mathcal{C} & \longrightarrow & X \times \operatorname{Spec}\mathbb{C}[\varepsilon] \\ \downarrow & & \\ \operatorname{Spec}\mathbb{C}[\varepsilon] & & \end{array}$$

such that $C_0 = C$ *and* $\mathcal{C}$ *is locally trivial at any point of* C *different from* $p_1, \dots, p_k$.

Moreover, setting $\Delta = \{t \in \mathbb{C} : |t| < \delta\}$, he proves the following.

PROPOSITION (10.18). *In case* $H^1(C, \widehat{N}_k) = 0$, *any element* s *of* $H^0(C, \widehat{N}_k)$ *arises, for any sufficiently small* δ, *from a flat family*

$$\begin{array}{ccc} \mathcal{C} & \longrightarrow & X \times \Delta \\ \downarrow{\scriptstyle \pi} & & \\ \Delta & & \end{array}$$

satisfying ii) and iii) *(but not necessarily* i)*) in Lemma (10.15), via the inclusion* $\operatorname{Spec}\mathbb{C}[\varepsilon] \hookrightarrow \Delta$ *given by* $t \mapsto \varepsilon$.

To prove (10.15), we consider the exact sequence

$$0 \to \widehat{N}_0 \to \widehat{N}_k \to \mathcal{T} \to 0 , \tag{10.19}$$

where $\mathcal{T}$ is supported on $\{p_1, \dots, p_k\}$. We will show that

$$H^1(C, \widehat{N}_0) \cong H^1(\widetilde{C}, N_\phi) = 0 . \tag{10.20}$$

To see this, recall that

$$N_\phi = \phi^*(\omega_X^{-1}) \otimes \omega_{\widetilde{C}} .$$

Let $\widetilde{\Gamma}$ be any component of $\widetilde{C}$ and set $\Gamma = \phi(\widetilde{\Gamma})$. Since, by assumption, $(\omega_X \cdot \Gamma) < 0$, we obtain

$$\begin{aligned} \deg\left(N_\phi|_{\widetilde{\Gamma}}\right) &= 2g(\widetilde{\Gamma}) - 2 - (\omega_X \cdot \Gamma) \\ &> 2g(\widetilde{\Gamma}) - 2 . \end{aligned}$$

From (10.20) and (10.19) we deduce that $H^1(C, \widehat{N}_k) = 0$ and that $H^0(C, \widehat{N}_k)$ surjects onto $H^0(C, \mathcal{T})$. Choose a section s of $\widehat{N}_k$ whose image in $H^0(C, \mathcal{T})$ does not vanish at any one of the points $p_1, \dots, p_k$. Let

$$\begin{array}{ccc} \mathcal{C} & \longrightarrow & X \times \Delta \\ \downarrow & & \\ \Delta & & \end{array}$$

be the family whose existence is guaranteed by (10.18). By the very choice of s, this family also satisfies property i) of Lemma (10.15). This concludes the proof of the lemma.

Q.E.D.

Following the classical terminology introduced by Severi, we shall call the nodes $p_{k+1}, \dots, p_h$ the *assigned nodes* of C, and the nodes $p_1, \dots, p_k$ the *virtually nonexistent nodes* of C. The reason for these names is that we allow the virtually nonexistent nodes to disappear under deformation, whereas we insist that the assigned nodes remain. We are now almost ready to prove Theorem (10.14). The only thing we need is a way of telling when the curves C_t constructed in Lemma (10.15) are irreducible. This is taken care of by the following simple observation of Severi.

REMARK (10.21). Let X, C, C_t, $p_1, \dots, p_h$ be as in Lemma (10.15). Let C' be the partial desingularization of C at the assigned nodes $p_{k+1}, \dots, p_h$. If C' is connected, then C_t is irreducible for $t \neq 0$.

This is essentially obvious. Blow up $\mathcal{C}$ along the curves traced out by the moving assigned nodes to obtain a flat family $\mathcal{C}' \to \Delta$ whose central fiber is C'. Thus C'_t is connected for every t. On the other hand, for $t \neq 0$, C'_t is the normalization of C_t.

We now proceed to prove (10.14). We consider three separate cases. Let $r = 2$; in this case we choose C to be the union of d general lines,

$$C = \ell_1 \cup \cdots \cup \ell_d .$$

We number the nodes of C so that $p_1, \dots, p_{d-1}$ are the intersections of ℓ_1 with the other lines in C. For any integer g such that

$$0 \leq g \leq \pi(d, 2) = \frac{(d-1)(d-2)}{2},$$

we set

$$k = d - 1 + g .$$

By Remark (10.21), when $t \neq 0$, the curve C_t, whose existence is guaranteed by Lemma (10.15), is an irreducible plane curve of degree d with exactly

$$\frac{d(d-1)}{2} - k = \frac{(d-1)(d-2)}{2} - g$$

nodes and no other singularity. New let us turn to the case of $r \geq 3$. In this case set

$$X = \mathbb{P}^1 \times \mathbb{P}^1 .$$

Let L_1, L_2 be curves belonging to the two rulings of X. We view X as embedded in $\mathbb{P}^r$ by means of the very ample linear system $\left|L_1 + \frac{r-1}{2} L_2\right|$. Set

$$m = \left[\frac{d-1}{r-1}\right], \quad n = d - 1 - \left(\frac{r-1}{2}\right)\left(\left[\frac{d-1}{r-1}\right] + 1\right).$$

Let C be the union of the diagonal of X, m distinct members of $|L_1|$, and n distinct members of $|L_2|$. We number the nodes of C so that $p_1, \ldots, p_{m+n}$ are the ones lying on the diagonal of X. For any integer g such that

$$0 \leq g \leq \pi(d,r) = m \cdot n ,$$

we set

$$k = m + n + g .$$

Arguing as in the previous case, we find an irreducible curve $C_t \in |C|$ of geometric genus g with exactly $\pi(d,r) - g$ nodes as singularities. The degree of $C_t \in |(m+1)L_1 + (n+1)L_2|$ is easily seen to be d. Moreover, the assumption that $d \geq r$ implies that $m + 1 \geq 2$, so that C_t cannot lie in a hyperplane of $\mathbb{P}^r$. The last case, r even and $r \geq 4$, can be handled in a similar manner and is left as an exercise. As a hint, we suggest the reader to try to construct the desired curve on the plane blown up at one point and embedded in $\mathbb{P}^r$ by means of the very ample linear system $\left|E + \frac{r}{2} L\right|$, where E is the exceptional divisor, and L is the proper transform of a line passing through the center of the blow-up.

Q.E.D.

A natural question to ask in connection with Theorem (10.14) is whether there exists a *smooth* connected nondegenerate curve in $\mathbb{P}^r$ of any given degree d and genus g with

$$0 \leq g \leq \pi(d,r),$$

when $r \geq 3$. In general the answer is no. Indeed, by the discussion at the end of Chapter I, any curve whose genus is close to the Castelnuovo bound necessarily lies on $\binom{r-1}{2}$ independent quadrics. When $r = 3$, such a curve must lie on a quadric Q in $\mathbb{P}^3$, and it is immediate to check that, for instance, there is no smooth curve of genus $\pi(d,3) - 1$ on Q.

11. The Hurwitz scheme and its irreducibility.

In this section we shall first recall a fundamental result due to Lüroth, Clebsch, and Hurwitz, concerning the irreducibility of $\mathcal{M}^1_{g,d}$. We shall closely follow Fulton's treatment [273]. Let C be a smooth curve of genus g. We recall that a ramified d-sheeted covering

$$f : C \to \mathbb{P}^1$$

is called *simple* if every ramification point of f has index equal to 2 and no two ramification points of f lie over the same point of $\mathbb{P}^1$. We shall also say that the corresponding g^1_d is *simple*. We denote by

$$R(f) \in \mathrm{Div}(C)$$

the ramification divisor of f and by

$$\Lambda(f) \in \mathrm{Div}(\mathbb{P}^1)$$

the branch locus of f, so that

$$\Lambda(f) = f_*\big(R(f)\big)$$

if f is simple. Let

$$w = 2d + 2g - 2$$

be the degree of the branch locus of f (w is the traditional symbol for the degree of the branch locus). Denote by $\mathbf{P}_w$ the open subset of the wth symmetric product of $\mathbb{P}^1$ consisting of unordered w-tuples of *distinct* points in $\mathbb{P}^1$. For any $A \in \mathbf{P}_w$, we let

$$H(d, A)$$

be the set of equivalence classes of d-sheeted ramified simple coverings

$$f : C \to \mathbb{P}^1 \tag{11.1}$$

such that $\Lambda(f) = A$. We recall that the covering (11.1) is said to be *equivalent* to a covering $f' : C' \to \mathbb{P}^1$ if there exists an isomorphism $\phi : C \to C'$ such that $f' \circ \phi = f$. We shall denote by $[f]$ the equivalence class of f.

We now recall Riemann's fundamental existence theorem. For this, we need to introduce a few definitions and some notation. If G and H are two groups, we set, as usual,

$$\mathrm{Hom}^{ext}(G, H) = \mathrm{Hom}(G, H)/\sim,$$

where $\lambda \sim \mu$ if there exists $h \in H$ such that $\mu = h\lambda h^{-1}$. We denote by $\mathfrak{S}_d$ the symmetric group on d letters. Now let $A = \{a_1, \dots, a_w\}$ be an element of $\mathbf{P}_w$ and choose a base point $y \in \mathbb{P}^1 - A$. We define a *standard system of generators* $\sigma_1, \dots, \sigma_w$ of $\pi_1(\mathbb{P}^1 \smallsetminus A, y)$ as in the following picture:

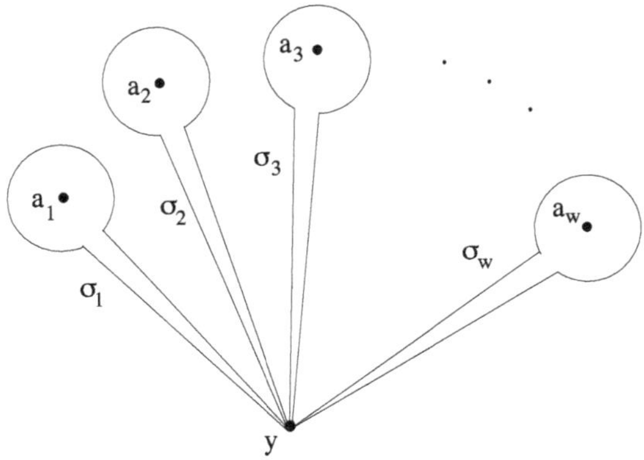

Figure 1.

We define the *monodromy representation*

$$\phi(f) : \pi_1(\mathbb{P}^1 \smallsetminus A, y) \to [\text{group of permutations of } f^{-1}(y)]$$

by mapping a loop σ in $\mathbb{P}^1 \smallsetminus A$ with base point at y to the permutation of $f^{-1}(y)$ obtained by associating to each point x of $f^{-1}(y)$ the end point of the unique lifting of σ^{-1} having initial point at x. Numbering the points of $f^{-1}(y)$ gives an identification between the group of permutations of $f^{-1}(y)$ and $\mathfrak{S}_d$, canonical up to inner automorphisms of $\mathfrak{S}_d$. Therefore, the monodromy representation induces a well-defined map

$$\phi : H(d, A) \to \mathrm{Hom}^{ext}(\pi_1(\mathbb{P}^1 \smallsetminus A, y), \mathfrak{S}_d), \tag{11.2}$$

obtained by associating to every element $[f]$ of $H(d, A)$ the class of $\phi(f)$. This being understood, we can state the following special instance of Riemann's existence theorem.

THEOREM (11.3) (RIEMANN'S EXISTENCE THEOREM). *The map (11.2) is injective. Moreover, the image of (11.2) consists of those classes which are induced by irreducible representations ξ such that*

$$t_i = \xi(\sigma_i)\,, \quad i = 1, \ldots, w\,,$$

is a transposition and $\prod t_i = 1$.

That t_i is a transposition is a reflection of the fact that we are dealing with simple coverings. In conclusion, via ϕ, we can identify $H(d, A)$ with the set

$$\mathcal{G}_{d,w} = \left[\begin{array}{c}\text{set of conjugacy classes } [t_1, \ldots, t_w] \text{ of } w\text{-tuples of} \\ \text{transpositions such that } \prod t_i = 1 \text{ and } t_1, \ldots, t_w \\ \text{generate a transitive subgroup of } \mathfrak{S}_d\end{array}\right].$$

We then consider the set

$$H(d,w) = \coprod_{A \in \mathbf{P}_w} H(d,A) .$$

Given an element $[f] \in H(d,A)$, we say that $[t_1, \ldots, t_w]$ is *the symbol of* $[f]$ *with respect to the basis* $\{\sigma_1, \ldots, \sigma_w\}$. There is a natural map

$$\Lambda : H(d,w) \to \mathbf{P}_w$$

given by

$$[f] \to \Lambda(f) .$$

Since the fibers of Λ may all be identified with the finite set $\mathcal{G}_{d,w}$, the set $H(d,w)$ can be equipped with a unique complex structure which makes $H(d,w)$ into a w-dimensional complex manifold and Λ into a topological covering. We shall call $H(d,w)$ the *Hurwitz space of type* (d,w). It follows from the construction that there is, over the Hurwitz space, a family of d-sheeted simple coverings, in the following sense. There exist a complex manifold X and smooth maps

$$\begin{array}{ccc} X & \xrightarrow{\ F\ } & \mathbb{P}^1 \\ {\scriptstyle\psi}\downarrow & & \\ H(d,w) & & \end{array}$$

such that, for each $s \in H(d,w)$, the fiber $\psi^{-1}(s) = X_s$ is a smooth curve. Moreover,

$$F_s = F\big|_{X_s} : X_s \to \mathbb{P}^1$$

is a simple d-sheeted ramified covering such that $[F_s] = s$. Clearly, by Riemann–Hurwitz's formula, X_s is a curve of genus g, where

$$w = 2g + 2d - 2 .$$

Therefore, there is a natural morphism

$$H(d,w) \to M_g \tag{11.4}$$

given by

$$s \mapsto [X_s] .$$

The fundamental result concerning $H(d,w)$ is the following.

THEOREM (11.5) (LÜROTH–CLEBSCH–HURWITZ). *$H(d,w)$ is connected.*

Proof. Consider the topological covering

$$\Lambda : H(d, w) \to \mathbf{P}_w$$

and let $A = \{a_1, \ldots, a_w\}$ be a point in $\mathbf{P}_w$. The proof of Theorem (11.5) consists in showing that $\pi_1(\mathbf{P}_w, A)$ acts transitively on the fiber

$$\Lambda^{-1}(A) = H(d, A).$$

To show this, let us consider loops

$$\Gamma_i, \quad i = 1, \ldots, w,$$

in $\mathbf{P}_w$ with endpoints at A of the form

$$\Gamma_i(t) = \{a_1, \ldots, a_{i-1}, \gamma_i(t), \gamma_i'(t), a_{i+2}, \ldots, a_w\},$$

where

$$\gamma_i, \gamma_i' : [0, 1] \to \mathbb{P}^1 \setminus \{a_1, \ldots, a_{i-1}, a_{i+2}, \ldots, a_w\}$$

are arcs such that

$$\gamma_i(0) = \gamma_i'(1) = a_i,$$
$$\gamma_i(1) = \gamma_i'(0) = a_{i+1},$$

as in Figure 2. It suffices to show that the subgroup Γ of $\pi_1(P, A)$ generated by the Γ_i acts transitively on $H(d, A)$. The following picture shows how to interpret the action of Γ_i on the set $\mathcal{G}_{d,w}$ to which $H(d, A)$ has been identified.

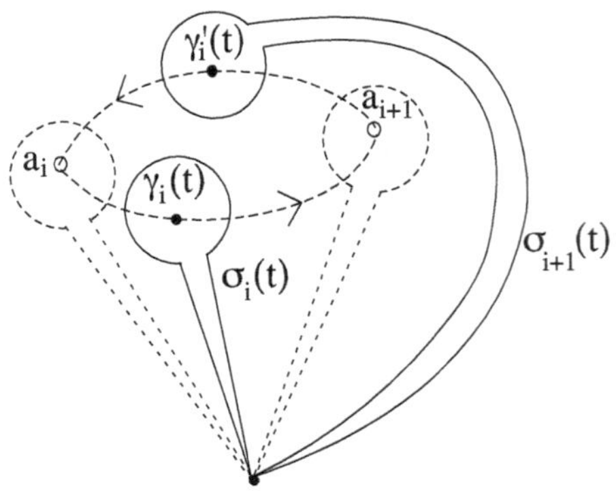

Figure 2. How Γ_i acts.

As t varies between 0 and 1, we have two varying loops $\sigma_i(t)$ and $\sigma_{i+1}(t)$ with

$$\sigma_i(0) = \sigma_i, \qquad \sigma_i(1) = \sigma_{i+1},$$
$$\sigma_{i+1}(0) = \sigma_{i+1}, \qquad \sigma_{i+1}(1) \sim \sigma_{i+1}\sigma_i\sigma_{i+1}^{-1}.$$

Thus, if we start at $t = 0$ with a d-sheeted covering having symbol $[t_1, \ldots, t_w]$ with respect to the basis $\{\sigma_1, \ldots, \sigma_i, \sigma_{i+1}, \ldots, \sigma_w\}$, we end up, at $t = 1$, with a d-sheeted covering having symbol $[t_1, \ldots, t_w]$ with respect to the basis $\{\sigma_1, \ldots, \sigma_{i-1}, \sigma_{i+1}(1), \sigma_{i+1}, \sigma_{i+2}, \ldots, \sigma_w\}$ or, equivalently, with a d-sheeted covering having symbol $[t_1, \ldots, t_{i-1}, t_i t_{i+1} t_i, t_i, t_{i+2}, \ldots, t_w]$ with respect to the basis $\{\sigma_1, \ldots, \sigma_i, \sigma_{i+1}, \ldots, \sigma_w\}$. Thus, the action of Γ_i on an element

$$[t_1, \ldots, t_w] \in \mathcal{G}_{d,w}$$

is as follows:

$$\Gamma_i \cdot [t_1, \ldots, t_w] = [t_1, \ldots, t_{i-1}, t_i t_{i+1} t_i, t_i, t_{i+2}, \ldots, t_w].$$

Thus,

$$\Gamma_i^{-1} \cdot [t_1, \ldots, t_w] = [t_1, \ldots, t_{i-1}, t_{i+1}, t_{i+1} t_i t_{i+1}, t_{i+2}, \ldots, t_w].$$

It is now a combinatorial exercise to show that the subgroup Γ of $\pi_1(P, A)$ acts transitively on $\mathcal{G}_{d,w}$. Following the treatment of Enriques [215] (Vol. III, Libro V, Cap. 1, page 26), we will show that the orbit of any element of $\mathcal{G}_{d,w}$ under the action of Γ contains the element

(11.6)
$$[\underbrace{(1\ 2), (1\ 2), \ldots, (1\ 2)}_{2g+2 \text{ times}}, (2\ 3), (2\ 3), (3\ 4), (3\ 4), \ldots, (d-1\ d), (d-1\ d)]\,,$$

where $(\alpha\ \beta)$ stands for the simple transposition that exchanges α with β. The proof will be broken up in several steps. Denote by G_1 the set of transpositions

$$(1\ \alpha)\,, \qquad 1 < \alpha\,,$$

and by G_1' its complement in the set of all simple transpositions. Let τ be an element of $\mathcal{G}_{d,w}$.

LEMMA (11.7). *There is an $h \geq 1$ such that the orbit of τ contains an element of the form*

$$[t_1, \ldots, t_w]$$

with

$$t_1, \ldots, t_h \in G_1\,, \qquad t_{h+1}, \ldots, t_w \in G_1'\,.$$

Such an h is necessarily even.

Proof. Write $\tau = [s_1, \ldots, s_w]$. Since the subgroup generated by the s_i is transitive, $s_j = (1\ \alpha)$ for some j. If $j > 1$, Γ_{j-1}^{-1} acts by moving s_j one place to the left without affecting the other s_i, except the one immediately preceding it. This kind of move can be applied repeatedly to push s_j all the way to the left. The result is $\tau' = [s_1', \ldots, s_w']$, where $s_1' = s_j = (1\ \alpha)$. Now we can apply the same procedure to $s_2', \ldots, s_w'$, and so on, until the desired configuration is reached. That h is even follows from the fact that $\prod t_i = 1$, and hence

$$t_1 \cdots t_h = 1,$$

since $t_1, \ldots, t_h$ are the only transpositions in τ involving 1.

Q.E.D.

LEMMA (11.8). *Let h be the minimum integer such that (11.7) holds. Then*

$$t_1 = \cdots = t_h .$$

Proof. We know that $h > 1$. If we had

$$t_{h-1} = (1\ \alpha), t_h = (1\ \beta), \quad \alpha \neq \beta ,$$

applying Γ_{h-1}^{-1} would yield

$$[t_1, \ldots, t_{h-2}, t_h, (\alpha\ \beta), \ldots],$$

contradicting the minimality of h. Arguing by induction, we will be done if we can show that, whenever $t_{h-k+1} = \cdots = t_h$, with $k \geq 2$, then $t_{h-k} = t_h$ as well. If k is even, applying $\Gamma_{h-1}^{-1} \cdots \Gamma_{h-k}^{-1}$ to $[t_1, \ldots, t_w]$ yields

$$[t_1, \ldots, t_{h-k-1}, t_{h-k+1}, \ldots t_h, t_{h-k}, t_{h+1}, \ldots, t_w] .$$

By the previous argument, $t_{h-k} = t_h$. If k is odd, applying $\Gamma_{h-2}^{-1} \cdots \Gamma_{h-k}^{-1}$ to $[t_1, \ldots, t_w]$ yields

$$[t_1, \ldots, t_{h-k-1}, t_{h-k+1}, \ldots t_{h-1}, t_{h-k}, t_h, \ldots, t_w] ,$$

and again one concludes that $t_{h-k} = t_h$.

Q.E.D.

Using (11.8) and conjugation, if necessary, we may assume that (11.7) holds with

$$t_1 = \cdots = t_h = (1\ 2) .$$

Obviously,

$$t_{h+1} \cdots t_w = 1 ,$$

and $t_{h+1}, \ldots, t_w$ act transitively on $\{2, \ldots, d\}$. It then follows by induction that the orbit of τ contains an element of the form $[s_1, \ldots, s_w]$ with

$$\begin{aligned} s_1 = \cdots &= s_h = (1\ 2) , \\ s_{h+1} = \cdots &= s_{h+h_2} = (2\ 3) , \\ &\vdots \\ s_{h+h_2+\cdots+h_{d-2}+1} = \cdots &= s_{h+h_2+\cdots+h_{d-1}} = (d-1\ d) , \end{aligned}$$

where $h, h_2, \ldots, h_{d-1}$ are even and different from zero. We now want to reduce the number of (2 3)'s, (3 4)'s, etc., down to two. Suppose for example that

$$[s_1, \ldots, s_w] = [\ldots, (1\ 2), (1\ 2), (2\ 3), (2\ 3), (2\ 3), (2\ 3), \ldots] .$$

Apply Γ_h^{-1} to get:

$$[\ldots,(1\ 2),(2\ 3),(1\ 3),(2\ 3),(2\ 3),(2\ 3),\ldots];$$

then Γ_{h+1}:

$$[\ldots,(1\ 2),(2\ 3),(1\ 2),(1\ 3),(2\ 3),(2\ 3),\ldots];$$

then Γ_{h+2}:

$$[\ldots,(1\ 2),(2\ 3),(1\ 2),(1\ 2),(1\ 3),(2\ 3),\ldots];$$

then $\Gamma_{h+1}^{-1}\Gamma_h^{-1}$:

$$[\ldots,(1\ 2),(1\ 2),(1\ 2),(2\ 3),(1\ 3),(2\ 3),\ldots];$$

then Γ_{h+2} to get, at long last:

$$[\ldots,(1\ 2),(1\ 2),(1\ 2),(1\ 2),(2\ 3),(2\ 3),\ldots].$$

By repeatedly applying this procedure we finally deduce that the orbit of τ contains the element (11.6), where the number of (1 2)'s must be $2g+2$ by the Riemann–Hurwitz formula. This concludes the proof of (11.5).

Q.E.D.

A graphical representation of a d-sheeted ramified simple covering of $\mathbb{P}^1$ of type (11.6) can be given as follows.

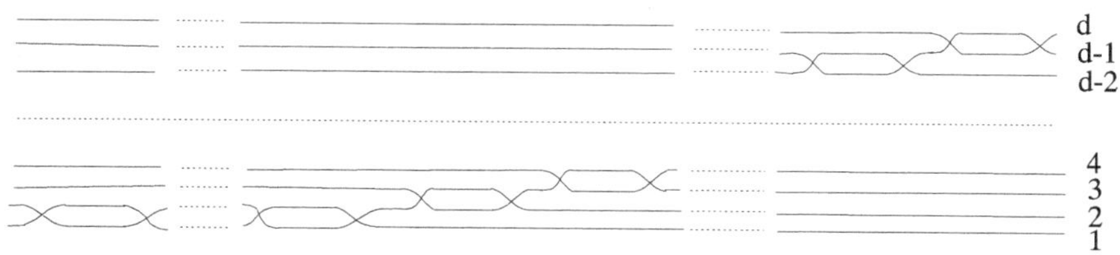

Figure 3.

A simple consequence of Theorem (11.5) is the following.

Corollary (11.9). *$M^1_{g,d}$ and M_g are irreducible.*

Proof. By the Existence Theorem (2.3) in Chapter VII,

$$M_g = M^1_{g,d}$$

as soon as $d \geq (g/2)+1$. Therefore, the second assertion is a special case of the first one, and we may assume that $d \leq (g/2)+1$. Let X be the image of the Hurwitz space $H(d, 2d+2g-2)$ in M_g via (11.4). Obviously,

$$X \subset M^1_{g,d}\,.$$

In view of Theorem (11.5), it suffices to show that $M^1_{g,d}$ is in fact the closure of X in M_g. We shall prove that *the general point of any component of $\mathcal{W}^1_d$ corresponds to a smooth curve C together with a complete, base-point-free, simple g^1_d*. That such a point corresponds to a (complete) g^1_d follows from the fact that, on any smooth curve C, no component of $W^1_d(C)$ is entirely contained in $W^2_d(C)$ (cf. Lemma (3.5), Chapter IV). The fact that such a g^1_d is base-point-free follows from a simple dimension count. In fact, we know (cf. Proposition (6.8)) that $\mathcal{W}^1_d$ is of pure dimension

$$\dim \mathcal{W}^1_d = 2d + 2g - 5\,,$$

whereas the dimension of the sublocus of $\mathcal{W}^1_d$ corresponding to g^1_d's with r base points is not greater than

$$\dim \mathcal{W}^1_{d-r} + r = 2d + 2g - 5 - r < \dim \mathcal{W}^1_d\,.$$

To show that the general g^1_d is simple, we proceed as follows. Let

$$\phi : C \to \mathbb{P}^1 \tag{11.10}$$

be the corresponding d-sheeted ramified covering. Suppose that ϕ has a ramification point p of index $h > 2$, so that in a neighborhood of p the map ϕ is given by

$$w = z^h.$$

Let

$$s \in H^0(C, N_\phi)$$

be a section of the normal sheaf to ϕ which vanishes to order exactly equal to 1 at p (this makes sense in the present context). Since

$$H^1(C, N_\phi) = 0\,,$$

s is the Horikawa class of an effective one-parameter deformation (cf. Theorem (8.15) or Remark (8.20))

$$\{\phi_t : C_t \to \mathbb{P}^1,\, t \in \mathbb{C},\, |t| < \delta\}.$$

In local coordinates near p, ϕ_t is given by

$$w = z^h + t(a_1 z + \ldots + a_{h-2} z^{h-2}) + [2]\,,$$

where $a_1 \neq 0$, and, as usual, $[2]$ stands for terms of order at least 2 in t. Therefore,

$$\frac{dw}{dz} = h z^{h-1} + t(a_1 + \ldots + (h-2) a_{h-2} z^{h-3}) + [2]\,.$$

If t is sufficiently small and $t \neq 0$, this function of z has zeros of order strictly less than $h-1$ in a neighborhood of $z=0$. This contradicts the assumption that (11.10) is general.

Summing up, we have shown so far that all the ramification points of ϕ have index equal to two. Suppose now that two of them, p_1 and p_2, lie over the same point of $\mathbb{P}^1$. Arguing as before, we are led to a contradiction by taking an effective one-parameter deformation of ϕ corresponding to a section of N_ϕ which vanishes at p_1 but not at p_2.

Q.E.D.

We finally remark that *any smooth curve C can be realized as a d-sheeted ramified simple covering of $\mathbb{P}^1$ if d is sufficiently large.* This can be seen as follows. Let

$$\phi : C \to \mathbb{P}^r, \quad r \geq 3,$$

be an embedding of C as a curve of degree d. For this, it suffices that $d \geq 2g+1$. Project generically C into a $\mathbb{P}^2$. The image curve Γ will be a plane curve of degree d having only ordinary double points (nodes). Since Γ has only a finite number of flexes and bitangents, a general projection of Γ to a $\mathbb{P}^1$ yields a representation of C as a d-sheeted ramified simple covering of $\mathbb{P}^1$.

12. Plane curves and g^1_d's.

In Section 6 we showed that, in genus $g \geq 2$,

$$\dim \mathcal{W}^1_d = 2d + 2g - 5 \quad \text{if } 2 \leq d < g+2$$

(cf. Proposition (6.8). Thus,

$$\dim M^1_{g,d} \leq 2d + 2g - 5\,. \tag{12.1}$$

Our main goal in this section is to show that equality holds in (12.1) when the right-hand side is less than or equal to $3g-3$, i.e., in case $d \leq g/2+1$. This will be achieved as follows. Let $p : \mathcal{C} \to B$ be a Kuranishi family of genus g smooth curves and set $\mathcal{W}^r_d = \mathcal{W}^r_d(p)$, according to the general conventions (3.16). Then it will suffice to show that there is a point ℓ in $\mathcal{W}^1_d - \mathcal{W}^2_d$ (which, by Proposition (6.8), is the smooth locus of $\mathcal{W}^1_d$) such that

$$d\eta : T_\ell(\mathcal{W}^1_d) \to T_{\eta(\ell)}(B) \tag{12.2}$$

is injective, where $\eta : \mathcal{W}^1_d \to B$ is the natural forgetful morphism. In fact, the moduli map $m : B \to M_g$ is finite-to-one, and we proved in Section 11 that $M^1_{g,d}$ is irreducible. We will then get the following result.

THEOREM (12.3). *If $2 \le d \le g/2+1$, then $M^1_{g,d}$ is irreducible of dimension*

$$\dim M^1_{g,d} = 2d + 2g - 5\,.$$

As we just pointed out, it suffices to find a point $\ell \in \mathcal{W}^1_d - \mathcal{W}^2_d$ such that (12.2) is injective. The point ℓ will correspond to a line bundle L of degree d over a smooth curve C of genus g such that $h^0(C,L) = 2$. The kernel of $d\eta$ is

$$\ker(d\eta) = T_L\big(W^1_d(C)\big).$$

By Proposition (4.2), Chapter IV,

$$T_L\big(W^1_d(C)\big) = (\mathrm{Im}\ \mu_0)^{\perp}.$$

Therefore, to show that this is zero, it suffices to show that, for L,

$$\begin{aligned}\dim(\ker\mu_0) &= -\rho \\ &= -(g - 2(g-d+1)) \\ &= g - 2d + 2\,.\end{aligned}$$

In view of the first part of Proposition (6.7), it then suffices to find a smooth curve C equipped with a degree d line bundle $L = \mathcal{O}(D)$ such that

i) $h^0\big(C, \mathcal{O}(D)\big) = 2$;
ii) $|D|$ has no base points;
iii) $h^0\big(C, \omega_C(-2D)\big) = g - 2d + 2$.

It has been proved by Beniamino Segre [618] that curves satisfying i), ii), and iii) can in fact be realized as plane curves of a very special kind. Here we will present a slightly modified version of Segre's constructions. In order to carry these out, it is necessary to use a beautiful result of Castelnuovo [108] about linear systems of curves on regular surfaces. Castelnuovo originally stated his theorem only for rational surfaces, but in fact the only consequence of rationality that is needed in the proof is the regularity of the surface.

We first need to introduce some notation and terminology. Let S be a smooth irreducible surface. Let Σ be a linear system on S without fixed components. We let

$$\phi_\Sigma : S \dashrightarrow S' \subset \mathbb{P}^r, \qquad r = \dim\Sigma\,, \ \ S' = \phi_\Sigma(S)\,,$$

be the rational map defined by Σ. We say that Σ is *irreducible* if its general member is an irreducible curve. Let Σ be an irreducible system, and $C \in \Sigma$ a general member. Then the *genus* $g(\Sigma)$ *of* Σ is defined to be the genus of the normalization of C. If Σ has s base points (both ordinary and infinitely near*) of multiplicities $\nu_1, \dots, \nu_s$, respectively, it is well known that

$$g(\Sigma) = \dim H^1(C, \mathcal{O}_C) - \sum_{i=1}^{s} \frac{\nu_i(\nu_i - 1)}{2}.$$

A positive divisor Γ on S is said to be *fundamental* for a linear system Σ if Γ is blown down by ϕ_Σ. Given a linear system Σ, a point $p \in S$ and a nonzero tangent vector $v \in T_p(S)$, we set

$$\text{(12.4)} \qquad \begin{aligned} \Sigma_p &= \{C \in \Sigma : p \in C\}, \\ \Sigma_{p,v} &= \{C \in \Sigma_p : v \in T_p(C)\}. \end{aligned}$$

Finally, given points $p_1, \dots, p_\delta$, we set

$$\text{(12.5)} \quad \Sigma_{2p_1,\dots,2p_\delta} = \{C \in \Sigma : C \text{ has multiplicity} \geq 2 \text{ at } p_i,\ i = 1, \dots, \delta\}.$$

In proving the following result we shall make use of standard material from the elementary theory of linear systems of curves on surfaces, such as Bertini's theorem and the Riemann–Roch formula. References for this are [213,610,690,55,52].

THEOREM (12.6) (Castelnuovo). *Let S be a smooth regular surface. Let Σ be an irreducible r-dimensional complete linear system on S, without fixed components and of genus g. Assume that there is a positive integer δ such that*

$$\delta \leq g \leq r - 2\delta - 1.$$

Let $p_1, \dots, p_\delta$ be general points on S. Then

i) $\dim \Sigma_{2p_1,\dots,2p_\delta} = r - 3\delta$;
ii) $\Sigma_{2p_1,\dots,2p_\delta}$ *is irreducible;*
iii) *The genus of $\Sigma_{2p_1,\dots,2p_\delta}$ is $g - \delta$.*

The intuitive motivation of the theorem is the following. Since it is three conditions for an irreducible curve to have a double point at a chosen point and since $r \geq 3\delta + 1$, one expects that there will be at least a pencil of curves of Σ having δ generically chosen double points. As a curve acquires a double point, either it becomes reducible, or its genus drops by one. Therefore, the condition $g \geq \delta$ is the obvious necessary condition to get an irreducible curve with δ nodes.

We first establish a number of lemmas.

LEMMA (12.7). *Let S be a smooth regular surface, and let Σ be a complete irreducible r-dimensional linear system of genus g on S. If $r \geq g + 2$, then ϕ_Σ is a birational map.*

---

*In classical terminology an infinitely near base point is an ordinary base point for the proper transform of Σ on a blowup of S.

Proof. Blowing up S, if necessary, we may assume that Σ has no base points. Let C be a general member of Σ. By Bertini's theorem, C is smooth. To prove the lemma, it suffices to show that the restriction of ϕ_Σ to C is not composed with an involution. Since S is regular, the sequence

$$0 \to H^0(S, \mathcal{O}_S) \to H^0(S, \mathcal{O}_S(C)) \to H^0(C, \mathcal{O}_C(C)) \to 0$$

is exact, so that

$$h^0(C, \mathcal{O}_C(C)) \geq r \geq g+2\,.$$

By the Riemann–Roch theorem, this implies that

$$(C)^{\cdot 2} \geq 2g+1\,. \tag{12.8}$$

An immediate consequence is that $h^1(C, \mathcal{O}_C(C)(-p-q)) = 0$ for any p, q in C, which shows that the restriction of ϕ_Σ to C is in fact an embedding.

Q.E.D.

LEMMA (12.9). *Let S, Σ, g, and r be as in the previous lemma. Let p be a general point on S, and $v \in T_p(S)$ a general nonzero tangent vector. Let Σ_p and $\Sigma_{p,v}$ be as in (12.4). Suppose that $r \geq g+3$. Then ϕ_{Σ_p} and $\phi_{\Sigma_{p,v}}$ are birational.*

Proof. Clearly, $\dim \Sigma_p = r-1$ and $\dim \Sigma_{p,v} = r-2$. Since Σ_p and $\Sigma_{p,v}$ can be viewed as complete linear series on suitable blow-ups of S, the result for Σ_p follows from Lemma (12.7), and so does the one for $\Sigma_{p,v}$ if $r > g+3$. There remains the case $r = g+3$. As in the proof of the previous lemma, we may assume that Σ has no base points and hence that a general member C of Σ is smooth. Let p be a general point of C, and let v be a nonzero tangent vector to C at p. Set $L = \mathcal{O}(C) \otimes \mathcal{O}_C$. By the Riemann–Roch theorem, $\deg L = 2g+2$. As we observed in the proof of Lemma (12.9), to show that $\phi_{\Sigma_{p,v}}$ is birational, it suffices to prove that its restriction to C is not composed with an involution. Since S is regular, $\Sigma_{p,v}$ cuts on C the linear system $|L(-2p)|$, which is composed with an involution exactly when C is hyperelliptic and

$$L(-2p) \cong \mathcal{O}(gD)\,, \tag{12.10}$$

where $|D|$ is the g^1_2 on C. On the other hand, (12.10) cannot hold for infinitely many p's in C. Therefore, $|L(-2p)|$ gives an embedding, by the generality of p.

Q.E.D.

LEMMA (12.11). *Let S be a smooth regular surface. Let Σ be an r-dimensional linear system on S without fixed divisors and such that ϕ_Σ*

is birational. Assume that Σ has positive genus. Let $\Sigma' \subset \Sigma$ be an $(r-1)$-dimensional linear subsystem all of whose members are reducible. Let Γ be the fixed divisor of Σ'. Set $\Sigma' = \Sigma'' + \Gamma$. Then

a) *Γ is a fundamental curve for Σ;*
b) *Σ'' is an irreducible system.*

Proof. After blowing up S, if necessary, we may assume that Σ has no base points. Let C be a general element of Σ. Since $\dim \Sigma' = r-1$, Σ is generated by Σ' and C. Since Σ has no base points, we must have $(\Gamma \cdot C) = 0$, proving a). Now suppose that a general member of Σ'' is reducible. By Bertini's theorem, there is then a pencil Λ such that every member of Σ'' is a sum of members of Λ. Since S is regular, all members of Λ are linearly equivalent. We now claim that $(F \cdot C) < 2$ if F is a member of Λ. Arguing by contradiction, suppose that this is not the case. Then, given a general point $p \in C$, there is a point $q \in C$ such that, if F is the unique member of Λ passing through p, then F cuts on C a divisor containing $p+q$. This means that every curve of Σ' passing through p also passes through q. But then the same is true for the linear system Σ, which is generated by Σ' and C. This contradiction shows that $(F \cdot C) = 1$. But this is also impossible, since F moves in a linear system of positive dimension and C has positive genus.

Q.E.D.

LEMMA (12.12). *Let S and Σ be as in the preceding lemma. Let p be a general point of S, and let Σ_p be as in (12.4). Then every positive divisor which is fundamental for Σ_p is also fundamental for Σ.*

Proof. As usual, we can assume that Σ has no base points. Let $\pi = \pi_p : \tilde{S} \to S$ be the blow-up map of S at p. Set $E = \pi^{-1}(p)$. Let C be a general member of Σ. We must show that, for any positive irreducible divisor Γ on $\tilde{S}$ with $(\Gamma \cdot (\pi^* C - E)) = 0$, one has $(\Gamma \cdot \pi^* C) = 0$. We first show that $(\Gamma \cdot E) < 2$. Suppose not. Then the divisor cut by Γ on E contains a divisor of the form $p_1 + p_2$. Since $((\pi^* C - E) \cdot \Gamma) = 0$, any member of $|\pi^* C - E|$ passing through p_1 contains Γ and hence also passes through p_2. This is absurd. In fact, since the point p is general on S, the linear system $|\pi^* C - E|$ separates points on E. If the conclusion of the lemma is false, there is an algebraic family $\{\Gamma_p\}_{p \in S}$ such that

$$(\Gamma_p \cdot E) = (\Gamma_p \cdot \pi_p^* C) = 1$$

and Γ_p is irreducible for general p. By construction, as p varies, the curves $\pi_p(\Gamma_p)$ are all algebraically equivalent and hence linearly equivalent, since S is regular. Also, $(\pi_p(\Gamma_p) \cdot C) = 1$. This is absurd since C is not rational.

Q.E.D.

Proof of Castelnuovo's theorem (12.6). Arguing by induction, we may as well assume that $\delta = 1$. By blowing up S, if necessary, we may also assume that Σ has no base points. Now let $p = p_1$ be a general point of S, and $v \in T_p(S)$ a general tangent vector. Let Σ_p, $\Sigma_{p,v}$, Σ_{2p} be as in (12.4) and (12.5). We have

$$\Sigma_p \supset \Sigma_{p,v} \supset \Sigma_{2p} .$$

Since p is a general point of S, we have

$$\dim \Sigma_p = r - 1, \tag{12.13}$$

$$\dim \Sigma_{p,v} = r - 2, \tag{12.14}$$

$$\Sigma_{p,v} \neq \Sigma_{p,v'} \text{ if } v \neq v'. \tag{12.15}$$

Since $\Sigma_{p,v} \supset \Sigma_{2p}$ for every v, (12.14) and (12.15) give $\dim \Sigma_{2p} \leq r - 3$. The reverse inequality is trivial. This proves i). We now prove ii). Suppose that Σ_{2p} is reducible. By i) and (12.14), Σ_{2p} has codimension 1 in $\Sigma_{p,v}$. Furthermore, since $r \geq g+3$ and p is general, $\phi_{\Sigma_{p,v}}$ is birational by Lemma (12.9). Therefore, by Lemma (12.11),

$$\Sigma_{2p} = (\Sigma_{2p})' + \Gamma,$$

where Γ is fundamental for $\Sigma_{p,v}$, and $(\Sigma_{2p})'$ is irreducible. This holds for every $v \in T_p(S)$. Therefore Γ is fundamental for Σ_p. By Lemma (12.12), Γ is then fundamental for Σ. In particular, $p \notin \Gamma$. This means that the curves of $(\Sigma_{2p})'$ have a double point (at least) at p. Let $\Sigma'' + \Gamma$ be the linear subsystem of Σ consisting of all curves of Σ that meet Γ. Since Σ'' contains $(\Sigma_{2p})'$, it is irreducible. Let $g' \leq g$ be the genus of Σ''. Then

$$\begin{aligned} \dim \Sigma'' &= r - 1 \geq g' + 2 , \\ \Sigma'' &\supset \Sigma''_{2p} \supset (\Sigma_{2p})' , \\ \dim (\Sigma_{2p})' &= r - 3 . \end{aligned}$$

By Lemma (12.7), $\phi_{\Sigma''}$ is birational. Since the curves of $(\Sigma_{2p})'$ have a double point at p, we get a contradiction by applying part i) to the system Σ''. This proves ii). Finally we prove iii). Let $\nu_1, \dots, \nu_s$ be the multiplicities of the base points of the irreducible system Σ_{2p}. Let C be a general element of Σ_{2p}. Let $\pi : \tilde{S} \to S$ be a blow-up of S for which the proper transform of Σ_{2p} has no base points. Let $\tilde{C}$ be the proper transform of C in $\tilde{S}$. Let $\tilde{g}$ be the genus of $\tilde{C}$. We must show that $\tilde{g} = g - 1$. We have

$$\begin{gathered} \tilde{g} = g - \frac{\sum \nu_i(\nu_i - 1)}{2} \leq g - 1 , \\ \dim |\tilde{C}| = \dim \Sigma_{2p} = r - 3 \geq g \geq \tilde{g} + 1. \end{gathered}$$

We also have

$$\tilde{C}^{\cdot 2} = C^{\cdot 2} - \sum \nu_i^2 .$$

By the Riemann–Roch theorem we then have

$$\begin{aligned} r - 3 = \dim |\tilde{C}| &= 1 + \tilde{C}^{\cdot 2} - \tilde{g} = 1 + C^{\cdot 2} - g - \frac{\sum \nu_i(\nu_i+1)}{2} \\ &= \dim |C| - \frac{\sum \nu_i(\nu_i+1)}{2} = r - \frac{\sum \nu_i(\nu_i+1)}{2} . \end{aligned}$$

This means that $\frac{\sum \nu_i(\nu_i+1)}{2} = 3$, proving that all the ν_i are equal to zero except one, which is equal to two.

Q.E.D

We now come to Beniamino Segre's existence theorem.

THEOREM (12.16) (B. Segre). *Let $g \geq 2$. For any integer d such that $2 \leq d \leq g/2 + 1$, there exists a smooth curve C of genus g carrying a complete g_d^1 $|D|$, without base points and such that*

$$\begin{aligned} h^0(C, \omega_C(-2D)) &= -(\textit{Brill–Noether number of the } g_d^1) \\ &= g - 2d + 2 . \end{aligned}$$

Proof. When $g = 2$, any C will do; it suffices to take as $|D|$ the canonical series. We therefore assume that $g \geq 3$. In this case C, or rather a birational model of it, will be constructed as a curve lying on a smooth quadric in $\mathbb{P}^3$, having nodes as singularities, and the g_d^1 will be cut out on C by one of the rulings of the quadric. Curves lying on quadrics have been encountered in Chapter III, Section 2, and provide a useful source of examples. Let Q be a smooth quadric in $\mathbb{P}^3$. Let $|L_1|$ and $|L_2|$ be the two rulings of Q. For every pair of integers m and n, consider the complete linear system

$$\Sigma_{m,n} = |mL_1 + nL_2|.$$

It is immediate to see that

$$\begin{aligned} \Sigma_{m,n} = \emptyset \text{ if } \min(m,n) < 0 , \\ \dim \Sigma_{m,0} = m, \dim \Sigma_{0,n} = n \text{ if } m \geq 0, n \geq 0 . \end{aligned}$$

Now suppose that $m > 0$, $n > 0$. It then follows from Bertini's theorem that $\Sigma_{m,n}$ is irreducible and that its general member is smooth. Recall that the canonical line bundle on Q is given by

$$\omega_Q = \mathcal{O}(-2L_1 - 2L_2).$$

Then, by adjunction, the genus of $\Sigma_{m,n}$ is

$$g_{m,n} = (m-1)(n-1).$$

Finally, by the Riemann–Roch theorem we have

$$\begin{aligned} \dim \Sigma_{m,n} &= mn + m + n \\ &= g_{m,n} + 2m + 2n - 1. \end{aligned} \tag{12.17}$$

We notice that, given any fixed integer $m \geq 2$, the intervals

$$I_{m,n} = [g_{m,n} - m - n + 1,\, g_{m,n}], \quad n \geq 1,$$

cover the half line $[1, \infty]$. This implies that, given integers g and d such that

$$2 \leq d \leq g/2 + 1\,,$$

there exists an integer $h \geq 4$ such that $g \in I_{d,h}$. Equivalently, there exist an integer $h \geq 4$ and an integer δ such that

$$0 \leq \delta \leq h + d - 1$$

and

$$g = g_{d,h} - \delta\,. \tag{12.18}$$

By (12.17), this implies that

$$\delta \leq g_{d,h} \leq \dim \Sigma_{d,h} - 2\delta - 1.$$

We may now apply Castelnuovo's theorem to the system $\Sigma_{d,h}$ and conclude that, given δ general points $p_1, \dots, p_\delta$ on Q, there exists an irreducible curve $\Gamma \in \Sigma_{d,h} = |dL_1 + hL_2|$ having δ nodes at $p_1, \dots, p_\delta$ and no other singularities. By construction, the genus of the normalization C of Γ is equal to $g = g_{d,h} - \delta$. The curve C comes naturally equipped with a g^1_d, namely the one cut out on Γ by the ruling $|L_2|$; let D be a divisor in the g^1_d. We will now compute $h^1(C, \mathcal{O}(\nu D))$ for each $\nu \geq 0$.

The residual series $|\omega_C(-\nu D)|$ is cut out on Γ by the (possibly empty) linear system $\Sigma_{d-2,h-\nu-2,\delta}$ consisting of those curves in $\Sigma_{d-2,h-\nu-2}$ which pass through $p_1, \dots, p_\delta$. Since $p_1, \dots, p_\delta$ are general points on Q, we have

$$\max(-1, \dim \Sigma_{d-2,h-\nu-2,\delta}) = \max(-1, \dim(\Sigma_{d-2,h-\nu-2}) - \delta). \tag{12.19}$$

From (12.17), (12.18), and (12.19) we conclude that

$$h^1(C, \mathcal{O}(\nu D)) = \begin{cases} \max(0, g - \nu d + \nu) & \text{if } 0 \leq \nu \leq h-2\,, \\ 0 & \text{if } \nu \geq h-1\,. \end{cases}$$

Since $2 \leq d \leq g/2+1$, we have, in particular,

$$h^1(C,\mathcal{O}(D)) = g-d+1\,,$$
$$h^1(C,\mathcal{O}(2D)) = g-2d+2\,.$$

The first equality tells us that the g^1_d is complete. The second one tells us that

$$h^0(C,\omega_C(-2D)) = -(\text{Brill–Noether number of the } g^1_d).$$

Q.E.D.

Returning to the discussion at the beginning of this section, this concludes the proof of Theorem (12.3). It may be amusing to remark that we proved that the morphism

$$m\circ\eta : \mathcal{W}^1_d \to M^1_{g,d}$$

is generically finite by going to a point of $M^1_{g,d}$ over which the fiber of $m\circ\eta$ is not finite (at least in some cases). In fact, referring to the proof of (12.16), notice that the two rulings of Q cut out on C a g^1_d and a g^1_h; it may very well happen that $h<d$. In this case the curve C has an "isolated" g^1_d and a continuous family of g^1_d's of the form $g^1_h + p_1 + \cdots + p_{d-h}$.

Segre's theorem implies that, when $d \leq g/2+1$, a smooth curve C which corresponds to a general point of $M^1_{g,d}$ carries only a finite number of g^1_d's. We shall now prove that in fact, when $d<g/2+1$, C carries a unique g^1_d. More precisely, we shall prove the following result (cf. [24]).

THEOREM (12.20). *Let d,g be integers such that*

$$2 \leq d < g/2+1\,.$$

Let C be a smooth curve of genus g corresponding to a general point of $M^1_{g,d}$. Then C carries a unique g^1_d. Moreover, if C carries a base-point-free g^1_h with $h<g/2+1$, then such a series is composed with the g^1_d.

Proof. We proceed by contradiction. Suppose that C carries a base-point-free g^1_h with $h<g/2+1$ which is not composed with the g^1_d. Let

$$f : C \to \mathbb{P}^1\,,$$
$$f' : C \to \mathbb{P}^1$$

be the morphisms corresponding to the g^1_d and g^1_h, respectively. Since C corresponds to a general point of $M^1_{g,d}$, we know that the g^1_d has no base points and, moreover, that f is a d-sheeted ramified simple covering (cf. the proof of Corollary (11.9)). Let

$$\phi : C \to \mathbb{P}^1 \times \mathbb{P}^1 = Q$$

be the morphism defined by $\phi = (f, f')$. We first show that ϕ *is not composed with an involution.* For this, we let $p_1 : Q \to \mathbb{P}^1$ be the projection onto the first factor so that $f = p_1 \circ \phi$. Set $\Gamma = \phi(C)$. Since f is simple, either Γ projects isomorphically onto $\mathbb{P}^1$ via p_1, or else ϕ is not composed with an involution. The first case cannot occur. If it did, we would have that $\phi(p) = \phi(q)$, and hence $f'(p) = f'(q)$ whenever $f(p) = f(q)$. Therefore, we would have $f' = \lambda \circ f$ for some ramified covering $\lambda : \mathbb{P}^1 \to \mathbb{P}^1$, and this would mean that the g^1_h is composed with the g^1_d, contrary to our assumption. Now consider the normal sheaf $N = N_\phi$ to the morphism ϕ and the basic exact sequence

$$0 \to \mathcal{K} \to N \to N' \to 0$$

associated to ϕ (cf. (9.1)). Recall (cf. (9.2)) that $h^1(C, N) = h^1(C, N')$. Since C corresponds to a general point of $M^1_{g,d}$, by Lemma (9.10) we have

$$h^0(C, N'_\phi) \geq \dim M^1_{g,d} + \dim(\mathrm{Aut}(Q)),$$

i.e.,

$$h^0(C, N'_\phi) \geq 2d + 2g + 1. \tag{12.21}$$

Since N'_ϕ is a line bundle, this implies that

$$h^1(C, N_\phi) = h^1(C, N'_\phi) = 0.$$

In turn, arguing as in the proof of (10.7), this implies that $\mathcal{K} = 0$, so that

$$N = N' \cong \det(\phi^*(T_Q)) \otimes \omega_C \cong \omega_C L^2 L'^2,$$

where $L = f^*\mathcal{O}(1)$, $L' = f'^*\mathcal{O}(1)$. Thus,

$$\deg N = 2g - 2 + 2d + 2h. \tag{12.22}$$

Using (12.21), (12.22), and the Riemann–Roch theorem, we get $2h - 2 \geq g$. This is a contradiction.

Q.E.D

13. Unirationality results.

In this section we will prove, among other things, the following classical result.

THEOREM (13.1). *M_g is unirational if $g \leq 10$.*

We recall that an irreducible algebraic variety X is said to be *unirational* if there exists a dominant rational map

$$g : \mathbb{P}^N \dashrightarrow X$$

for some N. As we mention in the bibliographical notes, at the time of writing this book, the space M_g is known to be unirational for $g \leq 14$. The situation in low genus contrasts with the one in high genus, where the following theorem, due first of all to Harris and Mumford [353], and then to Harris [348] and Eisenbud and Harris[204], holds.

THEOREM (13.2). *M_g is of general type when $g \geq 24$.*

We refer the reader to the bibliographical notes for a brief history of these results.

In order to prove Theorem (13.1), we come to another interesting aspect of Segre's theory of polygonal curves.

THEOREM (13.3) (B. Segre). *Let d and g be integers such that $3 \leq d \leq g$. Let C be a smooth curve of genus g equipped with a base-point-free g^1_d such that the corresponding d-sheeted ramified covering*

$$f : C \to \mathbb{P}^1$$

represents a general point of the Hurwitz space $H(d, w)$, where $w = 2g + 2d - 2$. Then, for any integer n such that

$$n \geq \frac{g + d}{2} + 1,$$

there exist a degree n plane curve Γ with an ordinary $(n - d)$-fold point p, $\delta = nd - \frac{d(d+1)}{2} - n + 1 - g$ nodes and no other singularities, and a birational map

$$\phi : C \to \Gamma \subset \mathbb{P}^2$$

such that the given g^1_d on C is cut out on Γ by the pencil of lines through p.

Following Segre, we will find it convenient to break up the proof of Theorem (13.3) into two propositions. The first result is really a mildly generalized version of Severi's existence theorem (10.13).

PROPOSITION (13.4). *Given integers g, d, n such that*

$$0 \leq g \leq nd - \frac{d(d+1)}{2} - n + 1,$$
$$2 < d < n,$$

there exists an irreducible plane curve Γ *of degree* n *and geometric genus* g *having an ordinary* $(n-d)$*-fold point* p*,*

$$\delta = nd - \frac{d(d+1)}{2} - n + 1 - g \tag{13.5}$$

nodes, and no other singularities. Moreover, if

$$n \geq \frac{g+d}{2} + 1\,, \tag{13.6}$$

it is possible to find a plane curve Γ *as above which has no adjoint* curve of degree* $n-4$ *having a point of multiplicity at least* $n-d$ *at* p*.*

Before proving (13.4) we shall state and prove a second proposition, which, combined with the previous one, actually yields a slightly strengthened version of Segre's Theorem.

PROPOSITION (13.7). *Let* g*,* d*,* n*,* δ *be as in Proposition (13.4). Suppose that (13.6) is satisfied. Let* V *be an irreducible component of the variety of irreducible plane curves of degree* n *and genus* g *with an ordinary* $(n-d)$*-fold point at* p *and* δ *nodes. Assume that* V *contains a point corresponding to a curve* Γ *which has all the properties listed in Proposition (13.4). Then projection from* p *defines a dominant rational map*

$$V \dashrightarrow H(d,w)\,,$$

where

$$w = 2g + 2d - 2\,.$$

Proof. Let X be the blow-up of $\mathbb{P}^2$ at p, and let

$$\sigma : X \to \mathbb{P}^1$$

be the projection from p. Let $\tilde{\Gamma}$ be the proper transform of Γ. Let

$$\phi : C \to X\,, \quad \tilde{\Gamma} = \phi(C),$$

*For a plane curve C having ordinary singularities, the classical term *adjoint curve* means a curve having multiplicity at least $\nu - 1$ at each ν-fold point of C. When C is irreducible, the adjoint curves of degree ℓ cut out on the normalization of C the complete linear system $|\omega((\ell - d + 3)D)|$, where D is a divisor cut out by a line. When C has arbitrary singularities we define adjoint curves in the same way but where infinitely near multiple points are included; if C is irreducible, the above result concerning the system cut out by adjoint curves is still valid.

be the normalization map for $\tilde{\Gamma}$. Since ϕ is unramified, we have that

$$N_\phi \cong \omega_C \otimes \phi^*(\omega_X)^{-1}.$$

Since $(\omega_X \cdot \tilde{\Gamma}) < 0$, we deduce that

$$H^1(C, N_\phi) = 0.$$

By Horikawa's theory there exists a local universal family of deformations of ϕ

$$\begin{array}{ccc} \mathcal{C} & \xrightarrow{\tilde{\phi}} & X \\ {\scriptstyle\pi}\big\downarrow & & \\ b_0 \in B & & \end{array} \qquad\qquad C = \pi^{-1}(b_0);\quad \phi = \tilde{\phi}|_C$$

where B is smooth, and $T_{b_0}(B) = H^0(C, N_\phi)$. Now consider the d-sheeted covering

$$f = \sigma \circ \phi : C \to \mathbb{P}^1.$$

Since the support of N_f is zero-dimensional,

$$H^1(C, N_f) = 0\,.$$

Again, we have a local universal family of deformations of f

$$\text{(13.8)} \qquad \begin{array}{ccc} \mathcal{C} & \longrightarrow & \mathbb{P}^1 \\ {\scriptstyle\lambda}\big\downarrow & & \\ a_0 \in A & & \end{array} \qquad\qquad C = \lambda^{-1}(a_0);\quad f = \tilde{f}|_C$$

where A is smooth, and $T_{a_0}(A) = H^0(C, N_f)$. Composing $\tilde{\phi}$ with σ yields a family of deformations of f parameterized by (B, b_0). By the universal property of (13.8), this family is induced (after shrinking B, if necessary) by a unique morphism

$$h : (B, b_0) \to (A, a_0).$$

The differential of h at b_0 is given by the homomorphism

$$dh : H^0(C, N_\phi) \to H^0(C, N_f)$$

induced by the corresponding map of sheaves in the commutative diagram

$$
\begin{array}{ccccccccc}
 & & & & 0 & & 0 & & \\
 & & & & \downarrow & & \downarrow & & \\
 & & & & L(\Delta) & = & L(\Delta) & & \\
 & & & & \downarrow & & \downarrow & & \\
0 & \longrightarrow & T_C & \longrightarrow & \phi^*(T_X) & \longrightarrow & N_\phi & \longrightarrow & 0 \\
 & & \| & & \downarrow{\scriptstyle d\sigma} & & \downarrow & & \\
0 & \longrightarrow & T_C & \longrightarrow & f^*(T_{\mathbb{P}^1}) & \longrightarrow & N_f & \longrightarrow & 0 \\
 & & & & \downarrow & & \downarrow & & \\
 & & & & 0 & & 0 & &
\end{array}
\tag{13.9}
$$

where L is the pullback to C of the hyperplane bundle of $\mathbb{P}^2$, and Δ is the pullback of the exceptional divisor on X. To see that the kernel of $d\sigma$ is actually $L(\Delta)$, it suffices to take determinants in the middle column of (13.9):

$$\ker(d\sigma) \cong f^*\omega_{\mathbb{P}^1} \otimes \phi^*\omega_X^{-1} \cong L^{-2}(2\Delta) \otimes L^3(-\Delta) \cong L(\Delta)\,.$$

We now come to the central point of the argument, namely to the proof that the homomorphism dh is surjective. It suffices to show that

$$h^1(C, L(\Delta)) = h^0(C, \omega_C \otimes L^{-1}(-\Delta)) = 0\,.$$

This is clear since $|\omega_C \otimes L^{-1}(-\Delta)|$ is cut out on Γ by adjoint curves of degree $n-4$ having a point of multiplicity at least $n-d$ at p and, by assumption, no such curve exists. We have thus proved that the morphism h is open. We have already remarked that a general point of A corresponds to a simple d-sheeted ramified covering of $\mathbb{P}^1$ (cf. the proof of (11.9)). This completes the proof of (13.7).

Q.E.D.

We now prove Proposition (13.4). Let $\ell_1, \dots, \ell_d$ be d general lines in $\mathbb{P}^2$, and let $\ell_{d+1}, \dots, \ell_n$ be $n-d$ general lines through p. Let

$$\tau : X \to \mathbb{P}^2$$

be the blow-up at p, and let $Y_1, \dots, Y_n$ be the proper transforms of $\ell_1, \dots, \ell_n$. Set

$$Y = Y_1 + \ldots + Y_n\,.$$

The divisor Y is reduced and has

$$\bar{\delta} = nd - \frac{d(d+1)}{2}$$

nodes. We number the nodes of Y in such a way that $p_1, \dots, p_{n-1}$ are the nodes lying on Y_1 and that

$$(13.10)\quad \{p_1, \dots, p_{3n-d-3}\} = \left\{ \begin{array}{c} \text{nodes belonging to } Y_1 \text{ or } Y_2, \text{ or nodes} \\ \text{of the form } Y_3 \cap Y_i,\ i = d+1, \dots, n \end{array} \right\}.$$

Set $k = n-1+g$ and notice that by assumption $\bar{\delta} \geq k$. Taking $p_{k+1}, \dots, p_{\bar{\delta}}$ to be the assigned nodes, by Lemma (10.15) and Remark (10.21) the linear system $|Y|$ contains an irreducible curve $\tilde{\Gamma}$ such that $\Gamma = \sigma(\tilde{\Gamma})$ is an irreducible plane curve of degree n and geometric genus g having an ordinary $(n-d)$-fold point at p,

$$\delta = \bar{\delta} - k$$

nodes, and no other singularity. So far (13.10) has not been used. In case assumption (13.6) is satisfied, we get

$$k \leq 3n - d - 3\,.$$

Set

$$q_i = \tau(p_i), \qquad i = 1, \dots, \bar{\delta}\,.$$

We wish to show that Γ can be constructed in such a way that it has no adjoint curves of degree $n-4$ with a point of multiplicity at least $n-d$ at p. By upper semicontinuity it suffices to show that there are no plane curves of degree $n-4$ passing through $p_{k+1}, \dots, p_{\bar{\delta}}$ and having a point of multiplicity at least $n-d$ at p. In fact, such a curve would contain n points of each of the lines ℓ_i, $i = 4, \dots, d$, and at least $n-1$ points of each of the lines $\ell_{d+1}, \dots, \ell_n$. It would then be a curve of degree $n-4$ containing $n-3$ distinct lines, which is absurd.

Q.E.D.

Remark (13.11). It is important to notice that when

$$2 < d < n\,,$$
$$\frac{g+d}{2} + 1 \leq n\,,$$
$$0 \leq g \leq nd - \frac{d(d+1)}{2} - n + 1\,,$$

Segre's argument actually shows that if

$$\delta = nd - \frac{d(d+1)}{2} - n + 1 - g$$

and $\{p, q_1, \dots, q_\delta\}$ is a general set of $\delta + 1$ points in $\mathbb{P}^2$, there is no curve of degree $n-4$ passing through $q_1, \dots, q_\delta$ and having a point of multiplicity at least $n-d$ at p.

We end this chapter by proving a unirationality theorem for some Hurwitz spaces. Our proof will be based on Segre's result (13.3) and on the following lemma, for a proof of which the reader is referred to [24].

LEMMA (13.12). *Let n, d, δ be positive integers such that*

$$n > d > 0\,,$$

$$\frac{(n-1)(n-2)}{2} - \frac{(n-d)(n-d-1)}{2} - \delta \geq 0\,,$$
$$\frac{n(n+3)}{2} - \frac{(n-d)(n-d+1)}{2} - 3\delta \geq 0.$$

Let $\{p, q_1, \ldots, q_\delta\}$ be a general set of $\delta + 1$ points in $\mathbb{P}^2$. Then there exists an irreducible plane curve of degree n having an ordinary $(n-d)$-fold point at p, nodes at $q_1, \ldots, q_\delta$, and no other singularity, with the exception of the case

$$n = 6, \quad d = 4, \quad \delta = 8\,.$$

In fact, the only sextic passing doubly through nine general points in $\mathbb{P}^2$ is a cubic counted twice.

The result we wish to prove is the following.

THEOREM (13.13). *Let d be an integer greater than or equal to 3. Then the Hurwitz space $H(d, 2d+2g-2)$ of simple, genus g, d-sheeted ramified coverings of $\mathbb{P}^1$ is unirational in each of the following cases:*

i) $d \leq 5,\ g \geq d-1$;
ii) $d = 6,\ 5 \leq g \leq 10$ *or* $g = 12$;
iii) $d = 7,\ g = 7\,.$

As a consequence, in each of these cases $M^1_{g,d}$ is unirational.

As the reader will certainly notice, some known cases of unirationality, as, for instance, the one of $H(2, 2g+2)$, are not covered by our statement. This is due to our particular method of proof.

To prove (13.13), let n be the minimum integer such that

$$n \geq \frac{g+d}{2} + 1$$

and set

$$\delta = nd - n - \frac{d(d+1)}{2} + 1 - g\,.$$

It is straightforward to verify that in each one of cases i)–iii) the assumptions of Lemma (13.12) are satisfied and that, moreover, the exception mentioned in the lemma cannot occur. Let $\{p, q_1, \ldots, q_\delta\}$ be a general set of $\delta + 1$ points of $\mathbb{P}^2$. Let Γ be the irreducible plane curve whose existence is guaranteed by Lemma (13.12). By Remark (13.11), Γ has no adjoint curves of degree $n-4$ with a point of multiplicity at least $n-d$ at p. Let V be the irreducible component of the variety of irreducible plane curves of degree n with an ordinary $(n-d)$-fold point at p and δ nodes which contains Γ. By (13.7) there is a dominant rational map from V onto $H(d, 2d+2g-2)$. On the other hand, by (13.12), there is a dominant rational map

$$f : V \dashrightarrow \mathrm{Sym}^{\delta}(\mathbb{P}^2)$$

given by

$$\Gamma \mapsto \{\text{nodes of } \Gamma\}.$$

The variety V is contained in the projective space $\mathbb{P}^N$ of plane curves of degree n, and the fibers of f are linear subspaces of this $\mathbb{P}^N$; therefore, V is unirational. This concludes the proof of (13.13).

Since, by the fundamental existence theorem (3.2) and by the proof of (11.9), a general curve of genus g can be represented as a simple d-sheeted ramified covering of $\mathbb{P}^1$ when

$$d \geq \frac{g+2}{2},$$

we get Theorem (13.1) from case ii) of Theorem (13.13).

14. Bibliographical notes and further reading.

For the general theory of the Picard functor, we refer the reader to Kleiman's beautiful Chapter 5 in [243] and its exhaustive bibliography. A natural problem, not treated in this book, is the one of compactifying the Picard varieties of stable curves. This has been the subject of intensive study starting with papers by D'Souza [157] and Altman and Kleiman [8,9], Oda and Seshadri [568], and Rego [605]. Among the many papers devoted to this problem, we may mention the ones by Caporaso [95], Pandharipande [583], Esteves [220], Esteves and Kleiman [222], Alexeev [6], and Melo [506]. A very useful overview on this subject and further bibliography can be found in Caporaso's paper [98].

Our treatment of Brill–Noether varieties over moduli and their tangent spaces is based on [23], [25], and [26]. What is modernly called Petri's conjecture is the statement by K. Petri quoted on p. 215 of our first volume. This statement was rediscovered by Arbarello and Sernesi [34]; this is where the μ_1 map, although not thus denoted, first appeared (see p. 215 in [34]).

The first result pointing to Petri's conjecture is the weaker statement to the effect that, for every line bundle on a general curve, the Brill–Noether number ρ is positive. This was proved by the third author and

Joe Harris in [319]. The first proof of Petri's conjecture is due to Gieseker [293]. Gieseker's argument was simplified by Eisenbud and Harris in [196]. The three papers [319], [293], and [196] are all based on a degeneration argument by considering families of smooth curves degenerating to either a g-noded rational curve [293] or to suitable stable curves possessing g elliptic tails [196].

The first proof of Petri's conjecture without degeneration to a singular curve is due to Lazarsfeld [464], who proved the theorem for curves lying on a K3 surface. For $g > 11$, these curves are far from being general curves in the sense of moduli. The idea that, in some sense, curves on a K3 surface behave like general curves has been very fruitful. In a different context it was used by Voisin in her breakthrough paper on Green's conjecture about syzygies of canonical curves [674] (see also [675] and [16]). Building on Voisin's remarks concerning the μ_1 map [672], Pareschi [588] considerably simplified Lazarsfeld's proof of Petri's conjecture; this is the proof we present in Section 7.

The link between the μ_1 map and first-order obstructions to deformations of triples (curve, line bundle, space of sections) was established in [23]. The first author proposed to study the higher μ's which are introduced in Exercises B-1–B-3. The link between the higher μ's and higher obstructions to deformations of triples (curve, line bundle, space of sections) was extensively studied by the three authors of this book without success. The question has been revived and studied by Clemens in [126,127].

There is a completely straightforward generalization of the μ_1 map. Given two line bundles L and L' on a smooth curve C, let $\mu_{0,L,L'} : H^0(C,L) \otimes H^0(C,L') \to H^0(C,LL')$ be the multiplication map and define

$$\mu_{1,L,L'} : \ker \mu_{0,L,L'} \to H^0(C,\omega_C LL') \tag{14.1}$$

by setting $\mu_{1,L,L'}(\sum_i s_i \otimes t_i) = \sum_i (s_i dt_i - t_i ds_i)$. This map was studied by Wahl in [678] for the case $L = L' = \omega_C$ and in [679] for the general case; it is often referred to as the *Gaussian map* or the *Wahl map*. Fundamental progress on the properties of this map were made by Ciliberto, Harris, and Miranda [122] and by Voisin [672] (see also [585], [569], [400], [123], [85], and [587]).

In our book we study only linear series on smooth curves. The idea of using degeneration to singular curves in order to prove properties of smooth curves goes back at least to Max Noether, Poincaré, and the Italian school of algebraic geometry. When dealing with linear series, there are many difficulties in making these arguments rigorous. In the case of one-dimensional linear series many of these difficulties can be circumvented by using admissible covers, which were first introduced by Beauville [54] (see also Knudsen and Mumford [425,426,427], Harris and Mumford [353], Chapter 3, Section G of Harris and Morrison [352],

Abramovich, Corti, and Vistoli [2], and our Section 5 in Chapter XVI). However, for higher-dimensional linear series, degeneration arguments are vastly more complicated. The first modern treatment is due to Kleiman, who uses Hilbert-scheme techniques to tackle the problem.

A different point of view is taken by Eisenbud and Harris in their theory of limit linear series [199]. Their method is very powerful and leads to a number of fundamental results, such as an alternative proof that M_g is of general type for $g \geq 24$, the existence of Weierstrass points with given semigroup, the monodromy of Weierstrass points over moduli, the one of zero-dimensional Brill–Noether varieties, and many more [198,202,195]. The theory of limit linear series has also been studied by Esteves and his collaborators, e.g., in [224] and [223] (see also [580] and [89]).

A third, natural approach to the study of linear series on singular curves is to define Brill–Noether subvarieties in the compactified Picard variety of, say, a stable curve, which is easier said than done. This construction, together with the first results on the Brill–Noether theory for stable curves, is given by Caporaso in [99].

Brill–Noether subvarieties of M_g, i.e., the subvarieties of M_g corresponding to curves carrying a linear series with negative Brill–Noether number ρ (or else having a special configuration of ramification points) have been extensively studied from the point of view of their existence, of their dimension, and of their cycle class. The case $\rho = -1$ is solved by Eisenbud and Harris in [205]. The case $\rho = -2$ is studied by Edidin in [188]. On a general curve one expects only finitely many linear series of given degree and dimension and with $\rho = 0$. Then the question of monodromy arises. This problem is addressed, and solved in the case of one-dimensional linear series, by Eisenbud and Harris in [200].

The computation of the cycle classes of Brill–Noether varieties in terms of tautological classes is of fundamental importance. The first step in this direction was taken by Mumford in [556], where he computed the cycle classes of the various strata of a flag of subvarieties of M_g defined in terms of Weierstrass points. Another instance of this type of computation can be found in the proofs by Harris and Mumford and by Eisenbud and Harris of their theorems on the Kodaira dimension of $\overline{M}_g$ (cf. [353] and [204]). Yet other instances of this type of computation appear in Diaz's work on the divisor of moduli space defined by exceptional Weierstrass points and in Cukierman's study of families of Weierstrass points (cf. [172], [152] and also Gatto's survey in [280]). Finally, Farkas in his study of the Gieseker–Petri divisor and Farkas and Popa in their study of the effective and ample cones of $\overline{M}_g$ perform a large number of very instructive computations of cycle classes of Brill–Noether type (see [249], [253], [252],[257], and [257]).

The Brill–Noether theory for vector bundles has been studied through the works of Narasimhan, Seshadri, Mukai, Le Potier, Hirschowitz, Newstead, Teixidor i Bigas, Ballico, and many others, leading to an explosive development (see, for example, [44], [45], [46], [68], [70], [82], [83], [84], [87], [211], [289], [364], [417], [461], [468], [470], [500], [507], [508], [537], [539], [649], [596], [628], [651], [656], [657], [658], [659], [660], [661], and [687]). A nice overview of the Brill–Noether theory for vector bundles on curves is given by Grzegorczyk and Teixidor i Bigas in [81]. For the general theory of moduli of vector bundles on a fixed curve, we refer the reader to Le Potier [468,469], Ramanan [600], Seshadri [628,629], and to Chapter 10 in Mukai's book [541].

Looijenga's vanishing theorem appears in [489]. The fact that $g-2$ is a bound for the dimension of a complete subvariety of M_g (cf. Theorem (6.4), Chapter XVII) was originally established by Diaz in series of papers centered around the geometry of the moduli spaces of curves carrying a special Weierstrass point [169,172,170,171,173]. A nice discussion of Diaz's theorem is in Chapter 6, section B of [352]. Complete subvarieties of the moduli space of principally polarized abelian varieties are studied in [234]. The reader can find a good survey of results on moduli of abelian varieties in the paper [282] by van der Geer and Oort.

Weierstrass points have attracted the attention of many authors throughout a long period of time. From the point of view of existence of Weierstrass points with a given gap sequence, after the work of Pinkham [595] and Rim and Vitulli [606] the best results have been obtained by Eisenbud and Harris in [202]. The dimension of the subloci of M_g defined in terms of Weierstrass points was first determined by Rauch [603]. A flag of irreducible Weierstrass subvarieties of M_g was introduced in [19] (see also [556], Section 7). Weierstrass points on moving curves, their limits on stable curves and the cycles they define in the moduli space are studied by Lax, Laufer, Diaz, Cukierman, Laksov, Thorup, Esteves, Gatto, and many others (see, for example, [463], [458], [455], [171], [170], [172], [173], [153], [456], [457], [279], [280], [223], [131], [281], [225], and [156]).

Horikawa's theory is developed in [367], [368], [369], and [370]. The theorem of de Franchis on the finiteness of the number of nonconstant morphisms from a fixed curve to another, possibly variable, curve of genus ≥ 2 first appeared in [269]; it had been previously shown by Castelnuovo [110] and Humbert [375,376] that such morphisms are rigid. Our treatment is based, with modifications, on the ones by Samuel [611] and Howard and Sommese [371].

The decomposability of the normal bundle to a projective curve C and its link to the curve C being a complete intersection is studied by Griffiths-Harris and Harris-Hulek in [320] and [350].

Severi's theorem on the existence of irreducible plane nodal curves of given degree and genus appears in [633], Anhang F, and is also discussed in Zariski's book [690]. The existence of irreducible plane curves with given number of nodes and cusps and their deformation theory is studied, among others, by Wahl [677], Tannenbaum [652,654], and Zariski [691]. Our discussion is based on [23], [24], and [25]. The theory culminated in Harris' proof of the long-standing Severi conjecture [349]. The geometry of the Severi variety has been studied by Diaz and Harris [175,176].

Curves carrying a theta characteristic of dimension $r \geq 1$ form a subvariety of M_g. These subvarieties and their dimensions have been studied, among others, by Teixidor i Bigas [655], Nagaraj [561], Colombo [128], Fontanari [266], and Farkas [249].

Brill–Noether varieties have been studied from the point of view of the Clifford index by Coppens and Martens [134,135,136,137,138], Coppens, Keem, and Martens [132,133], and Eisenbud, Lange, Martens, and Schreyer [206]. The Clifford index is at the center of yet another very important development in the theory of linear series on algebraic curves. This entire new field was initiated by Green and Lazarsfeld [312,313] and led to their well-known conjectures connecting the Clifford index of a smooth curve to the syzygetic resolution of its canonical ring via Koszul cohomology. In that direction we already mentioned Voisin's breakthrough papers [674,675], but for the vast literature on this subject, we refer the reader to the beautiful book by Aprodu and Nagel [14] and its exhaustive bibliography.

The Hurwitz scheme was introduced by Fulton [273] to give an algebraic proof of the irreducibility of the moduli space of curves of given genus. Our discussion is partly based on his paper. The combinatorial argument due to Lüroth and Clebsch is taken from the book [214] by Enriques and Chisini.

In general, there at least two main difficulties in studying the geometry of the Brill–Noether varieties $\mathcal{G}^r_d$, $\mathcal{W}^r_d$, and $M^r_{g,d}$. First of all, they are singular, but, most importantly, it is very hard to extend their definition, in a tractable way, to the case of stable curves. This difficulty is in fact the main motivation for Eisenbud and Harris' theory of limit linear series, which, in essence, shortcuts the problem of compactification.

However, when $r = 1$, the situation is more favorable. One advantage is that $\mathcal{G}^1_d$ is smooth, as we proved in Proposition (6.8). The other advantage is that, instead of working with $\mathcal{G}^1_d$ or $\mathcal{W}^1_d$, one may work with the Hurwitz schemes, which can be compactified using admissible covers. All of this makes the theory of one-dimensional linear series more manageable. Earlier in these bibliographical notes, we pointed out the many instances in which admissible covers proved to be an extremely useful tool, such as the papers by Beauville [54], Knudsen and Mumford [425,426,427], Harris and Mumford [353], Harris [348], Diaz [172], Harris and Morrison [351], and Abramovich, Corti, and Vistoli [2].

The (co)homology of Hurwitz spaces has been studied, among others, by Diaz and Edidin [174] and Kazarian and Lando [406].

Another way of compactifying the Hurwitz scheme is given by Ekedahl, Kazarian, Lando, and Vainshtein in [209]. In this paper, the four authors describe Hurwitz numbers, i.e., the number of ramified covers of the sphere with given degree and ramification, in terms of top intersections of the moduli space of curves. As we already observed in the bibliographical notes to Chapter XX, their work has been extremely influential and has been used, especially in the study of the intersection theory of $\overline{M}_{g,n}$, by many authors including Okounkov [570], Okounkov and Pandharipande [572,577], Shadrin [639], Zvonkine [695,694], Lando and Zvonkine [460], and Kazarian and Lando [407].

Another point of view from which one may study Hurwitz numbers is taken by Fantechi and Pandharipande [245] who use Gromov–Witten theory.

In the bibliographical notes to Chapter XVII we already mentioned the work of Goulden, Jackson, and Vakil [303,304,305], Bertram, Cavalieri, and Todorov [66], Cavalieri [113], and Liu and Xu [481]. All these papers involve, in one way or another, the Hurwitz scheme.

The idea for the proof of the unirationality of M_g for $g \leq 10$ is due to Severi [633] (see also Segre [619]). His argument was incomplete in more than one respect, and particularly because of the illicit (at that time) assumption of Harris' theorem on the irreducibility of the variety of curves of given degree and genus. A modern treatement of Severi's argument is given by Arbarello and Sernesi [35]. The proof given here is taken from [24], which was inspired by Segre [618]. As with many of his works, and certainly due to a change of style in the subject, Segre's paper contained fundamental ideas but was apparently overlooked.

The Bourbaki seminar [676] by Voisin and Farkas' paper [254] offer beautiful overviews of unirationality results for moduli spaces of curves. Many of these results are based on the so-called *Mukai's models* for M_g when $g \leq 9$ [534,536,538,540]. The unirationality of M_g has been established for $g \leq 14$, in genus 12 by Sernesi [622], in genera 11 and 13 by Chang and Ran [114], and in genus 14 by Verra [669] via an inventive argument which uses ideas close to one of Mukai [540]. A proof that M_{11} is uniruled is due to Mori and Mukai [517]. Bruno and Verra [90] prove that M_{15} is rationally connected. At the other extreme, in their breakthrough paper [353], Harris and Mumford showed that M_g is of general type for all odd $g \geq 25$; this was later extended to all $g \geq 24$ by Harris [348] and Eisenbud and Harris [204]. An alternative proof of these results is offered by Farkas [254], who uses the syzygetic resolution of the canonical ring of a curve and Koszul cohomology; in the same paper, Farkas proves that M_{22} is of general type as well.

At the time of this book's publication, there has been very interesting progress on the Kodaira dimension of moduli spaces. In [259], Farkas and

Verra prove that $\overline{\mathcal{C}}_{g,n}$ is unirational for $g < 10$ and $n < g$, that $\overline{\mathcal{C}}_{10,n}$ is uniruled for $n \neq 10$, that $\overline{\mathcal{C}}_{11,n}$ is uniruled for $n \neq 11$, that the Kodaira dimension of $\overline{\mathcal{C}}_{11,11}$ equals 19, and finally that the Kodaira dimension of $\overline{\mathcal{C}}_{g,g}$ equals $3g-3$ when $g \geq 12$.

In the same vein, Bini, Fontanari, and Viviani [71] prove that, if $(d+g-1$ and $2g-2$ are relatively prime, then the Kodaira dimension of $\mathbf{Pic}_g^d$ is equal to $-\infty$ for $4 \leq g \leq 9$, to 0 for $g = 10$, to 19 for $g = 11$, and to $3g-3$ for $g \geq 12$. All these results give a new, deeper meaning to the $g = 11$ threshold.

Genus g hyperplane sections of polarized K3 surfaces depend on $g+19$ parameters. A naive count of moduli predicts that the general curve of genus $g \leq 11$ should be realized as such a hyperplane section. As Mukai shows in [540], this is true, with the exception of the genus 10 case. In any event, for $g \geq 12$, a hyperplane section of a K3 surface is far from being a general curve of genus g. On the other hand, in Lazarsfeld's proof of Petri's conjecture we saw that hyperplane sections of K3 surfaces, from the point of view of special divisors, behave as curves with general moduli. This idea has been efficiently used by Voisin in her work on Green's conjecture. Curves on K3 surfaces are also used by Farkas and Popa [257] to give counter-examples to the slope conjecture. Good references and extensive bibliographies on this circle of ideas are again to be found in Voisin [676], Farkas [254], and Aprodu and Farkas [13].

15. Exercises.

A. Low genus

A-1. Do the Brill–Noether theory for genus $g \leq 5$.

B. Higher μ's

In this series of exercises we are introducing homomorphisms μ_k, $k \geq 0$, generalizing the homomorphisms μ_0 and μ_1.

B-1. Let L be a line bundle on a smooth curve C. Consider sections s_i of L and r_i of $\omega_C L^{-1}$. By covering C with coordinate disks U_α, with local coordinate z_α in each U_α, write $s_i = \{s_{i,\alpha}\}$ and $r_i = \{r_{i,\alpha}\}$ with $s_{i,\alpha}$ and $r_{i,\alpha}$ holomorphic in U_α and $i = 1, \ldots, l$. Suppose that

$$(15.1) \qquad \sum \left(\frac{d^h s_{i,\alpha}}{dz_\alpha^h} \right) r_{i,\alpha} = 0\,, \qquad h = 1, \ldots, k-1\,,$$

for every α. Show that the cochain

$$\left\{\left(\sum_i \left(\frac{d^k s_{i,\alpha}}{dz_\alpha^k}\right) r_{i,\alpha}\right) dz_\alpha^{k+1}\right\}$$

is, in fact, a section $\mu_k(\psi) \in H^0(C, \omega^{k+1})$ and that one gets an array of homomorphisms

$$\begin{array}{lccc} & H^0(C,L)\otimes H^0(C,\omega\otimes L^{-1}) & \xrightarrow{\mu_0} & H^0(C,\omega) \\ & \cup & & \\ & \ker\mu_0 & \xrightarrow{\mu_1} & H^0(C,\omega^2) \\ (15.2) & \cup & & \\ & \ker\mu_1 & \xrightarrow{\mu_2} & H^0(C,\omega^3) \\ & \cup & & \\ & \vdots & & \vdots \end{array}$$

B-2. Show that if (15.1) holds, then

$$\begin{aligned}\sum_i \left(\frac{d^k s_{i,\alpha}}{dz_\alpha^k}\right) r_{i,\alpha} &= -\sum_i \left(\frac{d^{k-1} s_{i,\alpha}}{dz_\alpha^k}\right) \frac{dr_{i,\alpha}}{dz_\alpha} = \cdots \\ &= (-1)^k \sum_i s_{i,\alpha} \frac{d^k r_{i,\alpha}}{dz_\alpha^k}\,.\end{aligned}$$

B-3. Looking at a suitable Wronskian, show that $\ker\mu_k = 0$ if $k \geq r$.

B-4. Let $\Delta : C \to C \times C$ be the diagonal morphism and set $D = \Delta(C)$. Denote by π_1, π_2 the projections of $C \times C$ onto the two factors and set

$$M = \pi_1^*(L) \otimes \pi_2^*(\omega \otimes L^{-1}).$$

Show that:
i) $\Delta^* M \simeq \omega$.
ii) The homomorphism

$$H^0(C,L)\otimes H^0(C,\omega\otimes L^{-1}) \cong H^0(C\times C, M) \to H^0(C,\omega)$$

induced by Δ is just μ_0.
iii) Show that the map μ_h can be identified, up to a multiplicative constant, with

$$H^0(C\times C, M(-hD)) \to H^0(C, \Delta^*(M(-hD))) \simeq H^0(C, \omega_C^{h+1})\,.$$

iv) Show that, under the identifications

$$H^0(C, L) \otimes H^0(C, \omega \otimes L^{-1}) \simeq H^0(C \times C, M)$$
$$H^0(C, \Delta^*(M(-kD))) \simeq H^0(C, \omega_C^{k+1}), \qquad k = 0, 1, 2, \dots,$$

$\ker \mu_{h-1} = H^0(C \times C, M(-hD))$, and μ_h is a nonzero multiple of the restriction map

$$\Delta^* : H^0(C \times C, M(-hD)) \to H^0(C, \Delta^* M(-hD)).$$

B-5. It is classical that curves on $C \times C$ can be interpreted as correspondences on C, and the exercises above suggest that we may take this point of view in analyzing the spaces $\ker \mu_h$. To this end, we set

$$C_{1,p} = \pi_1^{-1}(p) = \{p\} \times C,$$
$$C_{2,p} = \pi_2^{-1}(q) = C \times \{q\}$$

and agree to consider divisors on $C_{1,p}$, $C_{2,q}$ as divisors on $C \times C$. Consider a curve

$$T \in |M(-(k+1)D)|. \tag{15.3}$$

Show that, for general p and q, we have

$$\begin{cases} T \cdot C_{1,p} \in |L(-(k+1)p)|, \\ T \cdot C_{2,q} \in |\omega L^{-1}(-(k+1)q)|. \end{cases} \tag{15.4}$$

B-6. Show that the mapping

$$\mathbb{P}(\ker \mu_k) \to \left|\omega_C^{k+2}\right|$$

induced by μ_{k+1} sends each correspondence T satisfying (15.3) to the cycle of its united points, i.e., points p such that

$$p \in T \cdot C_{1,p}.$$

B-7. Consider a base-point-free pencil $|L|$ corresponding to a branched covering $f : C \to \mathbb{P}^1$. Define the symmetric correspondence $E = \{(p, q) : q \in f^{-1}(f(p)) - p\}$.

i) Show that

$$T = E + \sum_i C_{1,p_i}$$

is a sum of E plus "vertical" fibers.

ii) From the second condition in (15.4) we obtain

$$\sum_i p_i \in |\omega L^{-2}| . \tag{15.5}$$

Using exercise B-3 for $r = 1$, show that each member of the linear system $|M(-D)|$ consists of the fixed curve E plus the variable curve $\sum_i C_{1,p_i}$, where (15.5) is satisfied. Deduce that $\ker \mu_0 \simeq H^0(C, \omega L^{-2})$.

B-8. Now suppose $|L|$ is a base-point-free linear system of dimension 2 and degree d, and let

$$f : C \to \mathbb{P}^2$$

be the corresponding morphism.

i) Assume, for the moment, that f is an immersion. The exercise will consist in showing that

$$\ker \mu_1 = 0 .$$

Argue by contradiction and let $T \in |M(-2D)|$. The first of the two equations (15.4) reads $T \cdot C_{1,p} \in |L(-2p)|$. Define the tangential correspondence

$$E = \{(p, q) : q \in f^{-1}(\ell_p \cdot f(C)) - 2p\} .$$

Show, as before, that

$$T = E + \sum_i C_{1,p_i} .$$

The inverse correspondence to E is given by

$$\begin{aligned} E^{-1}(q) &= \{p \in C : q \in f^{-1}(\ell_p \cdot f(C)) - 2p\} \\ &= \text{ramification divisor of the composition} \\ &\quad\ \text{of } f \text{ with the projection from } q. \end{aligned}$$

Use the Riemann–Hurwitz formula,

$$E^{-1}(q) \in |\omega L^2(-2q)| ,$$

and the second condition in (15.4) to

$$\sum_i p_i \in |L^{-3}| , \tag{15.6}$$

which is absurd.

ii) Now, suppose that f is birational onto its image but is not necessarily an immersion. Let Z be the divisor of zeros of the differential of f. Show that (15.6) has to be replaced by

$$\sum p_i \in |L^{-3}(Z)|.$$

This is still absurd when

$$\deg(Z) < 3d\,. \tag{15.7}$$

Conclude that if (15.7) holds, $\mathcal{W}^2_d$ is smooth and of dimension $g + 3d - 9$ at $L \to C$.

C. Curves on quadrics

In this series of exercises we will give examples of linear systems for which $\ker \mu_1$ is not zero. We shall use the notation

$\alpha =$ codimension of $\mathcal{W}^r_d$ in Pic^d at $L \to C$,
$\beta = h^0(C, L) \cdot h^1(C, L)$,
$\gamma = \dim(\text{Image}\,\mu)$,

so that

$\beta =$ "expected" codimension of $\mathcal{W}^r_d$ at $L \to C$,
$\gamma =$ codimension of the Zariski tangent space to $\mathcal{W}^r_d$ at $L \to C$.

We have the following obvious inequalities:

$$\beta \geq \alpha \geq \gamma\,. \tag{15.8}$$

We shall consider smooth curves of type (m, n) on a smooth quadric $Q \subset \mathbb{P}^3$, i.e., curves C belonging to the linear system

$$|mL_1 + nL_2|,$$

where L_1 and L_2 are lines in the two rulings of Q.

C-1. Show that the genus g and the degree d of C are

$$d = m + n\,,$$
$$g = (m-1)(n-1)\,,$$

and that $\omega_C \cong \mathcal{O}_Q((m-2)L_1 + (n-2)L_2) \otimes \mathcal{O}_C\,.$

C-2. From now on, we shall assume that $m, n \geq 3$. Show that

$$\alpha \leq 3(m-1)(n-1) - 2m - 2n + 4\,, \tag{15.9}$$

the inequality being due to the fact that there might be curves of degree d and genus g not lying on a quadric.

C-3. Show that

$$E = \{(p,q) : q \in \mathcal{P}_p \cdot C - 2p\}$$

is a symmetric correspondence on $C \times C$ and that

$$E \in |\pi_1^* L \otimes \pi_2^* L(-2D)|.$$

On the other hand,

$$|M(-2D)| = |\pi_1^* L \otimes \pi_2^* (\omega L^{-1})(-2D)|.$$

Moreover, D is a fixed component of $|M(-2D)|$. It follows that, for $m, n \geq 3$,

$$\begin{aligned} \dim(\ker \mu_1) &= h^0(C, \omega L^{-2})^\vee \\ &= h^0(C, \mathcal{O}_Q((m-4)L_1 + (n-4)L_2) \otimes \mathcal{O}_C) = (m-3)(n-3). \end{aligned}$$

We may "explain" this equality as follows. Our curve C on Q may be considered as an abstract curve having a g^1_m and a g^1_n. The infinitesimal conditions imposed on the moduli of C to keep both these pencils are given by the vector space of quadratic differentials vanishing on the sum of the ramification divisors of the branched coverings of $\mathbb{P}^1$ corresponding to the g^1_m and to the g^1_n. This is just $H^0(C, \omega L^{-2})$.

C-4. Show that

$$\gamma = \beta - (m-3)(n-3) \qquad (15.10)$$

and

$$\begin{aligned} \beta &= 4(g - d + 3) \\ &= 4(m-1)(n-1) - 4m - 4n + 12\,. \end{aligned}$$

Using (15.8), (15.9), and (15.10), conclude that

$$\begin{aligned} &\gamma = \alpha \leq \beta\,, \\ &\alpha < \beta \qquad \text{if } m \geq 4,\ n \geq 4. \end{aligned}$$

This has the following interpretation: *let m, n be integers greater than 3. Set $g = (m-1)(n-1)$, $d = m+n$. The points of $\mathcal{W}^3_d$ that correspond to smooth curves of bidegree (m,n) lying on a smooth* *quadric in* $\mathbb{P}^3$ *fill up a dense open subset* U *of a component of* $\mathcal{W}_d^3$. *Moreover, every point* x *of* U *is a smooth point of* $\mathcal{W}_d^3$, *and*

$$\dim_x \mathcal{W}_d^3 = 3g - 3 + \rho + (m-3)(n-3)\,.$$

Observe that, in this example, the Brill–Noether number ρ is strictly negative, so that a general curve of genus $g = (m-1)(n-1)$ does not have a g_d^3.

D. Miscellaneous exercises on $\mathcal{W}_d^r$ and $M_{g,d}^r$.

D-1. Show that:

i) a general curve of genus 7 does not possess a g_6^2.
ii) the most general curve of genus 7 having a g_6^2 in fact possesses two distinct g_6^2's.
(Hint: consider a plane model of C).

D-2. Consider the locus $M_{10,8}^3$ of genus 10 curve possesing a g_8^3. Show that:

i) a general curve C in M_{10} has no g_8^3.
ii) the locus $M_{10,8}^3$ has two irreducible components Σ_1, Σ_2 of dimensions 19, 18. (Hint: use Castelnuovo's bound)
iii) For a general C in Σ_1, $\dim W_8^3(C) = 2$, and for a general C in Σ_2, $\dim W_8^3(C) = 1$.

REMARK (15.11). The preceding two exercises show that Proposition (6.7) is false for g_d^r's when $r \geq 2$.

D-3. Show that $M_{g,d}^1$ lies in the closure of $M_{g,d+1}^1$ for every g and d. (Hint: use (3.7)).

D-4. On the other hand, show that for $g \geq 3$, the locus of elliptic-hyperelliptic curves does not lie in the closure of the locus $M_{g,2}^1$ of hyperelliptic curves or viceversa.

D-5. For any d, $m >> d$, and $g >> md$, study the subvariety

$$\{(C, L^m) \mid (C, L) \in \mathcal{W}_d^1\} \subset \mathcal{W}_{dm}^m\,.$$

E. Some $M_{g,d}^r$'s.

In this sequence of exercises we will study some of the subvarieties $M_{g,d}^r$. In exercises 1–12 we consider the locus $M_{10,6}^2 \subset M_{10}$ of genus 10 curves having a g_6^2.

E-1. Let C be a curve of genus 10 and $\mathcal{D}$ a g^2_6 on C. Show that if $\phi_{\mathcal{D}} : C \to \mathbb{P}^2$ is birational onto its image, then $\mathcal{D}$ is base-point-free, and $C = \phi_{\mathcal{D}}(C)$ is a smooth plane sextic. Show also that if $\phi_{\mathcal{D}}$ fails to be birational, then either

i) $\mathcal{D}$ is base-point-free, and $\phi_{\mathcal{D}}$ is 2-1 onto a cubic curve;
ii) $\mathcal{D}$ is base-point-free, and $\phi_{\mathcal{D}}$ is 3-1 onto a conic;
iii) $\mathcal{D}$ has two base points, and $\phi_{\mathcal{D}}$ is 2-1 onto a conic;

E-2. Conclude that $M^2_{10,6}$ is the union of the loci
$\Sigma_1 = \{\text{smooth plane sextics}\}$,
$\Sigma_2 = \{\text{trigonal curveds}\}$,
$\Sigma_3 = \{\text{hyperelliptic curves}\}$,
$\Sigma_4 = \{\text{elliptic-hyperelliptic curves}\}$.

E-3. Show that the loci Σ_2, Σ_3, and Σ_4 are all disjoint. Show also that Σ_1 is disjoint from Σ_2 and Σ_3, and that Σ_1 is disjoint from Σ_4.

E-4. Show that the dimension of the loci Σ_i are:
$\dim \Sigma_1 = 19$,
$\dim \Sigma_2 = 21$,
$\dim \Sigma_3 = 19$,
$\dim \Sigma_4 = 18$.

E-5. Show that the loci Σ_3 and Σ_4 are closed, while $\overline{\Sigma}_2 = \Sigma_2 \cup \Sigma_3$.

E-6. Show that the closure $\overline{\Sigma}_1$ of Σ_1 contains some but not all trigonal curves. Can you say explicitely which trigonal curves lie in the closure of Σ_1? That is, describe $\overline{\Sigma}_1 \cap \Sigma_2$. (Answer (due to Lazarsfeld): $\overline{\Sigma}_1 \cap \Sigma_2$ consists of those curves C with a g^1_3 $\mathcal{E}$ such that $\mathcal{O}(6\mathcal{E}) = \omega_C$.)

E-7. Show similarly that $\overline{\Sigma}_1$ meets Σ_3 but does not contain it. Can you describe the locus $\Sigma_1 \cap \Sigma_3$?

E-8. Show that $\overline{\Sigma}_1 \supset \Sigma_4$.

E-9. Conclude from the above that $M^2_{10,6}$ has exactly two irreducible components $\overline{\Sigma}_1$ and $\overline{\Sigma}_2$. Note that for the latter of these two components, the g^2_6 on a general curve is not birational, while on the smaller component it is.

We now turn our attention to the variety $\mathcal{W}^6_2 = \mathcal{W}^2_{10,6}$.

E-10. Show that the fiber of the map $\mathcal{W}^6_2 \to M^2_{10,6}$ over a point $C \in \Sigma_1 \cup \Sigma_2$ is just one point; while for $C \in \Sigma_3$, it is two-dimensional, and for $C \in \Sigma_4$, it is one-dimensional.

E-11. Letting Ω_i be the inverse image of σ_i in $\mathcal{W}^6_2$, conclude that
$\dim \Omega_1 = 19$,
$\dim \Omega_2 = 21$,

$\dim \Omega_3 = 21$,
$\dim \Omega_4 = 19$,
and hence that the closures $\overline{\Omega}_i$ are all irreducible components of $\mathcal{W}^2_{10,6}$.

E-12. Show that

$$\overline{\Omega}_1 \cap \Omega_4 = \{(C, L) \ : \ C \in \Sigma_4 \,, \ L^3 \cong \omega_C\}$$

and

$$\overline{\Omega}_2 \cap \Omega_3 = \{(C, L) \ : \ |L| = 2g^1_2 + 2p\} \, .$$

To summarize some of the conclusions of the above exercises,

(15.12) $M^2_{10,6}$ *has 2 irreducible components, of dimesions 19 and 21*,

(15.13)
$\mathcal{W}^2_{10,6}$ *has 4 irreducible components, of dimesions 19, 21, 21, and 19*.

We now turn our attention on $M^3_{10,9}$ and $\mathcal{W}^3_9 = \mathcal{W}^3_{10,9}$.

E-13. Suppose that C is a curve of genus 10 and $\mathcal{D}$ a g^3_9 on C. Assuming that $\phi_{\mathcal{D}}$ is birational, show that the image curve $C_0 = \pi_{\mathcal{D}}(C)$ is either:

i) a smooth curve of type $(3, 6)$ on a quadric $Q \subset \mathbb{P}^3$,
ii) a curve of type $(3, 4)$ on a quadric $Q \subset \mathbb{P}^3$, having two nodes (or a specialization thereof).
iii) the smooth complete intersection of two cubic surfaces in $\mathbb{P}^3$.

(Hint: compute $h^0(C, \mathcal{O}(3))$ to conclude that C_0 lies on at least two cubic surfaces; ask whether C_0 lies on a quadric)

E-14. Suppose now that $\phi_{\mathcal{D}}$ fails to be birational. Show that if C is nonhyperelliptic, then either

i) $\mathcal{D}$ is base-point-free, and $\phi_{\mathcal{D}}$ is 3-1 onto a twisted cubic; or
ii) $\mathcal{D}$ has one base-point, and $\phi_{\mathcal{D}}$ is 2-1 onto an aliptic quartic curve.

Show that if C is hyperelliptic, then $\phi_{\mathcal{D}}$ is 2-1 onto either a twisted cubic, a rational quartic (i.e., a curve of type $(3, 1)$ on a quadric), or a singular curve of type $(2, 2)$ on a quadric if $\mathcal{D}$ is incomplete.

E-15. Conclude from the above that the locus $M^3_{10,9}$ is the union of the loci

$\Sigma_2 = \{$trigonal curves$\}$,
$\Sigma_3 = \{$hyperelliptic curves$\}$,
$\Sigma_4 = \{$elliptic-hyperellipticcurves$\}$,
$\Sigma_5 = \{$curves of type $(3, 6)$ on a quadric$\}$,
$\Sigma_6 = \{$curves of type $(4, 5)$ on a quadric$\}$,
$\Sigma_7 = \{$complete intersections of two cubics in $\mathbb{P}^3\}$.

(The notation is chosen to be consistent with the definitions of $\Sigma_i \subset M^2_{10,6}$.)

E-16. Count dimensions to show that
$\dim \Sigma_2 = 21$,
$\dim \Sigma_3 = 19$,
$\dim \Sigma_4 = 18$,
$\dim \Sigma_5 = 21$,
$\dim \Sigma_6 = 21$,
$\dim \Sigma_7 = 21$.

E-17. Show that if C is a general trigonal curve and $|D|$ the trigonal series on C, then $\mathcal{O}(6D) \neq \omega_C$. Show that $r(\omega(-4D)) \geq 1$ and that the map

$$\phi = \phi_D \times \phi_{\omega(-4D)} : C \to \mathbb{P}^1 \times \mathbb{P}^1 = Q$$

embeds C as a curve of type $(3,6)$ on a quadric. Conclude that Σ_5 is a dense open set in Σ_2.

E-18. Show that the remaining loci Σ_i are disjoint.

E-19. Show that a curve C of genus 10 with a g^3_9, denoted by $\mathcal{D}$, with $\mathcal{D} \subset |D|$, of type E-13, iii) above, then $\omega_C = \mathcal{O}_C(2D)$, that if $(C, \mathcal{D})$ is of type E-13, ii), then $\omega_C = \mathcal{O}_C(2D)$ if and only if the two nodes of C_0 lie on the same line of the ruling of Q meeting C_0 four times, and that if $(C, \mathcal{D})$ is of type E-13, i), that is, C is trigonal with a g^1_3, denoted by $\mathcal{E}$, then $\omega_C = \mathcal{O}_C(2D)$ if and only if $\omega_C = \mathcal{O}_C(6\mathcal{E})$.

E-20. Show that hyperelliptic curves and elliptic–hyperelliptic curves of genus 10 both possess semicanonical g^3_d's.

NOTATION (15.14). *We denote by $\mathcal{T}h^r_g$ the locus in M_g of curves possessing a theta characteristic of dimension r. Thus,*

$$\mathcal{T}h^r_g = \left\{ [C] \in M_g : \begin{array}{c} \text{there is a divisor } D \text{ on } C \text{ with} \\ \mathcal{O}_C(2D) \cong \omega_C \text{ and } \dim |D| \geq r+1 \end{array} \right\}.$$

E-21. Conclude from the previous two exercises that $\mathcal{T}h^3_{10} = \overline{\Sigma}_7$ and in particular that $\Sigma_3 \cup \Sigma_4 \subset \overline{\Sigma}_7$. Also use this to describe the intersections $\Sigma_2 \cap \overline{\Sigma}_7$ and $\Sigma_6 \cap \overline{\Sigma}_7$.

E-22. Conclude from he above that $M^3_{10,9}$ has exactly three irreducible components: $\overline{\Sigma}_3 = \overline{\Sigma}_5$, $\overline{\Sigma}_6$, and $\overline{\Sigma}_7$.

E-23. Suppose that $C \in \Sigma_2$ and that $|D|$ is the g^1_3 on C. Show that $W^3_9(C) = \{u(3D)\} \cup \{u(\omega(-3D))\}$.

E-24. Show that if $C \in \Sigma_6$, then $W^3_9(C)$ consists of exactly two points.

E-25. If Ω_i is the inverse image in $\mathcal{W}_9^3 = \mathcal{W}_{10,9}^3$ of Σ_i, show that

$\dim \Omega_2 = 21$,
$\dim \Omega_3 = 22$,
$\dim \Omega_4 = 20$,
$\dim \Omega_6 = 21$,
$\dim \Omega_7 = 21$.

Show moreover that Ω_2 has exactly two irreducible components, Ω'_2 and Ω''_2, while the other Ω_i are irreducible.

E-26. Conclude that $\Omega_4 \subset \overline{\Omega}_7$ (and hence an alternative proof that $\Sigma_4 \subset \overline{\Sigma}_7$).

E-27. Combining the above exercises, show that, in genus 10, the variety $\mathcal{W}_9^3$ has five irreducible components $\overline{\Omega}'_2$, $\overline{\Omega}''_2$, $\overline{\Omega}_3$, $\overline{\Omega}'_6$, and $\overline{\Omega}_7$, whose general points are given, respectively, as pairs

i) (C, L) : C has a g_3^1, $|D|$, and $L = \mathcal{O}_C(3D)$;
ii) (C, L) : C has a g_3^1, $|D|$, and $L = \omega_C(-3D)$;
iii) (C, L) : C has a g_2^1, $|D|$, and $L = \mathcal{O}_C(3D + p_1 + p_2 + p_3)$;
iv) (C, L) : ϕ_L is of type E-13, ii);
v) (C, L) : ϕ_L is of type E-13, iii).

Conclude, in summary, that

(15.15) In genus 10, $\mathcal{W}_9^3$ has 5 components of which:

- two map birationally to the same component of $M_{10,9}^3$,
- one maps 2-1 onto another component of $M_{10,9}^3$,
- one maps birationally to a thin component of $M_{10,9}^3$,
- two map birationally, with fiber dimensions 2 and 3, to subvarieties of the third component of $M_{10,9}^3$.

Finally, following the notation in Section 2, Chapter III, we let $g = \pi(d, r)$ and consider the locus $\Sigma_d^r \subset M_{g,d}^r$ of extremal curves.

E-28. In case $r = 3$, show that $\dim \Sigma_d^r = g + 2d - 7$ and that $\Sigma_d^r =$ is irreducible. (Note: this requires a little argument in case d is odd.) Observe that for $d \geq 8$, this dimension count violates the naive estimate.

E-29. In the case $d = 8$, $g = 9$, show that $M_{g,d}^r$ is irreducible but that $\mathcal{W}_d^r$ has three components of dimensions 17, 18, and 19.

E-30. Show that if $r \geq 4$, then Σ_d^r is irreducible of dimension

$$\dim \Sigma_d^r = \frac{(m+1)(m+2)}{2}(r-1) + (m+2)(r-3-\varepsilon) - 7\,.$$

E-31. Continuing the preceding exercise, in the case $\varepsilon = 0$, show that Σ_d^r has exactly two irreducible components Σ_1 and Σ_2 of dimensions

$$\dim \Sigma_1 = \frac{(m+1)(m+2)}{2}(r-1) + (m+2)(r-3) - 7\,,$$

$$\dim \Sigma_2 = \frac{m(m+1)}{2}(r-1) + 2(m+2) - 7\,.$$

E-32. In the case $r = 4$, $d = 10$ what is the intersection of Σ_1 and Σ_2? What is it in general?

F. Exercises on theta-characteristics and μ_0.

In this exercises we will use the terminology from Appendix B of Chapter VI.

F-1. Let L be a theta-characteristic (i.e., $L \cong \omega L^{-1}$) with $h^0(C, L) \geq 3$. Considering $\Lambda = \overset{2}{\wedge} H^0(C, L)$ as a subspace of $H^0(C, L) \otimes H^0(C, \omega L^{-1})$, show that $\Lambda \subset \ker \mu_0$ and $\Lambda \cap \ker \mu_1 = 0$. Show that $\mathcal{T}h_g^2$ has codimension 3 in M_g.

F-2. Show that for $g = 7, 8$, the locus $\mathcal{T}h_g^2$ is exactly the locus of hyperelliptic curves of genus g, and so $\mathcal{T}h_7^2$ has codimension 5, and $\mathcal{T}h_8^2$ has codimension 6.

F-3. Let C be a curve of genus 9 with a g_8^3 denoted by $\mathcal{D}$, and let $\mathcal{D} \subset |D|$. Show that if $\phi_{\mathcal{D}}$ is birational, the $\phi_{\mathcal{D}}$ embeds C in $\mathbb{P}^3$ as the intersection of a quadric and a quartic surface, and hence that $\omega_C \cong \mathcal{O}_C(2D)$.

F-4. Show that if $\phi_{\mathcal{D}}$ fails to be birational, then $\phi_{\mathcal{D}}$ either maps C in a 2-1 fashion onto a quartic or onto a twisted cubic. Conclude that C is either hyperelliptic or elliptic–hyperelliptic.

F-5. Show that the loci

$\Gamma_1 = \{\text{hyperelliptic curves}\}$,

$\Gamma_2 = \{\text{elliptic-hyperelliptic curves}\}$,

$\Gamma_3 = \{\text{intersections of a quadric and a quartic in } \mathbb{P}^3\}$

in M_g have dimensions 17, 16, and 18 respectively. Conclude that $\mathcal{T}h_9^3 = \overline{\Gamma}_3$ and in particular that $\mathcal{T}h_9^3$ is irreducible of codimension 6 in M_g.

F-6. Observe that Γ_1, $\Gamma_2 \subset \overline{\Gamma}_3$. Can you describe explicitly a family of curves C_λ in $\overline{\Gamma}_3$ specializing to a hyperelliptic curve C_0 in Γ_1?

F-7. Show that in genus 10, $\mathcal{T}h_{10}^3$ is exactly the closure of the locus of intersections of pairs of cubic surfaces in $\mathbb{P}^3$. Conclude that $\mathcal{T}h_{10}^3$ is irreducible of codimension 6 in M_{10}.

F-8. Let $\Gamma \subset M_{11}$ be the locus of curves C lying on a quartic surface $S \subset \mathbb{P}^3$ and such that $\mathcal{O}_S(C) = \mathcal{O}_S(2)(L_1 + L_2)$, where L_1 and L_2 are skew lines in $\mathbb{P}^3$ lying on S. Show that $\mathcal{T}h^3_{11}$ is irreducible of codimension 6 in M_{11}.

F-9. It may be conjectured, on the basis of the examples above, that for $g \geq 8$, $\mathcal{T}h^3_g$ has codimension 6 in M_g; and more generally that for any r and $g >> 0$, the subvariety $\mathcal{T}h^r_g$ has codimenion $\frac{r(r+1)}{2}$ in M_g. Is this true?

G. Miscellaneous exercises on the normal sheaf.

G-1. Let us go back to Exercises C-1–C-4 and compute directly the dimension of the kernel of μ_1 for a smooth curve C of type (m, n), $m, n \geq 3$, on a smooth quadric Q in $\mathbb{P}^3$.

i) Show that there is an exact sequence

$$0 \to A \to N \to B \to 0,$$

where

$$A = \mathcal{O}_C(mL_1 + nL_2), \qquad B = \mathcal{O}_C(2L_1 + 2L_2).$$

ii) Show that $h^1(C, A) = 0$.

iii) Show that

$$\begin{aligned} \dim \ker \mu_1 &= h^1(C, N) = h^1(C, B) \\ &= h^0(Q, \mathcal{O}((m-4)L_1 + (n-4)L_2)) \\ &= (m-3)(n-3). \end{aligned}$$

This is in accordance with Petri's statement, together with the well-known fact that the only *smooth* curves of genus $g \geq 3$ with general moduli lying on a smooth quadric in $\mathbb{P}^3$ are curves of type $(3, 3)$.

G-2. Reconsider Exercise B-8 from the point of view of the normal bundle. First of all, let

$$\phi : C \to \mathbb{P}^2$$

be a morphism corresponding to a g^2_d on C which is not composed with an involution. Denote by L the line bundle C corresponding to the g^2_d. Show that

$$(15.16) \qquad 3d > \deg Z \implies H^1(C, N_\phi) = 0.$$

In particular, if one wishes to prove directly Petri's statement (8.13) for g^2_d's, it suffices to show that (15.16) always holds when C is general; that is, it suffices to show that *any plane model of a general curve of genus g does not have too many cusps* (cf. (9.16)).

G-3. This exercise proves Petri's statement (8.13) when $r = 2$. Let C be a general curve of genus g. Consider a g^2_d on C corresponding to a nondegenerate morphism

$$\phi : C \to \mathbb{P}^2 .$$

Assume that

$$H^1(C, N_\phi) \neq 0 .$$

Construct a deformation

$$\begin{array}{ccc} \mathcal{C} & \xrightarrow{\ \widetilde{\phi}\ } & \mathbb{P}^2 \\ \big\downarrow{\scriptstyle p} & & \\ S & & \end{array}$$

inducing a commutative diagram

$$\begin{array}{ccc} T_s(S) & \xrightarrow{\qquad \rho_s \qquad} & H^1(C, T_C) \\ & {\scriptstyle h}\searrow \quad \nearrow{\scriptstyle \delta} & \\ & H^0(C, N_\phi) & \end{array}$$

with ρ surjective. Use Proposition (9.10) to reach a contradiction.

G-4. It is natural to ask if the arguments used in Exercise G-3 can be generalized to the case $r \geq 3$ The answer, as matters stand, is no. By squeezing hard the arguments used in Exercise G-3, show that, given a nondegenerate morphism $\phi : C \to \mathbb{P}^3$ with C general, one has $h^1(C, N_\phi) \leq 1$.

G-5. Looking at smooth curves of type (m, n) on a smooth quadric Q, show that the analogue of Theorem (10.1) for $\mathcal{W}^r_d$ is definitely false if $r > 2$.

G-6. Complete the proof of (10.14): r even and $r \geq 4$.

G-7. Construct singular plane curves with nonzero μ_1.

H. Mumford's example revisited

We go back to Mumford's example in Section 6 of Chapter IX. Keeping the notation of that section, we study curves lying on a smooth cubic surface $F \subset \mathbb{P}^3$. We denote by L one of the 27 lines on F and by X the hyperplane section, we fix integers n and m such that $n-1 > m \geq 2$, and we let C be a general (smooth) member of $|X^n L^m|$ of degree $d = 3n+m$ and genus $g = 3(n^2-n)/2 - (m^2-m)/2 + nm + 1$. Finally, we let N denote the normal bundle of C in $\mathbb{P}^3$.

H-1. **i)** Show that there is an exact sequence

$$0 \to A \to N \to B \to 0,$$

where $A = H^n L^m_{|C}$ and $B = H^3_{|C}$.

ii) Show that $h^1(C,A) = 0$ and $H^1(C,N) \cong H^1(C,B)$.

iii) Show that $h^1(C,N) = h^2(F, X^{3-n}L^{-m}) = h^0(F, X^{n-4}L^m)$.

Deduce that $h^1(C,N)$ is never zero and is equal to one only when $n=4$, $m=2$.

H-2. Show that $h^1(C,N) = g - 3d + 18$ if $n \geq m+3$ and $h^1(C,N) = g - 3d + 19$ if $n = m+2$.

H-3. Show that $h^0(C, X_{|C}) = 3$, so that $X_{|C}$ determines a point $w \in \mathcal{W}^3_d \smallsetminus \mathcal{W}^4_d$.

Set

$$\begin{aligned}
\alpha &= \text{codimension of } \mathcal{W}^3_d \text{ in } \mathbf{Pic}^d \text{ at } w\,,\\
\beta &= \text{expected dimension of } \mathcal{W}^3_d\\
&= 4g - 4d + 12\,,\\
\gamma &= \text{codimension of } T_w(\mathcal{W}^3_d) \text{ in } T_w(\mathbf{Pic}^d)\,,\\
\alpha' &= \text{codimension in } \mathbf{Pic}^d \text{ of the subvariety corresponding to the family}\\
&\quad \text{of all smooth curves belonging to } |H^n L^m| \text{ for some cubic surface } F,
\end{aligned}$$

and recall that $\gamma = \beta - h^1(C,N)$.

H-4. Show that

$$\gamma = \begin{cases} 3g - d - 6 \text{ if } n \geq m+3\,,\\ 3g - d - 7 \text{ if } n = m+2\,,\end{cases}$$
$$\beta = 4g - 4d + 12\,,$$
$$\alpha' = 3g - d - 6\,.$$

In particular, $\beta > \alpha'$ with the only exception $\beta = \alpha'$ in the case $n=4$, $m=2$.

H-5. Now restrict to Mumford's case: $n=4$, $m=2$. Show that when $g = 24$, there is a component Z of $\mathcal{W}^3_{14}$ whose codimension in $\mathbf{Pic}^d$ is 52, whereas, for any point $w \in \mathcal{W}^3_{14} \smallsetminus \mathcal{W}^4_{14}$, the codimension of $T_w(Z)$ in $T_w(\mathbf{Pic}^d)$ is equal 51. This means that Z is nowhere reduced.

In theses two series of exercises, we will give another approach to the proof of the unirationality of the Hurwitz spaces $H(3,b)$ and $H(4,w)$.

I. Unirationality of $H(3,w)$

I-1. Let C be a smooth curve of genus g, and $f : C \to \mathbb{P}^1$ a map of degree 3. Show that we have an inclusion

$$0 \to \mathcal{O}_{\mathbb{P}^1} \to f_*\mathcal{O}_C$$

and that the quotient sheaf E is locally free of rank 2. Show moreover that

$$c_1(E) = -g - 2.$$

I-2. Show that we have a natural map

$$\alpha : C \to \mathbb{P}E$$

and that this map is in fact an embedding—in other words, *every smooth trigonal curve may be embedded in a Hirzebruch surface* $\mathbb{F}_n$ *for some* $n \equiv g \pmod 2$*.*

I-3. If the surface $\mathbb{P}E$ in the preceding problem is the Hirzebruch surface $\mathbb{F}_n$, show that the image $\alpha(C) \subset \mathbb{P}E$ is a divisor of class

$$C \sim 3e_0 + \frac{g-3n+2}{2}f \;=\; 3e_\infty + \frac{g+3n+2}{2}f\,,$$

where f is the class of a fiber of $\mathbb{P}E \to \mathbb{P}^1$, and e_0 and e_∞ are the classes of sections of $\mathbb{P}E \to \mathbb{P}^1$ of self-intersection n and $-n$, respectively.

I-4. Now suppose the map $f : C \to \mathbb{P}^1$ above is a general 3-sheeted cover of genus g (equivalently, if $g \geq 4$, the curve C is a general trigonal curve of genus g). Show that $\mathbb{P}E \cong \mathbb{F}_0$ if g is even and $\mathbb{P}E \cong \mathbb{F}_1$ if g is odd. Show moreover that the locus of those simple covers for which $\mathbb{P}E \cong \mathbb{F}_n$ with $n > 1$ is codimension $n-1$ in the Hurwitz space $H(3, 2g+4)$ (if nonempty).

I-5. Now let $\mathbb{P}^N$ be the projective space parameterizing the linear system $|3e_0 + mf|$ on $\mathbb{F}_0$ (respectively, $\mathbb{F}_1$). Show that a general point in $\mathbb{P}^N$ corresponds to a smooth curve of genus g and deduce from the above that we have a dominant rational map $\mathbb{P}^N \to H(3,4m)$ (respectively, $H(3,4m+6)$) and hence that $H(3,b)$ is unirational for any (even) b.

J. Unirationality of $H(4,w)$

Here we give another approach to the unirationality of the Hurwitz spaces $H(4,b)$. The analysis is similar to that of the preceding series of exercises; this time we realize a general tetragonal curve as a complete intersection in a rational threefold.

J-1. Let C be a smooth curve of genus g, and $f : C \to \mathbb{P}^1$ a map of degree 4. As before, we have an inclusion

$$0 \to \mathcal{O}_{\mathbb{P}^1} \to f_*\mathcal{O}_C;$$

show that the quotient sheaf E is locally free of rank 3 and

$$c_1(E) = -g - 3.$$

Show as well that the natural map

$$\alpha : C \to \mathbb{P}E$$

is in fact an embedding.

J-2. For the time being, we will restrict ourselves to the case where g is divisible by 3 and write $g = 3k$. In this case, show that if the map $f : C \to \mathbb{P}^1$ above is a general 4-sheeted cover of genus g, then

$$E \cong \mathcal{O}_{\mathbb{P}^1}(-k-1)^{\oplus 3},$$

and in particular

$$\mathbb{P}E \cong \mathbb{P}^1 \times \mathbb{P}^2.$$

J-3. Continuing in this case (in particular, continuing under the assumption that $f : C \to \mathbb{P}^1$ is general), let ω and $\eta \in A^1(\mathbb{P}^1 \times \mathbb{P}^2)$ denote the pullbacks to $\mathbb{P}^1 \times \mathbb{P}^2$ of the hyperplane classes on $\mathbb{P}^2$ and $\mathbb{P}^1$, respectively; let E and D denote the corresponding divisors restricted to C. Show that

$$E = \omega_C - (k-1)D.$$

Deduce in particular that the degree of E is $2k+2$ and hence that the class of the curve C in $\mathbb{P}^1 \times \mathbb{P}^2$ is

$$[C] = 4\omega^2 + (2k+2)\omega\eta.$$

J-4. Now, suppose further that $g \equiv 3 \bmod 6$, so that k is odd; write $k+1$ as 2ℓ. Compare the dimensions of the linear series $|\mathcal{O}_{\mathbb{P}^1 \times \mathbb{P}^2}(\ell, 2)|$ and its restriction $|\mathcal{O}_C(\ell D + 2E)|$ to show that the curve $C \subset \mathbb{P}^1 \times \mathbb{P}^2$ lies on a pencil of divisors of type $(\ell, 2)$ on $\mathbb{P}^1 \times \mathbb{P}^2$; deduce moreover that C *is the base locus of this pencil.*

J-5. To finish this case, let $\mathbb{P}^N$ be the projective space parameterizing the linear system $|\mathcal{O}_{\mathbb{P}^1 \times \mathbb{P}^2}(\ell, 2)|$, and $\mathbb{G}(1, n)$ the Grassmannian parameterizing lines in $\mathbb{P}^N$. Show that for a general pencil $\{X_\lambda\} \subset |\mathcal{O}_{\mathbb{P}^1 \times \mathbb{P}^2}(\ell, 2)|$, the base locus $\cap X_\lambda$ is a smooth curve of genus g and deduce from the above that we have a dominant rational map $\mathbb{G}(1, n) \to H(4, 2g+6)$ and hence that $H(4, b)$ is unirational for any $b \equiv 0 \pmod{12}$.

J-6. Suppose now that $g \equiv 0 \pmod 6$, so that k is even; in this case write $k = 2\ell$. Show, analogously to Exercise 4, that the curve $C \subset \mathbb{P}^1 \times \mathbb{P}^2$ is a complete intersection of divisors of type $(\ell, 2)$ and $(\ell + 1, 2)$ on $\mathbb{P}^1 \times \mathbb{P}^2$; and deduce similarly that $H(4, b)$ is unirational for any $b \equiv 6 \pmod{12}$.

J-7. Finally, in case $g \equiv 1$ or $2 \pmod 3$ and $f : C \to \mathbb{P}^1$ is general, describe the vector bundle E and corresponding $\mathbb{P}^2$-bundle $\mathbb{P}E$; show that the curve $C \subset \mathbb{P}E$ is a complete intersection of two divisors (as in the case $g \equiv 0 \pmod 3$, the classes of these divisors may depend on the class of $g \pmod 6$), and deduce that $H(4, b)$ is unirational for all remaining congruence classes of $b \pmod{12}$.

Bibliography

1. Abikoff, William, *The real analytic theory of Teichmüller space*, Lecture Notes in Mathematics, 820, Springer-Verlag, Berlin, 1980.
2. Abramovich, D., Corti, A. and Vistoli, A., *Twisted bundles and admissible covers*, Comm. Algebra **31** (2003), 3547–3618.
3. Adem, A., Leida, J. and Ruan, Y., *Orbifolds and stringy topology*, Cambridge Tracts in Mathematics, 171, Cambridge University Press, New York, 2007.
4. Ahlfors, Lars V., *Lectures on quasiconformal mappings*, University Lecture Series, 38, American Mathematical Society, Providence, RI, 2006.
5. Ahlfors, Lars and Bers, Lipman, *Riemann's mapping theorem for variable metrics*, Ann. of Math. (2) **72** (1960), 385–404.
6. Alexeev, Valery, *Compactified Jacobians and Torelli map*, Publ. Res. Inst. Math. Sci. **40** (2004), 1241–1265.
7. Altman, Allen B. and Kleiman, Steven L., *Compactifying the Jacobian*, Bull. Amer. Math. Soc. **82** (1976), 947–949.
8. Altman, Allen B. and Kleiman, Steven L., *Compactifying the Picard scheme*, Adv. in Math. **35** (1980), 50–112.
9. Altman, Allen B. and Kleiman, Steven L., *Compactifying the Picard scheme. II*, Amer. J. Math. **101** (1979), 10–41.
10. Andreotti, Aldo, *On a theorem of Torelli*, Amer. J. Math. **80** (1958), 801–828.
11. Aprodu, Marian, *Brill–Noether theory for curves on $K3$ surfaces*, in Contemporary geometry and topology and related topics, Cluj Univ. Press, Cluj-Napoca, 2008, pp. 1–12.
12. Aprodu, Marian and Farkas, Gavril, *Koszul cohomology and applications to moduli*, arXiv:0811.3117v1 [math.AG].
13. Aprodu, Marian and Farkas, Gavril, *Green's conjecture for curves on arbitrary K3 surfaces*, arXiv:0911.5310 (2009).
14. Aprodu, Marian and Nagel, Jan, *Koszul cohomology and Algebraic Geometry*, University Lecture Series, 52, American Mathematical Society, Providence, RI, 2010.
15. Aprodu, Marian and Pacienza, Gianluca, *The Green conjecture for exceptional curves on a $K3$ surface*, Int. Math. Res. Not. IMRN **14** (2008), 8–25.

E. Arbarello et al., *Geometry of Algebraic Curves*,
Grundlehren der mathematischen Wissenschaften 268,
DOI 10.1007/978-3-540-69392-5,

16. Aprodu, Marian and Voisin, Claire, *Green–Lazarsfeld's conjecture for generic curves of large gonality*, C. R. Math. Acad. Sci. Paris **336** (2003), 335–339.
17. Arakelov, S. Ju., *Families of algebraic curves with fixed degeneracies*, Izv. Akad. Nauk SSSR Ser. Mat. **35** (1971), 1269–1293.
18. Arakelov, S. Ju., *An intersection theory for divisors on an arithmetic surface*, Izv. Akad. Nauk SSSR Ser. Mat. **38** (1974), 1179–1192.
19. Arbarello, Enrico, *Weierstrass points and moduli of curves*, Compositio Math. **29** (1974), 325–342.
20. Arbarello, Enrico, *On subvarieties of the moduli space of curves of genus g defined in terms of Weierstrass points*, Atti Accad. Naz. Lincei Mem. Cl. Sci. Fis. Mat. Natur. Sez. Ia (8) **15** (1978), 3–20.
21. Arbarello, Enrico, *Sketches of KdV*, in Symposium in Honor of C. H. Clemens (Salt Lake City, UT, 2000), Contemp. Math., Amer. Math. Soc., Providence, RI, 2002, pp. 9–69.
22. Arbarello, Enrico and Cornalba, Maurizio, *Su di una proprietà notevole dei morfismi di una curva a moduli generali in uno spazio proiettivo*, Rend. Sem. Mat. Univ. Politec. Torino **38** (1980), 87–99.
23. Arbarello, Enrico and Cornalba, Maurizio, *Su una congettura di Petri*, Comment. Math. Helv. **56** (1981), 1–38.
24. Arbarello, Enrico and Cornalba, Maurizio, *Footnotes to a paper of Beniamino Segre: "Sui moduli delle curve poligonali, e sopra un complemento al teorema di esistenza di Riemann" [Math. Ann.* **100** *(1928), 537–551]*, The number of g^1_d's on a general d-gonal curve and the unirationality of the Hurwitz spaces of 4-gonal and 5-gonal curves, Math. Ann. **256** (1981), 341–362.
25. Arbarello, Enrico and Cornalba, Maurizio, *A few remarks about the variety of irreducible plane curves of given degree and genus*, Ann. Sci. École Norm. Sup. (4) **16** (1983), 467–488.
26. Arbarello, Enrico and Cornalba, Maurizio, *The Picard groups of the moduli spaces of curves*, Topology **26** (1987), 153–171.
27. Arbarello, Enrico and Cornalba, Maurizio, *Combinatorial and algebro-geometric cohomology classes on the moduli spaces of curves*, J. Algebraic Geom. **5** (1996), 705–749.
28. Arbarello, Enrico and Cornalba, Maurizio, *Calculating cohomology groups of moduli spaces of curves via algebraic geometry*, Inst. Hautes Études Sci. Publ. Math. **88** (1998), 97–127.
29. Arbarello, Enrico and Cornalba, Maurizio, *Teichmüller space via Kuranishi families*, Ann. Sc. Norm. Super. Pisa Cl. Sci. (5) **8** (2009), 89–116.
30. Arbarello, Enrico and Cornalba, Maurizio, *Divisors in the moduli spaces of curves*, in Geometry of Riemann surfaces and their moduli spaces (Lizhen Ji, Scott Wolpert, Shing-Tung Yau, eds.), Surveys in Differential Geometry 14, International Press, Somerville, MA, 2010, pp. 1–22.

31. Arbarello, Enrico and Cornalba, Maurizio, *Jenkins-Strebel Differentials*, Rend. Lincei Mat. Appl. (9) **21** (2010), 115–157.
32. Arbarello, E., Cornalba, M., Griffiths, P., and Harris, J., *Geometry of algebraic curves, I*, Grundlehren der mathematischen Wissenschaften, 267, Springer-Verlag, New York, 1985.
33. Arbarello, E., De Concini, C., Kac, V., and Procesi, C., *Moduli spaces of curves and representation theory*, Comm. Math. Phys **117** (1988), 1–36.
34. Arbarello, Enrico and Sernesi, Edoardo, *Petri's approach to the study of the ideal associated to a special divisor*, Invent. Math. **49** (1978), 99–119.
35. Arbarello, Enrico and Sernesi, Edoardo, *The equation of a plane curve*, Duke Math. J. **46** (1979), 469–485.
36. Arsie, Alessandro and Vistoli, Angelo, *Stacks of cyclic covers of projective spaces*, Compos. Math. **140** (2004), 647–666.
37. Artin, Michael, *Algebraic spaces*, Yale University Press, New Haven, Conn., 1971.
38. Artin, Michael, *Théorèmes de représentabilité pour les espaces algébriques*, Les Presses de l'Université de Montréal, Montreal, Que., 1973, pp. 282. En collaboration avec A. Lascu et J.-F. Boutot, Séminaire de Mathématiques Supérieures, No. 44 (Été, 1970).
39. Atiyah, M. F., *Complex analytic connections in fibre bundles*, Trans. Amer. Math. Soc. **85** (1957), 181–207.
40. Baer, R., *Kurventypen auf Flächen*, J. Reine angew. Math. **156** (1927), 231–246.
41. Baer, R., *Isotopie von Kurven auf orientierbaren, geschlossenen Flächen und ihr Zusammenhang mit der topologischen Deformation der Flächen*, J. Reine angew. Math. **159** (1928), 101–111.
42. Baily, Walter L., Jr., *On the moduli of Jacobian varieties*, Ann. of Math. (2) **71** (1960), 303–314.
43. Baily, Walter L., Jr. and Borel, Armand, *Compactification of arithmetic quotients of bounded symmetric domains*, Ann. of Math. **84** (1966), 442–528.
44. Ballico, Edoardo, *Brill–Noether theory for vector bundles on projective curves*, Math. Proc. Cambridge Philos. Soc. **124** (1998), 483–499.
45. Ballico, Edoardo, *On the Brill–Noether theory of stable bundles on curves*, Int. Math. J. **4** (2003), 281–284.
46. Ballico, Edoardo, *Brill–Noether theory for stable vector bundles with fixed determinant on a smooth curve*, Int. J. Pure Appl. Math. **36** (2007), 413–415.
47. Bardelli, Fabio, *Lectures on stable curves*, in Lectures on Riemann surfaces (Trieste, 1987), World Sci. Publ., Teaneck, NJ, 1989, pp. 648–704.
48. Barja, Miguel Ángel, *On the slope of bielliptic fibrations*, Proc. Amer. Math. Soc. **129** (2001), 1899–1906.

49. Barja, Miguel Ángel and Stoppino, Lidia, *Linear stability of projected canonical curves with applications to the slope of fibred surfaces*, J. Math. Soc. Japan **60** (2008), 171–192.
50. Barja, Miguel Ángel and Stoppino, Lidia, *Slopes of trigonal fibred surfaces and of higher dimensional fibrations*. `arXiv:0811.3305v1 [math.AG]`.
51. Barja, Miguel Ángel and Zucconi, Francesco, *On the slope of fibred surfaces*, Nagoya Math. J. **164** (2001), 103–131.
52. Barth, W. P., Hulek, K., Peters, C. A. M., and Van de Ven, A., *Compact complex surfaces*, Springer-Verlag, Berlin, 2004, pp. xii+436. Ergebnisse der Mathematik und ihrer Grenzgebiete. 3. Folge. A Series of Modern Surveys in Mathematics, vol. 4, second edition.
53. Bayer, D., *The division algorithm and the Hilbert scheme*, Harvard thesis, Order number 82–22558, University Microfilms Int'l., Ann Arbor, Michigan, 1982.
54. Beauville, Arnaud, *Prym varieties and the Schottky problem*, Invent. Math. **41** (1977), 149–196.
55. Beauville, Arnaud, *Surfaces algébriques complexes*, Société Mathématique de France, Paris, 1978, pp. iii+172. Astérisque, No. 54.
56. Beauville, Arnaud, *La conjecture de Green générique (d'après C. Voisin)*, Astérisque **299** (2005), Exp. No. 924, vii, 1–14.
57. Behrend, K. and Fantechi, B., *The intrinsic normal cone*, Invent. Math. **128** (1997), 45–88.
58. Beilinson, A. A. and Manin, Yu. I., *The Mumford form and the Polyakov measure in string theory*, Comm. Math. Phys. **107** (1986), 359–376.
59. Beilinson, A. A., Manin, Yu. I., and Schechtman, V. V., *Sheaves of the Virasoro and Neveu–Schwarz algebras*, in K-theory, arithmetic and geometry (Moscow, 1984–1986), Lecture Notes in Math., Springer, Berlin, 1987, pp. 52–66.
60. Beilinson, A. A. and Schechtman, V. V., *Determinant bundles and Virasoro algebras*, Comm. Math. Phys. **118** (1988), 651–701.
61. Benedetti, R. and Petronio, C., *Lectures on Hyperbolic Geometry*, Universitext, Springer-Verlag, Berlin, 1992.
62. Bers, Lipman, *Spaces of Riemann Surfaces*, in Proc. Int. Congress of Math., Cambridge, 1958 (J. A. Todd, ed.), Cambridge Univ. Press, Cambridge, 1960.
63. Bers, Lipman, *Quasiconformal mappings and Teichmüller's theorem*, in Analytic functions, Princeton Univ. Press, Princeton, N.J., 1960, pp. 89–119.
64. Bers, Lipman, *On Moduli of Riemann surfaces*, summer Lectures Forschunginst. Math. 38. With supplemental chapters by C. J. Earle, I. Kra, M. Shishikura and J. H. Hubbard, Eidgenössische Technische Hochschule, Zurich, 1964.

65. Bers, Lipman, *Nielsen extensions of Riemann surfaces*, Ann. Acad. Sci. Fenn. Ser. A I Math. **2** (1976), 29–34.
66. Bertram, A., Cavalieri, R., and Todorov, G., *Evaluating tautological classes using only Hurwitz numbers*, Trans. Amer. Math. Soc. 360 (2008), 6103–6111.
67. Bertram, A., Ein, L., and Lazarsfeld, R., *Surjectivity of Gaussian maps for line bundles of large degree on curves*, in Algebraic geometry (Chicago, IL, 1989), Lecture Notes in Math., Springer, Berlin, 1991, pp. 15–25.
68. Bertram, A. and Feinberg, B., *On stable rank two bundles with canonical determinant and many sections*, in Algebraic geometry (Catania, 1993/Barcelona, 1994), Lecture Notes in Pure and Appl. Math., Dekker, Berlin, 1998, pp. 259–269.
69. Bessis, D., Itzykson, C., and Zuber, J. B., *Quantum field theory techniques in graphical enumeration*, Adv. in Appl. Math. **1** (1980), 109–157.
70. Bhosle, Usha N., *Brill–Noether theory on nodal curves*, Internat. J. Math. **18** (2007), 1133–1150.
71. Bini, G., Fontanari, C., and Viviani, F., *On the birational geometry of the universal Picard variety*, arXiv:1006.0722.
72. Bini, Gilberto and Harer, John, *Euler Characteristics of Moduli Spaces of Curves*, arXiv:math/0506083.
73. Birkar, C., Cascini, P., Hacon, C. D., and McKernan, J., *Existence of minimal models for varieties of log general type*, J. Amer. Math. Soc. **23** (2010), 405–468.
74. Birman, Joan S., *Braids, links, and mapping class groups*, Annals of Mathematics Studies 82, Princeton University Press, Princeton, New Jersey, 1975.
75. Biswas, I., Nag, S., and Sullivan, D., *Determinant bundles, Quillen metrics and Mumford isomorphisms over the universal commensurability Teichmüller space*, Acta Math. **176** (1996), 145–169.
76. Boggi, M. and Pikaart, M., *Galois covers of moduli of curves*, Compositio Math. **120** (2000), 171–191.
77. Borel, Armand, *Stable real cohomology of arithmetic groups*, Ann. Sci. École Norm. Sup. (4) **7** (1974), 235–272.
78. Borel, Armand, *Stable real cohomology of arithmetic groups. II*, in Manifolds and Lie groups (Notre Dame, Ind., 1980), Progr. Math., Birkhäuser, Boston, MA, 1981, pp. 21–55.
79. Bost, Jean-Benoît, *Semi-stability and heights of cycles*, Invent. Math. **118** (1994), 223–253.
80. Bowditch, B. H. and Epstein, D. B. A., *Natural triangulations associated to a surface*, Topology **27** (1988), 91–117.
81. Bradlow, S. B., Brambila-Paz, L., García-Prada, O., and Ramanan, S. (editors), *Moduli spaces and vector bundles*, London Mathematical

Society Lecture Note Series 359, Cambridge University Press, Cambridge, 2009.

82. Bradlow, S. B., Garcia-Prada, O., Muñoz, V., and Newstead, P. E., *Coherent systems and Brill–Noether theory*, Internat. J. Math. **14** (2003), 683–733.
83. Brambila-Paz, L., Grzegorczyk, I., and Newstead, P. E., *Geography of Brill–Noether loci for small slopes*, J. Algebraic Geom. **6** (1997), 645–669.
84. Brambila-Paz, L., Mercat, V., Newstead, P. E., and Ongay, F., *Nonemptiness of Brill–Noether loci*, Internat. J. Math. **11** (2000), 737–760.
85. Brawner, James N., *The Gaussian-Wahl map for trigonal curves*, Proc. Amer. Math. Soc. **123** (1995), 1357–1361.
86. Brézin, É. and Kazakov, V. A., *Exactly solvable field theories of closed strings*, Phys. Lett. B **236** (1990), 144–150.
87. Brivio, Sonia and Verra, Alessandro, *The Brill–Noether curve of a stable vector bundle on a genus two curve*, in Algebraic cycles and motives. Vol. 2, London Math. Soc. Lecture Note Ser., Cambridge Univ. Press, Cambridge, 2007, pp. 73–93.
88. Brown, Kenneth S., *Cohomology of groups*, Graduate Texts in Mathematics 87, Springer-Verlag, New York, 1994.
89. Bruno, Andrea, *Degenerations of linear series and binary curves*, PhD Thesis, Brandeis, 1999.
90. Bruno, Andrea and Verra, Alessandro, *M_{15} is rationally connected*, in Projective varieties with unexpected properties, Walter de Gruyter GmbH & Co. KG, Berlin, 2005, pp. 51–65.
91. Bryan, Jim and Pandharipande, Rahul, The local Gromov–Witten theory of curves, J. Amer. Math. Soc. 21 (2008), 101–136.
92. Brylinski, J.-L., *Propriétés de ramification à l'infini du groupe modulaire de Teichmüller*, Ann. Sci. École Norm. Sup. (4) **12** (1979), 295–333.
93. Buff, X., Fehrenbach, J., Lochak, P., Schneps, L., and Vogel, P., *Moduli spaces of curves, mapping class groups and field theory*, SMF/AMS Texts and Monographs, 9, American Mathematical Society, Providence, RI, 2003.
94. Canonaco, Alberto, *Introduction to algebraic stacks*, preliminary version, 2004.
95. Caporaso, Lucia, *A compactification of the universal Picard variety over the moduli space of stable curves*, J. Amer. Math. Soc. **7** (1994), 589–660.
96. Caporaso, Lucia, *Counting curves on surfaces: a guide to new techniques and results*, in European Congress of Mathematics, Vol. I (Budapest, 1996), Progr. Math., Birkhäuser, Basel, 1998.
97. Caporaso, Lucia, *Néron models and compactified Picard schemes over the moduli stack of stable curves*, Amer. J. Math. **130** (2008), 1–47.

98. Caporaso, Lucia, *Compactified Jacobians, Abel maps and theta divisors*, in Curves and abelian varieties, Contemp. Math. 465, Amer. Math. Soc., Providence, RI, 2008, pp. 1–23.
99. Caporaso, Lucia, *Brill–Noether theory of binary curves*, arXiv:0807.1484v3 [math.AG].
100. Caporaso, L., Casagrande, C., and Cornalba, M., *Moduli of roots of line bundles on curves*, Trans. Amer. Math. Soc. **359** (2007), 3733–3768.
101. Caporaso, L., Coelho, J., and Esteves, E., *Abel maps of Gorenstein curves*, Rend. Circ. Mat. Palermo (2) **57** (2008), 33–59.
102. Caporaso, Lucia and Harris, Joe, *Enumerating rational curves: the rational fibration method*, Compositio Math. **113** (1998), 209–236.
103. Caporaso, Lucia and Harris, Joe, *Counting plane curves of any genus*, Invent. Math. **131** (1998), 345–392.
104. Caporaso, Lucia and Sernesi, Edoardo, *Characterizing curves by their odd theta-characteristics*, J. Reine Angew. Math. **562** (2003), 101–135.
105. Caporaso, Lucia and Sernesi, Edoardo, *Recovering plane curves from their bitangents*, J. Algebraic Geom. **12** (2003), 225–244.
106. Cartan, Henri, *Quotient d'une variété analytique par un groupe discret d'automorphismes, Séminaire Henri Cartan, 6 (1953–1954), Exposé No. 12.*
107. Castelnuovo, Guido, *Ricerche di geometria sulle curve algebriche*, Atti R. Accad. Sci. Torino. Cl. Sci. Fis. Mat. Nat. **24** (1889), 346–373.
108. Castelnuovo, Guido, *Massima dimensione dei sistemi lineari di curve piane di dato genere*, Ann. Mat. Pura Appl. (2) **18** (1890), 119–128.
109. Castelnuovo, Guido, *Ricerche generali sopra i sistemi lineari di curve piane*, Mem. R. Accad. Sci. Torino. Cl. Sci. Fis. Mat. Nat. **42** (1891), 3–43.
110. Castelnuovo, Guido, *Sulla linearità delle involuzioni più volte infinite appartenenti ad una curva algebrica*, Atti R. Accad. Sci. Torino. Cl. Sci. Fis. Mat. Nat. **28** (1893), 727–738.
111. Catenacci, R., Cornalba, M., Martellini, M., and Reina, C., *Algebraic geometry and path integrals for closed strings*, Phys. Lett. B **172** (1986), 328–332.
112. Cattani, E., Kaplan, A., and Schmid, W., *Degeneration of Hodge structures*, Ann. of Math. (2) **123** (1986), 457–535.
113. Cavalieri, R., *Generating functions for Hurwitz–Hodge integrals*, Adv. Math. **218** (2008), 1419–1429.
114. Chang, M. C. and Ran, Z., *Unirationality of the moduli spaces of curves of genus* 11, 13 *(and* 12*)*, Invent. Math. **76** (1984), 41–54.
115. Chang, M. C. and Ran, Z., *Closed families of smooth space curves*, Duke Math. J. **52** (1985), 707–713.
116. Chang, M. C. and Ran, Z., *The Kodaira dimension of the moduli space of curves of genus* 15, J. Differential Geom. **24** (1986), 205–220.

117. Chang, M. C. and Ran, Z., *A note on the slope of subvarieties of* $\overline{M}_{15}$ $(6\frac{2}{3} \leq S_{15} \leq 6\frac{3}{4})$, in Mathematical aspects of string theory (San Diego, Calif., 1986), Adv. Ser. Math. Phys., World Sci. Publishing, Singapore, 1987, pp. 408–411.
118. Chang, M. C. and Ran, Z., *Deformations and smoothing of complete linear systems on reducible curves*, in Algebraic geometry, Bowdoin, 1985 (Brunswick, Maine, 1985), Proc. Sympos. Pure Math., Amer. Math. Soc., Providence, RI, 1987, pp. 63–75.
119. Chang, M. C. and Ran, Z., *On the slope and Kodaira dimension of* $\overline{M}_g$ *for small* g, J. Differential Geom. **34** (1991), 267–274.
120. Chang, M. C. and Ran, Z., *Dimension of families of space curves*, Compositio Math. **90** (1994), 53–57.
121. Chern, Shiing-shen, *An elementary proof of the existence of isothermal parameters on a surface*, Proc. Amer. Math. Soc. **6** (1955), 771–782.
122. Ciliberto, C., Harris, J., and Miranda, R., *On the surjectivity of the Wahl map*, Duke Math. J. **57** (1988), 829–858.
123. Ciliberto, C., Lopez, A. F., and Miranda, R., *On the Wahl map of plane nodal curves*, in Complex analysis and algebraic geometry, de Gruyter, Berlin, 2000, pp. 155–163.
124. Clebsch, A., *Zur Theorie der Riemann'schen Fläche*, Math. Ann. **6** (1873), 216–230.
125. Clemens, C. H., *Lectures 6–10*, in Seminar on degeneration of algebraic varieties (conducted by P.A. Griffiths), Institute for Advanced Study, Princeton, NJ, 1969, pp. 33–50.
126. Clemens, Herb, *A local proof of Petri's conjecture at the general curve*, J. Differential Geom. **54** (2000), 139–176.
127. Clemens, Herb, *Erratum: "A local proof of Petri's conjecture at the general curve" [J. Differential Geom.* **54** *(2000), no. 1, 139–176]*, J. Differential Geom. **77** (2007), 169–173.
128. Colombo, Elisabetta, *On curves with a theta-characteristic whose space of sections has dimension* 4, Math. Z. **215** (1994), 655–665.
129. Coolidge, Julian L., *Introduction to the theory of non-linear systems of plane algebraic curves*, Ann. of Math. (2) **31** (1930), 321–334.
130. Coppens, Marc, *Brill–Noether theory for non-special linear systems*, Compositio Math. **97** (1995), 17–27.
131. Coppens, Marc and Gatto, Letterio, *Limit Weierstrass schemes on stable curves with 2 irreducible components*, Atti Accad. Naz. Lincei Cl. Sci. Fis. Mat. Natur. Rend. Lincei (9) Mat. Appl. **12** (2001), 205–228.
132. Coppens, M., Keem, C., and Martens, G., *Primitive linear series on curves*, Manuscripta Math. **77** (1992), 237–264.
133. Coppens, M., Keem, C., and Martens, G., *The primitive length of a general* k*-gonal curve*, Indag. Math. (N.S.) **5** (1994), 145–159.
134. Coppens, Marc and Martens, Gerriet, *Secant spaces and Clifford's theorem*, Compositio Math. **78** (1991), 193–212.

135. Coppens, Marc and Martens, Gerriet, *Linear series on a general k-gonal curve*, Abh. Math. Sem. Univ. Hamburg **69** (1999), 347–371.
136. Coppens, Marc and Martens, Gerriet, *Linear series on 4-gonal curves*, **Math. Nachr.** (2000), 35–55.
137. Coppens, Marc and Martens, Gerriet, *On the varieties of special divisors*, Indag. Math. (N.S.) **13** (2002), 29–45.
138. Coppens, Marc and Martens, Gerriet, *Divisorial complete curves*, Arch. Math. (Basel) **86** (2006), 409–418.
139. Cornalba, Maurizio, *Systèmes pluricanoniques sur l'espace des modules des courbes et diviseurs de courbes k-gonales (d'après Harris et Mumford)*, Astérisque **121–122** (1985), 7–24.
140. Cornalba, Maurizio, *On the locus of curves with automorphisms*, Ann. Mat. Pura Appl. (4) **149** (1987), 135–151.
141. Cornalba, Maurizio and Harris, Joe, *Divisor classes associated to families of stable varieties, with applications to the moduli space of curves*, Ann. Sci. École Norm. Sup. (4) **21** (1988), 455–475.
142. Cornalba, Maurizio, *Moduli of curves and theta-characteristics*, in Lectures on Riemann surfaces (Trieste, 1987), World Sci. Publ., Teaneck, NJ, 1989, pp. World Sci. Publ., Teaneck, NJ.
143. Cornalba, Maurizio, *A remark on the Picard group of spin moduli space*, Atti Accad. Naz. Lincei Cl. Sci. Fis. Mat. Natur. Rend. Lincei (9) Mat. Appl. **2** (1991), 211–217.
144. Cornalba, M. D. T., *On the projectivity of the moduli spaces of curves*, J. Reine Angew. Math. **443** (1993), 11–20.
145. Cornalba, Maurizio, *A simple proof of the projectivity of Kontsevich's space of maps*, Istit. Lombardo Accad. Sci. Lett. Rend. A **129** (1995), 111–119 (1996).
146. Cornalba, Maurizio, *Cohomology of moduli spaces of stable curves*, in Proceedings of the International Congress of Mathematicians, Vol. II (Berlin, 1998), Documenta Mathematica, (1998), 249–257.
147. Cornalba, Maurizio, *The Picard group of the moduli stack of stable hyperelliptic curves*, Atti Accad. Naz. Lincei Cl. Sci. Fis. Mat. Natur. Rend. Lincei (9) Mat. Appl. **18** (2007), 109–115.
148. Cornalba, Maurizio, *Erratum: "On the locus of curves with automorphisms" [Ann. Mat. Pura Appl. (4)* **149** *(1987), 135–151; MR0932781]*, Ann. Mat. Pura Appl. (4) **187** (2008), 185–186.
149. Cornalba, Maurizio and Griffiths, Phillip A., *Some transcendental aspects of algebraic geometry*, in Algebraic geometry (Proc. Sympos. Pure Math., Vol. 29, Humboldt State Univ., Arcata, Calif., 1974), Amer. Math. Soc., Providence, R.I., 1975, pp. 3–110.
150. Cornalba, Maurizio and Stoppino, Lidia, *A sharp bound for the slope of double cover fibrations*, Michigan Math. J. **56** (2008), 551–561.
151. Courant, R. and Hilbert, D., *Methods of mathematical physics. Vol. II: Partial differential equations*, (Vol. II by R. Courant.), Interscience

Publishers (a division of John Wiley & Sons), New York-London, 1962.

152. Cukierman, Fernando, *Families of Weierstrass points*, Duke Math. J. **58** (1989), 317–346.

153. Cukierman, Fernando and Fong, Lung-Ying, *On higher Weierstrass points*, Duke Math. J. **62** (1991), 179–203.

154. Cukierman, Fernando and Ulmer, Douglas, *Curves of genus ten on $K3$ surfaces*, Compositio Math. **89** (1993), 81–90.

155. Carlson, J., Müller-Stach, S., and Peters, C., *Period mappings and period domains*, Cambridge Studies in Advanced Mathematics 85, Cambridge University Press, Cambridge, 2003.

156. Cumino, C., Esteves, E., and Gatto, L., *Limits of special Weierstrass points*, Int. Math. Res. Pap. IMRP **2** (2008), Art. ID rpn001, 65.

157. D'Souza, Cyril, *Compactification of generalised Jacobians*, Proc. Indian Acad. Sci. Sect. A Math. Sci. **88** (1979), 419–457.

158. Date, E., Kashiwara, M., Jimbo, M., and Miwa, T., *Transformation groups for soliton equations*, in Nonlinear integrable systems—classical theory and quantum theory (Kyoto, 1981), World Sci. Publishing, Singapore, 1983, pp. 39–119.

159. Dehn, Max, *Die Gruppe der Abbildungsklassen*, Acta Math. **69** (1938), 135–206. (English translation in: Max Dehn, Papers on Group Theory and Topology, Springer-Verlag, New York, 1987).

160. Deligne, Pierre, *Équations différentielles à points singuliers réguliers*, in Lecture Notes in Mathematics, Annals of Mathematics Studies, Springer-Verlag, Berlin, 1970, pp. iii+133.

161. Deligne, Pierre, *Théorie de Hodge. I*, in "Actes du Congrès International des Mathématiciens (Nice, 1970)", Tome 1, Gauthier-Villars, Paris, 1971, 425–430.

162. Deligne, Pierre, *Théorie de Hodge. II*, Inst. Hautes Études Sci. Publ. Math. **40** (1971), 5–57.

163. Deligne, Pierre, *Les constantes des équations fonctionnelles des fonctions L*, in Modular functions of one variable, II (Proc. Internat. Summer School, Univ. Antwerp, Antwerp, 1972) Lecture Notes in Math., Vol. 349, Springer, Berlin, 1973, pp. 501–597.

164. Deligne, Pierre, *La Formule de Dualité Globale*, in Théorie des Topos et Cohomologie Étale des Schemas (S.G.A. 4), tome 3, Lecture Notes in Mathematics 305, 1973, pp. 481–587.

165. Deligne, Pierre, *Théorie de Hodge. III*, Inst. Hautes Études Sci. Publ. Math. **44** (1974), 5–77.

166. Deligne, Pierre, *Le déterminant de la cohomologie*, in "Current trends in arithmetical algebraic geometry (Arcata, CA, 1985)", Contemp. Math. 67, Amer. Math. Soc., Providence, RI, 1987, pp. 93–177.

167. Deligne, Pierre and Mumford, David, *The irreducibility of the space of curves of given genus*, Publ. Math. IHES **36** (1969), 75–110.

168. Di Francesco, P., Itzykson, C., and Zuber, J.-B., *Polynomial averages in the Kontsevich model*, Comm. Math. Phys. **151** (1993), 193–219.
169. Diaz, Steven, A bound on the dimensions of complete subvarieties of $\mathcal{M}_g$, Duke Math. J. **51** (1984), 405–408.
170. Diaz, Steven, *Tangent spaces in moduli via deformations with applications to Weierstrass points*, Duke Math. J. **51** (1984), 905–902.
171. Diaz, Steven, *Moduli of curves with two exceptional Weierstrass points*, J. Differential Geom. **20** (1984), 471–478.
172. Diaz, Steven, *Exceptional Weierstrass points and the divisor on moduli that they define*, Mem. Amer. Math. Soc. **56** (1985), iv–69.
173. Diaz, Steven, *Deformations of exceptional Weierstrass points*, Proc. Amer. Math. Soc. **96** (1986), 7–10.
174. Diaz, Steven and Edidin, Dan, *Towards the homology of Hurwitz spaces*, J. Differential Geom. **46** (1996), 66–98.
175. Diaz, Steven and Harris, Joe, *Geometry of the Severi variety*, Trans. Amer. Math. Soc. **309** (1988), 1–34.
176. Diaz, Steven and Harris, Joe, *Geometry of Severi varieties, II. Independence of divisor classes and examples*, in Algebraic geometry (Sundance, UT, 1986), Lecture Notes in Math. 1311, Springer, Berlin, 1988, pp. 23–50.
177. Diaz, Steven and Harris, Joe, *Ideals associated to deformations of singular plane curves*, Trans. Amer. Math. Soc. **309** (1988), 433–468.
178. Dickey, L. A., *Soliton equations and Hamiltonian systems*, Second edition. Advanced Series in Mathematical Physics, 26, World Scientific Publishing Co. Inc., River Edge, NJ, 2003.
179. Dijkgraaf, Robbert, *Intersection theory, integrable hierarchies and topological field theory*, in New symmetry principles in quantum field theory (Cargèse, 1991), Plenum, New York, 1992, pp. 95–158.
180. Dijkgraaf, R., Faber, C., and van der Geer, G. (Editors), *The moduli space of curves. Papers from the conference held on Texel Island, April 1994*, Progress in Mathematics, 129, Birkhäuser Boston, Boston, MA, 1995, pp. xii+563.
181. Dijkgraaf, R., Verlinde, E., and Verlinde, H., *Loop equations and Virasoro constraints in nonperturbative two-dimensional quantum gravity*, Nucl. Phys. B **348** (1991), 435–456.
182. Dijkgraaf, Robbert and Witten, Edward, *Mean field theory, topological field theory, and multi-matrix models*, Nuclear Phys. B **342** (1990), 486–522.
183. Dolgachev, Igor, *Lectures on invariant theory*, London Mathematical Society Lecture Note Series 296, Cambridge University Press, Cambridge, 2003, pp. xvi+220.
184. Douglas, M. R., *Strings in less than one dimension and the generalized KdV hierarchies*, Phys. Lett. B **238** (1990), 176–180.
185. Douglas, M. R. and Shenker, S. H., *Strings in less than one dimension*, Nuclear Phys. B **335** (1990), 635–654.

186. Dubrovin, B. A., *Geometry of 2D topological field theories*, in Integrable systems and quantum groups (Montecatini Terme, 1993), Springer, 1996, pp. 120–348.

187. Edidin, Dan, *The codimension-two homology of the moduli space of stable curves is algebraic*, Duke Math. J. **67** (1992), 241–272.

188. Edidin, Dan, *Brill–Noether theory in codimension-two*, J. Algebraic Geom. **2** (1993), 25–67.

189. Edidin, Dan, *Picard groups of Severi varieties*, Comm. Algebra **22** (1994), 2073–2081.

190. Edidin, Dan, *Notes on the construction of the moduli space of curves*, in Recent progress in intersection theory (Bologna, 1997), Trends Math., Birkhäuser Boston, Boston, MA, 2000, pp. 85–113.

191. Edidin, Dan and Fulghesu, Damiano, *The integral Chow ring of the stack of at most 1-nodal rational curves*, Comm. Algebra **36** (2008), 581–594.

192. Edidin, Dan and Fulghesu, Damiano, *The integral Chow ring of the stack of hyperelliptic curves of even genus*, Math. Res. Lett. **16** (2009), 27–40.

193. Eguchi, T., Hori, K., and Xiong, C.-S., *Quantum cohomology and Virasoro algebra*, Phys. Lett. B **402** (1997), 71–80.

194. Eisenbud, David, *Commutative algebra*, With a view toward algebraic geometry, Springer-Verlag, New York, 1995, pp. xvi+785. Graduate Texts in Mathematics 150.

195. Eisenbud, D., Elkies, N., Harris, J., and Speiser, R., *On the Hurwitz scheme and its monodromy*, Compositio Math. **77** (1991), 95–117.

196. Eisenbud, David and Harris, Joe, *A simpler proof of the Gieseker-Petri theorem on special divisors*, Invent. Math. **74** (1983), 269–280.

197. Eisenbud, David and Harris, Joe, *Divisors on general curves and cuspidal rational curves*, Invent. Math. **74** (1983), 371–418.

198. Eisenbud, David and Harris, Joe, *Limit linear series, the irrationality of M_g, and other applications*, Bull. Amer. Math. Soc. (N.S.) **10** (1984), 277–280.

199. Eisenbud, David and Harris, Joe, *Limit linear series: basic theory*, Invent. Math. **85** (1986), 337–371.

200. Eisenbud, David and Harris, Joe, *Irreducibility and monodromy of some families of linear series*, Ann. Sci. École Norm. Sup. (4) **20** (1987), 65–87.

201. Eisenbud, David and Harris, Joe, *When ramification points meet*, Invent. Math. **87** (1987), 485–493.

202. Eisenbud, David and Harris, Joe, *Existence, decomposition, and limits of certain Weierstrass points*, Invent. Math. **87** (1987), 495–515.

203. Eisenbud, David and Harris, Joe, *The monodromy of Weierstrass points*, Invent. Math. **90** (1987), 333–341.

204. Eisenbud, David and Harris, Joe, *The Kodaira dimension of the moduli space of curves of genus* ≥ 23, Invent. Math. **90** (1987), 359–387.
205. Eisenbud, David and Harris, Joe, *Irreducibility of some families of linear series with Brill–Noether number* -1, Ann. Sci. École Norm. Sup. (4) **22** (1989), 33–53.
206. Eisenbud, D., Lange, H., Martens, G., and Schreyer, F.-O., *The Clifford dimension of a projective curve*, Compositio Math. **72** (1989), 173–204.
207. Ekedahl, Torsten, *Mumford's conjecture*, Institut Mittag-Leffler Report 68 (2006/2007).
208. Ekedahl, T., Lando, S., Shapiro, M., and Vainshtein, A., *On Hurwitz numbers and Hodge integrals*, C. R. Acad. Sci. Paris Sér. I Math. **328** (1999), 1175–1180.
209. Ekedahl, T., Lando, S., Shapiro, M., and Vainshtein, A., *Hurwitz numbers and intersections on moduli spaces of curves*, Invent. Math. **146** (2001), 297–327.
210. Eliashberg, Y., Galatius, S., and Mishachev, N., *Madsen–Weiss for geometrically minded topologists*, arXiv:0907.4226.
211. Ellia, Ph., Hirschowitz, A., and Manivel, L., *Problème de Brill–Noether pour les fibrés de Steiner. Applications aux courbes gauches*, Ann. Sci. École Norm. Sup. (4) **32** (1999), 835–857.
212. Enriques, Federigo, *Sulla proprietà caratteristica delle superficie algebriche irregolari*, Rendiconti della R. Accademia delle Scienze di Bologna **9** (1904), 5–13.
213. Enriques, Federigo, *Le Superficie Algebriche*, Nicola Zanichelli, Bologna, 1949, pp. xv+464.
214. Enriques, Federigo and Chisini, Oscar, *Lezioni sulla teoria geometrica delle equazioni e delle funzioni algebriche. 1. Vol. I, II*, Nicola Zanichelli Editore S.p.A., Bologna, 1985, Vol. I: xvi+398 pp.; Vol. II: 713. Reprint of the 1915 and 1918 editions.
215. Enriques, Federigo and Chisini, Oscar, *Lezioni sulla teoria geometrica delle equazioni e delle funzioni algebriche. 2. Vol. III, IV*, Nicola Zanichelli Editore S.p.A., Bologna, 1985, Vol. III: viii+594 pp.; Vol. IV: iv+274. Reprint of the 1924 and 1934 editions.
216. Epstein, D.B.A., *Curves on 2-manifolds and isotopies*, Acta Math. **115** (1966), 83–107.
217. Epstein, D. B. A. and Penner, R. C., *Euclidean decompositions of noncompact hyperbolic manifolds*, J. Differential Geom. **27** (1988), 67–80.
218. Esnault, Hélène and Viehweg, Eckart, *Logarithmic de Rham complexes and vanishing theorems*, Invent. Math. **86** (1986), 161–194.
219. Esteves, Eduardo, *Linear systems and ramification points on reducible nodal curves*, Mat. Contemp. **14** (1998), 21–35.

220. Esteves, Eduardo, *Compactifying the relative Jacobian over families of reduced curves*, Trans. Amer. Math. Soc. **353** (2001), 3045–3095.
221. Esteves, Eduardo and Gatto, Letterio, *A geometric interpretation and a new proof of a relation by Cornalba and Harris*, Comm. Algebra **31** (2003), 3753–3770.
222. Esteves, Eduardo and Kleiman, Steven, *The compactified Picard scheme of the compactified Jacobian*, Adv. Math. **198** (2005), 484–503.
223. Esteves, Eduardo and Medeiros, Nivaldo, *Limits of Weierstrass points in regular smoothings of curves with two components*, C. R. Acad. Sci. Paris Sér. I Math. **330** (2000), 873–878.
224. Esteves, Eduardo and Medeiros, Nivaldo, *Limit canonical systems on curves with two components*, Invent. Math. **149** (2002), 267–338.
225. Esteves, Eduardo and Salehyan, Parham, *Limit Weierstrass points on nodal reducible curves*, Trans. Amer. Math. Soc. **359** (2007), 5035–5056.
226. Faber, Carel, *Some results on the codimension-two Chow group of the moduli space of stable curves*, in Algebraic curves and projective geometry (Trento, 1988), Lecture Notes in Math. 1389, Springer, Berlin, 1989, pp. 71–81.
227. Faber, Carel, *Chow rings of moduli spaces of curves. I. Some results on the Chow ring of* $\overline{\mathcal{M}}_3$, Ann. of Math. (2) **132** (1990), 331–419.
228. Faber, Carel, *Chow rings of moduli spaces of curves. II. Some results on the Chow ring of* $\overline{\mathcal{M}}_4$, Ann. of Math. (2) **132** (1990), 421–449.
229. Faber, Carel, *Intersection-theoretical computations on* $\overline{\mathcal{M}}_g$, in Parameter spaces, Banach Polish Acad. Sci., Vieweg, Warsaw, 1994, pp. 71–81.
230. Faber, Carel, *A non vanishing result for the tautological ring of* M_g, *arXiv:math/9711219v1 [math.AG]*, 1997, pp. 1–4.
231. Faber, Carel, *A conjectural description of the tautological ring of the moduli space of curves*, in Moduli of curves and abelian varieties, Aspects Math., E33, Vieweg, Braunschweig, 1999, pp. 109–129.
232. Faber, Carel, *Hodge integrals, tautological classes and Gromov–Witten theory, info Sūrikaisekikenkyūsho Kōkyūroku, 1232*, in Proceedings of the Workshop "Algebraic Geometry and Integrable Systems related to String Theory" (Kyoto, 2000), 2001, 78–87.
233. Faber, Carel, *A remark on a conjecture of Hain and Looijenga*, `arXiv:0812.3631`.
234. Faber, Carel and van der Geer, Gerard, *Complete subvarieties of moduli spaces and the Prym map*, J. Reine Angew. Math. **573** (2004), 117–137.
235. Faber, Carel and Looijenga, Eduard, *Remarks on moduli of curves*, in Moduli of curves and abelian varieties, Aspects Math., E33, Vieweg, Braunschweig, 1999, pp. 23–45.

236. Faber, Carel and Pandharipande, Rahul, *Hodge integrals and Gromov–Witten theory*, Invent. Math. **139** (2000), 173–199.

237. Faber, Carel and Pandharipande, Rahul, *Logarithmic series and Hodge integrals in the tautological ring*, (With an appendix by Don Zagier; Dedicated to William Fulton on the occasion of his 60th birthday), Michigan Math. J. **48** (2000), 215–252.

238. Faber, Carel and Pandharipande, Rahul, *Hodge integrals, partition matrices, and the λ_g conjecture*, 1, Ann. of Math. (2) **157** (2003), 97–124.

239. Faber, Carel and Pandharipande, Rahul, *Relative maps and tautological classes*, 1, J. Eur. Math. Soc. (JEMS) **7** (2005), 13–49.

240. Faber, C., Shadrin, S., and Zvonkine, D., *Tautological relations and the r-spin Witten conjecture*, 2009 (arXiv:math/0612510).

241. Fantechi, Barbara, *Deformations of symmetric products of curves*, in Classification of algebraic varieties (L'Aquila, 1992), Contemp. Math., Amer. Math. Soc., Providence, RI, 1994, pp. 135–141.

242. Fantechi, Barbara, *Stacks for everybody*, in European Congress of Mathematics, Vol. I (Barcelona, 2000), Progr. Math., Birkhäuser, Basel, 2001.

243. Fantechi, B., Göttsche, L., Illusie, L., Kleiman, S., Nitsure, N., and Vistoli, A., *Fundamental algebraic geometry*, Mathematical Surveys and Monographs, 123, Mathematical Surveys and Monographs, Providence, RI, 2005, pp. x+339.

244. Fantechi, Barbara and Manetti, Marco, *Obstruction calculus for functors of Artin rings. I*, J. Algebra **202** (1998), 541–576.

245. Fantechi, Barbara and Pandharipande, Rahul, *Stable maps and branch divisors*, Compositio Math. **130** (2002), 345–364.

246. Fantechi, Barbara and Pardini, Rita, *On the Hilbert scheme of curves in higher-dimensional projective space*, Manuscripta Math. **90** (1996), 1–15.

247. Farkas, Gavril, *The geometry of the moduli space of curves of genus 23*, Math. Ann. **318** (2000), 43–65.

248. Farkas, Gavril, *Brill–Noether loci and the gonality stratification of $\mathcal{M}_g$*, J. Reine Angew. Math. **539** (2001), 185–200.

249. Farkas, Gavril, *Gaussian maps, Gieseker-Petri loci and large theta-characteristics*, J. Reine Angew. Math. **581** (2005), 151–173.

250. Farkas, Gavril, *Syzygies of curves and the effective cone of $\overline{\mathcal{M}}_g$*, Duke Math. J. **135** (2006), 53–98.

251. Farkas, Gavril, *The global geometry of the moduli space of curves*, in Algebraic geometry—Seattle 2005. Part 1, Proc. Sympos. Pure Math. 80 Amer. Math. Soc., Providence, RI, 2009, pp. 125–147.

252. Farkas, Gavril, *Koszul divisors on moduli spaces of curves*, Amer. J. Math. **131** (2009), 819–867.

253. Farkas, Gavril, *Rational maps between moduli spaces of curves and Gieseker-Petri divisors*, J. Algebraic Geom. **19** (2010), 243–284.

254. Farkas, Gavril, *Birational aspects of the geometry of M_g*, in Geometry of Riemann surfaces and their moduli spaces, (Lizhen Ji, Scott Wolpert, Shing-Tung Yau, eds.), Surveys in Differential Geometry 14, International Press, Somerville, MA, 2010.

255. Farkas, Gavril and Gibney, Angela, *The Mori cones of moduli spaces of pointed curves of small genus*, Trans. Amer. Math. Soc. **355** (2003), 1183–1199.

256. Farkas, G., Mustaţă, M., and Popa, M., *Divisors on $\mathcal{M}_{g,g+1}$ and the minimal resolution conjecture for points on canonical curves*, Ann. Sci. École Norm. Sup. (4) **36** (2003), 553–581.

257. Farkas, Gavril and Popa, Mihnea, *Effective divisors on $\overline{\mathcal{M}}_g$, curves on $K3$ surfaces, and the slope conjecture*, J. Algebraic Geom. **14** (2005), 241–267.

258. Farkas, Gavril and Verra, Alessandro, *The intermediate type of certain moduli spaces of curves*, arXiv:0910.3905.

259. Farkas, Gavril and Verra, Alessandro, *The classification of universal Jacobians over the moduli space of curves*, arXiv:1005.5354.

260. Farkas, Hershel M., *Special divisors and analytic subloci of Teichmueller space*, Amer. J. Math. **88** (1966), 881–901.

261. Farkas, Hershel M., *Special divisors on compact Riemann surfaces*, Proc. Amer. Math. Soc. **19** (1968), 315–318.

262. Fiorenza, Domenico and Murri, Riccardo, *Feynman diagrams via graphical calculus*, J. Knot Theory Ramifications **11** (2002), 1095–1131.

263. Fiorenza, Domenico and Murri, Riccardo, *Matrix integrals and Feynman diagrams in the Kontsevich model*, Adv. Theor. Math. Phys. **7** (2003), 525–576.

264. Fitting, H., *Die Determinantenideale eines Moduls*, Jahresber. Deutsche Math.-Ver. **46** (1936), 195–228.

265. Fogarty, J., *Invariant Theory*, W. Benjamin, New York, 1969.

266. Fontanari, Claudio, *Theta loci and deformation theory*, Atti Accad. Naz. Lincei Cl. Sci. Fis. Mat. Natur. Rend. Lincei (9) Mat. Appl. **13** (2002), 135–142.

267. Fontanari, Claudio and Looijenga, Eduard, *A perfect stratification of $\mathcal{M}_g$ for $g \leq 5$*, Geom. Dedicata **136** (2008), 133–143.

268. Ford, L.R., *Automorphic Functions*, Chelsea Publishing Company, New York, 1929 (1972).

269. de Franchis, M., *Un teorema sulle involuzioni irrazionali*, Rend. Circ. Mat. Palermo **36** (1913), 368.

270. Fricke, Robert and Klein, Felix, *Vorlesungen über die Theorie der automorphen Funktionen. Band 1: Die gruppentheoretischen Grundlagen. Band II: Die funktionentheoretischen Ausführungen und*

die Andwendungen, Bibliotheca Mathematica Teubneriana, Bände 3, 4, Johnson Reprint Corp., New York, 1965.

271. Friedman, Robert, *Hodge theory, degenerations and the global*, Torelli problem Harvard Thesis, 1981.

272. Fujiki, Akira and Nakano, Shigeo, *Supplement to "On the inverse of monoidal transformation"*, Publ. Res. Inst. Math. Sci. **7** (1971/72), 637–644.

273. Fulton, William, *Hurwitz schemes and irreducibility of moduli of algebraic curves*, Ann. of Math. (2) **90** (1969), 542–575.

274. Fulton, William, *Introduction to intersection theory in algebraic geometry*, second edition, CBMS Regional Conference Series in Mathematics, 54, Springer-Verlag, Published for the Conference Board of the Mathematical, Sciences, Washington, DC, 1984, pp. v+82.

275. Fulton, William, *Intersection Theory*, Second edition, Springer-Verlag, Berlin, 1998.

276. Fulton, William and Pandharipande, Rahul, *Notes on stable maps and quantum cohomology*, in Algebraic geometry—Santa Cruz 1995, Proc. Sympos. Pure Math., Amer. Math. Soc., Providence, RI, 1997, pp. 45–96.

277. Galatius, S., Tillmann, U., Madsen, I., and Weiss, M., *The homotopy type of the cobordism category*, Acta Math. **202** (2009), 195–239.

278. Gardiner, Friederick P., *Teichmüller theory and quadratic differentials*, Wiley-Interscience Publication, Wiley, New York, 1987.

279. Gatto, Letterio and Ponza, Fabrizio, *Derivatives of Wronskians with applications to families of special Weierstrass points*, Trans. Amer. Math. Soc. **351** (1999), Trans. Amer. Math. Soc.

280. Gatto, Letterio, *Intersection theory on moduli spaces of curves*, Monografías de Matemática 61, Instituto de Matemática Pura e Aplicada (IMPA), Rio de Janeiro, 2000, pp. viii+168.

281. Gatto, Letterio, *On the closure in $\overline{M}_g$ of smooth curves having a special Weierstrass point*, Math. Scand. **88** (2001), 41–71.

282. van der Geer, Gerard and Oort, Frans, *Moduli of abelian varieties: a short introduction and survey*, in Moduli of curves and abelian varieties, Aspects Math., E33, Vieweg, Braunschweig, 1999, pp. 1–21.

283. Gervais, S., *A finite presentation of the mapping class group of a punctured surface*, Topology **40** (2001), 703–725.

284. Getzler, Ezra, *Operads and moduli spaces of genus* 0 *Riemann surfaces*, in The moduli space of curves (Texel Island, 1994), Progr. Math., Birkhäuser Boston, Boston Mass., 1995, pp. 199–230.

285. Getzler, Ezra, *Topological recursion relations in genus* 2, in Integrable systems and algebraic geometry, Kobe/Kyoto, 1997, World Sci. Publ., River Edge, NJ, New York, 1998, pp. 73–106.

286. Getzler, Ezra, *The Virasoro conjecture for Gromov–Witten invariants*, in Algebraic geometry: Hirzebruch 70 (Warsaw, 1998), Amer. Math. Soc., Providence, RI, 1999, pp. 147–176.
287. Getzler, Ezra and Looijenga, Eduard, *The Hodge polynomial of* $\overline{M}_{3,1}$, arXiv:math/9910174 1999.
288. Getzler, Ezra and Pandharipande, Rahul, *Virasoro constraints and the Chern classes of the Hodge bundle*, Nuclear Phys. B **530** (1998), 701–714.
289. Ghione, Franco, *Un problème du type Brill–Noether pour les fibrés vectoriels*, in Algebraic geometry—open problems (Ravello, 1982), Lecture Notes in Math. 997, Springer, Berlin, 1983, pp. 197–209.
290. Gibney, Angela, *Numerical criteria for divisors on* M_g *to be ample*, arXiv:math/0312072v1 [math.AG].
291. Gibney, A., Keel, S., and Morrison, I., *Towards the ample cone of* $\overline{M}_{g,n}$, J. Amer. Math. Soc. **15** (2002).
292. Gieseker, David, *Global moduli for surfaces of general type*, Inv. Math **43** (1977), 233–282.
293. Gieseker, David, *Stable curves and special divisors: Petri's conjecture*, Invent. Math. **66** (1982), 251–275.
294. Gieseker, David, *Lectures on moduli of curves*, Tata Lecture Notes, 69, Springer-Verlag, New York, 1982.
295. Gieseker, David, *Geometric invariant theory and applications to moduli problems*, in Invariant Theory, LNM 996, Springer-Verlag, New York, 1983, pp. 45–73.
296. Gieseker, David, *A degeneration of the moduli space of stable bundles*, J. Diff. Geom. **19** (1984), 173–206.
297. Gieseker, David and Morrison, Ian, *Hilbert stability of rank two bundles on curves*, J. Diff. Geom. **19** (1984), 1–29.
298. Gillet, Henri, *Intersection theory on algebraic stacks and* Q*-varieties*, J. Pure Appl. Algebra **34** (1984), 193–240.
299. Givental, Alexander B., *Gromov–Witten invariants and quantization of quadratic Hamiltonians*, Mosc. Math. J. **1** (2001), 551–568, 645.
300. Givental, Alexander B., *Symplectic geometry of Frobenius structures*, in Frobenius manifolds, Aspects Math., E36, Vieweg, Wiesbaden, 2004, pp. 91–112.
301. Gorchinskiy, Sergey and Viviani, Filippo, *Picard group of moduli of hyperelliptic curves*, Math. Z. **258** (2008), 319–331.
302. Gotzmann, G., *Eine Bedingung für die Flachheit und das Hilbertpolynom eines graduierten Ringes*, Math. Zeit. **158** (1978), 61–70.
303. Goulden, I.P., Jackson, D.M., and Vakil, R., *The Gromov–Witten potential of a point, Hurwitz numbers, and Hodge integrals*, Proc. London Math. Soc. (3) **83** (2001), 563–581.
304. Goulden, I.P., Jackson, D.M., and Vakil, R., *Towards the geometry of double Hurwitz numbers*, Adv. Math. **198** (2005), 43–92.

305. Goulden, I.P., Jackson, D.M., and Vakil, R., The moduli space of curves, double Hurwitz numbers, and Faber's intersection number conjecture, arXiv:math/0611659 (2006), 1–45.
306. Graber, Tom and Pandharipande, Rahul, *Localization of virtual classes*, Inventiones Mathematicae **135** (1999), 487–518.
307. Graber, Tom and Pandharipande, Rahul, *Constructions of nontautological classes on moduli spaces of curves*, Michigan Math. J. **51** (2003), 93–109.
308. Graber, Tom and Vakil, Ravi, *On the tautological ring of* $\overline{\mathcal{M}}_{g,n}$, Turkish J. Math. **25** (2001), 237–243.
309. Graber, Tom and Vakil, Ravi, *Hodge integrals and Hurwitz numbers via virtual localization*, Compositio Math. **135** (2003), 25–36.
310. Graber, Tom and Vakil, Ravi, *Relative virtual localization and vanishing of tautological classes on moduli spaces of curves*, Duke Math. J. **130** (2005), 1–37.
311. Grauert, Hans, *Der Satz von Kuranishi für kompakte komplexe Räume*, Invent. Math. **25** (1974), 107–142.
312. Green, Mark L., *Quadrics of rank four in the ideal of a canonical curve*, Invent. Math. **75** (1984), 85–104.
313. Green, Mark and Lazarsfeld, Robert, *On the projective normality of complete linear series on an algebraic curve*, Invent. Math. **83** (1985), 73–90.
314. Green, Mark and Lazarsfeld, Robert, *Special divisors on curves on a* $K3$ *surface*, Invent. Math. **89** (1987), 357–370.
315. Griffiths, Phillip A., *Periods of integrals on algebraic manifolds. I. Construction and properties of the modular varieties*, Amer. J. Math. **90** (1968), 568–626.
316. Griffiths, Phillip A., *Periods of integrals on algebraic manifolds. II. Local study of the period mapping*, Amer. J. Math. **90** (1968), 805–865.
317. Griffiths, Phillip A., *Periods of integrals on algebraic manifolds: Summary of main results and discussion of open problems*, Bull. Amer. Math. Soc. **76** (1970), 228–296.
318. Griffiths, Phillip and Harris, Joseph, *Principles of Algebraic Geometry*, Wiley, New York, 1978.
319. Griffiths, Phillip and Harris, Joseph, *On the variety of special linear systems on a general algebraic curve*, Duke Math. J. **47** (1980), 233–272.
320. Griffiths, Phillip and Harris, Joseph, *Infinitesimal variations of Hodge structure. II. An infinitesimal invariant of Hodge classes*, Compositio Math. **50** (1983), 207–265.
321. Griffiths, Phillip and Morgan, John, *Rational homotopy theory and differential forms*, Progress in Mathematics, 16, Birkhäuser, Boston, Mass., 1981.

322. Griffiths, Phillip and Schmid, Wilfried, *Recent developments in Hodge theory: a discussion of techniques and results*, in "Discrete subgroups of Lie groups and applicatons to moduli (Internat. Colloq., Bombay, 1973)", Oxford Univ. Press, Bombay, 1975, pp. 31–127.

323. Gross, D. J. and Migdal, A. A., *A nonperturbative treatment of two-dimensional quantum gravity*, Nuclear Phys. B **340** (1990), 333–365.

324. Grothendieck, Alexander, *Éléments de géométrie algébrique (rédigés avec la collaboration de Jean Dieudonné): III. Étude cohomologique des faisceaux cohérents*, Première partie, I.H.É.S. Publications mathématiques **11** (1961), 5–167.

325. Grothendieck, Alexander, *Techniques de construction en géométrie analytique*, in Séminaire Cartan, 13ème année (1960–61), Exp. 7–17, W.A.Benjamin Inc., 1967.

326. Grothendieck, Alexander et al., *Séminaire de géometrie algébrique, 1: Revêtements étales et groupe fondamental*, LNM 224, Springer-Verlag, Heidelberg, 1971.

327. Grothendieck, Alexander, *Géométrie formelle et géométrie algébrique*, in Séminaire Bourbaki, Vol. 5, Exp. No. 182, Soc. Math. France, Paris, 1995, pp. 193–220, errata p. 390.

328. Grothendieck, Alexander, *Technique de descente et théorèmes d'existence en géométrie algébrique. II. Le théorème d'existence en théorie formelle des modules*, in Séminaire Bourbaki, Vol. 5, Exp. No. 195, Soc. Math. France, Paris, 1995, pp. 369–390.

329. Grothendieck, Alexander, *Techniques de construction et théorèmes d'existence en géometrie algébrique, III: Préschemas quotients*, Exposé 212, in Seminaire Bourbaki, 1960-61, Exposés 205–222, Soc. Math. France, Paris, 1995, pp. 99–118.

330. Grothendieck, Alexander, *Techniques de construction et théorèmes d'existence en géometrie algébrique, VI: les schemas de Hilbert*, Exposé 221, in Seminaire Bourbaki, 1960-61, Exposés 205–222, Soc. Math. France, Paris, 1995, pp. 249–276.

331. Grothendieck, Alexandre, *La longue marche à travers la théorie de Galois. Tome 1*, Université Montpellier II Département des Sciences Mathématiques, Montpellier, 1995.

332. Grothendieck, Alexandre, *Esquisse d'un programme. Geometric Galois actions, 1*, London Math. Soc. Lecture Note Ser. 242, Cambridge Univ. Press, Cambridge, 1997, pp. 5–48.

333. Haboush, William, *Reductive groups are geometrically reductive*, Ann. of Math. (2) **102** (1975), 67–83.

334. Hacking, Paul, *The moduli space of curves is rigid*, Algebra Number Theory **2** (2008), 809–818.

335. Hain, Richard and Looijenga, Eduard, *Mapping class groups and moduli spaces of curves*, in Proc. Sympos. Pure Math., LNM 996, Amer. Math. Soc., Providence, RI, 1997, pp. 45–73.

336. Halpern, Noemi, *A proof of the Collar Lemma*, Bull. London Math. Soc. **13** (1981), 141–144.
337. Harer, John L., *The second homology group of the mapping class group of an orientable surface*, Inv. Math. **72** (1983), 221–239.
338. Harer, John L., *Stability of the homology of the mapping class groups of orientable surfaces*, Ann. of Math. (2) **121** (1985), 215–251.
339. Harer, John L., *The cohomology of the moduli space of curves*, in C.I.M.E. notes: Montecatini, 1985.
340. Harer, John L., *The virtual cohomological dimension of the mapping class group of an orientable surface*, Invent. Math. **84** (1986), 157–176.
341. Harer, John L., *The third homology group of the moduli space of curves*, Duke Math. J. **63** (1991), 25–55.
342. Harer, John L., *Improved stability for the homology of the mapping class group of surfaces*, Duke University preprint (1993).
343. Harer, John L., *The rational Picard group of the moduli space of Riemann surfaces with spin structure*, in Mapping class groups and moduli spaces of Riemann surfaces (Göttingen, 1991/Seattle, WA, 1991), Contemp. Math., Amer. Math. Soc., Providence, RI, 1993, pp. 107–136.
344. Harer, John and Zagier, Don, *The Euler characteristic of the moduli space of curves*, Invent. Math. **85** (1986), 457–485.
345. Harris, Joe, *Galois groups of enumerative problems*, Duke Math. J. **46** (1979), 685–724.
346. Harris, Joe, *Theta-characteristics on algebraic curves*, Trans. Amer. Math. Soc. **271** (1982), 611–638.
347. Harris, Joe, *Families of smooth curves*, Duke Math. J. **51** (1984), 409–419.
348. Harris, Joe, *On the Kodaira dimension of the moduli space of curves. II. The even-genus case*, Invent. Math. **75** (1984), 437–466.
349. Harris, Joe, *On the Severi problem*, Inv. Math. **84** (1986), 445–461.
350. Harris, Joe and Hulek, Klaus, *On the normal bundle of curves on complete intersection surfaces*, Math. Ann. **264** (1983), 129–135.
351. Harris, Joe and Morrison, Ian, *Slopes of effective divisors on the moduli space of stable curves*, Invent. Math. **99** (1990), 321–355.
352. Harris, Joe and Morrison, Ian, *Moduli of curves*, Graduate Texts in Mathematics, 187, Springer-Verlag, New York, 1998, pp. xiv+366.
353. Harris, Joe and Mumford, David, *On the Kodaira dimension of the moduli space of curves*, Invent. Math. **67** (1982), 23–88.
354. Hartshorne, Robin, *Connectedness of the Hilbert Scheme*, Publ. Math. I.H.E.S. **29** (1966), 5–48.
355. Hartshorne, Robin, *Ample subvarieties of algebraic varieties*, Lecture Notes in Math., 156, Springer-Verlag, Berlin-Heidelberg-New York, 1970.

356. Hartshorne, Robin, *Algebraic Geometry*, Springer-Verlag, New York, 1977.
357. Hartshorne, Robin, *Deformation Theory*, Graduate Texts in Mathematics, 257, Springer-Verlag, New York, 2010, pp. vii+234.
358. Hassett, Brendan, *Classical and minimal models of the moduli space of curves of genus two*, in Geometric methods in algebra and number theory, Progr. Math. 235, Birkhäuser Boston, Boston, MA, 2005, pp. 169–192.
359. Hassett, Brendan and Hyeon, Donghoon, *Log canonical models for the moduli space of curves: the first divisorial contraction*, Trans. Amer. Math. Soc. **361** (2009), 4471–4489.
360. Hassett, Brendan and Hyeon, Donghoon, *Log minimal model program for the moduli space of stable curves: The first flip*, `arXiv:0806.3444v1 [math.AG]`.
361. Hassett, Brendan and Tschinkel, Yuri, *On the effective cone of the moduli space of pointed rational curves*, in Topology and geometry: commemorating SISTAG, Contemp. Math. 314, Amer. Math. Soc., Providence, RI, 2002, pp. 83–96.
362. Hernández, J. A. and Sanz, J., *Gérard–Sibuya's versus Majima's concept of asymptotic expansion in several variables*, J. Aust. Math. Soc. **71** (2001), 21–35.
363. Hilbert, David, *Hilbert's invariant theory papers*, in Lie Groups: History, Frontiers and Applications VIII, Math. Sci. Press, Brookline, 1978.
364. Hirschowitz, André, *Problèmes de Brill–Noether en rang supérieur*, C. R. Acad. Sci. Paris Sér. I Math. **307** (1988), 153–156.
365. Hirschowitz, A. and Ramanan, S., *New evidence for Green's conjecture on syzygies of canonical curves*, Ann. Sci. École Norm. Sup. (4) **31** (1998), 145–152.
366. Hirschowitz, André and Simpson, Carlos, *Descente pour les n-champs*, arXiv:math/9807049v3 [math.AG].
367. Horikawa, Eiji, *On deformations of holomorphic maps*, Proc. Japan Acad. **48** (1972), 52–55.
368. Horikawa, Eiji, *On deformations of holomorphic maps. I*, J. Math. Soc. Japan **25** (1973), 372–396.
369. Horikawa, Eiji, *On deformations of holomorphic maps. II*, J. Math. Soc. Japan **26** (1974), 647–667.
370. Horikawa, Eiji, *On deformations of holomorphic maps. III*, Math. Ann. **222** (1976), 275–282.
371. Howard, Alan and Sommese, Andrew J., *On the theorem of de Franchis*, Ann. Scuola Norm. Sup. Pisa Cl. Sci. (4) **10** (1983), 429–436.
372. Hubbard, John Hamal, *Teichmüller theory and applications to geometry, topology, and dynamics. Vol. 1*, Matrix Editions, Ithaca, NY, 2006.

373. Hubbard, J., Papadopol, P., and Veselov, V., *A compactification of Hénon mappings in* $\mathbb{C}^2$ *as dynamical systems*, Acta Math. **184** (2000), 203–270.

374. Hulek, K., Paranjape, K., and Ramanan, S., *On a conjecture on canonical curves*, J. Algebraic Geom. **1** (1992), 335–359.

375. Humbert, Georges, *Sur une classe de surfaces à génératrices rationnelles*, C. R. Acad. Sci. Paris **116** (1893), 1350–1352.

376. Humbert, Georges, *Sur quelques points de la théorie des courbes et des surfaces algébriques*, J. Math. Pures Appl. (4) **10** (1894), 169–201.

377. Humphries, Stephen, *Generators for the Mapping Class Group*, in Topology of low-dimensional manifolds, Lecture Notes in Math. 722, Springer, Berlin, 1979, pp. 44–47.

378. Hurwitz, Adolf, *Ueber Riemann'sche Flächen mit gegebenen Verzweigungspunkten*, Math. Ann. **39** (1891), 1–61.

379. Huybrechts, Daniel and Lehn, Manfred, *The geometry of moduli spaces of sheaves*, Aspects of Mathematics, E31, Friedr. Vieweg & Sohn, Braunschweig, 1997, pp. xiv+269.

380. Iarrobino, Anthony *Reducibility of the families of* 0*-dimensional schemes on a variety*, Invent. Math. **15** (1972), 72–77.

381. Iarrobino, Anthony, *Hilbert scheme of points: overview of last ten years*, in Algebraic geometry, Bowdoin, 1985 (Brunswick, Maine, 1985), Proc. Sympos. Pure Math. 46, Amer. Math. Soc., Providence, RI, 1987, pp. 297–320.

382. Ide, Manabu and Mukai, Shigeru, *Canonical curves of genus eight*, Proc. Japan Acad. Ser. A Math. Sci. **79** (2003), 59–64.

383. Igusa, Kiyoshi, *Combinatorial Miller–Morita–Mumford classes and Witten cycles*, Algebr. Geom. Topol. **4** (2004), 473–520.

384. Igusa, Kiyoshi, *Graph cohomology and Kontsevich cycles*, Topology **43** (2004), 1469–1510.

385. Imayoshi, Yôichi and Taniguchi, Masahiko, *An Introduction to Teichmüller Spaces*, Springer-Verlag, Berlin, 1992.

386. Ionel, Eleny-Nicoleta, *Topological recursive relations in* $H^{2g}(\mathbb{M}_{g,n})$, Invent. Math. **148** (2002), 627–658.

387. Ionel, Eleny-Nicoleta, *Relations in the tautological ring of* $\mathbb{M}_g$, Duke Math. J. **129** (2005), 157–186.

388. Itzykson, C. and Zuber, J. B., *The planar approximation. II*, J. Math. Phys. **21** (1980), 411–421.

389. Itzykson, C. and Zuber, J.-B., *Matrix integration and combinatorics of modular groups*, Comm. Math. Phys. **134** (1990), 197–207.

390. Itzykson, C. and Zuber, J.-B., *Combinatorics of the modular group. II. The Kontsevich integrals*, Internat. J. Modern Phys. A **7** (1992), 5661–5705.

391. Ivanov, Nikolai V., *Stabilization of the homology of Teichmüller modular groups*, Algebra i Analiz **1** (1989), 110–126.

392. Ivanov, Nikolai V., *On the homology stability for Teichmüller modular groups: closed surfaces and twisted coefficients*, in Mapping class groups and moduli spaces of Riemann surfaces (Göttingen, 1991/Seattle, WA, 1991), Contemp. Math., Amer. Math. Soc., Providence, RI, 1993, pp. 149–194.
393. Izadi, Elham, *The Chow ring of the moduli space of curves of genus 5*, in The moduli space of curves (Texel Island, 1994), Progr. Math., Birkhäuser Boston, Boston Mass., 1995, pp. 267–304.
394. Jambois, Thomas, *Lectures 4–5*, in Seminar on degeneration of algebraic varieties (conducted by P.A. Griffiths), Institute for Advanced Study, Princeton, NJ, 1969, pp. 16–32.
395. de Jong, A. Johan, *Smoothness, semi-stability and alterations*, Inst. Hautes Études Sci. Publ. Math. **83** (1996), 51–93.
396. de Jong, A. Johan, *Families of curves and alterations*, Ann. Inst. Fourier (Grenoble) **47** (1997), 599–621.
397. de Jong, A. Johan (maintainer), *The Stacks Project*, `http://math.columbia.edu/algebraic_geometry/stacks-git/`.
398. de Jong, A. Johan and Oort, Frans, *On extending families of curves*, J. Algebraic Geom. **6** (1997), 545–562.
399. Jost, Jürgen, *Compact Riemann Surfaces*, Universitext, Springer-Verlag, Berlin, 1997.
400. Kang, Eunju, *On the corank of a Wahl map on smooth curves with a plane model*, Kyushu J. Math. **50** (1996), 471–492.
401. Kac, V. G. and Raina, A. K., *Bombay lectures on highest weight representations of infinite-dimensional Lie algebras*, Advanced Series in Mathematical Physics, 2, World Scientific Publishing Co. Inc., Teaneck, NJ, 1987.
402. Kaku, M., *Strings, conformal fields, and M-Theory. Second edition*, Graduate texts in Contemporary Physics, Springer Verlag, New York, 2000.
403. Kas, Arnold, *On deformations of a certain type of irregular algebraic surface*, Amer. J. Math. **90** (1968), Amer. J. Math.
404. Kawamoto, N., Namikawa, Y., Tsuchiya, A., and Yamada, Y., *Geometric realization of conformal field theory on Riemann surfaces*, Comm. Math. Phys. **116** (1988), 247–308.
405. Kazakov, V. A., *The appearance of matter fields from quantum fluctuations of 2D-gravity*, Modern Phys. Lett. A **4** (1989), 2125–2139.
406. Kazarian, M. E. and Lando, S. K., *On intersection theory on Hurwitz spaces*, Izv. Ross. Akad. Nauk Ser. Mat. **68** (2004), 91–122.
407. Kazarian, M. E. and Lando, S. K., *An algebro-geometric proof of Witten's conjecture*, J. Amer. Math. Soc. **20** (2007), 1079–1089.
408. Keel, Seán, *Intersection theory of moduli space of stable n-pointed curves of genus zero*, Trans. Amer. Math. Soc. **330** (1992), 545–574.

409. Keel, Seán, *Basepoint freeness for nef and big line bundles in positive characteristic*, Ann. of Math. (2) **149** (1999), 253–286.
410. Keel, Seán and Mori, Shigefumi, *Quotients by groupoids*, Ann. of Math. (2) **145** (1997), 193–213.
411. Keen, Linda, *Collars on Riemann surfaces*, in Discontinuous Groups and Riemann Surfaces, Ann. of Math. Stud., Princeton Univ. Press, Princeton N.J., 1974, pp. 263–268.
412. Kempf, George, *On the geometry of a theorem of Riemann*, Ann. of Math. (2) **98** (1973), 178–185.
413. Kempf, George, *Instability in invariant theory*, Ann. of Math. (2) **108** (1978), 299–316.
414. Kempf, George R., *Deformations of symmetric products*, in Riemann surfaces and related topics: Proceedings of the 1978 Stony Brook Conference (State Univ. New York, Stony Brook, N.Y., 1978), Ann. of Math. Stud., Princeton Univ. Press, Princeton, N.J., 1981, pp. 319–341.
415. Khosla, Deepak, *Moduli spaces of curves with linear series and the slope conjecture*, arXiv:math/0608024v1 [math.AG].
416. Khosla, Deepak, *Tautological classes on moduli spaces of curves with linear series and a push-forward formula when* $\rho = 0$, arXiv:0704.1340v1 [math.AG].
417. King, A. D. and Newstead, P. E., *Moduli of Brill–Noether pairs on algebraic curves*, Internat. J. Math. **6** (1995), 733–748.
418. Kirwan, Frances, *Moduli spaces in algebraic geometry*, in Moduli spaces in mathematics and physics (Oxford, 1998), Hindawi Publ. Corp., Cairo, 2001, pp. 1–16.
419. Kirwan, Frances, *Cohomology of moduli spaces*, in Proceedings of the International Congress of Mathematicians, Vol. I (Beijing, 2002), Higher Ed. Press, Beijing, 2002.
420. Kleiman, Steven L., *r-special subschemes and an argument of Severi's*, Advances in Math. **22** (1976), 1–31.
421. Kleiman, Steven L., *The structure of the compactified Jacobian: a review and announcement*, in Geometry seminars, 1982–1983 (Bologna, 1982/1983), Univ. Stud. Bologna, Bologna, 1982, pp. 81–92.
422. Kleiman, Steven L. and Laksov, Dan, *On the existence of special divisors*, Amer. J. Math. **94** (1972), 431–436.
423. Kleiman, Steven L. and Laksov, Dan, *Another proof of the existence of special divisors*, Acta Math. **132** (1974), 163–176.
424. Kleppe Jan O., *The Hilbert-flag scheme, its properties and its connection with the Hilbert scheme. Applications to curves in 3-space*, Preprint Inst. of Math. Univ. Oslo, 1981.
425. Knudsen, Finn Faye and Mumford, David, *The projectivity of the moduli space of stable curves. I. Preliminaries on "det" and "Div"*, Math. Scand. **39** (1976), 19–55.

426. Knudsen, Finn Faye, *The projectivity of the moduli space of stable curves II. The stacks* $M_{g,n}$, Math. Scand. **52** (1983), 161–199.
427. Knudsen, Finn Faye, *The projectivity of the moduli space of stable curves III. The line bundles on* $M_{g,n}$*, and a proof of the projectivity of* $\overline{M}_{g,n}$ *in characteristic* 0, Math. Scand. **52** (1983), 200–212.
428. Knutson, Donald, *Algebraic spaces*, Springer-Verlag, Berlin, 1971, pp. vi+261. Lecture Notes in Mathematics, Vol. 203.
429. Kobayashi, Shoshichi, *Differential geometry of complex vector bundles*, Publications of the Mathematical Society of Japan, Volume 15, Kanô Memorial Lectures, 5, Princeton University Press, Princeton, NJ, 1987, pp. xii+305.
430. Kodaira, Kunihiko, *Characteristic linear systems of complete continuous systems*, Amer. J. Math. **78** (1956), 716–744.
431. Kodaira, Kunihiko, *A theorem of completeness of characteristic systems for analytic families of compact submanifolds of complex manifolds*, Ann. of Math. (2) **75** (1962), 146–162.
432. Kodaira, Kunihiko, *On the structure of compact complex analytic surfaces. I*, Amer. J. Math. **86** (1964), 751–798.
433. Kodaira, Kunihiko, *A certain type of irregular algebraic surfaces*, J. Analyse Math. **19** (1967), 207–215.
434. Kodaira, Kunihiko, *Complex manifolds and deformations of complex structures*, Springer-Verlag, New York, 1986.
435. Kodaira, K. and Spencer, D. C., *On deformations of complex analytic structures, I, II*, Ann. of Math. (2) **67** (1958), 328–466.
436. Kodaira, K. and Spencer, D. C., *A theorem of completeness of characteristic systems of complete continuous systems*, Amer. J. Math. **81** (1959), 477–500.
437. Kodaira, K. and Spencer, D. C., *On deformations of complex analytic structures, III. Stability theorems for complex structures*, Ann. of Math. (2) **71** (1960), 43–76.
438. Kodaira, K., Nirenberg, L., and Spencer, D. C., *On the existence of deformations of complex analytic structures*, Ann. of Math. (2) **68** (1958), 450–459.
439. Kollár, János, *Projectivity of complete moduli*, J. Differential Geom. **32** (1990), 235–268.
440. Kollár, János, *Rational curves on algebraic varieties*, Springer-Verlag, Berlin, New York, 1996.
441. Konno, Kazuhiro, *A lower bound of the slope of trigonal fibrations*, Internat. J. Math. **7** (1996), 19–27.
442. Konno, Kazuhiro, *Clifford index and the slope of fibered surfaces*, J. Algebraic Geom. **8** (1999), 207–220.
443. Kontsevich, M. L., *The Virasoro algebra and Teichmüller spaces*, Funktsional. Anal. i Prilozhen. **21** (1987), 78–79.
444. Kontsevich, Maxim, *Intersection theory on the moduli space of curves and the matrix Airy function*, Comm. Math. Phys. **147** (1992), 1–23.

445. Kontsevich, Maxim, *Feynman diagrams and low-dimensional topology*, in First European Congress of Mathematics, Vol. II (Paris, 1992), Progr. Math., Birkhäuser, Basel, 1994, pp. 97–121.
446. Kontsevich, Maxim, *Enumeration of rational curves via torus actions*, in The moduli space of curves (Texel Island, 1994), Progr. Math., 128, Birkhäuser Boston, Boston, MA, 1995, pp. 335–368.
447. Korn, A., *Zwei Anwendungen der Methode der sukzessiven Annäherungen*, Schwarz Festschrift, Berlin, 1914, 215–229.
448. Kouvidakis, Alexis, *The Picard group of the universal Picard varieties over the moduli space of curves*, J. Differential Geom. **34** (1991), 839–850.
449. Kouvidakis, Alexis, *Ample divisors in the moduli space of* 0*-genus pointed curves*, Comm. Algebra **21** (1993), 3165–3181.
450. Kouvidakis, Alexis, *Picard groups of Hilbert schemes of curves*, J. Algebraic Geom. **3** (1994), 671–684.
451. Kouvidakis, Alexis, *On some results of Morita and their application to questions of ampleness*, Math. Z. **241** (2002), 17–33.
452. Kuranishi, Masatake, *On the locally complete families of complex analytic structures*, Ann. of Math. (2) **75** (1962), 536–577.
453. Kuranishi, Masatake, *On deformations of compact complex structures*, in Proc. Internat. Congr. Mathematicians (Stockholm, 1962), Inst. Mittag-Leffler, Djursholm, 1963, pp. 357–359.
454. Kuranishi, Masatake, *New proof for the existence of locally complete families of complex structures*, in Proc. Conf. Complex Analysis (Minneapolis, 1964), Springer, Berlin, 1965, pp. 142–154.
455. Laksov, Dan, *Weierstrass points on curves*, in Young tableaux and Schur functors in algebra and geometry (Toruń, 1980), Astérisque, 87, Soc. Math. France, Paris, 1981, pp. 221–247.
456. Laksov, Dan and Thorup, Anders, *The Brill–Segre formula for families of curves*, in Enumerative algebraic geometry (Copenhagen, 1989), Contemp. Math., Amer. Math. Soc., Providence, RI, 1991, pp. 131–148.
457. Laksov, Dan and Thorup, Anders, *Weierstrass points and gap sequences for families of curves*, Ark. Mat. **32** (1994), 393–422.
458. Laufer, Henry B., *On generalized Weierstrass points and rings with no prime elements*, in Riemann surfaces and related topics: Proceedings of the 1978 Stony Brook Conference (State Univ. New York, Stony Brook, N.Y., 1978), Ann. of Math. Stud., 97, Princeton Univ. Press, Princeton, N.J., 1981, pp. 391–402.
459. Lando, S. K., *Ramified coverings of the two-dimensional sphere and intersection theory in spaces of meromorphic functions on algebraic curves*, Uspekhi Mat. Nauk **57** (2002), 29–98.
460. Lando, S. K. and Zvonkine, D., *Counting ramified converings and intersection theory on spaces of rational functions. I. Cohomology of Hurwitz spaces*, Mosc. Math. J. **7** (2007), 85–107.

461. Laumon, Gérard, *Fibrés vectoriels spéciaux*, Bull. Soc. Math. France **119** (1991), 97–119.
462. Laumon, Gérard and Moret-Bailly, Laurent, *Champs algébriques*, Springer-Verlag, Berlin, 2000, pp. xii+208. Ergebnisse der Mathematik und ihrer Grenzgebiete. 3. Folge. A Series of Modern Surveys in Mathematics, vol. 39.
463. Lax, R. F., *Weierstrass points of the universal curve*, Math. Ann. **216** (1975), 35–42.
464. Lazarsfeld, Robert, *Brill–Noether–Petri without degenerations*, J. Differential Geom. **23** (1986), 299–307.
465. Lazarsfeld, Robert, *A sampling of vector bundle techniques in the study of linear series*, in Lectures on Riemann surfaces (Trieste, 1987), World Sci. Publ., Teaneck, NJ, 1989, pp. 500–559.
466. Lazarsfeld, Robert, *Positivity in algebraic geometry. II*, Ergebnisse der Mathematik und ihrer Grenzgebiete. 3. Folge. 49, Springer-Verlag, Berlin, 2004.
467. Lee, Y.-P. and Vakil, R., *Algebraic structures on the topology of moduli spaces of curves and maps*, arXiv:0809.1879v1 [math.AG].
468. Le Potier, J., *Fibrés vectoriels sur les courbes algébriques*, Publications Mathématiques de l'Université Paris 7–Denis Diderot, Université Paris 7–Denis Diderot, U.F.R. de Mathématiques, Paris, 1995, pp. 167.
469. Le Potier, J., *Lectures on vector bundles*, Cambridge Studies in Advanced Mathematics, 54, Cambridge University Press, Cambridge, 1997.
470. Le Potier, J., *Problème de Brill–Noether et groupe de Picard de l'espace de modules des faisceaux semi-stables sur le plan projectif*, in Algebraic geometry (Catania, 1993/Barcelona, 1994), Lecture Notes in Pure and Appl. Math., Dekker, New York, 1998, pp. 75–90.
471. Lefschetz, Solomon, *L'analysis situs et la géométrie algébrique*, Gauthier-Villars, Paris, 1950, pp. vi+154.
472. Lehto, Olli, *Univalent functions and Teichmüller spaces*, Graduate Texts in Mathematics, 109, Springer-Verlag, New York, 1987.
473. Li, Jun and Tian, Gang, *A brief tour of GW invariants*, in Surveys in differential geometry: Differential geometry inspired by string theory, Surveys in differential geometry 5, Int. Press, Boston, MA, 1999, pp. 543–569.
474. Li, Yingchen, *A singularity theorem in Brill–Noether theory of higher rank*, Internat. J. Math. **4** (1993), 409–419.
475. Lichtenbaum, D. and Schlessinger, M., *The cotangent complex of a morphism*, Trans. A.M.S. **128** (1967), 41–70.
476. Lichtenstein, L., *Zur Theorie der konformen Abbildung nichtanalytischer, singularitätenfreier Flächenstücke auf ebene Gebiete*, Bull. Int. de l'Acad. Sci. Cracovie, Sér. A (1916), 192–217.

477. Lickorish, W. B. R., *A representation of orientable, combinatorial 3-manifolds*, Annals of Math. **76** (1962), 531–540.
478. Lickorish, W. B. R., *A finite set of generators for the homeotopy group of a 2-manifold*, Proc. Camb. Phil. Soc. **60** (1964), 769–778.
479. Lickorish, W. B. R., *Corrigendum: On the homeotopy group of a 2-manifold*, Proc. Camb. Phil. Soc. **62** (1966), 679–681.
480. Liu, Qing, *Algebraic Geometry and Arithmetic Curves*, Oxford University Press, New York, 2002.
481. Liu, K. and Xu, H., *A proof of the Faber intersection number conjecture*, arXiv:0803.2204v2 [math.AG].
482. Liu, Xiaobo and Tian, Gang, *Virasoro constraints for quantum cohomology*, J. Differential Geom. **50** (1998), 537–590.
483. Lochak, Pierre and Schneps, Leila, *The Grothendieck-Teichmüller group and automorphisms of braid groups*, in The Grothendieck theory of dessins d'enfants (Luminy, 1993), London Math. Soc. Lecture Note Ser., 200, Cambridge Univ. Press, Cambridge, 1994, pp. 323–358.
484. Logan, Adam, *The Kodaira dimension of moduli spaces of curves with marked points*, Amer. J. Math. **125** (2003), 105–138.
485. Looijenga, Eduard, *Cohomology of* $\mathcal{M}_3$ *and* $\mathcal{M}_{1,3}$, in Mapping class groups and moduli spaces of Riemann surfaces (Göttingen, 1991/Seattle, WA, 1991), Contemp. Math., Amer. Math. Soc., Providence, RI, 1993, pp. 205–228.
486. Looijenga, Eduard, *Intersection theory on Deligne–Mumford compactifications (after Witten and Kontsevich)*, Séminaire Bourbaki, Vol. 1992/93, Exp. No. 768, Astérisque **216** (1993), 187–212.
487. Looijenga, Eduard, *Smooth Deligne–Mumford compactifications by means of Prym level structures*, J. Algebraic Geom. **3** (1994), 283–293.
488. Looijenga, Eduard, *Cellular decompositions of compactified moduli spaces of pointed curves*, in The moduli space of curves (Texel Island, 1994), Progress in Mathematics 129, Birkhäuser Boston, Boston, MA, 1995, pp. 369–400.
489. Looijenga, Eduard, *On the tautological ring of* $\mathcal{M}_g$, Invent. Math. **121** (1995), 411–419.
490. Looijenga, Eduard, *Stable cohomology of the mapping class group with symplectic coefficients and of the universal Abel–Jacobi map*, J. Algebraic Geom. **5** (1996), 135–150.
491. Looijenga, Eduard, *A minicourse on moduli of curves, School on Algebraic Geometry (Trieste, 1999)*, ICTP Lect. Notes 1, Abdus Salam Int. Cent. Theoret. Phys., Trieste, 2000, 267–291.
492. Macaulay, F. R., *Some properties of enumeration in the theory of modular systems*, Proc. Lond. Math. Soc. **26** (1927), 531–555.
493. Madsen, Ib and Tillmann, Ulrike, *The stable mapping class group and* $Q(\mathbb{C}P^{\infty}_{+})$, Invent. Math. **145** (2001), 509–544.

494. Madsen, Ib and Weiss, Michael, *The stable mapping class group and stable homotopy theory*, in European Congress of Mathematics, Eur. Math. Soc., Zürich, Zürich, 2005, pp. 283–307.

495. Madsen, Ib and Weiss, Michael, *The stable moduli space of Riemann surfaces: Mumford's conjecture*, Ann. of Math. (2) **165** (2007), 843–941.

496. Manetti, Marco, *Lectures on deformations of complex manifolds (deformations from differential graded viewpoint)*, Rend. Mat. Appl. (7) **24** (2004), 1–183.

497. Manin, Yu. I., *Frobenius manifolds, quantum cohomology and moduli spaces*, Colloquium publications, 47, American Mathematical Society, Providence, RI, 1999.

498. Manin, Yu. I., *Generating functions in algebraic geometry and sums over trees*, in The moduli space of curves (Texel Island, 1994), Progr. Math. 129, Birkhäuser Boston, Boston, MA, 1995, pp. 401–417.

499. Manin, Yu. I., *Quantum strings and algebraic curves*, in Proceedings of the International Congress of Mathematicians, Vol. 1, 2 (Berkeley, Calif., 1986), Amer. Math. Soc., Providence, RI, 1987, pp. 1286–1295.

500. Markman, Eyal, *Brill–Noether duality for moduli spaces of sheaves on $K3$ surfaces*, J. Algebraic Geom. **10** (2001), 623–694.

501. Martin-Deschamps, Mireille and Perrin, Daniel, *Le schéma de Hilbert des courbes gauches localement Cohen–Macaulay n'est (presque) jamais réduit*, Ann. Sci. École Norm. Sup. (4) **26** (1996), 757–785.

502. Matelski, P., *A compactness theorem for Fuchsian groups of the second kind*, Duke Math. J. **43** (1976), 829–840.

503. Matsumura, Hideyuki, *Commutative algebra*, second edition, Mathematics Lecture Note Series 56, Benjamin/Cummings Publishing Co., Inc., Reading, Mass., 1980.

504. Matsuzaki, K., *An extension of the collar lemma*, RIMS Kokyuroku, 2003, 58–61.

505. Mayer, Alan L., *Compactification of the variety of moduli of curves*, Seminar on degeneration of algebraic varieties (conducted by P.A. Griffiths), Institute for Advanced Study, Princeton, NJ, 1969, pp. 6–15.

506. Melo, Margarida, *Compactified Picard stacks over $\overline{\mathcal{M}}_g$*, Math. Z. **263** (2009), 939–957.

507. Mercat, Vincent, *Le problème de Brill–Noether et le théorème de Teixidor*, Manuscripta Math. **98** (1999), 75–85.

508. Mercat, Vincent, *Le problème de Brill–Noether pour des fibrés stables de petite pente*, J. Reine Angew. Math. **506** (1999), 1–41.

509. Mestrano, Nicole, *Conjectures de Franchetta plongées*, C. R. Acad. Sci. Paris Sér. I Math. **301** (1985), 503–505.

510. Mestrano, Nicole, *Conjecture de Franchetta forte*, Inv. Math. **87** (1987), 365–376.

511. Mestrano, N. and Ramanan, S., *Poincaré bundles for families of curves*, J. Reine. Angew. Math. **362** (1985), 169–178.

512. Miller, Edward, *The homology of the mapping class group*, J. Diff. Geom. **24** (1986), 1–14.

513. Mirzakhani, Maryam, *Weil-Petersson volumes and intersection theory on the moduli space of curves*, J. Amer. Math. Soc. **20** (2007), 1–23.

514. Moerdijk, Ieke and Pronk, Dorette A., *Orbifolds, sheaves and groupoids*, K-Theory **12** (1997), 3–21.

515. Moishezon, Boris, *Stable branch curves and branch monodromies*, in Algebraic geometry, Univ. Ill. Chicago Circle, 1980, LNM 862, Springer-Verlag, New York, 1981, pp. 107–192.

516. Mondello, Gabriele, *Riemann surfaces with boundary and natural triangulations of the Teichmueller space*, arXiv:0804.0605 2008 (2008).

517. Mori, Shigefumi and Mukai, Shigeru, *The uniruledness of the moduli space of curves of genus* 11, in Algebraic geometry (Tokyo/Kyoto, 1982), Lecture Notes in Math., Springer, Berlin, 1983, pp. 334–353.

518. Morita, Shigeyuki, *Characteristic classes of surface bundles*, Bull. Amer. Math. Soc. (N.S.) **11** (1984), 386–388.

519. Morita, Shigeyuki, *Characteristic classes of surface bundles*, Invent. Math. **90** (1987), 551–577.

520. Morita, Shigeyuki, *Problems on the structure of the mapping class group of surfaces and the topology of the moduli space of curves*, in Topology, geometry and field theory, World Sci. Publ., River Edge, NJ, 1994, pp. 101–110.

521. Morita, Shigeyuki *Structure of the mapping class groups of surfaces: a survey and a prospect*, in Proceedings of the Kirbyfest (Berkeley, CA, 1998), Geom. Topol. Monogr. 2, Geom. Topol. Publ., Coventry, 1999, pp. 349–406.

522. Morita, Shigeyuki, *Geometry of characteristic classes*, Translated from the 1999 Japanese original, Iwanami Series in Modern Mathematics, American Mathematical Society, Providence, RI, 2001, pp. xiv+185.

523. Morita, Shigeyuki *Generators for the tautological algebra of the moduli space of curves*, Topology 4 (2003), 787–819.

524. Morita, Shigeyuki *Cohomological structure of the mapping class group and beyond*, in Problems on mapping class groups and related topics, Proc. Sympos. Pure Math., Amer. Math. Soc. Providence, RI 2006, pp. 329–354.

525. Morita, Shigeyuki, *Introduction to mapping class groups of surfaces and related groups*, in Handbook of Teichmüller theory. Vol. I, IRMA Lect. Math. Theor. Phys., Eur. Math. Soc., Zürich, 2007, pp. 353–386.

526. Moriwaki, Atsushi, *A sharp slope inequality for general stable fibrations of curves*, J. Reine Angew. Math. **480** (1996), 177–195.

527. Moriwaki, Atsushi, *Relative Bogomolov's inequality and the cone of positive divisors on the moduli space of stable curves*, J. Amer. Math. Soc. **11** (1998), 569–600.

528. Morrey, Charles B., Jr., *On the solutions of quasi-linear elliptic partial differential equations*, Trans. Amer. Math. Soc. **43** (1938), 126–166.

529. Morrison, Ian, *Projective stability of ruled surfaces*, Inv. Math. **50** (1980), 269–304.

530. Morrison, Ian, *Subvarieties of moduli spaces of curves: open problems from an algebro-geometric point of view*, in Mapping class groups and moduli spaces of Riemann surfaces (Göttingen, 1991/Seattle, WA, 1991), Contemp. Math., Amer. Math. Soc., Providence, RI, 1993, pp. 317–343.

531. Morrison, Ian, *GIT Constructions of Moduli Spaces of Stable Curves and Maps*, in Geometry of Riemann surfaces and their moduli spaces, (Lizhen Ji, Scott Wolpert, Shing-Tung Yau, eds.), Surveys in Differential Geometry 14, International Press, Somerville, MA, 2010.

532. Morrison, Ian, *Mori theory of moduli spaces of stable curves*, http://projectivepress.com/moduli/moristablecurves.pdf (2009).

533. Mukai, Shigeru, *Biregular classification of Fano 3-folds and Fano manifolds of coindex* 3, Proc. Nat. Acad. Sci. U.S.A. **86** (1989), 3000–3002.

534. Mukai, Shigeru, *Fano 3-folds*, in Complex projective geometry (Trieste, 1989/Bergen, 1989), London Math. Soc. Lecture Note Ser. 179, Cambridge Univ. Press, 1992 pp. 255–263.

535. Mukai, Shigeru, *Curves and symmetric spaces*, Proc. Japan Acad. Ser. A Math. Sci. **68** (1992), 7–10.

536. Mukai, Shigeru, *Curves and symmetric spaces. I*, Amer. J. Math. **117** (1995), 1627–1644.

537. Mukai, Shigeru, *Vector bundles and Brill–Noether theory*, in Current topics in complex algebraic geometry (Berkeley, CA, 1992/93), Math. Sci. Res. Inst. Publ., Cambridge Univ. Press, Cambridge, 1995, pp. 145–158.

538. Mukai, Shigeru, *Curves and $K3$ surfaces of genus eleven*, in Moduli of vector bundles (Sanda, 1994; Kyoto, 1994), Lecture Notes in Pure and Appl. Math., Dekker, New York, 1996, pp. 189–197.

539. Mukai, Shigeru, *Noncommutativizability of Brill–Noether theory and 3-dimensional Fano varieties*, Sugaku **49** (1997), 1–24.

540. Mukai, Shigeru, *Curves, $K3$ surfaces and Fano 3-folds of genus* ≤ 10, in Algebraic geometry and commutative algebra, Vol. I, Kinokuniya, Tokyo, 1998, pp. 357–377.

541. Mukai, Shigeru, *An introduction to invariants and moduli*, Cambridge Studies in Advanced Mathematics, 81, Cambridge University Press, Cambridge, 2003.
542. Mukai, Shigeru, *Polarized K3 surfaces of genus thirteen*, in Moduli spaces and arithmetic geometry, Adv. Stud. Pure Math., Math. Soc. Japan, Tokyo, 2006, pp. 315–326.
543. Mulase, Motohico, *Matrix integrals and integrable systems*, in Topology, geometry and field theory, World Sci. Publ., River Edge, NJ, 1994, pp. 111–127.
544. Mulase, Motohico, *Algebraic theory of the KP equations*, in Perspectives in mathematical physics, Conf. Proc. Lecture Notes Math. Phys., III, Int. Press, Cambridge, MA, 1994, pp. 151–217.
545. Mulase, Motohico, *Asymptotic analysis of a Hermitian matrix integral*, Internat. J. Math. **6** (1995), 881–892.
546. Mulase, Motohico and Safnuk, Brad, *Mirzakhani's recursion relations, Virasoro constraints and the KdV hierarchy*, Indian J. Math. **50** (2008), 189–218.
547. Mumford, David, *Further pathologies in algebraic geometry*, Am. J. Math. **84** (1962), 642–648.
548. Mumford, David, *Further comments on boundary points*, Lecture notes prepared in connection with the Summer Institute on Algebraic Geometry held at Woods Hole, MA, American Mathematical Society, 1964. (unpublished).
549. Mumford, David, *Picard groups of moduli problems*, in Arithemetic algebraic geometry (Purdue Univ. 1963), Harper and Row, New York, 1965, pp. 33–81.
550. Mumford, David, *Lectures on curves on an algebraic surface*, Princeton Mathematical Series 59, Princeton University Press, Princeton, New Jersey, 1966.
551. Mumford, David, *Abelian quotients of the Teichmüller modular group*, Abelian quotients of the Teichmüller modular group **18** (1967), 227–244.
552. Mumford, David, *Abelian Varieties*, Oxford University Press, Bombay, 1970.
553. Mumford, David, *The structure of the moduli spaces of curves and Abelian varieties*, in Actes du Congrès International des Mathématiciens, (Nice, 1970), Tome 1, Gauthier-Villars, Paris, 1971, pp. 457–465.
554. Mumford, David, *Theta characteristics of an algebraic curve*, Ann. Sci. École Norm. Sup. (4) **4** (1971), 181–192.
555. Mumford, David, *Stability of projective varieties*, L'Ens. Math. **23** (1977), 39–110.
556. Mumford, David, *Towards an enumerative geometry of the moduli space of curves*, Progress in Math. 36 (1983), 271–326, in Arithmetic and geometry, II, Birkhauser, Boston.

557. Mumford, David, *The red book of varieties and schemes*, Springer-Verlag, Berlin, 1999. Lecture Notes in Mathematics 1358, second expanded edition. Includes the Michigan lectures (1974) on curves and their Jacobians.

558. Mumford, D., Fogarty, J., and Kirwan, F., *Geometric invariant theory. Third edition*, Ergebnisse der Mathematik und ihrer Grenzgebiete (2), Springer-Verlag, Berlin, 1994, pp. xiv+292.

559. Mumford, David and Suominen, Kalevi, *Introduction to the theory of moduli*, in Algebraic geometry, Oslo 1970 (Proc. Fifth Nordic, Summer-School in Math.), Wolters-Noordhoff, Groningen, 1972, pp. 171–222.

560. Nag, Subhashis, *The complex analytic theory of Teichmüller spaces*, Canadian Mathematical Society Series of Monographs and Advanced Texts, John Wiley & Sons Inc., New York, 1988.

561. Nagaraj, Donihakkalu S., *On the moduli of curves with theta-characteristics*, Compositio Math. **75** (1990), 287–297.

562. Nagata, Masayoshi, *Lectures on Hilbert's fourteenth problem*, Tata Lectures 31, Springer-Verlag, New York, 1965.

563. Nagata, Masayoshi, *On the 14-th problem of Hilbert*, Am. J. Math. **81** (1959), 766–772.

564. Nakano, Shigeo, *On the inverse of monoidal transformation*, Publ. Res. Inst. Math. Sci. **6** (1970/71), 483–502.

565. Newstead, P. E., *Introduction to moduli problems and orbit spaces*, Tata Institute of Fundamental Research Lectures on Mathematics and Physics, 51, Tata Institute of Fundamental Research, Bombay, 1978.

566. Nielsen, Jakob, *Untersuchungen zur Topologie der geschlossenen zweiseitigen Flächen I*, Acta Math. **50** (1927), 189–358.

567. Northcott, D. G., *Finite free resolutions*, Cambridge University Press, Cambridge, 1976, pp. xii+271. Cambridge Tracts in Mathematics, No. 71.

568. Oda, Tadao and Seshadri, C. S., *Compactifications of the generalized Jacobian variety*, Trans. Amer. Math. Soc. **253** (1979), 1–90.

569. Ohbuchi, Akira, *On the higher Wahl maps*, Tsukuba J. Math. **18** (1994), 79–99.

570. Okounkov, Andrei, *Toda equations for Hurwitz numbers*, Math. Res. Lett. **7** (2000), 447–453.

571. Okounkov, Andrei, *Random trees and moduli of curves*, in Asymptotic combinatorics with applications to mathematical physics (St. Petersburg, 2001), LNM 1815, Springer-Verlag, Berlin, 2003, pp. 89–126.

572. Okounkov, Andrei and Pandharipande, Rahul, *Gromov–Witten theory, Hurwitz numbers, and Matrix models*, 2001, `http://www.math.princeton.edu/~rahulp/gwhm.ps`.

573. Okounkov, Andrei and Pandharipande, Rahul, *Hodge integrals and invariants of the unknot*, Geom. Topol. **8** (2004), 675–699.
574. Okounkov, Andrei and Pandharipande, Rahul, *Gromov–Witten theory, Hurwitz theory, and completed cycles*, Ann. of Math. (2) **163** (2006), 517–560.
575. Okounkov, Andrei and Pandharipande, Rahul, *The equivariant Gromov–Witten theory of* $\mathbb{P}^1$, Ann. of Math. (2) **163** (2006), 561–605.
576. Okounkov, Andrei and Pandharipande, Rahul, *Virasoro constraints for target curves*, Invent. Math. **163** (2006), 47–108.
577. Okounkov, Andrei and Pandharipande, Rahul, *Gromov–Witten theory, Hurwitz numbers, and matrix models*, in Algebraic geometry—Seattle 2005. Part 1, Proc. Sympos. Pure Math. 80, Amer. Math. Soc., Providence, RI, 2009, pp. 325–414.
578. Oort, Frans, *Coarse and fine moduli spaces of algebraic curves and polarized abelian varieties*, in Symposia Mathematica, Vol. XXIV (Sympos., INDAM, Rome, 1979), Academic Press, London, 1981, pp. 293–313.
579. Oort, Frans and Steenbrink, Joseph, *The local Torelli problem for algebraic curves*, in Journées de Géometrie Algébrique d'Angers (Angers, Juillet 1979), Sijthoff & Noordhoff, Alphen aan den Rijn, 1980, pp. 157–204.
580. Osserman, Brian, *A limit linear series moduli scheme*, Ann. Inst. Fourier (Grenoble) **56** (2006), 1165–1205.
581. Palamodov, Victor, *Deformations of complex spaces*, Russian Math. Surveys **31** (1976), 129–197.
582. Pandharipande, Rahul, *A Geometric Invariant Theory Compactification of* $M_{g,n}$ *via the Fulton–MacPherson Configuration Space*, arXiv:alg-geom/9505022v1.
583. Pandharipande, Rahul, *A compactification over* $\overline{M}_g$ *of the universal moduli space of slope-semistable vector bundles*, J. Amer. Math. Soc. **9** (1996), 425–471.
584. Pandharipande, Rahul, *Three questions in Gromov–Witten theory*, in Proceedings of the International Congress of Mathematicians, Vol. II (Beijing, 2002), Higher Ed. Press, Beijing, 2002, pp. 503–512.
585. Paoletti, Roberto, *Generalized Wahl maps and adjoint line bundles on a general curve*, Pacific J. Math. **168** (1995), 313–334.
586. Paranjape, Kapil and Ramanan, S., *On the canonical ring of a curve*, in Algebraic geometry and commutative algebra, Vol. II, Kinokuniya, Tokyo, 1988, pp. 503–516.
587. Pareschi, Giuseppe, *Gaussian maps and multiplication maps on certain projective varieties*, Compositio Math. **98** (1995), 219–268.
588. Pareschi, Giuseppe, *A proof of Lazarsfeld's theorem on curves on* $K3$ *surfaces*, J. Algebraic Geom. **4** (1995), 195–200.
589. Penner, R. C., *The decorated Teichmüller space of punctured surfaces*, Comm. Math. Phys. **113** (1987), 299–339.

590. Penner, R. C., *Perturbative series and the moduli space of Riemann surfaces*, J. Differential Geom. **27** (1988), 35–53.

591. Penner, R. C., *The Poincaré dual of the Weil-Petersson Kähler two-form*, Comm. Anal. Geom. **11** (1993), 43–69.

592. Piene, Ragni and Schlessinger, Michael, *On the Hilbert scheme compactification of the space of twisted cubics*, Amer. J. Math. **107** (1985), 761–774.

593. Pikaart, Martin, *An orbifold partition of* $\overline{M}_{n,g}$, in The moduli space of curves (Texel Island, 1994), Birkhäuser Boston, Boston, MA, 1995, pp. 467–482.

594. Pikaart, Martin and de Jong, A. Johan, *Moduli of curves with non-abelian level structure*, in The moduli space of curves (Texel Island, 1994), Birkhäuser Boston, Boston, MA, 1995, pp. 483–509.

595. Pinkham, Henry C., *Deformations of algebraic varieties with* G_m *action*, Astérisque, No. 20, Société Mathématique de France, Paris, 1974.

596. Polishchuk, Alexander, A_∞*-structures, Brill–Noether loci and the Fourier-Mukai transform*, Compos. Math. **140** (2004), 459–481.

597. Polishchuk, Alexander, *Witten's top Chern class on the moduli space of higher spin curves*, in Frobenius manifolds, Aspects Math., E36, Vieweg, 2004, pp. 253–264.

598. Polyakov, A. M., *Quantum geometry of fermionic strings*, Phys. Lett. B **103** (1981), 211–213.

599. Powell, Jerome, *Two theorems on the mapping class group of a surface*, Proc. Amer. Math. Soc. **68** (1978), 347–350.

600. Ramanan, S., *The moduli spaces of vector bundles over an algebraic curve*, Math. Ann. **200** (1973), 69–84.

601. Ramanan, S., *Curves and their canonical imbeddings*, in The Curves Seminar at Queen's, Vol. V (Kingston, ON, 1987–1988), Queen's Papers in Pure and Appl. Math. 80, Exp. No. A, 17, Queen's Univ., Kingston, ON, 1988.

602. Ratcliffe, J. G., *Foundation of Hyperbolic Manifolds*, Graduate Texts in Mathematics, Springer-Verlag, Berlin, 2006.

603. Rauch, H. E., *Weierstrass points, branch points, and moduli of Riemann surfaces*, Comm. Pure Appl. Math. **12** (1959), 543–560.

604. Rauch, H. E., *Addendum to "Weierstrass points, branch points, and moduli of Riemann surfaces"*, Comm. Pure Appl. Math. **13** (1960), 165.

605. Rego, C. J., *The compactified Jacobian*, Ann. Sci. École Norm. Sup. (4) **13** (1980), 211–223.

606. Rim, Dock Sang and Vitulli, Marie A., *Weierstrass points and monomial curves*, J. Algebra **48** (1977), 454–476.

607. Rosenlicht, Maxwell, *Toroidal algebraic groups*, Proc. Amer. Math. Soc. **12** (1961), 984–988.

608. Roth, Mike and Vakil, Ravi, *The affine stratification number and the moduli space of curves*, in Algebraic structures and moduli spaces, CRM Proc. Lecture Notes 38, Amer. Math. Soc., Providence, RI, 2004, pp. 213–227.

609. Ruan, Yongbin and Tian, Gang, *A mathematical theory of quantum cohomology (article)*, J. Differential Geom. **42** (1995), 259–367.

610. Members of the seminar of I. R. Šafarevič, *Algebraic surfaces*, Translated from the Russian by Susan Walker. Translation edited, with supplementary material, by K. Kodaira and D. C. Spencer. Proceedings of the Steklov Institute of Mathematics No. 75 (1965), American Mathematical Society, Providence, R.I., 1965, pp. ix+281.

611. Samuel, Pierre, *Lectures on old and new results on algebraic curves*, Notes by S. Anantharaman. Tata Institute of Fundamental Research Lectures on Mathematics, No. 36, Tata Institute of Fundamental Research, Bombay, 1966.

612. Satake, Ichirō, *On a generalization of the notion of manifold*, Proc. Nat. Acad. Sci. U.S.A. **42** (1956), 359–363.

613. Satake, Ichirō, *On the compactification of the Siegel space*, J. Indian. Math. Soc. **20** (1956), 259–281.

614. Satake, Ichirō, *The Gauss–Bonnet theorem for V-manifolds*, J. Math. Soc. Japan **9** (1957), 464–492.

615. Schlessinger, Michael, *Functors of Artin rings*, Trans. A.M.S. **130** (1968), 208–222.

616. Segal, Graeme, *The definition of conformal field theory*, in Topology, geometry and quantum field theory, London Math. Soc. Lecture Note Ser., Cambridge Univ. Press, Cambridge, 2004, pp. 421–577.

617. Segal, Graeme and Wilson, George, *Loop groups and equations of KdV type*, Inst. Hautes Études Sci. Publ. Math. **61** (1985), 5–65.

618. Segre, Beniamino, *Sui moduli delle curve poligonali, e sopra un complemento al teorema di esistenza di Riemann*, Math. Ann. **100** (1928), 537–551.

619. Segre, Beniamino, *Sui moduli delle curve algebriche*, Ann. Mat. Pura Appl. (4) **4** (1930), 71–102.

620. Segre, Beniamino, *Sulla completezza della serie caratteristica di un sistema continuo di curve irriducibili tracciate su di una superficie algebrica.* Rend. Circ. Mat. Palermo **55** (1931), 443–449.

621. Seifert, Herbert, *Bemerkungen zur stetigen Abbildung von Flächen*, Abh. Math. Semin. Hansische Univ. **12** (1937), 29–37.

622. Sernesi, Edoardo, *Unirationality of the variety of moduli of curves of genus twelve*, Ann. Scuola Norm. Sup. Pisa Cl. Sci. (4) **8** (1981), 405–439.

623. Sernesi, Edoardo, *On the existence of certain families of curves*, Invent. Math. **75** (1984), 25–57.

624. Sernesi, Edoardo, *Topics on families of projective schemes*, Queen's Papers in Pure and Applied Mathematics 73, Queen's University, Kingston, ON, 1986, pp. viii+203.
625. Sernesi, Edoardo, *Deformations of algebraic schemes*, Springer, Berlin-Heidelberg, 2006.
626. Serre, Jean-Pierre, *Géométrie algébrique et géométrie analytique*, Ann. Inst. Fourier **6** (1956), 1–42.
627. Serre, Jean-Pierre, *Rigidité du foncteur di Jacobi d'échelon $n \geq 3$*, in Séminaire Henri Cartan, 1960/61, Appendice à l'Exposé 17, Secrétariat mathématique, Paris, 1960/1961.
628. Seshadri, C. S., *Fibrés vectoriels sur les courbes algébriques*, Astérisque 96, Société Mathématique de France, Paris, 1982.
629. Seshadri, C. S., *Vector bundles on curves*, in Linear algebraic groups and their representations (Los Angeles, CA, 1992), Contemp. Math. 153, Amer. Math. Soc., Providence, RI, 1993, pp. 163–200.
630. Severi, Francesco, *Intorno alla costruzione dei sistemi completi non lineari che appartengono ad una superficie irregolare*, Rend. Circ. Mat. Palermo **20** (1905), 93–96.
631. Severi, Francesco, *Sulla classificazione delle curve algebriche e sul Teorema di esistenza di Riemann*, Rendiconti della R. Accad. Naz. Lincei **24** (1915), 877–888.
632. Severi, Francesco, *Sulla teoria degl'integrali semplici di 1ª specie appartenenti ad una superficie algebrica. V*, R. Acc. Lincei Rend. (5) **30** (1921), 296–301.
633. Severi, Francesco, *Vorlesungen über Algebraische Geometrie*, Teubner, Leipzig, 1921.
634. Severi, Francesco, *Sul teorema di esistenza di Riemann*, Rendiconti del circolo matematico di Palermo **46** (1922), 105–116.
635. Severi, Francesco, *La teoria generale dei sistemi continui di curve sopra una superficie algebrica*, Atti Accad. Italia. Mem. Cl. Sci. Fis. Mat. Nat. **12** (1942), 337–430.
636. Severi, Francesco, *Intorno ai sistemi continui di curve sopra una superficie algebrica*, Comment. Math. Helv. **15** (1943), 238–248.
637. Severi, Francesco, *Sul teorema fondamentale dei sistemi continui di curve sopra una superficie algebrica*, Ann. Mat. Pura Appl. (4) **23** (1944), 149–181.
638. Shadrin, S. V., *Geometry of meromorphic functions and intersections on moduli spaces of curves*, Int. Math. Res. Not. **38** (2003), 2051–2094.
639. Shadrin, S. V., *Polynomial Hurwitz numbers and intersections on* $\overline{M}_{0,k}$, Funktsional. Anal. i Prilozhen. **37** (2003), 92–94.
640. Simpson, Matthew, *On Log Canonical Models of the Moduli Space of Stable Pointed Curves*, arXiv:0709.4037.
641. Smith, Roy and Varley, Robert, *Deformations of singular points on theta divisors*, in Theta functions—Bowdoin 1987, Part 1 (Brunswick,

ME, 1987), Proc. Sympos. Pure Math. 49, Amer. Math. Soc., Providence, RI, 1989, pp. 571–579.
642. Smyth, David Ishii, *Towards a classification of modular compactifications of the moduli space of curves*, arXiv:0902.3690v1 [math.AG].
643. Soulé, Christophe, *Lectures on Arakelov geometry*, Cambridge Studies in Advanced Mathematics 33, Cambridge University Press, Cambridge, 1992. With the collaboration of D. Abramovich, J.-F. Burnol and J. Kramer.
644. Spanier, Edwin H., *Algebraic Topology*, Springer-Verlag, New York, 1981.
645. Springer, George, *Introduction to Riemann Surfaces*, AMS Chelsea Publishing, New York, 2002.
646. Stillwell, John, *The Dehn–Nielsen theorem*, in "Max Dehn, Papers on Group Theory and Topology", Springer-Verlag, New York, 1987, pp. 363–396.
647. Stoppino, Lidia, *Slope inequalities for fibred surfaces via GIT*, Osaka J. Math. **45** (2008), 1027–1041.
648. Strebel, Kurt, *Quadratic differentials*, Ergebnisse der Mathematik un ihrer Grenzgebiete (3) 5, Springer-Verlag, Berlin-Heidelberg, 1984.
649. Sundaram, Narasimhan, *Special divisors and vector bundles*, Tohoku Math. J. (2) **39** (1987), 175–213.
650. Talpo, Mattia and Vistoli, Angelo, *Deformation theory from the point of view of fibered categories*, arXiv:1006.0497.
651. Tan, Xiao-Jiang, *A model of Brill–Noether theory for rank two vector bundles and its Petri map*, Asian J. Math. **7** (2003), 539–550.
652. Tannenbaum, Allen, *Families of algebraic curves with nodes*, Compositio Math. **41** (1980), 107–126.
653. Tannenbaum, Allen, *Families of curves with nodes on $K3$ surfaces*, Math. Ann. **260** (1982), 239–253.
654. Tannenbaum, Allen, *On the classical characteristic linear series of plane curves with nodes and cuspidal points: two examples of Beniamino Segre*, Compositio Math. **51** (1984), 169–183.
655. Teixidor i Bigas, Montserrat, *Half-canonical series on algebraic curves*, Trans. Amer. Math. Soc. **302** (1987), 99–115.
656. Teixidor i Bigas, Montserrat, *Brill–Noether theory for stable vector bundles*, Duke Math. J. **62** (1991), 385–400.
657. Teixidor i Bigas, Montserrat, *Brill–Noether theory for vector bundles of rank* 2, Tohoku Math. J. (2) **43** (1991), 123–126.
658. Teixidor i Bigas, Montserrat, *On the Gieseker–Petri map for rank* 2 *vector bundles*, Manuscripta Math. **75** (1992), 375–382.
659. Teixidor i Bigas, Montserrat, *Rank two vector bundles with canonical determinant*, Math. Nachr. **265** (2004), 100–106.
660. Teixidor i Bigas, Montserrat, *Existence of vector bundles of rank two with sections*, Adv. Geom. **5** (2005), 37–47.

661. Teixidor i Bigas, Montserrat, *Petri map for rank two bundles with canonical determinant*, Compos. Math. **144** (2008), 705–720.
662. Tillmann, Ulrike, *On the homotopy of the stable mapping class group*, Invent. Math. **130** (1997), 257–275.
663. Tillmann, Ulrike, *Strings and the stable cohomology of mapping class groups*, in Proceedings of the International Congress of Mathematicians, Vol. II (Beijing, 2002), Higher Ed. Press, Beijing, 2002, pp. 447–456.
664. Tromba, Anthony J., *Teichmüller theory in Riemannian geometry*, Lectures in Mathematics ETH Zürich, Birkhäuser Verlag, Basel, 1992.
665. Vakil, Ravi, *A tool for stable reduction of curves on surfaces*, in Advances in algebraic geometry motivated by physics (Lowell, MA, 2000), Contemp. Math. 276, Amer. Math. Soc., Providence, RI, 2001, pp. 145–154.
666. Vakil, Ravi, *The moduli space of curves and its tautological ring*, Notices Amer. Math. Soc. **50** (2003), 647–658.
667. Vakil, Ravi, *Murphy's Law in algebraic geometry: Badly-behaved moduli deformation spaces*, Invent. Math. **164** (2006), 569–590.
668. Vakil, Ravi, *The moduli space of curves and Gromov–Witten theory*, in Enumerative invariants in algebraic geometry and string theory, Lecture Notes in Math. 1947, Springer, Berlin, 2008, pp. 143–198.
669. Verra, Alessandro, *The unirationality of the moduli spaces of curves of genus 14 or lower*, Compos. Math. **141** (2005), 1425–1444.
670. Viehweg, Eckart, *Quasi-projective moduli for polarized manifolds*, Ergebnisse der Mathematik und ihrer Grenzgebiete (3) 30, Springer, Berlin, 1995.
671. Vistoli, Angelo, *Intersection theory on algebraic stacks and on their moduli spaces*, Inv. Math. **97** (1989), 613–670.
672. Voisin, Claire, *Sur l'application de Wahl des courbes satisfaisant la condition de Brill–Noether–Petri*, Acta Math. **168** (1992), 249–272.
673. Voisin, Claire, *Déformation des syzygies et théorie de Brill–Noether*, Proc. London Math. Soc. (3) **67** (1993), 493–515.
674. Voisin, Claire, *Green's generic syzygy conjecture for curves of even genus lying on a $K3$ surface*, J. Eur. Math. Soc. **4** (2002), 363–404.
675. Voisin, Claire, *Green's canonical syzygy conjecture for generic curves of odd genus*, Compos. Math. **141** (2005), 1163–1190.
676. Voisin, Claire, *Géométrie des espaces de modules de courbes et de surfaces $K3$ (d'après Gritsenko–Hulek–Sankaran, Farkas–Popa, Mukai, Verra, et al.)*, Séminaire Bourbaki. Vol. 2006/2007, Exp. No. 981, x, Astérisque **317** (2008), 467–490.
677. Wahl, Jonathan M., *Deformations of planes curves with nodes and cusps*, Amer. J. Math. **96** (1974), 529–577.
678. Wahl, Jonathan M., *The Jacobian algebra of a graded Gorenstein singularity*, Duke Math. J. **55** (1987), 843–871.

679. Wahl, Jonathan, *Gaussian maps on algebraic curves*, J. Differential Geom. **32** (1990), 77–98.

680. Wahl, Jonathan, *Introduction to Gaussian maps on an algebraic curve*, in Complex projective geometry (Trieste, 1989/Bergen, 1989), London Math. Soc. Lecture Note Ser. 179, Cambridge Univ. Press, Cambridge, 1992, pp. 304–323.

681. Wajnryb, Bronislaw, *A simple presentation for the mapping class group of an orientable surface*, Israel J. Math. **45** (1983), 157–174.

682. Weyl, Hermann, *Classical groups, their invariants and representations*, Princeton Landmarks in Mathematics, Princeton University Press, Princeton, New Jersey, 1997.

683. Whitney, Hassler, *Geometric integration theory*, Princeton University Press, Princeton, New Jersey, 1957.

684. Witten, Edward, *Two-dimensional gravity and intersection theory on moduli space*, in Surveys in differential geometry vol. 1 (Cambridge, MA, 1990), Lehigh University, Bethlehem, PA, 1991, pp. 243–310.

685. Witten, Edward, *On the Kontsevich model and other models of two-dimensional gravity*, in Proceedings of the XXth International Conference on Differential Geometric Methods in Theoretical Physics, Vol. 1, 2 (New York, 1991), World Sci. Publ., River Edge, NJ, 1992, pp. 176–216.

686. Witten, Edward, *Algebraic geometry associated with matrix models of two-dimensional gravity*, in Topological methods in modern mathematics (Stony Brook, NY, 1991), Publish or Perish, Houston, TX, 1993, pp. 235–269.

687. Xiao, Jian Bo and Tan, Xiao Jiang, *Brill–Noether theory for rank two vector bundles generated by their sections*, Acta Math. Sin. (Engl. Ser.) **24** (2008), 1995–2010.

688. Xiao, Gang, *Fibered algebraic surfaces with low slope*, Math. Ann. **276** (1987), 449–466.

689. Yamaki, Kazuhiko, *A direct proof of Moriwaki's inequality for semistably fibered surfaces and its generalization*, J. Math. Kyoto Univ. **42** (2002), 485–508.

690. Zariski, Oscar, *Algebraic surfaces*, Springer-Verlag, Berlin, 1995, pp. xii+270. With appendices by S. S. Abhyankar, J. Lipman, and D. Mumford. Preface to the appendices by Mumford. Reprint of the second (1971) edition.

691. Zariski, Oscar, *Dimension-theoretic characterization of maximal irreducible algebraic systems of plane nodal curves of a given order n and with a given number d of nodes*, Amer. J. Math. **104** (1982), 209–226.

692. Zieschang, H., Vogt, E., and Coldewey, H.-D., *Flächen und ebene diskontinuierliche Gruppen*, Lecture Notes in Mathematics 122, Springer-Verlag, Berlin-Heidelberg-New York, 1970.

693. Zvonkin, Alexander, *Matrix integrals and map enumeration: an accessible introduction*, Math. Comput. Modelling **26** (1997), 281–304.
694. Zvonkine, Dimitri, *Enumeration of ramified coverings of the sphere and 2-dimensional gravity*, arXiv:math/0506248v1 [math.AG].
695. Zvonkine, Dimitri, *Counting ramified coverings and intersection theory on Hurwitz spaces. II. Local structure of Hurwitz spaces and combinatorial results*, Mosc. Math. J. **7** (2007), 135–162.

Index

E. Arbarello et al., *Geometry of Algebraic Curves*,
Grundlehren der mathematischen Wissenschaften 268,
DOI 10.1007/978-3-540-69392-5,

Printed in the USA
CPSIA information can be obtained
www.ICGtesting.com
'W060449010924
'CB00026B/99

9 783662 506202